T0349028

Mycorrhizal Symbiosis
Third Edition

Cover photographs

Top left.   Rhizoid of the mycoheterotrophic liverwort *Cryptothallus mirabilis* (Aneuraceae-Jungermanniales) showing hyphae of its fungal symbiont, a *Tulasnella* sp., growing towards the thallus. Rhizoid diameter 20 microns.

Top right.   Root of the mycoheterotrophic plant *Apteria aphylla* (Burmanniaceae) showing the clear apical zone lacking fungal colonization and the more mature distal portion with cells occupied by dense coils (yellow) of an arbuscular mycorrhizal (AM) fungus – a *Glomus* sp. Root diameter 0.5 mm.

Bottom left.   Seedling of the helleborine orchid *Epipactis gigantea* (Orchidaceae) recovered from a seed packet buried in nature, showing surface colonization by mycelia of its fungal symbiont – a *Tulasnella* sp. Maximum length of seedling 1 cm.

Bottom right.   Seedling of the mycoheterotrophic plant *Monotropa uniflora* (Monotropoideae-Ericales) recovered from a seed packet buried in nature with developing root system enveloped in mycelia of its fungal symbiont, *Russula decolorans*. Length of seedling 2.3 cm.

Photographs courtesy of M. Bidartondo.

Mycorrhizal Symbiosis
Third Edition

Sally E. Smith FAA

Soil and Land Systems, School of Earth and Environmental Sciences, Waite Campus, The University of Adelaide, Adelaide, Australia

and

David Read FRS

Department of Animal and Plant Sciences, University of Sheffield, Sheffield, UK

AMSTERDAM • BOSTON • HEIDELBERG • LONDON • NEW YORK • OXFORD
PARIS • SAN DIEGO • SAN FRANCISCO • SINGAPORE • SYDNEY • TOKYO
Academic Press is an imprint of Elsevier

Academic Press is an imprint of Elsevier
84 Theobald's Road, London WC1X 8RR, UK
Radarweg 29, PO Box 211, 1000 AE Amsterdam, The Netherlands
30 Corporate Drive, Suite 400, Burlington, MA 01803, USA
525 B Street, Suite 1900, San Diego, CA 92101-4495, USA

First edition 1983
Second edition 1997
Third edition 2008
Reprinted 2009

Notice
No responsibility is assumed by the publisher for any injury and/or damage to persons
or property as a matter of products liability, negligence or otherwise, or from any use
or operation of any methods, products, instructions or ideas contained in the material
herein. Because of rapid advances in the medical sciences, in particular, independent
verification of diagnoses and drug dosages should be made

British Library Cataloguing in Publication Data
A catalogue record for this book is available from the British Library

Library of Congress Cataloging-in-Publication Data
A catalog record for this book is available from the Library of Congress

ISBN–13: 978-0-1237-0526-6

For information on all Academic Press publications
visit our website at www.elsevierdirect.com

Printed and bound by CPI Group (UK) Ltd, Croydon, CR0 4YY

Transferred to Digital Print 2011

Working together to grow
libraries in developing countries

www.elsevier.com | www.bookaid.org | www.sabre.org

ELSEVIER   BOOK AID   Sabre Foundation
              International

Contents

The colour plate section appears at the back of the book

Preface

Nine years have elapsed since the publication of the second edition and we approached the preparation of this new edition with much trepidation. First, there has been a vast expansion of information and, secondly, much of this information has been derived from the application of new techniques of molecular biology, which had barely commenced at the time of preparation of the second edition. Now the task of writing is completed to the best of our abilities we can only agree with the statement made in 1983 (first edition) that it is 'almost impossible for one person or even two to keep up with all the experimental work and speculation on the subject of mycorrhizas'! We have done our best, but apologize to those readers who feel that our treatments of some areas lack sufficient coverage.

In this edition, we have retained part of the text from the second edition but, again, there has been so much new work that detailed reference to some research included in the first two editions has had to be reduced. We have, however, attempted to retain a feeling for the way the subject has developed and to highlight major contributions of early researchers and we urge current and future researchers to delve into early (pre-electronic) literature because it contains many ideas and much valuable information which should not be wasted or describes experiments that need not be repeated.

Again, it has been impossible to review all topics in detail and this third edition once more reflects our interests and provides personal views of the subject. Where the first edition explicitly avoided evolutionary discussions, on the grounds that insufficient information was available to make this profitable, we have capitalized on new information based on molecular phylogeny of fungi and of plants (introduced in the second edition), to reveal likely evolutionary pathways by which mycorrhizas have arisen. We have also maintained the emphasis on the extraradical mycelium, both with respect to development in all types of mycorrhizas and to function. In most areas of research, the structural and functional diversity among mycorrhizas has continued to be revealed. Accordingly, we highlight this aspect of mycorrhizal biology and show how it may be important in ecological situations.

The second edition was written when the very first results of molecular biological investigation of mycorrhizas were being published. In the intervening time there has been an explosion of such research, both targeted to increasing our understanding of well-known features of symbiosis, such as nutrient transfer, as well as to revealing 'unknown' aspects of the symbioses. The roles of mycorrhizas as the normal nutrient absorbing organs of the vast majority of plant species in nature is increasingly being revealed and appreciated by ecologists. Throughout the book we have emphasized the experimental approaches that have proved useful and that may need to be applied in the future.

We have retained the same general structure of the book, with three sections providing general accounts of the main types of mycorrhiza, including information on the identity of the symbionts, structure and development of the mycorrhizas formed by them, as well as their function and ecological significance. The fourth section is devoted to the functioning of mycorrhizas in broader contexts, in which we believe integration of ideas and information is essential for clear understanding. These chapters build on the material presented earlier, but so that they can be read separately we have deliberately included some repetition and have also tried to improve the cross-referencing in the text. We concentrate on the presence and activity of mycorrhizal symbioses in the major global biomes, on their roles in ecological interactions and on applications related to agriculture, horticulture and forestry. The essays in the first edition have served their purpose. Information on general aspects of nutrient transfers (second edition) has now been integrated as far as possible into chapters on the different mycorrhizal types. Readers interested in these topics will need to use the earlier editions.

Throughout the text we have tried to indicate where future experimental research might be directed, although we realize that our personal views may have introduced some bias, which will be recognized by those who know us.

Many friends and colleagues have helped us in countless ways, especially by discussions, by reading drafts of chapters and by supplying illustrations and results (some unpublished). We have not repeated the list from the second edition, but list alphabetically all those to whom we are particularly indebted in the preparation of this third edition: Susan Barker, Martin Bidartondo, Tom Bruns, Duncan Cameron, Tim Cavagnaro, Michel Chalot, Mark Chase, Sandy Dickson, Jeff Duckett, Evelina Facelli, Jose Facelli, Andrea Genre, Manuela Giovannetti, Emily Grace, Lisa Grubisha, Sarah Hambleton, Maria Harrison, Jan Jansa, Jonathan Leake, Francis Martin, Maria Manjarrez, Randy Molina, Hugues Massicotte, Joe Morton, Jesus Perez-Moreno, Larry Peterson, Arthur Schüßler, Andrew Smith, Lee Taylor, Sari Timonen, Jean-Patrick Toussaint, Andrew Smith, Michael Weiss and Katya Zimmer. We are grateful to them all for their critical and helpful comments, many of which have been incorporated into the text. However, we must take responsibility for all the ideas and views expressed whether good or bad; the errors are (once again) all ours.

We have also received invaluable help in preparation of the manuscript. Jayne Young typed many drafts of the chapters and produced the final manuscript; Rebecca Upson and Chris Read provided us with much-needed support in obtaining copyright clearances and cross-checking information and Glyn Woods provided assistance with production of many images.

We have used much previously published material in both tables and figures. The sources are acknowledged in each case, but we here wish to give special thanks to all the authors and publishers who generously allowed us to use their copyright material.

Last, but by no means least, our thanks and love go to the 'long-sufferers', Andrew Smith and Chris Read, who provided invaluable moral support which helped to see us through the many months of largely self-centred devotion to writing. They have had to go through this for a second time, so we are doubly grateful.

The book is again dedicated to the memory of Jack Harley, whose influence upon the development of our subject was so enormous. To expand the list of those who have made highly significant contributions to mycorrhizal research would have opened a Pandora's box. Who should we choose and what would happen if we inadvertently omitted crucial names? We hope that our attempts at historical treatment will make abundantly clear where credit for particular advances should be acknowledged.

Among Jack Harley's many contributions, two stand out as being of pre-eminent importance. Frustrated in his earlier career as an ecologist by much woolly thinking about the biology of mycorrhiza, Jack determined to subject the topic to rigorous physiological analysis. Over several decades and using as his research material the ectomycorrhizal roots of beech, he and his collaborators evaluated the basic processes whereby nutrients are exchanged between partners in the symbiosis. The second and arguably more important contribution arose out of his skill as a communicator. *The Biology of Mycorrhiza*, published in 1959, was the first ever attempt to synthesize the many and sometimes disparate strands of thought which had developed in over 100 years of research. With characteristic incisiveness he cut through much, often pedantic, debate to focus upon those questions which were in need of resolution. The work was of inestimable importance to many who, like ourselves, were struggling to take their first steps in research on this subject. These combined contributions provided impetus to a further expansion of research to the extent that, by 1983, a new and even more substantial volume, the first edition of *Mycorrhizal Symbiosis*, was required. One of us had the privilege to collaborate in that enterprise.

The influence of Jack Harley goes on. Although, sadly, he has not been here to assist us in updating the book through two editions, we have both been conscious of his legacy. Without him it is doubtful whether the subject would be in the pre-eminent position it enjoys today, increasingly recognized by physiologists and ecologists and now molecular biologists as being of central importance in plant and fungal biology.

INTRODUCTION

The interest in symbioses, noted in the introductions to both previous editions, has continued to expand and to encompass an increasing range of disciplines. The importance of symbioses between different prokaryotes in the evolution of eukaryotic cells is firmly established and there is increasing recognition that symbioses at the level of more complex organisms is the rule, rather than the exception. Compared with those at the cellular level, the mutually beneficial associations between identifiably different organisms, such as those between fungus and alga in lichens, plant and fungus in most mycorrhizas, alga and coelenterate in corals and bacteria and plant in nitrogen-fixing symbioses are of relatively recent origin, but they play exceedingly influential roles in the lives of plants and in the ecology of natural ecosystems. These roles are increasingly being recognized by biologists whose primary interest is not the symbioses themselves, but who find that for one reason or another account must be taken of the interactions between symbionts and their environments at molecular, cellular, whole plant and ecosystem levels.

The term 'symbiotismus' (symbiosis) was probably first used by Frank (1877) as a neutral term that did not imply parasitism, but was based simply on the regular coexistence of dissimilar organisms, such as is observed in lichens. de Bary (1887), who is often credited with the introduction of the term, certainly used it to signify the common life of parasite and host as well as of associations in which the organisms apparently helped each other. In time, the meaning of the terms symbiosis and parasitism changed. Symbiosis was used more and more for mutually beneficial associations between dissimilar organisms and parasite and parasitism came to be almost synonymous with pathogen and pathogenesis. de Bary had pointed out that there was every conceivable gradation between the parasite that quickly destroys its victim and those that 'further and support' their partners and, in recent years, biologists have come back to this view. In this book, we use 'symbiosis' in the broad sense originally developed by de Bary. Mutualistic symbioses are those in which both partners can benefit from the association, although there is unresolved discussion as to what actually constitutes 'benefit', increased 'fitness' or 'aptness' for particular environmental circumstances. Most mycorrhizal symbioses are now clearly recognized, together with lichens, corals and nitrogen-fixing systems, as being common and significant representatives of the mutualistic end of the symbiotic spectrum, whereas biotrophic pathogens, such as rusts and mildews, are parasitic symbioses. However, as will become clear, not all of those associations that are described as mycorrhizas have been shown to be mutualistic by experimental analysis of nutritional interactions or determination of fitness. Increasingly, it is recognized that, even in a single type of mycorrhiza, there may be a continuum of outcomes ranging from the mutualistic to the parasitic. Indeed, the relationship may change over the life of a single plant–fungus partnership.

If a symbiosis is mutualistic and based on bidirectional exchange of nutrients or other benefits, then the description of one partner as the 'host' seems inappropriate. Again, in this third edition we have adopted the terms 'plant' and 'fungus' to describe the partners in mycorrhizal symbioses, although we have retained such terms as 'host range' to describe the diversity of plant species with which a single fungus associates. We have also retained the term 'colonization', which replaced 'infection' because of forceful pleas by some that we should avoid any suggestion of negative effects, despite the diversity of responses.

de Bary (1887) believed that there was some degree of common life, i.e. of symbiosis, in all or almost all examples of association between plants and fungi. The associations are often classified as necrotrophic or biotrophic, depending on whether both symbionts remain alive, or whether the death of one was necessary before substances could be absorbed by the other. There is, however, a great range of behaviour between these two apparently clear-cut extremes, not only between different types of association but also at different times or under different environmental conditions in the same association. These variations in function, sometimes also apparent in changes in structural relationships between the symbionts, are seen in mycorrhizal associations as much as in other symbioses. The recognition of the considerable diversity of structure, development and function that exists even within a single mycorrhizal type was noted in the second edition. Research in the intervening years has greatly increased our knowledge of this diversity and begun to indicate its importance in nature. We take this opportunity to reiterate a point made forcibly in the Introduction to the first edition: 'It is clear that since these kinds of variation occur, much of the discussion based on classification and nomenclature is pedantic unless it is helpful in formulating clear questions which will lead to experiments from which answers will be obtained'. Indeed, the work of the last decade has revealed not only structural and developmental diversity, but an enormous range of physiological variation which has come to be termed the 'mutualism-parasitism continuum' (Johnson *et al.*, 1997). The significance of this diversity in determining plant interactions in nature is only just beginning to be revealed.

Although the term mycorrhiza implies the association of fungi with roots, relationships called mycorrhizal associations, which are involved in the absorption of nutrients from soil, are found between hyphal fungi and the underground organs of the gametophytes of many bryophytes and pteridophytes, as well as the roots of seed plants and the sporophytes of most pteridophytes. Recent work has highlighted and increased our knowledge of some of the lesser known groups and we have provided a new chapter on 'Fungal symbioses in the lower land plants'. Molecular techniques applied to these symbioses, as well as to better known groups, are revealing both which fungi are involved and the phylogenetic positions of both plant and fungal symbionts. In consequence, it has become possible to make evolutionary inferences that are grounded on evidence. We believe that Jack Harley, who was unsympathetic towards what he termed 'evolutionary speculation' would have recognized the worth of much of this new research.

Mycorrhizas, not roots, are the chief organs of nutrient uptake by land plants and recent work has amply confirmed the earlier observations (see Nicolson, 1975) that the earliest land plants, which had no true roots, were colonized by hyphal fungi that formed vesicles and arbuscules strikingly similar to modern arbuscular mycorrhizas.

It is now generally accepted that the colonization of the land was achieved by such symbiotic organisms, which were able to access nutrients unavailable to non-symbiotic individuals. We have not adopted new names for symbioses, recognized as 'mycorrhizal', in plants that do not have true roots and for which, therefore, the connotation mycorrhiza or 'fungus-root' may not be strictly appropriate. We do not consider the proliferation of new names to be helpful. Rather we regard research on functional integration of the symbionts and the significance of the symbioses in evolutionary and ecological contexts as much more relevant.

One of the important advances noted in the introduction of the second edition was an increasing emphasis on the structure, organization and function of the external mycelium. This trend has continued and has been linked to increasingly sophisticated and sensitive methods both to detect its presence and to uncover its roles in soil processes including nutrient mobilization and structural stabilization. Mycorrhizal fungi are specialized members of the vast population of microorganisms that colonize the rhizosphere. With a few exceptions, they are completely dependent on the plant for organic C. Being independent of the scarce and patchily distributed organic C resources in soil, they are likely to be in a good position to compete with saprotrophs in the mobilization of nitrogen, phosphate and other nutrients. Another vital difference between mycorrhizal associations and the general association of organisms with the root surface or rhizosphere lies in the closeness of the relationship and the recognizable and consistent structures formed. In mycorrhizas, there is always some penetration of the tissues of the root by the fungus, or a recognizable structure conforming to one of the common types, or both. A major difference between the mycorrhizal symbiosis and those symbioses caused by parasites which lead to disease is that the mycorrhizal condition is the normal state for most plants under most ecological conditions.

The way in which plants are potentially linked together in common mycelial networks formed by the external mycelium of their fungal symbionts has continued to fascinate researchers. Such links certainly exist and reflect the general lack of high specificity typical within the main mycorrhizal types. Nevertheless, evidence that the common networks form below-ground conduits that move nutrients from plant-to-plant in ecologically meaningful amounts, rather than pathways for internal redistribution, is still confined to rather few investigations. Physiological mechanisms that would facilitate interplant transfers have still not been elucidated. It can be said, however, that interplant transport of nutrients is better substantiated for ectomycorrhizal (ECM) than for arbuscular mycorrhizal (AM) symbioses. Indeed, the fungi forming ectomycorrhizas are, in some cases, capable of associating with achlorophyllous plants which derive all their nutritional requirements from them.

These mycoheterotrophic associations are mycorrhizal in the sense that they are normal (indeed essential) for the plants. They differ from other mycorrhizas in that C movement occurs in the same direction as mineral nutrients, so that strictly the plant is primarily parasitic on its fungal associate and secondarily dependent on another (photosynthetic) mycorrhizal partner. Considerable advances have been made in the last decade, such that it is now appropriate to discuss all such relationships in a single new chapter. Consequently, the structural and functional attributes of mycorrhizas in green orchids are considered separately from those orchids that remain achlorophyllous throughout their lives.

The second edition was written before the recent explosion of molecular biology, now applied to almost all aspects of biological investigation. For AM interactions, the parallels uncovered between mycorrhizal and rhizobial symbioses have provided enormous advantages to mycorrhizal researchers. In consequence, a huge amount of new information has been forthcoming, encompassing almost every aspect of AM biology. Major advances have also been made with ectomycorrhizas, including the recent release of a full genome sequence for the ECM fungus *Laccaria bicolor*. We have attempted to incorporate the new knowledge gained from use of these techniques as they shed light on different mycorrhizal types. At this stage of development of the research, there is a tendency to adopt experimental 'models', so that research has become focused on just a few, in the main easily grown, species of both plants and fungi. This sharply focused approach is understandable and is indeed normal in the initial stages of a new research direction. However, it must be recognized that the use of model species will make it impossible, in the short term, to address those aspects of diversity of symbiotic structure and function that have so clearly been revealed by other methods in the past decade. The need to address issues related to functional diversity must not be forgotten. Furthermore, as techniques become more sophisticated and also more demanding of resources, it is already apparent that simple measures of symbiotic development and effectiveness (such as plant growth, nutrient uptake and extent of colonization) are often not carried out – they are not high tech or trendy and they are time consuming. However, this basic information would add immeasurably to our ability to interpret the new results in terms of the biology of the plants and fungi in relation to the environmental conditions under which they are grown. In the longer term, it is the integration of molecular biological, biochemical and physiological information that will lead to real advances.

The same types of criticism have sometimes been levelled at physiologists. Their reductionist experimental approaches using single plants grown in pots has been viewed as lacking relevance to ecological situations. However, such experiments have played a significant part in revealing functional diversity of symbioses. Without this information we will be unlikely to understand mechanisms that underlie ecological interactions. We will not be able to open the 'underground black box' and reveal the processes to which mycorrhizas contribute in ecosystems. However, it is increasingly important for physiologists to design experiments that will contribute to answering questions that ecologists consider are worth asking. In other words, collaborative research in all areas is the only really effective way forward to understand complex symbiotic systems, including interactions between the partners and with their environments.

Classification of mycorrhizas

The classification of mycorrhizas adopted here has not changed significantly since the publication of the first edition and is shown in Table I.1 (taken from the second edition), again with further modifications which reflect advances in knowledge. As before, the classification aims to be descriptive and to emphasize problems in need of solutions, rather than to gloss over difficulties. The kinds of mycorrhiza are divided as before, on the basis of their fungal associates into those involving

Table 1.1 The characteristics of the important mycorrhizal types. The structural characters given relate to the mature state, not the developing or senescent states. Entries in brackets indicate rare conditions.

Kinds of mycorrhiza	Arbuscular mycorrhiza	Ectomycorrhiza	Ectendomycorrhiza	Arbutoid mycorrhiza	Monotropoid mycorrhiza	Ericoid mycorrhiza	Orchid mycorrhiza
Fungi septate	−	+	+	+	+	+	+
aseptate	+	−	−	−	−	−	−
Intracellular colonization	+	−	+	+	+	+	+
Fungal mantle	−	+	+ or −	+ or −	+	−	−
Hartig net	−	+	+	+	+	−	−
Achlorophylly	− (+)	−	−	−	+	−	+†
Fungal taxa	Glomero	Basidio/Asco (Glomero)	Basidio/Asco	Basidio	Basidio	Asco	Basidio
Plant taxa	Bryo Pterido Gymno Angio	Gymno Angio	Gymno Angio	Ericales	Monotropoideae	Ericales Bryo	Orchidales

† All orchids are achlorophyllous in the early seedling stages. Most orchid species are green as adults. The fungal taxa are abbreviated from Glomeromycota, Ascomycota and Basidiomycota; the plant taxa from Bryophyta, Pteridophyta, Gymnospermae and Angiospermae.

largely aseptate endophytes in the Glomeromycota and those formed by septate fungi in the Ascomycetes and Basidiomycetes.

The recognition of the Glomeromycota as a separate phylum of true fungi, rather than as zygomycetes (classified first in the order Endogonales and later in the Glomales), represents an important advance in the last decade. The extent of genetic variability within what were previously considered to be single taxa has been amply confirmed by the application of molecular genetic techniques to mycor-rhizal Ascomycota and Basidiomycota. The ascomycete, *Hymenoscyphus ericae*, long known to form ericoid mycorrhiza, has been shown to be an aggregate species within which one clade only, now named *Rhizoscyphus ericae*, is involved in forming this type of relationship. Some members of other clades within the *H. ericae* aggregate form ectomycorrhizas, while others are saprotrophic or mildly pathogenic. The extent of genetic and functional variability contained within a single genus of mycor-rhizal fungi is nowhere better demonstrated than in *Sebacina*, members of which have been shown to form ectomycorrhizas, mycorrhizas with green orchids and to be the symbionts of several fully mycoheterotrophic orchids. They also occur as symbiotic partners in some liverwort thalli and as associates, with as yet unknown function, of roots in the Ericaceae. Widespread and ecologically important ECM species such as *Pisolithus tinctorius*, which we suggested in the last edition of the book was likely to represent a species complex, have now been resolved as such. At least eleven species of *Pisolithus* are now recognized, there being the likelihood that others, including species endemic to particular areas, remain to be discovered. The molecular genetic advances, by enabling resolution of the identities of fungal species inhabiting individual roots of mycohetrotrophs, has also revealed the extent to which these plants selectively parasitize narrow clades within large basidiomycete genera such as *Rhizopogon* and *Russula*. The ecological implications of these previously unrecognized levels of specificity are considered in this edition.

In this edition, we have gone with the flow and adopted the now widespread use of 'arbuscular mycorrhiza' to describe those associations formed by members of the Glomeromycota. These symbioses were previously known as vesicular-arbuscular mycorrhizas (VAM) because some of the associations were characterized by the presence of both vesicles and arbuscules. Whereas it is true that not all associations formed by members of the Glomeromycota form vesicles, it is also true that some do not form arbuscules either. The term we have now accepted exemplifies the problems attendant on proliferation of names. We urge stability and simplicity in future.

The plant symbionts in mycorrhizas are so many and so taxonomically diverse that a primary classification on that basis would be wholly impractical. Those plants forming arbuscular mycorrhizas may belong to all phyla: Bryophyta, almost all groups of Pteridophyta, all groups of Gymnospermae and the majority of families in the Angiospermae. This is the most ancient mycorrhizal type, with fossils of *Aglaeophyton* from the Devonian containing both arbuscules and vesicles. In addition, members of the Glomeromycota associate not only with bryophytes, but also with a cyanobacterium, suggesting even earlier ancestral symbiotic forms. Fungi in this phylum are unculturable and it is normally assumed that they are wholly dependent on a photosynthetic plant. In consequence, the confirmation using molecular techniques that the mycorrhizas in roots of achlorophyllous members of the Burmanniaceae and Gentianaceae are formed by members of the Glomeromycota, raises fascinating questions about the nutrition of these fungi and indeed of their plant associates. Advances such as these are now discussed together in a separate chapter, encompassing all mycoheterotrophic species.

The septate fungi of the remaining kinds of mycorrhiza include members of almost all orders of basidiomycetes and many ascomycetes. The plant associates on which ecto-, ectendo- and ericoid mycorrhizas develop, are usually autotrophic trees, shrubs and, rarely, herbs. Arbutoid mycorrhizas are also formed by trees and shrubs, although some of the plants, such as species of *Pyrola*, are herbs and are often partially achlorophyllous. The closely related Monotropoideae are all achlorophyllous and are also herbaceous. Members of the Orchidaceae are all achlorophyllous at first, but most are photosynthetic as adults. The type of mycorrhiza formed can be influenced by the identity of both plant and fungus. For example, as mentioned above, the same fungus can form arbutoid (monotropoid) and ectomycorrhizas, or ecto- and ectendomycorrhizas or ecto- and orchid mycorrhizas, depending on the identity of the plant associate, so that there is a plexus of behaviour among the species of plant and the septate fungi with regard to mycorrhizal structures that they produce. Molecular approaches to taxonomy have clarified some previous problems. For convenience, we now group ectendo- and arbutoid mycorrhizas (including the chlorophyllous *Pyrola*) in a new chapter. We recognize that this separation is artificial and that mycorrhizas of *Pyrola* form a continuum with the phylogenetically related *Monotropa*, so that, like the green and achlorophyllous orchids, they could have been discussed together. However, we believe that the separate treatments we have now adopted will increase appreciation of the achlorophyllous, mycoheterotrophic forms as well as of green orchids, leading to increased understanding of common features as well as of differences.

In ectomycorrhizas, the fungus forms a structure called the mantle (or sheath) which encloses the rootlet. From it hyphae or rhizomorphs radiate outwards into the substrate. Hyphae also penetrate inwards between the cells of the root to form a complex intercellular system, which appears as a network of hyphae in section, called the Hartig net. There is little or no intracellular penetration. In a few plants, the development of the Hartig net is slight or absent and, in these, it is particularly important for experiments to confirm that these associations behave in a typically mycorrhizal manner, as for example in *Pisonia*.

In ectendomycorrhizas, the sheath may be reduced or absent; the Hartig net is usually well developed, but the hyphae penetrate into the cells of the plant. As already mentioned, the same species of fungus may form ectomycorrhizas on one species of plant and ectendomycorrhizas on others. Arbutoid mycorrhizas possess sheath, external hyphae and usually a well-developed Hartig net. In addition, there is extensive intracellular development of hyphal coils in the plant cells.

The ectomycorrhizas, ectendomycorrhizas and arbutoid mycorrhizas do have several features in common (see Table I.1) and it might be supposed that, as all the plant symbionts are photosynthetic, there are grounds for grouping them all together. However, this approach would hide problems that require investigation. The development of ectendomycorrhizas and the identity of the fungal symbionts that form them is understood much more clearly now than it was in 1983. In these, and in arbutoid mycorrhizas, more information on the physiology of the associations and on the factors that induce a single fungal species to produce different structures and different extents of intracellular penetration on different plants is now required. Nevertheless, the fungal associates appear to be able to access organic forms of nutrients, so that the symbioses expand the repertoire of nutrients available to the plants.

Two kinds of mycorrhiza are associated with plants that are totally achlorophyllous for the whole or part of their lives. They differ markedly in structure. The mycorrhizas formed on roots of members of the Monotropoideae are somewhat similar to the three kinds of mycorrhiza just considered, as they have a well-developed fungal sheath and Hartig net. In addition, a highly specialized haustorium-like structure (the fungal peg) penetrates the epidermal cells and goes through a complicated developmental pattern as the plant grows and flowers. The fungus also forms ectomycorrhizas on neighbouring autotrophic plants and the assumption is that organic carbon (C) is transferred to the monotropoid plant, although the mechanism of transfer and the role of the fungal peg still require experimental investigation. The experimental emphasis has been on organic C as a nutrient for the plant, but the very poor development of roots makes it highly likely that the fungus is important in mineral nutrition of these plants also.

In the Orchidaceae, the plants are partially or wholly achlorophyllous for some part of their life. They form mycorrhizas with basidiomycetes of various affinities. The division between orchids that are green for part of their lives and those that are wholly achlorophyllous is mirrored by the identities of their fungal associates. Whereas the fungal symbionts of green orchids are highly effective saprophytes broadly belonging to the form-genus *Rhizoctonia*, those of achlorophyllous orchids are more likely to be able to form ectomycorrhizas on autotrophic plants. For the green orchids, there is new evidence that the adult plants have some capability to provide the fungal symbionts with recent photosynthate for part of the life of the symbiosis, thus apparently reversing the direction of C flow between the partners. The mechanisms behind such bidirectional transfer of organic C have not yet been revealed, but it appears that, contrary to previous suppositions, there is the potential for mutualism.

In many autotrophic members of the Ericaceae and related families, the hair-like roots are enmeshed in an extensive weft of hyphae, which also penetrate the cells of the root. Normally no sheath is formed. The fungi certainly identified as forming ericoid mycorrhizas are all ascomycetes. Many ericaceous plants grow in habitats where most of the nutrients in the soil are in organic form and it is clear that the fungi have a considerable role in mobilizing these nutrients and making them available to the plant, again opening up a nutrient niche which is unavailable to non-mycorrhizal or arbuscular mycorrhizal plants.

These brief descriptions make it clear that the term mycorrhiza is used to describe many symbiotic associations between fungi and plants and it is important here to try to define what particular characteristics, in combination, make a 'mycorrhiza'. These are constancy of structure, development and presence under natural conditions. Each type of mycorrhiza is likely to have its own characteristic function and the term 'mycorrhiza' must certainly not be taken to imply that all types have the same function, or that every constant root–fungus association constitutes a 'mycorrhiza'.

Associations between fungi and roots are the norm in nature and, clearly, there must be some defining features of the mycorrhizal condition. At one level, mycorrhizal associations have distinctive structural attributes which are relatively easily recognized. Function is more difficult to ascribe, but some definition is required or any regular or common fungus–root association could indeed be described as mycorrhizal. A unifying feature is likely to be the role of the external hyphae in supplying soil-derived nutrients to the plant. In the second edition, we commented that such a definition required qualification, as some plants appeared to receive no *net* nutritional benefit from their fungal symbionts. Researchers rightly questioned how an apparently mutualistic symbiosis could persist in evolutionary terms and invoked benefits arising from improved drought and pest resistance. While these benefits are real, recent research on arbuscular mycorrhizas has also established that the fungal pathway of nutrient uptake is maintained and often contributes a large proportion of total uptake, even when there are no benefits in terms of growth. Furthermore, plants may benefit from this uptake in competitive situations, providing a rationale for evolutionary persistence. This is an example where reductive physiological experiments, coupled with molecular biology, have led to a better understanding of the widespread occurrence of mycorrhizas in nature.

The role of C nutrition as a defining feature in the symbiosis is more difficult to assess, because its source varies in the different symbioses. In most mycorrhizas, the plant is autotrophic and, in these, the symbiosis has the potential to be truly mutualistic with both partners deriving some nutritional benefit from the association. In others, the plant is clearly non-photosynthetic and it is doubtful if these symbioses are mutualistic with respect to bidirectional nutrient transfer, although future experiments may reveal other benefits that the fungus derives from the plant. Nevertheless, new work is revealing benefits that may be offset in time through the lives of the symbionts and hence have not been noticed in short-term pot experiments.

In the following chapters, these types of mycorrhizas will be discussed separately so as to emphasize what is known about each and the outstanding problems about them. After that, some general and comparative aspects of mycorrhizal symbiosis will be discussed. The topics we have chosen are different from those in earlier editions. Information on nutrient transfer between the symbionts has now been integrated into the chapters on different mycorrhizal types because it is increasingly appreciated that the sources and forms of the nutrients are different. The ecological importance of mycorrhizas is covered in two chapters. We now discuss the occurrence and importance of mycorrhizas of all types in different global biomes separately from details of the roles of mycorrhizas in particular ecological interactions. This expansion reflects both the importance of mycorrhizas in nature and the increased interest from ecologists in the roles that they play in plant–microbe interactions at all levels. The applications, actual and potential, of mycorrhizas in agriculture, horticulture and forestry are covered in a single chapter, which also includes consideration of the production of edible fungi both for high value markets and in subsistence production systems.

Section 1
Arbuscular mycorrhizas

1

The symbionts forming arbuscular mycorrhizas

Introduction

Arbuscular mycorrhizas (AM) are the most common mycorrhizal type. They are formed in an enormously wide variety of host plants by obligately symbiotic fungi which have recently been reclassified on the basis of DNA sequences into a separate fungal phylum, the Glomeromycota (Schüßler *et al.*, 2001). The plants include angiosperms, gymnosperms and the sporophytes of pteridophytes, all of which have roots, as well as the gametophytes of some hepatics and pteridophytes which do not (Read *et al.*, 2000; see Chapter 14). It seems highly likely that the fungi had their origins possibly over 1000 million years ago (predating current estimates of colonization of land) and that arbuscular mycorrhizal (AM) symbioses are also extremely ancient. Through their roles in nutrient uptake, AM fungi were probably important in the colonization of land by plants (Simon *et al.*, 1993; Remy *et al.*, 1994; Taylor *et al.*, 1995; Redecker *et al.*, 2000; Heckman *et al.*, 2001); they remain major determinants of plant interactions in ecosystems to the present day.

The name 'arbuscular' is derived from characteristic structures, the arbuscules (Figure 1.1a) which occur within the cortical cells of many plant roots and also some mycothalli colonized by AM fungi. Together with storage vesicles located within or between the cells, these structures have been considered diagnostic for AM symbioses. However, a rather wide range of intraradical structures formed by AM fungi is recognized (see Chapter 2 and Dickson, 2004), including well-developed intracellular hyphal coils, which sometimes occur in the absence of any arbuscules (Figure 1.1b). The variations in developmental pattern are determined by both plant and fungal partners, adding to the complexities of identifying a symbiosis as 'AM' on the basis of intraradical fungal morphology. The term vesicular-arbuscular mycorrhiza (VAM), which was in use for many decades, has been dropped in recognition that vesicles are formed by only 80% of AM fungi, but the name 'arbuscular' is currently retained, regardless of the structural diversity which is more and more widely appreciated.

An arbuscular mycorrhiza has three important components: the root itself, the fungal structures within and between the cells of the root and an extraradical mycelium in the soil. The last may be very extensive but does not form complex mycelial strands or rhizomorphs, nor any vegetative pseudoparenchymatous structures comparable to

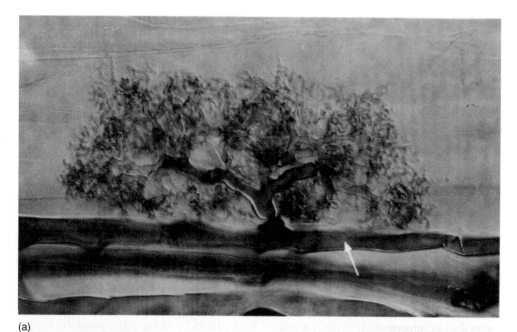

(a)

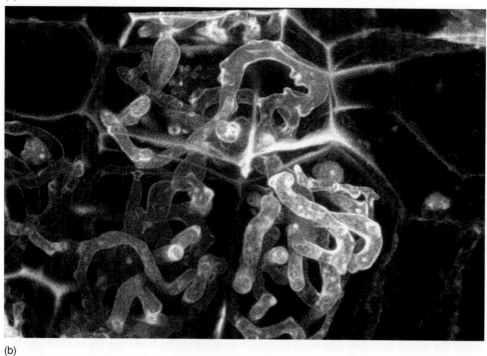

(b)

Figure 1.1 (a) A mature *Arum*-type arbuscule of *Glomus mosseae* within a cortical cell of *Allium porrum* (leek). The arbuscule has grown from a well-developed intercellular hypha (arrow). From Brundrett *et al.* (1984), with permission. (b) *Paris*-type intracellular coils of *Glomus intraradices* in cortical cells of *a* root of *Panax quinquifolius*, viewed by confocal microscopy. Note absence of intercellular hyphae. Reproduced from Peterson *et al.* (2004), with permission. See Chapter 2 for discussion of *Arum*- and *Paris*-type AM.

the fungal sheath typical of ectomycorrhizas (see Chapter 6). A few of the fungi do form sporocarps or clusters of spores with limited amounts of sterile mycelium. Because the characteristic fungal structures develop within the root and because changes in the rates of root growth and branching are discernible only by detailed comparison with non-mycorrhizal plants, it is not usually possible to tell if a root system is colonized without staining and microscopic examination or molecular probing for diagnostic DNA sequences.

Arbuscular mycorrhizas were first recognized and described in the last decades of the nineteenth century. Their widespread occurrence and common presence in plants of many phyla in most parts of the world, especially in the tropics, was realized very soon (Janse, 1897; Gallaud, 1905), but very little functional information was learnt about them until the mid-1950s. Almost all writings about the identity of the fungi until 1953 may be ignored, except for those of Peyronel (1923) who showed that hyphae of the endophyte could be traced to the sporocarps of species of fungi, then classified in the Endogonaceae, in the surrounding soil. Later, Butler (1939) in an influential review, agreed that the fungi called *Rhizophagus* were almost certainly imperfect members of the Endogonaceae, which then included the majority of fungi now transferred to the Glomeromycota. The work of Mosse (1953), which showed convincingly that mycorrhizal strawberry plants were colonized by a species of *Endogone* (later transferred to *Glomus*), may be said to have heralded the modern period. Soon Mosse, Baylis, Gerdemann, Nicolson and Daft and Nicolson greatly extended these early observations and demonstrated by inoculation that fungi in the Endogonaceae were symbiotic with many kinds of plants. Further information about the history of AM research is outlined in the highly readable 'A history of research on arbuscular mycorrhiza' by Koide and Mosse (2004).

A major milestone was reached in 1974 with a successful symposium on endomycorrhizas at which a number of key ideas were developed for the first time. Many of the papers presented at that meeting remain classics (Sanders *et al.*, 1975) and, together with the two previous editions of this book (Harley and Smith, 1983; Smith and Read, 1997), provide a general introduction to AM symbioses. While recent research has extended and confirmed the generalizations established by the early work, there has also been an increased appreciation of the genetic, structural and functional diversity to be found in AM fungi and the symbioses that they form and an important new emphasis on the cellular and molecular interactions between the symbionts and their roles in ecosystems.

In brief, AM fungi have been recognized as obligate symbionts of a very wide range of plant species. The symbioses are biotrophic and normally mutualistic, the long-term compatible interactions being based largely on bidirectional nutrient transfer between the symbionts, sometimes supplemented by other benefits such as drought and disease tolerance (see Chapters 5 and 6).

Arbuscular mycorrhizal fungi

General biology and development

The spores (Figure 1.2; see Colour Plate 1.1) formed by AM fungi are very large (up to 500 μm diameter), with abundant storage lipid, some carbohydrate and thick, resistant

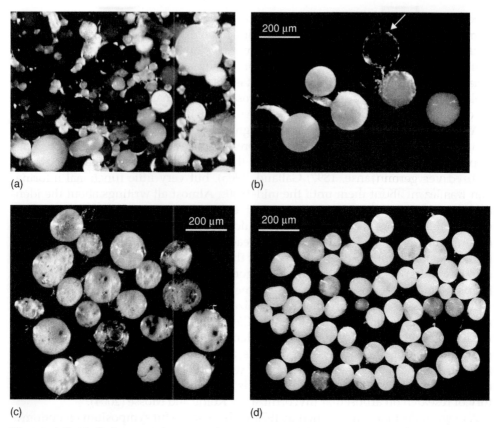

Figure 1.2 (a) Collection of spores of nine species of AM fungi isolated from a grassland that had developed from an abandoned agricultural site. Reproduced from Bever *et al.* (2001) Arbuscular mycorrhizal fungi: more diverse than meets the eye, and the ecological tale of why. *Bioscience* **51**, 923–931, Copyright, American Institute of Biological Sciences, with permission. (b) Spores of *Acaulospora laevis*. A translucent sporiferous saccule can be seen attached to one of the spores (arrowed). (c) A group of spores of *Glomus mosseae*. (d) A group of spores of *Gigaspora gigantea*. Images (b), (c) and (d) are reproduced courtesy of Joe Morton. See also Colour Plate 1.1.

walls which all contain chitin and, in some cases, β (1–3) glucan (Gianinazzi-Pearson *et al.*, 1994b; Lemoine *et al.*, 1995). Each spore contains a huge number of nuclei, with estimates ranging from 800 to about 35000 in different species (Hosny *et al.*, 1998). Most recent evidence indicates that, in *Glomus intraradices* and *G. etunicatum* and probably also *Scutellospora castanea*, all the nuclei are haploid as in other fungi, with no indication of variations in ploidy (Bianciotto *et al.*, 1995; Hijri and Sanders, 2004, 2005). The presumption is that this is the case with all AM fungi (but see below). DNA content per nucleus appears rather variable; amounts of DNA per spore and estimates of the number of nuclei give values of 1.7 and 3.4 pg for *Glomus versiforme* and *Scutellospora persica* (Viera and Glenn, 1990), whereas fluorimetry of the stained nuclei gives lower values of between 0.26 and 1.65 pg for a range of fungi from four genera (Bianciotto and Bonfante, 1992; Hosny *et al.*, 1998). The genome size of ~16.54 Mb (0.017 pg) in *G. intraradices* is low compared with other fungi but in some other AM species it is much larger, with a value of 1058.4Mb (1.1pg) reported for *S. pellucida* (Bianciotto and Bonfante, 1992; Hosny *et al.*, 1998). High values are unlikely to be the result of polyploidy, but rather of accumulation of repeated sequences (Hosny and Dulieu, 1994). Spores of many species, particularly in the Gigasporaceae, contain bacteria-like organisms (BLOs) as endosymbionts, revealed by electron microscopy and molecular analysis. Many are related to *Burkholdaria* and have been shown to possess functional bacterial genes including members of the *nif* operon, which is potentially involved in dinitrogen fixation. The BLOs can be vertically transmitted to new spore generations, but their functional significance in the life of AM fungi or in symbioses is yet to be resolved (Bianciotto *et al.*, 1996, 2004; Minerdi *et al.*, 2002).

As spores germinate, hyphal growth involves some nuclear division (Bianciotto and Bonfante, 1992; Bécard and Pfeffer, 1993; Bianciotto *et al.*, 1995), use of carbohydrate and lipid reserves (Bécard *et al.*, 1991; Bago *et al.*, 1999a) and production of limited amounts of branching, coenocytic mycelium, which is capable of anastomosis. In the absence of a host, plant growth eventually ceases, probably due to lack of signal molecules from the roots which have been shown to stimulate hyphal branching (Giovannetti *et al.*, 1993a; Buee *et al.*, 2000; Tamasloukht *et al.*, 2003; Akiyama *et al.*, 2005; Besserer *et al.*, 2006) (see Chapter 3). The growth arrest appears to be programmed, with controlled retraction of cytoplasm and nuclei and production of septa, allowing the spore and associated mycelium to retain long-term viability and the capacity to regerminate and colonize roots (Koske, 1981; Logi *et al.*, 1998). Once symbiosis with a host has been established, mycelial growth proceeds both within roots and in the soil and ultimately leads to the formation of new multinucleate spores terminally on the hyphae (see Chapters 2 and 3). Despite many efforts, AM fungi have generally proved to be unculturable and are unable to complete their life cycles without forming a symbiosis with an autotrophic plant. Nevertheless, one recent investigation (Hildebrandt *et al.*, 2006) reported the production of a small number of viable and infective spores by a *Glomus intraradices* isolate, co-cultured with two particular bacteria. This encouraging finding may lead to a better understanding of stimuli required for growth and sporulation.

The mycelium originating from a single spore is not likely to remain independent. Anastomosis between hyphae has been known for a century (Gallaud, 1905) and was observed in presymbiotic mycelium by Mosse (1959). Recently, detailed analysis has demonstrated anastomoses between hyphae originating either from spores or colonized roots, resulting in cytoplasmic continuity and nuclear migration (Giovannetti *et al.*, 1999, 2001, 2003, 2004; Giovannetti and Sbrana, 2001; de la Providencia *et al.*, 2005) (Figure 1.3a, b, c). The frequency of anastomosis and, hence, potential genetic exchange, is higher within a mycelium originating from a single spore than between mycelia emanating from different spores of the same isolate. Anastomosis between different geographic isolates of the same species or between different species has not been observed, but more work is needed with a wider range of fungi. Within *G. mosseae*, genetic distance revealed by vegetative incompatibility tests (Figure 1.3b) has been confirmed by total protein profiles and ITS RFLPs (Giovannetti *et al.*, 2003), suggesting that the tests may provide an assay to detect genetic relatedness among AM fungi. The occurrence of anastomoses has significant implications for understanding the genetics of AM fungi, as well as in establishing and maintaining mycelial networks involved in plant colonization and nutrient transfers (Jakobsen, 2004).

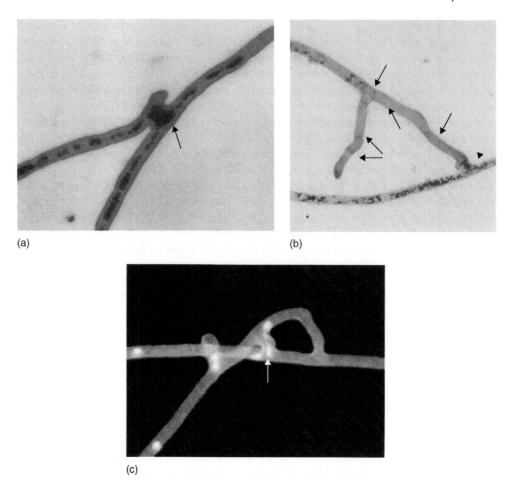

(a)   (b)

(c)

Figure 1.3 (a) Bright field micrograph of anastomosis (arrowed) between hyphae of the external mycelium of *Glomus mosseae*. Fusion of hyphal walls has occurred and cytoplasmic continuity is evidenced by staining indicating succinate dehydrogenase activity. From Giovannetti *et al.* (2001), The occurrence of anastomosis formation and nuclear exchange in intact arbuscular mycorrhizal networks. *New Phytologist* **151**, 717–724, with permission. (b) Bright field micrograph illustrating an incompatible interaction (arrowhead) between hyphae of *Glomus mosseae* originating from geographically different isolates. Hyphal contents have been withdrawn and septa (arrows) formed, as evidenced by absence succinate dehydrogenase activity. From Giovannetti *et al.* (2003), Genetic diversity of isolates of *Glomus mosseae* from different geographic areas detected by vegetative compatibility testing and biochemical and molecular analysis. *Applied and Environmental Microbiology* **69**, 616–624, with permission of the American Society for Microbiology. (c) Epifluorescence micrograph of anastomosis between two hyphae of *Glomus mosseae* following DAPI staining. Note presence of nuclei in the middle of the fusion bridge (arrow). From Giovannetti *et al.* (1999), Anastomosis formation and nuclear and protoplasmic exchange in arbuscular mycorrhizal fungi. *Applied and Environmental Microbiology* **65**, 5571–5575, with permission of the American Society for Microbiology.

Genetic organization

The genetic organization of AM fungi is not well understood, although rapid advances in unravelling long-standing problems are being made as new techniques clear the way to dealing with these 'difficult' organisms. The outline below can do no more than provide an overview of progress, acknowledging that much remains in doubt. As mentioned above, those AM fungi that have been studied are haploid; the suggestion that some may be polyploid (Pawlowska and Taylor, 2004) requires further investigation. Morphologically identifiable sexual structures have not been observed, leading to the assumption that the Glomeromycota are asexual and clonal. However, there is very high genetic diversity even in single isolates, as shown by the regular occurrence of several different ITS sequences in the nuclear rDNA of single multinucleate spores (Sanders *et al.*, 1995; Lloyd-MacGilp *et al.*, 1996; Lanfranco *et al.*, 1999; Antoniolli *et al.*, 2000; Pringle *et al.*, 2000; Rodriguez *et al.*, 2001; Jansa *et al.*, 2002b) as well as in AFLP patterns (Rosendahl and Taylor, 1997) and apparently in some genes involved in cellular function (Kuhn *et al.*, 2001; Sanders *et al.*, 2003; Pawlowska and Taylor, 2004). The way the variant sequences are arranged among the nuclei in a multinucleate spore or hypha is of considerable interest from the perspective of origin and transmission of genetic information to new generations of spores or different clonal mycelia. There are two possibilities; either the nuclei are genetically different and the organisms heterokaryotic or the nuclei are identical (homokaryotic), each containing all sequence variants (Figure 1.4).

There is evidence for both arrangements. The heterokaryotic model is supported by inheritance of spore shape in *Scutellospora pellucida* (Bever and Morton, 1999) and by the indication that nuclei in single spores of *S. castanea* contain different numbers of two divergent sequences in the ITS region of rDNA, as shown by fluorescent *in situ* hybridization (FISH) (Kuhn *et al.*, 2001). Furthermore, within-spore variation of a DNA polymerase-α (*POL1*-like sequence) in *G. etunicatum* seems unlikely to be explained by polyploidy (Hijri and Sanders, 2005). If these asexual organisms are heterokaryotic, the implication is that an assemblage of different nuclei, represent-

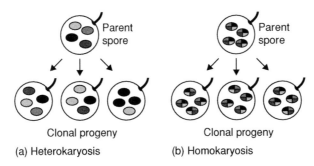

(a) Heterokaryosis   (b) Homokaryosis

Figure 1.4 Sorting of variants of a polymorphic marker in single-spore cultures of *Glomus* according to two models proposed to explain the organization of genetic variation in AM fungi. (a) Sorting under the heterokaryotic model; (b) sorting under the homokaryotic model. Variants of the polymorphic genetic marker contained in nuclei are represented by different shading. Reprinted by permission from MacMillan Publishers Ltd: Pawlowska and Taylor, *Nature* **427**, 733–737 (2004).

ing different genomes, is transmitted through the coenocytic mycelia and from one generation to the next. This would be an extremely unusual situation and hence requires critical investigation. Both Pawlowska and Taylor (2004) and Stukenbrock and Rosendahl (2005a) provide data which can only be aligned with the heterokaryotic model if frequencies of different nuclei are maintained in absolutely constant proportions during fungal development. This seems unlikely in a coenocytic and frequently anastomosing mycelium that may cover many meters (Rosendahl and Stukenbrock, 2004). Stukenbrock and Rosendahl (2005b) suggest that the conflicting evidence could be explained by differences between fungal species investigated or isolates and/or in the types of sequences (based on DNA or cDNA) used as markers to detect variation and heterokaryosis.

The genetic organization is important because the existence of multiple genomes (heterokaryosis) would be highly unusual and could pose problems for control of gene expression in the organism as a whole. The homokaryotic model requires refinement to explain the existence of marked genetic diversity in the absence of polyploidy. An understanding is relevant to determining the flow of genetic information within mycelia and between generations, the evolution of the group and also the mechanisms of control of gene expression which could be highly complex in a multigenomic organism.

If AM fungi are both ancient and asexual, then we need to ask how they avoid accumulation of large numbers of deleterious mutations (Judson and Normark, 1996). It may be that they are not asexual at all, but engage in cryptic sexual events or other processes with similar outcomes in 'cleaning up the genome', such as gene conversion or mitotic crossing over. Evidence for recombination has been sought and results indicate that it may be absent or extremely rare (Rosendahl and Taylor, 1997; Kuhn et al., 2001; Stukenbrock and Rosendahl, 2005a), but more frequent recombination is indicated by the analyses of Vandenkoornhuyse et al. (2001) and Gandolfi et al. (2003). Again, differences in methodology may underlie the disparate findings. It must be stressed that identification of recombination events does not tell us how often they occur or by what mechanism, or indeed how long ago. If AM fungi are completely asexual and recombination events are rare, then it follows that mutation (and possibly heterokaryosis) provides the main bases of the variation necessary to permit adaptation to environmental change and continuing evolution. This may not be a serious problem because the pressures for change are likely to be lower in mutualistic symbioses than in parasitic interactions, for which it is argued that the host and parasite need to recombine and evolve continually to keep pace with one another (Taylor et al., 1999). Indeed, Law and Lewis (1983) suggested that, in mutualistic symbioses, the endobiont (in this case the fungus) will evolve away from a sexual habit because the selection pressures will be to maintain similarity to, rather than difference from the parents. Superficially, this seems to fit with the situation in AM fungi, for which the important ecological niche for organic carbon (C) acquisition is the apoplast of the root cortex, in which homeostasis exercised by the plant will maintain reasonably constant environmental conditions. However, the idea has been criticized because it ignores the extraradical phase of the fungus that is subjected to fluctuating soil conditions; it includes the implicit assumptions that there is little diversity in the physiological outcomes of the symbiosis for either host or fungus; and it assumes that specificity in the interactions between AM fungi and their hosts is low. Any variations in these characters would lead to pressures selecting for more advantageous partnerships and hence pressures for recombination and the evolution of sex.

Rather little attention has been paid to the relationship between the genetic diversity described above and the phenotypic diversity which has also been shown to be high both between species of AM fungi and among isolates of a single species. One investigation showed that in a group of *G. intraradices* isolates from the same site the genetic diversity was greater than phenotypic diversity in hyphal growth and spore production expressed in monoxenic root organ culture (Koch *et al.*, 2004). It will be interesting to determine whether there is real genetic redundancy in terms of function in symbiosis with whole plants growing in soil or whether the findings are a consequence of the uniform and artificial culture environment. The critical importance of phenotypic diversity in symbiotic outcomes and consequences for plant interactions in ecosystems will be discussed in later chapters. In any event, the extent of both specificity and phenotypic diversity is probably much greater than previously realized, with important consequences for pressures leading to evolution.

Systematics and phylogeny

The difficulties of culturing AM fungi and of identifying vegetative stages of different taxa in soil or roots has meant that, until very recently, classification was based almost entirely on the development and wall structure of the spores. These can be collected from almost any soil and used to establish cultures on plant hosts ('pot cultures'), providing spores for further analysis or inoculum for experiments. Many cultures have deliberately been started from single spores to ensure that only single species were cultured. In consequence, isolates raised in this way are likely to have a much narrower genetic base than the species they are presumed to represent (Clapp *et al.*, 2002).

It was assumed until recently that AM fungi were most closely related to the Zygomycota because of their aseptate hyphae and, despite absence of typical sexual stages, this view was not challenged until the use of DNA sequences forced a re-evaluation of their relationships. The first Linnaean classification of the Endogonaceae (Zygomycota) made no attempt to relate taxonomy to the phylogeny of the group (Gerdemann and Trappe, 1974, 1975). Gerdemann and Trappe regarded their classification as a 'temporary solution to a difficult taxonomic problem', but it was important because it put the study of AM fungi on a firm basis, with descriptions of four genera, *Gigaspora, Acaulospora, Glomus* and *Sclerocystis*. The number was later increased to five, with the addition of *Scutellospora*. The classification allowed researchers to refer to the species of fungi used in their investigations, rather than using descriptive terms such as 'E3' and 'yellow vacuolate' as had been the common practice. Up to 1993, about 150 species, most widely distributed globally, had been described and, although there has been a major re-evaluation of the taxonomy and phylogeny of the group, rather few new species have been added. Considering the long evolutionary history, the paucity of species is surprising as it suggests remarkably little diversification. However, as we have seen, a single species can, in fact, encompass very high genetic and functional diversity.

The structure and development of the spores, particularly their walls, remain important components of taxonomic descriptions and phylogenetic analysis. The first approach to a classification representing phylogenetic relationships was made by Morton (1990a) using cladistic tools and assuming evolutionary significance for 27 characters. A new order, Glomales, was separated from the Endogonales (but still included in the Zygomycota) and defined as a monophyletic group containing only those fungi for which 'carbon is acquired obligately from their host plants via intraradical dichotomously branching arbuscules' (Morton and Benny, 1990). Although this was an attempt to define AM fungi as a separate group, it was widely recognized as a very restrictive definition because it makes a number of assumptions which have not been substantiated physiologically (Gianinazzi-Pearson *et al.*, 1991a; Smith and Smith, 1996) and it also created some practical difficulties. It was assumed that the arbuscule is a key unifying structure and that it is the sole site of C acquisition by the fungi. This is by no means certain, for there is no *a priori* reason why intercellular hyphae or intracellular coils should not be sites of organic C transfer (see Chapter 4). There are other difficulties: the definition requires that all descriptions of AM fungal species should be accompanied by evidence that the fungi can form arbuscules, which is problematic considering that, in some symbioses, hyphal coils predominate; fungi with typical spore development and vegetative morphology, but 'atypical' C transfer also exist (see Chapter 13); and the fungal partner in the unique cyanobacterial (*Nostoc*) symbiosis *Geosiphon pyriformis* is certainly related to AM fungi on the basis of DNA sequence information as well as spore morphology, but forms no structures similar to those in roots (Schüßler *et al.*, 1994; Schüßler, 2002). Setting aside these problems, the remaining characters used by Morton and Benny were based on spore characteristics that vary qualitatively and are stable and discrete. Vegetative structures (other than existence of arbuscules) were not used because of their developmental plasticity and variations within different host plants. Application of the cladistic approach to determination of evolutionary relationships yielded the phylogenetic tree shown in Figure 1.5a, which has served as a hypothesis for subsequent investigations.

The key features of the tree are the separation of the *Glomus/Sclerocystis* group from *Gigaspora/Scutellospora* and the existence of *Acaulospora/Entrophospora* as a line apparently diverging from *Glomus*. Initial DNA sequence analysis provided evidence confirming the 'glomalean' fungi as true fungi of monophyletic origin, divided into the same three families shown by the cladisitic approach. Lipid analysis also supported the existence of three families (Sancholle and Dalpé, 1993), but the carbohydrates in the fungal walls suggested a less clear-cut phylogeny and, significantly, called into question the relationships of AM fungi with the Zygomycota (Gianinazzi-Pearson *et al.*, 1994b; Lemoine *et al.*, 1995).

Development of molecular methods has transformed our knowledge of the evolution and phylogeny of AM fungi. Using SSU rRNA sequences, Simon *et al.* (1993)

Figure 1.5 Phylogenetic trees for AM fungi, derived from different types of information. (a) A cladogram showing taxonomic and phylogenetic divergence among genera of AM fungi, based on comparative developmental sequences of the spores. Courtesy Joe Morton. (b) A maximum likelihood tree based on molecular data from near full-length SSU rRNA gene sequences, which reflects the topology of different neighbour joining and parsimony bootstrap analyses, showing the phylogeny of the AM fungi (Glomeromycota). Courtesy Arthur Schüßler, modified from Schüßler (2002).

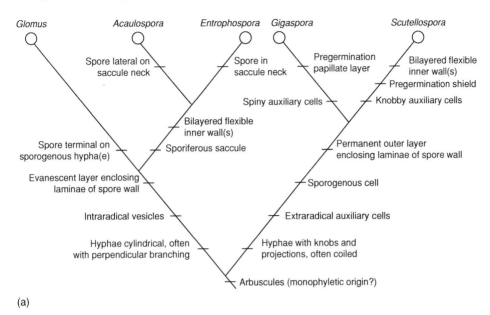

(a)

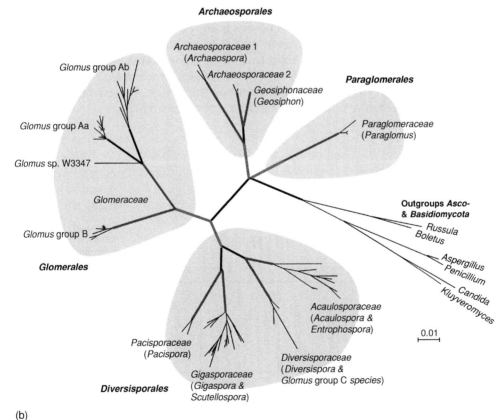

(b)

Figure 1.5 (Caption opposite)

dated the probable origin of the group to between 460 and 350 million years ago, which appeared to accord well with the fossil record of AM colonization in vascular land plants (Figure 1.6 and see below). Reappraisal now puts the timing much earlier, at between 1400 and 1200 million years ago, and the origin of the first land plants at around 900 million years ago, but there is still some uncertainty because of the absence of fossils with which to calibrate the molecular data (Heckman *et al.*, 2001; Schüßler, 2002). In any event, the discovery that the fungus *Geosiphon pyriformis* appears to be ancestral to other Glomeromycota suggests a role for these fungi in very early plant colonization of land, via symbiosis with cyanobacteria (Gehrig *et al.*, 1996).

A major advance in phylogeny was enabled by the availability of large numbers of sequences of SSU rRNA genes of a wide range of fungi, including many members of the Zygomycota and AM fungi, as well as the very well studied basidiomycetes and ascomycetes. The new data, taken with molecular, morphological and ecological characteristics, unequivocally show that AM fungi are quite different from other fungal groups and should be separated from them in a new monophyletic clade, given the status of phylum. According to SSU sequence phylogeny, this phylum, designated as Glomeromycota, probably diverged from the same common ancestor as the Ascomycota and Basidiomycota, but is not related to the Zygomycota (Schüßler *et al.*, 2001) (see Figure 1.5b). The monophyletic status of the Glomeromycota based on SSU sequences is supported by studies of β-tubulin, actin and elongation factor alpha gene phylogenies, although these suggest that the sister groups may not be ascomycetes and basidiomycetes, but chytridiomycetes or zygomycetes (*Mortierella*) (Helgason *et al.*, 2003; Corradi *et al.*, 2004). The confirmation of ascomycetes and basidiomycetes as the closest living relatives of the Glomeromycota therefore remains a subject for future research.

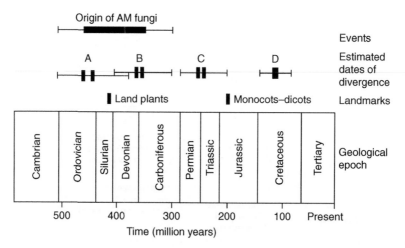

Figure 1.6 Estimated dates of origin and divergence of arbuscular mycorrhizal fungi. From Simon *et al.* (1993). Reprinted with permission from *Nature*, **363**, 67–69. Copyright McMillan Magazines Ltd.

At a finer scale, SSU sequence phylogeny supports the existence of four groups, given the status of order by Schüßler *et al.* (2001) (see Figure 1.5b). The Glomerales (formerly Glomales) as understood by Morton and Benny (1990) is now represented by two orders, the Glomerales containing two families Glomus group A and Glomus group B, and the Diversisporales containing Acaulosporaceae, Diversisporaceae, Gigasporaceae and Gerdemanniaceae (Schüßler *et al.*, 2001; Walker *et al.*, 2004). Two more ancestral lineages are also recognized, Archaeosporales, containing Archeosporaceae and Geosiphonaceae, and Paraglomerales, currently containing a single family, the Paraglomeraceae. These taxa are well supported by the SSU sequence analysis. Of the very well known groups, the Gigasporaceae is monophyletic and contains *Gigaspora* and *Scutellospora*, but the genus *Glomus* is not monophyletic, with members distributed in at least four families, Glomus group A, Glomus group B, Diversisporaceae (Schwarzott *et al.*, 2001) and Gerdemanniaceae (Walker *et al.*, 2004).

A species concept for AM fungi

The asexual and clonal nature of members of the Glomeromycota means that the Biological Species Concept is inappropriate and they should be regarded as phenetic or form species (Morton, 1990b; Walker, 1992). The increasing availability of SSU rDNA sequences means, as we have seen, that the classification and phylogeny based on spore characteristics (Morton *et al.*, 1992) can be supplemented by sequence data, providing a species with a range of 'sequence variants' that represent its molecular diversity and 'bar codes' by which it or its isolates could be identified. This information has already proved useful in ecological investigations of the distribution of AM fungi among root systems of different species and in different environments. The low number of species (~150) in the Glomeromycota based on spore characters superficially might suggest very low diversity in the group. However, it is becoming increasingly apparent that the high genetic diversity among different isolates of one phenetic species also extends to phenotypic diversity at the levels of development, function and symbiotic performance (Hart and Reader, 2002a; Koch *et al.*, 2004; Munkvold *et al.*, 2004). The genetic control of developmental variation is highlighted by different abilities of isolates of *Glomus intraradices* (separated on the basis of rDNA sequences as well as origin) to enter roots of a reduced mycorrhizal colonization (*rmc*) mutant of tomato. Most isolates are blocked at the root epidermis, but *G. intraradices* WFVAM23 is able to develop normally in the root cortex and forms a functional symbiosis (Gao *et al.*, 2001; Poulsen *et al.*, 2005; Jansa *et al.*, unpublished). If phenotypic diversity is high within a phenetic species, then we need to question the value of the species name in terms of understanding its evolution, characterizing it phenotypically and functionally and comparing different investigations purporting to use the same fungus. A major challenge for the future is to provide a workable species concept for the Glomeromycota and the tools to permit rapid identification and description of species or subspecific variants so that names have developmental and functional relevance and predictive capacity in relation to symbiosis. In this context, the variations in ability of AM fungi to anastomose within and between isolates and species, if linked with information on sequence similarities and symbiotic function, may be an avenue worth exploring.

The range of plants forming arbuscular mycorrhizas

The range of potential plant partners for AM fungi is extremely wide and has been responsible for the oft-quoted statement (Gerdemann, 1968) that 'the symbiosis is so ubiquitous that it is easier to list the plant families in which it is not known to occur than to compile a list of families in which it has been found'. This continues to hold good. Some members of most families of angiosperms and gymnosperms, together with sporophytes of ferns and lycopods develop arbuscular mycorrhizas. Additionally, the free-living gametophytes of pteridophytes, as well as those of some hepatics (liverworts, see Chapter 14), are often colonized by fungi now identified as members of the Glomeromycota, regardless of structural features of the colonization and photosynthetic capacity of the plant. One fungus from the Glomeromycota engulfs the cyanobacterium *Nostoc* to form the rare bladder-like symbiosis known as *Geosiphon piriformis* (Schüßler *et al.*, 1994) (Figure 1.7; see Colour Plate 1.2).

On the other hand, some plants characteristically do not form mycorrhizas of any type and it is possible to generalize that colonization is unlikely to occur in

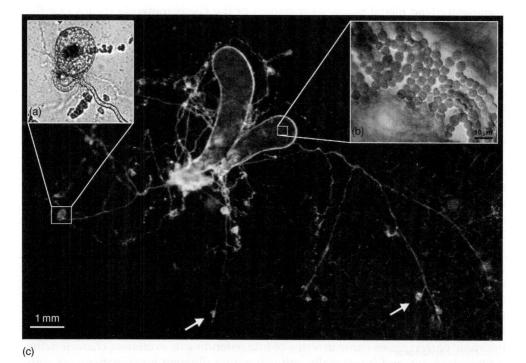

Figure 1.7 *Geosiphon pyriformis*, a glomeromycotan soil fungus forming endosymbiosis with a cyanobacterium and representing an *in-vitro* culture of a mycorrhiza-like association. (a) Spore and subtending hypha of *G. pyriformis*; (b) cyanobacterial partner; (c) bladders of *G. pyriformis*, with emanating external mycelium. From Schüßler and Wolf (2005), *Geosiphon pyriformis* – a glomeromycotan soil fungus forming endosymbiosis with Cyanobacteria. In *In Vitro Culture of Mycorrhizas*. Eds S Declerck, DG Strullu, JA Fortin pp. 272–289, with kind permission of Springer Science and Business Media. See also Colour Plate 1.2.

several major families, including Chenopodiaceae, Brassicaceae, Caryophyllaceae, Polygonaceae, Juncaceae and Proteaceae. However, even in these, colonization of the roots varying from both arbuscules and vesicles to scattered hyphae and perhaps vesicles, is sometimes observed (see below). The physiology of these inter-actions has rarely been explored and, in consequence, it is unwise to be too categori-cal about the functional characteristics of the symbioses. These may well vary from parasitic utilization of organic C, through neutral interactions to typical mutualistic symbioses (see Chapters 4 and 5).

Trappe (1987) produced a most valuable compilation of the incidence of all types of mycorrhizas within the angiosperms, taken from published material. Records of arbuscular mycorrhizas are to be found in all the orders from which plants have been examined and are about equally frequent in Dicotyledonae and Monocotyledonae (Table 1.1). He stressed that only about 3% of species had actu-ally been examined at that time and our knowledge of the mycorrhizal status of some taxa is still very poor, despite an increasing number of surveys. Nevertheless, the more we look the greater number of species turn out to be mycorrhizal, even in supposedly non-mycorrhizal families and in habitats such as wetlands, salt marshes and arid or disturbed environments, where the incidence of mycorrhizas has long been viewed as relatively unlikely (Bagyaraj *et al.*, 1979; Clayton and Bagyaraj, 1984; McGee, 1986; Khan and Belik, 1995; O'Connor *et al.*, 2001; Carvalho *et al.*, 2004). Harley and Harley (1987a, b) surveyed the literature on the incidence of

Table 1.1 Numbers and percentages of species of subclasses and classes of Angiospermae examined for mycorrhiza formation and percentage of examined species by type of mycorrhiza.

Taxon				Per cent with mycorrhizal types			
	Total species	Species examined	Per cent examined	AM only	Other	AM+ other	NM
Division Angiospermae	223 400	6507	3	50	15	5	18
Class Dicotyledonae	173 500	5020	3	50	14	6	17
Subclass							
Magnoliidae	12 000	270	2	66	3	4	17
Hamamelidae	3400	265	8	27	44	11	6
Caryophyllidae	11 000	317	3	14	4	2	59
Dilleniidae	25 000	792	3	33	29	7	20
Rosidae	62 100	1838	3	56	16	5	12
Asteridae	60 000	1538	3	63	2	5	15
Class Monocotyledonae	49 900	1487	3	49	18	2	21
Subclass							
Alismatidae	500	26	5	4	0	0	88
Arecidae	5600	61	1	56	3	3	30
Commelinidae	15 000	826	6	55	1	2	28
Zingiberidae	3800	28	1	71	4	0	11
Lilliidae	25 000	546	2	37	48	2	7

Based on Trappe, 1987. Plant classification according to Chronquist (1981). AM: arbuscular mycorrhiza; Other: mycorrhizas formed by ascomycetes and basidiomycetes; NM: non-mycorrhizal (non-host).

mycorrhizas at the species level in the very-well-studied British flora. In many families, over 40% and sometimes as high as 80% of the species had been investigated, often more than once. All families listed contained mycorrhizal species and, frequently, these constituted a very high proportion of the total. Similar trends are repeated in most surveys carried out more recently (Fitter and Moyersoen, 1996; Wang and Qiu, 2006). In many cases, species have been recorded as occurring in both mycorrhizal and non-mycorrhizal states and members of some plant families typically form mycorrhizas of types other than AM or, indeed, more than one type of mycorrhiza (Newman and Reddell, 1987; Wang and Qiu, 2006). The reasons for failure of a potentially mycorrhizal species to become colonized are many and include lack of inoculum of an appropriate fungus at the site, environmental conditions such as high nutrients, cold or waterlogging and seasonal variation in the development of the fungi in roots. Species which are sometimes, but not always, colonized are often referred to as 'facultatively mycorrhizal', to distinguish them from those 'obligately mycorrhizal'[1] species that are consistently colonized. It is best to avoid using these terms in relation to the extent to which plants respond to colonization, because responsiveness of a plant (whether positive or negative) is markedly influenced by the identity of the AM fungus colonizing the roots and by the environmental conditions (Chapter 4).

Arbuscular mycorrhizas are found in most herbaceous species that have been studied (see above for exceptions), but are by no means restricted to herbs. As long ago as 1897, Janse examined 46 species of tree in Java and found them all to have arbuscular mycorrhizas. More recent work, reviewed by Smits (1992), Janos (1987) and Wang and Qiu (2006), confirms the widespread occurrence of AM in trees. Even the Dipterocarpaceae, long thought to be exclusively ectomycorrhizal (Smith and Read, 1997), has recently been suggested to include as many as 75% of species that may form arbuscular mycorrhiza (Tawaraya *et al.*, 2003; Wang and Qiu, 2006), confirming the prevalence of AM symbioses in taxonomically diverse tropical forests, as well as in some temperate forest systems. Thus, Baylis (1961, 1962) stated that arbuscular mycorrhizas are ecologically the most important type of mycorrhiza in New Zealand forests and this can sometimes be true also of northern temperate broadleaf forests, even when the canopy dominants form ectomycorrhizas.

In general terms, arbuscular mycorrhizas are characteristically found in species-rich ecosystems, in contrast to ectomycorrhizas which predominate in forest ecosystems where only a few host species are present. Whereas the Pinaceae are predominantly ectomycorrhizal (ECM), all other conifer families are mainly AM, as are most other gymnosperms, all of which are woody. Although arbuscular mycorrhizas have often been ignored by foresters, they are characteristic of such valuable trees as *Acer*, *Araucaria*, *Podocarpus* and *Agathis*, as well as all the Cupressaceae, Taxodiaceae, Taxaceae, Cephalotaxaceae and the majority of tropical hardwoods. The importance of considering the appropriate mycorrhizal associates for trees used in reafforestation programmes in temperate and tropical ecosystems is now widely recognized

---

[1] We avoid the use of 'mycotrophic' which is both vague and potentially confusing. The term is used in at least two different ways: to indicate whether or not a plant becomes colonized by mycorrhizal fungi, which can be determined by microscopic investigation or molecular ecological techniques; and/or to indicate whether a plant is 'responsive' to colonization, which depends greatly on identity of the fungus involved.

(see Chapter 17). While most of the experimental work on arbuscular mycorrhizas has been done with herbs, some trees have also been used and include *Malus* (apple), *Citrus*, *Salix*, *Populus*, *Persea* (avocado), *Coffea*, *Araucaria*, *Khaya*, *Anacardium* (cashew), *Liquidambar*, *Acacia* and many others. It is certainly important to realize that arbuscular mycorrhizas may be significant in nutrient absorption and in nutrient cycling of woody species. Therefore, work with trees and other perennials, as well as in associated understorey plants, is very important both from an ecological point of view and from a need to consider forest and crop production. Indeed, although work with herbs allows greater control of conditions in growth rooms etc., the propagation of some woody species from cuttings may have great advantages in providing genetically uniform experimental material which may partly offset the lengthy growth periods necessary for the study of long-lived plants. The extensive work on *Citrus* mycorrhiza, by groups in California and Florida, exemplifies the experimental use of an economically important tree species with a view both to increase production and to understand the development and physiology of the AM symbiosis (Menge *et al.*, 1982; Graham and Syvertsen, 1985; Graham, 1986; Eissenstat *et al.*, 1993; Graham *et al.*, 1997). It is to be expected that the genome mapping project in *Populus*, which forms both ecto- and arbuscular mycorrhizas (Martin *et al.*, 2004; Strauss and Martin, 2004), will provide a rich avenue for future exploration of AM symbioses with woody plants.

Strong emphasis has been placed on angiosperms and gymnosperms largely because of their significance in ecosystems and in primary production, but AM symbioses are also characteristic of fern sporophytes, as was well recognized by Boullard (1951, 1958) and later by many others who have examined the group or included them in more general surveys (Cooper, 1976). Sporophytes of Lycopodiaceae and Psilotacae are now known to be AM and, although information on gametophytes is more scanty, those of most of these families, plus many thalloid liverworts are also characteristically AM (Read *et al.*, 2000; Winther and Friedman, 2007). The gametophyte stages provide a fascinating area for study because some are achlorophyllous and hence require an external source of organic C, possibly supplied by their AM fungal symbionts. Accordingly, these symbioses may have carbohydrate physiology in which the fungus supplies the plant with organic C (see Chapters 13 and 14).

Recent surveys worldwide have greatly increased the number of species known to form more than one kind of mycorrhiza (Wang and Qiu, 2006). Indeed, it is possible that, due to their ancient origin and significance during terrestrial plant evolution, AM fungi may have the ability to invade the underground organs of almost all land plants. It follows that failure of a plant to form arbuscular mycorrhizas or to form another type is likely to be a derived or secondary character and all plant genomes may carry evolutionary footprints of present or former AM status. Evidence from the phylogeny of host plants and the fossil record supports this conclusion (see below).

Phylogenetic relations of arbuscular mycorrhizal plants

The first significant attempt to examine the mycorrhizal status of plants in relation to their phylogeny was made by Trappe (1987) who, using only those angiosperm taxa for which the mycorrhizal status was known in at least 10% of the species and

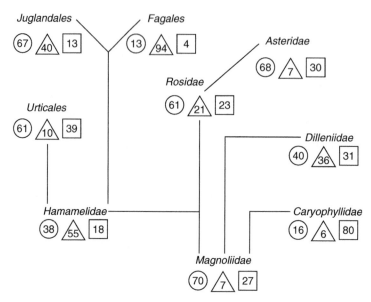

Figure 1.8 Phyologenetic dendrogram for the subclasses of Dicotyledonae, with percentage of species with arbuscular mycorrhizas (circles), asco- and basidiomycetous (ericoid, and ecto) mycorrhizas (squares), or no mycorrhizas (triangles). Many species have mycorrhizas in more than one category, so that percentages total more than 100. From Trappe (1987) in *Ecophysiology of VA Mycorrhizal Plants*, ed. GR Safir. Copyright CRC Press, Boca Raton, Florida.

the phylogenetic classifications of Cronquist (1981), prepared dendrograms which allow some preliminary evolutionary conclusions to be drawn (Figure 1.8).

More recent studies (Fitter and Moyersoen, 1996; Brundrett, 2002; Wang and Qiu, 2006), using the phylogenetic classification system of angiosperms based on molecular sequence data (Soltis *et al.*, 1992; Stevens, 2004) have come to very similar conclusions. Using reports from the literature, Wang and Qui surveyed 3617 species from 263 families for mycorrhizal status of any type. They show that, among land plants as a whole, 80% of species and 92% of families potentially form at least one type of mycorrhiza. For angiosperms, the figures are 84 and 94%, closely similar to numbers reported by Trappe (1987) (Tables 1.1 and 1.2). Arbuscular mycorrhizas appear as the ancestral type, occurring in the vast majority of land plant species and, very frequently, in all the early diverging lineages of the major clades. Mycorrhizal types other than AM and non-mycorrhizal species appear in lineages of more recent origin. Orchid mycorrhizas appear to have evolved only once, but ecto- and ericoid mycorrhizas probably had independent origins in many unrelated plant lineages. The same is true of both mycoheterotrophs (achlorophyllous plants in symbiosis with various fungal associates including fungi in the Glomeromycota, see Chapter 13) and non-mycorrhizal plants.

Non-mycorrhizal plants

Supposedly completely non-mycorrhizal plant families appear among mosses, ferns and many families of angiosperms distantly related to each other. Additionally,

Table 1.2 Mycorrhizal status of four major groups of land plants, indicating numbers and percentages of the families and species that have always (obligate), sometimes (facultative) and never (non-mycorrhizal) been observed to form mycorrhizas.

Group	Number of families/species surveyed	Number (percentage) of mycorrhizal families	Number (percentage) of obligate mycorrhizal species	Number (percentage) of facultative mycorrhizal species	Number (percentage) of non-mycorrhizal families	Number (percentage) of non-mycorrhizal species
Bryophyta	28/143	20 (71)	60 (42)	6 (4)	8 (29)	77 (54)
Pteridophyta	28/426	26 (93)	185 (43)	39 (9)	2 (7)	202 (47)
Gymnospermae	12/84	12 (100)	83 (99)	1 (1)	0 (0)	0 (0)
Angiospermae	195/2964	184 (94)	2141 (72)	396 (3)	11 (60)	427 (14)
Total	263/3617	242 (92)	2469 (68)	442 (12)	21 (8)	706 (20)

Modified from Wang and Qiu (2006). For more detailed information see the original publication. Angiosperm classification according to Soltis et al. (1992).

some families have both mycorrhizal and non-mycorrhizal members and, even in predominantly non-mycorrhizal groups, some mycorrhizal species do occur. As these groups include the Brassicaceae, Caryophyllaceae, Proteaceae and Cyperaceae, all traditionally thought of as wholly non-mycorrhizal, it is clear once again that we must not be too dogmatic about the mycorrhizal status of a species or group that has not actually been examined.

The independent (polyphyletic) origin of the non-mycorrhizal condition in different plant lineages suggests that the condition is secondary and that the cellular and physiological bases for it may be quite diverse. Although families containing large numbers of species in which mycorrhizal colonization is characteristically absent are relatively rare, they are worth studying in their own right, with respect to environmental conditions that may have led to loss of the symbiosis, mechanisms by which the fungi are excluded, as well as the means by which the plants acquire nutrients from soil and compete effectively with mycorrhizal plants in ecological situations (see Chapters 3, 15 and 16) (Lamont, 1981, 1982; Pate, 1994; Marschner, 1995; Lambers et al., 1998). It is thought that several factors may have led to loss of mycorrhizal status, including adaptation to aquatic habitats and growth in nutrient rich, extremely nutrient poor or disturbed environments. These plants often appear to have evolved compensatory mechanisms of effective nutrient uptake, such as fine roots with well developed root hairs, proteoid and dauciform roots and exudates that increase the solubility of P in the soil. However, as noted above, complete absence of mycorrhizas from either particular taxonomic or ecological groups of plants is not observed.

Fossil history of arbuscular mycorrhizas

Phylogenetic arguments about mycorrhizal origins are supported by the fossil record. The long fossil history of both spores of AM fungi and colonized absorbing structures is well recognized. Fossils resembling the spores of AM fungi date from as early as the Silurian and Ordovician (440–410 million years ago) (Redecker et al., 2000) and *Glomus*-like spores are also common within plant axes or decaying plant material from the famous and beautifully preserved Rynie chert flora (Kidston and Lang, 1921), which is approximately 400 million years old. This flora includes some of the earliest terrestrial tracheophyte fossils. Recent re-examination has revealed arbuscules, vesicles and intercellular hyphae within the protosteles of sporophytes of several fossil species including *Aglaeophyton major*, as well as a distinct tissue layer containing arbuscules in the gametophytes of the same plant (*Leonophyton rhynensis*), leaving no doubt that arbuscular mycorrhizas had evolved by that time (Remy et al., 1994; Taylor et al., 2004, 2005) (Figure 1.9a). The presence of AM structures in these early tracheophytes agrees with the supposition that arbuscular mycorrhizas were the first of all mycorrhizal types to evolve and that the AM condition is ancestral for land plants. Even earlier associations of Glomeromyota with plants are suggested by the cyanobacterial symbiosis *Geosiphon* and by the AM status of many present-day liverworts. These supposed links with early colonizers of land have led to the assumption that AM symbioses have continued in an unbroken line from the earliest, pre-tracheophyte land flora and that ascomycete or basidiomycete symbioses with modern liverworts came later. Some doubt has been cast on this simple view, based on structural features of colonization and the association of some modern liverworts with members of the relatively recently evolved Glomus group A (Selosse, 2005; see Chapter 14).

The significance of the fungal associates of the sporophyte fossils was noted by Kidston and Lang themselves (1921) and by Nicolson (1975). Subsequently Pirozynski and Dalpé (1989) provided a critical review of the geological history of the fungi. This shows continuous occurrence of *Glomus*-like structures into the Quaternary period and the occurrence of *Glomus*-like spores from silicified peat from the Triassic deposits in the Antarctic (Stubblefield *et al.*, 1987a, 1987b). Unfortunately, no reliable fossils of other genera in the Glomeromycota have been found which would shed more light on phylogenetic relationships within the group.

The fossil record of symbiotic organs in later Carboniferous deposits has revealed many gymnosperm fossils with arbuscular mycorrhizas. The best known and preserved is *Amyelon radicans*, which again resembled the arbuscular mycorrhizas of living gymnosperms (Nicolson, 1975). The Triassic flora from Antarctica (250–210 million years ago) has also yielded important evidence for the development of intraradical vegetative structures, including intercellular hyphae and arbuscules and well-developed intracellular coils (Stubblefield *et al.*, 1987a, 1987b; Phipps and Taylor, 1996). Beautifully preserved roots of *Antarcticycas*, containing both septate and aseptate hyphae and structures strongly resembling mycorrhizal arbuscules, vesicles and spores have been described by Stubblefield *et al.* (1987a). Sections and peels of these fossils are virtually indistinguishable from present-day mycorrhizal cycad roots (Figure 1.9b, c, d).

The emphasis has been on identifying the presence of arbuscules as easily recognized and supposedly defining features of arbuscular mycorrhizas and some variation in arbuscular structure has been noted. Those of *Aglaeophyton* are delicate structures with secondary branches of 1–2 μm diameter, similar to present-day arbuscules in *Arum*-type mycorrhizas (see Chapter 2), but different from the better preserved arbuscules found in *Antarcticycas* from the Triassic (compare Figures 1.1a and 1.9a and c). With the exception of the work of Phipps and Taylor (1996) on cycads, little attention has been given to identifying the diversity of structures, including coils, which are increasingly recognized as providing important intracellular interfaces between AM fungi and their hosts (*Paris*-type mycorrhizas) in modern plants, including the mycotrophic gametophytes of some hepatics and pteridophytes (see Chapters 2 and 14). Coils are of course harder to diagnose as 'AM', but would provide a significant avenue of research into structural diversity in the early land flora and might provide some evolutionary insights into development of different morphological types.

We cannot, of course, be sure about the physiology of these fossil arbuscular mycorrhizas but, if they functioned in a manner similar to present-day forms, their role in colonization of the land and in subsequent plant evolution is likely to have been considerable. The early land plant sporophytes did not have roots and their underground protosteles were poorly developed with respect to accessing nutrients from soil. The gametophytes had slightly swollen basal protocorms bearing rhizoids which, again, would have had limited access to the substratum. The soils colonized by early land plants were likely to have been poorly developed and deficient in available mineral nutrients. In consequence, it is likely that mutualistic symbioses with fungi in the Glomeromycota were important to the success of the autotrophic plants invading the terrestrial environment (Baylis, 1972; Nicolson, 1975; Pirozynski and Malloch, 1975).

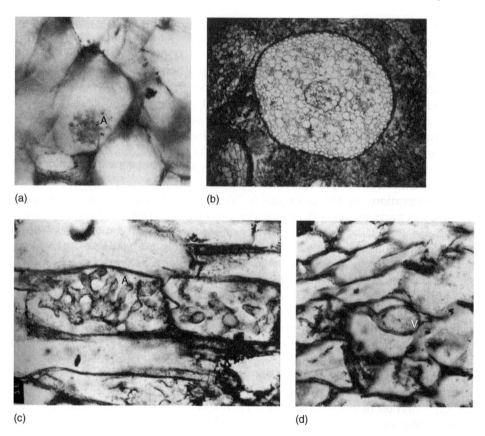

Figure 1.9 Fossil arbuscular mycorrhizas. (a) Arbuscule (A) in a cell of *Aglaeophyton* from the Devonian flora of the Rhynie Chert. From Remy *et al.* (1994). Copyright, National Academy of Sciences. (b) Transverse section of a mycorrhizal root of *Antarcticycas*, from the Triassic deposits of Antarctica. Note the colonized central cortex of the root. (c) and (d) Details of colonization in *Antarcticycas*. (c) Dichotomously branched arbuscule (A), with relatively robust branches; (d) vesicle (V). (b), (c), (d) from Stubblefield *et al.* (1987a; 1987b), with permission.

Fungus–plant specificity

So far, we have confined the discussion to the potential of different taxa of plants to form arbuscular mycorrhizas in field or experimental conditions without concerning ourselves greatly with questions about whether a species is always AM, how extensively the roots are colonized, or how far the plant may be dependent on the AM fungal symbionts for growth or reproductive success. These are complex issues which are important in discussions of cellular interactions and plant–fungus specificity and compatibility, as well as of ecology. Here we consider the range of potential partners that are available for plant and fungal symbionts. This aspect of specificity needs to be considered at both taxonomic and ecological levels. Taxonomic specificity indicates whether a given species of fungus can form an AM relationship with more than one species of plant or whether a given species of plant associates with more than one species of fungus. This can be extended to lower taxonomic levels, where subspecific genetic strains or isolates of a fungus may form AM attuned in some way to the species or subspecific genetic strains of the plant. Molecular tools will enable this issue to be explored.

The early investigators Magrou (1936), Stahl (1949) and Gerdemann (1955) reached the conclusion that there is no absolute specificity between taxa of AM fungi and taxa of potential host plants and until recently almost all research continued to support this view. Specificity appeared to be very low or absent at the level of species, but to be exerted at higher taxonomic levels, with generalizations possible about lack of mycorrhiza formation or the type of mycorrhiza formed in particular families, as discussed above. There is still agreement that much specificity is not qualitatively absolute, that it may be generally low at the level of currently described species and that it encompasses a continuum of variations in extent of colonization or effects on host plants.

Many observations support the existence of low or very low specificity and hence wide choice of partners in many plant–fungus combinations. The low number of AM fungal species (~150, based on current species concept) compared with the very large number of potential host species (perhaps 200 000 or 80–90% of terrestrial plants), means that each fungal species theoretically must have many hosts. Field observations (using both conventional and molecular detection methods) indicate that a single plant root system can contain many AM fungi and that different plant species at the same site often contain the same fungi or fungal sequence variants. ^{14}C can be transferred between roots of different plant species, indicating that the same fungus colonizes both. Finally, a very large number of experiments with a wide range of plant and fungal species have shown that an AM fungus isolated from one species of host plant can be expected, with reasonable confidence, to colonize any other species which has been shown to be capable of forming arbuscular mycorrhizas. Furthermore, it has been argued that AM fungi (like many soil-inhabiting fungi) have relatively ineffective dispersal mechanisms and will benefit from low specificity, enabling them to access organic C from a wide range of plant species. Additionally, mutualistic symbioses, like arbuscular mycorrhizas, are not expected to show the same specificity as pathogenic symbioses (Vanderplank, 1978; Law and Lewis, 1983; Smith and Douglas, 1987; Douglas, 1998). As Vanderplank said: 'Mutations to resistance in mycorrhizal plants are eliminated by selection because they are disadvantageous; and the elimination also eliminates a major source of specificity'.

However, these arguments for limited or low specificity can be questioned (Fitter, 2001). First, as we have seen, the species concept for AM fungi is inadequate and described species show remarkable intraspecific variation. A change in species concept to take this functional diversity into account could well increase the number of species and hence increase possibilities for specificity, although it is unlikely that the number of fungi would reach that of potential plant partners. Second, most fungi used in experimental tests are raised in pot cultures and, in consequence, have been selected for tolerance to disturbance and rapid colonization of many species of experimental plant; they are likely to be generalists, pre-selected for lack of specificity. Many fungi known and described from spores isolated from soil or detected as sequences in roots have never been successfully raised in pot culture, possibly because the appropriate host plant species or environmental requirements have not been met. Examples include *Entrophospora infrequens* from many sites (Rodriguez *et al.*, 2001) and *Scutellospora dipurpurescens* from Pretty Wood (a deciduous woodland in the UK) (Helgason *et al.*, 2002), although isolates of the latter fungus have been obtained in culture from other sites. Third, the arguments of Law and Lewis and of Vanderplank, fail to take into account diversity in symbiotic response both of fungi and plants, which could exert selection pressures leading to a narrowed choice of effective symbiotic partner. Douglas (1998) argues, however, that variation in benefit as a result of environmental heterogeneity, as well as unpredictable availability of the 'best' symbiont would counter selection pressures for high specificity. Recently, Kiers and van der Heijden (2006) have outlined four hypotheses put forward to explain the long term stability and evolutionary cooperation between AM fungi and their plant symbionts. They conclude, not surprisingly, that exchange of 'surplus resources' is of key importance. They consider that such exchange is 'enforced through sanctions by one or both parties. Importantly, such sanctions and the mechanisms by which they operate are as yet speculative.

Setting these arguments aside, we need to ask if there is direct evidence for any specific interactions between particular plant species and particular AM fungal species. Genotype-dependent variations in the extent of colonization have been observed in a number of species (Smith *et al.*, 1992; Peterson and Bradbury, 1995), but caution needs to be exercised in relating this to specificity, because quantitative variations in percent of the root length colonized are a function not only of the ability of a fungus to enter and spread in a root, but also of the rate of root growth which may itself be cultivar dependent. Furthermore, both fungal spread and root growth may be strongly influenced by environmental factors such as nutrient availability, temperature and light (see Chapter 2). In any event, per cent root length colonized is not a good predictor of 'benefit' likely to accrue to the plants.

Tighter specificity has been observed in a number of investigations. The ability of three legume species, *Medicago sativa*, *Hedysarum coronarium* and *Onobrychis viciaefolia*, to be colonized by four *Glomus* species was tested (Giovannetti and Hepper, 1985). Using two soils of different phosphorus (P) availability, the results showed that whereas *M. sativa* was extensively (though variably) colonized by all four fungi, there were considerable differences in colonization of the other two plant species. *H. coronatum* showed the most striking differences, being colonized to the same extent as *M. sativa* by *G. mosseae*, but scarcely or not at all by *G. caledonium* or by one of the isolates of *G. fasciculatum*. *H. coronarium* is certainly not a 'non-mycorrhizal' plant, but there is clearly some degree of selectivity in its receptiveness to different fungi, which has not been further investigated. Similarly, Douds *et al.* (1998) showed that, in soil-grown plants, *M. sativa* was a poor symbiont with *Gigaspora margarita* but formed developmentally and functionally normal mycorrhizas with *G. intraradices*. In monoxenic cultures, there was evidence of a hypersensitive reaction of *M. sativa* roots to *Gi. margarita* and a complete failure of the fungus to colonize, whereas colonization was again normal with *G. intraradices*. More recently, Helgason *et al.* (2002) studied the colonization of several woodland species, using a combination of conventional staining and molecular detection methods. The canopy dominant *Acer pseudoplatanus* was only colonized by *G. hoi* (to which it responded positively in a pot experiment in terms of growth and P uptake), but not by other fungi from the site, including *Glomus* sp. UY1225, *Acaulospora trappei* and *Scutellospora dipurpurescens*. *Acer* therefore appeared to show selectivity in choice of fungal partner, but all the fungi appeared to have wide host ranges among the herbaceous understorey plants. Similarly, Vandenkoornhuyse *et al.* (2003) studied the occurrence of 24 groups of related fungal sequences (called by them phylotypes) in the roots of co-occurring grassland species *Trifolium repens* and *Agrostis capillaris* and found that many of them did not occur in common, although others did.

Some of the most extreme examples of specificity are provided by plant mutants or genotypes of otherwise highly AM species. These have been identified in a number of plants, particularly among non-nodulating genotypes of legumes as well as in tomato (Peterson and Guinel, 2000; Marsh and Schultze, 2001). In the majority, colonization is blocked at the root epidermis or in the hypodermal (exodermal) cell layer, but in a few cases it proceeds further, resulting in formation of intercellular hyphae or even abnormal arbuscules in cortical cells. In each case, a recessive mutation in a single gene is enough to inhibit symbiotic development (see Chapter 3). In most instances, the mutations induce the same modified developmental pattern with all fungal species tested. However, the *rmc* mutation in tomato (Barker *et al.*, 1998a), is perceived differently by different AM fungal species. Some fungi are blocked at the epidermis, others are able to penetrate as far as the exodermis and one (a variant of *Glomus intraradices*) is capable of functional colonization of the cortex (Gao *et al.*, 2001; Poulsen *et al.*, 2005). Thus, the recessive *rmc* mutation allows AM fungi to express specificity, whereas the dominant *Rmc* allele does not. These findings show that the fungal genome is important in 'cross-talk' leading to colonization and also how specificity could evolve and plants become apparently 'selective' towards their fungal partners. As long as the specific symbioses provide benefits to both partners they would be subject to positive selection and hence persist in evolutionary terms. It will be of interest to extend the analysis of genes identified by mutation to other species where host specificity has been found, to gain a better understanding of the signalling and cellular processes that underlie these interactions.

Additional examples of extreme plant–fungus specificity have been identified recently in those mycoheterotrophic symbioses that involve fungi from the Glomeromycota. *Arachnites*, a monocotyledonous species in the Corsiaceae and *Voyria* and *Voyriella*, both dicotyledons in the Gentianaceae, all associate with limited but somewhat different ranges of fungal symbionts (Bidartondo *et al.*, 2002). This fits the emerging pattern of narrower specificity in mycoheterotrophic plants forming different types of mycorrhiza, compared with autotrophic relatives (Bidartondo, 2005). It may be that only a few fungal species or sequence variants have proved amenable to involvement in the necessary tripartite symbiosis between a mycoheterotroph and an autotrophic host, which would necessitate alterations in the flow of organic C (see Chapters 4 and 13).

There are relatively few known examples of AM fungi with very narrow ranges of potential symbionts among autotrophic plants, indicating that there may well be considerable selective advantages for them to keep their options open with respect to plants which provide suitable sources of organic C. Fungal host range may, nevertheless, occasionally be restricted for, in a survey of 19 species of host plant, *Glomus gerdemanni* formed arbuscular mycorrhizas only with *Eupatorium odoratum* (Graw *et al.*, 1979). Additionally, there is much indirect evidence from use of different plant species to trap AM fungi from field soil that plant identity has a strong influence on the sporulation of different fungi (Bever *et al.*, 1996; Jansa *et al.*, 2002a) and hence on one attribute related to fungal fitness.

These examples highlight the fact that specificity or selectivity in AM symbioses may well be somewhat higher than previously believed. It will be important for investigators to remain open to this possibility, focusing on subspecific sequence variants of fungal species in relation to their abilities to form symbioses with different plant species. It seems likely that a wide range of strategies will be revealed, ranging from the very tight specificity as exemplified by the mycoheterotrophs, to complete generalists among both plants and fungi. Generalists most certainly do exist, as shown by the fact that numerous research groups use one or a few species of plant on which to maintain pot cultures of a large number of fungal species. For example *Plantago lanceolata*, *Trifolium subterraneum* and *Sorghum sudanense* are used widely for the maintenance of pot cultures and become extensively colonized by a wide variety of fungi in the Glomeromycota, as shown in Table 1.3. The selective advantages of the different strategies and their consequences for plant and fungal interactions in ecosystems will be a fascinating area for future research.

Ecological considerations

Harley and Smith (1983) suggested that specificity might be closer under natural conditions than in artificial one-to-one tests of colonizing ability in pot experiments and they termed this 'ecological specificity'. There certainly are plant species that in the field may sometimes form arbuscular mycorrhizas and at other times not; the facultatively AM species referred to above. In many cases, the reasons for sporadic colonization have not been examined, but may be related to availability of inoculum, particularly in disturbed environments, as well as environmental conditions. The complexities of field environments, with simultaneous challenge of roots by many different potential AM colonizers, may also lead to quantitative variations in the extent of colonization by different fungi that would not be observed in simple one-to-one tests. Using only the broad distinction between colonization caused by the 'fine endophyte' (*Glomus tenuis*) and those caused by AM fungi with wider diameter hyphae, McGonigle and Fitter (1990) demonstrated that *Holcus lanatus* was apparently highly receptive to *G. tenuis*, which contributed over 60% of the colonized length, regardless of season. In contrast, the same fungus contributed 10% or less of the colonized length in *Ranunculus acris* and *Plantago lanceolata*, although the total percentages of the root length colonized were the same or higher than *Holcus*. In the same pasture, *Phleum pratense* was only slightly colonized by any AM fungi. Seasonal and site-related differences in the fungi colonizing woodland plants have been observed using molecular detection methods (Clapp *et al.*, 2002). Whether these cases represent variations in absolute specificity of the symbiosis is doubtful. They may equally result from variations among the fungi with respect to relative rates of colonization and competition with each other (Wilson, 1984; Vierheilig, 2004a; Jansa *et al.*, 2007), as well as variations in response to environmental variables (Chapter 2). What is important to discover is which fungi, or combinations of fungi, colonize the roots of which plants in the field and whether the symbioses thus formed are functionally similar with respect to bidirectional transfer of commodities (for example P for C) and fitness of the partners. The first question is beginning to be addressed as molecular methods of detection are developed. The second question is still very hard to answer in field-based investigations, but it is abundantly clear from pot experiments that different plant–fungus combinations exhibit enormous diversity with respect to symbiotic outcomes, not only between species but also among genotypes of plants and isolates or sequence variants of the fungi (Streitwolf-Engel *et al.*, 1997; van der Heijden *et al.*, 1998a, 1998b; Munkvold *et al.*, 2004; Smith SE *et al.*, 2004a).

Table 1.3 Species of fungi from the Glomeromycota confirmed to form arbuscular mycorrhizas with *Plantago lanceolata*, *Zea mays* and *Sorghum sudanense*. The lists are restricted to species of fungi that have been unequivocally identified; numerous other fungal species for which either the nomenclature or identification are uncertain are also associated with these plants. Gaps in the table do not imply that any species of fungus will not form an association with any particular species of plant.

Plant species	Fungal species				
	Scutellospora	Gigaspora	Acaulospora	Entrophospora	Glomus
P. lanceolata	*S. calospora*	*Gi. candida*	*A. delicata*		*G. clarum*
	S. castanea	*Gi. margarita*	*A. laevis*		*G. coronatum*
		Gi. rosea	*A. longula*		*G. dimorphicum*
			A. scrobiculata		*G. fasciculatum*
			A. spinosa		*G. fistulosum*
			A. trappei		*G. flavisporum*
					G. geosporum
					G. intraradices
					G. mosseae
					G. occultum
Z. mays	*S. aurigloba*	*Gi. candida*			*G. albidum*
	S. scutata	*Gi. margarita*			*G. fistulosum*
					G. fragilistratum
					G. geosporum
					G. mosseae
S. sudanense	*S. calospora*	*Gi. albida*	*A. delicata*	*E. colombiana*	*G. aggregatum*
	S. coralloidea	*Gi. descipiens*	*A. dilatata*	*E. infrequens*	*G. caledonium*
	S. dipurpurescens	*Gi. gigantea*	*A. gerdemannii*		*G. claroideum*
	S. erythropa	*Gi. margarita*	*A. lacunosa*		*G. clarum*
	S. fulgida	*Gi. rosea*	*A. laevis*		*G. clavispora*
	S. gregaria		*A. longula*		*G. constrictum*
	S. heterogama		*A. mellea*		*G. deserticola*
	S. pellucida		*A. morrowiae*		*G. diaphanum*
	S. persica		*A. scrobiculata*		*G. etunicatum*
	S. reticulata		*A. spinosa*		*G. fasciculatum*
	S. scutata		*A. trappei*		*G. fistulosum*
	S. verrucosa				*G. fragilistratum*
					G. geosporum
					G. intraradices
					G. invermaium
					G. lamellosum
					G. leptotichum
					G. manihotis
					G. mosseae
					G. occultum

Data of C. Walker and J.B. Morton.

Why are these questions important? If there is no absolute specificity, then the roots of all plants at a site may be linked together not simply in one common mycelial network (CMN) but by several separate mycelia, one for each fungal species. Links between different species of plant have been proved by direct observation (Newman *et al.*, 1994) and by transfer of ^{14}C between their roots (Grime *et al.*, 1987; Robinson and Fitter, 1999). They are also implied (but not proven to exist) by the presence of the same sequence variants within the roots of adjacent species. If some combinations of plants and fungi are absolutely specific, they essentially 'opt out' of networks that involve other species. As we have seen, there is little evidence for such a completely restrictive pattern. However, there is evidence that some plant species limit their access to only one network. This would be the case of *Acer* in Pretty Wood (see above), which apparently only associates with *Glomus hoi* and would only be connected to other plants at the site via the *G. hoi* network. What selective advantages this might have remains to be investigated, but might include restricting support in terms of C to the network formed by the fungal species that provides the greatest benefit (Helgason *et al.*, 2002).

The complexities are clearly enormous and the difficulties of experimentation, particularly in the field, extremely challenging. Up to ~30 AM fungal species have been found at one undisturbed field site, together with 50 plant species (Bever *et al.*, 2001). Although disturbed agricultural sites may have a single host species and much smaller AM fungal populations (Helgason *et al.*, 1998; Boddington and Dodd, 2000), this is not invariably the case (Johnson, 1993; Jansa *et al.*, 2002a). In consequence, multiple symbioses are likely, with the potential both for functional redundancy and/or complementarity.

Despite the complexity, it is encouraging that several different groups are beginning to address the issues both at theoretical and experimental levels in both temperate and tropical environments (Bever, 1999, 2002a, 2002b; Fitter, 2001; Bever *et al.*, 2002; Sanders, 2002; Kiers and van der Heijden, 2006).

Plants need to be adaptable, responding to local environments including the local availability of symbionts. A plant may be better off associating with an 'inferior' mutualist rather than none at all. In any event, a fungus that is inferior under one set of conditions may perform well under others, so that symbionts that always 'cheat' their partners may well not exist (see Johnson *et al.*, 1997). The advantages of maintaining flexible options in highly diverse ecosystems may well outweigh the apparent advantages of evolving specific and hence restrictive 'best friend' partnerships.

Conclusions

Arbuscular mycorrhizas are formed by members of all phyla of land plants and this mycorrhizal type is characterisitic of highly diverse ecosystems, containing many potential hosts. As currently described, the fungal symbionts appear to fall into only about 150 species in relatively few genera in the order Glomeromycota. The group is monophyletic and of very ancient origin, as indicated by the fossil record and by DNA sequences of living members of the Glomeromycota. The genetic organization of the fungi is the subject of considerable interest and experimentation which should soon lead to clarification of major uncertainties, such as the existence of recombination and heterokaryosis. Fungal isolates assigned to the same species show considerable functional as well as genetic diversity, so that current taxonomy does not necessarily have predictive value with respect to symbiotic performance. The limitations of current taxonomy also pose problems with respect to determining whether symbioses are specific or not. The symbioses are certainly ancient, probably evolved only once and most likely played an important role in colonization of the land by plants.

The number of species of present-day plants forming arbuscular mycorrhizas is very large and their diversity is considerable, not only in taxonomic position but also in life form and geographical distribution. Nearly all herbaceous plants, shrubs and trees of temperate and tropical habitats can form arbuscular mycorrhizas. Whereas most fungi are generalists, associating with a wide range of plants, there is increasing evidence for specificity or selectivity of some plant species for particular fungal symbionts. This is an important area which, together with increasing appreciation of functional diversity among plant–fungus combinations, has significant implications for roles of AM fungi in plant communities. Non-mycorrhizal plants and plants which form more than one type of mycorrhiza are found in a number of families, supporting the idea that loss of AM status or gain of another type of mycorrhiza has evolved many times, probably as a result of different selection pressures and based on different mechanisms.

2

Colonization of roots and anatomy of arbuscular mycorrhizas

Introduction

This chapter provides an account of the main characteristics of arbuscular mycorrhizal (AM) roots of different types and shows how they develop from sources of inoculum in the soil. The topics covered include:

1 the nature of the propagules in soil that initiate colonization, including spores, infected root fragments and mycelium
2 the development of structures within the root which characterize morphologically different types of arbuscular mycorrhiza and provide the interfaces between the symbionts across which signal molecules and nutrients are exchanged
3 the growth of the extraradical mycelium in soil and production of spores
4 the dynamic interactions between fungal colonization and root growth which lead to the development of AM root systems.

Details of the cellular and molecular interactions between plants and fungi before and during colonization and the importance of these in establishing and maintaining the symbiosis are considered only briefly here and will be discussed in greater detail in Chapter 3.

Sources of inoculum – overview

Colonization of roots by AM fungi can arise from three main sources of inoculum in soil: spores, infected root fragments and hyphae – collectively termed propagules. The large spores (Figure 1.2, see Colour Plate 1.1) with thick resistant walls and numerous nuclei are long-term survival structures with some capacity for dispersal by wind and water (Koske and Gemma, 1990; Friese and Allen, 1991; Gemma and Koske, 1992) and also by animals. Spores and sporocarps can survive passage through the guts of a number of different invertebrates, birds and mammals and may be dispersed locally through animal movements, although actual transmission has not been directly demonstrated in all cases (McIlveen and Cole, 1976; Daniels Hetrick, 1984; Reddell and Spain, 1991; McGee and Baczocha, 1994; Reddell *et al.*, 1997a). The distribution of spores and root fragments in soil is altered by the burrowing activities of both large and small animals and changes in mycorrhiza development associated with ant and gopher mounds have been documented (Koide and Mooney, 1987; Allen and McMahon, 1988; Friese and Allen, 1993).

For many years, it was assumed that spores were the most important propagules, possibly the only ones. Using wet sieving techniques (see below), spores of many fungal species were described and much was learned about their distribution and frequency in soil and roles in initiating colonization of roots. The spore assemblage at a site may be typically composed of around 20–50 species (Bever *et al.*, 2001; Fitter, 2005), encompassing different ages and different states of dormancy or quiescence (Tommerup, 1983). Little is known about the timing of germination and infectivity of such complex communities in soil, but germination of some species may be poor or occur rather slowly and variably, providing a reservoir of inoculum which persists for many years but may not always be important in early colonization of root systems (McGee, 1989; Braunberger *et al.*, 1994; Merryweather and Fitter, 1998a). AM fungi show varying abilities to colonize roots from different sources of inoculum. Klironomos and Hart (2002) tested the abilities of three types of inoculum (spores, fresh root fragments with adhering hyphae and hyphal fragments) from eight AM fungi to colonize roots of *Allium porrum*. They showed that *Glomus* and *Acaulospora* spp. colonized roots from all three inoculum sources, but *Scutellospora* and *Gigaspora* appear to depend entirely on spores. This investigation did not, however, test the potential of hyphae still attached to living plants to initiate colonization.

In many habitats, persistent hyphal networks in soil (Figures 2.1 and 2.3), together with root fragments (Figure 2.2), are the main means by which plants become colonized even when significant spore populations are also present (Hepper, 1981; Smith and Smith, 1981; Tommerup and Abbott, 1981; Birch, 1986; Jasper *et al.*, 1992; Merryweather and Fitter, 1998c). Consequently, as a seedling grows in an established community, it becomes linked into a complex underground network of mycelium of different fungal species and of roots growing from plants of different species and ages. Even in highly seasonal environments, where plants may be lacking for part of the year, there is evidence that mycelial networks persist in dry or frozen soil and play an important role in colonization of new generations of plants. From a fungal perspective, the survival and spread of networks to colonize new plants is a crucial scavenging process which provides ongoing sources of organic carbon (C) (Olsson *et al.*, 2002). Disruption of the network by disturbance can result in much reduced infectivity of the soil and lower rates of nutrient uptake by both

Figure 2.1 Variation in the hyphae forming the external mycelium of arbuscular mycorrhizas. (a) Camera lucida drawings showing thick-walled (H) and thin-walled (h) hyphal elements with many angular projections (arrowed) in the mycelium associated with the roots of *Dactylis glomerata*. Bars = 20 μm. From Nicolson (1959), with permission. (b) Diagrammatic representation of the types of hyphae and hyphal architecture associated with AM roots. Hyphae entering roots from root fragments or spores. Growth, branching and anastomosis of hyphae outside the roots gives rise to runner hyphae, bridge hyphae and hyphal networks. Intraradical colonization not shown. Redrawn from Friese and Allen (1991).

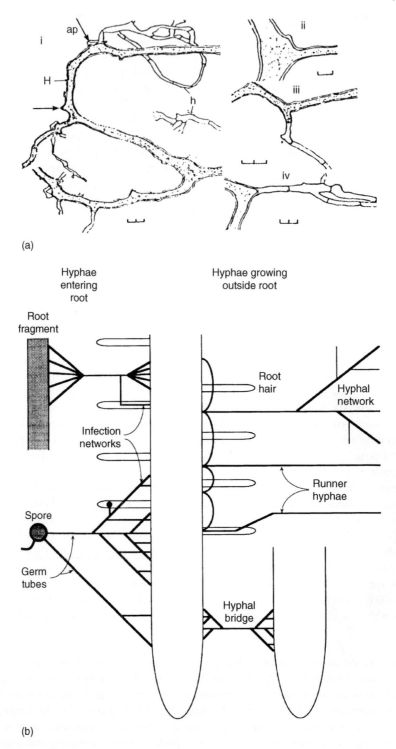

Figure 2.1 (Caption on p. 43)

crop and native plant species (Birch, 1986; McGonigle and Fitter, 1988a; Jasper *et al.*, 1989, 1991, 1992; Evans and Miller, 1990; Miller and Jastrow, 1992b; Miller and Lodge, 1997; Merryweather and Fitter, 1998c; McGonigle and Miller, 2000).

In practical terms, it is difficult to sort out the relative contributions of the different types of propagules to colonization of the root systems of plants growing in any particular field situation; that is to the 'infectivity' of the soil. The density of the spores in soil can be determined but, although this sometimes shows a correlation to the extent of root colonization, this is certainly not always the case. The relationship is complex because, although the extent of colonization may be related to the availability of spores as inoculum, it also influences the capacity of the mycorrhizal root system to produce new spores (see below). Furthermore, the spore assemblage is diverse with respect to species composition, viability and dormancy and sources of inoculum other than spores may play more important roles in the colonization of roots. Attempts to relate discrete and quantifiable spores and root fragments to colonization have been most successful in agricultural environments where the mycelial network has been disturbed. Not surprisingly, the rate of initiation of colonization and spread within a root system is correlated to the propagule density or infectivity of the soil (see below). The relatively robust propagules have been counted and weighed (Thompson, 1987) or collectively assayed by the most probable numbers (MPN) method, using trap plants to determine their presence or absence. The MPN method involves dilution and mixing of samples, which also destroys the hyphal

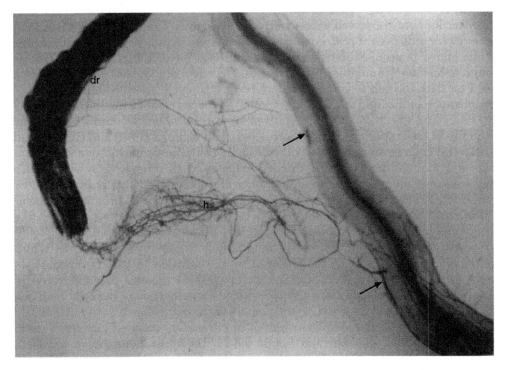

Figure 2.2 Growth of mycorrhizal hyphae (h) from a dead root fragment (dr) and initiation of colonization in a growing root of *Trifolium subterraneum* (arrowed). Photograph SE Smith.

network, but has proved useful to enumerate the reserve of infective, robust propagules in soil, as long as its constraints and wide confidence limits are appreciated
(Porter, 1979; Wilson and Trinick, 1982; An *et al.*, 1990; Jasper *et al.*, 1992). Assessment
of the infectivity of undisturbed soil, including the contribution of hyphal networks, is much more difficult, but can be achieved by bioassays using standard host
plants in undisturbed soil cores (Gianinazzi-Pearson *et al.*, 1985; Jasper *et al.*, 1989;
Braunberger *et al.*, 1994). Both MPN methods and bioassays have a great advantage because they use colonization of roots of trap plants to determine the presence
of viable propagules. Unfortunately, neither method can distinguish the relative
importance of the different types of propagule, nor the contributions of different
species of AM fungi to root colonization. This information is important because the
variations in specificity/selectivity and functional diversity among different plant–
fungus combinations have significant implications for community interactions.
Morphological variations in colonization patterns and molecular detection methods
are beginning to show which fungi colonize the roots of which plant species
but, as yet, quantification of the contributions of different species is by no means
straightforward.

   While an extraradical hyphal network associated with living plants has inde-
terminate growth and a long-term capacity to initiate colonization, the growth of
hyphae from spores or root fragments is very restricted. Saprotrophic growth in soil
does not occur. Hyphal extension following germination is at the expense of spore
reserves, although it serves to increase the chances of growth into the neighbourhood of roots. Although spores are probably widely dispersed by wind, at least in
low densities, other discrete propagules are likely to be moved only short distances.
Consequently, the fungi depend either on the growth and persistence of a mycelial
network or on dispersal of plant partners; lack of absolute specificity for particular
plant species (Chapter 1) will help to avoid spatial or temporal discontinuities in the
availability of sources of organic C. Plants which rely on the symbiosis for success
are also dependent on the capacity of the fungal propagules to survive in a wide
range of soil types and environmental conditions.

Spores as sources of inoculum and indicators of fungal assemblages

Occurrence of spores

Spores are the best-defined sources of inoculum and are the only propagules that
can currently be identified to species with any degree of certainty (see Chapter 1).
Consequently, they are of central importance in isolating AM fungal species, determining their distribution and establishing them in pot cultures for experimental or
identification purposes. Early work tended to ignore species diversity of AM fungal
assemblages, so that data were collected for overall spore densities in soil and per
cent root length colonized. With the recognition of marked genetic and functional
variability among AM fungi has come increased effort to describe the numbers of
species present and their relative contributions to communities. Information often
now includes density of spores in soil and their species diversity, increasingly supplemented by microscopic identification of fungal morphology in roots and molecular identification of the fungi (Abbott, 1982; Merryweather and Fitter, 1998a; Clapp

et al., 2002; Dickson, 2004). Nevertheless, as the examples given below illustrate, it is still extremely difficult to make worthwhile generalizations about factors influencing either spore assemblages or densities in soil.

The most commonly used method to isolate spores from soil is still 'wet-sieving and decanting' (Gerdemann, 1955; Gerdemann and Nicolson, 1963) and details plus modifications can be found in various handbooks and laboratory manuals (Schenck, 1982; Brundrett *et al.*, 1994, 1996; Norris *et al.*, 1994). Field-collected spores can be difficult to identify to species, so they are frequently used to establish pot cultures from which a new generation of spores is isolated and more certain identifications made. Field isolation is often supplemented by 'trap-culturing', a process in which one or several plant species are grown in the field soil for several months and spores later isolated and identified. In the majority of investigations, direct spore isolation gives a range of AM fungal species that is somewhat different from those producing spores in trap cultures, and different trap-plant species again give different but overlapping ranges of AM fungal species (Sanders and Fitter, 1992a; Bever *et al.*, 1996, 1997, 2001; Antoniolli, 1999; Jansa *et al.*, 2002a). Not only do these findings indicate higher diversity of AM fungal assemblages than is observed from direct spore counts but, importantly, they show the potential influence of plant species on fungal selectivity and success in sporulation. The numbers of species isolated often depends on the sampling effort (Bever *et al.*, 2001; Fitter, 2001, 2005), with mean values commonly around 20–25 species per site, but sometimes reaching 50 or more (Fitter, 2001). In an ongoing investigation of a mown grassland site, only 11 species were isolated as spores in the first year of work, but this value increased to 37 species over several years of repeated direct seasonal sampling and trap-culturing using three of the ~50 plant species present at the site (Bever *et al.*, 2001). Molecular methods broadly indicate the presence of similar numbers of taxa as spore isolations or morphological identification in roots, with eight to nine taxa detected in bluebell roots in a temperate woodland (Clapp *et al.*, 1995; Merryweather and Fitter, 1998c; Helgason *et al.*, 1999, 2002) and up to 30 sequences in roots of tree seedlings in a tropical forest (Husband *et al.*, 2002b) (see Chapter 15). Despite the general agreement between findings using different methods, there are examples of AM fungal sequences or morphotypes in roots that have never been detected as spores (Merryweather and Fitter, 1998a), underlining the significance of propagules other than spores in initiating colonization and emphasizing that spore isolation may give an inadequate picture of which AM fungi are actually important as symbionts.

Spore densities are highly variable. In some habitats they are not found in all seasons and even the seasonal maxima can be quite variable with reports ranging between 1 and 5 spores per gram of soil in woodland (McGee, 1986; Louis and Lim, 1987; Merryweather and Fitter, 1998a), 9–89 spores per gram at an agricultural site (Sutton and Barron, 1972) and a maximum of ~4 spores per cubic centimetre of sand in a Rhode Island sand-dune (Gemma and Koske, 1988).

More detailed analyses sometimes indicate that individual AM fungal species may differ in season of spore production. In one example, from a mown grassland in North Carolina, *Gi. gigantea* reached a maximum late in the year, but spore densities of *Scutellospora calospora* and *Acaulospora colossica* were highest in late spring (Bever *et al.*, 2001). Despite these findings, caution is needed because repeat sampling at the same woodland site investigated by Clapp *et al.* (1995) failed to confirm the seasonality of sporulation. In this case, peaks in spore density were not clearly associated with stage of plant development or any climatic or seasonal variable (Merryweather and Fitter, 1998b). Agricultural sites often have relatively high total spore densities with values of up to 120 spores per gram recorded (Johnson *et al.*, 1991, 1992; Jansa *et al.*, 2002a). In these examples, there were marked effects of site, season and crop species, but not of either tillage or long-term differences in P application. Disturbance due to agriculture can lead to changes in the composition of the AM spore assemblage as well as to reductions in diversity. A preponderance of *Glomus* (particularly *G. mosseae*) and reductions in other genera such as *Gigaspora*, *Scutellospora* and *Acaulospora* are frequently observed, with generalizations based on spore densities, supported by changes in DNA sequences detected (Johnson, 1993; Helgason *et al.*, 1998; Daniell *et al.*, 2001; Jansa *et al.*, 2002a). It is thought that vigorously sporing *Glomus* species are able to survive and maintain higher infectivity and spore production even if hyphal networks are disrupted, compared with AM fungi that are more dependent on hyphal spread to reach new plants.

In situations where density of spores is positively correlated with the extent of root colonization, both may increase during the growing season of annual plants. Decreases in density with depth of soil could well be associated with the associated decline in density of AM colonized roots (Hayman, 1970; Sutton and Barron, 1972; Jakobsen and Neilsen, 1983; Giovannetti, 1985). However, this relationship does not always hold and an inverse relationship between spore density and colonization in four perennial trees from lowland tropical rainforest has been observed (Louis and Lim, 1987) and, in some investigations, no correlation at all has been found between spore populations and infectivity (Powell, 1977).

Germination and hyphal growth

Germination of spores can take place in three ways; through germination shields, as in *Acaulospora* and *Scutellospora*, directly through the spore wall in *Gigaspora* and some *Glomus* species, or via regrowth through the hyphal attachment which is common in many other *Glomus* species (Siqueira *et al.*, 1985). The process does not require plant roots, although per cent germination is sometimes increased in their presence. Germination in soil or on agar media produces small amounts of presymbiotic mycelium, of the order of 20–30 mm per spore. There are indications that this mycelium may have slight capacities to utilize exogenous sugar and acetate (Bago *et al.*, 1999a) and possibly for saprotrophic growth in soil (Hepper and Warner, 1983) but, in the absence of roots or root exudates, the hyphae have very slow metabolic rates and all attempts at long-term culture have failed (Azcón-Aguilar *et al.*, 1999; Giovannetti, 2000; Bécard *et al.*, 2004). Presence of a root or root exudates stimulates growth and branching of the mycelium and apparently converts it into an 'infection ready' state (see Chapter 3), but ongoing growth does not occur and no new spores are produced unless successful colonization of a root system occurs; if the spore becomes detached, growth of the hyphae ceases.

The majority of studies of germination have used water or nutrient agar, variously modified, but an alternative and ecologically more relevant approach involves burying spores in soil packaged in such a way that they can be recovered. Erratic germination may be related to dormancy for, as Tommerup (1983) showed, spores of '*Gigaspora*' (*Scutellospora*) *calospora*, *Acaulospora laevis* and two *Glomus* species are dormant when first formed but after periods of storage dormancy is overcome. The spores then become quiescent and capable of germinating rapidly and fairly synchronously under appropriate conditions of moisture and temperature. This agrees with other investigations showing that periods of storage in dry soil or at low temperature increase the percentage germination, depending on the species (Sylvia and Schenck, 1983; Tommerup, 1984a; Gemma and Koske, 1988; Louis and Lim, 1988; Safir et al., 1990).

Hepper and Mosse (1975) and subsequently Hepper and co-workers (Hepper and Smith, 1976; Hepper, 1979, 1983a, 1983b, 1984a; Hepper and Jakobsen, 1983) carried out an extensive and systematic investigation of germination of a single species, *Glomus caledonium*. These and other investigations (Schenck et al., 1975; Green et al., 1976; Daniels and Duff, 1978; Daniels and Trappe, 1980) allow some very broad generalizations to be made (reviewed by Azcón-Aguilar et al., 1999; Giovannetti, 2000). Species of fungi vary in the optimum pH, soil matric potential and temperature for maximum germination and, although the effects of high concentrations of P and other mineral nutrients are variable, heavy metals (Zn, Mn and Cd), organic acids and a range of sugars can be inhibitory. High salinity reduces germination, probably via its effect on water potential (Juniper and Abbott, 1993, 2004). A few investigations suggest that glucose, root exudates or extracts from host species stimulate germination (Graham, 1982; Gianinazzi-Pearson et al., 1989; Vilarino and Sainz, 1997; Bécard et al., 2004), while others report negative effects or none at all (Schreiner and Koide, 1993a, 1993b, 1993c; Vierheilig and Ocampo, 1990a, 1990b).

Germination is sometimes, but not invariably, increased in the presence of microorganisms or decreased in sterile soil (Azcón-Aguilar et al., 1986a, 1986b; Mayo et al., 1986; Azcón, 1987; Wilson et al., 1988; Daniels Hetrick and Wilson, 1989; Xavier and Germida, 2003). Complex interactions between microbial activity and spore germination and mycelial growth are to be expected and possible mechanisms include removal of toxins or germination inhibitors, production of specific stimulatory compounds and the maintenance of elevated CO_2 concentrations, although the effects on germination itself may be minor (Le Tacon et al., 1983).

Germination and hyphal growth from spores has been intensively studied with the objective of determining why it is so limited in the absence of root colonization, as well as in the hope of producing axenic mycelium for use as inoculum. Early work focused on identifying an essential metabolite that might be lacking. Little understanding was gained, although it was consistently found that increased P concentration in the medium reduced hyphal growth, that the stimulatory effects of peptone could be traced to the lysine, cysteine and glycine components and that K (rather than Na) salts of sulphite and metabisulphate were more stimulatory than sulphate or thiosulphate. Various metabolic inhibitors like actinomycin, cycloheximide, ethidium bromide and so on, also failed to shed a great deal of light, except to indicate that there are no serious limits to DNA or protein synthesis during spore germination (Hepper, 1979, 1983a, 1983b, 1984a; Siqueira et al., 1982, 1985; Hepper and Jakobsen, 1983; Pons and Gianinazzi-Pearson, 1984). Nevertheless, the metabolic rate of spores and germ tubes appears to be low (Tamasloukht et al., 2003; Bécard et al., 2004). Stored lipid is transformed to carbohydrate and small amounts of glucose and acetate can be absorbed, but lipid synthesis does not take place during germination (Bago et al., 1999a).

The work on effects of plant exudates and volatiles has been much more enlightening (Koske and Gemma, 1992). As early as 1976, Powell (1976) noticed increased branching of hyphae in soil as they grew very close to roots. In axenic tests, soluble exudates or extracts from the roots of host species, as well as from cell cultures, stimulated the growth and branching of mycelium growing from spores (Graham, 1982; Carr et al., 1985; Elias and Safir, 1987; Vierheilig et al., 1998a; Buee et al., 2000), whereas exudates from non-hosts had no effect (Gianinazzi-Pearson et al., 1989; Schreiner and Koide, 1993b; Buee et al., 2000). High P supply to the roots from which exudates were collected had negative effects on hyphal responses (Tawaraya et al., 1996), possibly related to reduced accumulation of stimulatory compounds (Akiyama et al., 2002, 2005). Again in soil, Giovannetti et al. (1993b) showed complex hyphal branching patterns associated with mycelial growth from spores on roots of a number of host species (Fragaria, Helianthus, Oncimum, Lycopersicon and Triticum), but not on the non-hosts Brassica, Dianthus, Eruca or Lupinus. At the same time, a number of studies were initiated on the effects of various phenolic compounds produced by roots or seeds and known to influence symbiotic development between Rhizobium and Agrobacterium and their hosts. In summary, flavonoids have consistent stimulatory effects on the growth and branching of germ tubes of Gi. margarita and some Glomus species (Gianinazzi-Pearson et al., 1989; Tsai and Phillips, 1991; Bécard et al., 1992; Buee et al., 2000) and can also lead to increased colonization of roots by the fungi (Nair et al., 1991; Siqueira et al., 1991; Akiyama et al., 2002). There were some inconsistencies in the results from different groups using related compounds but, overall, they suggested that flavonoids (sometimes together with elevated CO_2) might be key signal molecules involved in stimulation of presymbiotic mycelium and mycorrhiza formation and possibly of directional growth towards roots (Vierheilig et al., 1998a). This view now appears to be incorrect because, although flavonoids have stimulatory effects, they are not essential for mycorrhiza formation. Bécard et al. (1995) showed clearly that colonization occurred in roots of species that did not produce flavonoids and that mutants of maize, deficient in chalcone synthase and hence unable to produce flavonoids, are colonized normally. It appears most probable that another lipophilic component of root exudates, termed branching factor (BF) and recently identified as a sesquiterpene, is implicated in enhanced mycelial growth, changes in morphogenesis and the processes that lead to them (Buee et al., 2000; Tamasloukht et al., 2003; Akiyama et al., 2005; Besserer et al., 2006). The compound is active at very low concentrations and is likely to be a signal molecule rather than a nutrient. In any event, the increased branching and anastomosis leads to the formation of an enlarged and interconnected presymbiotic hyphal network with increased likelihood of contacting a plant root to initiate colonization, as well as of retaining integrity and transport capacity if damaged.

Spores of some species appear to be adapted to survive situations where germination is not immediately followed by colonization of roots and establishment of a symbiotic relationship. Infectivity of spores of Acaulospora laevis and Glomus caledonium was retained in moist field soil for at least 4 weeks in the absence of suitable plants, but declined between 4 and 10 weeks (Tommerup, 1984b). Similarly, spore-based inoculum of G. intraradices retained infectivity for up to 3 weeks in moist soil at temperatures up to 38°C (Haugen and Smith, 1992). The basis is unknown for these species, but in G. caledonium a programmed growth arrest of the germ tubes occurs if no plant species becomes available for colonization (Logi et al., 1998). Furthermore, spores of Gi. gigantea are capable of producing a number of germ tubes successively if the earlier ones are cut off (Koske, 1981) and it is probable that nuclear division occurs, replacing the nuclei that migrated to the mycelium during the initial stages of growth (Bécard and Pfeffer, 1993). *Glomus epigaeus* produces secondary sporocarps in long-term storage without any intervening colonization of roots (Daniels and Menge, 1980).

Overall, spores appear to be well adapted to their roles as units able to initiate colonization in roots. They are capable of long-term survival in soil, but can germinate repeatedly in the absence of roots and maintain low metabolic activity. In the presence of roots and stimulated by root exudates, metabolic changes occur which result in increased branching, hyphal extension and root contact. Hyphal anastomosis between closely related presymbiotic mycelia will produce larger mycelia, again leading to increased chances of the fungal unit intercepting a suitable root, initiating colonization and hence accessing an ongoing supply of sugars.

Root fragments

Root fragments can be an important source of inoculum in many soils, but much less is known of their biology than that of spores. Regrowth of hyphae from infected root fragments has been observed frequently (Magrou, 1946; Hall, 1976; Powell, 1976; Hepper, 1984b) and the fragments have often been used to initiate colonization experimentally (Biermann and Linderman, 1983) (see Figure 2.2). We do not know how long vegetative hyphae survive in senescent or dead roots, but some results suggest that they may do so for at least 6 months in dry soil and that infectivity is not related to the presence of vesicles. Accordingly, the potential for hyphal regrowth and infectivity was retained both in *Scutellospora calospora* (a species that does not form vesicles) and in two *Glomus species* (Tommerup and Abbott, 1981). However, vesicles, like spores, do store large amounts of lipid and contain many nuclei which, together with their thick walls, suggests a function either as propagules or to support the regrowth of intercellular hyphae. Development of very thick walls by the intercellular hyphae of one AM fungus in old *Trifolium* roots was studied ultrastructurally (Lim *et al.*, 1983). The laminated walls resembled those of sclerotia and surround apparently fairly normal cytoplasmic contents. This may represent an adaptation to survival for long periods, but it is not known how widespread the phenomenon is nor whether the thick-walled hyphae are capable of subsequent germination.

Thompson (1987) found that with a long fallow period (up to 2 years without plants) numbers of both root fragments and spores in soil were low and so was soil infectivity; with short fallow the converse was true. The density and distribution of root fragments in pots is important in influencing the rate of colonization of *Trifolium subterraneum* seedlings and their location on main and lateral roots (Smith and Smith, 1981). Redistribution of root fragments by soil animals has an effect, not only on development of mycorrhizal roots, but also on plant distribution. Friese and Allen (1993) showed that harvester ants accumulate very large quantities of clipped root material (and AM spores) in their underground nests. After the nests were abandoned, plants of *Artemisia tridentata* and *Oryzopsis hymenoides* adjacent to the nests were colonized by AM fungi very rapidly and extensively. In some situations, the dead infected roots may be localized in such a way as to provide inoculum exactly where new roots will grow. This was suggested to be the situation in bluebell woods where new roots develop close to the dead roots remaining at the end of the previous season (Daft *et al.*, 1980). However, major soil disturbance reduced colonization considerably, indicating that the hyphal network as well as more robust dead roots were likely to have been an important source of inoculum in this case (Merryweather and Fitter, 1998c). The capacity of root fragments to maintain infectivity through periods of repeated wetting and drying is not clear, but may have considerable ecological significance in very seasonal habitats where intermittent summer rainfall wets the soil and induces hyphal growth, but is insufficient to support seedling establishment (Braunberger *et al.*, 1994).

Hyphal networks

There is no doubt that hyphal networks in soil linked to established plants are of key importance in the colonization of seedlings in perennial vegetation systems, a fact that has been appreciated for a very long time. The extensive development of mycelium was described very early in AM research (Peyronel, 1923) and the importance of hyphae in linking plants together clearly appreciated (Harley, 1991). Nicolson (1959) was one of the first to make a systematic investigation of the extraradical mycelium associated with grass roots from natural habitats. He described, as others have done since, 'the striking variation in diameter (2–27 μm) among hyphal filaments, with accompanying variation in wall thickness'. He showed how the thick-walled hyphae form the permanent bases for the short-lived hyphal complexes associated with roots. The main hyphae, which normally contain cytoplasm and nuclei, give rise to lateral systems of more and more finely branched and septate, lateral hyphae (see Figure 2.1a). The hierarchical branching pattern indicated by these observations is extremely difficult to see in soil, but is clearly apparent in monoxenic cultures of AM fungi (Bago *et al.*, 1998a) (Figure 2.3a) and in a two-dimensional system used to observe the anastomosing mycelium linking plants of different species together in soil (Giovannetti *et al.*, 2004) (Figure 2.3b).

The importance of living hyphal networks in initiating rapid colonization in seedlings and in mobilizing nutrients has deservedly received increasing emphasis in recent years (Read, 1992; Read *et al.*, 1985; Olsson *et al.*, 2002; Leake *et al.*, 2004a; Simard and Durall, 2004). Hyphal connections grow from plant to plant, forming bridges which can be simple or may branch as a root is approached to form multiple colonizing hyphae and appressoria (Friese and Allen, 1991) (see Figure 2.1b). Runner hyphae forming external loops along the surface of the root also initiate secondary colonization (see Figure 2.1b) (Peyronel, 1923; Cox and Sanders, 1974; Brundrett *et al.*, 1985; Friese and Allen, 1991; Wilson and Tommerup, 1992). Mycelium of *Glomus mosseae* has been shown to spread through soil at 3 mm/day to initiate colonization in soybeans (Camel *et al.*, 1991) and a maximum distance appears to be around 20–30 mm for a number of species (Warner and Mosse, 1983; Schubert *et al.*, 1987), although values up to 90 mm have been recorded. A mycelial 'infection front' can spread through a population of plants in sterilized soil with rates between 0.2 and 2.5 mm/day depending on host plant and fungal species (Powell, 1979; Scheltema *et al.*, 1987b). Spread is also influenced by the texture of the soil through which the hyphae grow and, as with other characters of AM fungi, there is interspecies diversity with respect to both distance the fungi can grow to colonize a new plant and ability to pass through narrow and tortuous soil pathways (Drew *et al.*, 2003, 2005). The AM fungal mycelial network can be very extensive and molecular fingerprinting has shown that a single AM fungal

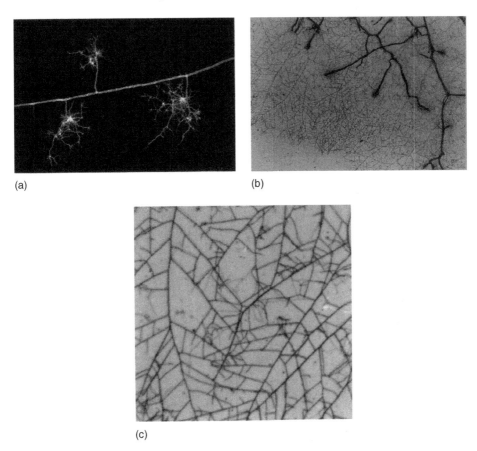

(a)   (b)

(c)

Figure 2.3 (a) External mycelium of *Glomus intraradices* in monoxenic culture with tomato roots, showing a runner hypha subtending branched absorbing structures (BAS), external to a root growing in monoxenic culture. Extended focus confocal image constructed of 40 optical slices taken at 1 μm intervals on the z-axis. Courtesy of Sandy Dickson and Alberto Bago. (b) Bright field image of intact extraradical mycelium of *Glomus mosseae* spreading from AM roots of *Prunus cerasifera*. Courtesy of Manuela Giovannetti. (c) Bright field image of detail of the network structure of the external mycelium of *Glomus mosseae*, showing marked interconnectivity due to frequent formation of ansastomoses. Courtesy of Manuela Giovannetti.

clone may possibly extend over 10 m between roots of *Hieraceum pilosella* in an undisturbed sand-dune system (Rosendahl and Stukenbrock, 2004). A fragile mycelium of this type can be disrupted by the activities of soil organisms, such as grazing collembolans and burrowing earthworms and also by agricultural practices. Disruption may lead to reduced infectivity of the fungi and also to reduced uptake of nutrients by the associated plants, indicating that the established mycelium has greater scavenging ability than newly developed mycelia associated with individual plants (McGonigle and Fitter, 1988a; Fitter and Sanders, 1992; Pattinson *et al.*, 1997).

The rapidity with which colonization occurs from an established network, supported by existing plants, is emphasized by experimental work using colonized 'nurse plants' to initiate colonization (Brundrett *et al.*, 1985; Rosewarne *et al.*, 1997). Colonization of transplanted seedlings was initiated by appressorium formation within 2 days, with subsequent overall colonization occurring extremely rapidly and reaching a maximum at around 10 days, with a peak in arbuscule development at 12 days (Rosewarne *et al.*, 1997) (Figure 2.4). These findings agree with observations of extremely rapid colonization of seedlings in undisturbed field situations, where there will be several different anastomosing networks belonging to different AM fungi (see Chapter 1). Some plant species will be associated with several of these networks, but others perhaps as few as one, depending on plant selectivity.

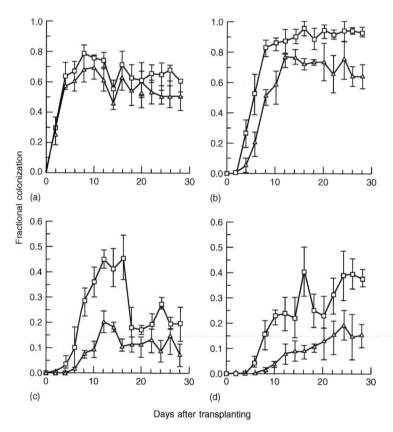

Figure 2.4 Colonization of *Solanum esculentum* (tomato) roots by *Glomus intraradices*. Tomatoes were raised in soil with low (□) or high (Δ) phosphate (P) and subsequently transplanted into 'nurse pots' containing 3-week-old mycorrhizal plants of *Allium porrum* growing in low P soil. Values are fraction of root length associated with (a) external hyphae, (b) internal hyphae, (c) arbuscules and (d) vesicles. Means and standard errors of means of four plants. Reprinted from *Mycological Research*, **101**. Rosewarne *et al.* (1997), Production of near-synchronous fungal colonization in tomato for developmental and molecular analyses of mycorrhiza. pp. 966–970.

Hyphal networks appear capable of surviving and retaining infectivity during periods when the vegetation with which they developed are either dormant or actually dead. Work in seasonally very dry and hot (Jasper *et al.*, 1989; McGee, 1989; Braunberger *et al.*, 1994) or cold (Addy *et al.*, 1994, 1997, 1998) environments indicates the importance of this survival to rapid colonization when conditions favourable to plant growth return. For some fungi, the maintenance of the infectivity of the network may depend on whether it has dried before sporulation commenced, so that the significance of spores and/or hyphae and the effects of disturbance will be diverse (Jasper *et al.*, 1993).

Morphology and anatomy of arbuscular mycorrhizas

Variations in morphology of colonization: *Arum*- and *Paris*-type mycorrhizas

It has already been emphasized that there are three important components of any AM root system, the root itself and two associated mycelial systems, one in the soil and the other within the root apoplast. These two fungal systems provide crucial interfaces involved in symbiotic nutrient uptake by the partner organisms and grow and develop in very different environments; the first is highly variable, influenced by soil heterogeneity, whereas the second is relatively constant and controlled through root homeostasis. The mycelium in soil is involved in searching for new plants and in acquisition of mineral nutrients that are used by both fungus and plant, whereas the intraradical interfaces are concerned with nutrient transfers between the symbionts (see Chapters 4 and 5). Here we describe the development of the critical interfaces in the root, emphasizing variations in structures formed by different fungi within different plant species, the existence and significance of which has only relatively recently been re-emphasized (Smith and Smith, 1996, 1997).

Descriptions and illustrations of the internal mycelium were made in the late nineteenth century (Janse, 1897) and beautiful drawings illustrating the details of fungal interactions with plant cells and tissues of the root were published by Gallaud (1905) (Figure 2.5a, c). Although Gallaud's interpretation of the significance of some of the structures has not stood the test of time, the main features are clearly recognizable. His observations indicate that AM roots can contain a wide range of structures, which he believed fell into one of two general anatomical groups. The type frequently regarded as a 'typical arbuscular mycorrhiza' and often described in the fast-growing root systems of crop plants, belongs to Gallaud's *Arum*-type. In these associations the fungus spreads relatively rapidly in the root cortex by intercellular hyphae which extend along well-developed intercellular air spaces. Short side branches penetrate the cortical cell walls and branch dichotomously in the cell lumen to produce characteristic highly branched arbuscules. Hyphal coils may be formed during penetration of the exodermal cell layers of the root, but they are not a major component of cortical colonization. Both one of Gallaud's illustrations and a photomicrograph of a typical *Arum*-type infection unit in *Allium porrum* are illustrated in Figures 2.5a, b and will be described in greater detail later. A single arbuscule is shown in Figure 1.1a.

Gallaud also recognized what he called *Paris*-type mycorrhizas, in which cortical colonization of roots is characterized by extensive development of intracellular

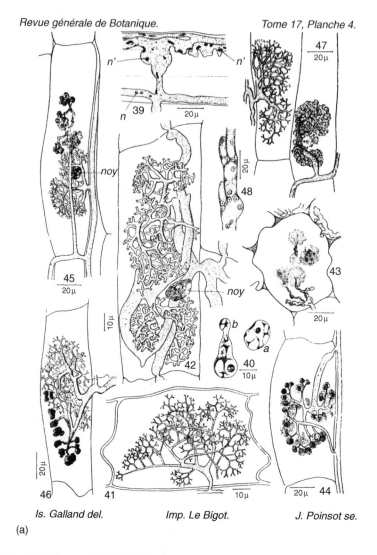

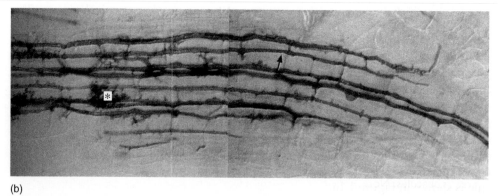

Figure 2.5 (Caption opposite)

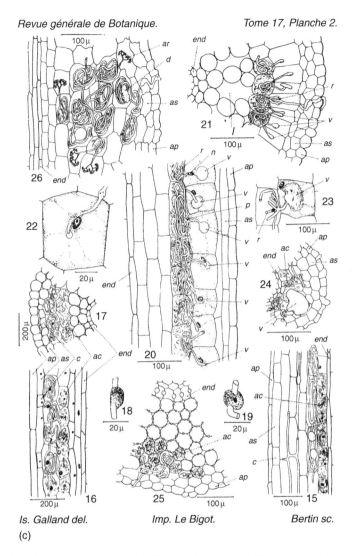

Revue générale de Botanique.   Tome 17, Planche 2.

Is. Galland del.   Imp. Le Bigot.   Bertin sc.

(c)

Figure 2.5 Variations in intraradical AM colonization. (a) Drawings of *Arum*-type mycorrhizas showing intracellular arbuscules originating from intercellular hyphae in a range of host plants (41, 45, 46, 47). Stages in the disintegration of arbuscules also shown (43, 44, 47). *Paris*-type arbusculate coil shown in 42. From Gallaud (1905). (b) A single, *Arum*-type infection unit of *Glomus versiforme* in a 10-day-old root of *Allium porrum*. Note the intercellular hyphae (arrowed) growing longitudinally between the root cortical cells. Branches may develop into arbuscules (*) or remain as short projections on the hyphae. From Brundrett *et al*. (1985), with permission. (c) Drawings of variations in *Paris*-type mycorrhizas, showing the prolific development of intracellular coils and relatively sparse development of arbuscules. From Gallaud (1905). (d) Arbuscular mycorrhizal development in American ginseng (*Panax americana*), showing highly developed coils in cortical cells, occasionally subtending sparse arbuscules. Note absence of intercellular hyphae; colonization spreads directly from cell to cell. From Peterson *et al*. (2004), with permission.

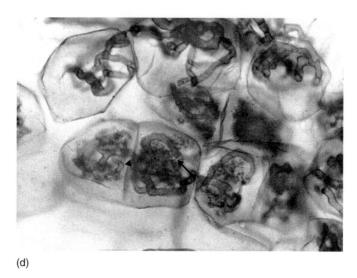

(d)

Figure 2.5 (Caption on p. 57)

coiled hyphae which spread directly from cell to cell (see Figure 2.5c). Arbuscule-like branches sometimes grow from these coils (Figure 2.5a, central panel-42), but there is very little, if any, intercellular growth. These two main AM types represent extremes along a developmental continuum which includes mycorrhizas character-ized by straight hyphae which spread longitudinally within, rather than between cortical cells, as well as situations where both inter- and intracellular hyphal devel-opment occurs (Dickson, 2004) (Figure 2.6). The control of development and func-tion of the different interfaces formed within roots has not been fully elucidated, but it must be recognized that *Arum*-type mycorrhizas are but a subset of the structural possibilities, so that functional inferences and generalizations based solely on them must be dangerous.

We do not know how common the *Paris*-type of colonization pattern really is, but increasingly surveys of the incidence of mycorrhizas record details of intraradical AM morphology and the emerging picture suggests that *Paris* types occur in a wide range of plants and habitats. Gallaud (1905) described them in the European wood-land plants *Paris, Parnassia* and *Colchicum* and they were subsequently depicted in members of the Gentianaceae (Jacquelinet-Jeanmougin and Gianinazzi-Pearson, 1983; McGee, 1985; Jacquelinet-Jeanmougin *et al.*, 1987) and in a number of wood-land species including *Erythronium, Trillium, Asarum* (Brundrett and Kendrick, 1990a, 1990b), *Taxus* (Strullu, 1978), *Acer saccharum* (Yawney and Schultz, 1990; Cooke *et al.*, 1993), *Liriodendron* (Gerdemann, 1965) and *Ginkgo* (Bonfante-Fasolo and Fontana, 1985). More recent surveys have confirmed the occurrence of *Paris*-type AM in a wide var-iety of plant species and habitats.

Despite these early observations, *Paris*-type AM have frequently been regarded as atypical and generally ignored. Smith and Smith (1997) surveyed the literature for information on the occurrence of both *Arum* and *Paris* types and found reports of *Paris*-type AM in many families of pteridophytes, gymnosperms and angiosperms. Coils rather than arbuscules also predominate in many heterotrophic plants such as

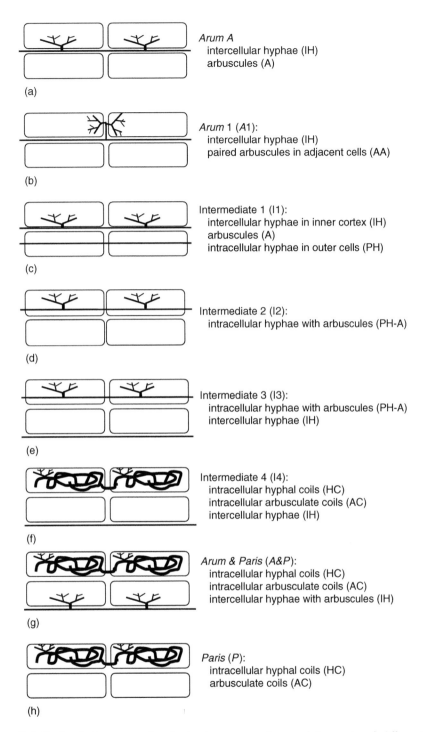

Figure 2.6 Stylized diagrams of colonization types observed in roots of different plant species, colonized by different AM fungi. (a) *Arum*-type; (b) *Arum*-type 2; (c–f) four different intermediate types; (g) *Arum*- and *Paris*-types in the same root system; (h) *Paris*-type. IH, intercellular hyphae; PH, intracellular hyphae; A, arbuscules; AA, paired arbuscules; HC, hyphal coils; AC, arbusculate coils; IH-A, arbuscules associated with intercellular hyphae; PH-A, arbuscules associated with intracellular hyphae. From Dickson (2004), with permission.

the gametophytes of *Psilotum* (Peterson *et al.*, 1981) and in the roots of some achloro-phyllous members of the Gentianaceae and Burmanniaceae (see Chapter 13). Initially, doubt was cast on the identity of the fungi but, increasingly, inoculation tests and molecular fingerprinting have confirmed them as members of the Glomeromycota. The increasing focus on structural diversity of mycorrhizas formed by AM fungi has led to an enormous increase in knowledge of the plant families known to form *Paris*-type AM (Dickson *et al.*, 2007). Overall, more families have been recorded forming *Paris*-types with intracellular coils than *Arum*-types with 'typical' arbuscules, empha-sizing that *Paris*-types are certainly not rare or unusual. Indeed, they may have been overlooked because of the overwhelming focus on arbuscules as supposedly defining features of the symbiosis and in fungal taxonomy.

Until the 1990s, little experimental emphasis had been placed on AM variants other than *Arum* types. Apart from developmental studies using light microscopy, a few details of the ultrastructure of *Gentiana*, *Ginkgo* and *Liriodendron* mycorrhizas (Kinden and Brown, 1975a, 1975b; Bonfante-Fasolo and Fontana, 1985; Jacquelinet-Jeanmougin *et al.*, 1987) and a small amount of experimental work on growth responses, we were relatively ignorant of the *Paris*-type mycorrhizas. Smith and Smith (1997) have provided a more extensive review of the early work than can be included here. The balance in emphasis is now being redressed by vegetation sur-veys and experimental investigations targeting development and function of *Paris*-type AM and other structural variants formed by AM fungi. The roles of plant and fungal genome in determining mycorrhizal morphology are likewise being inves-tigated. Accordingly, it is recognized that plant identity plays a part in determin-ing AM type and that it is possible to make generalizations with respect to which type a particular plant family or subfamily is likely to form (Smith and Smith, 1997). However, the fungal genome is also important. For example, tomato forms *Arum*-type AM with some fungi and *Paris*-AM with others (Cavagnaro *et al.*, 2001a). Some records of plant species with 'mixed' infections or families in which more than one type has been reported to occur may partially be explained by different or multiple AM fungal colonists (Smith and Smith, 1996; Cavagnaro *et al.*, 2001a).

Establishment of colonization

Precolonization events

Regardless of intraradical morphology, AM colonization of roots can be initiated from hyphae growing from any of the three sources of inoculum described earlier. Details of the colonization process were initially studied using spores, sporocarps or infected segments of root as inoculum, either in monoxenic cultures (Mosse and Phillips, 1971; Mosse and Hepper, 1975; Hepper, 1981; Bécard and Fortin, 1988; Bécard and Piché, 1989a) or on slides or membranes buried in soil (Powell, 1976; Giovannetti and Citernesi, 1993; Giovannetti *et al.*, 1993a, 1993b, 2000, 2004; Giovannetti, 1997). Observations of colonization from natural inoculum in soil have been made non-destructively in glass-sided boxes (Friese and Allen, 1991) or using repeated harvest times in 'nurse pots' containing mycorrhizal plants into which the plants of inter-est are transplanted once a hyphal network has developed (Brundrett *et al.*, 1985; Rosewarne *et al.*, 1997) (see Figure 2.4).

Primary colonization of roots from discrete propagules can be initiated from as far away as 13 mm, as shown by calculations of the effective width of the rhizosphere of *Trifolium subterraneum*, which increased linearly at 0.5 mm/day up to 12 days, suggesting that hyphae grew towards the root at this rate (Smith SE *et al.*, 1986). In a number of investigations in both soil and axenic systems, the main hypha (diameter 20–30 μm) approaching a root branches profusely (Figure 2.7a) and gives rise to a characteristic fan-shaped complex of lateral branches (diameter 2–7 μm), which may be septate and colonization of the root usually occurs from these narrow lateral hyphae (Figure 2.1b). Giovannetti *et al.* (1993a, 1993b) used an elegant and simple system of millipore membrane sandwiches to study the differential morphogenesis of hyphae in response to the presence of host and non-host roots, while preventing actual contact between them. They showed, as can be seen in Figure 2.7b, development of a densely branched hyphal network on the surface of the membrane immediately over the roots of host, but not non-host plants.

These results indicate the existence of exchange of signals in prepenetration stages but, as Friese and Allen (1991) emphasize, the formation of precolonization fans or obvious changes in hyphal morphogenesis do not always occur in soil and sometimes a relatively undifferentiated, thick-walled hypha infects the root directly, confirming some much earlier observations (Nicolson, 1959).

Contact and penetration

Hyphal contact with the root is usually followed by adhesion of the hypha to the root surface and, after about 2–3 days, the formation of swollen appressoria (Brundrett *et al.*, 1985; Bécard and Fortin, 1988; Giovannetti *et al.*, 1993b; Peterson and Bonfante, 1994; Rosewarne *et al.*, 1997; Giovannetti and Sbrana, 1998), followed by root penetration and formation of arbuscules around 2 days later (Brundrett *et al.*, 1985; Rosewarne *et al.*, 1997). Although not all AM fungi form well-defined appressoria under all conditions, morphogenetic changes on the surface of the root indicate that the fungus has recognized the presence of a potential host plant (see Figure 2.7b). The stimulus to form appressoria appears to be associated with epidermal cell walls (Nagahashi and Douds, 1997) and to be absent from many non-host plants and from artificial fibres (Giovannetti *et al.*, 1993a; Nagahashi and Douds, 1997; Giovannetti and Sbrana, 1998).

Early stages of colonization were studied at the ultrastructural level in *Allium porrum–Glomus versiforme* (Garriock *et al.*, 1989). The elongated and elliptical appressoria are aligned with their long axes parallel to the long axes of the epidermal cells (Figure 2.8c). There was little variation in either shape or position of the appressoria, but the length ranged from 16.8 to 79.8 μm. Wide diameter colonizing hyphae bearing small projections always develop from the appressoria and, in a high proportion of cases, penetrate both the adjacent epidermal cells (Figure 2.8d). Colonization by other fungi is less well documented, but it is clear that single hyphae originating from the appressorium are more common for some (Figure 2.8e).

Changes in the plant indicate recognition of fungal attachment. Garriock *et al.* (1989) observed the regular occurrence of slight wall thickenings on the epidermal cell adjacent to the penetrating hyphae and, even in the absence of thickening, the walls fluoresced strongly after staining with acriflavine-HCl for polysaccharides

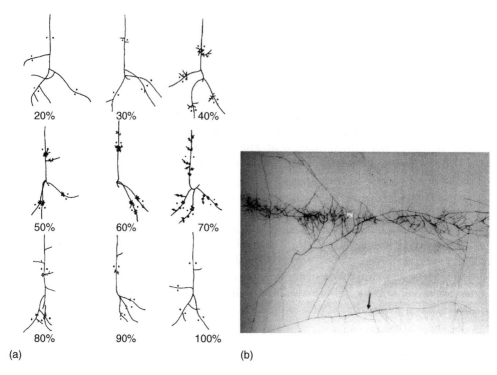

Figure 2.7 (a) Tracings (taken at ×20) of germ tubes and hyphae growing from spores of *Gigaspora gigantea*, 24 h after injections of various fractions isolated from exudates of roots of carrot (*Daucus carota*). The filtered carrot root exudates from 28-day-old root cultures were passed through a C18 SEPAK cartridge. The compounds bound to the cartridge were removed by successive washes with increasing concentrations of methanol (20–100%) collected and concentrated. No activity was found in material that did not bind to the cartridge (not shown) or in the 20 and 30% fractions. Greatest branching activity was observed in 50–70% fractions, with slight activity also found in 40 and 89% fractions. Reprinted from *Mycological Research*, 104, Nagahashi and Douds, (2000). (b) Differential hyphal morphogenesis elicited in *Glomus mosseae* by roots of host plants growing underneath a millipore membrane. Dense hyphae showing morphogenetic response above the roots of *Oncimum basilicum*. Non-elicited hyphae at some distance from the roots (arrowed). From Giovannetti *et al.* (1993a), with permission.

with vicinal hydroxyl groups. As there was no fluorescence with aniline blue or berberine sulphate, these thickenings probably did not contain either callose or lignin, a point confirmed by Harrison and Dixon (1994) for *G. versiforme* colonizing *Medicago*. However, *G. mosseae* did not induce such a response in *Pisum sativum* (Gollotte *et al.*, 1993) and *G. versiforme* caused no changes in synthesis of phenols in either *Allium* or *Ginkgo* (Codignola *et al.*, 1989). It is therefore not clear whether the variations are related to plant or fungal species but, in any event, the thickenings did not prevent the penetration of fungal hyphae through the walls.

Penetration of plant cell walls is always associated with narrowing of the hyphal diameter to form a peg, followed by expansion as the hypha enters the lumen of the cell (Figure 2.8e). The cell wall may sometimes bulge as the hyphae penetrate, indicating that pressure may play a part in the penetration process (Cox and Sanders, 1974; Holley and Peterson, 1979). However, changes in the middle lamella structure when intercellular spaces are colonized by hyphae indicate the involvement of fungal enzymes such as pectinases (Kinden and Brown, 1975b; Gianinazzi-Pearson et al., 1981b), a suggestion supported by biochemical evidence of their production by spores and external mycelium (García-Romera et al., 1990, 1991). Activities of other hydrolytic enzymes, such as cellulases and xyloglucanases, are also elevated in AM roots, although direct involvement in fungal penetration of plant cell walls has not been demonstrated (García-Garrido et al., 1992, 2000).

The penetration of the outer cell layers of the root (epidermis and exodermis) is a key point in colonization. Plant control of this step is indicated by changes in fungal morphology as hyphae pass through the cells and by the existence of a large number of AM-defective mutants which do not permit penetration to the root cortex (Giovannetti and Sbrana, 1998; Bonfante et al., 2000; Gao et al., 2001). Recent work showing the rapid development of a highly specialized prepenetration apparatus (PPA) that facilitates hyphal passage through the root epidermal cells (Genre et al., 2005) is described in Chapter 3.

In many plant species, including the very well-studied Allium, the layer of cells immediately beneath the epidermis is characterized by the presence of a casparian band on the tangential walls and, as the root matures, by increasing depositions of suberin on both tangential and radial walls (Shishkoff, 1986; Peterson, 1988; Enstone et al., 2002). In some species, this exodermal layer is dimorphic and, whereas the numerous long cells become rapidly suberized, in the short 'passage cells' suberization is delayed. The casparian band on the tangential walls offers no barrier to the entry of hyphae of AM fungi but the suberization on the radial walls appears to do so. Consequently, the fungus enters by the passage cells and always coils within them (Gallaud, 1905; Kinden and Brown, 1975a; Brundrett et al., 1985; Smith et al., 1989; Brundrett and Kendrick, 1990a, 1990b). The interaction of the fungi with the exodermal layer may be important both because it may affect the timing of susceptibility of the root (see below) and the control of the composition of the solution in the cor-tical apoplast, with consequences for nutrient transfer between the symbionts (Smith and Read, 1997).

Figure 2.8 Early stages of AM colonization of roots. (a, b) Precolonization branching of external hyphae (EH) and formation of appressoria (arrowed) by *Glomus monosporus* on the roots of *Trifolium subterraneum*. From Abbott (1982), with permission. (c) A single appressorium (*) of *Glomus versiforme* formed as a swollen branch hypha on the surface of a root of *Allium porrum*. External hyphal arrowed. From Garriock et al. (1989), with permission. (d) Infection branches (arrowed) developing from an appressorium (*) of *Glomus versiforme* on the surface of a root of *Allium porrum*. From Garriock et al. (1989), with permission. (e) Section of resin embedded tissue of *Allium porrum* showing an appressorium (*) of *Glomus versiforme* in contact with the epidermal cell (C) and the development of infection branches. From Garriock et al. (1989), with permission.

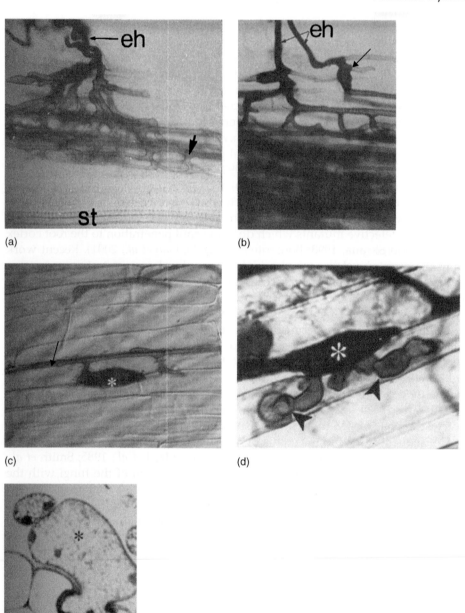

Figure 2.8 (Caption on p. 63)

Development of infection units in Arum-type mycorrhizas

Following the formation of an appressorium and penetration of the epidermis and exodermal cells, hyphal branches pass into the middle and inner cortex of the root and, in *Arum*-type mycorrhizas, grow longitudinally in the intercellular spaces. Thus the fungal hyphae develop in a fan-shaped way across the outer cells of the cortex (see Figures 2.8a, b), with a growing internal mycelium subtended by one or a few closely associated entry points on the epidermis. These independent colonies or mycelia were called 'infection units' (Cox and Sanders, 1974), a term which we retain. Each infection unit develops longitudinally and, to some extent, radially in the cortex of the root. Branches from the longitudinal hyphae give rise to arbuscules in the cells. Hence, the oldest arbuscules are closest to the site of penetration and the young and immature ones are progressively further away (see Figure 2.5b). The rates of growth of the intercellular hyphae have been estimated and, depending on the method, on the species of plant and on environmental conditions, range from 0.13 to 1.22 mm/day (Smith and Walker, 1981; Walker and Smith, 1984; Brundrett *et al.*, 1985; Tester *et al.*, 1986; Brundrett and Kendrick, 1990a; Bruce *et al.*, 1994). The maximum longitudinal extent of the infection units appears to be about 5–10 mm in each direction from a simple or complex entry point. Intercellular hyphae may branch, anastomose and form multihyphal cords in some particularly large intercellular spaces. With the electron microscope or appropriate staining, the hyphae can be seen to be multinucleate and they contain dense cytoplasm with numerous organelles, bacteria-like organisms (BLOs) and vacuoles, as well as glycogen and lipid reserves (Scannerini and Bonfante-Fasolo, 1983; Bonfante-Fasolo *et al.*, 1984; Peterson and Bonfante, 1994; Bago *et al.*, 1999b). The distribution of nuclei is fairly uniform (Bonfante-Fasolo *et al.*, 1987; Cooke *et al.*, 1987; Bécard and Pfeffer, 1993; Bianciotto and Bonfante, 1993) and the hyphae are long-lived compared with the arbuscules, at least in fast growing crop plants (Holley and Peterson, 1979; Smith and Dickson, 1991). Thus, in *Arum*-type mycorrhizas, the intercellular hyphae appear to provide the living and persistent communication pathway of a mycorrhizal infection unit whose functions include translocation of nutrients to and from the extraradical mycelium and possibly also transfer between the symbionts.

The characteristics of the infection units vary in different plant and fungal species. In some plants, such as *Allium*, arbuscules are scattered in all cortical cell layers (Figure 2.9a), whereas in others, such as *Medicago* and *Cucumis*, they develop as a dense layer in the cells of the inner cortex close to the vascular cylinder, with the outer cortical cell layers relatively free of intracellular fungal structures (Figure 2.9b).

Development and turnover of arbuscules

Arum-type arbuscules are usually relatively short-lived and their development, maturation and collapse has been investigated at both the light and electron microscope levels in many plant–fungus combinations, so that it is possible to generalize about the changes that occur in cells of both symbionts (Scannerini and Bonfante-Fasolo, 1983; Bonfante-Fasolo *et al.*, 1984; Peterson and Bonfante, 1994). Detailed studies of arbuscule development and degeneration in several plant species have been made, using morphometric techniques (Toth, 1992).

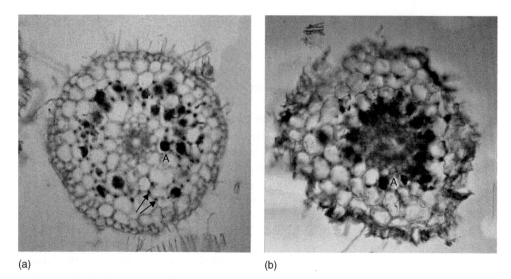

(a)   (b)

Figure 2.9 Arrangements of arbuscules in the root cortex. (a) Arbuscules (A) and intercellular hyphae (arrowed) distributed throughout the cortex of a root of *Allium porrum*. (b) Arbuscules (A) localized in a ring immediately adjacent to the endodermis in the inner root cortex of *Cucumis sativus* (cucumber). Courtesy Sandy Dickson.

When a hyphal branch penetrates the plant cell wall to form the main trunk of the arbuscule (Figure 2.10a, b, c) the plasma membrane is not penetrated, but is invaginated so that the invading hypha and all its branches remains enveloped by it and the fungal structure is always located outside the plant cell cytoplasm within an apoplastic compartment (Figure 2.10d, e). The plant membrane surrounding the arbuscule (periarbuscular membrane or PAM) is clearly modified functionally, although it retains staining properties and some activities similar to the peripheral plasma membrane of the cell from which it is derived (Dexheimer *et al.*, 1979, 1985). Specialization of the PAM is indicated by its reactions with monoclonal antibodies (Gianinazzi-Pearson *et al.*, 1990) and activities of enzymes, nutrient transporters and aquaporins (Gianinazzi-Pearson *et al.*, 2000; Rausch *et al.*, 2001; Harrison *et al.*, 2002; Glassop *et al.*, 2005; Porcel *et al.*, 2006; Aroca *et al.*, 2007). Such detailed investigations of the membrane surrounding intracellular coils have not been made, but there are clear indications of specialization in the distribution of mycorrhiza-inducible phosphate transporters around them (Karandashov *et al.*, 2004; Glassop *et al.*, 2005).

Together with the fungal plasma membrane in the arbuscule branches, the PAM delimits an interfacial zone (the interfacial matrix or apoplast) which appears highly specialized with respect to the molecules deposited within it and which has an important part to play in nutrient transfer between the symbionts (Smith SE and Smith, 1990; Bonfante-Fasolo *et al.*, 1992; Balestrini and Bonfante, 2005). At the base of the trunk of the arbuscule, a layer of material is laid down (see Figure 2.10d, e) which has similar chemical composition to the primary wall of the plant cell. It is thick at the base where it is continuous with that wall, but higher up the trunk hypha it becomes gradually thinner and appears absent from the finest branches of the arbuscule. A range of affinity probes, such as antibodies, lectins and enzymes,

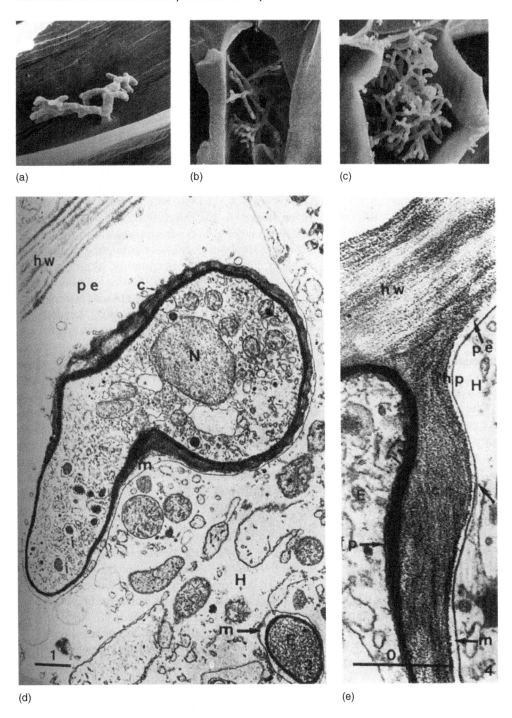

(a)

(b)

(c)

(d)

(e)

Figure 2.10 (Caption on p. 68)

has been employed to investigate the macromolecular composition of the walls of the symbionts and the interfacial apoplast. Wall deposition by both symbionts is clearly curtailed in the fine arbuscular branches. Structural molecules of plant origin, such as β (1–4) glucans, non-esterified polygalacturonans and hydroxyproline-rich glycoproteins (HGRP) are present, but are not polymerized as they are in a typical wall. Chitin is much reduced in the fine branches of the arbuscules, reflecting major changes in thickness and composition of the fungal walls as colonization proceeds.

Spores have thick, composite walls which contain a high proportion of chitin laid down in a complex helicoidal arrangement (Bonfante-Fasolo and Grippiolo, 1984; Bonfante-Fasolo, 1988; Grandmaison *et al.*, 1988), extraradical, intracellular coils and intercellular hyphae also have relatively thick walls (around 500nm) with laminated chitin fibrils, which may become even thicker in the intercellular phase as the roots themselves age and undergo secondary thickening. In contrast, the walls of the arbuscular branches show progressive thinning to about 50nm as the hyphae themselves are reduced in diameter to 1–2μm. No fibrillar structure is apparent, although chitin molecules have been detected, at least in *Glomus versiforme* (Bonfante-Fasolo *et al.*, 1992; Bonfante-Fasolo and Perotto, 1992). As shown in Figures 2.11a and b, prepared by freeze substitution, the PAM and fungal wall are closely adpressed. Very little is known of the solute composition of the interfacial apoplast, but it has a relatively low pH (Güttenberger, 2000) as would be predicted from activities of H^+-ATPases involved in membrane transport. Concentrations of other solutes such as P, sugars, amino acids, mono- and divalent cations have not been determined. Nevertheless, they are of key importance in transport between the symbionts and information is urgently needed.

The fact that arbuscules provide a considerable increase in surface area of contact between fungus and plant has led to the belief that they are involved in nutrient transfer and it is highly probable from other evidence that they are indeed the sites for movement of soil-derived nutrients, like P and Zn, to the plant. They are, however, not the only interfaces that may be involved. It is less clear that they are the sites of carbohydrate transfer to the fungus; arguments for the possible role of intercellular hyphae and *Paris*-type coils, as well as arbuscules will be discussed later. Here

---

Figure 2.10 Scanning electron microscopy of stages in the development of arbuscules of *Glomus mosseae* within cells of *Liriodendron tulipifera*. (a) Young arbuscule showing penetration point and dichotomous branching. (b) and (c) Later stages in arbuscule development showing how the hyphal branches come to fill the cell volume. From Kinden and Brown (1975), with permission. (d) Section through an arbuscule trunk which is giving rise to a branch. The interfacial matrix (m) can be seen to be an extension of the host periplasm (pe). The densely stained fungal wall is clearly distinguishable from the surrounding coating of fibrils and also from the host wall (hw). Features of fungal and host cytoplasm can be distinguished, both contain mitochondria and a nucleus. Bar=1μm. From Dexheimer *et al.* (1979), with permission. (e) Detail of a penetration point of a hypha (E) into a host cell (H). The host plasma membrane (hp) is invaginated by the arbuscular trunk hypha, so that the fungus is surrounded by a continuation of the host periplasm (pe), forming an interfacial matrix (m). Within the matrix an apposition layer (c) of fibrillar material has been laid down, which is continuous with the host wall (hw) and clearly distinguishable from the darkly stained fungal wall. Bar=0.5μm. From Dexheimer *et al.* (1979), with permission.

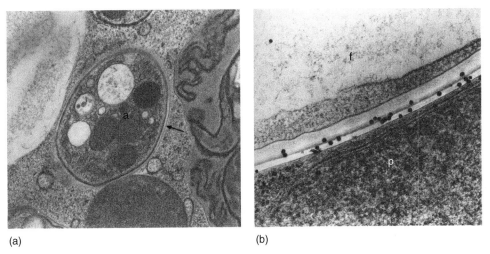

(a)   (b)

Figure 2.11 Detail of interfacial regions between AM fungi and root cortical cells in sections prepared by freeze substitution. (a) Mycorrhiza formed between *Gigaspora margarita* and *Trifolium*. The arbuscule branches (a) are surrounded by the invaginated host membrane, forming a periarbuscular membrane (arrowed). In this material the thin fungal wall is in close contact with the periarbuscular membrane (arrowed). Photograph courtesy Paola Bonfante. (b) Detail of the host–fungus interface after treatment with Wheat Germ Agglutinin-gold, to reveal N-acetylglucosamine residues (•) in the amorphous fungal wall. Plant cell, p; fungal hypha, f. Photograph courtesy Paola Bonfante.

it is important to point out that the calculations of Cox and Tinker (1976), together with new information on the presence of H^+-ATPases and phosphate transporters (Gianinazzi-Pearson *et al.*, 1991a, 2000; Rausch *et al.*, 2001; Harrison *et al.*, 2002; Karandashov and Bucher, 2005) (see Chapter 5), indicate that the collapse and presumed 'digestion' of arbuscules are unlikely to play a significant role in nutrient transfer from fungus to plant.

At the stage where an arbuscule is growing and reaching maturity in the intracellular apoplast, the mycorrhiza can be envisaged as an association between metabolically active fungal structures and living root cells (Dexheimer *et al.*, 1979). There is also very good evidence for increased physiological activity of the plant cells including increased mRNA (Chapter 3) and protein content (Delp *et al.*, 2003). The nucleus and nucleoli increase in size and the nucleus moves from a peripheral position close to the cell wall to take up a central position surrounded by the arbuscule branches. Major changes in cytoskeletal activity are implicated in these movements, as well as in the accommodation of the arbuscule itself within the cell (Bütehorn *et al.*, 1999; Timonen *et al.*, 2001; Timonen and Peterson, 2002; Delp *et al.*, 2003; Timonen and Smith, 2005). The volume of the cytoplasm, which contains a full complement of organelles, is also increased and the vacuoles appear to become fragmented. The increased size of the plant nuclei is associated with an increase in the amount of decondensed chromatin, indicating greater activity and delayed senescence and, in some examples, increase in ploidy (Berta *et al.*, 1988, 1990a, 1990b, 1991, 1993, 1995, 2000; Blair *et al.*, 1988).

The short life span of arbuscules differs quite markedly from intercellular hyphae and intracellular coils. Again, it was the early workers who first noted the way in

Table 2.1 The effect of development of arbuscules on the surface area of protoplasts, S(p) and surface:volume ratios of cells Sv (p,c) of different plant species colonized by *Glomus fasciculatus*.

Plant species	Cell volume	S(p)1 uninvaginated (μm^2)	Sv(p,c)2 uninvaginated cell ($\mu m^2/\mu m^3$)	S(p) invaginated cell (μm^2)	Sv(p,c) invaginated cell ($\mu m^2/\mu m^3$)	Area of arbuscular interface per cell (μm^2)
Wheat	212 874	23 416	0.11	83 012	0.39	59 605
Oats	221 875	22 188	0.10	73 219	0.33	51 031
Maize	177 328	19 506	0.11	79 798	0.45	60 292
Mean for grasses			0.11	78 679	0.39	56 978
Onion	340 468	34 074	0.10	64 688	0.19	30 614
Bean	146 574	16 123	0.11	49 835	0.34	33 712
Tomato	147 910	16 270	0.11	42 894	0.29	26 624
Mean for non-grasses			0.11	52 472	0.27	30 316

Values averaged over the whole cell cycle. Based on a table from Toth *et al.*, 1990.

which arbuscules grow in the cells and subsequently collapse to form clumps, while the plant cell remains alive (Gallaud, 1905) (see Figure 2.5a). During this phase, the fungal chromatin condenses and nuclear degeneration takes place (Balestrini *et al.*, 1992). The collapsed arbuscule material contains high concentrations of Ca, but composition otherwise differs little from turgid arbuscule branches (Ryan *et al.*, 2003). Rarely, a cell may become reinfected by the fungus and come to enclose several digestion clumps. Following the work of Cox and Tinker (1976), a number of elegant studies of arbuscule turnover have been carried out in different plant species using morphometric techniques applied to electron micrographs of arbuscules in different stages of growth and collapse (Toth and Toth, 1982; Alexander *et al.*, 1989; Toth *et al.*, 1990, 1991; Toth, 1992). Important points to note are that the surface:volume ratio of the plant cells is at first increased considerably, representing a 2–4-fold increase in plasma membrane area as a result of invagination by the arbuscules. Figure 2.12 shows that the species of host plant has a considerable effect on the per cent volume change and surface:volume ratio of plant protoplasts during development of arbuscules of *Glomus fasciculatum*. Arbuscules in grasses were generally larger than in the non-grasses, with consequent effects on the increase in volume of the plant cytoplasm and the surface area of interface between the symbionts (Table 2.1).

The duration of the arbuscular cycle also varies. In *Triticum aestivum*, formation of arbuscules took 2–3 days and the whole arbuscular cycle around 7 days, which agrees well with earlier estimates (Bevege and Bowen, 1975; Brundrett *et al.*, 1985) and appears to be typical of mycorrhizas of rapidly growing crop species (Alexander *et al.*, 1988). However, in slow growing woodland plants, the arbuscules appear not only to be much longer lived, but may also have wider and more robust branches (Brundrett and Kendrick, 1990a). We do not know why arbuscules have such short life spans in many plants. None of the proposed explanations are well supported by evidence, but (apart from the idea of 'digestion' being involved in nutrient transfer)

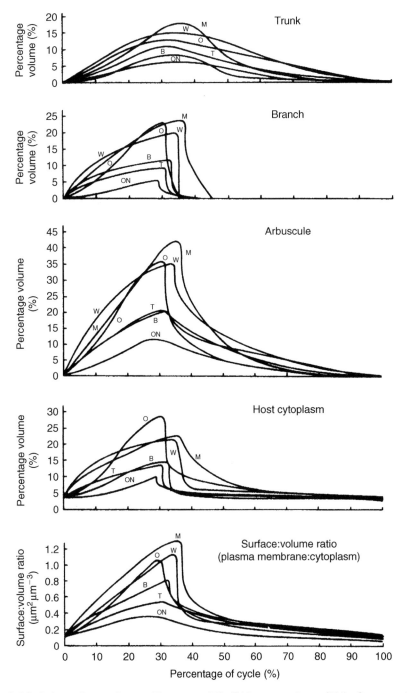

Figure 2.12 Arbuscular cycles in *Zea mays* (M), *Triticum aestivum* (W), *Oryza sativa* (O), *Phaseolus vulgaris* (B), *Lycopersicon esculentum* (T) and *Allium cepa* (ON). The volume fractions of the different structures are plotted as a percentage of the plant cell volume occupied by each feature and the surface/volume ratio of the cell is given in $\mu m^2 \ \mu m^{-3}$. From Alexander *et al.* (1989), with permission.

they include a manifestation of a host defence reaction against progressive fungal invasion and a response to extreme lowering of apoplastic pH, consequent on membrane transport events. Support for active host cell involvement is generally lacking (see Chapter 3) and a 'dead end' or programmed cell death of the fungus, which undergoes autolysis in what turns out to be the stressful environment of the cortical cell, may be the best explanation (Harley and Smith, 1983; Peterson and Bonfante, 1994). In support of this idea, each degenerating arbuscule is cut off from its subtending intercellular hyphae by a cross wall, indicating a likely stress response and preventing back-flow of nutrients contained in the arbuscule to the intercellular hypha (Dickson and Smith, 2001).

Intracellular development in Paris-type mycorrhizas

Much less is known of the development of infection units typical of intracellular development of *Paris*-type mycorrhizas. Although the general impression is that colonization is slower than in *Arum* types, the data of Dickson (2004) do not fully support this conclusion for she showed similar rates of overall development of *Arum*-, *Paris*- and intermediate arbuscular mycorrhizas in a range of host species. Nevertheless, infection units are frequently dense and compact, reflecting the way the fungi grow directly from one cortical cell to another, rather than extending through the relatively low-resistance pathways provided by intercellular spaces. The absence of intercellular development in some plants was first thought to be a consequence of lack of these spaces (Gallaud, 1905; Brundrett and Kendrick, 1990a, 1990b), but is certainly not the universal reason because *Paris* types can develop in root systems with extensive intercellular spaces (Imhof and Weber, 1997) and both *Arum*- and *Paris*-type morphologies develop in the same root system, most likely depending on the species of AM fungi (Smith and Smith, 1996). For example, tomato forms *Arum*-AM with one group of fungi, including *Glomus intraradices* and *G. mosseae*, but *Paris*-AM with *G. coronatum*, *Scutellospora calospora* and *Gigaspora margarita* (Cavagnaro *et al.*, 2001a). Mycorrhizal type was not closely correlated with hyphal diameter, as might be expected if relative dimensions of hyphae and intercellular spaces controlled which type could develop (Toussaint and Cavagnaro, unpublished).

Investigations of *Paris*-type development indicate that simple coils develop prior to the formation of any arbusculate branches (Cavagnaro *et al.*, 2001c) and that, as with arbuscules, the plant nucleus moves from a peripheral position in the plant cortical cell to a central position within the coil (Cavagnaro *et al.*, 2001b). In *Asphodelus* (one of the few species to be investigated in detail), simple hyphal coils were found in the outer cortical cell layers and arbusculate coils closer to the vascular cylinder (Cavagnaro *et al.*, 2001c). Confocal microscopy, coupled with 3D reconstruction, shows that the surface area of interface between coils and cortical cells can be as large as with arbuscules (Dickson and Kolesik, 1999), emphasizing that both types of structures considerably increase the area of interface across which nutrient transport could occur. Coils formed by *Glomus intraradices* in *A. fistulosus* are active with respect to succinate dehydrogenase and acid and alkaline phosphatase activity (van Aarle *et al.*, 2005). These authors suggest that both coils and arbusculate coils may be involved in P transfer to the plant, substantiating the observations of localization of H^+-ATPase activity to the plant membrane surrounding arbusculate coils and mycorrhiza-inducible phosphate transporters expressed in cortical cells containing coils (Marx *et al.*, 1982; Karandashov *et al.*, 2004). Formation of *Paris*-type AM can lead to positive plant responses, for Cavagnaro *et al.* (2003) showed improved P nutrition in *Asphodelus fistulosus*; all three lines of evidence implicate coils in mycorrhizal P transfer. All in all evidence is accumulating that supports the widespread occurrence and functional importance of patterns of mycorrhiza development that differ markedly from the well-known *Arum*-AM. However, there is an urgent need to undertake comparative studies which target other intracellular phases such as coils. If we are fully to understand the diversity of AM and their importance in ecosystems, it is crucial to extend investigations to encompass the full structural and functional variations that have now been rediscovered.

Formation of vesicles

As individual infection units of all types age, thick-walled vesicles may be formed, depending in part on the identity of the fungus; members of the Gigasporaceae (*Scutellospora* and *Gigaspora*) never develop vesicles, but instead produce auxiliary cells on the extraradical mycelium. Members of all other genera develop vesicles to varying degrees and in either intercellular or intracellular positions in the cortex (Abbott, 1982). Environmental conditions strongly affect vesicle development, so that at high P or low irradiance they are reduced in the same way as arbuscules. Vesicles are thick-walled structures of varying shapes from ovoid, irregularly lobed to box-like, depending on the species of fungus and where the vesicle is formed. They contain abundant lipid and numerous nuclei and it is likely that they are important storage organs, playing a significant role as propagules within root fragments, as already mentioned. Nevertheless, little is known of their biology particularly with respect to either germination or mobilization of the reserves.

Fungal biomass in roots

The earliest estimates of biomass of AM fungi associated with roots were based on determinations of root chitin content or volume of fungus (Harris and Paul, 1987; Toth *et al.*, 1991). Hepper (1977), using chitin content, obtained estimates of the dry weight of the fungus within the roots which ranged from 4% to 17% of the total dry weight. Bethlenfalvay *et al.* (1982a) estimated the chitin content of extraradical and internal hyphae of *Glomus fasciculatus* grown with *Glycine max* and found that both the internal mycelium and extraradical mycelium reached a maximum at about 8 weeks, when the fungus contributed 20% of the biomass of the mycorrhizal roots, of which about a third was external hyphae or spores. Later work generally confirms these estimates, although more information is needed on the contribution of the extraradical mycelium (Bethlenfalvay *et al.*, 1982a; Kucey and Paul, 1982; Toth *et al.*, 1991). Changes in AM fungal biomass can also be determined from changes in accumulation of the specific signature fatty acids (16:1ω5) (Olsson, 1999; Olsson *et al.*, 2002). Ergosterol has also been suggested as a quantitative marker for AM colonization (Hart and Reader, 2002a), but lack of specificity for AM fungi will limit the reliability of data, particularly from field samples where other fungi may be present (Olsson *et al.*, 2003). The relative contributions of plant and fungus to root activity

have also been estimated from rRNA accumulation, with somewhat variable conclusions (Maldonado-Mendoza *et al.*, 2002; Delp *et al.*, 2003; Isayenkov *et al.*, 2004).

Co-colonization of roots by different species of fungi

The diverse assemblages of AM fungal species found in soil and the low specificity of
the symbioses formed by them imply that a single root system will, in all probability,
be colonized by several different fungi. The extent to which each occupies and exploits
the root environment and also grows into the soil is of considerable interest with
respect to understanding the complex plant–fungus interactions in field situations.
Daft and Hogarth (1983) compared the competitiveness of four AM fungi, assessed
after multiple inoculation, in terms of production of spores. They showed that *G.
mosseae* suppressed spore production by *G. caledonium* in both maize and onion.
Wilson took a direct microscopic approach to understanding the way different fungi
compete for root occupancy (Wilson and Trinick, 1983; Wilson, 1984; Wilson and
Tommerup, 1992). Exploiting variations in developmental pattern of different AM
fungi in roots of *Trifolium subterraneum*, she studied the effects of different ratios of
inoculum of pairs of fungi on timing and extent of fungal colonization and concluded
that mixed inoculation usually resulted in similar values to colonization formed by
the same total inoculum level of the most infective fungus. However, competitive outcomes varied with the identity of the fungal competitors, some being more aggressive
(competitive) than others. This character was inversely proportional to the extent of
colonization at the final harvest (referred to as 'infectivity'). Some fungi appeared to
exclude others from the same sections of root, preventing 'dual occupancy'. However,
dual or multiple occupancy has frequently been observed. Detection of four AM fungal species by specific molecular fingerprinting in stained root fragments of *Allium*,
and analysis of changes in frequency of occurrence in the fragments, indicated that
Gigaspora rosea and *Scutellospora castanea* occurred more frequently when they were in
the presence of other fungi (van Tuinen *et al.*, 1998).

Jansa *et al.* (2007) attempted a more sophisticated quantification using real time
PCR to determine copy number of the ribosomal large subunit (LSU) genes of three
AM fungi inoculated singly or together in the roots of *Medicago* or *Allium*. As shown
in Figure 2.13, colonization of *Medicago* by single fungi varied between 50 and
95% of root length. *Glomus mosseae*, when coinoculated with other fungi, was the
best competitor, occupying around 75% of the root systems when applied at equal
inoculum densities with *G. intraradices* or *G. claroideum* or in a mixture of all three
AM fungi. *G. intraradices* also out competed *G. claroideum* in dual inoculation of
leek and *Medicago* and also following triple inoculation of leek. These examples
show that even in plant species that are well known to become rapidly and extensively colonized, AM fungi vary considerably in the amount of root they come to
occupy during competitive interactions. The possible mechanisms underlying such
differences were outlined by Wilson and Tommerup (1992). They discounted direct
competition for space on the grounds that roots were often not completely colonized and that co-occurrence of two species in the same small root segment was
often observed, but considered that different modes of root colonization might
contribute. Access to sugars was also suggested to influence the results of competition, supported by the results of Pearson *et al.* (1994) on soluble carbohydrate status of roots of *T. subterraneum*. Variations in sink-strength for organic C between

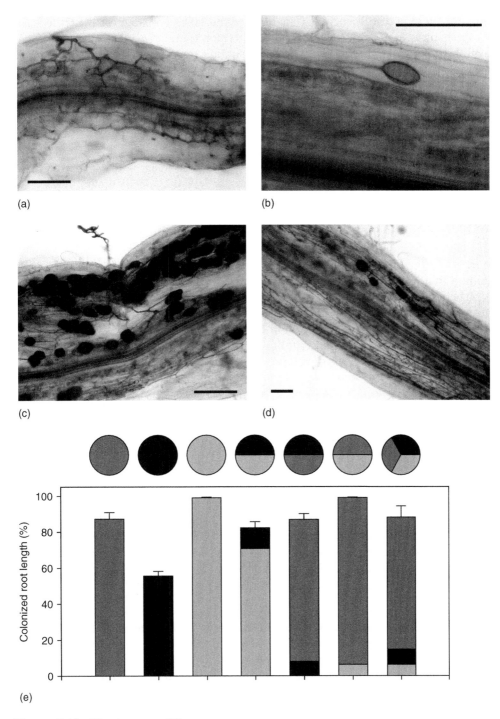

(a)

(b)

(c)

(d)

(e)

Figure 2.13 (Caption on p. 76)

different fungi and influenced by identity of the host, also suggest this is a feasible mechanism to explain outcomes of competition (Lerat *et al.*, 2003a). This mechanism, however, was discounted as an explanation for reduction in colonization of one half of a split root system of barley by prior colonization of the other half by *G. mosseae*, because sink-strength was not a function of extent of colonization (Vierheilig, 2004b). Here, the mechanism proposed was a direct 'phytoprotective' one, which additionally must have been mediated by shoot signals.

Growth of external hyphae and spore production

External hyphae

As noted above, once a fungus is established in the root and is growing vigorously in the soil, the external hyphae form an important source of inoculum for ongoing colonization of the same root system and of root systems of different plants. The extensive growth represents foraging behaviour of the fungi in search of new sources of organic C as well as soil-derived nutrients. Most researchers are of the opinion that mycelial development occurs by extension and branching of existing hyphae outside the roots; there is little or no direct microscopical evidence showing intraradical hyphae growing out from the root cortex by penetrating root cell walls. Ideally, electron microscopy is required to demonstrate passage of hyphae outward from the internal mycelium and to follow their route through different cell layers.

At later stages in symbiotic development, spores and auxiliary cells are formed on the external mycelium, sometimes at very high densities, so that transfer of organic C from the roots to soil can be considerable and energy is further expended in both uptake and translocation of nutrients as well as fungal growth and respiration. Development of the external mycelium represents considerable organic C flow to the soil, which may be distributed to sites well beyond the zone normally designated as the rhizosphere (Jakobsen *et al.*, 1992a), (Figure 2.14a).

Extensive growth of external mycelium does not begin until after the root has been penetrated by the invading fungal hyphae, but it is still not clear which stages of colonization are required. Mosse and Hepper (1975) noted considerable growth after the formation of appressoria. Hepper (1981) confirmed this and observed that growth outside the root could precede the formation of any arbuscules within the cells. This suggests that nutrients could be transferred across the interface from plant cells to intercellular hyphae and that transfer across the arbuscular interface need not be involved, a point supported by intercellular fungal growth in *Pisum*

Figure 2.13 Composition of AM fungal communities in *Medicago truncatula* roots 8 weeks after planting, as affected by competition among fungal species. Plants were inoculated with *Glomus mosseae*, *G. claroideum* and *G. intraradices*, singly or in combination. The overall per cent root length colonized was assessed by the magnified intersects method of McGonigle *et al.* (1990). Morphology of colonization by *Glomus mosseae* (a), *G. claroideum* (b), *G. intraradices* (c) and by a mixture of all three species (d). The contributions of the different AM fungal species were assessed by real-time PCR quantification of copy numbers of large ribosomal subunit gene with species-specific primers (e). The pie charts above each graph indicate composition of AM fungal inoculum, consisting of one, two, or three species. Dark grey, *Glomus mosseae*; black, *G. claroideum*, and light grey, *G. intraradices*. Images courtesy Jan Jansa.

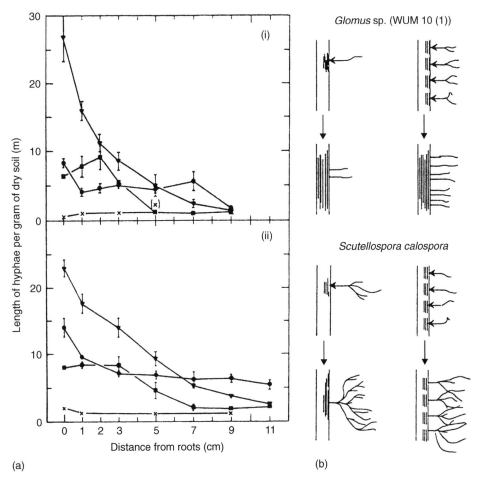

(a)   (b)

Figure 2.14 Development of external mycelium of AM fungi. (a) Length density of external hyphae spreading from mycorrhizal roots of *Trifolium subterraneum* after 28 days (top) and 47 days (bottom). Soil cores were sampled at increasing distances from a 'root compartment', up to 11 cm. Means and standard errors of means. ●, *Acaulospora laevis*; ■, *Glomus* sp.; ▼, *Scutellospora calospora*; ×, control. From Jakobsen *et al.* (1992a), with permission. (b) Suggested patterns of development of hyphae within and outside roots for two different AM fungi. From Abbott *et al.* (1992), with permission.

mutants which have highly reduced arbuscules (Gianinazzi Pearson *et al.*, 1994a). Nevertheless, Bécard and Piché (1989b) came to the conclusion, based on progressive observations of monoxenic root organ cultures, that formation of arbuscules is an absolute requirement for external hyphal growth. This conclusion is supported by the failure of a large number of AM fungi to develop any external mycelium in association with mycorrhiza defective mutants, even those in which some intraradical colonization has been shown to occur (Kling *et al.*, 1996; Poulsen *et al.*, 2005; Manjarrez-Martinez, 2007). Again, if arbuscules do provide either an essential signal for external hyphal growth or sites for fungal C acquisition, then it is a reasonable hypothesis that intracellular coils in *Paris*-AM will carry out the same roles.

Outside the root, main hyphae give rise to the characteristic branching systems illustrated in Figures 2.1 and 2.3. Friese and Allen (1991) measured the branches in glass-sided boxes and showed that up to eight orders were produced, which became progressively narrower in diameter with the finest (5th–8th orders) about $2\mu m$ in diameter. Similar observations have been made in monoxenic cultures (Bago et al., 1998a, 1998b). The finely branched hyphal fans are clearly well adapted to explor-ation of soil pores, where their diameters may change in response to pore diameter (Drew et al., 2003). Hyphae are consistently associated with organic mat-ter in soil where they proliferate in localities where mineralization of nutrients may be occurring (Nicolson, 1959; Hepper and Warner, 1983; St John et al., 1983a, 1983b). Several investigations indicate that growth of external mycelium of AM fungi is not stimulated by application of inorganic P to root-free compartments (Li et al., 1991a; Olsson and Wilhelmsson, 2000), but Cavagnaro et al. (2005) found that growth var-ied between different fungal species, with G. intraradices responding to high P by increased hyphal length. Variations between AM fungi have also been found in extent of proliferation with different distances from roots, in root free compartments and in soils of different pore-size and tortuosity (Jakobsen et al., 1992a; Smith FA et al., 2000; Smith SE, 2004a; Drew et al., 2003, 2005). The result of hyphal proliferation is the formation of a hyphal network that plays important roles, both in foraging for new sources of organic and inorganic nutrients and in stabilizing soil aggregates (Tisdall and Oades, 1979, 1982; Miller and Jastrow, 2002).

Because the external mycelium is important in so many ways, considerable effort is being made to overcome the difficulties of studying it. Early work involved pick-ing out the hyphae and weighing them or estimating their extent by the weight of soil adhering to roots. Immunofluorescence methods have been tested (Hahn et al., 1994, 2001; Hahn and Hock, 1994), but no applications have yet become routine. DNA-based methods currently have promise, particularly with respect to distin-guishing mycelium produced by different fungal species, but are not yet quantita-tive. Most of the published data on hyphal development have been obtained either using a membrane filter technique to capture hyphal fragments which are later measured by a grid intersect method (Jakobsen et al., 1992a) or, more recently, from quantitative determination of signature fatty acids such as 16.1ω5 in neutral and phospholipids in roots and soil (Olsson and Wilhelmsson, 2000; Olsson et al., 2002). These widely applied and destructive methods have been supplemented by vari-ous imaging techniques, including autoradiography following feeding $^{14}CO_2$ as a tracer to detect actively translocating hyphae in intact networks (Francis and Read, 1994) and phosphoimaging (Nielsen et al., 2002). Patterns of nutrient depletion (Li et al., 1991a), acquisition of isotopically labelled P (Jakobsen et al., 1992b; Smith et al., 2000; Jansa et al., 2003) and colonization of distant plants (Drew et al., 2005) have also provided information on the distribution of activities of the external mycelium in experimental systems.

The data show that, in general terms, hyphae proliferate vigorously in soil and grow considerable distances from colonized roots. Values of hyphal length den-sity (HLD) in pot experiments range from as low as 0.06 (Schubert et al., 1987) up to ~40 m/g soil (Sanders et al., 1977; Smith SE et al., 2004a) and measured distances from the root reach at least 15 cm from roots (Jansa et al., 2003). Considerable dif-ferences in HLDs are observed depending on fungal species, soil conditions and distance from the root (Jakobsen et al., 1992a; Hart and Reader, 2002b). Figure 2.14a shows the results of one of the first and most detailed experiments that demonstrated different patterns of hyphal development by different fungi. Such differences have been shown repeatedly, enabling Abbott *et al.* (1992) to make generalizations about types of mycelial development depicted in Figure 2.14b.

Some fungi, like *Scutellospora calospora*, consistently produce extensive external mycelium, which increases as the extent of colonization of the roots increases. With this species the HLD declined more or less linearly as distance from the root increased. Other fungi have a more limited capacity for mycelial growth, which is not necessarily related to extent of internal colonization (Abbott, 1982; Jakobsen *et al.*, 1992a; Smith *et al.*, 2000, 2004) (Figure 2.14b). In some instances there is a general correl-ation between the ability to absorb P and extent of spread and HLD of the external myc-elium (Jakobsen *et al.*, 1992b; Smith SE, *et al.*, 2004a), but this is certainly not always the case, consistent with dual roles in seeking out new plants as well as foraging for soil-derived nutrients. In a comparison of two fungi in symbiosis with *Medicago*, *S. calospora* was again shown to produce extensive external mycelium, far exceeding that produced by an isolate of *G. caledonium*, but was much less effective in acquiring and transporting ^{33}P from a distance (Smith *et al.*, 2000). Similar differences between fungal species and even isolates of the same species have now been observed in many investigations (Smith FA *et al.*, 2000; Smith SE, 2004a; Drew *et al.*, 2003; Munkvold *et al.*, 2004).

There is rather little information on effects of soil conditions on development of external hyphae. High P levels (Abbott *et al.*, 1984; de Miranda and Harris, 1994), changes in pH (van Aarle *et al.*, 2002a), texture (Drew *et al.*, 2003) or compaction of the growing medium (Schüepp *et al.*, 1987; Li *et al.*, 1997; Nadian *et al.*, 1998) and presence of roots (Smith SE, *et al.*, 2004a) have all been shown to reduce the length or biomass of hyphae, but the effects are certainly not sufficiently well established for generalizations to be made with confidence. As shown in Figure 2.15, *Gi. rosea* grew poorly in soil and failed to penetrate root-free compartments. *G. intraradices* grew similarly in presence and absence of roots, whereas *G. caledonium* proliferated

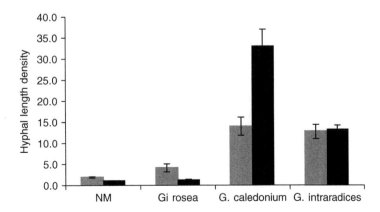

Figure 2.15 Effects of the presence of roots on development of external mycelium of three AM fungi in symbiosis with *Solanum lycopersicum*. Data for hyphal length densities in soil in root compartments (grey) and root-free, hyphal compartments (black). Data from Smith *et al.* (2004).

more in root-free hyphal compartments. Proliferation of hyphae in organic matter or organic amendments has also been noted (St John *et al.*, 1983a; Warner, 1984;
Ravnskov *et al.*, 1999; Gavito and Olsson, 2003). The effect of presence of roots in
reducing hyphal proliferation by some fungi requires further attention with respect
to competition for available P and it is worth noting that the majority of measurements of HLD have been carried out in pot systems using soil compartments from
which roots were excluded by mesh.

Measurements of mycelial development are much more difficult in field systems
than they are in pots. In the Serengeti Park, Tanzania, the length of mycelium associated with C4 grasses varied between 0.03 and 6.95 m/g soil and, in this habitat, was
negatively correlated with soil organic matter and nutrient status (McNaughton and
Oesterheld, 1990). Other measurements of HLD in field soils have been related to
soil structural stability and varied considerably with species of host plant used in
the experiments, with values associated with *Lolium perenne* and *Trifolium repens* of
13.9 and 3.1 m/g, respectively (Tisdall and Oades, 1979), which are well within the
ranges found in pot experiments.

Spore production

The external mycelium is important in the production of spores and translocates
relatively large amounts of carbohydrate and, more especially lipid, into them, adding considerably to the biomass of fungus outside the root in the soil. Sieverding
et al. (1989) found 28 spores/g (not an unusually high figure) associated with a cassava crop and calculated a biomass of up to 919 kg/ha which, as they pointed out,
was actually greater than some estimates of ectomycorrhizal fruit body production.
Values for spore densities in soil can therefore give an indication of variations in
fungal C deposition in soil, bearing in mind that these will be minimum estimates,
both because they omit hyphae and also because spores of different species contain
quite variable amounts of lipid (Madan *et al.*, 2002).

In pot experiments, spore production, as well as colonization of the root systems,
is affected by such factors as plant growth, fertilizer application, light intensity and
so on. Low irradiance or defoliation, which reduce photosynthesis and hence carbon
supply to the plants, have also been repeatedly shown to reduce sporulation (Furlan
and Fortin, 1977; Daft and El Giahmi, 1978). Application of increasing quantities of P
to plants of *Solanum lycopersicum* reduced not only the proportion of the root system
colonized by a *Glomus* species, but also the number of spores associated with each
plant after 84 days growth (Daft and Nicolson, 1972). This observation is supported
by the observation of lower spore production by several fungi when full Hoaglands
nutrient solution was applied to plants, compared with the same solution lacking P
(Douds and Schenck, 1990). Similar effects of P, and also of N, on spores produced
in field soils have also been found (Ross, 1971; Hayman *et al.*, 1975; Porter *et al.*, 1978;
Egerton-Warburton and Allen, 2000; Burrows and Pfleger, 2002). Low nutrient concentrations are conducive to high colonization, so there may be a link between extent
of intraradical colonization and spore production. This was certainly shown for two
species of *Acaulospora*, which required different critical lengths of intensely colonized
mycorrhizal root before sporulation commenced, and for '*Gigaspora*' (*Scutellospora*)
calospora in which numbers of spores formed at 119 days were closely correlated with
colonization of the roots of *Trifolium subterraneum* at 91 days (Scheltema *et al.*, 1987a;
Gazey *et al.*, 1992). However, poor correlation between spore production and extent of

colonization is also observed (Daniels Hetrick and Bloom, 1986). In the tropics, similar seasonal influences are less likely and the results of Louis and Lim (1987) showed no clear correlations between environmental factors and changes in spore populations.

An effect of the plant species on spore production has been observed both in pots and in the field. For example, Struble and Skipper (1988) found that four *Glomus* and one *Gigaspora* species always had poor spore production on soybean and that each of the fungi sporulated to different extents on *Zea*, *Paspalum* and *Sorghum*. *G. clarum* also sporulated better on maize and sorghum than on chick pea and on the first two plant species the process continued until plant growth reached a maximum (Simpson and Daft, 1990). None of the three *Glomus* species investigated by Daniels Hetrick and Bloom (1986) sporulated well on *Asparagus*, and again spore production was greater on *Sorghum* than on the dicotyledons *Tagetes*, *Trifolium* or *Lycopersicon* (*Solanum*). Using co-occurring plants from a grassland, Bever and co-workers (Bever *et al.*, 1996; Bever, 2002a) showed differential effects of plant species on spore production. These variations with different hosts exemplify differences in symbiotic effectiveness and fitness from a fungal perspective.

The extent of colonization of roots

Per cent colonization
The fraction or percentage of the root length colonized by AM fungi following staining remains the most widely used method of assessment. It is a relative measure which is recognized as having limitations, some of which are discussed below. Preparation of roots for light microscopy involves clearing the tissue (usually with 10% KOH) and after washing, staining with trypan blue (Phillips and Hayman, 1970), acid fuchsin (Merryweather and Fitter, 1991) or chlorazole black (Brundrett *et al.*, 1984). Variations in the effectiveness of stains as well as their toxicity has led to an ongoing search for improvements. Trypan blue is probably still the most commonly used, but is now registered as a carcinogen. Aniline blue (or cotton blue, as it is often called) is reported to be equally effective (Grace and Stribley, 1991) and is non-hazardous, while the other two stains can give excellent results, particularly with epifluorescence or Nomarsky optics (Merryweather and Fitter, 1991; Brundrett *et al.*, 1994, 1996). Furthermore, acid fuchsin is highly effective for distinguishing fungal tissue within plant roots under confocal microscopy (Dickson and Kolesik, 1999). A new and very simple stain with considerable promise for routine bright field microscopy is 'ink and vinegar' (Vierheilig *et al.*, 1998b); it gives excellent results with most AM fungi in many hosts tested so far and may overcome some of the problems of background staining of roots encountered in the past (Dickson *et al.*, unpublished). Vital staining on fresh material to detect the activity of different enzymes such as succinate dehydrogenase and alkaline and acid phosphatases has also been used to determine the proportion of the fungus that is active (Smith and Gianinazzi-Pearson, 1990; Smith and Dickson, 1991; Tisserant *et al.*, 1993; van Aarle *et al.*, 2002b, 2005). Problems with all these methods include differential tissue penetration by the chemicals (particularly when fungal walls are thick and impermeable as they are in vesicles) and background staining in plant tissues. Nevertheless, their use has added considerably to our knowledge of the proportion of fungal material that is active at any one time in soil or roots and to turnover of different structures (Dickson and Smith, 2001; Dickson *et al.*, 2003).

The most widely used method for determination of per cent root length colonized is a modification of a grid intersect method devised to measure root length (Newman, 1966). Giovannetti and Mosse (1980) compared this method with others and concluded that the standard error of estimates of per cent colonization was lower than for any of the methods based on mounting root pieces on slides. The grid intersect method has the additional advantage of permitting relatively large root samples to be scored and, if weighed subsamples of root are used, can give the length to weight ratio and hence provide very important data for the total root length of the plants. Furthermore, it can be adapted to measure lengths of hyphae on membrane filters as mentioned above.

In addition to per cent colonization, the importance of determining the characteristics of colonization, that is the intensity of colonization of the cortex and the extent of development of intraradical structures (intercellular hyphae, arbuscules, coils and vesicles) is increasingly recognized. This is most usually done by applying the grid intersect method at high enough magnifications to visualize the different structures (McGonigle et al., 1990; Cavagnaro et al., 2001c) or by a semi-subjective ranking method using root segments observed at high magnification (Trouvelot et al., 1986).

Progress of colonization in root systems

Colonization of a root system by AM fungi is a dynamic process, in which both root and fungal components grow and develop. The root grows apically by cell division, elongation and differentiation and it initiates lateral roots. At the same time, the fungus initiates both primary and secondary infection units, which grow and colonize the root cortex. The rate at which the root system becomes colonized (and hence the per cent colonization) is influenced not only by the rate of formation of the infection units and their rate of growth, but also by the rate of growth of the root system (Sutton, 1973; Smith and Walker, 1981; Sanders and Sheikh, 1983). This is not the place to discuss environmental effects on root growth and development, but clearly nutrient availability, temperature, soil compaction and so on all have indirect effects on both apical extension and branching which will affect values of per cent colonization.

Direct effects of AM colonization on root growth have been investigated in several host species and, although some work indicates that there are no changes in rates of apical extension or initiation of branches (Buwalda et al., 1984), it is well established that changes do take place even if the colonization of the root by AM fungi is relatively slight. Berta et al. (1990a, 1991) demonstrated that the rate of growth of root apices slowed down after colonization by a Glomus species and that this was associated with a decrease in the mitotic index because of extensions of G1, S and metaphase and marked reductions in the duration of G2. At the same time, an increase in initiation of lateral roots was observed, presumably stimulated by the loss of activity of the apices of the main adventitious roots (Barker et al., 1998b). A major effect of G. mosseae on root growth of a maize mutant severely impaired in root development (lrt-1) is to stimulate production of a new type of lateral roots which, together with P uptake via external mycelium, restored the growth of the mutant in low P soil (Paszkowski and Boller, 2002). The mechanisms underlying these changes in root development are obscure and clearly operate at a distance from the actual sites of colonization but, as Koske and Gemma (1992) have pointed out, the increased branching of roots, as well as that of hyphae would increase the chances of encounters between roots and infective hyphae.

It is very difficult to determine direct effects of environmental variables on fungal processes using per cent colonization of the root system, unless it has already been ascertained that there are no environmental effects on root growth itself. This would be a very unusual situation. Any factor that influences the rate of root growth, such as temperature, plant nutrition or elevated CO_2 will inevitably influence per cent colonization. Moreover, differences in per cent colonization of different species or different genotypes of the same species of plant may be determined by differences in their rates of root growth and cannot safely be interpreted as differences in susceptibility of the root systems to AM colonization. This point is frequently overlooked, but is of critical importance in determining effects of environmental conditions, comparisons of different plant species or cultivars and in screening plant germ plasm for variations in colonization. Single harvest measurements of per cent colonization, unaccompanied by measurements of plant, and especially root growth, are of limited usefulness. An appreciation of the dynamics of colonization requires multiple harvest experimentation, including measurements of root growth and root:shoot ratio. Having said this, there is no doubt that per cent colonization is a very convenient parameter to use. It can be measured on appropriately stained, representative subsamples of a root system and large numbers of samples can be processed relatively quickly and results of widely different investigations can be compared with reasonable safety. It is worth noting, however, that other parameters, such as fungal biomass, colonized root length per plant or per unit weight or volume of soil may be more appropriate for some investigations. As always, the method chosen should be dictated by the question being addressed.

A graph of per cent of the root length colonized against time typically has a sigmoid form (Figure 2.16). Similar curves have been obtained for single AM fungi in pot culture with different plant species, as well as for mixed assemblages from field soil. The key elements of the relationship are the lag phase before colonization is detectable, a phase of rapid increase in per cent colonization during which fungal spread exceeds the rate of root growth and a plateau phase in which spread of the fungus and growth of the root are constant relative to each other, although both remain active. In some investigations there may also be a late decline, in which the rate of root growth exceeds the rate of fungal spread. Multiple sampling is clearly required to obtain data on the progress of colonization in both field and pot investigations. It cannot be emphasized too strongly that such detailed information is critical for interpretation of data on such parameters as growth and P uptake. Furthermore, and as Fitter *et al.* (2000) have emphasized, the typical sigmoid curve shows what is happening during the establishment phase of both plants and fungi typically from discrete propagules in soil. Its closest parallel in the field is to be found in annual crops. In undisturbed ecosystems, the situation will be modified by the existence of mycorrhizal networks and by seasonal patterns of root growth.

The sigmoid relationships are also usually based on staining with non-vital stains and consequently take no account of possible death of either fungal hyphae or root cells. A number of studies using different vital stains have all demonstrated that, while in young plants all hyphae and arbuscules are active, as the plant and fungus age a smaller and smaller percentage of the intraradical mycelium remains alive

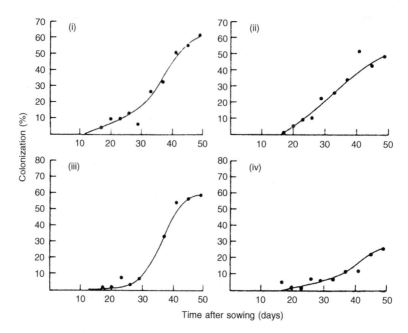

Figure 2.16 The progress of colonization in roots of *Allium cepa* inoculated with four different mycorrhizal fungi: (i) *Glomus mosseae*; (ii) *Glomus macrocarpus*; (iii) *Gigaspora (Scutellospora) calospora*; (iv) *Glomus microcarpus*. From Sanders *et al.* (1977), with permission.

(Smith and Dickson, 1991; Toth *et al.*, 1991). Smith and Dickson (1991) used image analysis of sectioned roots to demonstrate that the decline in fungal activity is associated with a reduction in density of arbuscules, while the intercellular hyphae remain alive for longer. Different 'vital stains' do not necessarily give the same picture of fungal activity, so that care in interpretation and comparison of data is needed (Tisserant *et al.*, 1993). Close correlations between per cent colonization and physiological effects on plant growth and P uptake are not by any means always observed (Fitter and Merryweather, 1992). Reasons include progressive loss of activity of the fungus and change in the quality or intensity of colonization, as well as involvement of the fungi in P uptake even when there are no net effects on plant growth or P content (see Chapter 5). In any event, if colonization is only determined at a final destructive harvest, then a strong correlation with plant response is even less likely, because early, rapid colonization (short lag phase and rapid spread) have been shown to be important in influencing nutrient absorption and growth, rather than the final plateau value of per cent colonization.

Quite large differences in per cent colonization are frequently observed with different plant–fungus combinations and are often related to differences in rates of root growth and susceptibility of the plants, as well as to different fungal strategies in root colonization. Again, low colonization is not necessarily an indicator of a small effect of the symbiosis on plant growth. Plants like cereals, with rapid rates of root growth, tend to have lower plateau values for per cent colonization than those with slower root growth, like clover or leek. In a comparison of the interactions between three AM fungi and three plant species, flax had lower per cent colonization by two fungi than either medic or tomato, but it showed an equal or greater response to them in terms of plant growth (Smith *et al.*, 2004). Comparisons of fungi are more difficult because it is hard to standardize inoculum, but where that has been done it is clear that major differences in rates of spread do exist (Daniels *et al.*, 1981; Bowen, 1987; O'Connor, 1994; Nadian *et al.*, 1998; Hart and Reader, 2002a, 2002c; Smith SE *et al.*, 2004a; Jansa *et al.*, 2007).

Influence of environmental factors on per cent colonization

There are also important environmental influences on the extent of colonization, of which the most extensively investigated are density of inoculum, temperature, availability of nutrients (particularly P), light and elevated CO_2 (eCO_2). High propagule densities, including the existence of infective mycelial networks, reduce the length of the lag phase in the curve of per cent colonization versus time (Figure 2.17a) and may also be associated with a rapid spread of the fungus within the roots. After primary colonization from propagules has taken place, the growth of the extraradical mycelium gives rise to increased fungal colonization of the soil and also results in the formation of secondary infection units, which increase the number of connections between the internal fungal structures and the external mycelium (Wilson and Tommerup, 1992). Experimentally, it is almost impossible to distinguish between primary and secondary entry points and data on factors affecting their formation are nearly always combined. Low numbers of propagules in field soils (e.g. in eroded sites) may result in low levels of colonization so that the need to evaluate the infectivity or propagule density in soil in order to predict outcomes in terms of P uptake and growth of crops or in restoration programmes is widely recognized (Reeves *et al.*, 1979; Jasper *et al.*, 1988; Allen, 1989; Abbott and Robson, 1991; Miller and Jastrow, 1992a).

The effects of temperature on the rate and extent of colonization are complex, the responses varying with both fungus and plant. In experimental systems there is usually an increase in per cent colonization between 10° and 30°C, but some plant–fungus combinations can develop normally at much lower or much higher temperatures (Bowen, 1987). The occurrence of well-developed AM colonization in biomes from the tropics to the sub-Antarctic argues for considerable adaptability (Janos, 1980; Frenot *et al.*, 2005). Gavito *et al.* (2005) investigated the effects of temperatures between 6° and 24°C on growth of three fungi in symbiosis with transformed carrot roots. They showed direct effects of temperature on fungal colonization of roots and growth of the external mycelium, as well as indirect effects mediated by differences in root growth. The direct effect appeared to be, at least in part, the result of reduced C translocation to the fungi at low temperature, measured directly with ^{13}C-labelled glucose. The majority of pot experiments in soil have been carried out above 15°C, but many plants, both wild and cultivated, grow and develop mycorrhizas at lower soil temperatures in temperate regions. Baon (1994) found that barley failed to become colonized by *Glomus etunicatum* when root temperatures were held at 10°C, although it became colonized at 15°C. However, barley cultivars grown in field soil from Montana, USA became colonized at 11°C, leading to the suggestion that the local fungi were cold tolerant (Grey, 1991). Some of the effects can certainly be related to seasonality of growth of plants in particular communities for Bentivenga and Hetrick (1992) showed that mycorrhizal activity was greatest in cool season grasses during their growing period when temperatures were relatively low. Similarly, Daft *et al.* (1980) commented that bluebell roots became colonized during their winter growth period when the soil temperatures were around 5°C. Later, Merryweather and Fitter (1998b) confirmed this and showed almost exclusive colonizaton by *Scutellospora dipurpurescens* during this period, although other fungi colonized later in the season. Less cold-tolerant plants show different responses and there are again variations with fungal identity, so that Martin and Stutz (2004) showed differences between AM fungal species in colonization ability of *Capsicum* between 20° and 25°C, compared with 32°–38°C, which were associated with differences in growth response. Variations in temperature can also have differential effects on intraradical and extraradical colonization. Colonization in roots of pea occurred at both 10° and 15°C and almost doubled with the increase in temperature, but no external mycelium was formed at 10°C, leading to marked differences in the mycorrhizal contribution to P uptake (Gavito *et al.*, 2003).

Generalizations are obviously difficult and more work is required which is specifically directed at understanding the effects of temperature on the biology of propagule survival, germination and colonization of roots as well as the functions of the mycorrhizas in particular communities. This, as several authors have recently emphasized, will be very important in understanding plant–fungus interactions and ecosystem changes in response to global warming (Rillig *et al.*, 2002a, 2002b; Gavito *et al.*, 2003; Heinemeyer *et al.*, 2004, 2006).

It is frequently stated that high soil P concentrations markedly reduce or eliminate mycorrhizal colonization, with the implication that the plant is in control and 'the fungi are no longer needed to enhance nutrient uptake'. This phytocentric view is a serious oversimplification and certainly not true in many situations. Furthermore, most of the work has been done in the context of the manipulation of AM fungi in agricultural crop production using fertilizer applications that result in soil P levels hugely exceeding natural environments. While it is well known that P availability does influence per cent colonization and the formation of arbuscules, the magnitude of the effect is strongly influenced by host species and other environmental factors, particularly irradiance. A typical response curve of P addition versus per cent colonization is shown in Figure 2.17b. As can be seen, very low P availability may actually inhibit colonization so that small additions result in increased values (Tinker, 1975a; Bolan *et al.*, 1984; de Miranda and Harris, 1994). Further additions of P frequently result in reductions and the sensitivity of the response appears to differ markedly between different host plants. Two examples illustrate this, although many others are to be found in the literature. Baon *et al.* (1992) found that, whereas wheat, barley and rye grown in soil with 5 mg/kg bicarbonate extractable P had up to 40% of their root length colonized, additions of 5 mg/kg P markedly reduced colonization and 30 or 60 mg/kg effectively eliminated it. Furthermore, different cultivars of barley were not only colonized to different extents by *Glomus intraradices*, but the extent of colonization was differently sensitive to P addition (Baon *et al.*, 1993). On the other hand, in *Trifolium subterraneum*, application of enough P to achieve maximum growth only reduced colonization from 74 to 53% (Oliver *et al.*, 1983). Irradiance interacts strongly with P, with marked reductions in per cent colonization following P application to *Allium cepa* which were only apparent at low irradiance (Graham *et al.*, 1982a; Son and Smith, 1988; Smith and Gianinazzi-Pearson, 1990) (Figure 2.18).

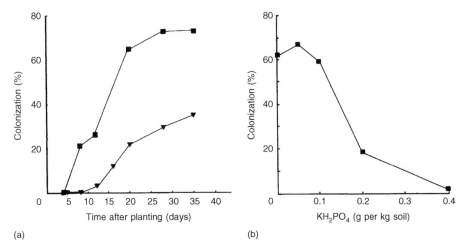

Figure 2.17 (a) Effects of differences in propagule density on the progress of colonization in *Trifolium subterraneum*. Plants were grown in non-sterile soil with a propagule density (determined by the most probable numbers method) of 4.0 propagules per gram (■) or in soil diluted with steamed sand to provide a propagule density of 0.4 per gram (▼). From Smith and Smith (1981), with permission. (b) The effect of additions of KH_2PO_4 to soil, on the percentage of the root length of *Allium cepa* colonized by AM fungi after 8 weeks growth. Redrawn from Sanders and Tinker (1983).

Part of the effect of increased P supply is most certainly mediated by increased root growth. Bruce *et al.* (1994) showed that P applied to mycorrhizal *Cucumis* plants increased rates of initiation and of apical extension of lateral roots. However, these effects did not occur early enough to provide a complete explanation for reduced per cent colonization which was first observed at 10 days. More recent work suggests that initial colonization, particularly from spores, may well be influenced by effects of increased P in reducing the production of hyphal branching factor (Tawaraya *et al.*, 1998) and hence initiation of entry points (see Chapter 3). In the investigation of Bruce *et al.* (1994), growth of infection units was also reduced by added P from as early as day 8 and, although the length of the lag phase of colonization was unaffected, the rate of production of entry points was very much lower in the presence of added P from day 20 onwards. This was probably due to reduced secondary colonization, for negative effects of high soil P on the development of external mycelium have sometimes been observed (Abbott *et al.*, 1984). Effects of P on arbuscule development are somewhat variable. Sanders and Tinker (1973) and many subsequent observers (Hayman, 1974; Graham *et al.*, 1982a; Amijee *et al.*, 1989; Smith and Gianinazzi-Pearson, 1990; Braunberger *et al.*, 1991; Bruce *et al.*, 1994) have demonstrated reductions in the densities of arbuscules in roots, but although Abbott and Robson (1979) observed lower per cent colonization in *T. subterraneum*, *Erodium botrys* and *Lolium rigidum*, they saw no effects on arbuscule development. Again, plant and fungal identity are likely to underlie this variability.

The effects of P operate on the fungi at several steps in the colonization process. High P concentrations in the medium reduce the growth of germ tubes and hence

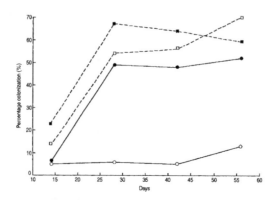

Figure 2.18 Effects of irradiance and additions of P on AM colonization of roots of *Allium cepa* in low P soil. High light, low P, ■; high light, added P, ●; low light, low P, □; low light, added P, ○. Note the combined effect of added P and low light in reducing the percentage of the root length colonized. Data of Chris Son and Sally Smith (see Son and Smith, 1988).

may have a direct effect on initiation of colonization. High P also reduces the production of root exudates that stimulate branching of hyphae approaching roots, as well as growth of the extraradical mycelium, with consequences for both primary and secondary colonization. High P status of roots appears to have a direct effect on intraradical fungal growth rate and development of extraradical mycelium, possibly associated with reduction in carbohydrate supplies available for the fungus (Jasper *et al.*, 1979; Graham *et al.*, 1981).

Following early work which indicated reductions in colonization at low light and fuelled speculation on the relationships with carbohydrate supply, the effects of light and defoliation have continued to be investigated and it is clear that per cent colonization is almost always reduced in situations where supply of sugars might also be expected to be lower. Daft and El Giahmi (1978) found that both defoliation and either shading or short day length reduced per cent colonization and numbers of secondary spores produced by two *Glomus* species in a variety of host plants. Hayman (1974) found that colonization was higher at higher light intensities and that this was correlated to sugar concentrations in the roots. The reduction in colonization at low irradiance is more marked if P supply is high (Hayman, 1974; Daft and El Giahmi, 1978; Graham, 1982; Son and Smith, 1988; Smith and Gianinazzi-Pearson, 1990; Thomson *et al.*, 1990a) and there have been a number of investigations of the way light and P supply interact with the pools of soluble carbohydrate in roots and the amount and composition of root exudates, with the aim of correlating these with mycorrhizal colonization.

Light and P do not necessarily operate in the same way in reducing per cent colonization, for Tester *et al.* (1986) found no effect of reduced irradiance on the rate of growth of infection units in *Trifolium*, whereas this parameter was influenced by P supply in *Allium* and *Cucumis* (Amijee *et al.*, 1989; Bruce *et al.*, 1994). In summary, while it is clear that both low irradiance and high P reduce colonization, the mechanism(s) by which they do so remain to be elucidated. Results of different investigations were not consistent and it seems unlikely that gross changes in carbohydrate pools play a major role in controlling fungal development.

Interest in global change (see Chapter 15) has led to a large number of experiments on the effects of eCO$_2$ on AM colonization. The outcomes have also been reviewed extensively (Hodge, 1996; Diaz, 1996; Treseder and Allen, 2000; Rillig *et al.*, 2002a; Staddon *et al.*, 1998, 2002; Fitter *et al.*, 2004; Alberton *et al.*, 2005), so that it is only necessary to provide a summary here. In separate experiments, per cent AM colonization has been shown either to increase or to remain unchanged with eCO$_2$; there are very few reports of a decrease. Growth of plants is also frequently increased, so interpretation of results in terms of direct influence on AM fungal growth *per se* is extremely difficult and requires multiple-harvest experiments in which both plant and AM fungal growth rates are assessed. The inadequacy of using per cent colonization to evaluate effects has been highlighted by Alberton *et al.* (2005). In a meta-analysis of a large number of different studies involving AM and ectomycorrhizal symbioses (see also Chapter 8), they concluded that both partners responded positively to eCO$_2$; plants by 25% and AM fungi by 21%, suggesting that there is no major change in C allocation between them. Differential effects on symbioses formed by different plant and fungal partners have scarcely been examined, but might be expected because individual AM fungal species probably do have different organic C requirements. Indications that eCO$_2$ alters AM fungal community structure supports this idea. However, Klironomos *et al.* (2005) showed that the effects were greatest when [CO$_2$] was elevated in a single step (as has been usual in most investigations) rather than gradually increased over 21 plant generations. The implication is that symbiotic communities adapt to gradual changes.

Conclusions

The propagules that can initiate AM colonization have been identified as spores, root fragments and hyphae, present either as fragments or (more importantly) as established mycelia in soil. These complex networks link the roots of plants of the same and different species and when these are growing the infectivity of the mycelium is very high. The networks appear to be able to survive in both dry and cold conditions, which is probably very important in initiating colonization early in the following season. The processes leading to the colonization of roots and the way in which AM fungi develop in the root systems is well understood in a few plant species. The outcome of the colonization process is that the fungus comes to occupy two different apoplastic compartments in the root, the intercellular spaces between cortical cells and a more specialized intracellular apoplast, surrounding arbuscules or coils. These interfaces play critical roles in bidirectional nutrient transfers between the symbionts (see Chapters 4 and 5). The morphological structures formed by the fungi, particularly in the root cortex, are highly varied and include both 'classic' arbuscules as well as less well known and understood intracellular coils and arbusculate coils. There are thus several interfaces between the symbionts which show different specializations and are likely to have different functions. Complexity is increased by colonization of a single root system by several different fungal species which may compete for occupancy and each forms a different suite of interfaces.

Outside the root, AM fungi form extensive mycelia, which undergo differentiation, so that different types of hyphae potentially perform different functions in colonization, nutrient acquisition and possibly survival. The foraging behaviour is now recognized to involve both search for new plants as sources of organic C and soil for mineral nutrients. Although variations in colonization patterns were described early in this century, the comparative study of *Arum*- and *Paris*-type mycorrhizas and other morphological variants has only recently been revived. Increased surveys of plant species, together with ongoing structural, developmental and physiological investigations are required to understand these diverse associations in field situations, as well as for managing mycorrhizas in primary production (see Chapter 17). The complexities of field situations, where many different species of fungi and plants coexist, together with other soil organisms are appreciated but have scarcely been investigated and will provide very significant challenges.

3

Genetic, cellular and molecular interactions in the establishment of arbuscular mycorrhizas

Introduction

It has long been appreciated that development of arbuscular mycorrhizas (AM) involves a well synchronized sequence of events, during which morphogenetic changes to both fungus and plant take place, supporting the maintenance of a compatible, biotrophic symbiosis (see Chapter 2). As we suggested in the second edition of this book, arbuscular mycorrhizas might represent a state of basic compatibility between plants and fungi (whether mutualistic or pathogenic), which underlies more taxon-specific resistance mechanisms. It is encouraging that plant pathologists are renewing interest in control of compatibility, so advances in understanding arbuscular mycorrhizas may well inform their researches (Heath, 1981; Paszkowski, 2006; Mellersh and Parniske, 2006).

The processes involved in establishing AM interactions have long been predicted to be closely programmed and to involve signalling between partners and coordinated gene expression, but details were rather slow to be revealed. On the plant side, some key steps in the colonization programme were recognized relatively early, but the fungi have been more difficult to study because they are obligate symbionts with non-synchronous development in root tissues and their genetics is so poorly understood (see Chapters 1 and 2). In 1997, molecular genetic methods were just beginning to be used, so that the second edition predicted that '… rapid advances are to be expected in our knowledge of the molecular-genetic regulation of the symbiosis, as new approaches and methods are applied'. This prediction has been amply borne out and research is currently at a very exciting stage, with details of both plant and fungal roles in establishing partnerships rapidly being revealed. Advances have been facilitated by recognition of common symbiotic pathways in nitrogen fixing symbioses (NFS) between legumes and rhizobia and arbuscular mycorrhizas. Although NFS involve a considerably smaller number of plant species than arbuscular mycorrhizas and evolved much more recently, they have proved more tractable to research, partly because of the ease of independent culture of the bacterial symbionts. The details of NFS signalling and genetic programming are still

Table 3.1 Phenotypic stages in the normal development of AM fungi during colonization of typical plant host species. (For more detail see text.)

Developmental stage	Fungal response/stimulus	Plant response/stimulus	Reference
Spore germination	Increased germination	Some evidence for stimuli produced by roots	Elfstrand et al., 2005; David-Schwartz et al., 2001; Besserer et al., 2006
Preinfection growth of hyphae	Unidentified diffusible factor produced by fungi (MYC factor)	Induces gene expression (MtENOD11-GUSA) in roots of M. truncatula	Kosuta et al., 2003
	Growth inhibited by exudates from pmi tomato mutant M161	Signal transduction genes in plant activated	Weidman et al., 2004; Gadkar et al., 2003
	H^+-ATPase (GmPMA1) highly expressed in preinfection stages and downregulated in symbiosis		Requena et al., 2003
Preinfection branching	Fungus branches profusely	Presence of plant required	Powell, 1976; Glenn et al., 1988; Giovannetti et al., 1993 a,b
	Respiratory activity increased	Exudates/strigolactones produced by hosts	Buee et al., 2000; Tamasloukht et al., 2003; Akiyama et al., 2005; Besserer et al., 2006
Appressorium formation	Swollen structure, often in groove between plant cells	Stimulus not known (not thigmotropic)	See Chapter 2
		Plant wall thickenings variable	Albrecht et al., 1998; Roussel et al., 2001
		Genes upregulated/specifically expressed	Bilou et al., 2000; Brechenmacher et al., 2004; Weidman et al., 2004
	Strong induction of H^+-ATPase (GmHA5)		Requena et al., 2003

Structure	Fungal activity	Observation	Reference
Epidermal 'opening'	Separation of epidermal cells allowing fungus to pass between	Observed in *M. truncatula*. *MtSym15* required	Novero et al., 2002; Demchenko et al., 2004
Penetration	Hypha narrows, forming infection peg Pressure and hydrolytic enzymes produced by fungus		Garcia-Romera et al. 1990, 1991 See Chapter 2
		Formation of prepenetration apparatus Increased DAPI fluorescence in plant nuclei Expression of *MtENOD11*	Genre et al., 2005 Sgorbati et al., 1993 Chabaud et al., 2002
Coils in exodermal passage cells	Fungal growth through non-suberized cells, often forming coils	Possible response to plant wall composition	Bonfante Fasolo, 1988
Intercellular hypha	Hyphae grow between cells in cortical parenchyma	Present in *Arum*-type plant species	Chapter 2
Coils in cortical parenchyma (*Paris*-types)		Some plant control as in *Paris*-type plant species	Chapter 2
Arbuscules (*Arum*-types)	Intracellular hyphal branching Reduced fungal wall synthesis Possible increase in permeability to P	Nuclei take up central position in coil AM-inducible P transporters expressed in plant plasma mambrane Invagination of plant plasma membrane	Cavagnaro et al., 2001b Karandashov et al., 2004 Chapters 2 and 5
		Plant nuclei take up central position between arbuscule branches Alterations in cytoskeletal activity Accumulation of HRGP in interface	Balestrini et al., 1992 Timonen and Peterson, 2002 Bonfante Fasolo et al., 1992

(Continued)

Table 3.1 (Continued)

Developmental stage	Fungal response/stimulus	Plant response/stimulus	Reference
		Transitory defence responses	Harrison and Dixon, 1994 Salzer et al., 2000; Bonanomi et al., 2001; Elfstrand et al., 2005
	Arbuscular cycle	Changes in plastid organization, and accumulation of carotenoid degradation products	Fester et al., 2001, 2002a,b
Periarbuscular membranes		Plant H$^+$- ATPase gene upregulated, H$^+$-ATPase activity increased	Gianinazzi-Pearson et al., 1991, 2000; Requena et al., 2003; Murphy et al., 1997
		Plant Pht1 transporters induced in periarbuscular and pericoil membranes	Rausch et al., 2001; Harrison et al., 2002; Pazksowski et al., 2002; Karandashov et al., 2004; Glassop et al., 2005
External hyphae	Rapid growth only after colonization	Nature of host influence unknown but sugar availability presumably important	Chapters 2 and 5
	GvPT, GiPT and GmosPT expression; Ginmyc1 and Ginhb1 only expressed in external mycelium; Ginmyc2 expressed in external mycelium and within roots		Harrison and Van Buuren, 1995; Maldonado Mendoza et al., 2002; Requena et al., 2003; Benedetto et al., 2005 Delp et al., 2003

better understood than those of AM symbioses and, accordingly, are capable of being exploited to understand the more ancient of the two. Rapidly advancing molecular genetic techniques, adoption of model plant species, particularly the legumes *Medicago truncatula* and *Lotus japonicus*, and their inclusion in genome sequencing projects, has greatly facilitated research into both symbioses. Furthermore, advances in techniques with the capacity to analyse very small samples have facilitated identification of key metabolites and signalling molecules. However, it cannot be overstated that these advances have only been made possible because research was accompanied by very careful microscopical studies of the phenotypes of the interactions between symbionts at the cellular level, accompanied in some cases by detailed functional information. Imaging of cytological changes in living cells now provides a means of following plant–fungus interactions as they occur.

This chapter will highlight some of the recent advances, focusing in particular on the identification of key control steps in colonization, cytological modifications, particularly with respect to the cytoskeleton, genetic approaches, including use of mutants, changes in gene transcription and signalling between the symbionts as the AM fungi invade roots. Several reviews have been published over the last 15 years which provide different perspectives on these complex interactions and show how the subject has developed (Koide and Schreiner, 1992; Gianinazzi-Pearson, 1996; Gianinazzi-Pearson *et al.*, 1996; Kapulnik *et al.*, 1996; Harrison, 1999, 2005; Gadkar *et al.*, 2001; Kistner and Parniske, 2002; Timonen and Peterson, 2002; Bécard *et al.*, 2004; Vierheilig, 2004b; Genre and Bonfante, 2005; Oldroyd *et al.*, 2005; Smith *et al.*, 2006).

Key steps in colonization

The changes that occur in each organism during normal AM colonization indicate that there are a number of 'switches' during the establishment of the biotrophic and compatible interaction which require the exchange of signals leading to changes in gene expression and developmental pattern. Identification of control points has mainly been based on morphological and cytological changes in the development of plant and fungus, supplemented increasingly by physiological or molecular information and studies of mutants deficient in AM colonization (Gianinazzi, 1991; Bonfante-Fasolo and Perotto, 1992; Giovannetti *et al.*, 1994; Smith, 1995; Giovannetti and Sbrana, 1998). As outlined in Chapter 2, from a fungal perspective, the main features of the sequence are spore germination, growth and morphogenesis of the presymbiotic mycelium, formation of appressoria, penetration of the root epidermis and exodermis, colonization of the root cortex and growth of the extraradical, symbiotic mycelium (Table 3.1). Finally, production of viable and infective spores completes the fungal life cycle.

Reception of the fungi in the root also involves changes in the plant cells, encompassing wall alterations, nuclear movements, alterations in cytoskeletal activity and membrane proliferation and modification, including the formation of a complex prepenetration apparatus (PPA; Figure 3.1; see Colour Plate 3.1) (Genre *et al.*, 2005). At all stages of development, the fungi remain in the plant apoplast, so that the symbiotic interfaces are bounded by both fungal and plant plasma membranes and contain varying amounts of fungal- and plant-derived wall material, as well as specialized depositions of various types. The interfacial apoplastic compartments are certainly

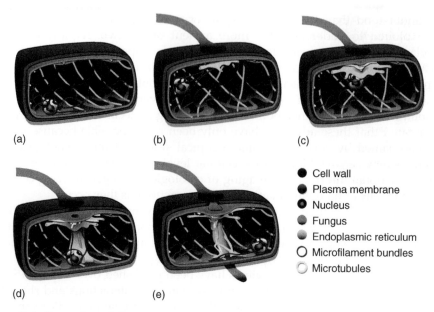

(a)   (b)   (c)

● Cell wall
● Plasma membrane
◉ Nucleus
● Fungus
◖ Endoplasmic reticulum
○ Microfilament bundles
○ Microtubules

(d)   (e)

Figure 3.1 Formation of the prepenetration apparatus (PPA) that facilitates passage of AM fungi through root epidermal cells. (a) An epidermal cell prior to fungal contact, showing peripheral position of the plant cell nucleus and arrangement of cytoskeletal elements and endoplasmic reticulum (ER). (b) Appressorium formation on the outer surface of the host cell results in important rearrangements of microtubules, microfilament bundles and ER, associated with initial nuclear movement toward the surface appressorium. (c, d) Assembly of the transient PPA within the cytoplasmic column and subsequent transcellular nuclear migration. (e) AM infection hypha penetrates the cell wall and crosses the epidermal cell through the apoplastic compartment constructed within the cytoplasmic column. Coding of different structures as in legend. Courtesy Andrea Genre. See also Colour Plate 3.1.

involved in nutrient transfers and presumably also in transmission of signals and stimuli that induce developmental changes at different steps in colonization.

Cytological changes during root colonization

Epidermal wall responses to invasion vary with plant–fungus combination, thickening in some (Figure 3.2a) and indicating recognition of fungal contact. In normal colonization events, the appositions contain no special compounds related to defence against pathogens. Major cytological events have recently been revealed that are initiated before any plant wall penetration by an AM fungus occurs (Genre *et al.*, 2005). Contact between the fungal appressorium and plant epidermal cell wall in *M. truncatula* stimulates the epidermal cell nucleus to move rapidly (<2h) from close to the cell wall and to take up a position immediately below the appressorium (see Figure 3.1a,b and Colour Plate 3.1). Cortical microtubules are rapidly reorganized and thick actin bundles radiate from the nucleus towards the site of fungal contact. Ongoing

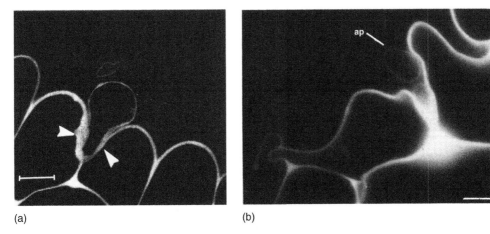

(a)   (b)

Figure 3.2 Cellular reactions during epidermal penetration of roots by AM fungi. (a) Acriflavine-positive wall thickenings in the epidermal cells of *Allium porrum*, in response to normal colonization by *Glomus versiforme*. Bar = 10 μm. From Garriock *et al.* (1989), with permission. (b) Callose-containing wall thickenings in the epidermal cells of myc^{-1} mutants of *Pisum sativum* cv Frisson, in response to colonization. Bar = 10 μm. From Gollotte *et al.* (1993), with permission.

cytoskeletal and endoplasmic reticulum (ER) reorganization leads to the formation of a doughnut-shaped structure below the site of appressorium contact (ACS) (see Figure 3.1c and Colour Plate 3.1). Now the plant nucleus moves again, at 15–20 μm/h, towards the underlying cell layer. Migration is associated with assembly of a complex cytoplasmic column that links the migrating nucleus to the site beneath the ACS. The column is an assemblage of microtubules and microfilament bundles and a dense region of ER cisternae, forming a hollow tube joining the nucleus to the ACS (see Figure 3.1d and Colour Plate 3.1). Following establishment of the column, an infection hypha penetrates the outer epidermal cell wall immediately above the doughnut-formation and grows through the epidermal cell, following precisely the path laid down by the column (see Figure 3.1e and Colour Plate 3.1). The column appears to provide a new apoplastic compartment, in which the invading hypha is contained. It seems possible that the prepenetration apparatus (PPA) has features in common with the infection threads that lead rhizobial cells through the root hair lumen to the legume root cortex. However, parallel studies on legumes have not yet been reported.

As penetration and colonization of the epidermal and exodermal cells proceeds, the plant responds in a number of ways, which again vary in different plant–fungus combinations. Nuclear size and DAPI fluorescence increase, possibly indicating genome reduplication or increased rates of gene transcription (Berta *et al.*, 1991; Sgorbati *et al.*, 1993). Penetration of the outer tangential walls of the epidermal cells follows appressorium formation and is marked by narrowing of the invading hypha to form an infection peg which penetrates the epidermis or underlying exodermis, before expanding again in the cell lumen. In *M. truncatula*, this key step

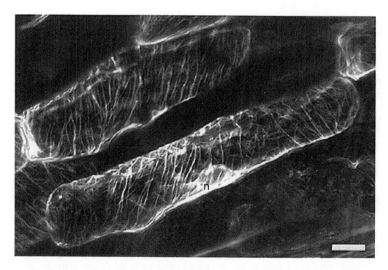

Figure 3.3 Arrangement of microtubules inside a parenchyma cell of *Nicotiana tabacum*, prior to colonization by AM fungi. Microtubule spirals are evident in the cortical cytoplasm. Other microtubule bundles envelop the nucleus (n), which remains in a peripheral position adjacent to the cell wall. From Genre and Bonfante (1998), with permission.

is accompanied by separation of the radial walls of two adjacent epidermal cells, allowing the fungus to grow between them. Fungal penetration of plant cell walls appears to require localized production of wall-degrading hydrolytic enzymes by the fungus, as well as hydrostatic pressure exerted by the hyphal tip (Table 3.1, see Chapter 2).

Cortical colonization of the root involves the formation of intercellular and intracellular hyphae, intracellular coils and/or arbuscules. Plant–fungus interactions have been studied by following changes in the deposition of molecules in the intercellular and intracellular interfaces, including fungal and plant walls (Bonfante-Fasolo, 1988; Lemoine *et al.*, 1995) and by the striking changes in fungal morphology during arbuscule and coil formation. In both arbuscule- and coil-containing cells, the plant nuclei migrate from the periphery of the cortical cells (Figure 3.3) to take up a central position enveloped by the fungal structure (Figure 3.4a, b; Figure 3.5) (Balestrini *et al.*, 1992; Cavagnaro *et al.*, 2001b). They increase in size and the chromatin decondenses, as shown by antibody labelling (Balestrini *et al.*, 1992). At this stage, actin filaments run along the plasma membrane and round the nucleus. Microtubular spirals run round the cells (Genre and Bonfante, 1998). Marked changes in cytoskeletal organization accompany both arbuscule and coil formation within cortical cells (Figure 3.5; see Colour Plate 3.2). Microfilaments come to surround each arbuscule branch, the plant cell nucleus and even collapsed arbuscules (Figure 3.5; see Colour Plate 3.2) (Genre and Bonfante, 1997, 1998, 1999; Armstrong and Peterson, 2002; Timonen and Peterson, 2002). Future work on PPA-like structures that seem likely to be formed during invasion of cortical cells by arbuscules or coils will be fascinating but technically challenging.

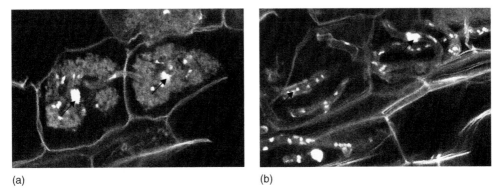

(a)   (b)

Figure 3.4 Positioning of the plant cell nucleus in root cortical cells of *Asphodelus fistulosus* with *Paris*-type AM colonization by *Glomus coronatum*. Extended focus confocal images of cells containing (a) an arbusculate coil and (b) hyphal coils. Roots were stained with acid fuchsin and a series of 2 μm optical sections in the z axis were obtained and used for quantification of size and position of the nuclei. Plant nuclei in colonized cells were significantly larger than those in non-colonized cells and were further from the nearest cell wall (i.e. they were positioned centrally). From Cavagnaro *et al.* (2001), with permission.

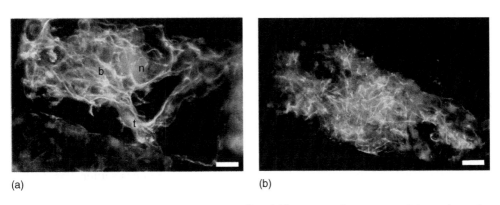

(a)   (b)

Figure 3.5 Microtubule arrangement in cells of *Nicotiana tabacum* containing arbuscules of *Gigaspora margarita*. (a) Bundles can be seen running through the arbuscule branches (b), along the arbuscular trunk (t) and around the nucleus (n). (b) Shorter bundles are visible among the arbuscule branches, connecting them together. Bar = 10 μm. From Genre and Bonfante (1998), with permission. See also Colour Plate 3.2.

Recent work has highlighted the marked changes in plastid behaviour and associated carotenoid metabolism that occur during the arbuscular cycle in many plant species. Fester *et al.* (2001) demonstrated that invasion of root cortical cells and development of arbuscules is accompanied by the formation of an extended plastid network that comes to cover mature arbuscules. Later, as arbuscules degenerate, the network disappears and carotenoid degeneration products (including 'mycorradicin') accumulate (Fester *et al.*, 2002a, 2002b; Hans *et al.*, 2004). These are the yellow pigments that have so often been observed in highly colonized AM roots (Becker and Gerdemann, 1977). The significance of these changes remains to be revealed.

Much less is known about changes in the fungal cytoskeleton, but major differences in microtubule organization were not observed between extraradical mycelium, intercellular hyphae and arbuscules in tomato colonized by *G. intraradices* (Timonen *et al.*, 2001). However, changes in genes encoding cytoskeletal elements have been detected as colonization progresses (Delp *et al.*, 2003).

Genetic approaches to understanding AM colonization – interactions of AM fungi with non-hosts and mutants

Considerable progress has been made in revealing the genetically controlled check points facilitating or restricting symbiotic development and identifying the plant and fungal genes that are involved. Early attempts to identify which steps in colonization are under genetic control were hindered by the absence of naturally occurring, incompatible interactions between AM fungi and non-host plant species that are closely related to normal AM hosts. Exclusion of AM fungi by roots is usually common to most members of major, non-host taxa such as the Brassicaceae and Caryophyllaceae (Gianinazzi-Pearson, 1984), so that comparisons had to be made between distantly related plants and rather little progress was made. The model species *Arabidopsis thaliana* is unavailable for AM studies, due to lack of AM colonization in this non-host. However, since 1989, when the first AM mutant plants were identified (Duc *et al.*, 1989), increasing numbers of mutants of normal host plants which block AM fungal development at various stages of colonization have been identified and have provided keys to defining the symbiotic check points (Figure 3.6) (Peterson and Bradbury, 1995; Marsh and Schultze, 2001; Barker *et al.*, 2002). Genetic analyses of the kinds that have been undertaken have some limitations. They are capable of identifying stages in symbiotic development where a critical plant function is encoded by a single gene, but where multiple genes are required the approaches are less likely to be successful. Furthermore, mutants blocking early stages in colonization are likely to be phenotypically non-mycorrhizal for later stages also. The work has centred on plant responses and genes rather than on the fungi for a number of reasons. First, the developmental sequence in *Arum*-type mycorrhizas is well known and comparisons of different species and cultivars indicate that the plant genotype can affect the extent of colonization and the response. Secondly, plant genomes are increasingly well understood and several AM hosts have been adopted as models and, in the case of *Oryza sativa* (rice), the genome completely sequenced. Thirdly, the plants can be grown with or without AM inoculum, so that their development and gene expression can be studied in both AM and non-mycorrhizal states.

The first mutant plants to be identified with abnormal AM phenotypes were among non-nodulating genotypes of the legumes *Pisum sativum* and *Vicia faba* (Duc *et al.*, 1989). In all instances, the so-called myc⁻ plants blocked colonization at appressorium formation (Figures 3.6, 3.7a), so that the fungi failed to penetrate the roots (now referred to as Pen⁻). Epidermal wall reactions are strongly enhanced in these mutants, so that well-defined wall thickenings with increased deposition of phenolics and β-1,3-glucans (possibly callose) occur beneath the appressoria (Gollotte *et al.*, 1993) and hyphae fail to enter the cells (see Figure 3.2b). Subsequent screening of 66 nodulation mutants in *P. sativum* yielded a second AM phenotype, in

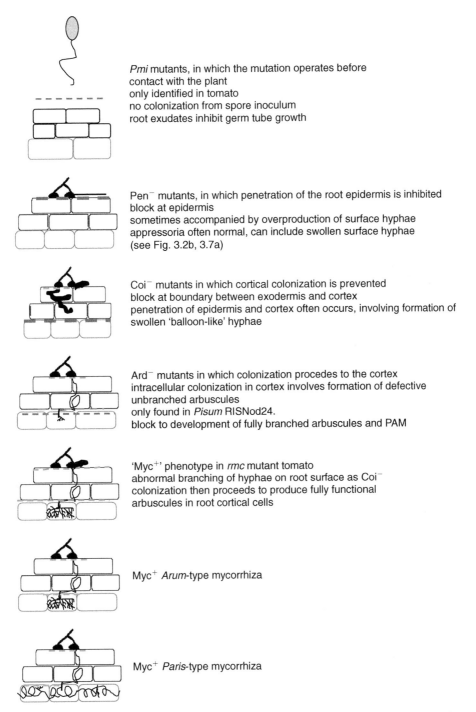

Pmi mutants, in which the mutation operates before contact with the plant
only identified in tomato
no colonization from spore inoculum
root exudates inhibit germ tube growth

Pen⁻ mutants, in which penetration of the root epidermis is inhibited
block at epidermis
sometimes accompanied by overproduction of surface hyphae
appressoria often normal, can include swollen surface hyphae
(see Fig. 3.2b, 3.7a)

Coi⁻ mutants in which cortical colonization is prevented
block at boundary between exodermis and cortex
penetration of epidermis and cortex often occurs, involving formation of
swollen 'balloon-like' hyphae

Ard⁻ mutants in which colonization procedes to the cortex
intracellular colonization in cortex involves formation of defective
unbranched arbuscules
only found in *Pisum* RISNod24.
block to development of fully branched arbuscules and PAM

'Myc⁺' phenotype in *rmc* mutant tomato
abnormal branching of hyphae on root surface as Coi⁻
colonization then proceeds to produce fully functional
arbuscules in root cortical cells

Myc⁺ *Arum*-type mycorrhiza

Myc⁺ *Paris*-type mycorrhiza

Figure 3.6 Diagrammatic representations and brief descriptions of mutant and wild-type mycorrhizal phenotypes, together with commonly used abbreviations. Roots are represented as having three tissue types (top to bottom in each cartoon) epidermis, exodermis with suberized radial walls and cortical parenchyma. The main points at which colonization appears to be blocked are indicated by dashed horizontal lines. Developed from the concept of Marsh and Schulze (2001).

which the fungus is able to form appressoria, penetrate the epidermis and to grow intercellularly in the cortex, but in which arbuscule formation is much reduced and those arbuscules that are formed have very few branches (ard⁻, for arbuscules defective, Figure 3.6) (Gianinazzi-Pearson *et al.*, 1991b). These findings provided the impetus for screening programmes in other species and a foundation for recent advances. AM-defective mutants have now been identified in the legumes *Medicago truncatula*, *M. sativa*, *Lotus japonicus*, *Glycine max* and *Phaseolus vulgaris* (as well as *P. sativa* and *V. faba*) and in the non-legumes *Solanum* (*Lycopersicon*) *esculentum* and *Zea mays*. All mutations identified so far are, with one exception, recessive, so that loss, or at least change of function, is expected to underlie the changed phenotype. The maize mutant *Pram1* facilitates faster colonization by *Glomus mosseae*, but how this dominant and hence 'gain of function' mutation operates remains to be elucidated (Paszkowski *et al.*, 2006).

For almost all the legume mutants reported so far, screening for AM phenotypes was undertaken on nodulation-defective genotypes. Subsequent genetic analysis revealed that nodulation and AM development were at least partially under the control of the same genes, leading to the recognition that the NFS pathway almost certainly evolved from the more ancient AM pathway (Gianinazzi-Pearson and Denarie, 1997; van Rhijn *et al.*, 1997). Recent evidence, obtained mainly from *M. truncatula* and *L. japonicus*, has confirmed the existence of at least seven common symbiosis genes, as well as additional distinct AM- and NFS-specific genes (Harrison, 2005; Kistner *et al.*, 2005). Not all nodulation-defective mutants have altered phenotypes with the AM fungi tested so far, indicating divergence in the NFS and AM pathways. The assumption is that all AM-forming plant species will, regardless of taxonomic position, contain similar suites of genes that facilitate colonization, but only some of these appear to have been taken over by the legumes. Much more work on suitable non-legumes, such as rice, maize and tomato, is required to reveal details of steps that are not common to the two symbioses. Indeed, the ancient origin of AM suggests that many of these genes may also be found in non-hosts. Evolution of the non-host state is not likely to have involved simultaneous loss of all genes involved in mycorrhiza development and is more probably based on presence of one or a few genes conferring 'resistance', or on loss of 'switch' or receptor genes that normally open the way to the colonization programme. Given the origins of non-hosts in different phylogenetic lineages of plants (see Chapter 1), several different mechanisms conferring failure of colonization by AM fungi are likely to be found. These have not yet been investigated by molecular-genetic methods, but *Arabidopsis thaliana* would provide an excellent subject for such work.

The existence of mutants with different colonization patterns (see Figure 3.6) has confirmed the key control steps in colonization, some of which appear to involve operation of more than one plant gene. The earliest known step occurs before any contact is made between the symbionts, so that the *pmi* mutants in tomato (M20 and M161 in *S. lycopersicum* cv Microtom) reduce colonization from spores of *Glomus intraradices* (David-Schwartz *et al.*, 2001, 2003). Rather surprisingly, the mutation does not prevent colonization from more complex inoculum containing root fragments and hyphae as well as spores, apparently indicating that different stimulatory or signalling processes operate on the different types of propagule. Exudates from roots of M161 inhibit the preinfection growth of germ tubes of *Gigaspora gigantea* and *G. intraradices* (Gadkar *et al.*, 2003), suggesting that the mutation alters the composition of a signalling compound or produces an inhibitor. The latter option is less likely because the mutation is recessive. More work is required to clarify and extend these findings, particularly now that plant signal compounds have been identified.

The majority of *Pisum* (Pen⁻) mutants do not affect appressorium formation, which is normal both with respect to frequency and morphology (Giovannetti *et al.*, 1993b; Gollotte *et al.*, 1993). However, a number of plant genes do influence this stage, because sometimes appressoria are not formed at all (Gao *et al.*, 2001; Paszkowski *et al.*, 2006) or increased numbers of abnormally shaped appressoria are produced (Figure 3.7b,d) (Bradbury *et al.*, 1991; Barker *et al.*, 1998a; Gao *et al.*, 2001). Variable responses by different AM fungi may contribute to the diverse

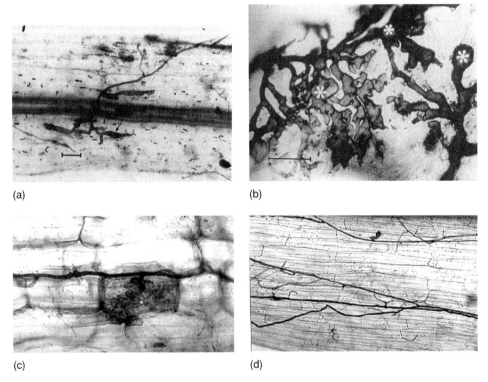

(a)   (b)

(c)   (d)

Figure 3.7 Phenotypes of mutants with altered patterns of mycorrhizal colonization. (a) The first mutant to be identified: Myc⁻¹ mutant (P6) of *Pisum sativum* cv Frisson, showing normal appressorium formation (apr⁺, arrowed), but absence of penetration into the root tissues. Bar = 50 μm. From Duc *et al.* (1989), with permission. (b) Hypertrophied appressoria formed on 'non-mycorrhizal' genotypes of *Medicago sativa*. Bar = 50 μm. From Bradbury *et al.* (1993), with permission. (c) Coi⁻ phenotype in the association between *Solanum lycopersicum* mutant *rmc* and *Scutellospora calospora*. The fungus frequently enters epidermal and hypodermal cells, where the hyphae swell. Cortical colonization never occurs. From Gao *et al.* (2001), with permission. (d) Pen⁻ phenotype in the association between *Solanum lycopersicum* mutant *rmc* and *Glomus intraradices*. Extensive superficial hyphae grown on the root surface, but few appressoria are formed and penetration of root cells is extremely rare. From Gao *et al.* (2001), with permission.

phenotypes observed. Overproduction of appressoria may be a fungal response to failure of tissue colonization (Bradbury *et al.*, 1993), indicating ongoing attempts to breach defences. Alternatively, the swollen structures may not be true appressoria, but rather equivalent to the much branched hyphae produced by AM fungi when actual contact with the root is prevented by artificial membranes, as in the studies by Giovannetti and co-workers (see Chapter 2). In these cases, the implication is that the Pen⁻ genotypes continue to produce compounds that stimulate AM hyphal growth and branching, but do not provide the appropriate stimulus for the formation of appressoria. Such a stimulus has not been unequivocally identified, but appears specific to cell walls of epidermal cells (Nagahashi and Douds, 1997).

Failure of AM fungi to gain access to the root epidermal cells is common to a number of different mutants in several plant species, indicating a further check point in the interaction. In Pen⁻ *Pisum* mutants, cytoskeletal reorganization and changes in nuclear size occur to a lesser extent than in wild-type plants (Berta *et al.*, 1991; Sgorbati *et al.*, 1993). In the mutant *Ljsym4–2* (Pen⁻), the cytoskeleton becomes disorganized both before and during penetration attempts. Cells lose actin and tubulin, cytoplasm becomes disorganized and cells die (Bonfante *et al.*, 2000; Genre and Bonfante, 2002). At the same time, the invading hypha may swell, forming the balloon-like structures observed in interactions with several plant species (see Figure 3.7c), but eventually also dies. The PPA that facilitates hyphal invasion of epidermal cells is not formed in *Mtdmi2* mutants (Genre *et al.*, 2005). In *L. japonicus*, mutation of *Ljsym-15*, one of the common SYM genes, completely prevents *G. intraradices* from penetrating between epidermal cells, entering the exodermis or cortex or developing intracellularly. However, this fungus responds to the presence of roots by producing extensive surface hyphae and slight swelling of appressoria (Demchenko *et al.*, 2004; Kistner *et al.*, 2005), a pattern which is also found in some other Pen⁻ mutants, such as *M. truncatula dmi2* (Calantzis *et al.*, 2001). Again in *L. japonicus*, mutations in the common SYM genes SYMRK (*Ljsym2/Ljsym21*), CASTOR (*Ljsym4*), POLLUX (*Ljsym22*), *Ljsym3*, *Ljsym6* and *Ljsym24*, permit slightly deeper penetration, so that the same fungus forms balloon-like swellings in the epidermis and exodermis (Wegel *et al.*, 1998; Novero *et al.*, 2002; Demchenko *et al.*, 2004; Kistner*et al.*, 2005). However, general and rapid penetration to the cortex is usually prevented (Coi⁻), probably by death of the plant cells and/or deployment of defence responses as shown for CASTOR (*Ljsym4*) mutations (Bonfante *et al.*, 2000) and some tomato *rmc* interactions (Gao *et al.*, 2004), respectively. Despite these general effects in outer root cell layers, AM fungi colonizing mutants do sometimes gain access to the cortex and form arbuscules, although more slowly than in wild-type interactions. The only mutations which have consistently been shown completely to prevent any cortical colonization and arbuscule formation (arb⁻) are CASTOR (*Ljsym4*) and *Ljsym15*.

Although little is known about *Ljsym15* beyond the effect described above, there is much more information about the function of other genes, gained in part by parallel studies in *M. truncatula* and *P. sativum*, capitalizing on knowledge of their roles in the NFS pathway (Figure 3.8). A group of at least seven common SYM genes is strongly implicated in early AM signalling, including a receptor-like kinase identified from several legumes (NORK, SYMRK/DMI 2/PsSYM19), which operates upstream of a channel protein (DMI1/POLLUX/CASTOR) and a calcium-calmodulin protein kinase (DMI3, PsSYM19 and PsSYM30) (Harrison, 2005; Kistner

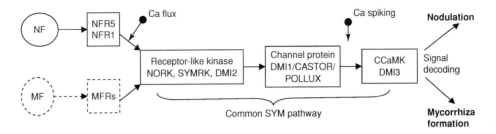

Figure 3.8 Diagrammatic representation of proposed signalling pathways in arbuscular mycorrhizal and nitrogen fixing symbioses, indicating features of the common SYM pathway. Putative 'myc factors' (MF) and myc factor receptors (MFR) are indicated as boxes with dashed boundaries. The importance of changes in calcium status have been demonstrated for NFS, but not for AM interactions. After Oldroyd *et al.* (2005).

et al., 2005). In NFS, these genes are involved in changes in calcium status in root hairs (calcium flux and calcium spiking) in response to signalling by nod factors produced by the bacterium.

By analogy with NFS, involvement of these genes suggests that calcium may act as a crucial second messenger in signalling in the early stages of AM symbiosis and that 'myc factors' produced by AM fungi are likely to be important in triggering responses (Ané *et al.*, 2004; Levy *et al.*, 2004; Bersoult *et al.*, 2005; Harrison, 2005; Oldroyd *et al.*, 2005). However, the plastid location of CASTOR and POLLUX protein products is unexplained, although the importance of plastids as calcium stores has been suggested (Imaizumi-Anraku *et al.*, 2005). Plastids are also strongly implicated in alterations of carotenoid metabolism that occur during the arbuscular cycle. How the putative calcium signals are received and decoded to promote AM development remains to be seen. The common SYM genes operate downstream of nod factor receptor genes (*NFR1* and *NFR5*) in the NFS pathway. Mutations in NFR genes do not affect AM colonization, indicating that they are not involved in reception of putative 'myc factors' (see below). The existence of 'myc factors' and receptors for them is implied by changes in expression of other mycorrhiza-related genes, but we have no idea of their identities as yet.

Once an AM fungus has reached the root cortex it may proliferate in intercellular spaces producing simple hyphae which subtend intracellular arbuscules (*Arum*-type AM), or it may grow directly from cell to cell forming intracellular hyphal coils and arbusculate coils (*Paris*-type AM) (see Chapter 2). Only *Arum*-type interactions have been investigated in detail with mutants. In these, arbuscule formation in *L. japonicus* is prevented by mutations in CASTOR (*Ljsym4*). The *Ljsym4–1* allele of this gene often permits intercellular cortical colonization, but arbuscules are never formed, suggesting that mutations act to prevent intracellular development either in the cortex or (in the case of *Ljsym4–2*) in the epidermis or exodermis. The *RisNod24* mutant in *P. sativum* permits intracellular colonization of root cortical cells, but the fungus fails to produce fully developed arbuscules (ard⁻) and growth of extraradical mycelium is also impaired (Gianinazzi-Pearson *et al.*, 1996; Kling *et al.*, 1996). This developmental difference from wild-type *P. sativum* interactions has been exploited to identify genes with arbuscule-related expression profiles (Grunwald *et al.*, 2004). Plant genetic control of arbuscule formation is also implied by results showing that development is enhanced in supernodulating mutants of *M. truncatula* and *P. sativum* (Morandi *et al.*, 2005) and in the *Pram1* mutant in maize (Paszkowski *et al.*, 2006). In the work on legumes, considerable care was taken to minimize the possibility that changes in root growth in relation to fungal development were the basis of observed increases in colonization (see Chapter 2). A link between super-nodulation and AM colonization is suggested by work on soybean, which also indicated a possible hormonal basis for the effects (Meixner *et al.*, 2005).

Increasingly mutants with defective AM phenotypes are being created or identified by such strategies as RNAi (RNA interference) and TILLING (target-induced local lesions in genomes). Knockout or knockdown mutations involving the AM-regulated phosphate transporters *MtPT4* (from *Medicago truncatula*), *LePT4* (*Solanum lycopersicum*) and *LjPT3* (*Lotus japonicus*) have been used primarily to investigate their roles in phosphate transfer from fungus to plant (Nagy *et al.*, 2005; Maeda *et al.*, 2006; Javot *et al.*, 2007; see Chapter 5). However, in both *M. sativa* and *L. japonicus* the mutations affect fungal colonization of roots, particularly the development of arbuscules. As yet the mechanisms underlying the intriguing effects can only be speculative.

The role of the fungal partners in cross-talk supporting AM development has received much less attention. Few analyses of plant mutant phenotypes have involved more than one or two species of AM fungi, the most commonly used being *Glomus intraradices* and *G. mosseae*. However, work with tomato (*S. esculentum*) using a large number of different fungi has shown that the mutant *rmc* can support at least three developmental phenotypes, including pen$^-$ and coi$^-$ (no cortical colonization) and a fully functional myc$^+$ symbiosis. The first two are markedly similar to phenotypes controlled by different plant mutations in *L. japonicus* and *M. truncatula*, with the fungi blocked either at the surface (see Figure 3.7d) or forming swollen structures within epidermal and exodermal cells (see Figure 3.7c) (Barker *et al.*, 1998a; Gao *et al.*, 2001; M. Manjarrez, unpublished). Although tomato forms both *Arum*- and *Paris*-AM (see Chapter 2), there is no correlation between wild-type and mutant phenotypes developed with different fungi. The myc$^+$ phenotype in this otherwise AM-defective mutant has so far only been clearly identified with a single isolate of *G. intraradices* WFVAM23 which shows abnormal penetration structures and slow cortical development but fully developed arbuscules, expression of mycorrhiza-inducible P transporters and P delivery to the plant via external mycelium (Gao *et al.*, 2001; Poulsen *et al.*, 2005). However, delayed but normal colonization following development of abnormally thick appressoria has been observed in several *M. truncatula* mutants, including *Mtdmi2*. The functionality of these delayed arbuscular mycorrhizas has not yet been investigated.

In combination, these findings support the existence of control steps in particular root cell layers. Differential abilities of the fungi to traverse the check points highlight the potential for a degree of 'specificity' and complexity in AM symbioses that mirrors the situation in NFS. The earliest steps in communication appear to be separate in the two symbioses (see Figure 3.8), but there is the potential for parallel lines of communication leading to the common receptors.

Signalling between symbionts in AM symbioses

Evidence of the importance of exchange of signals between AM symbionts during mycorrhiza establishment is increasing rapidly and has been shown to operate at several different stages. Although most research indicates that plant exudates are not essential for AM spore germination, there are a few reports of stimulatory effects of root exudates of AM host species (Graham, 1982; Gianinazzi-Pearson *et al.*, 1989; David-Schwartz *et al.*, 2001) and negative effects of exudates of non-hosts such as Brassicas (Tommerup, 1984c; Vierheilig and Ocampo, 1990a; Schreiner and Koide, 1993a, 1993b). The active compounds have not been identified, but may include flavonoids, CO_2 and strigolactones (Gianinazzi-Pearson *et al.*, 1989; Bécard *et al.*, 1992, 1995; Chabot *et al.*, 1992; Besserer *et al.*, 2006).

Once germination has occurred, there is unequivocal evidence for the role of signals produced by roots of a wide range of host species in stimulating preinfection growth and branching of fungal germ tubes (see Chapter 2, Figure 2.7a, Table 3.1). A branching factor (BF) from *L. japonicus* has recently been purified and identified as a strigolactone, 5-deoxy-strigol (Akiyama *et al.*, 2005). This compound belongs to a group of sesquiterpene lactones that had previously been shown to stimulate seed germination of the root parasites *Striga* and *Orobanche*. Potential parallels between AM symbioses and the *Striga* system were noted by Koide and Schreiner (1992) and the intriguing new finding suggests that communication between these plant parasites and their hosts may be another example of piracy of a pre-exisiting AM strategy. In any event, not only 5-deoxy-strigol, but also other naturally occurring strigolactones and a synthetic analogue are active in stimulating branching of germ tubes of *Gi. margarita* at very low, although variable concentrations. It therefore seems quite possible that an array of related compounds may be involved in plant–AM fungus signalling and that variations in both activity and sensitivity could play a part in differences in selectivity.

Research on the effects of root exudates of host species from several families, as well as of purified compounds, have revealed that extremely low concentrations are required to induce fungal branching and concurrent nuclear division; exudates from non-hosts in the Brassicaceae and Chenopodiaceae had no effect (Buee *et al.*, 2000). Both exudates of carrot root organ cultures and strigolactones themselves induce transcription of fungal genes involved in respiratory pathways and changes in mitochondrial activity within 0.5–1.0h of application, followed by physiological changes in terms of O_2 consumption and reduction of tetrazolium salts between 1.5 and 3h and changes in morphology and orientation of mitochondria in about 4h. The first morphological changes in branching were observed about 5h after application of the exudates (Tamasloukht *et al.*, 2003; Besserer *et al.*, 2006). Other work also points to significant changes in the fungi in response to root exudates or extracts, including very rapid decrease in membrane potential (Ayling *et al.*, 2000), a rather slower increase in cytoplasmic pH (Jolicoeur *et al.*, 1998) and marked changes in actin filaments in the newly-formed finely branched hyphae (Åström *et al.*, 1994). Membrane transport processes are implicated in these changes and may not be fully operational in germ tubes growing from spores. Some early work investigating this possibility showed that activity of H^+-ATPases on fungal plasma membranes apparently depended on colonization (Lei *et al.*, 1991; Thomson *et al.*, 1990b). More recently, Requena *et al.* (2003) found a change in expression of genes encoding fungal H^+-ATPases, with strong induction of *GmHA5* as appressoria develop, and concurrent downregulation of *GmPMA1* which is highly expressed in preinfection hyphae. Branching appears to follow marked modifications in metabolic acitivity and be associated with conversion to an 'infection-ready' state. There is also an intruiging link between the effects of increased plant P supply in concurrent reduction of production of BF and colonization (Tawaraya *et al.*, 1996). It is tempting to speculate that these physiological and morphological changes are accompanied by the production of 'myc factors' that signal the presence of a fungus that has the capacity to colonize roots (Figure 3.8).

'Myc factors' produced by AM fungi have not been identified, but evidence for their existence is accumulating and indicates that they may operate at several steps during colonization. Using a *pMtENOD11*–GUSA reporter construct, Kosuta *et al.* (2003) demonstrated that diffusible factors from preinfection mycelia of four AM fungi induced upregulation of the gene in the cortex of wild-type *M. truncatula*. Similar positive results were observed with three nod⁻/myc⁻ mutants (*dmi1*, *dmi2* and *dmi3*) in which AM colonization is halted at the stage of appressorium formation. All three mutants are blocked in their response to nod factors produced by rhizobia, so the results provide important evidence that AM and NFS pathways do not completely overlap (see Figure 3.8).

Appressorium formation on the root epidermis is the next step that is identified by clear changes in fungal morphogenesis (Giovannetti and Sbrana, 1998). The stimulus has again not been clearly identified, but is likely to be associated with the topography of or chemical signals from the root epidermis (see Chapter 2). The resulting contact induces the formation of a PPA in wild-type *M. truncatula*, but not in *dmi2* or *dmi3* mutants. Tissue penetration and formation and function of characteristic intracellular structures, particularly arbuscules, involves ongoing communication. Localized induction of H^+-ATPases, P transporters and other genes occurs in cortical cells containing arbuscules or intracellular coils (Gianinazzi-Pearson *et al.*, 2000; Rausch *et al.*, 2001; Harrison *et al.*, 2002; Karandashov *et al.*, 2004; Glassop *et al.*, 2005) (see Chapter 5). In some cases, communication appears not to require tissue penetration, because cytoskeletal changes have been observed in root cortical cells that have not actually been invaded by fungi but are adjacent either to intercellular hyphae or to cells with intracellular fungal structures (Blancaflor *et al.*, 2001). Likewise, two genes (*MtScP*, a serine carboxy peptidase, and *MtGst*, a glutathione S transferase) are expressed in non-colonized cells adjacent to infected regions (Liu *et al.*, 2003; Wulf *et al.*, 2003). These findings not only imply signalling, which could involve molecules produced by the fungus *per se* or by the colonized cells, but also indicates that the cytoskeleton plays an active role in facilitating intracellular invasion. Signalling over longer distances in the plant is indicated by changes in expression of *Mt4*, a gene which is downregulated by AM colonization and phosphate (P). Using a split root system, Burleigh and Harrison (1999) showed reduction of expression both in the colonized and non-colonized halves of the root, suggesting long-distance signalling via the shoot. Formation of appressoria on the epidermis of a Pen⁻ mutant of *M. sativa* was sufficient to generate the signal. Long-distance signal transmission in roots is also indicated by partial autoregulation of the extent of AM colonization mediated via the shoot (Vierheilig, 2004a; Meixner *et al.*, 2005) and by transmission of systemic acquired resistance (SIR) from *G. mosseae*-colonized regions of roots towards *Phytophthora parasitica* in tomato (Cordier *et al.*, 1998). The nature and origin of the signals is not known in any of these examples.

Changes in gene transcription during AM colonization

The plant side

The hypertrophy of plant nuclei, increased staining with DAPI and susceptibility to degradation by DNAase were early taken as evidence indicating that colonization induces higher rates of gene transcription in AM roots, leading to increases in accumulation of mRNA on a fresh weight basis (Schellenbaum *et al.*, 1993; Franken and Gnädinger, 1994; Murphy *et al.*, 1997). Such changes are to be expected given the greatly increased activity of colonized cells associated with proliferation of organelles, changes in plastid behaviour and other cytoplasmic changes apparent from electron micrographs and visualization of changes in the plant cytoskeleton. Alterations in expression of particular genes during AM development were first explored in both targeted and non-targeted ways using techniques such as differential screening of cDNA libraries and differential display PCR (Lapopin and Franken, 2000), coupled with real-time RT-PCR, *in situ* hybridization and other methods to determine timing and tissue localization of expression. Targeted approaches were designed to increase understanding of AM symbiotic activities that could be predicted on the basis of other evidence to be important aspects of plant–fungus interactions. These included exploration of plant defence-related responses, membrane transport processes and overlaps with nodulation in NFS. The non-targeted approaches were designed to uncover changes in gene expression that could not easily be predicted from preceding work and to identify AM-specific genes. Both lines of research have led to major advances and some surprises (see Table 3.1), which are discussed below and (in the case of nutrient transport) in Chapters 4 and 5. Among the early investigations, many genes identified to be regulated in AM symbioses have turned out to be regulated by other factors as well, such as P, nodulation or pathogen attack; few AM-related genes were identified that might be involved in specific control of symbiotic development. However, some of the early screening of cDNA libraries may have missed symbiosis-specific members of multiple gene families because the probes used lacked the necessary specificity.

Genes whose expression is altered at specific steps in colonization have been identified, including a number of appressorium-related genes that may be involved in signalling as well as other cellular functions (Weidmann *et al.*, 2004). Several early nodulation genes are induced in AM roots of legumes, including *MtENOD11*, which is implicated in responses to AM fungal signals (Kosuta *et al.*, 2003), as well as during arbuscule formation (van Rhijn *et al.*, 1997; Albrecht *et al.*, 1998; Journet *et al.*, 2001), in line with the overlaps in programming already discussed. Also at the arbuscule stage, there is transitory expression of genes from families first identified in responses to plant pathogens, but evidence is lacking that the function of the genes is related to defence against AM fungi (Harrison and Dixon, 1994; Salzer *et al.*, 2000; Bonanomi *et al.*, 2001; Brechenmacher *et al.*, 2004; Elfstrand *et al.*, 2005). AM-inducible transporters and H^+-ATPases have been localized to plant membranes surrounding coils or arbuscules in cortical cells (Rausch *et al.*, 2001; Gianinazzi-Pearson *et al.*, 2000; Harrison *et al.*, 2002; Paszkowski and Boller, 2002; Karandashov *et al.*, 2004; Glassop *et al.*, 2005). Grunwald *et al.* (2004) have identified a further seven genes with arbuscule-related expression patterns, following suppressive subtractive hybridization (SSH) using wild-type and ard⁻ *P. sativum RISNod24*. Apart from genes involved in basic cell functions which would be expected to be upregulated in these active cells, one gene was related to cell wall modifications and also expressed in rhizobium-induced infection threads, one was involved in cell cycle regulation and there were several stress-related genes. One of the surprises has been the expression in AM roots of leghaemoglobin genes, normally seen in NFS nodules and generally considered to be important in O_2 delivery to bacteroids (Fedorova *et al.*, 2002; Vieweg *et al.*, 2004, 2005). The role of leghaemoglobin in arbuscule-containing cells is suggested to be in scavenging NO and hence suppression of defence processes, which also occurs in these cells (Uchiumi *et al.*, 2002; Vieweg *et al.*, 2004).

A broader and more comprehensive picture of transcriptional changes during AM symbiosis is beginning to be obtained from larger scale investigations, based on SSH, analysis of macro- and microarrays and other techniques that indicate differential expression of large numbers of genes (Gianinazzi-Pearson and Brechenmacher, 2004). Among the goals has been identification of AM-specific genes and also genes common between AM symbioses and NFS, P nutrition and pathogenic fungal interactions. A general picture is beginning to emerge, but the detailed roles of individual AM-regulated genes in symbiotic processes awaits more detailed analysis. Results confirm that considerable transcriptional reprogramming takes place and provide pointers to avenues of research that might not have been readily predicted and are worth exploring. With the adoption of *M. truncatula* and *L. japonicus* as model legumes, increasingly rich resources for studies of symbiotic gene expression are being opened up. Although neither species has been completely sequenced, an enormous amount of information is available. So far large-scale, transcriptomics approaches have mainly centred on *M. truncatula* (Journet *et al.*, 2002; Liu *et al.*, 2003, 2004; Wulf *et al.*, 2003; Hohnjec *et al.*, 2005), but studies on *L. japonicus* (Kistner *et al.*, 2005) and rice (Güimil *et al.*, 2005) have also commenced. In general terms, between 1 and 4% of the plant genes expressed appear to be differentially regulated on AM symbiosis, mostly involving increases rather than decreases in transcription. Journet *et al.* (2002) analysed the available *M. truncatula* EST data set and identified a group of genes predicted with some confidence to be upregulated in AM libraries and therefore involved in AM symbiotic programmes. Of the genes to which some function could be ascribed, many were related to protein synthesis and processing, primary metabolism, response to abiotic stimuli and defence-related activities. Acknowledged difficulties inherent in the *in silico* approach made unequivocal conclusions difficult, but the analysis did predict changes in expression of P transporter and membrane intrinsic protein (MIP) genes that had already been shown by other methods to be regulated in AM roots, providing confidence in the approach. Subsequent investigations have used various combinations of *in silico* analyses, microarrays and SSH, coupled with quantitative real-time RT-PCR and/or tissue expression analyses to identify AM-related genes. In all cases, considerable numbers of such genes were identified, including many that had not previously been associated with AM symbioses and are therefore potentially AM-specific or related to unsuspected features, as well as groups of genes to which functions have been ascribed. Not surprisingly, genes encoding ion and sugar transporters have been consistently shown to be upregulated, particularly the AM-inducible P transporters *MtPT4* and *OsPT11*, but also including a second AM-inducible P transporter from rice (*OsPT13*) and nitrate, ammonium and sugar transporters from *M. truncatula*. Other commonly-identified groups include lectins, signalling- and defence-related genes, as well as the common symbiosis genes.

Only one investigation so far has directly compared transcriptional profiles of a plant (*M. truncatula*) inoculated with two different AM fungi (Hohnjec *et al.*, 2005). There was major overlap between the profiles as might be expected, but also some interesting differences including the extent of downregulation of the P transporters *MtPT1* and *MtPT2*. This observation may have been related to differences in the contributions of direct and AM P uptake pathways but, unfortunately, the paper did not report plant growth or P responses that might have shed light on the physiological significance of the differences in gene expression. As transcriptional profiling is increasingly undertaken, concurrent physiological characterization of plant material by standard methods should become routine to provide an essential functional context for interpretation of transcriptional changes.

The only non-legume to be subjected to AM transcriptional profiling so far is rice. Güimil *et al.* (2005) have taken advantage of the complete genome sequence to compare expression of genes in AM interactions with those in two interactions with pathogenic fungi, also taking into account differences that might be influenced by P nutrition. They found considerable overlap between the three fungal programmes, with more in common between AM and the hemi-biotrophic *Magnoporthe grisea* than the necrotrophic *Fusarium moniliforme*. The overlapping genes were suggested to be involved in supporting plant–fungus compatibility. This may prove to be a very interesting line of enquiry and lead to a better understanding of mechanisms supporting basic compatibility in fungal symbioses generally. Predictably, some commonalities in response to P supply and AM colonization were also observed. An innovative aspect of this work was the demonstration that 34% of the AM-associated genes identified in rice were also associated with AM symbioses in dicotyledons, confirming the prediction that responses are conserved between these two major angiosperm classes. However, there is also evidence for some divergent evolution among angiosperms. For example, the promoter regions of AM-inducible Pht1 transporters genes from the dicotyledons potato (*StPT3*) and medic (*MtPT4*) drove GUS expression in Eudicots in the Asteridae and Rosideae, but the promoter from rice (*OsPT11*) failed to do so (Karandashov *et al.*, 2004; Karandashov and Bucher, 2005). Putative regulatory elements in the promoter regions of different Pht1 transporter genes, as well as the AM-inducible gene *MtGst*, have been identified and some appear as likely candidates specifying activity in roots and/or arbuscular mycorrhizas.

The fungal side

Knowledge of regulation of gene expression in AM fungi is much more limited. A genome sequencing project for *G. intraradices* is underway, which will greatly expand resources (Martin *et al.*, 2004). As with plants, both targeted and non-targeted approaches have been applied. The targeted approaches have identified expression of genes involved in membrane transport and mineral nutrition, including the phosphate transporters *GvPT*, *GiPT* and *GmosPT* (Harrison and van Buuren, 1995; Maldonado-Mendoza *et al.*, 2001; Benedetto *et al.*, 2005), H^+-ATPases (Ferrol *et al.*, 2000; Requena *et al.*, 2003), nitrate reductase (Kaldorf *et al.*, 1998; Hildebrandt *et al.*, 2002), acid and alkaline phosphatases (Aono *et al.*, 2004; Ezawa *et al.*, 2005), carbon metabolism (Franken *et al.*, 1997; Harrier *et al.*, 1998; Lammers *et al.*, 2001) and cytoskeletal activity (Bonfante *et al.*, 1996; Bütehorn *et al.*, 1999; Delp *et al.*, 2003). Non-targeted approaches at various scales have, as with the plant partners, identified genes with no known functions as well as genes whose functions can be tentatively attributed to general metabolism, signalling, cytoskeletal activity and the cell cycle (Sawaki and Saito, 2001; Jun *et al.*, 2002). Differential regulation of genes at different stages of symbiosis has been identified for some of these genes, including H^+-ATPases (Requena *et al.*, 2003), a superoxide dismutase (Lanfranco *et al.*, 2005) and a number of genes with potentially regulatory or signalling functions (Requena *et al.*, 1999, 2002; Delp *et al.*, 2003). Random sequencing of inserts from a cDNA library of germinating *G. intraradices* spores yielded ESTs potentially encoding proteins with a wide range of functions, including once again transport and metabolism, cytoskeletal activity and cell cycle and nucleus related activities. Using SSH to prepare a cDNA library enriched in genes upregulated during appressorium formation, Breuninger and Requena (2004) found that 63% had no known homologues and therefore might be specific for AM symbiosis. They also found genes common to other plant–microbe interactions including, tantalizingly, some associated with Ca^{2+}-related signalling, as also found for nodulation. It seems likely that an explosion of information will shortly be forthcoming, followed by the challenge of integrating the new data to understand the contributions of the fungal partner to development and function of AM symbioses.

Proteomics approaches

Gene expression is only one step to determining how plant and fungal processes are programmed during AM symbiosis. In the final analysis, uncovering what happens to synthesis and localization of proteins and to metabolic regulation and whole plant physiology will be required. Protein profiling is an important step in this direction and has been attempted in a number of investigations. Analysis of soluble protein composition (Pacovsky, 1989; Arines *et al.*, 1993; Schellenbaum *et al.*, 1993; Dumas-Gaudot *et al.*, 2004) revealed production of new polypeptides in AM roots, some of which were certainly fungal, but others were plant proteins related to particular stages of colonization. Recently, more comprehensive proteomics approaches are being adopted, taking advantage of highly sensitive analytical tools (Bestel-Corre *et al.*, 2004; Canovas *et al.*, 2004). Although major advances have not yet been made, the research is positioned to complement both genetic and well established physiological approaches in the immediate future.

Effects of AM colonization on plant defence responses

The reasons why AM fungi, which have many similarities with damaging plant pathogens, fail to trigger major defence responses in plants has intrigued researchers for many years. The fungi certainly do not grow in an unrestricted fashion in roots, but follow closely programmed paths through epidermal, subepidermal and cortical cells, never invading the stele. Comparisons of plant defence responses to biotrophic parasitic and AM fungi has yielded some coherent information. It is important to recognize, however, that although compatible interactions in these two types of biotrophic symbioses are superficially similar, there are significant differences. In AM interactions with wild-type host plants the only sign of loss of biotrophic status is the regular degeneration of arbuscules (see Chapter 2); it is the fungus and not the plant cell which collapses and dies. In contrast, in biotrophic symbioses involving fungal parasites the plant cells die, either rapidly in hypersensitive, resistant responses or more slowly in compatible interactions. AM associations are consequently most closely analogous to compatible biotrophic interactions, but care must be taken in drawing direct comparisons between them. Nevertheless, AM symbioses may represent a state of basic compatibility between plant and fungus which probably underlies all the more taxon-specific resistance mechanisms. Having evolved so long ago, the AM recognition and signalling systems may well predate evolution of specific mechanisms of resistance to plant disease that have been so well studied in angiosperms. Overlaps observed between genetic programming in rice interacting with an AM fungus and with two root-infecting pathogens adds some weight to this suggestion (Güimil et al., 2005). Progress on signalling in AM symbioses, together with identification of receptor-like molecules, can be expected to permit direct comparisons with programmes involving both resistance and compatible responses triggered by biotrophic pathogens.

Nevertheless, plant mutants with AM colonization compromised in various ways have not been shown to exclude or reduce infection by any of the root-infecting parasites tested thus far. These include root-knot nematodes, *Rhizoctonia solani*, binucleate rhizoctonias, *Fusarium oxysporum* and bulb and potato aphid (David-Schwarz et al., 2001; Morandi et al., 2002; Barker et al., 2005; Gao et al., 2006). In contrast, the *rmc* mutant of tomato has been shown to be more susceptible to *F. oxysporum* f. lycopersici and to permit more extensive reproduction of *Meliodogyne incognita* (Barker et al., 2005). Tantalizing suggestions of commonalities in genetic programming between AM colonization and root-knot nematode development have yet to be confirmed (Tahiri-Alaoui and Antoniw, 1996). Furthermore, mutations in the common symbiosis genes in *Lotus japonicus* had no effect on infection of shoots by the biotrophic pathogen *Uromyces loti* (Mellersh and Parniske, 2006).

In AM symbioses, some defence responses are apparently mobilized for a short period but are later suppressed. In typical AM interactions, there are no major changes in synthesis of lignin or callose in the plant cells. Metabolism of secondary metabolites, including flavonoids, is altered, but to a much lesser extent than in response to pathogen attack. Enzymes of the phenylpropanoid pathway have been investigated at levels of transcription and activity, particularly in legumes. During AM development, early stimulation of both transcription and activity of phenylalanine ammonia lyase (PAL), chalcone synthase (CHS) and isoflavone reductase (IFR) occurs in *Medicago* and *Phaseolus*. In *Medicago*, an increase in transcription of chalcone isomerase (CHI) has also been reported (Harrison and Dixon, 1993; Lambais and Mehdy, 1993; Volpin et al., 1994, 1995). The increase in activity of PAL and CHS is transitory in *M. sativa* (Volpin et al., 1994) but, in *M. truncatula*, accumulation of PAL and CHS transcripts is maintained at 1.75 and 2.25 fold respectively above uninoculated controls for some weeks, while IFR transcripts decline. These increases are small compared with plant–pathogen interactions, but are consistent. The pattern of transcription of IFR is correlated with changes in accumulation of the phytoalexin medicarpin and appears to indicate the suppression of a defence response in mycorrhizal plants, which is not observed in the nod⁻myc⁻ *M. sativa* genotype (Harrison and Dixon, 1993). In *Glycine max*, small increases in glyceollin production have been observed following inoculation with *Glomus* spp., but these are either slow to develop or not significantly greater than the uninoculated controls (Morandi *et al.*, 1984; Wyss *et al.*, 1991). It must be remembered that the values are averaged over the different cells types in the roots, not all of which show increases in levels of transcripts. It remains to be seen whether AM fungi induce the same gene-family members as pathogens and, indeed, whether the genes have similar functions.

In *Arum*-type interactions in *M. truncatula* and *Phaseolus vulgaris, in situ* hybridization shows that the increased levels of defence-related mRNAs are confined to cells containing arbuscules, possibly indicating a localized stress response or a mechanism controlling, or even enabling, intracellular invasion (Harrison and Dixon, 1994; Blee and Anderson, 1996). No increases associated with intercellular hyphae were observed. Arbuscule-related expression also indicates upregulation of a number of genes involved in defence and stress responses. These findings may help to explain the short life span of arbuscules, but details of the triggers and processes that lead to collapse are not clear. *Paris*-AM have been less well studied. However, a comparison of defence-related gene expression in both *Arum*- and *Paris*- types formed by different fungi in tomato indicated a substantially greater increase in expression in *Paris*-type AM formed by *Scutellospora calospora*, compared with *Arum*-types formed by two *G. intraradices* isolates (Gao *et al.*, 2004). This finding led to the suggestion that the large number of cell wall penetrations in *Paris*-AM increases the release of cell wall fragments capable of acting as endogenous elicitors. Higher defence-related gene expression in *Paris*-AM was not associated with lower root colonization, indicating that the gene products do not control invasion of the cortex, nor is their suppression a prerequisite for successful AM colonization. This finding supports previous work showing that constitutive expression of several defence-related genes had no effect on the progress or final extent of AM colonization in tobacco (Vierheilig *et al.*, 1995).

Localization of hydroxyproline-rich glycoproteins (HPRG) in the arbuscular interface also suggests mobilization of small and localized defence responses (Bonfante-Fasolo *et al.*, 1992). However, although Franken and Gnädinger (1994) observed consistent increases in accumulation of transcripts of a gene coding for HRGP during mycorrhizal colonization of *Petroselenium crispum*, preliminary *in situ* hybridization analysis showed that the tissues involved were the stele and root apex and not the cells colonized by arbuscules. In this non-legume, which has been extensively used to study plant–pathogen interactions, there were no major changes in the accumulation of phenolic compounds or transcripts from *PAL, CHS* or 4-coumarate:coA ligase (*4CL*) genes in AM plants, or genes encoding other enzymes which respond to treatment with elicitors from *Phytophthora megasperma* f. sp. *glycinea*, including peroxidase.

Chitinases and β-1,3-glucanases are also implicated in responses of plants to parasites, while peroxidase is important in the final stages of lignin deposition, again as a defence response. Chitinase and peroxidase sometimes show transitory increases in activity in roots colonized by AM fungi (Spanu and Bonfante-Fasolo, 1988; Spanu *et al.*, 1989; Lambais and Mehdy, 1993; Vierheilig *et al.*, 1994; Volpin *et al.*, 1994; Salzer*et al.*, 2000; Bonanomi *et al.*, 2001; Elfstrand *et al.*, 2005), whereas activities of β-1, 3-glucanases are either unaffected or suppressed, compared with uninoculated controls (Lambais and Mehdy, 1993; Vierheilig *et al.*, 1994). Again, activities of these enzymes do not appear to be related to the control of colonization, because the progress of colonization is normal in transgenic *Nicotiana* plants expressing different forms of chitinase. Spanu *et al.* (1989), using gold-labelled antibodies to bean chitinase, showed that in *Allium porrum* these enzymes are localized in plant vacuoles and intercellular spaces of both AM and non-mycorrhizal plants and were never found bound to the walls of *G. versiforme*. Indeed, the lack of reaction may be because the chitin in the fungal walls is rendered inaccessible to the enzymes by the presence of other wall components. In the non-hosts *Brassica*, *Spinacea* and *Lupinus*, increases in chitinase and β-1, 3-glucanase activity, as well as ethylene production, have been observed in response to the presence of AM fungi, but the effects were all weak or transitory and did not markedly differ from responses of AM host plants (Vierheilig *et al.*, 1994). In the myc^{-1} mutant of *Pisum*, the defence responses appear to be stronger, accompanied by production of phenolics and callose, and to prevent the fungi from entering the root (Gollotte *et al.*, 1993). The occurrence of a pathogenesis related protein (Pbr1) in the wall thickenings of myc^{-1} mutants indicates that other features of host defence against pathogens are mobilized (Gollotte *et al.*, 1994). Variations in defence-related responses in the tomato *rmc* mutant challenged by different AM fungi indicated low gene expression in the Pen$^-$ interaction with *G. intraradices*, but much stronger responses in the Coi$^-$ phenotype with *S. calospora* which is restricted to epidermal and hypodermal layers. In *L. japonicus LjSym 4* mutants, the block at this step is associated with plant cell death, reminiscent of a hypersensitive response (Bonfante *et al.*, 2000).

In conclusion, rather than inducing major deployment of defence responses, AM colonization apparently provokes minor, transitory defence responses which are followed by general suppression. This is consistent with the persistent state of compatibility which arbuscular mycorrhizas represent. There is no evidence from the plant species so far tested that failure of colonization by AM fungi in non-host plants involves mechanisms similar to those which are mobilized in plants resistant to attack by pathogens. Nevertheless, the small changes observed may be sufficient to confer some pre-immunity on the plants which reduces the effects of subsequent attack by damaging pathogens (Cordier *et al.*, 1998; Barker *et al.*, 2005). Such effects might confer benefits in both natural ecosysems and agricultural and horticultural production. There appears to be a general assumption that AM symbioses have evolved from parasitic symbioses. The rationale for this is unclear, especially as ECM fungi have most obviously evolved from saprophytes (see Chapter 6). Exploration of this possibility for AM fungi is difficult because the Glomeromycota have no close relatives. Nevertheless, a change of conceptual framework might prove illuminating.

Conclusions

Research in the last 10 years has revolutionized our understanding of genetic programming and control of AM colonization, although much detail remains to be revealed. We know that the development of arbuscular mycorrhizas is under the control of plant and fungal genes, which act in a coordinated manner to produce the characteristic, biotrophic and compatible interaction in AM plants. It is possible to describe the colonization process in precise terms and with the help of increasing numbers of mutants, identify key regulatory steps and some of the genes that control them. Signalling between the symbionts, particularly prior to contact, is now confirmed and the identification of the molecules involved is likely to lead to much increased understanding of both the receptors with which the signals interact and of the roles of signals and receptors in facilitating integrated symbiotic development, as well as in conferring plant–fungus selectivity or specificity. Commonalities with NFS have been fully established and a common genetic programme identified in model legumes, with some functional attributes identified. Major advances have also been made in showing how gene expression is altered in both symbionts; future developments will surely come in understanding what roles the genes play, as well as downstream changes in protein synthesis, metabolism and physiology. One persistent gap in knowledge is how the symbiosis enables AM fungi to complete their life cycles. The answer may lie in the plant signals that result in changes to carbohydrate and lipid metabolism.

Research is at the stage where common features of AM symbiosis are being sought and a generalized picture developed. In future, it will be essential to appreciate the considerable diversity that has been revealed with respect both to structural aspects of colonization and to function and link the information to gene expression. Ongoing research will need to give increased attention to AM mutations that are not also involved in nodulation and to model plant species other than legumes. In this way processes and genes that are unique to AM symbiosis are more likely to be identified and aspects of symbiotic programming that lead to variations in AM structure, function and specificity will be revealed. Understanding control of structurally and functionally different AM symbioses at genetic and cellular levels will provide insights into their evolution and hence the present-day interactions in natural ecosystems.

4

Growth and carbon economy of arbuscular mycorrhizal symbionts

Introduction

This chapter is the first of two on the physiological interactions between the symbionts in arbuscular mycorrhizal (AM) plants. It will cover the basis of the symbiotic relationships in terms of reciprocal transfer of organic carbon (C) and soil-derived nutrients between the symbionts and will also describe the C economy of the symbioses and the growth responses of the plants to colonization by AM fungi. New work on C metabolism in AM fungi is presented and the importance of mycelial links between plants is considered, in so far as they affect C allocation among members of a group of AM plants. Interactions between AM fungi and non-photosynthetic (mycoheterotrophic) plants will be discussed briefly, because of their unusual method of organic C nutrition, but are covered in greater detail in Chapter 13.

The symbioses between AM fungi and autotrophic plants are generally regarded as mutualistic, with the main basis of mutualism assumed to be bidirectional transfer of nutrients. With the exception of a few achlorophyllous species, AM plants are autotrophic and, although normally colonized by AM fungi in the field, many are capable of satisfactory growth in the absence of colonization, as long as mineral nutrient supplies are adequate. These species are considered to be facultative symbionts. Others, viewed as obligate symbionts, are less capable of acquiring nutrients when non-mycorrhizal and are much more dependent on the symbiosis for survival. There is increasing recognition of wide diversity in the extent to which plants depend on, or respond to, AM fungi and the consequent continuum of symbiotic interactions, grading from mutualism to parasitism and strongly influenced by identities of the partners and environmental conditions (Johnson *et al.*, 1997; Jones and Smith, 2004). In contrast, the fungi are all ecologically obligate symbionts. There is no good evidence that any AM fungi have significant saprotrophic ability and the limited capacity of their propagules to produce mycelium in the absence of symbiosis is based on mobilization of reserves from spores or other propagules. In consequence, AM fungi depend on recent photosynthate supplied by the autotroph and, as will be discussed later, utilize a considerable proportion of the assimilated C. The development of external mycelium of the fungal symbiont in the soil ensures that it has access to soil-derived nutrients, some of which are passed to the host plant.

The simple view of mutualism, based largely on plant growth measured as vegetative biomass or whole plant nutrient uptake, takes no account of changes in the quantitative balance of nutrient transfer between symbionts at different times as the plants develop over days, weeks or years, nor of the fact that nutritional interactions within an uneven-aged plant community, composed of different species, may be very complex indeed. Furthermore, it ignores non-nutritional features of the symbioses. The more stable physicochemical conditions in the root apoplast (e.g. water potential, solute concentrations, pH), may be an advantageous habitat for the fungus compared to the soil environment, quite apart from organic C supply. For plants, there is increasing evidence that AM colonization may increase resistance to pathogens and insect herbivores, and tolerance of water deficit. It has been argued that increased fitness (rather than biomass or nutrient uptake) is, for both partners, the most appropriate basis on which to determine whether an association is mutualistic. Whereas the fungi cannot complete their life cycles without plants, the consequences of AM symbiosis for plant fitness over complete life histories, have only been investigated in a very few instances. Consequently, it may be difficult to demonstrate that a symbiosis is mutualistic, especially in natural ecosystems, and the term 'mutualism' should be used rather cautiously (see Jones and Smith, 2004). What is important is that we appreciate that considerable diversity in responsiveness is the norm and seek to understand how the symbioses work in nature.

The issues that will particularly be addressed in this chapter concern the organic C transfer from plant to fungus; C metabolism of AM fungi, an area in which major advances have been made in the last 10 years; the growth of AM plants, particularly how they allocate photosynthate to growth of roots and shoots and to the development of the symbiotic fungus; and variations in plant responsiveness in different plant–fungus combinations. Understanding these physiological relationships, including growth responses, requires some understanding of the role played by AM fungi in mineral nutrition, as well as their requirements for organic C. This topic will be covered briefly here, but details of the integration of fungal and plant nutrient uptake processes and the transfer of soil-derived nutrients between symbionts are discussed in Chapter 5.

Effects of AM colonization on plant nutrition and growth – overview

The relationship between the development of arbuscular mycorrhizas and increased growth of the host was recognized by Asai (1944) in his studies of AM colonization and nodulation in a large number of legumes. He concluded that colonization was important both in plant growth and in the development of nodules. Subsequently, a large number of experiments by many investigators were carried out which demonstrate that, in many plants, colonization is followed by considerable stimulation of growth. This early work has been extensively reviewed, with particular emphasis on the importance of arbuscular mycorrhizas in increasing plant P nutrition (Gerdemann, 1968, 1975; Mosse, 1973; Tinker, 1975a, 1975b; Gianinazzi-Pearson and Gianinazzi, 1983; Hayman, 1983; Koide, 1991a; Smith, 1980; Smith and Gianinazzi-Pearson, 1988).

Pioneering work on the potential significance of mycorrhizas in plant nutrition was carried out by Mosse (1957) on apples, Baylis (1959, 1967) on *Griselinia* and other New Zealand plants and Gerdemann (1965) on *Liquidambar* and maize. Subsequently, Daft and Nicolson (1966, 1969a, 1969b, 1972) and Hayman and Mosse (1971, 1972; Mosse and Hayman, 1971) independently investigated the basis for the growth responses in a number of plant species, particularly with respect to soil conditions and fungal inoculum density. They demonstrated that development of AM roots and their effects on plant growth are greater in soils of low or imbalanced nutrient status, particularly if P is in short supply, and made valuable advances in interpretation of the mechanisms of these effects, which are discussed in Chapter 5.

The C economy of AM plants needs to be considered in the context of the effects of AM colonization on mineral nutrition and the relative costs of fungal C use, in relation to benefits derived from increased nutrient uptake. It is well established that AM roots are often more efficient in nutrient acquisition, per unit length, than non-colonized roots. Most of the evidence for this comes from pot experiments conducted with AM-responsive plants, in the controlled conditions of glasshouse or growth room, but some field studies have given similar results (Jakobsen, 1986, 1987; Dunne and Fitter, 1989; Merryweather and Fitter, 1995a, 1995b). Furthermore, even when there is no increase in nutrient uptake efficiency in tissue concentrations or in total plant nutrient content, it is now established that the fungal partner can make a considerable contribution to nutrient uptake. The largest effect of AM formation is on P nutrition. Not only is P required by both symbionts in relatively large amounts, but it is poorly mobile in soil and occurs in very low concentrations in the soil solution, being rapidly fixed as iron, aluminium or calcium phosphate or immobilized in the microbial biomass. There is also convincing evidence for increased uptake of zinc (Zn), which is also poorly mobile and is deficient in some soils, and of copper (Cu). Only relatively recently has attention turned to N which, in addition to its organic forms, occurs as either poorly mobile ammonium or nitrate; the latter is highly mobile in moist soil, but not in dry soil. Evidence is accumulating that AM fungi have the potential to play a considerable part in plant N nutrition and that uptake of both ammonium and nitrate can be increased in mycorrhizal plants (see Chapter 5).

AM plants have two potential pathways of nutrient uptake, directly from the soil or via an AM fungal symbiont. The AM pathway depends on three essential processes: uptake of the nutrients by the fungal mycelium in the soil; translocation for some distance within the hyphae to the intraradical fungal structures (hyphae, arbuscules and coils) within the roots; and transfer to the plant cells across the complex interface between the symbionts (Figure 4.1). The fungal mycelium in soil can absorb nutrients beyond the zone depleted through uptake by the roots themselves, so that they increase the effectiveness with which the soil volume is exploited. Furthermore, the soil pores that can be penetrated by hyphae are perhaps an order of magnitude smaller than those available to roots (Table 4.1). Consequently, the effects of AM colonization on P nutrition are often large and may have indirect effects on other aspects of plant metabolism, so that direct effects of the symbiosis on other nutrients are masked. For many years, and despite suggestions to the contrary (e.g. Jakobsen, 1995), it was assumed that uptake of P via the AM pathway always supplemented continuing direct uptake via root hairs and epidermis. This may certainly occur in some cases, but it is becoming increasingly apparent from experiments with isotopically labelled P, that the AM uptake pathway

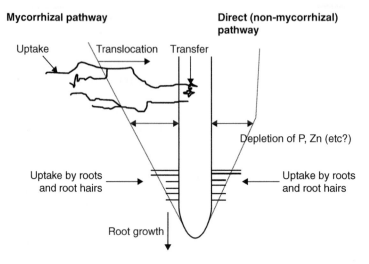

Figure 4.1 Diagrammatic representation of potential pathways of nutrient acquisition from soil in an arbuscular mycorrhizal root. The mycorrhizal pathway involves nutrient uptake by the external mycelium of an AM fungus, translocation through the hyphae to fungal structures in roots and transfer across symbiotic interfaces to the plant root cells. The direct pathway involves uptake by root hairs and epidermis. Depletion of relatively immobile nutrients in soil, such as P, following rapid uptake via either pathway is also indicated.

Table 4.1 Representative measurements of dimensions and spheres of influence of roots and hyphae of associated arbuscular mycorrhizal (AM) fungi growing in soil.

Feature	Hyphae	Roots
Diameter (μm)	2–10	>300
Specific length (m/g soil)	2–40	<0.10[1]
Radius of influence (m from the root)	0.25	<0.01[2]
Inter-hyphal or inter-root distance (μm)	~130	2000

[1] excluding root hairs; [2] including root hairs.

is probably the major route by which nutrients are absorbed, not only when large benefits of the symbiosis are apparent, but also when there are no increases in whole-plant P content. The implication is that uptake via the direct pathway can be greatly reduced or eliminated in many AM plants, due to P depletion in the rhizosphere and/or to downregulation of P transporter expression in root hairs and epidermis. This will be discussed further in Chapter 5, but the fact that even non-responsive plants may rely on their symbiotic partners for delivery, certainly of P and possibly other major and minor nutrients, needs to be appreciated in the later discussion of net costs and benefits of AM symbiosis.

When the availability in soil is low, non-mycorrhizal root systems may be unable to absorb P effectively and the plants become P deficient and grow poorly. AM colonization and P uptake lead to relief of this nutrient stress and, in consequence, plant growth is increased. This is the well-known mycorrhizal growth response (the 'big and little plant effect') which has been demonstrated for an enormous number of species mainly in pot experiments. In addition to increased growth rate, responsive AM

Table 4.2 Effects of AM colonization and P nutrition on growth and P concentrations of roots and shoots of *Trifolium subterraneum* (31–35 days old). Means of three replicate determinations, with standard errors of means in parentheses.

Added P (mmol/kg)	Colonization (%)	DW (mg/plant)			R/S	P concentration (µg/mg DW)		Response (%)
		Root	Shoot	Total		Root	Shoot	
0	0	36 (6)	39 (4)	75 (9)	0.92	0.53	0.79	
0	74	51 (1)	58 (3)	109 (2)	0.88	1.34	1.98	45
0.2	0	57 (4)	63 (6)	120 (10)	0.90	0.75	1.02	
0.2	72	57 (3)	90 (3)	147 (5)	0.63	2.12	2.83	22
0.4	0	70 (8)	97 (6)	172 (14)	0.72	1.20	1.29	
0.4	63	60 (2)	104 (3)	164 (5)	0.58	3.00	3.11	0
0.67	0	87 (11)	132 (8)	218 (20)	0.66	1.77	1.57	
0.67	53	67 (1)	120 (2)	187 (2)	0.56	2.33	2.83	−14

Data from Oliver *et al.* 1983 and unpublished). Per cent response (calculated from fresh weight data) = $100 \left(\dfrac{M - NM}{NM} \right)$.

plants sometimes have higher tissue P concentrations than non-mycorrhizal plants grown in soil of the same P status and allocate a lower proportion of total plant weight to roots (Table 4.2). As soil P is increased, the growth increase of responsive plants to AM colonization declines, so that if sufficient P is available to support near maximum growth of NM plants, the colonized plants may actually grow less than non-mycorrhizal ones. However, although there is then a negative growth response to colonization, the fungus continues to have an effect on the physiology of the plants. The extent of this depends on the sensitivity of the symbiosis to P supply (see Chapter 2), but the AM roots continue to function as effective nutrient absorbing organs, the root: shoot ratio is reduced and the plants accumulate P in luxury amounts (see Smith and Gianinazzi-Pearson, 1988). Nutrient uptake by the fungus continues, but does not lead to a positive growth response because other factors, including the rate of C acquisition via photosynthesis, limit the rate of growth. Indeed, negative growth responses in AM plants can be induced by low irradiance and are most commonly observed in experiments carried out in glasshouses in winter or in growth rooms with poor light sources, a further indication that there may be a delicate balance between the benefits to be gained from symbiosis in nutrient acquisition and the costs incurred by supporting the heterotrophic fungal symbiont (Buwalda and Goh, 1982; Bethlenfalvay *et al.*, 1983; Bethlenfalvay and Pacovsky, 1983; Koide, 1985a; Modjo and Hendrix, 1986; Smith FA *et al.*, 1986; Modjo *et al.*, 1987; Son and Smith, 1988).

Increased growth has been demonstrated for a very wide variety of plant species including many crop plants and trees, and is often manifest as increased biomass of roots as well as shoots, reduced root/shoot ratio and increased tissue P concentrations (Table 4.2). In a few plant species, increased flower production and yield has also been demonstrated. Nodulation and N fixation in AM legumes and dually colonized actinorrhizal plants are also increased and have been shown to result in increased tissue N concentrations in some experiments (Barea and Azcón-Aguilar, 1983). Indeed, so many pot experiments have been carried out with such a wide variety of plant species that most of them cannot be cited. The majority of the positive effects on growth can be attributed, directly or indirectly, to improved mineral nutrition and, in facultatively AM plants, similar changes have often been shown in response to application of fertilizer in the absence of AM colonization. However, it must be stressed here that in discussing C balance in AM plants and C use by the fungal partner, it is essential to distinguish direct mycorrhizal effects from those which would inevitably follow any changes in size, form or nutrient content and could be independent of colonization itself. Although this point has been made repeatedly, a large number of investigations continue to ignore its importance.

Emphasis on positive plant growth responses, particularly in much early work, meant that data indicating negative effects were ignored or even suppressed. In consequence, the diversity of responses and the 'mutualism-parasitism con-tinuum' (Johnson et al., 1997) went unrecognized for many years. The responsive-ness of plants to AM colonization varies enormously between species and cultivars and is markedly influenced by nutrient supply. This responsiveness (sometimes referred to as 'mycorrhizal dependency') is expressed as the per cent difference in dry matter between AM and non-mycorrhizal plants grown in the same soil (Table 4.2). Responsiveness is, as Gerdemann (1975) emphasized, strongly affected by the nutrient status of the soil and by the irradiance. In other words, both the capacity of a plant to absorb nutrients independently and its capacity to support a hetero-trophic symbiont with 'excess' photosynthate are important and may be genetically and environmentally influenced. Current work is increasingly directed to revealing underlying symbiotic processes and in addressing questions relating to the ecologi-cal and evolutionary benefits of AM symbiosis both for obligately AM species and for species that do not apparently benefit from the association in terms of increased growth or nutrient acquisition. In most cases, the benefits, or lack of them, have been identified in highly simplified pot experiments taking no account of complex interac-tions which exist in the field. Elucidation requires both detailed information on sites, mechanisms and amounts of organic C transfer to the fungus, as well as knowledge of the physiological bases of variations in plant responses to colonization.

C transfer to the fungal partner

Interfaces between fungus and plant

Transfer of hexoses from plant to fungus, as well as mineral nutrient transfer in the opposite direction, occurs across symbiotic interfaces. These may be intercellular, where hyphae grow in the intercellular spaces of the root cortex, or intracellular involving arbuscules, hyphal coils or arbusculate coils. The structure and compos-ition of these interfaces vary in the details of specialization, surface area, amounts of wall material and membrane modifications (see Chapter 2), but all have the same basic design. As shown in Figure 4.2, they all involve plasma membranes of both partners, separated by an apoplastic interfacial compartment which contains vary-ing amounts of wall or wall-like material. In consequence, the intraradical phase of the fungus is always located in a more or less specialized apoplastic compartment in the plant. There is no cytoplasmic continuity between the symbionts and transfer of nutrients in either direction requires efflux from one partner followed by uptake by the other, with control potentially operating at either step. However, there is no *a priori* reason to suppose that direct exchange of organic C and mineral nutrients occurs bidirectionally across the same interface(s). Membrane transport processes at the interfaces must operate by the same 'rules' as apply elsewhere in plants, dictated by (electro) chemical potential gradients of individual solutes between compartments and the requirements to maintain charge balance and cytoplasmic pH. Thus, transfer of uncharged hexoses from plant to fungus (Shachar-Hill *et al.*, 1995) could be passive (facilitated diffusion), with the direction dictated by concentration gradients across the interface maintained by differential production and consumption of the solute. Alternatively, passive efflux from the plant might be followed by active proton co-transport into the fungus, requiring maintenance of proton motive force (PMF) across the fungal membrane. At this stage, there is no evidence that allows us to distinguish between the two hypotheses, but rapid conversion of hexoses to specific fungal metabolites would maintain a concentration gradient in favour of the fungus.

Which interfaces are involved in C transfer from plant to fungus remains an open question. There is a general assumption that arbuscules are involved in *Arum*-type AM, but definitive evidence is still lacking and it has been suggested that the interface between the intercellular hyphae and the root cortical cells could also be important. Evidence to support the latter idea includes the longevity of the hyphae (Smith and Dickson, 1991; Tisserant *et al.*, 1993) and their persistent continuity with the external mycelium, even when frequency of arbuscules declines as the plants age. The presence of H^+-ATPases on hyphal plasma membranes indicates that these are energized, capable of generating the PMF necessary to support proton cotransport of sugars absorbed from the intercellular apoplast (Gianinazzi-Pearson *et al.*, 1991a, 2000) (Figure 4.3). Furthermore, circumstantial evidence that arbuscules may not be

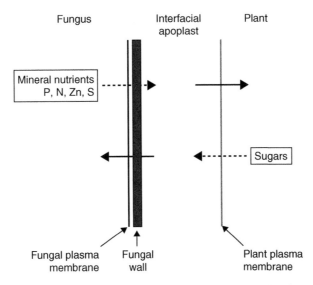

Figure 4.2 Diagrammatic representation of a generalized mycorrhizal interface, indicating the key structural components of fungal plasma membrane, fungal wall material, interfacial apoplast and plant plasma membrane. The arrows indicate directions of movement of soil-derived nutrients from fungus to plant and sugars from plant to fungus. Efflux steps are indicated by dotted arrows and uptake steps by solid arrows.

the sole site of C acquisition from the host comes from the observations of Mosse and Hepper (1975), who showed that extraradical mycelium began to grow as soon as intercellular hyphae began to colonize the root cortex and before arbuscules were formed. Some transitory proliferation of external mycelium also occurs in the Pen$^+$ Coi$^-$ interaction between the *rmc* mutant of tomato (see Chapter 3) and *Scutellospora calospora*. However, the fungus was unable to complete its life cycle and produce fully developed spores indicating that, although some C transfer via the reduced interface is possible, this is certainly low compared with the normal myc$^+$ interactions (Manjarrez-Martinez, 2007).

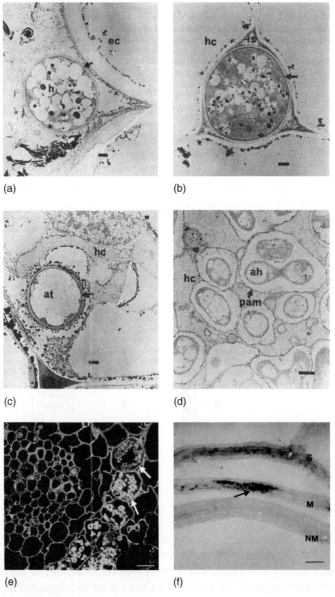

(a) (b) (c) (d) (e) (f)

Figure 4.3 (Caption opposite)

The distribution of H^+-ATPases in the arbuscular interface is consistent with the role of arbuscules in polarized P transfer to the plant, rather than bidirectional transfer of both C and P (Gianinazzi-Pearson et al., 1991a, 2000) and the site of C transfer in Paris-type AM, with few or no intercellular hyphae, must be presumed to be the intracellular coils (Smith and Smith, 1996). More work is still required to elucidate the pathway(s) and mechanisms of organic C transfer to the fungi.

AM fungal carbon metabolism

AM fungi are completely dependent on their plant partners for organic C and are unable to complete their life cycles without forming a symbiosis. Evidence for the dependence of AM fungi on plant-derived C is both direct and indirect. The indirect evidence is provided by the greatly increased growth of extraradical mycelium and production of spores, which take place once colonization is established, and by the influence of irradiance on the colonization of the roots (see Chapter 2). Direct evidence of C transfer was first obtained from $^{14}CO_2$-labelling experiments, with several groups showing that there is rapid translocation of ^{14}C-labelled photosynthate to root systems of AM plants and that some of this C passes into intracellular fungal structures and into hyphae outside the root (Ho and Trappe, 1973; Bevege and Bowen, 1975; Cox et al., 1975; Hirrel and Gerdemann, 1979; Francis and Read, 1984). The experiments showed that C transfer certainly takes place and that lipids are a major organic ^{14}C pool in AM fungi (Cooper and Lösel, 1978), but did not provide quantitative information on the amounts or identity of compounds transferred. Two approaches have contributed to considerable progress in this area. First, the isotopic ^{14}C labelling approach has been extended and, while confirming the earlier findings, has provided much more complete data on the quantities of material transferred from plant to fungus. Second, application of ^{13}C and ^1H nuclear magnetic resonance (NMR) spectroscopy, associated with HPLC and GC-MS, has revolutionized our understanding of the pathways of C metabolism and trafficking in AM roots

---

Figure 4.3 (a–d) Transmission electron micrographs of details of the arbuscular mycorrhizal interaction between *Allium cepa* and *Glomus intraradices*. Sections of roots fixed and stained to demonstrate ATPase activity by the deposition of lead phosphate which appears as fine electron-dense precipitates. (a), (b), (c) no inhibitors; (d) in the presence of molybdate, which inhibits non-specific phosphatases but not H^+-ATPases. (a), Extraradical hypha at the root surface; (b), intercellular hypha; (c), arbuscular trunk hypha in a cortical cell; (d), fine arbuscular branches within a cortical cell. Arrows indicate the presence of ATPase activity on fungal (a–c) and plant (d) membranes. Bars, 0.5 μm. The plate is reproduced from Smith and Smith, 1996, with permission. Origins of figures as follows, (a)–(c), V. Gianinazzi-Pearson, unpublished; (d), from Gianinazzi-Pearson et al. 1991. (e) Transverse section of a root of *Nicotiana tabacum* colonized by *Glomus fasciculatum* showing light microscopic immunolocalization of H^+-ATPase (bright silver signal) in arbuscule-containing root cortical cells (solid arrows). No signal can be seen in non-colonized cells or in the stele (open arrows). Bar = 15 μm. (f) GUS activity in arbuscule containing cells of a recently colonized root segment (M) of *N. tabacum* transformed with a *pma4-gusA* reporter construct (blue staining arrowed). No GUS activity is evident in non-mycorrhizal roots (NM) or in roots containing senescent arbuscules (S). Bar = 0.5 mm. See also Color Plate 4.1.

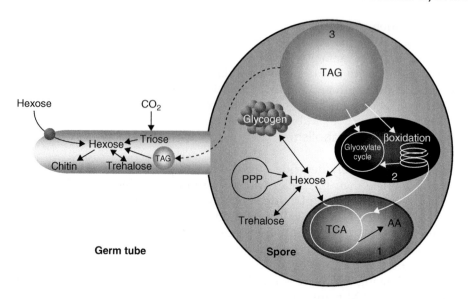

Figure 4.4 Proposed scheme for pathways of carbon (C) metabolism and trafficking in asymbiotic spores and germ tubes of an arbuscular mycorrhizal fungus. Broken arrows indicate transport between different C pools. (1) Mitochondrion, (2) glyoxysome and (3) lipid body. Reproduced from Bago *et al. Plant Physiology*, **124**, 949–957 (2000), with permission of the American Society of Plant Biologists.

(Pfeffer *et al.*, 2001). Much of this work was done using monoxenic cultures of AM fungi with transformed carrot roots. Although this biological system is highly artificial, it has proved tractable to following the fate of ^{13}C-labelled substrates in roots and external mycelium and, in particular, to tracking labelling positions, hence gaining critical information on the biochemical pathways operating both in germinating spores and in symbiosis. It will be important to verify findings, particularly the quantitative aspects, in more realistic whole-plant systems with normal source-sink relationships, shoot signalling and hormonal balance.

There are major differences between presymbiotic and symbiotic phases of fungal development. AM spores and fungal germlings, grown in the absence of any stimulation by root exudates, rapidly utilize stored trehalose and lipids to support growth (Figure 4.4). High rates of lipid breakdown to form hexoses and trehalose are associated with considerable dark CO_2 fixation, consistent with maintenance of tricarboxylic acid cycle (TCA) intermediates. Both enzyme assays and ^{13}C labelling patterns are consistent with operation of glycolysis, TCA cycle and pentose phosphate pathway (PPP). During this presymbiotic stage, the fungus can absorb small amounts of hexose or acetate, both of which are metabolized to trehalose, consistent with trehalose turnover. Notably, there is no evidence for lipid synthesis and this may be a significant feature distinguishing presymbiotic and symbiotic fungal development (Beilby, 1980; Beilby and Kidby, 1980; Amijee and Stribley, 1987; Bécard *et al.*, 1991; Schubert *et al.*, 1992; Bago *et al.*, 1999a, 2000).

Considerable changes in fungal C metabolism occur as symbiosis with roots develops (Figure 4.5). The intraradical fungal phase has a high capacity to absorb hexoses; glucose to a greater extent than fructose (Shachar-Hill *et al.*, 1995). There are few

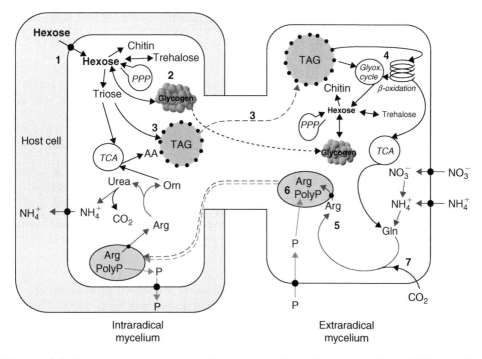

Figure 4.5 A proposed scheme describing metabolism of organic carbon (C), N and P in AM symbioses. Pathways and directions of fluxes are based on data from NMR findings. Names of metabolic pathways are in italics. The numbers refer to results of NMR experiments: (1) the uptake of hexose by intraradical parts of the fungus; (2) the formation of glycogen and trehalose from hexose within the host root; (3) the site of synthesis for storage lipids (TAG) is within the intraradical fungal tissue; this lipid is then exported to the extraradical mycelium; (4) the glyoxylate cycle is active in transforming exported lipid into gluconeogenic precursors for hexose synthesis in the extraradical mycelium; (5) arginine is actively synthesized by AM fungi by the usual metabolic pathway; (6) arginine may be associated with polyP in mycorrhizal vacuoles; (7) dark fixation of CO_2 is active in AM fungi. Reproduced from Pfeffer *et al. New Phytologist*, **150**, 543–553 (2001).

data for the rate of transfer of sugars across the interface(s). However, the values of C use by *Glomus fasciculatum* associated with *Cucumis* (Jakobsen and Rosendahl, 1990), together with information on mycorrhizal development (*G. mosseae*) in the same host (Smith and Dickson, 1991) can be used to estimate the C flux (as glucose) across the interface, assuming that hyphae alone or arbuscules or both are involved in the transfer process (Table 4.3). If it is assumed that hyphae are the only site of transfer, fluxes of C could be around 28 nmol/m²/s; if arbuscules are involved, either solely or together with intercellular hyphae, then the fluxes would be around 5.5 or 4.5 nmol C/m²/s, respectively. These estimates remain preliminary, but they show that combining measurement of C allocation to the fungus by ^{14}C labelling with determination of the development of arbuscules and intercellular hyphae has potential for obtaining accurate data for fluxes in single experiments.

Proposed pathways of C trafficking in the symbiotic fungus is shown in Figure 4.5. Hexoses transferred to the fungus are rapidly incorporated into trehalose and

Table 4.3 Fluxes of C (as glucose equivalents) from *Cucumis sativus* to *Glomus fasciculatum*.

C use by *G. fasciculatum* (mg C/plant/day)	7.4
Equivalent glucose transfer (nmol/s)	1.16
Length AM root per plant (m)	36
Hyphal interface (m²/m)	1.2×10^{-3}
Hyphal interface per plant (m²)	41.4×10^{-3}
Arbuscular interface (m²/m)	5.9×10^{-3}
Arbuscular interface per plant (m²)	212×10^{-3}
Total interface per plant (m³)	253×10^{-3}
Flux (nmol/m²/s)	
hyphae only	28
arbuscules only	5.5
whole interface	4.5

Calculated from data for C use by the fungus and per cent colonization of the roots from Jakobsen and Rosendahl (1990). Data for development of intercellular hyphae and arbuscules (by *G. mosseae*) in the same host from Smith and Dickson (1991).

glycogen, almost certainly providing a buffer against excessive glucose accumulation in the cytoplasm (Shachar-Hill *et al.*, 1995; Solaiman and Saito, 1997). There is also major synthesis of lipid in the intraradical mycelium which, together with glycogen, is exported to the external mycelium (Pfeffer *et al.*, 1999; Bago *et al.*, 2002a, 2002b, 2003). At this stage, the mechanism of hexose transfer across the symbiotic interface(s) is not known. However, it can be presumed that the plant delivers sucrose to AM roots and that sucrose hydrolysis by plant enzymes maintains a pool of hexoses available for transfer. Several investigations have shown increases in patterns of expression of both invertase and sucrose synthase genes in AM roots, consistent with this idea (Blee and Anderson, 2002; Ravnskov *et al.*, 2003), although expression was not straightforwardly related to the sizes of carbohydrate pools. Activity of alkaline invertase in AM soybean was consistently higher than non-mycorrhizal, so this enzyme may play a part in providing hexoses for AM fungi (Schubert *et al.*, 2004). Nevertheless, rapid conversion of sucrose to hexoses in the plant, and of hexoses to lipid and trehalose in the fungi, would help to maintain a concentration gradient in favour of continuing facilitated diffusion across the interface (Smith *et al.*, 1969; Patrick, 1989; Smith FA and Smith, 1990; Smith SE and Smith, 1990). Increased expression of a plant hexose transporter in AM roots (Harrison, 1996) could be related either to export of sugars from plant cells, or to reabsorption of hexose from the interfacial apoplast, possibly as a mechanism to control the C drain to the fungal symbiont.

Lipid arriving in the external mycelium is progressively broken down and converted to hexoses and trehalose. There is again evidence for considerable activity of glycolysis, the glyoxylate cycle and PPP, the latter presumably involved in the production of reducing equivalents and pentoses for nucleic acid synthesis. The external mycelium has little or no capacity to utilize exogenous sugars for catabolism, storage or transfer to the plant, nor does it synthesize any lipid. C supplies for this critical phase of the symbiosis come entirely as lipid and, to a lesser extent, glycogen from the intraradical phase. The lipid moves in the cytoplasmic stream as lipid bodies, which have a central triacylglycerol (TAG) core, surrounded by a phospholipid monolayer in which stabilizing proteins are inserted. Visualization of size and movement of lipid bodies suggests that they are progressively consumed as they pass outwards from the plant, but labelling patterns indicate some return flow, consistent with transport via bidirectional cytoplasmic streaming. The amounts of C moved are considerable, with estimates of 0.26 μg TAG/h per runner hypha of *G. intraradices* and 1.34 μg/h per runner hypha of *Gi. margarita*, sufficient to support observed spore production in the monoxenic culture system employed in the experiments (Bago *et al.*, 2002b). Dependence of AM fungi on the apparently very inefficient system of synthesizing lipid from sugars in the intraradical phase, exporting it to the external mycelium, and then resynthesizing carbohydrate from it is rather surprising. Bago *et al.* (2002b) suggest that lipid transport is a way of mobilizing a large amount of organic C in a form that does not induce osmotic stress and making sure that it reaches all parts of the symbiotic fungus. The process also argues that availability of organic C is not a major constraint to AM fungal growth and metabolism and that 'economies' for an organism continuously tapped into a photosynthesizing plant may not be generally required. At late stages of fungal development, large quantities of lipid are stored in AM roots, particularly in vesicles, as well as in spores in the soil.

Although some C cycling as lipid between intraradical and external phases of fungal development have been observed, no evidence for C transfer from the intraradical fungus back into the plant has been obtained. Evidence has been sought by feeding labelled substrates (acetate, glycerol and CO_2) to the external mycelium and also by using two monoxenic root cultures, linked by a common mycelial network (CMN) and supplying labelled substrates to one of them. In neither case was any labelling of sucrose (a characteristic plant sugar) observed (Pfeffer *et al.*, 2004). Significantly, as there is no photosynthetic tissue in the monoxenic system, the transfer of labelled C to the plant via photosynthetic fixation of labelled CO_2 was precluded. Furthermore, although transfer of N from fungus to plant has now been demonstrated in the monoxenic system, it occurs in inorganic form (Govindarajulu *et al.*, 2005; Jin *et al.*, 2005) (see Chapter 5). This is an important finding because it is no longer realistic to postulate that transfer of organic N from fungus to plant might facilitate C movement between plants linked in a CMN (Smith FA and Smith, 1990).

Quantities of organic carbon used by the fungal symbionts

Fungal biomass associated with roots has been estimated at between 3 and 20% of root weight (Harris and Paul, 1987; Douds *et al.*, 2000), but most early estimates did not include external hyphae or spores. These add to the biomass very considerably, contributing up to 90% of AM fungal biomass in some systems (Sieverding *et al.*, 1989; Olsson *et al.*, 1999). In an investigation with 26-day-old *Cucumis* with 95% of its root length colonized by *Glomus fasiculatum*, the dry weight of external hyphae was calculated at 2.6% of root dry weight (Jakobsen and Rosendahl, 1990). Spore production would not have contributed much to the overall C allocation in these young plants but, at later stages of development, spores, as was noted in Chapter 2, can be produced at very high densities and represent considerable further C allocation to the fungal symbiont.

Balance sheets of the C allocation in AM and non-mycorrhizal plants, taking into account their P nutrition and growth, have given a reasonably consistent picture of the proportion of plant photosynthate used by the fungal symbiont both in growth and respiration. By feeding $^{14}CO_2$ to AM and non-mycorrhizal plants matched for shoot P concentration or to a single plant with a split (AM and non-mycorrhizal) root system and determining the distribution of label in different fractions after a chase period, it has been calculated that AM roots of a range of herbaceous and woody plants receive about 4–20% more of the total photosynthate than non-mycorrhizal roots (Koch and Johnson, 1984; Harris and Paul, 1987; Douds *et al.*, 1988, 2000; Jakobsen and Rosendahl, 1990; Eissenstat *et al.*, 1993). C is deployed in growth of the intra- and extraradical mycelium and in respiration to support both growth and maintenance, representing a considerable increase in C flux to the soil. At this stage there is little indication of the reasons for the variations in the estimates, but they are likely to include species of plant and fungus, fungal biomass and rate of colonization, as well as the metabolic activity of the fungus.

Tables 4.4 and 4.5 (Jakobsen and Rosendahl, 1990) show data for ^{14}C incorporation into the AM roots of young *Cucumis* plants and the external AM mycelium

Table 4.4 C allocation in *Cucumis sativus*, colonized by *Glomus fasciculatus*. The plants were grown in pots divided into compartments by mesh, in order to separate hyphae from roots. Uptake of C and distribution of ^{14}C in 26-day-old AM *Cucumis sativus* 80h after labelling of shoots with $^{14}CO_2$ for 16h.

C uptake	
Total (mg C/day)	$37.0 \pm 0.5^*$
Specific (mg C/dm^2/h)	1.36 ± 0.03
^{14}C distribution (%)	
Shoot	54.1 ± 0.6
Shoot respiration	2.5
Root	13.2 ± 0.8
External AM hyphae[†]	0.8 ± 0.1
Soil organic C	2.3 ± 0.1
Below-ground respiration	27.0 ± 1.1
Ratio ^{14}C lost from roots:^{14}C translocated to roots	0.70 ± 0.09

Data from Jakobsen and Rosendahl, 1991. [*]Means ± SE of five plants. [†]Hyphal densities assumed to be similar in hyphal compartments (HC) and root compartments (RC).

Table 4.5 C allocation in *Cucumis sativus*, colonized by *Glomus fasciculatus*. The plants were grown in pots divided into compartments by mesh, in order to separate hyphae from roots. Length, dry weight and C incorporation of hyphae in hyphal compartment (HC) and length, AM colonization and C incorporation of roots in root compartment of AM *Cucumis sativus* 80h after labelling of shoots with $^{14}CO_2$ for 16h.

Hyphae in HC	
Length (cm/g dry soil)[*]	2708 ± 206[†]
Diameter (μm)	2.6 ± 0.1
Dry weight (μg/g dry soil)[‡]	34 ± 3
C incorporation	
Total (μg C/plant/day)	125 ± 14
Specific (μg C/mg dry wt/day)	41 ± 3
Roots	
Total length (cm/g dry soil)	24 ± 1
AM length (cm/g dry soil)	23 ± 1
C incorporation	
Total (μg C/plant/day)	4965 ± 301
Specific (μg C/mg dry wt/day)	17 ± 1

[*]Data corrected for hyphal counts in HC of NM plants. [†]Means ± SE of five plants. [‡]Dry wt = biovolume × 0.23 (Bakken and Olsen, 1983). The compartments contained 150g soil so that absolute lengths of colonized root can be calculated (3450 cm per plant).

of *Glomus fasciculatum* associated with it. The C incorporation (μg C/plant/day) into extraradical hyphae in this case was 6% of the incorporation by roots and the specific ^{14}C incorporation (μg C/mg dry weight/day) was 2.4 times that of the roots. The hyphae constituted 26% of the extraradical organic ^{14}C, emphasizing their importance in transfer of organic matter to soil (Table 4.4). Using a number of reasonable assumptions about fungal growth, specific incorporation into intraradical and external mycelium and intraradical fungal biomass, Jakobsen and Rosendahl (1990) calculated that the fungus could have been using as much as 20% of the total $^{14}CO_2$ fixed by the plant, equivalent to 7.4 mg ^{14}C per day. These plants were only 26 days old and were already extensively (95%) colonized, so that it is likely that most of the fungal tissue was alive and active (Smith and Dickson, 1991), possibly explaining the high C use. The fungal contribution to below-ground respiration was not determined separately, but a recent field study has shown that between 5 and 8% of C lost by pasture plants was respired by the external AM hyphae in soil (Johnson *et al.*, 2002b). More recently, Heinemeyer *et al.* (2006) showed by ^{13}C-labelling that the C demand of the extraradical mycelium of *Glomus*, associated with *Plantago lanceolata* was less than 1% of net photosynthesis. Respiration by hyphae was strongly influenced by irradiance, but effects of temperature were slight.

It has been suggested that a major potential transfer to soil is in the form of 'glomalin', supposed to be a glycoprotein produced in large amounts by AM fungi (e.g. Wright and Upadhyaya, 1998). However, recent work suggests that glomalin is most probably a heat shock-related protein (Gadkar and Rillig, 2006). It remains very firmly bound to fungal walls and enters soil only after hyphal death and breakdown, so is unlikely to contribute to C budgets in short-term experiments (Driver *et al.*, 2005). In any event, the large amounts of C used by AM fungi may, as discussed below, represent a cost to the plant in maintaining the symbiosis if the C could otherwise have been utilized in increasing growth or fitness. However, the costs may well be offset in a number of ways such as increases in rates of photosynthesis, reductions in root growth or alterations in root/shoot ratio.

Different fungal species appear to use different proportions of the total photosynthate. With *Cucumis* as the plant symbiont, the proportion of total fixed ^{14}C respired below ground was 16.3, 17.3 and 26.2% when the roots were colonized by *Glomus caledonium*, *Glomus* spp. WUM 10(1) and *Scutellospora calospora* respectively. Non-mycorrhizal control roots respired only 9.7% (Pearson and Jakobsen, 1993a). The transfer of ^{32}P to the plant via the fungi was studied in the same experiment and showed that *S. calospora* was the least efficient fungus and *G. caledonium* the most efficient in terms of hyphal ^{14}C use per unit of ^{32}P transported. The underlying reasons for these differences have not yet been determined, but could involve different colonization patterns, where efficiency might be related to a higher proportion of fungal biomass in arbuscules (involved in P transfer), compared with intercellular and external hyphae (involved in P uptake and in C transfer and distribution). Alternatively, there may be variations in symbiotic function at the level of rates of transfer across the interfaces or differences in metabolic activity. In this context, further exploration of the observation that N supply to monoxenic root organ cultures influences C allocation to *Glomus intraradices*, as well as expresson of AM fungal P transporter genes, may provide some clues (Olsson *et al.*, 2005). Further investigation of variations in fungal efficiency in different plant fungus combinations and under different environmental conditions will also provide valuable insights into the functional diversity of symbioses.

Cost-benefit analysis

C fixed by the plant during vegetative growth is allocated above ground to photo-synthetic tissue and below ground to nutrient absorbing roots and arbuscular mycorrhizas. Both these investments are essential for the continuing growth of the plant. However, it has frequently been suggested that the C used by the fungus represents a considerable cost to the plant, which may or may not be offset by a benefit in terms of nutrient uptake (Stribley *et al.*, 1980a, 1980b; Fitter, 1991). AM plants may sometimes be limited by C rather than nutrients, with direct evidence coming from the negative growth responses under low irradiance (see above) and indirect evidence from increased nutrient concentrations in tissues and/or increased fresh weight/dry weight ratios (Stribley *et al.*, 1980a, 1980b; Snellgrove *et al.*, 1982; Smith *et al.*, 1986; Son and Smith, 1988).

However, cost or benefit of a symbiosis determined as dry weight difference (or responsiveness) are estimates of the net effect and do not provide information on the gross effects, including total C fixed as well as the actual consumption of C by the fungal symbiont which, as shown above, can be up to 20% of the total fixed C. This consumption will be a cost only if the C involved could otherwise have been deployed by the plant in increasing biomass or other feature related to fitness. Furthermore, large contributions of AM fungi to plant nutrition, even when there is no net benefit in plant growth or fitness, complicate the rationale for this type of analysis.

Koide and Elliott (1989) were among the first to advocate the use of a common currency (C) for the 'economic' analysis of the efficiency of AM symbiosis. They defined the cost of the symbiosis as the C expended to support the symbiosis, gross benefit as the extra C fixed as a consequence of AM colonization and net benefit as gross benefit minus cost. The efficiency of the symbiosis was defined as net benefit/gross benefit. Tinker *et al.* (1994) extended the approach and provided a useful account of appropriate methods for making cost-benefit analyses under different conditions. Analysis of the C economy is valuable in determining the direct influence of an AM fungus on a plant, the amount of photosynthate used in fungal growth and respiration, and the relative efficiencies of roots and fungi under particular soil P conditions. It is important to extend the use of this approach to compare the efficiencies of different plant–fungus combinations and particularly to determine the C-use efficiencies of different fungi in slow and fast growing plants. The data may be valuable in identifying model systems for experiments to determine what characteristics of the fungi or plants lead to efficient or inefficient symbioses and, possibly, in selecting fungi for horticultural and agricultural applications. So far, few systematic investigations of this sort have been made, so that comparisons are mainly based on the easier, but somewhat less useful, measures of AM responsiveness or dependency.

In simple terms, AM colonization of a root system and the involvement of the fungus in energy- and substrate-requiring activities like nutrient uptake, vegetative growth and spore production, would certainly represent a cost to the plant if there were no way of compensating for it. This is the situation in plant–pathogen associations. In AM plants, the involvement of the fungus in nutrient acquisition may directly or indirectly increase the ability of the plants to fix CO_2 and, consequently, the 'expense' of the fungus is offset (Allen *et al.*, 1981; Snellgrove *et al.*, 1982; Koch and Johnson, 1984; Brown and Bethlenfalvay, 1988; Fredeen and Terry, 1988; Wang *et al.*, 1989; Eissenstat *et al.*, 1993; Schwob *et al.*, 1998; Wright *et al.*, 1998b).

Comparing AM *Citrus* plants grown in low P conditions with non-mycorrhizal, P-supplemented plants, Eissenstat *et al.* (1993) showed that the AM plants had higher photosynthetic rates than the non-mycorrhizal, and that a marked increase occurred during the period (7–8 weeks after planting) when fungal colonization was particularly rapid and P concentration in the tissues increasing. The AM plants in low P soil were highly dependent (in the strict sense of the word) on the symbiosis, so that, although the fungus used a considerable proportion of the fixed C, this was clearly compensated for by the improved nutrient uptake which resulted in a higher rate of photosynthesis and overall increased growth. Increased sink strength of AM roots has been shown to be associated with increased rates of photosynthesis and with increased activity of sucrose-degrading enzymes (Ravnskov *et al.*, 2003). AM roots typically, though not invariably, have lower sugar and starch concentrations and it appears that the 'extra' C produced in photosynthesis is not used in biomass production but is utilized by the fungal symbionts, illustrating internal compensation in support of the symbiosis which was independent of P or N nutrition and plant size (Wright *et al.*, 1998a, 1998b). Variations in sink strength mediated by identity of both fungus and plant have been observed (Lerat *et al.*, 2003a, 2003b). Increased photosynthesis and/or changes in C allocation may also be mediated via increased availability of inorganic P in the leaves (Sivak and Walker, 1986) and hence be affected by increased P uptake mediated by either by AM fungi or by increased P supply (Eissenstat *et al.*, 1993; Peng *et al.*, 1993; Fay *et al.*, 1996; Grimoldi *et al.*, 2005).

When the rate of photosynthesis and growth is limited by nutrient availability, the C cost of producing roots may be significant. In contrast, the cost of producing a given length of hyphae (around 2–10 μm diameter) is about two orders of magnitude less than for roots (>300 μm diameter), so that under these severely limited conditions, a plant may be able to 'afford' hyphae but not roots (see Table 4.1). However, at high P, the total cost of maintaining a mycorrhiza may be greater than an uncolonized root (Eissenstat *et al.*, 1993; Graham and Abbott, 2000). In *Citrus*, AM effects on the per cent below-ground recovery of ^{14}C (fed as ^{14}CO$_2$) that were independent of P were a reduction in the fibrous root component at early stages of growth, and increases in the soil component (including hyphae) and in below-ground respiration (Eissenstat *et al.*, 1993). Overall, the authors concluded that at equivalent P status AM plants had a lower efficiency of C production (change in whole plant C gain/change in C expended below ground) and were, surprisingly, somewhat less efficient than non-mycorrhizal plants in P acquisition.

Koide *et al.* (1999, 2000) have suggested that a more meaningful approach to costing investment in AM symbiosis may be, for a P limited plant, the expenditure of P itself in obtaining more. They tested the use of the parameter P efficiency index (PEI or dP/dt (1/P)) on two AM responsive species and a non-host species and showed that AM inoculation increased PEI of AM lettuce and abutilon, but not the non-host *Beta*. Smith (2000) showed how this growth analysis approach could be expanded to emphasize effects on P uptake of those plant and fungal features likely to contribute to high AM responsiveness that are listed in Table 4.6. Analyses such as these have highlighted effects of both AM colonization and P nutrition on C allocation and growth. However, they are inevitably based on values for biomass and total P uptake and therefore do not take into account reductions in the contribution

Table 4.6 Plant and fungal factors that may be associated with high plant responsiveness to AM colonization.

Fungus	Interface	Root	Plant
External hyphae	Fast development	Short length	Constant shoot [P]
High growth rate	Large area of contact	Low root/shoot ratio	Fluctuation in non-structural carbohydrates
High extension into soil	High longevity	Little branching	
High nutrient influx capacity	High nutrient flux to roots	Large diameter	
High nutrient translocation		Few/short root hairs	
Low C drain		Selectively flexible root/shoot ratio	
Internal hyphae		Inability to modify rhizosphere	
High growth rate		Low nutrient influx capacity	
Fast nutrient delivery to interfaces			

of the direct P uptake in AM plants and hence to the real magnitude of AM effects on P uptake. The recognition that there are significant changes in the pathway of P uptake in all AM plants regardless of their responsiveness has completely changed the way that costs and benefits of the symbiosis should be considered.

In summary, if P is readily available in soil at any point in time, then non-AM roots can frequently absorb it more efficiently than AM roots because of the high C cost of maintaining the fungal symbiont if there are no compensatory changes in rates of photosynthesis or biomass allocation within the plants. However, if P is in short supply or only transitorily available, then non-mycorrhizal roots cannot absorb it at rates that will support photosynthesis and growth. Investment of C in rapid growth of fine roots with long root hairs that help to overcome problems of development of depletion zones, is not possible (see Chapter 5) and the plant is dependent on the fungus because AM roots can obtain a resource that roots alone cannot. Even facultatively mycorrhizal plant species that are able to acquire P efficiently when non-mycorrhizal may receive considerable proportions of P via AM symbionts and may or may not exhibit net increases in P uptake or growth.

Variations in AM responsiveness of plants

Although positive benefits in terms of plant growth and/or total P uptake are generally highlighted, some plants are apparently unresponsive to AM colonization, even in soil of low P status. An enormously wide range of responsiveness, usually determined in pot experiments in both wild species and crops, is increasingly being recognized and investigated (Wilson and Hartnett, 1998; Klironomos, 2003; Tawaraya, 2003). An example of such variation is shown in Figure 4.6 (Klironomos,

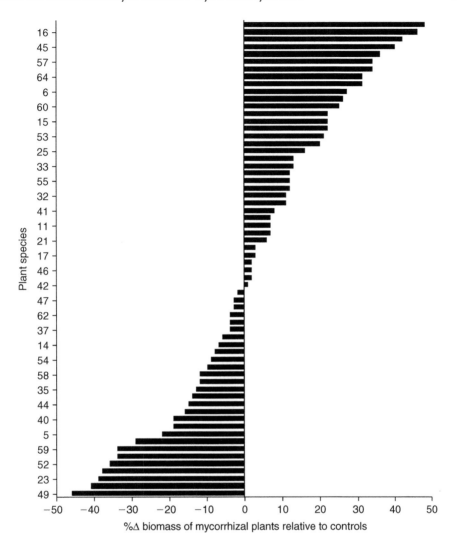

Figure 4.6 The influence of *Glomus etunicatum* on the growth of 64 plant species from the same site and grown in the same soil as that from which the AM fungus originated. Data are the per cent change in biomass of AM plants, compared to non-mycorrhizal controls. Mean 95% confidence interval was 17%. Identities of the plant species are given in Table 1 of the reference. Reproduced from Klironomos, *Ecology*, **84**, 2292–2301 (2003), with permission of The Ecological Society of America.

2003). Sixty-four plant species from an old field community in Canada were grown separately with an isolate of *Glomus etunicatum* from the same site and their growth compared with non-inoculated plants in the same soil. Growth responses in terms of whole-plant biomass varied between the species from −46 to +48%. The data highlight the fact that quite different AM strategies can be present among coexisting plant species. Unfortunately, in this investigation no attempts were made to relate responsiveness to changes in P uptake, root characteristics or variations in fungal colonization in roots or hyphal growth in soil. Understanding why plant species respond differently and the importance of such variation in ecological situations will only come from more detailed analyses, some of which are considered later.

Responsiveness also varies considerably with changes in nutrient availability in soil and this not only provides insights into mechanisms underlying the variations, but also affects the design of experiments aimed at determining the amounts of C allocated to AM fungi and the economics of the symbiosis in terms of bidirectional transfer of P and C. Measurements of the rate of photosynthesis and allocation of photosynthate in the growing plants need to take into account the effects of the symbiosis on P nutrition, so that comparisons should be made between AM and non-mycorrhizal plants that are as similar as possible for the rate of growth and nutrient status (matched plants). In other words, if the aim is to determine whether a fungus has any direct effects on the physiology of a plant, it is not useful to compare a small nutrient deficient non-mycorrhizal plant with a large, nutrient-sufficient mycorrhizal one, because the indirect effects of improved P uptake on the AM plants will be impossible to distinguish from the direct effects of the fungus itself. This means that AM and non-mycorrhizal plants need to be grown in soil with different P concentrations and ideally a full P response curve for the non-mycorrhizal plants obtained, so that the most appropriate comparisons with AM plants grown in either low or high P soil can be made (Stribley et al., 1980a; Eissenstat et al., 1993; Peng et al., 1993; Graham and Eissenstat, 1994; Tinker et al., 1994; Wright et al., 1998b). An alternative is to employ plants with split root systems, with and without AM colonization or supplied with different amounts of P (Koch and Johnson, 1984; Douds et al., 1988; Lerat et al., 2003a).

The ontogeny of the plants and the rate of colonization of the roots also need to be considered in the design of experiments; ideally multiple harvests at different stages of plant development are required to gain a thorough understanding of the symbiotic relationships, including relative growth rate and nutrient uptake rates. The biomass of the fungus and relative frequency of arbuscules, hyphae, vesicles as well as development of extraradical mycelium in soil are all important in revealing mechanisms underlying responses, but it remains true that few investigations have provided data for all these parameters (Harris and Paul, 1987; Douds et al., 2000).

Variations in responsiveness occur in both cultivated and wild plants, including woody species, grasses and forbs, annuals and perennials (Hetrick et al., 1991, 1992; Wilson and Hartnett, 1998; Tawaraya et al., 2001; Klironomos, 2003; Tawaraya, 2003). Plant characters that are generally accepted as associated with high responsiveness are shown in Table 4.6. They include features which are associated with inefficient nutrient uptake and include short roots and slow root growth, thick (magnolioid) roots with few and short root hairs, as well as inherently low nutrient uptake capacities and lack of mechanisms that mobilize nutrients from unavailable forms (Smith, 2000; Smith et al., 2001; Jakobsen et al., 2002). Inherent rate of growth and selectively flexible root:shoot ratio may sometimes be important in allowing a plant to balance uptake between roots and AM hyphae depending on conditions, and hence take advantage of AM nutrient uptake. Koide (1991a) has argued that another crucial determinant of responsiveness is the P deficit of the species or cultivar. This is essentially the difference between P demand for growth and P uptake to meet that demand, with the latter parameter influenced by P supply in the soil, as well as by mechanisms such as rate of root growth and mycorrhiza formation which influence efficiency of uptake. Again, the relative contributions of the direct and AM uptake pathways in meeting the deficit need to be incorporated into these analyses.

The work on variation in responsiveness of *Citrus* and related species has included consideration of C use by the colonizing fungus as well as extent and rate of colonization and AM responsiveness in low and high P soils. In these plants, the rate of spread of the fungus rather than the maximum extent of colonization appeared to be important in C use and its effect on responsiveness. *Citrus* species of low AM responsiveness became colonized more slowly than highly responsive species, leading to the suggestion that this character would have been favoured during evolution because of the considerable drain on photosynthate that the fungus exerts during rapid colonization of the roots (Graham *et al.*, 1991, 1997; Graham and Eissenstat, 1994). The suggestion is supported by the fact that non-responsive citrus varieties regulate non-structural carbohydrates more tightly than responsive cultivars and hence seem likely to be able to restrict development of AM fungi in their roots (Graham and Eissenstat, 1994; Jifon *et al.*, 2002). In highly responsive species, which would be expected to be characteristic of sites low in available P, C expenditure would be offset by increased nutrient acquisition and would not have been disadvantageous in evolutionary terms. In these highly responsive plants, the rate and extent of colonization is correlated with limited branching of the roots and slow root growth (Graham and Syvertsen, 1985) and, in contrast to the responsive barley cultivar Shannon (Baon *et al.*, 1993) (Figure 4.7), is not markedly reduced by high P supply, compared with less responsive species (Eissenstat *et al.*, 1993; Peng *et al.*, 1993).

Evolution and possible effects of plant breeding programmes on plant responsiveness have been followed in several crops. A comparison of barley cultivars with respect to responsiveness of P supply and AM colonization, showed that these characters are related (Baon *et al.*, 1993) (Figure 4.7a). The study also demonstrated differences in extent to which soil P status influenced AM colonization (Figure 4.7b). Comparison of 22 land races of wheat (*Triticum aestivum*) with 22 high yielding varieties showed that the land races were significantly more responsive to colonization in shoot and ear production (Manske, 1989). In this investigation, growth responses were not related to either per cent colonization or root length, but the picture is confused by the fact that some land races had greater root length than the high yielding varieties and, in consequence, longer colonized root length. P fertilization reduced per cent colonization in both groups and there were no positive growth responses to colonization under these conditions. Similarly, responsiveness of modern wheat cultivars, in terms of P uptake, was less than for old varieties (Zhu *et al.*, 2001), but the same was not true for a land race (Sahara) and improved variety (Clipper) of barley, because the latter was both more responsive to and obtained a higher proportion of P from its fungal symbiont (Zhu *et al.*, 2003). Ancestors of wheat carrying the D genome (e.g. *Aegilops squarrosa* or *Triticum tauschii*) showed much greater variation in responsiveness than genotypes carrying A or B genomes, indicating that the D genome may be the source of such variation (Kapulnik and Kushnir, 1992; Hetrick *et al.*, 1996).

Responses determined in pot experiments with single species give only limited insight into natural ecosystems, because interpretation of the advantages of AM colonization in terms of responsiveness, growth rate and P demand are complicated by the age structures and species diversity of the plant and fungal assemblages in the field. They are also complicated by the recognition that AM fungi can play considerable roles in plant nutrition and provide other benefits that are not apparent from

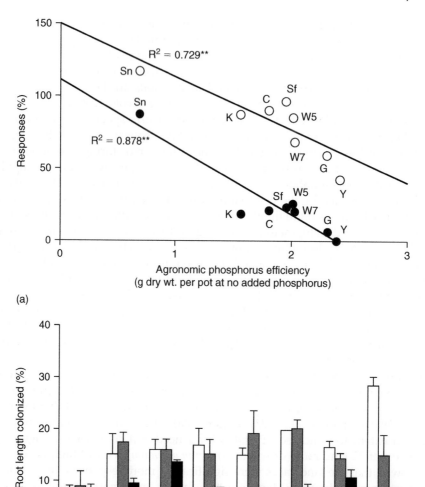

(a)

(b)

Figure 4.7 Interactions of arbuscular mycorrhizal fungi with cultivars of *Hordeum vulgare*, differing in P efficiency. From Baon *et al.* (1993), with permission of Kluwer Academic Publishers. (a) The relationship between agronomic P efficiency of barley (*Hordeum vulgare*) and response to colonization by *Glomus etunicatum* or P fertilization with 10 mg/kg P. Cultivars: Sn, Shannon; K, Kaniere; C, O'Connor; Sf, Skiff; W5, WI 2539; W7, WI 2767; G, Galleon; Y, Yagan. ●, mycorrhizal response; ○, phosphorus response. Regressions significant at 0.01 level of probability. (b) Effects of P fertilization on per cent mycorrhizal colonization of cultivars of *H. vulgare* by *Glomus etunicatum*. Bars are standard errors of the means of determinations from three replicate plants. P additions (mg/kg): □, zero; ▓, 10; ■, 20.

determinations of whole-plant responses. The nutritional effects have barely been explored, but may confer advantages in competitive situations (see Chapters 15 and 16). Indeed, advantages are to be expected because it is otherwise hard to understand how the AM condition could have persisted in evolutionary terms in such a large number of facultatively AM plants that are apparently not responsive to colonization and hence appear to be supporting essentially parasitic symbionts.

In natural ecosystems, unrelated plant species with different nutrient uptake strategies and different AM responsiveness occur together. One of the most comprehensive investigations, involving both pot and field experiments, has been in tallgrass prairies. Experiments involving almost 100 plant species and both natural and artificial inoculation in pots showed that warm-season perennial C4 grasses and perennial forbs are, with very few exceptions, highly responsive to AM colonization (in terms of dry weight accumulation) to the extent that some are obligately mycorrhizal. In contrast, cool-season perennial C3 grasses and annual and biennial forbs were both less responsive and somewhat less colonized (Hetrick *et al.*, 1989, 1991; Bentivenga and Hetrick, 1992; Wilson and Hartnett, 1998). The responsive warm-season grasses generally had less fibrous root systems, lower specific root length and more plastic architecture than the cool season grasses and their degree of branching was reduced when they became colonized, all characters regularly identified with high responsiveness (see Table 4.6). Annual and biennial species, with lower responsiveness, are characteristic of disturbed patches in which interspecific competition is relatively low. In such situations, investment in finer roots with high turnover rates appears to be the more advantageous strategy, but facultative AM status presumably allows this type of plant to overcome occasional nutrient deficiencies and competitive stresses.

The pot experiments were linked with microcosm and field studies in which AM colonization was effectively suppressed by regular application of the fungicide benomyl. This treatment resulted in decreases in abundance of dominant and highly AM responsive grasses and increases in abundance of subordinate, less responsive grasses and forbs (Wilson and Hartnett, 1997; Hartnett and Wilson, 1999; Smith *et al.*, 1999), so that it was clear that activities of AM fungi had major impacts on competitive interactions that were at least partially mediated by plant responsiveness. For the plants in this community, pot experiments were able to predict the likely competitive outcomes following AM suppression in the field, despite the fact that AM fungi were active in roots of non-responsive, as well as responsive, plants (Bentivenga and Hetrick, 1992). Simplistic interpretations almost certainly ignore very complex underground interactions because studies of ^{32}P uptake via the AM pathway show that AM fungi make very significant contributions to P uptake, not only in the responsive warm season grass *Andropogon gerardii* at both low and high temperatures, but also in the cool season *Bromus inermis*, when it was grown at low temperatures (19°C) (Hetrick *et al.*, 1994b).

The variations in P transfer at different temperatures highlights the fact that AM responsiveness may also vary with seasonal requirements or the stage of growth of the plant, particularly with respect to periods of high P demand. Establishment of seedlings of fast growing ephemerals may well require high rates of P uptake before the root systems are well developed. Concentrations of P in seeds do influence plant establishment and dependence on AM colonization (Allsopp and Stock, 1992a). Rapid AM colonization in small-seeded plants may be important in providing adequate

P for plant growth, even though the high C demand of the fungus may lead to transitory growth depressions (Smith, 1980), unless C requirements are met by other plants supporting a common mycelial network (CMN). P may also be required in large amounts at other stages of growth. For example, *Fragaria* has very high P demands during fruiting (Dunne and Fitter, 1989) during which time P uptake occurs at rates that might be strongly influenced by mycorrhiza activity. Moreover, *Hyacinthoides non-scripta* (the English Bluebell) has a very limited root system which grows at a season when there is no above-ground tissue. The plant is highly dependent on AM colonization for P absorption at this stage (Merryweather and Fitter, 1995a).

Luxury P accumulation following AM colonization in situations where P supply does not (at least temporarily) limit productivity, may be a significant advantage for wild plants, providing a stored pool which can be mobilized in periods of high requirement or transitory deficit in soil. This was shown in the tallgrass prairie plants discussed above. *A. gerardii* (responsive) took up more P and grew more, maintaining a relatively constant tissue P concentration. In contrast, the *B. inermis* (non-responsive) accumulated higher concentrations of P (Hetrick *et al.*, 1994b). The role of AM fungal P uptake as a mechanism to ensure that transitory peaks of nutrient availability are exploited to the maximum extent has been largely ignored. This stems from the early observations that AM fungi do not, unlike ectomycorrhizal (ECM) fungi, store large quantities of P. However, the same eventual outcome for the plant could result from luxury accumulation in roots, leaves and seeds and there is evidence that AM plants 'protect' their reproductive structures from severe P deficiency by altering patterns of P allocation (Koide and Lu, 1992; Merryweather and Fitter, 1996; Li *et al.*, 2005). Another outcome of high P uptake in non-responsive plants may include capture of soil P, making it unavailable to competitors (see Chapter 5).

Much of the work on variations in plant responsiveness has employed either single species of AM fungi or mixed inoculum derived from field assemblages. Increasingly, experiments are now revealing marked effects of different AM fungal species and isolates on plant responses. Using combinations of 10 plant species and 10 AM fungi, all derived from the same old field site, Klironomos (2003) has demonstrated that the per cent change in plant biomass of a single species following inoculation can be markedly positive, markedly negative or neutral depending on the fungus (Figure 4.8). The direction of change with particular fungi was not consistent between the various plant species, underscoring the huge functional diversity to be expected in field situations. Again, no attempt was made to determine which plant or fungal characteristics underpinned the observed diversity in responses. However, some other investigations have attempted to address these issues. Graham and Abbott (2000) highlighted the importance of fungal colonization strategies, showing that rapidly colonizing, aggressive fungi were able to colonize wheat roots extensively at both low and high P supply, but that there were major differences in growth response. Only two aggressive AM fungal colonizers increased wheat biomass at low P and none of the fungi tested gave a positive growth response at high P. However, low or negative growth responses have also been shown to occur in plants with low colonization (e.g. Bethlenfalvay *et al.*, 1982a, 1982b; Modjo and Hendrix, 1986), and hence low fungal C cost. The mechanisms underlying the effects are only just beginning to be explored (Smith FA, unpublished).

The importance of patterns of hyphal growth in soil in determining effects on plant P uptake is highlighted in a number of investigations (Jakobsen *et al.*, 1992a; Smith FA,

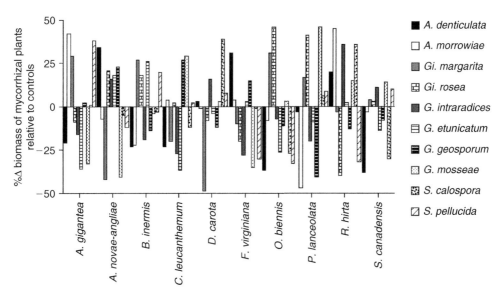

Figure 4.8 The influence of different AM fungi (see legend) on the growth of different plant species. Bars indicate change in biomass of AM plants compared to non-mycorrhizal plants grown in the same soil. Mean confidence interval for the treatments is 18%. Reproduced

et al., 2000; Smith SE, 2004a) mainly focusing on individual fungal species. Again using cucumber, marked variations in plant responses depending on fungal symbiont were observed (Munkvold *et al.*, 2004). Internal root colonization was generally high, but patterns of fungal development in soil varied not only between, but also within species. Whereas there was little intraspecific variation in rates of P uptake per unit hyphal length, the total amounts of P delivered to the plants also varied within and between species. These authors emphasized that even fungal communities of low species diversity might contain intraspecific AM fungal variants with considerable functional diversity.

All in all there is increasing recognition of very high diversity in the outcomes of interactions between different plant species and their AM symbionts. The complexities revealed indicate that although it is possible to make broad generalizations about determinants of plant responsiveness or fungal effectiveness, in terms of increases in growth or P uptake, exceptions are always to be found. Most research has involved simple pot experiments, so that our knowledge of the way plants respond to colonization in natural ecosystems is still scanty (see Chapters 15 and 16). Only more research to determine mechanisms underlying the symbiotic interactions in communities of plants and their symbionts will provide the answers.

AM mycelial links between plants: importance in carbon allocation in a plant community

The existence and potential importance of CMNs linking plants of the same or different species is now widely appreciated (Read *et al.*, 1985; Newman, 1988; Robinson and Fitter, 1999). It is likely that the networks function differently to those involving ECM plants and the following discussion relates only to AM communities. External AM hyphae are important as sources of inoculum and probably account to a very large extent for the rapid colonization of root systems in many undisturbed habitats, particularly when conditions favour perennial vegetation (see Chapter 2). C allocation among plants could be influenced by the hyphal links in a number of ways and it is consequently important to gain as much information as possible about how the links function. It has been suggested that the CMNs are involved in net transfer of organic C from plant to plant. Evidence for this remains controversial, but the controversy does not diminish the possible importance of links in modifying plant interactions. There seems to be no doubt that a seedling growing up in a community of AM plants would become very rapidly linked into and acquire inorganic nutrients from a CMN which had developed at the expense of photosynthate from already established plants. This interaction could provide a considerable saving of organic C to seedlings and result in a higher chance of successful establishment for small-seeded species, where shade, for example, might limit their photosynthetic capacity beneath the canopy. Growth depressions are observed at the seedling stages of experiments where all plants are the same age (Smith, 1980) and are likely to be the result of C utilization as the fungus enters into the stage of rapid root colonization, but before it has a major effect on P uptake. If the hyphae colonizing the roots were supported (even temporarily) by photosynthate from an established plant, rather than by limited reserves in discrete propagules, a considerable benefit might be gained. At later stages of growth, a successful seedling might be expected to become a 'supporter' of the CMN in terms of C supply, as well as a 'user' in terms of P uptake. Evidence that the C cost of colonization of a 'user' can be borne by a 'supporter' comes from experiments with *Centaurium*, which are only successfully colonized if they grow with an AM companion plant (McGee, 1985; Grime *et al.*, 1987), from the very rapid colonization observed in grassland communities (Birch, 1986) and from a number of experimental systems using older 'nurse' or 'supporter' plants to initiate colonization in seedlings (Brundrett *et al.*, 1985; Rosewarne *et al.*, 1997).

Sharing the C cost of maintaining a CMN does not necessarily depend on transfer of fixed C from one plant to another. All that is required is that the 'supporter' provides the majority of the organic C to the CMN. AM effects on both frequency and diversity of species in microcosm experiments (Grime *et al.*, 1987; Zabinski *et al.*, 2002) can be explained on the basis of competition if the species vary in the amount of C they provide to the system and in their responses to colonization in terms of nutrient uptake and growth; they do not depend on net organic C movement from dominant to subordinate plants (see Chapter 16). Some experiments in which donor plants linked into a CMN were fed $^{14}CO_2$ have indicated limited transport of C to receiver plants connected to the same network (Hirrel and Gerdemann, 1979; Francis and Read, 1984; Grime *et al.*, 1987; Fitter *et al.*, 1998a). However, the final destinations and quantitative importance of such transfers has generated considerable discussion and are still not fully clear. The questions that need to be answered are: does organic C move between plants in the CMN, or is there transfer directly through soil or via fixation of CO_2 released by below-ground respiration? If C does move in the CMN is it retained in the fungal compartment of the receiver roots or is it transferred across the symbiotic interface to the plant, and thence to the shoots? Additionally, does transfer from donor to receiver represent net transfer in amounts which could significantly affect plant growth?

Newman (1988) explained that, even in cases where small amounts of ^{14}C appear in shoots, this simply suggests the existence of a pathway for transfer (possibly along the hyphae of the CMN and across the fungus–plant interface). It does not show that net C transfer between plants occurs and, with one possible exception (Lerat *et al.*, 2002), no experiments have yet shown unequivocally that net transfer of C from an AM fungus to autotrophic plant takes place via the CMN. Neither shade nor clipping, which might be expected to increase the driving force for such source-sink transfer of C, increased the labelling in shoots of the receivers, although they did in their roots (Read *et al.*, 1985; Newman, 1988; Fitter *et al.*, 1998a). Robinson and Fitter (1999) reviewed many of the papers addressing these questions and pointed out that, in most experiments, transfer by pathways other than the CMN remained a possibility and that in a large number of cases almost all the ^{14}C remained in the AM roots and did not appear in shoots; the likelihood was that it remained in the intraradical fungal structures. Although Fitter *et al.* (1998a) found that large amounts of C were transferred in the CMN, there was no evidence for transfer to shoots of receiver plants and considerable evidence that C was in fact accumulated in fungal vesicles within roots, representing fungal storage rather than support for nutrient uptake by the CMN. As mentioned above, no labelled C was incorporated into plant metabolites in a monoxenic system that precluded photosynthetic fixation of CO_2 and hence return of labelled C to plant shoots (Pfeffer *et al.*, 2004). All in all, it seems unlikely that autotrophic AM plants regularly share ecologically meaningful quantities of photosynthate via CMNs. Rather, C storage by the fungi and support of the CMN by larger or more photosynthetically active individuals, with consequent changes in competition, probably underlie the effects of AM connections observed in communities of autotrophic plants.

If the potential receiver plants are non-photosynthetic, the situation is certainly different. Some achlorophyllous members of the Polygalaceae, Gentianaceae and Burmanniaceae are associated with AM fungi which form mycorrhizas characterized by aseptate hyphae and extensive intracellular coils (see Chapters 1 and 13). It has been assumed that these plants are mycoheterotrophs and obtain all their organic C via the fungus. The implication is that this must come from an autotrophic 'donor' plant via a CMN, because AM fungi have no saprotrophic capacity (Leake, 2004). The mycoheterotrophs are epiparasites on the autotrophic plants and net transfer of organic C from fungus to the heterotrophic plant must occur. At present, there is no experimental evidence to support this idea or information on underlying mechanisms, but it is clearly an area which is wide open for research and should prove amenable to NMR-based studies if an appropriate experimental system could be devised (see Chapter 13).

Conclusions

AM fungi are dependent on an organic C supply from a photosynthetic partner. Between 4 and 20% of net photosynthate is transferred to the fungus and used in both production of both vegetative and reproductive structures, and in respiration to support growth and maintenance, including nutrient uptake. Hexoses are absorbed by the intraradical fungal structures, rapidly converted to lipid and glycogen and exported to the external mycelium which itself has little or no capacity either for organic C uptake from the environment or for lipid synthesis.

There appears to be considerable variation in the C expended by different AM fungi in the transfer of P to the plant and, at this stage, it is not known whether the differences are due to variations in patterns or rates of colonization or hyphal extension in soil, or to more subtle differences in membrane transport capacity or respiratory efficiency. More data are needed in all these areas. Expenditure of fixed C by the plant to maintain the fungal symbiont can be regarded as an investment, resulting in greater efficiency of nutrient acquisition when nutrient availability in the soil is low, and sometimes in luxury accumulation of nutrients for later use. Very large variations in plant responses to AM colonization are increasingly being identified, both in crops and wild plants. The magnitude of the responses is influenced by both environmental factors (e.g. nutrient availability and irradiance) and biotic factors (such as the identity of fungal symbionts), as well as by community interactions. Attempts to make cost-benefit analyses of the symbiosis have proved difficult, because of the complex ways in which both AM colonization and plant growth respond to and interact with mineral nutrition.

It is generally accepted that plants of the same and different species may be linked into CMNs. Organic C support for a CMN may not be equally shared between the plants which the networks link together. The result is that some plants may effectively support others by reducing the C drain on them, at least temporarily. There is no unequivocal evidence at present that this support involves net transfer of C from one plant to another via a CMN. The way in which mycoheterotrophs associated with AM fungi obtain organic C is at present unknown, although it seems likely that epiparasitism via mycelial links may be involved.

5

Mineral nutrition, toxic element accumulation and water relations of arbuscular mycorrhizal plants

Introduction

Mineral nutrition of arbuscular mycorrhizal (AM) plants has received more emphasis and been the subject of more research than any other aspect of the symbiosis. The trend has continued in recent years, but with changes in approach. It is generally appreciated that the influence of AM fungi on mineral nutrition of plants is an essential contributor to mutualistic biotrophy (Lewis, 1973) and that there are marked differences in outcome of symbioses between different plant species and AM fungi. Functional characterization of interactions has moved well beyond comparisons of non-mycorrhizal and AM plants, demonstrating effects of colonization on whole plant nutrition and growth (see Chapter 4). At the same time, there has been recognition that many plant–fungus combinations do not result in improved mineral nutrition at the whole plant level, at least in pot experiments. However, acceptance that AM fungi make major contributions to nutrient uptake even in these symbioses has been somewhat slow. Emphasis on the significance of AM fungal development is switching from extent of internal colonization of the roots to spread and longevity of the external mycelium in soil. The role of this mycelium in delivering nutrients to plants has been followed in compartmented systems, using isotopic tracers or by measuring depletion profiles in soil. These approaches have led to much improved quantitative information on the amounts of nutrients reaching plants via AM fungal symbionts. At the cellular level, details of fungal phosphorus (P) and nitrogen (N) metabolism are being revealed which, together with information on the effects of AM symbiosis on expression of both fungal and plant nutrient transporters, is beginning to lead to a coherent general picture of how fungal and plant nutrient acquisition processes are integrated. Nevertheless, much remains to be elucidated in detail, particularly with respect to aspects of the functional diversity among different symbiotic combinations.

In early work on mineral nutrition, the main emphasis, as shown in Chapter 4, was the large positive effects of AM colonization on growth and nutrition of many responsive species. In the 1960s, a number of general reviews were written which covered the occurrence of colonized plants and anatomy of AM roots and also addressed the problems of experimentation that arise from difficulties in identification of the fungi involved in the symbiosis and production of satisfactory inoculum (Baylis, 1962; Nicolson, 1967; Gerdemann, 1968). During that early period, details of the effects of the symbiosis on mineral nutrition were poorly understood, although some clues had begun to appear and key experiments on P nutrition were carried out. In 1957, Mosse published the results of an experiment with apple seedlings which clearly demonstrated increased amounts of K, Fe and Cu per unit weight of tissue in AM plants, compared with uninoculated controls. Later, several researchers (Gerdemann, 1964; Daft and Nicolson, 1966; Baylis, 1967) established that tissue concentrations of P (which were not measured by Mosse) were sometimes higher in AM plants. Mechanism(s) contributing to this effect in positively responsive plants were quickly addressed and the first papers showing increased uptake of P on the basis of root length soon appeared (Sanders and Tinker, 1971, 1973), around the same time as the first demonstrations of differential effects of different fungal species (Hayman and Mosse, 1971; Mosse and Hayman, 1971).

Consequently, by 1973, Mosse was able to comment that, even though the fungi had not been grown in pure culture, the methodological problems had been largely overcome and that there had been a change in emphasis of the research towards effects on plant growth and P uptake. Important reviews followed that introduced concepts from soil chemistry to mycorrhizal studies and have influenced the research to the present day (Tinker, 1975a, 1975b, 1978). This early work established that AM roots of responsive plants take up P, Zn and probably Cu and ammonium from soil more efficiently than non-mycorrhizal root systems and that the extraradical hyphae play an essential part in increasing the volume of soil effectively available for acquisition of these nutrients. There was a general assumption (now recognized as incorrect) that AM fungi did not contribute to nutrition of non-responsive plants and, accordingly, other benefits of the symbiosis were invoked to explain their evolutionary persistence.

Research then extended to cover effects of the symbiosis on many aspects of the physiology of the plants (Smith, 1980; Harley and Smith, 1983; Smith and Gianinazzi-Pearson, 1988; Koide, 1993; Jakobsen, 1995) and a vast array of books and reviews has continued to appear, reflecting the interest of researchers in many basic and applied fields. Moreover, the importance of arbuscular mycorrhizas, rather than non-mycorrhizal roots, as the normal nutrient-absorbing organs of most plant species was clearly recognized and their significance addressed in more general contexts.

It is the purpose of this chapter to present an overview of the current understanding of the role of arbuscular mycorrhizas in the mineral nutrition of plants and it is impossible to quote more than a small proportion of the relevant papers. The discussion centres on the mechanisms that have been suggested to account for the AM effects and the experimental evidence that supports them. The focus is on P and N uptake, as well as on the uptake of other nutrients and toxic elements for which there is now unequivocal evidence of AM fungal involvement. In addition, the effects of arbuscular mycorrhizas on the water relations of plants are discussed. The roles of arbuscular mycorrhizas in mineral nutrition of plants growing in the field are discussed in Chapters 15 and 16.

Phosphorus availability in soil

The development of research on AM roles in nutrient uptake has been inextricably linked to knowledge of soil chemistry, particularly in relation to pools and availability of P. The amount and form of P in soil and the factors affecting its availability are important in determining the way in which AM fungi influence uptake by plants (Tinker, 1975b; Bolan, 1991; Comerford, 1998; Tinker and Nye, 2000). P is required in relatively large amounts but is often poorly available in soil. It is absorbed by cells as inorganic orthophosphate ions (specifically, $H_2PO_4^-$) from the soil solution where it is present at very low concentrations (typically $0.5-10\mu M$), controlled mainly by soil chemical reactions and, to a lesser extent, by biological processes (Schachtman *et al.*, 1998). P supplies may also be patchily distributed both in space and in time, affecting their availability to and use by plants and microorganisms (Lodge *et al.*, 1994; Robinson, 1994).

P in soil can be broadly categorized as inorganic (Pi) or organic (Po). Pi may be held very firmly in crystal lattices of largely insoluble forms, such as various Ca, Fe and Al phosphates and may also be chemically bonded to the surface of clay minerals. Some of this P exchanges very slowly with the soil solution and comprises a non-labile pool, that is unavailable to plants. Less tightly bound (or labile) P exchanges relatively rapidly with the soil solution and is viewed as being in isotopic equilibrium with it. It is this pool of labile P which is regarded as being available to plants, although different chemical extraction methods to determine its size do not necessarily reflect what the plants actually absorb (see below). Pi is most readily available in soil around pH 6.5. At lower pH the decreasing solubility of Fe and Al phosphates controls the solution concentration, whereas at higher pH decreasing solubility of Ca phosphates becomes important. Localized changes in rhizosphere pH may play a role in altering the availability of these different sources of P. Furthermore, the production of chelating compounds, such as organic anions (e.g. citrate and oxalate) increases the availability of P from some sources (Marschner, 1995; Comerford, 1998).

The predominant forms of Po that can be extracted from soil have long been thought to be inositol phosphates (phytate), phospholipids and nucleic acids. However, doubt has now been cast on the identification of phytate as the major source of Po in soil and it may actually only contribute about 5% to this pool (Smernik and Dougherty, 2007). Experiments demonstrating the roles of mycorrhizal fungi in hydrolysis of phytate may therefore be of limited relevance. Nevertheless, the conversion of all types of Po to inorganic form and consequent availability to plants depends on hydrolysis by microorganisms. Po does not seem to provide a major pool directly available to non-mycorrhizal plants, although there is some evidence for the activity of phosphatases on the surface of roots that could affect hydrolysis. Some AM fungi appear to have significant capacity to hydrolyse some sources of Po or to influence phosphatase production by plants, but the quantitative significance to plant nutrition remains to be explored. It should however be noted that both ectomycorrhizal (ECM) and ericoid mycorrhizal (ERM) fungi make very significant use of organic sources of both P and N, a point of divergence between these mycorrhizal types and arbuscular mycorrhizas (see Chapters 9, 10 and 11).

Soluble Pi entering the soil following hydrolysis (mineralization) or application of fertilizer results in localized and short-term increases in the concentration of orthophosphate in the soil solution. However, much of it is removed from solution by 'fixation', which is rapid at first and continues for a long time without reaching equilibrium. Fixation involves sorption of ions on soil surfaces, precipitation of mineral phosphates and use by organisms, resulting in immobilization of P in the biomass. These fixation processes have important effects on the concentration of orthophosphate in the soil solution and its movement in soil and hence the uptake by roots and microorganisms, including AM fungi.

It is quite apparent that mass flow of solution in soil is unable to supply orthophosphate to roots at rates that can account for the amount of P absorbed and it is taken as axiomatic that orthophosphate reaches absorbing surfaces by diffusion (Tinker and Nye, 2000). Furthermore, the rate of diffusion of P ions is several orders of magnitude lower in soil (10^{-8}–10^{-11} cm^2/s) than in free water (10^{-5} cm^2/s) and will vary with the P content of soil, the buffering capacity and the tortuosity of the diffusion pathway. Slow diffusion of orthophosphate in the soil solution, contrasted with rapid absorption, results in the development of depletion zones around roots (see Figure 4.1). Uptake is limited by the rate of diffusive movement of orthophosphate into these depletion zones, rather than by the rate of the transport across living membranes into the root (i.e. the absorbing capacity). The longer a segment of root remains actively absorbing from soil at a rate greater than that of movement of orthophosphate to it, the wider will be the depletion zone, with consequent reduction in the rate of arrival of P at the absorbing surface and hence of uptake by the plant.

It is possible to make approximate calculations of the inflow (uptake per unit length per unit time) into non-mycorrhizal roots to be expected once a 'zero concentration sink' has developed round the roots. The magnitude of this inflow depends on a number of soil factors including the size of the labile pool and buffering capacity, which affect the solution concentration in the bulk soil, together with characteristics which influence the diffusion coefficient. These include the solution concentration, the pH, redox potential and ionic strength of the soil solution as well as the water content and other factors that affect the tortuosity of the diffusion path. Compared with these soil factors, the effects of root radius and root absorbing capacity (influenced by K_m and V_{max} of the uptake system) on rate of uptake are relatively small. However, the smaller the diameter of the absorbing structure the smaller will be the depletion effect and the greater the importance of absorbing capacity. Root hairs effectively extend the diameter of the absorbing surface of the root beyond the depletion zone and they have significant effects on P uptake, especially if they are long (Claassen and Barber, 1976; Clarkson, 1985; Schweiger and Jakobsen, 1999a; Jakobsen et al., 2005a). Hyphae of AM fungi extend into soil far beyond the root surface or root-hair zone (up to 25 cm has been measured). Their small diameter (similar to or even less than root hairs, see Table 4.1) means that marked depletion zones do not develop around them and, furthermore, they are able to grow into soil pores that roots with much larger diameters are unable to access. This means that they are able to exploit solution-filled soil pores at considerably lower soil water potentials than roots and hence absorb P from drier soils.

Effects of arbuscular mycorrhizal colonization on plant phosphorus nutrition

Effects of AM colonization in enhancing P uptake and growth of many plants was discussed in Chapter 4. Indirect evidence that AM roots can be more efficient

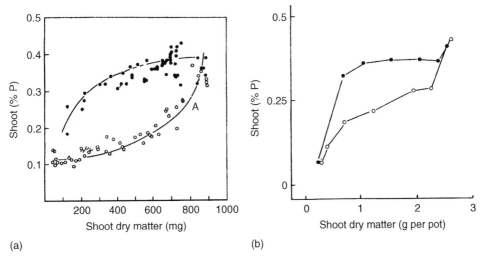

Figure 5.1 The relationship between P concentration in the shoots (per cent dry weight) and dry weight of shoots in: (○) non-mycorrhizal plants, and (●) plants colonized by *Glomus mosseae*. (a) *Allium porrum* (leek) grown on ten γ-irradiated soils differing in initial P content and receiving five different levels of added P to give the 50 soil treatments. Redrawn from Stribley *et al.* (1980). (b) *Trifolium subterraneum* grown in soil with the addition of superphosphate. Redrawn from Pairunan *et al.* (1980).

in nutrient uptake than non-mycorrhizal roots came from the observations that responsive AM plants are both larger and contain higher concentrations of P in their tissues than uncolonized controls. The explanation first suggested was that AM colonization increased efficiency of absorption by roots. However, this is not the only possible explanation for the increases. Increased total root length or efficiency in AM plants would certainly contribute to increased total uptake, but would not necessarily lead to elevated tissue concentration. If growth keeps pace with P uptake, tissue concentrations remain constant, for they are dependent upon the relative rates of uptake and growth. If tissue concentrations rise (as shown in Table 4.2 and Figure 5.1), some factor other than P must be limiting growth (Pairunan *et al.*, 1980; Stribley *et al.*, 1980b). As shown in Chapter 4, elevated concentrations may result from increased organic C use in AM plants, leading to C-limitation and 'luxury' accumulation of P. Whatever the mechanism, the elevated tissue concentrations of P and other nutrients in responsive AM plants certainly alerted early investigators to the possible role of arbuscular mycorrhizas in plant nutrition.

More direct evidence of increased efficiency of P absorption was obtained by expressing uptake on the basis of the amount of absorbing tissue. The contribution of the AM fungi was determined by subtracting values for P uptake by non-mycorrhizal plants from those obtained from AM plants ('subtraction method'). Most results have been expressed per unit root length (inflow, mol/m/s) or root weight (specific absorption rate; mol/g/s). Inflow gives a realistic basis for comparison of uptake of immobile, diffusion-limited ions such as orthophosphate, because linear extension of the root system into undepleted soil is more important in determining uptake than the surface area presented to a zone of soil in which nutrients are at very low

Table 5.1 Inflow of P into mycorrhizal (M) and non-mycorrhizal (NM) roots of *Allium cepa* and the calculated contribution of hyphae (H) to inflow in the mycorrhizal roots for two experiments.

Expt	Colonization	Inflow (mol P/m/s $\times 10^{-12}$)				Flux in hyphae (mol P $\times 10^{-4}/m^2/s$)
	(%)	M	NM	Hyphae	Hyphae colonized regions	Mean for 2 expts
A	50	13.0	4.2	8.8	17.6	3.8
B	45	11.5	3.2	8.3	18.5	

From Sanders and Tinker 1973. The colonized regions of the mycorrhizal roots had about 600 entry points per m and entry point hyphae were 15 μm in diameter, with a central lumen of 10 μm. The cross-sectional area of hyphae via which P entered the roots was therefore approximately $4.7 \times 10^{-8} m^2/m$. These values, together with the per cent root length colonized by the fungus (a *Glomus* spp.) were used to calculate the flux of P translocated into the roots via the hyphae.

concentrations. This approach to determining fungal contribution to uptake is limited to situations where the AM plants have higher total P contents than non-mycorrhizal controls and the assumption is made that colonization itself has no effect on direct nutrient uptake via root epidermis and root hairs, so that the contributions of direct and AM uptake pathways (see Figure 4.1) are additive. As already mentioned in Chapter 4, this is certainly an oversimplification. Data are accumulating that the AM pathway plays a major role not only when plants respond positively to colonization, but also when there are no differences in growth and nutrient uptake between AM and non-mycorrhizal treatments. Nevertheless, the early calculations of nutrient uptake efficiency placed investigations on a sound quantitative basis that could be linked to the abilities of both roots and hyphae to extract P from soils of different P availability.

The first demonstration of increased inflow of P in AM roots was in highly responsive *Allium cepa* colonized by *Glomus* spp. (Sanders and Tinker, 1971, 1973). Values for AM roots were on average about three to four times greater than into non-mycorrhizal roots and Sanders and Tinker calculated that the AM fungi contributed about 70% of the P absorbed by the AM plants, on the assumption that direct uptake by AM and non-mycorrhizal roots was similar, because rates of direct uptake would be diffusion limited (Table 5.1).

The absolute magnitudes of the inflows measured in different experiments varied quite considerably, presumably as a function of soil P status, plant species and AM fungus. An important early investigation showed that the mean inflows of P to whole root systems of the highly responsive plant *A. cepa* colonized by three different AM fungi were much higher than inflows into non-mycorrhizal roots (Sanders *et al.*, 1977). A fourth fungus colonized poorly and had no effect on inflow compared with the controls. Those fungi that increased inflow above the control values produced a fairly constant amount of external mycelium per metre of colonized root, so that it appeared possible to conclude that, once the fungus has become established, the rate of P uptake by hyphae was related to the length of the external mycelium (Sanders *et al.*, 1977). Several subsequent investigations have also shown a general correlation between extensive external mycelium and high AM uptake (Graham *et al.*, 1982b; Munkvold *et al.*, 2004; Smith SE *et al.*, 2004a). However, fungi producing extensive mycelium are not always effective symbionts (Smith *et al.*, 2000) and, in these cases, it is presumed that hyphal length density may not reflect the actual spread of hyphae away from roots or that colonization of the soil is related to foraging for new host plants rather than nutrient uptake (see Chapter 2).

Estimates of the fungal contribution to inflow must be regarded as minimum values because they are based on the assumption that direct uptake via root epidermis and root hairs is always maintained at the same rate regardless of AM colonization. We know that this is not the case from direct evidence of the contribution of the AM uptake pathway. Evidence for reduced direct uptake capacity has been accumulating for many years. Physiological experiments have shown that K_m and V_{max} of uptake systems in roots are influenced by internal P concentration, so that absorbing power is reduced in roots of high P status, at least in well stirred solutions (Lefebvre and Glass, 1982; Elliott *et al.*, 1984; Jungk *et al.*, 1990; Schachtman *et al.*, 1998; Smith FW *et al.*, 2003a), most likely because of downregulation of epidermal P transporter expression (see below). Reductions in uptake capacity might also be induced directly if AM colonization *per se* reduces expression of P transporters in epidermis and root hairs. In any event, $^{32/33}$P uptake studies have now demonstrated that direct uptake is often reduced in AM plants (see below).

A positive correlation between calculated P uptake via an AM fungus and per cent root length colonized has been observed in some investigations, but certainly not invariably (McGonigle, 1988; McGonigle and Fitter, 1988b; Fitter and Merryweather, 1992; Sanders and Fitter, 1992b; Smith SE *et al.*, 2004a). Explanations for the variations include progressive death of the fungus within the root as the plants age, reduction in the contribution of arbuscules to the colonized length and/or death or destruction of the extraradical hyphae. In fact, the extent and distribution of the external mycelium may be more important in limiting P uptake via the fungi than area of interface between the symbionts (determined from per cent colonization). In natural ecosystems, the involvement of arbuscular mycorrhizas in mineral nutrition has not always been apparent from comparisons of uptake by AM and non-mycorrhizal plants. A good correlation between P uptake and colonization was found in *Hyacinthoides non-scripta* (Merryweather and Fitter, 1995a, 1995b) and *Ranunculus adoneus* (Mullen and Schmidt, 1993), but this was not the case for the grass *Vulpia ciliata* (West *et al.*, 1993b) and a number of grassland species (McGonigle and Fitter, 1988b; Sanders and Fitter, 1992b). Part of the explanation may be changes in availability of P resulting from techniques applied to eliminate or reduce AM colonization. Correlations may also have been based on inaccurate data, because of the general underestimation of fungal contribution to uptake using the subtraction method.

As the diversity of whole plant responses began to be generally acknowledged (Johnson *et al.*, 1997 and see Chapter 4), considerable interest began to be generated in the possible selective advantages of AM colonization for plants that do not show increases in either total plant P or efficiency of P uptake (non-responsive or negatively responsive). Additional benefits of AM colonization were uncovered, such as improved water relations (see below) and reduced disease expression (see Chapter 16). However, by the early 1990s, an important development in methodology involving the use of pots separated into root and hyphal compartments (HCs) using mesh began to provide direct evidence of the activity of the AM uptake

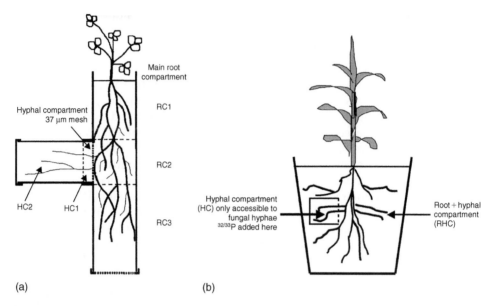

Figure 5.2 Compartmented pots used for investigations of the effects of extraradical mycelium on nutrient uptake by plants. The pots are divided into compartments using mesh, to
provide a root and hyphal compartment (RHC) and a compartment from which roots but
not hyphae are excluded (HC). (a) 'Cross-pot' devised by Jakobsen using plumbers pipe;
roots of the plant are excluded from the large hyphal compartment by 37 μm mesh. (b)
Compartmented pot in which the HC is a small tube buried in the main root hyphal compartment and separated from it by mesh. Redrawn from Smith SE et al. 2003c.

pathway (Figures 5.2 and 5.3). Using such systems, the hyphal distribution in soil
and AM fungal contribution to uptake by plants or depletion of nutrients in soil
have been studied in the absence of the roots themselves. The depletion of bicarbo-
nate- or water-extractable P, hyphal distribution (Figure 5.3) or the uptake of ^{32}P or
^{33}P from hyphal compartments (Figures 5.4 and 5.5) have been measured in different
experiments. By the early 1990s, there were a number of direct demonstrations of
the major contributions of the AM uptake pathway, not only in responsive *Trifolium*
spp. and *Andropogon gerardii* (Li *et al.*, 1991a, 1991b; Jakobsen *et al.*, 1992a, 1992b;
Hetrick *et al.*, 1994a), but also in relatively non-responsive wheat, *Bromus inermis*
and *Agropyron desertorum* (Hetrick *et al.*, 1994a, 1996; Cui and Caldwell, 1996a). Later
research has confirmed these general findings in a range of plant species (Hartnett
and Wilson, 2002; Zhu *et al.*, 2003). The results clearly demonstrate the major contribution of hyphal uptake of P from soil and translocation through the hyphae to AM
plants of varying responsiveness.

The real significance of the findings in confirming that the AM pathway delivers
significant amounts of P regardless of responsiveness, and in demonstrating that
direct plant P uptake must be decreased in non-responsive plants, was slow to be
generally recognized. However, the findings have been confirmed and quantitative determinations of AM fungal contributions indicate that they are large, can be
as high as 100%, and are not closely related to per cent internal root colonization

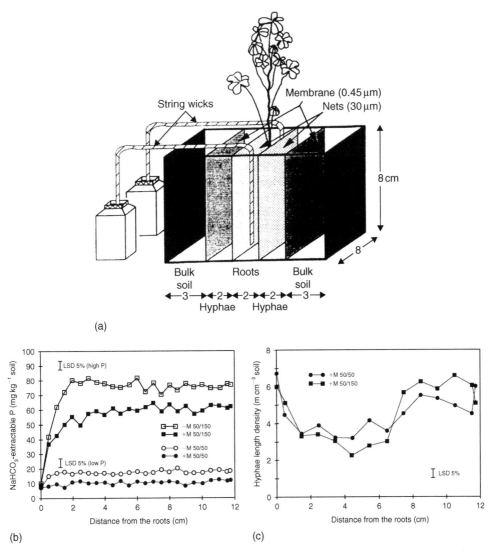

Figure 5.3 Compartmented pot and data on P depletion and hyphal spread obtained following growth of AM or non-mycorrhizal *Trifolium repens*. Reproduced from Li *et al.* (1991a), with permission of Kluwer Academic Publishers. (a) Compartmented pot in which a small root hyphal compartment containing the plant is separated from a large hyphal compartment by 30 μm mesh. The outermost compartments (representing bulk soil) are further separated by 0.45 μm membranes which exclude both roots and hyphae. (b) Depletion profiles of NaHCO₃-extractable (Olsen method) P in the hyphal compartments grown at two P levels. In both cases the AM treatments (solid symbols) have depleted the soil to a greater extent than the non-mycorrhizal treatments (open symbols). (c) Hyphal length density at different distances from the boundary of the root compartment, at the same two P levels used in (a). Note that there were small effects of P on hyphal development, so that the different depletion profiles cannot be attributed to different hyphal densities.

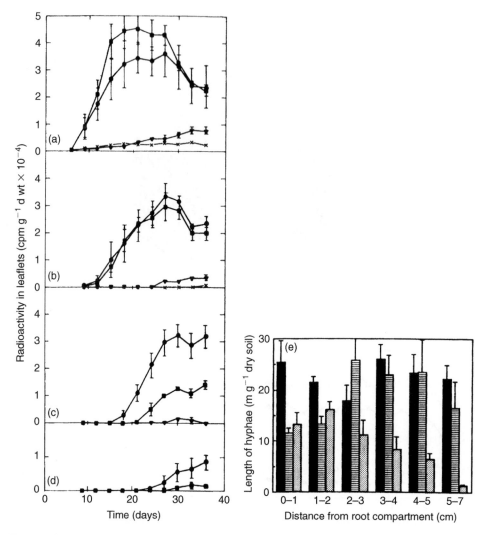

Figure 5.4 (a)–(d). Time course of appearance of radioactivity in young leaflets of *Trifolium subterraneum* in association with *Acaulospora laevis*, ●; *Glomus* spp., ■ and *Scutellospora calospora*, ▼ or non-mycorrhizal, x. Distances (cm) between the ^{32}P-labelled soil and the root compartment were 0, (a), 1.0 (b), 2.5 (c) or 4.5 (d). Bars are standard errors of means. (e) Length of hyphae in soil sections from the hyphal compartments of the experimental units with *Trifolium subterraneum* in association with *Acaulospora laevis*, ■ *Glomus* spp., ▤ or *Scutellospora calospora*, ▨. Values were corrected for background hyphae. From Jakobsen *et al.* (1992), with permission.

nor increases in P uptake (Smith SE *et al.*, 2003c, 2004a) (Figure 5.5). Taken with new information on expression of both plant and fungal P transporters, the emerging picture is that AM colonization markedly changes the pathway of P acquisition and that the changes must be taken into account in interpretation of data for total P uptake by plants of different responsiveness across the mutualism–parasitism continuum and in competitive interactions in field situations (see Javot *et al.*, 2007).

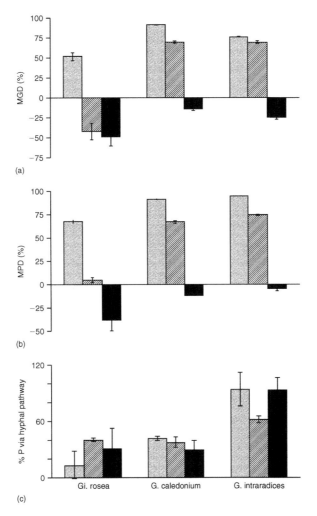

Figure 5.5 Diversity of outcomes of symbiosis between three plant species and three AM fungi. (a) Mycorrhizal growth dependency (MGD), (b) mycorrhizal P dependency (MPD) and (c) per cent P entering plants via the AM fungal pathway in *Linum usitatissimum* (stippled bars), *Medicago truncatula* (cross-hatched bars) and *Solanum lycopersicum* (black bars) after 7 weeks growth, colonized by three different AM fungi as indicated. Means and standard errors of means of 3 replicate pots. Data from Smith SE *et al.*, 2003c and 2004a.

Although Rhodes and Gerdemann (1975) demonstrated the ability of hyphae of AM fungi to absorb ^{32}P from some distance away from the roots (7 cm), they provided no information on actual depletion of soil P. Early observations suggested that hyphal depletion might only extend a few millimetres from the root surface (Owusu-Bennoah and Wild, 1979), but the data in Figure 5.3 clearly demonstrate the marked effect of hyphal spread on the depletion of P in soil with two levels of P applied in the hyphal compartment (Li *et al.*, 1991b). The per cent colonization of the roots and the hyphal length density in the hyphal compartment differed little between the P treatments (76–79% and around $5.0 \, \text{m/cm}^3$ soil, respectively, see Figure 5.3c) and, assuming all the hyphae were alive, values of P inflow to the hyphae of 3.3 and $4.3 \times 10^{-13} \, \text{mol P/m/s}$ for the two P treatments were calculated and are very close to the value of $2.25 \times 10^{-13} \, \text{mol P/m/s}$ that can be derived from the data of Sanders and Tinker (1973) making the same assumptions (see Table 5.1). Depletion of P, and hence hyphal uptake, extended at least 12 cm into the hyphal compartment (the maximum possible in the experimental system) and had a significant effect on the AM plants, which had both higher tissue P concentrations and higher dry weights than the non-mycorrhizal controls. Furthermore, the proportional contribution of the hyphae to total uptake was slightly higher with the higher P supply, emphasizing that localized patches of high P in soil (in this case the soil in the hyphal compartment) can be effectively exploited by the AM mycelium. Differential exploitation of nutrient-rich patches by AM fungi has been confirmed in several investigations (Cui and Caldwell, 1996b; Cavagnaro et al., 2005) and may have the potential strongly to influence plant population structure (Facelli and Facelli, 2002). This is an important point because increasing patchyness of soil is generated by plant and fungal uptake, as well as by nutrient-rich inputs. Furthermore, although the effects of AM colonization on uptake are frequently greatest when soil P supply is low, significant effects may be apparent even when levels are adequate for near maximal plant growth, particularly during later stages of plant development when depletion zones would be large and much of the P initially available in the pots had been absorbed (Smith, 1982; Son and Smith, 1988; Dunne and Fitter, 1989; Li et al., 2005).

The extent of hyphal development in soil is quite variable, depending on fungal species and also soil conditions (see Chapter 2). Suggestions that some functional complementarity may exist between AM fungal species with different mycelial spread colonizing the same root systems and between AM fungi and root hairs have been made (Koide, 2000). It is certainly true that many plants with fine, rapidly growing root systems and dense or long root hairs are less responsive to AM colonization than those lacking these attributes. A recent study comparing wild-type barley with long root hairs with a root-hairless mutant, has confirmed a large AM contribution to the efficiency of P uptake when root hairs are lacking, with major differences in effects of different AM fungal species (Jakobsen et al., 2005a).

Differences between AM fungal species and even isolates of the same species in respect of both hyphal proliferation in soil and P uptake and delivery are increasingly being revealed. In an early example (Jakobsen et al., 1992b), hyphal uptake and translocation of ^{32}P into Trifolium subterraneum from a source localized in the HC at different distances (0, 1.0, 2.5, 4.0 and 7.0 cm) from the root plus hyphal compartment (RHC) were followed for up to 37 days (see Figure 5.4). Hyphae of a Glomus spp. (WUM 10(1)) were relatively dense close to the roots and this fungus transferred most ^{32}P to the plants when the P source was similarly close. In contrast, Acaulospora laevis had a higher hyphal length density between 2 and 5 cm from the root than closer to it and absorbed ^{32}P effectively from more distant placements. Scutellospora calospora had relatively low hyphal length densities and did not transfer much ^{32}P to the plants from any distance. However, hyphae of this fungus contained around four times more ^{32}P than the others and the conclusion was that, although the S. calospora was able to absorb and translocate ^{32}P, transfer to the plants occurred at a low rate and consequently P accumulated in the hyphae. Only a relatively small total amount of P needs be absorbed and retained by the hyphae for this effect to be apparent. Subsequent experiments have generally confirmed that fungi with poor hyphal development have a poor capacity for P delivery (Smith FA *et al.*, 2000; Smith SE, 2004a; Munkvold *et al.*, 2004). However, extensive external mycelium is not always accompanied by high P translocation and transfer, reflecting roles of the hyphae in foraging for C as well as absorbing P (see Chapter 2).

The reasons for the low rate of transfer in some plant–fungus combinations have not been clearly determined, but might relate to low density of active arbuscules or coils (Smith and Dickson, 1991) or to inherently low transfer fluxes across the symbiotic interface(s). The relative effectiveness of the same isolates of *Glomus* spp. and *S. calospora* used by Jakobsen *et al.* (1992b) was confirmed in dual labelling experiments in which ^{33}P was supplied to roots plus hyphae in an RHC and ^{32}P to hyphae alone in an HC (Pearson and Jakobsen, 1993b). However, hyphal uptake from the HC by a third fungus, *Glomus caledonium*, was as great as uptake from RHC, leading to the suggestion that root uptake was completely inhibited by the presence of the fungus, by a then unknown mechanism (Pearson and Jakobsen, 1993b; Jakobsen, 1995). It now seems likely that this was an early demonstration of the way that the AM uptake pathway takes over from direct uptake as AM colonization develops.

More precise quantification of the AM contribution to uptake has confirmed the basic finding that AM fungi make highly significant contributions to P uptake regardless of the overall responsiveness of the plants. Using the pot system depicted in Figure 5.2b, and measurements of the specific activity of ^{32}P in the plant-available P pool in soil (which had not been included previously). Smith SE *et al.* (2003c, 2004a) compared the outcome of symbiosis between three plant species of varying responsiveness (*Linum*, *Medicago* and tomato) and three fungi (*Gigaspora rosea*, *Glomus caledonium* and *G. intraradices*). All fungi colonized roots effectively and extensively, and the two *Glomus* spp. developed considerable external mycelium both in the presence of roots and in root-free HCs. However, *Gi. rosea* formed little external mycelium regardless of plant symbiont. *Linum* responded positively to all fungi in growth and P uptake and *Medicago* to *Glomus caledonium* and *G. intraradices*, but not *Gi. rosea*. Tomato showed no positive responses to any of the fungi (see Figure 5.5a, b). Regardless of these overall plant responses, transfer of ^{32}P from the hyphal compartments to the plants showed that the fungal pathway made large contributions to P uptake in five of the nine plant fungus combinations and significant contributions in a further two. *G. intraradices* delivered close to 100% of the P to all three plant species, including non-responsive tomato (see Figure 5.5c). The contribution of the fungal pathway was again not related to per cent root length colonized or to growth or P responses. The low contribution of *Gi. rosea* was probably related to poor development of external hyphae in soil. The conclusions confirmed the results of Pearson and Jakobsen (1993b) with cucumber, which showed that AM colonization can result in complete inactivation of the direct P uptake pathway via root hairs and epidermis, and that lack of a positive plant response does not mean that the AM fungal uptake pathway is making no contribution to P acquisition. The results emphasize once again that calculations of AM contributions to P uptake from total plant P using the subtraction method will often be highly inaccurate and can provide little useful information for non-responsive plants.

The effectiveness of a symbiosis in terms of P transfer is, as the data in Figure 5.5 show, dependent on the identity of both symbiotic partners. *Gi. rosea* only contributed significantly to the P nutrition of flax. Similar results indicating marked variations in 'functional compatibility', regardless of compatibility at the level of colonization *per se*, have been shown for different symbiotic partners (Ravnskov and Jakobsen, 1995; Burleigh *et al.*, 2002). There have been few direct investigations of the operation of the AM uptake pathway using tracers in the field. However, Schweiger and Jakobsen (2000a) clearly demonstrated the contribution of native AM fungi to P uptake by wheat, even in the absence of growth or P responses.

Pathways of nutrient uptake in AM plants

Overview

As already indicated, an AM root has two possible pathways by which it can absorb P and other nutrients from soil: the direct pathway, through root hairs and epidermis, as in non-mycorrhizal plants and the AM pathway which involves uptake by fungal hyphae in soil, rapid translocation over long distances in the extraradical mycelium, and ultimate delivery to intraradical fungal structures and transfer across symbiotic interface(s) in the root cortex (see Figure 4.1). As described in Chapter 4, the interfaces are bounded by fungal and plant plasma membranes, separating an interfacial apoplastic compartment, so that transfer of nutrients requires both efflux from the fungus and uptake by the plant (see Figure 4.2). The mechanisms contributing to uptake, translocation and transfer of nutrients via the AM pathway and its relative contribution to plant nutrient acquisition are the subject of active research.

Roles of extraradical hyphae in P uptake

Extraradical hyphae of AM fungi grow into soil, absorb orthophosphate from the soil solution and translocate P to the roots. These processes, together with subsequent transfer from fungus to plant, are much faster than diffusion through the soil. Consequently, hyphal transfer via the AM pathway overcomes the reductions in rate of plant P uptake which result from the slow P diffusion and development of depletion zones around the roots. The production of hyphae involves a smaller expenditure of C per unit length or per unit absorbing area than the production of roots (Tinker, 1975b) and their small diameter reduces the development of depletion zones and also allows them to penetrate soil pores that are inaccessible to roots, effectively increasing the volume of soil solution available for uptake. Rapid spread of hyphae many centimetres away from roots and turnover of exploratory hyphae maintains a large surface for absorption. Schweiger and Jakobsen (1999a) calculated that kinetic parameters (K_m and V_{max}) of P uptake systems will have a very much larger effect on P uptake by hyphae (with narrow diameters) than by roots (with larger diameters), in agreement with predictions (see above). Considerable variation exists in the extent of development of external mycelium by different AM fungi (see Chapter 2) and there is limited evidence that hyphae of some AM fungi proliferate differentially in root-free soil and in both organic and inorganic nutrient-rich patches, indicating abilities to sense environmental signals that may lead to more effective P acquisition (Maldonado-Mendoza *et al.*, 2001; Olsson *et al.*, 2002; Smith SE *et al.*, 2004a; Cavagnaro *et al.*, 2005).

Orthophosphate is probably absorbed as $H_2PO_4^-$ and uptake must be energy-requiring because it occurs against a large electrochemical potential gradient between soil solution and cytoplasm (Schachtman *et al.*, 1998). Physiological measurements indicate the operation of dual P uptake systems (high and low affinity) in AM fungi (Thomson *et al.*, 1990b), as in cells of higher plants and saprotrophic fungi. Phosphate transporter genes belonging to the Pht1 family have been cloned from *Glomus versiforme* (*GvPT*), *G. intraradices* (*GiPT*) and *G. mosseae* (*GmosPT*) (Harrison and van Buuren, 1995; Maldonado-Mendoza *et al.*, 2001; Benedetto *et al.*, 2005). They appear to be high affinity transporters; all are strongly expressed in the external mycelium but less so in intraradical structures, consistent with a proposed role in P uptake from the low concentrations in the soil solution. Estimates of K_m values of around $18\,\mu M$ for *G. versiforme* are rather high considering the normal solution concentration of P in the soil solution and very much higher than values obtained from physiological measurements (around $0.17\,\mu M$; Schweiger and Jakobsen, 1999a). Expression of both *GiPT* and *GmosPT* in the external mycelium is influenced by P concentrations in the solution and by the P status of the mycorrhiza, suggesting that both internal and external sensing mechanisms influence expression. In the case of *GmosPT*, consistent weak expression in intraradical structures was observed, independent of P supply. External hyphae of AM fungi also have active H^+-ATPases on the plasma membrane which would be capable of generating the required proton motive force to drive H^+–phosphate co-transport (Smith and Smith, 1996; Lei *et al.*, 1991) (see Figure 4.3a). Several AM fungal H^+-ATPase genes have been cloned and shown to be differentially regulated by stage of colonization and by phosphate supply, but a clear picture of their roles in nutrient uptake or transfer to the plant has not yet been obtained (Ferrol *et al.*, 2000; Requena *et al.*, 2003).

The work on P uptake by germ tubes of *Gigaspora margarita* remains somewhat contradictory. Although Thomson *et al.* (1990b) were able to demonstrate the existence of two P uptake systems in 14-day-old germlings, Lei *et al.* (1991) found no evidence of ^{32}P accumulation in germinating spores using autoradiography, unless they had been stimulated by the presence of root exudates (see Chapter 2). The unstimulated germ tubes showed no H^+-ATPase activity on their plasma membranes, while stimulated germ tubes did. Growth of the stimulated germ tubes was inhibited by application of the H^+-ATPase inhibitor diethyl stilbestrol (DES) (Lei *et al.*, 1991). These findings support the significance of stimulation of germ tubes by plant-derived signals to convert them to an 'infection-ready' state and may indicate that an inability to accumulate nutrients might be one factor accounting for the failure of AM fungi to grow extensively in the absence of a host plant (see Chapter 3). Further work is required to confirm this suggestion, particularly as the two research groups used material of different ages.

There have still been rather few measurements of the rate of uptake of P by the mycelium of any AM fungus in association with the plant. Table 5.2 shows values of inflow for a range of AM fungi associated with different plants, determined by different methods. Estimates range from $\sim 1.1 \times 10^{-15}$ to $4.3 \times 10^{-13}\,mol/m/s$; the wide range probably reflecting differences in methodology, the AM fungi and plants used, as well as the available P concentration in soil, which will have a marked

Table 5.2 Inflow of P to hyphae from soil determined in three different pot systems using plant–fungus combinations showing different plant responsiveness.

Plant–fungus	R or NR	Inflow (mol/m/s)	Reference
Allium cepa/Glomus[1]	R	2.25×10^{-13}	Sanders and Tinker, 1973; see Tinker, 1975
Trifolium repens/ G. mosseae[2]	R	$3.3–4.3 \times 10^{-13}$	Li et al., 1991
Linum usitatissimum/ G. caledonium[3]	R	$0.1–1.1 \times 10^{-14}$	Smith et al., unpublished
L. usitatissimum/ G. intraradices[3]	R	$1.1–4.3 \times 10^{-14}$	Smith et al., unpublished
Medicago truncatula/ G. caledonium[3]	R	$1.1–2.2 \times 10^{-14}$	Smith et al., unpublished
M. truncatula/ G. intraradices[3]	R	$0.6–2.6 \times 10^{-14}$	Smith et al., unpublished
Solanum esculentum G. caledonium[3]	NR	$0.1–0.9 \times 10^{-14}$	Smith et al., unpublished
S. esculentum G. intraradices[3]	NR	$0.6–3.2 \times 10^{-14}$	Smith et al., unpublished
T. subterraneum/ G. intraradices[3]	R; data for shoots only	2×10^{-15}	Schweiger and Jakobsen, 1999

[1]AM and NM plants grown in pots. P uptake via AM pathway determined by the subtraction method (see text).
[2]Hyphal uptake determined from uptake by AM and NM plants grown in compartmented pots. AM plants had access to a larger volume of soil that NM plants.
[3]Uptake via the AM pathway determined from transport of ^{32}P in from a hyphal compartment. See Smith et al., 2004. R, responsive, NR non-responsive.

influence on rates of P uptake (Munkvold *et al.*, 2004; Smith SE *et al.*, 2004a). Uptake on a hyphal surface area basis has been measured even less often. Sukarno *et al.* (1993) estimated the surface area of hyphae of a *Glomus* species as $12 \times 10^{-6} m^2/m$ hyphal length and used this value to derive uptake rates of $20–40 \times 10^{-9} mol/m^2/s$. These values are quite high, but of the same order of magnitude as rates of uptake by other types of cells.

Calculations of the relative rates of uptake by external hyphae and of transfer from fungus to plant showed that uptake per unit area of absorbing hyphae was about 10 times slower than transfer per unit area of symbiotic interface (Sukarno *et al.*, 1996). The difference emphasizes that the external mycelium must absorb nutrients over a large surface area in soil to supply P through the entry points to the interfacial regions and thence to the plant.

Use of sources of P by AM fungi that are unavailable to roots

There has been considerable interest in whether AM fungi have any ability to access P from the non-labile fractions in soil. This question has been addressed by assuming that only the labile Pi would exchange with ^{32}P added to soil and that it was the labile pool from which non-mycorrhizal plants absorbed P via the direct uptake pathway. Although AM onions, rye grass and soybeans took up more total P from soil with the labile fraction labelled than non-mycorrhizal plants, there was no difference in specific activity of ^{32}P in the two groups of plants, suggesting that AM plants had no access to non-labile P sources (Sanders and Tinker, 1971; Hayman and Mosse, 1972; Mosse *et al.*, 1973; Powell, 1975; Pichot and Binh, 1976; Gianinazzi-Pearson *et al.*, 1981a). Furthermore, heating soil to provide a range of concentrations of 'fixed' P gave no evidence of significant difference between AM and non-mycorrhizal plants in accessing P in the different fractions (Barrow *et al.*, 1977). One investigation on potatoes apparently showed that AM plants could take up fixed P (Swaminathan, 1979), but the general conclusion is that there are no major differences in pools of P accessible to the two groups of plants (Bolan, 1991). Nevertheless, the range of soil types for which P availability to AM fungi has been tested remains relatively low and confirmation of the conclusions with soils of different pH, P-fixing capacity and dominant forms of fixed P is warranted.

Despite the conclusion, reached many years ago, that plants and AM fungi use the same sources of soil P, the suggestion that AM roots might be able to exploit sources of P in soil not normally available to plants has continued to be explored. These include relatively insoluble forms of Pi, such as rock phosphate (RP) and Fe and Al phosphates, as well as sources of Po such as phytate. It has certainly been shown that growth of AM plants does respond to the application of Fe phosphates, RP or tricalcium P, whereas these fertilizers had much smaller effects on the growth of non-mycorrhizal plants at the rates of application used (Murdoch *et al.*, 1967; Pairunan *et al.*, 1980; Bolan *et al.*, 1987). Similar results have been obtained following application of insoluble P fertilizers for a variety of host plants, usually in soils of low pH. Although detailed investigations over a full P response curve indicated that there were no absolute differences in availability (in accord with the general conclusions, above), at moderate and realistic levels of application, and at any level of P equivalent to the range of superphosphate applications to crops (0–0.8 g P/kg soil), AM plants were more effective at extracting P from RP (Pairunan *et al.*, 1980). The results with Fe-phosphates were also interpreted in terms of more effective spatial exploitation of soil by AM hyphae (Bolan *et al.*, 1987).

The mechanisms underlying increased uptake might depend upon hyphal exploitation of the soil volume, synergistic action between AM fungi and P-solubilizing microorganisms, and the possible excretion by hyphae of H^+ that lowers pH or organic anions with chelating ability (Smith, 1980; Comerford, 1998). There is no direct experimental support for production of chelating agents by hyphae of AM fungi, but reductions in soil pH (up to 1.0 unit) in HCs have been shown concurrently with P depletion in a cambisol, when P was supplied as $Ca(H_2PO_4)_2$ and N as $(NH_4)_2SO_4$ (Li *et al.*, 1991b) (Figure 5.6). The mechanism by which the pH was reduced is likely to have been via the extrusion of H^+, following ammonium assimilation by the hyphae (Raven and Smith, 1976; Smith, 1980; Bago and Azcón-Aguilar, 1997), but other mechanisms, such as increased CO_2 production, may also have contributed (Li *et al.*, 1991b).

Work on the possibility that enzymes from roots or AM hyphae hydrolyse sources of Po is receiving increasing emphasis. An early investigation (Mosse and Phillips, 1971) found that phytates were satisfactory sources of P for plant and AM fungal growth in agar cultures and that calcium phytate stimulated fungal growth.

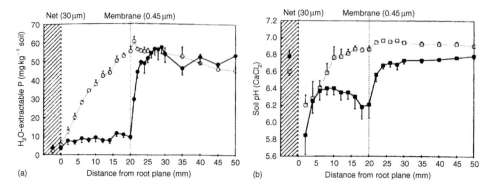

Figure 5.6 Depletion profiles of (a) H₂O-extractable P and (b) soil pH in the hyphal and bulk soil compartments (refer to Figure 5.3a) from non-mycorrhizal (open symbols) and mycorrhizal (closed symbols) *Trifolium repens*, grown in a cambisol. Bars represent standard errors of means. From Li *et al.* (1991b), with permission.

Subsequently, several investigations indicated increased capacity of AM plants to access Po of various availabilities (Gianinazzi-Pearson *et al.*, 1981a; Jayachandran *et al.*, 1992; Joner, 1994; Joner and Jakobsen, 1994; Koide and Kabir, 2000), but it has proved difficult to be sure that the effects were due to hydrolysis by enzymes produced by the AM fungi themselves, rather than indirect effects of enhanced microbial activity in soil (Joner and Johansen, 2000). Phosphatases produced by external hyphae have been shown to have an acidic pH optimum, but most are internal or wall-bound, with almost no soluble external activity detectable. Nevertheless, hydrolysis of artificial Po substrates by AM fungi has been demonstrated in monoxenic culture (Koide and Kabir, 2000), supporting the idea that increases in rhizosphere activity could be mediated by the fungi (Dodd *et al.*, 1987; Tarafdar and Marschner, 1994). However, enzymes could also be of plant origin. An acid phosphatase (ACPase), whose activity is increased in AM roots, has been identified from *Tagetes patula* (Ezawa and Yoshida, 1994). It is related to plant purple acid phosphatases and is excreted into the rhizosphere. The gene encoding this enzyme (*TpPAP1*) is upregulated by AM colonization of the roots and downregulated by increased P supply (Ezawa *et al.*, 2005), leading the authors to suggest that the AM fungus induced the plant to activate a response to low P availability. Any relationship between production of the enzyme and P mobilization and uptake remains to be revealed.

The idea that extraradical hyphae of AM fungi might effectively increase the 'competitive ability' of an AM root system, *vis à vis* soil microorganisms, in acquiring P from the soil solution and thus circumvent the problems of immobilization of P in the biomass has been canvassed (Linderman, 1992). The dependence of AM fungi on recent photosynthate from the plant means that their activity would not be affected by the availability of organic C substrates in soil or their C:P ratio, giving them considerable advantages over saprotrophic microorganisms. Barea *et al.* (1975) were among the first to investigate interactions between P-solubilizing bacteria and AM inoculation in the mobilization of P from rock phosphate (RP). They observed positive, synergistic effects in growth and P uptake by *Zea mays* and *Lavendula spica* that were significant in some soil–plant combinations. Similarly, the potential of a P-solubilizing fungus *Penicillium bilaji* to increase the availability of RP to *Triticum aestivum* and *Phaseolus vulgaris* depended on AM activity (Kucey, 1987; Kucey and Janzen, 1987). Jayachandran *et al.* (1989) used three soils and showed that if strong Fe chelating agents were added, the P released from Fe-phosphates was available to AM, but not to non-mycorrhizal plants. They found no evidence for production of chelating agents by the AM fungi themselves and concluded that the outcome was the result of effective exploitation of the soil and competition with resident microflora. This conclusion is supported by additional work showing that *Glomus* spp. (WUM 10 (1)) and *G. caledonium* were both capable of intercepting Pi released during mineralization of Po by microorganisms and preventing immobilization in the biomass or sorption on clay minerals (Joner and Jakobsen, 1994). The proliferation of external hyphae in soil organic matter and in dying roots (St John *et al.*, 1983a; Warner, 1984; Ritz and Newman, 1985; Newman, 1988; Eason and Newman, 1990; Eason *et al.*, 1991; Olsson *et al.*, 2002) would be an appropriate strategy for the operation of this competitive effect.

More work is needed to understand the interactions between the role of the soil microflora and microfauna in both mineralizing and immobilizing soil P and the potential capacity of AM fungi to short-circuit this aspect of nutrient cycling. The interactions between AM colonization and populations of particular functional groups of organisms, both in the rhizosphere and further from the root where hyphae may proliferate, will be discussed further in Chapter 16.

Fungal P metabolism

Orthophosphate absorbed from the soil solution into the fungal cytoplasm is utilized first for synthesis of key molecules for cellular functions. Subsequently, P is transferred to acidic compartments (pH ~5.6), presumed to be vacuoles, where both orthophosphate and polyphosphate (polyP) accumulate in varying amounts. P storage in vacuoles has been most clearly demonstrated by *in vivo* NMR spectroscopy from the pH-dependent chemical shifts of P-containing metabolites (Rasmussen *et al.*, 2000).

PolyP is a linear polymer of variable numbers of orthophosphate residues, linked by high energy phosphoanhydride bonds. It is synthesized by a wide range of microorganisms and fungi and the most likely function is as a storage molecule, buffering the cytoplasmic orthophosphate concentration within physiologically acceptable limits and reducing osmotic stress (Harold, 1966; Beever and Burns, 1980). PolyP accumulation has also been suggested to act as an energy store (Beever and Burns, 1980), but this is now seen as unlikely in AM fungi because polyphosphate glucokinase-type activity, which would provide a route for glucose phosphorylation using the high energy bonds in polyP, is much lower than hexokinase which utilizes ATP (Ezawa *et al.*, 2001).

Metabolism of polyP in AM fungi has received considerable attention following early observation of accumulation, detected by staining or electron microscopy, and the suggestion that insoluble polyP existed in 'granules' that might be transported over long distances in external mycelium by cytoplasmic streaming (Cox and Tinker, 1976; Callow *et al.*, 1978; Cox *et al.*, 1980). Both chemical extraction and *in vivo* NMR spectroscopy indicate that much of the polyP in AM fungi is of relatively short chain length (up to about ~17 mer) (Solaiman *et al.*, 1999) and, hence, not in granular form. Some recent work suggests that the granules may be artefacts of specimen preparation

(Orlovich and Ashford, 1993; Ezawa *et al.*, 2002), but 'granules' have been seen in living hyphae of ECM fungi (Bücking and Heyser, 1999; see Chapter 10). Studies of polyP metabolism are beset by methodological problems which have hindered clear demonstration of occurrence and distribution in hyphae, ranges of chain lengths and turnover. Although these matters are still unresolved, there seems no doubt that polyP, together with Pi, does play a significant storage and translocatory role in AM hyphae and more work with a range of techniques is required to sort out the details.

The amount of polyP in extraradical and intraradical mycelium varies considerably, from undetectable to relatively large proportions of the total P, and it can be synthesized very rapidly when P is supplied to P-deprived AM roots. In one investigation involving *Tagetes patula* colonized by *Archaeospora leptoticha*, the percentage of extraradical hyphal length showing metachromatic staining with toluidine blue O (to visualize polyP) increased from 25 to 44% from 0 to 1 h after addition of relatively high P concentration (1 mM), reaching a maximum of 50% by 3 h (Ezawa *et al.*, 2004). PolyP accumulation (assessed using a highly sensitive polyphosphate kinase assay (Ohtomo *et al.*, 2004)) was very rapid (46.4 ± 15.1 nmol/min/mg), doubling between 1 and 3 h and reaching 10.0 μmol Pi equivalents/mg protein. In this investigation, some of the accumulated polyP was shown by electrophoresis to have chain lengths greater than 300 Pi units (Ezawa *et al.*, 2004) and would therefore not be visible by *in vivo* solution NMR which has an upper limit of $\sim$75 units, because of the mobility of the atomic nuclei. The high rates of accumulation and long polymers may well indicate an ability of AM fungi to capitalize rapidly on spatially or temporally localized sources of P in soil. PolyP chain lengths in external mycelium are sometimes longer than in intraradical mycelium, suggesting hydrolysis associated with P transfer to plants (Solaiman *et al.*, 1999), but variations due to differences in fungal symbionts and in P supply are also likely, so that it is premature to generalize.

Information on pathways of P metabolism in AM fungi still does not give a very coherent picture. AM roots of onion have been shown to contain higher activities of both exopolyphosphatase (PPX) and endopolyphosphatase (PPN) than non-mycorrhizal roots, but no polyphosphatase activities were detected in external hyphae. Polyphosphate kinase (PPK) was apparently also detectable in AM roots and external hyphae, but not in non-mycorrhizal roots, while polyphosphate glucokinase (PPGK) was only found in the external hyphae (Capaccio and Callow, 1982). Neither PPK nor PPGK have been detected in any other eukaryotes and more recent investigations have failed to confirm their presence in two AM fungi (Ezawa *et al.*, 2002), so their roles in AM polyP turnover must now be questioned. Indeed, the way in which polyP is synthesized in vacuoles remains enigmatic (Ezawa *et al.*, 2002), but will be an important area for future research. Hydrolysis of polyP at sites of P utilization or transfer to the plant probably occurs through the combined activities of PPX, PPN and non-specific acid phosphatases (ACPases). At least two PPX-type enzymes have been identified in AM fungi, with different pH optima, K_m values and specificity for different substrate chain lengths; they are also differently expressed between intraradical and extraradical mycelium, possibly contributing to the differences in polyP chain length observed in the two fungal compartments.

Other phosphatase enzymes have also been implicated in P metabolism in AM fungi. 'Mycorrhiza-specific' alkaline phosphatases (ALPases) received considerable attention following cytochemical detection in the vacuoles of the mature arbuscules and intercellular hyphae (Gianinazzi *et al.*, 1979), as well as in hyphal coils of

Paris-type AM (van Aarle *et al.*, 2005). Circumstantial evidence that activity is linked to the presence of arbuscules and transfer of P to the plant (Gianinazzi-Pearson and Gianinazzi, 1978) led to the suggestion that this enzyme might be a useful marker for efficient P metabolism in the fungi. Early observations also indicated that activity was restricted to intraradical AM fungal structures and did not occur in the external phase. Activity in germ tubes was restricted to the tip region and there was apparently a lag in development of activity within the root, which is not seen for succinate dehydrogenase (SDH) activity (Tisserant *et al.*, 1993). However, ALPases were subsequently also demonstrated in external mycelium (Dodd, 1994; van Aarle *et al.*, 2002b; Aono *et al.*, 2004) and although (together with SDH) they may be good cytochemical markers for fungal activity, their role in P metabolism remains unclear. Larsen *et al.* (1996) showed that whereas the fungicide benlate inhibited P uptake and transfer to plants via AM hyphae it did not affect ALPase activity, calling into question the relevance of activity as a marker for P efficiency in this instance. Furthermore, substrate specificities and inhibitor studies of ALPase suggest that it is unlikely to be able to hydrolyse pyrophosphate compounds (including polyP), but it may be involved in sugar metabolism. Acid phosphatases (ACPases) show little substrate specificity and appear more likely candidates for involvement in polyP hydrolysis, particularly as the pH of fungal vacuoles (~5.6) where polyP is located, would not favour activity of ALPases (Ezawa *et al.*, 1999, 2002).

Translocation of P by AM fungi

The importance of hyphal translocation in soil in delivering P (and other nutrients) to plants was demonstrated directly as early as 1973 (Hattingh *et al.*, 1973; Hattingh, 1975; Rhodes and Gerdemann, 1975, 1978a, 1978c). Using compartmented perspex chambers, ^{32}P, ^{35}S and ^{45}Ca were injected at different distances from AM onion plants and the appearance of the tracer followed. ^{32}P was translocated by hyphae of *Glomus mosseae* and *G. 'fasciculatus'* up to 7 cm through soil to roots of *Allium cepa*. When hyphae between the source of ^{32}P and the root were cut, no translocation to the root was observed. Extensions of this approach have, as described above, been increasingly used to compare effectiveness of P delivery, particularly among different plant–fungus combinations.

Motile vacuoles containing polyP may be involved in long-distance translocation of P along AM hyphae, replacing the earlier idea of movement of granules of polyP by cytoplasmic streaming (Smith and Read, 1997). Unfortunately, polyP with chain lengths longer than about 75 residues is not detectable by solution NMR, so that this otherwise valuable technique cannot be applied to determining the role of solid phase molecules in P dynamics. The matter is also complicated by the fact that AM fungi, as well as other fungi including those forming ectomycorrhizas, appear to have two vacuole systems. In addition to spherical vacuoles, pleiomorphic vacuolar tubules have been observed, both by *in vivo* staining and electron microscopy (Shepherd *et al.*, 1993a, 1993b; Rees *et al.*, 1994; Uetake *et al.*, 2002). In ectomycorrhizal (ECM) fungi, the tubules are found in extraradical hyphae and in the fungal sheath (Allaway and Ashford, 2001) and have been shown to contain polyP; pulsation results in transfer of their contents over short distances. Transfer over the long distances that would be required for translocation in external mycelium is still an open question (Ashford and Orlovich, 1994). A similar tubular vacuole system in AM fungi, whose movement is not closely related to cytoplasmic streaming, has been observed in germ tubes as well as extraradical and intercellular hyphae of the AM fungus *Gigaspora margarita* and could be involved in P translocation (Uetake *et al.*, 2002). Effects of temperature and the cytoskeletal inhibitor cytochalasin B, shown to inhibit translocation (Cooper and Tinker, 1981), would be expected to act similarly on any motile system, be it tubular vacuoles or cytoplasmic streaming and so cannot distinguish between these possibilities.

P translocation in *G. mosseae* to *Trifolium subterraneum*, determined from ^{32}P movement, was shown to be between 2 and 20×10^{-6} mol/m^2/s (Pearson and Tinker, 1975; Cooper and Tinker, 1978, 1981). Higher values have been obtained both for entry-point hyphae in soil-grown plants (3.8×10^{-4} mol/m^2/s) and for runner hyphae in monoxenic cultures (1.3×10^{-3} mol/m^2/s) (Sanders and Tinker, 1973; Nielsen *et al.*, 2002). A feature of results obtained using tracers is a lag that frequently occurs after application of ^{32}P to established external mycelium, either in split plate agar systems or hyphal compartments in soil, before a steady rate of appearance in the host is established (Cooper and Tinker, 1978; Johansen *et al.*, 1993a; Nielsen *et al.*, 2002). Suggestions that the lag is due to a delay between arbuscule formation and degeneration and hence transfer to plant cells (Cox and Tinker, 1976) can be discounted, for the lag occurs not only in appearance of ^{32}P in the shoots but also in roots containing fungal structures (Cooper and Tinker, 1978). It seems likely that the lag is due to two factors: a delay due to the time taken for the tracer to travel between the source and the site of detection and a delay caused by the equilibration of tracer (^{32}P) with ^{31}P already present in the translocation pathway. Thus, although translocation of total phosphate (^{31}P+^{32}P) occurs at a constant rate, this will not be apparent from measurements of radioactivity until the specific activity of ^{32}P is uniform throughout the translocation pathway.

Transfer of P from fungus to plant

It is generally accepted that the considerable transfer of P from fungus to plant involves membrane transport steps at living interfaces, which are comprised of the membranes of both symbionts and an apoplastic region between them (see Chapters 2 and 4). Early workers assumed that degeneration (digestion) of arbuscules was the mechanism that made fungal P available to the plants. Cox and Tinker (1976) calculated that arbuscules do not contain sufficient P for their 'digestion' to be a credible mechanism for P transfer, but more recent measurements suggest that their estimates of arbuscular P content may have been rather low and some researchers consistently maintain that arbuscule breakdown may contribute to transfer (F.E. Sanders, personal communication). Nevertheless, it is generally assumed that the site of transfer is the mature arbuscular interface in *Arum*-type mycorrhizas and intracellular coils in *Paris*-types. Distribution of membrane-bound H$^+$-ATPases and AM-inducible P transporters certainly supports this view (see below), but there is still no definitive evidence that would exclude intercellular hyphae from involvement in transfer.

There is no doubt that the interface between the symbionts in cells colonized by arbuscules would provide, as Cox and Tinker (1976) emphasized, a relatively

Table 5.3 Transfer of P to the plant across the symbiotic interface, assuming the arbuscular interface alone is responsible for transfer.

Plant–fungus	Transfer flux (mol/m^2/s)	Reference
Allium cepa/Glomus mosseae	13×10^{-9}	Cox and Tinker, 1976
Allium cepa/Glomus WUM 16	4–29×10^{-9}	Sukarno *et al.*, 1996
Allium porrum/Glomus mosseae	2.0–3.2×10^{-9}	Smith *et al.*, 1994
Allium porrum/Glomus sp WUM 16	5.0–12.8×10^{-9}	Smith *et al.*, 1994

Hyphal contribution to inflow calculated from total P uptake in mycorrhizal and non-mycorrhizal plants, using the subtraction method. Area of interface determined from numbers of arbuscules and invagination of the plant plasma membrane (see Cox and Tinker, 1976; Toth etc.).

large surface area across which P and other nutrients also could be transferred by membrane transport. They measured the area of interface in such cells and calculated the flux required to support transfer of P from fungus to plant, based on total P accumulated in AM and non-mycorrhizal plants. P flux via arbuscules of *G. mosseae* to *A. cepa* was 13×10^{-9} mol/m^2/s. Similar calculations have now been made for *Glomus coronatum* and *G. mosseae* colonizing *A. porrum* (leek) and found to be of the same order (Smith *et al.*, 1994) (Table 5.3). No equivalent calculations are available for *Paris*-type AM, but coils do provide an area of interface that can be as large as arbuscules (Dickson and Kolesik, 1999), AM-inducible P transporters are expressed in coil-containing cells (Karandashov *et al.*, 2004) and it is known that these mycorrhizas can improve P nutrition of the plants that form them (Cavagnaro *et al.*, 2003; Smith SE *et al.*, 2004a).

The magnitude of the transfer fluxes is of the same order as uptake of P by free-living plant and fungal cells and very much larger than measured rates of efflux (Beever and Burns, 1980; Elliott *et al.*, 1984). As transfer across a symbiotic interface involves both efflux and uptake operating in series (see Figure 4.2), the two processes must occur at equal rates. P transfer from fungus to plant far exceeds the rate which might be expected if the process were dependent on 'normal' efflux (Smith *et al.*, 1994), so the inevitable conclusion is that P transfer from fungus to plant involves special modifications to increase the efflux from the fungus and probably also to suppress fungal reabsorption (uptake) from the apoplast, as this would negate the efflux (Smith *et al.*, 1995; Schachtman *et al.*, 1998). Mechanisms of efflux are not of wide interest in free-living organisms, so work on P metabolism in yeast is not a very helpful parallel for AM fungi in this case. Special mechanisms promoting efflux in symbiotic systems need to be sought and probably include mobilization of P reserves such as polyP and liberation from the vacuolar pool, followed by efflux to the interfacial apoplast (Ezawa *et al.*, 2002). There is a little evidence that P efflux from intraradical fungal structures, separated from roots by digestion, is enhanced by exogenously supplied glucose and deoxyglucose (Solaiman and Saito, 2001; Bücking and Shachar-Hill, 2005). Low expression of the AM fungal high-affinity P transporter, *GvPT*, in intraradical structures would certainly prevent fungal reabsorption (Harrison and van Buuren, 1995). However, *GmosPT* is consistently expressed at a low level within roots, a finding that led Benedetto *et al.* (2005) to suggest that the fungus may actually control the amounts of P made available to the plant by reabsorbing P released to the apoplast.

Expression of P transporters in AM fungi and roots

The last decade has seen an enormous explosion of knowledge of the families of transporter proteins that facilitate P transfer through plant and fungal membranes (Schachtman *et al.*, 1998; Raghothama, 1999; Smith FW *et al.*, 2003a). Investigations of the extent and sites of expression of genes that encode members of the Pht1 family of transporters show that colonization of roots by AM fungi induces very significant modifications. P transporters have been cloned from three AM fungi and shown to be expressed in the external mycelium, consistent with active P uptake from the soil solution (see above). In plants, P transporters have been identified that are either expressed only in AM roots (AM-specific) or which show markedly increased expression (AM-inducible). So far, these transporters have been identified in the eudicots *Solanum tuberosum* and *S. lycopersicon* (Solanaceae), *Medicago truncatula* and *Lotus japonicus* (Fabaceae) and *Populus* (Betulaceae) (see Martin, 2007) and in the monocots *Oryza sativa, Triticum aestivum, Hordeum vulgare* and *Zea mays* (Poaceae) (Rausch *et al.*, 2001; Harrison *et al.*, 2002; Paszkowski *et al.*, 2002; Glassop *et al.*, 2005, 2007; Güimil *et al.*, 2005; Nagy *et al.*, 2005a, 2005b; Maeda *et al.*, 2006). This wide distribution suggests that similar transporters are likely to occur in all potentially AM plants. Three different AM-inducible transporters have been identified in the solanaceous species, two in *O. sativa* and one in each of the members of the Fabaceae and in *T. aestivum, H. vulgare* and *Z. mays*. In *Populus*, different members of the Pht1 family were upregulated depending on whether the roots were colonized by AM or ECM fungi (Martin, 2007). Where tissue localization has been investigated, mRNA expression is confined to root cortical cells containing *Arum*-type arbuscules (Figure 5.7a; see Colour Plate 5.1a) or *Paris*-type coils (Figure 5.7b; see Colour Plate 5.1b) (Rausch *et al.*, 2001; Harrison *et al.*, 2002; Karandashov *et al.*, 2004; Glassop *et al.*, 2005; Maeda *et al.*, 2006) and, in one case, immunolocalization has confirmed that the protein is confined to the plasma membrane of mature arbuscules (Harrison *et al.*, 2002) (Figure 5.8; see Colour Plate 5.2) and disappears from cells containing collapsing arbuscules.

Increased expression of both *LePT3* and *LePT4* in tomato was shown to depend on colonization of the root cortex by AM fungi and was accompanied by physiological demonstration of the operation of the AM P uptake pathway using ^{32}P (Poulsen *et al.*, 2005). None of these events took place in roots of tomato mutants lacking colonization. At the individual cell level, activity of the *StPT3* promoter in monoxenic potato roots appears to require the presence of an intraradical fungal structure (arbuscule or coil), because no activity was detected in adjacent non-colonized cells. Activity was also temporally regulated, so that different parts of the same infection unit showed different levels of promoter activity (Karandashov *et al.*, 2004). The same work confirmed that *StPT3* promoter activity was only induced by members of the Glomeromycota, and not by any other root-colonizing fungi, whether pathogenic or neutral. Coupled with differential expression of plasma membrane H^+-ATPases in AM roots and localization of both activity and protein to the periarbuscular membranes (Gianinazzi-Pearson *et al.*, 1991a, 2000; Murphy *et al.*, 1997) (see Figure 4.3 and Colour Plate 4.1), these findings are completely consistent with the roles of intracellular arbuscules and coils as the sites of transfer of P from fungus to plant during the operation of the AM uptake pathway.

The functional significance of multiple AM-inducible P transporter genes in a single species is not yet clear. It appears that some (like *MtPT4, LePT4* and *OsPT11*)

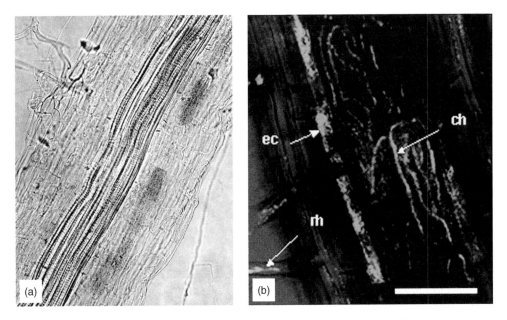

Figure 5.7 Cell-type-specific promoter activity of mycorrhiza-induced Pht1 transporters. (a) Histochemical staining for GUS activity in *Medicago truncatula* roots carrying an *MtPT4* promoter-UidA fusion. Blue staining indicates GUS activity associated with *Arum*-type arbuscules of *Glomus versiforme*. Image courtesy Maria Harrison. (b) Confocal image of hairy root of *Solanum tuberosum*, transformed with an *StPT3* promoter-Fluorescent Timer chimeric gene. Green fluorescence originating from Fluorescent-Timer in cells colonized by *Paris*-type coils formed by *Gigaspora margarita*. rh, root hairs; ec, epidermal cells; ch, coiled hyphae. Reproduced from Karandashov *et al.*, (2002) *Proceedings of the National Academy of Sciences of the United States of America*, **101**, 6285–6290. Copyright National Academy of Sciences, USA See also Colour Plate 5.1.

are only expressed when roots are colonized by AM fungi (AM-specific), whereas others (such as *StPT3, LjPT3*) are expressed at a low level in uncolonized roots, but are upregulated by AM colonization. Knockout of *MtPT4* completely prevents AM phosphate uptake by *M. truncatula* and there do not appear to be any other AM-regulated P transporters that take over this role (Javot *et al.*, 2007). Mutants created by RNAi or identified by TILLING showed markedly deficient arbuscule development, premature degeneration of intraradical mycelium and lack of extraradical mycelium. The knockdown mutant of *LjPT3* showed both reduced arbuscule development and reduced AM P uptake by *L. japonicus* (Maeda *et al.*, 2006). These genes appear to have complex or multiple functions that have not yet been elucidated. Speculation that P delivered via the AM pathway is essential for C transfer and proper arbuscule development appears somewhat premature at this stage. In contrast, knockout of the *LePT4* gene did not prevent operation of the AM uptake pathway in tomato and arbuscule development was apparently normal (Nagy *et al.*, 2005a). In this case there appears to be some functional overlap (redundancy) of the roles of the transporters, but it is possible that future investigations will reveal subtle differences in operation, such as differences in K_m to cope with different

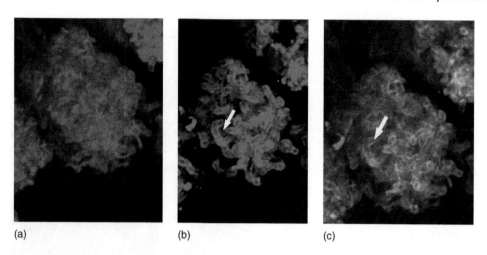

(a)   (b)   (c)

Figure 5.8 Confocal images showing immunolocalization of MtPT4 in roots of *Medicago truncatula* colonized by *Glomus versiforme*. The roots were probed with anitbodies specific for MtPT4 and these were visualized with a secondary antibody conjugated with AlexaFluor488 (green fluorescence). Roots were counterstained with WGA-Texas red to visualize fungal structures (red fluorescence). (a) WGA-Texas red visualization of arbuscules; (b) green fluorescence showing immunolocalization of MtPT4 surrounding branches of the arbuscule (arrowed); (c) merged images showing both red and green fluorescence. Images courtesy Maria Harrison. See also Colour Plate 5.2.

P concentrations in the interfacial apoplast or differential induction by different AM fungi. At the evolutionary level, it appears that signal perception and transduction leading to expression of AM-inducible P transporters are conserved within the eudicot species, so that the *StPt3* promoter was able to direct AM-related expression not only in *S. tuberosum*, but also *Petunia hybrida, Daucus carota, M. truncatula* and *L. japonicus*. The *MtPT4* promoter also drove expression in *S. tuberosum*. However, regulatory mechanisms appear not to be conserved between the monocot rice and eudicots, because the *OsPT11* promoter was unable to direct AM-inducible expression in either *S. tuberosum* or *M. truncatula* root organ cultures (Karandashov *et al.*, 2004). Phylogenetic tree analysis does, however, show high protein sequence similarity between the AM-inducible transporters OsPT11, MtPT4, StPT4/5 and LePT4/5, with HvPT8 and OsPT13 sequences also more similar to these than to other members of the Pht1 family in the different plants. This suggests that most AM-inducible transporters identified thus far arose from the same ancient ancestral protein. However, the StPT3/LePT3 proteins group more closely to StPt1 and LePt1 which show more general expression patterns, suggesting a separate evolutionary gain of AM inducibility in these genes (Karandashov *et al.*, 2004).

Physiological evidence indicates that operation of the AM P uptake pathway is often accompanied by reduction in P absorbed directly by root hairs and epidermis. The reduction could in part be due to the depletion of P in the rhizosphere (see above), but reduced expression of Pht1 transporters in root epidermal cells and particularly in the root hairs could also contribute to the effect. For example, *MtPt1* from *M. truncatula* (Chiou *et al.*, 2001), *HvPt1* and *HvPt2* from *H. vulgare* (Glassop *et al.*, 2005) and *StPT1* and *StPT2* from *S. tuberosum* (Rausch *et al.*, 2001) are sometimes downregulated when roots are colonized by AM fungi. Expression of these genes is frequently P sensitive, so that increased P concentrations in AM plants could bring about the observed reduction. Alternatively, it is possible that direct fungal signalling may be involved. Both mechanisms may operate concurrently and more work on regulation of expression by different factors is required.

The K_ms of AM-inducible transporters are of considerable interest in relation to the concentrations of orthophosphate that exist in the interfacial apoplasts. Estimates of the K_ms made in heterologous yeast systems generally suggest that they are high (low affinity). However, the Km of *HvPT8*, determined from measurements of P flux into barley roots overexpressing the gene, appears to be much lower (E. Grace, unpublished). Some variation in affinity may in fact be advantageous in respect of the efficiency with which different AM fungi deliver P to the interfaces. It will therefore be of considerable interest to compare the kinetic properties of the multiple AM-inducible transporters from single species, as well as determining the stimuli that regulate them.

Nitrogen nutrition

Nitrogen in soil

Organic nitrogen (No) dominates soil N pools, with immobilization and mineralization consequently highly dependent on activities of soil microorganisms (McNeill and Unkovich, 2006). Microorganisms also bring about conversions such as nitrification and denitrification and the important conversions of gaseous N_2 to ammonium via biological N_2 fixation. Plants and microorganisms (including AM fungi) can absorb both nitrate and ammonium from the soil solution, as well as some soluble forms of No. Both major inorganic N sources are regarded as relatively mobile in soil, transported to roots by mass flow in the soil solution. Major N depletion zones are not usually considered to be a serious limitation to uptake. Nevertheless, ammonium is less mobile than nitrate and movement of both can be restricted in dry soil (Tinker and Nye, 2000). Accordingly, there is potential for external mycelium of AM fungi to play a role in soil-to-plant N transfers and attention needs to be paid to the quantitative contribution of this pathway, bearing in mind that plants require about 10 times more N than P.

Arbuscular mycorrhizal effects on nodulation and N_2 fixation

Increased N concentrations have been reported in AM plants. Of course, where the plants are also symbiotic with N_2-fixing bacteria or actinomycetes, this can be attributed to increased rates of N_2 fixation induced secondarily, rather than to direct uptake of N compounds from the soil. The assimilation of N_2 in rhizobial root nodules is certainly increased when plants growing in low P soils are also colonized by AM fungi. This effect was probably first observed by Asai (1944) who made detailed observations of growth, nodulation and mycorrhizal status of a large number of legumes. Subsequently, nodulation and N_2 fixation by AM and non-mycorrhizal legumes (Smith and Daft, 1977; Abbott and Robson, 1978; Abbott *et al.*, 1979; Bethlenfalvay *et al.*, 1982a, 1982b, 1997; Bethlenfalvay and Newton, 1991; Bethlenfalvay, 1992a), as well as plants nodulated by *Frankia* (Rose, 1980; Rose and Youngberg, 1981; Gardner, 1986; Reddell *et al.*, 1997b; Wheeler *et al.*, 2000; Duponnois *et al.*, 2003), have been the subject of many experiments. In most cases, improved nodulation and N_2 fixation in AM plants appears to be the result of relief from P stress and possibly uptake of some essential micronutrients, which result in both a general improvement in growth and indirect effects upon the N_2-fixing system. The differences between AM and non-mycorrhizal plants usually disappear if the latter are supplied with a readily available P source. More detailed information can be obtained from the many reviews of this area (Bowen and Smith, 1981; Barea and Azcón-Aguilar, 1983; Bethlenfalvay and Newton, 1991; Barea *et al.*, 1992; Bethlenfalvay, 1992a). The fact that the effects are, in the main, not directly attributable to the AM fungi themselves should not be allowed to detract from the interest of these tripartite symbioses which may be very important in agriculture, natural ecosystems and in revegetation programmes, where nutrients are in short supply in soil (Barea and Azcón-Aguilar, 1983; Lamont, 1984; Pate, 1994).

Uptake of N from soil

Hyphae of AM fungi, as well as roots, are able to absorb both ammonium and nitrate. However, in contrast to the situation with other mycorrhizal types, the role of extraradical hyphae of AM fungi in mineralization of organic forms of N from soil remains somewhat equivocal. Despite the observations of Hodge *et al.* (2001), suggesting that AM fungi both enhance organic N decomposition and plant N capture, other work does not substantiate these conclusions (Frey and Schüepp, 1993; Hawkins *et al.*, 2000; Hodge, 2001). Ames *et al.* (1983) showed that, although ^{15}N supplied to extraradical hyphae in organic form reached AM plants of *Apium graveolens* (celery), transfer required a considerable period of time. They assumed that mineralization by soil microflora was an essential step in making the organic N available and that this caused the delay. Even though Hawkins *et al.* (2000) did show increased ^{15}N uptake from ^{15}N-glycine by AM plants, the transfer was very small and insufficient to influence the N status of the plants. On the other hand, there is increasing evidence that AM fungi are involved in uptake and transfer of inorganic N, although it is still not clear whether this always occurs in amounts that are significant for whole plant nutrition.

The first results on roles of AM fungi in inorganic N uptake and transfer were somewhat inconclusive. Haines and Best (1976) reported that loss of ammonium, nitrate and nitrite from soil by leaching with water was retarded when plants of *Liquidambar styraciflua* were AM with *Glomus mosseae*. Unfortunately, the root systems of the AM plants were considerably larger than the non-mycorrhizal plants, so that the results did not indicate unequivocally that the AM fungi themselves were involved. Increased inflow of N to roots of AM *Trifolium subterraneum* supplied with inorganic N has been observed, but the results of different experiments were not consistent (Smith *et al.*, 1986a, 1986b) and uptake of N by the hyphae did not seem to play an important role in net plant N nutrition because increased uptake per plant was not observed. Recent work exploring the ability of native AM fungi to influence N nutrition of several old-field perennials also failed to show enhancement of total N uptake from any of the inorganic or organic sources tested (Reynolds *et al.*, 2005).

Despite these findings, evidence is accumulating that the AM pathway makes considerable contributions to plant N uptake from soil, regardless of total N uptake and N responses. Again, the use of mesh-compartmented pots has played an important part. Ames et al. (1983) supplied ($^{15}NH_4$)$_2SO_4$ to the HC and observed considerable ^{15}N transfer to the AM plants. The amount was correlated with the per cent colonization of the roots of A. graveolens by G. mosseae, with the hyphal length density in the HC and with the number of hyphal crossings of the mesh. Using the number of crossings and assuming that all these hyphae were alive, Ames et al. (1983) were able to calculate a translocation flux for N of $7.42 \times 10^{-4} \, mol \, N/m^2/s$. This is roughly the same as the flux of P through entry-point hyphae calculated by Sanders and Tinker (1973) and, considering that the N requirement of plants is about 10 times that of P, the value seems low. A similar approach has been used with different plant–fungus combinations and the data have confirmed hyphal ^{15}N transfer to AM plants of several species (including Cucumis sativus, Trifolium subterraneum, Zea mays and Solanum esculentum) when $^{15}NH_4$ or $^{15}NO_3$ was supplied in the HC colonized by hyphae of AM fungi (Johansen et al., 1992, 1993a; Frey and Schüepp, 1993; Hawkins et al., 2000; Mader et al., 2000) (Figure 5.9). In several of these investigations, the ^{15}N in the HC became significantly depleted (Johansen et al., 1992, 1993b, 1994; Frey and Schüepp, 1993; Johansen, 1999). However, rather few of these demonstrations of N translocation and transfer have been associated with increased plant N content or growth. Johansen et al. (1992) suggested that the small physical size of the experimental systems might have contributed to this and it is certainly also true that differences between species of fungi in accessing ^{15}N have frequently been found to be related to differences in extent and distribution of hyphae in the HC, which is often also influenced by soil conditions as well as N supply itself. Nevertheless, Mader et al. (2000) showed increases in N concentration in AM tomato that were attributed to

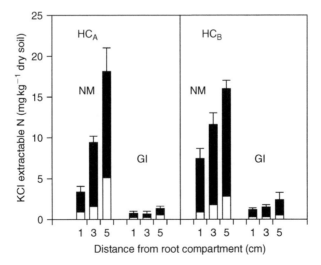

Figure 5.9 Effect of mycorrhizal hyphae of *Glomus intraradices* on depletion of KCl-extractable NH_4^+ and NO_3^- in the hyphal compartment of plants of *Cucumis sativus* fertilized (HC$_A$) or not (HC$_B$) with added N. Mycorrhizal plants, GI; non-mycorrhizal plants, NM. Bars are standard errors of the means of 4 replicates. From Johansen et al. (1992), with permission.

the ability of external hyphae to access N-enriched hyphal compartments. They calculated that up to 42% of plant N had been absorbed via the AM pathway. Similarly high transfer (around 30% of N in the roots) has been observed in monoxenic cultures of carrot roots with *G. intraradices* (Govindarajulu *et al.*, 2005). There is inconsistency in reports of the importance of the hyphal pathway for uptake of nitrate versus ammonium. In some investigations N from both sources was transferred to the plants equally, whereas in others either nitrate or ammonium was preferentially transferred. In one investigation, no difference in ^{15}N enrichment between AM (*Glomus fasciculatum*) and non-mycorrhizal plants of *Lactuca sativa* supplied with $^{15}NO_3$ under well-watered conditions was observed. However, in dry soil, the enrichment of the AM plants was four times higher, probably reflecting the much lower mobility of nitrate in dry soil and consequently reduced uptake via the direct pathway (Tobar *et al.*, 1994). Regardless of net effects of AM colonization on whole plant N status, the accumulating evidence points to significant 'hidden' N transfer via the AM uptake pathway.

AM effects on N nutrition have been studied under field conditions and the potential for increased uptake of N from soil, as well as the P mediated effects on N_2 fixation, have been demonstrated in *Hedysarum coronarium* (Barea *et al.*, 1987). In mixed plantings, a twofold increase in ^{15}N transfer from soybean to maize was observed in mycorrhizal plots, together with a relative increase in productivity of maize (Hamel and Smith, 1991). The type of mycorrhiza formed by plants also appeared to have an influence on their capacity to use nitrate in a *Banksia* woodland. At recently burnt sites nitrate predominated over ammonium and was stored and used most effectively by non-mycorrhizal and AM herbaceous plants. Woody species, including some that could be hosts to both AM and ECM fungi, appeared to use a wider range of N sources (Pate *et al.*, 1993). There is also an indication that $\delta^{15}N$ values may differ in AM and non-mycorrhizal plants, due to unknown fractionation processes (Handley *et al.*, 1993) and this might account for the variability in $\delta^{15}N$ values in plants of different mycorrhizal status and life form in the *Banksia* woodland. It might also have contributed to the problems encountered by Hamel and Smith (1991) in using natural abundance methods to elucidate the AM interactions between soybean and maize.

Nitrogen assimilation in AM roots

Details of the mechanisms involved in absorption, translocation and transfer of N to plants are beginning to be revealed. Early work focused on activities of potentially important enzymes in AM roots, but it is now recognized that the external mycelium is at least as important and that activities of intraradical and external hyphae are likely to differ. As ammonium is frequently present at very low concentrations in soil, it is thought that assimilation depends on the activity of glutamine synthetase (GS) and glutamate synthase (GOGAT), rather than glutamate dehydrogenase (GDH), because of the higher affinity of GS for ammonium (Miflin and Lea, 1976). The external mycelium certainly has GS activity (Toussaint *et al.*, 2004), as does fungal tissue separated from AM roots (Smith *et al.*, 1985). Assimilatory nitrate reductase activity was detected, albeit at very low levels, in isolated spores of two *Glomus* spp. (Ho and Trappe, 1975). Subsequently, genes encoding assimilatory nitrate reductase have been detected in AM fungi (Kaldorf *et al.*, 1994, 1998) and shown to have different expression patterns from that of the maize used as the plant symbiont. Accumulation of mRNA transcripts of the maize gene was lower in roots and shoots of AM plants and enzyme activity in shoots was also lower. However, the fungal gene was expressed in AM roots and nitrite formation was mainly NADPH-dependent. These findings are consistent with the fact that most fungi that reduce nitrate have an NADPH-dependent nitrate reductase. However, in other experiments, NADPH-dependent nitrate reductase activity in AM roots of *Trifolium subterraneum* could not be demonstrated, although there was substantial NADH-dependent activity, more characteristic of plants than fungi (Oliver *et al.*, 1983). Thus, both plant and fungal nitrate reductases could conceivably be involved in assimilation of nitrate in AM plants.

Following absorption and initial assimilation of either nitrate or ammonium, new evidence indicates substantial accumulation of amino acids and particularly arginine in the extraradical mycelium of *G. intraradices* in monoxenic cultures with carrot. The pattern of ^{15}N labelling is consistent with nitrate and nitrite reduction (when nitrate is available), followed by assimilation of ammonium via GS/GOGAT, asparagine synthase and the urea cycle (Govindarajulu *et al.*, 2005; Jin *et al.*, 2005). Most of the N is rapidly incorporated into arginine, which is the dominant amino acid in the extraradical mycelium (Johansen *et al.*, 1996; Govindarajulu *et al.*, 2005; Jin *et al.*, 2005) and shown to be transferred between hyphal and root compartments in monoxenic cultures. A surprising finding, based on labelling patterns in plant and fungal metabolites following feeding of ^{13}C-acetate or ^{15}N-arginine to the extra-radical mycelium, is that N is transferred from fungus to plant as inorganic ammonium and not an amino acid as previously suggested. Thus, the arginine delivered to the intraradical mycelium is broken down and, whereas the ammonium released is transferred to the plant, the other breakdown products are apparently recycled in the fungal tissue (Figure 5.10). At present, the roles of plant and fungal ammonium transporters in transfer to the plant are unclear, as are the nature and significance of processes involved in recycling products of arginine breakdown. Transfer of ammonium to the plant also has important consequences for expression of plant enzymes required for reassimilation to amino acids, as well as for pH regulation mechanisms to cope with the excess protons generated during the assimilation.

Uptake of other nutrients

Copper, zinc and other micronutrients

After some years, during which the importance of arbuscular mycorrhizas in micro-nutrient uptake received only limited emphasis, there is now consistent evidence that the efficiency of uptake of both Zn and Cu is increased in AM plants. Some of the earliest work showed an increase in concentration of Cu in AM apple seedlings (Mosse, 1957) and, subsequently, similar results were obtained in such diverse species as *Zea mays* (Daft *et al.*, 1975), *Avena sativa* (Gnekow and Marschner, 1989), *Phaseolus vulgaris* (Kucey and Janzen, 1987), *Allium porrum* (Gildon and Tinker, 1983) and *Trifolium repens* (Li *et al.*, 1991c). Again, using a compartmented pot system, Li *et al.* (1991c), demonstrated hyphal uptake and translocation of Cu to *T. repens*. The fungi not only contributed up to 62% of the total Cu uptake, but uptake via the AM pathway was independent of the effects of colonization on P nutrition (see below).

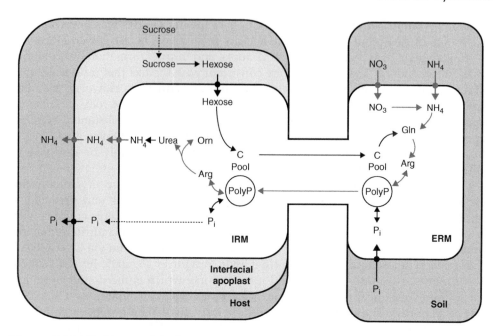

Figure 5.10 Working model of nitrogen uptake, metabolism and transfer in the AM symbiosis. The pathway of nitrogen movement is shown by grey arrows. Uptake of inorganic N from soil is followed by assimilation into amino acids, synthesis of arginine, translocation of arginine from the extraradical mycelium (ERM) to fungal structures within the root (IRM), breakdown of arginine to release ammonium and efflux of ammonium into the interfacial apoplast and subsequent uptake by plant cells. Bold arrows show active uptake processes, dashed arrows show passive processes. Reproduced from Jin et al. (2005) New Phytologist, **168**, 687–696, with permission.

Increased Cu uptake in AM plants has also been confirmed for a number of plant–fungus combinations (Killham and Firestone, 1983; Manjunath and Habte, 1988), although transfer from the fungus to the plant is often very small (Manjunath and Habte, 1988) and the mechanisms controlling transfer have not been investigated.

As with P, the mobility of Zn in soils is very low and its uptake by organisms is diffusion-limited. Consequently, similar effects of AM colonization on whole-plant uptake were expected, but proved difficult to demonstrate because of strong P–Zn interactions. It was shown at an early stage that AM colonization increased uptake of ^{65}Zn by Araucaria roots (Bowen et al., 1974) and that ^{65}Zn was translocated along hyphae of G. mosseae into T. repens growing in an agar plate system (Cooper and Tinker, 1978). The rate of ^{65}Zn translocation was 2.1×10^{-8} mol/m^2/s, considerably lower than the rate of P translocation, but probably adequate given the lower requirement of plants for this micronutrient. The tracer studies gave little quantitative information on the transfer of Zn to the plants and effects on Zn nutrition, but Zn deficiency symptoms in peach disappeared as arbuscular mycorrhizas developed (Gilmore, 1971). Subsequently, a number of studies have shown unequivocally that Zn uptake via the AM uptake pathway is important and can alleviate Zn deficiency in several species in both pot and field experiments (Manjunath and Habte, 1988; Evans and Miller, 1988; Lu and Miller, 1989; Faber et al., 1990; Thompson, 1990;

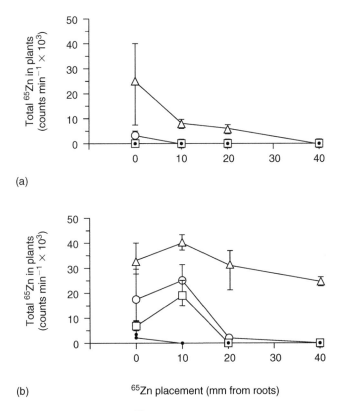

(a)

(b)

^{65}Zn placement (mm from roots)

Figure 5.11 Effect of placement of ^{65}Zn at different distances from the roots of non-mycorrhizal and mycorrhizal clover plants on the ^{65}Zn activity of the whole plants at 21 days (a) and 35 days (b). Non-mycorrhizal plants, ●; *Acaulospora laevis*, △; *Glomus sp*, ○; and *Scutellospora calospora*, □. Bars indicate standard errors of means. From Bukert and Robson (1994), with permission.

Wellings *et al.*, 1991; Bürkert and Robson, 1994). AM involvement in Zn nutrition has been implicated in the negative effects of both tillage (Evans and Miller, 1988; Fairchild and Miller, 1988) and long fallow (Thompson, 1987, 1990; Wellings *et al.*, 1991) on crop growth. In symbiosis with *Trifolium subterraneum*, *Acaulospora laevis* was shown to obtain ^{65}Zn from distances up to 40 cm from the plants, whereas *Glomus* spp. (WUM 10(1)) and *Scutellospora calospora* (WUM 12(2)) were less effective. As with P uptake, the distribution and length density of hyphae in soil were important contributors to the differences between the fungi (Bürkert and Robson, 1994) (Figure 5.11). These findings were largely confirmed by Jansa *et al.* (2003), who demonstrated concurrent transfer of ^{32}P and ^{65}Zn to maize, via *G. intraradices* from up to 14 cm. Not unexpectedly, Zn transfer was again much lower than P transfer, but there was a strong linear correlation between the amounts of P and Zn transported, substantiating the importance of hyphal length density in soil in the acquisition of both nutrients.

Surprisingly, AM colonization is also implicated in reducing Zn accumulation and hence plant toxicity in soils with high Zn content (Li and Christie, 2001; Zhu *et al.*, 2001; Burleigh *et al.*, 2003). The mechanism is unclear, but is presumably different from that promoting Zn uptake under conditions of deficiency. In at least one investigation (Li and Christie, 2001), the protective effect was independent of indirect effects due to increased growth and P uptake in AM plants (see below).

Few data exist on the molecular and physiological mechanisms underlying either increased or decreased Zn accumulation. A Zn transporter cloned from *G. intraradices* has characteristics that suggest a role in protecting the fungus from Zn stress, rather than in uptake of an essential nutrient (Gonzalez-Guerrero *et al.*, 2005). Similarly, an investigation of a plant Zn transporter from *Medicago truncatula* (*MtZIP2*) showed that it is localized to the plasma membrane and that its expression is downregulated by AM colonization, but strongly upregulated by increasing Zn supply (Burleigh *et al.*, 2003). This pattern suggests that it is not involved in the AM pathway for Zn uptake, but its role has not been clarified.

Uptake of other micronutrients via external hyphae is not so well established (Marschner and Dell, 1994) and, in fact, the uptake of Mn is most commonly reduced in AM plants. This effect was attributed to lower Mn^{+IV} reducing potential in the rhizosphere of AM plants, probably because the populations of microorganisms responsible were lower (Kothari *et al.*, 1991).

Interactions between P fertilization and deficiencies of trace elements are well known in several species of typically AM plants (Wallace *et al.*, 1978). In general, when the availability of P is increased, P uptake and plant growth also increase. Concentrations of Cu and Zn in the tissues fall, sometimes to levels at which deficiency symptoms become apparent. AM colonization has been shown to affect these interactions (Lambert *et al.*, 1979; Timmer and Leyden, 1980), so that at moderate levels of P fertilization deficiency symptoms are alleviated because AM fungi increase uptake of the trace elements and tissue concentrations rise. At very high levels of P, AM colonization itself may be reduced (see Chapter 2) with consequent reductions in AM uptake and reappearance of the deficiency symptoms. Interactions such as these may be involved in some cases of alleviation of Zn toxicity in polluted sites (Dueck *et al.*, 1986). If the sites are P deficient, then AM P uptake could result in increased growth and dilution of Zn in the tissues. However, as shown above, this cannot be the explanation in all cases.

Potassium

Analyses of K concentrations in plant tissues have occasionally indicated increases in K uptake in AM plants, which might be expected considering the relative immobility of this ion in soil (Mosse, 1957; Holevas, 1966; Possingham and Groot Obbink, 1971; Huang *et al.*, 1985). However, in the majority of investigations, K was found to be at lower concentrations in the tissues of AM than in those of non-mycorrhizal plants. Extrapolation from tissue concentrations can be dangerous, as we have seen for other nutrients, because of the simultaneous effects of P nutrition on growth. Smith *et al.* (1981) observed elevated concentrations of K in shoots (but not roots) of AM *Trifolium subterraneum* when plants were grown on P-deficient soils. If sufficient P was supplied to soil to remove any AM growth response, then K concentrations in both groups of plants were very similar. This suggests an indirect effect of arbuscular mycorrhizas on K uptake in P-deficient plants, similar to effects observed with sulphate (Rhodes and Gerdemann, 1978b). However, K depletion in HCs colonized by hyphae of *Glomus mosseae* accompanied by increased accumulation in associated

AM *Agropyron repens* has been observed (George *et al.*, 1992), so that direct effects are also possible.

Accumulation of K is strongly influenced by the form of N available (nitrate or ammonium), as well as by other cations, particularly Na^+. It might also be influenced by the synthesis and storage of polyP, so that carefully designed experiments to investigate the influence of AM colonization on K nutrition need to take all these potentially confounding factors into account.

Toxic elements

Some essential elements are required in very small quantities but, when accumulated at high concentrations in plants, may become toxic. Consequently, heavy metal toxicity may derive from excessive uptake of Zn, Cu, Fe and Co as well as from other elements and ions which are normally regarded as toxic (e.g. Pb, Cd, Cr, Ni, Ti, Ba and As). General aspects of the interactions between fungi, including AM fungi, and these metals have been reviewed (Gadd, 1993; Meharg, 2003). Results of different AM investigations are not always consistent, with interpretations often made difficult because of concurrent effects of colonization on plant nutrition and growth. Large effects of arbuscular mycorrhizas in increasing accumulation of Cu, Ni, Pb and Zn have been found (Killham and Firestone, 1983), but many reports show the converse or show that whereas total accumulation was increased, the actual concentrations and hence potentially toxic effects were reduced (Weissenhorn *et al.*, 1995; Karagiannidis and Hadjisavva-Zinoviadi, 1998; Joner and Leyval, 2001; Andrade *et al.*, 2003, 2004; Janouskova *et al.*, 2005). It is obvious that interactions with P nutrition and growth, together with other aspects of AM physiology, must be taken into account in studies of accumulation and tolerance of toxic elements. Tissue dilution of a toxic element can occur as a consequence of improved P nutrition and increased plant growth, even in situations where the uptake per plant is actually increased (El-Kherbawy *et al.*, 1989). Only measurements of inflow, use of plants matched for growth and P nutrition, or compartmented systems can determine whether the hyphae of AM fungi are directly involved in uptake of toxic elements and their transfer to plants. Interpretations of mechanisms underlying effects have sometimes been made difficult when test soils contain high levels of several contaminants and have also been coloured by the rationale of experiments. Thus, increases in total uptake have been seen as potential benefits for phytoremediation programmes and reductions in concentrations as benefits in decreasing entry of toxic metals into food chains.

External hyphae of AM fungi have been shown, in compartmented systems, to be capable of absorbing and translocating not only Cu and Zn, but also Cd, Ni and U. Effects on uptake of other metals, such as Pb, have been inferred from changes in whole plant uptake. Lee and George (2005) used a compartmented pot system (see Figure 5.3a) to supply metals and/or P to external mycelium of *G. mosseae* growing with cucumber. They found that concentrations of Zn and Cu, as well as P, were elevated in AM plants, demonstrating once again the role of hyphae in metal transfers. Both Cd and Ni concentrations were reduced in shoots of AM plants, suggesting reduced transfer from roots to shoots. However, interpretations of the data were rendered somewhat difficult by the marked increases in growth of AM, compared with non-mycorrhizal plants.

Uptake by AM fungi often results in increased concentrations of elements in roots, but decreased transfer to shoots (Joner and Leyval, 1997; Rufyikiri *et al.*, 2004; Chen *et al.*, 2005a). One investigation of U and As uptake by the As hyperaccumulator *Pteris vittata* from contaminated soil (Chen *et al.*, 2005b) showed that AM colonization increased U accumulation in roots; transfer to shoots was also slightly increased, although overall the amounts transferred were low. This contrasted with earlier findings with barley, where root to shoot transfer was reduced in AM plants (Chen *et al.*, 2005a). In *P. vittata*, AM colonization had no effect on As concentrations, even though plant growth depressions might have 'concentrated' the metalloid in the tissues. Total As in AM plants was therefore lower than or the same as non-mycorrhizal plants, depending on plant dry weights.

Mechanisms to explain effects of AM fungi in reducing shoot accumulation have been reviewed by Meharg (2003) and include surface binding by the hyphae (Joner *et al.*, 2000), changes in bioavailability in soil (Citterio *et al.*, 2003) and the possibility that metals may be sequestered in the hyphae and hence not transferred to the plants (Christie *et al.*, 2004). Electron energy loss spectroscopy of AM roots of *Pteridium aquilinum* showed greater accumulation of Cd, Ti and Ba in the fungal structures than in the root cells themselves (Turnau *et al.*, 1993). It was suggested that sequestration of the metals by polyP in the fungus might have been important in minimizing transfer to the plant, but this requires confirmation. No increases in metal transport by *G. mosseae* were observed as P supply to the mycelium and P transfer to plants increased. This suggests that there is no direct link between P and metal transfers (Lee and George, 2005). Using a compartmented pot system, Joner and Leyval (1997) showed that, although ^{106}Cd uptake by roots alone was not significantly influenced by the AM status of the plant, the AM fungus took up considerable ^{106}Cd from an HC. However, much of the Cd remained in the roots and again was probably sequestered in the fungal tissues, preventing transfer to the plant. Similarly, external mycelium of *G. lamellosum* can take up and translocate ^{137}Cs in a monoxenic culture system and also apparently sequestered the radionuclide in the hyphae (Declerck *et al.*, 2003).

The possibility that toxic element accumulation might result in damage to the fungi and consequently have significant negative effects on AM mediation of P or Zn uptake has also been addressed. Rufyikiri *et al.* (2004) showed that U appears to have no effect on AM colonization of *T. subterraneum* in spiked soil, but both Cd and Pb, as well as contaminated soil containing several different toxic metals, have been reported to have negative effects on colonization and/or mycelial growth or spore production (Joner and Leyval, 2001; Andrade *et al.*, 2003, 2004; Tullio *et al.*, 2003; Janouskova and Vosatka, 2005). Not unexpectedly, variations between different AM fungal species and sources of the inoculum have been observed and there is evidence that prolonged exposure to Cd and other toxic elements can result in the development of tolerance in *Glomus* spp., but again the mechanism is not known (see Weissenhorn *et al.*, 1994; Tullio *et al.*, 2003).

The significance of arsenic (As) as a major contaminant of water and soil is increasingly being recognized and efforts made to reduce transfer to human food chains (Meharg, 2004). Arsenate tolerance in plants is most interesting because it appears to depend on modifications of the P uptake system (which also transports arsenate) and hence the root absorbing capacity (Meharg, 1994). Reduced As uptake in some non-mycorrhizal plants (e.g. *Holcus lanatus*) involves constitutive downregulation of expression of high-affinity P transporters in roots (Meharg and Hartley-Whitaker, 2002) and P uptake by tolerant lines appears to occur via low-affinity transporters, at least in solution culture (Meharg and MacNair, 1992). However, these transporters are unlikely to be of much significance in soil, where high-affinity uptake is essential to absorb P from solution concentrations that rarely exceed $10\,\mu$M. Tolerant genotypes of *H. lanatus* are more highly colonized by AM fungi in the field than non-tolerant genotypes, leading to the suggestion that the tolerant plants are dependent on the fungi for P uptake and consequent success (Meharg *et al.*, 1994). Furthermore, survival of As-tolerant plants in contaminated soil points to the possible involvement of AM symbiosis in compensating for loss of high-affinity P uptake systems and poor root growth as a consequence of As toxicity. If correct, this explains some observations that As tolerance interacts with AM effects on plants and that AM plants can be more tolerant to As than non-mycorrhizal plants (Meharg *et al.*, 1994; Fitter *et al.*, 1998b; Wright *et al.*, 2000; Gonzalez-Chavez *et al.*, 2002; Liu *et al.*, 2005a, 2005b).

Interplant transfer of nutrients

Hyphal links between plants offer potential pathways for the movement of soil-derived nutrients, just as they do for plant-derived C and could play important roles in interplant and interspecies competition and redistribution of nutrients in ecosystems. Consequently, the technical problems in determining whether observed tracer movement of nutrients is likely to be nutritionally or ecologically significant need to be overcome (Newman, 1988; Miller and Allen, 1992).

The survival of AM hyphae within dead or dying roots could lead to rapid uptake and transfer of nutrients released by autolysis or by the activity of microorganisms to plants linked by a common mycelial network (CMN). Transfer of mineral nutrients from one living plant to another via a CMN is more problematic. There is certainly evidence for the transfer of trace amounts of ^{32}P and ^{15}N from 'donor' to 'receiver' plants (Whittingham and Read, 1982; Ritz and Newman, 1985; van Kessel *et al.*, 1985; Newman and Ritz, 1986; Haystead *et al.*, 1988; Hamel and Smith, 1991; Johansen and Jensen, 1996), but the amounts were often small and results did not show net movement of the nutrient in question. Francis *et al.* (1986) did not use tracers, but instead supplied either nutrient solution or water to one half of the root system of the donor plants (either *Plantago lanceolata* or *Festuca ovina*). Receivers were grown with the other half of the root system, either inoculated with AM fungi or uninoculated. AM receivers (*P. lanceolata* or *F. ovina*) responded positively in both growth and nutrient content when nutrients were supplied to the donors, but the non-host *Arabis hirsuta* and non-mycorrhizal *P. lanceolata* and *F. ovina* did not, apparently indicating net transfer of nutrients from the larger nutrient-sufficient donors to the receivers via the CMN. However, Newman (1988) suggested that these results could be explained not by direct transfer but by changes in the competitiveness of the donor plants as a result of differences in nutrient supply to them, with large nutrient-sufficient donors competing less strongly with the 'receivers'.

Newman's suggestion is supported by data of Ocampo (1986) indicating that competition could have been the dominant effect in nutrient redistribution between large and small plants of *Sorghum vulgare*. Eissenstat (1990) re-examined this question using both nutrient applications and ^{32}P and ^{15}N in a compartmented pot system. By measuring the ratio of tracer in the receiver to tracer in the donor, he concluded that P transfer between individuals of *P. lanceolata* was increased following P fertilization, while N transfer was unaffected. In quantitative terms, the transfer of P was too small to have any effect on plant growth, whereas the transfer of N was about 10-fold higher, as would be expected from the relative requirements of plants for these nutrients, and could have had an effect on the receivers in very nutrient deficient soils. His overall conclusion was again that alteration in the competitive balance, rather than direct transfer was the most important effect. This conclusion is further supported by the results of Johansen *et al.* (1992), who found that although external hyphae of *Glomus intraradices* effectively depleted total N from soil in a hyphal compartment (HC) and transferred it to plants of *Cucumis sativus*, very little ^{15}N was transferred via the plant to a second hyphal compartment. This work was followed up in an experiment investigating transfer of both N and P between barley and pea (Johansen and Jensen, 1996). The compartmented system was devised to prevent roots intermingling and hence eliminate direct transfer and prevent competition. Transfer of P and N in both directions was assessed using long-term labelling of donor plant roots with either ^{15}N or ^{33}P. Transfer was very low between intact plants and occurred bidirectionally, so that net transfer was insignificant compared with the needs of the plant. If the shoots of the donor plants were cut off, transfer was slightly increased, most probably due to mineralized N and P becoming available to AM hyphae as the roots decomposed. Indeed, root turnover and consequent nutrient release could explain the small transfer between intact plants. The combined results indicate that net P or N transfer between plants via a CMN is probably ecologically negligible, but that nutrient cycling from decaying plant tissues may be short-circuited. Furthermore, our knowledge of nutrient transfer across symbiotic interfaces suggests that movement of N and P is strongly polarized from fungus to plant. No mechanisms supporting bidirectional movement of soil-derived mineral nutrients across living interfaces have yet been revealed (Smith and Smith, 1990a, 1990b). The lack of convincing data supporting net transfer of nutrients between plants, underlines the fact that very carefully designed experiments are required in this area.

Water relations

Mosse and Hayman (1971) observed that AM *Allium cepa* did not wilt when transplanted, but that non-mycorrhizal plants did. Subsequently, several similar observations have been made (Busse and Ellis, 1985; Huang *et al.*, 1985) and there is no doubt that AM colonization does affect the water relations of plants (Augé, 2001; Ruiz-Lozano, 2003). As with other aspects of the physiology of AM plants, it is relevant to distinguish direct effects of fungal colonization from indirect effects resulting from changes in plant size or P status. The subject is complex and there are many inconsistencies in the literature, not all of which can be easily explained (Nelsen, 1987; Fitter, 1988; Koide, 1993). The relationships between AM colonization and plant water relations were comprehensively reviewed by Augé (2001), who concluded that AM effects on water relations included direct effects, as well as effects that were strictly related to changes in plant nutrition and size. In field situations, size effects can be important in plant survival and success, particularly in species that avoid rather than tolerate drought. Augé also highlighted the fact that we know very little about variations in water relations consequent on different plant–fungus combinations. However, some recent work has included a number of different AM fungi in investigations of responses to water stress. Considerable diversity in outcome has been revealed (Marulanda *et al.*, 2003; Aroca *et al.*, 2007), which may go some way to explaining apparently contradictory results in earlier investigations.

The influence of AM colonization on plant water relations was first investigated systematically with soybean (Safir *et al.*, 1971, 1972). Results showed that AM plants had lower resistances to water transport than non-mycorrhizal plants and, in this instance, it appeared that most of the difference was attributable to changes in root resistance, for shoot resistances were small and did not differ in the two groups of plants. The conclusion was that the effect was probably due to improved nutrition, because the differences could be eliminated if nutrients were supplied or fungicide applied (Safir *et al.*, 1972). Transpiration rates of AM plants are generally higher than those for non-mycorrhizal plants (Allen *et al.*, 1981; Allen, 1982; Nelsen and Safir, 1982; Huang *et al.*, 1985; Koide, 1985b; Fitter, 1988). As Koide (1993) showed, the discussions and inconsistencies centre on whether the increases are due to increased stomatal conductance or whether decreased resistance to water transport in the below-ground system is also important, as suggested by the data of Safir (Safir *et al.*, 1971, 1972). Levy and Krikun (1980) used *Citrus jambhiri* and *Glomus* spp. with growth and fertilizer conditions that permitted the comparison of AM and non-mycorrhizal plants of similar size and growth rate. The major effect of AM colonization was an increase in transpirational flux and stomatal conductance, both during stress and recovery. There were apparently no differences between the two groups of plant in terms of resistance of the root to water movement. Similar conclusions were reached with several different plant species (Allen *et al.*, 1981; Allen, 1982; Huang *et al.*, 1985; Koide, 1985b; Fitter, 1988). Again, transpiration rates were increased in AM plants, while stomatal resistances were greatly reduced. Both Koide (1985b) and Fitter (1988) came to the conclusion that the high stomatal resistance in P deficient non-mycorrhizal plants was a nutritional effect and Figure 5.12 shows the good correlation between stomatal conductance and leaf P concentration in *T. pratense*, regardless of whether the differences in P nutrition were induced by fertilization or AM colonization. However, there have been suggestions that stomatal behaviour is influenced by those hormonal changes in the plant that certainly occur as a result of either changes in P nutrition or AM colonization *per se* (Allen *et al.*, 1980, 1982; Allen, 1982; Dixon *et al.*, 1987; Baas and Kuiper, 1989; Danneberg *et al.*, 1993; Druge and Schonbeck, 1993).

In *Bouteloua gracilis*, the changes in stomatal conductance could not be explained in terms of differences in gross anatomy or morphology and there were no changes in size of mesophyll cells, bundle sheath cells or stomata, nor was the stomatal density altered following AM colonization. Resistance to water transport was not separated into contributions of root and shoot by Allen *et al.* (1981), but they did comment that increased branching of the roots in AM plants could lead to substantial increases in root surface area without changes in root biomass and that this might reduce the root resistance to water uptake. However, Koide (1985b) measured the hydraulic resistance to water transport between soil and leaf and between soil and stem below the leaf (root plus stem resistance) over a range of transpiration rates in well watered soils. When AM and non-mycorrhizal plants of the same size

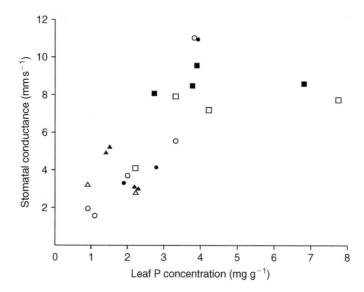

Figure 5.12 Relationship between stomatal conductance and tissue P concentration, compiled by Fitter (1988). Symbols represent *Rosa* (△), *Helianthus*, (○) and *Trifolium* (□). Mycorrhizal plants, solid symbols and non-mycorrhizal plants, open symbols. Reprinted from the *Journal of Experimental Botany* by permission of Oxford University Press.

and root length were compared, there were no differences in hydraulic properties and the effects of arbuscular mycorrhizas could be entirely explained on the basis of decreased stomatal resistance. Increased stomatal conductivity and increased transpiration rates in *Leucaena* were accompanied by higher xylem pressure potentials and the stomata responded more rapidly to changes in humidity. The AM plants lowered the water potential of the soil in the pots, indicating much higher water uptake than into the non-mycorrhizal plants (Huang *et al.*, 1985). The conclusion in this case was that higher water use in AM plants was offset by increased C gain, because the stomata remained open for a greater part of the day. Large root systems and rapid stomatal response were also important in the overall water economy.

The results of a number of investigations have, however, suggested that AM colonization can reduce the hydraulic resistance to water uptake in the roots (Hardie and Leyton, 1981; Nelsen and Safir, 1982; Graham and Syvertsen, 1984; Bildusas *et al.*, 1986; Aroca *et al.*, 2007), although this appears to be a less common outcome. On a whole plant basis, this effect could be mediated by increases in the size or branching of the root system. Hardie and Leyton (1981) could only attribute part of the decrease in hydraulic resistance of the root systems of *Trifolium pratense* colonized by *Glomus mosseae* to this effect. They therefore concluded that hyphal growth in the soil was important in reducing root resistance to water flow and this point is discussed further below. However, Koide (1993) has emphasized that, because root hydraulic conductivity is not linearly related to root system size (Ficus and Markhart, 1979), it is not valid to compare values obtained from root systems of different sizes, even if the results are expressed per unit root length. Consequently, only data from matched AM and non-mycorrhizal plants should be used for comparison and, when this requirement was met, either no effects were observed (Graham and Syvertsen, 1984; Koide, 1985b; Graham *et al.*, 1987) or there was a decrease in hydraulic conductivity (Levy *et al.*, 1983). Koide (1985b) further noted that apparent effects of AM colonization on the root hydraulic conductivity could have been recorded because this parameter varies with transpirational flux, which is normally higher in AM plants. Studies of plasma membrane aquaporins (PIPs), which facilitate water flux through the membrane, have indicated that expression is often lower in AM plants and that such plants responded to drought by further down-regulation (Porcel *et al.* 2006; Ruiz-Lozano *et al.* 2006). These findings suggest effects of AM colonization operate at cell membrane level, as well as at tissue and whole plant levels. Again, more work is required.

The possible roles of hyphae of AM fungi in water uptake and transfer to plants requires examination, despite the fact that there is little unequivocal evidence that would implicate these processes in water relations of AM plants. Hardie (1985) studied transpirational flux in non-mycorrhizal and AM *Allium porrum* and *T. pratense* that were not nutrient limited. AM clover had slightly higher transpirational fluxes and lower stomatal resistances than the non-mycorrhizal controls. Removal of external hyphae did not affect the stomatal resistance, but appeared to reduce the transpirational flux, although the differences were not significant. She used the maximum reduction in transpirational flux caused by hyphal removal and transplanting (9.9×10^{-9} l/m root length/s) to calculate that damage to forty hyphal entry points per metre of colonized root would account for this reduction if entry points had the same water flux as *Phycomyces blakesleeanus*. This was a bold assumption, given the lack of the significance of the data. However, there is some evidence for depletion of soil water, both in single compartment pots and in HC of compartmented systems, implying hyphal uptake and transport of water (Faber *et al.*, 1991; Marulanda *et al.*, 2003). Faber calculated an apparent water flux through hyphae of 375×10^{-9} l/h per hypha crossing the mesh, or using internal radius of hyphae as 5 μm, 1.32 l/m^2/h. Kothari *et al.* (1990) also examined the question using *Zea mays*. In this case, the difference in transpiration rates between AM and non-mycorrhizal plants was considerable and could largely be attributed to increased leaf area. However, root lengths in AM plants were reduced by 31% and in consequence the water uptake per unit length of root was much higher than in non-mycorrhizal plants ($\Delta 7.3 \times 10^{-9}$ l/m/s); a value similar to that found by Hardie (1985). Assuming 60% colonization of the root system and 1000 entry points (of 10 μm diameter) per metre of colonized length (Kothari *et al.*, 1990), the cross-sectional area of hyphae through which the water would enter can be estimated as 4.7×10^{-8} m^2/m and the flux of water as 0.15 l/m^2/s, which is an order of magnitude less than the estimate of Faber *et al.* (1991). Kothari *et al.* (1990) calculated that if P was delivered in the hyphae by mass flow of a solution of concentration 16 mmol/l, the inflow should have been 10×10^{-10} mol/m/s. This is one or two orders of magnitude higher than the actual inflow, which was in the range 6.2–22.0×10^{-12} mol/m/s. They concluded that the apparent bulk flow of water in hyphae did not seem to make an important contribution to P inflow. The conclusion that mass flow of solution in the hyphae did not occur to any great extent was supported by the fact that consumption of water from the hyphal compartments in their experiments was negligible. Furthermore, George *et al.* (1992) confirmed these findings and showed no difference in water depletion in a hyphal compartment, regardless of whether the plants were well watered or water stressed, or whether the hyphae were cut.

Recently, differential effects of different fungal species on soil drying have been highlighted, which go some way to explaining the diverse findings. Using *Lactuca sativa*, well matched for growth between AM and non-mycorrhizal treatments, it was shown that six different AM fungi had quite different effects on soil drying and that the differences were related to development of external mycelium in soil (Marulanda *et al.*, 2003). The results were interpreted in terms of differences in water uptake and transfer by the hyphae. However, there is a possibility that hyphal effects on soil properties contribute to the results. Using wild-type *Phaseolus vulgaris* that forms normal arbuscular mycorrhizas and an AM deficient mutant that does not, Augé *et al.* (2004) were able to show that AM effects on stomatal conductance and drought tolerance could be partitioned into those that were the result of actual colonization of the roots and those that followed from growth of the external mycelium in soil. A noteworthy feature was that mutant (non-mycorrhizal) plants grown in 'mycorrhizal' soil took longer to reach stomatal closure during drought than those grown in 'non-mycorrhizal' soil. This investigation provides confirmation of the importance of external AM hyphae in influencing soil structure and hence water relations, as well as a partial explanation for the relationship between hyphal length density and soil drying found by Marulanda *et al.* (2003).

AM colonization may thus have effects on drought tolerance that are not directly related to plant water relations. Available water is higher in well structured soils, so that the well-documented effects of AM colonization on soil structure and aggregate stability would be likely to play an important role (Tisdall and Oades, 1982; Miller and Jastrow, 2002b; Rillig and Mummey, 2006). AM fungi also probably influence soil conductivity, in the same way as root hairs, by prevention of formation of air gaps and maintaining contact between roots and water-filled pores, as well as more general effects on structure. Changes in soil moisture characteristic curves, as well as improved structure in soils well colonized by external AM mycelium, compared with 'non-mycorrhizal' soils have been documented (Augé, 2001, 2004). Nutrients, as we have seen, become less and less available as soil dries because of the increasing tortuosity of the diffusion path. Growth of non-mycorrhizal plants is likely to be increasingly limited by nutrient availability under drought conditions and reduced root growth would limit the accessibility of water. Hyphal contribution to uptake of nutrients in AM plants would then become more and more important. Indeed, Neumann and George (2004) showed that AM sorghum was much better able to access P from dry soil than non-mycorrhizal plants. Part of the effect they observed can be attributed to higher soil moisture as a consequence of hydraulic lift in which hyphae as well as roots are involved in redistributing soil water along water potential gradients in the soil.

Conclusions

There is excellent evidence demonstrating that external hyphae of AM fungi absorb non-mobile nutrients (P, Zn, Cu) from soil and translocate them rapidly to the plants, thus overcoming problems of depletion in the rhizosphere which arise as a consequence of uptake by roots. Transfer across the symbiotic interface often results in increased nutrient acquisition by the plant. These processes act in series and, with efficient fungi producing extensive external mycelium, lead to depletion of nutrients in the soil well beyond the rhizosphere. The contribution of AM uptake to whole plant nutrition has been shown to be significant not only in plants that respond positively to colonization, but also in non-responsive or negatively responsive ones. Molecular and physiological mechanisms underlying the integration of plant and fungal uptake pathways are increasingly being revealed. The new work showing how fungal and plant P uptake mechanisms are integrated at the physiological and molecular levels should change the way we think of the roles of mycorrhizas in plant nutrition, not only in simple pot experiments but also in field situations. It must be recognized that the symbiotic uptake plays a significant role much more frequently than has been thought in the past and that interpretation of experiments must take potentially 'hidden' AM nutrient uptake into account. N nutrition is a case in point. New work has revealed operation of both AM uptake of N, as well as underlying mechanisms, but quantitative significance remains to be demonstrated.

The sources of P (and other nutrients) available to the fungi are less clear. The soil solution, in equilibrium with the so-called labile fraction, must be the primary source. Hyphae are able to penetrate soil pores inaccessible to roots and may also be able to compete effectively with soil-inhabiting microorganisms for recently mineralized nutrients. Rapid removal of P from solution at sites of dissolution will accelerate the use of P that exchanges rapidly with the solution. Evidence that AM fungi directly access sources of P or N unavailable to plants is rather slim, but work continues particularly in relation to mobilization of organic nutrient sources by fungal enzymes. It also appears that localized alterations in pH might play a role in increased P mobilization in microsites. Production of chelating agents that would increase the availability of Fe or Al phosphates has not been demonstrated, despite the apparent differences in accessibility of these P sources to AM and non-mycorrhizal plants.

The interactions between AM colonization and accumulation of heavy metals and other toxic elements is an area of considerable interest in relation both to production of safe food and bioremediation programmes. A number of different mechanisms may be involved, including tissue dilution of the toxic element due to interactions with P nutrition and growth, sequestration of the toxic metal in the fungus and development of tolerance by the fungi.

Water relations of plants are modified in various ways by AM interactions. AM and non-mycorrhizal plants grown in soil well colonized by AM hyphae often have higher stomatal conductance and higher drought resistance. Effects on root hydraulic properties have been observed less often. The mechanisms are difficult to sort out, but direct effects as well as those related to changes in nutritional status and to changes in soil properties have been demonstrated. There is no clear evidence for actual water transport via the fungal hyphae, but this remains a possibility.

Section 2
Ectomycorrhizas

6

Structure and development of ectomycorrhizal roots

Introduction

An ectomycorrhizal (ECM) root is characterized by the presence of three structural components: a sheath or mantle of fungal tissue which encloses the root, a labyrinthine inward growth of hyphae between the epidermal and cortical cells called the Hartig net, and an outwardly growing system of hyphal elements (the extraradical or external mycelium) which form essential connections both with the soil and with the sporocarps of the fungi forming the ectomycorrhizas.

Almost all of the plants upon which ectomycorrhizas develop are woody perennials. The anatomical structures of the mantle and of the emanating mycelia, are stable at least at the level of the fungal genus and are increasingly used to facilitate characterization of ectomycorrhizas (Agerer, 1987–2002; Ingleby *et al.*, 1990; Goodman *et al.*, 1998). This type of organ is also clearly distinguishable from all other types of mycorrhiza on the basis of the absence of intracellular penetration by the fungus. In the event of penetration of healthy root cells by an ECM-forming fungus, whether from the Hartig net or from the hyphae of the sheath, the structure is referred to as an ectendomycorrhiza (see below and Chapter 7). The identity of the plant can influence the outcome. Some fungi, for example the ascomycete *Wilcoxina mikolae*, routinely produce ectendomycorrhizas on young plants of *Pinus* and *Larix* in nursery soils while forming ectomycorrhizas on *Abies, Picea* and *Tsuga* (Mikola, 1988). However, most ECM fungi are capable of forming intracellular penetrations in senescent parts of the root axis, or when the nutrient balance of the association is disturbed. In these circumstances, the fungus appears to be behaving in a weakly pathogenic manner. Many fungi that produce typical ectomycorrhizas on members of the Pinaceae and Fagaceae form extensive intracellular growths as well as Hartig net and sheath in certain ericaceous hosts. This type of colonization is recognized as being of the distinct 'arbutoid' category (see Chapter 7).

While the presence of the three structural elements signifies an ectomycorrhiza, there may be considerable variation in the extent to which Hartig net, mantle and extraradical mycelium develop. Indeed, structures that have only a patchy mantle, as in some Asteraceae (Warcup, 1980), that lack a mantle altogether, as in *Pinus* spp. colonized by *Tricholoma matsutake* (Ogawa, 1985; Yamada *et al.*, 2006), or lack a Hartig net, as in the roots of *Pisonia grandis* (Ashford and Allaway, 1982), have been referred to as 'ectomycorrhiza'. The danger in broadening the category to this extent is that relationships between structure and function established by study of 'typical' forms may break down. It is noteworthy in this context that no evidence for positive growth response to colonization was reported in any of the above studies. Of the many studies in which symbioses between *T. matsutake* and its hosts have been experimentally established, only one (Guerin-Laguette *et al.*, 2004) has shown stimulation of seedling growth.

Phylogenetic studies described below strongly suggest that the ECM habit arose independently on a number of occasions and over a considerable time span, as fungal saprotrophs formed symbiotic partnerships with autotrophs. In such circumstances, the range of structural and functional relationships observed within the ECM consortium may reflect different stages in the evolutionary transition from the saprotrophic towards mutualistic lifestyles. In view of the dynamic evolutionary state of the ECM condition, as well as the diversity of ecological niches and of plant–fungus combinations in which it occurs, it should not be surprising that a wide range of structural variants is found. However, it is important not to lose sight of the fact that the vast majority of associations defined as ectomycorrhizal conform to the basic pattern in which a mantle, Hartig net and some, if only seasonal, development of extraradical mycelium are all present, while intracellular penetration is scarce. These well-conserved features appear to have been favoured because they confer particular functional attributes that further distinguish them from types not having all of these defining characteristics.

Taxonomic, evolutionary and geographic aspects of the ectomycorrhizal symbiosis

The plants forming ectomycorrhizal associations

Table 6.1 lists examples of families and genera within which ECM colonization has been reported. Some are exclusively ECM, but others may also form arbuscular mycorrhizas and, indeed, this may be the typical mycorrhizal type for the taxon (see Chapter 1). While a relatively small number, probably around 3% (Meyer, 1973), of phanerogams (seed plants) are ECM, their global importance is greatly increased by their disproportionate occupancy of the terrestrial land surface and their economic value as the main producers of timber. Thus the Pinaceae, members of which form the major component of the vast boreal forests of the northern hemisphere, the Fagaceae, dominants or co-dominants of the northern and some southern hemisphere temperate forests, as well as tropical forests of South-East Asia, are predominantly ECM species. The occurrence and importance of ECM trees in the wet tropics has been consistently underestimated. In the Dipterocarpaceae alone, there are over 500 species, all of which are frequently ECM (Alexander and Högberg, 1986; Taylor and Alexander, 2005) (see Chapters 1 and 15). Tropical caesalpinoid legumes are also characteristically ECM (Alexander, 1989a, 1989b), most members of the tribe Amherstieae and some genera such as *Afzelia, Intsia* and *Eperua* in the tribe Detareae have this type of association. Groves, as well as extensive mono-dominant stands, of caesalpinoid genera such as *Tetraberlinia, Microberlinia* and *Jubernaldia* occur in

Table 6.1 Genera reported to contain at least one species on which ectomycorrhiza has been described.

Family	Genus	Family	Genus
Aceraceae	B *Acer*	Hammamelidaceae	*Parrotia*
Betulaceae	B *Alnus*	Juglandaceae	B *Carya*
	B *Betula*		B *Juglans*
	B *Carpinus*		*Pterocarya*
	B *Corylus*	Caesalpinoideae	*Afzelia*
	B *Ostrya*		*Aldina*
	B *Ostryopsis*		*Anthonota*
Bignoniaceae	*Jacaranda*		*Bauhinia*
Caprifoliaceae	B *Sambucus*		*Brachystegia*
Casuarinaceae	B *Casuarina*		*Cassia*
	B *Allocasuarina*		*Eperua*
Cistaceae	B *Helianthemum*		*Gilbertiodendron*
	B *Cistus*		*Julbernardia*
Compositae	B *Lactuca* (Mycelis)*		*Monopetalanthus*
Cyperaceae	B *Kobresia**		*Paramacrolobium*
Dipterocarpaceae	B *Anisoptera*		*Swartzia*
	B *Balanocarpus*	Mimosoideae	*Acacia*
	B *Cotylelobium*	Papilionoideae	*Brachysema*
	B *Dipterocarpus*		*Chorizema*
	B *Dryobalanops*		*Daviesia*
	B *Hopea*		*Dillwynia*
	B *Monotes*		*Eutaxia*
	B *Shorea*		B *Gompholobium*
	B *Valica*		B *Hardenbergia*
Elaeagnaceae	*Shepherdia*		*Jacksonia*
Epacridaceae	*Astroloma*		*Kennedya*
Ericaceae	*Arbutus*		B *Mirbelia*
	Arctostaphylos		B *Oxylobium*
	Chimaphila		*Platylobium*
	Gaultheria		*Pultenaea*
	Kalmia		B *Robinia*
	Ledum		B *Vicia*
	Leucothoe		B *Viminaria*
	Rhododendron	Myricaceae	*Comptonia*
	Vaccinium		*Myrica*
Euphorbiaceae	*Poranthera*	Myrtaceae	B *Angophora*
	B *Uapaca*		B *Callistemon*
Fagaceae	B *Castanea*		B *Campomanesia*
	B *Castanopsis*		B *Eucalyptus*
	B *Fagus*		B *Leptospermum*
	B *Lithocarpus*		B *Melaleuca*
	B *Nothofagus*		B *Tristania*
	B *Pasania*	Nyctaginaceae	B *Neea*
	B *Quercus*		B *Torrubia*
	B *Trigonobalus*		B *Pisonia*
Gentianaceae	*Bartonia*	Oleaceae	B *Fraxinus*
Goodenaceae	*Brunonia**	Platanaceae	B *Platanus*
	B *Goodenia**	Polygalaceae	B *Comeosperma*

(Continued)

Table 6.1 (Continued)

Family	Genus	Family	Genus
Polygonaceae	*Coccoloba*	Sterculiaceae	B *Lasiopetalum*
	*Polygonum**		*Thomasia*
Rhamnaceae	*Cryptandra*	Stylidiaceae	B *Stylidium*
	Pomaderris	Thymeliaceae	B *Pimelia*
	Rhamnus	Tiliaceae	B *Tilia*
	Spyridium	Ulmaceae	B *Ulmus*
	Trymalium		*Celtis*
Rosaceae	*Chaembatia*	Vitaceae	B *Vitis*
	Cirocarpus	Cupresseceae	B *Cupressus*
	B *Crataegus*		B *Juniperus*
	B *Dryas*	Pinaceae	*Abies*
	B *Malus*		*Cathaya*
	B *Prunus*		B *Cedrus*
	B *Pyrus*		*Keteleeria*
	B *Rosa*		*Larix*
	B *Sorbus*		*Picea*
Salicaceae	B *Populus*		*Pinus*
	B *Salix*		*Pseudolarix*
Sapindaceae	*Allophylus*		*Pseudotsuga*
	Nephelium		*Tsuga*
Sapotaceae	*Glycoxylon*	Gnetaceae	*Gnetum*

Modified from Harley and Smith (1983). This list cannot pretend to be exhaustive but illustrates the wide range of families and genera of Angiospermae and Gymnospermae in which ectomycorrhizas have been observed. A record of the presence of ectomycorrhizal individuals in a genus does not mean that all species are or may be ectomycorrhizal, nor does it mean ectomycorrhizal colonization is necessarily consistently or even normally present in any species of that genus. Those marked * are herbaceous and those marked B may form both ecto- and arbuscular mycorrhizas, with the latter in many cases being the most common mycorrhizal type observed.

rainforests of the Guineo-Congolan basin (Newbery *et al.*, 1988) (see Chapters 15 and 16). Woodlands of the miombo type formed by the prominent caesalpinoid ECM genera *Brachystegia* and *Isoberlinia* cover vaste areas of dry savannahs of East and South-Central Africa (Högberg, 1982; Högberg and Pearce, 1986; Alexander and Högberg, 1986; Alexander, 1989a; Taylor and Alexander, 2005).

In contrast to the Caesalpinoideae, the subfamilies Mimosoideae and Papilionoideae appear, with a very few exceptions that need to be confirmed, to be made up of AM species. There are reports of ECM colonization in species such as *Acacia* (Mimosoideae) (Warcup, 1985; McGee, 1986) even though neither typical Hartig net nor mantle development is present. Again, since this genus is more widely associated with AM fungi, there seems little to be gained by classifying them as ECM.

Some genera of shrubs and a very small number of herbaceous species of angiosperm are routinely found to be ECM. Of these, the shrubs *Dryas* (Rosaceae) and *Helianthemum* (Cistaceae) are of particular ecological significance. Among the herbaceous species, the dicotyledenous herb *Polygonum viviparum* and the cyperaceous monocot *Kobresia myosuroides* have typical ECM short roots with sheath and Hartig net. Neither of these plants would normally be colonized by AM fungi.

One further category of woody plants is of interest because it shows the facultative ability to be both AM and ECM. Members of the Myrtaceae (Lapeyrie and Chilvers,

1985; Jones *et al.*, 1998) and Salicaceae (Lodge and Wentworth, 1990; van der Heijden and Vosatka, 1999; van der Heijden, 2000; van der Heijden and Kuyper, 2001a) fall into this group as do some dipterocarps (see Chapter 1). Intensive study of the dune shrub *Salix repens* showed that the two types of colonization co-occurred across 16 different dune habitat types, but that ECM colonization was always more prolific than that by AM fungi. Pot studies comparing the effects of ECM with AM colonization showed that mycorrhizal benefits expressed in terms of shoot yield were generally larger in the former case, but that even small amounts of AM colonization were still beneficial (van der Heijden and Kuyper, 2001a). Similar conclusions were reached by Jones *et al.* (1998) working with seedlings of the myrtaceous tree *Eucalyptus coccifera* (see Chapter 8).

Our limited understanding of the evolutionary history of ECM plants is based upon evidence derived from fossils, from our understanding of the geographic and phylogenetic origins of the plant groups forming this type of symbiosis, and from molecular clock data. There is a very restricted amount of fossil evidence. Fifty million year-old cherts containing well preserved ectomycorrhizas typical of *Pinus* spp. have provided the first unequivocal evidence for the occurrence of this type of symbiosis in the fossil record (Le Page *et al.*, 1997). Small roots, each having a diameter of 3–5 mm, bear coralloid clusters of attenuated and thickened, dichotomously branched rootlets (Figure 6.1a, b). Transverse sections reveal the presence of a Hartig net-like fungal tissue between the cells of the cortex and up to the position of the endodermis (Figure 6.1c). Only traces of mantle tissue remain and extraradical hyphae are scarce. Those that are present lack clamp connections. Among the small number of fungal associates of extant *Pinus* spp. that produce such coralloid clusters of rootlets and lack clamp connections are members of the hypogeous genus *Rhizopogon* which has close affinities with *Suillus* (see below). If, as proposed by Le Page *et al.* (1997), these fossil mycorrhizas were formed by *Rhizopogon*, the findings would support the view of Trappe (1987) that hypogeous basidiomycetes originated more than fifty million years ago (mya). They also provide support for molecular clock-based dating (see below) of the origin of the epigeous forms that were ancestral to *Rhizopogon*, concurrent with the earliest fossils of the genus *Pinus* around 130 mya (Axelrod, 1986).

The fossil evidence has been supplemented by analyses of the likely origins and radiation patterns of those Angiosperm plant families that today are almost exclusively ECM. Studies of one such family, the Dipterocarpaceae, have been particularly instructive. Dipterocarps are of gondwanan origin (Ashton, 1982). While they currently predominate in the rainforests of South-East Asia, they are also found in South America and in Madagascar, where a related family, the Sarcolaenaceae with which they share a common ancestor, is ECM (Ducousso *et al.*, 2004). These observations place the likely origin of ectomycorrhizas in the South-East Asian dipterocarp clades at a time before the separation of the Indian land mass from the Madagascan-East African continent around 88 mya. Since relict populations of the only South American dipterocarp genus, *Pakaraimaea*, are also ECM (Moyersoen, 2006), the same reasoning leads to the conclusion that the symbiosis predates the separation of South America and Africa some 135 mya. Similar arguments may apply to the Fagaceae, another almost exclusively ECM family of gondwanan origin, with the important genera *Nothofagus* and *Quercus* in southern and northern hemisphere forests, respectively. Members of the Fagaceae also co-occur with dipterocarps in Asia. Together the distributions provide geographic links between the two hemispheres. A very approximate date of 130-plus mya for the origin of ECM plants is broadly

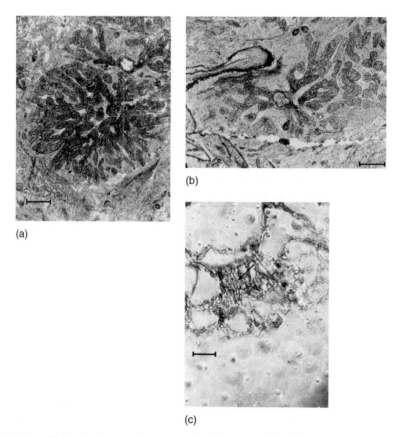

Figure 6.1 Fossil ectomycorrhizas of a *Pinus* spp. recovered from the Princeton cherts of Eocene age in British Columbia. (a) Coralloid clusters of ectomycorrhizal roots. Bar = 1 mm. (b) Individual rootlets showing the dichotomous branching characteristic of *Pinus* mycorrhizas. Bar = 1 mm. (c) Transverse section showing Hartig net-like fungal tissue (arrowed) surrounding root cells and showing evidence of labyrinthine structure. Bar = 20 μm. From Le Page *et al.* (1997), with permission.

consistent with evidence from molecular clock data (see below) and is in line with the dates currently proposed for the origin of the basidiomycetous fungi that form most ECM associations (Berbee and Taylor, 2001).

However, it places the origin of angiosperm ECM at a time that is very close to that of all flowering plants (Willis and McElwain, 2002). While this is well before the oldest fossils that are directly attributable to extant lineages of ECM-forming angiosperms (see below), it is consistent with dates for the possible origin of the gymnosperm lineage Pinaceae. As reported above, pines for which fossil evidence confirms a gondwanan origin are thought likely to have been ECM from this time and hence may have been a repository of fungal lineages capable of forming this type of symbiosis on the earliest angiosperms.

Alternative explanations for the origins of ECM in angiosperms have been considered (Alexander, 2006). It is possible, for example, that the ancestors of ECM dipterocarps, beeches and legumes, were not themselves ECM prior to the break-up of Gondwana but that they carried with them a genetic propensity to form this type of symbiosis that was expressed later in their separated continental habitats. A further proposal, and one that would seriously undermine the c 135 mya date for possible origins of the angiosperm ECM, is that the disjunct distribution of supposedly gondwanan species is attributable to transoceanic transport of propagules after the African and American continental plates had separated (Pennington *et al.*, 2004). This view is consistent with some dated molecular phylogenies which, in the dipterocarps for example, indicate an origin between 14 and 28 mya that clearly would preclude the involvement of continental drift in their distribution (Wikström *et al.*, 2001). While such dates have been contested (Givnish and Renner, 2004), it is evident that uncertainties concerning the origins of ECM in these critical taxa will remain until debates over the roles played by tectonism and transoceanic transport in their distibution have been resolved.

The fungi forming ectomycorrhizal associations

Basidiomycetes as ectomycorrhizal fungi

Despite having a uniform and distinctive suite of structural features, ECM symbioses are characterized by the great diversity of fungi involved in their formation. Molina *et al.* (1992) estimated that between 5000 and 6000 species of fungi form ectomycorrhizas. Our knowledge of the identities of these fungi has improved progressively over time as more rigorous methods have been applied to their study. Early observations using sporocarp occurrence as an indicator of ECM status were largely replaced, first by methods based on morphological analysis (e.g. Agerer, 1987–2002) and later by molecular techniques. Records of plant–sporocarp association have provided a useful basis on which to commence consideration of issues relating to host range and specificity of the fungi (Table 6.2). However, there is increasing recognition of the diversity, both structural and functional, seen within previously designated ECM fungal genera and species. Some of these developments are discussed below, where major emphasis is placed upon advances facilitated by the newer molecular approaches. For the most part, these have been applied directly to the study of ECM roots, removing the uncertainties associated with study of plant–sporocarp associations. Because the vast majority of ECM fungi are in the Basidiomycota, the present treatment is largely devoted to this group of fungi. However, in recognition of our increasing awareness of the occurrence of ascomycetes as root symbionts, these are considered in a later section.

Early molecular studies (Gardes *et al.*, 1991b; Henrion *et al.*, 1992; Gardes and Bruns, 1993; Karen *et al.*, 1997) matched restriction fragment length polymorphism (RFLP) patterns of the Internal Transcribed Spacer (ITS) region of unknown mycorrhizas to known fungi. However, difficulties were experienced in comparing RFLP patterns between studies because of the inaccuracies of fragment-size estimates and the considerable intraspecific variation often found. Greater accuracy in size estimates can be achieved by labelling the terminal fragments and resolving them with an automated sequencer. This TRFLP approach also simplifies the patterns produced and enables the analysis of complex mixtures of ECM fungi. The method has been used to examine ECM fungi in soil (Dickie *et al.*, 2002; Koide *et al.*, 2005), but the TRFLP method can produce artefacts in such complex communities (Avis *et al.*, 2006). The use of specific

Table 6.2 Examples of ectomycorrhizal fungal species with little host restriction (broad host range) by fruiting habit, class, family, and genus.

Habit, class, family	Genus	Species
Epigeous habit		
Basidiomycotina		
Amanitaceae	*Amanita*	*aspera, fulva, gemmata, inaurata, muscaria, pantherina, phalloides, rubescens, solitaria, spissa, strobiliformis, vaginata, verna, virosa*
Astraeaceae	*Astraeus*	*hygrometricus, pteridus*
Boletaceae	*Boletus*	*appendiculatus, calopus, edulis, erythropus, luridus, minatioolivaceus, pulverulentus, regius*
	Gyroporus	*castaneus, cyanescens*
	Phylloporus	*rhodoxanthus*
	Pulveroboletus	*ravenlii*
	Tylopilus	*chromapes, felleus, gracilis, porphyrosporus*
	Xerocomus	*armeniacus, badius, chrysenteron, rubellus, spadiceus, subtomentosus, truncatus*
Cantharellaceae	*Cantharellus*	*cibarius, infundibuliformis, tubiformis*
Clavariaceae	*Ramaria*	*aurea, botrytis, flava, formosa, mairei, subbotrytis*
Corticiaceae	*Byssocorticium*	*atrovirens*
	Byssoporia	*sublutea*
	Piloderma	*byssinium, croceum, sulphureum*
Cortinariaceae	*Cortinarius*	*acutus, anomalus, bicolor, bivelus everneus, hemitrichus, leucophanes, mucosus, multiformis, obtusus, phrygianus, saniosus*
	Dermocybe	*anthracina, cinnamomea, malicoria, palustris, phoenicea*
	Hebeloma	*crustuliniforme, cylindrosporum, hiemale, longicaudum, mesophaeum, minus, pumilum, sinapizans*
	Inocybe	*asterospora, bongardii, brunnea, cincinnata, dulcamara, fastigata, jurana, lacera, lanuginella, petiginosa, terrigena, umbrina*
	Rozites	*caperata*
Hydnaceae	*Dentinum*	*repandum*
	Hydnellum	*velutinum*
	Hydnum	*imbricatum, rufescens, scabrosum*
Hygrophoraceae	*Hygrophorus*	*capreolarius, camarophyllus, chrysodon, discoideus, hypothejus, karstenii, marzulus, pudorinus*
Paxillaceae	*Paxillus*	*involutus*
Pisolithaceae	*Pisolithus*	*tinctorius*
Polyporaceae	*Albatrellus*	*cristatus*
Russulaceae	*Lactarius*	*decipiens, fuliginosus, helvus, necator, piperatus, repraesentaneus, rufus, scrobiculatus, spinosulus, uvidus, vellereus, volemus*
	Russula	*aeruginea, albonigra, amoena, anthracina, cyanoxantha, densifolia, emetica, foetens, heterophylla, lutea, nigricans, ochroleuca, odorata, olivacea, paludosa, palumbina, parazurea, vesca, virescens, xerampelina*
Sclerodermataceae	*Scleroderma*	*bovista, cepa, citrinum, laeve, polyrhizum, verrucosum*
Strobilomyceataceae	*Boletellus*	*betula, chrysenteroides*
	Strobilomyces	*floccopus*
Thelephoraceae	*Thelephora*	*anthocephala, atrocitrina, penicillata, terrestris*
Tricholomataceae	*Laccaria*	*amethystina, bicolor, laccata, montana, proxima*
	Tricholoma	*caligatum, columbetta, flavobrunneum, flavovirens, myomyces, saponaceum, sulphureum*

(Continued)

Table 6.2 (Continued)

Habit, class, family	Genus	Species
Hypogeous habit		
Ascomycotina		
Balsamiaceae	*Balsamia*	*magnata, platyspora, vulgaris*
Elaphomycetaceae	*Cenococcum*	*geophilum*
	Elaphomyces	*anthracinus, granulatus, muricatus, mutabilis, reticulatus, variegatus*
Geneaceae	*Genabea*	*cerebriformis*
	Genea	*gardneri, harknessii, intermedia*
Helvellaceae	*Hydnotrya*	*tulasnei*
Pezizaceae	*Pachyphloeus*	*citrinus, ligericus, melanoxanthus*
Terfeziaceae	*Choiromyces*	*alveolatus, venosus*
Tuberaceae	*Tuber*	*aestivum, borchii, brumale, californicum, excavatum, melanosporum, puberulum, rapaeodorum, rufum*
Basidiomycotina		
Cortinariaceae	*Hymenogaster*	*bulliardii, calosporus, citrinus, decorus, lilacinus, luteus, olivaceus, populetorum, tener, vulgaris*
Hysterangiaceae	*Hysterangium*	*membranaceum*
Leucogastraceae	*Leucogaster*	*nudus*
Melanogastraceae	*Melanogaster*	*ambiguus, broomeianus, euryspermus, intermedius, tuberiformis, variegatus*
Russulaceae	*Elasmomyces*	*mattirolianus*
	Zelleromyces	*stephensii*
Sclerodermataceae	*Scleroderma*	*hypogaeum*
Strobilomycetaceae	*Gautieria*	*graveolens, mexicana, otthii*
Zygomycotina		
Endogonaceae	*Endogone*	*lactiflua*

Modified from Molina *et al.*, 1992.

DNA probes for individual taxa or genotypes has been explored (Marmeisse *et al.*, 1992a; Bruns and Gardes, 1993), but development has been slow and probes have not yet been used for large surveys. Such probes do, however, have considerable potential when used in conjunction with microarray technologies, as shown by their use for characterization of bacterial communities (Wilson *et al.*, 2002).

Direct sequence analysis has become the main tool in more recent studies. By extracting DNA from herbarium specimens and from cultures of identified fungi, Bruns *et al.* (1998) assembled a sequence database for a small piece of the mitochondrial LSU rRNA gene from 80 genera of the hymenomycete lineage (Basidiomycota). This data-set included many of the most widespread ECM genera. Phylogenetic studies using the nuclear LSU rRNA have now produced a much more extensive database for ECM and saprotrophic basidiomycetes (Moncalvo *et al.*, 2002) and the accumulation of ITS sequences from phylogenetic and ecological studies has enabled a higher resolution within many ECM groups. The UNITE database (http://unite.ut.ee/) contains a well curated set of ITS sequences from ECM fungi that are connected to vouchered specimens (Koljalg *et al.*, 2005).

Phylogenetic studies have enabled a better understanding of the range of fungi involved in forming ectomycorrhizas while, at the same time, providing valuable insights concerning the likely evolutionary relationships among them. The Bruns *et al.* (1998) analysis (Figure 6.2) resolved 18 distinct groups of ECM fungi within the

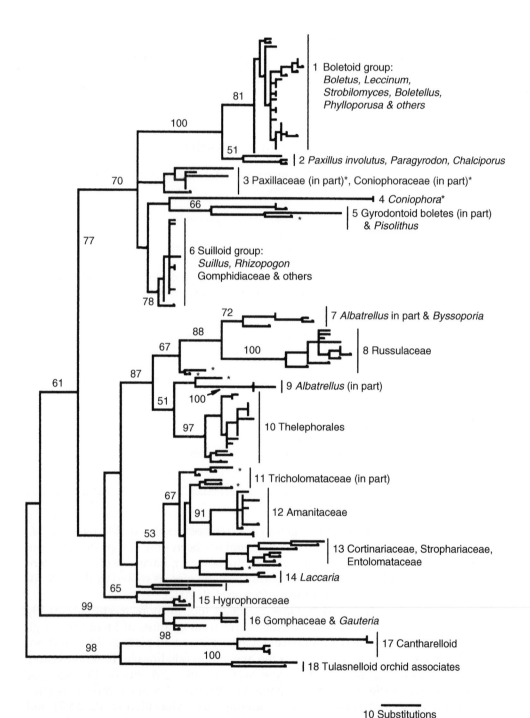

Figure 6.2 A neighbour-joining tree indicating relationships between some of the major groups of ECM forming fungi in the Basidiomycota. The tree is based upon a sequence database obtained using a small region of the mitochondrial large subunit and rRNA gene. It is constructed using patristic distances generated from PAUP and branches supported by more than 50% of bootstrap replicates are indicated. Modified from Bruns et al. (1998).

hymenomycete lineage. A feature of the tree is that several of the groups, particularly the Boletoid (Group 1), Suilloid (Group 6), Russulaceae (Group 8), Thelephorales (Group 10) and Amanitaceae (Group 12) (each of which were represented by large and diverse ranges of taxa in the analysis) exhibited very short within-group branch lengths relative to other branches within the tree. Short branches, which arise because numbers of nucleotide changes are small, are indicative both of closeness of the within-group relationships and also of relatively recent evolutionary radiation.

It is also evident (Figure 6.2) that saprotrophic and ECM taxa are intermingled throughout the tree. For example, the Boletales (Group 2), wood-decaying species in the Coniophoraceae (Groups 3 and 4) and Paxillaceae (Group 3) appear to be

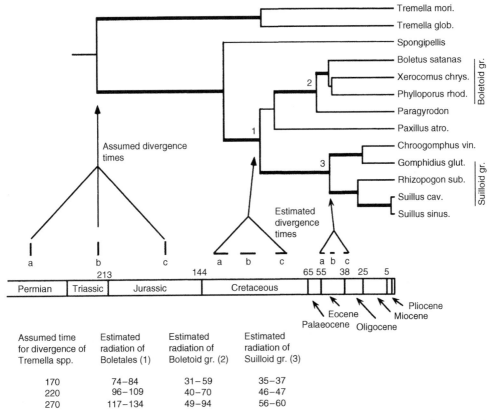

Figure 6.3 Estimated divergence times for the Boletales (1), boletoid group (2) and suilloid group (3) based on the maximum likelihood analysis of nuclear small subunit rRNA gene sequences. Berbee and Taylor's (1993) estimated divergence of 220 mya (b) for the node indicated (*) is assumed. Estimated times are given graphically for the Boletales and the suilloid group and in tabular form for all three lineages. Range of estimates is derived by allowing ±50 mya (a & c) variation for the Berbee and Taylor date and through analysis of 12 other topologies that differ slightly from the one shown. All 12 trees shared the internal branches indicated in bold and were not significantly different from each other based on Kishino and Hasewaga (1989) tests. Epochs of the Tertiary: P, Palaeocene; E, Eocene; O, Oligocene; M, Miocene; unmarked, Pliocene. Modified from Bruns et al. (1998).

close relatives of the Boletoid and Suilloid groups. Similarly, more detailed analysis of the same tree (Bruns *et al.*, 1998) revealed that basal to the Russulaceae and Thelephoraceae clades are three wood-decaying taxa, including the economically important *Heterobasidion annosum*. These and other close relationships between ECM fungi and saprotrophs are supported by several more recent studies, based upon additional genes and greatly expanded taxon sampling (Hibbett and Donoghue, 1995; Binder and Hibbett, 2006; James *et al.*, 2006). Collectively, these examples demonstrate that the switch from a presumably ancestral saprotrophic habit to a symbiotic mode occurred convergently and on a number of separate occasions.

Phylogenetic analyses by Hibbett *et al.* (2000) confirmed that ECM fungi have evolved repeatedly from saprotrophic precursors, but have also suggested that multiple reversals from ECM to the free-living condition have taken place. This conclusion has been disputed by Bruns and Sheffarson (2004), on the basis that several of the assumptions of the method employed by Hibbett *et al.* (2000) are either questionable or invalid. In the case of only one fungus, *Lentaria byssoides*, did the latter authors claim that there was 'unambiguous' evidence for reversal from the ECM to the saprotrophic condition. Even in this case, the fungus, while fruiting on woody substrates, also formed ectomycorrhizas. A later, more extensive, sampling of taxa related to *Lentaria* has confirmed that this genus is a basal relative of presumed ECM taxa rather than derived from them (Humpert *et al.*, 2001). Bruns and Shefferson (2004) argue on two grounds that gains of the ECM habit are more likely to have occurred than losses, over evolutionary time. First, evidence indicates that unambiguous gains of ECM habit have taken place in the fungi at large, as well as from basal clades of saprotrophs across a broad range of basidiomycetous fungi. There are no unambiguous cases of a non-mycorrhizal taxon that is nested well within a strongly supported clade of ECM species. Secondly, loss of a complex trait like the ECM habit might be expected to occur easily but, in order to be successful, it would have to be accompanied by restoration of saprotrophic capabilities. The convergent evolution of the ECM habit, which seems to have occurred in radiations, may have been facilitated by the presence of open niches as the geographic extent of potentially ECM taxa of autotrophs expanded.

On the assumption that the ECM habit is derived, it is of interest to consider the possible time-scale for the divergence of ECM fungi from their saprotrophic ancestors. Events can be placed in the context of our, so far restricted, understanding of the time-scales for evolution of the fungi as a whole. Based on the occurrence of AM-type fossils and of molecular clock data, the origin of the Glomeromycota is placed at least 450 mya and possibly very much earlier (see Chapter 1 and Figure 1.6). Separation of agarics and boletes from the jelly fungi (*Tremella* spp.) is estimated at 220 mya (Berbee and Taylor, 1993). With 220 (±50) my as an assumed divergence time for these basal groups, Bruns *et al.* (1998) used molecular clock data to estimate the likely origin of the Boletales and subsequent divergence of two distinct ECM groups (the boletoid and suilloid groups) (see Figure 6.3). The estimated median time for the first origin of the Boletales is the mid-Cretaceous 100 mya. The boletoid group then originates in the early Eocene 58 mya and the suilloid group in the mid-Eocene 45–50 mya. The last date broadly coincides with that estimated by Le Page *et al.* (1997) for the earliest known fossil ectomycorrhizas (see Figure 6.1) which, as reported above, are thought to have had suilloid affinities. These dates are in line with the suggestion of Bruns *et al.* (1998) that radiation of these groups has occurred relatively recently.

Plant macrofossil assemblages suggest that radiation of the early ECM fungal lineages was co-associated with the emergence of their autotrophic partners. As indicated above, members of the Pinaceae, Dipterocarpaceae and Fagaceae almost certainly originated in Gondwana in the course of the Cretaceous (144–65 mya) and radiated with their ECM fungal partners thereafter. The Eocene–Oligocene transition (about 37 mya) was a period of climatic cooling, during which temperate forests dominated by obligately ECM trees of the families Fagaceae and Pinaceae became widespread in the northern hemisphere (Tallis, 1991). This may therefore have been a period of particularly active fungal speciation (Bruns *et al.*, 1998).

Resolution of the phylogenetic status of selected Basidiomycete fungi

In addition to the provision of insights on evolutionary issues, molecular approaches have contributed to a deeper understanding of the genome of ECM basidiomycetes and to the resolution of a number of taxonomic uncertainties within key genera. Selected here as examples of these advances are studies of the full genome sequence of *Laccaria bicolor* (Martin, 2007), analyses of the phylogeography of *Pisolithus* species (Martin *et al.*, 2002), a revised phylogeny of the hypogeous genus *Rhizopogon* (Grubisha *et al.*, 2002) (Figure 6.4) and a phylogenetic analysis of the genus *Sebacina*, some members of which, in addition to forming ECM, have been implicated as symbiotic associates of ericaceous plants (see Chapter 11), green and mycoheterotrophic orchids (see Chapters 12 and 13) and liverworts (see Chapter 14).

The first whole genome sequence of an ECM fungus was publicly released in July 2006 (http://mycor.nancy.inra.fr/IMGC/LaccariaGenome/). The hypogeous basidiomycete *L. bicolor* was selected for analysis because it is both widely distributed as an ECM symbiont of many autotrophs in nature and is tractable as a laboratory organism. Its genome of 65 Mb is much larger than that of other basidiomycetes whose genomic sequences have been published. A striking result of the analysis is that 50% of the genes identified are of unknown function. The organization of the genome in *L. bicolor* is considerably more complex than that of ascomycetes, such as *Saccharomyces cereviseae* for which the sequence is available, but is similar to those of *Cryptococcus neoformans* (Loftus *et al.*, 2005) and the lignocellulose-degrading *Phanerochaete chrysosporium* (Martinez *et al.*, 2004), both of which are also basidiomycetes. A total of about 20 000 genes encoding proteins were identified, including proteases, lipases, phytases and around 315 involved in carbohydrate metabolism, among which were glucanases, cellulases and chitinases, which indicate a diverse catabolic repertoire probably enabling the degradation of polymers found in organic horizons of soil and a limited saprotrophic capacity. Comparisons of this ECM genome with those of saprotrophic and pathogenic fungi is already proving instructive. It has emerged that, in comparison with the saprotrophs C. *neoformans* and *P. chryosporium*, most of the increase in size of gene families in *L. bicolor* has occurred in genes predicted to have roles in signal transduction pathways, with the encoded proteins having significant roles in infection processes and in setting up dialogue between the partners that leads to functional ECM symbiosis (Martin, 2007). Further evaluations of these *L. bicolor* sequences will contribute a better understanding of the contrasts and similarities between ECM and saprotrophic modes of nutrition.

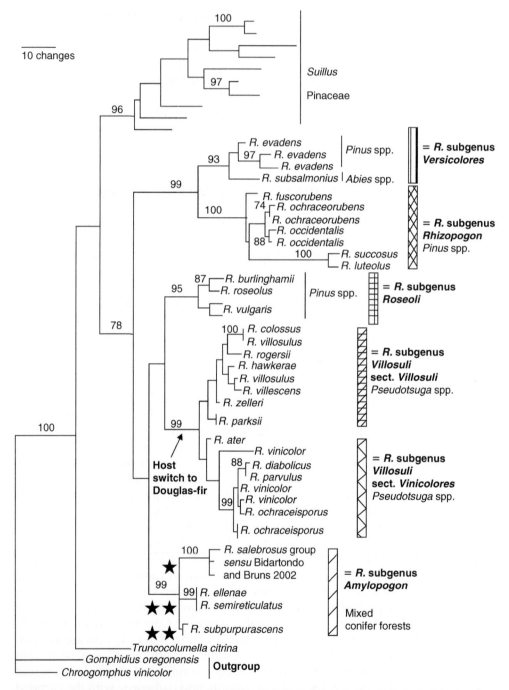

Figure 6.4 Phylogram (with bootstrap values indicated at respective nodes) showing current understanding of relationships of species in sections of the genus *Rhizopogon*. Individual sections appear to be associated with either pines or Douglas fir, with the exception of subgenus *Amylopogon* that has broad host range. In *Amylopogon*, *R. salebrosus* associates specifically with the monotropoid mycoheterotroph *Pterospora andromedea* (★) and *R. ellenae* and *R. subpurpurascens* (★ ★) associate specifically with another monotropoid species *Sarcodes sanguinea* (see Chapter 13). From Grubisha *et al.* (2002).

The globally distributed ECM genus *Pisolithus*, in the past widely treated as consisting of the single species *P. tinctorius* (Cairney, 2002), is now known to encompass at least 11 species. Martin *et al.* (2002) sequenced the rDNA of the ITS region in 102 lineages of the fungus collected worldwide and also analysed 48 additional Gen Bank accessions. The type species, *P. tinctorius*, was confirmed to be widely distributed in the Holarctic, particularly in association with *Pinus* and *Quercus*. Probably as a result of the introduction of exotic pines inoculated with this fungus (see Chapter 17), *P. tinctorius* is now also widely present in other biogeographic regions. However, several other *Pisolithus* lineages, including *P. aurantioscabrosus*, were identified as occurring in restricted areas associated with endemic plants such as *Afzelia* spp. in eastern Africa. *P. albus*, *P. marmoratus* and *P. microcarpus* are associated with the Australasian hosts *Eucalyptus* and *Acacia* spp., but have also been distributed worldwide with these plants. In contrast, two additional unnamed species are restricted to Australia. This study revealed that evolutionary lineages within *Pisolithus* are closely related to the biogeographical origin of their hosts, which brings the likelihood that a large number of endemic species of *Pisolithus* remain to be discovered in association with endemic autotrophic partners.

Rhizopogon, in the order Boletales, is the largest genus of hypogeous basidiomycetes and contains more than 100 species. It has a worldwide distribution as an ECM fungus and associates almost exclusively with members of the Pinaceae (see Colour Plate 6.1e,f), particularly members of the genera *Pinus* and *Pseudotsuga*. Despite this cosmopolitan range, most of the species are found in forests dominated by these genera in the Pacific Northwest of the USA (Molina *et al.*, 1999). The genus *Rhizopogon* was divided by Smith and Zeller (1966) into four sections, *Amylopogon*, *Fulviglobae*, *Rhizopogon* and *Villosuli*, on the basis of morphological characteristics. Subsequent maximum parsimony analysis of ITS sequences of 27 *Rhizopogon* and 10 *Suillus* species selected, respectively, to test sectional and phylogenetic relationships, showed first that the section *Rhizopogon* was not monophyletic. The analysis (see Figure 6.4) resolved six sections, the subgenera *Versicolores* and *Roseoli* being clearly distinguishable from *Rhizopogon* (Grubisha *et al.*, 2002). Resolution of the taxonomic affinities of this genus is significant, not only because of the important and often highly specific associations which its members have within the Pinaceae, but also because scattered through the genus are species which form highly specialized relationships with mycoheterotrophs in the Monotropaceae (see Chapter 13).

In addition to expanding our knowledge of the interrelationships of fungi long known to be ECM, DNA-based technologies have revealed the presence of large cohorts of fungi hitherto not recognized as being of ECM habit. Prominent among these are members of the heterobasidomycete order Sebacinales (Weiss *et al.*, 2004) and a range of resupinate homobasidomycete genera, most notably *Tomentella* and *Tomentellopsis* (Koljalg *et al.*, 2000). Members of the Sebacinales are now known to form ectomycorrhizas (see Colour Plate 6.1d) on a wide range of tree hosts with global distributions (Glen *et al.*, 2002; Selosse *et al.*, 2002a; Tedersoo *et al.*, 2003; Urban *et al.*, 2003). They have also been identified as the fungal symbionts in some orchid mycorrhizas (McKendrick *et al.*, 2002; Selosse *et al.*, 2002a; Taylor *et al.*, 2003; see Chapter 12), of a newly described association in some liverworts (Kottke *et al.*, 2003; see Chapter 14) and as associates of ericoid mycorrhizal (ERM) roots (Allen *et al.*, 2003; see Chapter 11). The functional roles of *Sebacina* spp. as associates of liverwort thalli and ericaceous roots remains to be

evaluated. The order Sebacinales occupies a basal position among the basidiomycetes and has a close phylogenetic affiliation with *Geastrum* spp. It has been
hypothesized (Taylor *et al.*, 2003; Weiss *et al.*, 2004) that the common ancestor of the
Geastrum/Sebacinales clade or even of the whole Hymenomycete group was ECM.

Estimates of phylogenetic relationships within the Sebacinales by Weiss *et al.* (2004),
based upon Markov chain Monte Carlo analyses (MCMC) (Figure 6.5) show not only
that the order is divisible into two subgroups A and B, but also that the two groups
are only distantly related to one another. Group A contains all the sequences obtained
from sporocarps from ectomycorrhizas and from the orchids *Neottia, Epipactis* and
Hexalectris (see Chapter 13). These fungi have so far not been brought into axenic culture. In contrast, some members of subgroup B, which were referred to as *S. vermifera*
in the past, are culturable. The original isolates were obtained from roots of a green
orchid (Warcup and Talbot, 1967; Warcup, 1988; see Chapter 12) and assigned to the
form genus *Rhizoctonia*. Group B also contains the three liverwort isolates described
by Kottke *et al.* (2003) and those detected using molecular methods in ericaceous
roots (Allen *et al.*, 2003; Selosse *et al.*, 2007). Whereas Warcup (1988) claimed to have
synthesized ectomycorrhizas on myrtaceous hosts with his *S. vermifera* isolates, there
are no records of the formation of this type of mycorrhiza by Group B *Sebacina* spp.
in nature. On the basis of their phylogenetic analysis, Weiss *et al.* (2004) stress that
Group B is a species complex. The strains identified by Warcup and Talbot (1967) as
Sebacina vermifera have recently been transferred to the genus *Serendipita* (Roberts, 1993)
as *S. vermifera* (Roberts, 1999). This, as more phylogenetic information emerges, may
prove to be the first of many such revisions in the *Sebacina* complex.

These advances pave the way for a large expansion of knowledge which will reveal
key features of the symbiotic lifestyle in comparison with both wood-destroying
saprotrophs and damaging pathogens. We may also expect increased insight into
ECM fungal phylogeny, linked to biogeographic information and hence to biodiversity and function of ecosystems.

Figure 6.5 Phylogenetic relationships in Sebacinales. The phylogram is inferred from heuristic maximum likelihood analysis from an alignment of nuclear DNA sequences coding for
the D1-D2 region of the large ribosomal subunit (nucLSU) using a general time-reversible
model of nucleotide substitution, assuming a percentage of invariable sites. Γ-distributed
substitution rates at the remaining sites. Branch support values are given as non-paramet-
ric maximum likelihood bootstrap values/posterior probabilities as estimated from Bayesian
Markov chain Monte Carlo analyses. Support values below 50% are omitted or denoted with
an asterisk. The trees were rooted with *Auricularia auricular-judae*. The two main subgroups
are denoted A and B. Ectomycorrhizal members are only known from group A, whereas
Sebacinales observed in ericoid, cavendishoid or jungermannoid tissues all belong to group
B. Some orchid mycorrhizal Sebacinales belonging to group A can form tripartite associations involving ectomycorrhizas. Basidiome-forming Sebacinales are included in group A; the
only teleomorphic specimens known from group B have all been assigned to the *S. vermifera* complex. Acronyms used for mycorrhizal types are: JMM, jungermannoid mycorrhiza
(*sensu* Kottke *et al.*, 2003); ORM, orchid mycorrhiza. Acronyms used for origins of sequences:
A, Austria; AUS, Australia; CAN, Canada; CHN, P.R. China; ECU, Ecuador; EST, Estonia; FRA,
France; GER, Germany; IND, India; MEX, Mexico; NOR, Norway; SPA, Spain. From Weiss *et al.*,
(2004), see also Chapters 11, 12 and 14.

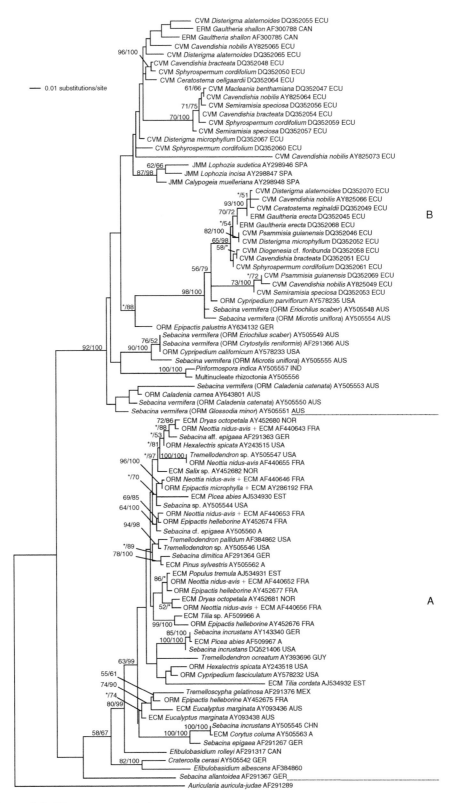

Figure 6.5 (Caption opposite)

Ascomycetes as ectomycorrhizal fungi

The application of molecular techniques has enabled improved understanding of the phylogenetic positions of the ascomycetous fungi that have been considered to produce ECM. Lobuglio *et al.* (1996) concluded, on the basis of 18S rDNA analysis, that the ECM condition arose independently on at least four occasions in the Ascomycota; in the loculoascomycetes (*Cenococcum geophilum*), the plectomycetes (*Elaphomyces*), the discomycetes (*Tuber*) and in a group encompassing *Phialophora* (syn *Cadophora*) *finlandia*, known to be part of the *Hymenoscyphus* aggregate (see Chapters 7 and 11).

It has long been known that some of the most commonly occurring ECM fungi are ascomycetes. The most prominent of these, *Cenococcum geophilum*, has both a worldwide geographic distribution and an enormous range of autotrophic partners (Ferdinandsen and Winge, 1925; Mikola, 1948; Molina and Trappe, 1982). Where it occurs, it can be the dominant fungus present (e.g. Koide *et al.*, 2005). Molecular analyses place *C. geophilum* in a basal intermediate lineage between the orders Pleosporales and Dothidiales of the loculoascomycetes (Lobuglio *et al.*, 1996; Lobuglio and Taylor, 2002). Its widespread occurrence and the formation of mantle and Hartig net, show that it is, indeed, an ECM fungus, but we still know far too little of its functional role in plant communities. While some positive impacts on plant performance have been demonstrated (e.g. Marx and Zak, 1965; Kropp *et al.*, 1985; Rousseau *et al.*, 1994; see Table 10.5), many reports indicate neutral or inhibitory effects of colonization (Mejstrik and Krause, 1973; Marx *et al.*, 1978; Dixon *et al.*, 1984; Dickie *et al.*, 2002). It is worthy of note that, at the end of his extensive monograph on the ecology–physiology of *C. geophilum*, Mikola (1948) concluded that: 'the question as to whether *Cenococcum* is as beneficial to its host plant as other mycorrhizal fungi, must be considered as undecided so far'. The situation remains much the same to this day! In view of extent of the occurrence of this fungus as a root associate, there is a particular need to scrutinize the nature of its relationships with its autotrophic partners and to determine, under a range of environmental conditions, its position on the mutualism–parasitism continuum (see Chapter 4). In combination, the recognition of the drought resistance of *C. geophilum* (Coleman *et al.*, 1989) and the observation that its representation in the community of ECM symbionts increases under dry conditions (Worley and Hacskaylo, 1959; Pigott, 1982), suggest that this fungus will become a greater contributor to the root mycoflora in a warming world.

Perhaps the best known ascomycetous ECM fungi are in the truffle genus *Tuber*, which are members of the discomycete order Pezizales (Percudani *et al.*, 1999). While originally having a natural occurrence in three distinct areas, Europe, South-East Asia and North America, some members of the genus, notably the gastronomically important black truffle (*T. melanosporum*), have been exported for use in truffle orchards so now have a worldwide distribution (see Chapter 17). *Tuber* spp. (of which there are about 100 species) have broad host ranges, but are largely restricted in their ecological distributions to soils in the pH range 7–8 (Pacioni and Comandini, 1999). Because members of the genus are difficult to grow axenically, information concerning their functional roles is limited. However, both nursery (Pirazzi and Di Gregario, 1987) and field (Hall *et al.*, 1994) studies have demonstrated increases of plant size and weight as a result of inoculation. The antagonistic impacts of *Tuber* mycelium on herbaceous vegetation, revealed in the form of scorched rings of dead plants around the colonized trees, so-called 'brulé', can be expected to be indirectly advantageous to the ECM plant by weakening competition for nutrients and water.

Several other members of the Pezizales have been considered to be ECM (Maia *et al.*, 1996). Among these are *Genea* (Jacuks *et al.*, 1998), *Geopixis* (Vralstad *et al.*, 1998), *Humaria* (Ingleby *et al.*, 1990), *Sphaerosporella* (Danielson, 1984) and *Wilcoxina* (Ursic and Peterson, 1997). Of these, *Wilcoxina* spp. are better known to form ectendomy-corrhizas (see Chapter 7). Structurally, the ectomycorrhizas formed by these fungi have been characterized as having thin pseudoparenchymatous mantles with few mycelial connections to the soil and an absence of rhizomorphs. Agerer (2001) referred to the mycelium as being of the 'short distance exploration type' (see Figure 6.25), as distinct from the medium or long-distance exploration types seen in the majority of basidiomycetous mycorrhizas. On the basis of sequencing studies of root tips in Northern European forests, Tedersoo *et al.* (2006) recently added *Geopora, Trichophaea, Helvella, Pachyphloeus, Peziza* and *Sarcosphaera* to the list of suggested ECM formers. Members of the Pezizales comprised 3–13% of the taxa in the four communities examined and were present on 4.5–6.1% of the root tips. In contrast, in a post-fire situation of the kind well known to stimulate the activities of pezizal-ean fungal communities, Fujimura *et al.* (2005) found no root tips colonized by these fungi. The physiological properties of some ECM members of the Pezizales have been investigated in axenic culture (El Abyad and Webster, 1968) that have been reported to be ECM. Following an analysis of the functional status of a wide range pezizaean fungi, Egger and Paden (1986) concluded that covered a continuum from parasitic to mycorrhizal. There remains the need for more re-synthesis experiments to confirm the extent to which these fungi function as ECM symbionts.

Specificity in ectomycorrhizal symbioses

Since almost all species of ECM plants associate with large numbers of distantly related fungi, the plants can be considered to show a low level of specificity. Trappe (1977) estimated that, over its natural geographic range, a single plant species may associate with thousands of fungal species. Likewise, at a local scale, tens of ECM fungi can be found on single trees or on small pure stands of a given tree species (Bruns, 1995). It has been argued that the lack of specificity shown by the plants will be advantageous not only because it increases the chance that the roots of seedlings will find appropriate colonists, but also because association with fungal species of different physiological attributes may provide access to a broader range of nutri-ent pools (Molina *et al.*, 1992). Examples of plants that show greater specificity are rare but *Alnus* spp., with specialist symbionts in the genus *Alpova*, are often quoted in this context (Molina, 1979; Miller *et al.*, 1991; Molina *et al.*, 1992). However, since *Alpova* spp. occur in several distantly related lineages, *Alnus* does not actually show high specificity. Rather, the genus associates with a set of fungi that are themselves specialized on this tree genus. There remains the possibility that the tropical tree *Pisonia grandis* may be exclusively associated with fungi in the Thelephoraceae (Chambers *et al.*, 1998), but more extensive studies are required to confirm this.

While many of the fungal partners of ECM symbioses also lack specificity, it is becoming increasingly clear that there are greater levels of specialization among the fungi than was hitherto believed. The application of molecular tools to enable secure identification of fungi forming mycorrhizas on individual root tips has confirmed that, while generalist fungi can be present on a high proportion of the roots of co-associated tree species (Horton and Bruns, 1998), there are a number of ecologically important fungal taxa which may be present on smaller numbers of roots and which show specialization towards particular species of autotroph. Considerable attention has been devoted to the suilloid group which is a monophyletic lineage of fungi comprising *Suillus, Rhizopogon, Truncocolumella, Gomphidius* and *Chroogomphus* specialized on members of the Pinaceae (Kretzer *et al.*, 1996; Bruns *et al.*, 1998; Kretzer and Bruns, 1999). Most suilloid species are entirely restricted to single plant genera or species groups in the Pinaceae and individual clades within both *Suillus* and *Rhizopogon* frequently exhibit these narrow host association patterns. Thus, *Suillus pungens* occurs exclusively with *Pinus muricata* and *P. radiata*, two North American species with small and scattered natural ranges in California. *S. pungens* fruits prolifically while occupying only a small fraction of the ECM root assemblage (Gardes and Bruns, 1996; Bruns *et al.*, 2002). Likewise, *Rhizopogon salebrosus* (=*R. subcaerulescens*) and *R. occidentalis* (=*R. ochraceorubens*) are restricted to the same pine forests in which, again, they produce abundant sporocarps while accounting for only a small proportion of the ECM root biomass (Gardes and Bruns, 1996; Horton and Bruns, 1998; Stendell *et al.*, 1999; Taylor and Bruns, 1999).

The possible advantages and disadvantages to both plant and fungus of specialization of the kind seen in the suilloid group must be considered. Molina *et al.* (1992) hypothesized that advantages might accrue to the plant because the risk that organic C allocated to its fungal partner would be accessible to other plants forming ectomycorrhizas with it via a common mycelial network (CMN), so-called facultative epiparasitism (Bruns *et al.*, 2002), would be reduced. The quantitative significance of C movement between plant species through CMNs of generalist fungi remains to be evaluated (see Chapters 8 and 16), but the potential for functional discrimination between specialist and generalist fungi is evident. If avoidance of facultative epiparasitism were a factor selecting in favour of specialization, then specificity would be expected to be more prevalent in stands made up of mixed rather than single species. For this reason, comparative analyses of the extent of specialization in stands with different extents of interspecific mixing of plants would be valuable.

A further possibility is that plants favour specialist fungi because they have lower C demands. Reduced C costs could accrue if the fungi either produced lower reproductive and vegetative biomass or, through saprotrophy, had separate access to exogenous sources of organic C. However, there is little evidence to support these possibilities. The specialist fungi are characterized by intensive fruiting associated with relatively small amounts of colonization (Gardes and Bruns, 1996), which is suggestive of high rather than low C demand. In contrast to generalists like many *Russula* species (Agerer, 1995; Redecker *et al.*, 2001), they produce vigorous and extensive mycelial systems (Bonello *et al.*, 1998; Molina *et al.*, 1999) which must carry high C costs for both production and maintenance. The notion that the C economies of *Suillus* and *Rhizopogon* species might be subsidized by saprotrophy is not supported by their carbon isotope signatures which are similar to those of other pine-associated ECM fungi and distinct from those of saprotrophic fungi (Högberg *et al.*, 1999). A remaining possibility is that more effective exploitation of an otherwise growth-limiting resource such as N or P provides a greater return per unit of

C investment by the plant. On the basis that the suilloid species *S. bovinus* and *R. rose-olus* provide pines with access to organic sources of N (Chapter 9), they were included in the category 'protein fungi' (Abuzinadah and Read, 1986b). Detailed investigations of the relative N mobilizing capabilities of specialist and generalist ECM fungi occurring in the same ecosystem will be required to resolve these questions.

The observation that pioneer tree species have a greater preponderance of specialist fungi than those occurring in late successional forests had been made earlier (Kropp and Trappe, 1982). Molina *et al.* (1992) suggested that such specificity might both increase the chance of early contact with an appropriate fungus and then provide invigoration of its mycelia by allocation of C in the absence of competing sinks.

Genetics of ectomycorrhizal fungi

There has long been an awareness of the extent of interspecific variability among ECM fungi in the structure and function of the mycorrhizas they form. Studies have demonstrated, however, that in some of the most widely occurring fungi, such as *Pisolithus tinctorius* (Lamhamedi *et al.*, 1990; Lamhamedi and Fortin, 1991; Burgess *et al.*, 1995), *Laccaria bicolor* (Kropp *et al.*, 1987; Wong *et al.*, 1989; Wong and Fortin, 1990) and *Hebeloma cylindrosporium* (Debaud *et al.*, 1986; Marmeisse *et al.*, 1992b), the magnitude of intraspecific variability can be as great as that between species.

Most basidiomycetes are heterothallic, with very complex genetic control of mating, there being several thousand mating types determined by one to four multi-allelic loci (Aα, Aβ, Bα, Bβ) (Kües and Casselton, 1992; Casselton and Kües, 1994; Debaud *et al.*, 1995). A sexually sterile monokaryotic mycelium is produced from germination of a basidiospore in which meiosis has occurred. In ECM basidiomycetes, such monokaryons are normally unable to produce fully developed ectomycorrhizas (see below), but the mating of two monokaryotic mycelia enables the resulting dikaryon both to produce fruit bodies and to form mycorrhizas. Two mating systems are recognized. In bipolar forms, which constitute approximately 25% of heterothallic species, compatibility between homokaryons is controlled by multiple alleles at a single locus, A. The remaining 75% are tetrapolar species in which there are two unlinked mating-type loci, A and B, again with multiple alleles. In the tetrapolar system, homokaryons are compatible with each other when they have different alleles at both mating-type loci. ECM basidiomycetes are much harder to work with than the saprotrophs that have been used to reveal these genetic systems. However, mating types of the bipolar kind have been identified in three *Suillus* species (Fries and Neumann, 1990; Fries and Sun, 1992), while *H. cylindrosporum* (Debaud *et al.*, 1986) and *Pisolithus tinctorius* (Lamhamedi *et al.*, 1990) are of the tetrapolar type. Pairings between monokaryons of *H. cylindrosporum* that were derived from the progeny of six wild dikaryotic strains of disjunct geographical distribution, have demonstrated the occurrence of multiple alleles at the A and B mating-type loci in this species (Debaud *et al.*, 1986).

Within the genus *Laccaria*, a complex pattern of mating systems exists. The most important ECM species, *L. amethystea, L. bicolor, L. laccata* and *L. proxima*, have tetrapolar mating systems (Fries and Mueller, 1984), in which all four mating types are rarely found in the progeny of a dikaryon. Doudrick *et al.* (1990) showed the presence of a large number of alleles in *L. laccata* var *moelleri*. By pairing isolates obtained from different regions of North America, they estimated the outbreeding efficiency of this system to be 88%. However, there is also evidence that, in *L. laccata*, genes other than those determining mating types can restrict pairings between homokaryons from morphologically similar strains. Thus, Fries (1983) found two incompatible groups of the species in a restricted area of Sweden, and Mueller (1991) detected three such groups in North America that were also incompatible with the isolates of Fries. *L. bicolor* also shows evidence of incompatibility groups (Kropp and Fortin, 1988; Doudrick and Anderson, 1989). Such analyses highlight the inadequacy of our understanding of the 'species' as a unit and emphasize the need to characterize and describe the origin of isolates used in any experimental study.

Intraspecific genetic variation can be expressed at the physiological level in the form of differences in growth, production of enzymes or of auxins (Gay and Debaud, 1987), or at the level of ECM infectivity or aggressiveness (Wong and Fortin, 1990; Burgess *et al.*, 1994). Gay and Debaud (1987) measured rates of production of indole acetic acid (IAA) by different species of *Hebeloma* and compared the observed rates with those of wild strains, of monokaryons produced by germination of spores from a single fruit body (sib-monokaryons) and of dikaryons synthesized from the monokaryons of *H. cylindrosporum* (Figure 6.6). Intraspecific variation in auxin production was as large as between species. Analyses of differences in growth and of enzymes such as glutamate dehydrogenase (Wagner *et al.*, 1988), nitrate reductase (Wagner *et al.*, 1989) and acid phosphatase (Meysselle *et al.*, 1991) reveal relatively broad variability (Table 6.3). Similar differences in expression of the latter enzyme have been observed in dikaryotic progenies of controlled matings in *L. laccata* (Kropp, 1990).

Debaud *et al.* (1995) calculated that up to 36% of the progeny of a dikaryon are more efficient in terms of enzyme production than the parents, and point out the obvious potential which this presents for designing programmes to select improved mycelia. The problem faced by any such programme is that phenotypic expression is strongly influenced by environmental conditions. Gay *et al.* (1993) measured the growth of 50 synthesized dikaryons of *H. cylindrosporum* under a range of widely used culture conditions and showed that the expression of genetic diversity in terms of yield was strongly influenced by nutritional conditions (Table 6.4). It is evident from such studies that a genotype selected for performance under one set of conditions will not necessarily be useful in other circumstances. Variations in growth within a range of *H. cylindrosporum* genotypes was less than variation in enzyme expression, probably because growth is controlled by more genes than is expression of activity of a single enzyme. The interactive nature of gene expression in control of growth leads to the generation of phenotypes in which performance is ultimately less distinctive than might be predicted from preliminary screening of progeny.

Some insights into the processes of genetic control of ECM development have been obtained by challenging potential host plants with fungal strains of known genotype. Lamhamedi *et al.* (1990) examined the ability of 28 sib-monokaryotic and 78 reconstituted dikaryotic strains of *P. tinctorius* to form mycorrhizas with *Pinus*. While a few of the monokaryons showed limited ability to colonize roots, fully developed mycorrhizas and growth promotion of the plant were obtained only with dikaryons. The ability to produce rhizomorphs was also restricted to dikaryotic strains (Lamhamedi and Fortin, 1991). Strong influences of the environment upon the expression of genetic potential were indicated by the observation that development of rhizomorphs by a given dikaryotic strain was far more extensive over soil

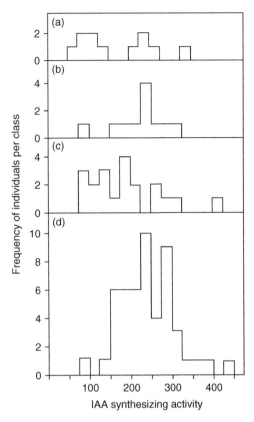

Figure 6.6 Specific IAA synthesizing activity of *Hebeloma*. (a) Different *Hebeloma* species, (b) different wild strains of *H. cylindrosporum*, (c) sib-monokaryons being the progeny of the HC1 dikaryotic strain of *H. cylindrosporum*, (d) the dikaryons synthesized from all possible fusions between these monokaryons. Specific activity expressed as nmol IAA synthesizing protein/mg/h. From Gay and Debaud (1987), with permission.

than over cellulose sheets. Here, it is interesting to speculate that the expression of hydrophobin genes may be involved. These have been detected in the population of symbiosis related (SR) proteins in *Eucalyptus* mycorrhizas (Martin and Tagu, 1995). Hydrophobins are also implicated in formation of hyphal aggregates and fruit bodies of the wood-decay fungus *Schizophyllum commune* in which some of the genes are only expressed by the dikaryons (see Wessels, 1992).

Using sib-selected homokaryotic cultures of *L. bicolor*, Kropp *et al.* (1987) demonstrated that monokaryotic strains could form ectomycorrhizas, at least under laboratory conditions. While some of the monokaryons were strongly mycorrhizal, others quickly lost the ability to colonize roots of *Pinus banksiana* and, in some cases, no colonization took place. Wong *et al.* (1989), also using strains of *L. bicolor*, showed that a given genotype could proceed only to a particular stage of ECM development, arrest being seen at defined points in the differentiation processes. Some strains did not respond to the presence of the root, others produced only a surface weft of hyphae, while others produced a Hartig net but no mantle. The strains could also be distinguished by the rate of development of these structures and the morphology

Table 6.3 Range of intraspecifc variation of different properties of the ectomycorrhizal fungus *Hebeloma cylindrosporum*.

Character	11 Wild dikaryotic mycelia		HCI parental dikaryon	20 Sib-monokaryons		50 synthesized dikaryons		Synthesized dikaryons showing a higher activity than that of the parental dikaryon HCI (%)
	Range of variation	Average		Range of variation	Average	Range of variation	Average	
Growth in the presence ammonium (mg protein per culture)	4.5–7.0	5.3	6.6	5.9–8.0	6.9	4.1–7.8	5.8	14
GDH activity (nkat/mg protein)	1.5–11.6	7.78	9.14	0.19–9.0	2.8	4.0–19.8	8.5	36
Growth in the presence of nitrate (mg protein per culture)	9.1–16.1	12.0	16.1	6.7–35.0	14.2	1.1–23.6	9.7	12
Nitrate reductase activity (mmol NO_2 synthesized/mg protein/h)	201–700	345	397	51–510	211	72–689	344	28
Growth on a P-poor medium (mg protein per culture)	0.10–0.15	0.12	0.14	0.11–0.20	0.17	0.16–0.28	0.20	100
Acid phosphatase activity (nmol p-nitrophenyl phosphate hydrolysed/mg protein/min)	49–676	238	676	85–791	352	186–756	454	8
IAA synthesizing activity (nmol IAA/mg protein/h)	85–321	22	279	108–561	220	94–437	241	30

GDH: Glutamate dehydrogenase; IAA: indole-3-acetic acid. From Debaud et al., 1995.

Table 6.4 Estimates of the components of the variance recorded in 50 synthesized dikaryons, which were progeny of the *Hebeloma cylindrosporum* HC1 strain.

Item	Compatibility group	Phenotypic variance	Environmental variance[a]	Genetic variance	
				Interactive variance[a]	Parental variance[a]
Growth in the presence of ammonium (mg protein per culture)	A1B2 × A2B1	100	22.6	33.4	44.0
	A1B1 × A2B2	100	15.5	51.5	32.9
GDH activity (nkat/mg protein)	A1B2 × A2B1	100	14.1	42.7	43.0
	A1B1 × A2B2	100	6.0	40.0	53.9
Growth in the presence of nitrate (mg protein per culture)	A1B2 × A2B1	100	6.9	79.0	14.1
	A1B1 × A2B2	100	8.3	7.2	19.5
Nitrate reductase activity (nmol NO$_2$ synthesized/mg protein/h)	A1B2 × A2B1	100	13.3	85.6	1.0
	A1B1 × A2B2	100	–	–	–
IAA synthesizing activity (nmol IAA/mg protein/h)	A1B2 × A2B1	100	14.3	43.3	42.4
	A1B1 × A2B2	100	57.8	26.7	15.3

GDH: Glutamate dehydrogenase; IAA: indole-3-acetic acid. [a]Expressed as a percentage of the phenotypic variance.
Data from Gay and Debaud, 1987; Wagner et al., 1988, 1989; Gay et al., 1993; Debaud et al., 1995.

of the fungus in the Hartig net (Wong *et al.*, 1990). Observations of this kind suggest that each stage of the processes of ECM development is under separate genetic control, completion of the sequence being dependent upon a cascade of gene expression in both organisms. The differences within fungal species in their ability to express physiological or morphogenetic attributes can be ascribed to genetic differences, most of which are features of DNA sequences of the nuclear genome. The use of variants should permit the description of colonization as a sequence of well-defined steps which can be reasonably presumed to be under the influence of particular fungal genes. The studies of Wong *et al.* (1990) and Burgess *et al.* (1994, 1995) provide excellent starting points for analysis in *Pinus* and *Eucalyptus* respectively. Further progress in understanding the control of these processes, which has been dependent upon isolating and cloning the genes coding for each attribute, is now being made.

The formation of ectomycorrhizas

The early stages of development of an ectomycorrhiza will obviously be influenced by the source of the inoculum. Two distinct patterns can be envisaged. In nature on established trees, first, second and further orders of lateral roots all having restricted potential for extensive growth, are produced, usually seasonally, from the axis of the long roots of unlimited growth. These laterals become colonized either from the Hartig net present on the long root as in the case of *Pinus* spp. (Robertson, 1954; Wilcox, 1968a, 1968b) or from the inner mantle of the subtending long root, as in the case of *Eucalyptus* (Massicotte *et al.*, 1987b, 1987c). In both of these circumstances, the colonizing fungus will be the same as that forming mycorrhizas on the parent root. In contrast, where the lateral root emerges through a portion of the uncolonized long root, or where a seed germinates to produce a completely new root system, the potential arises for colonization by different fungi from propagules in the soil. In these cases, the processes of ECM formation are determined by phenomena such as recognition, compatibility and inoculum potential. The diversity of ECM types and fungal species observed in nature on a single tree reflects the complexity of temporospatial events determining the colonization process.

In order to elucidate such complex events, most research to date has employed simplified systems in which plants, usually seedlings, are challenged by a single fungal species in monoxenic culture. By sequential analysis, some of the major events in development of the symbiosis from approach to contact and differentiation have been evaluated.

Precolonization events

It is well known that plant roots release compounds into their immediate environment and the importance of such exudates, mainly as nutrients for the general microbial population, is evidenced by enhancement of this community in the rhizosphere. The challenge remains to determine which, if any, of the array of compounds so far identified is sufficiently specific in its effects to exert selective impacts and so enable communication with potentially symbiotic partners. That specific root exudates may be involved in the ECM establishment situation is suggested by experiments using compatible and incompatible isolates of *Pisolithus tinctorius* and

Paxillus involutus (Horan and Chilvers, 1990). By interposing a permeable membrane between plant and putative symbiont, it was shown that hyphae of compatible symbionts were attracted towards the membrane while those of the incompatible species were not. The likelihood is that specific signal molecules are in very much lower concentration than those which exert nutritional effects. There are parallels in the flavonol compounds involved in signalling in *Agrobacterium* and *Rhizobium* interactions with plants (see Peters and Verma, 1990) and with the recently identified strigolactones that have been shown to have profound influences on metabolism of presymbiotic AM fungi (see Chapter 3).

Debate continues on the chemical nature of such attractants in ECM systems. It is known that hormones, including cytokinins (Gogala, 1991) and indole-3-acetic acid (IAA) (Gay and Debaud, 1987), can influence hyphal branching and growth, but again there is little evidence for specificity. However, exudates of plant roots can stimulate germination of spores of ECM species (Fries, 1987; Ali and Jackson, 1988), but often such effects can also be obtained using roots of non-hosts (Wong and Fortin, 1990). It is known that in the vicinity of a host root hyphal morphology changes, there being increased branching and growth. In eucalypt ECM, host-produced metabolites such as the flavonol rutin and the cytokinin zeatin, strongly modified hyphal branching patterns and were identified as triggers or 'branching factors' (Lagrange *et al.*, 2001). The increased mycelial ramification enhances the probability of contact with the root.

Exposure of the fungus *Pisolithus microcarpus* to zeatin stimulates accumulation of metabolites in its hyphae (Beguiristain *et al.*, 1995). One of these, hypaphorine (a tryptophan derivative), is known to be produced in large amounts by this fungus during ECM development (Beguiristain and Lapeyrie, 1997). It was suggested that competitive antagonism between endogenous IAA produced by the plant and hypophorine secreted by the fungus must be involved in regulation of development of the symbiosis (Ditengou *et al.*, 2000; Ditengou and Lapeyrie, 2000; Jambois *et al.*, 2005). These studies again highlight the likely role of IAA in control of ECM formation, demonstrated by the production of the multiseriate Hartig net (Figure 6.7) (Gea *et al.*, 1994; Tranvan *et al.*, 2000).

In order to determine whether auxin-regulated plant genes could play a role in the establishment of ECMs, Charvet-Candela *et al.* (2002a) screened a cDNA library from auxin-treated roots of *Pinus pinaster* and isolated several auxin-inducible cDNA clones. This screen led to the identification of two auxin-inducible *P. pinaster* genes that were shown to be upregulated on establishment of the ECM symbiosis (Charvet-Candela *et al.*, 2002b; Reddy *et al.*, 2003). One of these, *Pp-iaa88*, encodes a putative transcription factor that may play a role in triggering a cascade of molecular events in pine roots leading to the formation of mycorrhizas.

The demonstration (Garbaye, 1994) that the rate and extent of ECM formation can be enhanced under some circumstances by the presence of fluorescent pseudomonads, so-called 'mycorrhiza helper bacteria' (see Chapter 16), indicates the possibility that trigger compounds are released by organisms other than ECM fungi.

Contact between fungus and root – molecular events

Techniques in which seedlings are placed in contact with a developed mycelial inoculum and then harvested sequentially in the hours or days following contact

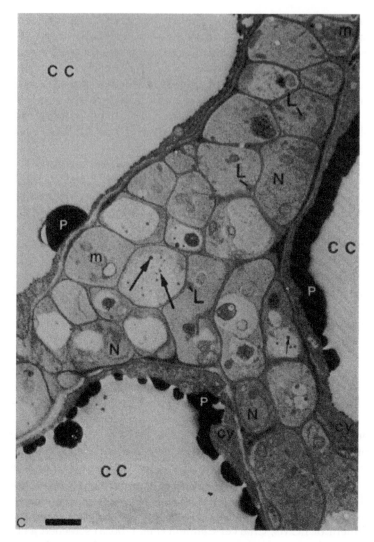

Figure 6.7 Extensive development of a multiseriate Hartig net in a mycorrhiza formed between *Pinus pinaster* and a mutant of *Hebeloma cylindrosporum* overproducing IAA (see text). Cortical cells (CC). Bar=2 μm. From Gea *et al.* (1994), with permission.

and formation of ectomycorrhizas, have enabled elucidation of the responses of the partners at the molecular level. The symbiosis between *Eucalyptus* (*E. globulus* and *E. grandis*) and *Pisolithus* spp. has been widely used for such studies. The paper-sandwich method (Horan *et al.*, 1988) yields synchronous colonization of lateral roots over a period of days and the cellophane-over-agar method (Malajczuk *et al.*, 1990) enables the same sequence of events to be completed on the primary root within hours (Table 6.5).

Establishment of the symbiosis must be under the control of genes of both partners and changes of gene expression, revealed in the form of changes of protein biosynthesis and mRNA populations, have provided valuable insights into the

Table 6.5 Sequence of the formation of *Eucalyptus* ectomycorrhiza obtained using the cellophane-over-agar technique (see text) after Malajczuk *et al.* (1990).

Time	Developmental stage	Anatomical features
0–12 h	Preinfection	Hyphal contact with the root
12–24 h	Symbiotic initiation	Fungal attachment to the epidermis
24–48 h	Fungal colonization	Initial layers of mantle
		Hyphal penetration between epidermal cells
48–96 h	Symbiotic differentiation	Rapid buildup of mantle hyphae
		Hartig net proliferation
96 h–7 days	Symbiotic function	Mantle well developed and tightly appressed to epidermal cells
		End of Hartig net growth

events occurring at the subcellular level. Hilbert and Martin (1988) established the *E. globulus–P. tinctorius* association in paper sandwiches and used two-dimensional gel electrophoresis (2-D PAGE) to compare the protein profiles of ECM roots harvested over several weeks of development with those of uncolonized roots and of free-living mycelium. In the established mycorrhizas, they observed a large downregulation of those polypeptides present in the free-living partners and of a total of 520 polypeptides found, only 10 were unique to the symbiotic condition. These symbiosis-related (SR) proteins were termed 'ectomycorrhizins'.

By harvesting the primary roots of seedlings grown in the colonized and uncolonized condition, Hilbert *et al.* (1991) subsequently showed that major changes of polypeptide synthesis occurred within hours of colonization by the fungus. They detected the accumulation of seven ectomycorrhizins during the early stages of ECM establishment. At the same time, there were marked decreases in several plant and fungal polypeptides, the loss being referred to as 'polypeptide cleansing'. Studies of the pattern of incorporation of labelled ^{35}S-methionine into protein (Hilbert *et al.*, 1991) showed that, whereas there was extensive downregulation of polypeptides resulting from inhibition of synthesis, intense labelling of ectomycorrhizins occurred. This is indicative of acceleration of current biosynthesis of SR proteins during symbiotic development. Such changes were observed only when roots were challenged by compatible races of *P. tinctorius*.

The most important changes observed by Hilbert *et al.* (1991) were in a group of fungal acidic polypeptides which increased as the symbiosis became established up to 4 days after contact. The effects were most marked in the presence of the highly aggressive isolate, H2144, and also observed with the moderately aggressive one, H441. A strongly upregulated fungal polypeptide apparently accumulated in the fungal walls of *E. globulus–P. tinctorius* mycorrhizas was shown to be composed of isoforms of an acidic polypeptide which accumulates during the early stages of colonization (Martin and Tagu, 1995). This protein, (32kD-CWP, see Figure 6.8a), subsequently recognized as symbiosis-related acid polypeptide 32 (SRAP 32), is now known to be encoded by part of a multi-gene family of fungal cell wall proteins (Laurent *et al.*, 1999). That this was only one of many changes in gene expression associated with mycorrhiza formation in the *E. globulus–Pisolithus* system was demonstrated by comparative analysis of the mRNAs accumulated in free-living mycelium of the fungal symbiont and in mycorrhizas formed by it (Figure 6.8b) (Tagu *et al.*, 1993).

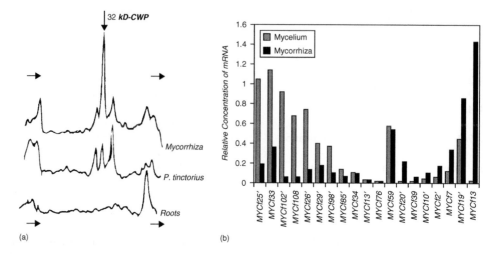

Figure 6.8 (a) Cell wall polypeptides in ectomycorrhizas formed between *Eucalyptus globulus* and *Pisolithus tinctorius* 441, non-colonized *Eucalyptus* roots and cultured *Pisolithus tinctorius* 441. Densitograms of 1-D SDS-PAGE from cell wall proteins (CWP). Note the increased accumulation of a fungal band (32-kDa CWP) in mycorrhizas (arrowed). Data of De Carvalho and Martin, unpublished. From Martin and Tagu (1995), with permission. (b) Changes in gene expression during the formation of ectomycorrhizas between *Eucalyptus globulus* and *Pisolithus tinctorius* 441. Free living mycelium, hatched bars and ectomycorrhizas stippled bars. From Tagu *et al.* (1993), with permission.

A further gene family detected and characterized from SR proteins in these early molecular studies were the hydrophobins, among which HydPt-2 and HydPt-3 were markedly upregulated as the *E. globulus–Pisolithus* symbiosis developed (Tagu *et al.*, 1996, 2001a, 2002). The likely importance of these molecules in cellular recognition processes and in the water relations of ECM plants has become increasingly clear and is considered further below.

Using a different *Eucalyptus* species, *E. grandis*, Burgess *et al.* (1993, 1994, 1995) investigated aspects of fungal specificity and possible function of SR protein production by screening the impacts of different *P. tinctorius* isolates on protein synthesis in ECM roots. Three isolates were used: one very aggressive, one moderately so, and another incapable of forming mycorrhiza with this eucalypt species. Only during colonization by the most aggressive isolate, which was also shown to enhance growth of the test plants, was there a large upregulation of fungal protein biosynthesis, which occurred within 4 days of contact with the root. The plant showed a marked inhibition of polypeptide synthesis. Early studies also revealed changes in polypeptide profiles in mycorrhizas formed between *Betula pendula* and *Paxillus involutus* (Simoneau *et al.*, 1993) and between *Pinus resinosa* and *P. involutus* (Duchesne *et al.*, 1989b). However, SR proteins were not detected in the *Picea–Amanita* symbiosis examined by Güttenberger and Hampp (1992) and Güttenberger (1995), though there were quantitative changes in tissue protein content associated with ECM formation.

Progress towards a more complete understanding of the molecular events associated with ECM formation has been enabled by the application of DNA microarray technology. Transcription profiling has now been achieved in a number of plant–fungus symbioses and the likely functions of many of the expressed SR genes are becoming clearer. A pioneering study (Voiblet *et al.*, 2001) used cDNA arrays to compare gene expression in 4-day-old *Eucalyptus–Pisolithus* mycorrhizas with those of non-mycorrhizal seedlings of the same age, and of the fungus grown axenically. There were significant differences in the expression levels of 17% of the 850 genes represented on the array. Transcription profiles subsequently obtained from a number of different plant–fungus combinations including *Tilia–Tuber* (Polidori *et al.*, 2002), *Pinus–Laccaria* (Podila *et al.*, 2002) and *Betula–Paxillus* (Johansson *et al.*, 2004) have confirmed similar changes. In the latter study for example, 10% of the 2284 plant and fungus genes were differentially regulated in the symbiosis.

Application of transcriptional profiling to plant–fungus associations from the time of first contact between the partners to establishment of the fully developed mycorrhiza has provided further advances. Two recent studies, one using the *E. globulus–P. microcarpus* (Duplessis *et al.*, 2005), and the other the *B. pendula–P. involutus* (Le Quere *et al.*, 2005) model systems, have described global patterns of gene expression over a period of 20–21 days. In each case up- or downregulation was determined relative to levels of expression in free-living mycelia of the respective fungal or non-symbiotic plant partner. Duplessis *et al.* (2005) observed five major and distinct temporal patterns of induction or repression, each associated with a different group of genes that were responsive at early, mid or late stages of ECM formation (Figure 6.9). At day 4, the differentially expressed genes of *P. microcarpus* included those coding for cell wall SR proteins like SRAP32 and hydrophobins (HydPt-2 and HydPt-3) which are candidate markers for SR changes in the cell wall. These declined markedly towards day 7 (Pattern I, Figure 6.9). Increased expression of hydrophobin and SRAP transcripts during the early stages of ECM formation is again suggestive of a direct participation of the corresponding proteins in morphogenetic events related to the adhesion to root surfaces. Several fungal transcripts involved in primary metabolism, such as hexokinase, NAD-malate dehydrogenase, aspartate aminotransferase and NADH dehydrogenase were increasingly expressed between days 4 and 7 (Pattern II, Figure 6.9), suggesting that C transfer between plant and mycobiont was already taking place. Increased C flux would explain the observed increases in expression of genes associated with glycolysis, the TCA cycle and respiration. Several cellular functions were induced at 12 days (Pattern III, Figure 6.9). These included protein synthesis, mitochondrial activity and the appearance of signalling pathway components. Transcripts involved in amino acid metabolism also showed their highest expression at this stage. Pattern IV (Figure 6.9) demonstrates a decline in a number of functions which are assumed to be primarily associated with maintenance of the mycelium in pure culture and which are downregulated as the symbiosis develops.

From days 7 to 12, transcripts coding for various homologues of stress- and defence-related proteins, such as metallothionein-like proteins, a hypersensitive response-induced protein and a pathogenesis-related (PR) protein were observed. The increased levels of these compounds suggest that colonized cells mount some resistance to the initial phases of colonization, but that these diminish towards 21 days (Pattern I, Figure 6.9) as the relationship develops in a compatible manner. At this stage, transcripts coding for defence-related proteins were upregulated in the plant tissues. The genes that were transcriptionally responsive in the mid and later stages of ECM formation coded for enzymes associated with amino acid and protein biosynthesis.

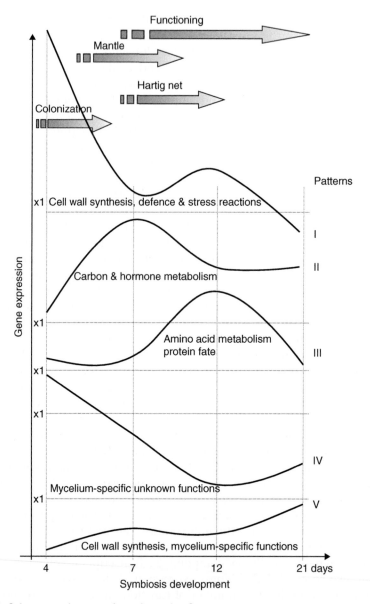

Figure 6.9 Schematic drawing describing the five major gene expression patterns of plant and fungal genes during development of the *Eucalyptus–Pisolithus* mycorrhiza, see text. Data from Duplessis *et al.* (2005).

Transcripts involved in protein synthesis are also upregulated in the plant 12 days after contact. These include those associated with the ubiquitin/proteosome pathway, serine carboxypeptidase synthesis and primary events in N and C metabolism.

The study of Le Quere *et al.* (2005), employing the *Betula–Paxillus* system, demonstrated remarkable similarities in the expression patterns to those observed by Duplessis *et al.* (2005), despite the differences of both plant and fungal symbionts.

An upregulation in expression of genes involved in fungal C metabolism coincident with mantle and Hartig net formation, and the initiation of a plant defence programme similar to that described in pathogenic interactions, was again observed. In the *Betula* symbiosis, differential expression also involved several genes that are widely implicated as determinants of tissue water balance. These include aquaporins, which influence the permeability of membranes to water, and dehydrins, which contribute to the provision of tolerance to water stress (see Chapter 10).

There have now been two analyses of gene expression in the more mature situation with mycorrhizas supporting fully developed extramatrical mycelial systems (Morel *et al.*, 2005; Wright *et al.*, 2005). They are more closely related to the real-world situation than the earlier studies which viewed colonized roots as isolated compartments. Both again employed the *Betula–Paxillus* system. While much emphasis has been placed in recent times on the likely importance of the external mycelial systems of ECM roots (Leake *et al.*, 2004a), these are the first investigations to examine aspects of their function at the molecular level using intact systems. In combination, they provide strong indications of functional specialization between these morphologically distinct components of the system (see also Chapter 9).

In their analysis, Morel *et al.* (2005) identified 65 genes that were differentially expressed between external mycelium and ECM roots. Genes coding for urea and spermine transporters were upregulated four- and sixfold, respectively in the external mycelium whereas, in the ECM roots, a putative phosphatidyl serine decarboxylase (Psd), which may contribute to membrane remodelling during ECM formation, was upregulated 24-fold. Wright *et al.* (2005) generated comparative global gene expression profiles of ECM root tips, rhizomorphs and the distal mycelium of *P. involutus* as it colonized ammonium-enriched organic matter. Statistical analysis revealed that 337 of 1075 fungal genes were differentially regulated between these three tissue types. While the profiles of the mycelium and rhizomorphs were similar to each other, they were distinctly different from those of the mycorrhizal tips. Clusters of genes exhibiting distinct expression patterns within specific tissues were identified. As in the Morel *et al.* (2005) study, those putatively implicated in the nitrogen cycle, most notably amino acid and urea metabolism, were relatively strongly expressed in the mycelia and rhizomorphs. Genes involved in the provision of C skeletons for ammonium assimilation and in the glyoxylate cycle were also highly expressed in these two compartments. A tubulin gene homologue, expressed particularly strongly in rhizomorphs, is of interest because actin is implicated in the motile tubular vacuole system that may function in long-distance transport through ectomycorrhizal hyphae (Ashford and Allaway, 2002; see Chapter 10). Genes implicated in glycolysis/gluconeogenesis, glycerol metabolism and amino sugar synthesis are among those selectively upregulated in the root tips.

Events at the whole root level

Root hairs proliferate behind the apices of growing uncolonized roots and provide a large surface area for potential contact with any ECM mycelium in surrounding soil. It appears that, on making contact with root hairs, ECM hyphae can alter their orientation of growth to the surface of the root and partially envelope the hairs (Massicotte *et al.*, 1989; Thomson *et al.*, 1989). When hyphae first make contact with the root surface, they may show morphological changes prior to the production of mantle or Hartig net. These include increased branching and fusion of hyphal tips (Jacobs *et al.*, 1989). The mechanisms that enable attachment of compatible hyphae to the roots and provide discrimination against the incompatible remain unclear. That they are likely to be subtle is indicated by the marked differences in compatibility revealed when hosts are challenged with different isolates of ECM fungus (Malajczuk *et al.*, 1990; Lei *et al.*, 1991; Burgess *et al.*, 1994). There is evidence that, among the plant defence processes, increased deposition of phenolic materials at the point of contact with incompatible species or strains may be involved (Malajczuk *et al.*, 1984), but this is likely to be a late manifestation of many interacting effects, the most critical of which will be occurring at the molecular level. Albrecht *et al.* (1994) found that the induction of chitinases and peroxidases in *Eucalyptus* was related to the aggressiveness of the fungal strain, with only good colonizers inducing a strong response. This calls into question the widely accepted roles of these enzymes in plant defence.

Ultrastructural studies of the plant–fungus interface of ectomycorrhiza have demonstrated the presence of fibrillar material, probably made up of glycoproteins, extending from the fungal wall towards that of the plant (Piché *et al.*, 1983a, 1983b; Lei *et al.*, 1990a, 1990b, 1991). A layer of extracellular fibrillar polymers is present on the surfaces of free-living mycelium of *Laccaria bicolor* (Lei *et al.*, 1991) and *Pisolithus tinctorius* (Lei *et al.*, 1990a) before contact with a root and it seems likely that reorientation of these towards the surface of the plant cell is one of the important initial steps in ECM formation. Giollant *et al.* (1993) detected binding sites on the root surface of spruce (*Picea*) for lectins isolated from the hyphal walls of *Lactarius deterrimus*. These were different from those of *L. deliciosus* in symbiosis with *Pinus*.

There is the possibility that receptor sites are present on both partners in the symbiosis, but that they are masked by unreactive materials. Lapeyrie and Mengden (1993) found that, in the case of the fungal wall, the masking compounds could be removed by the enzymes laminarase or protease, raising the possibility that recognition was through a programmed process of enzyme release, leading to the exposure of the receptor sites on the contiguous surfaces of both partners and establishment of the symbiosis.

Establishment of the composite ECM structure

Use of the pouch system of Fortin *et al.* (1980) enabled detailed studies of the development of mycorrhizas formed by symbionts with roots of known age. Among the plant–fungus combinations studied were those between *Alnus crispa* and *Alpova diplophoeus* (Massicotte *et al.*, 1986), *Eucalyptus pilularis* and *Pisolithus tinctorius* (Massicotte *et al.*, 1987a, 1987b) and *Betula alleghanensis* and *P. tinctorius* (Massicotte *et al.*, 1990).

In the case of the *A. crispa–A. diplophoeus* association, colonization of first-order laterals by rhizomorphs leads to ECM formation within 2–4 days of contact. Fungal hyphae first contact the growing root at a position immediately proximal to the root cap, from which point they grow both basipetally and acropetally to produce a covering of hyphae that keeps up with root elongation. This stage is reached within 24–48 hours of initial contact and, at the same time, a swelling of the root appears proximal to the root cap. The final morphology of each mycorrhiza is dependent upon the stage of lateral root outgrowth at which colonization is initiated. The swollen tip is a feature of a mycorrhiza formed after a lateral has elongated, whereas colonization at an early stage of outgrowth results in a mycorrhiza of uniform thickness.

Longitudinal sections revealed that hyphae of the inner mantle start to penetrate between cells of the root cap immediately behind the apex and, within a very short distance proximal to this, penetrate between epidermal cells to form the Hartig net. The Hartig net develops in an acropetal direction but never penetrates beyond the outer tangential wall of the first layer of cortical cells. The epidermal cells enveloped by hyphae show only slight radial elongation. A combination of light and electron microscopy (EM) has elicited a schematic composite drawing of the *Alnus–Alpova* mycorrhiza (Figure 6.10) which reveals the progressive development as the symbiosis matures behind the apex. The diagram indicates the proliferation of the hyphal walls of the Hartig net produced by repeated branching in the intercellular position (Figure 6.10c, d), the production, typical of this particular association, of ingrowths on the epidermal cell walls (Figure 6.10d), and the concentration of rough endoplasmic reticulum (ER) and mitochondria in fungal and epidermal cells (Figure 6.10d), the latter becoming progressively more vacuolate as they mature (Figure 6.10e).

The development of the *Eucalyptus–Pisolithus* mycorrhizas (Massicotte *et al.*, 1987a, 1987b) follows a broadly similar pattern to that of *Alnus*. The final morphology of the individual mycorrhiza again depends on the stage of lateral root elongation at the time of colonization. Some differences are, however, observed. In *Eucalyptus*, there is a well-defined zone behind the apex in which Hartig net formation is lacking, the so-called apposition or pre-Hartig net zone (Figure 6.11). The epidermal cells of *Eucalyptus* show a rapid response to the presence of ECM fungi as considerable radial enlargement, which is not accompanied by normal apical elongation. Such changes of epidermal cell development are thought to be specifically induced by fungal colonization since, as shown previously by Chilvers and Pryor (1965), they are not seen in non-mycorrhizal roots. A further feature which appears to distinguish the *E. pilularis–P. tinctorius* association is that the plant–fungus interface in the Hartig net is relatively simple. There are no fungus-induced wall ingrowths in the root cells involved in forming the interface and only a small number of labyrinthine branches in the fungal tissue. Simplicity in the interface at the cellular level may be compensated for by the precocious development of lateral roots to form extensive clusters and tubercles, which would have the effect of increasing the total surface area of interface per root system (Chilvers and Gust, 1982; Dell *et al.*, 1994).

The developmental sequence in the *Eucalyptus–Pisolithus* mycorrhizas has been examined in detail by Massicotte *et al.* (1987a, 1987b). Longitudinal sections viewed under the light microscope reveal that there are inherent anatomical differences between primary roots and first-order laterals, even in the absence of ECM colonization (Figure 6.12). The primary roots (Figure 6.12a) are pointed and have a complex organization with a root cap consisting of a distinct columella, an extensive meristem, a cortex of three or four layers, a differentiated stele and a single-layered epidermis. In contrast, first-order laterals (Figure 6.12b) have a rounder, blunt apex, a reduced apical meristem and a subapical construction consisting of three or four layers of cortical cells and a poorly developed stele. Cortical cells and tracheary elements are seen to mature closer to the apex in these laterals than in the primary root (Figure 6.12a). The distinctive structure of the two types of roots is a basic feature of heterorhizic systems and has been described also in *Fagus* (Clowes, 1951; Warren-Wilson and Harley, 1983) and in *Pinus* (Hatch and Doak, 1933; Wilcox, 1964, 1968a).

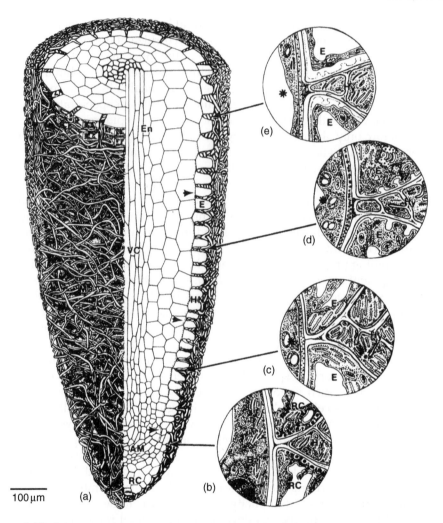

Figure 6.10 Schematic drawings of an ectomycorrhiza formed between *Alnus crispa* and *Alpova diplophloeus*. (a) Three-dimensional drawing of a mycorrhizal root, showing the mantle (Ma), paraepidemal Hartig net (HN) and epidermal cells (E), which become progressively more radially enlarged from the apex back. (b) Root cap region. Hyphae (Hy) have begun to penetrate between the cells of the root cap (RC). The densely cytoplasmic epidermal cell (E) is also shown. (c) Hartig net region. Fungal hyphae (Hy) with rough endoplasmic reticulum and labyrinthine wall branching are penetrating between the epidermal cells (E). (d) Mature Hartig net region. Fungal hyphae (Hy) have reached as far as the modified wall of the exodermis (*). Epidermal cells (E) show wall modifications, including ingrowths. (e) Older Hartig net region. Fungal hyphae (Hy) show reduced numbers of cysternae of endoplasmic reticulum and mitochondria. Epidermal cells (E) show modifications including deposition of wall material, vacuolation and a decrease in number of mitochondria. The hypodermal cell (*) is also more highly vacuolated. From Massicotte *et al.* (1986), with permission.

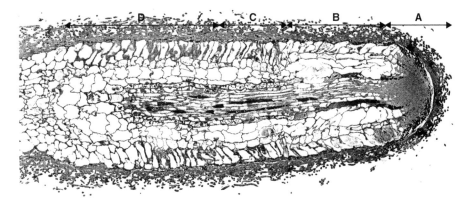

Figure 6.11 Light microscopy of mycorrhiza formed between *Eucalyptus pilularis* and *Pisolithus tinctorius*. Longitudinal section showing zonation. Zone A, root cap meristem; zone B, apposition (pre-Hartig net); zone C, young Hartig net; zone D, older Hartig net. Bar = 100 μm. From Massicotte *et al.* (1987b), with permission.

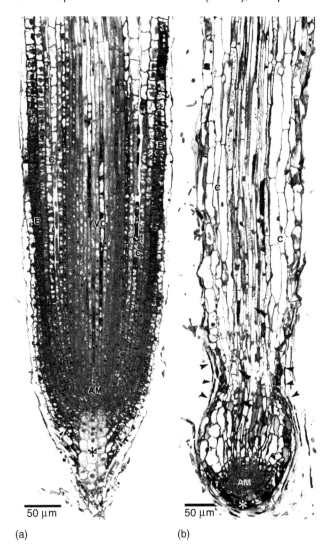

(a)

(b)

Figure 6.12 (Caption on p. 228)

The important functional difference between the two types of structure is that, while the primary root has the capacity for continuous growth, the laterals do not and may abort if not colonized by an ECM fungus.

In the case of *Eucalyptus* seedlings growing in pouches with inoculum of *P. tinctorius*, first-order laterals become colonized by hyphae or rhizomorphs extending from the plug of inoculum. A thin mantle was formed over the lateral, sometimes within 12–24 hours of contact. In longitudinal sections of such a root, four zones, each representing a distinct type of colonization, were recognized (zones A, B, C, D, see Figure 6.11). In zone A, a thin but continuous mantle is seen to form over the blunt meristematic area. In zone B, the hyphae do not penetrate between the cortical cells, but remain in contact with the surface. This is referred to as the apposition or pre-Hartig net zone. In zone C, the Hartig net begins to form and the cells of the epidermis begin to show radial extension. In the mature Hartig net zone, the epidermal cells are fully extended in a radial direction and all are surrounded by the net. Second-order root primordia appear in the pericycle of first-order roots very close to the meristem (Figure 6.13a). Here the endodermis may already have differentiated and de-differentiation occurs at the sites of lateral root initiation. At this level, hyphal penetration between cells of the epidermis of the parent long root is commencing (Figure 6.13a, b). Transverse sections taken proximally to this (Figure 6.13c) show stages of emergence of the second-order laterals into the cortex of the parent root and eventually into the epidermis which is pushed out (Figure 6.13d). At higher magnification (Figure 6.13e), early stages of differentiation of a root cap are seen around the meristem of the primordium, but there is still no sign of fungal penetration of its tissues. Soon after this an apical meristem and vascular cylinder are seen (Figure 6.14a) and longitudinal sections (Figure 6.14b, c, d) show proliferation of hyphae over the apex of the emerging root, radial extension of epidermal cells and stages in the formation of the Hartig net.

Transmission electron microscopy (TEM) enables the interaction between fungus and plant to be visualized in each of the four zones shown in Figure 6.11 (Massicotte *et al.*, 1987b). In zone A (Figure 6.15a), a thick mantle made up of densely cytoplasmic hyphae is seen and some of the hyphae have penetrated between the root cap cells despite the presence, between the cap and the mantle, of an elaborate electron-dense layer. In the young Hartig net zone (C), densely cytoplasmic hyphae penetrate between vacuolate epidermal cells up to the hypodermis (Figure 6.15b). At higher magnification, the epidermal cells can be seen to contain a diffuse electron dense matrix or peripheral deposits of electron dense materials (Figure 6.15c). Hyphae of the Hartig net are densely cytoplasmic (Figure 6.15d) and contain numerous lipid droplets as well as mitochondria. The walls of the epidermal cells do not produce ingrowths in this association, which thus appears to be of a simpler kind than that seen in *Alnus, Pisonia* (Allaway *et al.*, 1985) or *Pinus* (Duddridge and Read, 1984a, 1984b). In zone C, contiguous walls between epidermal and exodermal cells show

---

Figure 6.12 Longitudinal sections of non-mycorrhizal roots of *Eucalyptus pilularis*. (a) Primary root, showing well-developed root cap and apical meristem (AM), giving rise to well-defined tissue layers. (b) A young lateral root, showing limited root cap, reduced apical meristem and subapical constriction (arrowed). Bars = 50 μm. From Massicotte *et al.* (1987a), with permission.

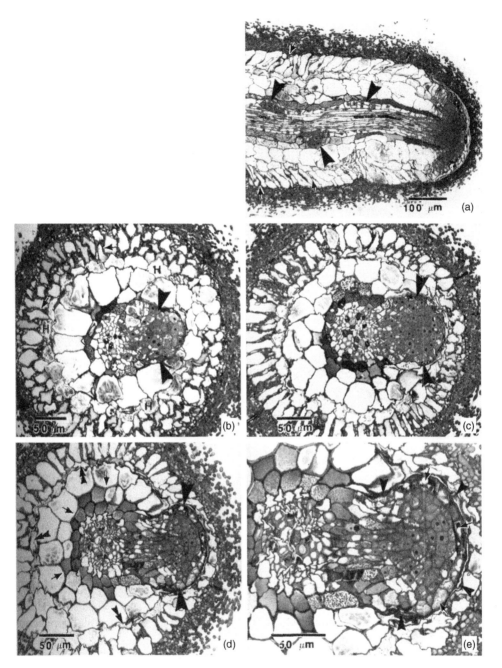

Figure 6.13 Sections of mycorrhizal first-order lateral roots of *Eucalyptus pilularis*. (a) Longitudinal section showing lateral root primordia (arrowheads). Surface hyphae are present, but hyphal penetration between the cortical cells has only just begun (arrow). Bar = 100 μm. (b) Transverse section, showing a young lateral root primordium (arrowheads). Hartig net (arrowed) is completely formed and the exodermis (H) is collapsing. Bar = 50 μm. (c) Transverse section showing growth of the lateral primordium up to the exodermis. Bar = 50 μm. (d) Lateral primordium has completely penetrated the cortex and pushed out the epidermis. Wall thickenings in the cortex are becoming evident (arrowheads). Bar = 50 μm. (e) Higher magnification of the lateral root primordium shown in (d) surrounded by collapsed cells (arrowheads) and showing initiation of root cap (arrows). Bar = 50 μm. From Massicotte *et al.* (1987a), with permission.

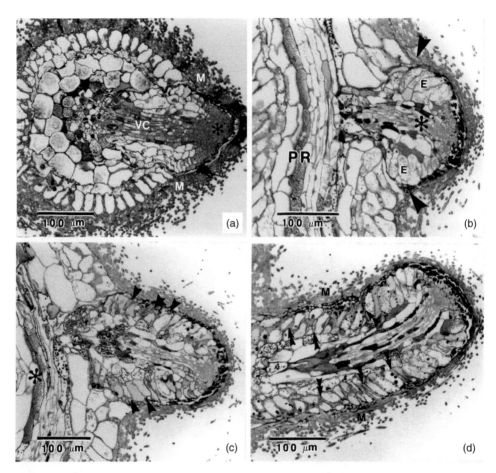

Figure 6.14 Development of lateral root primordia in *Eucalyptus pilularis*. (a) Transverse section of a primordium protruding into the mantle (M). There is no hyphal penetration of tissues. (b) Longitudinal section of primary root (PR) and primordium (*). Aggregations of hyphae at either side of the primordium (arrowheads) and radial enlargement of epidermal cells (E) are evident. (c) Later stage in which hyphal penetration between epidermal cells of the young lateral is evident (arrowheads). (d) Longitudinal section of a fully emerged lateral showing paraepidermal Hartig net (double arrowheads) and mantle (M). Bars = 100 μm. From Massicotte *et al.* (1987a), with permission.

Figure 6.15 Transmission electron microscopy of zones of a mycorrhizal root of *Eucalyptus pilularis* (refer to Figure 6.11). (a) Root cap meristem zone A, showing thick mantle (M) and penetration of hyphae (H) between the cells of the root cap (RC). Bar = 5 μm. (b) Young Hartig net zone B. Cytoplasmic hyphae (arrowed) are present between the vacuolated epidermal cells (E), up to the hypodermis (*). Bar = 5 μm. (c) Higher magnification of region similar to (b). Bar = 1.0 μm. (d) Hyphae between the epidermal cells (E), the walls of which (arrows) do not show structural modifications. Bar = 1.0 μm. (e) Contiguous walls between an epidermal (E) and hypodermal (*) cell, showing suberin lamellae (arrowheads) in the hypodermal wall and plasmodesmata (arrows) between the cells. Bar = 0.1 μm. From Massicotte *et al.* (1987c), with permission. Electron microscopy of mycorrhizas formed between *Eucalyptus pilularis* and *Pisolithus tinctorius* in the older Hartig net zone D.

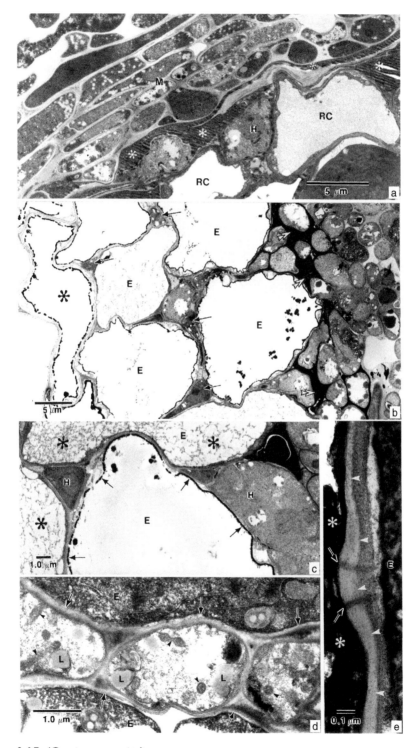

Figure 6.15 (Caption opposite)

suberin lamellae in the exodermal cell walls (Figure 6.15) and plasmodesmatal connections between the cells.

TEM micrographs of the mature Hartig net zone D (Figure 6.16a) show that penetration by now vacuolate *Pisolithus* hyphae between the epidermal cells is far more extensive. They still, however, do not pass into the exodermis. There are extensive formations of suberin lamellae in the exodermal cell wall (Figure 6.16b). In some

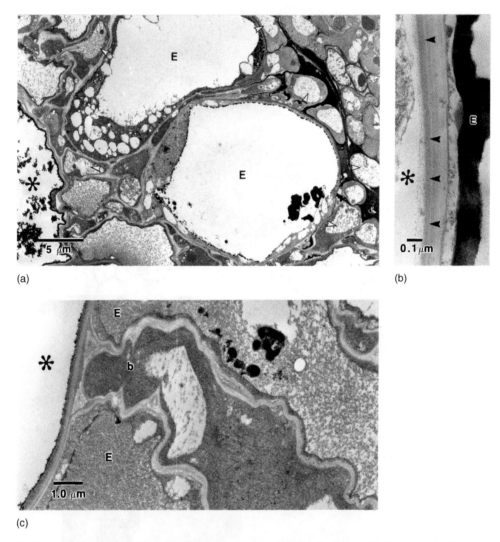

(a)

(b)

(c)

Figure 6.16 (a) A well-developed Hartig net (*Pisolithus tinctorius*) between epidermal cells (E) of *Eucalyptus pilularis*, showing vacuolation of the hyphae in the inner mantle and Hartig net (arrowheads). The Hartig net does not pass the exodermis (*). Bar=5 μm. (b) Contiguous walls between epidermal (E) and exodermal (*) cells showing duberin lamellae (arrowheads) but no plasmodesmata. Bar=0.1 μm. (c) Hartig net adjacent to the exodermis (*), where the hyphae appear to disrupt the middle lamella between the epidermal and exodermal cells. Bar=1.0 μm. From Massicotte *et al.* (1987c), with permission.

cases, hyphae that have reached the exodermis appear to disrupt the middle lamellae (Figure 6.16c). Hyphae of the inner mantle in this region are nucleate, contain many mitochondria and have an interhyphal matrix made up of electron dense and electron lucent materials.

When *Betula alleghaniensis* was challenged by the same strain of *P. tinctorius* in pouches, mycorrhizas were formed on first-order laterals within 4–10 days. Prior to mantle formation, preferential growth of hyphae was observed between root hair papillae in the subapical region of the root (Figure 6.17a). The mantle becomes progressively thicker (Figure 6.17b) until eventually root hairs are no longer visible (Figure 6.17c). A pre-Hartig net zone was again detectable and where this Hartig net is present, it is, like that of *Alnus*, of the para-epidermal kind. A feature of this ECM association is that, while there is no ingrowth of the epidermal cell walls adjacent to the Hartig net, the fungus branches prolifically to form a very complex system of

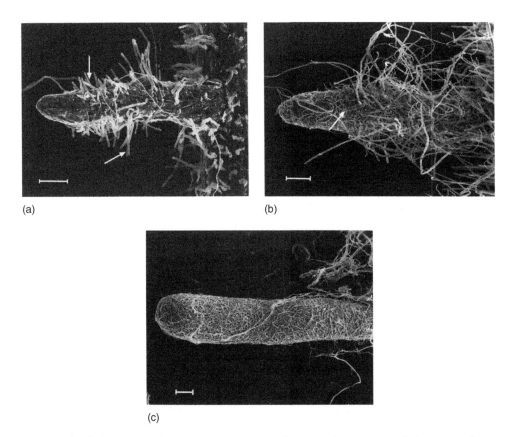

(a)   (b)

(c)

Figure 6.17 Scanning electron microscopy of mycorrhizas formed between *Betula alleghaniensis* and *Pisolithus tinctorius*. (a) Early stages of mycorrhiza formation, in which a few hyphae are present on the root surface and numerous root hairs (arrowed) are evident. (b) A thin mantle (arrowed) has been formed on the root surface. (c) A compact mantle covers the root and root hairs are no longer evident. Bars = 100 µm. From Massicotte *et al.* (1990), with permission.

labyrinthine growths in which the spaces between the fungal branches are barely wide enough to accommodate the elongated mitochondria.

Tuberculate mycorrhizas

Many of the plant species which produce individual ectomycorrhizal roots exposed to surrounding soil, also support compound structures in which clusters of such roots, each with its own mantle and Hartig net, are enclosed in a globular rind, sometimes referred to as a peridium, of fungal tissues. The often densely packed complex of roots can have a diameter of up to 20 mm and is termed a tuberculate mycorrhiza (see Colour Plate 6.1f). Analysis of the fungal rind in a *Eucalyptus* tubercle reveals a two-layered structure, the outermost being 0.15–0.20 mm thick and composed of densely packed hyphae cemented together by a matrix of carbohydrate and lipid-rich material to yield a tissue likely to be highly impermeable (Dell *et al.*, 1990). Similar anatomical features have been revealed in studies of *Pinus* (Randall and Grand, 1986), *Pseudotsuga* (Zak, 1971; Massicotte *et al.*, 1992) and *Engelhardtia* (Juglandaceae) (Haug *et al.*, 1991). Since the peridium appears effectively to isolate the roots within it from soil nutrient pools, the function of the whole structure is unclear. However, extensive systems of fungal rhizomorphs are often seen radiating outwards from young tubercles into surrounding soil. Zak (1971) described rhizomorphs associated with tubercles of Douglas fir, the outer tissues of which were continuous with those of the rind, while the central core of hyphae entered the tubercle to form connections with the individual roots within the tubercle. Dell *et al.* (1990) also observed branching rhizomorphs within tuberculate systems of *Eucalyptus*. These observations suggest that the roots contained within the tubercles have a nutrient storage rather than an absorptive function. A further possibility is that a reduced partial pressure of oxygen within these sealed compartments is conducive to the support of N_2 fixation by free-living bacterial communities. *In situ* assays of acetylene reduction by tuberculate roots formed by *Suillus tomentosus* on *Pinus contorta*, have recently provided indirect evidence for significant quantities of N_2 fixation in these systems (Paul *et al.*, 2007). There is clearly a need to confirm these observations using ^{15}N to provide direct measurements of fixation rates.

The structure of the mature Hartig net

Relationships between plant and fungus at the tissue level

The zone of contact between the symbionts is of key importance in ECM function. In this zone, the Hartig net is produced by hyphae that penetrate between the outer cells of the root axis. This penetration is normally from the inner mantle (see Figures 6.10, 6.11, 6.13). It can occasionally, for example in the case of *Picea abies* (Nylund and Unestam, 1982), occur as soon as hyphae reach the root surface and hence before the mantle is formed. The depth of this penetration differs in angiosperms and gymnosperms. In the majority of angiosperms, penetration is confined to the epidermal layer, so forming what is referred to as an 'epidermal' Hartig net (Godbout and Fortin, 1983). Examples of this type are seen in *Eucalyptus* (see Figures 6.11, 6.16), *Alnus* (see Figure 6.10) and *Betula* (Figure 6.18). Within the epidermal type,

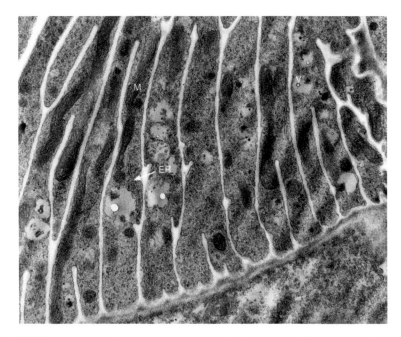

Figure 6.18 Transmission electron microscopy of the mature Hartig net of a mycorrhiza formed between *Betula alleghaniensis* and *Pisolithus tinctorius*. Multibranched hyphae, showing mitochondria (M), endoplasmic reticulum (arrowed, ER) and vacuoles (V). From Massicotte *et al.* (1989b), with permission.

two variants are recognized: the so-called 'para-epidermal' type, in which there is a partial encircling of the epidermal cell as described earlier in *Alnus* and the 'peri-epidermal' structure, in which hyphae encircle the whole cell (Godbout and Fortin, 1983). Of the two types, the former appears to be most common, though Godbout and Fortin (1985) report that in *Populus tremuloides* the peri-epidermal state develops from the para-epidermal under certain culture conditions, indicating that there is possibly a continuum in which the final state is determined by age or environmental conditions.

In ECM gymnosperms, the Hartig net typically penetrates beyond the epidermis to enclose several layers of cortical cells (Figure 6.19), sometimes extending even to the endodermis. This type of structure, best described as a 'cortical Hartig net', is also seen in a few genera of angiosperms such as *Cistus* (Giovannetti and Fontana, 1982) and *Dryas* (Alexander and Bigg, 1981; Debaud *et al.*, 1981) and is occasionally reported in genera, for example *Populus*, which normally have a net of the more superficial kind.

It is of interest that the radial elongation of epidermal cells appears to be restricted to those mycorrhizas which have an epidermal Hartig net. When hyphae penetrate the cortex, whether in the gymnosperm hosts or in angiosperm genera such as *Dryas*, *Cistus* and *Populus*, no such radial elongation is observed. This suggests that increased surface contact between the symbionts is achieved by penetration of the fungus in the cortical type and by extension of the plant cell wall in the epidermal type.

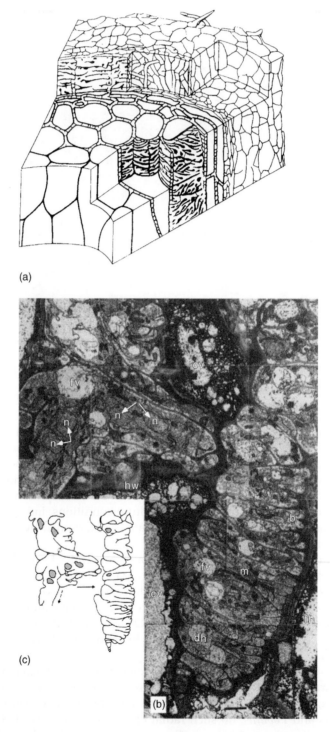

(a)

(c)

(b)

Figure 6.19 (Caption opposite)

Microscopic structure

Intercellular penetration induces profound morphogenetic change in the ECM hyphae. A repeatedly lobed fan-like hyphal front advances across the radial surfaces of the plant cell. A structure of this kind was observed under the light microscope by Mangin (1910) who referred to the fungal lobes as 'palmetti'. It is now generally accepted that the lobes are a product of repeated and prolific hyphal branching producing a labyrinthine structure (see Figures 6.18 and 6.19). In addition to proliferation of the hyphal structures, in some mycorrhizas the presence of the fungus can induce the plant to form wall ingrowths into the adjacent epidermal cells. These have been observed in *Alnus–Alpova* mycorrhizas (see Figure 6.10) (Massicotte *et al.*, 1986) and were apparently induced in *Pinus sylvestris–Suillus bovinus* mycorrhizas by the presence of exogenous C sources (Duddridge and Read, 1984c). In the case of mycorrhizas formed on roots of *Pisonia grandis*, wall ingrowths are produced in the absence of a Hartig net (Ashford and Allaway, 1982; Ashford *et al.*, 1988; Cairney *et al.*, 1994). Whether or not wall ingrowths are produced by the cells of the plant, the prolific branching of the fungus as it encircles the epidermal or cortical cells yields a structure of immensely enlarged surface area. The hyphal walls are so closely associated with those of the plant that the two appear to be fused into a joint structure which has been termed the 'contact zone' (Strullu and Gerault, 1977) or 'involving layer' (Scannerini, 1968; Duddridge and Read, 1984c).

The detailed structure of a typical mature Hartig net is revealed in an ultra-thin tangential section of a mycorrhiza of *Picea abies* formed by *Amanita muscaria* and viewed under TEM (see Figure 6.19b) (Kottke and Oberwinkler, 1987). The main growth direction of the hyphae, indicated by an arrow (see Figure 6.19c), is transverse to the axis of the root, the ultimate finger-like branches being extremely fine and packed together so that there is little or no space between them. These parts are densely cytoplasmic, non-vacuolate and contain large numbers of mitochondria which, together with the extensive ER, appear to be stretched in the direction of hyphal growth. In more proximal parts, the Hartig net hyphae are seen to be of larger diameter and some are vacuolate. Since more than two nuclei can sometimes be observed to occur in a non-septate part of the mycelium in this type of mycorrhiza (see Figure 6.19b), Kottke and Oberwinkler (1987) regarded the structure as having a coenocytic construction. However, Massicotte *et al.* (1989) found only one or two nuclei in the Hartig net compartments of *B. alleghanensis* mycorrhizas and a similar situation

---

Figure 6.19 Development of the Hartig net. (a) Block diagram showing typical structure of the Hartig net in different sectional aspects and of a pseudoparenchymatous mantle. The main growth direction of the hyphae in the Hartig net is transverse to the root axis. (b) Transmission electron microscopy of a mycorrhiza formed between *Picea abies* and *Amanita muscaria*. Ultrathin section through the intercellular space and several cortical cells showing fully developed, mature Hartig net. Extensive branching leads to the formation of narrower and narrower hyphae (fh). Numerous mitochondria (m) and nuclei can be seen. The presence of two dikaryons (arrowed) indicates the coenocytic nature of the tissue. Bar = 2 μm. (c) Outline of the fungal hyphae in (b). Main growth of the hyphae is in the direction of the full arrow. Dolipore septum and dikaryons are marked. From Kottke and Oberwinkler (1987), with permission.

was reported in association of *A. rubra* with *A. diplophloeus* (Massicotte *et al.*, 1989). More investigations of nuclear migration patterns and mitotic events are required to enable the nuclear organization of the Hartig net to be evaluated.

The similarities between the elaborate structures produced by the proliferation of hyphae in the Hartig net and those seen in transfer cells which increase the surface area for exchange of solutes in many physiologically active plant tissues has been recognized (Duddridge and Read, 1984c; Kottke and Oberwinkler, 1987).

Cellular interactions in Hartig net formation

Restriction of Hartig net formation to a specific zone closely proximal to the root cap (zones C and D, see Figure 6.11) is indicative of the fact that intercellular penetration can occur only at a specific stage of differentiation of the epidermal cell. The conventional view is that this penetration is achieved by mechanical means (Foster and Marks, 1966; Nylund and Unestam, 1982; Piché *et al.*, 1983b; Duddridge and Read, 1984c), although it has been suggested (Duddridge and Read, 1984c; Nylund, 1987) that, in conifers, the walls of cortical cells in the zone susceptible to penetration have a higher pectin cellulose ratio than fully mature walls which, at this early stage in their development, might make them more susceptible to penetration by the fungus.

The possibility of involvement of fungal enzymes in the penetration process is suggested by observation of lysis of the middle lamella of epidermal cells in advance of the hyphal tips in the *A. crispa–A. diplophloeus* mycorrhiza (Massicotte *et al.*, 1986). Likewise, disruption of the middle lamella was seen in the epidermal cells of *E. pilularis* being colonized by *P. tinctorius* (Massicotte *et al.*, 1987). Further support for enzymic involvement in the process of Hartig net formation comes from analysis of the mature interface between the partners, in which adjacent plant and fungal walls become indistinguishably fused to form the interfacial matrix. Duddridge and Read (1984c) interpreted this fusion as a loss of integrity of the plant wall, a view supported by Nylund (1987), who showed by histochemical means that pectic and polysaccharide components of the plant wall and of the interfacial matrix were similar.

Harley (1985) interpreted these temporospatially related events in terms of interference by the fungus with the processes of deposition of the plant primary wall and middle lamella of the plant cell. He proposed that there may be impairment of the supply of precursors for wall assembly, possibly as a result of co-polymerization of fungal proteins with those plant enzymes responsible for assembly of the plant wall. The possibility of direct attack by fungal enzymes upon the plant cell wall cannot be discounted. The potential of ECM fungi to express enzymes necessary to degrade non-lignified walls under some circumstances is demonstrated by their ability to penetrate epidermal cells in the production of arbutoid and ectendomycorrhizas (see Chapter 7). Indeed, in ECM roots themselves, the fungi are widely observed to penetrate the epidermal or cortical cells of the plant in older parts of the root (Nylund *et al.*, 1982; Downes *et al.*, 1992) and this is very likely to require the production of wall-degrading enzymes. A continuum may thus be envisaged in the zones shown in Figure 6.11 from a situation in zones A and B where epidermal cells are resistant to penetration, to a balanced interaction in zones C and D. The developmental sequence facilitates the production of a joint structural entity, probably involved in nutrient exchange, leading finally to a situation where the fungus breaks through the wall into the now moribund cell. According to Downes *et al.* (1992), this sequence of events may occur over a period of about 80 days. By this stage, the sheath may still be present as a moribund structure or it may be progressively lost (Abras *et al.*, 1988).

The mycorrhizal mantle

Whereas the Hartig net forms the most extensive interface between fungus and plant, its biomass in most ectomycorrhizas is small relative to that of the overlying mantle. By separating the mantle from the core of selected mycorrhizas of *Fagus*, Harley and McCready (1952b; see Figure 8.2) were able to calculate that 40% of the weight of the colonized root was due to the fungus. This value does not, of course, include the weight of the Hartig net and has been widely, and often loosely, used by later workers even though they were studying other plant–fungus associations growing under different conditions. Vogt *et al.* (1982, 1991) reported values similar to those obtained for *Fagus* from *Abies amabilis* growing in sub-alpine forest. However, they observed that, in the low altitude forest of *Pseudotsuga menziesii*, the sheath constituted only 20% of the root weight. Bearing in mind that Harley and McCready were specifically collecting relatively large ectomycorrhizas with fleshy mantles probably formed by *Lactarius subdulcis*, it is reasonable to assume that 40% may be a high value for fungal weight and that estimates between 20 and 40% would be more commonplace.

Use of mantle structure for classification of ectomycorrhizal types

The need for a system of classification of ECM types that enables identification of the fungal associates has long been recognized. Apart from the intrinsic interest in the question of diversity of species present on a given plant, a rational system for identification and selection of defined types of the symbiosis is an essential prerequisite for rigorous classification of the functional differences between them. Early attempts to define ECM types (Melin, 1927; Dominik, 1969), which were based upon differences of macroscopic and microscopic characteristics, did little to facilitate identification of the fungi involved. Trappe (1967) emphasized the need to identify fungal partners and suggested various hyphal and other microscopic characteristics that could be used. In an analysis of *Pseudotsuga* mycorrhizas, Zak (1971, 1973) used gross morphological characters and emphasized colour as an important distinguishing feature. Chilvers (1968a, 1968b) added descriptions of mantle structure based upon surface characteristics. All of these systems offered only limited scope for the determination of the fungal partners involved in the association.

There are circumstances in which it is possible to identify the causal organism by careful exposure of hyphal connections between sporophore and ECM mantle. However, in many genera, such as *Lactarius, Russula* and *Inocybe*, tracing of this kind is very difficult. The ephemeral nature of fruit bodies and the fact that many important ECM types are not as yet attributable to a given species of fungus are features which have further advanced the need for a method of classification based upon stable, easily recognizable characteristics of the fungal mantle and its associated mycelial structures.

One such system has been developed by Agerer and the results are assembled in an 'Atlas' of ECM types (Agerer 1987–2002). Classification requires preliminary recording of colour, using daylight quality film, and of morphology of the colonized root and its emanating hyphae (see Colour Plate 6.1a, b, c). This level of analysis, or 'morphotyping', widely used in the past as the sole method of description, is supplemented by anatomical characterization of the mantle, using Nomarski interference contrast microscopy. While the morphology of ECM roots is largely under the control of the plant, the construction of the mantle is a well-conserved feature apparently largely determined by the fungi. The known types of ECM mantle, as revealed by simple scrapings of the surface, can be divided into two main groups. In the first, the hyphae can be discerned as individual structures which form a loose plectenchymatous or prosenchymatous (*sensu* Chilvers, 1968) construction while, in the second, they lose their identity as individual structures being packed, normally as irregularly shaped cells, in diagnostic patterns that produce a pseudoparenchymatous structure. Agerer (1991a, 1995) recognizes nine types of plectenchymatous (a–i) and seven (k–q) of pseudoparenchymatous construction (Figure 6.20 and Table 6.6). The assembled morphological and anatomical details of a given type are presented in the form of colour plates showing the characteristic appearance of whole ECM roots, and of half-tone plates which selectively demonstrate features of the mantle, revealed by scraping and sectioning, and of emanating hyphae (Agerer 1987–1993).

Following the investigation of Brand and Agerer (1986) and Brand (1991), some of the most important ECM types of *Fagus* are described in the Atlas and can be distinguished to the level of the fungal species. Thus, for example, the robust and apparently smooth mycorrhizal type which was selectively collected by Harley and co-workers for use in studies of nutrient absorption is described (Colour Plate 6.1b; Figures 6.21 and 6.22) and the fungus responsible for its formation, *Lactarius subdulcis*, has been identified. Colour Plate 6.1b (see Colour Plate section) reflects the overall appearance of the roots colonized by *L. subdulcis*. Views of the surface and of varying depths in the mantle (Figures 6.21 and 6.22), the latter as scrapings, obtained using Nomarski optics, reveal its pseudoparenchymatous nature, the characteristic angularity of its individual cells and the presence of laticiferous hyphae which are a diagnostic feature of the genus *Lactarius*.

While the specific identity of this type of mycorrhiza was originally obtained by tracing connections between fruit bodies of *L. subdulcis* and the root, there are many types in which, to date, no such connections have been established. Gronbach and Agerer (1986) propose that in those cases where the identity of the fungus forming a widely distributed and well-characterized ECM type is not known, a binomial system of nomenclature can usefully be retained. In such cases, the name is made up of the genus of the host plant and a characterizing epithet, e.g. *Picierhiza bicolorata* (see Colour Plate 6.1c). The major features of the gross morphology and of the anatomy of the mantle and emanating hyphae are again described. When the identity of the fungal partner is discovered, the artificial name of the ectomycorrhiza can be replaced by that of the fungus.

Comparable progress has been made with characterization and identification of the mycorrhizas of *Picea* (Agerer, 1986, 1987). One of the most widely occurring ECM types on *Picea abies* in Europe is that produced by *Russula ochroleuca*. A combination of features, including the presence of bright yellow-green patches revealed by colour microscopy and of angular cells in the mantle packed with yellow granules enables identification (see Colour Plate 6.1a). Structural features of the mantle and Hartig net provide further information to assist in identification.

The structure and exploration strategies of rhizomorphs

While the bulky tissue of the ECM mantles may provide structures suitable for nutrient storage and, through its intimate contact with the root surface, play key roles in control of nutrient transfer between fungus and plant, it does little to increase the surface area of the colonized root in contact with the soil (see below). This function is served by the extraradical mycelium which, as single hyphae or linear aggregates of such hyphae, extend from the mantle. Structural attributes of these mycelial systems are of additional importance, because singly and collectively their constituent hyphae form the connection between mantle and soil, and so provide the pathways for nutrient exchange. Recognition of the critical role played by the components of ECM systems constitutes an important change of emphasis in research on the symbiosis in recent years (Read, 1984, 1992; Agerer, 2006).

The simplest level of organization seen in ascomycete associates such as *Cenococcum geophilum* and *Tuber* spp., and in some of the more widespread basidiomycetes, is one in which hyphae emanating from the mantle retain their individuality, growing as single elements into the soil. The majority of basidiomycetous associates of ECM roots do, however, produce structures in which hyphae aggregate and grow in parallel to some extent, as they leave the mantle, so forming a composite linear organ. Some confusion has arisen over nomenclature applied to these multi-hyphal linear aggregates and a number of terms have been used to describe them, including bundle, cord, strand, rope and rhizomorph. The term rhizomorph has the combined advantages that it appears to have been the first used to describe this type of structure and that it highlights their root-like morphology. Earlier reluctance to employ this term for ECM organs was based upon the view that rhizomorphs were structures which extend, by the action of initials, in a well-defined apical meristem (Garrett, 1963; Motta, 1969), in contrast to the hyphal aggregates of ECM fungi which have loose, apically spreading modes of growth. It is now known that rhizomorphs of *Armillaria* spread, not as a result of meristematic activity, but as a looser front of apically extending hyphae which interdigitate to produce a structure which superficially resembles a meristem (Rayner *et al.*, 1985). There is thus a continuum of structural differentiation (Figure 6.23) from apically dominant organs (Figure 6.23a), such as those of *Armillaria* at one extreme, to the apically diffuse and spreading systems typical of many ECM fungi, at the other (Figure 6.23e). In view of this, Cairney *et al.* (1991) made the sensible recommendation that the term rhizomorph be used to describe all linear aggregates of hyphae.

Agerer (1991a, 1995) recognizes six categories of rhizomorph structure (Figure 6.24). The simplest structures are undifferentiated aggregations of loosely woven hyphae of equal diameter, the whole having an ill-defined margin (type a, Figure 6.24). A still undifferentiated but more compactly arranged structure (type b, Figure 6.24) with a smooth margin is produced by *Laccaria* and *Lactarius* spp. (Table 6.7). Some enlargement of central hyphae within the aggregate yields a somewhat more complex structure (type c, Figure 6.24). Considerable increases in complexity are seen in types d, e and f (Figure 6.24) where internal hyphae have much enlarged diameters. In type d, the thicker hyphae are randomly distributed, while in the most

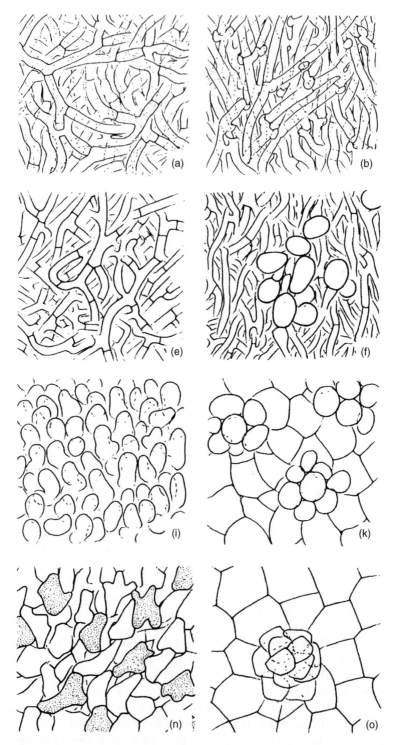

Figure 6.20 Schematic drawings of surface plan views of plectenchymatous (a–i) pseu-doparenchymatous (k–q) mantles, based on surface scrapings. From Agerer (1991a), with permission.

Figure 6.20 Continued.

Table 6.6 Examples of ectomycorrhizal fungal species which show the types of mantle construction depicted in Figure 6.20.

Plectenchymatous	Pseudoparenchytamous
a *Amanita muscaria, Boletus edulis, Scleroderma citrinum*	k *Russula fellea*
b *Cortinarius armillatus, Hebeloma crustuliniforme, Laccaria laccata,*	
Paxillus involutus, Pisolithus tinctorius	l *Tuber aestivum*
c *Lactarius glyciosmus, L. pubescens*	m *T. melanosporum, Lactarius rufus*
d *Russula aeruginea, Thelephora terrestris*	n *Russula emetica*
e *Rhizopogon luteolus*	o *Russula ochroleuca*
f *Boletinus cavipes, Leccinum scabrum*	p *Lactarius subdulcis*
g *Cenococcum geophilum*	q *Lactarius pallidus*
h *Russula xerampelina*	
i *Lactarius picinus*	

The letters a–q refer to Figure 6.20. Data from Agerer (1991a).

highly organized types (e and f), there is a central core of thick hyphae, those in type f showing dissolution of transverse septa.

The well-defined distinctions in the extent of structural differentiation between rhizomorph types are also reflected in the extents of their proliferation through the soil. Agerer (2001) described five distinct 'exploration strategies' (Figure 6.25), each representing a progressively greater extension of the hyphae or hyphal aggregates and their associated mycelial networks into the surrounding growth medium. Strategies 1 and 2 are 'contact' and 'short distance' respectively, and are seen in fungi which do not form rhizomorphs. A few *Russula* and *Lactarius* spp. fall into this category. Most of the ECM-forming ascomycetous fungi, notably *Cenococcum*, show the short distance strategy. Strategy 3, 'medium distance' exploration, has three variants (Figure 6.25). These exhibit different levels of organization in largely undifferentiated rhizomorphs of the structural type a (see Figure 6.24). Among the variants, Strategy 3b encompasses those fungi like *Gautieria, Hysterangium* and *Hydnellum* that produce hyphal 'mats'. In contrast, Strategy 4 (Figure 6.25), 'long distance' exploration, involves the production of robust rhizomorphs of the differentiated kind exemplified by structural type f (see Figure 6.24). Many important ECM fungi are in this category, including members of the genera *Boletus, Paxillus, Pisolithus, Rhizopogon* and *Suillus*. Strategy 5 (Figure 6.25), referred to as 'pick-a-back' exploration, is found most widely in the Gomphidiaceae, members of which produce contact or short distance hyphae that penetrate and pass along the rhizomorphs of other ECM fungi. The extent to which these architectural aspects of exploration strategy are reflected in functional terms remains to be investigated in detail, but the recognition that the anatomy and architecture of extraradical mycelial systems may determine the potential for nutrient capture itself represents an advance. Agerer (2006) provides a detailed review of the relationships between the taxonomic groupings of ECM fungi and the structural features of their hyphal mantles, rhizomorphs and mycelial systems.

The process of rhizomorph construction

On artificial media, such as agar gels or liquids, the mycelia of ECM fungi show a dense 'fluffy' but largely undifferentiated growth form, which is unrepresentative

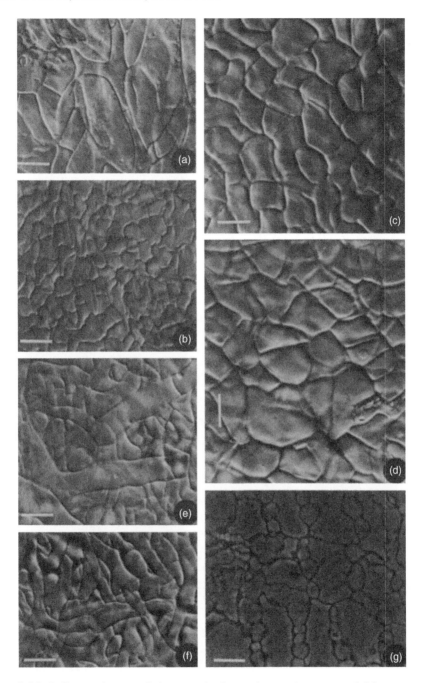

Figure 6.21 Different layers of the mantle formed in a *Lactarius subdulcis* mycorrhiza. (a) Hyphal reticulum forming mantle surface. (b) Surface view of the very tip of the mycorrhiza. (c) Plan view of outer pseudoparenchymatous layer of the mantle. (d) As (c), but in an older region of the mantle. (e) Plan view of inner, plectenchymatous mantle, showing lactiferous hyphae. (f) Inner surface of the mantle. (g) Tangential section through the outermost cortical cells and Hartig net. Bars = 10 μm. From Agerer *et al.* (1987–2002), with permission.

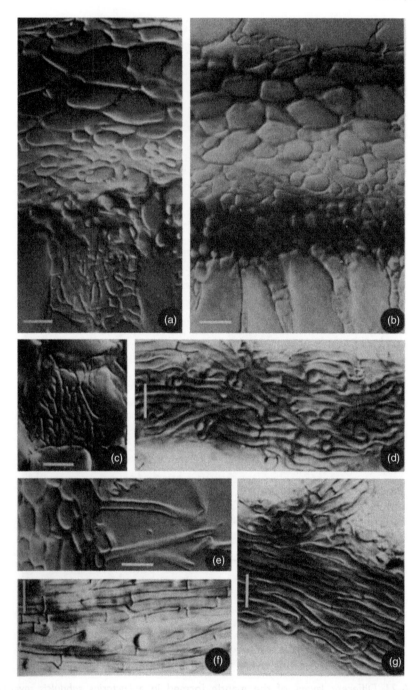

Figure 6.22 Characteristics of a *Lactarius subdulcis* mycorrhiza. (a) Cross-section of the Hartig net in plan view. (b) Longitudinal section, showing young emanating hyphae. (c) Cross-section of the Hartig net in plan view. (d) Rhizomorph, close to mantle. (e) Cross-section of emanating hyphae. (f) Young rhizomorphs, showing anastomoses. (g) Branching rhizomorph. Bars = 10 μm. From Agerer *et al.* (1987–2002), with permission.

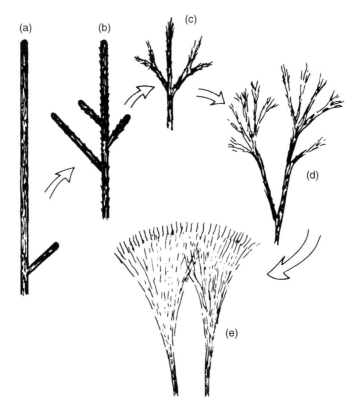

Figure 6.23 Stages in the development of types of rhizomorph from (a) systems of the *Armillaria* kind with organized apices, to those typical of most ectomycorrhizal fungi (d and e), in which a loose and usually fan-shaped hyphal front explores the medium. From Rayner *et al.* (1985), with permission.

of that seen in nature when the fungi are growing in association with plants. In heterogeneous natural substrates, a considerable morphogenetic plasticity is revealed both at the leading edge of the hyphal front and in the maturing regions behind the front. Hyphae making contact with hitherto uncolonized roots initiate mycorrhiza formation. By this mechanism, seedlings for example, become integrated into the mycelial network (Figure 6.26). When hyphae of the advancing front make contact with particles or aggregates, they can be induced to branch repeatedly, to increase in diameter by up to fourfold and to develop thick gelatinous walls. Such morphogenetic changes have been described in association with mineral and organic materials (Read *et al.*, 1985; Ponge, 1990; Agerer, 1992). Intensive hyphal proliferation can be induced by introducing organic substrates of a particular quality to sparsely growing mycelial systems (Figure 6.27, see also Colour Plates 6.2c, d, 6.3b) (Read, 1991a; Unestam, 1991; Bending and Read, 1995a; Read *et al.*, 2004). There is evidence (see Chapters 9, 10 and 15) that these proliferations effect nutrient mobilization from the added substrate.

Behind the leading edge of the mycelial fan, hyphae growing in parallel approach each other more and more closely to form the linear aggregates which eventually

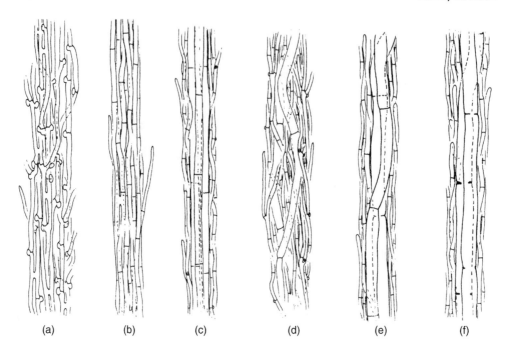

(a)   (b)   (c)   (d)   (e)   (f)

Figure 6.24 Schematic drawings of different types of rhizomorph formed by ectomycor-rhizal fungi. See text for descriptions. From Agerer (1991a), with permission.

Table 6.7 Examples of ectomycorrhizal species having different types of rhizomorph construction.

a	*Cortinarius obtusus, Dermocybe cinnamomea, Tricholoma sulfureum*
b	*Laccaria amethystina, Lactarius deterrimus, Lactarius vellereus*
c	*Gomphidius glutinosus, Thelephora terrestris*
d	*Cortinarius hercynicus, Cortinarius variecolor*
e	*Tricholoma saponaceum*
f	*Leccinum scabrum, Paxillus involutus, Scleroderma citrinium, Suillus bovinus, Xerocomus chrysenteron*

The letters a–f refer to Figure 6.24. From Agerer (1991a).

achieve dimensions of sufficient magnitude to be visible as rhizomorphs. Construc-tion of, and cohesion within, the rhizomorph is achieved by a number of mecha-nisms which can be seen individually or in combination to produce structures of the kinds shown in Figures 6.23 and 6.24. Among these, inter-hyphal bridges formed by anastomoses, adhesion between hyphae facilitated by gelatinous wall materials and the production of backwardly and forwardly growing ramifications have been iden-tified by Agerer (1992) as being the most important. Bridges may provide open con-nections between the attached hyphae or may become secondarily 'closed' by the formation of clamps or simple septa. Since the septa of higher fungi have a central pore, such 'closure' does not necessarily prevent either cytoplasmic continuity or transfer of materials. However, complete closure of the pore may be advantageous under some circumstances, in that it would facilitate isolation of any segments of

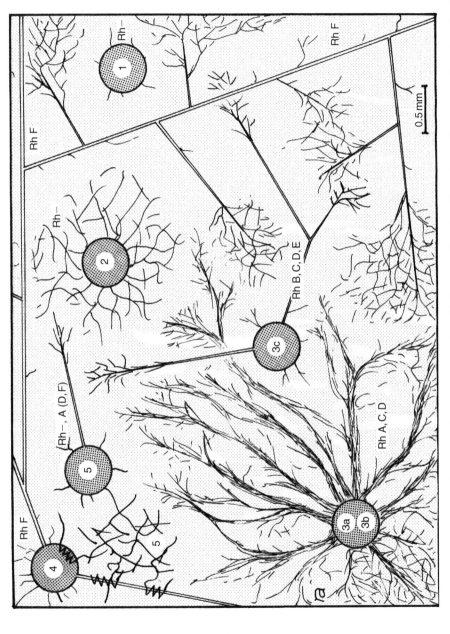

Figure 6.25 Schematic drawings of different exploration types. 1 = Contact exploration; 2 = short distance exploration; 3 a, b = medium distance fringe exploration; 3c = medium distance smooth exploration; 4 = long distance exploration; 5 = pick-a-back exploration. All figures are to scale. Rh = rhizomorph; – = rhizomorph lacking; A–F = organization types of rhizomorphs. From Agerer (2001).

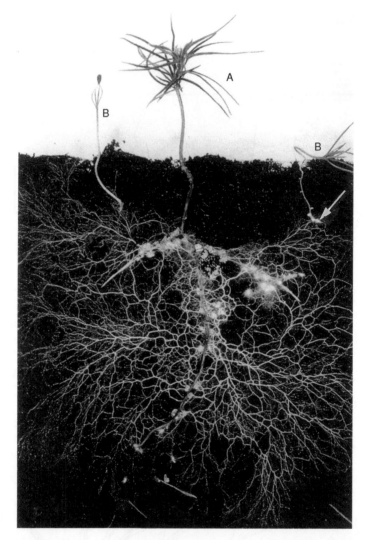

Figure 6.26 Development of extensive mycelium and rhizomorphs by *Suillus bovinus* in symbiosis with a seedling of *Pinus sylvestris* (A). Note the fan-like spread of the extraradical fungal system, with which the germinating seedlings (B) have already formed mycorrhizas (arrowed). From Read (1991c), with permission. See also Colour Plate 6.3a.

the rhizomorph, in which, for whatever reasons, cytoplasmic continuity has been disrupted.

The production of backward or forward growing elements, often from clamp connections on the main hyphae, has been known to be a prominent feature of construction of rhizomorphs since the studies of Butler (1958) of the ontogeny of the wood-decay fungus *Serpula lacrymans*. These secondary elements, referred to as 'tendril hyphae', ensheath the larger central structures. A similar type of construction is seen also in the rhizoidal rhizomorphs of the moss *Polytrichum juniperinum* (Agerer,

1991c). Its independent origins in the plant and fungal kingdoms suggest favourable attributes probably including increased mechanical strength. The tendrils ensheath the main hyphae from which they have developed, become densely cytoplasmic and produce an extra-hyphal matrix that helps to cement the structure together. In the most complex mycorrhizal rhizomorphs, such as those produced by *Suillus bovinus* (Duddridge *et al.*, 1980; Read, 1984) and *Leccinum scabrum* (Brand, 1989), these thinner structures encase a group of central hyphae of much larger diameter. The central elements lose their cytoplasm and show progressive breakdown of transverse septa, to form pipe-like structures having some superficial resemblance both in their axial location and tubular conformation to the xylem vessels of plants. For this reason, they have been referred to as 'vessel hyphae'.

Microscopy reveals that fully differentiated vessel hyphae first appear in the rhizomorph well behind the advancing mycelial front. They occupy the full length of the rhizomorph but do not enter the mantle of ECM root. The junction between rhizomorph and mantle surface is normally formed by loose aggregates of undifferentiated hyphae, which may fan out to produce a delta-like confluence with the ensheathing system. Analysis of the zone of conjunction by light microscopy indicates that direct continuity between any differentiated components of the rhizomorph and the mantle is lacking. Agerer (1990) described ramification of the vessel-like hyphae of *S. bovinus* as the rhizomorph approached the mantle surface, the narrower elements arising from it crossing the transition zone as a delta- or fan-like structure to intertwine with the mantle hyphae. In the less well differentiated rhizomorphs of *Sarcodon imbricatus*, the slightly thicker central hyphae again ramify proximally, and their ultimate branches grow into the inner mantle where they lie in close contact with the root surface (Agerer, 1991b). There is evidence of elaborate anastomosis between the mantle hyphae and those entering from the rhizomorph. The basic structure thus seems to be one which can give polarized transport between the mantle and soil, and dispersed distribution within it.

At the distal end of the rhizomorph in soil, a fan-like arrangement of fine hyphae is again seen, the pattern of distribution of the individual elements apparently being ultimately determined by the nature of the resources they encounter. The structure of a typical mantle–rhizomorph–hyphal front pathway is summarized in Colour Plate 6.2a–d and Colour Plate 6.3b (see Colour Plate section). These features are of key importance both in scavenging resources from the soil and distributing these and organic C derived from the root throughout the fungal biomass. The ramifying hyphal system also rapidly colonizes roots of seedlings as they penetrate the soil after germination, so incorporating these juveniles into the common mycelial network (see Colour Plate 6.3a).

Development of the extraradical mycelium system

In his early experiments on ectomycorrhiza, Hatch (1937) stressed the importance of plant architecture, attributing greater nutrient absorption associated with ECM formation to the increased branching of the roots which arose as a result of colonization. An indication of the relatively great importance of mycelial architecture was provided by Rousseau *et al.* (1994) who compared the increases of absorptive area gained by branching of roots with that provided by the mycelium of the fungi

C. geophilum and *P. tinctorius* as they colonized *Pinus taeda* (see Table 10.5). They found a fourfold increase of root surface area arising from colonization by *P. tinctorius*, but no increase in area in seedlings colonized by *C. geophilum*. The major impacts upon both absorptive surface area and mineral nutrient acquisition by the plant were seen to be associated with the mycelial characteristics. Colonization by *P. tinctorius* yielded a 47-fold increase in surface area of mycelium and a significant enhancement of P capture; *C. geophilium* gave a 28-fold increase of surface area but no increase of P uptake relative to that seen in non-mycorrhizal plants. Since *P. tinctorius* and *C. geophilum* occupy extreme ends of the spectrum of mycelial exploration strategy (Strategies 4 and 2, respectively) (Agerer, 2001), this experiment goes some way towards validation of a structure–function relationship in the mycelial systems involved. The extent to which such differences are expressed in nature will clearly be determined by soil nutrient status. Effects are likely to be maximized under conditions of impoverishment.

The basic processes of growth and differentiation of the extraradical mycelia of rhizomorph-forming ECM fungi have been described and quantified by observing their development in transparent observation chambers containing non-sterile natural substrates (Figure 6.27) (Brownlee *et al.*, 1983; Read *et al.*, 1985; Finlay and Read, 1986). Under controlled conditions of temperature (15°C day, 10°C night) and day-length (18h), fungi such as *Suillus bovinus* extend from colonized roots of *Pinus* spp. as an undifferentiated fan-like front of hyphae at a constant rate of 2–4mm/day. Coutts and Nicoll (1990a) measured the extension rates of *Thelephora terrestris* and *Laccaria laccata* growing from roots of *Picea sitchensis* through a growing season, under natural conditions of temperature and day length. A mean extension rate of 3mm/day was recorded for *T. terrestris* mycelium in July, while *L. laccata* achieved 2mm/day (Figure 6.28). Between June and November the fans had extended from colonized roots to reach a distance of 0.24m. The mycelia of *T. terrestris* continued to grow, albeit at a much slower rate of 0.44mm/day throughout the winter.

During the growing season, the primary roots of spruce grew at a slightly faster rate than the advancing hyphal front (see Figure 6.28, compare c and d) but, because the mean distance from the tip of such a root to the youngest short root to emerge from its axis is normally greater than that to the hyphal front, such laterals are rapidly colonized as they break through the cortex into the domain occupied by the fungus. Whereas extension growth of roots ceases during the winter months, the mycelium of *T. terrestris* continued to extend from December to March. In contrast, the mycelium of *L. laccata* was not evident after October (Coutts and Nicoll, 1990a). When the mycelial network of *T. terrestris* was subjected to waterlogging in winter, the undifferentiated parts of the network senesced, but the rhizomorphs retained their structure and thus provided a skeletal framework from which new hyphae could be regenerated when soil conditions ameliorated (Coutts and Nicoll, 1990b).

The extent of development of extraradical mycelia has been quantified in several studies (Read and Boyd, 1986; Jones *et al.*, 1990; Francis and Read, 1994; Rousseau *et al.*, 1994) enabling their contribution to the potential absorbing surface of the plant to be estimated. Read and Boyd (1986) determined hyphal lengths of a number ECM fungi after they had grown for three months from seedlings of *Pinus sylvestris* across non-sterile unfertilized peat. Values ranging from 2000 to 8000m/m of colonized root were obtained; *Pisolithus tinctorius* growing for eight weeks from seedlings of *P. taeda*, into sterilized and fertilized quartz sand had hyphal lengths of 504m/m root

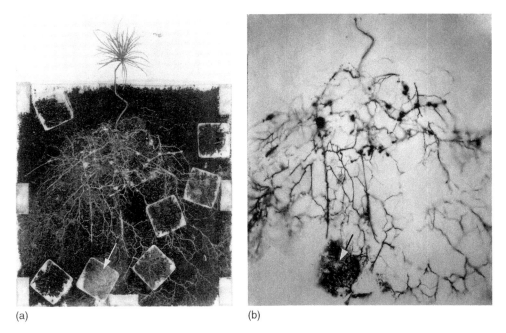

(a)   (b)

Figure 6.27 Development of the mycelium of *Suillus bovinus* growing from a mycorrhizal plant in an observation chamber. (a) Dispersed colonization of the peat substrate and intensive proliferation of mycelium in trays containing FH horizon litter (arrowed). (b) Autoradiograph of the same chamber after feeding $^{14}CO_2$ to the shoot. Note the intense labelling in the densely colonized tray (arrowed). From Bending and Read (1995a), with permission.

(Rousseau *et al.*, 1994). This value is similar to the 289 m/m and 308 m/m reported by Jones *et al.* (1991) for *L. proxima* and *T. terrestris*, respectively after growth for 12 weeks, again in a sterilized and fertilized sandy matrix, but from *Salix viminalis*.

Differences between experiments are to be expected when measuring hyphal lengths, especially where substrate conditions range over extremes from peat to sand. It is arguable that the high values obtained in organic matter of low fertility are likely to be more representative of those seen in many soils in nature. Whatever their cause, however, the differences in hyphal length recorded between experiments cannot disguise the main feature of these results which is the enormous increase in potential absorption area which the root system gains from the presence of the extraradical mycelium. It is difficult to emphasize the importance of this component sufficiently.

This area of study has been advanced by the development of techniques enabling the quantification of mycelial length in the field (Wallender *et al.*, 2001; Nilsson and Wallander, 2003; Hagerberg *et al.*, 2003). The most widely used of these involves burial of sand-filled mesh bags in the rooting zone of forest soils. These have apertures which exclude roots but permit the ingrowth of hyphae. Sand is employed so as to inhibit proliferation of saprotrophic fungi and because of the difficulty in quantifying hyphal biomass in a matrix of forest mor-humus. Even using this substrate, it remains a prerequisite of the experiment that distinction can be made between the hyphae of ECM and any saprotrophic fungi which may enter the bags.

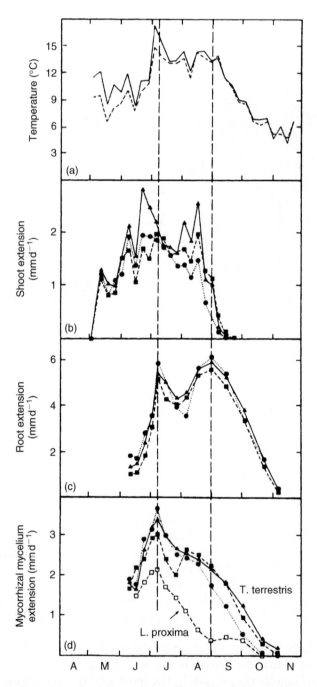

Figure 6.28 Cumulative growth of three clones (■, ▲, ●) of *Picea sitchensis* and their mycorrhizal fungi during 1987. In (d) the open symbols are *Laccaria proxima* and the closed symbols are *Thelephora terrestris*. (a) Seasonal variations in temperature; (b) shoot extension; (c) root extension; (d) mycelial extension. From Coutts and Nicholl (1990a), with permission.

Working in a mixed stand of pine (*P. sylvestris*) and spruce (*P. abies*), Wallender *et al.* (2001) obtained discrimination between these functional groups using two independent methods. The first involved comparison of the mycelial contents of bags placed in undisturbed forest with those in plots from which ECM mycelia were removed by trenching to exclude roots. The second involved mass spectrometric determination of carbon isotope signatures of harvested mycelia. This is based upon awareness that, because ECM and saprotrophic mycelial systems are sustained by different organic C sources, these groups have distinctive ^{13}C signatures (Högberg *et al.*, 1999). Both methods indicated that ~85% of the mycelia occupying bags placed in undisturbed forest were of the ECM type. Using a molecular approach to the characterization of mycelia colonizing sand-filled mesh bags in a beech forest, Kjoller (2006) subsequently reported that 83% of recovered sequences were of ECM fungi.

Wallander *et al.* (2001) employed quantitative assays using the fungus-specific biomarker ergosterol (Nylund and Wallander, 1992) and the phospholipid fatty acid (PFLA) 18:2ω6,9 (Olsson, 1999), to determine the biomass of the ingrowing ECM mycelium. An annual mycelial production of 125–200 kg/ha was estimated, this value being close to that of the standing crop of fine roots from which the mycelia extended. Assuming, following Rousseau *et al.* (1994), a mean hyphal length to mass ratio of 0.83 m/μg, this level of productivity would yield a potential absorbing surface area of 70–112 m²/m² of forest surface. On an area basis, the total ECM mycelial biomass (i.e. extraradical mycelia plus ECM mantles) was estimated to be 700–900 kg/ha. The Wallander *et al.* (2001) study confirmed the observation of Coutts and Nicholl (1990a) that, in boreal coniferous ecosystems, most of the mycelial growth occurs in association with fine root production between July and October.

Recent developments suggest that even the high levels of ECM mycelial productivity indicated by studies using sand-filled mesh bags may represent underestimates. Hendricks *et al.* (2006) carried out comparative analyses of mycelium production in bags filled with sand and those containing soil derived from a pine-forest floor. While confirming that in both types of bag the mycelia were largely of ECM origin, they also showed that hyphal production in the natural soil was ~300% higher than in the corresponding sand-filled bags. Clearly, while further studies using soils of differing quality are desirable, it appears that estimates of ECM hyphal length and biomass obtained under field conditions to date are likely to be conservative.

Högberg and Högberg (2002) calculated the contribution of ECM fungal mycelium to the microbial biomass of a pine-dominated boreal forest soil. They measured the decline of microbial C and N following large-scale tree girdling in mid-August (see Chapter 8) and, on the assumption that this reflected largely the impact of cessation of assimilate supply to the mycorrhizal component of the microflora, they concluded that the ECM mycelium constituted at least one third of the total microbial biomass and that it contributed, together with its associated roots, half of the dissolve organic C in the soil.

Molecular-genetic approaches to characterization of the distribution of ectomycorrhizal mycelia and roots in nature

A combination of somatic incompatibility testing, DNA sequence and microsatellite marker studies have enabled an increase in our knowledge of the distribution

of ECM fungal genotypes in forest soils. In the heterokaryotic phase, basidiomycete mycelia that have originated from the same genet are able to fuse, while those
strains that are not genetically identically reject each other, thereby maintaining
their genetic integrity. Fries (1987) demonstrated variation between ECM fungi in
the extent to which they show somatic incompatibility. *Paxillus involutus, Pisolithus
tinctorius* and *Thelephora terrestris* do not show the phenomenon, *Suillus luteus* and
S. variegatus show it to varying extents, whereas in *Amanita muscaria, Hebeloma mesophaeum* and *Laccaria proxima*, heterokaryons from different genets are routinely
incompatible. By recording the presence or absence of rejection in pairings of isolates from fruit bodies of *S. bovinus* collected in *Pinus sylvestris* stands of different
ages in Sweden, Dahlberg and Stenlid (1990) were able to determine the relationships between the numbers of genets present, the area occupied by each and the
age of the trees with which they were associated. The area occupied by any single
genet and the number present per unit area both decreased with the age of the trees.
In 10–20-year-old stands, there were 700–900 genets/ha and the distance between
the outermost compatible fruit body of each genet was 4–5 m. In 100-year-old
stands, there were only 30–100 genets/ha, the diameter of each being of the order of
14–20 m. Genets of such a size encompass a number of trees, all of which, therefore,
are likely to be interlinked by the same genetically distinct mycelium.

Analyses of DNA sequences (Anderson *et al.*, 2001; Sawyer *et al.*, 2003; Guidot
et al., 2004) or of microsatellite markers (Dunham *et al.*, 2003; Kretzer *et al.*, 2004)
have the advantage that they circumvent the need for the time-consuming procedures of isolation and culture in order to establish genet identity. They can also be
applied to species or isolates that do not show incompatibility in culture. These
studies have shown that genet sizes vary widely between species and confirm that
they are strongly influenced by disturbance and by age of the forest supporting
their growth. Individual genets may extend over large contiguous areas, some of
which may occupy over $50 \, m^2$ (Anderson *et al.*, 2001; Sawyer *et al.*, 2003). However,
the extent to which such mycelial networks remain completely interlinked over
such wide areas remains to be revealed. There are likely to be intergeneric and interspecific differences in longevity and extents of mycelial genets. Thus, it appears that
most genets of *H. cylindrosporum* turn over rapidly relative to those of *S. bovinus*,
there being a well-developed propensity for the former to be replaced annually
through spore germination (Guidot *et al.*, 2004).

At a finer scale, the application of T-RFLP and ITS rDNA techniques has enabled
species-specific discrimination between the distribution of mycorrhizal roots and
that of their associated extramatrical mycelium down a weakly podzolized soil profile (Genney *et al.*, 2006). Spatial segregation of the two components was mapped in
seven different co-occurring mycorrhizal fungi, several of which were shown to produce their mycorrhizas in different soil layers from their extramatrical mycelium.
Observations at this level of resolution are likely to provide important clues concerning functional aspects of nutrient foraging in nature.

Longevity of ectomycorrhizal roots

The initiation, maturation and senescence of individual mycorrhizal roots and of
whole ECM root systems are dynamic processes and the developmental changes

involved in them have impacts at spatial scales ranging from nutrient transfer at the fungus–plant interface to biomass turnover at the ecosystem level. Perhaps not surprisingly, the processes occurring at the different spatial scales do so over different time-scales.

Thus, older studies have indicated that, while the functioning life span of the fungus–root interface may be of the order of days (Downes *et al.*, 1992), the structural elements of fine roots may persist in soils for years (Orlov, 1957, 1960). These temporal differences, which have considerable implications both for nutrient exchanges at the level of the individual tree and for ecosystem C cycling, have complicated both the understanding and the definition of ECM fine root longevity (Högberg and Read, 2006). In view of these complexities, it is appropriate to differentiate between functional longevity at the cellular scale and the longevity of whole ECM fine root.

Downes *et al.* (1992) carried out ultrastructural analyses of plant–fungus interface in ECM roots of spruce colonized either by *Tylospora fibrillosa* or *Paxillus involutus*. These indicate that the life span of this critical part of the nutrient exchange pathway is only a few days. The process of senescence of cortical cells invested by the Hartig net was initiated within 2–3 days of contact between the partners. Death was observed to proceed from proximal to distal regions of the root so that functional longevity of the short roots was ultimately determined by the length of time over which the meristem continued to add new cortical cells at the distal position. Senescent meristematic cells were seen in 60–85-day-old mycorrhizas of *T. fibrillosa* and in those of *P. involutus* at 25–50 days. It was confirmed, using fluorescein diacetate as a vital stain, that physiological activity in the regions of the plant–fungus interface declined markedly in mycorrhizas over 85 days old. The longevity estimate at this scale is complicated by the fact that some apparently senescent meristems can periodically produce new bursts of growth, each of which yields a young turgid cortex. This activity gives rise to 'beaded' mycorrhizas (Thomson *et al.*, 1990a). According to Downes *et al.* (1992), degeneration of the stele, which occurred proximal to the meristem, took place much later and, since fine roots were initiated in the pericycle, the plant retained the potential to produce new roots for a considerable period after the interface with the fungus had broken down. Lei and Dexheimer (1988) explored the distribution of membrane-bound ATPases at the interface of 32-day-old *Pinus sylvestris–Laccaria laccata* mycorrhizas in the Hartig-net region of which there were already a mix of senescent and living cells. A high enzyme activity was found only in association with living interfaces, again indicating that the exchange surfaces had a functional life span of the order of days.

Anatomical studies of the fine roots of 100-year-old spruce (*Picea abies*) (Orlov, 1957, 1960) throw into sharp relief the contrast between the functional life span of the mycorrhizal interface and the longevity of the entire root organ. Based largely on appearance of the tissues of the stele, Orlov calculated that over 40% of these roots retained viability for more than three years. Values of this order are in accord with those achieved by recent analyses based upon direct measurment of below-ground C allocation (Andrews *et al.*, 1999; Matamala *et al.*, 2003). In a large-scale FACE (Free Air Carbon dioxide Enrichment) facility, which enabled $^{13}CO_2$-labelling and tracing of products of photosynthesis, estimated turnover times for the finest roots of *Pinus taeda* were around 4 years. Similar values are produced by calculations based upon decay of the ^{14}C signal left in the environment as a legacy of 1960s nuclear weapons tests. These so-called 'bomb-carbon' estimates indicate life spans for fine roots of temperate forests to be of the order of 3–18 years (Gaudinski, 2000, 2001; see also Chapter 7). Direct estimates, involving burial of detached ECM roots of pinyon pine (*Pinus edulis*) in their natural habitat have confirmed that very little decomposition occurs over at least two years (Langley *et al.*, 2006).

Measurements of the kind described above cast considerable doubt on estimates derived largely from studies based on soil coring techniques indicating turnover times for ECM fine roots of spruce and pine of the order of months rather than years (Persson, 1979; Vogt *et al.*, 1981; Alexander and Fairley, 1983; Santantonio and Santantonio, 1987; Ruess *et al.*, 2003). According to the direct measurments of C allocation and turnover described earlier, the latter estimates, derived largely from observations of loss of root turgor and darkening of the root apex, represent considerable underestimates of the longevity of these structural units in soil. In view of the very large biomass of the fine root component of forest ecosystems, the apparent discrepancy between the two sets of observations has considerable implications for calculations of global C budgets (see Chapter 7). Current evidence indicates that the plant–fungus interfaces of ECM fine roots have a functional longevity of the order of days or weeks, while the longevity of the entire unit as a structural component of soil organic matter is of the order of years rather than months.

It is important to bear in mind that, even though the life of an individual short root as a functional mycorrhiza may be short, the long roots of ECM trees continue to extend through the growing season and from their axes new short roots can be continuously initiated.

Succession of mycorrhizas and mycorrhizal fungi

Community-level studies

Based upon changes in the spatiotemporal appearance of sporocarps around birch, Last and his group (Mason *et al.*, 1982, 1983; Last *et al.*, 1983; Deacon *et al.*, 1983; Dighton and Mason, 1985) concluded that species of *Hebeloma, Inocybe* and *Laccaria*, which were the first to 'fruit' around young trees, could be categorized as 'early stage' fungi and distinguished from those in the genera *Lactarius, Leccinum* and *Russula*, which appeared later and so were placed in the 'late stage' category (Figure 6.29). While having the benefit of simplicity, this scheme was criticized on the grounds that so-called 'early stage' fungi were not restricted to young trees (Arnolds, 1991) and that the observations were made under ecologically unrealistic circumstances involving trees planted into agricultural soils (Read, 1991a; Molina *et al.*, 1992; Newton, 1992). Justification of the latter criticism was provided by Fleming (1983, 1984) who showed that when birch seedlings developed, as they would most frequently in nature, under mature parental stands, they were predominantly colonized by fungi classified as 'late stage'. It was pointed out that, since many of the early fungal colonists persisted as 'multi stage' components of the ECM fungal community, rigid application of the essentially dichotomous scheme of Last *et al.* (1987) was inappropriate (Danielson and Pruden, 1989).

Despite difficulties arising from the use of the 'early' and 'late' stage terminology, these studies served to show that discrimination occurs in nature between a group of fungi capable of colonizing roots of trees establishing in virgin or disturbed habitats

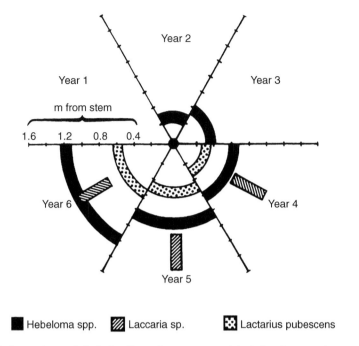

Figure 6.29 Sucession of fruit bodies of ectomycorrhizal fungi around trees of *Betula*, showing the mean distance of production from the trunk during the 6 years after planting. From Last *et al.* (1983), with permission.

and another group which colonizes seedlings by spread from adult roots in forests. This was first demonstrated by Fox (1986) who showed convincingly that the important biological characteristic of the fungi colonizing plants in the unforested environment was the ability to establish mycorrhizas from spores (Table 6.8). Subsequent studies have confirmed that the fundamental differences between fungi propagating largely from spores and those spreading primarily by mycelial extension are of both biological and ecological significance (Nara and Hogetsu, 2004; Nara, 2006a; see Chapter 16).

Whatever the merits of any classification based upon production of sporocarps, it is evident that it can provide only an indirect indication of the population structure of fungi forming ectomycorrhiza on the roots themselves. This point was made in a study of ECM assemblages of *Picea sitchensis* plantations (Taylor and Alexander, 1989) which showed that, despite the presence of a distinctive flora of ECM hymenomycetes as epigeous sporocarps, over 70% of ECM roots were colonized by the fungus *Tylospora fibrillosa* which was not represented in the sporophore assemblage. A study by Visser (1995) was also enlightening in this respect. She investigated changes of ECM morphotypes on roots, as well as sporocarp production in *P. banksiana* stands of a wide range of age cohorts (6, 41, 65 and 121 years) which had established by natural regeneration on pine sites after fire. Assessment based upon analysis of both sporocarps and ECM morphotypes suggested that there was a sequence of ECM fungi with stand age and that the succession was broadly similar, in terms of the identity of the fungi, regardless of the methods used. Early-stage fungi, such as *Thelephora terrestris* and those of the E-strain type were followed, but not completely

Table 6.8 Ability of ectomycorrhizal fungi to form mycorrhizas from freshly harvested spores on birch seedlings grown for usually 12–16 weeks in unsterile rooting media at 18°C; composite results from several tests.

Fungi that established mycorrhizas		Fungi that did not form mycorrhizas	
a*	*Hebeloma crustuliniforme*	e*	*Amanita muscaria*
a	*H. leucosarx*	b	*Cortinarius bulbosus*
a	*H. sacchariolens*	a	*C. debilitus*
abd	*Inocybe geophila*	a	*Elaphomyces muricatus*
abe	*I. lacera*	a	*Lactarius blennius*
a	*I. lanuginella*	a	*L. pubescens*
ab	*Laccaria proxima*	a	*L. rufus*
a	*L. tortilis*	bd	*L. spinosulus*
ce	*Paxillus involutus*	ac	*L. turpis*
		a	*L. vietus*
		ae	*Leccinum roseofractum*
		ac	*L. scabrum*
		b	*Russula cyanoxantha*
		a	*R. grisea*
		ae	*Scleroderma citrinum*
		a	*Suillus luteus*

*Condition of tests: (a) spores in brown earth; (b) spores in vermiculite-peat; (c) spores in coal spoil; (d) crushed fruit bodies in brown earth; (e) coal spoil samples from beneath fruit bodies. Data from Fox (1986).

replaced, by species of the late-stage category, among which *Cortinarius* spp., *Lactarius* spp., *Russula* spp. and *Tricholoma* spp. were strongly represented. In addition, there was a population of fungi which did not dominate at any particular stage, including *Inocybe* spp., *Suillus brevipes* and *Cenococcum geophilum*. In so far as evidence for complete extinction of early colonizing species was weak, it is arguable that succession, as a strict species-replacement process, was not observed. Rather, an increase in complexity of species composition of the community develops with age. The species-abundance distribution in the 6-year-old stand resembled a geometric series, typical of that seen in plant and animal communities of low species richness, dominated by 'r' selected species (May, 1981). The community increased in complexity within 6–41 years, as more species were added, so shifting the species abundance from geometric to lognormal. The lognormal abundance is representative of a more stable community, where a relatively large number of species is more equally distributed in complex interactive situations.

Whereas results of Dighton *et al.* (1986) and Last *et al.* (1987) indicated a decline in ECM fungal species diversity after canopy closure in forest stands between 20 and 30 years, the study of Visser (1995) over the much greater age range showed a marked increase of species richness as stands aged from 6 and 41 years. Thereafter, both the structure and composition of the ECM fungal community had largely stabilized (Figure 6.30). All of the abundant root morphotypes present at 41 years also occurred at 65 and 112 years.

It is important to consider the mechanisms driving the sequential development of the flora of ECM fungi up to the time stability is achieved. Some (e.g. Dighton and

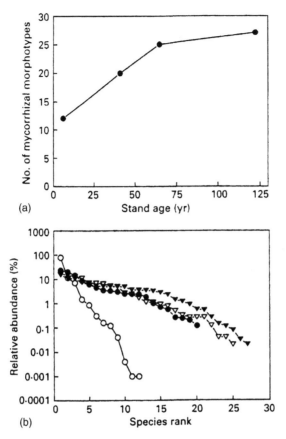

Figure 6.30 Species richness (a) and species relative abundance (b) for ectomycorrhizal fungi colonizing root tips of *Pinus banksiana* in an age sequence of *P. banksiana* stands. ○ = 6 years; ● = 41 years; ▽ = 65 years; ▼ = 122 years. From Visser (1995), with permission.

Mason, 1985; Last *et al.*, 1987) have suggested relationships between accumulation of organic matter on the forest floor and increasing diversity. It is logical to postulate that changes in the quality and quantity of substrates in which the symbiosis develops will lead to selection favouring fungi with the functional attributes enabling them to obtain essential nutrients from different resources. In the case of the *Pinus banksiana* stands studied by Visser (1995), however, the possibility of such a relationship was largely discounted because ECM roots of this plant, in common with those of some other *Pinus* spp., were thought to proliferate almost exclusively in the mineral soil just below the organic horizon. Changing patterns of C allocation associated with tree age were considered by Visser to provide the most likely driving force for shifts in the fungal communities, though the additional possibility that the nature of leachates from thickening organic horizons might change sufficiently to contribute to such shifts was acknowledged. There is a need here to explore the pattern of foraging of the extraradical mycelium as well as distribution of roots. Even where, as in the case of *P. banksiana*, ECM roots themselves appear to be localized in the mineral soil, the fact that the majority of the fungal colonists form rhizomorphs, indicates that they have the potential to explore the overlying litter resource for nutrients. If, as laboratory studies suggest, preferential foraging in organic substrates does occur, changes in the physicochemical nature of the organic horizon could still influence the vigour and, hence, inoculum potential, of the colonizing fungi even though the colonized roots were in a different substrate.

Molecular approaches to the study of ectomycorrhizal communities

Working in the Pinacaeae-dominated forests of western North America, Bruns and co-workers have clearly demonstrated the importance of spores as a source of ECM inoculum in successional post-fire environments (Horton *et al.*, 1998; Baar *et al.*, 1999; Grogan *et al.*, 2000). One- to five-month-old pine seedlings re-establishing after a wild fire which destroyed mature trees were exclusively colonized by suilloid fungi (*Rhizopogon subcaerulescens, R. ochraceorubens* and *Suillus pungens*), whereas those of the same age regenerating in unburned forest supported a wider range of fungi among which *Russula* and *Amanita* species were common (Horton *et al.*, 1998). Interestingly, seedlings of the same age regenerating in a nearby site previously dominated by non-ECM shrubs were also colonized exclusively by these suilloid fungi. Since there is little likelihood that the fungi will be present as mycelial units in the latter environment, this pattern strongly suggests that the suilloid fungi survive fire as spore banks in these soils. The ability of *Rhizopogon* species to survive disturbance as resistant propagules was further confirmed by experiments in which ECM community structures were compared on roots of mature trees of *Pinus muricata* and those of seedlings grown on previously air-dried soil collected from the same forest (Figure 6.31) (Taylor and Bruns, 1999). The lack of correlation between the two communities is striking, there being only one fungus, the thelephoroid *Tomentella sublilicina*, capable of forming ECM under both circumstances. The dominant position of *R. subcaerulescens* in the resistant propagule community was again evident.

Studies at the regional scale across California have thrown new light on the population biology of *Rhizopogon* communities occurring in geographically dispersed and discontinuous stands of *Pinus* spp. from coastal (Kjøller and Bruns, 2003) to inland montane (Rusca *et al.*, 2006) sites. When pots of natural soil from disjunct stands were sown with pine seedlings as 'bait' plants to facilitate selection of ECM-forming fungi, *Rhizopogon* was observed to be the major ECM-forming fungus in almost all sites, confirming its dominance of the spore bank in these forests. By sequencing the ITS region of *Rhizopogon* spp. isolated from ECM roots formed on the bioassay plants and comparing these sequences with those derived from identified sporocarps, the extent of the distribution of distinct genotypes of these fungi within and between stands was evaluated. At the within-site level, dominant *Rhizoctonia* clades were recognizable, these being uniformly distributed through sampling plots of the site. The considerable longevity of spores of this genus is likely to contribute to within-site homogeneity of distibution of a given genotype. In contrast, and perhaps not surprisingly in view of the discontinuity of host distribution and the hypogeous nature of its sporocarps, considerable evidence for genetic isolation was observed at the between-site level.

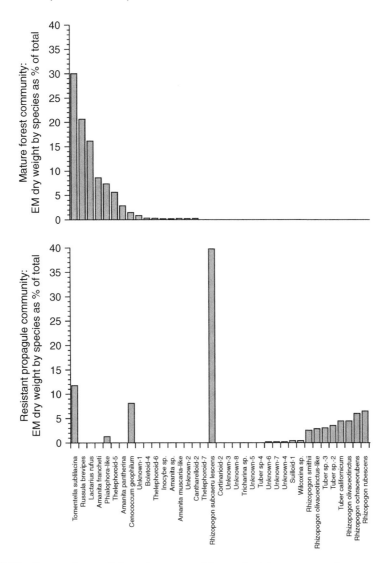

Figure 6.31 Comparison of fungal communities colonizing mature forest roots *versus* bioassay seedlings. The proportion of the total mycorrhizal biomass (dry weight) in each community accounted for by each fungus is shown. The predominance of *Rhizopogon sub-caerulescens* as a mycorrhiza-forming fungus in the spore bank community is evident. From Taylor and Bruns (1999).

At the broadest geographical scale, the *Amylopogon* and *Occidentalis* clades of the genus were more common at the coastal sites (Kjøller and Bruns, 2003), while the *Ellenae* and *Arctostaphyli* clades predominated in the Eastern Sierras (Rusca *et al.*, 2006). More locally, at the inter-site level in the coastal study, a striking case of geographic diversification was seen in the subgenus *Amylopogon*, distinct clades of which were found in sites separated by ~200 km. These sites had a low similarity index in respect of fungal species. The importance of geographical distance as a determinant of

genetic isolation was suggested by the observation that the two sites in closer proximity to one another (75 km) shared most of their sequence groups and had a high similarity index. Analysis of clades of *R. salebrosus* apparently shared between sites in the
closest proximity (~25 km) using more sensitive AFLP-based methods revealed that,
even in these cases, there was evidence of separation of the fungi into site-specific
groups. Thus, geographic isolation appears to be the driving force in diversification of
Rhizopogon. Nonetheless, soil and climatic factors cannot be excluded as contributors
to the observed patterns of diversification. Since each of the five pine species used in
the bioassay programme in the eastern Sierra study were capable of acting as hosts to
all the *Rhizopogon* species identified in the montane soils, it was concluded that intra-
generic host specificity played little role in determination of spore bank composition.

Isolation may be maintained in these hypogeous fungi by a combination of the
relative ineffectiveness of aerial distribution of their spores and the short distance
of spore transport by the main vectors which are thought to be small mammals
(Johnson, 1996) or deer (Ashkannejhad and Horton, 2006).

As we have already mentioned, molecular tools have introduced new possibilities for identification of fungi involved in ECM formation from small amounts of
vegetative material. Application of these techniques at the community level has
provided new insights into the occurrence and succession of fungi in their vegetative stages, which can be usefully compared with information derived from studies
based upon 'morphotypes' or sporocarps (Gardes and Bruns, 1996). Over a 4-year
period, Gardes and Bruns (1996) collected and identified all sporocarps of ECM
basidiomycetes produced above ground in plots established in a stand of *Pinus
muricata*. They simultaneously sampled mycorrhizas occurring in soil cores under
some of the most frequently occurring sporocarps, such as those of *Amanita franchetii* and *Suillus pungens* and, after initially separating them on the basis of morphotype, attempted identification using PCR-based molecular tools. Nearly all
of the morphotypes could be identified using these approaches, at least to generic
level. In general, despite the sporocarp-biased sampling, correspondence between
identity of fungi producing the sporocarp and the predominant ECM type was not
good. In cores taken both under *A. franchetii* and *S. pungens* (Figure 6.32a), the dominant ECM types were produced by *Russula amoenolens*, a *Boletus* type and *Tomentella
sublilacina* and *R. xerampalina*, whether the results were expressed as frequency of
occurrence (Figure 6.32b) or as a percentage of total ECM tips present (Figure 6.32c).
While the small, resupinate sporocarps of the thelephoroid fungus *T. sublilacina*
could have been overlooked, those of the *Russula* and *Boletus* type are conspicuous,
and the absence of a relationship between their occurrence as tips and appearance
on fruit bodies is striking. Likewise, the failure of *S. pungens* to appear as anything but a rare component both in terms of frequency or occurrence below ground
despite its ability to fruit prolifically is of considerable interest.

The reasons for the discrepancy between frequency and abundance of sporocarps
and mycorrhizas are not clear. Gardes and Bruns (1996) speculated that, in the case
of *S. pungens*, which has a narrow host range extending only to *P. muricata* and *P.
radiata*, there may be a particular efficiency in C transfer between roots and sporocarps enabling production of the latter, despite the small resource base. There is a
possibility, conversely, that the process of C transfer between ECM tips and mycelium of *R. amoenolens* is inefficient. An alternative explanation at the community
level is that those species, such as *R. amoenolens*, which fruit so rarely are made up

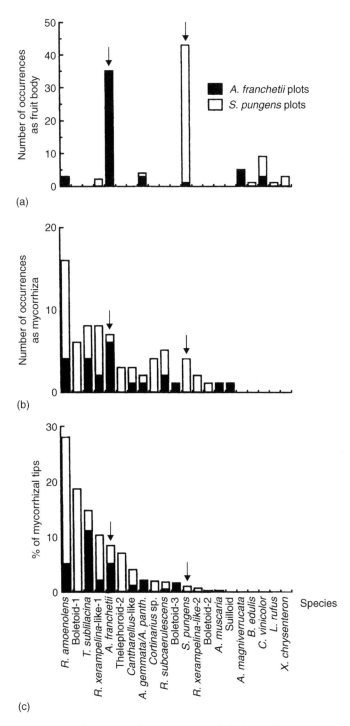

Figure 6.32 Relationship between the presence of fruit bodies and the frequency and abundance of mycorrhizas in plots containing *Suillus pungens* (open bars) and *Amanita francheti* (closed bars). Plots dominated by *A. francetii* or *S. pungens* are indicated by arrows. (a) Frequency of occurrence of fruit bodies of species within 1 m of the soil cores taken below the fruit bodies. (b) Frequency of occurrence as mycorrhizas. (c) Abundance of species based on the percentage of mycorrhizal root tips. From Gardes and Bruns (1996), with permission.

of a large number of genotypes, the clonal sizes of which are too small to obtain sufficient C to support sporocarp production.

Whatever the genetic and physiological bases for the lack of correspondence between sporocarp and mycorrhiza formation, the study of Gardes and Bruns (1996) is enlightening in a number of ways. It draws attention (as did that of Taylor and Alexander (1990) using conventional methodologies) to the danger inherent in attempts to determine below-ground population structures from patterns of sporocarp production. Further, and perhaps of greater long-term importance for the understanding of processes at the population level, it indicates that molecular tools have the power to provide the rapid and unequivocal identification of mycorrhizal types, upon which accurate determination of population structure depends.

Succession and replacement of fungi on roots and root systems

Replacement on individual roots: interactive replacement

The mechanisms by which species of fungi replace each other on a root system will play important roles in any community interactions and successions in the field. Once established on a short root, the mycorrhiza formed by a given fungus is resistant to replacement. Coexistence of two or more associations of ECM fungi can occasionally be observed on individual roots, but this is the exception rather than the rule, although dual associations have been reported to occur on as many as 1–4% of roots of *Pinus radiata* (Marks and Foster, 1967). In *P. banksiana*, Danielson and Visser (1989) observed the overgrowth of the mantle formed by early-stage fungi, particularly in the distal part of the short root, by fungi of the *Suillus* type. They refer to this process as one of interactive replacement and contrast it with the far more commonly seen phenomenon, so-called non-interactive replacement, in which certain species die, apparently spontaneously, after a period of occupancy of the short roots.

There is evidence that interactive replacement is more likely to occur as roots resume growth after dormancy. Wilcox (1968) observed that the root axis of *Pinus* sometimes extended in advance of regrowth of the fungal mantle, so that its uncolonized apex was exposed in soil to other fungal symbionts. Similarly, *Betula* roots initially forming mycorrhizas with *Hebeloma* spp. became colonized by *Lactarius pubescens* after dormancy (Fleming, 1985). Clearly, in view of the restricted longevity of an ECM association on a single root (see above), there is a possibility that interactive replacement is facilitated by senescence of the original colonist. There appears, as yet, to be no experimental evidence for replacement on this basis.

Replacement on whole root systems

In the majority of cases, the assemblage of fungal species on root systems appears to change in composition by a process of non-interactive replacement in which one population dies, the short roots either dying with the fungi or persisting in the non-mycorrhizal condition (Danielson and Visser, 1989), while a new suite of species colonizes the next generation of short roots. A relationship between age of the plant and its root system and the susceptibility to colonization by different fungi is suggested by observations such as those of Fleming et al. (1984). They showed that, whereas some fungi such as *Hebeloma* spp., colonize container-grown plants of *Betula* in the first year, others, for example *Lactarius pubescens*, do so only in the second year despite the persistent presence of inoculum of both fungi. This situation mirrors that seen in the field in newly afforested sites where *Hebeloma* spp. are recorded as early-stage and *L. pubescens* as being among the first of the late-stage species. By growing 2-year-old saplings of birch in transparent troughs containing soil pre-inoculated with *H. crustuliniforme* and *L. pubescens*, Gibson and Deacon (1988) were able to follow the sequence of development of mycorrhizas formed by the two fungi as the plants aged. They found that, whereas *H. crustuliniforme* could form mycorrhizas in any region of the root system, *L. pubescens* did so only on short roots emerging in the oldest parts of the system. Since the soil was of uniform quality throughout the length of the trough, it is logical to postulate that some age-related feature of the plant is the primary factor determining susceptibility to colonization.

Among the various features of the root system that have been considered as possibly contributing to age-related differences in susceptibility to colonization, the supply of C in sufficient quantities to sustain the relatively robust and highly differentiated mycelial systems of late-stage fungi emerges as being of probable importance. Gibson and Deacon (1990) compared the ability of a number of early- and late-stage fungi to grow in pure culture and to produce mycorrhizas on *Betula*, using agar media containing different concentrations of glucose. Whereas representatives of the early-stage category grew continuously across agar in the absence of glucose, the late-stage fungi, including *L. pubescens*, required glucose at a concentration of at least 0.1% to sustain growth. Similarly, while early-stage fungi colonized *Betula* roots in the absence of exogenous glucose, most of the late-stage fungi failed to do so. Clearly, these experimental conditions are unrealistic. They nonetheless demonstrate that ECM fungi depend to different extents upon their environment for C and raise the possibility that those in the 'late-stage' category colonize laterals in older parts of the root system because these provide superior access to endogenous sources of organic C.

The possibility of different patterns of C allocation or of C turnover in different regions of sapling root systems is amenable to investigation using $^{14}CO_2$, but the necessary experimental analyses appear not, so far, to have been carried out. Circumstantial evidence in favour of a C-related explanation for the observed colonization sequences comes from studies using transparent observation chambers (Finlay and Read, 1986c). In these, late-stage fungi are seen to have the ability to colonize the short roots in any part of a seedling root system, provided they are growing from an established plant. The pattern is repeated when seedlings are sown in the field under established trees already colonized by late-stage fungi (Fleming, 1985). To the extent that their inoculum potential is closely related to the size of the food base from which they were growing, fungi of the late-stage types responded to C availability in the same way as their rhizomorph-producing saprotrophic counterparts such as *Serpula lacrymans*, *Phanerochaete laevis* and *Armillaria mellea*. The larger the base, the more aggressively they could forage for new resources.

Conclusions

Evidence gained from a combination of molecular phylogenetic, biogeographical and palaeontological approaches is pointing to a common origin of ECM fungi and their autotrophic partners in the early to mid-Cretaceous period before the break-up of Pangea. Considerable genetic diversity within a number of important ECM fungi that were previously thought to be single taxa has been revealed. In addition, the full genome sequence of an ECM fungus has been determined for the first time. Building on many excellent structural studies, the molecular revolution has begun to show how the early development cross-talk between the symbionts takes place. In particular, advances have been made towards an understanding of the molecular events associated with the formation of mycorrhizas at the level of the individual root tip. Transcriptional profiling has enabled the identities of some of the genes that are upregulated during formation of the symbiosis to be elucidated. Among others, those involved in modification of cell wall structure, in glycolysis, the TCA cycle and respiratory activity have been identified.

Work on the characterization of the extraradical mycelial systems of ECM roots has progressed. New methods enabling quantification of its biomass in the field have been developed and, using molecular methods that now facilitate the identification of small lengths of ECM hyphae, information on selective niche exploitation is being obtained. It has been shown that spore banks can play an important role in facilitating ECM colonization, particularly after disturbance events such as fire.

Genetic profiling of fungal communities associated with fine roots has further confirmed that there is little correspondence between production of sporocarps and representation of a given fungus as a root symbionts and has identified fungi with hypogeous or resupinate fruit bodies as important components of many mycorrhizal fine root communities.

7

Ectendo- and arbutoid mycorrhizas

Introduction

Mycorrhizas with many of the characteristics of ectomycorrhizas (ECM), but also exhibiting a high degree of intracellular penetration, have been described on numerous occasions in the last century and in various species of tree and shrub. Two categories of this kind of structure are recognized here. The first, termed ectendomycorrhiza (Mikola, 1965; Laiho, 1965; Egger and Fortin, 1990; Yu *et al.*, 2001a), occurs primarily on *Pinus* and *Larix* and is distinguished by the fact that, in addition to a usually thin fungal mantle and well-developed Hartig net of the ECM type, the epidermal and cortical cells are occupied by intracellular hyphae. The fungal symbionts forming ectendomycorrhizas are also taxonomically distinctive. Originally described by Mikola (1965) as 'E-strain' fungi, it is now known that most are members of the pezizalean ascomycete genus *Wilcoxina* (Egger, 1996). Since most ECM fungi, as well as some that are weakly pathogenic, are capable of producing intracellular penetrations, especially in senescent parts of fine roots, there has been confusion concerning the status of root symbioses of these intracellular kinds. Melin (1917) recognized fungal symbionts, many of which had dark septate (DS) hyphae that had deleterious effects upon plant performance. Among these he described *Mycelium radicis atrovirens* (MRA) α and β, that formed intracellular penetrations that were disadvantageous to their hosts. He referred to these as 'pseudomycorrhizal' fungi. The application of molecular methods has enabled a clearer understanding of the taxonomic position and status of these fungi. Like *Wilcoxina*, they are shown to be Ascomycetes, but are representatives of genera including *Phialocephala* and *Phialophora* which occupy distinctive taxonomic positions within the Ascomycota (Gams, 1963; Ahlich and Sieber, 1996; Menkis *et al.*, 2004; see also Chapter 11). These fungi are described in this chapter. It is emphasized that while DS fungi do not form ectendomycorrhizas as defined above, many of them do penetrate cells of the root. They occupy that portion of the mutualism–parasitism continuum in which the colonization of the roots can have negative impacts on plant performance.

The second category of ectendo-type mycorrhiza, referred to as arbutoid, is found in the ericaceous genera *Arbutus* and *Arctostaphylos* and in several genera of the ericaceous subfamily Pyrolae. It is distinguished from the first category by the restriction of intracellular penetration to the epidermal layers of the root and by the involvement of a distinct suite of largely basidiomycetous fungi more normally found as ECM symbionts of trees. On the basis of similarities of structure, some (e.g. Zak, 1974) have considered structures of the arbutoid kind to be simply variants of ectendomycorrhizas. However, since there are consistent differences between the two types in terms both of structure and the fungal taxa involved in their formation, we here follow Peterson *et al.* (2004) in considering them as separate categories of mycorrhiza.

Ectendomycorrhizas

Occurrence and structure

Laiho and Mikola (1964) examined the initiation of mycorrhizal colonization in *Pinus sylvestris* and *Picea abies* seedlings. *Picea* developed normal ectomycorrhizas, but on *Pinus* a coarse Hartig net, intracellular colonization and a thin or absent fungal mantle were formed. These ectendomycorrhizas were initially very abundant and invariably present on *Pinus* in nursery beds of soil of agricultural origin, but were very infrequent indeed on *Picea*. The earliest sign of colonization was the formation of a Hartig net which followed closely behind the apical meristem as the root grew. Behind this, intracellular penetration increased in intensity towards the older part of the root, so that the cells became almost filled with coils of septate hyphae, each up to 15 μm wide. These ectendomycorrhizal roots persisted for at least a year, with few signs of hyphal degeneration. As Mikola (1965) wrote: 'Intracellular hyphae do not injure the cortical cells; both plant cells and intracellular hyphae were observed to live at least one year after commencement of colonization; even the nuclei of such heavily colonized cortical cells were clearly visible in stained preparations'.

Scales and Peterson (1991a) made detailed examination of the structure and development of the ectendomycorrhiza of *Pinus banksiana* formed by *Wilcoxina mikolae* var. *mikolae*. Emergent short roots become covered with hyphae which appear to be embedded in a matrix material (Figure 7.1a). The hyphae in this matrix are highly branched (Figure 7.1b). The sheath forms first behind the apex (Figure 7.1c) where it develops between protruding root hairs. At this stage, only a few hyphae of narrow diameter traverse the root apex (Figure 7.1c). In the mature structure (Figures 7.1d, 7.2), the apex is completely ensheathed except in those cases where the lateral roots grow very rapidly. As shown in Figure 7.2a, a uniseriate Hartig net begins to form just behind the apex. This penetrates between the epidermal and outer cortical cells at first, but eventually extends to the inner cortex adjacent to the endodermis. The Hartig net has a labyrinthine structure (Figure 7.2b), typical of that seen in ectomycorrhiza. One or two cells distal to those in which the earliest Hartig net formation is observed, intracellular penetration occurs (Figure 7.2a). The hyphae, having penetrated the cell, branch repeatedly (Figure 7.3).

All aspects of the structure and development of this ectendomycorrhiza, with the exception of intracellular penetration, appear to be similar to those seen in an ectomycorrhiza. When the same fungus was used to inoculate *Picea mariana* and *Betula alleghaniensis* (Scales and Peterson, 1991b), ectomycorrhizas typical of those formed by other fungi were produced. A well-developed Hartig net with

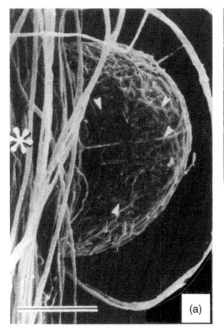
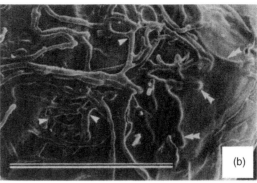
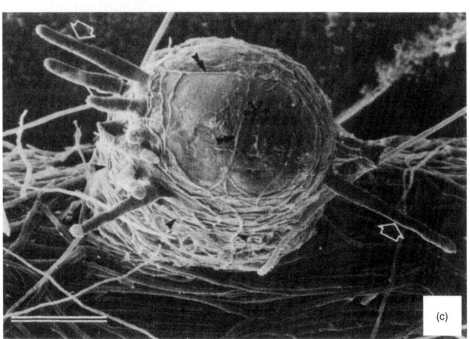

Figure 7.1 Scanning electron microscopy of short roots of *Pinus banksiana–Wilcoxina mikolae* var *mikolae* mycorrhizas. (a) Young emergent short root covered with hyphae (arrowed), that appear to be partially embedded in matrix material (*). (b) Subapical region of a mycorrhizal short root, showing highly branched hyphae (arrowheads) embedded in matrix material (double arrowheads) on the surface of the root. (c) Short root showing apical root hairs (large arrows) and hyphae (arrowheads) contacting the root surface in the colonization zone. Hyphae of small diameter (double arrowheads) traverse the root surface. (d) Apical region of a monopodial mycorrhizal roots, showing the mantle (*). Bars = 100μm. From Scales and Peterson (1991a), with permission.

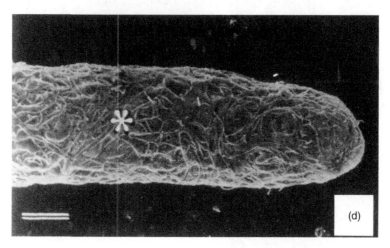

Figure 7.1 (Caption on p. 271)

labyrinthine growth occurred in both types, but neither showed any evidence of intracellular penetration. These descriptions show that the contention sometimes made that the cortical cells are dead and devoid of cytoplasm in EECM roots is not generally true. Neither Mikola (1965) nor Scales and Peterson (1991a) saw any signs of hyphal 'digestion'.

Mikola concluded that the kind of ectendomycorrhizas formed by E-strain fungi (see below) were confined to young, 1–3-year-old *Pinus* in nursery soil over a wide range of soil fertility, acidity and humus content and that their intensity of formation was not greatly dependent on light intensity. The longevity of this kind of mycorrhiza therefore equalled that of typical ectomycorrhizas with no intracellular colonization (one year or more), and was many times longer than non-mycorrhizal roots of similar position in the root system.

Laiho (1965) made a wide survey of ectendomycorrhizas in nurseries and forests in Europe and America and experimentally synthesized mycorrhizas on a number of species of tree using the E-strain fungi isolated from them. He encountered one form of ectendomycorrhiza only, in which the roots were inhabited by coarse septate mycelium which formed a strong Hartig net and intracellular colonization. He concluded that E-strain fungi were present in all ectendomycorrhizas that he examined in detail, but that there must be other causative fungi, for E-strain fungi would only colonize species of *Pinus* and *Larix*, whereas some ectendomycorrhizas undoubtedly occurred on species of *Picea* and other genera. He stressed, however, that there was a possibility of error because senescent mycorrhizas often had intracellular colonization and he quoted Mikola's (1948) observation that *Cenococcum* could penetrate the cortical cell walls of plants growing in unsuitable conditions. This point is even more strongly reinforced by the observation that it is quite usual for the senescent cortical cells to be invaded (Harley, 1936; Atkinson, 1975; Nylund, 1981). Laiho was impressed by the absence of a deleterious effect of EECM colonization on *Pinus* seedlings and concluded that it was a 'balanced symbiosis' and that it gave place to ectomycorrhizas as the seedling aged, especially in woodland conditions.

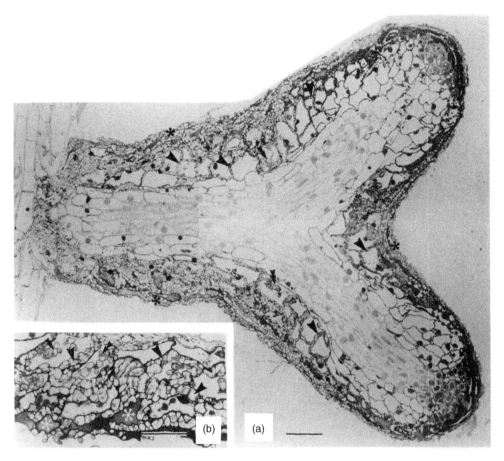

Figure 7.2 Light microscopy of dichotomous short roots of *Pinus banksiana–Wilcoxina mikolae* var *mikolae* mycorrhizas. (a) Longitudinal section showing Hartig net (arrowheads), sheath (*) and intracellular hyphae (double arrowheads). Bar = 100 μm. (b) Higher magnification of an area of the main axis of the root seen in (a). Hyphae in both Hartig net (arrowheads) and intracellular hyphae (double arrowheads) are quite vacuolate. Bar = 50 μm. From Scales and Peterson (1991a), with permission.

Wilcox (1968b, 1971) described similar ectendomycorrhizas on young plants of *P. resinosa* in which coarse hyphae form the Hartig net in both long and short roots. These hyphae tend to spiral round in the cell wall following the angle of the cellulose fibrils. They penetrate the cell wall through pit areas and also by means of rather complicated appressoria and grow into the lumen of the cell. The hyphae seem to be stimulated to spread intercellularly behind the apex of the long roots and around developing laterals, but do not invade the meristems. They may be active around the lateral initials and colonize the young roots as they pass through the cortex. Although ectendomycorrhizas are often present on most of the short roots of young seedlings in nursery soils, they are later replaced by ectomycorrhizas formed by slender hyphae, probably of another fungus.

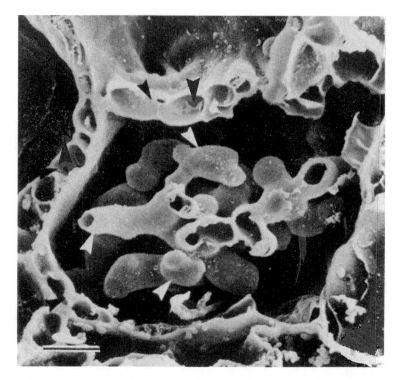

Figure 7.3 Scanning electron microscopy of a single cortical cell of a mycorrhiza formed between *Pinus banksiana* and *Wilcoxina mikolae* var *mikolae*. It can be seen that the intracellular hyphae (arrowheads) are branched and that the Hartig net (black arrowheads) is unisereate. Bar = 100 μm. From Scales and Peterson (1991a), with permission.

Ectendomycorrhizal fungi

The fungi involved in the formation of ectendomycorrhizas were first examined in detail by Mikola (1965) and Laiho (1965) who isolated more than 150 strains from *Pinus* seedlings growing in Finnish nurseries. On grounds of morphology of the mycelia growing on nutrient agar, Mikola considered that all of these strains belonged to the same species which he called the 'E-strain' fungus. The mycelia are typically brown and consist of two hyphal forms, one dark with verrucose walls and one thinner, smooth walled and hyaline. The diameter of the hyphae range from 2 to 12 μm. The taxonomic position of the fungus was unclear because no fruiting structures were found. Wilcox *et al.* (1974) isolated a similar fungus from seedlings of *P. resinosa* growing in nursery soils of the USA and observed the production of chlamydospores in culture. The chlamydospores of E-strain fungi have a distinctive crenulate surface and dimensions in the range 100–200 μm. A spore of this kind isolated from soil in a pine plantation in Iowa was named *Complexipes moniliformis* by Walker (1979) and placed in the Endogonaceae. However, this was later challenged (Danielson, 1982; Thomas and Jackson, 1982; Mosse *et al.*, 1981). The septal pores of vegetative hyphae of fungi producing these spores are of the simple type (Thomas and Jackson, 1982) and the fungus is resistant to benomyl (Danielson, 1982), both being features of fungi with ascomycetous affinities.

Danielson (1982) postulated that E-strain fungi may belong to the order Pezizales. This was confirmed when ascocarps of an operculate discomycete in the genus *Tricharina*, then called *T. mikolae*, were produced on a pot culture of soil supporting EECM seedlings of *P. resinosa* (Yang and Wilcox, 1984). The soil had been inoculated with *Complexipes*-type chlamydospores collected in an Oregon Douglas fir nursery. Later, Yang and Korf (1985) delineated the new genus *Wilcoxina* for E-strain fungi, *T. mikolae* being subsumed into *Wilcoxina*, as *W. mikolae*. Analyses of RFLPs in the nuclear (Egger and Fortin, 1990) and mitochondrial (Egger *et al.*, 1991) genomes of E-strain isolates have confirmed that most can be referred to *Wilcoxina* and that the majority of strains can be assigned to two taxa, *W. mikolae* and *W. rehmii*, each of which has a distinct habitat preference. *W. mikolae* is the chlamydospore-producing fungus and is predominantly found in disturbed mineral soils such as those of nurseries, whereas *W. rehmii* occurs in peaty soils and does not produce chlamydospores. Egger (1996) confirmed that *Wilcoxina* and *Tricharina* should be maintained as separate genera and that *Wilcoxina* taxa formed a distinct group, including the non-mycorrhizal species *W. alaskana*.

Ectendomycorrhiza-like structures have been reported in the epiphytic ericaceous plant *Cavendishia nobilis* var *capitata* (Setaro *et al.*, 2006). This is a member of the Andean clade of the subfamily Vacciniodeae within the Ericaceae (see Chapter 11). A well-developed mantle and Hartig net were present, together with intracellular penetration of the epidermal layer of root cells. Setaro *et al.* (2006) propose the term 'Cavendishioid' for this type of ectendomycorrhiza, but the structural differences described do not seem to us to justify creation of a new mycorrhizal category at this stage. Analyses of fungal DNA sequences revealed the presence of a mixture of *Sebacina* spp. and ascomycetes of the Leotiomycetes. It was not determined whether either or both of the fungal taxa were involved in formation of the mycorrhiza. All sebacinalean sequences from *C. nobilis* clustered in the subgroup *Sebacina* B (see Figure 6.5, Chapter 6), together with isolates of *S. vermifera* obtained from Australian orchids (Warcup, 1988), *Piriformospora indica*, sequences from *Gaultheria shallon* (Allen *et al.*, 2003; see Chapter 11) and from jungermannioid liverworts (Kottke *et al.*, 2003; see Chapter 14).

The occurrence, taxonomic status and function of dark septate (DS) fungi

The presence, both on the surface and within roots, of fungi other than those yielding the definitive mycorrhizal structures of arbuscular mycorrhizas (AM), ectomycorrhizal (ECM), ericoid mycorrhiza (ERM) and orchid associations, has long been recognized. In the course of his extensive analyses of roots of boreal forest trees, Melin (1923, 1925) observed the very frequent occurrence of fungi with dark-coloured mycelia, some of which he isolated and cultured. Since these could be recovered from roots even after extensive surface sterilization, he concluded that they were inhabitants of the root tissues rather than root-associated soil fungi. The most common of these isolates was referred to as *Mycelium radicis atrovirens* (MRA), while others were considered to be *Rhizoctonia* spp. These fungi are traditionally discussed in parallel with EECM fungi. We retain this arrangement, but point out that there are strong links to the fungi forming pseudomycorrhizas (see Chapter 6).

Melin (1923) recognized two types of MRA which he designated α and β, the former being characterized by the production of microsclerotia and being by far the more common. Some MRA α isolates had the ability to produce thin hyphal mantles on root surfaces, whereas others penetrated the tissues. They were described by Melin (1925) as being 'pseudomycorrhizal' and were shown to be weakly pathogenic on their hosts. The less common MRA β was shown to be a virulent pathogen. At the same time, Peyronel (1924) observed dark hyphae in roots that also contained fungi of the AM type, referring to these as 'dual infections'. He considered the fungi to be '*Rhizoctonia*-like' and recorded their occurrence in the roots of 135 species of angiosperm. Since Peyronel's description of dual infections, it has been recognized that, while these types of root colonization could co-occur in plant roots with those produced by mycorrhizal fungi, they were also extremely widespread as sole occupants of roots.

Because fungi of these kinds failed to produce sexual stages and in recognition of the fact that, apart from the normally dark colouration of their hyphae, they had few other defining features, Haselwandter and Read (1982) referred to them collectively as 'dark septate' (DS) fungi. Others, emphasizing their frequent and apparently benign presence within host tissues, have described the same types of fungi as 'endophytes' (Currah *et al.*, 1987), 'septate endophytes' (O'Dell *et al.*, 1993) or 'dark septate endophytes' (DSE) (Jumponnen and Trappe, 1998). Reviewing the occurrence of DS fungi through the plant kingdom, Jumponnen and Trappe (1998) recorded reports of their presence in nearly 600 species, representing 320 genera and 100 plant families.

Whereas the designations DS and DSE fungi are now widely adopted, it is necessary to recognize not only that many DS fungi produce hyaline hyphae, particularly when present within plant tissues, but also that some fungi in this broad category normally consist entirely of pale coloured, white or hyaline mycelia both in the roots and in culture (Haselwandter and Read, 1982; Newsham, 1999; Barrow and Aaltonen, 2001; Yu *et al.*, 2001a; Barrow, 2003; Hambleton and Sigler, 2005). These hyaline forms are extremely difficult to detect by standard methods of staining and light microscopy and so are almost certainly widely overlooked in studies of 'DS' fungi. The status of these fungi is discussed in Chapter 11.

The taxonomic status of DS fungi

Gams (1963) succeeded in inducing some isolates of MRA α to produce conidia in culture and recognized them to be members of the genus *Phialocephala*. He emphasized that MRA α probably consisted of a number of different genotypes.

Understanding was further advanced by Wang and Wilcox (1985) who isolated numerous strains of MRA-type fungi from coniferous roots and, after incubating them for periods of six months to a year at low temperature (5°C), induced sporulation in a few isolates. These cultures were shown to represent three different genera and species of phialide-producing hyphomycete, namely, *Phialocephala fortinii* (Figure 7.4a), *Chloridium pauciflorum* (Figure 7.4b) and *Phialophora* (now *Cadophora finlandia*) (Figure 7.4c). A fourth DS fungus, *Leptodontidium orchidicola*, which produces minute conidia from the tips of conidiogenous cells but lacks phialides, has since been added to the list of the most widely occurring DSE fungi (Fernando and

Currah, 1995, 1996). On the basis of nuclear DNA sequence analyses, all four genera are now recognized as being ascomycetes of the order Helotiales. Within this order, *P. fortinii* is considered to have affinities with mollisioid taxa in the Dermateaceae (Wilson *et al.*, 2004). Analyses of ITS sequence data of *Cadophora finlandica* have

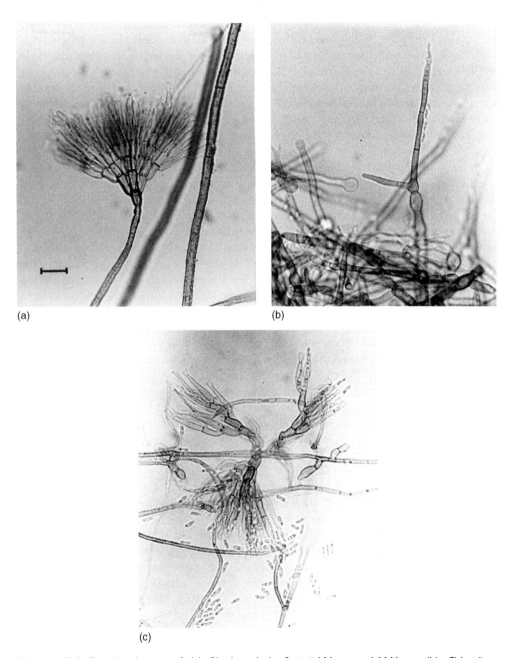

(a)

(b)

(c)

Figure 7.4 Conidiophores of (a) *Phialocephala fortinii* Wang and Wilcox; (b) *Chloridium pauciflorum* Wang and Wilcox; (c) *Phialophora finlandia* Wang and Wilcox (=*Cadophora finlandica* Harrington and New). Bars = 10 μm. Photograph courtesy J. Wang.

grouped this fungus within the *Hymenoscyphus ericae*-aggregate of the Helotiales (Vrålstadt *et al.*, 2002a; Hambleton and Sigler, 2005, see Chapter 11). *L. orchidicola* is also resolved on the basis of ITS analysis as being a member of the Dermateaceae within the Helotiales (Wilson *et al.*, 2004).

In-depth analyses of the taxonomic status and occurrence of *P. fortinii* have confirmed the views of Melin (1923) and Gams (1963) that fungi in what were known as the MRA α group are both extremely widely distributed and genetically heterogeneous. The *P. fortinii* component of MRA α occurs as the dominant root endophyte in families such as Poaceae, Caryophyllaceae, Rosaceae and Asteraceae in sub-alpine sites of western Canada (Stoyke *et al.*, 1992) and with a wide range of hosts, among which it is the dominant DS fungus in Pinaceae in Europe (Ahlich and Sieber, 1996). Thus, *P. fortinii* is confirmed to be neither host- nor site-specific, though its occurrence may be favoured in cold-stressed environments and it appears to have preference for some plant families.

On the basis of cultural characteristics, Ahlich and Sieber (1996) recognized four morphological types of *P. fortinii*. Subsequent population genetic studies in Europe (Menkis *et al.*, 2004; Queloz *et al*, 2005; Grunig *et al.*, 2006; Sieber and Grunig, 2006) and Canada (Piercey *et al.*, 2004) have confirmed that *P. fortinii senu lato* is composed of several cryptic species. Sieber and Grunig (2006) estimated that at least eight such species of *P. fortinii* occurred in Europe and one of these was recently described as *Acephala applanata* (Grunig and Sieber, 2005). These cryptic species can occur sympatrically in the same forest site and can even co-occur in the same root fragment (Grunig *et al.*, 2006).

Different geneotypes of *P. fortinii* have been identified as having the ability to form ectomycorrhizas (Fernando and Currah, 1996; Kaldorf *et al.*, 2004), as associates of healthy or decaying roots (Queloz *et al.*, 2005) or as a cause of soft-rot (Sieber, 2002). The fungi have also now been recorded above ground as occupants of the healthy wood of *Pinus sylvestris* stems (Menkis *et al.*, 2004). Studies of the related *Phialocephala* species, *P. dimorphospora* (Menkis *et al.*, 2004) suggest that it occurs mostly in tissues of *Picea abies* and, again, it is not restricted to roots. There appear to be no reports of its occurrence as a mycorrhizal fungus. Of the remaining widely occurring DS associates of woody roots, *Cadophora* (=*Phialophora finlandica* (Harrington and McNew comb, nov.)), was shown by Wilcox and Wang (1985, 1987a, 1987b) to form ectomycorrhizas with some coniferous hosts and ectendomycorrhizas with others (see Chapter 6). *C. finlandica* has now been identified as the fungus involved in the formation of the widely occurring ECM morphotype *Picierhiza bicolorata* (see Colour Plate 6.1c and Figure 11.12) and has also been shown to have the potential to form hyphal coils of the ERM type in roots of *Vaccinium* (see Chapter 11).

As the name suggests, *L. orchidicola* was first isolated from an orchid, although it was not shown to be an orchid mycorrhizal fungus (Currah *et al.*, 1987). It has since been recorded as an associate of other herbaceous plants, such as *Carex* spp. (Fernando and Currah, 1995) and *Ranunculus adoneus* (Schadt *et al.*, 2001; Harrington and McNew, 2003). The likely phylogenetic relationships between the major genera of DS fungi have been resolved by Menkis *et al.* (2004).

Functional roles of DS fungi

On the basis of an in depth analysis of the very large pre-1985 literature dealing with fungi of the DS type, Summerbell (2005) concluded that a potential for pathogenicity had been established 'beyond doubt' in fungi attributed historically to MRA α (now considered as *P. fortinii sensu lato*). However, the circumstances under which pathogenicity was expressed were often artificial. Some concluded that these fungi became aggressive only in forest nurseries which had been deprived by fumigation of their ECM inoculum (Richard and Fortin, 1974). Others, notably Robertson (1954), in a careful analysis of distribution of the fungus within individual long and short mycorrhizal roots of *Pinus sylvestris*, showed that the fungus occurred preferentially in naturally senescent tissues rather than in active mycorrhizas. Detailed aetiological studies of plant responses to invasion by *P. fortinii* have been few, but inoculation of *Asparagus* roots by the fungus gave rise to irregular wall thickenings adjacent to the hyphae which are suggestive of a weak defence reaction (Yu *et al.*, 2001b).

Most analyses of the modern era confirm the conclusions of the pioneer studies (Jumponnen, 2001; Mandyam and Jumponnen, 2005). Inoculation of pot cultures of a wide range of woody and herbaceous plants with *P. fortinii* has shown either no impact or negative effects upon plant growth (Wilcox and Wang, 1987; Stoyke and Currah, 1991, 1993; Currah *et al.*, 1993; O'Dell *et al.*, 1993).

Since the evidence is overwhelmingly in favour of the notion that DS fungi occur as benign or malignant associates of roots, it is appropriate to consider them as facultative biotrophs (Jumponnen and Trappe, 1998) rather than as mycorrhizal fungi. However, there are occasions when plants may benefit directly from their presence and, in these cases, the DS fungi concerned appear to be functioning towards the mycorrhizal end of the mutualism–parasitism continuum (Francis and Read, 1995; Johnson *et al.*, 1997).

Enhancement of P capture has been shown to arise from colonization of alpine *Carex* spp. (Haselwandter and Read, 1982) and *Pinus contorta* (Jumponnen *et al.*, 1998) by *P. fortinii* (see Chapter 10). Another fungus, *Aspergillus ustus*, regarded by Barrow and Osuna (2002) as being of the DS type, was shown to enhance P mobilization from rock and tricalcium phosphates and to increase transfer of P to the shrub *Atriplex canescens* in semi-arid grasslands.

Assays of polymer utilization by *P. fortinii* and *C. finlandica* (Caldwell *et al.*, 2000) suggest that these fungi have well-developed saprotrophic capability. Among the polymeric C sources exploited by both fungi were cellulose, starch and fatty acid esters. Protein and RNA were hydrolysed as potential organic N and N and P sources, respectively. Upson (2006) has shown that these capabilities can benefit plants growing in nutrient deficient soils (see Chapter 15). *P. fortinii* produces the hydroxamate siderophore, ferricrocin (Bartholdy *et al.*, 2001), which can be expected to facilitate iron capture for its autotrophic partners.

There remains the possibility that, by forming benign associations with significant portions of root system, DS fungi provide protection against virulent pathogens. Such protective effects have been demonstrated in AM colonized plants (see Chapter 16), but the appropriate experiments appear not to have been attempted using DS colonized plants.

Arbutoid mycorrhizas

Mycorrhizas in *Arbutus* and *Arctostaphylos*

The fungi mycorrhizal with *Arbutus* and other plants in the Arbutoideae were long believed to be basidiomycetes because of the structural similarities between ECM and arbutoid mycorrhizas. This has been confirmed both by synthesis experiments and by the descriptions of dolipore septa in fungi associated with mycorrhizas of *Arctostaphylos* (Duddridge, 1980; Scannerini and Bonfante-Fasolo, 1983; Read, 1983). The work of Zak (1973, 1974, 1976a, 1976b) who both traced mycelium and performed synthesis experiments, showed that mycorrhizas in *Arbutus menziesii* and *Arctostaphylos uva-ursi* are formed by fungi which also form ectomycorrhizas. The fungi involved included *Hebeloma crustuliniforme*, *Laccaria laccata*, *Lactarius sanguifluus*, *Poria terrestris* var. *subluteus*, *Rhizopogon vinicolor*, *Pisolithus tinctorius*, *Poria terrestris*, *Thelephora terrestris*, *Piloderma bicolor* and *Cenococcum geophilum*. Similarly, Molina and Trappe (1982a) tested the ability of 28 ECM fungi to form mycorrhizas with *Arbutus menziesii* and *Arctostaphylos uva-ursi* in pure culture. All but three produced arbutoid mycorrhizas with both species. The conclusion here must be that the plant plays an important part in regulating the development of mycorrhizas, with the consequence that different structures are produced in different taxa of plant.

There is no evidence which will allow comparison of function in arbutoid and ectomycorrhizas, but the assumption is that they operate similarly. The plants are all woody and photosynthetic and, since mycorrhizas are the common form of absorbing organ of members of the Arbutoideae, an important and ecologically significant group of species, the symbiosis must be assumed to be of selective advantage. This is even more likely because the sheath on the roots, as in ectomycorrhizas, may not only have a storage function, but also separates the plant from the soil. Hence, the fungus calls the tune in absorption by the short roots, and everything absorbed by them must pass through it. It seems extremely likely that the mycelium and rhizomorphs in soil are important in nutrient scavenging.

Massicotte *et al.* (1993) carried out a detailed analysis of the structure and histochemistry of arbutoid mycorrhizas, synthesized in growth pouches between *Arbutus menziesii* and the basidiomycetes *Pisolithus tinctorius* and *Piloderma bicolor*. The morphology of the mycorrhizas was strongly influenced by the identity of the fungal symbiont. In the case of plants colonized by *P. tinctorius*, repeated pinnate branching of first and second-order lateral roots produced a complex structure (Figure 7.5), similar to that originally described by Rivett (1924) in *Arbutus unedo*. This pattern of branching, which has been observed in associations between *Arbutus* spp. and a range of other fungi (Molina and Trappe, 1982a, 1982b; Giovannetti and Lioi, 1990), appears to arise from precocious initiation of individual lateral roots, rather than by dichotomy of the apical meristem of the root, as is typically seen in ectomycorrhizas of *Pinus* (Piché *et al.*, 1982; see Chapter 6). Each of the rootlets colonized by *P. tinctorius* is ensheathed in a well-developed mantle, from the outer layer of which an extensive system of rhizomorphs develops. This is a feature previously recorded in associations between *Arbutus* and several other species ECM fungi (Zak, 1976b; Molina and Trappe, 1982b).

Mycorrhizas formed by *P. bicolor*, in contrast, were largely unbranched and had a thin or non-existent mantle, sparse surface hyphae being embedded in mucilage in a manner similar to that seen in ectendomycorrhizas formed in *Pinus* by *Wilcoxina* spp. (Piché *et al.*, 1986; Scales and Peterson, 1991a).

A longitudinal section of the pinnately branched type of mycorrhiza formed by *P. tinctorius* (Figure 7.6) reveals a thick mantle, intercellular development of mycelium to produce a Hartig net and penetration of some epidermal cells by fungal hyphae which proliferate to form dense hyphal complexes (Massicotte *et al.*, 1993). The combined

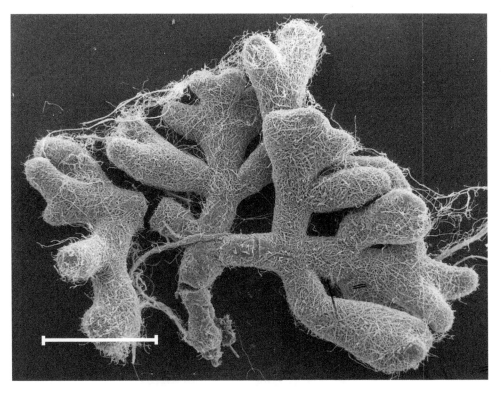

Figure 7.5 Scanning electron microscopy of mycorrhizal rootlets formed between *Arbutus menziesii* and *Pisolithus tinctorius*, showing three-dimensional pinnate branching. Bar = 500 μm. From Massicotte *et al.* (1993), with permission.

presence of mantle, Hartig net and intracellular proliferation are diagnostic features of arbutoid mycorrhizas which can only be revealed by anatomical investigation. There are reports (e.g. Largent *et al.*, 1980) that *Arbutus* spp. are ECM, but these are based only on superficial recognition of the presence of a mantle. Clearly, in the absence of more detailed structural analyses such reports must be regarded with suspicion.

It was observed by Rivett (1924) and subsequently confirmed (Fusconi and Bonfante-Fasolo, 1984; Münzenberger, 1991; Massicotte *et al.*, 1993) that the Hartig net in *Arbutus* is of the para-epidermal kind typically found in ectomycorrhizas in the majority of angiosperms (Brundrett *et al.*, 1990; Chapter 6). Massicotte *et al.* (1993) propose that deeper penetration may be prevented by deposition of suberin lamellae and a Casparian strip in radial walls of the outer tier of cortical cells, so forming an exodermis. The epidermal cell walls contain phenolic substances but no suberin and clearly do not inhibit fungal penetration. The physiological activity and potential storage role of the fungal tissue is indicated by the presence of glycogen rosettes and of polyphospate (polyP) (Ling Lee *et al.*, 1975).

Mycorrhizas in Pyrolae

Molecular phylogenetic analyses have led to taxonomic revisions involving loss of the family Pyrolaceae and transfer of its constituent genera to the tribe Pyrolae in the order Ericales (Kron *et al.*, 2002). These analyses reveal that the nearest relatives of the Pyrolae within this order are the tribes Monotropae and Pterosporae (see Chapter 13). The Pyrolae encompasses four genera viz: *Pyrola* (35 spp., e.g. *P. minor* – see Colour Plate 7.1a), *Chimaphila* (7 spp.), *Orthilia* (1 spp. *O. secunda* – see Colour Plate 7.1b) and *Moneses* (1 spp.). In contrast to the Monotropae and Pterosporae, all members of which are fully mycoheterotrophic, only one member of the Pyrolae, *P. aphylla*, is achlorophyllous and hence putatively a full mycoheterotroph *sensu* Leake (1994; see Chapter 13). All members of the tribe, henceforward referred to as 'pyroloid plants', are restricted in nature to habitats dominated by woody species with ECM associations. Below ground, pyroloid plants typically produce extensive white rhizome systems from which darker coloured branched root axes arise. Light microscopic examination of regions immediately distal to the root apices reveals dense intracellular mycorrhizal colonization of the epidermal cells of the roots (see Colour Plate 7.1c, d).

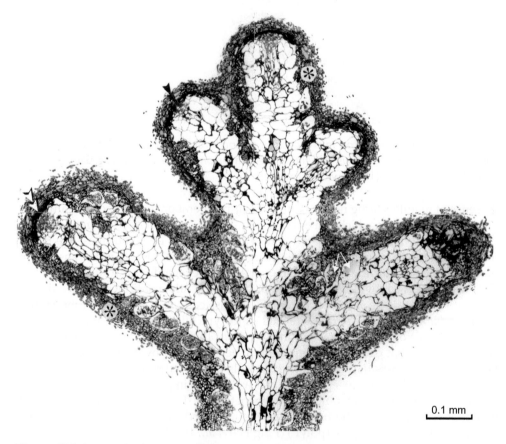

Figure 7.6 Longitudinal section (light microscopy) of a pinnate mycorrhiza formed by *Pisolithus tinctorius* on *Arbutus menziesii*. The individual branches have flattened apices (double arrowheads) and are covered by a thick mantle (*). Many of the epidermal cells are colonized by intracellular hyphae (arrows). Bar = 0.1 mm. From Massicotte *et al.* (1993), with permission.

Seeds of pyroloid plants are minute. They have greatly reduced embryos and hence are similar in structure to those of Monotropae, Pterosporae and most Orchidaceae, which are described as dust seeds (see Chapters 12 and 13). Asymbiotic germination and early development of *P. rotundifolia* and *P. secunda* were studied *in vitro* by Lihnell (1942), who described the formation of colourless branched axes supporting 'root-like' structures (procaulomes). These structures developed slowly in culture for 3 years without further differentiation. Lihnell reported that only few seeds germinated. Symbiotic germination of *P. secunda* and *O. secunda* has recently been observed in nature when seed of these species was sown in packets under stands of pyroloid plants in beech forest (Zimmer *et al.*, 2007). While amounts of germination were low, those seedlings observed had produced root axes within one year, the epidermal cells of which were heavily colonized by fungi producing mycorrhizal structures very similar to those seen in adult root systems (see Colour Plate 7.1e, f).

Lihnell (1942) isolated four fungi from the mycorrhizas of *P. rotundifolia* and one from *P. secunda*. From his descriptions, these appear to have produced sterile mycelia, at least some of which were of the DS kind. Fungi of this kind did not form any association with procaulomes produced aseptically. While Lück (1940) had isolated a clamp-bearing basidiomycete from *Pyrola,* only recently, with the use of molecular methods, has the extent of colonization by fungi of this kind been revealed (Tedersoo *et al.*, 2007; Zimmer *et al.*, 2007). Analyses of DNA confirm the observation of Lihnell that fungi of the DS kind proliferate in the vicinity of pyroloid roots. Among these, *Phialocephala fortinii* (Tedersoo *et al.*, 2007), as well as *Phialophora* and *Cadophora* spp. (Zimmer *et al.*, 2007; see Table 7.1) have been recorded. However, probably of greater significance is that the molecular analyses have identified a number of basidiomycetes, all of which are familiar as ECM forming fungi (Table 7.1). Tedersoo *et al.* (2007) found some evidence for a preferential association between *P. chlorantha* and *Tricholoma* spp. Though mycorrhizal syntheses have yet to be achieved, these studies point strongly to the possibility that pyroloid plants are, as in the case of *Arbutus* and *Arctostaphylos*, co-associated with the ectomycorrhizal symbionts of their non-ericalean partners.

Table 7.1 Mycorrhizal fungi detected in roots of pyroloid species from six German sites and one N Californian site (C6).

Pyroloid species	Site	n	Mycorrhizal fungi
Orthilia secunda	G1	5	**Piloderma spp.** (1), Ascomycete spp.* (1)
Orthilia secunda	G2	5	**Sebacina spp.** (1), Pezizales spp.* (2)
Orthilia secunda	G3	5	**Piloderma spp.** (2), **Wilcoxina spp.** (1)
Pyrola chlorantha	G4	5	**Russula spp.** (1)
Chimaphila umbellate	G5	5	**Tomentella spp.** (1), Phialophora spp.* (1)
Pyrola minor	G6	5	**Laccaria spp.** (1)
Pyrola picta	C6	5	**Cortinarius spp.** (1), **Hebeloma spp.** (1), **Piloderma spp.** (2), **Rhizopogon spp.** (2), **Russula spp.** (3), **Thelephoraceae spp.** (2), **Tomentella spp.** (2), Helotiales spp.* (1), Phialocephala spp. (1), **Wilcoxina spp.** (1), Cadophora spp. (1), Leptodontidium spp. (1)

From Zimmer *et al.* (2007). Obligate ectomycorrhizal lineages are shown in bold. Lineages that contain some ectomycorrhizal taxa are indicated by an asterisk. *n* is the number of individuals from which root samples were taken. Numbers in parentheses after each fungus indicate the number of plants with which it was associated.

Evidence in support of the possibility that their mycorrhizal fungi contribute to the N and C economies of the Pyrolae is emerging from comparative analyses of the δ^{15}N and ^{13}C isotope signatures of these plants (Tedersoo *et al.*, 2007; Zimmer *et al.*, 2007). These studies demonstrate that the δ^{15}N signatures of pyroloid plant tissues are considerably enriched relative to those of co-associated plants that are either non-mycorrhizal or colonized by AM fungi, but are less enriched than those of fully mycoheterotrophic plants. Since the high levels of δ^{15}N and ^{13}C enrichment seen in full mycoheterotrophs are acknowledged to be attributable to their use of fungus-derived sources of these elements (see Chapters 12 and 13), the values recorded in the Pyrolae are indicative of partial mycoheterotrophy in these plants. Zimmer *et al.* (2007) showed significant enrichment of δ^{15}N in *O. secunda* (Figure 7.7a, b, c), *P. chlorantha* (Figure 7.7d), *P. minor* (Figure 7.7f) and *Chimaphila umbellata* (Figure 7.7e) relative to the non-pyroloid autotrophs. Their tissues were less δ^{15}N enriched than those of co-occurring full mycoheterotrophs in the genera *Neottia* and *Monotropa*. Using a two-source isotope mixing model (see Chapter 13) Zimmer *et al.* (2007) calculated that, across a range of habitat types and light environments, between 30 and 80% of tissue N of these pyroloid plants was derived from their mycorrhizal fungi (Table 7.2).

In the study of Zimmer *et al.* (2007), analyses of δ^{13}C indicated that the contribution of mycoheterotrophy to the plant C economies was considerably less than that of N. Indeed, δ^{13}C values were, in contrast to those of fully mycoheterotrophic plants, generally not significantly different from those of other neighbouring autotrophs (Figure 7.7). In only one species, *O. secunda*, and at one deeply shaded site, was there a significant mycoheterotrophic C gain (Table 7.2). In the case of this element, the data of Zimmer *et al.* (2007) and Tedersoo *et al.* (2007) are discrepant. In the latter study, in addition to the significant mycoheterotrophic N gains, *O. secunda*, *P. chlorantha*, *P. rotundifolia* and *Chimaphila umbellata* were reported to acquire between 10 and 68% of their C from fungal sources. Clearly, further analyses will be required to determine whether this discrepancy reflects real biological distinction between the plants examined or, as seems more likely, a difference arising from the methods of calculating the relative levels of isotopic enrichment. Whatever

Table 7.2 Percentage of nitrogen ($N_{df} \pm 1$ SD) and carbon ($C_{df} \pm 1$ SD) derived from mycorrhizal fungi in the leaves of pyroloids as calculated from δ values and mean enrichment factors, using a linear two-source isotopic mixing model

Pyroloid/orchid species	Site	Relative irradiance [%]	N_{df} [%] ± 1 SD	C_{df} [%] ± 1 SD
Orthilia secunda	G1	1	56 ± 21**	28 ± 12**
Orthilia secunda	G2	100	62 ± 15**	0
Orthilia secunda	G3	15	59 ± 27**	0
Pyrola chlorantha	G4	15	74 ± 29**	0
Chimaphila umbellata	G5	15	40 ± 5***	0
Pyrola minor	G6	25	30 ± 9**	0
Pyrola picta	C3		83 ± 26**	0
Pyrola picta	C4		70 ± 11***	0

Significance levels for deviations from zero, based on a Student's t-test are indicated by asterisks: **$P<0.01$, ***$P<0.001$. Relative irradiance (%) was measured under the forest canopy at sites G1–G6. Sites: G = Germany; C = California. Relative irradiance %. Modified from Zimmer *et al.* (2007).

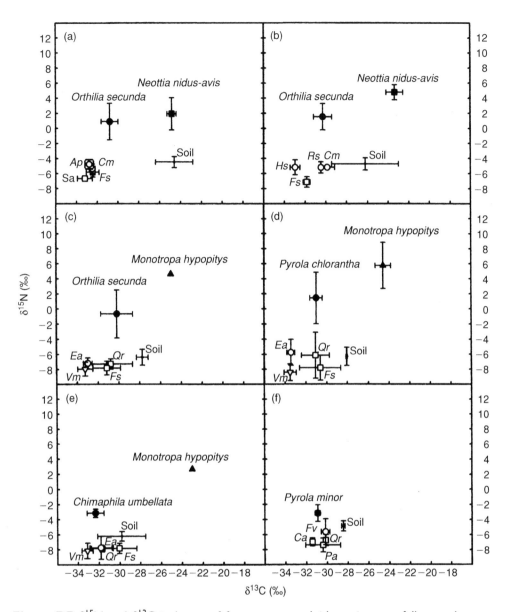

Figure 7.7 $\delta^{15}N$ and $\delta^{13}C$ in leaves of four green pyroloid species, two fully mycohetero-trophic species, 12 fully autotrophic reference species and in soil samples collected from six forest sites in NE Bavaria (Germany): (a) *Fagus sylvatica* forest G1; (b) *Fagus sylvatica* forest clearance G2; (c), (d) and (e) mixed *Pinus sylvestris* forest G3–5; (f) mixed conifer forest G6. Symbols represent a classification according to functional groups: ectomycorrhizal reference plants (□), arbuscular mycorrhizal or non-mycorrhizal reference plants (open circles), ericoid mycorrhizal reference plants (∇), green pyroloids (●), mycoheterotrophic orchids (■), mycoheterotrophic monotropoids (▲). Error bars are missing if $n < 3$. Plant species abbreviations: Ap = *Acer pseudoplatanus*, Ca = *Corylus avellana*, Cm = *Convallaria majalis*, Ea = *Epilobium angustifolium*, Fs = *Fagus sylvatica*, Fv = *Fragaria vesca*, Hs = *Hieracium sylvaticum*, Pa = *Picea abies*, Qr = *Quercus robur*, Rs = *Rubus saxatilis*, Sa = *Sorbus aucuparia*, Vm = *Vaccinium myrtillus*. Mean ± 1 SD values. From Zimmer et al. (2007).

the cause of these differences, the likelihood emerges from the Tedersoo *et al.* and Zimmer *et al.* studies that there is a close physical and nutritional interdependence between pyroloid plants, their mycorrhizal symbionts and their neighbouring autotrophs. Indeed, hitherto largely unrecognized mycoheterotrophic intereractions may explain the apparently obligate restriction of these plants to forest or shrub communities dominated by ECM trees.

Ultrastructure of mycorrhizal roots in Pyrolae

Careful ultrastructural analysis by Robertson and Robertson (1985) of development of the arbutoid mycorrhizas of a range of *Pyrola* species has enabled us to visualize the sequence of events involved in the formation and subsequent decline of the arbutoid association at the cellular level. The mycorrhiza is initiated a few millimetres behind the root tip, where hyphae from the surface weft (Figure 7.8) penetrate between the radial walls of epidermal cells to form a Hartig net. In *Pyrola* this, again, is of the para-epidermal type (Figure 7.9) associated with labyrinthine developments of the hyphal walls (Figure 7.10). The walls of these hyphae are more electron dense then those that penetrate the cells (Figures 7.10 and 7.11). As hyphae enter the epidermal cells from the Hartig net, they are surrounded by invaginated plasma membrane of the plant cell and by a matrix of material that is continuous

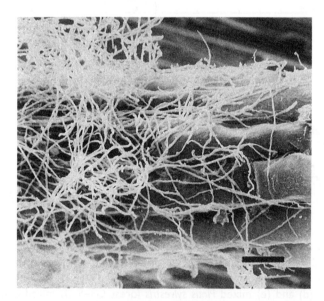

Figure 7.8 Mycorrhiza formation in *Pyrola* (*Orthilia*) *secunda*. A surface weft of fungal hyphae covers the root a short distance behind the root tip, but there is no organized sheath. Bar=50 μm. From Robertson and Robertson (1985), with permission.

Figure 7.9 Mycorrhiza formation by *Hysterangium separabile* on *Pyrola secunda*. Section showing sparse development of mycelium on the suface of the root (*), para-epidermal Hartig net (arrowed, HN) and extensive intracellular colonization of the epidermal cells, but no penetration into the root cortex. Bar=50 μm. From Robertson and Robertson (1985), with permission.

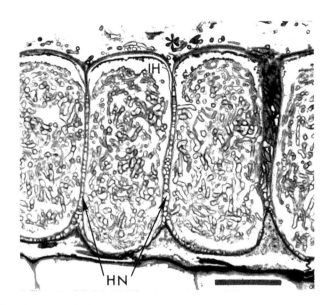

Figure 7.9 (Caption opposite)

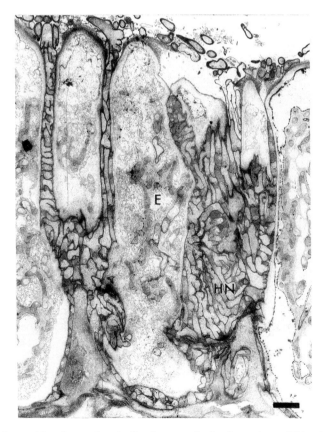

Figure 7.10 Mycorrhiza formation in *Pyrola secunda*. Surface view of Hartig net (HN) formation, showing anastomoses and labyrinthine developments in the hyphal walls. Bar = 10 μm. From Robertson and Robertson (1985), with permission.

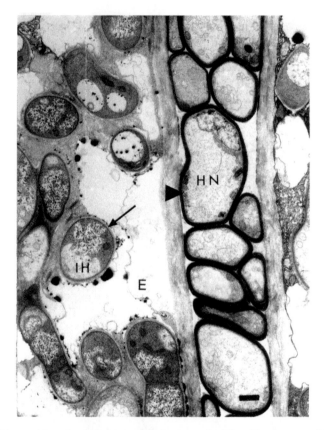

Figure 7.11 Mycorrhiza formation in *Pyrola secunda*. The Hartig net (HN) surrounds the epidermal cells (E). Note the differences in wall thickness between the hyphae of the Hartig net (arrowhead) and those colonizing the epidermal cells (arrow). Bar=1.0 μm. From Robertson and Robertson (1985), with permission.

with the plant cell wall (Figure 7.12). This matrix is considerably thicker (around 100 nm) than the wall of the hypha itself, which attains a thickness of only 25–35 nm (Figure 7.13). Tannin-like deposits occur along the plant membrane surrounding the hyphae and on the tonoplast (Figure 7.14). Despite this apparent resistance

Figure 7.12 Intracellular hyphal development in an epidermal cell of *Pyrola secunda*, showing penetration by a hypha from the Hartig net and invagination of the plasma membrane of the plant cell (arrowed). Bar=1.0 μm. Inset: Light micrograph of epidermal cell showing hyphal invasion from the Hartig net and extensive intracellular colonization. From Robertson and Robertson (1985), with permission.

Figure 7.13 Development of granular interfacial matrix (arrowed) surrounding the intracellular hyphae in *Pyrola secunda*. Note the relatively thin hyphal walls and extensive development of organelles in both fungal hypha and plant cells. Bar = 1.0 μm. From Robertson and Robertson (1985), with permission.

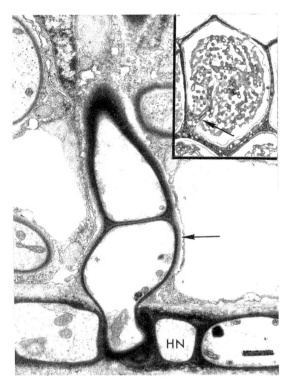

Figure 7.12 (Caption opposite)

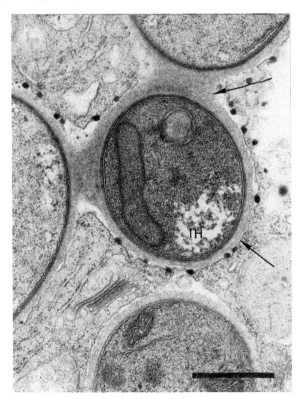

Figure 7.13 (Caption opposite)

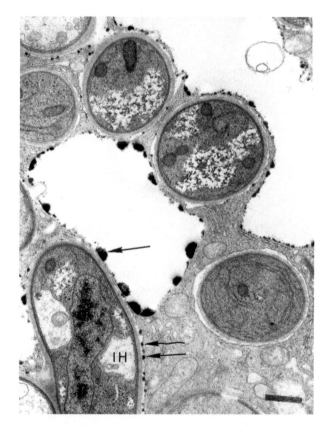

Figure 7.14 Intracellular fungal development in *Pyrola minor*, showing extensive tannin deposits on plant plasma membrane and tonoplast (arrowed). Note development of organelles in both symbionts. Bar = 1.0 μm. From Robertson and Robertson (1985), with permission.

response, the fungus grows extensively within the cells, forming complexes which, in TS, appear as hyphal profiles (Figures 7.9 and 7.12).

At the mature stage, the plant cytoplasm in colonized cells is packed with organelles including mitochondria, endoplasmic reticulum, ribosomes, dictyosomes and plastids of various kinds (see Figures 7.13 and 7.14). The cytoplasm of the intracellular hyphae is even more dense and, in addition to abundant mitochondria and nuclei, can have membranous sheets and multivesicular bodies (see Figure 7.13). The work of Robertson and Robertson (1985) also revealed intraspecific differences in the structural organization of the interfacial matrix in *Pyrola* species. In *P. minor*, the matrix is a homogeneous granular layer between fungal wall and plant plasma membrane (see Figure 7.13), whereas in *P. secunda*, it has a characteristic honeycomb appearance (Figure 7.15a, b).

The process of senescence of the association is similar to that observed in the ericoid mycorrhizas of *Rhododendron* (Duddridge and Read, 1982b; see Chapter 11). It begins in a localized region of a cell (Figure 7.16), with degeneration of cytoplasm of the plant and loss of integrity of its organelles. Again, as in *Rhododendron*, the fungal hyphae and surrounding matrix materials have a normal appearance at this stage.

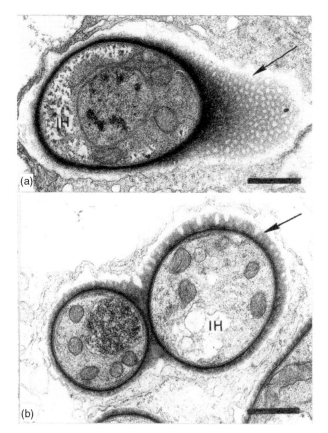

Figure 7.15 (a) and (b). Honeycomb structure of the interfacial matrix (arrowed) in *Pyrola secunda*. Bars = 1.0 μm. From Robertson and Robertson (1985), with permission.

Degeneration of the hyphae routinely occurs only after disorganization of the plant cytoplasm (Figures 7.17 and 7.18), though Robertson and Robertson (1985) occasionally observed collapsed hyphae embedded in apparently healthy plant cytoplasm. It thus seems clear that in arbutoid, as in ericoid, mycorrhizas degeneration of the plant cell frequently precedes that of the fungus, a situation which again contradicts earlier observations, based upon light microscopy, that 'digestion' of the fungus takes place as the association matures.

Figure 7.16 Two colonized epidermal cells of *Pyrola*, showing para-epidermal Hartig net (arrowhead) and mature intracellular colonization. Localized degeneration of the plant cytoplasm surrounding living hyphae is indicated by arrows. Bar = 1.0 μm. From Robertson and Robertson (1985), with permission.

Figure 7.17 Advanced stage of senescence, showing degeneration of plant cytoplasm around living hyphae in *Pyrola secunda*. Bar = 1.0 μm. From Robertson and Robertson (1985), with permission.

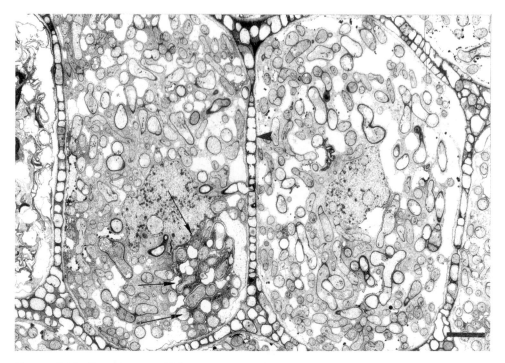

Figure 7.16 (Caption on p. 291)

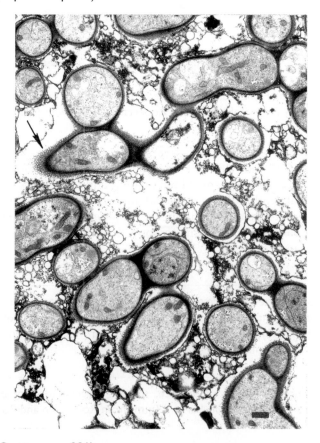

Figure 7.17 (Caption on p. 291)

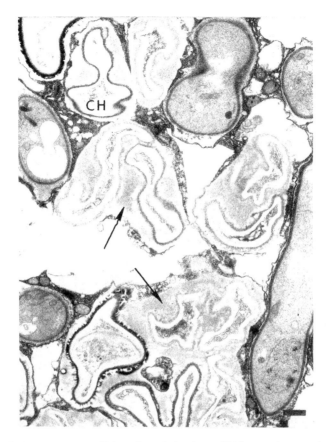

Figure 7.18 Collapsed intracellular fungal hyphae (CH) in the epidermis of *Pyrola*. Bar = 1.0 μm. From Robertson and Robertson (1985), with permission.

Conclusions

It is clear that ectendomycorrhizas are not aberrant or minor parasitic infections which occur in the absence of ECM fungi. They are symbiotic and potentially mutualistic associations which occur frequently as the early colonization state of seedlings of some conifers. Little is known of the function of this type of mycorrhiza, but it may be productive to make comparative estimates of the drain of organic C in relation to mineral nutrient absorption, to see if ectendomycorrhizas are in fact less C demanding than the ectomycorrhizas that develop on older plants which have reached their full photosynthetic potential.

One interesting question raised by Wilcox, relates to the consistent formation of ectendomycorrhizas on *Pinus* and ectomycorrhizas on *Picea*, by the same E-strain fungal isolates. The findings, which indicate an element of plant control over mycorrhiza development, are not unique to E-strain fungi.

The extremely widespread occurrence of dark septate fungi as associates of roots, a feature already recognized in the early literature, continues to cause interest. Thanks to the application of molecular methods, much progress has been made towards accurate identification of the taxa involved. The question of the function of these associations remains to be answered unequivocally, but the vast majority of studies point to these organisms as being facultative biotrophs rather than mycorrhizal symbionts.

Arbutus, *Arctostaphylos* and *Pyrola* form arbutoid mycorrhizas with the same fungi that form ectomycorrhizas in conifers. The emerging evidence from stable isotope analysis indicates that, for Pyrolae at least, the sharing of symbionts with ECM co-associates facilitates mycoheterotrophic supply of N and possibly also of C to the arbutoid symbiosis. Such observations go some way to explaining the apparently obligate coexistence of Pyrolae and ECM trees in nature. Several features of the Pyrolae, notably the extreme reductions of their seeds and the occurrence within the tribe of a fully achlorophyllous species, suggest that, in an evolutionary context, these plants have a status that is intermediate between full autotrophy and the full mycoheterotrophy of their close relatives in the Monotropae (see Chapter 13).

8

Growth and carbon allocation of ectomycorrhizal symbionts

Introduction

Frank (1894) conducted experiments on the effects of ectomycorrhizal (ECM) colonization on the growth of seedlings of *Pinus*. Those which were grown in unsterilized soil developed mycorrhizas and grew faster than those in sterilized soil. Although this experimental design is faulty, because heat sterilization may release both nutrients and toxic substances, the results have in principle been repeatedly confirmed. Refinements aimed at obviating the effects of heat sterilization have been used and the results can be accepted with confidence. In addition, observations on the establishment of exotic species of ECM trees in many parts of the world have shown that artificial inoculation is usually essential to success, although non-mycorrhizal seedlings may be grown if provided with adequate fertilizers (see Chapter 17).

Surprisingly, while there have been innumerable reports of growth enhancement as a result of ECM colonization, relatively few reports have provided unequivocal evidence, in the form of nutrient response curves, of the benefits accruing to a plant from colonization when its growth is limited by a single nutrient. Some such studies are now available (e.g. Heinrich and Patrick, 1986; Bougher *et al.*, 1990; Jones *et al.*, 1990, 1998).

The study of Bougher *et al.* (1990) is particularly instructive in the overall context of growth response. *Eucalyptus diversicolor* plants, whose small seeds have little initial nutrient reserve, were grown in a sandy soil of low organic content which was given mild sterilization unlikely to produce unnatural side effects. Furthermore, the experiments examined responses to inoculation with four fungi (two isolates, A and B, of *Descolea maculata* and one each of *Laccaria laccata* and *Pisolithus tinctorius*) and to one element, P, the deficiency of which was known to be a major factor limiting growth in soils supporting the plant in nature. Ecological relevance was thus added to good experimental design and the results provide a useful example of the type of growth response that might be expected in the field.

The growth response curves of non-mycorrhizal plants in relation to soil P were of the sigmoid type, as seen typically in other coarse-rooted plants (Bolan *et al.*, 1983; Figure 8.1a). In these, there is a threshold P concentration in soil below which growth will not occur. Colonization by *D. maculata* A and B and *L. laccata* modified the sigmoidal growth curve by removing the threshold effect and allowing plants to

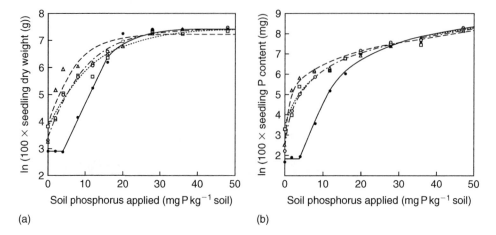

(a)   (b)

Figure 8.1 Growth and P uptake of *Eucalyptus diversicolor* in response to P application and mycorrhizal colonization by two strains of *Descolea maculata*, A (□) and B (○) or by *Laccaria laccata* (△), compared with uninoculated controls (●). (a) Dry weight; (b) P content. From Bougher *et al.* (1990), with permission.

grow in extremely nutrient-poor soil. This effect is believed to result from the ability of extraradical mycelium to exploit nutrients, in this case P, beyond depletion zones surrounding the root (see Chapter 5). The fungi may also have a lower threshold concentration for absorption of the element, or may have the ability to release P associated with complex substrates which would add to their ability to increase nutrient uptake (see Chapters 9 and 10).

In the range of P supply in which a growth response to ECM colonization occurred (2–12 mg P/kg soil) fungal isolates differed in their ability to promote seedling growth (Figure 8.1a). Dry weight of seedlings inoculated with *L. laccata* was significantly greater than that of plants colonized by *D. maculata* at 2 and 4 mg P/kg soil. At the latter concentration, biomass of *L. laccata* seedlings was nine times that of seedlings ECM with *D. maculata* and 21 times non-mycorrhizal plants. The pattern of P accumulation by plants in relation to soil P supply was similar to that of dry weight (Figure 8.1b) and again ECM colonization removed the threshold seen in uncolonized plants. However, seedling dry weight reached a maximum at around 28 mg P/kg soil, whereas P content increased linearly above this level of P application. ECM plants, in addition to showing no threshold, had higher tissue P concentrations than non-mycorrhizal plants even with no added P.

This example, in addition to demonstrating that ECM colonization has the potential to enhance growth, shows that it can fundamentally change the nature of growth response curves of plants at low nutrient concentrations. It further emphasizes that there may be considerable differences both between and within fungal species in their ability to acquire nutrients and promote growth. This biologically important observation also indicates that response curves comparing the performance of different ECM fungi can be used to predict more accurately their potential for application and effectiveness for nursery and afforestation practices (see Chapter 17). However, it must be borne in mind that this selected example describes responses to P when other nutrients were supplied in adequate amounts. Depending upon

Table 8.1 Growth of *Picea sitchensis* in axenic culture with *Lactarius rufus* over 14 weeks.

	Control	Inoculated	Significance of difference P<
Mycorrhizal colonization (%)	0	58.2 ± 9.9	
Shoot height (cm)	5.8	9.2	0.05
Shoot dry weight (mg)	47.1	100.2	0.01
Root dry weight (mg)	25.5	74.0	0.001
Total dry weight (mg)	72.6	174.2	0.001
Root–shoot ratio	0.59	1.76	NS
Number of lateral buds	4.4	5.9	NS
Length of lateral roots (cm)	98.2	219.0	0.01
Total number of root tips	60.8	236.7	0.05
Short root/cm lateral roots	0.64	1.17	NS
Short roots/mg weight	2.43	3.31	NS

Data from Alexander (1981). Substrate: vermiculite-peat with nutrient medium. The roots were kept below 25°C in a greenhouse. Midday irradiance 25W/m² approximately. NS, not significant.

the soil or ecosystem in which a plant occurs, other elements, in particular N (see Chapter 9), may be more important as growth limiting factors, and that often P and N may be co-limiting. Experiments investigating dose response curves in these different situations would be extremely instructive.

The difficulties of carrying out experiments with trees under controlled conditions have precluded extensive research except during a small part of their early life. In consequence, comparisons between ECM plants and controls have been made over one or a few seasons at most. Much work has also been done in open beds of soil, sometimes inoculated with soil, humus, or chopped mycorrhizas, with controls treated with sterilized inoculum (examples are given by Harley, 1969). In some cases, the soil has been sterilized by drenching or fumigation before planting. The published conclusions are simple: ECM seedlings are usually taller and have larger root systems, both shoots and roots are of greater dry weight but the ratios of root weight to shoot weight are very frequently smaller.

The change in the root:shoot ratio needs further investigation, for irrespective of mycorrhiza formation, the ratio diminishes during the early growth of the young seedling and is lower in soils of high nutrient availability, especially of N, as well as being sometimes reduced in conditions where photosynthesis is reduced. Although a lower ratio might be expected in mycorrhizal plants, both because of a greater ability to absorb nutrients and an increased concentration of nutrients in their tissues, it would be of great value to know more about the extent to which greater size and age themselves contribute to carbon (C) allocation in the seedlings. Alexander (1981) grew *Picea sitchensis* seedlings in monoxenic culture with *Lactarius rufus* for 14 weeks. The results are shown in Table 8.1. The usual increases of height and dry weight of both shoots and roots were observed, but the root:shoot ratio was greater (although below $P = 0.05$ significance) in ECM plants. In discussing this, Alexander observed that, if the fungal sheath were to comprise 20% of the weight of the roots, which is not impossible, the slight rise in root:shoot ratio might be explained by the fungal colonization directly. Bougher *et al.* (1990) also provided data relating root: shoot ratio of *E. diversicolor* to total seedling dry weight, which showed no differences in response to the fungi they employed and further, no differences between

ECM and non-mycorrhizal plants. This again emphasizes that effects on root:shoot ratio may be related less to colonization than to plant size.

Other observations of Alexander (1981) seem to confirm the much older results (e.g. McComb, 1938) that colonization increases the total number of short roots per seedling. In this case, the form of the root system changes, for there are more short roots both per centimetre of root length and per milligram of root weight on the ECM seedlings. There is a great need to disentangle the normal changes in form due to increase in size, those dependent on changes in nutrient supply and absorption, and those arising from ECM colonization and its physiological consequences. Moreover, although ECM colonization is known to increase growth rates of plants supplied with suboptimal levels of a major nutrient such as P (Bougher et al., 1990), relatively few studies have examined the consequences of ECM development in plants receiving optimal supplies of nutrients. On theoretical grounds it is evident that under these circumstances, the benefits of colonization in the form of increased nutrient capture would be eliminated, yet the costs to the plant in terms of C supplied to the fungus for growth and respiration (quantified below) would remain. In fact, several studies (Molina and Chamard, 1983; Ingestad et al., 1986; Rousseau and Reid, 1991) indicate that, while ECM plants suffered little or no growth reduction at adequate or supra-optimal nutrient supply, such reductions were experienced under lower nutrient regimes. A possible explanation for this apparent anomaly was sought by Rousseau and Reid (1991). They point out that plants in the higher nutrient regimes generally have lower root:shoot ratios relative to those under low regimes. By definition, plants with lower root:shoot ratio have more tissues that are able to gain rather than consume C. Therefore, even if equal quantities of root tissue were colonized in seedlings with different root:shoot ratios, those with the higher ratio would have proportionately more mycorrhizal development per plant than their counterparts with a low ratio of root to shoot. Thus, if an ECM fungus constituted 20% of the root system of a seedling with a 1:1 ratio of shoot to root, it would make up 10% of the dry weight; in cases where that ratio was 3:1, the weight of fungus would contribute only 5%. The relative C costs of supporting the fungus in a seedling with a 1:1 shoot root ratio could thus be double that of a seedling with a 3:1 ratio, even though the fungal biomass was the same in both cases. The data show that, in reality, the extent of C costs imposed by the fungus under high nutrient regimes may be further reduced by inhibition of development of the extraradical mycelium (Jones et al., 1990; Wallander and Nylund, 1992; and see Chapters 4 and 5).

Results which show no effect of ECM colonization or even a decrease of growth rate may be found scattered through the literature, but are not often emphasized. They are important for two reasons. First, they indicate that ECM colonization does not invariably increase growth of the plant in size or weight, and that there is frequently an initial phase, sometimes inexplicably prolonged when the growth of ECM plants is similar or slower than non-mycorrhizal ones. Secondly, they often show that different combinations of plant and fungus are differently effective in growth, as already discussed for *Eucalyptus diversicolor*. Although different species of fungus were compared on a single species of plant, there is good evidence that different strains of a single fungal species also differ in their effects on a plant. Table 8.2 gives one example from the work of Marx (1979b) of the effect of different strains of *Pisolithus tinctorius* on *Quercus rubra* seedlings. Variability of this kind is exactly what would be expected from the genetic, physiological and biochemical variability

Table 8.2 Growth of *Quercus rubra* seedlings with different strains of *Pisolithus tinctorius*.

Fungal strain	Height (cm)	Fresh weight (g)	Percentage of ectomycorrhizal short roots
138	13.3	8.5[a]	72[a]
136	13.5	6.5[c]	15[b]
145	11.5	7.0[b]	1[c]
Control	12.5	6.7[c]	0

Data from Marx (1979b). Day temperature 28°C, night 22°C. Day length 14.5 h. Growth period four months. Height differences insignificant at $P = 0.05$. Within other columns common letters denote insignificance. Percentage of ectomycorrhizas estimated visually.

of ECM fungi. Similarly, different biotypes of a plant may react differently with a single strain of a fungal species (Marx and Bryan, 1971; Marx, 1979b), although research on this aspect has not gone far. Clearly, further information on both these topics is essential to the full elucidation of the physiological relationships between the symbionts and to the determination of the most efficient combinations of fungus and plant for the production of forest crops. Experimentation should aim at describing in detail the structure and function of ECM which appear to be associated with greater or lesser growth rate. Information concerning Hartig net development, sheath thickness and, particularly, the quantity and extent of extraradical mycelium needs to be collected from plants grown in near-natural conditions, in order to calculate the C demand of the fungi (see below). Some advances in this area have been discussed in Chapter 6. Equally important are physiological properties of the fungal strains, in particular economic coefficients such as mass of C used per mass of mycelium formed. By such analyses the recognition of effective symbionts could be put on a firm basis, as has been done for single plant–fungus combinations by Jones *et al.* (1991) and Rygiewicz and Anderson (1994; and see below).

In all this work, the growth of the plant was the main preoccupation of the experimenters because they have been concerned, however distantly, with its ecology or its use for human benefit. From the point of view of understanding the physiology of symbiosis, much more knowledge is needed concerning the physiological activities of the fungi in symbiosis and in culture and factors that affect them.

The decreases in dry weight of the plant which sometimes follow colonization appear to be exactly similar to those observed with arbuscular mycorrhizal (AM) plants (see Chapter 4). A decrease in growth might be expected in a system where there is mutual dependency of one symbiont on the other for C compounds and for nutrients essential to support growth and photosynthesis. Decreases in growth rate would be expected in conditions of low irradiance that limit the rate of photosynthesis, but not the intensity of mycorrhizal colonization, or where the supply of soil-derived nutrients is adequate for growth but is not high enough to decrease the intensity of colonization. Similar examples where colonization results in no increase or sometimes a decrease in growth rate of the plant are also found in ericoid mycorrhizas (see Chapter 11).

Having discussed in general terms the growth of ECM plants, normally regarded as being a process of C accumulation, it is now appropriate to consider details of C requirements of the fungi before returning later to examine C distribution in intact ECM symbioses.

Carbon supplies for ectomycorrhizal fungi

Unlike the glomeromycotan AM fungi, many of the fungi of ectomycorrhizas have been isolated into culture and the physiology of their growth has been studied. However, the results have been disappointing. Much has been learned about factors which affect mycelial growth but little about intermediary metabolism or biochemistry. No positive characteristic that might explain the ECM habit has been discovered. Few of the early experiments even considered intermediary metabolism, respiration, secretion of metabolites or similar activities except in the sphere of the production of auxin-like substances. More recently, production of enzymes, organic anions, siderophores and metallotheionins has been highlighted, especially in relation to their roles in nutrition and heavy metal tolerance (see Chapters 9 and 10).

Frank (1894) assumed from the first that the source of organic C for ECM fungi was the photosynthetic plant, although he recognized that these might be supplemented by supplies from soil. Much later, Melin (1925) began to examine experimentally the requirements of the fungi for organic C. He found that most strains had little ability to grow on complex polymers, such as might be found in litter and humus, and that they could not use lignin or cellulose. They therefore seemed to be dependent on simple sugars such as might be produced by or released from the roots with which they were associated. Rommell (1938, 1939a) showed that a number of ECM fungal species depended upon association with living roots to produce fruit bodies, a reasonable expectation if they depended on the roots for C supplies. The positive relationship between irradiance and carbohydrate concentration in the root system and the intensity of mycorrhiza development in *Picea* and *Pinus* shown by Bjorkman's experiments (1942–1956) further appear to be consistent with this view.

From a survey of much work on growth of ECM fungi in culture, Harley and Smith (1983) concluded that most of the ECM fungi have at most, a limited ability to use lignin and cellulose as substrates for growth. Despite reports of slight lignase and cellulase activity, the ability to degrade polymers of this kind is much less than that of wood decomposing fungi or even of some ericoid mycorrhizal fungi (Trojanowski *et al.*, 1984; Haselwandter *et al.*, 1990; Cairney and Burke, 1994; see Chapter 11). Abilities vary within and between species in the ease with which starch, glycogen and insulin, the simpler oligosaccharides and the disaccharides sucrose and trehalose are used. The monosaccharides glucose, mannose and fructose are usually good sources of C for growth, whereas pectic substances can be used for growth by some ECM fungi but not others. This information has proved important in studies of the mechanisms of transfer of C from plant to fungus (see below).

The potential to use different sources of C clearly varies between fungal species and strains, and could possibly be related to fungal survival in soil. However, cellulose- and pectin-degrading abilities might be related not to use of litter as a source of organic C, but rather to penetration of root tissues, where enzyme production need only be localized and the degradation linked to softening the cell walls during Hartig net development. Much lower activity would be required for penetration than for releasing monosaccharides from complex polymers in amounts likely to affect the growth of the fungi. This point was clearly appreciated by Lindeberg and Lindeberg (1977) and research directed towards clarifying the extent of production of enzymes within plant roots and their roles in colonization processes is discussed in Chapter 6. The abilities of the fungi when free-living to use organic C-sources in soil has only limited relevance to nutritional interactions of the associated symbionts.

Björkman's conclusions (1970) that high internal carbohydrate supply in the roots favoured fungal colonization, although subject to more recent criticism (Nylund, 1988), seem to have been upheld by Marx *et al.* (1977). However, the fact that the root system is more susceptible to colonization when it contains high concentrations of soluble carbohydrates (such as sucrose) is not in itself a proof that the fungal symbiont derives its supply of C in whole or in part from the plant. Proof that this did occur was given by Melin and Nilsson (1957) who fed $^{14}CO_2$ to the leaves of axenically-grown *Pinus sylvestris* seedlings in combination with either *Suillus variegatus* or *Rhizopogon roseolus*. Photosynthetic products were translocated through the seedlings and were found in the roots and in their fungal mantles. Although this experiment demonstrated very clearly that the products of photosynthesis were translocated directly and rapidly to the fungal symbiont in ectomycorrhizas, it did not show anything about the quantities involved. Nor did it prove that all C compounds in the fungus were derived from the plant. Moreover, the controls in these experiments were decapitated ECM plants which also accumulated ^{14}C in their tissues. The quantities at the end of the experiment amounted to 6%, 11% and 15% of that in the intact photosynthesizing plants in stem, non-colonized roots and mycorrhizas, respectively. Lewis (1963) pointed out that this was explicable in terms of dark fixation of CO_2 which commonly occurs in plant material.

Much was learned about the C metabolism of ectomycorrhizas of *Fagus* using excised roots and some of this work is certainly relevant to a discussion of the supply of organic C to the fungal sheath by the plant (Harley and Jennings, 1958; Lewis, 1963; Lewis and Harley, 1965a, 1965b, 1965c). The monosaccharides glucose and fructose are readily absorbed by excised mycorrhizas from aerated solutions, but glucose is selected preferentially from mixtures. In mycorrhizas, sucrose is hydrolysed by an invertase attached to the plant wall and glucose preferentially absorbed from the products. The rate of absorption of hexoses is temperature- and oxygen-dependent and inhibited by metabolic inhibitors of the cytochrome oxidase pathway and of oxidative phosphorylation. The analysis of the tissues after uptake shows that the change in concentration of glucose and other monosaccharides is small and absorbed hexoses are rapidly converted to other compounds. Among the storage carbohydrates, mycorrhizas contain those characteristic of both the plant and the fungus (Table 8.3). Glucose and fructose are common to both. Trehalose, mannitol and glycogen are fungal, whereas sucrose and starch are from the plant. It may be assumed that sucrose is a plant sugar. The typical fungal carbohydrates, trehalose and mannitol, are absorbed by non-mycorrhizal roots at rates only one-tenth and one-twentieth of that of mycorrhizas (Table 8.4). The disaccharides, sucrose and probably trehalose, are hydrolysed before absorption and glucose and fructose have different destinations in the tissues. Since they competed in uptake it was thought that they probably had a common transporter on the plasma membrane. However, after absorption glucose was mainly converted to trehalose and fructose to mannitol among the sol-uble carbohydrates of the ECM mantle.

Table 8.3 Carbohydrate content of excised mycorrhizas and changes after storage for 20 h at 20°C in water or in 0.5% (w/v) solutions of glucose, trehalose, sucrose, fructose or mannitol.

	Initial sample	Changes after storage in					
		Water	Mannitol	Fructose	Sucrose	Trehalose	Glucose
Total soluble	14.49	−5.36	−5.66	−1.53	−0.10	+2.02	+2.91
Total reducing sugars	4.86	−1.02	−0.22	+0.34	−0.43	−0.69	+0.06
Sucrose	5.07	−2.34	−3.20	−0.99	−0.80	−0.06	+0.28
Trehalose	4.74	−2.00	−2.24	−0.88	+1.13	+2.77	+2.57
Mannitol	Nil	Nil	+7.69	+6.36	+3.86	+2.27	+3.86
Insoluble (glucose units)	25.94	−2.84	+3.87	+6.83	+5.20	+8.10	+11.14
Total carbohydrate present	40.63	32.43	38.84	45.93	45.73	50.75	54.68

Values as mg per g fresh weight (= 150 mg dry weight approximately). Data from Lewis and Harley (1965a).

Table 8.4 Relative rate of uptake of carbohydrate from solution by mycorrhizal and non-mycorrhizal roots of *Fagus*.

Sugar supplied	Ratio of uptake rate mycorrhizal:non-mycorrhizal
Glucose	3.1
Fructose	2.7
Sucrose	2.3
Trehalose	11.8, 10.8
Mannitol	19.6, 23.6

Data from Lewis and Harley (1965b, c).

Lewis and Harley (1965c) conducted experiments in which they fed the cut stumps of *Fagus* mycorrhizas with ^{14}C-sucrose by placing agar blocks on them (Figure 8.2) in order to stimulate C supply in intact roots. The analysis of the apical region of the mycorrhizas after separation of the sheath and core tissue by dissection (Figure 8.2c) showed movement of C into the sheath. Between 55% and 76% of the ^{14}C that had been translocated through the plant tissues to the tip region and not released as CO_2 was found in the fungal sheath, where it was present mainly as trehalose, mannitol and glycogen.

The uptake and destination of monosaccharides has been confirmed in intact systems (Söderström *et al.*, 1988). The inhibitory effect of fructose on glucose absorption was confirmed and has been suggested, together with regulation of apoplastic invertase activity, to be a factor important in the control of transfer of C from plant to fungus (Salzer and Hager, 1993).

Amanita muscaria, Hebeloma crustuliniforme and *Pisolithus tinctorius* are unable to use sucrose in culture because they lack a wall-bound invertase that would enable them to hydrolyse the disaccharide to glucose and fructose (Taber and Taber, 1987; Salzer and Hager, 1991; Schaeffer *et al.*, 1995). Protoplasts of *A. muscaria* can absorb both these monosaccharides, but whereas the uptake of fructose is strongly inhibited by glucose, the converse was not found and nor did sucrose inhibit uptake of either

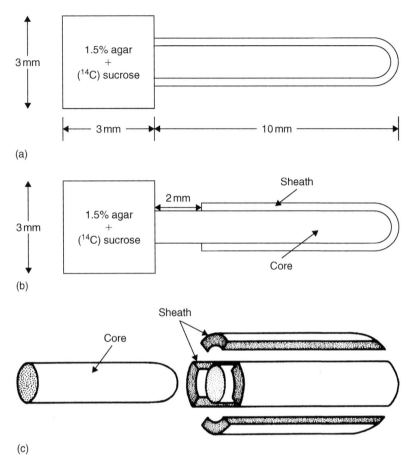

Figure 8.2 Schematic drawings to show the method of feeding mycorrhizal roots with ^{14}C-labelled sugars in agar blocks in studies of ^{14}C transfer from plant to fungus (see text). (a) The agar block abuts on both sheath and core; (b) a collar of sheath tissue has been removed to prevent contact between the fungus and the agar block, so that direct transfer of ^{14}C-sucrose to the fungus is also prevented; (c) dissection of the fungal sheath from the root core at the end of the feeding period. Modified from Lewis and Harley (1965c), with permission.

monosaccharide. The uptake system had a much higher affinity for glucose than for fructose (K_m 1.25 and 11.3 mM and V_{max} 18 and 30 pmoles per 10^6 protoplasts per minute, for glucose and fructose respectively; Chen and Hampp, 1993). It appears that the only way that *A. muscaria* can use sucrose is in symbiosis when the disaccharide is hydrolysed by an apoplastic or wall-bound invertase derived from the plant (Salzer and Hager, 1991).

Two hexose transporter genes have now been identified in *A. muscaria* (Nehls *et al.*, 1998). Of these, *AmMst1* and *AmMst2* encode proteins of 520 and 519 amino acids, respectively. *AmMst1* has a high sequence homology with a hexose transporter from the basidiomycotan fungus *Ruomyces fabae*, while *AmMst2* has closest homology to transporter *Stl1* from the ascomycete *Saccharomyces cerevisae*.

When expressed in yeast, *AmMst1* had K_m values of 0.4 mM for glucose but 4.0 mM for fructose (Wiese *et al.*, 2000). The much higher affinity for glucose is reflected in culture studies which show that hyphal uptake of glucose is strongly favoured, even in the presence of large excesses of fructose (Nehls *et al.*, 2001a). At glucose concentrations above 0.5 mM, uptake of fructose is inhibited. The expression of these genes is upregulated by a threshold response mechanism determined by the external concentration of monosaccharides (Nehls *et al.*, 1998). Up to a monosaccharide concentration of 2 mM both of the transporters show a basal level of expression, but concentrations above this threshold trigger a fourfold increase in transcript accumulation. This increase takes place slowly over a period of up to 24 hours.

In contrast to the *AmMst1* gene that exhibits slow sugar-dependent enhancement of expression, Nehls *et al.* (1999) identified a further gene in *A. muscaria*, *AmPAL*, that showed rapid sugar-dependent repression. Transcripts of this gene encode a key enzyme of secondary metabolism, phenylalanine ammonia lyase (PAL), which is involved in phenol biosynthesis. Transcripts accumulated to high levels in free-living mycelia grown at low glucose concentrations and it was suggested that they may be involved in production of defence-related compounds, protecting hyphae from fungivores as they extend through the soil.

Working on the same fungus growing in ECM association with spruce, Nehls *et al.* (1998) found an increase of *AmMst1* expression similar to that found in free-living mycelia when grown at elevated monosccharide concentrations. It was suggested that both the extended lag phase for enhanced *AmMst1* expression and its threshold response to elevated monosaccharide concentrations are hexose-regulated adaptations to the homeostatic conditions found at the fungus–plant interface. Here the hexose-enriched ambient environment would contrast strongly with that of the soil which is largely free of available monosaccharides. However, not only *AmMst1* but also *AmPAL* was strongly expressed in intact mycorrhizas. Since these genes were differentially expressed in a hexose-dependent manner in pure culture, it could only be suggested that the regulation of the genes was affected by their location in the mycorrhizal system. To address this question, ectomycorrhizas were dissected and gene expression was investigated separately in the hyphal, fungal mantle and Hartig net compartments (Nehls *et al.*, 2001b). As in pure culture, *AmMst1* was expressed only at the basal level in hyphae of the fungal mantle. In contrast, *AmPAL* showed a high transcript level in this part of the mycorrhiza. In Hartig net hyphae the opposite expression pattern was observed. As for hyphae in pure culture, in the presence of high external hexose, the transcript level of *AmMstl* was enhanced sixfold, while that of *AmPAL* was barely detectable. The opposing patterns of gene expression in hyphae of the fungal sheath and Hartig net (the latter resembling that of genes in pure culture situations) led Nehls *et al.* (2001b) to postulate that there must be a hexose gradient between the apoplast of the root–fungus interface (i.e. the Hartig net where the hexose concentration must be assumed to be above 2 mM) and the fungal sheath (Figure 8.3). The mechanisms that maintain gradients remain to be elucidated.

The question of the extent to which, if at all, the plant partner may be able to control the movement of sugars to the fungus has been little studied. However, a recent analysis of monosaccharide transporter gene expression in roots of *Populus tremula* showed that one of three such transporters, *PttMST3.1*, showed a 12-fold increase

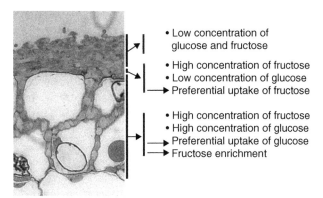

- Low concentration of glucose and fructose

- High concentration of fructose
- Low concentration of glucose
→ Preferential uptake of fructose

- High concentration of fructose
- High concentration of glucose
→ Preferential uptake of glucose
→ Fructose enrichment

Figure 8.3 Hypothetical model showing the spatial distribution of hexose uptake by the fungal hyphae in ectomycorrhizas. Sucrose hydrolysis in the apoplast of the Hartig net results in high glucose and fructose concentrations. In this compartment, glucose is preferentially taken up since the uptake of fructose is inhibited (by glucose concentrations above 0.5 M). In the innermost one or two layers of the fungal sheath, glucose concentration is low due to efficient uptake by fungal hyphae of the Hartig net. Thus, mainly fructose is taken up. In the apoplast of other layers of the fungal sheath, glucose as well as fructose concentrations are low due to the efficient hexose uptake by hyphae of the Hartig net and the inner layers of the sheath. From Nehls *et al.* (2001b).

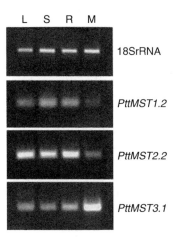

L   S   R   M

18SrRNA

PttMST1.2

PttMST2.2

PttMST3.1

Figure 8.4 Expression profiles of *Populus tremula* × *tremuloides* monosaccharide transporter genes *PttMST1.2*, *PttMST2.2* and *PttMS3.1* in leaves (L), stems (S), non-mycorrhizal (R) and mycorrhizas (M), showing upregulation of *PttMST3.1* in mycorrhizal roots. From Grunze *et al.* (2004).

of transcript accumulation in ECM relative to non-mycorrhizal roots (Figure 8.4; Grunze *et al.*, 2004). Upregulation in response to the presence of the fungus suggests that root cells are able to compete with fungal hyphae for hexoses present in the common apoplast of the Hartig net. On the basis of these observations, Grunze *et al.* (2004) suggested that (through its ability to control expression of its own hexose transporters) an autotroph could restrict the flux of sugars towards the fungus under circumstances in which mineral nutrient supply from the heterotroph was restricted. By such mechanisms the plant would be in a position to regulate the carbon balance of the symbiosis.

The characteristics of acid invertases have been investigated in *Picea abies* in relation to their possible role in nutrient supply to ECM fungi. Using cell-suspension cultures, Salzer and Hager (1993) showed that there were two important acid invertases associated with the apoplast. Both had relatively low pH optima, with the ionically bound form showing high activity between pH 3.5 and 4.5 and the tightly bound form having a sharp optimum at pH 4.5. Above pH 6 neither form showed significant activity. Both invertases had relatively high K_ms with respect to sucrose (16 and 8.6 mM for the ionically and tightly bound forms, respectively) and were competitively inhibited by fructose but not by glucose. These authors suggested that control of carbohydrate supply might be regulated by the fungus, via changes in apoplastic pH and uptake of fructose. Using whole *Picea abies* roots, Schaeffer *et al.* (1995) showed an acid invertase with a pH optimum of around 4.0 and a K_m (sucrose) of 5.7 mM, which is rather less than for the suspension cultures. By comparing different segments of ECM and non-mycorrhizal long roots it was shown that, although overall the activity of invertase was lower in ECM roots (in agreement with findings of Lewis and Harley, 1965b), when the weight of fungal tissue was taken into account, invertase production by the plant was unaffected by fungal colonization. Around 75% of the invertase activity was associated with cells of the root cortex rather than the stele and was thus in the tissue accessible to the fungal symbiont. Interestingly, Schaeffer *et al.* (1995) could find no evidence for increase in plant invertase activity associated with fungal colonization and they highlight the difference between this situation and that in some parasitic symbioses.

Figure 8.5 shows a hypothetical scheme for sugar transfer from plant to fungus, based on current knowledge. The overall picture is one in which, following sucrose efflux to the apoplast and hydrolysis by plant invertase, glucose would be preferentially absorbed by the fungus via the monosaccharide transporter *AmMst1*. Fructose released by the invertase would exert feedback inhibition on invertase activity, whereas glucose would inhibit fructose uptake by the fungus. As the concentration of glucose fell, the fungus would absorb fructose, releasing invertase inhibition and permitting further sucrose hydrolysis. One important gap appears in this reasoning and that is the possible need to reduce or eliminate uptake by the plant of both sucrose and the hexoses from the interfacial apoplast. Here again proton-symports are likely to operate and net flux in the direction of the fungus could only be maintained by a mechanism favouring net efflux (as sucrose) from the plant and net uptake (of monosaccharides) by the fungus. Perhaps fructose also competes with reabsorption of glucose by the plant or perhaps plant hexose transporters are downregulated. This area clearly requires continued investigation. Monosaccharide uptake by the fungus might be passive, with a concentration gradient maintained by metabolic conversion within the fungus. However, it seems just as likely that active proton co-transport would drive the inward flux, especially as low apoplastic pH controlled by the membrane-bound H^+-ATPase (of either organism) is necessary to maintain invertase activity. ATPase activity has been demonstrated cytochemically on both fungal and plant plasma membranes in the Hartig net zone of *Pinus sylvestris–Laccaria laccata* mycorrhizas and shown by its DES sensitivity to be, in all probability, an H^+-ATPase (Lei and Dexheimer, 1988).

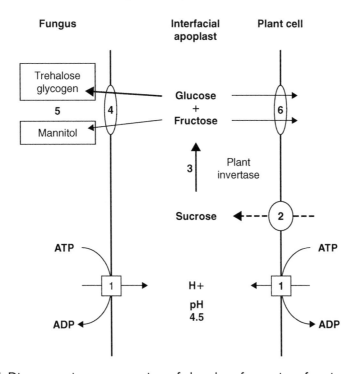

Figure 8.5 Diagrammatic representation of the plant–fungus interface in an ectomycorrhiza, showing processes likely to be important in sugar transfers between plant and fungus. (1) Activity of both plant and fungal H^+-ATPases transfers protons to the interfacial apoplast, lowering the pH and creating proton motive force necessary for active transport. (2) Sucrose, delivered to the roots is exported from the plant cells and hydrolysed by a plant wall-bound invertase (acid pH optimum) to glucose and fructose (3). (4) Glucose is preferentially absorbed by the fungal partner via a hexose transporter (*AmMst1* and 2 in *Amanita muscaria*), which may be a proton hexose symporter. Fructose can also be absorbed, but is outcompeted by glucose. (5) Glucose is used by the fungus to synthesize large amounts of glycogen and lesser amounts of trehalose. Fructose is converted to mannitol. (6) Upregulation of a plant monosaccharide transporter *PttMST3.3* in ECM poplar roots suggests that this plant may have a means of reabsorbing some of the sugar lost to the apoplast (see Figure 8.4) (Grunze *et al.*, 2004). However, no such upregulation of plant sugar transporter activity was found in *Betula* (Wright *et al.*, 2000).

The importance of maintaining a net flux of organic C in favour of the fungus has been the subject of considerable discussion. Since trehalose and mannitol were only very slowly absorbed by non-mycorrhizal roots, it was suggested that they, with glycogen, constituted a sink in the ECM mantle into which carbohydrates were accumulated in a form not readily available to the adjacent plant tissues even if they were present in the apoplast. This idea was elaborated in a review by Smith *et al.* (1969) in which they consider allied features of carbohydrate movement between the partners in lichens, pathogenic associations and other symbioses. The hypothesis of a 'biochemical valve' ensuring movement in the direction of the fungus has not yet been effectively superseded.

Although sugars from the plant certainly move to the fungus, movement of organic C is not exclusively in that direction. Harley (1964) showed, again using excised roots, that the rate of dark fixation of CO_2 by the mycorrhizas of *Fagus* was greatly increased if NH_4^+ was being absorbed. Using ^{14}C-bicarbonate he was able to show, and Carrodus (1967) later confirmed, that the main destination of the increased fixation was into glutamine. Since this method provided a simple way of labelling glutamine, Reid and Lewis (see Lewis, 1976) used it to observe the movement of glutamine from the fungal sheath to the plant, demonstrating how C compounds may return to the plant as the C-skeletons of amino compounds. Subsequently, the pathways for assimilation of inorganic N and the mobilization of organic N sources in soil received considerable attention (see Chapter 9). There is as yet little evidence on the extent to which C-skeletons of the organic N sources utilized by the fungi contribute to the C budgets of the plants. Abuzinadah and Read (1989a) calculated that, in ECM seedlings of *Betula pendula* supplied with N exclusively in the organic form, as much as 8% of total plant C could be derived from the fungus. The movement of C and N compounds synthesized in the fungus certainly provides a route for 'cycling' of organic C between the symbionts. The relative fluxes in the two directions have not been measured and this will certainly be an important challenge in assessing the potential of the 'cycling' of organic N through the interface to support net interplant transfer of C (see Chapters 9 and 15).

Carbon distribution in intact plant–fungus systems

In their study of distribution of ^{14}C-labelled photosynthetic products between ECM and non-mycorrhizal roots on the same plants of *Pinus radiata*, Bevege *et al.* (1975) found that 15 times more C was allocated to the colonized roots, 45–50% of the label being in trehalose and 1–22% in mannitol. A similar study using individual plants of *Eucalyptus pilularis*, some roots of which were colonized by *Pisolithus tinctorius*, showed that the ratio of ^{14}C-accumulation in ECM versus non-mycorrhizal roots was 18:1 (Cairney *et al.*, 1989). The ability of these ectomycorrhizas to attract photosynthate was greatest soon after their formation and there was a progressive reduction in the amount translocated to them with age so that 90 days after inoculation all translocation had ceased. Cairney and Alexander (1992) compared allocation of ^{14}C to younger and older mycorrhizas of *Picea sitchensis* formed by *Tylospora fibrillosa* growing on peat. Although some ^{14}C-labelled compounds were translocated to older mycorrhizas in all plants, the ratio of activity in young to older mycorrhizas, which was initially around 2:1, became progressively greater as the whole root system aged, reaching 54:1 38 weeks after transfer of the newly colonized seedlings to the peat substrate.

In the studies quoted above, the extent of onward transport of fixed C from root to extraradical mycelium was not determined. The importance of this mycelium as a potential sink for fixed C is evident from studies in which intact systems develop over semi-natural substrates (see Chapters 6, 9 and 15). Autoradiographic techniques have shown clearly that ^{14}C fed to assimilating shoots of *Pinus* is rapidly transported to the extraradical network (see Figures 6.27a, b and Colour Plate 6.2a, b) which, in addition to exploring the soil, provides interconnection between individual mycorrhizal short roots in the same and on adjacent compatible plants (Finlay and Read, 1986a) (Figure 6.26). Thus the fungus supplies a series of potential pathways for the flow of C from structures of small or decreasing sink strength, such as old roots, to younger tissues.

A question of fundamental importance concerns the amount of C allocated by the plant to support the growth and maintenance of its ECM fungi. The first studies of respiratory activity specifically of the mycelial phase of the ECM association (Söderström and Read, 1987) demonstrated that approximately 30% of total respiration was attributable to the ECM mycelium. This respiration was shown to be highly dependent upon the supply of current assimilate and, if mycelial connections to the roots were severed, there was a reduction in respiration rate of at least 50% within 24 hours (Figure 8.6). A similar dependence upon current assimilate for production of sporophores has been demonstrated in an association of *Pinus strobus* with *Laccaria bicolor* (Lamhamedi *et al.*, 1994) (Figure 8.7).

Rygiewicz and Anderson (1994) fed *P. ponderosa* seedlings, either colonized by *Hebeloma crustutiniforme* or non-mycorrhizal, with $^{14}CO_2$ and examined the distribution of label in different fractions after incubating the plants in microcosms for 72 hours. Both sets of seedlings retained around 40% of labelled C in their shoots, but

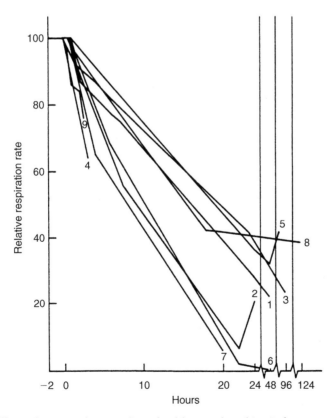

Figure 8.6 Effect of cutting the mycelium (and hence detaching it from a source of carbohydrate from the plant) on respiration of ectomycorrhizal mycelium. The curves represent the relative respiration rate (per cent of value before cutting) for nine separate combinations of plant and fungal species. From Söderström and Read (1987), with permission.

the presence of the fungus (representing only 5% of total seedling weight) increased the below-ground allocation of ^{14}C by 23% relative to that in non-mycorrhizal plants. Dry matter allocation to roots and mycorrhizas was similar, the difference being due to increased respiration by ECM roots and mycelium. The greater allocation below ground led to a small reduction of seedling biomass in the ECM plants.

Jones *et al.* (1991) considered C allocation to mycorrhizas as a cost to the plant. They examined efficiency in terms of amount of P taken up per unit of C expended below ground, using cuttings of *Salix viminalis* either colonized by *Thelephora terrestris* or grown in the non-mycorrhizal condition. Efficiency was calculated according to the formula $\Delta P/\Delta C_b$ of Koide and Elliott (1989) (see Chapter 4), where ΔP is the amount of P taken up by the plants over a defined interval, and ΔC_b is the total amount of C allocated to the below-ground system during the same interval. C_b includes the C incorporated into root or fungal tissue, lost in respiration and deposited in the soil. C_b was calculated for both ECM and non-mycorrhizal plants, which had been pulse-labelled prior to each harvest, for the intervals up to the first harvest (0–50 days) and from the first harvest to the final harvest (50–98 days).

Two methods were used to obtain C_b.

In the first method, it was calculated as:

$$C_{b(Pn)} = P_n \frac{\%C_{BG}}{100 - \%C_{SR}} \, 57\,600$$

where $C_{b(Pn)}$ is the amount of C allocated below ground in mMol C in a 24 hour period; P_n is the net rate of photosynthesis as mmol C/s for a whole shoot system;

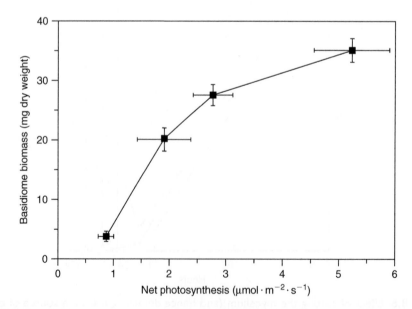

Figure 8.7 The influence of rate of net photosynthesis of *Pinus strobus* seedlings on the biomass of fruit bodies produced by associated *Laccaria bicolor* after 20 days. Bars are standard errors of means. From Lamhamedi *et al.* (1994), with permission.

%C_{BG} is the percentage of the total $^{14}CO_2$ absorbed allocated below ground over a 9 day period; %C_{SR} is the percentage of the absorbed $^{14}CO_2$ released as shoot respiration; and 57 600 is the length of the daily light period in seconds. The term 100–%C_{SR} is a correction for the fact that P_n is exclusive of shoot respiration. A curve was then constructed of $C_{b(Pn)}$ against time and the resulting equation integrated over the interval between each harvest to give the total amount of C allocated below ground over these intervals ($\Delta C_{b(Pn)}$).

In the second method, shoot weight was used as an integrated measure of the amount of C deposited in tissue over a harvest interval. The relationship between C deposited in shoot tissue and C allocated below-ground as determined by the ^{14}C labelling was then used to calculate ΔC_b directly:

$$\Delta C_{b(Wt)} = \Delta W_s \frac{\%C_{BG}}{\%C_{ST}}$$

where ΔW_s is the mean increase in shoot weight over the interval and %C_{ST} and %C_{BG} are the mean percentages of the ^{14}C fixed allocated to shoot tissue, and to the below-ground compartments, respectively. For the first interval, %C_{ST} and %C_{BG} from the first harvest were used. For the second interval, a weighted (i.e. based on the length of time between harvests) average of the %C_{ST} and %C_{BG} value for harvests two, three and four were used. Dry weight was converted to g C using a correction factor of 0.5 g C/g dry weight.

The amount of C allocated below ground was consistently greater in ECM than non-mycorrhizal plants throughout the experimental period (Figure 8.8).

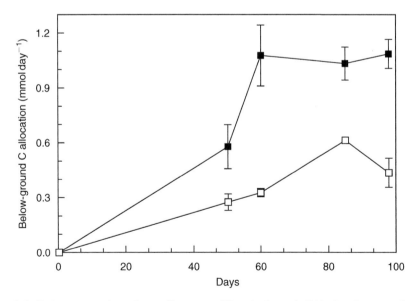

Figure 8.8 Below-ground carbon allocation ($C_{b(Pn)}$) (mmol C/day) of mycorrhizal (■) and non-mycorrhizal (□) *Salix viminalis* during a 98-day growth period. Values are means of 3 replicates at 50, 60 and 85 days and 4 (M) or 5(NM) at 98 days ± standard error of means. From Jones *et al.* (1991), with permission.

Table 8.5 P uptake, C translocated below ground and P acquisition efficiency of mycorrhizal *Salix viminalis* over a 98-day growth period (see text).

Mycorrhizal treatment	Growth period (days)	$\Delta P(\mu mol)$	$\Delta C_{b(Pn)}$ † (mmol)	$\Delta C_{b(Wt)}$ ‡ (mmol)	$\Delta P/\Delta C_{b(Pn)}$ † (μmol P/mmol C)	$\Delta P/\Delta C_{b(Wt)}$ ‡ (μmol P/mmol C)
Mycorrhizal	0–50	50.1	14.2	17.0	3.52	2.94
Non-mycorrhizal	0–50	17.1	5.7	9.7	2.97	1.75
Mycorrhizal	50–98	25.6	46.6	69.8	0.55	0.37
Non-mycorrhizal	50–98	17.9	22.0	21.2	0.82	0.84

From Jones *et al.* (1991). ΔC_b is the amount of C translocated below ground during a given interval, including that deposited in root and fungal tissues, that respired by roots and soil and that deposited in the soil. ΔP is the amount of P taken up during a given interval. † Values were calculated using photosynthetic measurements. ‡ Values were calculated using weight measurements. Data based on the 3–5 plants labelled with ^{14}C prior to each harvest.

When the data shown in Figure. 8.6 were integrated, it was shown (Table 8.5) that over the interval up to the first harvest ECM plants allocated 2.5 times more C ($\Delta C_{b(Pn)}$) below ground than did those not colonized by *T. terrestris*. Expressed on a weight basis, C allocation ($\Delta C_{b(Wt)}$) was 1.75 times greater in ECM plants. Since colonized plants absorbed three times as much P from soil over the same period, they had a higher P use efficiency according to both methods of calculation. Because, over the second half of the experiment the difference in P uptake between ECM and non-mycorrhizal plants was less, the efficiency of P acquisition was higher in the non-mycorrhizal than ECM plants. Cost-benefit analysis shows the costs, expressed as C required to produce and maintain the nutrient absorbing structures, are lowest at the earliest stage of growth while the benefits, in the form of P acquisition, are highest. This is a pattern which Jones *et al.* (1991) suggest will match the requirements of field-grown perennial plants where, in temperate and boreal climates, maximum demands for nutrients may be experienced in spring.

Jones *et al.* (1998) subsequently analysed the growth, C and P relations of *Eucalyptus coccifera* when colonized by ECM or AM fungi, comparing the performance of the plants with these mycorrhizal associates and without mycorrhizas. Growth promotion was significantly greater in the ECM than in the AM or uncolonized states but, when the plants were fed with $^{14}CO_2$ after 89 days growth no significant differences were detected between treatments in the quantities of C allocated below ground. This is surprising in view of the fact that the ECM plants produced three to seven times more extraradical mycelium and acquired significantly more P than their AM counterparts (see also Chapter 10). The authors propose that failure to detect effects of ECM on below-ground C allocation was attributable to the small size and low root weight fractions (0.21–0.27) of their young plants. In an earlier experiment, differences in below-ground C allocation between *Salix viminalis* grown in the ECM and AM conditions increased with time but were not significant until the root weight fraction of the ECM plants exceeded 0.35 (Durall *et al.*, 1994).

The seasonality of C allocation is generally overlooked. The literature (e.g. Shiroya *et al.*, 1966; Gordon and Larson, 1968; Ursino *et al.*, 1968; Glerum and Balatinecz, 1980) suggests that the main surge of C allocation to below-ground systems occurs not in spring but towards the end of the growing season after stem elongation and bud set are complete. This is also the time at which the main flush of root growth occurs in many of the tree species which are characteristically ECM (Lyr and Hoffman, 1967) and it coincides with the late summer flush of fruit body production by epigeous ECM-forming mushrooms. Since these are, as shown above, dependent upon current assimilate for their development, it is implicit in such observations that this is also a time during which the extraradical mycelium must be particularly active. There is evidence (Langlois and Fortin, 1984) for a distinct seasonality in the pattern of absorption of P (see Chapter 10) by ECM roots of *Abies balsamea*, with maximum rates again observed in August, after bud set. This feature may also be a reflection of greater below-ground C allocation. A requirement for enhanced nutrient inflow in the autumn would be expected in boreal and temperate systems because winter-hardening involves significant augmentation of cellular components, in particular membrane phospholipids (Siminovitch *et al.*, 1975) which are remobilized to support the spring flush of above-ground growth.

Environmental factors, in particular availability of soil moisture, may be superimposed on the inherent seasonality of ECM processes, but it is important to bear in mind, whether carrying out experiments under constant conditions of day length and irradiance in the laboratory or collecting ECM roots in the field, that the results obtained at any one time may not reflect accurately the situation as it prevails in the field. There is scope for much more work on seasonality in function in ECM systems, though it appears likely that the main phase of activity is at the end rather than the beginning of the growing season as surmised by Jones *et al.* (1991).

Non-nutritional effects upon carbon assimilation

Formation of ECM symbioses can influence the C balance of the plant via a number of interrelated processes but, in particular, through its effects on net photosynthetic rate and mineral nutrition. These influences can be detected both in the leaves which are the C sources and in the roots and their symbionts which are the C sinks. The impacts of ECM colonization can be detected in the source leaves. The regulation of sucrose synthesis in the leaf cytosol mainly occurs through the activities of sucrose phosphate synthase (SPS) and fructose 2,6-bisphosphatase (FBPase), the latter being inhibited by an effector metabolite, fructose 2,6-bisphosphate (F26BP) (Quick and Schaffer, 1996). Loewe *et al.* (2000) reported increased activation of SPS and decreased levels of the inhibitor FBPase in ECM seedlings of spruce (*Picea abies*). This is indicative of an increased capacity for sucrose synthesis in source tissues of ECM plants. This effect of ECM formation is almost certainly linked to the role of the fungal partners as sinks for assimilates produced in the source tissues. The importance of sink strength in determining the rates of C assimilation in leaves is now well established (Herold, 1980; Gifford and Evans, 1981; Sonnewald *et al.*, 1994; Quick and Shaffer, 1996). In the context of mycorrhizas, the assumption is that, through their ability to maintain a net flux of C in their favour (Lewis and Harley, 1965a; Nehls *et al.*, 2001) and to increase the overall C demand of the root (Finlay and Söderström, 1992), the fungal partners exert direct impacts upon C assimilation by the plant. It remains necessary to determine the relative importance of enhanced provision of nutrients by the ECM fungi and the increased sink strength arising from their presence in determining the observed shifts of plant carbon balance.

Whereas several earlier studies showed that photosynthetic rates are enhanced in ECM plants relative to those grown in the non-mycorrhizal condition (Reid *et al.*, 1983; Nylund and Wallander, 1989; Loewe *et al.*, 2000), the confounding influence of changed mineral nutrient balance in the foliage of colonized plants makes it impossible to discriminate between the effects of nutritional and non-nutritional factors in these experiments. The necessary discrimination can be achieved by adding nutrients to non-mycorrhizal plants so as to compensate for the gains normally associated with colonization by the fungi. Experiments using plants nutritionally matched in this way (Dosskey *et al.*, 1990; Rousseau and Reid, 1990; Conjeaud *et al.*, 1996; Wright *et al.*, 2000) have confirmed that ECM sink strength alone can have major impacts both upon assimilatory activities in the source tissues and upon fluxes to the roots. Rousseau and Reid (1990) grew ECM and non-mycorrhizal *Pinus taeda* seedlings that were nutritionally matched to varying extents by exogenous supply of P to the uncolonized plants. The photosynthetic rates of the ECM plants, some of which were lightly, some moderately and some heavily colonized by the fungus, were then compared with those of uncolonized plants from each of the P treatments, before harvests were taken to determine tissue P status.

Relationships between mycorrhiza development, net photosynthetic rate and foliar P (Figure 8.9) indicate that, with low and medium colonization, assimilation

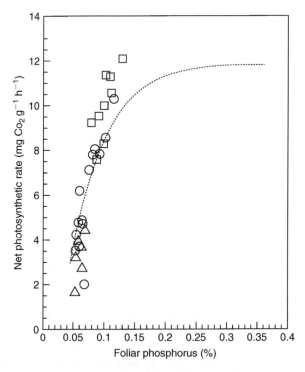

Figure 8.9 Net rate of photosynthesis (mg CO_2/g/h) versus foliar P concentration (percent dry weight) for ectomycorrhizal seedlings. The different symbols indicate seedlings with different concentrations of glucosamine (mg/g) derived from the fungal symbiont associated with their roots: △, 0–7.99 mg; ○, 8–15.99 mg; □, 16–24 mg. The dashed line shows the regression obtained from non-mycorrhizal seedlings. From Rousseau and Reid (1990), with permission.

of seedlings responded to increasing foliar P concentration in the same way as uncolonized plants, suggesting that the rate was responding to increased P accumulation. However, at high levels of colonization net rates of photosynthesis in ECM plants were significantly greater (up to 17%) than those seen in non-mycorrhizal plants of the same foliar P concentration. Clearly, this suggests an effect of colonization upon photosynthesis not related to P status. The likelihood that the increased sink strength imposed by the highest level of colonization was responsible for the enhancement of assimilation was stressed by Rousseau and Reid (1990), though alternative explanations, including changes of hormone balance induced by the fungus, were also discussed. The possibility that nutrients other than P had become limiting for the highly colonized plants, despite the apparently adequate nutrient supplies, also deserves attention.

Increased rates of photosynthesis in ECM plants can certainly occur independently of nutrient enrichment (Dosskey *et al.*, 1990). Colonization of *Pseudotsuga menziesii* by *Rhizopogon vinicolor* led to significant increases of net photosynthesis, whereas two other fungi, *Hebeloma crustuliniforme* and *Laccaria laccata* had no effect. Despite the increased net assimilation rate, *Rhizopogon*-colonized seedlings were smaller than the controls. The same type of result was observed in seedlings of *Pinus pinaster* when colonized by *H. cylindrosporum* (Conjeaud *et al.*, 1996). Despite significant increases of photosynthetic rate over non-mycorrhizal plants, with the same or even greater tissue N and P concentrations, there was an accompanying 35% reduction of growth. In these cases, stimulation of photosynthetic rate arising from increases in sink strength are not sufficient to compensate for the costs of production of mycorrhizas and associated extraradical mycelium. A similar effect was observed by Wright *et al.* (2000) using *Betula pendula* colonized by *Paxillus involutus*. Analysis of diurnal carbon budgets of ECM and nutritionally matched non-mycorrhizal plants revealed that the net amount of C assimilated per unit dry weight was 29% higher in the colonized than in the uncolonized plants (Table 8.6). This further confirms the important role of sink strength in determining C assimilation rates. However, because more C was respired by the root systems of ECM than non-mycorrhizal plants, the net C gain in both sets of plants was similar. Overall, the total biomass of the ECM plants was again significantly lower than that of their

Table 8.6 Diurnal carbon budgets of non-mycorrhizal (NM) birch plants and birch plants ectomycorrhizal with *Paxillus involutus* (M).

	CO_2 exchange			
	μmol CO_2/plant/day		μmol CO_2/g DW/day	
	Non-mycorrhizal	Mycorrhizal	Non-mycorrhizal	Mycorrhizal
Net CO_2 assimilation	690 ± 158	444 ± 28	2233 ± 394	2870 ± 437
Dark shoot respiration	123 ± 18	79 ± 3*	355 ± 21	516 ± 95
Root system respiration	225 ± 67	252 ± 43	593 ± 33	1029 ± 136**
Carbon gain of whole plant (NM) and plant and fungus (M)	342 ± 94	113 ± 44	1394 ± 257	1325 ± 326

From Wright *et al.* (2000). Results are expressed either per unit dry weight (μmol CO_2/g DW/day) or on a per plant basis (μmol CO_2/plant/day). Values are means ± SE ($n = 4$). The data for each parameter for M and NM plants were compared by one-way ANOVA. *$P < 0.05$, **$P < 0.01$.

uncolonized counterparts, indicating that the increases in C assimilation were insufficient to compensate for the demands of both the plant and fungal partners.

Community level patterns of carbon allocation

Rommell (1939b) provided probably the first estimate of the C demand of ECM fungi under field conditions, based upon fruit body production and showed that the C required was 10% of that invested by the plants in annual timber production. The calculation assumed a fruit body yield of 180 kg/ha/year which some workers have suggested is a high value. Fogel and Trappe (1978) quote values of dry weight of epigeous fruit bodies for a number of forest types in the Northern Hemisphere in the range 3 to 180 kg/ha/year. However, Vogt et al. (1982) recorded 380 kg/ha/year of hypogeous and 30 kg/ha/year of epigeous fruit bodies in a 180-year-old stand of Abies amabilis. In addition, Cenococcum sclerotia yield 2700 kg/ha/year in these forests. By combining all such values with estimated fungal biomass in mantles and Hartig nets, Vogt et al. (1982) estimated that 15% of net primary production was allocated to the fungi, a value which still did not take account of vegetative mycelium in the soil, nor the possibility of increased C allocation due to turnover of the ECM rootlets.

It is a prerequisite for determination of the carbon costs of ECM symbioses that the annual production of vegetative mycelium of the system be calculated. It is, of course, difficult to determine the biomass of ECM hyphae growing as they do in mixed populations with other fungi in forest soils. However, some estimates are now available. On the basis of regression relationships between mycelial respiration and biomass, Finlay and Söderström (1989) estimated that in a Swedish pine forest soil dominated by Lactarius rufus there were 200 m/g soil of ECM hyphae. This figure is remarkably similar to that calculated by Read and Boyd (1986) for mycelial systems of Suillus bovinus growing in peat in observation chambers (see Chapter 6) and is not dissimilar to values of 100–700 m/g presented earlier by Söderström (1979) for lengths of active hyphae (stained with fluoroscein diacetate) in the FH horizon of Swedish forests. When converted to fungal biomass, the value of 200 m/g dry soil is equivalent to 3.5 kg of living mycelium per ha/year. Assuming a turnover time of one week throughout a five-month growing season, the hyphal production would be 70 kg/ha/year. This is more than double the estimated dry matter yield of fruit bodies of L. rufus which, in Swedish forests, is considered to be around 30 kg/ha/year (Richardson, 1970). Based on results obtained using the mesh bag ingrowth method (Chaper 6), Wallander et al. (2001) calculated an annual production of ECM mycelium of 125–200 kg/ha/year in a mixed conifer forest and 590 kg/ha/year in a spruce forest (Wallander et al., 2001, 2004). This represents 80% of the total ECM fungal biomass. Nevertheless, according to a more recent assessment (Hendricks et al., 2006), even these high values may be considerable underestimates.

Determination of total ECM biomass in the system also requires estimates of fine root production and turnover of fine roots, all of which are likely to be mycorrhizal in boreal forest systems (Taylor et al., 2000). Unfortunately, considerable discrepancies have arisen between estimates of these parameters. Some studies based upon soil coring methods (e.g. Persson, 1978; Vogt et al., 1982; Gill and Jackson, 2000) indicate that the population of fine roots (<2 mm diam) turns over several times per year in boreal and temperate forests. The high values obtained by these approaches uggest that fine root production and turnover is the major short-term fate of plant C that is allocated below ground. In contrast, results obtained from recently developed direct methods of measuring C allocation have led to these estimates being challenged (Högberg and Read, 2006). In a Free Air Carbon Dioxide Enrichment (FACE) facility (Andrews *et al.*, 1999), which enabled $^{13}CO_2$-labelling of photosynethetic products, much slower rates of fine root turnover were suggested, the median value being about 4 years. Measurements using the ^{14}C signature left as a legacy of nuclear weapons testing (Gaudinski *et al.*, 2000, 2001) support those based on the FACE technology, suggesting fine-root turnover times of 3–18 years.

New observations on the life span of fungal components of the mycorrhiza indicate that, in contrast to the fine roots, they have a short life span. Using the bomb-^{14}C method, Hobbie *et al.* (2002) showed the ages of ECM sporophores to be of the order of weeks, which is consistent with visual observations. A further feature indicating that the production of fine roots cannot be the major sink for photosynthate allocated below ground is the decline of soil respiratory CO_2 flux, by 56% within 14 days, following girdling of trees late in the growing season (Figure 8.10; Högberg *et al.*, 2001). This was considered to be a conservative estimate of the respiratory contribution of mycorrhizas because root starch reserves were shown to be used following curtailment of C supplies from the canopy. Together with almost immediate cessation of sporophore production of the ECM fungi following girdling (Table 8.7), the findings show that, at this time of year, the major short-term sink for current assimilate allocation below ground is the fungal mycelium and its associated reproductive structures. Godbold *et al.* (2006) have shown that the external ECM mycelium of popal can provide the dominant pathway by which C enters the soil organic matter pool, far exceeding the inputs from leaf litter and fine root turnover. These observations are consistent with the losses of respiratory activity seen following hyphal severance

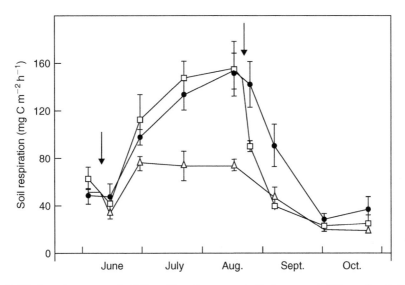

Figure 8.10 Respiratory soil CO_2 efflux from ungirdled control (●) early girdled (△) and late girdled (□) plots of *Pinus sylvestris*. Arrows indicated times of early (June) and late (August) girdling. After the early girdling, soil respiration was reduced by ~54% within one month, and after the late girdling it was reduced by 37% within 5 days. Högberg *et al.* (2001).

Table 8.7 The numbers of species and of sporocarps and their dry weights collected from early (June) and late (August) girdled and from control (ungirdled) plots in a 50-year-old stand of *Pinus sylvestris* at Aheden, Sweden.

Treatment	No. of species	No. of sporocarps	Biomass (g dry wt)
Control	11 ± 3	252 ± 116	84.4 ± 24.4
Early girdling	1 ± 1 (72)	4 ± 4 (72)	0.5 ± 0.5 (72)
Late girdling	10 ± 1 (2–3)	108 ± 48 (2–3)	40.9 ± 8.5 (2–3)

From Högberg *et al.* (2001). Figures in brackets are the number of days since the girdling treatment. Sporocarp production was virtually eliminated 72 days after early girdling and more than halved within 2–3 days of late girdling.

(Söderström and Read, 1987; see Figure 8.6) and the collapse of sporocarp production following canopy shading (Lamhamedi *et al.*, 1994; see Figure 8.7).

The girdling study revealed two further features that emphasize the importance of seasonality in determining the C balance of the ECM community. First, the decline of respiration at 5 days following girdling early in June was, at 27%, considerably less than that in September. This is probably in part attributable to the fact that temperate forest conifers preferentially allocate C to support shoot, needle and bud expansion growth early in the season (Hansen *et al.*, 1997). Another contributory factor is that, at this time of year, the roots contain significantly more starch as a potential respiratory substrate (Högberg *et al.*, 2001). Second, the observation that the highest levels of respiratory activity were recorded in August and September, well after the period of maximum solar irradiance and highest temperature in June and July, eliminate these factors as being the primary drivers of the respiratory events. Högberg and Högberg (2002) showed a decline of microbial biomass in the girdled plots of around 32% 1–3 months after the treatment was applied and calculated that one third of the total soil microbial biomass was attributable to the ECM mycelial component. These authors stress that the possible contribution of the ECM biomass to soil respiration is likely to be underestimated because respiration of the heterotrophic community was probably stimulated by substrate released from the decomposing ECM structures. In a follow-up study two years following girdling, it was shown that respiratory output in the treated plots had reduced by 65% (Bhupinderpal *et al.*, 2003). This can be expected to be a more precise measure of the contribution of the mycorrhizal roots and mycelium to soil respiration because, by this stage, the alternative substrates released as a result of the girdling activity, and hitherto being used to support the heterotrophic community, would have been exhausted.

Two scenarios emerge for below-ground C allocation. The first predicts that fine roots are a major and rapidly turning over destination for current assimilate. The second shows that the roots are a less important sink and that the mycelial and fruiting components of the ECM systems are the primary C sinks, their production and activities being tightly coupled to current assimilate supply. Calculations based upon the first scenario indicate that fine root production in pine forest of the kind studied by Finlay and Söderström (1989) is of the order of 2030 kg/ha/year (Persson, 1978). With the conservative estimate that 90% of these roots are ECM and that 40% of an ECM root is fungal (Harley and McCready, 1952b; Vogt *et al.*, 1982), 730 kg of fungal sheath would be produced per year. Thus, total annual fungal production (mycelium = 70 kg, sporophores = 30 kg, sheath = 730 kg, see above) would be 830 kg/ha. If the C content of the fungal material is taken to be 40% and the respiratory efficiency is 60%, the C demand of ECM fungi in the forest would be 830 kg C/ha/year. Photosynthetic production in the same forest has been estimated at 5800 kg C/ha/year (Linder and Axelsson, 1982). On this basis, the ECM fungi use around 15% of assimilated C. This value, as Finlay and Söderström (1992) point out, is strikingly similar to those presented by Vogt *et al.* (1982) and, indeed, by Rommell (1939b).

Clearly, these values may need to be revised in the light of the suggestion that fine root and hence fungal mantle production and turnover are lower, perhaps by at least four times, than those indicated by the coring studies. Set against the possible reductions in estimates of C allocation required by a lower value of fine root production, are the observations that both mycelial (Wallander *et al.*, 2001, 2004) and fruit body production (Högberg *et al.*, 2001) in these types of ECM forest can be considerably greater than earlier estimates suggest. Accordingly, while values for production of biomass of mantles may need to be reduced by four times, those for mycelial biomass and fruit body production could require sixfold and fourfold increases, respectively. Repeating the earlier calculation assuming the same C conversion and efficiency factors the values, based on 172 kg/ha/year for fine root turnover, 420 kg/ha/year for mycelium and 12 kg/ha/year for fruit bodies is equivalent to a carbon allocation of 722 kg/ha/year. This would still be around 12.5% of annual primary production of the stand.

Conclusions

The construction of nutrient response curves has gone some way to providing an understanding of the influence of ECM colonization on plant growth. When plants face a shortage of a mineral element such as P, there is a threshold of availability below which growth will not occur. By enhancing access to the growth-limiting nutrient, colonization can significantly reduce the threshold. The extent of plant response varies with the fungus and its pattern has been shown to be distinctive at inter- and even at intraspecific level, both in the effectiveness with which the nutrient resource is exploited and in the C demands imposed upon the plant. It is now recognized that relationships between effectiveness of resource exploitation and extent of C demand can usefully be explored by means of cost-benefit analysis. Costs and benefits of ECM associations are again shown to differ with fungus and experimental circumstances.

Further progress has been made towards understanding of the molecular-biochemical mechanisms involved in C transfer from autotroph to heterotroph. While not changing our perceptions concerning the nature and localization of the biochemistry of transfer at the interface between the partners, the new research has begun to identify specific sugar transporters and to elucidate some of the under-lying mechanisms that control the transfer.

We are now better aware of the extent of coupling between the flow of current assimilate, the activities of the extraradical mycelium and the production of sporocarps. It had been calculated from laboratory experiments that between 10 and 20% of photosynthate C was allocated to sustain the vegetative mycelium of ECM fungi, but major advances have now been achieved by extending these studies to the field.

Elimination of photosynthate supply by tree stem girdling in pine, leads rapidly to major reductions of soil respiratory CO_2 release and almost instantaneously to cessation of sporocarp production. Calculations based upon this type of field experiment indicate that ECM mycelia constitute at least one third and probably a good deal more of total soil microbial biomass. As a major sink for plant C these components of the mutualism directly influence assimilation rates in the leaves, giving the potential for positive feedbacks on allocation of C below ground. There is now a need for field experiments involving manipulation of source-sink relations to be carried out in different types of forest in order to determine the relative importance of mycorrhizal communities as C sinks in a range of ecosystems.

9

Nitrogen mobilization and nutrition in ectomycorrhizal plants

Introduction

Constraints upon plant growth imposed by low availability of N are a characteristic feature of many ecosystems dominated by ectomycorrhizal (ECM) plants. Thus, in the extensive boreal (Tamm, 1991) and temperate (Ellenberg, 1988) forests of the northern hemisphere the accessibility of N is the most important determinant of productivity. Our understanding of the N dynamics of these systems has been improved by the recognition that the forms in which the element may be available to ECM plants in forests are diverse. Despite the early suggestion of Frank (1894) that ECM colonization may provide plants with access to organic forms of N, the view that mineral N was the only important source of the element predominated during much of the following century. Re-emphasis of the potential importance of organic N sources for the nutrition of mycorrhizal plants came from studies first of ericoid mycorrhiza (ERM) (Bajwa *et al.*, 1985; see Chapter 11) and then of ECM symbioses (Abuzinadah and Read, 1986a, 1986b; Read, 1991a; Bending and Read, 1995a, 1995b; Chalot and Brun, 1998; Emmerton *et al.*, 2001a, 2001b; Nasholm and Persson, 2001). By stressing that the activities of the fungal symbionts might provide plants with access to forms of N which would otherwise be completely unavailable to them, these studies led to a resurgence of interest in the roles of organic N in plant nutrition. The result has been that there is now a more balanced appreciation of the spectrum of N sources likely to be involved in the nutrition of ECM plants and of the environmental circumstances under which they are likely to be the predominant available forms. Attention is now being given to the fundamental molecular and biochemical events associated with the mobilization, assimilation and transport of all potentially available forms of N and the improved understanding of the environmental constraints is helping to place these processes into particular ecological contexts.

Use of N by ectomycorrhizal fungi in pure culture

Inorganic N sources

The ECM fungi that have been successfully grown in axenic culture are similar to other fungi in the kinds of N compounds that they can use for growth. The early workers (Norkrans, 1950; Rawald, 1963; Lundeberg, 1970) found a range of relative abilities to use ammonium and nitrate among the species and strains of ECM and saprotrophic fungi that they studied (see Harley and Smith, 1983 for details). Nothing was discovered which distinguished ECM fungi as a group and later work confirmed these observations (France and Reid, 1984; Littke *et al.*, 1984; Genetet *et al.*, 1984; Plassard *et al.*, 1991). Most of them grow fastest on ammonium and some can use nitrate but others not. Experimental work on N absorption is beset with difficulties concerning pH change, which can be very significant in the unbuffered media often employed. Absorption of ammonium results in a marked lowering of the pH, which may be followed by a sharp cessation of growth. The form of N present in the medium is strongly affected by pH. The pKa for protonation of ammonia is 9.25, with the result that ammonium rather than ammonia will be the predominant form in most culture media and under most growth conditions. This has implications for the membrane transport processes likely to be involved in uptake. Absorption of nitrate, as well as the release of ammonium from amides or other readily hydrolysed compounds, causes an increase in pH which is usually slower and may have a less marked effect on growth than the reductions in pH. In spite of these problems, there is not much doubt of the broad results, although there is much variation between different species and strains of ECM fungi in requirements for inorganic N.

There is striking variability both between (France and Reid, 1984; Plassard *et al.*, 1986; Anderson *et al.*, 1999) and within (Ho and Trappe, 1987; Anderson *et al.*, 1999) ECM fungal species with respect to their abilities to use nitrate. Among those ECM fungi that most readily use nitrate are members of genera like *Hebeloma* which, as pioneer occupants of disturbed habitats, are most likely to encounter this form of N (Marmeisse *et al.*, 1998). The nitrate nutrition of *H. cylindrosporum*, some strains of which achieve a 10-fold greater biomass on nitrate than on ammonium, has been extensively investigated (Plassard *et al.*, 1991, 1994). The Km for nitrate uptake in this fungus of $67\,\mu M$ (Plassard *et al.*, 1994), compared with $12\,\mu M$ for *Rhizopogon roseolus* (Gobert and Plassard, 2002), indicates the large interspecific differences between fungi in their affinity for nitrate. In the latter fungus, nitrate uptake occurred at the same rate regardless of whether the mycelium had been pretreated in $0.05\,mM$ nitrate or not, suggesting that no inducible nitrate transporter exists in this organism. Jargeat *et al.* (2003) isolated a gene (*NRT2*) coding for a nitrate transporter from *H. cylindrosporum*. The polypeptide which it encodes is characterized by 12 transmembrane domains that are suggested to consist of a long possibly intracellular loop and a short C-terminal tail. Transcription of the *NRT2* gene was repressed by ammonium and stimulated under conditions of low exogenous supplies of both nitrate and simple sources of organic N.

Progress has been made towards characterization of the molecular events and physiological processes associated with nitrate assimilation. Genes encoding proteins involved are induced by nitrate and subject to N catabolite repression.

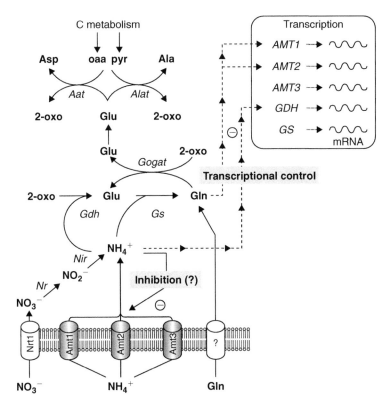

Figure 9.1 A model describing the regulation of nitrogen transport and assimilation in *Hebeloma cylindrosporum*. This ectomycorrhizal fungus is able to use nitrate, ammonium and amino acids as nitrogen sources. Under low ammonium status, the genes *AMT1*, *AMT2*, *AMT3*, *GDH* (*GDHA*) and *GS* (*GLNA*) are transcribed, which results in elevated ammonium uptake and metabolism capacities. Under ammonium excess, *AMT1*, *AMT2* and *GDHA* are efficiently repressed, which results in reduced ammonium assimilatory capacities. Under these conditions, *AMT3* and *GLNA* would ensure the maintenance of a basal level of ammonium assimilation. *AMT1* and *AMT2* transcript levels are controlled through the effect of intracellular glutamine, whereas the *GDHA* and *NAR1* mRNA levels are controlled by ammonium (dotted lines). Ammonium uptake activity may be controlled by intracellular NH_4^+ through a direct effect (dotted lines). *2-oxo* 2-oxoglutarate, *oaa* oxaloacetate, *pyr* pyruvate, *GOGAT glutamate synthase, Aat* aspartate aminotransferase, *Alat* alanine aminotransferase, *NR* nitrate reductase, *NiR* nitrite reductase, *Nrt1* nitrate transporter, *GAP1* general amino acid transporter. From Javelle *et al.* (2003b), with permission.

Two nitrate reductase genes have been cloned in *H. cylindrosporum* (Jargeat *et al.*, 2000). One (*nar 1*) has been transcribed and shown to code for a 908 amino acid polypeptide, while the other (*nar 2*) is considered to be an ancestral but non-functional duplication of *nar 1*. Above a basal level, transcription of *nar1* is repressed by high concentrations of ammonium, but strongly induced in the presence of nitrate or under low total N conditions. Transcription also occurs when organic sources such as glycine constitute the sole sources of N in the bathing medium, indicating that an exogenous supply of nitrate is not necessarily required to induce transcription of the gene. *Nar1* gene expression in *H. cylindrosporum* thus appears to be regulated primarily by the availability of ammonium. These authors propose that the lack of a nitrate induction system may represent an adaptation to the forest soil environment in which many different N sources are present all at low concentrations.

Since ammonium is the major source of mineral N in most forest soils (Read, 1991a; Marschner and Dell, 1994), it is perhaps not surprising that the majority of ECM fungi preferentially use the ammonium ion when grown in culture. The kinetics and energetics of ammonium uptake have been examined in *Paxillus involutus* (Javelle *et al.*, 1999). K_m and V_{max} were 180 and 380 µM, respectively. Experiments carried out using both short-term (hours) and long-term (days) incubation indicated that ammonium uptake was strongly regulated by the presence of the organic N sources glutamate and leucine. This observation is of ecological interest in view of the frequent co-occurrence of amino compounds and ammonium in soils supporting ECM plants (Read *et al.*, 2004).

In common with other fungi, most of the N from ammonium is assimulated by ECM fungi into the amide group of glutamine by the glutamine synthetase (GS-GOGAT) pathway, or as the amino group of glutamate through the glutamate dehydrogenase (GDH) pathway. While both of these routes can be present, the relative importance of each differs according to fungal species.

The genes coding for ammonium uptake and transport have now been cloned and this has enabled progress towards understanding of both their function and regulation. The ammonium transporter gene *AMT1* has been cloned from *Tuber borchii* (Montanini *et al.*, 2002) and *H. cylindrosporum* (Javelle *et al.*, 2001, 2003a). This is a high-affinity transporter which is expressed only under conditions of N deficiency (Javelle *et al.*, 2003a). Further transporters, *AMT2* and *AMT3* have also been characterized from the latter fungus (Javelle *et al.*, 2001). One of these, *AMT3*, is a low-affinity transporter which is highly expressed but not highly regulated. In addition to the transporters, genes encoding both GS (*GLNA*) and one encoding GDH (*GDHA*) (see below) have also been cloned and characterized (Javelle *et al.*, 2003b, 2004). This work has enabled the construction of a model describing the uptake and regulation of ammonium transport in *H. cylindrosporum* (see Figure 9.1). When exposed to low levels of ammonium, the regulatory genes *AMT1*, *AMT2*, *AMT3*, *GLNA* and *GDHA* are transcribed, resulting in increased capacity for ammonium uptake and metabolism. Under conditions of elevated ammonium availability, *AMT1*, *AMT2* and *GDHA* are effectively repressed, reducing the ability of the mycelium to assimilate ammonium. In the latter situation, it is expected that the activities of *AMT3* and *GLNA* would ensure a basal level of ammonium assimilation. The model proposes that the transcript levels of *AMT1* and *AMT2* are controlled through the effects of intracellular glutamine, whereas the *GDHA* mRNA accumulation is controlled by ammonium (dotted lines in Figure 9.1).

The high-affinity *AMP1* ammonium transporter is a member of a gene family, *MEP2*, which is involved in mycelial proliferation in fungi and is essential for hyphal differentiation in yeast (*Saccharomyces cerevisiae*) (Madhani and Fink, 1998). On the basis that incorporation of the *AMP1* and 2 genes of *H. crustuliniforme* into *MEP2*-deficient *S. cerevisiae* restored the ability of the yeast to produce pseudohyphae, Javelle *et al.* (2003b) have proposed that these ammonium transporters may also act as ammonium sensors which stimulate hyphal proliferation in response to low levels of ammonium enrichment. Observations such as these indicate that genes identified as having a specific function in the context of nutrition may also have other roles, in this case as triggers of morphogenetic change. The possibility that expression of the two attributes takes place in concert to facilitate ammonium capture is particularly interesting.

Analysis of the major pathways of N metabolism in *Cenococcum geophilum* (Genetet *et al.*, 1984; Martin, 1985; Martin *et al.*, 1988a), *Hebeloma cylindrosporum* (Chalot *et al.*, 1991) and *Laccaria laccata* (Brun *et al.*, 1992) indicates that the GS pathway of ammonium assimilation predominates in these fungi. It was shown that GS represented about 3% of the total soluble protein pool of *L. laccata*, a value considerably greater than that for GDH, which accounted for only 0.15%. Using ^{15}N-labelled ammonium as tracer, up to 40% of assimilated ^{15}N was found in the amide group. The GS of *L. laccata* is known to have a very high affinity for ammonium, further suggesting that, in this fungus at least, GS is the main route of ammonium assimilation (Brun *et al.*, 1992). This may also be the case in many ECM fungi and would be consistent with earlier work showing that, in the presence of ammonium, dark fixation of $^{14}CO_2$ results in preferential incorporation of label into glutamine by excised *Fagus* mycorrhizas (Harley, 1964; Carrodus, 1967).

The gene, *GNLA*, encoding GS in *H. cylindrosporum*, has been cloned and characterized (Javelle *et al.*, 2003a). A single mRNA of 1.2 kb was detected. Transfer of the fungus from an ammonium-containing to an N-free medium resulted in an increase of GS activity. However, when the mycelium was re-supplied with ammonium, GLNA transcripts remained unchanged or decreased only slowly, suggesting that GS in *H. cylindrosporum*, while being highly expressed, is not highly regulated.

The role of GDH should not be overlooked. In *H. crustuliniforme*, for example, assimilation of ammonium appears to be mainly via the GDH pathway (Quoreshi *et al.*, 1995). Two forms of this enzyme are recognized, one NAD (EC 1.4.1.2)- and the other NADP (EC 1.4.1.4)-dependent. The former has been ascribed a catabolic role, whereas the latter is considered to be involved in glutamate biosynthesis. Both have been found and characterized in the ECM fungus *L. laccata* (Dell *et al.*, 1989; Ahmad *et al.*, 1990; Brun *et al.*, 1992; Botton and Chalot, 1995; Garnier *et al.*, 1997). Its properties are similar to those reported for NADP-GDH enzymes of *Neurospora crassa* and yeasts (Stewart *et al.*, 1980). When GS is inhibited by methionine sulphoximine (MSX), glutamate, alanine and aspartate accumulate in mycelium of ECM fungi, confirming the presence and operation of the GDH pathway. These results suggest that ammonium assimilation, in these three fungi at least, is achieved by parallel action of GDH and GS-GOGAT (see also Chalot *et al.*, 1994a, 1994b). The need for examination of other fungi is highlighted by the observation that, in *Pisolithus tinctorius*, GS activity is low (Ahmad *et al.*, 1990) and that MSX blocked the synthesis of other amino acids suggesting the operation of the GS-GOGAT pathway. An NADH-dependent GOGAT has been detected in *L. bicolor* (Vezina *et al.*, 1989), but the instability of this enzyme renders characterization difficult and its status in ECM fungi remains uncertain.

The gene (*GDHA*) encoding NADP-GDH has been cloned and characterized from *L. bicolor* (Lorillou *et al.*, 1996), *H. cylindrosporum* (Javelle *et al.*, 2003a), and *Tuber borchii* (Vallorani *et al.*, 2002). Quantification of mRNA accumulation using a cDNA probe encoding the *L. bicolor* NADP-GDH confirmed that the growth of mycelia on nitrate and on N-free media resulted in an increased accumulation of transcripts encoding

NADP-GDH (Lorillou *et al.*, 1996). In the case of *H. cylindrosporum*, transfer of the fungus from 3 mM ammonium to N-free medium resulted in a 12-fold increase in the GDH transcript level, corresponding to a similar increase of enzyme activity. In contrast, feeding the mycelium with ammonium resulted in a rapid decrease of transcripts encoding GDH, which correlated with a decline in GDH-specific activity.

The studies outlined above provide an overview of the genes involved and aspects of the physiological events involved in assimilation of mineral N sources. However, they are heavily biased towards a very small number of ECM fungal species and analyses across a broader range of fungi, particularly those characteristic of distinctive soil types or ecosystems, are highly desirable.

Organic N sources

The ability of some ECM fungi to use organic N sources has been appreciated for many years (see Harley and Smith, 1983). Lundeberg (1970), in his cultural studies of a number of species and strains, identified several members of the genus *Suillus* that grew better on asparagine and glycine than they did on inorganic N. Again, there is clearly a good deal of variability between different species and even between strains of the same species. Laiho (1970) examined the N-source preferences of a number of strains of *Paxillus involutus*. All were able to use casein hydrolysate, peptone and a mixture of amino acids and all isolates grew well on both glutamate and arginine. Ability to grow on other amino acids as sole N source was low and variable. There was no case of an organic N source being very readily used by one strain and being totally useless to another.

To test the ability of ECM fungi to use organic N compounds in forest soil organic matter, Lundeberg (1970) prepared humus agar in which the concentrations of inorganic N compounds were reduced and the organic-N was labelled with ^{15}N. He allowed fungi to grow from an inoculum which straddled the interface between the N-free glucose agar and the humus agar. None of the ECM fungi absorbed organic N from the humus in significant quantities, although five other saprotrophic fungi which produced some or all of the hydrolytic enzymes, cellulase, pectinase, proteinase and laccase, were able to do so. These results agree with the findings of Mosca and Fontana (1975), on the use of protein-N by *Suillus luteus* and together such data have been taken to support the tentative conclusion that ECM fungi may not be very effective at acquiring organic N and must compete with other organisms for inorganic N mineralized by free-living members of the soil microflora.

Awareness of the possible importance of amino acids as primary sources of N for ECM fungi and for the plants which they colonize has been heightened by the recognition that these can constitute a significant pool, particularly in acid organic soils (Nemeth *et al.*, 1987; Abuarghub and Read, 1988a; Kielland, 1997), leading to re-examination of the role of ECM fungi in their mobilization. Some amino acids and amides, including glutamine, glutamate and alanine, which appear to occur in significant amounts in soil solution, are readily assimilated by ECM fungi in pure culture and, at equivalent concentrations of N and C, they can support yields as large as those obtained on ammonium as sole N source (Abuzindah and Read, 1988a; Finlay *et al.*, 1992; Keller, 1996; Anderson *et al.*, 1999) (Table 9.1).

Using *Paxillus involutus*, Chalot *et al.* (1994a, 1994b) demonstrated, by means of tracer and enzyme inhibition techniques, that glutamate, glutamine and alanine

Table 9.1 Yields of three ectomycorrhizal fungi, *Suillus bovinus, Amanita muscaria* and *Hebeloma crustuliniforme*, when grown with a range of mineral or amino N sources at a concentration of 60 mg N/l and at the same C:N ratio.

Nitrogen source	Wt of N source added (g)	Wt of glucose added (g)	Dry weight yields (mg) after 30 days		
			S. bovinus (mean ± SE)	A. muscaria (mean ± SE)	H. crustuliniforme (mean ± SE)
Mineral N.					
Ca (NO$_3$)$_2$ 4H$_2$O	0.504	3.004	17.7 ± 1.5	4.7 ± 0.3	12.6 ± 1.2
(NH$_4$)$_2$ SO$_4$	0.284	3.004	30.7 ± 1.2	26 ± 1.2	25.3 ± 1.4
Acidic amino acids					
L-Aspartic acid	0.572	2.372	31.3 ± 2.6	34.3 ± 2.0	23.0 ± 0.9
L-Glutamic acid	0.632	2.216	32.0 ± 1.0	36.3 ± 1.5	24.9 ± 1.3
Basic amino acids					
L-Arginine	0.224	2.764	33.6 ± 1.8	34.0 ± 2.5	15.6 ± 0.7
L-Lysine	0.392	2.528	17 ± 0.6	18.0 ± 0.1	5.9 ± 0.5
L-Histidine	0.300	2.688	5.0 ± 0.0	14.7 ± 1.2	4.2 ± 0.4
Neutral amino acids					
L-Alanine	0.380	2.528	31.6 ± 0.3	21.7 ± 1.2	20.4 ± 1.3
L-Asparagine	0.284	2.688	29 ± 1.5	23.3 ± 1.9	19.0 ± 0.7
L-Cysteine	0.520	2.528	7.7 ± 0.3	3.3 ± 0.8	6.0 ± 0.5
L-Cystine	0.516	2.528	6.0 ± 0.6	19.7 ± 0.7	5.6 ± 0.2
L-Glutamine	0.312	2.608	29.3 ± 1.4	36.3 ± 0.3	18.1 ± 0.7
L-Methionine	0.640	2.216	7.7 ± 0.3	2.3 ± 0.3	3.5 ± 0.2
Glycine	0.324	2.688	5.7 ± 0.3	22.7 ± 1.4	9.9 ± 0.4
L-Phenylalanine	0.708	1.588	11.3 ± 0.7	4.7 ± 0.7	4.3 ± 0.7
L-Hydroxy-L-proline	0.560	2.528	3.7 ± 0.3	4.7 ± 0.3	4.2 ± 0.5
L-Isoleucine	0.564	2.216	22.3 ± 2.3	7.3 ± 0.9	11.7 ± 0.3
L-Leucine	0.564	2.216	23.0 ± 3.2	7.3 ± 0.3	8.1 ± 0.4
L-Proline	0.492	2.216	4.0 ± 0.6	6.3 ± 0.9	4.4 ± 0.3
L-Serine	0.452	2.528	27.7 ± 0.9	18.7 ± 0.7	16.3 ± 1.2
L-Threonine	0.512	2.372	4.7 ± 0.3	6.7 ± 0.3	7.6 ± 0.3
L-Tryptophane	0.436	2.136	2.0 ± 0.0	2.7 ± 0.7	7 ± 0.1
L-Tyrosine	0.776	1.588	1.7 ± 0.3	5.3 ± 0.9	8.2 ± 1.0
L-Valine	0.300	2.220	17 ± 1.2	11.3 ± 0.9	18.9 ± 3.1
Minus N	–	3.004	7.3 ± 0.7	4.3 ± 0.3	6.2 ± 0.2

From Abuzinadah and Read (1988).

were absorbed intact and incorporated respectively by GS, GOGAT and alanine aminotransferase (AIAT) into the various assimilation pathways. Here the N sources would supplement the free amino acid pools, which represent important sinks for C in mycelia fed with mineral sources of N. It has been shown, for example, in *Cenococcum geophilum* and *Sphaerosporella brunnea* that 16–40% of C, fed as 1-^{13}C glucose to the mycelium, entered the amino acid pools (Martin and Canet, 1986; Martin *et al.*, 1988b). Exogenous supply of amino acids and amides would therefore be expected to supplement the organic C, as well as the N economy of ECM fungi.

In studies of assimilation of glutamine, glutamate and alanine carried out over a range of pHs, with and without the presence of ammonium, nitrate and glucose,

Chalot *et al.* (1995) showed that amino acid absorption had a distinctly acid pH optimum (Figure 9.2) and that neither ammonium, at concentrations from 0.05 to 0.5 mM, the low range (similar to those in soil), nor glucose, had an impact upon uptake. Over a period of 5–6 weeks, uptake of amino acids decreased by a factor of 4–10 as the mycelium aged, while the size of endogenous pools progressively increased. Extrapolation of information on pool turnover in pure cultures to symbiotic systems must be done with caution, because of the absence of a sink for assimilated N compounds which in ECM would be provided either by ongoing growth of the fungus in soil or by the associated plant.

It is widely accepted that amino acids are transported by proton symport mechanisms. Analysis of the energetics and specificity of a general amino acid transporter in *P. involutus* (Chalot *et al.*, 1996) showed that, in the case of glutamate, the concentration-dependent uptake failed to obey simple Michaelis Menten kinetics. Rather, there were two separate components of uptake, one of which was saturable and carrier mediated, the other a non-saturable diffusion-like process. The pH-dependence of amino acid uptake of this fungus was again demonstrated, it being optimal between pH 3.9 and 4.3 in the cases of glutamine and glutamate, and between 3.9 and 5.0 for alanine and aspartate. Both pH dependence and susceptibility of the uptake processes to inhibitors such as 2,4 dinitrophenol (DNP) are consistent with a proton symport mechanism for amino acid uptake by *P. involutus*. Competition

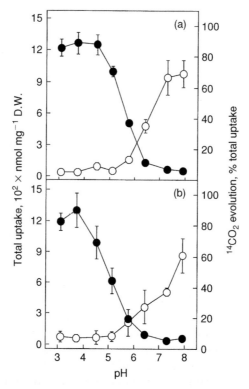

Figure 9.2 Effect of external pH on the uptake (●) and respiration (○) of L-glutamate and L-glutamine by the ectomycorrhizal fungus *Paxillus involutus*. Data are means of three replicate determinations ± standard errors of the means. From Chalot *et al.* (1995), with permission.

studies in this fungus (Chalot *et al.*, 2002, 2006) suggest a broad substrate recognition that is consistent with the role of a general amino acid permease.

A gene encoding an amino acid importer (*AmAAP1*) of *Amanita muscaria* has been isolated and the activities of its transcripts characterized by heterologous expression in yeast (Nehls *et al.*, 1999). This study confirmed the protein to be a high-affinity, general amino acid permease with a K_m of 22 μm for histidine and up to 100 μm for proline. A low constitutive expression of the transporter was detected in the presence of amino acids. By contrast, under conditions of N starvation, or in the presence of nitrate or phenylalanine, neither of which is used as an N source by *A. muscaria*, expression of the gene was considerably enhanced. It was concluded that in *A. muscaria*, as in yeast and *Aspergillus nidulans*, gene expression of amino acid transporters is regulated at the transcriptional level by N repression (Nehls *et al.*, 1999). *AmAAP1* was shown to have a higher affinity for basic than for acidic or neutral amino acids, which may reflect a preponderance of basic amino compounds in acidic forest soils. Nehls *et al.* (1999) raise the possibility that, in addition to having a role in amino acid uptake for nutrition, the enhancement of expression under conditions of N starvation may prevent amino acid loss by leakage from hyphae when suitable N sources are lacking. In a complementary study, Wipf *et al.* (2002) isolated an amino acid transporter (*HcBap1*) from *H. cylindrosporum* by functional complementation of a transporter-deficient yeast strain.

Whereas some of the amino acids present in the soil solution are undoubtedly derived from the free pools of these molecules in living plant and microbial tissues, others are likely to be a product of the cleavage of polymeric peptides and proteins associated by decomposer organisms in soil. Since the polymers constitute the bulk of the N in many ECM forest soils (see Chapter 15), it is relevant to determine whether they too might be accessible to ECM fungi. Abuzinadah and Read (1986a) showed that *Suillus bovinus* (Figure 9.3b), *Rhizopogon roseolus* and *Pisolithus tinctorius* readily used a series of alanine peptides of increasing chain length from the di- to the penta-peptide. Hexa-alanine was also used, but more slowly. *Laccaria laccata*, in contrast, made poor growth on all these substrates (Figure 9.3a).

In early experiments, Chalot and Brun (1998) reported that uptake of ^{14}C-labelled peptides was slower than that of amino acids. Later, Benjdia *et al.* (2006) demonstrated that *H. cylindrosporum* was capable of taking up di- and tripeptides and of growing on them as sole N sources. Using an *H. cylindrosporum* cDNA library, two peptide transporters (*HcPTR2A* and *B*) were isolated by means of yeast functional complementation and these were shown to mediate dipeptide uptake. *HcPTR2A* was involved in high-efficiency peptide uptake under conditions of limited N availability, whereas *HcPTR2B* was expressed constitutively.

The capacity of ECM fungi to use proteins as N sources has been known for some time (Melin, 1925; Lundeberg, 1970), but the extent of fungal proteolytic capability was not fully appreciated. Using soluble proteins of animal and plant origin, Abuzinadah and Read (1986a, 1986b) screened eight ECM fungi over a wide range of pH conditions for the ability to exploit these polymers as sole sources of N. On both bovine serum albumin (BSA, MW 67000, N content 16%) and gliadin (MW approx. 30000, N content 14%) some ECM fungi produced yields as large or even larger than those obtained with ammonium as sole N source. These, including *Amanita muscaria*, *Cenococcum geophilum*, *Paxillus involutus*, *R. roseolus*, *S. bovinus* and *H. crustuliniforme*, were referred to as 'protein fungi'. In contrast, *L. laccata* and

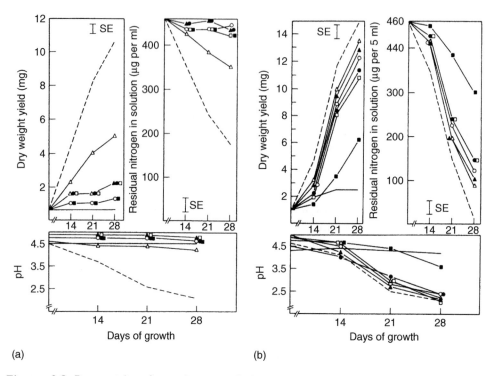

Figure 9.3 Dry weight of mycelium, residual N and pH of the culture medium (a) of *Laccaria laccata* and (b) *Suillus bovinus* grown in liquid medium containing ----- ammonium; △ alanine; a range of alanine peptides of chain length 2–6 (▲ A2; ○ A3; ● A4; □ A5; ■ A6) or —, lacking N. Data are means of three replicates ± standard errors of the means. From Abuzinadah and Read (1986a), with permission.

Lactarius rufus, the so-called 'non-protein' fungi, produced only small yields on these organic N substrates. *P. tinctorius* was in an intermediate category. Subsequent examination of *Pisolithus* (Anderson *et al.*, 1999) showed a generally low level of ability to use BSA, but also considerable inter- and intraspecific variation.

El-Badaoui and Botton (1989) isolated a protein-rich fraction from forest litter and observed that it induced greater proteolytic activity in *A. rubescens*, *C. geophilum* and *H. crustuliniforme* than did BSA or gelatin. There is obviously variation in the accessibility of commercial proteins for fungal growth. For example, Hutchison (1990) screened a very large number of fungi for their ability to breakdown gelatin to its constituent amino acids but failed to detect activity even in species such as *H. crustuliniforme* and *P. involutus* which are known to be able to hydrolyse BSA or gliadin. The extracellular acid proteinase of *H. crustuliniforme* has been purified and characterized (Zhu *et al.*, 1990). The enzyme was most stable and had greatest activity over the pH range 2–5. Protein and some individual amino acids, notably glycine (Table 9.2), induced activity when they were supplied as sole N sources. Enzyme production was not affected by the addition of ammonium at 3.2 mM, but was repressed at higher concentrations. Production required the presence of a simple C source and was not repressed by glucose between 0.5 and 2.0% (Zhu *et al.*, 1994). Consequently, it appears that, as in the case of the ERM fungus *Rhizoscyphus ericae* (Leake and

Table 9.2 Effect of various nitrogen sources on growth and proteinase production of *H. crustuliniforme* after 20 days of growth[a].

Nitrogen sources[b]	Conc.	Dry weight (mg/plate)	Proteinase activity		Final pH
			(units/mg dry wt)	(% of control)	
Ammonium	3.2 mM	21.3 ± 1.5	32.4 ± 2.1	Control	2.8
Asparagine	10 mM	36.1 ± 3.4	16.7 ± 1.2	52	5.4
Glycine	10 mM	10.4 ± 1.6	74.7 ± 3.4	230	5.3
Glutamine	10 mM	89.2 ± 5.6	8.6 ± 0.9	27	6.5
Casein hydrolysate	0.04%	37.4 ± 2.7	9.6 ± 2.1	30	6.4
Casein	0.04%	36.1 ± 3.6	18.4 ± 2.7	57	4.9
Gelatin	0.04%	44.7 ± 3.2	60.1 ± 3.2	185	4.8
Gliadin	0.04%	25.7 ± 2.1	52.9 ± 4.6	163	4.1
BSA	0.04%	28.6 ± 2.4	70.8 ± 5.2	219	4.5

From Zhu *et al.* 1994. [a] Data are presented as mean ± standard deviation of five replicates. [b] Nitrogen sources were added to the basal medium containing 1% glucose.

Read, 1991; see Chapter 11), the acid proteinase of *H. crustuliniforme* is largely regulated by induction. It may be repressed by some forms of N, but catabolite repression in general does not seem to play an important part. The observation that amino acids supplied singly can act as inducers is important, in view of current recognition that soils of the kinds in which ECM roots grow can have sizeable pools of these compounds (Abuarghub and Read, 1988a; Kielland *et al.*, 1994).

Nehls *et al.* (2001c) isolated and characterized two aspartic proteases which are released in a pH-dependent manner by mycelia of *Amanita muscaria*. AmProt1 with a molecular mass of approx. 45 kDa is mainly produced up to pH 5.4, with an optimum around pH 3.0, whereas the excretion of AmProt2, which had a mass of 90 kDa, was only detectable at pHs between 5.4 and 6.3 and had a somewhat higher pH optimum for activity. One cDNA clone presumed to encode AmProt1 was identified and also shown to be expressed in a pH-dependent manner. C- and N-limitation significantly enhanced AmProt1 expression. However, whereas N starvation alone increased expression of the gene by three to fourfold, the absence of a C source increased transcript levels by a factor of 12 independently of the presence of N in the bathing medium. Thus, the endogenous C levels in ECM hyphae can be expected to exert strong regulatory controls on levels of proteolytic activity.

H. crustuliniforme, when grown on protein as sole N source shows a prolonged lag phase, followed by a phase of more rapid growth during which proteolysis results in accumulation of amino acids in the medium (Read *et al.*, 1989) (Figure 9.4). These are subsequently assimilated, but only after the protein concentration has been markedly reduced. At no stage during growth is there evidence of ammonium release. Indeed, ammonium ions are detectable in the medium only when the amino compounds have themselves been virtually exhausted. By this stage, C-starvation of the fungus is likely to have led to deamination of the residual compounds and release and utilization of the C-skeletons. Such starvation and the release of ammonium associated with it, may be both artefacts of the pure culture environment and unrepresentative of the symbiotic condition in which C supply from the plant is assured. Furthermore, the failure of ammonium to appear in the medium at earlier stages of growth could

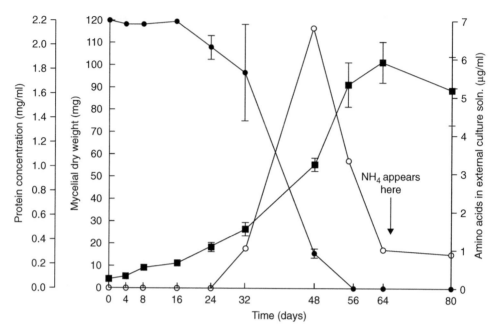

Figure 9.4 The relationship between protein utilization, mycelial biomass production and amino acid release when *Hebeloma crustuliniforme* is grown in liquid culture with protein as the sole N source. Note that ammonium does not appear in the medium until all the protein has been used and the fungus therefore starved of C. ● protein concentration; ○ amino acid concentration; ■ mycelial dry weight. Data are standard errors of means. From Read *et al.* (1989), with permission.

reflect the relative rates of deamination of protein (if that occurs) and uptake of ammonium which would be likely to be rapid in actively growing mycelium.

The availability of proteins in soil depends on their structure and solubility and on their interactions with other soil components. Thus, the relatively insoluble gelatin is less available in culture than the soluble BSA but, in soil, all proteins might be rendered unavailable by tanning reactions with phenolic compounds, by ionic reactions with soil organic matter and clays, or by physical occlusion in the soil pores. The success of ECM fungi in obtaining these less-available resources might very well be increased, compared with saprotrophic fungi because, in symbiosis, they are not dependent on soil organic C. For this reason, if for no other, it is clearly essential to extend studies of the abilities of the fungi using intact symbiotic systems in which the normal pathways of organic C supply would be operating.

Use of N by mycorrhizal roots and intact plants

Inorganic N sources

In a comparative analysis of the abilities of ECM and non-mycorrhizal roots of *Picea abies* to take up nitrate and ammonium, Eltrop and Marschner (1996) reported that neither the mycorrhizal status of the plants nor the form of N supplied greatly influenced

Table 9.3 Kinetic uptake parameters (substrate affinity K_m in µM, and maximal uptake rate V_{max} in µmol/g f.wt/h, after Lineweaver-Burk) for ammonium and nitrate in intact root systems of 22-week-old non-mycorrhizal and mycorrhizal Norway spruce seedlings.

	Ammonium		Nitrate	
	K_m	V_{max}	K_m	V_{max}
Supply of N as NH₄NO₃				
Non-mycorrhizal	227	0.38	214	0.28
Laccaria laccata	229	0.55	207	0.33
Paxillus involutus	222	0.47	228	0.32
Pisolithus tinctorius	236	0.53	367	0.51
Supply of N as (NH₄)₂SO₄ or KNO₃				
Non-mycorrhizal	567	0.96	98	0.46
Pisolithus tinctorius	334	1.24	263	0.97

Nitrogen was supplied either as NH_4NO_3 or as $(NH_4)_2SO_4$ or KNO_3 at a total concentration of 800 µm. Data from Eltrop and Marschner (1996).

Table 9.4 Kinetic parameters for methylammonium and amino acid uptake by mycorrhizal beech roots calculated by non-linear regression analysis in a substrate concentration range of 0.001–0.25 mol/m.

Fungus	Compound	V_{max}	K_m
Lactarius subdulcis	Methylammonium	48.5 ± 6.2	220 ± 62
	Glutamine	12.4 ± 0.9	79 ± 13
	Glycine	22.8 ± 4.2	197 ± 67
	Glutamic acid	10.1 ± 1.3	115 ± 33
Russula ochroleuca	Methylammonium	27.9 ± 7.2	555 ± 191
	Glutamine	1.4 ± 0.1	86 ± 20
	Glycine	3.1 ± 0.3	233 ± 35
Xerocomus chrysenteron	Methylammonium	40.8 ± 8.0	441 ± 122
	Glutamine	16.5 ± 1.5	80 ± 18
	Glycine	18.7 ± 2.5	84 ± 26

Values for V_{max} (µmol/g DW/h) and K_m (mmol/m³) (±SE of the fitted parameters) were calculated by non-linear curve fitting of the experimental data to the Michaelis-Menten equation [$v = (V_{max} \times [S])/(K_m + [S])$] in a substrate concentration range of 0.001–0.25 mol/m³. Sampling site: Hathersage, Derbyshire, UK. From Wallenda and Read (1999).

the kinetic properties of the uptake process (Table 9.3). K_mS for uptake of nitrate were slightly lower than those for ammonium, suggesting a higher affinity for the former. Caution must, however, be exercised when generalizing from such studies because it was subsequently shown that uptake rates of methyl ammonium, as an analogue for ammonium, are strongly fungus-dependent roots colonized by *Lactarius subdulcis* having higher affinities for the analogue than those supporting *Russula ochroleuca* (Wallenda and Read, 1999) (Table 9.4). This variation might explain some of the large variability seen in K_m for methyl ammonium uptake observed by Kielland (1994) in field collected ECM roots of *Salix* and *Betula* (Table 9.5).

Table 9.5 Kinetic parameters for methylamine (ammonium) uptake by excised roots of major plant species from four Arctic tundra communities.

Community	Species	Mycorrhizal type	Half saturation constant, K_m (μmol/l)	Maximum uptake capacity, V_{max} (μmol/g/h)
DH	*Betula nana*	ECM	67	5.6
WM	*Carex aquatilis*	NM	153	17.8
	Eriophorum angustifolium	NM	935	17.8
TT	*Betula nana*	ECM	3821	142.8
	Carex bigelowii	NM	78	5.4
	Eriophorum vaginatum	NM	242	13.7
	Ledum palustre	ERM	256	7.6
	Salix pulchra	ECM	1197	75.9
ST	*Betula nana*	ECM	663	31.7
	Salix pulchra	ECM	6717	272.8

Kinetic constants were calculated from four replicate measurements at each of four solution concentrations. Abbreviations: DH = dry heath, ECM = ectomycorrhizal, ERM = ericoid, NM = non-mycorrhizal, TT = tussock tundra, ST = shrub tundra and WM = wet meadow. From Kielland (1994).

The fate of ^{15}N from labelled nitrate and ammonium fed to external mycelium attached to intact ECM plants was studied by Finlay *et al.* (1988, 1989) and by Ek *et al.* (1994a). Glutamine was identified as a major sink for absorbed N, with alanine, arginine and aspartate-asparagine also important (Finlay *et al.*, 1992; Martin and Botton, 1993; Botton and Chalot, 1995). The contribution of arginine to the pool of amino acids and amides differs with fungal species and the suggestion has been made that this may relate to variations in accumulation in vacuoles, where arginine might play a role in stabilizing polyphosphate (polyP) (Finlay *et al.*, 1992; see Chapter 10). The data obtained with ^{15}N on intact ECM plants are consistent with earlier findings using ^{14}C with excised roots, confirming that glutamine is most likely to be the major form in which N is translocated in the mycelium and transferred across the symbiotic interface to the plant (Harley, 1964; Carrodus, 1967; Reid and Lewis, in Lewis, 1976). There is no doubt that transfer occurs very rapidly in the intact systems (Finlay *et al.*, 1988, 1989). ^{15}N-labelling experiments have shown that ammonium is preferentially absorbed from ammonium nitrate by *Paxillus involutus* in association with *Betula pendula* and also that ammonium inhibited ^{15}NO$_3$ assimilation in the external hyphae (Ek *et al.*, 1994a). Again, the ammonium was assimilated into glutamine at the uptake site, providing evidence for the importance of this amide in translocation.

France and Reid (1983) provided a conceptual model of the mechanisms thought to be important for assimilation of ammonium in ECM roots, in which the pathways involving GS, GOGAT and GDH were all present. Subsequent studies (Martin and Botton, 1993; Botton and Chalot, 1995) have revealed a diversity of pathways in which the plant exercises considerable influence over the activities of the fungi in the symbiotic condition, as well as important differences between plant–fungus combinations (Figure 9.5). Whereas the GDH-GS pathway appears to predominate in the fungi, ammonium assimilation in higher plants is known to occur primarily via

GS-GOGAT (Robinson *et al.*, 1991; Oaks, 1994; Miflin and Lea, 1976). If fungal and higher plant cells have distinctive pathways of inorganic N assimilation, it is important to determine how these are controlled in the ECM condition. In some cases, the fungal enzymes are downregulated in proximity to plant tissues, a feature that gave rise to the concept of metabolic zonation (Martin *et al.*, 1992) while, in others, activity is relatively unaffected. Thus, in *Picea* mycorrhizas formed by *Hebeloma*, NADP-GDH activity was highest in the extraradical hyphae and mantle, but reduced in the Hartig net (Dell *et al.*, 1989). In line with this, the quantity of GDH polypeptide, revealed by immunogold labelling, decreased progressively from the peripheral cells of the mantle to the Hartig net (Chalot *et al.*, 1990a). In *Fagus* mycorrhizas, on the other hand, both the activity (Dell *et al.*, 1989) and the amount (Chalot *et al.*, 1990a) of GDH were strongly suppressed throughout the mantle whether it was formed by *Hebeloma* or by *Cenococcum geophilum* or *P. involutus*.

A similar picture was revealed in *Fagus* roots with *Lactarius*-type ECM with successive incorporation of $^{15}NH_4$ into the amido-N of glutamine, glutamate and alanine, consistent with GS activity (Martin *et al.*, 1986). The importance of this pathway was confirmed by incubation of roots in MSX which inhibited ^{15}N-incorporation into glutamine and glutamate by 90%. Supporting evidence for the role of GS-GOGAT was obtained by use of the GOGAT inhibitor azaserine which completely blocked glutamate synthesis from ammonium. Such observations are suggestive of a complete downregulation of fungal GDH in some ECM types, although there remains the possibility of expression of some activity in the mycelium as it grows away from the root. In addition, N sources (ammonium versus nitrate) and concentration (N-rich versus N-starved) regulate NADP-GDH biosynthesis and activity through the alteration of GDH mRNA accumulation in *L. laccata* S238 (Lorillou *et al.*, 1996), suggesting the potential for regulation by N availability in soil.

Further evidence for the role of ammonium concentration in controlling pathways of ammonium assimilation in the fungi (albeit from studies in pure culture) comes from the use of NMR spectroscopy to monitor ^{15}N-labelling, after feeding *L. biolor* with $^{15}NH_4$ (Martin *et al.*, 1994). Rapidly growing mycelium assimilated $^{15}NH_4$ into glutamine via GS, which was probably the main route. However, when this pathway was inhibited by MSX, GDH activity became apparent. In stationary-phase growth with low concentrations of ammonium in the medium, both pathways operated.

The pathways of ammonium assimilation and the distribution of the enzymes involved have important implications for the mechanisms of transport of N across the fungus–root interface. Martin and Botton (1993) recognized three basic patterns of N incorporation. In *Fagus*, GS is localized in the fungal mantle and GOGAT in the root. In *Picea*, GDH and GS occur in extraradical mycelium and mantle respectively, with asparagine synthetase (AS) in the root. A third pattern (Figure 9.5) is representative of ECM formed by *Pisolithus tinctorius*, in which the GS-GOGAT pathway appears to operate in this fungus (Figure 9.5). It should be noted that, although Figure 9.5 shows transfer of hexoses from plant to fungus, this is an oversimplification because (as shown in Chapter 8) transfer of organic C probably involves sucrose efflux to the apoplast and its hydrolysis to hexoses before uptake by the fungus.

The question of uptake and transport of solutes can best be addressed in systems in which the sources of nutrient supply (the soil) and the ultimate sinks (the transpiring shoots) are intact. Enhancement of ammonium uptake was demonstrated when intact seedlings of *Tsuga heterophylla*, *Picea sitchensis* and *Pseudotsuga menziesii* were colonized by *H. crustuliniforme* compared with non-mycorrhizal seedlings, over a pH range of 3–7 (Rygiewitz *et al.*, 1984). Significantly greater rates of uptake were maintained in all three species when the plants were mycorrhizal. However, absolute amounts absorbed were always lower in both ECM and non-mycorrhizal plants under acidic conditions. Finlay *et al.* (1988) fed $^{15}NH_4$ to the distal parts of mycelial systems of *Rhizopogon roseolus*, *Suillus bovinus*, *Paxillus involutus* and *Pisolithus tinctorius* growing across peat from colonized roots of *Pinus sylvestris*. Measurements of the distribution of ^{15}N after 72 hours showed high labelling in glutamate-glutamine, aspartate-asparagine and alanine in all plant–fungus combinations, except that involving *P. involutus* which showed no labelling in aspartate-asparagine. Within this time period, 5–50% of the amino acids in the shoot were also labelled, indicating the very rapid operation of the translocation and transfer processes. Within the plant, despite large differences in the size of the different amino acid pools, the levels of ^{15}N-enrichment were similar, indicating that equilibrium between the pools, dependent upon activities of amino transferases, was rapidly achieved.

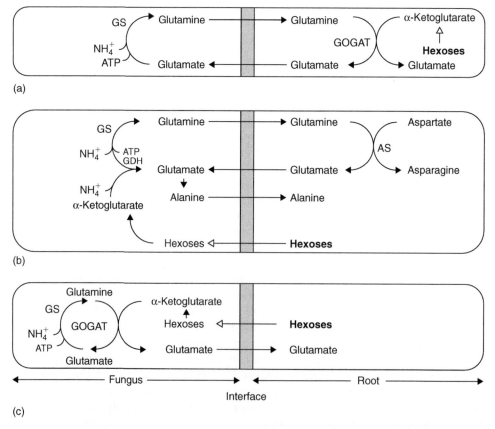

Figure 9.5 Possible localization of enzymes of ammonium assimilation in different types of ectomycorrhizas. (a) *Fagus*: glutamine synthetase (GS) in the fungus and glutamate synthase (GOGAT) in the root. (b) *Picea*: NADP-glutamate dehydrogenase (GDH) and GS in the fungus and asparagine synthetase (AS) in the root. (c) *Pisolithus* mycorrhizas: GS and GOGAT in the fungus. Modified from Martin and Botton (1993).

An alternative to the view that ammonium assimilation takes place primarily by GS-dependent glutamine accumulation has been proposed by Chalot *et al.* (2006). This suggests (Figure 9.6) that N taken up as ammonium may be translocated and transferred across the ECM interface, either as ammonium or uncharged ammonia. Evidence in support of this possibility comes from the description of a root-specific high-affinity ammonium importer (*PttAMT1.2*) in aspen (*Populus tremula*), the expression of which was greatly increased in an N-independent manner upon ECM formation (Selle *et al.*, 2005). Strong downregulation of genes coding for GS in fungal cells in close contact to the root (Wright *et al.*, 2005) and decreases of the activity of enzymes involved in N assimilation during ECM colonization (Blaudez *et al.*, 1998) would contribute to the maintenance of ammonium concentrations at the fungus–plant interface that are necessary to achieve transfer.

The nature of any possible ammonium transfer mechanisms remains a matter of conjecture (Selle *et al.*, 2005; Chalot *et al.*, 2006). Diffusion of ammonia across the plasma membrane into the interfacial apoplast of the Hartig net is a possibility. If this occurred, the low ambient pH of the apoplast (Nenninger and Heyser, 1998) would lead to protonation of ammonia and prevention of its return by diffusion. The N-dependent repression of the ammonium importer gene of the fungal partner observed by Javelle *et al.* (2003a) could also be expected to contribute to inhibition of back-flow. Ammonium efflux proteins of the kind reported in yeast (Guaragnella and Butow, 2003) could contribute to ammonium export at the fungus–plant interface. Tarkka *et al.* (2006) report that genes encoding proteins homologous to those in yeast are expressed in *Amanita muscaria*. The possible involvement of non-specific channels, such as aquaporins, and of voltage dependent cation channels in ammonium import by ECM plants have also been suggested (Chalot *et al.*, 2006).

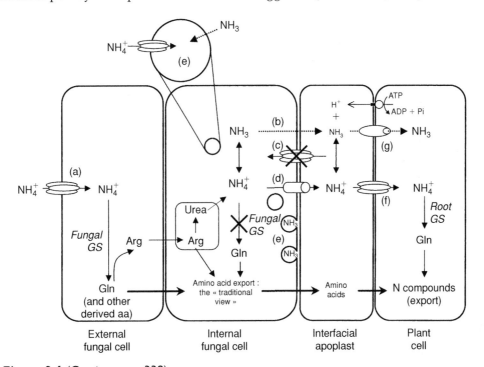

Figure 9.6 (Caption on p. 338)

The possibility that ammonium may be at least one of the forms of N transferred across the ECM interface has been raised at the same time as a similar scenario has been revealed for AM symbioses (see Chapter 5). For both associations it now needs to be appreciated that transfer of ammonium or ammonia will incur potential problems relating to pH regulation and charge balance (Raven and Smith, 1976). Sorting out how the plants meet these challenges will be an important topic in future research.

Studies of the impact of ECM colonization upon nitrate assimilation and growth of the plant made clear the differences between patterns seen in the axenic and symbiotic systems. Scheromm et al. (1990a), using the isolate of H. crustuliniforme shown in pure culture to assimilate nitrate preferentially, could find no direct effect of the fungus upon either uptake or reduction of nitrate by mycorrhizal Pinus pinaster. Using the same fungus, Rygiewitz et al. (1984) observed no differences in nitrate uptake between colonized and uncolonized plants of Picea sitchensis or Tsuga heterophylla. Increases in uptake of nitrate were, however, observed in colonized Pseudotsuga seedlings. pH was the major factor determining uptake of nitrate in both categories of plant, it being increased as pH was increased.

Figure 9.6 Possible ammonium transfer mechanisms (a–g) across different mycorrhizal interfaces as identified by Chalot et al. (2006). (a) After soil-derived ammonium is taken up by plasma membrane-based ammonium transporters (Amts) of the external hyphae, specific septal pores will ensure symplastic continuity and cell-to-cell transport of ammonia. N transfer between fungal and root cells will occur at the interfacial apoplast, either as organic N compounds according to the 'traditional view' (bottom horizontal arrows), or as ammonia. Transport mechanisms for organic nitrogen, which have yet to be revealed, would be of crucial importance if N is taken up by the fungus in organic form. (b) At ammonium concentrations in the millimolar range measured in the fungal cells, ammonia concentration in the cytosol can be far higher than in the apoplast, thus generating a favourable gradient for passive diffusion of ammonia. Such a model requires apoplastic concentrations to be kept low, potentially by an ammonia acid-trapping mechanism provided by the low pH thought to occur at the mycorrhizal interface and owing to the increased activity of H^+-ATPase along the plant membrane flanking the interface. An excretion mechanism based predominantly on ammonia diffusion is questionable because of the lack of overall control of the process. A large array of membrane proteins could fulfil the function of transferring ammonia from the fungal cytosol to the interfacial apoplast. (c) It is anticipated that the number of Amts would be decreased at the plasma membrane in ECM mantle hyphae compared with those in the extraradical hyphae. (d) Transcriptomic studies in various ECM associations have identified fungal efflux systems that could be promising candidates for the release of ammonia at the fungal membrane. (e) Alternatively, diffusion of ammonia or active transport of ammonium via an Amt-mediated system into acidified-vesicles would ensure compartmentation of excess ammonia. The ammonia-loaded vesicles could then move via microtubules to the symbiotic membrane where vesicles would fuse with the plasma membrane and release ammonia into the interfacial apoplast. (f) The further transfer of ammonia from the apoplast to the plant host cytoplasm might involve plant Amts. (g) Non-specific channels such as aqua-ammoniaporins or voltage-dependent cation systems might also contribute to ammonia transport from the interfacial apoplast to the plant cell cytoplasm. Abbreviations: Arg, arginine; Gln, glutamine; GS, glutamine synthetase. From Chalot et al. (2006).

Table 9.6 Levels of ^{15}N enrichment (atom % excess) in free amino acids in ectomycorrhizal systems of *Fagus sylvatica* infected with the fungus *Paxillus involutus* and supplied with ^{15}N-labelled ammonium or nitrate.

N source	Mycelium		Mycorrhizal tips		Roots		Shoots	
	NO_3^-	NH_4^+	NO_3^-	NH_4^+	NO_3^-	NH_4^+	NO_3^-	NH_4^+
Amino acid								
Ala	30.8	57.9	25.9	38.6	6.4	13.0	0.0	0.0
Gly	11.5	24.7	4.7	7.9	0.0	1.9	0.0	0.4
Thr	2.5	8.0	1.1	1.6	0.8	2.4	0.0	0.0
Ser	8.6	26.6	3.1	41.3	0.9	6.5	0.1	2.0
Leu	0.0	6.7	0.0	3.7	2.7	7.1	0.0	0.5
Ileu	3.5	5.1	0.9	6.6	5.5	7.8	0.0	0.0
Gaba	30.8	61.5	22.0	29.8	5.1	16.2	0.0	1.0
Pip	0.0	0.0	11.9	1.3	0.0	0.3	0.0	0.0
Asx	30.0	53.9	14.0	17.7	4.6	16.9	1.4	1.3
Glx	37.7	68.5	25.3	51.1	6.4	21.0	0.0	2.1
Lys	0.9	5.8	1.2	3.5	1.3	3.1	0.0	0.0
Tyr	9.9	10.7	2.6	5.3	1.0	13.2	0.0	9.5
Arg	1.2	0.0	6.8	11.1	0.4	4.1	0.0	0.0
No. I	38.1	69.6	28.5	56.1	6.6	23.3	0.0	1.4
Mean	14.7	28.5	10.6	19.7	2.9	9.8	0.1	2.1

From Finlay *et al.* (1989). Values represent means of duplicate growth chambers. No. I represents an unidentified amino acid. Pip = pipecolinic acid.

Table 9.7 Levels of ^{15}N enrichment (atom % excess) in protein-incorporated amino acids in ectomycorrhizal systems of *Fagus sylvatica* infected with the fungus *Paxillus involutus* and supplied with ^{15}N-labelled ammonium or nitrate.

N source	Mycelium		Mycorrhizal tips		Roots		Shoots	
	NO_3^-	NH_4^+	NO_3^-	NH_4^+	NO_3^-	NH_4^+	NO_3^-	NH_4^+
Amino acid								
Ala	1.1	4.7	3.8	6.8	1.6	4.1	0.3	0.3
Gly	0.2	1.6	3.1	3.4	1.5	3.0	0.2	0.0
Val	1.2	2.8	2.6	3.4	1.5	2.4	0.0	0.2
Ser	1.3	1.9	2.9	2.9	1.6	2.3	0.2	0.0
Leu	1.8	4.6	4.4	5.3	1.8	3.2	0.1	0.0
Ileu	1.1	3.6	3.7	4.8	1.7	3.1	0.0	0.0
Pro	0.6	3.3	1.9	2.8	1.2	1.9	0.0	0.0
Asx	2.1	5.4	5.0	6.2	2.0	4.5	0.4	0.0
Phe	1.0	2.8	3.2	3.7	2.3	3.5	0.1	0.0
Glx	3.5	9.3	6.0	8.3	2.2	4.4	0.2	0.0
Lys	1.3	1.1	1.4	1.5	0.7	1.1	0.3	0.0
Tyr	1.0	0.1	0.8	1.5	0.5	1.6	0.5	0.0
Arg	3.2	1.4	2.9	1.3	1.2	3.4	1.3	1.4
Mean	1.6	2.7	2.8	3.4	1.4	2.6	0.3	0.2

From Finlay *et al.* (1989). Values represent means of duplicate growth chambers.

Finlay *et al.* (1989) fed $^{15}NO_3$ to the extraradical mycelium of *Paxillus involutus* associated with *Fagus sylvatica* and compared the patterns of uptake, assimilation and transport of the label with those obtained in systems fed with $^{15}NH_4$. ^{15}N was taken up from both sources, incorporated into a range of free amino acids and transported to the shoots (Table 9.6). However, the amounts of enrichment of most free and protein-bound amino acids were usually greater in the systems fed with $^{15}NH_4$, than in those fed with nitrate. N assimilated from nitrate was only 62% of that obtained from ammonium (Table 9.7). Interesting though these results are, the immense genetic variability, even within races of *P. involutus* itself (Laiho, 1970) with respect to nitrate utilization must be borne in mind. The possibility remains that, where roots of trees such as *Fagus* grow in nitrifying environments, selection may favour colonization by fungi which preferentially use nitrate.

Organic N sources

Since Melin and Nilsson (1953a) demonstrated uptake and transfer of ^{15}N-glutamate by *Suillus granulatus* in ECM association with *Pinus*, there have been surprisingly few studies of amino acid utilization by intact ECM plants. Alexander (1983) fed aspartic acid and serine at concentrations of 0.5 mM to ECM and non-mycorrhizal *Picea sitchensis*. Whereas both N sources depressed growth of non-mycorrhizal seedlings, the inhibitory effect of aspartic acid did not occur in ECM plants and that of serine was reduced. Clearly, at high concentrations, toxicity of amino acids can be a problem for both fungus and plant. Accumulation of alanine, arginine and aspartic acid in ECM roots of *Pseudotsuga menziesii* and *Tsuga heterophylla* has been demonstrated in short-term uptake studies using colonized and uncolonized plants (Sangwanit and Bledsoe, 1987). Wallenda and Read (1999) carried out a comparative analysis of the uptake kinetics of the amino acids glutamine and glycine by a range of ECM types excised from pine, spruce and beech. All types took up amino acids via high-affinity transport systems. The kinetic parameters (Table 9.8) indicated that ECM roots have similar or even higher affinities (lower K_m values) for glutamine and glycine than for methylammonium. This confirms the potential of these organic forms to contribute significantly to total N uptake by ECM plants. The data further demonstrated that, as in the case of ammonium (see above), the considerable differences in the kinetic properties of amino acid uptake seen within a single species of plant are determined by the fungal partner.

It has been demonstrated that ECM colonization of birch (*B. pendula*) by *P. involutus* led to major changes in the metabolic fate of exogenously supplied amino acids (Blaudez *et al.*, 2001). The uptake of ^{14}C glutamate was increased by up to eight times, especially in the early stages of ECM formation. In addition, it was shown that glutamine was the major ^{14}C sink in ECM roots. In contrast, citrulline and insol-uble compounds were the major compounds to be labelled in non-mycorrhizal roots.

When grown with alanine or its peptides as sole N sources over a longer period of time, *Betula* showed striking responses to ECM colonization (Abuzinadah and Read, 1989a). In the absence of colonization, birch seedlings appeared to have no ability to use alanine and were N deficient. In contrast, when colonized by *Hebeloma crustuliniforme*, *Amanita muscaria* or *P. involutus*, the plants grew vigorously and their tissues contained N concentrations of the kind seen in healthy plants. When the plants were supplied with peptides of alanine of chain lengths from two to six units, again,

Table 9.8 Kinetic parameters for amino acid uptake by mycorrhizal roots collected along a European North/South gradient.

Tree species	Glutamine		Glycine	
	V_{max}	K_m	V_{max}	K_m
	Lactarius subdulcis			
Fagus sylvatica	13.3 ± 1.0	81 ± 14	18.1 ± 1.5	105 ± 18
Fagus sylvatica	11.6 ± 1.2	77 ± 18	14.5 ± 1.1	102 ± 16
Fagus sylvatica	12.4 ± 0.9	79 ± 13	22.8 ± 4.2	197 ± 67
	Russula ochroleuca			
Fagus sylvatica	5.3 ± 0.6	79 ± 21	4.4 ± 0.9	127 ± 49
Picea abies	n.a.	n.a.	4.8 ± 0.6	135 ± 30
Fagus sylvatica	1.4 ± 0.1	86 ± 20	3.1 ± 0.3	233 ± 35
	Unidentified morphotypes			
Pinus sylvestris	11.6 ± 1.6	130 ± 38	8.3 ± 1.0	50 ± 17
Picea abies	7.7 ± 1.4	36 ± 22	10.7 ± 1.8	42 ± 20
Picea abies	24.7 ± 3.1	104 ± 29	n.a.	n.a.
Picea abies	24.0 ± 6.7	111 ± 59	34.1 ± 5.6	113 ± 38
Fagus sylvatica	14.7 ± 1.6	19 ± 8	17.9 ± 1.7	21 ± 7

Values for V_{max} (μmol/g DW/h) and K_m (mmol/m^3) ($\pm$SE of the fitted parameters) were calculated by non-linear curve fitting of the experimental data to the Michaelis-Menten equation [$v = (V_{max} \times [S])/(K_m + [S])$] in a substrate concentration range of 0.001–0.25 mol/m^3; n.a. = not available. From Wallenda and Read (1999).

no growth was observed without colonization. However, growth and N concentration were greatly increased in ECM plants. There were some differences in the effectiveness of the three fungi, *H. crustuliniforme* giving higher yields and N contents than *A. muscaria* which was, in turn, more effective than *P. involutus*. Yields and N contents were, in general, higher on the peptides than on the amino acids and, in plants colonized by the two most effective fungi, the high values were obtained with the largest peptide units (Figure 9.7).

Plants colonized by those ECM fungi which can hydrolyse BSA and gliadin show increases in growth and N content when grown with these proteins as sole N source. Thus, yields of *Pinus contorta* colonized by *R. roseolus* or *S. bovinus* were significantly higher than those of non-mycorrhizal plants supplied with BSA and were similar to those grown with ammonium at the same N concentration. Colonization by *P. tinctorius*, shown in pure culture to have lower proteolytic ability, gave little access to the substrate and the plants grew less well (Abuzinadah *et al.*, 1986; Figure 9.8a, b). The 'protein fungus' *H. crustuliniforme* gave similar responses in *B. pendula*, *P. sitchensis* and *P. contorta*, while none of these species had access to protein N without colonization

Figure 9.7 Growth (G) and nitrogen contents (N) of roots (open bar), shoots (diagonally hatched bars) and whole plants (vertically hatched bars) of *Betula pendula* colonized by *Hebeloma crustuliniforme* (HC), *Amanita muscaria* (AM), *Paxillus involutus* (PI), or non-colonized (NM) grown with different amino acids and peptides as sole N sources for 75 days. Vertical bars represent least significant difference (LSD) for roots (R), shoots (Sh) and whole plants (Pl). (a) Alanine; (b) di-alanine; (c) penta-alanine; (d) hexa-alanine. From Abuzinadah and Read (1986b), with permission.

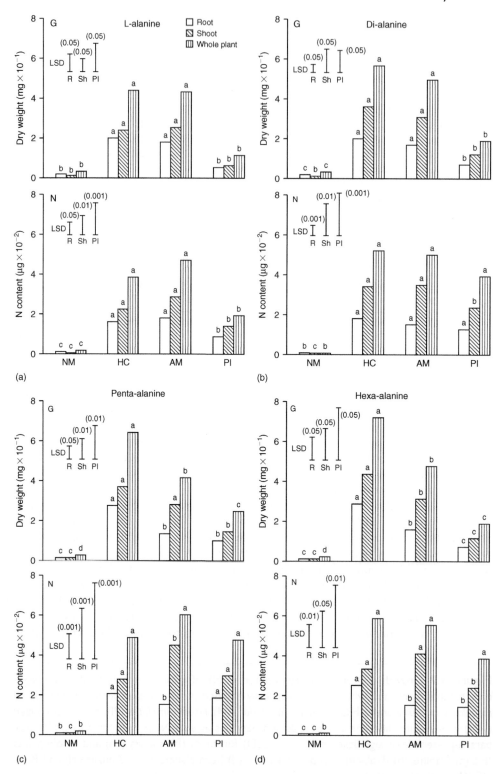

Figure 9.7 (Caption on p. 341)

(Abuzinadah and Read, 1986b). ECM colonization also had a major impact upon N utilization by *Eucalyptus* seedlings (Turnbull *et al.*, 1995). Both *E. grandis* and *E. maculata* used amino acids and protein when they were ECM, an ability lacking in non-mycorrhizal seedlings. Both species grew more than non-mycorrhizal plants on protein as well as on the amino acids arginine, asparagine and histidine, when colonized by an *Elaphomyces* spp. (Figure 9.9a). Although both species used protein less readily when colonized by a *Pisolithus* spp. (Figure 9.9b), the ECM plants were still significantly larger that their non-colonized counterparts. When *Betula* was grown monoxenically with selected mycorrhizal or saprotrophic fungi, colonization by *A. muscaria* and *H. crustuliniforme* gave highest yields and N content on protein N, while *P. involutus* was somewhat less effective as a symbiont (Abuzinadah and Read, 1988). In the presence of a seed-borne saprotroph, *Ulocladium botrytis,* or an ERM fungus, *Rhizoscyphus ericae,* no growth responses or net increases of N content of the *Betula* were observed. In contrast, inoculation with the saprotroph *Oidiodendron griseum,* while giving lower plant yields and total N contents than *A. muscaria* or *H. crustuliniforme,* did facilitate accumulation of N in the tissues of *Betula,* presumably because N became available via mineralization (Abuzinadah and Read, 1989b).

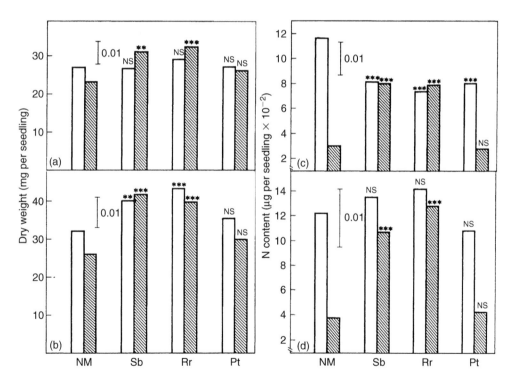

Figure 9.8 Dry weights (a and b) and nitrogen contents (c and d) of *Pinus contorta* colonized by *Suillus bovinus* (Sb), *Rhizopogon roseolus* (Rr) or *Pisolithus tinctorius* (Pt), or non-mycorrhizal (NM) after 40 (a and c) or 80 (b and d) days growth on ammonium (open bars) or bovine serum albumin (hatched bars) as sole N source. Vertical bars represent least significant difference (LSD) and asterisks indicate significant differences between mycorrhizal and non-mycorrhizal plants within each treatment. * $P < 0.05$; ** $P < 0.01$; *** $P < 0.001$; NS, not significant. From Abuzinadah *et al.* (1986), with permission.

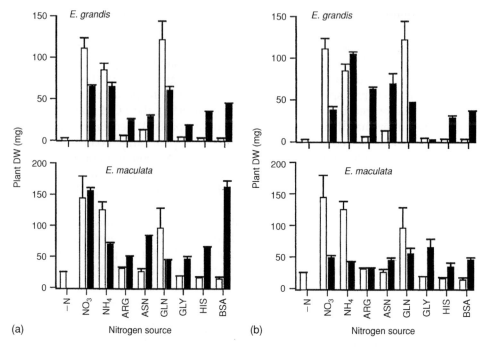

Figure 9.9 Growth of seedlings of *Eucalyptus grandis* or *E. maculata* in agriculture on a range of inorganic and organic N sources, either non-mycorrhizal (open bars) or ectomycorrhizal (closed bars) with (a) *Elaphomyces* spp. or (b) *Pisolithus* spp. Data are means of 12 replicate plants, ± standard errors of means. From Turnbull *et al.* (1995), with permission.

These results highlight one of the difficulties in extrapolating from experiments carried out under monoxenic conditions. In this case, a non-mycorrhizal saprotroph presumably converted protein to a form which could be absorbed and assimilated by the plant. In a mixed population of heterotrophs, such products are less likely to accumulate. Clearly, the extent of benefit accruing to the plant from its fungal associates will depend upon a combination of factors which include plant–fungus compatibility, the ability of the symbiont to compete with other soil or rhizosphere organisms for a resource and, having gained access to a substrate, its capacity to mobilize the nutrients which it contains. These factors can only be realistically evaluated in the presence of mixed natural populations of symbionts and saprotrophs, although under these conditions the mechanisms operating will be hard to unravel.

Dighton *et al.* (1987) approached this problem by examining the breakdown of the N-containing substrates powdered animal hide and chitin by ECM plants of *P. contorta*, grown in the presence or absence of the basidiomycete decomposer *Mycena galopus*. When roots of the plant were colonized by the ECM fungi *H. crustuliniforme* or *Suillus luteus*, significantly greater degradation of these substrates occurred, together with enhanced plant growth. In the presence of the saprotroph, the effectiveness of the ECM fungi in acquiring N was much reduced. However, interpretation of these results is complicated by the addition of readily available C to the medium, thus increasing the C:N ratio and probably enhancing, to an unrealistic extent, the ability of the saprotroph to compete for N in the organic substrates. Indeed, one of the advantages which all mycorrhizal fungi may have is that (in symbiosis) they are not dependent for C on organic substrates in soil. In consequence, the C:N (or C:P etc.) ratio of the substrates will be considerably less important for them than for saprotrophs.

Ultimately evaluation of the role of ECM fungi in mobilization of N under natural conditions can be achieved only by supplying colonized plants with the substrates which they would normally encounter in the field, in the presence of a natural population of soil microorganisms. Bending and Read (1995a) grew plants of *Pinus sylvestris* colonized by *S. bovinus* or *Thelephora terrestris* on non-sterile peat in transparent observation chambers. Organic matter freshly collected from the fermentation horizon (FH) of a pine forest was supplied in weighed aliquots on plastic trays. On reaching such introduced organic materials, the mycelia of the ECM fungi typically proliferate intensively to form dense 'patches' over the substrates (see Colour Plate 6.2c, d). After a standardized period of 'occupation' of the organic matter, the total N content of the material was measured and the quantities of N mineralized during the incubation period determined. There was no significant loss of N in the controls, whereas in the organic matter colonized by both fungi there was a significant (23%) depletion of N in material colonized by *S. bovinus* and 13% in that colonized by *T. terrestris* (Figure 9.10). These values are similar to those obtained by Entry *et al.* (1991b) who measured decline of N concentration in litter of *Pseudotsuga menziesii* colonized by mycelia of the mat-forming mycorrhizal fungus *Hysterangium setchelli* in the field. Over one year they observed losses of 32% of initial N from the litter. In an experiment involving supply of litter obtained from different tree species and using *Paxillus involutus* as the mycobiont of *Betula pendula*, it was confirmed that ECM colonization of the litter resulted in improved plant yield, but patterns of exploitation of the individual nutrients were different from those reported by Bending and Read (1995a) using different ECM fungal partners (Perez-Moreno and Read, 2000). In only one litter type was the loss of N more than 10% and, whereas whole seedling N content was increased by colonization, tissue N concentrations were not significantly different in ECM and non-mycorrhizal plants. In this experiment, litter P was exploited by the ECM plants more effectively than N. The different patterns of nutrient exploitation in the two experiments probably arose at least in part from the use of different fungal symbionts. Thus, apart from the broad conclusion that ECM colonization facilitates mobilization and removal of nutrients from organic residues of the FH, further generalizations from observations involving particular plant–fungal partnerships should be avoided.

Experiments examining plant litter as a potential source of N for ECM plants have been followed by studies in which individual components of typical boreal forest mycorrhizospheres are provided as substrates. Among these, pollen (Perez-Moreno and Read, 2001a), seeds (Tibbett and Sanders, 2002), nematodes (Perez-Moreno and Rea, 2001b) and collembolans (Klironomos and Hart, 2001) have been selected on the basis that, during parts of the growing season at least, the cellular contents of these materials represent large and rapidly turning over potential sources of nutrients (see Chapter 15). In the case of pollen, the N contained in weighed aliquots of the substrate was reduced by 76% when supplied to plants grown in symbiosis with *P. involutus*, whereas in non-mycorrhizal systems only 42% was removed, presumably by saprotrophs, over the same period. Whereas a large proportion of the exported N was found in the ECM plants, the non-mycorrhizal plants gained relatively small amounts

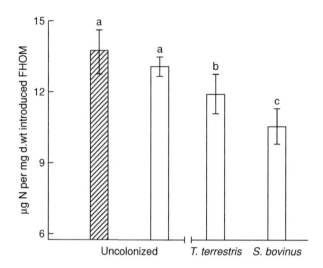

Figure 9.10 Utilization of N from litter collected from the FH horizon of forest soil. The hatched bar shows N content at the start of the experiment and open bars show N content after incubation, uncolonized or colonized by mycelium of *Thelephora terrestris* or *Suillus bovinus*, growing in symbiosis with *Pinus sylvestris*. FHOM = fermentation horizon organic matter. Treatments with different letters are significantly different at $P < 0.01$. From Bending and Read (1995a), with permission.

of the nutrient. When air-dried nematode cadavers were supplied as sole additional sources of N in the same way, 68% of the N was removed in the ECM systems and only 37% in those lacking a fungal symbiont (Perez-Moreno and Read, 2001b).

As yet we know little about the metabolic processes involved in the mobilization of N in naturally occurring organic residues. Their colonization by mycorrhizal fungi both in the field (Griffiths and Caldwell, 1992) and in microcosms (Bending and Read, 1995b) leads to significant increases of proteolytic activity. These observations, coupled with those showing proteolytic capability of the fungi in axenic culture and on colonized plants under aseptic conditions, are strongly suggestive of a direct role of ECM associations in mobilizing N from natural substrates, but do not provide definitive evidence of this. An alternative explanation is that the presence of ECM mycelium in the substrates somehow facilitates the activities of saprotrophs. What is evident is that the ECM fungi, by intensive colonization of the substrate and provision of organized mycelial aggregates connecting resource deposits to roots, are, in many cases, extremely effective as scavengers for, and transporters of, nitrogenous materials.

Conclusions

The ease with which many ECM fungi can be grown in axenic culture has enabled extensive screening of their abilities to use different forms of N. Most species readily use ammonium, nitrate and some simple organic-N compounds, although there are differences at both the inter- and intraspecific levels. Much new information has been gained concerning the biochemistry of N assimilation, in particular in relation to the enzymes involved in assimilation of ammonium which, in many ECM

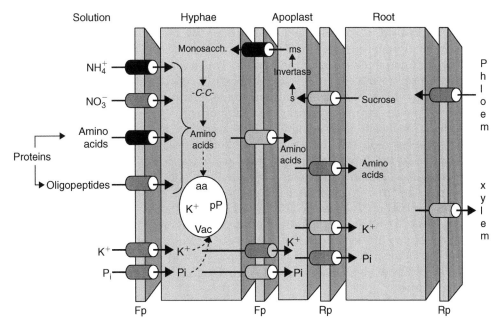

Figure 9.11 Diagrammatic representation of the current understanding of the location and function of nitrogen transporters in ectomycorrhizal tissues. Black cylinders represent structures in which at least one member of the transporter family has been fully characterized by functional complementation in a deficient strain of yeast. Dark grey cylinders represent putative transporters in which candidate genes have been identified as expressed sequence tags. Pale grey cylinders represent hypothetical transporters. The transporters putatively involved in carbohydrate transport, carboxylation of N compounds and in transfer of phosphorus (P) and potassium (K^+) ions are also shown. Fp, fungal plasma membrane; Rp, root plasma membrane; aa, amino acids; pP, polyphosphate; Vac, vacuole; s, sucrose; ms, monosaccharide; cc, carboxylation. Modified from Chalot *et al.* (2002), with permission.

fungi, appears to be the preferred source of inorganic N. These studies have been extended to allow analysis of N assimilation by the fungi when grown in symbiosis with plants and, importantly, to investigate the influence of the plants themselves upon the pattern of events. Progress has been made towards characterization of the processes of nitrate and ammonium uptake and transport at the molecular levels. This has enabled a better understanding of the mechanisms and pathways whereby uptake of mineral N sources takes place in ECM symbioses.

Increasingly, it has been recognized that much of the N contained in the superficial layer of soil occupied by ECM roots is in organic form and that some ECM fungi have access to these more complex N sources. Emphasis has turned on the one hand to characterization of the primary sources of organic N in forest soils, with consideration of the extent to which these are accessible to ECM fungi and, on the other, to consideration of the molecular and biochemical events involved in the uptake, assimilation and transport of organic N compounds.

The molecular studies, mostly using *Amanita muscaria* (Nehls *et al.*, 1999) or *Hebeloma* spp. (Javelle *et al.*, 2001; Wipf *et al.*, 2002; Selle *et al.*, 2005; Wright *et al.*, 2005;

Benjdia *et al.*, 2006), by revealing N dependent expression profiles of N importer genes, have enabled us to envisage how N uptake by the fungi and transfer to the plant may occur in nature (Figure 9.11). Expression of these genes is likely to be regulated by the internal N status of the hyphae. This will be relatively reduced when they are exposed to the essentially low ambient N concentrations of the soil solution, but higher in the N-accumulating mantle tissues surrounding the root. Under these circumstances strong expression of N importers in the soil compartment can be expected. Conversely, their activities will be repressed in the hyphae of the Hartig net adjacent to the root. This repression would inhibit reabsorption of N compounds by the fungus at the fungus–plant interface and help to facilitate the export of N either in the form of amino acid or ammonia, to the plant.

It is important to recognize that the pathways and processes involved in the import and transfer of N across the ECM interfaces are intrinsically linked to those by which carbon is transferred in the reverse direction (Figure 9.11 and see Chapter 8). In addition, the ability to capture and assimilate N will be influenced by the availability of the other macronutrients, phosphorus (P) and potassium (K). The processes whereby these elements are acquired and transported are considered in the following chapter.

10

Phosphorus and base cation nutrition, heavy metal accumulation and water relations of ectomycorrhizal plants

Introduction

The early view, effectively expounded by Frank (1894), that ectomycorrhizas were especially important in absorption of nitrogen (N) held sway for much of the early part of the twentieth century. Its prominence was weakened by the seminal paper of Hatch (1937), who demonstrated that mycorrhizal colonization in *Pinus strobus* led to increased concentrations of phosphorus (P) and potassium (K), as well as of N (Table 10.1). Hatch was of the opinion that the importance of ectomycorrhizas lay in their ability to increase the uptake of any nutrient in short supply. This broad view of ectomycorrhizal (ECM) function was also espoused by others, who carried out analyses of *P. strobus* (Mitchell *et al.*, 1937; Finn, 1942). However, some contemporary studies (e.g. McComb, 1938; McComb and Griffith, 1946; Stone, 1950) indicated that absorption of P was more enhanced than that of other nutrients. The publication of these findings was followed within a few years by the first commercial production of ^{32}P, which greatly improved the precision with which movement of P could be traced in biological systems. Its availability led to a significant shift of emphasis towards study of P nutrition of ECM plants and studies of ^{32}P uptake were carried out by Kramer and Wilbur (1949), Harley and McCready (1950) and Melin and Nilsson (1950). These set a trend of work on P uptake and nutrition of ectomycorrhizas which was performed almost to the exclusion of work on uptake of the other elements until the 1980s (see Harley and Smith, 1983).

While these pioneers laid the foundation for much elegant research on the uptake of P by whole plants, the broader view of ECM function espoused by Hatch (1937) and by Harley and Smith (1983) is the one that more closely reflects the cumulative results of research. There are, as discussed in Chapter 15, many circumstances in nature where P deficiency is clearly the primary limitation on productivity of plants

Table 10.1 Growth and specific nutrient uptake of nitrogen, phosphorus, potassium by *Pinus strobus* seedlings.

Degree of mycorrhizal infection	Dry weight (mg)	Root weight (mg)	Nitrogen		Phosphorus		Potassium	
			T	SA	T	SA	T	SA
Mycorrhizal (+)	448	180	5.39	0.030	0.849	0.0047	3.47	0.019
	361	170	4.62	0.027	0.729	0.0042	2.57	0.015
Uninfected (0)	300	174	3.16	0.013	0.229	0.0013	1.04	0.006
	361	182	2.51	0.018	0.268	0.0015	1.94	0.011
	301	152	2.40	0.016	0.211	0.0014	1.17	0.008

Data from Hatch (1937). T, total absorbed mg. SA, specific absorption mg/mg root dry weight.

and in which, therefore, ECM colonization may be of particular importance for P nutrition. These are the systems in which further work on P nutrition is still necessary. In those even more widespread areas where elements other than P limit plant growth, work on these nutrients should be emphasized. Whereas the balance has been redressed in this regard in the case of N (see Chapter 9), until recently, relatively few have followed Hatch in considering the possible role of ECM colonization in the capture of K and analyses of the other nutritionally important base cations, magnesium (Mg) or calcium (Ca), have been equally scarce. Fortunately, as described later in this chapter, there has now been some upsurge of interest in these elements. Increasing use of the stable isotopes of potassium (^{41}K), magnesium (^{25}Mg) and calcium (^{44}Ca) to trace the movement of these ions through extraradical mycelia and across the ECM mantle is providing new insights.

Uptake of P by excised ECM and non-mycorrhizal roots

Much of the detailed experimental work on the mechanism of uptake of nutrients, particularly P, by ectomycorrhizas has been done with excised roots in which both the external mycelial system in the soil and the throughput of water are eliminated. The overwhelming reason for using excised mycorrhizas for investigating certain aspects of mycorrhizal physiology is that uniform samples can be obtained for studying specific aspects of the uptake processes. Whole root systems are composed of mycorrhizas, uncolonized primary roots and secondarily thickened axes in different proportions so that it is almost impossible to conclude anything about detailed mechanisms in ECM roots from them. It is easy to obtain large numbers of similar ECM roots from the surface layers of forest soil, wash and prepare them with no great effort. Of course, excised mycorrhizas and non-mycorrhizal roots can also be obtained from aseptically grown plants in the laboratory, but the labour of providing them in sufficient quantity for experimental work on a large scale is daunting.

The criticisms levelled at the use of excised mycorrhizas are those applicable to all experiments with excised roots, i.e. that the transpiration stream is eliminated and the tissue may become starved of organic C during the experiment. In the case of ectomycorrhizas, there is the additional problem that the root is detached from what is increasingly seen to be the critical absorbing system, the extraradical mycelium. Setting these problems aside for the moment, the factors that affect the rate of absorption of nutrients by excised ectomycorrhizas are similar to those which affect the rate of absorption by most plant material including roots. This was an important contribution from the work of Harley's group using *Fagus* roots most probably colonized by *Lactarius subdulcis* (see Colour Plate 6.1b) (e.g. Harley and McCready, 1952a, 1952b; Harley *et al.*, 1953, 1954, 1956; Harley and Jennings, 1958; Harley and Wilson, 1959; Carrodus, 1966 and see Harley and Smith, 1983) and served to focus attention on the fact that nutrient absorption by mycorrhizas was more relevant to the natural situation than work with non-colonized roots. In the ensuing account, the aim will be to consider the manner in which the absorption physiology of ectomycorrhizas differs from that of non-colonized roots. The work with excised mycorrhizas was predicated on the idea that the mantle tissue that covers the root resembled the extraradical mycelium and that uptake characteristics of the mantle could be extrapolated in developing ideas of nutrient uptake by hyphae from soil. As we have seen from discussions of the changes in enzyme activities in mycelium and mantle (see Chapter 6), this assumption must be viewed very much as a first approximation.

The development of the fungal mantle, as well as the activity of the extraradical mycelium, has important impacts on the duration of active absorption by a root system. The most active non-colonized root apices differ from mycorrhizas in that they are dividing and growing. It is well known that rate of uptake in the tip region of a growing root is much greater than that in the region behind it. This is not generally true of ECM apices. McCready (unpublished) found that the uptake of P by *Fagus* mycorrhizas did not change greatly over distances of 12 cm. However, the rate of absorption of ECM and non-mycorrhizal roots may be very different as shown in Table 10.2 (Harley and McCready, 1950; Bowen and Theodorou, 1967). The differences were also emphasized by the autoradiographs of Kramer and Wilbur (1949) and also of Harley and McCready (1950) that show intense P accumulation in the mycorrhizas and in the extreme apices of non-colonized roots, but not elsewhere. Bowen (1968) scanned the long roots of *P. radiata* with a Geiger counter, following feeding of ^{32}P and showed that, whereas the most active region of accumulation in uncolonized roots was at the apex and in the positions of the apices of developing short roots, that of long root bearing mycorrhizas was at its apex and more particularly at the positions of the ECM rootlets.

Harley and McCready (1952a) showed, using ^{32}P, that the exposure of excised roots to a bathing solution resulted in greater accumulation of P in the fungal mantle. By dissecting the fungal layer from the plant tissue (the core; see Figure 8.2c), they were able to estimate the relative quantities accumulated in the two symbionts. This method was later used for more detailed analyses, in particular of C (see Chapter 8) as well as P fractions and other nutrients. In the case of P, about 90% of that absorbed was found in the sheath after uptake from low concentrations. This observation has since been often confirmed. Harley and McCready (1952a) verified that the accumulation was not dependent upon, nor influenced by, excision. They compared the distribution of P between mantle and root (the core) tissues in mycorrhizas attached to adult trees in the forest and those detached. Comparisons were made on three occasions: when the trees were leafless, when developing their leaves and when in full leaf. On all occasions, there was a great accumulation of P in the sheath (Table 10.3), although the extent to which P was absorbed within rather than adsorbed onto the surface of the sheath was not determined.

These observations of Harley and McCready (1952a) extended only up to July and suggested little seasonal variation in the rate of removal of P from solution. It

Table 10.2 Comparative uptake rates of phosphate by excised mycorrhizas and non-mycorrhizal roots.

Authors and host	Fully infected	Uninfected	Sheath poorly developed
Harley and McCready (1950)	5.18	0.88	4.76
Fagus sylvatica	6.68	0.75	1.20
	1.97	0.42	0.62
	2.72	0.61	–
	1.69	0.72	2.13
Bowen and Theodorou (1967)	7.5	3.5	–
Pinus radiata	15.5		
	15.0	5.5	5.5

The values for each host are relative to one another.

Table 10.3 Estimates of the proportion of phosphorus which accumulates in the fungal mantle of _Fagus_ mycorrhizas when attached to the parent root system or when excised. Experiments in Bagley Wood with roots of adult trees at three seasons. Mycorrhizas in aerated phosphate solution pH 5.5 at ambient temperature.

Condition of mycorrhizas	Attached				Detached			
Date	31 March	11 May	23 July		31 March	11 May	23 July	
mM KH$_2$^{32}PO$_4$	0.074	0.32	0.16	1.6	0.074	0.32	0.16	1.6
Mean percentage in sheath	88	88	90	87	91	89	85	91
Range	74–96	83–94	89–94	86–93	85–96	79–93	83–89	91–93

Data from Harley and McCready (1952a).

has since been shown that the ectomycorrhizal roots of _Abies balsamea_ (Langlois and Fortin, 1984) and _Picea sitchensis_ show a distinct seasonality in their ability to absorb P, maximum rates being achieved in the late summer or autumn after completion of shoot extension growth. Further, it must be borne in mind that the patterns of accumulation described by Harley and McCready (1952a, 1952b) were in all cases obtained using ECM roots from which the extraradical mycelium had become detached. When the kinetics of P absorption by these roots are compared with those seen in intact mycorrhizal systems (van Tichelen and Colpaert, 2000), it is observed that the rates achieved by the excised roots are considerably lower (Figure 10.1 and see below). While differences of experimental conditions and in the physiology of the mycobionts involved must contribute to such effects, there is every likelihood that the removal of the mycelial network strongly influences the observed rates of P uptake.

The results of Harley and McCready (1952a, 1952b) were obtained by dissecting roots following a period of exposure to solutions containing orthophosphate. Two features must be noted. First, as external solution is applied to the mantle, accumulation can appear to take place as P is adsorbed onto its surface or as it passes through the fungal tissue to the core within. Second, there may be a real accumulation depending upon the species of fungus and its activity. Garrec and Gay (1978) analysed the mycorrhizas of _Pinus halepensis_ using an electron probe and concluded that P is mainly accumulated in the fungal mantle and Hartig net region and is

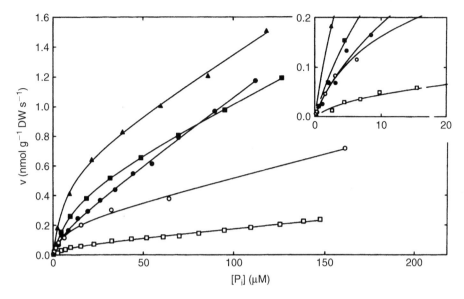

Figure 10.1 Representative Michaelis-Menten plot showing the net orthophosphate (P_i) uptake in intact mycorrhizal and non-mycorrhizal *Pinus sylvestris* seedlings as a function of solution P_i concentrations (0.2–160 μM). The inset figure shows the relationship at the lower P_i concentrations most representative for soil solution. Measurements were performed 9 weeks after inoculation and data points were fitted to a two-phase Michaelis-Menten equation. ▲ *Paxillus involutus*; ■ *Suillus bovinus*; ● *Thelephora terrestris*; □ non-mycorrhizal. For comparison data obtained with excised *Fagus sylvatica–Lactarius subdulcis* mycorrhiza (○) are also plotted (Harley and McCready, 1952b; with permission). From van Tichelen and Colpaert (2000).

lower in the plant tissue. Since the mantle, whether it accumulates nutrients permanently or acts as a temporary store, separates the plant tissue from the soil, the mechanism of the passage of substances through it and factors affecting their rate of transfer to the plant require investigation.

The absorption of orthophosphate into excised mycorrhizas of *Fagus* results in its immediate incorporation into nucleotides and sugars (Harley and Loughman, 1963). Both separated fungal mantle and plant core exhibit rapid incorporation of applied ^{32}P but, in intact mycorrhizas, the core tissue receives less P and appears more sluggish in incorporating orthophosphate into other compounds, with only about 10–20% of that absorbed passing steadily into the core tissues through the mantle. By studying the time course of esterification of the P entering the root core, Harley and Loughman were able to show that the labelling of orthophosphate represented 100% of the radioactivity in that tissue initially and that proportion fell, as nucleotides, sugar phosphates and other fractions became labelled. Since both the mantle and the plant tissues showed the same labelling pattern of soluble P compounds if allowed separately to absorb orthophosphate from radioactive solutions, it was concluded that in the intact excised mycorrhizas orthophosphate was the form which passed from the mantle to the plant when low concentrations were applied externally.

Harley and McCready (1952b) and Harley *et al.* (1958) studied the possible routes by which P might pass through the mantle from the external solution to the plant.

They showed first that the mantle prevented the plant from absorbing P at its maximum possible rate, except when very high concentrations were present in the solution. They concluded that from low P concentrations, such as might be expected in the soil, diffusive movement through the mantle did not take place at a significant rate. It might, however, occur at concentrations above about 1 mM, which are totally unrealistic ecologically. The P passing through the sheath to the plant root did not equilibrate with a large part of the P in the fungus. If it did so, as Harley *et al.* (1954) showed, a lag phase in the arrival of P in the core tissue would be expected and the quantity of P in the pathway to the core would be related to the length of the lag phase. Using low external P concentrations, the lag phase was exceedingly short, so that the quantity of P in the mantle with which the passing inorganic P equilibrated was small – of the order of 0.017 μg P per 100 mg dry weight of mycorrhizas. This is extremely low compared with the amount of P present in the sheath and it was concluded that orthophosphate is incorporated first into the metabolic pools of the fungal symplast in the sheath and that these constitute a small proportion of the total P, as they do in other roots (Crossett and Loughman, 1966).

Following uptake, ECM fungi, both in culture and in symbiosis, synthesize polyphosphate (polyP), which is stored in the vacuoles, thus maintaining relatively low cytoplasmic Pi concentrations. In ectomycorrhizas, some of the P in the extraradical mycelium and the sheath is certainly present as polyP. This was pointed out by Ashford *et al.* (1975) using *Eucalyptus* and subsequently shown to be true of ectomycorrhizal *Pinus radiata*, of the arbutoid mycorrhizas of *Arbutus unedo* and of the arbuscular mycorrhizas of *Liquidambar styraciflua* (Ling Lee *et al.*, 1975). Using the same cytochemical methods as Ling Lee *et al.* (1975), Chilvers and Harley (1980) described particles believed to be polyP in the sheath of *Fagus* mycorrhizas. The number and size of particles increased during P absorption at rates similar to the rate of absorption of P from similar concentrations and similar factors affected their formation as affected P uptake. At this stage it was assumed that polyP was stored as long-chain, insoluble granules, a view that has subsequently been re-evaluated (see below). The formation of polyP in *Fagus* mycorrhizas was further examined by Harley and McCready (1981) using the method of extraction and precipitation described by Aitchison and Butt (1973). Assuming that there was little hydrolysis in the extraction and that the precipitation with BaCl₂ was complete, it was concluded that a large amount of the P accumulated in the mantle tissue is polyP. More recent work has confirmed the location of the polyP to be in the fungal vacuoles (Ashford *et al.*, 1994; Ashford and Orlovich, 1994; Gerlitz and Gerlitz, 1997; Bücking and Heyser, 1999), the pH values of which are inherently low (Rost *et al.*, 1995). Much early work suggested that the polyP occurred as granules, stabilized by Ca^{2+} or arginine (e.g. Ashford *et al.*, 1975; Ling-Lee *et al.*, 1975). Hypotheses were formulated involving these polyP granules in both storage and translocation in fungal hyphae and in the mantle tissues of ectomycorrhizas. More recently, it has been suggested that the granules may be artefacts of specimen preparation (Orlovich and Ashford, 1993; see Figure 10.2) and hypotheses depending on their behaviour must be re-evaluated.

There were early indications from NMR spectroscopy (BC Loughman, personal communication, see Harley and Smith, 1983) that at least part of the polyP in the fungal mantle of ECM of *Fagus* occurs as relatively short-chain molecules. The data of Martin *et al.* (1985), again using ^{31}P NMR, also indicated a large, soluble polyP fraction in the mycelium of *Cenococcum geophilum* and *Hebeloma crustuliniforme*.

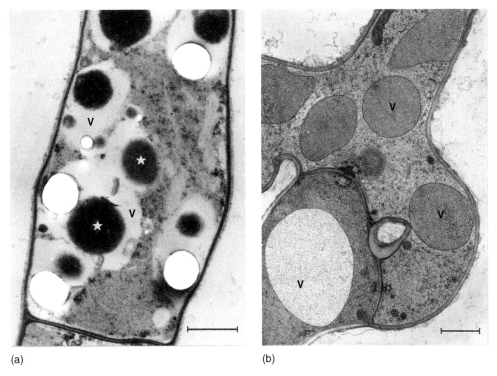

(a)   (b)

Figure 10.2 Effect of fixation methods on the preservation and distribution of polyphosphate in *Pisolithus tinctorius*. (a) Transmission electron micrograph of a glutaraldehyde-fixed hypha embedded in Spurr's resin and triple stained. Note the spherical opaque granules (★) in vacuoles (V). (b) Transmission electron micrograph of a stained ultrathin section of part of a freeze-substituted hypha near a clamp connection. The vacuoles (V) contain electron-opaque material which is evenly dispersed. There are no discrete granules. Bars = 1 μm. From Orlovich and Ashford (1993), with permission.

PolyP present either in chains longer than about 70 orthophosphate residues or as insoluble precipitated material (granules) cannot be detected by NMR spectroscopy. Importantly, the behaviour of polyP in the fungal mycelium revealed by NMR was similar to that shown by counting the (possibly artefactual) granules in excised mycorrhizal roots (Chilvers and Harley, 1980; Strullu *et al.*, 1981a, 1981b, 1982). PolyP accumulation varied with different stages of growth, being low when growth was rapid in young mycelia and linear in the early and late stages of the stationary phase when P in the medium was relatively abundant compared with N. When the mycelium was transferred to low P medium, the NMR spectra indicated mobilization of polyP, rapidly in *H. crustuliniforme* and more slowly in *C. geophilum*. Martin *et al.* (1985) discussed the apparently conflicting evidence on the form of polyP, concluding that the spin-lattice relaxation times of the ^{31}P nuclei in both fungi were consistent with a single pool of relatively fluid poly-P, possibly in the form of 'macromolecular aggregates'. They certainly found no evidence for granules and considered the investigation of purified granules by NMR spectroscopy to be 'particularly urgent'! Orlovich and Ashford (1993) obtained data from anhydrous freeze substituted material, indicating that, in *P. tinctorius*, polyP is uniformly distributed in the fungal vacuoles and is stabilized by K^+ (Figure 10.3). PolyP of about 15 ortho-phosphate units was extracted from the mycelium and identified by chromatography, gel electrophoresis and ^{31}P NMR. This investigation also illustrated the formation of granules stabilized by Ca^{2+} during chemical fixation, explaining the widespread observation of these 'structures'. However, X-ray microanalytical analysis of thin

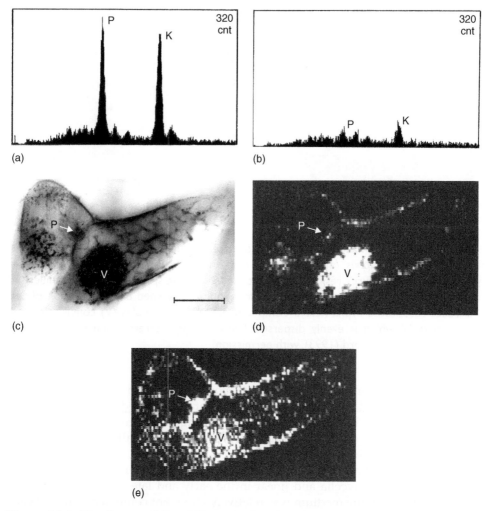

Figure 10.3 Distribution of P and K in a hypha of *Pisolithus tinctorius*. (a) and (b) Energy-dispersive X-ray spectra from the analysis of an unstained, freeze-substituted hypha cut dry at 1 μm. The full vertical scale, measured in X-ray counts, is at the top right of each spectrum. Vacuole showing large peaks for P and K shown in (a) and cytoplasm, showing relatively small peaks for P and K, shown in (b). (c) Transmission electron micrograph of a hypha near a clamp connection. Note the large vacuole (V). Bar = 2 μm. (d) and (e) X-ray maps of the hypha in (c), showing the similar distribution of P in (d) and K in (e) in the vacuole. From Orlovich and Ashford (1993), with permission.

sections of tissues prepared without such fixation (Bücking and Heyser, 1999) indicates that polyP granules are visible in the hyphae of ECM fungi. The matter of the state (fluid or solid) in which polyP occurs is therefore still a matter for debate.

PolyP is located in fungal vacuoles and in the tubular cysternae of ECM fungi (Orlovich and Ashford, 1993; Ashford *et al.*, 1994), all of which compartments are acidic. In consequence, the polyP carries a strong negative charge which must be balanced by association with cations. Positively charged nitrogen compounds, such as the basic amino acid arginine, may play a role as neutralizing agents, but microanalytical studies have provided evidence of the direct association of vacuolar P with K^+ and Mg^{2+} ions (Orlovich and Ashford, 1993; Bücking and Heyser, 1999).

Ashford (1998) has hypothesized that the fungal vacuoles occur as tubular systems through which the longitudinal translocation of polyP takes place (Figure 10.4). If this is the case, strong coupling of polyP with K and Mg would inevitably involve co-transport of these cations with P in the direction of the plant (see below). The compartmentation of polyP would enable a great part of the P in the mantle to be separated from the labile P that can move to the plant tissue.

The mechanisms of transfer of P from the ECM partner across the interfacial apoplast into the plant roots continues to be a subject of experiment. Harley and Loughman (1963), through short-term labelling experiments with excised roots of *Fagus*, showed that ^{32}P-labelled orthophosphate passed from fungus to host. The electrochemical potential difference between the Hartig net and apoplast will favour passive efflux from the fungus into the apoplast. However, rates of P loss by fungal hyphae are generally rather low, so that mechanisms which promote efflux and, at the same time, reduce retrieval or reabsorption, seem likely to operate at the interface. Uptake by the plant symbiont must be active, as the cells accumulate P against a strong electrochemical potential gradient. Recent work has identified transporter genes that may be implicated in P uptake by the fungal partner from soil, as well as plant genes potentially involved in transfer at the symbiotic interface. Several low- and high-affinity transporters have been identified in both *Laccaria bicolour* and *Hebeloma cylindrosporum*, but it is not yet known what roles they may play in P uptake from soil or redistribution in fungal tissues. On the plant side, 13 homologues of high-affinity Pi transporters in the Pht1 family have been identified in *Populus*. One of these is preferentially expressed in ECM root tips. It is presumed, but not yet demonstrated, that expression will be localized in the plant cells adjacent to the Hartig net (Martin, 2007). There have been suggestions that the supply of P to plants may be quantitatively linked to loss of sugars to the interfacial apoplast. However, as yet no mechanisms that relate influx of P to efflux of sugars have been identified.

Bücking and Heyser (2000) showed that the translocation of P across the ECM interface was regulated by the orthophosphate concentration in the cytoplasm of the Hartig net and by the efflux into the interfacial apoplast. These authors stress that the relationships between P supply to the mycorrhiza and transfer to the host are strongly dependent upon fungal species. Whereas in mycorrhizas formed by *Suillus bovinus* an increased P supply to *Pinus sylvestris* roots had no effect on translocation of the element across the ECM interface to the plant, a supply-dependent effect was observed in mycorrhizas formed by *Pisolithus tinctorius*. Clearly, caution must be exercised when attempting to generalize from patterns seen in one mycorrhizal system to all others.

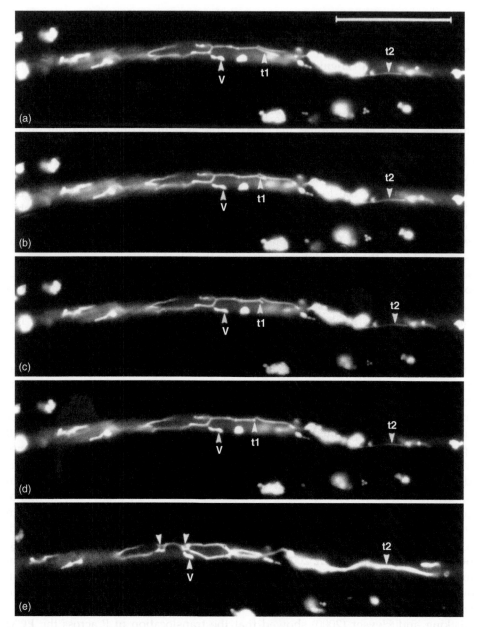

Figure 10.4 Time-lapse photomicrographs (a)–(e) taken at 4 second intervals in the same focal plane showing movements of the tubular reticulum in the apical cell of a hypha of *Pisolithus tinctorius* following loading with carboxyfluorescein. A vacuole (v) remains connected to the reticulum via a narrow fluoresent bridge throughout. The sequence shows consecutive frames of the retraction of a tubule (t1) that constitutes a branch of the reticulum. At some branch point (arrowheads) a plaque-like ring structure is seen surrounding a non-fluorescent area. The tubule (t2), which remained very fine initially (a) to (d) becomes obliterated (e) by a broader tubule that is confluent with the fluorescence of the vacuole cluster. From Shepherd *et al.* (1993b), with permission.

Phosphate absorption in intact plants

Following uptake, the slow transfer in excised roots is consistent with data from experiments on whole plants of *Pinus radiata* (Morrison, 1957a, 1962a), in which seedlings were grown at two levels of P in pots in the greenhouse for 17 weeks. With high P, the non-mycorrhizal plants grew somewhat better than the ECM ones, but the reverse was true with low P. Plants of each kind were then grown with fresh supplies of P, labelled with ^{32}P. In all experiments the movement of ^{32}P to the shoot tip of the non-mycorrhizal plants was rapid at first, but later decreased in rate and almost ceased, but could be increased again by further addition of P to the soil. Movement to the shoot tips of ECM plants, although much slower than the initial rate in non-mycorrhizal plants, continued steadily for weeks and was little affected by further additions of labelled P to the soil. The accumulated ^{32}P in the shoot tips of ECM plants eventually exceeded that of the non-mycorrhizal controls. If ECM and non-mycorrhizal plants were deprived of P after a period of uptake of ^{32}P, radio-activity continued to pass to the shoots of ECM plants for three weeks, but ceased to move to those of non-mycorrhizal plants after only a short period. This behaviour is explicable in terms of the accumulation of P in the fungal mantle, coupled with a steady rate of transfer to the plant when P is available. When P supplies are deficient, mobilization of the stored P in the mantle occurs. This type of study is instructive in terms of distribution of P in the plant, but tells us little about the processes of P capture from soil by the extraradical mycelium.

Stone (1950) compared two samples of seedlings of *P. radiata* with very different development of extraradical hyphae. Those with a more extensive system absorbed ^{32}P from the soil faster and translocated a greater quantity to their needles. Similarly, Melin and Nilsson (1950) showed that ^{32}P orthophosphate fed to the extraradical mycelium of *P. sylvestris* was translocated to the root by the hyphae and thence through the plant to the needles. Skinner and Bowen (1974a, 1974b), using *P. radiata* and *Rhizopogon luteolus*, confirmed the transport of P in ECM rhizomorphs. P absorption by rhizomorphs was inhibited by cyanide and was temperature dependent and subsequent translocation occurred over distances of up to 12 cm. However, there were large differences in extent of mycelial growth in the soil between strains of fungus and between samples of the same fungal isolate in different soil conditions (Skinner and Bowen, 1974b). Experiments such as these emphasize the need for the extent of production of extraradical hyphae and rhizomorphs to be fully described in experiments on the efficacy of different combinations of fungal strain and plant genotype. The differences between the absorptive capabilities of mycobionts has been emphasized in studies using *P. sylvestris* seedlings grown in the non-mycorrhizal condition or with a number of commonly occurring ECM fungi (Colpaert *et al.*, 1999; van Tichelen and Colpaert, 2000). At an external Pi concentration of 10 μM, ECM seedlings colonized by *Thelephora terrestris* and *Paxillus involutus* achieved P uptake rates that were, respectively, 2.5 and 8.7 times higher than those of their non-mycorrhizal counterparts. Positive correlations were found between P uptake rates and the biomass of the external mycelium of each mycobiont, measured by ergosterol assay. In detailed kinetic analyses, van Tichelen and Colpaert (2000) further emphasized the importance of mycobiont effects as determinants of the ability of *P. sylvestris* to capture P. Net P uptake was dependent upon concentration and was governed by Michaelis-Menten kinetics. However, a dual uptake process consisting of high- and low-affinity systems operating simultaneously was revealed. Representative Michaelis-Menten plots (see Figure 10.1) confirmed the large differences between mycobionts in net uptake of P and again demonstrated the effectiveness of *P. involutus* in this regard. Comparisons of P uptake rates in four mycobionts of *P. sylvestris* with those calculated from the data of Harley and McCready (1952a) for excised *F. sylvatica-Lactarius subdulcis* roots (see Figure 10.1 inset; Table 10.4) indicate that, whereas the high- and low-affinity systems are present in both excised and intact roots, the V_{max} values, as well as the overall P uptake rates, are higher in three of the pine mycobionts.

Finlay and Read (1986b) used autoradiography to examine the uptake of ^{32}P by the extraradical mycelium of *Suillus bovinus* and its transport to seedlings of *Pinus* spp. which were interlinked by the fungus (Figure 10.5a). A seedling of *P. sylvestris* colonized by the fungus was first introduced to an observation chamber containing non-sterile peat and the system was incubated for a period sufficient to enable the mycelium of *S. bovinus* to colonize both the peat and series of previously uncolonized seedlings of *P. contorta*. ^{32}P-orthophosphate fed at about 30cm from the seedling roots was absorbed by the fungus over a period of 72h and was translocated throughout the peat and to the ECM roots of all the plants interlinked by the fungus. ^{32}P accumulated in ECM roots in a pattern that would be predicted from the studies of excised roots described above. In some seedlings, onward transfer of ^{32}P to the shoots took place in the same period. Distribution of ^{32}P in the extraradical mycelium in the peat was irregular. It was clear that the rhizomorphs provided the main pathways for long-distance translocation, but there was also directional transport toward the actively growing hyphae at the advancing mycelial front and into patches of dense mycelium (Figure 10.5b).

The extraradical component of ectomycorrhizas is extremely important in colonizing the soil and may play a role similar to that described by arbuscular mycorrhizal (AM) hyphae (see Chapters 5 and 6). However, both the extent of development of

Table 10.4 Kinetic parameters of the high-affinity uptake system in intact mycorrhizal and non-mycorrhizal *P. sylvestris* root systems.

Harvest	Inoculation treatment	No. of plants	No. of points	Goodness of fit (mean) R	Deviation from model (Runs test) P	$K_m(\mu M)$	V_{max} (nmol/ g/s)
Week 7	*Paxillus involutus*	3	13	0.998	NS (0.73)	3.5	0.57
	Suillus bovinus	3	15	0.999	NS (0.52)	7.5	0.49
	Thelephora terrestris	3	20	0.997	NS (0.81)	8.7	0.13
Week 9	*Paxillus involutus*	3	9	0.999	NS (0.89)	5.9	0.62
	Suillus bovinus	3	10	0.999	NS (0.64)	10.2	0.52
	Thelephora terrestris	3	16	0.999	NS (0.81)	7.3	0.15
	Non-mycorrhizal	4	17	0.995	NS (0.11)	12.1	0.08
	Lactarius subdulcis		12	0.998	NS (0.99)	6.4	0.21

Data from van Tichelen and Colpaert (2000). NS, not significant. The K_m and V_{max} values were calculated by iterative fitting of the data to Equation (3) (sum of the two Michaelis-Menten terms); mean and range (between brackets) are shown. The average number of points used for this procedure is provided as well as the number of plants analysed. Kinetic parameters obtained with excised *F. sylvatica–L. subdulcis* mycorrhizas are included for comparison (calculated for the 0.2–160 μM P_i range, from Harley and McCready, 1952b).

the ECM mycelium and the abilities of some of the fungi to utilize sparingly soluble organic P sources are generally greater in ECM than in AM symbiosis (Marschner, 1995; George and Marschner, 1996; see Chapter 5). The hyphae not only extend beyond any zone depleted of nutrients near the surface of the mantle, but also may readily be extended or replaced with a small expense of C and nutrients per unit area of absorbing surface. Furthermore, the hyphae proliferate in microsites in the soil and exploit the resources in them (see Chapter 6). Laboratory studies (Wallander and Nylund, 1992; Ekblad *et al.*, 1995; Jentschke *et al.*, 2001a) have shown that proliferation of the mycelial system of ECM fungi can be greatly stimulated by the addition of P to otherwise P-deficient substrates. These observations, taken in association with those of Häussling and Marschner (1989) showing a linear relation-ship between the lengths of extraradical mycelial systems and phosphatase activity (Figure 10.6), are indicative of a close coupling between P supply to the plant and the extent of foraging for the element in soil.

Rousseau *et al.* (1994) quantified the difference in potential absorbing surface area between seedlings of *P. taeda* colonized by the fungi *Pisolithus tinctorius* and

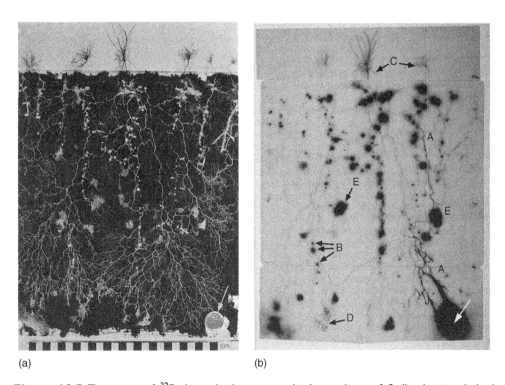

(a)   (b)

Figure 10.5 Transport of ^{32}P through the extraradical mycelium of *Suillus bovinus*, linked to seedlings of *Pinus sylvestris* and *P. contorta*. (a) Root observation chamber showing the mycelial connections between the plants and the site of feeding with ^{32}P in half-strength Melin-Norkrans medium (arrowed). (b) Autoradiograph of the same chamber showing the distribution of ^{32}P after 82 hours. Label has accumulated in the rhizomorphs (A), myc-orrhizal roots (B) and the shoots (C). There is also some accumulation in the advancing mycelial front (D). From Finlay and Read (1986b), with permission.

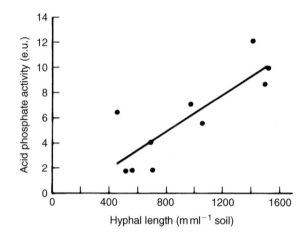

Figure 10.6 The relationship between length of external hyphae and phosphatase activity (in enzymes units, EU) in the humus layer in which mycorrhizal plants of *Picea abies* were growing. From Häussling and Marschner (1989), with permission.

Cenococcum geophilum and those that grew in the non-mycorrhizal condition. They also examined the differential effect of the two fungi on P uptake by the plants (Table 10.5). Whereas *P. tinctorius* stimulated some increase in branching of root tips and hence surface area of fine roots, *C. geophilum* did not do so. The major impact of the fungi on the area available for absorption was provided by the extraradical hyphae which led to an increase of approximately 40-fold in the case of *P. tinctorius* and 25-fold for *C. geophilum* in relation to the non-colonized controls. The greater lengths, and consequently surface areas, of the mycelia were associated with significant increases in P uptake by both fungi and with greater shoot weight in the case of plants ECM with *P. tinctorius*. While there appears to be a correlation here between hyphal development and P uptake, studies of this kind do not conclusively demonstrate a causal relationship between the two.

To determine such a relationship, P uptake must be measured in terms of inflow (uptake per unit length of root per unit time) or specific uptake rate (uptake per unit weight per unit time). That the effects of the fungus can be large has been appreciated for some time and, as Bowen (1973) pointed out, estimates of the specific uptake rate can be calculated from the data of Hatch and others (see Table 10.1) and the results demonstrate the greater uptake that follows from the improved exploitation of the soil with respect to poorly mobile P, even though the values must be underestimates because the contents of nutrients in the seed are not known.

Two studies, one of *Eucalyptus pilularis* colonized by a fungus of uncertain identity but probably *Cenococcum geophilum* (Heinrich and Patrick, 1986) and another of *Salix viminalis* colonized by *Thelephora terrestris* (Jones *et al.*, 1991) have shown unambiguously that ECM colonization of intact root systems increases P inflow significantly. Heinrich and Patrick (1986) established relationships between numbers of *Cenococcum*-type ectomycorrhizas and both seedling dry weight (Figure 10.7a) and total seedling P content (Figure 10.7b). Significant correlations were observed for both relationships. In the case of *Salix*, the inflows of P to ECM roots were almost three times higher than to those that were uncolonized. The substantial increase in

Table 10.5 A comparison of plant and fungal parameters for *Pinus taeda* seedlings colonized by *Pisolithus tinctorius* (Pt), *Cenococcum geophilum* (Cg) or left uncolonized (control).

	Pt	Cg	Control	P*
Mycorrhizal infection (%)	69.5 a	66.5 a	0.0 b	<0.0001
Shoot weight (g)	1.09 a	0.830 b	0.710 b	0.015
Foliar P conc. (g)	0.066 a	0.043 b	0.034 b	<0.0001
Shoot P content (mg)	0.669 a	0.340 b	0.238 c	<0.0001
Fine-root diameter (mm)	0.477 b	0.573 c	0.299 a	<0.0001
Root-tip ratio	3.72 b	1.39 a	1.55 a	<0.0001
Area fine-root (mm^2)	4.02 b	1.49 a	1.30 a	<0.0001
Area (mm^2/g soil)				
Hyphae	33.8 a	28.1 a	1.50 b	<0.0001
Rhizomorphs	13.6 a	0.00 b	0.00 b	0.0012
Total	47.4 a	28.1 b	1.50 c	<0.0001
Length (m/g soil)				
Hyphae	6.42 a	2.80 b	0.28 c	<0.0001
Rhizomorphs	0.36 a	0.00 b	0.00 b	0.0011
Total	6.78 a	2.80 b	0.28 c	<0.0001
Dry weight (μg/g soil)				
Hyphae	4.98 a	7.85 b	0.22 c	<0.0001
Rhizomorphs	14.3 a	0.00 b	0.00 c	<0.0001
Total	19.3 a	7.85 b	0.22 c	<0.0001
Hyphal diameter (μm)	1.60 a	3.18 b	–	<0.0001

Data from Rousseau *et al.* (1994). Values within a row having the same letter are not statistically different (Duncan's, $P < 0.05$). P, Probability values from one-way ANOVA between inoculation treatments.

P supply supported a twofold increase of growth (Jones *et al.*, 1991). Subsequently, Jones *et al.* (1998) determined the relative abilities of ECM and AM fungi to enhance P uptake rates of *Eucalyptus coccifera*. Although the two ECM symbionts, *T. terrestris* and *Laccaria bicolor*, differed in their effectiveness in P acquisition for the plant, they were both significantly more effective than the two species of AM fungi, *G. caledonium* and *G. mosseae*. Overall, P inflows to ECM plants were increased by 3.8 times and to AM plants by between 2.0 and 2.7 times, relative to those seen in the non-mycorrhizal condition.

There are studies indicating that hyphal development in soil can be a poor indicator of mycorrhizal effectiveness. Thomson *et al.* (1994) found that, while those fungi which were most effective in increasing the uptake of P and growth in *Eucalyptus globulus* were also those that colonized the roots most extensively, P uptake correlated poorly with hyphal length. Thus, the fungus most effective in increasing plant growth, *Descocolea maculata*, formed the smallest amount of external hyphae per metre of colonized root, while isolates of *L. laccata* developed more external hyphae per metre of root than other fungi, without any apparent additional benefit to the plant. Observations such as these indicate the need for caution in generalizing about the role of hyphal length in P uptake. Clearly, other factors, among which the viability and physiological characteristics of the mycelium and the compatibility between fungus and plant, as well as in exploring the soil for new plant sources of organic C may be important and should be taken into account.

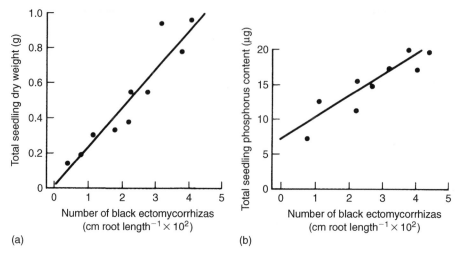

Figure 10.7 Growth and P uptake of *Eucalyptus pilularis* colonized by *Cenococcum geophilum* 176 days after sowing. (a) Relationship between number of black ectomycorrhizas and total seedling dry weight. (b) Relationship between number of black ectomycorrhizas and phosphate content of the seedlings. From Heinrich and Patrick (1986), with permission.

A number of other studies of responses of colonized and uncolonized plants to changing concentrations of P have now been published. That of Bougher *et al.* (1990) on *Eucalyptus* has already been described (see Chapter 8). *Pinus taeda* responded positively to inoculation with four fungi (*Pisolithus tinctorius, Rhizopogon roseolus, Scleroderma aurantium* and *T. terrestris*) when growing in Piedmont forest soil at all P concentrations supplied, relative to uninoculated controls after ten months (Ford *et al.*, 1985). Responses to *S. aurantium* were much greater than to the other fungi. This was a pot study in a glasshouse which, considered in isolation, could be taken to indicate that similar responses to P and inoculation might be expected in the field. However, the authors point out that attempts to improve growth of *P. taeda* in plantations established on these soils by application of P fertilizers had failed because N was in fact the limiting nutrient in the field. The growth responses to P achieved in pots occurred because N limitation had been removed by N fertilization at the start of the experiment. The experiment was thus instructive in terms of the potential of ECM colonization to improve growth, but also emphasizes that the significance of the roles of the fungi will be strongly influenced by field conditions (see Chapter 17).

Sources and mobilization of P in soil

Some discussion of the nature and availability of P sources in soil has been provided in Chapter 5. A large part of the P contained in the surface horizons of forest soils, where most ectomycorrhizas are localized, is present in organic forms (Dalal, 1977; Harrison, 1983). These can occur as phosphomonoesters such as inositol hexaphosphate, often referred to as 'phytate', or as phosphodiesters, among which nucleic acids and phospholipids are likely to be important. In this context the recent indications that 'phytate' may have been misidentified as a major organic P source in soil should be borne in mind and the relevance of studies with pure phytate questioned

(Smernik and Dougherty, 2007, see Chapter 5). Some of the monoesters, though important constituents of living cells, may have only a short life span in soil because endogenous phosphomonoesterases will attack them in the course of cell breakdown (see Beever and Burns, 1980). Others are clearly more resistant to breakdown, since a significant proportion of the organic P in acidic soils may be present as inositol penta- and hexa-phosphates (Cosgrove, 1967; McKercher and Anderson, 1968). The phosphomonoesterases are easily studied and have been detected in most ECM fungi but, in future, it will be important to investigate other (hitherto neglected) enzymes that can degrade organic P.

Evidence for the ability of ECM fungi to produce phosphomonoesterase has taken two forms, one indirect involving studies of growth on phytate supplied as sole P source and the other direct and dependent upon measurement of phosphatase activity using paranitrophenyl phosphate (PNPP) as the substrate. Growth of *Suillus granulatus*, *S. luteus*, *Cenococcum geophilum* and *Rhizopogon roseolus* takes place with phytates of Ca and P as sole sources of P (Theodorou, 1968), though it was shown that Fe phytate, likely to be of greater quantitative significance in acid organic soil, was little used. *R. luteolus* was shown to produce two types of phytase (Theodorou, 1971). Bartlett and Lewis (1973) examined the surface phosphatase activity of mycorrhizas of *Fagus* and also showed the presence of more than one phosphatase, because the activity had a double pH optimum and hydrolysed a range of P compounds including inorganic pyrophosphates and organic compounds especially inositol phosphates. They emphasized that phosphatase of such an activity on the surface of the fungal component of ectomycorrhizas might result in the immediate recycling of the phosphates present in the fallen litter back into the mycorrhizal system. Williamson and Alexander (1975) also examined *Fagus* mycorrhizas. They found that acid phosphatase was present throughout the fungal tissue and was not associated with contaminating microflora to any significant extent. They agreed with Bartlett and Lewis that more than one phosphatase enzyme was present and that each had different characteristics. Alexander and Hardy (1981) showed that mycorrhizas of *Picea sitchensis* possessed surface phosphatase activity that was inversely correlated with the concentration of extractable inorganic P in the soil. In this respect, the work is reminiscent of that of Calleja *et al.* (1980) who showed that the phosphatase activities of four species of ECM fungi were higher in the absence of soluble P in the culture medium. The effects of such environmental variables as pH, temperature and substrate concentration on the activities of acid phosphatases have now been examined in a number of ECM fungi (Antibus *et al.*, 1986) and it has again been shown (Antibus *et al.*, 1992) that enzyme activity and phytate utilization are greatest at low concentration of inorganic P. Colpaert *et al.* (1997) could find no evidence of phytase activity in the mycelia of *Thelephora terrestris* or *S. bovinus* symbiotic with *P. sylvestris* when they were grown in the presence of phytate. Further, the fungi were unable to obtain P from phytate fixed on HPLC resin. These workers were unable to find support for the hypothesis that phytate is a useful P source for ECM plants.

Dinnelaker and Marschner (1992) demonstrated that phosphomonoesterase activity was greater in the ECM roots of spruce and in the rhizomorphs of *T. terrestris* than in non-mycorrhizal roots. The localization of these enzyme activities on the surfaces of fungal hyphae necessitates close juxtaposition with appropriate P sources. The preferential proliferation of the extraradical mycelium in P-enriched substrates, whether they be hyphal mats (Griffiths and Caldwell, 1992) or mesh bags (Hagerberg *et al.*, 2003) in the field, or litter 'patches' in microcosms (Bending and Read, 1995a; Perez-Moreno and Read, 2000) (see Figure 6.27a, b), will facilitate this physical proximity.

To date, studies of what are likely to be the more important phosphodiesterase activities have been few. Griffiths and Caldwell (1992) found that the mat-forming ectomycorrhizal fungi *Gautieria monticola* and *Hysterangium gardneri*, together with an unidentified *Chondrogaster* species, were capable of hydrolysing the major phosphodiester RNA and it has been shown by Leake (unpublished) that *S. bovinus*, one of the fungi known to produce dense mycelial patches (see Figure 6.27a, b) can use DNA as sole source of P. There is, of course, the likelihood that both phosphomono- and diesters contained in senescent organic residues will be sequestered, along with nitrogenous components, in more complex aromatic and aliphatic macromolecules. Evidence is emerging that some ECM fungi can produce enzymes capable of hydrolysing these 'protected' substrates (see Chapters 15 and 16).

Interest in the ability of ECM fungi to gain access to P sources sequestered in mineral materials has been increased by the recognition that, in some soils, notably those of a podsolic nature, the mycobiont can proliferate in horizons below the superficial organic layers (see Chapter 6). Earlier studies (Stone, 1950; Bowen and Theodorou, 1967) showed that some ECM fungi had the ability to bring P into solution from rock phosphate. There was debate about the mechanisms involved in this release. Acidification of the local environment may be important. In *in vitro* experiments, Lapeyrie *et al.* (1991) concluded that acidification alone could explain release of P from calcium and iron phosphates. When the fungi were grown with ammonium or nitrate as N sources, significant solubilization of P occurred only in the acidic environments produced by ammonium assimilation. More recent studies of the same type (Mahmood *et al.*, 2002) have confirmed the importance of acidification in P release from mineral sources and Rosling *et al.* (2004b) have developed methods enabling quantification of substrate acidification in these systems.

Studies in soils supporting ECM fungal mats (Cromack *et al.*, 1979; Griffiths *et al.*, 1994) implicated a combination of acidification and organic anion production in the processes of P release from these more complex environments. While pH is significantly lower in the mat soils, there are also elevated levels of dissolved organic carbon (DOC) and calcium oxalate. Scanning electron microscopy of minerals in mat soils showed intense chemical weathering which was attributed to oxalate attack in the immediate vicinity of the ECM hyphae (Cromack *et al.*, 1979). While direct evidence for the involvement of the ECM fungi in these processes was not provided, the notion of 'ectomycorrhizal weathering' (rock eating) became established and a considerable literature on the subject has subsequently arisen (van Breemen *et al.*, 2000a, 2000b; Landeweert *et al.*, 2001; Hoffland *et al.*, 2004; Wallander, 2006). There is the suggestion, still unsubstantiated (see Chapter 15), that ECM fungi are responsible for the production of tunnels through mineral particles in podsolic soils (Smits, 2006). While it may inevitably be difficult to distinguish between the P mobilizing activities of ECM fungi from those of the general microflora in natural soil environments, some progress has been made towards evaluation of their roles using defined P minerals in *in vitro*, pot and field experiments.

Some ECM fungi can release P from the potentially important inorganic source of soil P, apatite. Wallander (2000) grew pine seedlings with and without ECM fungi in peat systems to which apatite was supplied as sole additional P source in root-free compartments. Two of the fungi, *S. variegatus* and an unidentified species, had a significant positive influence on the dissolution of apatite and the seedlings colonized by these fungi produced significantly more biomass than those which were either non-mycorrhizal or poorly colonized by a different isolate of *S. variegatus* (Figure 10.8). A budgeting approach indicated that, after 210 days of exposure to the fungi, ~1% of the apatite had been degraded. There were positive correlations between the amounts of oxalate present in the root-free compartments and P concentrations of their soil solutions, indicating that mobilization of P by the fungi may have been achieved by release of the organic acid.

Considerable differences have been observed in the abilities of ECM fungi to release oxalate (Ahonen-Jonnarth *et al.*, 2000; Casarin *et al.*, 2004) and those fungi like *Rhizopogon roseolus* and *Suillus* spp. with the highest rates of oxalate release are the most effective at freeing P from insoluble inorganic sources (Wallander, 2000; Casarin *et al.*, 2004). Van Schöll *et al.* (2006a) have shown that the production of oxalate and of another low molecular weight organic anion, malonate, by both ECM and non-mycorrhizal seedlings of pine was significantly increased under conditions of P deficiency. More oxalate was produced by seedlings colonized by *Paxillus involutus* than by non-mycorrhizal seedlings, though the latter produced a greater total quantity of organic anions.

Using the mesh-bag method (see Chapter 6), Hagerberg *et al.* (2003) confirmed that growth of ECM fungi in the field could be stimulated by the addition of apatite, but only if the soil was of inherently low P status. Under low P conditions, the quantity of ECM mycelium in apatite-containing bags was 50% greater than in control bags and there was a threefold increase in the number of ECM root tips immediately outside the bags. Such effects could be attributed either to increased allocation

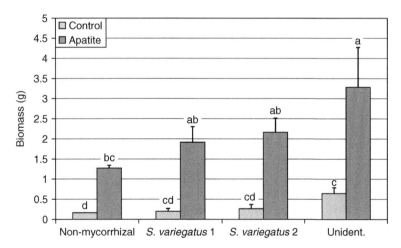

Figure 10.8 Biomass of non-mycorrhizal seedlings and seedlings colonized by one of three different ECM fungi, grown with or without apatite as P source. Bars indicate SE. Different letters indicate statistically different values using two-way ANOVA and LSD to separate the means ($P > 0.5$). Values were log-transformed to equalize the variances of the mean weights of seedlings grown with and without apatite. The P value for effects of EM colonization was 0.001 and for mineral addition $P = 0.000$. The interaction between EM colonization and mineral addition was not significant. From Wallander (2000).

of organic C to each of the fungal species initially present in the soil or to a change in the fungal flora, induced by the greater P availability, in the direction of species which produce more prolific external mycelia. The proliferation of root tips in proximity to the mesh bags suggests that the additional P supply had a strong influence on C allocation by the autotroph. It can be hypothesized that a considerable proportion of the additional C allocated to these resource-enriched areas will be used in the production of oxalate. The ECM-induced dissolution of apatite was 40% greater in the P-deficient soil than in soil adequately supplied with P.

Mobilization, uptake and translocation of potassium

High intracellular concentrations of K are required to maintain activity of enzymes involved in intermediary metabolism, biosynthesis and membrane transport processes. The K^+ also contributes to the osmotic potential of the cell. Hatch (1937) demonstrated that ECM colonization could enhance the capture of K by plants (see Table 10.1); subsequently, Harley (1978) stressed that the sheath of ectomycorrhizas acts as an important storage organ for many nutrients derived from soil, including K. It was shown that ~67% of the K^+ absorbed by excised ECM rootlets of *Fagus sylvatica* was retained in the fungal sheath (Edmonds and Harley, unpublished; see Harley and Smith, 1983). Despite these findings, interest in this element has been restricted until recently. Its high coefficient of diffusion in soil should make K^+ more accessible at absorbing surfaces than P and there has been little to suggest that growth of ECM systems is limited by its availability (Tamm, 1985). However, combinations of high biomass removal and leaching losses arising from anthropogenic soil acidification have led to fears that K could become growth limiting in some environments (Barkman and Sverdrup, 1996; Uebel and Heinsdorf, 1997; Jonsson *et al.*, 2003). Feedbacks from progressively reduced K concentrations in foliage can be expected to lead to reduced availability of the element in soil organic matter. For this reason, there has been renewed interest in the processes whereby K is released from its parent minerals in soils and in determination of the extent to which ECM fungi are involved in its mobilization and transport to the plant.

Among the most important K-containing minerals in soil are the micas phlogopite, biotite and vermiculite, and the silicate mineral, muscovite. It has been shown that, in axenic culture, *Paxillus involutus* can release K from phlogopite and in so doing convert it to vermiculite (Paris *et al.*, 1996). The process was, at least in part, facilitated by exudation of oxalate. Pot experiments comparing the ability of beech either colonized by *Laccaria laccata* or in the non-mycorrhizal condition to obtain K from phlogopite revealed increased K release in ECM treatments (Leyval and Berthelin, 1989) and showed that dual inoculation with *Agrobacterium* further enhanced the effect, apparently largely by acidifying the substrate (Leyval and Berthelin, 1991). In a pot experiment, van Schöll *et al.* (2006b) grew *Pinus sylvestris* either in the non-mycorrhizal condition or with each of three fungal symbionts (*P. involutus*, *Piloderma croceum* and *S. bovinus*) and supplied the plants with muscovite as the sole source of K. Both mineral and non-mineral pools of K were quantified after 27 weeks. Seedlings colonized by *P. involutus* showed almost twofold greater release of K from muscovite than that achieved by the other ECM fungi or by the non-mycorrhizal plants (Figure 10.9). This K mobilization resulted in increased K content of roots

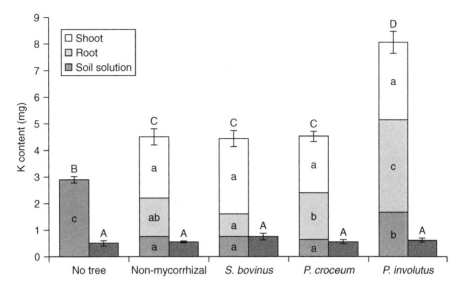

Figure 10.9 Non-mineral-bound K in the soil solution in non-mycorrhizal seedlings of *Pinus sylvestris* and in seedlings mycorrhizal with *Suillus bovinus*, *Piloderma croceum* and *Paxillus involutus* after growth in pots with muscovite as sole source of K. Dark grey bars denote the non-mineral-bound K in pots without muscovite addition. Mean of six replicates. Error bars represent the standard error of the non-mineral-bound K. Bars with the same letter are not significantly different ($P < 0.05$); capital letters above bars denote total amount of non-mineral-bound K in pot. From van Schöll *et al.* (2006b).

and adhering mycelia of the plants. The superior ability of *P. involutus* to release K from muscovite may be attributable to a prolific exudation of oxalate. In a further experiment (van Schöll *et al.*, 2006b), it was shown that, under K-deficient conditions, seedlings ECM with this fungus alone responded by increased production of oxalate. This observation is in line with that of Paris *et al.* (1996).

There is evidence from studies of soils heavily colonized by ECM fungi in the field that K-containing minerals are weathered to a greater extent in the presence of the associated mycelia (Arocena *et al.*, 1999; Arocena and Glowa, 2000). These processes led to a greater concentration of K in soil solution. One of the fungi present in the field, *P. croceum*, which is known to release oxalate in large quantities (Arocena *et al.*, 2001), was shown in laboratory studies to release K from a variety of minerals (Glowa *et al.*, 2003).

There are some discrepancies with regard to observations of the responses of ECM mycelia to the localized addition of K-containing minerals. In field studies, insertion of mesh bags containing biotite did not stimulate mycelial proliferation in the vicinity of the mineral (Hagerberg *et al.*, 2003). This finding is in marked contrast to the situation when the P-containing mineral apatite was added. Conversely, in peat microcosms to which pure potassium feldspar or quartz were added in localized patches, selective proliferation of the mycelia of *Hebeloma crustuliniforme* and *Piloderma fallax* was observed to occur on the feldspar (Rosling *et al.*, 2004b). The differences between the results of the two studies may be attributed to the K status of the surrounding substrates, there being a likelihood that K was more freely available in the forest environment than in the peat of the microcosms. Alternatively, known differences in the responses of plants to deficiencies of soil P and K may be involved. In non-mycorrhizal birch seedlings, P deficiencies lead to increased allocation of photosynthate to root production, whereas those of K (and Mg) have the opposite effect (Ericsson, 1995), possibly because, in the latter case, there is a tighter coupling between deficiency of the element and C-fixing ability. The same differential effects of the two deficiencies were observed in laboratory studies of ECM pine seedlings (Wallander and Nylund, 1992; Ekblad *et al.*, 1995). Deficiency of K, in contrast to that of P (see above), led to reduced allocation of assimilate to ECM mycelia. Results such as these suggest that the abilities of ECM systems to respond to K deficiency, either by more effective scavenging of available ions or by enhanced weathering of minerals, will be less than they are where P is in limiting supply.

Uptake and translocation of K through ECM hyphae have been demonstrated using Rb^{86} as an isotopic analogue of K (Finlay, 1992) and by mass balance approaches (Jentschke *et al.*, 2001a). Both studies show fluxes of K through the hyphae to be similar in rate to those of P. In the latter study, translocation of K through hyphae of *P. involutus* to ECM spruce plants occurred only when supplementary P had been supplied to the P-deficient system in a root-free compartment. Both this observation and that relating to fluxes of both K and P are consistent with the notion of co-transport of the two elements as K^+ polyP (see above). Hyphal acquisition of K was estimated by Jentschke *et al.* (2001a) to contribute 6% to total plant uptake of the element. Final calculated and measured K contents of P-fed ECM plants were significantly greater than those in the non-mycorrhizal plants (Table 10.6) confirming that hyphal transport of K can contribute in an important way to whole plant K budgets. Plassard *et al.* (2002) emphasized that K uptake capabilities differed between fungi. Whereas colonization of *Pinus pinaster* by *Rhizopogon roseolus* facilitated a marked increase in uptake of K^+, no such effects were produced when the plants were colonized by *Hebeloma cylindrosporum*.

Several investigations have explored the extent to which the ECM sheath forms an apoplastic barrier to the entry of solutes, including K to the root tissues. The

Table 10.6 Initial K content, uptake from nutrient solutions and final K contents of mycorrhizal (with *Paxillus involutus*) or non-mycorrhizal Norway spruce (*Picea abies*) seedlings grown in a two-compartment culture system with or without P addition to the hyphal compartment.

Treatment		Initial K content (mg per vessel) week 0	K uptake week 0 – week 11		Final K content (mg per vessel) week 11	
			Plant compartment	Hyphal compartment	Calculated	Measured
Non-mycorrhizal	−P	4.4a	22.7b	−4.3b	22.8b	24.4b
	+P	3.3a	26.2ab	−5.4b	24.1b	25.6b
P. involutus	−P	3.2a	27.4ab	−5.6b	25.0b	28.9ab
	+P	4.0a	32.6a	1.9a	38.5a	37.3a

Data from Jentschke *et al.* (2001a).

apoplastic phase of the fungal mantle of some ECM is impermeable and hence must be a barrier to nutrient transfer between soil and root (Ashford *et al.*, 1988, 1989; Bücking *et al.*, 2002). Ashford *et al.* (1989) showed that entry of the apoplastic marker cellufluor to the roots of *Eucalyptus pilularis* is prevented by the fungal mantle of *P. tinctorius* unless the rootlets were deliberately damaged to remove the outer, unwettable region of the mantle. Bücking *et al.* (2002) investigated the extent to which the ECM mantle formed an apoplastic barrier to the entry of K^+ into the tissues of the roots. Using the stable isotope ^{41}K as a tracer they showed that transfer of K^+ to the root cortex of pine was significantly slower in the presence of mantles formed by *P. tinctorius* and by *S. bovinus* than in non-mycorrhizal roots. The retardation of flow was attributed to hydrophobin production on the mantle surfaces. Time-dependent differences between the fungal species in terms of the patterns of isotope transfer were attributed to distinctive patterns of hyrophobin production on the mantle surface.

The presence of the impermeable layer means that all solutes reach the root cells via the fungal symplast of the mantle, first by translocation in the external mycelium and subsequently by efflux to the interfacial apoplast in the Hartig net region. Conversely, solutes from the root cells effluxing to the apoplast must pass to the fungal symplast and could not 'leak' to the soil via the apoplast of the mantle. The impermeable layers thus offer an opportunity for control of conditions and solute concentrations in the cortical apoplast and Hartig net region, where transport between the ECM symbionts must occur. As Ashford *et al.* (1989) point out, maximum efficiency requires that material must not be allowed to escape from the 'exchange compartment'. The initial work on apoplastic impermeability was carried out with *Pisonia grandis*, in the mycorrhizas of which there is no Hartig net. A similar, but slightly differently organized, exchange compartment could exist where the Hartig net penetrates as far as the outer layer cortical cells; in this case, the inner boundary would be the endodermis, which again provides a block to apoplastic transfer.

The fungal mantle has not been reported to be impermeable in all investigations and the discrepancies may relate to different plant–fungus combinations as well as to different experimental methods. The external mycelium of some species of fungi is also covered with non-wettable material (Unestam, 1991). As yet, it is not clear whether this is relevant to long-distance translocation within the rhizomorphs or whether it might be important in preventing desiccation as has been suggested for the aerial hyphae of some saprophytic fungi, or indeed whether both these attributes are important.

Release, uptake and transport of magnesium

Magnesium (Mg) is a mobile element in soil and, like K, is generally considered to be a non-limiting nutrient in ECM forest soils. However, the anthropogenic factors that are leading to a reduction of K stocks in these environments are also depleting those of Mg. The inability of ECM trees to replenish Mg lost in litter fall has been implicated as a direct cause of decline in European forests suffering from soil acidification (Schulze, 1989).

Van Schöll *et al.* (2006a) examined the ability of ECM fungi to release Mg from the mineral hornblende. In contrast to their observation that *P. involutus* was able to release K from muscovite (see above), they observed no weathering of hornblende by any of the fungi examined. This may have been because the oxalate-producing capabilities of the fungus, which are stimulated by Mg deficiency (van Schöll *et al.*, 2006b), were inhibited by a rise of pH to over 7.00 caused by addition of the hornblende. Whereas the ability of *P. involutus* to release K from hornblende appears to be small, the fungus has been shown to facilitate access of its ECM associate, *P. sylvestris* to Mg when supplied in root-free compartments (van Schöll, 2006). Hyphae of *P. involutus* proliferate in these compartments in response to the addition of Mg, either as $MgSO_4$ or $Mg_3(PO_4)_2$ (Figure 10.10). The Mg content of the seedlings was also increased (Figure 10.11) irrespective of whether P was supplied as an additional nutrient to the side chamber. This result conflicts with that of Jentschke *et al.* (2001a) who found that hyphal transport of Mg in the same plant–fungus association occurred only under conditions in which supplementary P supplies were provided to the root-free compartment. The difference may be resolved by the fact that, in the study of van Schöll *et al.*, access to Mg was the only factor limiting growth, whereas in that of Jentschke *et al.*, P was limiting and, in the absence of supplementation by this element, hyphal growth and foraging were severely restricted. Jentschke *et al.* (2000, 2001a) used the stable isotope ^{25}Mg to quantify fluxes through hyphae of *P. involutus*. While the values for this element were the lowest of those measured (Table 10.7), fungal translocations were estimated to have contributed approximately 4% of total plant uptake of Mg. Significantly higher concentrations of ^{25}Mg were found in mycorrhizal than in non-mycorrhizal seedlings after the 6 weeks labelling period (Figure 10.12).

It has been shown (Bücking *et al.*, 2002) that, as in the case of K, ECM colonization of pine seedlings significantly reduced the apoplastic transfer of Mg to the root cortex. In roots supporting a fungal mantle, even after an exposure of 72 h to a solution labelled with ^{25}Mg, at least one third of the apoplastic Mg content of the

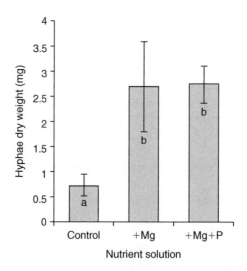

Figure 10.10 Yield of *Paxillus involutus* showing stimulus to hyphal proliferation (expressed as dry weight of mycelium) in root-free compartments to which magnesium (Mg) was added, either alone or with phosphorus (P). Bars with the same letter are not significantly different ($P < 0.05$). Mean of 4 with standard error. From van Schöll (2006).

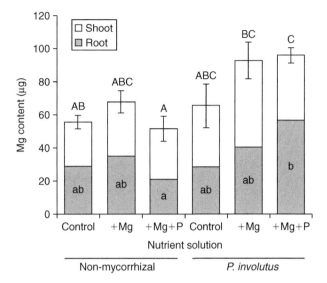

Figure 10.11 Mg content (μg) in shoot and root of *P. sylvestris*, either non-mycorrhizal or colonized by the ectomycorrhizal fungus *P. involutus*. Bars with the same letter are not significantly different ($P < 0.05$); capital letters refer to total Mg content of seedling. Mean of 4–5, standard error for total Mg content. From van Schöll (2006).

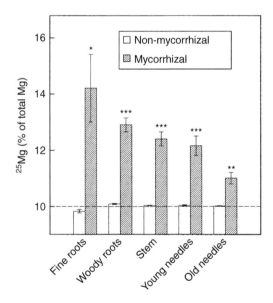

Figure 10.12 Concentration of the ^{25}Mg label (expressed as percentage ^{25}Mg of total Mg) in tissues of non-mycorrhizal and mycorrhizal Norway spruce seedlings after labelling the hyphal compartment in the culture system with ^{25}Mg for 6 weeks. Broken line indicates natural abundance of ^{25}Mg (10.0%). Values are mean values of 4 replicate pots. Small bars indicate standard error. Significance levels for differences between mycorrhizal and non-mycorrhizal seedlings: ★, $P \geqslant 0.01$; ★★, $P \geqslant 0.001$; ★★★, $P \geqslant 0.0001$. From Jentschke *et al.* (2000).

Table 10.7 Hyphal translocation of N, P, K and Mg in *Picea abies* seedlings in mycorrhizal association with *Paxillus involutus* during an 11-week experimental period.

Treatment	Hyphal translocation (μm per vessel)			
	Nitrogen	Phosphorus	Potassium	Magnesium
−P	70	nd	nd	nd
+P	660	120	50	7

Data of Jentschke *et al.* (2001a). P and K translocation amounts are from mass balance calculations. N and Mg translocation were determined by stable isotope labelling. Translocation of Mg was corrected for the shorter time period of ^{25}Mg labelling (6 weeks), assuming that the translocation rates were constant over the experiment. nd, not detected.

root cortex had still not exchanged with the externally applied label. In contrast, clear labelling of the apoplastic Mg and complete exchange of the internal and external pools occurred in non-mycorrhizal roots within minutes. These results appear to contrast with those of Kuhn *et al.* (2000) who found that levels of apoplastic ^{25}Mg equilibrated rapidly with those of a bathing medium in ECM roots of spruce. However, sections of spruce roots used in the latter study indicate that, while a Hartig net was well developed, virtually no fungal mantle was present. The absence of the barrier identified by Ashford *et al.* (1989) and Bücking *et al.* (2002) would explain the apparent discrepancy.

Mobilization, uptake and transport of calcium

In contrast to K and Mg, calcium (Ca) is primarily located in cell walls. Its intracellular role as a signalling molecule requires maintenance of extremely low concentrations in the cytosol. Until recently, it has been widely assumed that, as in the cases of K and Mg, supplies of Ca in soils of most ECM ecosystems were sufficient to eliminate the likelihood of deficiency. However, as with those elements, combinations of soil acidification and repeated crop harvesting are now seen to threaten soil Ca stocks in some areas (Yanai *et al.*, 2005). Unfortunately, little is known about the roles of ECM fungi in Ca mobilization and capture. Melin and Nilsson (1955) showed, using the isotope ^{45}Ca, that ECM mycelium could absorb and transport Ca to pine seedlings and mycelial transfer of Ca was confirmed by Jentschke *et al.* (2000). However, Lamhamedi *et al.* (1992) reported reduced levels of Ca in roots of *P. pinaster* plants in ECM association with *P. tinctorius* relative to non-mycorrhizal controls. Similarly van Schöll *et al.* (2005) observed no differences in Ca content of *P. sylvestris* needles when plants were grown in a semi-hydroponic system in the ECM and non-mycorrhizal condition. Bücking and Heyser (2000), using X-ray microanalysis of tissues, found that colonization of *P. sylvestris* by *S. bovinus* and *P. tinctorius* reduced the amounts of Ca detectable in roots, especially in the apoplasts of cortical cells. However, this effect was not observed with *P. involutus* as the mycobiont. An analysis of Ca uptake under different conditions of N supply (Jentschke *et al.*, 2001b) indicated the dynamic nature of the balance between anion and cation uptake.

When N was supplied as ammonium to ECM plants, uptake of Ca (and Mg) was less negatively influenced than in the case of non-mycorrhizal plants. It was concluded that acidification of the whole root compartment of the uncolonized plants led to reduced Ca (and Mg) uptake. In the case of mycorrhizal plants, acidification was localized in the hyphosphere and was less extreme, leading the authors to conclude that the mycelium exerted a buffering effect which ameliorated the negative impacts of ammonium supply on cation uptake.

Effects of ectomycorrhizal colonization on resistance to metal ion toxicity

ECM plants successfully dominate many natural environments where soil acidity and base cation leaching result in exposure to elevated levels of metals. That this exposure has led to some inherent constitutive tolerance to metal pollution in their ECM root systems is shown by the observation that, in many parts of the world, ECM plants successfully colonize mine spoils contaminated with mixtures of metal ions (Meharg and Cairney, 1999). There are reports of enhanced metal tolerance in ECM plants exposed to aluminium (Al) (Cumming and Weinstein, 1990; Hentschel et al., 1993; Schier and McQuattie, 1996; Lux and Cumming, 2001; Ahonen-Jonnarth et al., 2003), cadmium (Cd) (Jentschke et al., 1999), copper (Cu) (van Tichelen et al., 1999), lead (Pb) (Marschner et al., 1996), nickel (Ni) (Jones and Hutchinson, 1986, 1988a, 1988b) and zinc (Zn) (Brown and Wilkins, 1985). However, as emphasized elsewhere (Meharg and Cairney, 1999; Meharg, 2003), in many of these studies tolerance was assessed in terms of growth improvement in the colonized plant and the extent to which the effects were attributable to an enhanced nutrient supply was not resolved.

Evidence that long-term exposure to metals in nature can lead to the selection of constitutively resistant strains of mycobionts was provided by Colpaert and van Assche (1987). Whereas strains of *Suillus bovinus* isolated from Zn contaminated soils could grow in the presence of $1000\,\mu g/g$ Zn, those from uncontaminated sites produced little or no growth at Zn concentrations above $100\,\mu g/g$. Zn-resistant strains of this fungus conferred significantly more tolerance upon plants of *Pinus sylvestris* than did non-resistant strains. In contrast, Denney and Wilkins (1987) could find no evidence that strains of *Paxillus involutus* isolated from Zn-polluted sites had greater ameliorative effects upon birch than strains from unpolluted sites.

The role of fungal sensitivity to a pollutant as a determinant of the potential for amelioration was emphasized by Jones and Hutchinson (1986). Colonization of *Betula papyrifera* seedlings by *Laccaria proxima* or *Lactarius hibbardae* alleviated Ni toxicity at $32\,\mu M$ Ni, but the effect was lost at $64\,\mu M$ because, at this concentration, the fungus failed to grow. Similarly, in the case of *Picea* seedlings exposed to Cd, enhanced resistance to the metal at low concentration was lost as amounts increased to the point at which fungal growth was inhibited (Jentschke et al., 1999).

Where metal tolerance is conferred, it may operate at several spatially distinct locations along the mycelium-root-shoot pathway. These have been described by Bellion et al. (2006) and involve one or a mixture of routes (Figure 10.13), including extracellular binding on the extramatrical mycelium or fungal mantle by excreted ligands, surface sequestration by binding to the fungal cell wall in the mycelium or mantle, enhancement of efflux from the fungal cell, or by a metallothionein (MT),

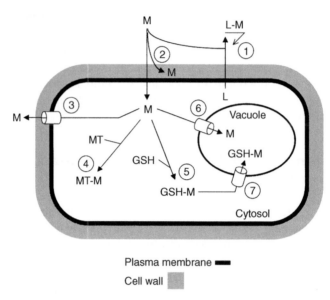

Figure 10.13 Schematic representation of cellular mechanisms potentially involved in metal tolerance by ECM fungi: ① Extracellular chelation by excreted ligands (L); ② cell wall binding; ③ enhanced efflux; ④ intracellular chelation by metallothionein (MT); ⑤ intracellular chelation by glutathione (GSH); ⑥ subcellular compartmentation (vacuole or other internal compartments); ⑦ vacuolar compartmentation of GSH-M complex. M = metal-ion. From Bellion *et al.* (2006).

intracellular chelation by glutathione (GSH), subcellular compartmentation in the cytosol or vacuole, or vacuolar sequestration as GSH complex.

Exclusion of metals by sequestration in the fungal component of the symbiosis contributes to avoidance of toxicity to the plant. The first contact between ECM fungus and a metal is likely to be made by the extramatrical mycelium as it explores the soil. Support for the view that the mycelium provides an important barrier comes from observations that fungi producing the largest quantity of external mycelium provide the greatest resistance to the plant. This has been shown in the cases of Zn (Colpaert and van Assche, 1992) and Cd (Colpaert and van Assche, 1993). Binding of Pb (Jentschke *et al.*, 1991) and Cu (Meharg, 2003) to hyphal surfaces has been demonstrated and there is a possibility that extracellular slime on hyphal walls provides binding sites for metals (Denny and Wilkins, 1987; Tam, 1995). The hyphal surface provides a large number of potential binding sites in the form of free carboxyl, amino, hydroxyl and phosphate groups. A Cu-binding protein exuded into the growth medium by isolates of *L. laccata* and *P. involutus* was considered by Howe *et al.* (1997) to be 'metallothionein-like', but could not be positively identified as such. Subsequently, Courbot *et al.* (2004) have reported a metallothionein-like compound in *P. involutus*. This finding is supported by the presence of a metallothionein sequence, homologous to a known metallothionein from *Agaricus bisporus*, in a cDNA array analysis of *P. involutus* exposed to Cd (Jacob *et al.*, 2004). The expression of this metallothionein gene was examined in the same fungus (Bellion *et al.*, 2006) and correlations were observed between transcript accumulation and exposure to metals. Metallothioneins can be retained in the cytosol or released to provide extra-cellular chelation. Courbot *et al.* (2004) showed that, in contrast to metallothioneins, there was a complete lack of phytochelatins in *P. involutus*.

The possibility that exudation of organic anions can lead to chelation of potentially toxic ions has been more extensively investigated. Bellion *et al.* (2006) report experiments showing that exposure to oxalic acid reduced Cd uptake in *P. involutus* by 85%. Of seven organic anions examined, oxalate provided by far the most effective metal binding activity. Ahonen-Jonnarth *et al.* (2000) showed that pine seedlings mycorrhizal with *Suillus variegatus* and *Rhizopogon roseolus* responded to Al exposure by strongly increasing release of the effective Al-chelator oxalate. Subsequently, it has been reported (van Schöll *et al.*, 2006a) that, while oxalate production was greatest in seedlings suffering Mg and P deficiency, no exudation occurred in the absence of Al ions. Large differences between fungal symbionts were observed in their ability to release oxalate. Indeed, differences in Al-induced organic anion exudation among seedlings colonized by different ECM fungi were as big as or bigger than those between non-mycorrhizal and ECM seedlings. Much more needs to be learned about the relative importance of the different processes involved in the exclusion of toxic metals and we still know little about the duration over which sequestered metals can be immobilized in the hyphosphere.

Depending upon the extent of its hydrophobicity, some extracellular localization of metals can also be expected to occur on, or in, the ECM mantle. Using X-ray microanalysis of cryosections, Frey *et al.* (2000) showed that two distinct mechanisms were involved in the binding of Zn and Cd ions in ECM roots. In the case of Cd, extracellular complexation occurred primarily in the Hartig net and in the cell walls of root cortical cells, indicating that transfer from fungus to plant occurred readily and that the pathway was primarily apoplastic. In contrast, Zn accumulated mainly in the cell walls and cytoplasm of the mantle hyphae, there being less transfer to the plant. A wide range of potentially toxic elements were shown to accumulate in the fungal mantles of *Pinus* roots growing on contaminated soil (Turnau *et al.*, 2002).

There have been relatively few analyses of the response of mycobionts to accumulation of metals. Metals absorbed by the external hyphae, mantle or Hartig net represent a threat to cellular metabolism. Oxidative stress is one such effect of entry and the activities of a number of enzymes released in response to this threat have been investigated after exposure of *P. involutus* mycelium to Cd ions (Ott *et al.*, 2002). The work confirmed that superoxide dismutase (SOD), catalase, glutathione reductase and glutathione peroxidase were active in the mycelium of the fungus. Ott *et al.* (2002) concluded that *P. involutus* is able to detoxify high concentrations of Cd by a strong induction of glutathione synthesis, accompanied by a rapid sulphur-dependent transport of Cd into the vacuole. Glutathione reductase activity was increased by exposure to low levels of Cd. A SOD gene was shown to be upregulated and to be under post-translational control following exposure of *P. involutus* to Cd ions (Jacob *et al.*, 2001). Using a desorption method with ^{109}Cd, Blaudez *et al.* (2000) showed that 20% of the Cd added to cultures of *P. involutus* appeared in the hyphal cytosol, while 30% was transported to the vacuole. Vacuolar accumulation of Cd was suggested to be an essential detoxification mechanism in this fungus. Enhancement of efflux or downregulation of genes involved in uptake of metals may be effective in providing avoidance of toxicity in some cases (Adriaensen *et al.*, 2005).

Searches of GenBank for sequences from a range of ECM fungi have revealed the presence of numerous expressed sequence tags (ESTs) or open reading frames that encode for proteins known in the yeast *Saccharomyces cerevisieae* to be involved in metal tolerance pathways (Table 10.8) (Bellion *et al.*, 2006). The demonstration that particular gene products are involved in increasing metal tolerance remains to be achieved in almost all cases and transformation systems enabling overexpression or disruption of target genes in ECM are urgently required in order to test their roles in contaminated soils.

Effects of ECM colonization on plant water relations

There has long been an interest in the possibility that ECM colonization might improve the water economy of trees and there are some reports that the symbioses can provide a measure of drought resistance at intermediate levels of water stress (Cromer, 1935; Zerova, 1955; Goss, 1960). However, as few of the early observations were accompanied by measurements of tissue water balance, it remains a possibility that such benefits could be attributable to nutritional rather than hydrological effects.

Advances were achieved by studies in which tissue water potentials were compared in ECM and non-mycorrhizal plants exposed to the same controlled drought treatments. Dixon *et al.* (1983) showed that the pre-dawn water potentials of container-grown *Quercus velutina* exposed to modest drought were higher in plants ECM with *P. tinctorius*, than in those lacking colonization. Similar effects of ECM colonization by this fungus were observed with *Pinus virginiana* (Walker *et al.*, 1982), *P. taeda* (Walker *et al.*, 1989) and *P. halapensis* (Morte *et al.*, 2001). Even in cases such as these involving monitoring of tissue water balance, it is not possible to discriminate between direct impacts of ECM formation on the uptake and transport of water and indirect effects arising from nutritional differences between colonized and uncolonized plants.

Comparative analyses of the water transport pathways of ECM and non-mycorrhizal plants provide a better view of the relative hydraulic conductances. In the case of *Ulmus americana*, ECM colonization leads to significant increases of apoplastic water transport and root hydraulic conductivity, these effects being maintained across a temperature range (Figure 10.14) (Muhsin and Zwiazek, 2002). On the basis that transmembrane water transport is mediated by aquaporin proteins (Maurel and Chrispeels, 2001), these authors hypothesized that application of a mercuric protein inhibitor would enable the relative importance of transmembrane and apoplastic transport to be determined. Since the application of the inhibitor had a lower impact on conductance in ECM than non-mycorrhizal roots, it was concluded that reduced conductivity at lower temperatures in the symbiotic condition were mediated more by changed water viscosity than by impacts on the membrane channels themselves. Nonetheless, a combination of apoplastic and membrane-based processes were considered to contribute to the increase in hydraulic conductance seen in the ECM plants.

The importance of aquaporins as determinants of water flow in ECM plants has been highlighted by elegant studies involving a combination of molecular and physiological analyses using *Populus* hybrids and the ECM fungus *Amanita muscaria* (Marjanovic *et al.*, 2005a, 2005b). It was first shown that, as in the case of *Ulmus* (see above), ECM roots had significantly higher hydraulic conductances, in this case

Table 10.8 Putative proteins from ectomycorrhizal fungi similar to proteins belonging to yeast metal tolerance pathways

Mechanism	Pathway	Function	Organism	GenBank accession no
Transcription factors	YAP1-like	Regulation of genes involved in oxidative stress tolerance and metal resistance	*Tuber borchii*	CN488390
	ZAP1-like	Regulation of zinc transporters	*Paxillus involutus*	CN072154
Transport systems involved in metal tolerance and homeostasis	Metal efflux into organelles	Cation diffusion facilitator	*Hebeloma cylindrosporum*	CK993155
		Cd-conjugate ABC transporter	*Hebeloma cylindrosporum* *Pisolithus microcarpus*	CK995083, CK992826 CB10722
		Metal-transporting ATPase	*Hebeloma cylindrosporum* *Tuber borchii*	CK992318, CK994170 AF487323
	Metal influx	Manganese transporter	*Hebeloma cylindrosporum*	CK995213, CK992324, CK995203
		Copper transporter Iron transporter	*Tuber borchii*	CN487781
Intracellular metal binding	Metal delivery to other proteins	Metallochaperone	*Paxillus involutus*	AAT91247, AAT31333, AAT91334 AAT91335, AAT91336, CD273262 CD273746, CD273829, CD275306 CD274894
			Hebeloma cylindrosporum	BU964154
	Cu and Cd binding	Metallothionein γ-glutamylcysteine synthetase	*Paxillus involutus* *Hebeloma cylindrosporum*	AAS19463 CK995328
	Glutathione synthesis	Glutathione synthetase	*Paxillus involutus* *Paxillus involutus*	CD273087 BG141319

(Continued)

Table 10.8 (Continued)

Mechanism	Pathway	Function	Organism	GenBank accession no
Protection against metal-induced oxidative stress	Regulation of cell redox homeostasis	Thioredoxin	Paxillus involutus	AAS19462, CD275083, CD275423, CD276018
				CK995145, CK995656
			Hebeloma cylindrosporum	BM26656, CN48764, CN48812
			Tuber borchii	CB011224, BF942541
			Pisolithus microcarpus	CB012066
			Laccaria bicolor	BM266155
		Glutaredoxin	Tuber borchii	BF942586
			Pisolithus microcarpus	CB10230, CB010243
			Laccaria bicolor	CB10617
		Catalase	Laccaria bicolor	
	Removal of reactive-oxygen species	Superoxide dismutase	Tuber borchii	BM266201
			Paxillus involutus	AD25353, AQ064502, AW064510
			Tuber borchii	BM266232
			Laccaria bicolor	CB010250, CB010696
			Hebeloma cylindrosporum	CK994166, CK991636, CK993733, CK992059, CK992841, CK994504, CK995143, CK991818, CK994684, CK994504, CK994795, CK991819

Data from Bellion et al. (2006). Selected protein sequences identified in Saccharomyces cerevisiae being involved in metal tolerance pathway were used to search for expression sequence tags or open reading frames from ectomycorrhizal fungi encoding putative proteins similar to them. Searches were made by TBLASTn or BLASTp in the NCBI database ($P = <0.05$).

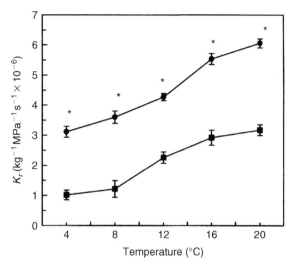

Figure 10.14 Root hydraulic conductance (K_r) of mycorrhizal (●) and non-mycorrhizal (■) seedlings *Ulmus americana* exposed to decreasing temperatures. Mean $(n = 8)$ ±SE are shown.* indicate statistically significant difference $(P = 0.05)$ between mycorrhizal and non-mycorrhizal roots. From Muhsin and Zwiasek (2002).

greater by up to 40% (Figure 10.15), than their non-mycorrhizal counterparts. Since the root systems of ECM plants had a surface area lower by 17% than that of the non-mycorrhizal plants, the real difference in hydraulic conductance between the two systems was larger, equivalent to an increase of 57% in water transport capacity (Marjanovic *et al.*, 2005a). Seven genes coding for aquaporins were isolated from a poplar ectomycorrhizal cDNA library and it was shown that four of them were preferentially expressed in roots prior to ECM formation. After colonization by *A. muscaria*, three showed an increased transcript accumulation including two (*PttPIP1.1* and *PttPIP2.5*), which are the most commonly expressed aquaporins in roots (Figure 10.16). When expressed in *Xenopus* eggs, these genes were confirmed to have the capability to transport water. The high expression, in particular of *PttPIP2.5*, associated with ECM colonization, suggests that these aquaporins play a key role in facilitating the increased water transport capacity seen in the ECM plants. It was subsequently shown (Marjanovic *et al.*, 2005b) that the expression of two aquaporin genes was more pronounced under conditions of drought in ECM poplar plants, indicating that the symbiosis may improve the capacity to transport water during periods of reduced supply.

Indirect evidence in support of the view that the extraradical mycelia might serve as conduits for water supply to plants was obtained by cutting the rhizomorphs of *Suillus bovinus* which were growing from a colonized plant into moist soil (Boyd *et al.*, 1986). Severance of the connections led to an almost instantaneous decline in plant transpiration. The importance of the external mycelial phase for water absorption was further emphasized by Lamhamedi *et al.* (1992) who examined the ability of a number of genetically distinct dikaryons of *P. tinctorius* to influence xylem water potential of *P. pinaster* growing under moderate drought. Significant correlations were found between plant water potential, total root system resistance

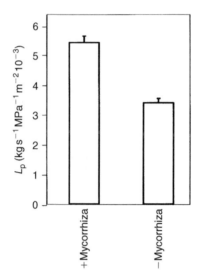

Figure 10.15 Measurements of the root hydraulic conductivity of mycorrhizal and non-mycorrhizal poplar root systems. Root hydraulic conductance was measured with a high-pressure flow meter in the excised undisturbed root systems of mycorrhizal and non-mycorrhizal poplar plants and used (together with the size of the root systems) to calculate the root hydraulic conductivity (L_p). From Marjanovic et al. (2005a).

and both the extension growth and rhizomorph diameters of the different fungal strains (Figure 10.17). Those genotypes that produced the most extensive systems of thick rhizomorphs enabled their plant partners to sustain the highest xylem water potential when soil water potential was low.

It is well known that soil water potentials can be reduced, particularly in surface horizons containing a high proportion of the ECM roots and mycelia, as water is extracted during the day by the demands of the transpiration flux. Predictably, this drying can have adverse effects upon the activities of the ECM symbiosis (Nilsen et al., 1998; Swaty et al., 1998) as well as on nutrient availability. However, after stomatal closure at dusk, the accumulated tissue water deficits are sufficient under some circumstances to enable roots to absorb and raise water from deep soil layers in a process termed 'hydraulic lift' (Caldwell and Richards, 1989). It has been hypothesized (Caldwell et al., 1998; Horton and Hart, 1998) that recharging of the water resources of the surface layers in this way could facilitate retention of mycorrhizal activity in what would otherwise be dry soil. In a test of this hypothesis, Querejeta et al. (2003) constructed compartmented microcosms enabling water with coloured tracers to be supplied to tap roots of droughted ECM live oaks (Quercus agrifolia) at dusk or dawn. Microscopic analysis of fine roots and their associated AM and ECM symbionts in the uppermost compartment at dawn, following feeding of tracer to the basal units at dusk, revealed extensive labelling of both root tissues and external mycelium in the surface soil. After feeding at dawn, no such upward movement was detected. Because the tracers used were membrane-impermeable, it was concluded that the pathway for water transport must be apoplastic, rather than by a process of leakage from plant tissues that would require reabsorption across fungal membranes.

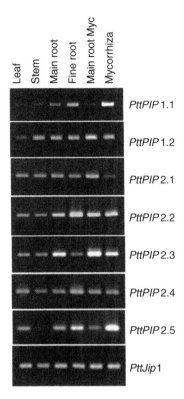

Figure 10.16 Expression profiles of *Populus tremula* × *tremuloides* plasma membrane proteins (*PttPIPs*) (i.e. aquaporins) in poplar organs and ectomycorrhizas. Total RNA was isolated from leaves, stems and roots (separated into fine and main roots) of non-mycorrhizal and mycorrhizal plants. Upregulation of *PttPIP1.1* and *PttPIP2.5* is observed in the mycorrhizal roots. From Marjanovic *et al.* (2005a).

Redistribution of dye through rhizomorphs of ECM fungi over distances of several centimetres from the root was observed.

Clearly, the potential implications of these observations for the functioning of both mycorrhizas and the wider ecosystem are large. It can be envisaged that hydraulic lift would be particularly advantageous in systems subjected to surface drying during the growing season. There is now a need to explore the extent to which this phenomenon occurs in nature and to quantify its effects in terms both of water supply to the symbionts and facilitation of nutrient mobilizing potentials.

Water shortages in soil can be expected to have influences upon the fungal community as well as upon the plants. There is ample evidence that drought can cause major changes both in the amount of ECM colonization present (Nilsen *et al.*, 1998; Bell and Adams, 2004), in the structure of the fungal communities (Swaty *et al.*, 1998, 2004; Shi *et al.*, 2002) and their physiological activities (Jany *et al.*, 2003; Bell and Adams, 2004). These impacts are likely to have lasting effects, especially if they arise as a result of loss of species that function most effectively under normal conditions of moisture supply. In those circumstances where hydraulic lift occurs, the process may provide some buffering of the fungal community against the effects

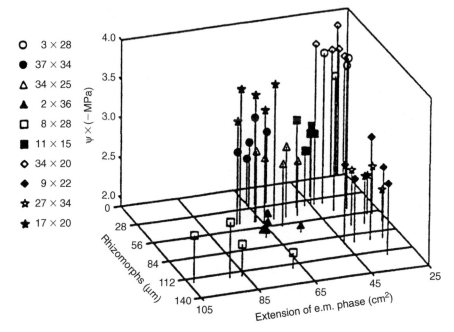

Figure 10.17 Effects of the diameter of rhizomorphs (μm) and of extension rates of the extraradical mycelial systems (cm^2) of a range of dikaryotic isolates (designated by number on left of figure) of *Pisolithus tinctorius*, upon the xylem water potential (ψ_{xylem}) of *Pinus pinaster* seedlings. From Lamhamedi *et al.* (1992), with permission.

of modest drought, but is unlikely to prevent the lethal impacts of prolonged or severe drought. Further, if hydraulic lift increases soil moisture content, effects both on activities of soil saprotrophs and on nutrient availability may also contribute to beneficial effects.

The differential ability of ECM fungi to withstand reductions of water potential when grown in pure culture has been adequately established (Mexal and Reid, 1973; Theodorou, 1978). Coleman *et al.* (1989) observed that all ECM fungi examined showed reduced growth as water potential decreased. Some, like *Laccaria laccata*, failing to grow below -1 MPa, whereas others, such as *Suillus granulatus* and *Cenococcum geophilum*, were able to grow at -3 MPa. These patterns of relative drought sensitivity are reflected to some extent in nature where drought is reported to promote increases of colonization by fungi like *C. geophilum* (Mikola, 1948). Pigott (1982) reported that colonization of *Tilia cordata* by this fungus enabled roots to survive in soils to water potentials as low as -5.5 MPa. However, in the absence of experimental analyses of the effects of *C. geophilum* on plant water relations, we are still unable to establish whether the increased presence of this fungus under conditions of drought makes any contribution to improving plant performance.

Conclusions

Early work emphasized the role of the fungal sheath in the processes of absorption, storage and transfer of P and laid the basis for our understanding of the physiology of nutrient transfer between fungus and plant in the ECM symbioses. Most of these studies were carried out using excised roots, but emphasis has moved more recently towards consideration of intact systems in which the effectiveness of different fungi in capture, transport and transfer of P to the plant has been examined in soil. The relationship between P acquisition and growth is now much more clearly understood and a number of studies have provided estimates of P inflow in intact ECM systems. These are of the same order of magnitude as those seen in AM systems. The role of the extraradical mycelium in exploring the soil and facilitating the mobilization of P from complex sources, both inorganic and organic, has been emphasized. The success of combined laboratory and field-based studies in elucidating the mechanisms and significance of increased P capture is clear. Both production of phosphatases, which release P from organic sources, and production of protons and organic anions, which can accelerate processes of chemical weathering, appear to be important in soils that have major mineral components. There is now a need to characterize more precisely the chemical nature of the main sources of P used by ECM plants in nature and to investigate the relative effectiveness of different species and races of fungal symbionts in providing access to them. Considerations of P uptake cannot be made in isolation. In many natural environments, P and N occur together in organic substrates and their release requires a complex suite of enzyme activities, some of them involving prior breakdown of polymeric carbon sources. We know little of the relative abilities of ECM and non-mycorrhizal microbial communities to achieve this breakdown, or of any competitive interactions that may take place between them.

We are beginning to recognize that, in addition to P and N limitation in forest ecosystems, other key elements (notably K, Mg and Ca), the availability of which has hitherto been largely taken for granted, may soon be in short supply. Clearly, we know a great deal less about the role of ECM fungi in the release, absorption and transfer of these elements and much more work needs to be done. There is much to suggest that through their abilities, both locally to secrete low molecular weight organic anions and more widely to release protons, these organisms may be critically involved in important biogeochemical weathering processes, as well as in the transfer of released cations to the plant. However, we still know too little about the relative contributions of ECM and saprotrophic communities to the overall mineral weathering budget to draw firm conclusions with regard to the role of the symbiotic community in the biogeochemistry of ectomycorrhizal environments dominated by ECM plants. Likewise it is difficult, on the basis of the small number of experiments carried out to date, to determine whether and under what circumstances ECM colonization can improve plant access to base cations. It also emerges not only that ECM colonization can significantly change both the processes of transfer to, and the balance of cations within, the tissues of the plants, but also that the mycobionts differ greatly among themselves in their propensities to influence cation supply to the root.

Recent work has extended knowledge of ECM symbioses beyond plant nutrition, to encompass aspects of responses to stresses. It is clear that ECM fungi have the ability to confer tolerance to both heavy metals and drought. In both cases, part of the beneficial effects may stem from improved nutrition, but it is also clear that direct effects conferred by activities of the fungal partners are implicated and that there is considerable diversity in responses, dependent on the identity of symbiotic partners. The significance both of enhanced tolerance and of symbiont-dependent effects in field situations requires ongoing investigation.

Section 3
Ericoid, orchid and mycoheterotrophic mycorrhizas

11

Ericoid mycorrhizas

Introduction

Until the 1990s, relationships between the plants comprising the order Ericales were addressed using traditional evolutionary taxonomic approaches largely involving the use of morphological characters (Cronquist, 1981). On this basis, the Ericaceae, Epacridaceae and Empetraceae were recognized as distinct families of the order. More recently, the combined application of morphological and phylogenetic methodologies, the latter making use of nuclear and chloroplast DNA sequence data, has enabled a thorough re-evaluation of relationships within the entire ericoid clade (Kron *et al.*, 2002). In this analysis (Figure 11.1), plants of what were traditionally seen as members of the families Epacridaceae and Empetraceae lose their family status and are placed in tribes, the Styphelioideae and Ericoideae, respectively, within the broadened Ericaceae. In order to avoid confusion, members of the Styphelioideae are subsequently referred to in this text as 'epacrids'.

Whereas the ericoid tribe have representatives which occur as dominant plants over vast areas of the northern hemisphere and in southern Africa, epacrids are of particular importance in Australasia. Throughout their geographic ranges, plants of these tribes, in the case of the ericoids often in association with members of the Vaccinioideae, are major components of distinctive heathland ecosystems (see Chapter 15). A further structural feature of the major tribes of the Ericaceae, one not used by, but supportive of, the phylogeny proposed by Kron *et al.* (2002), is that they have uniquely specialized distal roots, the epidermal cells of which are invaded by ascomycetous fungi to form 'ericoid mycorrhiza' (ERM) (Figures 11.2 and 11.3a, b). Only the least well supported branches of the consensus tree (see Figure 11.1), comprising Arbutoideae, Monotropoideae and Enkianthoideae, consist of plants which lack ericoid mycorrhiza, these characteristically forming symbioses of the arbutoid (see Chapter 7), monotropoid (see Chapter 13) and arbuscular mycorrhizal (AM) types, respectively (Abe, 2005). *Clethra*, a related member of the order Ericales but lying outside the Ericaceae (see Figure 11.1) is also reported to be an AM-forming genus (Kubota *et al.*, 2001). It is emerging that the mycorrhizas formed by at least some of the epiphytic members of the Ericaceae may be structurally distinguishable from typical ERM. Rains *et al.* (2003) report the presence of rudimentary mantle and Hartig net in *Cavendishia melastomoides* and *Disterigma humboldtii*. Intensive study of another epiphytic *Cavendishia* spp., *C. nobilis*, revealed a well-developed mantle and

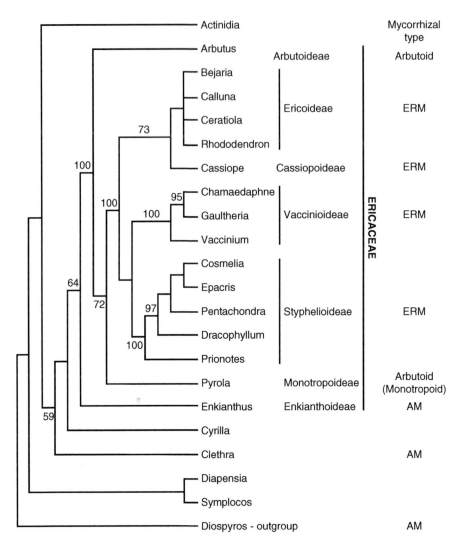

Figure 11.1 Relationships between phylogeny and mycorrhizal types within the order Ericales. The cladogram consists of a strict consensus of three trees produced by combining analyses of the 18s, rbcL and matK genomes of 22 taxa following Kron *et al.* (2002). The ericoid type of mycorrhiza (ERM) (right), is restricted to those tribes of the family Ericaceae that are most strongly supported by the consensus tree.

Hartig net as well as intracellular penetration, producing a structure reminiscent of an ectendomycorrhiza (Setaro *et al.*, 2006; see Chapter 7). These observations suggest that ectendomycorrhiza may be widespread in this Andean clade of the Vaccinoideae.

By combining fossil records with molecular clock estimates for the origins of the Ascomycota, Cullings (1996) indicated that the ericoid mycorrhizal (ERM) condition may have arisen by the early Cretaceous, around 140 million years ago. This view is broadly consistent with the proposal that the ancestral Ericales originated in, and radiated from, southern Gondwana during the mid- to late Cretaceous (Specht, 1979; Dettmann, 1992).

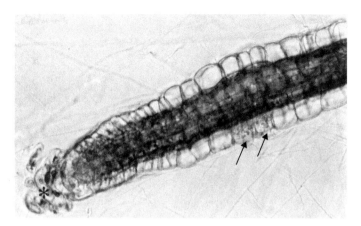

Figure 11.2 Light micrograph of the apical region of a hair root of *Calluna vulgaris*, showing the root cap (*) and the commencement of mycorrhiza formation in cells distal to the root apex (arrowed). Photograph DJ Read.

Much controversy surrounded the early studies of ericoid mycorrhizas, especially with respect to the identity of the fungal partner and, in particular, its distribution within the tissues of the plant. These arguments distracted attention from more important questions about the role of mycorrhizas in the growth, nutrition and ecology of the plants. The controversies have been discussed elsewhere (Harley, 1969; Read, 1983) and will only be briefly referred to later in this chapter. The first systematic studies of the ERM root were largely devoted to isolation of fungi from the hyphal complexes within the epidermal cells (Friesleben, 1935; Bain, 1937; Burgeff, 1961; Nieuwdorp, 1969; Pearson and Read, 1973a). These workers were at pains to point out that, while the fungi which they obtained from the hyphal complexes always yielded slow growing usually dark-coloured and sterile mycelia, a wide range of colony morphologies was evident among their cultured isolates. Access to molecular techniques has recently enabled improved evaluation of the taxonomic status of some of these isolates and this, in turn, has led to reappraisal of the status of some of the most prominent taxa. Most notably, sequence analysis of the ITS1 and 5.8S-ITS2 regions of rDNA in the genus Helotiales by Zhang and Zhuang (2004) has led to a major shift of nomenclature. The fungus strains used in nearly all studies of ERM function, previously known as *Hymenoscyphus ericae* (Read) Korf and Kernan, have been transferred to a new genus, *Rhizoscyphus* W.Y. Zhuang and Korf. The new combination *Rhizoscyphus ericae* (Read) W.Y. Zhuang and Korf is used throughout this text where reference is made to isolates known to be mycorrhizal with ericoid plants. However, in accordance with others (e.g. Hambleton and Sigler, 2005), the name '*Hymenoscyphus* aggregate' is retained where reference is made to the broader genetic consortium which includes fungi not known to form ericoid mycorrhiza.

Sadly, progress towards better understanding of the functional roles of the fungal symbionts has not kept pace with that obtained in the domain of phylogenetics. Nonetheless, there is increasing evidence that some of the genetic diversity revealed in molecular studies is reflected in diversity of function. This chapter will place emphasis upon structural, taxonomic and some functional aspects of ericoid mycorrhizas. Their possible ecological significance is reviewed in Chapter 15.

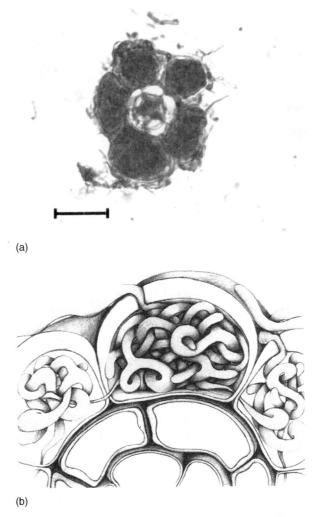

(a)

(b)

Figure 11.3 (a) Transverse section of a mycorrhizal root of *Calluna vulgaris*, showing the epidermis of one cell layer fully colonized by fungal hyphae. Bar = 100 μm. Photograph DJ Read. (b) Diagram of epidermal cells showing entry of fungal hyphae through thickened outer epidermal walls and intracellular hyphal complexes. From Peterson *et al.* (2004).

Roots of ericoid plants are delicate structures, the anatomy of which shows considerable uniformity across all tribes of the Ericaceae which are characterized by this type of mycorrhiza (Peterson *et al.*, 1980; Berta and Bonfante-Fasolo, 1983; Ashford *et al.*, 1996; Allaway and Ashford, 1996; Bell *et al.*, 1996; Read, 1996; Cairney and Ashford, 2002). Their most distinctive feature is the absence of root hairs. Because of their narrow diameters, which range from 100 μm to <50 μm in distal regions, such structures are referred to as 'hair roots'. Hair roots have a monarch stele containing only a single file of very small tracheids and sieve elements within a ring of pericycle cells. Surrounding the vascular elements is a two-layered cortex consisting of an endodermis and a suberized exodermis. The outermost cell layer, the epidermis, provides both the interface with the soil and it is exclusively these cells that are colonized by ERM fungi. They are often inflated and, when they are colonized by fungal endophytes, their volume is fully occupied by hyphal complexes (see Figure 11.3a, b). The root apical meristem, a small group of undifferentiated cells, is protected distally by a root cap which is invested in mucilage (see Figure 11.2). The mucilage extends backwards from the apex as a thin but discontinuous layer over the outer surface of the root.

The epidermal layer of the hair root is an ephemeral structure, which disappears in older roots. Suberized and thickened cells derived from the two cortical layers then come to form the outer surfaces of the root (Peterson et al., 1980; Allaway and Ashford, 1996). Since mycorrhizal colonization is restricted to expanded epidermal cells, these maturation processes define the 'window of opportunity' for formation of the symbiosis in both space and time. Secondary thickening provides for longevity in the more mature part of the roots, whereas the unthickened hair roots, which are of determinate growth, appear to be periodically shed. We lack precise information concerning the patterns of hair root turnover in nature, though Kerley and Read (1995) observed that, in moist heaths of the northern hemisphere, populations of colonized hair roots were present throughout the year. In the heaths of Australia, particularly those with a Mediterranean climate yielding hot dry summers, hair root length has been shown to decline rapidly as the soil dries out, but to increase progressively in both length and extent of fungal colonization as soils are re-wetted in the autumn (Hutton et al., 1994; Bell and Pate, 1996). Ashford et al. (1996) describe an interesting situation in the epacrid species *Lysinema ciliatum*, where apparently specialized thick-walled cells of the epidermis, which are readily detached from the root surface, become preferentially colonized by ERM fungi. It is suggested that these cells may act as resistant propagules, able to survive dry conditions and colonize new hair roots as they emerge after rain.

The colonization process

The hyphae of ERM fungi form a loose network over the zone of the hair root which contains a mature epidermis (Figure 11.4). As the root apex extends, new epidermal cells are differentiated and the hyphae at the leading edge of the network advance to colonize them. Slowing of root growth, for whatever reason, can enable the hyphal network to reach the root apex, but normally there is a zone of differentiating cells immediately behind the meristem which is free of fungal hyphae.

The molecular basis of the processes of recognition between fungi forming this network over ericoid roots is so far unknown, but ultrastructural and cytochemical observations, most of them using strains of R. ericae, provided insights. Some strains of R. ericae typically produce an extracellular fibrillar sheath which is rich in polysaccharides (Bonfante-Fasolo and Gianinazzi Pearson, 1982). This sheath is more strongly developed in infective (Figure 11.5a) than non-infective (Figure 11.5b) strains (Gianinazzi-Pearson and Bonfante-Fasolo, 1986) and it has been suggested (Gianinazzi-Pearson et al., 1986) that it may anchor the fungus to the plant as the first step in mycorrhiza formation (Figure 11.6). Using the gold-labelled lectin Concanavalin A (Con A) as a cytochemical marker for mannose or glucose residues, Bonfante-Fasolo et al. (1987b) observed an abundance of these compounds when an infective strain was in association with a plant root. While the fibrillar sheath may

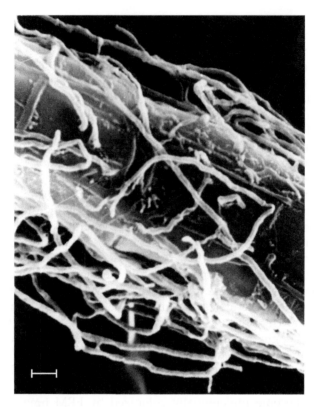

Figure 11.4 Scanning electron micrograph of a hair root of *Rhododendron* colonized by loose wefts of fungal hyphae which are penetrating the outer epidermal wall. Bar=10 μm. From Duddridge and Read (1982b), with permission.

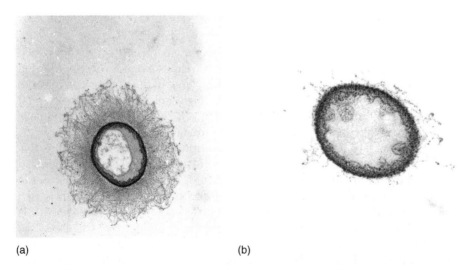

(a)   (b)

Figure 11.5 Development of surface fibrils on hyphae of *Rhizoscyphus ericae* in the presence of roots of *Calluna vulgaris*. (a) Strongly compatible isolate, producing many fibrils. (b) Sparse production of fibrils by weakly compatible isolate. Diameter of hyphae approx. 3 μm. Photographs courtesy V Gianinazzi-Pearson.

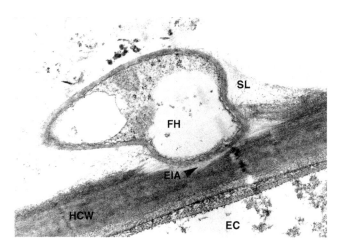

Figure 11.6 Transmission electron micrograph of a hypha of *Rhizoscyphus ericae* (FH) attached to the surface of a root epidermal cell (EC) of *Rhododendron ponticum*. Note the presence of fibrils (SL) and an electron-lucent zone (EIA) at the site of penetration of the cell wall (HCW). Photograph courtesy V. Gianinazzi-Pearson.

assist in attachment, which is clearly a prerequisite for further colonization, doubt remains about the extent to which it is involved in recognition of host as distinct from non-host plants because strains with a typical fibrillar sheath formed attachments to and penetrated roots of the non-host *Trifolium pratense* (Bonfante-Fasolo *et al.*, 1984). Differences between the responses of hosts and non-host roots to challenge by the fungus were seen only in the intracellular situation where, in the case of one ericoid plant, the fibrillar material was lost as the hyphae became invested by plant plasma membrane, while in non-host it was retained over a period during which organization of the invaded cell broke down.

Penetration of the plant wall

From the hyphae of the surface network branches emerge at right angles to penetrate the epidermal cells (see Figures 11.4, 11.6, 11.7). Typically, there is a single penetration point per cell, but multiple entry points are sometimes observed (Read and Stribley, 1975). Each entry may be preceded by the production of an appressorium but, as this structure is not always found, it is clearly not a prerequisite for successful colonization.

The mechanism of penetration is not fully understood. There is evidence from the use of the PATAG test for localization of polysaccharide that the carbohydrate-rich fibrils which ensheath the hyphae and, although present at the point of attachment to the surface of the epidermal cells, disappear as the hyphal tip penetrates the outer wall. An electron lucent zone becomes evident where the fungus enters the inner wall layer of the plant (see Figure 11.6; Duddridge and Read, 1982a; Bonfante-Fasolo and Gianinazzi-Pearson, 1982). This might represent dissolution of the plant cell walls by fungal enzymes. When grown in pure culture, *R. ericae* has the ability to use a range of plant cell wall-related mono-, di- and polysaccharides, including carboxymethyl cellulose (CMC), as sole sources of C (Pearson and Read, 1975; Varma

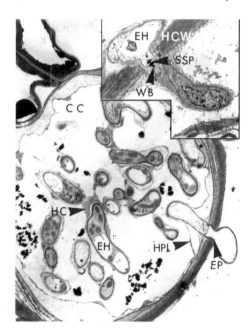

Figure 11.7 Newly colonized cell of a *Rhododendron* seedling showing point of penetration (EP) by a hypha of *Rhizoscyphus ericae*. Invagination of the plasma membrane of the plant cell is shown (HPL). Plant cytoplasm (HC), intracellular fungal hypha (EH). Inset: higher magnification of fungal penetration through the plant cell wall showing a simple septum in the hypha (SSP) and Woronin body (WB). From Duddridge and Read (1982b), with permission.

and Bonfante, 1994). Two polygalacturonases were identified when the fungus was grown with pectin as sole C source (Perotto *et al.*, 1993) and activity of both β 1–4 and β 1–3 glucanase was detected in cultures supplied with CMC or sterile root segments (Varma and Bonfante, 1994). Since these enzymes are involved in penetration of cell walls by a number of plant pathogenic fungi (Hahn *et al.*, 1989), their production by *R. ericae*, combined with the ultrastructural evidence indicating weakening of the plant cell wall in advance of the penetrating fungal hypha, strongly suggests the involvement of hydrolase enzymes in the penetration process. There is also the likelihood, discussed later, that these enzymes might be deployed in soil organic matter, resulting in release of the nutrients contained within it.

Features of intracellular colonization

Electron microscopy of mature infection units (Nieuwdorp, 1969; Bonfante-Fasolo and Gianinazzi-Pearson, 1979; Peterson *et al.*, 1980) has shown that colonizing hyphae retain a discrete structural integrity within the plant cell. In addition, there is some deterioration in the appearance of plant cytoplasm when there is no loss of integrity in the fungus. In a detailed analysis of the sequence of events from colonization of the plant cell through to collapse of the association, it was confirmed that breakdown begins with deterioration of the plant rather than the fungal tissue (Duddridge and Read, 1982b). Seedlings of *Rhododendron ponticum* were grown

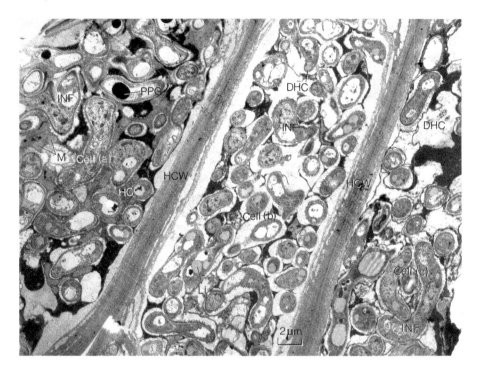

Figure 11.8 Transmission electron micrograph of a root of *Rhododendron ponticum*, showing three adjacent cells in longitudinal section and intracellular colonization of different ages. (a) Healthy, mature colonization by hyphae that fill the cell; (b) plant cytoplasm is degenerating, but the intracellular hyphae appear healthy; (c) cytoplasmic degeneration in both plant and fungus. DHC, degenerating plant cytoplasm; PPG, polyphosphate granule; LI, lipid droplet; HCW, plant cell wall; INF, intracellular fungal hypha. Bar = 2 μm. From Duddridge and Read (1982b), with permission.

either in soil partially sterilized by γ-irradiation and then inoculated with the ERM symbiont, or in soil freshly collected from underneath *Rhododendron* bushes. Surface colonization of roots occurred in irradiated soil after three weeks and in natural soil after four weeks, with penetration of the cortical cells following immediately (see Figures 11.6, 11.7). Once within the cell, the fungal hyphae proliferate extensively. The plant plasma membrane invaginates to envelope each branch of the invading fungus, but is separated from the fungal cell wall by a thin electron-lucent layer, the so-called interfacial matrix (see Figure 11.7). This contains flocculent, electron dense pectic material (Duddridge, 1980). The SEM and TEM studies have confirmed that colonization occurs through the outer wall of the cortical cell, so that each cell is an individual infection unit. Consequently, even adjacent cells may have fungal complexes which are of different ages (Figure 11.8). When each complex is mature, the volume of the plant cell is almost completely occupied by fungal hyphae and little or no vacuolar volume is apparent. Outside the electron-lucent zone the hyphae are ensheathed by plant cytoplasm which is packed with rough endoplasmic reticulum and mitochondria, suggesting that, at this stage, the unit is the site of considerable physiological activity (Figure 11.9).

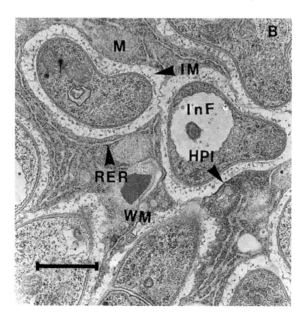

Figure 11.9 Early stage of degeneration of mycorrhizal colonization in roots of *Rhododendron ponticum*. The plant cells contain abundant rough endoplasmic reticulum (RER). M, mitochondria; IM, interfacial matrix; HPI, plant plasma membrane; InF, intracellular hypha; WM, wall material. Bar = 1 μm. From Duddridge and Read (1982b), with permission.

The first sign of breakdown observed by Duddridge and Read (1982b) was a loss of structural integrity of organelles in the plant cell, particularly the mitochondria. Plant cytoplasm then degenerates and the electron-lucent area between the plant plasma membrane and fungal cell wall becomes progressively wider. The integrity of the plant plasma membrane is finally lost and most of the plant organelles degenerate before fungal deterioration occurs (Figure 11.10). Evidence of deterioration in the fungal hyphae is seen as an increase in the size of the vacuoles only in the later stages of plant degeneration. Final breakdown of the fungus is not complete until after the plant cell loses it integrity. At the end of the breakdown process, therefore, the cell is empty apart from the debris of earlier fungal occupation. In the studies of Duddridge and Read (1982a, 1982b), the first indication of breakdown was observed eight weeks after inoculation and hence about four weeks after cellular penetration in inoculated soil and about 11 weeks after planting into natural soil. The differences of timing are probably attributable to differences in the vigour of the partners in the two systems, both fungus and plant being more active in the irradiated soil.

The events revealed by TEM are thus distinct from those described by light microscopy. Since breakdown occurs first in plant cytoplasm, nutrient transfer either from fungus to plant or in the reverse direction must take place during the few weeks after colonization, when both partners have full structural and hence presumably physiological integrity. This means that the active life span of the individual colonized cells is not more than five or six weeks. In addition, the notion of digestion or lysis of the fungus by the plant is not tenable.

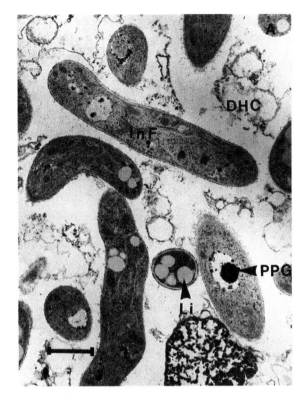

Figure 11.10 Late stage in the degeneration of mycorrhizal colonization in roots of *Rhododendron ponticum*. The plant cytoplasm (DHC) has almost completely degenerated, whereas the fungal hyphae (InF) are apparently still alive. Lipid droplets (Li) and polyphosphate granules (PPG) are apparent. Bar = 1 μm. From Duddridge and Read (1982b), with permission.

Ultrastructural analysis has also provided information concerning the taxonomic status of fungi associated with ericoid roots (see below).

The fungi forming ericoid mycorrhizas

The earliest attempts to determine the identity of the fungi involved in the formation of ericoid mycorrhizas were surrounded by controversy. There were claims (Rayner, 1915, 1927) that roots were colonized by a species of *Phoma*, the mycelium of which extended from the roots, through the shoots and into the floral organs where it reached the seed coat. This so-called 'systemic infection' was considered to enable the radicle of the germinating seed to become colonized as it passed through the seed coat. The view was expressed that subsequent development of the seedling was dependent upon this transfer of colonization (Rayner, 1915; Addoms and Mounce, 1931). At the time, some workers, notably Knudson (1929, 1933) in America and Christoph (1921) in Europe, took issue with this conclusion, showing that systemic infection was neither normal nor an essential prerequisite for seedling development, which could readily be obtained under aseptic conditions. If nothing else, these controversies exposed the need for rigour in isolation of ERM fungi from roots and for re-inoculation to ensure that a typical mycorrhiza was produced. These simple prerequisites for verification of the status of microbial colonists had earlier been clearly expounded in the form of Kochs postulates (Koch, 1912), but while they were widely followed in plant pathology, their application to studies of mycorrhizal colonization, especially of ericaceous plants, was largely overlooked by Rayner and her colleagues.

Problems with interpretation of results obtained from isolation studies persist. While most of the early workers did indeed target the intracellular hyphal complex which is the diagnostic feature of the ERM symbiosis, it has become common practice simply to take segments of hair root, normally several millimetres in length, and to culture fungi emerging from any part of the tissue (e.g. Berch et al., 2002; Allen et al., 2003; Midgley et al., 2004a). Such approaches will yield a mixture of ECM and non-mycorrhizal fungi, but the origin of any one of them in the root piece remains in doubt. While such methods are likely to overestimate the number of fungi regarded as being mycorrhizal, exposure of delicate hair root segments to sterilants, most commonly solutions of hypochlorite or hydrogen peroxide, will lead to a selective reduction in their number. It has long been known that chemical sterilants, even when applied at low concentrations and for short durations, can kill internal as well as external fungal occupants of hair roots (McNabb, 1961; Singh, 1965). Their use may account for the low rates of recovery of fungi reported in some studies (Berch et al., 2002). On the basis of their experiences in attempting to isolate mycelial fungi forming ectomycorrhizas, Harley and Waid (1955) recommended a less drastic approach to surface sterilization involving serial washing of roots in sterile water. When applied to ericoid hair roots this facilitated excellent recovery of ericoid endophytes (Pearson and Read, 1973a; Singh, 1974) which can be observed as they grow directly out of colonized cells (Figure 11.11).

Doak (1928) was the first to report isolation of an endophyte from an ericoid mycorrhizal root, in this case of *Vaccinium*, and to confirm its mycorrhizal status by back-inoculation. This type of experiment was repeated by Bain (1937) using a number of North American plants, as well as by Freisleben (1933, 1934, 1936), Burgeff (1961) and Pearson and Read (1973a) in Europe and McNabb (1961) and Reed (1987) in Australasia. Systematic observations of this kind revealed that the fungi isolated from ericoid hyphal complexes, and hence likely to be mycorrhizal, were predominantly slow growing, asexual and of dark colouration. Slow growing white or pale grey isolates were also occasionally reported (Melin, 1925; Bain, 1937).

One typically dark isolate obtained by Pearson and Read (1973a) eventually produced apothecia in culture, enabling its taxonomic placement, first in the genus *Pezizella* as *P. ericae* (Read, 1974), later in *Hymenoscyphus*, as *H. ericae* (Read) Korf and Kernan (Kernan and Finocchio, 1983) and subsequently in *Rhizoscyphus* as *R. ericae* (Read) Zhang and Zhuang (Zhang and Zhuang, 2004). This fungus has occasionally been placed in the order Leotiales but, in strict terms, while it is a member of the broader Class Leotiomycetes, both the genus *Rhizoscyphus* and the broader *Hymenoscyphus*-aggregate are members of the order Helotiales. Teleomorphs identical with those described by Read (1974) have since been reported from ericoid isolates in France (Vegh et al., 1979) and Canada (Hambleton et al., 1999), but apothecia have still not been observed in nature. A high proportion of the dark slow growing isolates obtained by Pearson and Read (1973a), including that yielding teleomorphs,

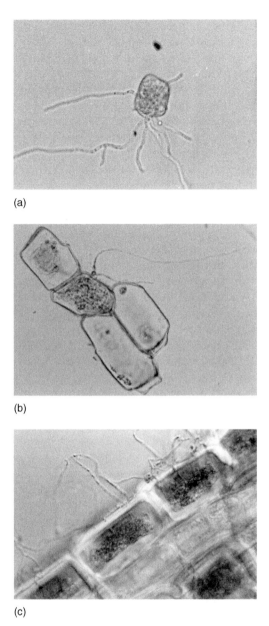

(a)

(b)

(c)

Figure 11.11 Emergence of hyphae of *Rhizoscyphus ericae* from colonized epidermal cells of *Calluna vulgaris* after serial washing, maceration and culture on water agar. (a) Isolated single epidermal cell with hyphae emerging after 7 days incubation. (b) Hyphae emerging from colonized cell situated in group of uncolonized individuals. (c) Hyphae emerging from epidermal cells of intact root segment. Photographs DJ Read.

segmented in agar to produce arthroconidia. An anamorph of this type was subsequently identified as *Scytalidium vaccinii* by Dalpé *et al.* (1989). On the basis of phenological and genetic similarities, Egger and Sigler (1993) later confirmed that *S. vaccinii* was an anamorph of *R. ericae*.

Attention has turned more recently to questions concerning the broader phylogenetic interrelationships of the slow growing fungi which can be isolated from ericoid roots and confirmed as ERM fungi, but which also have wider occupancy in the soil domain. Particular interest has surrounded the relationships within the broader *Hymenoscyphus* grouping which includes *R. ericae*. With the growing amount of available DNA sequence data has come the increasing consensus that the genus *Hymenoscyphus* contains an aggregate of genetically related fungi. Vrålstad *et al.* (2000) demonstrated that the fungus producing the ectomycorrhizal morphotype *Piceirhiza bicolorata* (*sensu* Agerer *et al.*, 1987–2002; see Chapter 6, Colour Plate 6.1c), in addition to having similar cultural characteristics, suggested 95% sequence identity in the ITS1 region to *R. ericae*. An inferred ITS1 phylogeny suggested that a single major evolutionary lineage of the fungus forming *P. bicolorata* embraced *R. ericae* in a 100% bootstrap-supported clade. This not only challenged the view that *R. ericae* is genetically well defined but also exposed the possibility that within the aggregate there may be genotypes capable of producing ECM as well as ERM colonization. Subsequent work involving comparative phylogenetic analysis of root isolates obtained from ERM and ECM hosts, together with those from culture collections (Vrålstad *et al.*, 2002a), strengthened the notion that *R. ericae* is part of a broader and closely related assemblage of helotialean ascomycetes. A large number of ITS sequences with broad affinities to Helotiales were obtained. Of these, 75% grouped within the *Hymenoscyphus* aggregate. Construction of a most parsimonious tree (MPT) from the data set revealed four major clades that were robust to alternative alignments. Type cultures of *R. ericae* and *S. vaccinii* clustered in Clade 3. Clade 1 also contained some ERM isolates while Clades 2 and 4 were largely made up of isolates from ECM roots. The most recent genetic analysis of the *H. ericae* aggregate, again primarily based upon ITS data (Hambleton and Sigler, 2005), has largely confirmed but has also broadened Vrålstad's phylogeny. A detailed MPT analysis (Figure 11.12) again resolved four major clades which are similar in terms of their constituent taxa to those described by Vrålstad *et al.* (2002a).

From a functional standpoint, the critical question concerns the mycorrhizal status of members of the *H. ericae* aggregate. When Vrålstad *et al.* (2002b) used cross inoculation experiments to test the ability of *H. ericae* strains from ECM or ERM roots to form these types of mycorrhiza on aseptically grown host plants, they observed that isolates from ECM roots were unable to form ERM associations and *vice versa*. Hambleton and Sigler (2005) also observed that isolates from ericoid roots produced only ericoid mycorrhiza. Such observations provide additional support for the view that Clade 3 of Vrålstad *et al.* (2002b) and of Hambleton and Sigler (2005) (see Figure 11.12) should be distinguished on functional as well as genetic grounds and designated *R. ericae* within the broader *H. ericae* aggregate.

The Hambleton and Sigler (2005) study, in addition to examining a wide range of slow growing dark sterile (DS) fungi of ERM and ECM origin, also included isolates of the paler types, having white to grey colouration in culture, that were reported in the older literature (see above). These were described as 'the variable white taxon' (VWT) by Hambleton and Currah (1997). On the basis of their morphological and

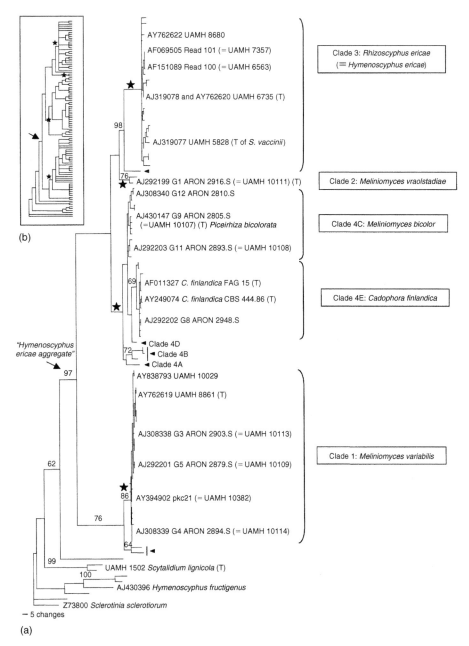

Figure 11.12 Phylogenetic affinities within the *Hymenoscyphus ericae* aggregate as revealed by ITS gene sequence analysis. The cladogram is one of 5000 most parsimonious trees (MPT's) from an aborted analysis of the ITS/1 and 5.8S/ITS2 sequences of cultures deposited in GenBank. All ingroup sequences were derived from cultures isolated from roots except for four derived from DNA extracted directly from roots. (a) A majority of taxa within the *H. ericae* aggregate were resolved into four clades indicated by stars adjacent to the relevant nodes, these clades being named in the blocks on the right of the figure. *Rhizoscyphus ericae* (Clade 3) contains almost all of the proven ERM fungi. (b) Strict consensus of all 5000 MPT's with taxon names removed. Stars indicate the same four clades as in (a). Modified by Sarah Hambleton from Hambleton and Sigler (2005).

ITS-based characterization of these isolates, Hambleton and Sigler (2005) created the new anamorphic genus *Meliniomyces* (Hambleton and Sigler, anam. gen. nov) for the two VWT's. Two species, *M. vraolstadiae* and *M. bicolor* form parts of Clades 2 and 4 respectively within the *H. ericae* aggregate, while the other, *M. variabilis*, most strains of which appear to be non-mycorrhizal, makes up Clade 1 (Figure 11.12).

Also within Clade 4 are fungi in the genus *Cadophora* (= *Phialophora*) which are known to produce ecto- and ectendomycorrhizas (see Chapters 5 and 7). In the only case of its kind so far reported, Villareal-Ruiz *et al.* (2004) observed that a member of this genus, *C. finlandica* (= *P. finlandia*), originally isolated from a *P. bicolor* ectomycorrhiza, was capable of forming both ericoid mycorrhiza on *Vaccinium* and ectomycorrhiza on pine when the two plant species were grown in dual culture. The possible ecological significance of this observation is discussed later (see Chapter 15).

Phylogenetic studies of the status of isolates obtained from roots of Australian epacrids have again revealed that, within a group of largely sterile slow growing darkly pigmented ascomycetous fungi, there is considerable genetic diversity (Hutton *et al.*, 1994; McLean *et al.*, 1999; Chambers *et al.*, 2000; Midgley *et al.*, 2002, 2004b; Bougoure and Cairney, 2005a). While it seems clear that most of these fungi are Helotiales, evidence that some are attributable to the *H. ericae* aggregate has now been produced by Midgley *et al.* (2002) from studies of *Woolsia pungens* and by Bougoure and Cairney (2005b) for the Australian endemic *Rhododendron lochiae*. A neighbour-joining tree, based upon ITS sequence data from northern hemisphere ericoid and Australian epacrid isolates (Sharples *et al.*, 2000a), separated ERM fungal isolates into two strongly supported clades. Whereas one clade included endophytes from the epacrids *W. pungens* and *Epacris impressa* as well as all the *Hymenoscyphus* isolates from non-epacrid ericaceous plants of the northern hemisphere, the other encompassed most of the epacrid endophytes described at that time which were regarded as having affinities with Helotiales while being outside the *Hymenoscyphus* aggregate. There is clearly a need for more precise evaluation of the taxonomic status of Australasian epacrid isolates. It is noteworthy that, to date, there are no molecular analyses of epacrids endophytes from the Western Australian sandplain heaths or from epacrids occurring outside continental Australia.

Another ascomycetous fungus frequently isolated from hyphal complexes of ERM roots in both northern and southern hemispheres is *Oidiodendron* (Ascomycota, Onygenales) (Burgeff, 1961; Dalpé, 1986; Douglas *et al.*, 1989; Hambleton *et al.*, 1998; Monreal *et al.*, 1999; Berch *et al.*, 2002; Allen *et al.*, 2003; Bougoure and Cairney, 2005a, 2005b; Addy *et al.*, 2005). While a number of species have been reported, many of these appear to be conspecific with *O. maius* (Hambleton *et al.*, 1998). This fungus has been confirmed to be mycorrhizal in re-inoculation experiments on *Gaultheria shallon* in Canada (Monreal *et al.*, 1999), as well as *Calluna vulgaris* (Leake and Read, 1991), *Rhododendron* spp. (Douglas *et al.*, 1989) and *Erica arborea* (Bergero *et al.*, 2000) in Europe. Isolates of *Oidiodendron* obtained by Bougoure and Cairney (2005b) from roots of *R. lochiae* produced typical ericoid hyphal complexes in *Vaccinium macrocarpon*.

Among other culturable ascomycetous genera which have been detected in ericoid roots by molecular methods and shown to form hyphal complexes in host cells are *Capronia* spp. (Chaetothyriales) (Bergero *et al.*, 2000; Berch *et al.*, 2002; Allen *et al.*, 2003) and *Acremonium strictum* (Xiao and Berch, 1995). The functional status of the former is unknown, but *Capronia* spp. have been reported to be mycoparasites (Untereiner and Malloch, 1999). The occurrence in ericoid roots of *A. strictum*, which is widely known as a shoot endophyte of angiosperms, was reported following sequence-based searches, but its presence has not been confirmed by subsequent workers.

A number of fungi reported to occur in ERM roots are so far unculturable, which again makes interpretation of their status difficult. Among these, basidiomycetes of the genus *Sebacina*, most notably *S. vermifera*, are prominent. It seems that most of the *Sebacina* spp. present in ericaceous roots can be assigned to Subgroup B in the broad phylogeny of the genus (see Figure 6.5, Chapter 6). Allen *et al.* (2003) found *Sebacina*-like DNA in 11 of 15 mycorrhizal roots of *G. shallon* and, since 92 of the 156 cloned DNAs from these roots were of the *Sebacina* type, they concluded that these fungi were regular associates of this plant in its natural habitat. Ultrastructural studies have revealed that hyphae with basidiomycete-type septal pores are occasionally present in ericoid epidermal cells. Whereas dolipore septa have been observed primarily in dead cells (Duddridge and Read, 1982b), they have been seen also in living cells of *C. vulgaris* (Bonfante Fasolo, 1980), *Pieris* (Peterson *et al.*, 1980) and *Dracophyllum* (Allen *et al.*, 1989). Septal pores diagnostic of the order Auriculariales, which encompasses *Sebacina*, have been reported (Bonfante-Fasolo and Gianinanzzi-Pearson, 1979). Since sebacinoid fungi are known to be mycorrhizal with some orchid (Warcup, 1988; McKendrick *et al.*, 2002a; Selosse *et al.*, 2002a; see Chapter 12) and ECM plants (Weiss and Oberwinkler, 2001; Glen *et al.*, 2002; Selosse *et al.*, 2002b; see Chapter 6), their status in ERM roots requires to be investigated further. Selosse *et al.* (2007) claimed that a number of sebacinoid fungi of Subgroup B formed ERM, but their definition of the symbiosis as 'a morphogenetic process uniting roots and soil fungi' does not conform to that employed here (see Introduction). Anatomical and cultural studies of hyphal complexes in those ericoid roots shown by DNA analysis to harbour *Sebacina*-like genotypes are required. The report that an orchid-derived strain of *S. vermifera* can be cultured (Warcup, 1988) suggests that direct culturing approaches would be worthwhile.

While molecular studies have alerted us to the presence of some hitherto unrecorded fungi like *Sebacina* in ericoid hair roots, by not detecting others which were previously claimed to be mycorrhizal, they have contributed to the resolution of some issues. The frequent occurrence of fruit bodies of the basidiomycete *Clavaria argillacea* in soil surrounding ericaceous plants in nature (Gimingham, 1960; Seviour *et al.*, 1973; Moore-Parkhurst and Englander, 1982), lead some to the view that this fungus may be a ERM symbiont. Serological (Seviour *et al.*, 1973), immunocytochemical (Mueller *et al.*, 1986) and nutrient transfer techniques (Moore-Parkhurst and Englander, 1982) indeed demonstrated that this fungus can form close associations with ericoid roots, but the failure of sequencing studies to detect clavarioid genotypes within roots castes further doubt on the possibility that the relationship is mycorrhizal.

Functional aspects of ericoid mycorrhizas

The soils that support ericaceous vegetation are characteristically extremely poor in available nutrients and it is logical to expect nutritional benefits to arise from ERM colonization. Experimental analysis of ERM effects has, however, been made difficult by the fact that sterilization of those organic soils on which northern hemisphere ericoid plants typically grow may release nutrients or toxins. Using autoclave-sterilized peat to generate non-mycorrhizal controls, Freisleben (1936) concluded that the main effect of ERM fungi was to detoxify the peat. Subsequent experiments also using autoclaved soil (Brook, 1952; Morrison, 1957b) provided some indication that ERM colonization might improve plant nutrition, but problems arising from toxicity and the uncertain nature of the inoculum made interpretations of their results difficult. Progress has been made using two distinct approaches to the assessment of function, one of which involves examination of the physiological attributes of ERM fungi in pure culture, while the other uses comparative analysis of the responses of ERM and non-mycorrhizal plants grown under defined experimental conditions. The two approaches are dealt with separately below.

Physiological attributes of ericoid mycorrhizal fungi in pure culture

Several features, including culturability, widespread global occurrence and the early establishment of taxonomic identity, have led to the selection of *R. ericae*, or its close relatives, as the test organisms of choice in the majority of pure culture studies. These have included a number that have confirmed the ability of *R. ericae* to assimilate, as would most fungi, simple sugars (Pearson and Read, 1975), ionic forms of the minerals N and P (Pearson and Read, 1975; Bajwa and Read, 1986) and amino acids (Figure 11.13).

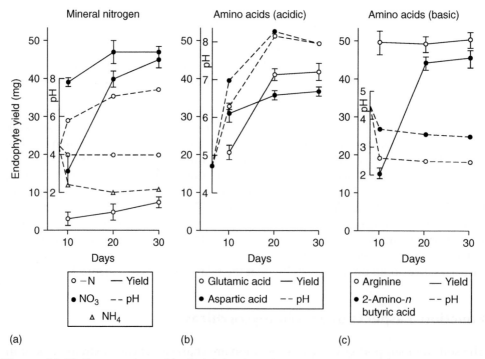

Figure 11.13 Dry weight yields (solid lines) and changes in pH (dashed lines) of the medium after growth of *Rhyzoscyphus ericae* on nitrate (NO$_3$), ammonium (NH$_4$) or a series of acidic and basic amino acids as sole sources of N. From Bajwa and Read (1986), with permission.

Some intraspecific differences in patterns of N utilization by *R. ericae* have now been recognized (Cairney *et al.*, 2000; Grellet *et al.*, 2005). The abilities of *R. ericae* strains to use nitrate, ammonium and glutamine are, not surprisingly, influenced by the availability of organic C in the medium (Grellet *et al.*, 2005). When C supply was high, growth differences between strains were explained by the total amount of N taken up, whereas under C-limiting conditions, strain differences were more closely linked to their N use efficiencies which may imply intraspecific differences in N metabolism. When Grellet *et al.* (2005) screened the same strains as those employed by Cairney *et al.* (2000) and under identical culture conditions, some differences in yield were observed. Since time over which the isolates had been maintained in culture was the only feature differing in the two sets of experiments, Grellet *et al.* (2005) emphasized the need to recognize age as a factor influencing the ability of these fungi to use N and probably other nutrients as well.

More enlightening from an eco-physiological standpoint have been those studies which have investigated the ability of ERM fungi to attack substrates of the kind prevalent in the characteristically acidic mor-humus soils which they normally occupy in association with their autotrophic partners. These degradative processes have been classified as being of two types; those that might facilitate the destruction of polymeric carbon sources hence contributing to the exposure of essential nutrients as well as to decomposition processes, and those that would be involved in providing direct access to the nutrients themselves (Leake and Read, 1997; Cairney and Burke, 1998; Read and Perez-Moreno, 2003). Polymeric carbon sources in soil include those produced by the autotrophs which, in the case of many ericaceous communities, are predominantly contributed by the ERM plants themselves, and those produced by the microbial and mesofaunal populations. Again, in acidic mor-humus, fungi and arthropods dominate. Since, in addition to normal cell wall components, ericaceous plants are characterized by the production of high levels of monomeric and polymeric phenolic compounds (Jalal *et al.*, 1982, 1983), the ability to metabolize these will contribute both to detoxification and to release of co-precipitated N compounds.

The ability of *R. ericae* to degrade pectin, cellulose, cellobiose and hemicellulose, which are the major structural components of plant cell walls, and to oxidize phenolic acids including tannins has been demonstrated in a number of reports (Table 11.1). Uncertainty still surrounds the extent, if any, of ligninolytical capability in *R. ericae*. Haselwandter *et al.* (1990) provided respirometric evidence for the release of $^{14}CO_2$ from aromatic dehydrogenated polymers representing structural components of lignin. Subsequently, Burke and Cairney (1998a) proposed that the fungus might facilitate degradation of lignin by the action of the hydroxyl or Fenton radical, which is produced from peroxidase (H_2O_2) in the presence of iron (Fe II). The hydroxyl radical is known to be the reactive intermediate produced by brown rot fungi in the course of lignin degradation. The production of a wall-bound peroxidase is a known prerequisite for the proven ability of *R. ericae* to oxidize pyrogallol (Bending and Read, 1997). Cairney and Burke (1998) argued that, since hydroxyl radicals are not sufficiently reactive to catalyse the cleavage of β-1 or β-O-4 bonds, such a mechanism would mediate only partial lignin degradation. This is consistent with the low rates of substrate breakdown reported by Haselwandter *et al.* (1990) and with the process being analogous to that involved in brown rather than white rot of woody material, the latter involving complete degrading of the polymer. Since the nature and extent of its ligninolytic activity will influence both the competitive

Table 11.1 Extracellular enzymes known to be produced by ericoid mycorrhizal fungi which would be expected to provide the ability to degrade structural components of plant litters in heathland and other ericaceous plant communities, thereby affecting decomposition processes and 'unmasking' of nutrients to facilitate attack upon nitrogen- (protein degradation) and phosphorus-containing (organic phosphorus) polymers.

Process	Substrate	Enzyme	Reference
Plant cell wall degradation	Pectin	Polygalacturonase	Perotto et al., 1990, 1997
	Cellulose	Cellulase	Varma and Bonfante, 1994; Burke and Cairney, 1997a
	Cellobiose	Cellobiohydrolase	Bending and Read, 1996a; Burke and Cairney, 1997a, 1998a
	Hemicellulose	Xylanase	Burke and Cairney, 1998b; Cairney and Burke, 1998
		β-Xylosidase	Bending and Read, 1996a; Burke and Cairney, 1997a, 1997b
		β-D-Mannosidase	Burke and Cairney, 1997a
		β-D-Galactosidase	Burke and Cairney, 1997a
		β-D-Arabinosidase	Burke and Cairney, 1997a
		β-1,3-Glucanase	Burke and Cairney, 1997a
Fungal cell wall degradation	Chitin	Chitinase	Leake and Read, 1990c; Mitchell et al., 1992, 1997; Kerley and Read, 1995, 1997
Oxidation of phenolic acids and tannins	Polyphenols	Polyphenol oxidase	Varma and Bonfante, 1994
		Laccase	Bending and Read, 1996b, 1997
		Catechol oxidase	Bending and Read, 1996a, 1997; Burke and Cairney, 1998
Hydrolysis of lignin	Lignin	Lignase	Burke and Cairney, 1998a; Haselwandter et al., 1990*
Degradation of protein-phyphenol complexes	Polyphenols	Polyphenol oxidases	Leake and Read, 1989b, 1990d; Bending and Read, 1996b
Degradation of nitrogen-containing polymers	Protein	Acid proteinase	Bajwa et al., 1985; Leake and Read, 1990b, 1991; Ryan and Alexander, 1992; Chen et al., 1999*; Xiao and Berch, 1999*
Organic phosphorus breakdown		Acid phosphatase	Lemoine et al., 1992
		Phosphodiesterase	Leake and Miles, 1996: Myers and Leake, 1996

*Results are based upon indirect method of observation, for example, presence of appropriate gene or growth promotion in test organism supplied with substrate. For additional results of earlier studies see Leake and Read (1997).

ability of *R. ericae* in the soil and the C balance of its ERM relationships, it is important that these questions be addressed as soon as possible.

It is acknowledged that structural components of fungal mycelia contribute in a major way to the organic C pools of mor-humus soils (Bååth and Söderström, 1979). Among these materials, chitin, which constitutes between 10 and 40% of hyphal cell walls (Michalenko *et al.*, 1976) is prominent contributing almost half of the total N in the F horizon of a *Calluna* heathland soil (Kerley and Read, 1997).

R. ericae is readily able to mobilize the N contained in the two major constitutent hexosamines of chitin, N- acetylglucosamine and N-acetylgalactosamine (Kerley and Read, 1995), as well as to degrade the pure chitin (Leake and Read, 1990c; Mitchell *et al.*, 1992; Kerley and Read, 1995). Purified hyphal wall fractions derived from pure cultures of ERM and ECM fungi were able to support growth of *R. ericae* when supplied in growth media as sole N sources (Kerley and Read, 1997) (Table 11.1).

These observations have implications for the C nutrition of ERM fungi, for the C and mineral nutrition of the plant and for decomposition processes in heathland soils. They are not, however, separable in terms of their physiological and ecological importance from another suite of degradative capabilities of the ericoid endophytes that involve the direct attack upon polymeric forms of molecules containing potential mineral nutrients. Among these, the organic monomers and polymers containing the elements N and P, which most limit plant growth on heathland soils, are of greatest interest.

That monomeric forms of organic N are readily assimilable by ERM endophytes (Bajwa and Read, 1985; Cairney *et al.*, 2000; Grellet *et al.*, 2005) (see Figure 11.13) is not surprising since most fungi readily use amino acids both as C and N sources, some preferentially so (Jennings, 1995). The ability to gain access to polymeric sources of N, whether in the form of peptides, pure proteins or of protein-polyphenol complexes (see Table 11.1) is of greater interest, particularly in an ecological context because autotrophs are unlikely to have any access to N contained in such macromolecules (see also Chapter 15). When supplied as sole N sources to *R. ericae*, the tripeptide glutathione and peptides in the form of alanine units of 2–6 amino acid residues, were all readily used as sole C and N sources by the fungus (Bajwa and Read, 1985). Their assimilation was generally slower the longer the chain length, but ultimate yields were the same on all peptides and were equivalent to those obtained on the organic monomers or on ammonium. Experiments in which the pure proteins bovine serum albumen (BSA-molecular weight 67 000) and the plant proteins gliadin (MW 27 000) and zein (MW 40 000) were supplied to *R. ericae* as sole sources of N and C (Bajwa *et al.*, 1985) indicated that the strains used could readily sustain biomass production on these substrates (Table 11.1). Four ERM fungal isolates from roots of the epacrid *Woolsia pungens* produced as much or more biomass yield on BSA as did a northern hemisphere isolate of *R. ericae* (Chen *et al.*, 1999). Subsequent studies have demonstrated considerable intraspecific variation in biomass production by strains of *R. ericae* grown on BSA (Cairney *et al.*, 2000). It is now known that the protease involved in breakdown of these molecules is an extracellular acid carboxy proteinase (EC 3.4.23.6), the production and activity of which is regulated by pH (Leake and Read, 1989a, 1990a). While maximal production and activity are observed between pH 2 and 5, both are practically eliminated above ph 6 (Figure 11.14). The ecological significance of these attributes is discussed later (see Chapter 15).

Proteolytic activity can be induced by the presence of protein itself, or by hydrolysates of protein (Leake and Read, 1990b, 1991) and is increased by the presence of ammonium. When pure protein was supplied as sole source of N for the fungus, both total and specific activities of the proteinase were strongly repressed by the presence of glucose. This observation led Leake and Read (1991) to suggest that *R. ericae* employs a 'Noah's Ark' strategy of enzyme regulation (Burns, 1986). In this, an emissary extracellular proteinase is produced under conditions of derepression which, in the presence of appropriate substrates, leads to the release of the preliminary

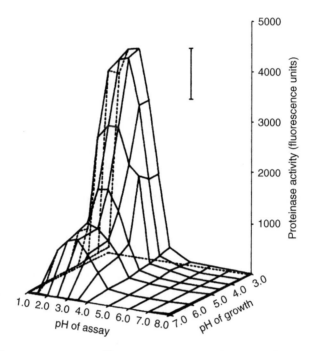

Figure 11.14 Proteinase activity (fluorescence units released in 3 hours, corrected for residual protein) in culture filtrates of *Rhizoscyphus ericae* in relation to the pH of the medium and pH at which the assay was performed. Values are means of four replicates. Vertical bar = LSD. $P < 0.05$. From Leake and Read (1990a), with permission.

Table 11.2 Nitrogen content and yield and ^{15}N excess of shoots of mycorrhizal (M), non-mycorrhizal (NM), and non-mycorrhizal saprophyte-inoculated (SAP) plants of *Vaccinium macrocarpon* after six months growth on $^{15}NH_4$-labelled soil.

Growth stage	N content (% oven-dry weight)	Yield (mg oven-dry weight)	Total N (mg/plant)	^{15}N excess (atom %)
Sterile seedlings	0.94	4.23	0.04	0
Plants 6 months after inoculation				
Mycorrhizal (M)	1.20	30.32+	0.36+	15.38+
Non-mycorrhizal (NM)	0.98	20.97+	0.21+	20.03+
Inoculated with				
Trichoderma spp. ⎱SAP	0.94	18.80	0.18	ND
Aspergillus spp. ⎰	0.82	16.20	0.18	ND

From Stribley and Read (1974). Note: Each figure represents a mean of 14 plants, except for sterile seedlings, where 30 plants were analysed; +, figures significantly different between M and NM categories at $P < 0.001$; ND, not determined.

products of protein hydrolysis as reporter molecules that induce full enzyme production. The rapid release of proteinase by *R. ericae* when the fungus is exposed to protein suggests that this strategy may be very effective. Whereas conditions of derepression would be expected to prevail in the typical soil environment of the ericoid hair root, the relatively high pH and sugar content of the intracellular environment would be expected to ensure that little or no release of this proteinase would take place in the colonized epidermal cells (Leake and Read, 1991).

While the potential role of proteolytic activity in the mobilization of N from organic substrates is evident, it has to be recognized that much of the organic N, particularly in mor-humus, will be present in more recalcitrant forms (Kuiters and Denneman, 1987; Bending and Read, 1996a, 1996b). However, it has been hypothesized that the ability of *R. ericae* to metabolize a number of phenolic acids as well as more elaborate tannic substrates (Leake and Read, 1989b, 1990d; Bending and Read, 1996a, 1996b, 1997) may provide the fungus with some access to protein-N when it is co-precipitated with phenolic compounds. Bending and Read (1996b) confirmed that the fungus had sufficient access to N complexed in this way with tannic acid to support biomass production. They observed that the removal of tannic acid by laccase and catechol oxidase activity (see Table 11.1) led to the production and polymerization of quinones in the growth medium. Metabolism of tannic acid with production of quinones would not only explain enhanced access of *R. ericae* to tannin-bound proteins but the quinones would be expected to contribute significantly to mor-humus formation in heathland soils.

Physiological attributes of the ericoid mycorrhizal symbiosis

While the plethora of results arising from studies of the ERM fungi grown axenically are indicative of likely functional roles in the symbiotic condition, the need to confirm that any such attributes are expressed in the plant–fungus system remains paramount. Further, in order that observations on the physiology of the symbiosis can be interpreted in an ecological context, it is necessary carefully to define the substrates in which the functions are being examined. This is particularly the case with the ericoid symbiosis, characterized as it is in nature by preferential occurrence across a narrowly defined suite of soil conditions (Specht, 1979; Read, 1991a). Progress has been made by using a range of systems from those that are simple and readily defined such as hydroponic sand culture, to heathland soil itself which is more realistic but correspondingly more difficult to handle. Use of irradiation to sterilize soil has helped because it does not release toxins, but there can be problems of excessive nutrient release, especially of ammonium (Stribley *et al.*, 1975). Using small quantities of irradiated heathland soil and strains of *R. ericae* isolated by Pearson and Read (1973a), it was shown in early experiments that ERM colonization consistently led to increased biomass production in seedlings of *Calluna* and *Vaccinium*, these responses being accompanied by enhanced concentrations not only of P, but particularly of N, in the plant tissues (Read and Stribley, 1973). Observations of this kind, which had parallels with the effects of ECM fungi on trees, led to a more intensive analysis of the nutritional role of the ericoid fungal symbionts when grown with their autotrophic partners.

The possibility that ERM colonization might provide the plant with access to organic sources of N emerged from a study using small quantities of irradiated heathland soil upon which *Vaccinium* plants were grown with and without colonization by *R. ericae* and supplied with ^{15}N-labelled ammonium, which was then thought to be the only significant source of N in this soil (Stribley and Read, 1974b). Colonized plants showed a stimulation of growth and N concentration, but a lower ^{15}N enrichment than their non-mycorrhizal counterparts (Table 11.2). Since little of the ^{15}N had been incorporated into the organic constituents, dilution of label in the colonized plants was attributed to them having access to N in the soil organic matter. This experiment led to others that were designed to determine the possible sources of this organic N. Amino acids were obvious candidates, though only later (see Chapters 8 and 15) was it found that potentially significant pools of these were present in heathland soil.

Subsequently, ERM and non-mycorrhizal plants were grown in sand culture to which the amino acids alanine, aspartic acid, glutamic acid, glutamine or glycine were added individually as sole N sources (Stribley and Read, 1980). Yields of colonized plants were comparable with those obtained on ammonium, but most of the amino compounds were little used by non-mycorrhizal plants, whether they were grown in sterile or non-sterile conditions (Stribley and Read, 1980). Clearly, a proportion of the N contained in both amino acids and their amides shown earlier to be assimilated by *R. ericae* (see Figure 11.13) in monoxenic systems, is readily transferred to ERM plants. Electrophysiological studies (Sokolovski *et al.*, 2002) have confirmed that the presence of *R. ericae* in epidermal cells of *C. vulgaris* facilitates a significant increase in amino acid influx. Transport capacity for aparagine, histidine, ornithine and lysine were particularly strongly enhanced in colonized cells. These data suggest that ERM colonization triggers a derepression of a high affinity amino acid uptake system in *C. vulgaris* hair roots. This would improve the effectiveness of uptake of these molecules from soil solution.

The extent to which ERM colonization facilitates access of ericoid plants to polymeric sources of N has been extensively studied. When ERM and non-mycorrhizal plants of *V. corymbosum* were grown with peptides as their sole N source, the colonized plants had higher N contents in all cases and higher yields than their non-mycorrhizal counterparts on all but two of the peptide sources (Bajwa and Read, 1985). Similar experiments employing BSA as sole N source showed significantly greater dry matter yields (Figure 11.15a) and tissue N contents (Figure 11.15b) in the

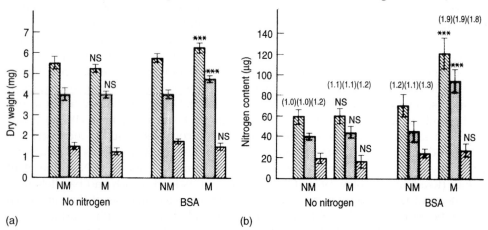

(a)   (b)

Figure 11.15 Partitioning of (a) dry weight and (b) N content in whole plant (left bar), shoot (centre bar) and root (right bar) of plants of mycorrhizal (M) and non-mycorrhizal (NM) *Vaccinium macrocarpon* grown with BSA or with no nitrogen for 30 days. Final N concentrations in parentheses. Significant differences between M and NM plants within the nutrient regimes at $P < 0.001$ (***); NS, no significant difference. Values are means and standard errors of means of 24 replicate plants. From Bajwa *et al.* (1985), with permission.

ERM plants (Bajwa *et al.*, 1985). *Gaultheria shallon* produced greater biomass on BSA when grown with *R. ericae*, *O. maius* and with two other ERM strains which were distinct from *R. ericae* but members of the Helotiales (Xiao and Berch, 1999).

The question of the extent to which mycorrhizal colonization may provide access to protein N when it is present as a synthetically produced protein-tannic acid complex has been addressed by growing *R. ericae* with this as sole N source (Bending and Read, 1996b). *R. ericae*, together with an *Oidiodendron* isolate, probably *O. maius*, was equivalent or greater on the complexed N source than when N was supplied either as pure protein or as ammonium. Read and Kerley (1995) showed that dead mycelium or purified cell wall fraction of *H. ericae* could act as sources of N for mycorrhizal, but not non-mycorrhizal, plants of *Vaccinium macrocarpon*. N was released from both materials in sufficient quantity to sustain growth of the ERM plants whether it was supplied as entire mycelium (Figure 11.16a) or purified hyphal walls (Figure 11.16b). Since separate assays of both substrates revealed that mineral N was scarcely detectable, it was concluded that mobilization of organic N was achieved by a combination of proteolytic and chitinolytic activity.

The residues of the ericaceous plants themselves will inevitably form a major component of heathland soil, particularly in what are often pure stands of these sclerophyllous shrubs. Kerley and Read (1998) investigated the extent to which the ERM symbiosis might be directly involved in mobilization of N from these residues. ERM and non-mycorrhizal plants of *V. macrocarpon* were grown aseptically and, after killing their tissues by drying, the sterile necromass was compartmented into shoot, ERM root and non-mycorrhizal root fractions, each of which were employed as a substrate containing defined quantities of the sole N source for living plants grown with and without *R. ericae*. Plants grown in the ERM condition alone had access to N contained in these substrates. They showed significantly greater biomass than their non-mycorrhizal counterparts on all tissue fractions, but exploited the ERM root necromass particularly effectively. After 60 days of growth, 76% of the N contained in this fraction was recovered in the test plants (Table 11.3). Obviously, substrates produced and killed under aseptic conditions are not precise surrogates for those to be found in nature, but such experiments indicate that ericoid ERM systems will have some ability to carry out both nutrient mobilizing and decomposer activities in substrates which they themselves have produced. The ecological implications of this observation are discussed later (see Chapter 15).

Comparisons of P concentrations in tissues of ericaceous plants grown with or without ERM colonization indicate that the fungi can increase access to this element as well as to N (Read and Stribley, 1973; Mitchell and Read, 1981). In many heathland soils, the main sources of P will again be the organic residues in which the roots proliferate. Phytates have been identified as being quantitatively the most important sources of organic P in soils (Cosgrove, 1967), though it is increasingly recognized that the phosphodiesters, such as nucleic acids, may be present in significant amounts and are more labile than the monoesters (Griffiths and Caldwell, 1992; see Chapter 9).

Under the acidic conditions prevailing in heathland soil, phosphomonoesters are likely to be complexed with iron and aluminium to form Fe- or Al-phytates. These have been shown to be accessible to ericoid endophytes isolated from *Vaccinium* and *Rhododendron* (Mitchell and Read, 1981) and considerable progress has been made towards characterization of the phosphomonoesterase enzymes involved in release of P from these sources. It was established some time ago that activity was greatest

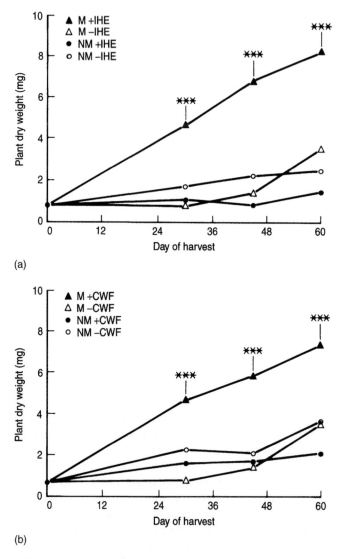

Figure 11.16 Use of hyphal material as sole N source by *Vaccinium macrocarpon* mycorrhizal with *Rhyzoscyphus ericae* (M) or non-mycorrhizal (NM). (a) Growth of plants with (+IHE) or without (−IHE) necromass of *R. ericae* as sole N source. (b) Growth of plants with (+CWF) and without (−CWF) pure cell wall fraction of *R. ericae* as sole N source. Symbols as shown. Vertical bars represent least significant differences at $P < 0.05$. ★★★ indicates result of one-way ANOVA of M+ substrate versus NM+ substrate: $P < 0.001$. From Read and Kerley (1995), with permission.

under conditions of low external P concentration (Pearson and Read, 1975). The production of two isoenzymes has subsequently been demonstrated (Straker and Mitchell, 1986; Straker *et al.*, 1989), the activities of both of which are stimulated in the presence of Fe^{3+} at low concentration. The low molecular weight form of the enzyme that appears to be produced selectively under conditions of low P supply

Table 11.3 Percentage of available substrate N (shoot, non-mycorrhizal root and mycorrhizal root necromass of *Vaccinium macrocarpon*) present in M and NM plants of *Vaccinium macrocarpon* grown in sterile culture for 60 days (percentages adjusted for initial seed N content).

Substrate	Available substrate N (mg)	Substrate N present in M plant (%)	Substrate N present in NM plant (%)
Shoot	0.26	40	4.6
NM root	0.26	43	1.9
M root	0.26	76	4.2

From Kerley and Read (1998).

(Lemoine *et al.*, 1992) is known to be able to hydrolyse such compounds as ATP, ADP and AMP (Straker and Mitchell, 1986).

In a comparative analysis of the extracellular acid phosphatase production of ERM endophytes isolated from a number of plant species, Straker and Mitchell (1986) found highest activity in a fungal isolate from the South African species *Erica hispidula*. Soluble orthophosphate made available by mineralization could subsequently be absorbed by low- and high-affinity uptake systems depending on the concentration (Straker and Mitchell, 1987). By ultrastructural localization using immunogold labelling, Straker *et al.* (1989) demonstrated that the high molecular weight phosphatase was exclusively associated with the walls and septa of living hyphae. There is some evidence that activity of this wall-bound enzyme is at its greatest in hyphae close to the root and that it is much reduced as distance from the plant increases (Gianinazzi-Pearson *et al.*, 1986). This could be due to lower concentrations of P in the vicinity of the root surface or to control of enzyme activity being exerted by the plant. In any event, maximal expression of activity in the rhizosphere could, especially in view of the low specificity shown by acid phosphatases towards phosphomonoesters, provide the plant with access to P from a range of organic sources.

There is increasing recognition of the quantitative importance of phosphodiesters in acid organic soils. Leake and Miles (1996) have investigated phosphodiesterase production by *R. ericae*, together with the ability of the fungus to use DNA as a sole source of P. The fungus grew well on this compound and achieved greater mycelial dry weight than on orthophosphate (Figure 11.17). At least part of the enzyme activity was attributed to the exonuclease 5′ nucleotide diesterase. The pH optimum for the activity was between 4.0 and 5.5. In a comparative study, Chen *et al.* (1999) showed that ERM isolates from roots of the epacrid *Woolsia pungens* and northern hemisphere strains of *R. ericae* were both readily able to use DNA and inositol hexaphosphate as P sources.

In addition to their involvement in the capture and transport of the major nutrients N and P, ERM fungi may play a role in acquisition of other elements by the plants. Their influence upon Fe nutrition has been most intensively studied. The chemical form and hence availability of this element, can change drastically in heathland soils where redox conditions fluctuate with seasonal changes in water balance. Shaw *et al.* (1990) emphasized that, under such circumstances, regulation of supply of the element across a wide range of concentrations was likely to be more important than acquisition or exclusion as individual processes. *R. ericae* was able to sustain productivity over a range of external Fe concentrations, typical of those to

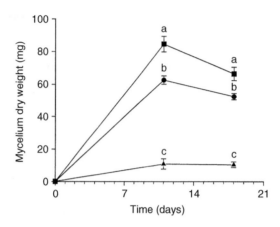

Figure 11.17 Growth of mycelium of *Rhyzoscyphus ericae* on DNA (■) or orthophosphate (5 mM P) (●) compared with growth with no P supply (▲). Vertical bars are standard errors of the mean and letter codes (a–c) indicate significant differences between means at each harvest derived from Tukey's test. From Miles and Leake (1996), with permission.

be expected in heathland soils. At the highest external Fe concentration (144 μg/ml), the concentration in the mycelium was raised considerably. This suggests that the fungus may retain the potential to regulate plant Fe consumption, even when exposed to soil solutions which are enriched in the element. Regulation appears to be achieved by processes which control the affinity of the fungus for Fe. Both in pure culture and in colonized roots of *Vaccinium macrocarpon* or *Calluna vulgaris*, *R. ericae* showed an extraordinarily high affinity for iron at low concentrations (2.0 μg/ml), which was progressively reduced as the external concentration of Fe increased. The high affinity appears to be explained by the production of iron-specific siderophores (Schuler and Haselwandter, 1988; Federspiel *et al.*, 1991), some of which have now been chemically characterized. The predominant siderophore in *R. ericae* and *O. griseum* is ferricrocin, while in a related endophyte isolated from the calcicolous shrub *Rhodothamnus chamaecistus* it was fusigen (Haselwandter *et al.*, 1992; Haselwandter, 1996). Such siderophores may be involved in facilitating the significant increases in specific absorption rate of iron seen in ERM plants of *Calluna* grown in the presence of calcium salts (Leake *et al.*, 1990a) and may be particularly important in those plants like *R. chamaecistus* which typically occur in relatively calcium-rich environments.

R. ericae also appears to play a role in determining the response of colonized plants to non-essential (and potentially toxic) elements. It has a remarkable tolerance of the presence of Al in solution, no inhibition of mycelial yield being observed in cultures contain 800 mg/l of Al^{3+} (Burt *et al.*, 1986). Cairney *et al.* (2001) examined the effects of Cd, Cu and Zn on biomass production of 13 ascomycetous ERM isolates of *Woolsia pungens* and compared these with the responses of *R. ericae* grown under the same conditions. Concentrations that effectively inhibited growth of the ERM fungi by 50% (EC 50) were similar, indicating that the impacts on plant metal resistance attributable to ERM endophytes of the northern and southern hemispheres are also likely to be similar.

A close link has been established between arsenate tolerance of ericaceous plants growing on As-contaminated china-clay spoils and their ericoid endophytes. Here, the fungi play a role not only in reducing exposure of the autotroph to As, but also contribute to the maintenance of P supply to the plants. As an analogue of phosphate, arsenate is known to be transported across the plasma membrane via the phosphate co-transporter in many plants and fungi (Meharg and Macnair, 1992). Sharples *et al.* (2000b, 2000c) showed that isolates of *R. ericae* obtained from mine sites readily accumulate As, but that they have an enhanced capacity to reduce the arsenate to arsenite. The arsenite then effluxes from the hypha. This enables the fungus actively to accumulate P, thus avoiding deficiency of this element, while avoiding As toxicity. The As tolerance was not shown by *R. ericae* isolates from uncontaminated heathland soils indicating that there had been selection for resistance in the presence of the toxin (Sharples *et al.*, 2000c).

The resistance of *R. ericae* and of another ERM fungus, *O. maius*, to Zn is also adaptive (Martino *et al.*, 2000). This group showed that strains of *O. maius* obtained from sites polluted by Zn differed from those of uncontaminated sites in their lower propensity to mobilize the metal from insoluble zinc oxide and zinc phosphate (Martino *et al.*, 2003). In contrast to the ERM fungus from the contaminated site, that from the site not exposed to Zn contamination readily mobilized the metal from both salts, doing so by a process involving excretion of citric and malic acid. Since such increases of bioavailability of the Zn would be disadvantageous in polluted envir-onments, it was concluded that the relatively low rates of mobilization reflected an adaptive strategy in the strain from the polluted site.

Gibson and Mitchell (2004) investigated the ability of four ERM mycobionts to solubilize P supplied as zinc phosphate to culture media containing a range of N sources and different concentrations of P and C. All the fungi released P from the source irrespective of type of N added, and there was no effect of increasing orthophosphate concentration of the bathing medium. No solubilization occurred in the absence of glucose. Under conditions found to be optimal for solubilization of zinc phosphate, all but one of the fungi was able to solubilize tricalcium phosphate while no solubilization was observed in media supplemented with aluminium phosphate, iron phosphate or copper phosphate.

In studies of the impacts of pollutant Cu on the growth, phosphomonoesterase (PMEase) and phosphodiesterase (PDEase) activities of isolates of *R. ericae* collected from Cu-polluted and unpolluted sites, Gibson and Mitchell (2005) observed that, whereas addition of Cu to the growth medium had no impact on yields of the strains from mine spoil sites, there was a significant reduction of growth in those from unpolluted sites. In the fungi from polluted soil, the activities of PDEase were stimulated by addition of 0.25 mM Cu in the growth medium, whereas these effects were not seen in isolates from unpolluted sites. In the fungi from Cu-contaminated soil, wall-bound phosphatase activity was not inhibited at Cu concentrations up to 5.0 mM. These results confirm earlier observations (Bradley *et al.*, 1981, 1982) that isolates of *R. ericae* obtained from Cu contaminated sites can develop considerable constitutive resistance to the pollutant.

Conclusions

The application of molecular techniques has improved our understanding of the inter-relationships both of the plants which form ericoid mycorrhizas and of the fungi involved in the associations. Cultural characteristics had revealed that, while there were apparently unifying features among ERM fungal isolates, there were also identifiable differences between them. Phylogenies based upon analysis of ITS sequences of some of the most commonly occurring isolates have confirmed a good deal of genetic diversity among fungi which were once referred to the single species *R. ericae.*

This fungus is seen as part of a larger *Hymenoscyphus ericae* aggregate, some members of which may form ectomycorrhizas. There is now evidence for the occurrence of members of this aggregate, and specifically of *R. ericae*, in epacrid plants of Australia. However, it is clear that, while most epacrid isolates are ascomycetes of the order Helotiales, large numbers of them may not fall within the *H. ericae* aggregate. DNA sequences of basidiomycetous fungi, notably *Sebacina* spp., have been identified as being present in ericoid hair roots but their location, status and function within these structures remains to be determined. *Sebacina* spp. of subgroup B have been identified in structures reminiscent of ectendomycorrhiza in epiphytic members of the Vaccinioideae in the Andes.

Detailed analysis by light, transmission and scanning electron microscopy has gone some way to elucidating the major events associated with attachment, penetration and internal proliferation of fungi in the epidermal cells of the root. Cytohistochemical and molecular tools are now being deployed to enable localization of some of the processes involved in nutrient transfer. These methods promise further exciting increases in our understanding of these fundamentally important events.

Experimental examination of the nutrition of ericoid and epacrid endophytes growing in axenic culture and of the enzymes involved in the mobilization of structural and nutritient containing polymers have revealed that these fungi possess a wide range of saprotrophic capabilities. A number of simple and complex substrates are attacked, providing access to, and mobilization of, N and P. When grown in symbiosis with ericaceous plants, the fungi facilitate transfer of the nutrients originally contained in the polymers to the plant. Progress has been made both towards characterization of the fungal exoenzymes involved in nutrient mobilization and in understanding of the environmental conditions leading to their induction and repression.

In addition to the direct nutritional benefits arising from colonization, there are other advantages that are likely to be of ecological significance. The ability of the fungal partner to sequester and, in some cases, to metabolize metal ions that are otherwise toxic to the plant appears to be important. This combination of nutritional and non-nutritional attributes of ERM associations of members of the Ericaceae contributes significantly to the ability of these plants to grow in contaminated mine spoils as well as in their natural heathland habitats.

12

The mycorrhizas of green orchids

Introduction

With recent estimates ranging from 20000 to 35000 species, the Orchidaceae is one of the largest and most diverse families in the plant kingdom (Cribb *et al.*, 2003). The family is a member of the monocot order Asparagales (Chase *et al.*, 2003), but within it the orchids occupy an isolated position (Chase, 2004). Though an ancient group, probably arising before the break-up of Pangaea about 100 million years ago, the Orchidaceae is nonetheless one of the most highly and uniquely modified of all angiosperm families and it is still undergoing rapid diversification and speciation (Chase, 2001). The family can be divided into five monophyletic subfamilies, the Apostasioideae, Cypripedioideae, Vanillioideae, Orchidioideae and the Epidendrioideae (Figure 12.1). Of these, the apostasioids are considered to be the basal lineage and the epidendroids, which are by far the largest group, the most advanced.

The vast majority of orchid species are green and hence at least putatively autotrophic in the above-ground stages of their life cycles. These forms are considered in this chapter. Consideration of the mycorrhizal relationships of those orchids which lack chlorophyll as adults, being entirely dependent upon fungi throughout their lives (the fully mycoheterotrophic species, *sensu* Leake, 2004), follows in Chapter 13. However, as already indicated in our treatment of the Pyrolaceae (see Chapter 7), the presence of chlorophyll may not be sufficient to confirm full autotrophy, there being now good evidence that some green plants in a number of families, including Orchidaceae, are partially mycoheterotrophic (Gebauer and Meyer, 2003; Bidartondo *et al.*, 2004; Selosse *et al.*, 2004). These so-called 'mixotrophic' orchid species, which include members of the genera *Cephalanthera* and *Epipactis*, are considered along with the full mycoheterotrophs in Chapter 13. We stress, however, that the change from full autotrophy to full mycoheterotrophy occurs along a continuum in the Orchidaceae and that our separation of the two groups in this way is to an extent a matter of convenience. It is, nonetheless, worthy of note that, whereas the supposedly fully autotrophic orchids considered here largely associate with fungi of the saprotrophic form-genus *Rhizoctonia*, the emerging evidence indicates that mixotrophic and fully mycoheterotrophic forms associate predominantly with ectomycorrhizal (ECM) fungi (see Chapter 13).

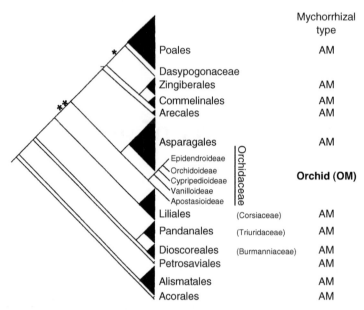

Figure 12.1 Cladogram, based on combined analysis of the 18s, *rbcL* and *matK* genomes, showing relationships between orders of the monocotyledonous families and indicating the position of the family Orchidaceae relative to that of the mycoheterotrophic families Corsiaceae, Triuridaceae and Burmanniaceae. Subfamily relationships within the Orchidaceae are also shown. Mycorrhizas of the orchid type (OM) are seen to be restricted entirely to the family Orchidaceae. * indicates a node with 50–74% bootstrap support, ** indicates 75–89% support, no asterisk indicates 90–100% support. Modified from Chase *et al.* (2006).

Green orchids have a cosmopolitan distribution, being found in most types of terrestrial habitat from high latitude bogs, heaths and forests to deserts and tropical rainforests. In addition to the terrestrial habit, a large number of species, particularly in the tropics, are epiphytic, and some are lianes. Despite the immense diversity, particularly of their floral morphologies, these plants have specialized unifying characteristics which may reflect their isolated taxonomic position. Two of these, seed morphology and mycorrhizal habit, are almost certainly interdependent. The seeds of almost all orchid species, often referred to as 'dust seeds', are extremely small (0.3–14 µg) consisting of minute undifferentiated embryos that, in the absence of an endosperm, have few reserves (Burgeff, 1936; Arditti and Ghani, 2000). The lack of reserves to support early seedling development makes all orchids in nature dependent upon the provision of nutrients by mycorrhizal fungi. The fungi, which facilitate the process known as 'symbiotic germination' produce intracellular coils (called pelotons) in the embryos of developing seedlings and in the rhizomes or roots of adult plants. These structures are sufficiently distinctive in appearance and pattern of distribution within orchid tissues that they form the major defining characteristics of orchid mycorrhiza. These are found in orchids from the basal apostasioid lineage (Kristiansen *et al.*, 2004) to the most advanced epidendroid forms. Apart from unpublished reports of morphologically similar forms of colonization in the most closely related but still distant family Boryaceae in the Asparagales (see Chase *et al.*, 2003), there appear to be no records of this type of mycorrhiza elsewhere in the plant kingdom. Nevertheless, caution must be exerted to avoid confusion between the robust pelotons formed in orchid cells and the superficially similar *Paris*-type configuration formed by arbuscular mycorrhizal (AM) fungi in the roots of a large number of plant species (see Chapter 2).

Symbiotic germination and subsequent development of orchid embryos to form a swollen protocorm are only achieved if the colonizing fungus has access to a soluble or insoluble source of carbohydrate which it translocates to the plant. In the orchids under consideration in this chapter, the fungus-driven mode of nutrition is retained until the appearance above ground of green leaves, at which stage they are thought to be fully autotrophic.

The first large-scale surveys of temperate and tropical orchid species (Wahrlich, 1886; Janse, 1897) revealed the regular occurrence of fungal colonization in roots of adult plants. However, it was Bernard (e.g. Bernard, 1899; for a more complete list see Smith and Read, 1997) and Burgeff (1909) who first described pelotons as being the distinctive structures of orchid mycorrhizas and carried out the earliest systematic isolations from these hyphal components of orchid roots. They also made the first attempts at classification of the fungi obtained and, through studies of symbiotic and asymbiotic germination, provided the earliest insights into the possible functions of the symbiosis in the development of orchid seedlings. These pioneers established that the fungi routinely isolated from orchid pelotons were basiodiomycetes which could be ascribed to the anamorph genus *Rhizoctonia*. Members of this genus have often subsequently been treated as though they belonged to a coherent taxonomic entity. However, the establishment, initially by Warcup and Talbot (see below), that a range of teleomorphic stages could be induced in cultures of '*Rhizoctonia*', led to recognition, subsequently confirmed by analysis of ribosomal DNA, that the form genus was polyphyletic. The application of molecular phylogenetic techniques has also confirmed the involvement of fungi other than those of the *Rhizoctonia* type in forming orchid mycorrhiza (Taylor *et al.*, 2002), but it is appropriate to make the assumption here that the majority of fully autotrophic orchid species are colonized by fungi of the *Rhizoctonia* type. Hanne Rasmussen (1995) has written an excellent review of the extensive researches on the conditions required for germination of orchids with autecological accounts of important temperate zone species.

As was the case with the early history of research on ericoid mycorrhizas (see Chapter 11), the first part of the twentieth century saw controversy about whether or not symbiosis was a prerequisite for germination and complete development of green orchids (see Arditti, 1992). It was established that orchid seeds could be routinely germinated in the absence of their symbionts provided that a supply of sugars was in the growth medium (Burgeff, 1936). Using this approach, a number of species were grown to maturity and flowering in a non-mycorrhizal condition (Bultel, 1926; Knudson, 1930). A combination of asymbiotic germination and vegetative propagation today form the basis of the economically important horticultural production of orchid plants that takes place on an industrial scale in many countries.

From the perspective of the biology and conservation of wild orchids, the important consideration is that, because soils lack carbon (C) sources of a kind that can be readily used by autotrophs, symbiotic germination must be the norm in nature.

As wild orchid populations are increasingly threatened by habitat destruction and eutrophication, the need to understand the roles played by mycorrhizal fungi in the processes of germination and sustenance of these plants at all developmental stages becomes ever greater (Dixon *et al.*, 2003).

The fungi forming mycorrhizas in green orchids

Isolation and identity

Bernard (1904a) was the first to record regular isolations of the *Rhizoctonia*-like fungi from orchid root tissues (Figure 12.2). Methods employed for isolation include plating fragments of surface-sterilized root on nutrient agar. These inevitably bring the attendant problems of contamination and difficulties of relating the fungi to actual mycorrhiza-forming species which were seen in studies of ericoid mycorrhizas (see Chapter 11). However, perhaps because most of the *Rhizoctonia*-like fungi forming pelotons in orchid roots are more vigorous in their growth than ericoid endophytes, there were soon a very large number of reports involving careful separation and plating of individual fungal pelotons. In many cases, the isolated fungi have been shown in reinoculation tests to produce pelotons, satisfying the requirements of Koch's postulates and providing considerable confidence that they are, indeed, mycorrhizal. Again, as in the case of ericoid mycorrhiza, there are a large number of papers that report simply the isolation of fungi from roots without defining details of their origin or localization.

Some of these (e.g. Bayman *et al.*, 1997) identify their intention to make broad surveys of the fungal endophyte flora of orchid roots, which can indeed be diverse, while acknowledging that many of the isolates obtained are likely to be non-mycorrhizal. Because they were unable to detect sufficient pelotons for isolation purposes in the roots of tropical epiphytic orchids, Otero *et al.* (2002) isolated *Rhizoctonia*-type fungi from surface sterilized root pieces of nine orchid species from

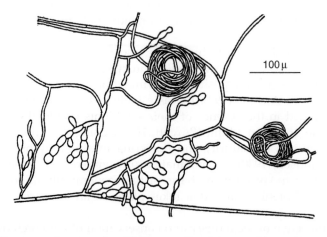

Figure 12.2 Drawing of a typical orchid mycorrhizal fungus, *Rhizoctonia repens* (*Tulasnella calospora*), isolated in pure culture. From Bernard (1909).

Puerto Rico and then subjected them to molecular systematic analysis. Most of the endophytes obtained were closely related to each other and formed a well supported group within the *Rhizoctonia*-type *Ceratobasidium* (see below). Among these was a strain very closely related to the widespread associate of temperate orchids *C. cornigerum*, which had previously been isolated from pelotons of a tropical orchid by Richardson *et al.* (1993). Clearly, there is a strong likelihood that some, if not all of these *Ceratobasidium* isolates are mycorrhizal, but studies involving resynthesis of associations and growth analysis are desirable, in particular because the roles of mycorrhizal colonization in tropical orchids have been hitherto neglected.

In addition to the readily culturable endophytes, there are also reports of the presence in orchid roots of fungi that cannot be isolated. Thus, for example, despite microscopic evidence for the presence of peloton-forming fungi in the roots of *Cypripedium acaule* and *Malaxis monophyllos*, Zelmer *et al.* (1996) were unable to isolate the fungi involved. Among these recalcitrant fungi are those that, on the basis that they produce clamp connections, are assumed to be basidiomycetes unrelated to *Rhizoctonia*. These fungi may be the sole occupants of the roots, may form pelotons in addition to those of *Rhizoctonia*-type fungi, or may even co-habit as peloton-forming fungi in the same cells as the rhizoctonias (Zelmer *et al.*, 1996). The emerging potential of molecular methods to enable identification of fungi at the level of single pelotons (see below) promises to enable elucidation of many of these uncertainties. The occurrence of fungi unrelated to those of the *Rhizoctonia*-type, and which are often unculturable, is more prevalent in fully mycoheterotrophic orchids and their possible functional attributes are discussed in that context (see Chapter 13).

Because of their widespread occurrence as peloton-forming orchid associates, much interest has been shown in the taxonomic status of rhizoctonias. Early studies emphasized morphological and microscopic aspects of their sclerotia and mycelia (Bernard, 1909). It was accepted that the hyphae generally lacked clamp connections and that many strains were characterized by the production of chains of swollen monilioid cells (see Figure 12.2). Later, hyphae of mononucleate, binucleate and multinucleate types were recognized. Among these, the mononucleate types have been shown to be pathogenic in forest nurseries (Hietala, 1997) and only a few have been reported as mycorrhizal associates of orchids. Otero *et al.* (2002) observed that 66% of the 108 *Rhizoctonia*-like fungi that they obtained from tropical epiphytic orchids were uninucleate, but these were not isolated from pelotons. Most multinucleate types are plant pathogens or soil saprophytes (Sneh *et al.*, 1991). In contrast, the vast majority of the *Rhizoctonia*-type isolates obtained from orchid mycorrhizas are 'binucleate rhizoctonias' (BNRs) (Currah *et al.*, 1997). These are also not restricted to orchids. Some can be present in the roots of non-orchid hosts, the growth of which can be stimulated (Hietela and Sen, 1996; Sen *et al.*, 1999) or depressed (Mazzola *et al.*, 1996) by the presence of BNR.

By inducing teleomorph development on their anamorphic strains of *Rhizoctonia*-type orchid isolates in culture, Warcup and Talbot (1967, 1970, 1980) established not only that the form-genus was essentially polyphyletic, but also that its members consisted of distantly related basidiomycete taxa. The teleomorph genera *Ceratobasidium* and *Thanatephorus* (Family Ceratobasidiaceae), *Tulasnella* (Family Tulasnellaceae) and *Sebacina* (Family Sebacinaceae) were described. This differentiation has been amply confirmed in more recent times by ultrastructural analysis of septal pores (Moore, 1987; Andersen, 1996; Muller *et al.*, 1998) and by molecular phylogenetic studies based upon DNA sequencing (Andersen, 1996; Taylor *et al.*, 2002; Ma *et al.*, 2003; McKormick *et al.*, 2004; Shefferson *et al.*, 2005). Analysis of both nuclear (Figure 12.3) and mitochondrial large subunit sequences from an array of Basidiomycetes, including a range of orchid isolates, effectively

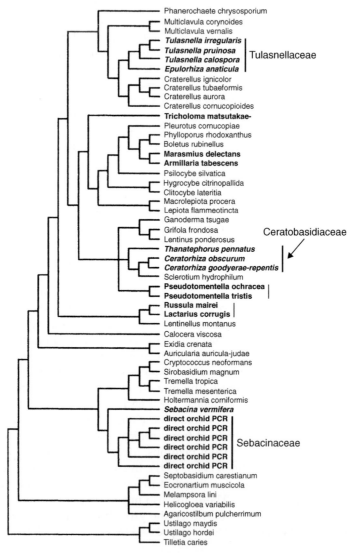

Figure 12.3 Strict consensus maximum-likelihood tree identifying the phylogenetic positions of orchid mycorrhiza-forming fungi of the genus *Rhizoctonia sensu lato* (indicated in bold italics) within the Basidiomycota. Placement in the families Tulasnellaceae, Ceratobasidiaceae or Sebacinaceae is based on the features of their teleomorphs. The data set included 96 taxa and 451 bases from the 5' end of the nuclear 28S ribosomal gene. For ease of presentation, numerous taxa (mainly of the Agaricales) were pruned from the tree after the search was complete. The Ustilaginiomycetes were designated as a monophyletic outgroup. The tree was constructed by D.L. Taylor and is modified from Taylor *et al.* (2002).

illustrates the large phylogenetic distances between the Orders Ceratobasidiales, Tulasnellales and Sebacinales, which make up the three major clades of orchid associated *Rhizoctonia* fungi (Taylor *et al.*, 2002). These are the orders in which most, if not all, the fungi associating with green orchids belong (Table 12.1).

Large phylogenetic distances can also occur between fungi placed within a single order of *Rhizoctonia*-type fungi. This has been strikingly demonstrated in studies involving DNA sequence analysis in the Sebacinales (Weiss *et al.*, 2004; see Chapter 6). Following the observations of Warcup (1981, 1988) that a member of this order, *Sebacina vermifera*, formed mycorrhizas with several green orchids species in Australia, there have been increasing numbers of reports of the presence of *Sebacina*-like taxa forming root-fungus symbioses across the plant kingdom. These range from endophytic colonization in the Hepaticae (Kottke *et al.*, 2003; see Chapter 14), the Ericaceae (Allen *et al.* 2003; see Chapter 11) and in fully mycoheterotrophic orchids (McKendrick *et al.*, 2002; Selosse *et al.*, 2002a; Taylor *et al.*, 2003; see Chapter 13), to ECM formation on trees (Glen *et al.*, 2002; Selosse *et al.*, 2002a, 2002b; see Chapter 6). Current understanding of the phylogeny of the Sebacinales (see Figure 6.5) suggests that those of its members that are associated with green orchids, including *S. vermifera*, should be placed in Group B which is distantly related to the ECM species (Weiss *et al.*, 2004). It also appears that members of Group B belong to the *Rhizoctonia* form genus (see Figures 6.5 and 12.3).

There has been some progress towards understanding the relationships between *Rhizoctonia* isolates that do not produce teleomorphs and hence in resolving their likely taxonomic status. In addition to septal pore characteristics, somatic compatibility tests have been revealing in this context. Since only hyphae of closely related genotypes will fuse and anastomose, strains that are compatible in this way can be placed together in Anastomosis Groups (AGs) (Sneh *et al.*, 1991; Carling, 1996). Most isolates of the widespread endophyte genus *Ceratorhiza* fall into AG-C, those of *Moniliopsis* into AG-6 and those of *Epulorhiza* into Rr1 and 2. By facilitating recognition of genetic relationships within anamorphic isolates, these groupings have enabled some advances in the study of specificity in the orchid mycorrhizal symbiosis (Ramsay *et al.*, 1987).

Table 12.1 Teleomorph-anamorph relationships in *Rhizoctonia*-type fungi.

Order	Teleomorph		Common orchid symbiont	Anamorph	Common orchid symbiont
	Family	Genus			
Ceratobasidiales	Ceratobasidaceae	Ceratobasidium	C. cornigerum	Ceratorhiza (Moore, 1987)	C. goodyerae-repentis
		Thanatephorus		Moniliopsis (Moore, 1987)	T. cucumeris
Tulasnellales	Tulasnellaceae	Tulasnella	T. calospora	Epulorhiza (Moore, 1987)	E. repens
Exidiales	Sebacinaceae	Sebacina (Oberwinker) = Serendipita (Roberts, 1993, 1999)	S. vermifera	Rhizoctonia (Warcup and Talbot, 1967)	

Fungal specificity

If orchid seeds are spread on a moist substratum, the undifferentiated embryos absorb water, swell slightly and may burst the testa, sometimes even producing epidermal hairs (Figures 12.4 and 12.5). Further development of the embryo can normally be obtained *in vitro* only if it receives an exogenous supply of sugars. In some species, other sources of organic C, growth factors and vitamins may be required. This type of development in the absence of a fungus is known as asymbiotic germination and can, in some species, lead to the development of mature plants (see below). In nature, however, continuing supplies of soluble sugars and other essential growth factors are not available from the soil and the conclusion reached by Bernard (1904c), that a fungus is required for successful germination, remains valid. Fungal colonization yields symbiotic germination, a process which is required both to stimulate gluconeogenesis and to provide ongoing nutritional support before photosynthesis commences.

From their extensive studies of orchid seed germination, both Bernard (1909a) and Burgeff (1911) concluded that, because the number of *Rhizoctonia*-type fungi that were effective in promoting symbiotic germination and seedling growth *in vitro* was small, there must be a high level of specificity in the relationships. Subsequent studies, of which there have been a great many (see Table 12.2 for a selective summary), have confirmed that the predominant mycorrhizal associates of terrestrial green orchid seedlings are members of the families Ceratobasidiaceae, Tulasnellaceae and Sebacinaceae.

Orchid species differ greatly in the extent to which they show specificity in *in vitro* germination tests and there are likely to be differences between levels of specificity shown *in vitro* and in the field. These differences are exemplified by studies of two eastern Australian orchids. In one, *Microtis parviflora*, Perkins *et al.* (1995) observed little specificity *in vitro* but the presence of only two fungi, both *Epulorhiza* spp., was detected in protocorms and plants grown in the field. Even greater specificity was shown both *in vitro* and in the field by another species, *Pterostylis acuminata* (Perkins and McGee, 1995), in which a single *Rhizoctonia* isolate was capable of germinating seeds and promoting the growth of protocorms *in vitro*, this fungus also being the only one recovered from pelotons of field-grown adult plants. Following Masuhara and Katsuya (1994), the distinction was drawn between the potential specificity seen in *in vitro* tests and the ecological specificity revealed by field studies.

The germination event itself is also not the only indicator of fungal compatibility. In some orchid species it is clear that a number of fungal strains isolated from an adult plant can stimulate germination *in vitro* but the extent to which they enhance subsequent growth of the same species may vary greatly according to the strain used even over the short term of four weeks (Rasmussen and Whigham, 1998). The need to assess impacts of fungi over a longer term was highlighted by Zelmer and Currah (1995). They isolated three fungi, *Ceratorhiza goodyerae-repentis* (≡ *Ceratobasidium cornigerum*) and two *Epulorhiza* species from mycorrhizas of field-grown adult plants of *Spiranthes lacera*, but only the *C. goodyerae-repentis* was recovered from protocorms that developed at the same site. On culture media, seeds of *S. lacera* germinated asymbiotically but their development did not progress much beyond the testa-splitting stage (see Figure 12.4). While both the adult and

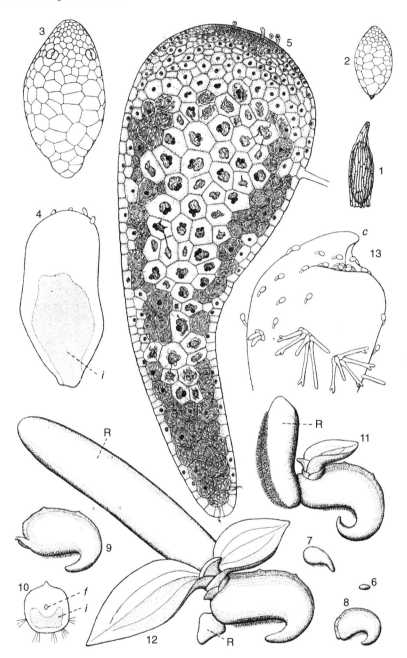

Figure 12.4 Stages in the development of *Phalaenopsis*, showing the minute seeds (1), undifferentiated embryos (2 and 3) and mycorrhizal protocorm (5). Rhizoids produced by germinating protocorms are shown (10 and 13). From Bernard (1909).

protocorm-derived *C. goodyerae-repentis* strains were capable of supporting further development of *S. lacera* seedlings *in vitro*, only the strain obtained from field-grown protocorms was capable of sustaining development of seedlings beyond the formation of an apical meristem.

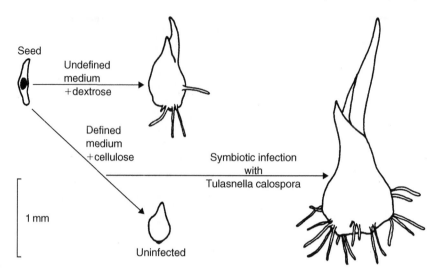

Figure 12.5 Development of *Spathiglottis plicata* after 5 weeks growth on two carbon sources, in relation to colonization by *Tulasnella calospora*. From Hadley (1969), with permission.

McCormick *et al.* (2004) analysed ITS and mtLSU sequences to resolve the identities of fungi colonizing protocorms and adults of three North American green orchid species. The fungi isolated from both stages of *Goodyera pubescens* were not genetically distinguishable. The same was true of *Liparis lilifolia*. However, in *Tipularia discolor*, different fungi were detected in the protocorm and adult stages. On the basis of ITS sequences, isolates from the adults appeared to include at least four distinctly different groups of tulasnelloid fungi. Some of these were closely related to the associates of *G. pubescens*, some to associates of *L. liliflora*, whereas two other groups of distantly related tulasnelloid fungi were specific to *T. discolor*. There were also differences between the fungi associated with protocorms and adults in *T. discolor*. Two fungi were the dominant colonists of protocorms, but a much greater variety was found in mature plants. In laboratory germination tests, all fungi isolated from adult *G. pubescens* were capable of supporting seed germination of that species, whereas only those isolates of *T. discolor* that were genetically very similar to those isolated from *G. pubescens* induced germination in the latter. All isolates from *L. liliflora* adults supported germination of this species.

A number of points emerge. In many orchid species, initial expansion of the embryo and cracking of the testa may occur *in vitro* without the involvement of a fungal symbiont. However, in the absence of exogenous sugar supplies, fungal colonization appears to be essential for further development of the protocorm to produce first seedlings and then adult plants. The levels of specificity shown in orchid–fungus relationships at this stage varies greatly between orchid species and it is likely that there is a continuum from those which are able to form stable symbioses with a large number of fungal associates to those that are compatible with a more restricted range of fungi.

Gaku Masuhara and co-workers carried out some of the first experimental studies of seed germination in the field. In a classic experiment involving comparative

Table 12.2 Selected studies describing fungal associates and specificity in green orchids.

Taxon	Samples	Origin	Fungal identification methods[1]	Reference	Identified fungi[2]	Trophic group[3]
Amerorchis rotundifolia	19 adults	Field	In planta + isolation: vegetative morphology	Zelmer et al., 1996	Epulorhiza (7), Moniliopsis (3).	Unknown
Acianthus reniformis	26 adults	Field	Isolation: morphology of vegetative and sexual stages	Warcup, 1981	25 of 26 isolates were Sebacina vermifera	Unknown
Acianthus caudatus	12 adults	Field	Isolation: morphology of vegetative and sexual stages	Warcup, 1981	Sebacina vermifera (2), Tulasnella cruciata (10)	Unknown
Acianthus exsertus	12 adults	Field	Isolation: morphology of vegetative and sexual stages	Warcup, 1981	Tulasnella calospora	Unknown
Acianthus pusillus	8 adults	Field	Isolation nuc LSU	Bougure et al., 2005c	Epulorhiza repens	Unknown
20 Caladenia spp.	98 adults	Field	Isolation: morphology of vegetative and sexual stages	Warcup, 1971	108 of 110 isolates were Sebacina vermifera	Unknown
Caladenia carnea	7 adults	Field	Isolation nuc LSU	Bougure et al., 2005c	Sebacina vermifera	Unknown
Calypso bulbosa	? adults	Field	In planta + isolation: morphology of vegetative and sexual stages	Currah et al., 1988	Rhizoctonia spp., Rhizoctonia anaticula, Thanatephorus pennatus, unidentified clamped fungi	Unknown
Cypripedium calceolus	2 adults	Field	In planta nuc LSU mt LSU	Shefferson et al., 2005	Tulasnellaceae (2), Thelephoraceae (1)	Unknown (Ectomycorrhizal?)
Cypripedium californicum	10 adults	Field	In planta nuc LSU mt LSU	Shefferson et al., 2005	Tulasnellaceae (3), Sebacinaceae (2), Ceratobasidaceae (3)	Unknown
Cypripedium candidum	6 adults	Field	nuc LSU mt LSU	Shefferson et al., 2005	Tulasnellaceae (2), Thelephoraceae (1)	Unknown (Ectomycorrhizal?)
Cypripedium fasciculatum	16 adults	File	In planta nuc LSU mt LSU	Shefferson et al., 2005	Tulasnellaceae (14), Sebacinaceae (1)	Unknown

(Continued)

Table 12.2 (Continued)

Taxon	Samples	Origin	Fungal identification methods[1]	Reference	Identified fungi[2]	Trophic group[3]
Cypripedium montanum	10 adults	Field	*In planta* nuc LSU mt LSU	Shefferson et al., 2005	*Tulasnellaceae* (10), *Thelephoraceae* (1)	Unknown (Ectomycorrhizal?)
Cypripedium parviflorum	10 adults	Field	nuc LSU mt LSU	Shefferson et al., 2005	*Tulasnellaceae* (6), *Sebacinaceae* (1), *Russula* (1)	Unknown (Ectomycorrhizal?)
Dactylorhiza purpurella	21 adults	Field	Isolation: vegetative morphology	Harvais and Hadley, 1967b	A variety of unidentified *Rhizoctonia* spp., *R. repens, R. solani* and other fungi	Unknown
Dactylorhiza purpurella	Seedlings	Lab.	*In vitro* germination tests	Harvais and Hadley, 1967b	Unidentified *Rhizoctonia* spp., *R. repens* and *R. solani* (including pathogenic strains)	Unknown
Dactylorhiza majalis	4 adults	Field	Single pelotons mtLSU	Kristiansen et al., 2001	*Tulasnella*	Unknown
5 *Diuris* spp.	28 adults	Field	Isolation: morphology of vegetative and sexual stages	Warcup, 1971	*Tulasnella calospora*	Unknown
Goodyera repens	Seedlings	Lab.	*In vitro* germination tests	Hadley, 1970	*Ceratobasidium cornigerum* (1/3), *Ceratobasidium* spp. (1/2), *Thanatephorus cucumeris* (3/9), *Rhizoctonia* spp. (2/4)	Unknown
Goodyera oblongifolia	8 adults	Field	*In Planta* + isolation: vegetative morphology	Zelmer et al., 1996	*Epulorhiza* (1), *Ceratorhiza* (34), *Moniliopsis* (3)	Unknown

Orchid species	Sample	Setting	Method	Reference	Fungi	Function
Goodyera pubescens	? adults	Field	*In planta* mt LSU	McCormick et al., 2004	*Tulasnellaceae*	Unknown
Liparis liliflora	? adults	Field	*In planta* mt LSU	McCormick et al., 2004	*Tulasnellaceae* (distinct from *Tipula* and *Goodyera pubescens*)	Unknown
Microtis parviflora	18 adults + 72 seedlings	Field	Isolation: vegetative morphology	Perkins et al., 1995	2 *Epulorhiza* spp. were isolated from seedlings and adults	Unknown
Microtis parviflora	Seedlings	Lab.	*In vitro* germination tests	Perkins et al., 1995	*Epulorhiza repens*, *Epulorhiza* spp., 3 *Ceratorhiza* spp.	N/A
Neuwiedia veratrifolia	Adults	Field	*In planta* + single mtLSU	Kristiansen et al., 2004	*Tulasnella*, *Thanatephorus*	Unknown
Platanthera hyperborea	13 adults	Field	*In planta* + isolation: vegetative morphology	Zelmer et al., 1996	*Epulorhiza* (7), *Ceratorhiza* (7), *Moniliopsis* (5)	Unknown
Platanthera hyperborea	15 seedling packets	Field	*In planta* + isolation: vegetative morphology	Zelmer et al., 1996	*Epulorhiza* (3), *Ceratorhiza* (1), unknown clamped fungus (4), unidentified (1)	Unknown
Platanthera leucophaea	? adults	Field	Isolation: vegetative morphology	Curtis, 1939	*Rhizoctonia robusta*, *R. sclerotica*, *R. stahlii*, *R. subtilis*[4]	Saprotrophs
Platanthera obtusata	14 adults	Field	Isolation: vegetative morphology	Currah et al., 1990	*Epulorhiza anaticula*, *Ceratorhiza goodyerae-repentis*, *Sistotrema* spp.	Unknown, Saprotrophs

(Continued)

Table 12.2 (Continued)

Taxon	Samples	Origin	Fungal identification methods[1]	Reference	Identified fungi[2]	Trophic group[3]
Pterostylis barbata	6 adults	Field	Isolation: vegetative morphology, anastomosis grouping	Ramsay et al., 1987	Binucleate *Rhizoctonia* spp., mostly anastomosis group P3	Unknown
Pterostylis obtusata	9 adults	Field	Isolation nuc LSU	Bougure et al., 2005c	*Ceratobasidium* spp.	Unknown
Pterostylis longifolia	9 adults	Field	Isolation nuc LSU	Bougure et al., 2005c	*Ceratobasidium* spp.	Unknown
Spiranthes sinensis	37 adults	Field	Isolation: vegetative morphology	Terashita, 1982	*Rhizoctonia repens* (32), *Rhizoctonia solani* (16)	Unknown
Spiranthes sinensis	Seedling	Lab.	*In vitro* germination tests	Masuhara et al., 1993	22 out of 23 *Rhizoctonia* tester strains, including binucleate and multinucleate isolates	N/A
Spiranthes sinensis	18 adults + 27 seedlings	Field	Isolation: vegetative morphology, anastomosis grouping	Masuhara and Katsuya, 1994	All plants contained *Epulorhiza repens*, 2 also had *R. solani*	Unknown
Thelymitra pauciflora	1 adult	Field	Isolation nuc LSU	Bougure et al., 2005c	*Epulorhiza* spp.	Unknown
Tipula discolor	? adults	Field	*In planta* mt LSU	McCormick et al., 2004	Tulasnellaceae	Unknown
Tolumnia variegata	Adult epiphyte	Field	*In planta* nucLSU	Otero et al., 2002	*Ceratobasidium,* uninucleate and binucleate *Rhizoctonias*	Unknown

Updated from Taylor et al. (2002). [1] Isolation refers to the culturing of fungi from pelotons or from whole tissue sections. *In planta* refers to the identification of fungi by direct methods that do not require fungal isolation. *Ex planta* refers to direct observations of fungal morphology or hyphal connections from plants to identifiable fungal structures (fruit bodies or ectomycorrhizal roots). For the molecular methods, mtLSU refers to fungal mitochondrial large subunit ribosomal gene sequences in the region described by Bruns et al. (1998). ITS-RFLP refers to restriction digests of the fungal nuclear internal transcribed spacer region. [2] Fungal taxa are as given in the cited publication; note that different authors have used different nomenclatures. For *Rhizoctonia* fungi, various names may refer to the same taxon (e.g. *Rhizoctonia repens*, *Epulorhiza repens* and *Tulasnella calospora*). In the case of laboratory seed germination tests, only fungi that produced compatible interactions are listed. Fractions in parentheses show the number of compatible strains over the total number of strains tested for a given fungal species. [3] In some cases we infer the trophic category of a fungus from other sources. However, if the authors have made a definitive statement concerning the trophic niche, their conclusion is given preference. [4] It is not clear how the *Rhizoctonia* epithets used or coined by Curtis fit into modern *Rhizoctonia* taxonomy.

analysis of *in vitro* versus field germination of the green orchid *Spiranthes sinensis*, they showed clearly that the expression of specificity under laboratory conditions was strikingly different from that seen in the field. Seeds of the orchid were challenged *in vitro* by 23 *Rhizoctonia* tester strains representing multinucleate and binucleate anastomosis groups, most of which were members of the *Ceratobasidium/Thanatephorus* clade, but including a member of the *Tulasnella* clade, *Epulorhiza* (=*Rhizoctonia*) *repens*. Symbiotic germination was obtained with all but one of the strains (Masuhara *et al.*, 1993). In contrast, seeds sown in mesh packets at a field site were stimulated to germinate in 26 out of 27 cases only by the *E. repens* strain (Masuhara and Katsuya, 1994). A similar predominance of *E. repens* was found in adults from the same site. When baits for the fungi in the form of herbaceous stems were buried across the field plots, fungi representing seven *Rhizoctonia* AGs were found to be widely distributed in the soil. Among these, one AG predominated yet was never recovered from *Spiranthes* plants. These isolates induced germination of *S. sinensis* seed *in vitro* despite the fact that they were not associated with the plants in the field. *E. repens* was only occasionally found in the baits. These results bring into sharp relief the distinction between 'potential' and 'ecological' specificity.

It is not known to what extent fungi other than those of the *Rhizoctonia*-type can initiate or maintain symbiotic germination of green orchids in the field. Zelmer *et al.* (1996) observed that clamp-forming basidiomycetes formed pelotons in protocorms of three *Cypripedium* species, two *Platanthera* spp. and in *Spiranthes romanzoffiana*. In some of these orchids, clamp-bearing hyphae were present in some sites but not at others. These fungi were not culturable. Since hyphae with clamp connections were never seen in adults, it was speculated that, in some species at least, fungal symbionts present in and perhaps responsible for germination of the seeds, may be distinct from those found in the adult phases of growth.

A recent molecular analysis of the fungal associates of *Cypripedium* spp. illustrates the power of the new techniques to resolve hitherto unanswerable questions of this kind concerning symbiont identity. Through direct PCR amplification of fungal genes in mycorrhizal tissue, Shefferson *et al.* (2005) identified the primary symbionts of 58 *Cypripedium* plants representing seven species (Table 12.3). Phylogenetic analysis revealed that the great majority of the fungi forming mycorrhizas in these plants are members of narrow clades within the fungal family Tulasnellaceae. Members of the Sebacinaceae and Ceratobasidiaceae occur rarely and the ascomycete *Phialophora* (= *Cadophora*) occurs as an occasional endophyte.

Knowledge of the identity of partners able to initiate and sustain symbiotic development is of primary importance for conservation programmes. The further use of molecular techniques will enhance resolution of some of the taxonomic uncertainties. Accordingly, the ability to analyse fungal rDNA sequences at the level of the individual peloton (Kristiansen *et al.*, 2001) promises to enable more rigorous evaluation of the identities of these partnerships at the below-ground stage.

Nutritional characteristics of the fungi

There is some diversity as to the source of carbon used by *Rhizoctonia*-type orchid fungi. Most are relatively fast-growing saprotrophs which, in culture, can use complex polymers such as starch, pectin and cellulose and occasionally lignin, as well as

Table 12.3 Fungal root endophytes in *Cypripedium*. Fungal groups represent closest-related taxa from nucLSU and mtLSU phylogenies of Basidiomycota and Ascomycota, and from NCBI BLAST results. Each row lists the corresponding numbers of plants associated with each fungal group. Numbers of plants yielding multiple fungi, and hence requiring cloning of PCR products, are listed in parentheses. Total numbers of plants of each species yielding PCR product are listed in the final row and total numbers of *Cypripedium* plants yielding each fungal group are listed in the final column. Column sums may be greater than totals listed in the final row, because some plants contained multiple fungi and are listed multiple times per column.

Fungal group	calceolus	californicum	candidum	fasciculatum	guttatum	montanum	parviflorum	Total plants
Tulasnellaceae	2(0)	3(1)	2(1)	14(7)	0(2)	10(7)	6(4)	39
Sebacinaceae	–	2(2)	–	1(1)	–	–	1(1)	4
Ceratobasidiaceae	–	3(0)	–	–	–	–	–	3
Thelephoraceae	–	–	1(1)	–	–	1(1)	–	2
Russula	–	–	–	–	–	–	1(1)	1
Agaricales	–	–	–	–	–	–	1(1)	1
Phialophora	–	3(3)	4(4)	3(3)	–	4(4)	4(4)	18
Glomus	–	1(1)	–	–	–	–	1(1)	2
Total plants	2	10	6	16	2	12	10	

From Shefferson *et al.* (2005).

soluble sugars (see Burgeff, 1936; Perombelon and Hadley, 1965; Smith, 1966; Harley, 1969; Hadley and Ong, 1978). These fungi also have unspecialized requirements for nutrients other than organic C. Most can use a wide range of N compounds, at least in pure culture (Hollander in Burgeff, 1936; see Arditti, 1979, 1992) and it seems likely that organic sources of N, and possibly also P, are important for these fungi growing in soil, although direct investigations have not been carried out. The significance in nature of the effects of B vitamins, yeast extract and root exudates which have been observed in culture has not been followed up, although it may have relevance to the horticultural production of orchids by symbiotic methods (Vermeulen, 1946; de Silva and Wood, 1964; Perombelon and Hadley, 1965; Hadley and Ong, 1978).

Seed and protocorm development

Asymbiotic

All orchids pass through a phase where they are non-photosynthetic and dependent on an external supply of nutrients, including organic C. The minute seeds (see Figure 12.4) contain very small amounts of high-energy protein and lipid and very little sugar (Harrison, 1977; Arditti, 1979; Manning and van Standen, 1987; Richardson *et al.*, 1992; Arditti and Ghani, 2000). Some species also contain small numbers of starch grains (Hadley and Williamson, 1971; Purves and Hadley, 1975). If the seeds are spread on a moist substratum, the undifferentiated embryos (see Figures 12.4 and 12.5) absorb water, swell slightly and may burst the testa and sometimes produce epidermal hairs. The embryo does not develop further unless it receives an exogenous supply of carbohydrate or is infected by a compatible mycorrhizal fungus (Figures 12.4, 12.5 and 12.6). In some species, vitamins or growth factors are also required for the embryos to develop asymbiotically. Lipid reserves are not mobilized, probably because of the absence of glyoxysomes and the very low concentrations of endogenous sugars (Harrison, 1977; Manning and van Standen, 1987). Starch is not hydrolysed in those species containing it and, like the lipid and protein, does not support ongoing protocorm development. If sugars (such as glucose, fructose, sucrose, maltose and trehalose; see Smith, 1973; Ernst, 1967; Ernst *et al.*, 1971) are supplied, together with vitamins for those that require them, then many species will develop further and, although growth is slow, mature plants can be obtained. The seeds of *Disa*, *Disperis*, *Cymbidium* and *Huttonia*, when supplied with sucrose synthesized small amounts of starch, developed glyoxysomes and began to use endogenous lipid (Manning and van Standen, 1987). In nature, continuing supplies of soluble sugars, vitamins, amino acids and growth factors are not available from the soil and the conclusion reached by Bernard (1904c), that the fungus is required for successful germination, remains valid. Fungal colonization is required both to stimulate gluconeogenesis and mobilization of reserves and to provide ongoing nutritional support, before photosynthesis commences.

Mycorrhizal colonization of protocorms

Most of the experimental work on fungal colonization of orchid tissue has been carried out using protocorms in monoxenic culture. The interaction between embryo

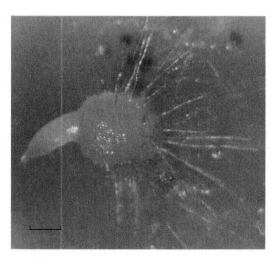

Figure 12.6 Light microscopy of orchid mycorrhizas: protocorm of *Dactylorhiza sambucina* inoculated with *Ceratobasidium cereale*. Bar = 1 mm. From Smreciu and Currah (1989), with permission.

and fungus can have three basic outcomes: a mycorrhizal interaction, with the formation of pelotons (see Figure 12.7a) which are later lysed; a parasitic interaction, in which the orchid cells are invaded by relatively disorganized hyphal growth and death of the protocorm results; and a resistant, rejection response in which the fungus is excluded from the orchid tissues (Figure 12.8) (Hadley, 1970; Beyrle *et al.*, 1995). These interactions may occur simultaneously in a population of protocorms, emphasizing the dynamic and relatively unstable nature of the fungus–plant association. The proportion in any category is influenced by nutritional and environmental factors, at least under laboratory culture conditions.

Figure 12.7 Ultrastructure of orchid mycorrhiza. (a) Scanning electron micrograph of pelotons (p) of *Rhizoctonia* in a protocorm of *Orchis morio*. Bar = 100 μm. From Beyrle *et al.* (1995), with permission. (b) Hyphae of *Rhizoctonia* in the host cytoplasm of a cell of *Dactylorhiza purpurella*. Both the living hyphae (h) and dead collapsed hyphae (dh) are surrounded by an encasement layer (e), which is probably of host origin. Paramural bodies (pb) occur in the interfacial matrix between the host plasma membrane and the encasement layer. Host cytoplasm contains endoplasmic reticulum (er) and a crystal (x). Host vacuole is also indicated. From Hadley (1975), with permission. (c) Inset: detail of the endoplasmic reticulum and encasement layer. From Hadley (1975), with permission. (d) Penetration of a host cell wall (w) by fungal hypha (h) in a normal mycorrhizal interaction between *Goodyera repens* and *Ceratobasidium cereale*. Note the dissolution of the host wall (arrowed). Bar = 1.0 μm. From Peterson and Currah (1990), with permission. (e) Resistant response of *Orchis morio* to penetration by *Rhizoctonia* under culture conditions in which the fungus is excluded and fails to form a mycorrhiza. Note the thickening of the plant wall (arrowed) beneath the penetration peg (arrowed). Fungal hypha (h); host wall (w). Bar = 5.0 μm. From Beyrle *et al.* (1995), with permission.

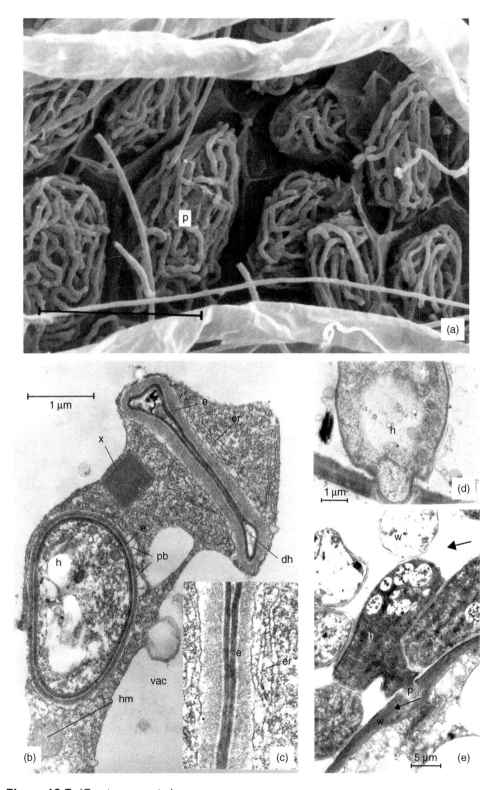

Figure 12.7 (Caption opposite)

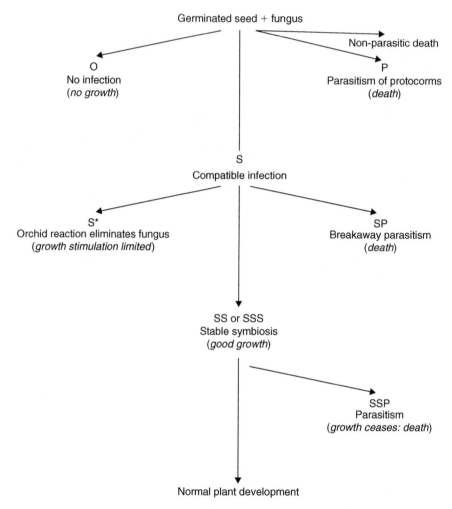

Figure 12.8 Scheme of symbiotic development in orchids. O, non-colonized, no growth; S, SS, SSS, greater numbers of protocorms develop; S*, initial colonization followed by hypersensitive reaction or complete fungal disintegration; P, fungus becomes an aggressive parasite, killing cells. SP, SSP, no infection, but protocorms die for other reasons. Modified from Hadley (1970).

The imbibed and swollen embryos are rapidly colonized by hyphae of suitable fungi. The most usual route appears to be via the suspensor but, in some species, the fungus enters via epidermal hairs (Williamson and Hadley, 1970; Hadley, 1982; Clements, 1988; Peterson and Currah, 1990). There is no good evidence for specific mechanisms of attraction between the symbionts or directional growth of hyphae prior to penetration (Clements, 1988). As it penetrates the cell of the embryo, the infecting hypha invaginates the plasma membrane of the orchid cell and becomes surrounded by a thin layer of cytoplasm (see Figure 12.7b) which remains healthy and continues to show protoplasmic streaming (Williamson and Hadley, 1970). In electron micrographs, the orchid cells appear physiologically active at all stages of colonization and contain numerous mitochondria, well-developed endoplasmic reticulum, Golgi and vacuoles of various sizes (Hadley *et al.*, 1971; Strullu and Gourret, 1974; Dexheimer and Serrigny, 1983). The nuclei of both colonized and non-colonized cells of a mycorrhizal protocorm are obviously hypertrophied (e.g. Burgeff, 1932) and are reported to have higher DNA contents than those of uninfected ones (Alvarez, 1968; Williamson and Hadley, 1969; Williamson, 1970). Another important physiological change is the disappearance of starch grains from the colonized cells and often from other cells in the mycorrhizal protocorm. This is clearly associated with the presence of the fungus, because it does not happen in axenically grown protocorms (e.g. Burgeff, 1959; Peterson and Currah, 1990; Beyrle *et al.*, 1995).

Within the protocorm, the fungus spreads from cell to cell so that the basal region becomes extensively colonized (Figures 12.4 and 12.9). Hyphae penetrating the plant cell wall narrow and do not distort the wall, nor induce thickening (Peterson and Currah, 1990; Beyrle *et al.*, 1995; and see Figure 12.7d), suggesting that localized hydrolysis rather than pressure is important in penetration. Growth and anastomosis of the intracellular hyphae result in the formation of complex pelotons which very much increase the interfacial area between the symbionts (see Figure 12.7a). In recently colonized cells, few vacuoles are present in the hyphae but, in mature coils, the cytoplasm of the fungus becomes vacuolated and contains nuclei, mitochondria, ribosomes, lipid globules and glycogen rosettes, but little ER (Hadley *et al.*, 1971; Strullu and Gourret, 1974; Hadley, 1975; Peterson and Currah, 1990).

The walls of the fungus do not appear to undergo changes during primary colonization and development of the pelotons (see Figure 12.7b, c). They are described as being formed of two layers: an electron-dense inner layer and an outer less dense more 'flocculent' layer which is always present and of variable thickness (Hadley *et al.*, 1971; Strullu and Gourret, 1974). It was at first thought that the outer layer was of fungal origin, but Hadley (1975) concluded that it was comparable to the interfacial matrix in other plant–fungus interactions. The composition of the walls has not been analysed as thoroughly as in AM symbioses. However, Peterson and Currah (1990) suggest that changes in the reaction of walls to cellufluor (see below) indicate that chitin staining is normally blocked, but that during lysis of the pelotons the blocking material is removed, resulting in a very strong reaction. They pointed out that many polysaccharides, including chitin, react strongly with cellufluor and did not agree with the conclusions of Barroso and Pais (1985) that the matrix material was cellulose. A thorough study of the interface region has been carried out by Peterson *et al.* (1996). They employed high-affinity techniques used previously to characterize the interfacial matrix between arbuscules and root cell cytoplasm of arbuscular mycorrhizas and the hyphal complexes of epidermal cells of ericoid mycorrhizas. The walls of peloton hyphae gave a positive reaction for β-1,3 glucans. These compounds are present in the interfacial matrix of arbuscular mycorrhizas but absent from those of ericoid mycorrhiza (ERM) cells (Perotto *et al.*, 1995). Labelling of the interfacial matrix around the peloton hyphae also indicated the presence of β-1,3 glucans as well of cellulose and pectins, indicating that the hyphae may either utilize these components or that they are being synthesized in this position.

There have been conflicting reports about the effects of intracellular penetration on microtubule (MT) formation. Dearnaley and McGee (1996) showed that intracellular penetration of *Microtis parviflora* protocorms led to loss of MTs. However, in studies of *S. sinensis*, Uetake *et al.* (1997) observed that, while colonization by the

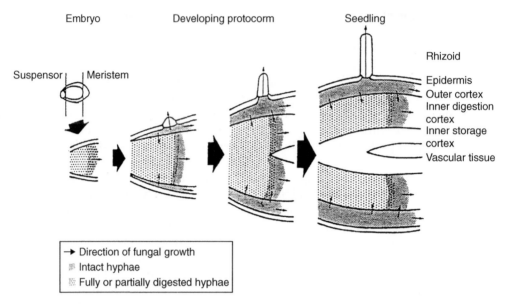

Figure 12.9 A schematic diagram showing the cell layers found in the protocorms and regions of fungal colonization. From Clements (1988), with permission.

fungal symbiont altered the arrangement of MTs, they were always retained and became closely associated with healthy peloton hyphae, with the cell nucleus and with degenerating pelotons. It was suggested that those associated with the functioning hyphae may be involved in the formation of the peri-fungal membrane. It seems highly likely that the plant cytoskeleton plays an essential role in cellular reorganization as the fungal symbiont becomes established intracellularly. Much more detailed examination of the roles of the cytoskeleton in this regard have been carried out with arbuscular mycorrhizas (see Chapter 3).

Hadley and his colleagues (Hadley *et al.*, 1971; Hadley, 1975) observed that the fungal plasma membrane was invaginated in places to form vesicles and tubules, while the fungal wall bore protuberances adjacent to the plant (*Dactylorhiza purpurella*) plasma membrane. Both these features increase the surface area across which transfer of nutrients might take place. However, Peterson and Currah (1990) did not find similar structures in protocorms of *Goodyera repens* colonized by *Ceratobasidium cereale*. Other modifications in the interfacial region include the presence of ATPases on both plant and fungal membranes, with the plant enzyme remaining active surrounding the collapsing pelotons. Neutral phosphatase activity in the interface and on the plant membrane surrounding active pelotons may be involved with the synthesis of interfacial matrix by the plant (Serrigny and Dexheimer, 1985).

The intracellular pelotons have a limited life even in the mycorrhizal interaction. Some of the earliest workers (Wahrlich, 1886; Janse, 1897) noted the characteristic clumping of the hyphal coils, which was thought to be the result of defensive phagocytosis by Bernard (1905, 1909b). Electron microscopy has revealed that, during lysis (sometimes called digestion), the hyphal contents become disorganized and the walls take on flattened or angular profiles. In the final stages, the wall material clumps together, forming an irregular mass (see Figures 12.4 and 12.7b). The plasma membrane of the plant remains intact and is separated from the clumped fungal material by an electron-lucent layer which has been shown to contain callose, pectins and a small amount of cellulose (Peterson and Currah, 1990; Peterson and Bonfante, 1994). During this process, the plant cells remain alive and active and may be recolonized by hyphae either apparently surviving the lytic process or invading from an adjacent cell (Burgeff, 1936; Burges, 1939; Strullu and Gourret, 1974). In this respect, the interaction is similar to arbuscular mycorrhizas (see Chapter 2) and different from ericoid mycorrhizas (see Chapter 11).

The timing of colonization, development of pelotons and subsequent lysis have been followed in a number of orchid–fungus combinations. In culture, fungal penetration occurs within a very few days of the symbionts coming into contact. For *Goodyera repens*, Mollison (1943) recorded that suspensors were invaded by *Rhizoctonia goodyerae-repentis* (*Ceratobasidium cornigerum*) within 5 days, pelotons were formed within 7 days and had lysed by 11 days. The symbiotic stimulus to protocorm growth occurred before lysis of pelotons. She also noted that if the protocorms were already imbibed, the process occurred more quickly. Hadley and Williamson (1971) followed colonization in *Dactylorhiza purpurella* continuously for several days, using an inverted slide technique (Figure 12.10). There was considerable variation within the population but, on average, colonization via the epidermal hairs occurred within 14.5 hours of the first contact and pelotons were formed by 29 hours. The number of pelotons and the number of lysed pelotons increased linearly with time. Lysis of pelotons was observed by 30–40 hours and continuous observation of a single, peloton-containing cell showed that lysis took less than 24 hours. The growth of protocorms (measured as volume) was followed at the same time as colonization and, although the number of pelotons was not linearly related to the rate of growth (Figure 12.10a–d), there was evidence that the mycorrhizal growth stimulus preceded lysis of pelotons, thus confirming Mollison's (1943) observations.

The causes of the hyphal collapse and lysis are still not really known. Many workers have believed it to be the result of activity of the cells of the orchid and a manifestation of a defence reaction against fungal invasion (e.g. Burges, 1939), or a means by which the orchid cell causes a release of nutrients from the fungus. Evidence still does not clearly distinguish between these two hypotheses but, on balance, the former seems the most likely, especially in view of the results cited above showing that a growth stimulus occurs before any lysis is apparent (see Figure 12.10). Williamson (1973) showed that acid phosphatase activity, often thought to be a marker for lysosome activity, increases in cells where hyphal collapse is taking place. Subsequently, it was shown that this activity was never found in the interface of the plant with young active hyphae but, even in the same cell, was markedly increased around old, highly vacuolated hyphae. These old hyphae themselves contained acid phosphatases which disappeared when the hyphae finally collapsed (Dexheimer and Serrigny, 1983; Serrigny and Dexheimer, 1986). Increased activities of oxidase systems (e.g. polyphenol oxidase, catalase, ascorbic acid oxidase) have been observed during lysis, but their role in the process was not clear (Blakeman *et al.*, 1976; Pais and Barroso, 1983). Orchids also produce fungitoxic phytoalexins such as orchinol, lorroglossol and hircinol (see below and Stoessl and Arditti, 1984). These compounds are mainly found in tubers where they are believed to be involved in controlling or excluding fungal colonization. The presence of orchinol has also been

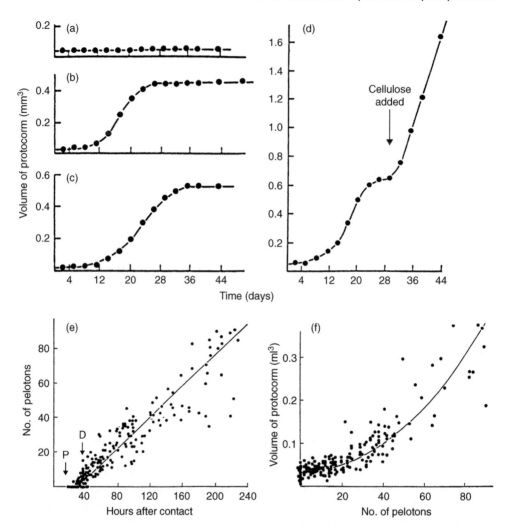

Figure 12.10 Development of the interaction between *Dactylorhiza purpurella* and *Rhizoctonia* spp. in inverted slide cultures. (a) Growth of protocorms when grown on water agar in the absence of a fungal symbiont. (b) Growth on water agar following inoculation. (c) Growth on Pfeffer's medium plus 0.1% glucose, following inoculation. (d) Growth on Pfeffer's medium plus cellulose, following inoculation. (e) Relationship between time after contact and number of pelotons formed. Mean first penetration time (P) = 14.5 hours; first digested pelotons observed (D) = 30–40 hours. (f) Relationship between the number of pelotons formed and volume of protocorms obtained. From Hadley and Williamson (1971), with permission.

detected in protocorms of *Orchis morio*, so that a role in the more dynamic stages of plant–fungus interaction is possible (Beyrle *et al.*, 1995).

Once colonization is established in a mycorrhizal (rather than parasitic) pattern, growth of protocorms proceeds rapidly with different phases of colonization restricted to well-defined regions. The protocorms differentiate into distinct tissues, including uncolonized storage cortex, vascular tissue, shoot meristem and (in the potentially photosynthetic orchids used) young leaves, as well as basal colonized regions containing both living and lysed pelotons (see Figures 12.4 and 12.10). Fungal hyphae connect the internal protocorm to the substrate, with the only entry/ exit points being the suspensor or the epidermal hairs (see Clements, 1988). The growth of mycorrhizal protocorms is generally much faster than asymbiotic proto-corms raised under the same conditions. For example, plantlets of *Goodyera repens*, colonized by *Ceratobasidium cornigerum*, reached a height of 5–10 cm in 12 months on cellulose agar, whereas asymbiotic protocorms were only 2–3 cm tall (Alexander and Hadley, 1984; see Figure 12.5).

The sequence and timing of different stages of development have rarely been fol-lowed under natural conditions. However, preliminary observations using seeds of several terrestrial species buried in mesh bags, suggest that colonization prob-ably occurs in spring–summer and may be much slower than in laboratory culture (Rasmussen and Wigham, 1994).

Mycorrhizas in adult orchids

Structural aspects

Adult orchids usually have mycorrhizal roots or tubers, although the extent of col-onization is variable. Temperate and tropical species of terrestrial orchids are usu-ally heavily colonized (see Hadley, 1982; Goh *et al.*, 1992; Figure 12.11). Roots are frequently produced anew each season and rapidly become mycorrhizal, with the fungus entering from the soil. In *Ophrys*, this coincides with the production of new tubers in autumn (Bernard, 1901). Roots of *Goodyera repens* are also colonized as soon as they are produced (4 mm long) and no difference in the extent of colonization at different seasons has been observed (Mollison, 1943; Alexander and Alexander, 1984). In *Bletilla striata*, root growth began in early summer (May) and colonization by living, undigested fungal pelotons was observed to be highest in young roots. As the rate of root growth increased and the season progressed, both active and total colonization declined. Seasonal isolations showed that *Rhizoctonia repens* was the most common endophyte of *B. striata* at the site surveyed (Masuhara *et al.*, 1988).

A few investigations have compared colonization in different orchid tissues. In the evergreen *G. repens*, colonization of rhizomes (up to 90%) occurred independently of that of roots and was at a maximum in winter (November to April) when the rate of elongation was lowest (Alexander and Alexander, 1984). *Spiranthes sinensis* var *amoena* produces true roots in autumn which, as in the species described above, become rapidly colonized up to a maximum the following summer. During flower-ing, the amount of living fungus declined and the roots decomposed. In contrast, tuberous roots were produced in spring and, although anatomically similar to the true roots, were only locally colonized. The fungus did not spread and the extent of colonization was always less than for the true roots (Masuhara and Katsuya, 1992). Five patterns of colonization have been found in Western Australian species from arid environments (Ramsay *et al.*, 1986). In species with stem and root tubers, the fungus was found in these structures throughout the year, although the extent of colonization was lowest during the summer period of aestivation. In other species,

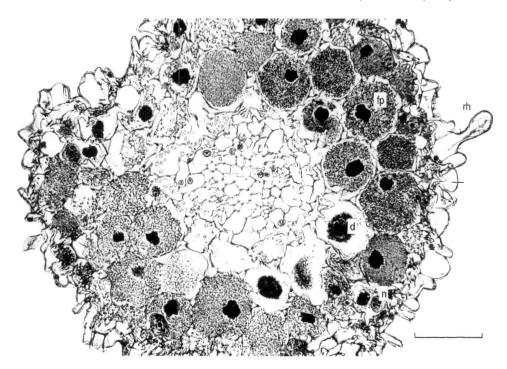

Figure 12.11 Transverse section across a mycorrhizal root of *Cypripedium arietinum* showing the presence of intact (fp) and degenerate (d) pelotons in the same areas of cortical tissue. Figure courtesy of R.L. Peterson.

seasonally produced stem collars and roots were colonized anew each season (as in temperate species), with a peak coinciding with the period of maximum vegetative development (Figure 12.12).

In epiphytic orchids, colonization appears, with a few exceptions, to be relatively sparse and sporadic compared with terrestrial species. Not all plants examined contain typical mycorrhizal structures and colonization, where present, was often limited to the regions of the roots adjacent to the support (Ruinen, 1953; Hadley and Williamson, 1972; Bermudes and Benzing, 1989; Lesica and Antibus, 1990; Richardson *et al.*, 1993).

At the cellular level, fungal colonization of tissues of adult orchids appears generally similar to protocorms. A fungal hypha penetrates the epidermis and enters the cells of the cortical parenchyma, forming characteristic and complex hyphal coils. Penetration of the suberized exodermis, which is often well developed in orchids, is via the short passage cells (Janse, 1897; Mejstrik, 1970; Esnault *et al.*, 1994). The fungi spread either as a result of repeated colonization from the soil or

Figure 12.12 The five categories of colonization type found in Western Australian orchids. Positions of characteristic colonization zones in roots, rhizomes or stems are indicated by arrows. From Ramsay *et al.* (1986).

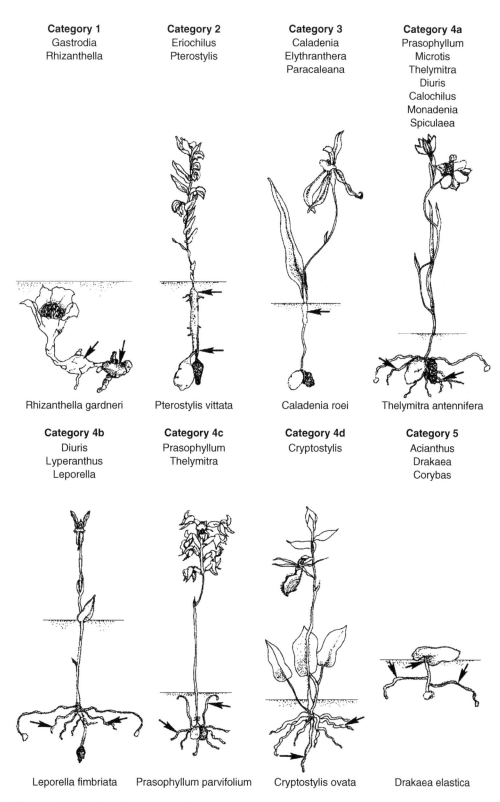

Category 1
Gastrodia
Rhizanthella

Category 2
Eriochilus
Pterostylis

Category 3
Caladenia
Elythranthera
Paracaleana

Category 4a
Prasophyllum
Microtis
Thelymitra
Diuris
Calochilus
Monadenia
Spiculaea

Rhizanthella gardneri

Pterostylis vittata

Caladenia roei

Thelymitra antennifera

Category 4b
Diuris
Lyperanthus
Leporella

Category 4c
Prasophyllum
Thelymitra

Category 4d
Cryptostylis

Category 5
Acianthus
Drakaea
Corybas

Leporella fimbriata

Prasophyllum parvifolium

Cryptostylis ovata

Drakaea elastica

Figure 12.12 (Caption opposite)

by hyphae penetrating from cell to cell in the root cortex. More than one species of fungus can form pelotons in a root or even within the same cell (Warcup, 1971) and within the cortex the cells may be uncolonized or contain active pelotons or clumps of degenerating hyphae in different proportions (see Figures 12.7a, b and 12.11).

Transfer of nutrients between symbionts

Overview

As made clear in the earlier discussion, the non-photosynthetic protocorm stages of orchids are dependent upon exogenous supplies of sugars and other nutrients supplied in nature by their fungal symbionts. A change in direction of transfer of organic C potentially occurs as the seedlings of green orchids develop photosynthetic capacity. It has been presumed that this physiological transition is of general occurrence but, until recently, few experiments had investigated it. As some green terrestrial orchids normally grow in shaded environments, there remains the possibility that organic C, as well as N and P and other nutrients might continue to be supplied by the fungi (see Chapter 13).

In the sections that follow, we separate what is known about protocorms from information on orchid seedlings and adult plants that have green tissues and are presumed to be, at least partially, autotrophic. It must, however, be recognized that the transition is likely to be gradual and a sharp distinction is quite artificial.

Protocorm nutrition

Carbon

The increase in growth following mycorrhizal colonization of protocorms is based on the fungi supplying organic C to the orchid. Much of the work on symbiotic growth of orchid protocorms has employed soluble sugars in the medium that can be used by both fungus and protocorm. However, growth responses to mycorrhiza establishment are regularly obtained on oatmeal-, starch- or cellulose-containing media, where hydrolysis by the fungus would be a prerequisite for use by either symbiont and on which no growth of asymbiotic protocorms occurs. Indeed, establishment of a mycorrhizal interaction and continued protocorm growth is more certain if the fungus is provided with cellulose than if presented with a medium rich in soluble carbohydrate (see Figure 12.5). In the latter case, a parasitic interaction is more likely to occur (Smith, 1966; Harvais and Hadley, 1967b; Hadley, 1969).

As with other mycorrhizal systems, translocation of nutrients is a prerequisite for effective delivery to the symbiotic partner. The first evidence of translocating ability in orchid endophytes was provided by Beau (1920) using a split-plate system which separated organic substrates for the fungus from the colonized protocorms. Growth only occurred if an intact mycelium was present. If the hyphae connecting the protocorms to the nutrient supply were cut, growth ceased. Later, Smith (1966) used the experimental system illustrated in Figure 12.13 to show that *R. solani* (*Thanetephorus cucumeris*) was not only capable of hydrolysing cellulose and absorbing its products, but also of translocating them to orchid seedlings in sufficient quantities for growth to occur. The results are given in Table 12.4. Harley and Smith (1983) used data for the growth of protocorms of *Dactylorhiza purpurella* and assumptions about the number of hyphal connections with a cellulose substrate to calculate an approximate (hexose) flux of $2.5 \times 10^{-3}/mol/m^2/s$. This flux is considerably higher than the fluxes of inorganic nutrients in AM mycorrhizal hyphae (see Chapter 5).

The translocating ability of orchid mycorrhizal fungi and delivery of sugars to symbiotic protocorms of *D. purpurella* and *Goodyera repens* was confirmed using ^{14}C-glucose (Smith, 1967; Purves and Hadley, 1975; Alexander and Hadley, 1985; see Figure 12.8). Both alcohol-soluble and insoluble fractions of the seedlings became labelled in *D. purpurella* and chromatographic analysis of the soluble fractions showed that uncolonized orchid tissues contained sucrose, glucose and fructose, whereas the fungi contain trehalose accompanied by glucose and occasionally by mannitol, but no sucrose. Seedlings fed with ^{14}C-glucose via the fungus in split-plates became labelled in the fungal sugars and also in the orchid sugar, sucrose (Figure 12.14). Changes in the pattern of labelling with time in *D. purpurella* are shown in Figure 12.15 and indicate that the fungal sugar, trehalose, is the most heavily labelled in the early samples but as time elapses sucrose becomes proportionally more heavily labelled as trehalose labelling declines (Smith, 1967).

There is thus good evidence that carbohydrate is translocated in the fungus and that during or following transfer to the orchid cells it is converted to sucrose. Since trehalose can support asymbiotic growth of the seedlings of several orchids, it is

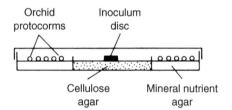

Figure 12.13 Modified Petri dish used to study translocation to orchid seedlings. Seeds were sown on mineral nutrient agar in the outer dish. The fungus was inoculated onto the cellulose-containing medium in the inner dish; it grew over the barrier into the outer compartment and formed mycorrhizas with the seedlings. From Smith (1966).

Table 12.4 Growth of seedlings of *Dactylorhiza purpurella* in 14 weeks on substrates with or without cellulose in the presence of absence of *Thanatephorus cucumeris* at 22.5°C, in the dark.

	Not inoculated	Inoculated	
		+cellulose	−cellulose
Number of seedlings[a]	14	226	324
Length (μm)[b]	248 ± 9	1170 ± 119	800 ± 72
Width (μm)[b]	206 ± 10	692 ± 44	519 ± 39

Data from Smith (1966). [a]seedlings remaining healthy out of approximately 800. [b] ± standard error.

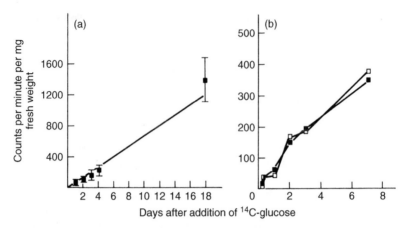

Figure 12.14 Translocation of ^{14}C-labelled compounds by mycorrhizal fungi into orchid protocorms after the fungi had been supplied with ^{14}C-glucose on split plates. (a) *Goodyera repens* colonized by *Rhizoctonia goodyerae-repentis*. Redrawn from Purves and Hadley (1975). (b) *Dactylorhiza purpurella* colonized by *R. solani*. Alcohol soluble fraction, ■; alcohol insoluble fraction, □. Redrawn from Smith (1967).

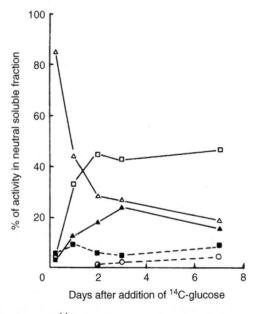

Figure 12.15 Distribution of ^{14}C in the components of the neutral ethanol-soluble fraction of mycorrhizal protocorms of *Dactylorhiza purpurella* after the mycorrhizal fungus, *Rhizoctonia solani*, had been supplied with ^{14}C-glucose on split plates. △, trehalose; □, sucrose; ■, glucose; ▲, mannitol; ○, fructose. Redrawn from Smith (1967).

possible that it is transferred directly from the fungus, but hydrolysis to glucose before absorption by the action of a fungal or orchid trehalase cannot be ruled out and might occur in the symbiotic interface within the protocorms. Trehalose is a suitable source of carbohydrate for germination of *D. purpurella* (Smith, 1973), *G. repens*

(Purves and Hadley, 1975), *Bletilla hyacintha*, *Dendrobium* spp. and *Phalaenopsis* spp. (Ernst, 1967; Ernst *et al.*, 1971; Smith, 1973). It can also be absorbed and metabolized by the leaves of *B. hyacintha*. The labelling patterns following absorption of ^{14}C-trehalose were identical with those following ^{14}C-glucose absorption (Smith and Smith, 1973). Ernst *et al.* (1971) observed low concentrations of glucose in the medium following incubation of trehalose with asymbiotic *Phalaenopsis* seedlings, possibly indicating that hydrolysis precedes absorption by the protocorms. Mannitol is less likely to be important as a carbohydrate for translocation and transfer. It occurs in only a proportion of the fungi examined and is suitable for the asymbiotic germination of only a few species of orchid. The leaves of *B. hyacintha* absorb and accumulate mannitol but do not metabolize it; nor is it suitable for the germination of the seeds of this species (Smith, 1973; Smith and Smith, 1973).

Vitamins and growth factors synthesized by the fungi may also be important or essential for the growth of the seedlings of some orchids and there is reasonable circumstantial evidence that the requirements of seedlings in asymbiotic culture are provided by the fungus in symbiotic systems. However, direct evidence for their transfer from fungus to host is lacking.

Mineral nutrients

The experiments of Hollander and Burgeff (see Burgeff, 1936) indicate major mycorrhizal effects on N nutrition. The increase in weight of *Cymbidium* seedlings on polypodium fibre was about 10-fold in three months, whereas the increase in N content was 25-fold. More direct evidence for uptake and translocation of N is not available, although split-plate techniques using ^{15}N-sources (both organic and inorganic) would be relatively simple to carry out. ^{32}P is translocated by the fungi and accumulates in protocorms of *D. purpurella* in split plates (Smith, 1967). However, again, quantitative information is lacking.

Transfer of nutrients from fungus to orchid in adult plants

Carbon, nitrogen and phosphorus

Early workers argued that the C transfer from fungus to protocorm seen to occur in heterotrophic stages of the orchid life cycle was reversed when the plants became photosynthetic (Burgeff, 1936; Scott, 1969). These assertions were unsupported by experimental evidence but appeared to be justified by a later study (Alexander and Hadley, 1985) which investigated the movement of ^{14}C, supplied as an insoluble source cellulose, from mycorrhizal mycelium to associated green plantlets of *Goodyera repens*. Conversely, the small number of experimental investigations of C transfer from autotrophic orchid to fungus had indicated that the heterotroph received little or no organic C from their plant partners. Hadley and Purves (1974) were unable to detect any movement of C to the fungus after exposing green shoots of *G. repens* to ^{14}CO$_2$. Similarly, Alexander and Hadley (1985) found that after one week less than 0.5% of the total ^{14}CO$_2$ fed to shoots of *G. repens* was released into the substrate, either to the mycorrhizal mycelium or as exudate. Conversely, the autotrophic plants, in contrast to the protocorms, failed to obtain C from their fungal partners, even after prolonged periods of darkness designed to enhance plant C demand. On the basis of these results, Alexander and Hadley (1985) also concluded that the orchid mycorrhiza underwent a physiological change which stopped the movement of C from fungus to plant once the orchid became photosynthetic. These studies, together with others which were unpublished but gave similar results (see Purves and Hadley, 1975), were limited by the sensitivity of the methods then available.

A new study (Cameron et al., 2006) has demonstrated significant C flow from plants of G. repens to its extensive extraradical mycelial network (Figure 12.16). Around 3% of the label from ^{14}C fed as a pulse of ^{14}CO$_2$ to the shoots was transferred to the fungus over a 72 h labelling period (Figure 12.17). As this value does not include ^{14}C respired by the fungus, it is likely to represent a significant underestimate of total C flow from fungus to plant. Despite such losses, the residual value of 3% for allocated C is within the range of estimates reported for the fungi in some AM symbioses (Leake et al., 2006; see Chapter 4).

Transfer of C and other nutrients in the reverse direction, from fungus to plant, has also been re-examined. Cameron et al. (2006) showed that when glycine was supplied to mycorrhizal mycelium, either as double-labelled (^{13}C-^{15}N) glycine (Figure 12.18a) or as ^{14}C-labelled glycine (Figure 12.18b), significant quantities of labelled C were transferred across a diffusion barrier to photosynthetic G. repens plantlets. This experiment clearly demonstrated that the pathway for C transfer from fungus to plant is retained into adulthood in this autotrophic orchid. It leaves open the possibility that G. repens, as in the case of some other green orchids that characteristically inhabit deeply shaded habitats, may be a partial mycoheterotroph in nature (see Chapter 13). It remains to be ascertained whether the contrasting results obtained in this experiment and that of Alexander and Hadley (1985) are attributable to differences of the substrate employed, to environmental conditions under which the experiments were carried out, or to different sensitivities of the methods.

The experiment of Cameron et al. (2006) also provides the first direct experimental evidence for transfer of N from fungus to autotrophic orchid (Figure 12.18a), although this was inferred by some early work reported by Burgeff (1936). The total amount of ^{15}N recovered in plant and fungal biomass on the receiver (plant) side of the diffusion barrier was more than double that of ^{13}C, despite the fact that the double-labelled glycine contained three times more C than N. This difference was attributed to respiration of C in the mycelial compartment of the growth chambers. The finding implies that glycine is assimilated by the fungus and that transfer of N and C across the interface is likely to be independently regulated. The total amount of ^{15}N recovered in biomass after 72 h exposure to the isotope decreased from fungus (78%) to root (20%) to shoot (3%). Establishment of a fungal pathway for N acquisition in orchids is important not just in the light of their poorly developed root systems but also because many orchids, both of the autotrophic and mycoheterotrophic kinds, have high foliar N contents and hence high N requirements (Gebauer and Meyer, 2003).

There have also been few analyses of the role of orchid mycorrhiza in P nutrition of adult orchids. Using ^{32}P-orthophosphate, Alexander et al. (1984) provided strong circumstantial evidence that the presence of orchid mycorrhizas enhanced the uptake of P in G. repens by demonstrating that application of the fungicide thiobenzidole significantly suppressed P acquisition by colonized plants. However, although ^{32}P transfer from fungus to orchid protocorms was shown many years ago (Smith, 1967), direct evidence of P transfer to an adult plant through mycorrhizal mycelium was

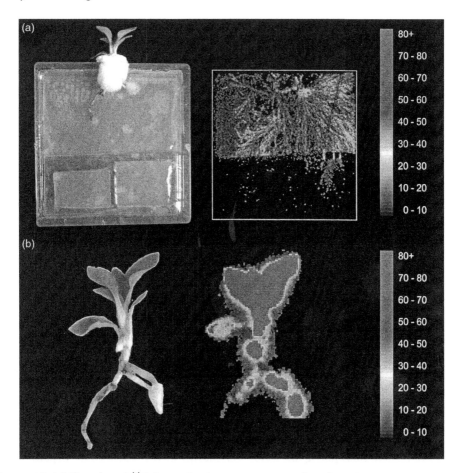

Figure 12.16 Transfer of ^{14}C from *Goodyera repens* to its fungal symbiont microcosm used for exposure of plants to $^{14}CO_2$ (left panel) and grey-scale digital autoradiographs obtained after 72 h exposure of the shoot showing (a) ^{14}C allocation through external mycorrhizal mycelium, into the lower right-hand block of agar through bridging hyphae and the absence of ^{14}C transfer to mycelium in the left-hand agar block where hyphal connections to the plant were severed before labelling (right panel). (b) Assimilation of $^{14}CO_2$ and its allocation to roots and rhizomes of *G. repens*. The shaded scale indicates the number of counts detected in pixel areas of $0.25\,mm^2$ in a period of 60 minutes. From Cameron *et al.* (2006).

lacking until recently. A study in which ^{33}P was supplied to the mycorrhizal mycelial network of *G. repens*, again with a diffusion barrier between the source and mycorrhizal plants, has now confirmed that significant quantities of P can be transported through the mycelium (Figure 12.19). After 7 days, 6.3% of the total ^{33}P transported over the barrier was recovered in the plant–fungus system (Cameron *et al.*, 2007).

Mechanisms of transfer

Although lysis of the fungus in orchid mycorrhiza is generally regarded as a manifestation of defence of the host against invasion, it has frequently been assumed that

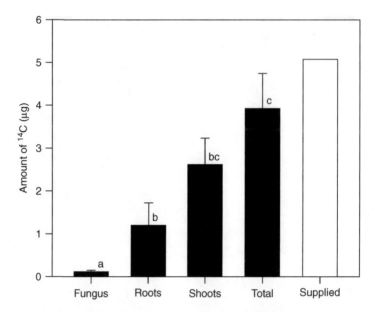

Figure 12.17 Amount of ^{14}C allocated to the biomass components: external mycelium compartment (fungus), roots, shoots and their sum (total biomass) from $^{14}CO_2$ supplied to shoots. The amount of ^{14}C originally supplied is shown on the right. Error bars represent + 1 standard error. Bars sharing the same letter are not significantly different ($P > 0.05$, 1-way ANOVA (Log_{10} transformed data): d.f. = 3,23; F = 29.82; $P < 0.001$). From Cameron et al. (2006).

it is also important in the transfer of nutrients. There is no clear evidence against this hypothesis, except that the growth response to colonization appears to start before any lysis of pelotons is observed (see Figure 12.10). At present, it seems more likely that nutrients are transferred across the intact membranes of fungus and orchid, with potential for separate control of movement of different nutrients, which is clearly required as shown by the work of Alexander and Hadley (1984, 1985) on *G. repens* at different stages of development, as well as by the more recent work of Cameron (2006). However, Serrigny and Dexheimer (1985) showed that DES-sensitive ATPases are active on the plasma membranes of both fungus and plant in the peloton interface and on the plant plasma membranes surrounding the lysed clumps of fungus, but not on the uninvaginated plasma membrane. This distribution might be taken as evidence that nutrients are absorbed by the plant from both intact and digested pelotons and further work is clearly required to sort this out.

Host–fungus interactions in the protocorm – mycorrhizal or not?

Until very recently, it appeared that interactions between green orchids and their fungal associates could not, realistically, be viewed as mutualistic. Prior to the work of Cameron et al. (2006), no experimental evidence had been produced that the fungi gained any nutritional benefit from the association. Even now it is not clear under what conditions net transfer of photosynthate to orchid fungi takes place, or how widespread it is among the many thousands of green orchid species that grow in

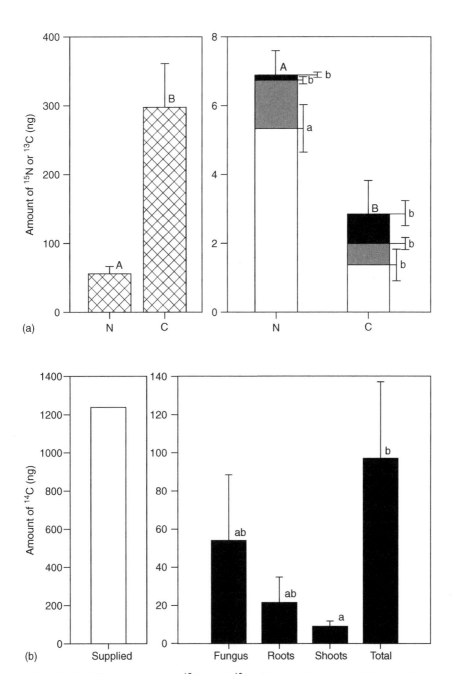

Figure 12.18 (a) The amount of ^{15}N and ^{13}C (a) supplied as double-labelled glycine source (hatched bars) and (b) the total ^{15}N and ^{13}C transferred into plant and fungal biomass, and its component parts: in (a) external mycelium compartment (□), into roots (▨) and shoots (■). Error bars represent + 1 standard error. Bars sharing the same letters are not significantly different ($P > 0.05$, 2-way ANOVA), N = 3–4. Upper case letter codes refer to the bar total, lower case letter codes refer to the component parts within each bar. From Cameron *et al.* (2006). (b) Amount of ^{14}C supplied in ^{14}C glycine (a) and transferred into the biomass components (b): external mycelium compartment, roots, shoots and total biomass. Error bars represent + 1 standard error. Bars sharing the same letter are not significantly different ($P > 0.05$). N = 4–5. From Cameron *et al.* (2006).

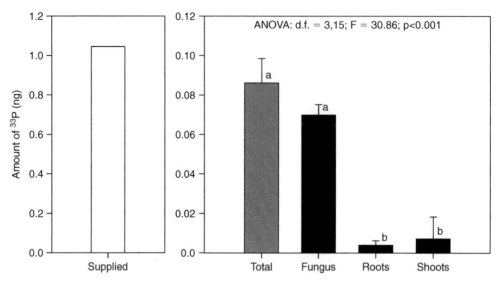

Figure 12.19 The amount of ^{33}P supplied (left panel) and mean total ^{33}P transferred across the diffusion barrier by the mycorrhizal mycelial network (lightly shaded bar) and the distribution of this between fungus, roots and shoots (black bars). Error bars represent $+$ 1 standard error, N = 5 microcosms. Bars sharing the same letter are not significantly different ($P > 0.05$, 1-way ANOVA: d.f. = 3,15; F = 30.86; $P < 0.001$). From Cameron *et al.* (2007).

diverse habitats. The demonstration does, however, indicate the potential for mutualism, offset in time between the symbionts.

Mycorrhizal symbioses in green orchids also appear to lack the long-term compatibility that is so characteristic of other types. Indeed, the outcome of the interaction between potentially mycorrhizal fungi and orchid protocorms is quite variable. In any population raised under apparently controlled conditions with a single endophyte, there may be actively growing mycorrhizal protocorms and also protocorms completely invaded and parasitized by the fungus or from which the fungus has been excluded (rejection). In the last two situations, little or no growth of protocorms occurs. Figure 12.8 shows a scheme of possible pathways of fungal interaction with orchid protocorms drawn up by Hadley (1970) after experiments with 32 strains of 10 species of fungus (isolated from 5 north temperate and 3 tropical species of orchid) with 10 species of protocorm in culture. Similar variations in response have been found for Australian orchids by Warcup (e.g. 1975) and Clements (1981, 1982b). Manipulation of culture conditions, especially N and C composition and concentration in the medium, can alter the proportions of protocorms in the different categories (Hadley, 1970; Beyrle *et al.*, 1991, 1995) and can be used to analyse the interactions, although the very artificial conditions are unlikely to relate to the situation in soil. Rejection of the fungus is accompanied by thickening of cell walls (see Figure 12.7e) and deposition of phenolics, while uncontrolled parasitism is characterized by soft rot, with the hyphae ramifying through the tissues and no evidence of mobilization of structural defence responses (Masuhara and Katsuya, 1991; Beyrle *et al.*, 1995). To add to the complexity, mycorrhizal protocorms may develop normally for some weeks but then succumb to what Hadley (1970) referred to as 'breakaway parasitism', in which the fungus changes its behaviour and ramifies through the tissues.

The proportion of protocorms in each class depends upon the strain of fungus and the species of orchid, as well as upon the environmental conditions. Little is known of what controls the balance between the two organisms, but the symbiosis is clearly much less stable than other types of mycorrhizal interaction and can be likened to a situation where attack by a potential pathogen is controlled by defence responses in the plant. Rhizoctonias, similar to orchid symbionts, produce cellulases and pectinases which may be deployed in colonization of plant tissues. The amounts of pectinases produced in culture by different species of *Rhizoctonia* are not related to their pathogenicity towards seedlings of *Dactylorhiza purpurella* (Perombelon and Hadley, 1965). A considerable range of cellulolytic ability has also been found (Smith, 1966; Hadley, 1969; and see Burgeff, 1936), so that control of activity in mycorrhizal interactions must occur, perhaps via end-product repression in tissues with relatively high hexose content compared with non-mycorrhizal orchid tissue (Purves and Hadley, 1975). During parasitism, the cell walls are degraded (Beyrle *et al.*, 1995), suggesting that control of hydrolytic activity no longer operates. In the mycorrhizal interactions, limited hydrolytic activity in the tissues might be significant in permitting penetration of cell walls (as in Figure 12.7d) and in releasing small quantities of oligosaccharides which would elicit defence responses in the plant and lead to control of invasion.

Evidence for the deployment of defence responses by orchids is quite extensive and dates from the initial observations of Bernard (1909b) who suggested that the absence of fungal infection from the tubers of some orchids and the resistance of some seeds to fungal attack was due to the presence within them of an antifungal principle. He showed that tubers of *Loroglossum* contained a substance which was toxic to many orchid endophytes including *Rhizoctonia repens* (*Tulasnella calospora*) but not to a strain of *R. solani*. Subsequently, it has been confirmed that orchids produce dihydroxyphenanthrene phytoalexins such as hircinol, loroglossol and orchinol, which inhibit the growth of mycorrhizal fungi and many other fungi and bacteria (see Arditti, 1979; Stoessl and Arditti, 1984). It is not completely clear whether these phytoalexins are present before contact with the fungi, but Gaumann and his colleagues (see Nuesch, 1963) showed that orchinol is formed in tubers of *Orchis militaris* when incubated with *Rhizoctonia repens*, not only by the cells immediately in contact with the fungus, but by those up to 12 mm away.

It seems certain that these substances play a part in controlling or restricting fungal invasion, particularly in tubers where they are produced in high concentrations. However, their role in the fine-tuning of the interactions in protocorms and roots is not at all clear and for many years neither orchinol nor related phytoalexins had been detected in protocorms. Beyrle *et al.* (1995) showed that low concentrations of orchinol were present in asymbiotic protocorms of *Orchis morio* and that levels increased on contact with a mycorrhizal *Rhizoctonia*. Neither the amount of orchinol synthesized nor the activity of phenylalanine ammonia lyase (PAL, a key enzyme in the synthetic pathway for phytoalexins) was related to the type of symbiotic response. In contrast, Reinecke and Kindle (1994), working with *Phalaenopsis*, detected no 9,10-dihydrophenanthrenes in young sterile plantlets, but contact with *Botrytis cinerea* and a *Rhizoctonia* caused synthesis of phytoalexins, including hircinol, and concomitant rises in PAL and bibenzyl synthase activity. Unfortunately, the type of symbiotic interaction with the *Rhizoctonia* was not recorded, so the picture remains incomplete.

Production of hydrolytic enzymes by the plant may be important in lysis of the fungus. The possible involvement of acid phosphatases and of 1,-3 β glucanases and chitinases has already been mentioned, but their activities in the different symbiotic responses has not been closely analysed. As the last two are recognized as patho-genesis-related (PR) proteins, produced in many plant–pathogen interactions and under conditions of stress, they may have considerable interest in orchid symbioses. It is quite clear that the dynamic and variable nature of the interactions between protocorms of orchids and their endophytic fungi sets them apart from the reactions in other mycorrhizal associations. Compared with AM systems, for example, in which host defence responses are transitory (see Chapter 3), the situation in orchids looks like a battlefield in which attack and defence mechanisms are mobilized by both partners. There is clearly scope for further work which would provide a clearer picture of gene expression and protein accumulation in asymbiotic protocorms and in the different symbiotic interactions.

Conclusions

Although the widespread occurrence of mycorrhizas in adult orchids has been rec-ognized for well over a century, it remains true that much of the work on orchid mycorrhizas has been directed towards understanding the interactions taking place during the early stages of seedling germination and growth under laboratory cul-ture conditions. All orchids have a relatively prolonged heterotrophic stage during germination and early growth. Although the seeds of many species can be induced to germinate and grow in the absence of a mycorrhizal fungus, as long as sugars and various growth factors are supplied, this does not occur in nature. During the het-erotrophic phase, green orchids (as well as the full mycoheterotrophs, see Chapter 13) are dependent on carbohydrate translocated to them via the mycelium of a myc-orrhizal fungus. In green orchids, the source of the carbohydrate is most frequently dead organic matter supplied by the saprotrophs. The fungi probably also supply N and P and other nutrients to both seedlings and adult plants, but this aspect of the physiology of the association should be investigated with a wider range of plants and, again, under ecologically relevant conditions. Recent work has shown that some organic C transfer from fungus to plant is maintained into the adult phase of at least one green orchid, providing evidence for some ongoing dependence on the fungus for C, as well as N and P.

Interestingly, it has also been demonstrated that the products of photosynthe-sis may also pass to the fungi. This finding provides limited evidence in support of the contention of Burgeff (1936) and Curtis (1939) that the direction of C trans-fer is reversed in adult orchids. This may be taken as evidence that the symbioses can be mutualistic, although benefits to plant and fungus are offset temporally over the whole life span of a green orchid. Nevertheless, the situation is highly complex and likely to be influenced by environmental conditions, particularly irradiance. It will be necessary to explore the relative magnitude of C transfer in both directions in order to evaluate the net transfer occurring under different conditions and in different orchid–fungus combinations. Only then will it become clearer where individual orchid–fungus partnerships sit on the mutualism–parasitism continuum and, at the latter extreme, which organism is the parasite.

In recent years, mycorrhizal associations in orchids have been relatively neglected, compared with the enormous amount of research on arbuscular and ectomycorrhizas. We hope this chapter has shown how a combination of conventional and molecular methods has the potential to unravel some of the complexities of host–fungus interactions at both physiological and taxonomic levels and provide data relevant to ecological situations. Such information will become increasingly important in orchid conservation, as populations decline as a result of habitat loss and anthropogenic changes in nutritional status of soils.

13

Mycorrhizas in achlorophyllous plants (mycoheterotrophs)

Introduction

The loss of chlorophyll and hence of the ability for autotrophic carbon fixation through-out the life cycle is seen in two distinct categories of plant (Cummings and Welsch-meyer, 1998). In one, the heterotrophic existence is sustained by direct physical attachment to autotrophs which enables the transfer of organic C through haustorium-like structures. These are the parasitic plants and they are restricted in their distri-bution to the dicotyledenous lineage of the plant kingdom. In the other category, which is the subject of this chapter, heterotrophy is sustained by forming associations with fungi, most, but not all of which are co-associated with the roots of neighbour-ing autotrophs. In these cases the plants occur as epiparasites on the fungal partners of the plants rather than upon the autotrophs themselves. Plants sustained through-out their lives by parasitizing fungi in this way are referred to as mycoheterotrophs (Leake, 1994). They are found in unrelated families across the plant kingdom from the liverworts (Brypophyta), through the ferns (Pteridophyta) to the Angiosperms (Figure 13.1). While there is considerable morphological diversity among the 400 species of mycoheterotroph recognized among the Angiosperms (Figure 13.2), these plants, in addition to their fungus-dependent habit, have a number of structural and developmental features in common. Mycoheterotrophs are characterized by loss of leaves, reduced vascularization and a sometimes extreme reduction in the extent of their root systems. In the majority of species, their seeds are also greatly reduced in size. Termed 'dust seeds' these consist of a tiny embryo of only a few cells and a min-imal volume of storage tissues. A remarkable expression of this reduction is seen in the gentianaceous mycoheterotroph *Voyria tenella* which, with an embryo consisting of three cells and seed diameter of less than 50 μm, is one of the smallest in the plant kingdom (Maas and Ruyters, 1986) (see Colour Plate 13.1a). Very early colonization by an appropriate fungus is therefore a prerequisite for embryo development in these seeds. Studies employing the seed packet technique (see below) have demonstrated that, in the tropical rainforest environments inhabited by *V. tenella*, progression from germination to flowering can be achieved within one year (see Colour Plate 13.1b–e).

Probably on the basis of the observation that achlorophyllous plants of these kinds normally live in richly organic forest soils they were long regarded as 'saprophytes';

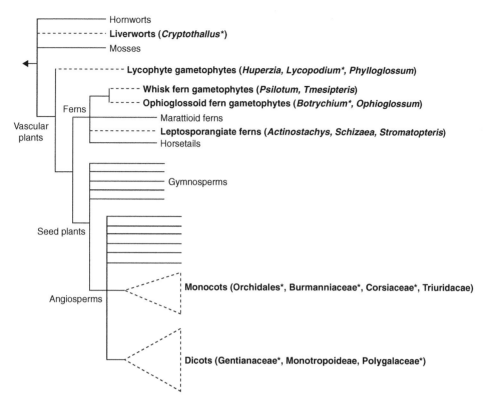

Figure 13.1 Lineages of plants that have evolved mycoheterotrophy (dashed branches). Lineages with full mycoheterotrophs are shown in bold (c. 500 spp.). The rest contain initially mycoheterotrophic plants only (c. 20 000 spp.). The exceptions are *Buxbaumia* and *Schizaea* which may be full mycoheterotrophs and *Parasitaxus* which may be a direct root parasite. All direct plant–plant parasites (not shown) fall in the dicots. Asterisks indicate lineages for which some molecular identification of mycorrhizal fungi has been carried out. Adapted from Bidartando (2005).

indeed, many modern botanical reference works continue to place mycoheterotrophs in this category (e.g. Preston *et al.*, 2002; Smith N *et al.*, 2004b). This is despite pioneering work on some such plants, most notably in the Monotropoideae by Kamienski (1882), whose seminal paper describing links between ectomycorrhizal (ECM) roots of trees and the roots of *Monotropa hypopitys* has recently been republished in translation (Berch *et al.*, 2005). While subsequent studies (Romell, 1939a; Bjorkman, 1960; Martin, 1985) went some way towards confirming the observations of Kamienski, compelling evidence from molecular studies (see below) has now established beyond doubt that monotropoid roots associate exclusively with the ECM fungal symbionts of autotrophs. As a result of these advances, what has been referred to as 'the myth of saprophytism' (Leake, 2005) can finally be debunked. The long history of sometimes rancorous argument over the structural and functional attributes of what we now know to be mycoheterotrophs has been reviewed elsewhere (Furmann and Trappe, 1971; Trappe and Berch, 1985; Leake, 1994; Bidartondo, 2005).

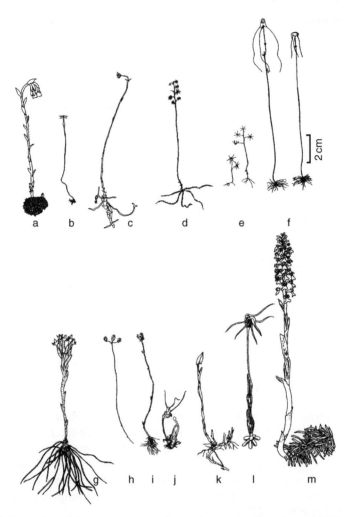

Figure 13.2 Examples of the main groups of mycoheterotrophic plants: (a) Monotropoideae: *Monotropa uniflora*; (b) Gentianaceae: *Voyria chiona*; (c) Petrosaviaceae: *Petrosavia*; (d) Triuridaceae: *Sciaphila albescens*; (e) *Peltophyllum lutea*, male and female plants; (f) *Triuris alata*, male and female plants; (g) Burmanniaceae: *Campylosiphon purpurascens*; (h) *Burmannia tenella*; (i) *Gymnosiphon brachycephalus*; (j) *Thismia saulensis*; (k) Corsiaceae: *Corsia ornata*; (l) *Arachnites uniflora*; (m) Orchidaceae: *Neottia nidus-avis*. From Leake (1994).

The order Orchidales contains the largest number of mycoheterotrophic species. These include, in addition to completely achlorophyllous and hence fully myco-heterotrophic species, a number of chlorophyll-containing species which, none-theless, receive some C from their fungal partners and so are regarded as being partially mycoheterotrophic or 'mixotrophic'. Both fully and partially mycoheterotrophic orchid species are considered in this chapter. All members of the ericaceous subfamily Monotropoideae are mycoheterotrophic, as are those of the Triuridaceae, Petrosaviaceae and Corsiaceae. The large family Burmanniaceae contains a mix of autotrophic and mycoheterotrophic species as well a number which are almost certainly mixotrophic. The Gentianaceae consists largely of autotrophic species. Of the around 1200 species (in 74 genera) in the family, only approximately 30 are mycoheterotrophic and, again, there are chlorophyll-containing genera (e.g. *Obolaria* and *Bartonia*) that are likely to be partially mycoheterotrophic.

Progress towards an accurate assessment of the nutritional relationships of these plants was provided by Rommell (1939a) who hypothesized that members of the Monotropoideae, subsequently referred to here as 'monotropes', were epiparasites on woody plants that associated with the same fungi, then by Bjorkman (1960) who advanced the concept of epiparasitism. He demonstrated that movement of radiolabelled C and P from spruce to *M. hypopitys* was greater than that to neighbouring autotrophs. Subsequent work, described below, has provided support for the notion that most mycoheterotrophs are indeed epiparasitic, but has emphasized that they directly parasitize the fungal component of the tripartite autotroph–fungus–mycoheterotroph relationship (Bidartondo, 2005).

There was much uncertainty until recently concerning the identities of the fungal partners of the disparate plant taxa to which mycoheterotrophic plants belong. The early descriptions of mycorrhizas in mycohetreotrophs of the families Burmanniaceae (Groom, 1895; Janse, 1897), Polygalaceae (Pijl, 1934) and Gentianaceae (Knöbel and Weber, 1988) show a range of intracellular structures including aseptate hyphae, hyphal coils, arbuscules and vesicles, which are reminiscent of arbuscular mycorrhizal fungal colonization. Since autotrophic members of the families Gentianaceae and Polygalaceae are known normally to support AM colonization, records of the occurrence of these fungi in their heterotrophic relatives were perhaps to be expected. However, the occurrence of AM fungi in Burmanniaceae, Corsiaceae and Triuridaceae was initially less predictable in view of the original taxonomic placement of these families in the Orchidales (see for example Cronquist, 1981; Leake, 1994). Recent revisions of phylogenetic relationships within monocotyledonous plants, based upon plastid nucleotide sequences have demonstrated that these families lie, respectively, in the Liliales, Dioscoreales and Pandanales, each of which are predominantly made up of plants with AM colonization (see Figure 12.1, Chapter 12). Structural and molecular analyses described below are now largely supportive of the view that mycoheterotrophic members of the families Burmanniaceae, Corsiaceae and Triuridaceae associate with AM fungi and the same is now known to be true of the subterranean achlorophyllous gametophytes of the Lycopodiales (the lycophytes) (Winther and Friedman, 2007b), of *Tmesipteris* and *Psilotum* (the whisk ferns) and of the ophioglossoid ferns *Botrychium* (Winther and Friedman, 2007a) and *Ophioglossum* (see Figure 13.1) (see also Chapter 14).

Large numbers of basidiomycetous fungi, most of which are known also to form ECM, have been claimed, usually on the basis of casual observation, to be involved in the formation of mycorrhizas in the Monotropoideae. For this reason, a view that specificity in the subfamily was low became established (Castellano and Trappe, 1985). Again, molecular methods have transformed our understanding of mycotrophy in these plants. While confirming the basidiomycete and ECM affinities of the fungi involved, analyses of nuclear and ribosomal DNA have exposed considerable, and occasionally extreme, levels of specificity in the fungus–plant relationship (Bidartondo and Bruns, 2005).

In the case of the fully mycoheterotrophic orchids, early research suggested, as it did in the Monotropoideae, that specificity might be low. This view arose in part because claims of mycorrhizal status have been made for fungi that were either not isolated from, or not re-inoculated to produce, mycorrhizas. Burgeff (1932), for example, reported the isolation of numerous fungi from roots of the achlorophyllous species *Neottia nidus-avis*, *Corallorhiza trifida* and *Epipogium nutans* which, on current evidence, would appear to have been saprotrophs. In contrast, the early report (Kusano, 1911a, 1911b), of a mycorrhizal association between the mycoheterotrophic orchid *Gastrodia elata* and strains of the pathogenic fungus *Armillaria mellea* has been confirmed and manipulations of this association now form the basis of an extensively used system for cropping this medicinally important orchid in Korea (Kim and Ko, 1995) and China (Xu and Guo, 2000).

As in the Monotropoideae, the application of molecular techniques, by providing the first reliable identification of many of the fungi involved in sustaining mycoheterotrophic orchids, has exposed a previously unexpected level of specificity. Parallels with the Monotropoideae occur also in terms of symbiont sharing between the mycoheterotrophic orchids and ECM trees and shrubs.

Fungal associations and specificity in monotropoid mycorrhizas

Wallace (1975) recognized 10 genera within the tribe Monotropoideae which he classified as a subfamily of the Ericaceae. The first accurate descriptions of the fungal partners of monotropes were provided by Jean-Francois Martin who, on the basis of meticulous microscopic analysis, concluded first that *M. hypopitys* was mycorrhizal with members of the genus *Tricholoma* (Martin, 1985). He then showed that members of its distant relative *M. uniflora* (examined as herbarium specimens!) were associated with *Russula* spp. (Martin, 1986). The notions of fungal specificity in these plants, espoused by Martin, have been confirmed and greatly extended by later workers using DNA-based methods (Figure 13.3) (Cullings *et al.*, 1996; Bidartondo and Bruns, 2001; Bidartondo *et al.*, 2002, 2003). As a result, understanding of relationships between the plants in the subfamily, with their sister groups and with their fungal partners has been transformed.

Parsimony analysis of nuclear DNA sequence data (Figure 13.3) indicates that members of the Montropoideae share a recent common ancestry with autotrophic members of the other ericaceous tribes Arbutoideae and Pyroloideae (Bidartondo and Bruns, 2001, 2002; Bidartondo, 2005) (see also Chapter 7). Within the Monotropoideae three lineages are recognized, the probably basal *Pterospora-Sarcodes-Pleuricospora* clade, a clade involving *Monotropa uniflora* and *Monotropastrum* and a further clade including *Pityopus*, the derived *Allotropa* and *Monotropa hypopitys*. The latter occurs as two distinct races, Eurasian and North American.

Parallel studies of fungal DNA sequences have enabled precise identification of the fungi which stimulate germination of monotropes in the laboratory (Bruns and Read, 2000) and in the field (Leake *et al.*, 2004b; Bidartondo and Bruns, 2005) and which form monotropoid mycorrhizas on adult plants (Cullings *et al.*, 1996; Kretzer *et al.*, 2000; McKendrick *et al.*, 2000a; Bidartondo and Bruns, 2001, 2002, 2005; Young *et al.*, 2002; Leake *et al.*, 2004; Yokoyama *et al.*, 2005). In so doing they have not only confirmed the deductions based upon microscopy of Martin (1985, 1986) but revealed unprecedented levels of specificity across the whole of the Monotropaceae. Examples of the taxa examined, the sample sizes and methodologies employed in these studies are shown in Table 13.1.

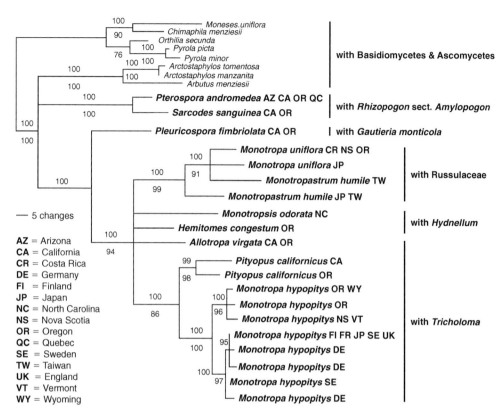

Figure 13.3 Consensus phylogram of relationships within Monotropoideae and with their most recent common ancestors in the Arbutoideae and Pyrolae, based on analysis of nuclear DNA, and showing the associations (on the right) between the different monotrope lineages and particular fungal families, genera or species. Plants were selected to maximize geographic coverage and habitat diversity. Simplified from Bidartondo (2005).

Host–symbiont specialization is seen at the level of the fungal genus in some monotropes and at that of the species in others. *In vitro* tests (Bruns and Read, 2000) on seeds of *Sarcodes sanguinea* (see inset Colour Plate 13.2a) and *Pterospora andromedea* showed that *Rhizopogon* species alone were capable of initiating germination of these plants. All *Rhizopogon* strains isolated from either species were capable of stimulating germination in both *Sarcodes* and *Pterospora*. Their seeds failed to germinate in the presence of six other genera of basidiomycetes. Since germination could be stimulated through cellophane and without direct contact between fungus and seed, it was concluded that a diffusible or volatile compound unique to *Rhizopogon* was involved in the process. The *in vitro* tests revealed that, despite the narrowness of the suite of fungi capable of stimulating germination, this range was still greater than seen in association with mature plants in the field (see below). A further finding of likely biological significance was that seeds collected over a number of years, some of them having over-wintered in the field, could be stimulated to germinate by the appropriate fungus, thus indicating a potential for long-term dormancy.

Table 13.1 Selected studies decribing the mycorrhizal symbionts of plants in the mycoheterotrophic tribe Monotropoideae.

Monotrope species	Type of sample	Origin	Fungal identification method[1]	Reference	Identified fungi[2]	Trophic group[3]
Allotropa virgata	18 adults	Field	In planta: molecular mtLSU, ITS	Bidartondo and Bruns, 2001	Tricholoma magnivelare	Ectomycorrhizal
Allotropa virgata	37 soil cores near adults	Field	In planta: ex planta: mycorrhiza morphology + molecular (ITS RFLPs)	Lefevre et al., 1998	Tricholoma magnivelare	Ectomycorrhizal
Hemitomes congestum	7 adults	Field	In planta: molecular mtLSU, ITS	Bidartondo and Bruns, 200	Hydnellum spp.	Ectomycorrhizal
Monotropa uniflora	23 adults	Field[4]	In planta: morphology	Martin, 1986	Russula species, and unidentified Russulaceae	Ectomycorrhizal
Monotropa uniflora	6 adults	Field	In planta: molecular ML5–6 sequences	Cullings et al., 1996	Russulaceae	Ectomycorrhizal
Monotropa uniflora	35 adults	Field	In planta: molecular mtLSU, ITS	Bidartondo and Bruns, 2001	Russulaceae	Ectomycorrhizal
Monotropa uniflora	15 adults	Field	In planta: morphology + molecular ITS	Young et al., 2002	Russulaceae	Ectomycorrhizal
Monotropa uniflora	Seedlings	Field	In planta: molecular ITS sequences	Bidartondo and Bruns, 2005	Russulaceae	Ectomycorrhizal
Monotropa hypopitys	11 adults	Field	In planta + culture	Martin, 1985	Several species of Tricholoma	Ectomycorrhizal
Monotropa hypopitys	9 adults	Field	In planta: molecular ML5–6 sequences	Cullings et al., 1996	Suilloid group (including Rhizopogon)	Ectomycorrhizal
Monotropa hypopitys	43 adults	Field	In planta: molecular mtLSU, ITS	Bidartondo and Bruns, 2001	Tricholoma spp.	Ectomycorrhizal
Monotropa hypopitys	3 adults, seedlings	Field	In planta: molecular mtLSU, ITS	Leake et al., 2004b	Tricholoma terreum, T. cingulatum	Ectomycorrhizal

Species	Count	Setting	Methods	Reference	Fungal associate	Type
Monotropastrum sp.	? adults	Field	In planta + ex planta: vegetative morphology and hyphal connections	Kasuya et al., 1995	Unknown yellow, clamped Basidiomycete	Ectomycorrhizal
Monotropastrum humile	2 adults	Field	In planta: molecular mtLSU, ITS	Bidartondo and Bruns, 2001	Russula spp.	Ectomycorrhizal
Monotropastrum humile	1 adult	Field	In planta: molecular mtLSU	Yokoyama et al., 2005	Russulaceae	Ectomycorrhizal
Monotropastrum humile var. glaberrimum	2 adults	Field	In planta: molecular mtLSU	Yokoyama et al., 2005	Thelephoraceae	Ectomycorrhizal
Monotropsis odorata	2 adults	Field	In planta: molecular mtLSU, ITS	Bidartondo and Bruns, 2001	Hydnellum spp.	Ectomycorrhizal
Pityopus californicus	12 adults	Field	In planta: molecular mtLSU, ITS	Bidartondo and Bruns, 2001	Tricholoma spp.	Ectomycorrhizal
Pityopus californicus	Seedlings	Field	In planta: molecular ITS	Bidartondo and Bruns, 2005	Tricholoma myomyces.	Ectomycorrhizal
Pleuricospora fimbriolata	42 adults	Field	In planta: molecular mtLSU, ITS	Bidartondo and Bruns, 2001	Gautieria monticola	Ectomycorrhizal
Pterospora andromedea	31 adults	Field	In planta: molecular (ITS RFLPs, ML5–6 sequences)	Cullings et al., 1996	Rhizopogon subcaerulescens group	Ectomycorrhizal
Pterospora andromedea	Seedlings	Lab.	In vitro germination tests	Bruns and Read, 2000	2 closely related Rhizopogon spp.	Ectomycorrhizal

(Continued)

Table 13.1 (Continued)

Monotrope species	Type of sample	Origin	Fungal identification method[1]	Reference	Identified fungi[2]	Trophic group[3]
Pterospora andromedea	77 adults	Field	In planta: molecular mtLSU	Bidartondo and Bruns, 2001	Rhizopogon salebrosus, R. arctostaphylli	Ectomycorrhizal
Pterospora andromedea	Seedlings	Field	In planta: molecular ITS	Bidartondo and Bruns, 2005	Rhizopogon salebrosus, R. arctostaphylli	Ectomycorrhizal
Sarcodes sanguinea	57 adults	Field	In planta + isolation: molecular ITS	Kretzer et al., 2000	Rhizopogon ellenae	Ectomycorrhizal
Sarcodes sanguinea	Seedlings	Lab.	In vitro germination tests	Bruns and Read, 2000	2 closely related Rhizopogon spp.	Ectomycorrhizal
Sarcodes sanguinea	12 adults	Field	In planta: molecular ML5–6	Cullings et al., 1996	Cantharellaceae, Rhizopogon, Suillus, unknown fungus	Ectomycorrhizal
Sarcodes sanguinea	93 adults	Field	In planta + molecular mtLSU, ITS	Bidartondo and Bruns, 2001	Rhizopogon ellenae, R. subpurpurascens	Ectomycorrhizal
Sarcodes sanguinea	Seedlings	Field	In planta + molecular ITS	Bidartondo and Bruns, 2005	Rhizopogon ellenae, R. subpurpurascens	Ectomycorrhizal

Updated from Taylor et al. (2002). [1] Isolation refers to the culturing of fungi from pelotons or from whole tissue sections. In planta refers to the identification of fungi by direct methods that do not require fungal isolation. Ex planta refers to direct observations of fungal morphology or hyphal connections from plants to identifiable fungal structures (fruit bodies or ectomycorrhizal roots). For the molecular methods, ML5–6 refers to fungal mitochondrial large subunit ribosomal gene sequences in the region described by Bruns et al. (1998). TSOP refers to 'taxon-specific oligonucleotide probe' and ITS-RFLP refers to restriction digests of the fungal nuclear internal transcribed spacer region. [2] Fungal taxa are as given in the cited publication; note that different authors have used different nomenclatures. In the case of laboratory seed germination tests, only fungi that produced compatible interactions are listed. [3] In some cases we infer the trophic category of a fungus from other sources. However, if the authors have made a definitive statement concerning the trophic niche, their conclusion is given preference. [4] Based on study of herbarium specimens.

Francke (1934), in the first experiments ever to employ buried packets of seed, obtained limited amounts of symbiotic seedling development in *M. hypopitys* both *in vitro* and in the field, but was unable to identify the fungus involved. Much later, Leake *et al.* (2004b), using essentially the same techniques, sowed seed packets close to and at distance from putative autotrophic co-associates of *M. hypopitys* (*Salix repens* and *Pinus sylvestris*). Packets, recovered sequentially over two years, yielded *M. hypopitys* seedlings showing progressively more advanced developmental stages. Inflorescence buds were formed by 20 months (see Colour Plate 13.2e). No germination was obtained in packets placed at distance from the ectomycorrhizal autotroph *Salix repens* but, when sown in the presence of these shrubs, greater amounts of germination and more advanced development were observed in packets placed close to adult *Monotropa* plants. Germination of *M. hypopitys* in the field was exclusively associated with *Tricholoma* species and, as predicted by Martin (1985), adult plants were also exclusively associated with fungi of this genus.

Application of a sophisticated version of the seed burial technique involving use of packets with multiple compartments each containing seed from different geographical sources, has demonstrated that, from the time of germination, monotropoid species show considerable regional specialization upon their particular fungal partners (Table 13.2) (Bidartondo and Bruns, 2005). Seed of *M. uniflora* collected in the eastern USA was shown to germinate in a Californian site (Mount Tamalpais) known to contain *Russula brevipes*, despite the fact that no monotrope plants occurred naturally in the area. *Monotropa hypopitys*, *M. uniflora* (see Colour Plate 13.2f), *Pityopus californicus*, *Pterospora andromedea* (see Colour Plate 13.2d), *Sarcodes sanguinea* and *Allotropa virgata* all germinated only where they were transplanted to areas supporting the fungi of the type known to be associated with the maternal source of the seed.

Examination of adult *M. hypopitys* selected from Swedish, Eurasian and North American clades of the plant confirms that specialization on the genus *Tricholoma* spp. is the norm for this species (Bidartondo and Bruns, 2001). Analyses of ribosomal protein (rps) and nuclear DNA (nrDNA) phylogenies within and between clades found high levels of specificity in all cases. With one possible exception, examples of fungi being shared between two or more of the clades were not found. The exclusively Swedish clade is associated with *T. columbetta*, *T. saponaceum* or *T. portentosum*, the Eurasian clade with *T. terreum* and *T. cingulatum* and the North American lineage associates with *T. sejunctum* and *T. flavovirens*. The possible exception arises in the case of *T. portentosum* which has also been identified as an associate of *M. hypopitys* in some of its North American locations. If *T. portentosum* is indeed the same species in European and North America localities, this apparently widespread fungus could have enabled the circumboreal expansion of *M. hypopitys*. The geographic mosaic of specificity of the kind observed by Bidartondo and Bruns (2001) in *M. hypopitys* is an attribute seen in many parasites and is derived from the requirement to specialize on narrow host lineages, in this case species of *Tricholoma* (Thompson, 1994).

A simpler example of a geographical mosaic is found in *Monotropa uniflora*. Adults of this exclusively North American species associate only with a narrow range of species within the family Russulaceae (Bidartondo and Bruns, 2005). In nature, single *M. uniflora* plants associate only with *Russula brevipes* or its very close relative even when other members of this family occur nearby (Figure 13.4).

Adult plants of *Pterospora* and *Sarcodes* in the basal lineage of the Monotropoideae (see Figure 13.3) specialize on a distinct and narrow subgroup, *Amylopogon*, of fungi

Table 13.2 Seeds of Monotropoideae were collected from plants with known mycorrhizal fungi at five sites (horizontal) and buried near plants with known mycorrhizal fungi at six sites (vertical) in the combinations indicated by grey cells. Mount Tamalpais, California, was the only site without any Monotropoideae plants, but it is known to harbor *Russula brevipes*, an associate of *Monotropa uniflora*. Germination was observed in the combinations indicated by a check mark (✓). At Eel Creek, Oregon, the only combination where germination was observed was Pa/RH (data not shown).

		Seed source (plant/mycorrhizal fungus)													
		Dinkey Creek, CA				SERC, MD			Del Norte County, CA		Perkins Creek, OR				
Seeds buried at ↓	Seeds buried near ↓	Pa/RHa	Pa/RHs	Ss/RHe	Pf/Gm	Mu/Rd	Mu/Rn	Mh/Ts	Mu/Rb	Mu/Rv	Ave/Tm	Mh/Tf	Mh/Tp	Pc/Tmy	Mu/Rb
Dinkey	Pa/RHa	✓	✓	✓											
	Pa/RHs	✓	✓	✓											
	Ss/RHe	✓	✓	✓											
	Pf/Gm														
SERC	Mu/Rn					✓	✓								✓
	Mu/Rd					✓	✓								▦
	Mh/Ts							▦							▦
Del Norte	Mu/Rb								✓	✓					
	Mu/Rv								✓	✓					
Perkins	Ave/Tm										▦				▦
	Mh/Tf										▦	✓			▦
	Mh/Tp										▦		✓		▦
	Pc/Tmy										▦			▦	▦
	Mu/Rb										▦				✓
Eel	Mh/Tl										▦				▦
	Ave/Tm										▦				▦
	Pc/Tfo										▦				▦
	Pa/RHs		✓												
Tamalpais	Rb					✓									✓

Plant lineage abbreviations: Av, *Allotropa virgata*; Mh, *Monotropa hypopithys*; Mu, *Monotropa uniflora*; Pa, *Pterospora andromedea*; Pc, *Pityopus californicus*; Ss, *Sarcodes sanguinea*. Fungal lineage abbreviations: Gm, *Gautieria monticola*; Rb, *Russula brevipes*; Rd, *Russula decolorans*; RHa, *Rhizopogon arctostaphyli*; RHe, *Rhizopogon ellenae*; RHs, *Rhizopogon salebrosus*; Rn, *Russula nitida*; Rv, *Russula vesca*; Tf, *Tricholoma flavovirens*; Tfo, *Tricholoma focale*; Tl, *Tricholoma luteo-maculosum*; Tm, *Tricholoma magnivelare*; Tmy, *Tricholoma myomyces*; Tp, *Tricholoma portentosum*; Ts, *Tricholoma sejunctum*. From Bidartondo and Bruns (2005).

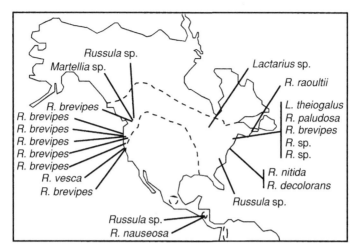

Figure 13.4 An example of a geographic mosaic of mycorrhizal specificity in the Monotropoideae. Fungal lineages associated with individual plants of *Monotropa uniflora* at various locations (each straight line corresponds to one location) throughout its geographic range (delimited by dashed lines). All fungi detected are members of the Russulaceae, one of the most diverse groups of obligate ectomycorrhizal fungi, with several hundred species in North America and typically several species at any one location. Single *M. uniflora* plants were not found to form mycorrhizas with more than one fungal lineage, even when neighbouring *M. uniflora* were mycorrhizal with other Russulaceae or if mushrooms and tree mycorrhizas of other Russulaceae were nearby. From Bidartondo (2005).

within the large and diverse genus *Rhizopogon* (see Figure 6.4) (Bidartondo and Bruns, 2001, 2002). Further, even when the natural ranges of *P. andromedea* and *S. sarcodes* overlap to the extent that individuals of the species coexist at the same site, they specialize on different *Rhizopogon* species; *P. andromedea* associates most frequently with *R. salebrosus* or *R. arctostaphyli* and *S. sanguinea* with *R. ellenae* (see Table 13.2). Bidartondo and Bruns (2001) analysed 93 adults of *S. sanguinea* and 77 of *P. andromedea* growing intermixed within a few metres of each other in the Sierra Nevada and could find absolutely no overlap between their fungal symbionts. Within the same monotrope lineage, *Pleuricospora fimbriolata* associates exclusively with *Gautieria* but, again, only with a small subset, the *G. monticola* group, within the large genus (Bidartondo and Bruns, 2002).

In combination, the *in vitro* (Bruns and Read, 2000) and seed packet (Bidartondo and Bruns, 2002, 2005) studies provide clues as to how plant genotypic diversification at local or regional scales is correlated with divergence of mycorrhizal specificity among closely related fungi. The *in vitro* studies show that germination of *Sarcodes* and *Pterospora* is triggered by diffusible or volatile chemical cues produced by closely related *Rhizopogon* species that colonize mature plants. The seed packet work with a broader set of monotropes and fungi indicates that germination is similarly stimulated only by fungi closely related to those associated with adult plants. These observations suggest that jumps to distantly related fungi are effectively constrained during germination by the absence of recognizable fungal cues. There is, nonetheless,

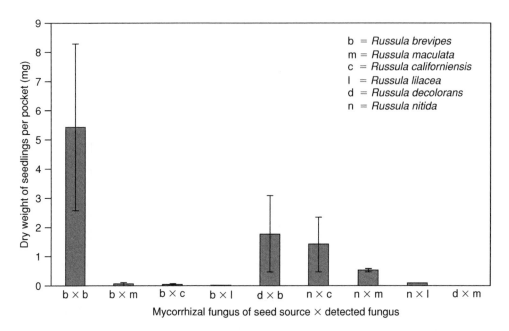

Figure 13.5 The extent of germination and of development of seedlings of *Monotropa uniflora*, sown in compartmented seed packets, at Mount Tamalpais, California, a site outside the natural range of the plant. The development is expressed in relation to the identity of the fungus associated with the maternal plants from which the seed was derived and that of the fungi initiating the germination. Thus, for example, b × b = *Russula brevipes* as maternal fungus × *R. brevipes* as the fungus which initiated germination. The first four bars correspond to seeds from Oregon and last five to seeds from Maryland. While some fungi other than those associated with maternal source plants were capable of stimulating germination, the extent of their development was much reduced, suggesting some physiological incompatibility. From Bidartondo and Bruns (2005).

some flexibility in the formation of partnerships during germination since closely related species can beinvolved in the process. In addition to a somewhat expanded receptiveness seen *in vitro* relative to that seen in nature, Bidartondo and Bruns (2005) recorded several associations between plants and fungi in seedlings of *M. uniflora* that were not seen in naturally occurring adults. However, these were often seen to result in a reduction of vigour of the seedlings concerned (Figure 13.5). There is a strong possibility that selection will operate against these less effective liaisons and that the weakly performing seedlings will not survive to produce seed. This would explain the situation seen in adult *Monotropa uniflora*, *Pterospora andromedea* or *Sarcodes sanguinea* in which in each case, adults are found to be colonized by only one of a number of closely related members of the genera *Russula* or *Rhizopogon*.

The structure of monotropoid mycorrhizas

Monotropa hypopitys (see Colour Plate 13.2c) and its relatives such as *Pterospora andromedea* (see Colour Plate 13.2b) and *Sarcodes sanguinea* (see Colour Plate 13.2a)

Figure 13.6 Excavated plant of *Pterospora*, showing the highly developed 'root ball' (arrowed). From Robertson and Robertson (1982), with permission.

frequently grow in forests of pine and other conifers, whose roots and mycorrhizas are closely associated with those of the monotropoid plants in tight complexes known as 'root-balls' (Figure 13.6). The plants overwinter underground and from the 'root' system flowering scapes develop. Adventitious buds develop on the apices of some of the roots of first and second order and the flowering shoots formed from them grow above the ground, mature and senesce over a period of months. Some of these plants reach considerable size, those of *S. sanguinea* achieving weights of several kilograms and of *P. andromedea* heights of 2 m (Bidartondo, 2005). The amount of organic C required for this development must be considerable.

The distinctive nature of the anatomy of mycorrhizas in Monotropoideae was first recognized by MacDougal (1899) who described the occurrence of peculiar peg-like structures which he called 'haustoria' in the outer cells of the *Monotropa* root. This structure was subsequently recognized as the characteristic feature of *Monotropa* mycorrhizas (Francke, 1934).

It was envisaged that the 'haustorium' (now known as the 'peg') arose as a result of encapsulation of the invading fungal hypha by the cell wall of the plant. Such a process by itself, however, would not be unique since similar phenomena are commonly observed in orchids for which the term 'ptyophagus' mycorrhiza was used (Burgeff, 1932; see below) and in ericoid mycorrhizas (Burgeff, 1961). The distinctive nature of the peg in Monotropoideae has been revealed by ultrastructural analysis (Lutz and Sjolund, 1973; Duddridge and Read, 1982a; Robertson and Robertson, 1982), which demonstrates that the intrusive structure is not a true haustorium. The peg is of sufficiently elaborate and specialized construction to justify a differentiation between this type of mycorrhiza for which the term 'monotropoid' was proposed by Duddridge and Read (1982a) and the arbutoid type (see Chapter 7) in which extensive internal proliferation of the fungus occurs.

The fungal mantle surrounding the roots of monotropes consists of a multilayered and compact sheath in which the boundaries between the layers are (or sometimes are) demarcated by tannin deposits (Figure 13.7). In *Monotropa* (Duddridge and Read, 1982a) and *Pterospora* (Robertson and Robertson, 1982), the sheath encloses the root apex, while in *Sarcodes sanguinea* (Robertson and Robertson, 1982), the apex remains free. In all three genera, a Hartig net surrounds the outer epidermal layer of relatively small cells but does not penetrate into the underlying cortex. There is no evidence of labyrinthine development in the Hartig net as seen in ECM (see Chapter 6).

From the Hartig net single hyphae grow into the epidermal cells, the walls of which often appear to extend around the fungus, or at least to deposit a very substantial interfacial matrix between the two organisms. This composite structure is referred to as the fungal peg. Elaborations in the form of ingrowths of the wall of the epidermal cell surrounding the hypha occur in the intracellular position. Robertson and Robertson (1982) point out that in *P. andromedea* and particularly in *S. sanguinea*, the intracellular penetration of the hyphae occurs in a precise manner. Fungi always enter

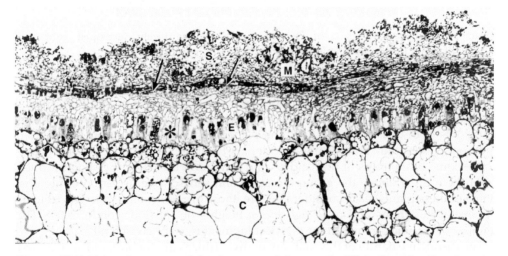

Figure 13.7 Light microscopy of development of the mantle (M) in *Sarcodes*, showing tannin deposits (arrowed) and hyphal penetration between the epidermal cells (E) and large cortical cells (C) containing starch grains. Region marked (*) is shown in higher magnification in Figure 13.8. From Robertson and Robertson (1982), with permission.

the cell through that radial wall of the epidermal cell which is orientated towards the tip of the root. In *Sarcodes*, the point of entry occurs near the base of the epidermal cell (Figure 13.8). As intrusion takes place, ingrowths develop from the encapsulating wall of the peg into the plant cytoplasm and become progressively more elaborate, their appearance differing according to their position along its length. Near the base of the peg, these ingrowths are sac-like and filled with granular contents (Figure 13.9), whereas closer to the tip they are thinner and, in *Pterospora* (Figure 13.10) and *Sarcodes* (Figure 13.11), have a long ribbon-like conformation. Both the basal and apical protuberances are extensions of the cell wall surrounding the peg. Since this wall is essentially fused with that of the fungus throughout its length, it is not possible in the absence of cytochemical data to be precise about the origin of the materials from which it is built. However, the effect of the proliferation is a massive increase of surface area within the epidermal cell, producing a structure which Duddridge and Read (1982a) recognized to be analogous to a 'transfer cell'. Cells of this type are widely distributed in the plant kingdom and are particularly associated with tissues that are involved in active nutrient exchange (see Gunning and Robards, 1976).

The final stage of development of the monotropoid mycorrhizas involves the opening of the tip of the peg and what appears to be the release from it of the fungal contents. The fungal material enters the epidermal cell, but is contained by a membranous sac derived from the plant plasma membrane which extends, balloon-like, into the cytoplasm of the epidermal cell (Figures 13.12). These latter events were observed under the light microscope by Francke (1934) who considered them to be indicative of 'digestion' of the fungus. However, while the process is rapid and has been referred to as 'bursting' of the peg, it appears to be a controlled event during which fungal and plant structures retain their integrity. During the event, an osmiophilic neck-band (Figure 13.13) develops around the open tip of the peg. Duddridge

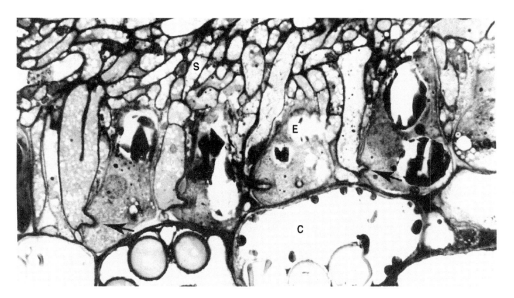

Figure 13.8 Detail of fungal colonization of the epidermal cells (E) of *Sarcodes* (refer to Figure 13.7) showing the development of fungal pegs (arrows). From Robertson and Robertson (1982), with permission.

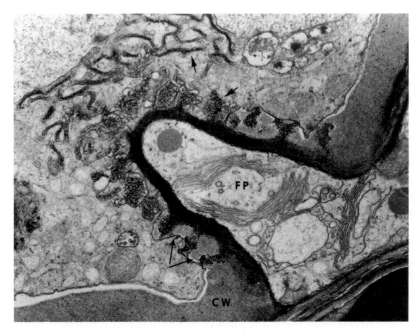

Figure 13.9 Development of the fungal peg (FP) in *Pterospora,* showing the numerous protruberances (arrows). From Robertson and Robertson (1982), with permission.

Figure 13.10 Development of the fungal peg in *Pterospora,* showing the long ribbon-like protruberances (arrows). From Robertson and Robertson (1982), with permission.

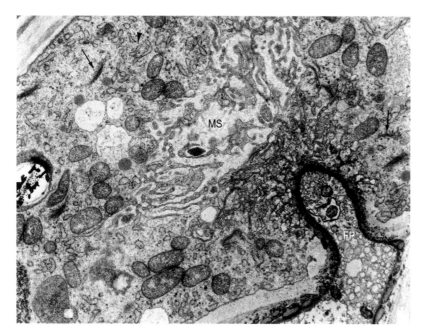

Figure 13.11 Epidermal cell of *Sarcodes*, showing fungal peg (FP) entering the wall. Note the membranous sac (MS) and mitochondria, golgi bodies and endoplasmic reticulum in the plant cytoplasm. From Robertson and Robertson (1982), with permission.

and Read (1982a) suggested that this may prevent back-flow of nutrients through the wall of the peg. This type of ring is seen around the haustorial neck of some biotrophic pathogens such as *Erysiphe* (Gil and Gay, 1977) and *Puccinia* (Heath and Heath, 1975) and is believed to create an apoplastic seal at the plant–fungus junction (see Smith SE and Smith, 1990). It was suggested (Duddridge and Read, 1982a), that 'bursting' of the fungal peg involves transfer of materials from fungus to plant, but as yet there is no evidence of this. If the membrane of the plant retains its integrity, then 'normal' processes of membrane transport would be expected to be involved in any such transfer. In *Pterospora* and *Sarcodes*, the sac appears to be a stable and well organized structure clearly delimited by the plasma membrane of the epidermal cell. It contains linear arrays of fibrils which have the dimensions of microtubules (inset Figure 13.13) and its membrane is, in places, greatly invaginated to produce what in section appear as small membrane-bounded vesicles in the cytoplasm.

Duddridge and Read (1982a) attempted to throw light on the functional aspects of the structures observed by relating differentiation of the fungal peg to the different stages of shoot development in *Monotropa* through the growing season. The maximum period of peg formation coincides with elongation of the flowering scape in June, while 'bursting' occurs as a late phenomenon at the time of seed-set from July onwards. They suggested that in the extension phase the epidermal proliferations act as transfer cells facilitating steady supply of nutrients to support the extension of the scape. There is clearly a possibility that the 'bursting' event provides a late surge of transfer of residual materials which could be used for seed production before the scape senesces. However, all such suppositions require experimental investigation.

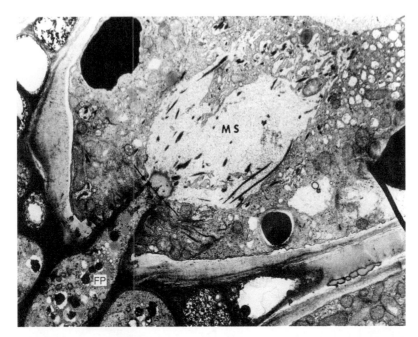

Figure 13.12 Epidermal cell of *Pterospora*, showing a median section of the fungal peg (FP), the tip of which has opened to form a membranous sac (MS). Note the osmiophilic ring at the tip of the peg (arrowed). From Robertson and Robertson (1982), with permission.

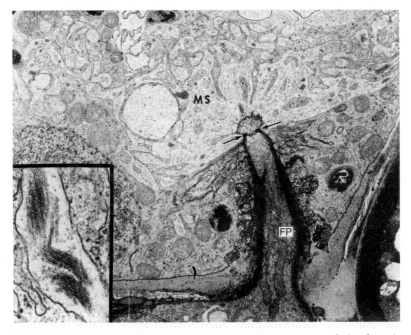

Figure 13.13 Epidermal cell of *Sarcodes*, showing median section of the fungal peg (FP). The membranous sac (MS) appears to be surrounded by the plant plasma membrane and the osmiophilic ring at the tip of the peg is again evident. Inset: higher magnification showing linear array of microtubule-like inclusions of the membranous sac. From Robertson and Robertson (1982), with permission.

Fungal associations and specificity in mycoheterotrophic orchids

As in the case of monotrope systems, molecular analysis has enhanced our under-standing of the phylogenetic distribution of fully mycoheterotrophic orchids and greatly improved knowledge of the mycorrhizal status of these plants. Genera containing achlorophyllous or partially achlorophyllous species and individ-uals are found in the tribes Epidendroideae, Orchidoideae and Vanilloideae of the Orchidaceae (see Figure 12.1). Of these, a cluster (*Cephalanthera, Epipactis, Limodorum, Neottia*) in the epidendroid subtribe Neottiae have been the subject of recent inten-sive investigation. A selective summary of the results of recent experimental analy-ses of the fungal associations of these plants is provided in Table 13.3. Remarkable parallels have been revealed with the Monotropoideae both in terms of high levels of specificity and of selective and preferential specialization on fungi that are also ECM associates of co-occurring trees or shrubs. The first definitive results came from analysis of adults of the achlorophyllous species *Cephalanthera austinae* (Taylor and Bruns, 1997) in which typical orchid pelotons, but with clamp connections indica-tive of a non-*Rhizoctonia* type fungus (Figure 13.14a), were shown to be produced by fungi in the Thelephoraceae. Races of this fungus with identical ITS sequence pat-terns formed typical ectomycorrhizas on adjacent tree roots (Figure 13.14b).

Recognition of the ectomycorrhizal association in *C. austinae* was important because this is the only fully achlorophyllous species in the genus *Cephalanthera*, other mem-bers of which are green and hence putatively photosynthetic in their adult stages. Subsequent analyses of normally green members of this genus (*C. damasonium, C. rubra*) revealed that they too are predominantly associated with thelephoroid fungi, though other ectomycorrhizal genera, notably *Inocybe* and *Hymenogaster*, can also be present (Bidartondo *et al.*, 2004). In a comparison of fungi isolated from pelotons of green photosynthetic and largely non-photosynthetic albino individuals of *C. damaso-nium*, the same thelephoroid fungi were found in both, with the additional presence of *Hymenogaster* in the green and of *Cortinarius* in the acholorophyllous individuals (Julou *et al.*, 2005). As in the case of *C. austinae*, Julou *et al.* (2005) found that identical strains of *Thelephora* formed ectomycorrhiza on nearby autotrophic trees and pelotons in the orchid roots. All *Cephalanthera* species so far studied associate in this way with autotrophic trees and they do so under the sometimes deep shade of their natural for-est environments. It therefore seems likely that access by way of their shared fungal symbionts to photosynthate of the ECM co-associates may be key to their survival. Physiological evidence in support of this contention is reported below. That there may be 'costs' involved in adopting the fully mycoheterotrophic habit is indicated by the observation of Julou *et al.* (2005) that albino plants were, at the same age, invariably of lower biomass than adjacent green individuals (Table 13.4).

The genus *Epipactis*, also within the Neottiae, contains species of open habitats (e.g. *E. palustris*), which retain fungal symbionts of the *Rhizoctonia*-type (*Ceratobasidium*, sebacinoid, tulasnelloid) and those of more shaded habitats (e.g. *E. helleborine, E. distans* and *E. atrorubens*), which again associate with ECM fungi (Bidartondo *et al.*, 2004). Among these are the ascomycetous species *Tuber* and *Wilcoxina*. In a study of chlorophyllous and achlorophyllous individuals of *E. microphylla*, Selosse *et al.* (2004) showed that plants of both types were colonized by ectomycorrhizal fungi, 78% of investigated peloton-containing root pieces being colonized by *Tuber* spp. No Rhizoctonias were recovered. Four of the ectomycorrhizal taxa were found to be

Table 13.3 Selected studies describing the mycorrhizal symbionts of achlorophyllous orchids.

Orchid species	Type of sample	Origin	Fungal identification methods[1]	Reference	Identified fungi[2]	Trophic group[3]
Cephalanthera austinae	26 adults	Field	In planta + isolation: molecular (ITS RFLPs, ITS sequences, ML5–6 sequences)	Taylor and Bruns, 1997	14 species spanning the Thelephora-Tomentella group	Ectomycorrhizal
Corallorhiza maculata	9 adults	Field	Isolation: vegetative morphology	Zelmer et al., 1996	7 Moniliopsis isolates	Parasitic
Corallorhiza maculata	104 adults	Field	In planta: molecular (ITS RFLPs, ML5–6 sequences)	Taylor and Bruns, 1997, 1999b	20 species spanning much of the Russulaceae	Ectomycorrhizal
Corallorhiza mertensiana	27 adults	Field	In planta: molecular (ITS RFLPs, ML5–6 sequences)	Taylor and Bruns, 1999b	3 closely related species in the Russulaceae	Ectomycorrhizal
Corallorhiza striata	8 adults	Field	In planta + isolation: molecular (ITS RFLPs, ML5–6 sequencing)	Taylor, 1997	A narrow clade within the Thelephora-Tomentella group	Ectomycorrhizal
Corallorhiza trifida	18+ adults	Field	In planta + isolation: vegetative morphology	Zelmer and Currah, 1995	Unknown yellow, clamped Basidiomycete	Ectomycorrhizal
Corallorhiza trifida	4 adults + 24 seedlings	Field	In planta + isolation: molecular (ITS RFLPs and ITS sequences)	McKendrick et al., 2000b	7 ITS RFLP types, all in the Thelephora-Tomentella group.	Ectomycorrhizal
Epipogeum roseum	Adults	Field	In planta + isolation: molecular (ITS sequences)	Yamato et al., 2005	Coprinaceae cf Coprinus disseminatus, Psathyrella spp.	Saprotrophs
Fulophia zollingeri	12 adults	Field	In planta (ITS sequences)	Ogura-Tsujita and Yukawa, 2008	Psathyrella cf candolleana	Saprotrophic
Galeola altissima ≡ Erythrorchis ochobiensis	Seedling	Lab.	In vitro germination tests	Umata, 1995, 1998a,b	Erythromyces crocicreas, Ganoderma australe, Loweporus tephroporus, Microporus affinus, Phellinus spp.	Wood decay saprotrophs
Galeola septentrionalis	? adults	Field	In planta + isolation: vegetative morphology	Hamada, 1939	Armillaria mellea	Parasitic + saptrotrophic
Galeola septentrionalis	? adults	Field	In planta + isolation: vegetative morphology	Terashita and Chuman, 1987	Armillaria tabescens	Parasitic + saptrotrophic

Species		Methods	Reference	Fungal taxa	Trophic niche
Gastrodia cunninghamii	? adults	Field	*In planta + ex planta*: vegetative morphology and hyphal tracing	Campbell, 1962	*Armillaria mellea*	Parasitic + saptrotrophic
Gastrodia elata	? adults	Field	*In planta + ex planta*: vegetative morphology. Isolation	Lan et al., 1994	*Armillaria mellea*	Parasitic + saptrotrophic
Gastrodia elata	Seedlings + adults	Field	*In planta + ex planta*: isolation	Xu and Mu, 1990	*Mycena osmundicola* (seedling) *Armillaria mellea* (adults)	Parasitic + saptrotrophic
Hexalectris spicata	Adults	Field	*In planta*: molecular (ITS RFLP's ML5–6 sequencing)	Taylor et al., 2003	Sebacinaceae – *Sebacina* spp.	Ectomycorrhizal
Neottia nidus-avis	8 adults + 7 seedlings	Field	*In planta*: molecular (ITS RFLPs, ITS sequences, 28S sequences)	McKendrick et al., 2002	*Sebacina* spp.	Ectomycorrhizal
Neottia nidus-avis		Field	*In planta*: molecular (ITS RFLP's ITS sequences)	Selosse et al., 2002a	*Sebacina* spp.	Ectomycorrhizal
Rhizanthella gardneri	1 adult	Field	Isolation: morphology of vegetative and sexual stages	Warcup, 1985, 1991	The single isolate obtained was named *Thanatephorus gardneri*	Ectomycorrhizal
Rhizanthella gardneri	Adult	Field	Molecular sequencing and morphological characterization	Bougure et al., 2006	*Thanatephorus*	Ectomycorrhizal

Updated from Taylor et al. (2002). [1]Isolation refers to the culturing of fungi from pelotons or from whole tissue sections. *In planta* refers to the identification of fungi by direct methods that do not require fungal isolation. *Ex planta* refers to direct observations of fungal morphology or hyphal connections from plants to identifiable fungal structures (fruit bodies or ectomycorrhizal roots). For the molecular methods, ML5–6 refers to fungal mitochondrial large subunit ribosomal gene sequences in the region described by Bruns et al. (1998). TSOP refers to 'taxon-specific oligonucleotide probe' and ITS-RFLP refers to restriction digests of the fungal nuclear internal transcribed spacer region. [2]Fungal taxa are as given in the cited publication; note that different authors have used different nomenclatures. In the case of laboratory seed germination tests, only fungi that produced compatible interactions are listed. [3]In some cases we infer the trophic category of a fungus from other sources. However, if the authors have made a definitive statement concerning the trophic niche, their conclusion is given preference.

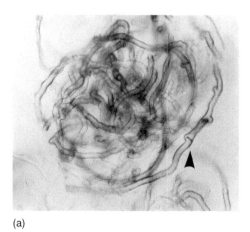

(a)   (b)

Figure 13.14 The first demonstration of specific association between the fungal symbiont of a fully mycoheterotrophic orchid and its co-associated autotrophic tree species. (a) Peloton formed by a member of the Thelephoraceae in a cortical cell of a root of the achlorophyllous orchid *Cephalanthera austinae*, showing clamp connection (arrowed). (b) Typical branched ectomycorrhizal roots of a co-associated coniferous tree formed by the same thelephoroid fungus. From Taylor and Bruns, *Proceedings of the National Academy of Sciences*, 94, 1997, with permission.

colonizing surrounding trees. Further studies will be required to determine whether shade-tolerant species of *Epipactis* are unique in terms of their specialization on Ascomycetes and on *Tuber* as orchid mycorrhizal associates.

The two genera of the sub-tribe Neottieae that have progressed furthest towards full achlorophylly, *Limodorum* and *Neottia*, conform to the pattern of fidelity to narrow ranges of ectomycorrhizal fungi. *L. abortivum* retains sufficient photosynthetic pigments in leaves and stems to enable activity of photosystem II, albeit at low efficiency (Girlanda *et al.*, 2006). Nonetheless, it and two other species (*L. trabutianum* and *L. brulloi*) associate exclusively with ECM fungi of surrounding trees, most of them being ascribed to the genus *Russula*. *Neottia nidus-avis* also shows tight specificity, but with *Sebacina* species of the Group A category (see Figure 6.5) which form ectomycorrhiza on surrounding shrubs and trees (McKendrick *et al.*, 2002; Selosse *et al.*, 2002a). Colonization by *Sebacina* is seen consistently in field grown plants from the earliest stages of seedling development to adulthood. Since germination of *Neottia* seeds fails to occur in the absence of ECM co-associates of the orchid, it is highly likely that the plant is dependent upon this fungus throughout its life cycle (McKendrick *et al.*, 2002).

Two fully mycoheterotrophic orchid genera of the more derived upper epidendroid clade, *Corallorhiza* and *Hexalectris*, have been shown also to conform to the pattern of high specificity. In the case of the former, 18 fungal ITS RFLP types were identified from *C. maculata*, all grouped within the Russulaceae (Taylor and Bruns, 1997). Sampling of populations and genotypes of *C. maculata* and the very closely related *C. mertensiana* over a wide range of habitat (forest) types, geographic areas and elevations in the USA (Taylor and Bruns, 1999) indicated that, while colonization was exclusively by Russulaceae in both genera, a combination of genetic and local factors played a strong part in determining a finer level specificity of the association. Different *Russula* species colonized the roots of a given orchid species in oak than in

Table 13.4 Comparison of green, albino and hypogeous *Cephalanthera damasonium* individuals for photosynthetic abilities, early growth and fungal diversity, as estimated by ITS sequencing on isolated pelotons or after cloning.

	Green individuals	Albino individuals	Hypogeous individuals
Total chlorophyll (μg/mg f. wt)	3.15 ± 0.50 ($n = 4$) a	0.029 ± 0.0004 ($n = 4$) b	–
Net assimilation (μmol $CO_2/m^2/s$)*	-0.10 ± 0.35 ($n = 5$) a	-0.065 ± 0.39 ($n = 6$) a	–
Stomatal conductance (mol $H_2O/m^2/s$)*	0.017 ± 0.003 ($n = 5$) a	0.036 ± 0.019 ($n = 6$) b	–
Quantum efficiency of PSII ($\Delta F/F_m$)*	0.728 ± 0.052 ($n = 6$) a	0.053 ± 0.041 ($n = 6$) b	–
Early growth (cm on 12 May 2004)	14.5 ± 1.9 ($n = 9$) a	9.2 ± 0.3 ($n = 5$) b	–
Fungal identified in isolated pelotons:	Hymenogaster spp. 1	Cortinariaceae spp. 1	Hymenogaster spp. 1
	Thelephoraceae spp. 2	Thelephoraceae spp. 2	
	Thelephoraceae spp. 3	Thelephoraceae spp. 3	
	Thelephoraceae spp. 1		
Fungi identified by cloning:			
Ascomycetes			
ECM or mycorrhizal	4	8	4
nonmycorrhizal	5	5	4
Basidiomycetes			
ECM Thelephoraceae	2	5	2
other ECM fungi	2	2	2
Ceratobasidium sp.	–	1	–
Fungi specific to the phenotype	3 out of 13	12 out of 21	5 out of 12
Basidiomycetes specific to the phenotype	1 out of 4	6 out of 8	2 out of 4

From Julou et al. (2005). Mean ± SD; values followed by different letters differ significantly between green and albino individuals according to a Student test, $P < 0.05$. *measured at PAR = 15 μmol/m²/s (at 20°C and 40% air humidity), representative of the site conditions.

conifer forest, indicating that the orchid selected compatible fungal genotypes out of the species-rich local population of the family. Because the geographical range of *C. mertensiana* is almost entirely within that of *C. maculata* and they occur together as intermixed individuals at many sites, habitat factors could be excluded when examining genetic influences on specificity. There was no overlap in *Russula* species between these orchids showing that, as in the case of the monotropes discussed above, plant genetic differences at the level of sibling species simultaneously control the acceptance and exclusion of closely related fungi. Even more strikingly, two co-occurring but distinct colour phenotypes of *C. maculata* were each colonized by distinct *Russula* species. The fact that neither species in the same genus nor varieties within a species share the same fungal association and that the separation has a strong genetic component indicates that the specialization is both dynamic and of recent origin.

The fungal symbionts of seedlings and adults of *C. trifida* (see Colour Plate 13.3a) collected from populations in Britain, Austria and two localities in the USA have been shown by ITS RFLP analysis to belong exclusively to the *Thelephora-Tomentella* complex of the ECM family Thelephoraceae (McKendrick *et al.*, 2000a). Burial of seeds of *C. trifida* in different ECM community types revealed that germination occurred within 7–8 months in all cases, indicating the presence of the appropriate thellephoroid strains of fungal symbiont. The most vigorously developed seedlings associated with the ECM shrub *Salix repens* had produced a shoot bud within 14 months (see Colour Plate 13.3b, c) and developmental studies indicated that without seed packet constraints flowering would occur within a further 12 months. These developmental chronologies are considerably faster than earlier estimates which have suggested that 3 to 4 years was a period required to reach maturity (Fuchs and Ziegenspeck, 1924, 1926).

Evidence for divergence of mycorrhizal specificity in closely related orchid genotypes is also seen in *Hexalectris* (Taylor *et al.*, 2003). However, all members of the species complex comprising this genus are exclusively associated with fungi of *Sebacina* Group A (see Figure 6.5). These fungi fall into six ITS RFLP types. Four of them were found in samples of *H. spicata* var *spicata*, a distinct type was found exclusively in *H. spicata* var *arizonica* and a further type exclusively in *H. revoluta*. Since, like *C. maculata* and *C. mertensiana*, two of these *Hexalectris* species (*H. spicata*-var *arizonica* and *H. revoluta*) co-occur at the same site, the specificity is likely to be attributable to genetic rather than geographic selection. In parallel with the *Corallorhiza-Russula* association, the evidence for differences in *Sebacina*-like associates between two coexisting floral forms suggests that mycorrhizal specificity may have diverged in concert with recent phylogenetic divergence in the orchid lineage.

In addition to the large majority of mycoheterotrophic orchids that associate with ECM fungi, there are a small number that have long been known to associate with pathogenic and decomposer fungi. One such plant is *Gastrodia elata*, the tubers of which have been harvested for medicinal purposes in China for over 2000 years (Xu and Guo, 2000). Research on the mycorrhizal relations of *G. elata* has enabled the orchid to be brought into cultivation (see Chapter 17) and has revealed aspects of its fungus-dependent growth cycle (Figure 13.15). The observation of Kusano (1911b) that plants of *Gastrodia elata* were mycorrhizal with the wood-rotting pathogen *Armillaria mellea* has been confirmed but shown to be incomplete. It is now known that while *A. mellea* is the natural symbiont of roots and tubers of the adult orchid, this fungus is parasitic on its young seedlings (Xu and Mu, 1990; Xu and Guo, 2000).

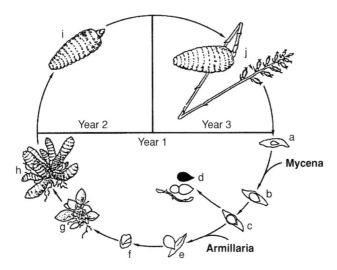

Figure 13.15 Life cycle of *Gastrodia elata* from seed (a) through primary colonization by *Mycena* (b, c, d) and subsequent development of those protocorms in which the *Mycena* is replaced by *Armillaria* (e–j). In these protocorms, flowering can take place in year 3. Modified from Xu and Guo (2000).

Experiments in which seeds were enveloped in decomposing leaves, thus simulating forest floor conditions, revealed that another fungus, *Mycena osmundicola*, was required to stimulate and sustain the early periods of germination. Exposure of seedlings at this stage to *A. mellea* led to their death. It is not yet clear whether *M. osmundicola* is the only fungus able to stimulate germination, but there is good evidence that it is incapable of sustaining development of the orchid beyond the protocorm stage. For such progression to occur *M. osmundicola* must be replaced by *A. mellea*, late in protocorm development (Figure 13.15e). It can be assumed that this replacement provides the orchid with access to the greater reserves of organic C contained in the massive woody substrates exploited by *A. mellea*. However, while there is some evidence of structural specialization in the mycorrhiza at the cellular level (see below), there remains a complete absence of information concerning the physiology of this fascinating mycorrhizal relationship.

Late protocorms invaded by *A. mellea* expand to form a primary stem tuber and this in turn buds off further tubers each of progressively larger size (Figure 13.15g, h). As the tubers proliferate, carrying with them fresh inoclum of *A. mellea*, the primary tubers decay and are consumed by the fungus. Approximately 3 years after germination tubers large enough to yield flowering spikes are produced (Figure 13.15i, j). Commercial vegetative production of this orchid in Korea is achieved by placing freshly harvested small tubers into holes drilled in buried oak logs that have been previously inoculated with *A. mellea* (Kim and Ko, 1995).

Galeola altissima is an achlorophyllous liane in which the mycorrhizal associate is another wood-destroying fungus, *Erythromyces crocicreas* (Umata, 1995). Additional fungi subsequently reported to be present as mycorrhizal associates in young plants of this orchid include the wood-rotting *Ganoderma australe* and the edible shiitake mushroom *Lentinus edodes* (Umata, 1998b). The extent and nature of the contribution of these fungi to the development and nutrition of the orchid remains to be elucidated.

Yamato *et al.* (2005) collected roots of *Epipogium roseum* from a number of sites in Japan and isolated fungi grown from individual pelotons. By sequencing the nuclear rDNA of these isolates they showed the mycorrhizal symbionts to be members of the Coprinaceae, which is an exclusively saprotrophic member of homobasidiomycetes. While some of these fungi were assigned to the genus *Psathyrella*, others were in *Coprinus*, some being particularly close to *C. disseminatus*. In a subsequent study, seeds of *E. roseum* were sown in a sterilized mixture of sawdust and volcanic soil, inoculated with a *C. disseminatus*-like isolate and incubated under laboratory conditions (Yagame *et al.*, 2007). While amounts of germination were small, some seeds developed in the presence of the fungus, producing plants which reached the flowering stage within 27 weeks. This enables the first chronological analysis of development of a mycoheterotrophic orchid from germination to flowering (see Colour Plate 13.4a–c). Protocorms were produced within 6 weeks of fungal inoculation. These produced lengthy filiform stolons which lacked fungal colonization (see Colour Plate 13.4a). Relatively robust peloton-containing rhizomes, some of them coralloid in form, developed on the stolons and these, in turn, produced swollen starch-filled tubers (see Colour Plate 13.4a inset) by 16 weeks. At this stage, the rhizomes decomposed leaving the tubers from which, in some cases by 18 weeks, inflorescence spikes emerged. By week 26, some of these bore fully developed flowers (see Colour Plate 13.4c and inset). While the developmental processes described by Yagame *et al.* (2007) may proceed more rapidly under laboratory conditions than in nature, the observations provide a striking indication of the effectiveness with which C can be mobilized from a complex source, in this case sawdust, in mycoheterotrophic symbioses.

Structural aspects of mycorrhiza in mycoheterotrophic orchids

There are two types of fungal colonization in fully mycoheterotrophic orchids. In the majority of species, hyphal pelotons reminiscent of those seen in green orchids (see Chapter 12) are produced. These coils develop extensively in the outer cortical cells of the roots but are only rarely found in the epidermal layer. Burgeff (1932) observed stages during which the intracellular hyphal coils became clumped in amorphous masses and believed that they were 'digested' in a process he described as tolypophagy. This kind of colonization has been reported in *Corallorhiza trifida* (Weber, 1981), *Neottia nidus-avis* (Barmicheva, 1989), *Galeola septentrionalis* (Terashita, 1985) and *Cephalanthera austinae* (Taylor and Bruns, 1997) (see Figure 13.14a). Stages in the autolysis of the intracellular hyphae of *Neottia* have been revealed by ultrastructural analysis (Dorr and Kollman, 1969; Barmicheva, 1989).

In the second type of colonization, described by Burgeff (1932) as ptyophagy, hyphal penetration of cells is restricted in the mid-cortex by the formation of encapsulating ingrowths of the plant cell wall. Ptyophagy has been described at the light microscope level in *Gastrodia*, *Galeola*, *Zeuxine* and *Cystorchis* (Burgeff, 1932). While pelotons may still be produced, they are usually less dense than in tolypophagous mycorrhizas and are restricted to the outer cortex. The main feature of the ptyophagus type is that the growth of hyphae entering cells of the middle cortex is arrested by the formation of an encapsulating ingrowth of the host cell wall. This produces a structure that is remarkably reminiscent of the 'peg' which is the defining feature of monotropoid mycorrhizas (see above). Ultrastuctural analysis of this structure in roots of *Galeola elata* colonized by *A. mellea* emphasize these similarities. Wang *et al.* (1997) describe longitudinally running canals between the outer and inner cortex of *G. elata* through which undifferentiated rhizomorph-like bundles of hyphae pass. The outer cortical cells, referred to as 'host' cells, are penetrated by hyphae from the bundle which form persistent intracellular coils. In contrast, hyphal penetration of the walls of the inner cortical cells, which Wang refers to as 'digestion cells', is arrested and the structure is invested by an interfacial plasma membrane densely packed with electron translucent lysosomal vesicles (Figure 13.16b). In what Wang *et al.* (1997) assume to be stages in the process of hyphal digestion, a radiating system of electron dense endocytotic vesicles proliferate in the host cytoplasm surrounding the hyphal walls (Figure 13.16a, b). These are believed to contain products of hyphal breakdown (Figure 13.16c, d). The difference between the peg-like structure seen in the analyses of Wang *et al.* and the monotropoid peg is that in the orchid there was no evidence of either bursting or sac formation at the hyphal tip. However, such

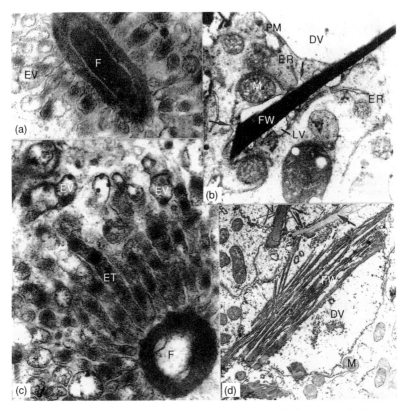

Figure 13.16 Sections showing intracellular ultrastructural features of the plant fungal interfaces in colonized cells of a mycorrhizal root of *Gastrodia elata*. (a) Endocytic vesicles (EV) forming around hyphae (F) in early stage of disintegration. Magnification × 24 000. (b) Lysosomal vesicles (LV) and endoplasmic reticulum (ER) fusing with perifungal plasmalemma to produce a large digestion vesicle (DV) around the remains of fungal wall (FW) material. Magnification × 30 000. (c) A radiating system of endocytic tubes (ET) forming around hyphae (F) and cut off endocytic vesicles (EV). Magnification × 26 000. (d) Digestion vacuole (DV) with remnants of fungal wall (FW) material. Magnification × 15 000. From Wang *et al.* (1997).

'bursting' has been observed in other orchids (Burgeff, 1936), so it is possible that Wang *et al.* (1997) recorded an earlier stage in the developmental process.

On the basis of light microscope observations, Burgeff (1936) described lysis of the hypal tip and release of its contents as the defining feature of 'ptyophagy'. If, as seems possible, tip lysis is ultimately shown to be the final stage in the process of hyphal collapse in these mycoheterotrophic orchids, the structural parallelism between them and the monotropes at this level will be remarkable.

It should be stressed at this point that the concepts of 'tolypophagy', 'ptyophagy' and 'digestion' advanced by Burgeff (1936), and since employed by others, are based entirely upon anatomical observations. Despite the fact that these terms carry with them the implication that the events observed involve nutrient release to the plant, there is no evidence that any such release is of significance for the C nutrition of the mycoheterotroph. Indeed, the more likely but as yet unproven scenario is one in which the critical processes of C transfer occur, as they appear to do in AM (see Chapter 4) and ECM (see Chapter 8) associations, across newly formed and hence physiologically active symbiotic interfaces between the partners. In this case the distinguishing feature of the C-transfer pathway in mycoheterotrophic plants would be that the direction of C movement is from fungus to plant rather than in the opposite direction. According to this interpretation, the various 'digestion' events observed by microscopy are analogous to processes of degeneration of the arbuscule or of the Hartig net seen in ageing AM and ECM associations, respectively.

Mycoheterotrophic plants with arbuscular mycorrhizas

AM colonization occurs in fully mycoheterotrophic representatives of the dicotyledon families Polygalaceae (1 genus, *Salomonia*) and Gentianaceae (4 genera), in the monocot families Triuridaceae (6 genera), Petrosaviaceae (1 genus), and Corsiaceae, all members of which are fully mycoheterotrophic, and in the Burmanniaceae (14 genera, not all fully mycoheterotrophic). The placement of Burmanniaceae and Corsiaceae in the Orchidales in older systems of classification (e.g. Cronquist, 1981) left the implication that this type of mycorrhiza is anomalous, but the issue is resolved by the new molecular phylogenies which show these families to be in different dicot orders and hence distantly removed from the Orchidaceae (see Figure 12.1). Earlier suggestions that the structures seen in the roots of these families were of the AM kind (Groom, 1895; Janse, 1897; Pijl, 1934; Knöbel and Weber, 1988; Schmid and Oberwinkler, 1994, 1996) have since been confirmed by molecular (Bidartondo *et al.*, 2002; Franke *et al.*, 2006) and structural analyses (Imhof, 1997, 1998, 1999a, 1999b, 1999c, 2001, 2003, 2006; Imhof and Weber, 1997, 2000; Yamato, 2001; Franke, 2002).

Molecular analysis of the gentianaceous genera *Voyria* and *Voyriella* from tropical rainforests of South America show them to be associated with a closely related set of *Glomus* species of Group A (see Chapter 1), with a single occurrence of a *Gigaspora* spp. (Bidartondo *et al.*, 2002). In the same study, the monocot *Arachnitis uniflora* (Corsiaceae) collected from three subantarctic forest sites in Argentina was also shown to be colonized by AM fungi. Again, specialization upon an extremely narrow lineage was detected. Since the *Glomus* species found in *Arachnitis* also formed mycorrhiza in three adjacent autotrophic plants representing three different families, the likelihood that the autotrophs were a distal source of C for the mycoheterotroph was evident.

Four mycoheterotrophic members of the genus *Afrothismia* (Burmanniaceae) were shown to be colonized by distinct clades of *Glomus* group A (Franke *et al.*, 2006).

The achlorophyllous gametophytes of the ophioglossoid ferns *Botrychium lanceolatum* and *B. crenulatum* also appear to be colonized by narrow lineages within the *Glomus* group A (Winther and Friedman, 2007a). One of these lineages had 18S sequences that were most closely related to those of the *Glomus* strain isolated from *Arachnitis uniflora* by Bidartondo *et al.* (2002), while others were very similar to the *Glomus* identified in roots of *Voyria corymbosa* in the same study. This suggests that multiple taxonomically isolated clades of glomeromycotan fungi have been independently recruited to support the mycoheterotrophic habit. Winther and Friedman (2007a) showed that these strains were capable of forming mycorrhizas simultaneously with the achlorophyllous gametophyte and the leafy sporophyte generations of both *Botrychium* species, indicating a potential for the direct supply of assimilates from the autotrophic to the mycoheterotrophic stages of the life cycle. One lineage was also present as an AM symbiont of an unrelated co-associated angiosperm which suggests that any such C supply may not occur exclusively from con-specific plants. It has now been shown (Winther and Friedman, 2007b) that the mycoheterotrophic gametophytes of some lycophytes and the sporophytes which arise from them are also colonized by narrow lineages of AM fungi with *Glomus* group A (see Figure 1.5).

Detailed structural analyses of representative members of these mycoheterotrophic families have revealed distinctive patterns of fungal colonization, root structure and hyphal morphology. *Burmannia tenella* (Burmanniaceae) produces a stellate root system consisting of 0.7–2 mm thick succulent but brittle roots which reach lengths of only 3 cm (Imhof, 1999c). They have an epidermal layer, a cortical parenchyma of about 10 cells in thickness, an endodermis and a very reduced vascular cylinder. Tangential sections reveal that the epidermal cells are colonized by thick coiled hyphae (Figure 13.17a). Most of the deeper cortical cells are very heavily colonized by intracellularly coiled hyphae, some of which produce arbuscule-like structures (Figure 13.17b), while others contain intercalary vesicles (Figure 13.17c). The epidermal layer remains free of colonization and very few fungal entry points are seen associated with it. This type of colonization is described by Imhof (1999c) as a *Paris*-type arbuscular mycorrhiza (see Chapter 2). *Afrothismia gesnerioides* illustrates even further reduction of the root system which consists of ovoid root tubercles (Figure 13.18a) from which may extend short filiform extensions lacking a vascular cylinder (Imhof, 2006). Two morphologically distinct types of hyphae, assumed to be representative of different fungal species, are involved in colonization of these roots. One, fungus A, is aseptate and believed to be of the AM type, the other, fungus B, has septate hyphae. The septate fungus colonizes the central cylinder of these structures at an early stage of their development (Figure 13.18b), while fungus A proliferates later and in the cortical tissues.

Figure 13.17 Tangential sections through the mycorrhizal roots of *Burmannia tenella*. (a) Section showing epidermal cells (uppermost layer) penetrated by thick aseptate hyphae which grow inwards to produce arbuscule-like structures and occasional vesicles in the lower cortical cells. (b) Close-up of an arbuscule-containing cell surrounded by epidermal cells with thick hyphae. (c) Intracellular vesicle together with a portion of an arbuscule and thick hyphae in a cortical cell. From Imhof (1999c).

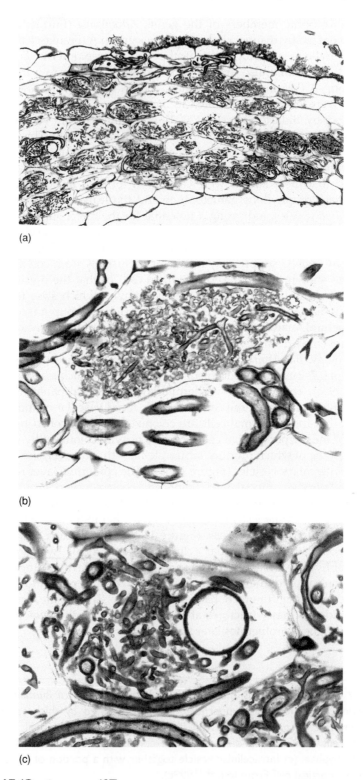

(a)

(b)

(c)

Figure 13.17 (Caption on p. 487)

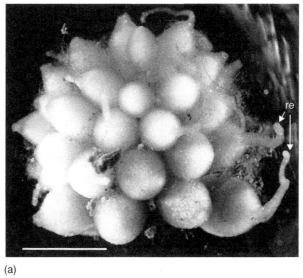

(a)

(b)

Figure 13.18 Root structure in *Afrothismia gesnerioides*. (a) Rhizome tip densely covered with root tubercles, having short filiform root extensions (re). Scale bar = 250 μm. (b) Longitudinal section of a root initial already being colonized close to the central cylinder (cc) by the septate fungus B. From Imhof (2006).

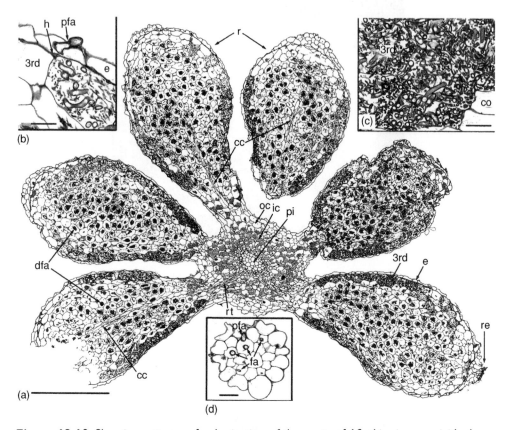

Figure 13.19 Showing patterns of colonization of the roots of *Afrothismia gesnerioides* by an aseptate fungus (fungus A). (a) Synoptic view of rhizome (in cross-section) and root tubercles (r) (in longitudinal section) showing uncolonized epidermal cells (e), heavy colonization of cells of the third layer below the epidermis (3rd) and clumped hyphal masses (dfa) in the deeper cortical cells indicating degeneration. (b) Highly magnified section through the root epidermis showing penetration by thick hypha (pfa), collapsed exodermis (h) and arbuscule formation in the third epidermal layer. Scale bar = 50 μm. (c) Tangential section of the third epidermal layer showing cells with dense coils of aseptate fungus A and uncolonized inner cortical cells (co). Scale bar = 50 μm. (d) Transverse section of filiform root extension showing penetration (pfa) and longitudinal growth of hyphae of fungus A (fa) in cross-section. Scale bar = 50 μm. From Imhof (2006).

A section across a group of root tubercles (Figure 13.19a) reveals the extensive intracellular occupation by fungal symbionts. These are seen to coil extensively in the third cortical layer of tissue below the epidermis of each root tubercle (Figure 13.19b, c). The presence of a less intensive form of colonization is visible in the filiform extensions (Figure 13.19d) (Imhof, 2006).

The fungus has aseptate hyphae believed to be of the AM type, but these retain their structural integrity to different extents in different zones. Straight and persistent hyphae are present in the epidermal cells (Figure 13.20a), coiled but still persistent hyphae occur in the third cell layer beneath the epidermis (Figure 13.20b, c), whereas the bulk of the deeper root cortical tissue is occupied by coiled hyphae undergoing degeneration in the so-called 'digestion zone' (Figure 13.20d). Again, the mycorrhiza

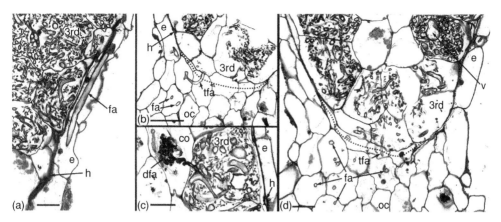

Figure 13.20 High magnification views of colonization of *Afrothismia gesnerioides* by fungus A. (a) Tangential section showing uncoiled hyphae in the epidermis (fa), collapsed exodermal layer (h) and dense proliferation of hyphae in the third epidermal layer. Scale bar $= 50\,\mu$m. (b) Section through connecting region between outer rhizome cortex and root. A hypha within the rhizome epidermis branches to directly penetrate into the root cortex (tfa). Scale bar $= 50\,\mu$m. (c) Showing intact hyphal colonization of third layer and degeneration of coils to produce amorphous clumps (dfa) in the cortical layer beneath. Scale bar $= 50\,\mu$m. (d) Section through tissues connecting outer rhizome cortex (oc) and root. The uncoiled hyphae growing in the outer cortex penetrate the root tissue (tfa) and proliferate forming coils in the third epidermal layer. Dotted line indicates demarcation of rhizome and root tissue. Scale bar $= 50\,\mu$m. From Imhof (2006).

is interpreted as being *Paris*-type AM. The presence of a second fungus with septate hyphae but of otherwise unknown affinity was also recorded in this plant. It was restricted to zones of tissue unoccupied by the aseptate hyphae. *Sciaphila polygyna* (Triuridaceae) also produces a stellate root system with individual roots up to 1.4 mm thick and 1 cm long radiating from a central point (Imhof, 2003). These roots show a remarkable bilateral symmetry when viewed in cross-section (Figures 13.21a, b, c, d) and, as in the burmanniaceous roots described above, distinct compartmentation of zones is recognizable. The third cell layer below the epidermis contains loose coils of aseptate hyphae from the dorsal to the lateral side of the root, in contrast to the extremely dense coils of thin hyphae in the ventral position (Figure 13.21a, b). In this layer, the hyphae produce vesicle-like swellings, in particular dorsal areas and retain their integrity (Figure 13.21b). The fourth cell layer is anatomically heteromorphic (Figure 13.21a, b) consisting of scattered 'giant cells' which reach up to $320\times130\,\mu$m in the ventral part and in which hyphae degenerate (Figure 13.21c). Prior to collapse, the hyphae swell markedly in diameter and then degenerate to form amorphous clumps. Imhof (2003) proposes that the function of the third root layer is to enable healthy development of the fungus before it grows into the giant cells for 'digestion'.

Many achlorophyllous gentians in the genus *Voyria* have stellate or coralloid root systems in which the axes of the individual roots are extremely short but a range of root morphologies and colonization patterns is seen. Thus, *V. truncatula* has some elongate roots with almost classical *Paris*-type AM (Imhof and Weber, 1997), while *V. tenella* (Imhof, 1997), *V. obconica* (Imhof and Weber, 2000) and *V. flavescens* (Franke, 2002) produce short roots (Colour Plate 13.1d and Figure 13.22) in which the

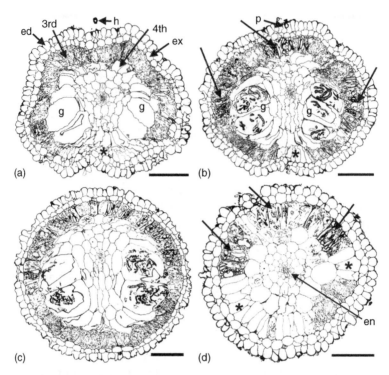

Figure 13.21 Transverse sections of the root of *Sciaphila polygyna* (Triuridaceae) taken through progressively more mature parts of the root. (a) Section behind tip showing distinct layers, *viz* uncolonized epidermis (ed) and exodermis (ex), third epidermal layer (3rd) heavily colonized by aseptate fungal hyphae in the dorsal region of the root and less so in the ventral region, uncolonized fourth layer (4th) containing some giant cells (g). Scale bar = 250 μm. (b) Section in middle part of root showing penetration point (p), thick hyphae and vesicle-like structures in the dorsal and lateral parts of the third layer (arrows) and extension of colonization into the ventral cells of the third layer. Thick hyphae, some of which form amorphous clumps, are seen in the giant cells (g) of the fourth layer. Scale bar = 250 μm. (c) Further maturation with complete colonization of the ventral cells of the third layer and structural differentiation between vacuolate hyphae in the dorsal cells and non-vacuolate hyphae in the ventral cells. Scale bar = 250 μm. (d) Close to the base of the root, dorsiventrality is less clear. Thick hyphae with vesicle-like swellings are visible in the dorsal cells of the third layer (arrows). En=endodermis. Scale bar=250 μm. From Imhof (2003).

distribution of mycorrhizal hypae is compartmented in a manner reminiscent of that seen in the Burmanniaceae (above). In *V. truncata*, aseptate hyphae extensively penetrate the epidermis and grow inwards to produce a densely colonized root cortical parenchyma. A contrasting pattern is seen in *V. tenella* (Imhof, 1997). Here there are few penetration points and those hyphae that gain access to the root grow directly towards the inner cortical layers where they spread along the narrow central cylinder immediately outside the endodermis. From this interior position they then turn back to grow in an outward direction (Figure 13.23), ramifying and producing intracellular coils throughout the central cortex. Eventually they swell and degenerate into what may be 'digestion' events.

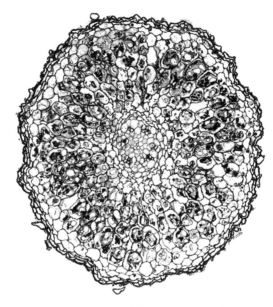

Figure 13.22 Transverse section of root of *Voyria obconica*, showing outer epidermal layers largely free of fungal colonization, dense colonization of the middle cortical zone and relatively light colonization in the inner cortex. From Imhof and Weber (2000).

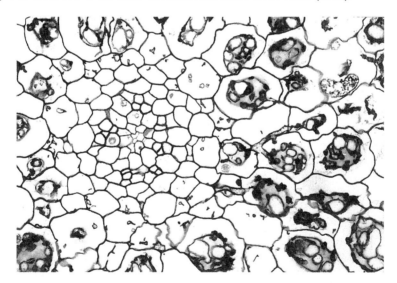

Figure 13.23 Transverse section of root of *Voyria tenella*, taken through transition zone between inner and outer cortex tissues, showing growth of intracellular hyphae from inner cortical cells into those of the outer cortex with marked swelling of hyphal contents. From Imhof (1997).

Stimulation of exogenous development of ectomycorrhizal mycelial systems around mycoheterotroph roots has been reported (Bidartondo *et al.*, 2004, see above), but the massive internal proliferation of mycelium seen in Burmanniaceae, Triuridaceae and Gentianaceae appears to be a unique feature of these AM-colonized plants. While it is legitimate to speculate, as does Imhof, about the functional role of these mycelial masses, it has to be said that we have no knowledge either of the mechanisms that stimulate the fungal development, or of the nature of the functional consequences for the plant.

Functions of mycoheterotrophic mycorrhizas

Kamienski (1881) hypothesized that *Monotropa* shared a common symbiotic fungus with the forest trees near which it grew. He also believed that it might be nourished, not saprophytically, but through the common mycelium from the neighbouring trees. This view was not widely accepted. Kerner (1894) was one of the first to recognize groups of plants that were parasitic upon fungi. His observation that 'we cannot easily familiarize ourselves with the idea of a flowering plant draining the mycelium of a fungus of nutriment' encapsulated the conceptual difficulties experienced by many of his contemporaries when considering the nutrient relations of these plants.

The first experimental test of Kamienski's hypothesis was carried out by Björkman (1960). He first separated *Monotropa* plants from the tree roots by metal sheets and observed that they grew poorly compared with attached plants. Later, ^{14}C-labelled glucose and ^{32}P-labelled orthophosphate were found to be translocated in five days from the *Picea* and *Pinus* trees into which they had been injected to the tissues of *Monotropa* growing close by. The distance between the trees and *Monotropa* plants was 1–2m and young developing plants became more radioactive than old mature plants. Björkman confirmed the view of Kamienski that the fungus infecting the tree roots and *Monotropa* was probably of the same mycelium. Other plants in the neighbourhood, such as *Calluna vulgaris*, *Vaccinium vitis-idaea* and *V. myrtillus*, did not become labelled from either ^{14}C-glucose or ^{32}P-orthophosphate injected into *Picea* and *Pinus* and this observation was regarded as confirmation of the need for hyphal connections.

In a similar but reversed experiment briefly reported by Furman (1966), *Monotropa* plants were injected with ^{32}P, which was transported to neighbouring *Quercus* and other trees which were ECM and (surprisingly) to the AM plant *Acer*. Repeated attempts to confirm transfer of ^{14}C after feeding ^{14}CO$_2$ to *Salix repens* associated with *M. hypopitys* failed to confirm these findings (Duddridge *et al.*, unpublished data), but raise the interesting possibility that the mycoheterotroph may be supported not by current photosynthate (which would be labelled during ^{14}CO$_2$ feeding), but from stored carbohydrate. Since in *M. hypopitys* and probably in other monotropes, the flowering axis of the current year is produced by extension growth from a bud laid down in the previous year (Duddridge and Read, 1982a), there is every likelihood that stored C indeed provides a substantial proportion of its assimilate requirement. Clearly, to resolve these questions movements of isotopic tracers would need to be followed over more than one growing season. Such new experiments, in addition to exploring temporal and spatial aspects of mycoheterotroph C supply, should also investigate the processes whereby N and P are acquired by the fungal partners and tranferred to the plants. The fascinating patterns of N isotope discrimination revealed by recent studies of mycoheterotrophs (Gebauer and Meyer, 2003; Zimmer *et al.*, 2007) raise many important questions which can only be resolved through analyses of the sources of N exploited by the fungal symbionts and of the processes by which this element is transferred to the mycoheterotroph. Whatever the route, there is no doubt that acquisition of these nutrients via symbiotic fungi would compensate for the very poor root development normally seen in these types of plant.

Since these earlier studies, the nutrient relations of mycoheterotrophic plants have not been subject to the detailed analytical evaluation that they deserve. However, some experiments involving the mycoheterotrophic orchid *Corallorhiza trifida* have provided fresh insights. It was first shown in a field experiment that seed of this plant germinated only when sown in association with the ECM autotrophs *Betula pendula* or *Salix repens* (McKendrick *et al.*, 2000a, and see above). *C. trifida* seedlings were recovered from seed packets sown in these experiments, weighed, then transferred to microcosms containing soil from the field site and non-mycorrhizal plants of *B. pendula*, *S. repens* or *Pinus sylvestris* (McKendrick *et al.*, 2000b). Roots of *B. pendula* and *S. repens* growing in the immediate vicinity of the introduced orchid produced ectomycorrhizas (see Colour Plate 13.3d, e, f). That the ECM of birch and willow were formed by the mycorrhizal symbiont of the orchid was demonstrated by microscopic observation of the development of hyphal linkages between the partners (see Colour Plate 13.3e). When the mycoheterotroph seedlings were connected in this way to the autotrophs they gained weight by between 6 and 14% over a 25–28-week period, whereas in microcosms supporting *P. sylvestris* with which no mycorrhizal linkages were formed or in those lacking autotrophs, they lost 13% of their weight over the same period (Figure 13.24). These observations demonstrate that connections to ECM autotrophs are required to facilitate the C transfer required for development of the orchid and that there is some specificity with regard to the ability to form these linkages.

In the course of the 28-week incubation, additional seedlings of *C. trifida* appeared in the microcosms containing *B. pendula* and *S. repens* but not in those supporting

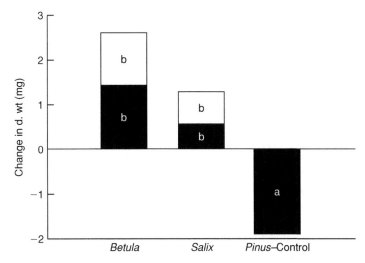

Figure 13.24 The change in dry weight of original *Corallorhiza* plants (black bars) and of recruits (open bars) which grew in association with *Betula* or *Salix* or in control microcosms in which they did not establish shared mycorrhizal associations with an autotroph (*Pinus* – control). Bars sharing the same letter are not significantly different. From McKendrick *et al.* (2000a).

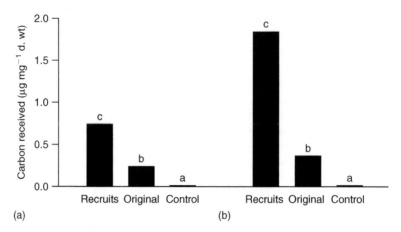

Figure 13.25 The amount of carbon received by recruits, original plants and control plants of *Corallorhiza* in microcosms containing (a) *Betula* and (b) *Salix*. In each case the differences between the three categories of plant are significant ($P < 0.05$, Tukey test). From McKendrick *et al.* (2000a).

P. sylvestris. These new seedlings, which were referred to as 'recruits', and which had developed from seed which had failed to germinate in the packets, also became linked to autotrophs and developed at rates similar to those of the originally sown seedlings. Digital autoradiography and scintillation counting revealed that when shoots of *B. pendula* and *S. repens* were supplied with $^{14}CO_2$, direct transfer of C took place from the autotrophs to both original and recruit seedlings of the mycoheterotroph in all cases where the associates had become connected by the shared fungal symbiont. Orchid seedlings lacking these connections, which were introduced to the microcosms as controls immediately before isotope feeding, failed to assimilate significant amounts of C (Figure 13.25). This study yields the first experimental confirmation that growth of a mycoheterotrophic orchid could be sustained by supply of C received directly from an autotrophic partner through linked fungal mycelium. In a subsequent experiment, in which the mycoheterotrophic liverwort *Cryptothallus mirabilis* (see Colour Plate 14.1c) was grown with *B. pendula*, it was again shown by pulse labelling the autotroph with $^{14}CO_2$, that the shared fungal symbiont, in this case a *Tulasnella* species (see Chapter 14), facilitated transfer of C to the mycoheterotroph (Figure 13.26) (Bidartondo *et al.*, 2003).

Indirect methods of analysis involving determination of the natural abundances of ^{15}N and ^{13}C have thrown light on the possible sources of these nutrients. Trudell *et al.* (2003) demonstrated that the $\delta^{15}N$ signatures of the monotropes *Hemitomes congestum*, *Allotropa virgata*, *Monotropa hypopitys* and *M. uniflora* differed from those of all non-mycoheterotrophic plants examined in Olympic National Park, Washington. Their signatures were most like those of co-associated ECM fungi with which their $\delta^{15}N$ values were strongly positively correlated. In contrast, the ^{13}C signatures of these plants were similar to those of their co-associated autotrophs. These authors speculated that the elevated ^{15}N values might reflect mass flow of mycelial contents through the monotropoid 'peg', the enrichment being enhanced by selective exclusion in this process of chitinous fungal wall material which is known to be strongly depleted in ^{15}N (Taylor AFS *et al.*, 1997). They proposed that, if sac bursting enabled

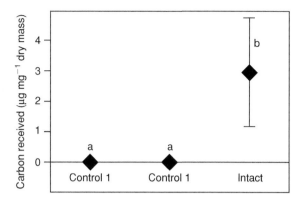

Figure 13.26 The concentration of carbon received 21 days after a $^{14}CO_2$ pulse by individual thalli of *Cryptothallus* grown with *Betula pendula*. The thalli had *Tulsanella* hyphal connections to the roots of *Betula* which were forming ectomycorrhizas. In control 1, hyphal connections between thalli and roots were severed immediately before $^{14}CO_2$ exposure. In control 2, thalli were introduced to the chambers immediately before $^{14}CO_2$ exposure. Bars show standard error of the mean. Different letters indicate significant difference ($P = 0.05$, Tukey test). From Bidartondo *et al.* (2003).

the mass transfer of fungal C, the absence of metabolic fractionation might explain the contrasting lack of discrimination in the ^{13}C signature.

The same isotopic distinction has also been shown between mycoheterotrophic orchids and co-associated autotrophs (Gebauer and Meyer, 2003). However, in this case both $\delta^{15}N$ (Figure 13.27) and $\delta^{13}C$ (Figure 13.28) of the fully mycoheterotrophic orchids *Neottia nidus-avis* and *Limodorum abortivum* were significantly more enriched than in the co-occurring non-orchids. The large differences in signatures of both isotopes seen in *N. nidus-avis* and *L. abortivum* provides evidence for the utilization of N and C sources that are distinct from those used by the autotrophs.

Despite speculation such as that of Trudell *et al.* (2003, see above), concerning the physiological basis for the distinctive $\delta^{15}N$ enrichments seen in mycoheterotroph, we remain lamentably ignorant of the nature of the N transfer pathways and processes involved in the production of these signatures. When similar enrichments are seen in the sporocarps of ECM fungi, some of which co-associate with mycoheterotrophs, it is concluded that they arise because the fungi selectively exploit recalcitrant organic sources of soil N that are themselves characteristically $\delta^{15}N$ enriched (see Chapter 15). On the same basis, it is logical to assume that the enrichment of $\delta^{15}N$ seen in mycoheterotrophic plants occurs because their N is derived, by way of the fungi, from the recalcitrant soil sources. The exploitation of N sources that are predominantly organic must in turn influence the $\delta^{13}C$ signatures of mycoheterotrophic tissues. While it has sometimes been assumed that almost all of their C is derived from co-associated autotrophs by way of the fungal partners, the complications involved in such partially heterotrophic patterns of C gain need to be taken into account. The application of isotope mixing models (see below) has gone some way to resolving the relative importance of different C and N sources used by mycoheterotrophic plants in nature.

Demonstration that the putatively photosynthetic *Cephalanthera damasonium*, along with its green helleborine relatives *C. rubra*, *Epipactis helleborine* and *E. atrorubens*, had N

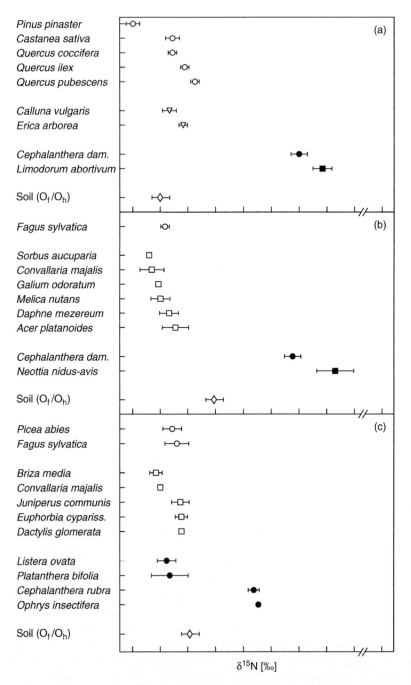

Figure 13.27 Mean values of $\delta^{15}N \pm 1$ SE in leaves of all individual plant species and in soil samples collected from four forest sites in Europe. (a) *Pinus pinaster* forest (Thezan, S. France); (b) *Fagus sylvatica* forest V1; (c) *Pinus sylvestris* forest V2; (d) *Pinus sylvestris* forest V3 (all Veldensteiner Forest, NE Bavaria, Germany). Symbols represent classification according to functional groups: ECM (O); ERM (∇); AM/NON ($\square$); AO (Autotrophic Orchids) (●); MHO (Mycoheterotrophic Orchid) (■); soil ($\diamond$). Error bars are missing if smaller than the symbols or if $n < 3$. From Gebauer and Meyer (2003).

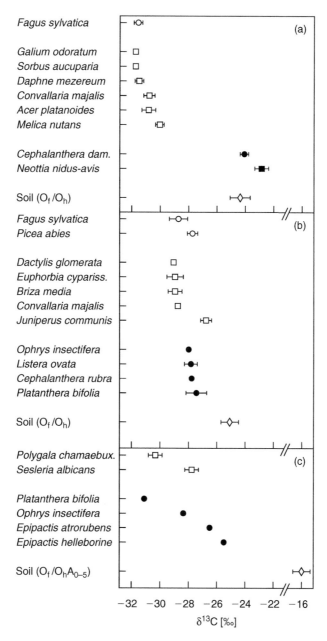

Figure 13.28 Mean values of $\delta^{13}C \pm 1$ SE in leaves of all individual plant species and in soil samples collected from three forest sites in Veldensteiner Forest, NE Bavaria, Germany. (a) *Fagus sylvatica* forest V1; (b) *Pinus sylvestris* forest V2; (c) *Pinus sylvestris* forest V3. Symbols represent classification according to functional groups: ECM (O); AM/NON (□); AO (●); MHO (■); soil (◇). Error bars are missing if smaller than the symbols, or if $n < 3$. From Gebauer and Meyer (2003).

and C signatures that were enriched relative to those of other, sometimes co-ocurring, green orchids (Gebauer and Meyer, 2003) (see Figures 13.27, 13.28), indicated the likelihood that these plants too receive considerable proportions of their C and N from heterotrophic sources. This view was reinforced by the observation that, as in the case of their fully mycoheterotrophic relative *C. austinae* (Taylor and Bruns, 1997, see above), these species were associated with ECM rather than the *Rhizoctonia*-like fungi normally found in autotrophic orchids (see Chapter 12) (Bidartondo *et al.*, 2004). This helps to explain both their access to photosynthetically derived C and contributes to an explanation of the relatively high shade tolerance shown by these species.

Using a linear two-source isotope mixing model, Gebauer and Meyer (2003) estimated that up to 85% of the C acquired by these green and hence putatively photosynthetic orchids was derived from their fungal partners. The same method, subsequently applied to leaves of these species in conjunction with parallel molecular determinations of the fungal symbionts in roots of the same individuals, indicated that the plants from shaded habitats received some of their C by way of their ECM fungal colonists. In contrast, *Epipactis palustris*, a helleborine from open habitats which was subsequently shown to be associated with fungi of the *Rhizoctonia*-type, showed no evidence of such partial mycoheterotrophy. In this species a small N gain from the fungal partner, similar to that seen in the autotrophic orchid *Dactylorhiza majalis* was detected (Gebauer and Meyer, 2003; Bidartondo *et al.*, 2004).

Recently, Zimmer *et al.* (2007) carried out a comparative analysis of the ^{15}N and ^{13}C contents of monotropes, *Pyrola* spp. and mixed AM or non-mycorrhizal autotrophs in Californian forest communities (Figure 13.29). At the three sites where they were analysed, *Sarcodes sanguinea* (Figure 13.29a, b, d) and *Pterospora andromedea* (Figure 13.29c, d, f) showed significant enrichment of both δ^{15}N and δ^{13}C relative to other plants in the communities. The achlorophyllous orchid *Corallorrhiza maculata* had isotope signatures similar to those of the monotropes (Figure 13.29a, b). These results confirm that, as in the case of the mycoheterotrophic orchids examined in European forests (Gebauer and Meyer, 2003; Bidartondo *et al.*, 2004), these plants are strongly mycoheterotrophic. Most of the *Pyrola* spp. examined occupied an intermediate position and appear to be partially mycoheterotrophic, at least for N (see Chapter 7), the exception being the achlorophyllous *P. aphylla* (Figure 13.29f) which, as might be predicted, has isotope signatures indicative of full mycoheterotrophy. Sample sizes of this plant were too small to enable statistical verification of its nutritional mode.

Isotope mixing models must be regarded with some caution because the net contributions of C and N that they predict to be gained through mycoheterotrophy are themselves derived from basal values of isotopic enrichment of co-occurring putatively fully autotrophic and fully mycoheterotrophic plants. These data, especially in the latter category, can vary considerably. Nonetheless, they provide a guide as to the likely proportions of the nutrients derived from the two sources and enable hypotheses that can be directly evaluated. It is on the combined basis of their association with what, for green orchids, is a novel functional group of mycorrhizal fungi and their distinctive mixotrophic C nutrition that most of the helleborine orchids are considered in this rather than in the previous chapter.

While further sampling is desirable, it has become apparent that with the transition from open to shaded forest habitats, fully autotrophic orchid species colonized by *Rhizoctonia*-type fungi of relatively low specificity are replaced by others in which a symbiont switch has occurred involving not only a different functional

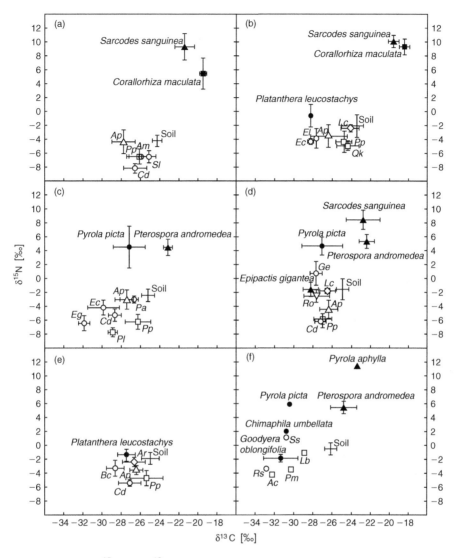

Figure 13.29 Mean δ^{15}N and δ^{13}C signatures of tissues of four distinct functional groups of plants and of soil collected from five forest sites in California. The functional groups are first, full mycoheterotrophs, *Sarcodes sanguinea*, *Pterospora andromedea* (Monotropoids), *Corallorhiza maculata* (Orchid), *Pyrola aphylla* (Pyroloid). Secondly, two green Pyroloid species, *Pyrola picta* and *Chimaphila umbellata*. Thirdly, three green orchids, *Goodyera oblongifolia*, *Epipactis gigantea*, *Platanthera leucostachys*. Lastly, 22 fully autotrophic reference species, names abbreviated as follows: Ac = *Abies concolor*; AM = *Acer macrophyllum*; Ap = *Arctostaphyllos patula*; Ar = *Alnus rhombifolius*; Bc = *Bromus carinatus*; Cd = *Calocedrus decurrens*; Ee = *Elymus elymoides*; Eg = *Elymus glaucus*; Ec (c) = *Epilobium canum*; Ec (b) = *Eriophyllum confertifolium*; Ge = *Glyceria elata*; Lb = *Linnea borealis*; Lc = *Lotus crassifolius*; Pa = *Pteridium aquilinum*; Pl = *Pinus lambertiana*; Pm = *Pseudotsuga menziesii*; Pp = *Pinus ponderosa*; Qk = *Quercus kellogii*; Rs = *Ribes* spp.; Ro = *Rhododendron occidentale*; Ss = *Smilacina* spp.; *Stipa lettermanii*. Error bars are missing if smaller than the symbols, or if $n < 3$. From Zimmer et al. (2007).

group of fungi, the ECMs, but also a much greater level of specificity. Members of the Monotropoideae, all of which frequent deeply shaded habitats, show tight specificity with the same functional group of fungi. They differ from the orchids only in so far as they appear to lack autotrophic or putatively autotrophic relatives. However, members of the closely related tribe Pyrolae, which also associate exclusively with ECM fungi, show a gradation from complete autotrophy to mycoheterotrophy (see Chapter 7), indicating that evolutionary trends towards dependence upon resources acquired from fungal symbionts seen in Orchidales, have their parallels in the Ericales.

It is of interest to consider why the shift towards increasing dependence upon mycoheterotrophy should lead to greater specificity in the fungus–plant relationship. Having lost photosynthetic ability, these plants are entirely dependent upon parasitism of their fungal symbionts as a source of C. In the first instance, the switch from *Rhizoctonia*-type saprophytes to ECM fungi may have been driven by the inability of the former to provide C in sufficient quantities to sustain the development of plants beyond a protocorm-like stage. Since, to date, there are no recorded examples of orchid species that switch within their life cycles from *Rhizoctonia* to ECM fungi, the evolutionary stages through which such transitional phases presumably passed must remain a matter of speculation. Nevertheless, the switch yields success in terms of plant fitness that is demonstrated by the fact that mycoheterotrophs are capable of completing their life cycles in habitats so dark that autotrophs are often completely excluded.

The basis of extreme specificity in mycoheterotrophic plants is of interest in particular because it would appear to carry with it the disadvantage of a reduced potential for a seed to make contact with a suitable fungus. Not only are mycoheterotrophs placed at one extreme of the specificity continuum, they are also probably the only completely parasitic plants in the 'mutualism–parasitism' continuum. As we have described, they are directly parasitic on their fungal symbionts and it is with the latter that the specific relationship occurs. The mycoheterotrophs sometimes, but not always, show some specificity towards their secondary 'hosts', the autotrophic plants. It is notable that among other fungus–plant symbioses it is the parasitic interactions that show specificity. These are most clearly exemplified in the biotrophic parasites (pathogens) of shoots, where close genetic control of specificity has been repeatedly revealed. Close specificity has evolved where fungus and plant are engaged in a dynamic evolutionary process in which the plant evolves resistance to exclude the parasite, while the fungus evolves to overcome the resistance. This results in extreme specificity, often at the level of race of parasite and variety of plant.

We can speculate that the specificity in mycorrhizal fungus–mycoheterotroph symbioses is the result of similar evolutionary processes but, in these cases, it is the plant that is acting as a parasite. It is possible that those fungi that associate with mycoheterotrophs are either those that have been unable to evolve mechanisms to resist the plants or that suffer less from the association because they are highly efficient at obtaining the 'extra' organic C that support of a mycoheterotroph requires.

Nothing is known of the genetics of interactions between mycoheterotroph and parasitized mycorrhizal fungus, so that anything we might say in that regard would be wholly speculative. A starting point might be to investigate in detail the interactions between mycoheterotrophs, the fungi with which they associate and close relatives of the latter at very early stages of the relationships. Some work in this direction has already been done. We know that signal molecules are produced by 'suitable' fungal symbionts and that these stimulate germination of mycoheterotrophs.

'Unsuitable' fungi do not produce such compounds or amounts are insufficient to trigger a seed response. However, this may be only one part of the molecular dialogue between the symbionts. Whereas, recognition of such cues by the seeds would be an advantage for the plants, their production would seem to be a disadvantage to the fungi. Again, it is possible to speculate that the fungal molecules may be produced in response to a primary signal emanating from the seed. What is needed here is detailed microscopy of both symbionts as the interactions develop. It might be revealing to investigate cellular responses using vital staining of various types, as results would indicate whether any resistance or hypersensitive-like responses occurred in non-compatible interactions.

Furthermore, until we know more of the qualitative and quantitative aspects of nutrient transfers, and the extent to which these are or can be controlled by fungal symbionts, we will not be able to determine what the real costs to the fungi are of maintaining mycoheterotrophs. Related to this is the issue of movement of organic C from fungus to plant, a direction which, despite the possibility of some organic N movements to autotrophs in other types of mycorrhizal symbiosis is atypical for the majority of associations. Such movement clearly takes place and the mechanisms that underpin it presumably require changes in expression of nutrient, particularly sugar, transporters at the interface. It has been suggested that, in many mycorrhizal associations involving autotrophs, the plant retains an ability to reabsorb organic C lost to the interfacial apoplast, thus minimizing transfer to the fungal partners (see Chapters 4 and 8). It is exactly this type of mechanism that can be envisaged as supporting uptake of organic C by the mycoheterotroph. The significant question may relate to the mechanisms that induce loss of hard-won organic C by the fungal associate. Additionally, it will be important to elucidate the presumably highly complex mechanisms underlying mixotrophy and also the ability of mycorrhizas of green orchids to change the direction of organic C flow between symbionts at temporal scales that may be diurnal, seasonal or over the whole life span of the plant (see Chapter 12).

In any event, one of the outcomes of a specific relationship is that the chances of finding an appropriate fungal partner are reduced. Clearly, the production of very large numbers of seeds which is typical of this group of plants, accompanied by prolonged seed dormancy, are attributes that will increase the chances of contact between symbionts with the potential to set up appropriate associations. A complex system of signals is also likely to be even more important than they are in other, mutualistic, mycorrhizal types.

Conclusions

Technological developments have enabled significant advances in our understanding of the phylogenetic relationships between mycoheterotrophic plants and their fungal symbionts. Analyses of DNA of the respective partners in a wide range of alliances in the Monotropaceae and Orchidaceae have revealed remarkable levels of specificity in the associations. In these families, the mycoheterotrophs selectively form mycorrhizas with basidiomycetous fungi. In contrast to the tight specificity seen at the level of the individual symbiont partnerships, the fungi involved in the different associations are representative of a very wide range of totally unrelated taxa. The affinities of those fungi associated with mycoheterotrophs have been revealed by analyses of the nuclear ribosomal large subunit, enabling Lee Taylor to construct a phylogenetic tree spanning the most widely involved group, the Hymenomycetes (Basidiomycota) (Figure 13.30). Not only are they genetically distinct, but also they occupy different functional groups. While the majority of the parasitized fungi are of the ectomycorrhizal habit, a significant number are wood decomposers and some may be soil saprophytes.

Some mycoheterotrophic plants associate with AM fungi and early results indicate the possibility that the strains involved are selected from a narrow range of genotypes within this group, but further research is necessary to confirm this. In the cases of associations between mycoheterotrophs and both ECM and AM fungi, the evidence indicates that these plants have subverted the bipartite mutualism between autotroph and fungus by parasitizing the heterotrophic component of the mycorrhiza. A relatively small number of experiments involving feeding of ^{14}C-labelled compounds to autotrophs, or comparative analysis of stable isotope signatures of mycoheterotrophs, lend strong support to earlier suppositions that this subversion provides access to photoassimilates produced by the co-associated autotrophs. This invasion of a normally bipartite mutualism has been seen as a form of 'cheating', since there are no obvious benefits either to the fungal or autotrophic partners of the symbiosis. A form of cheating it may be, but since the fitness costs to the fungus or indeed to the invariably much larger autotroph in the tripartite association appear to be small, this may be a factor contributing to the stability of partnerships in these complex symbioses.

The nutritional relationships of those orchids that parasitize wood decay fungi are even less well understood, though it seems reasonable to assume that the exceptionally large biomass achieved by some representatives of this group are a reflection of the considerable volumes of C available to their fungal associates in decomposing woody substrates. Experimental analyses of these systems is urgently required. Recent advances enabling the *in vitro* culture of one such orchid from seed to flowering

---

Figure 13.30 Phylogenetic affinities of Hymenomycete (Basidiomycota) fungi associated with mycoheterotrophs as revealed in a tree based upon the nuclear ribosomal large subunit. The MOR Homobasidiomycete alignment of Hibbett *et al.* (2005) was downloaded from http://mor.clarku.edu/ and reduced to the core set of taxa. This data-set encompasses most recognized major clades within the Basidiomycetes (Hibbett *et al.*, 2005). Several sequences were added to fill out representation of mycoheterotroph-associated lineages and numerous taxa were removed due to space constraints. Models of sequence evolution were evaluated using ModelTest 3.7 (Posada and Crandall, 1998), and the best model according to Akaike Information Criteria was GTR+G+I. A complete heuristic search utilizing the HKY model and the genetic search algorithm was conducted in Metapiga 1.0 (Lemmon and Milinkovitch, 2002). The most likely tree from this search was then utilized as basis for a further incomplete heuristic search in PAUP*4.0b10 utilizing the GTR+G+I model. *Calocera cornea*, *Christiansenia pallida*, *Dacrymyces chrysospermus* and *Tremella foliacea* (Dacrymycetales and Tremellales) were used as the outgroup and deleted from the figure for space. Fungal genera reported with mycoheterotrophs are shown; note that in most cases, species represented in GenBank were used in place of species actually reported to occur with a given plant species. Fungal taxa with an asterisk are not believed to be the primary fungal symbiont of the orchid indicated. Adapted from Taylor *et al.* (2002).

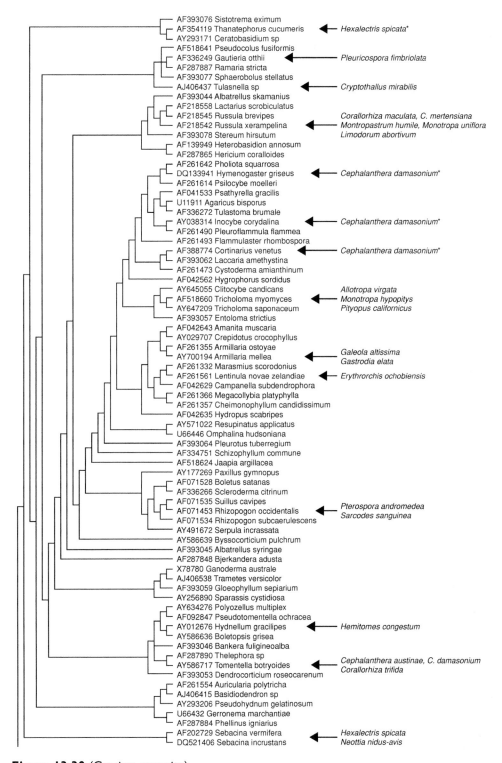

Figure 13.30 (Caption opposite)

using sawdust as a sole C source reveals the potential for experimental analysis of the nutrition of these plants.

Extraordinary levels of specificity of the kind exhibited by mycoheterotrophic associations are typical of those seen in a wide range of host–parasite relationships and are thought to arise from the need of the parasite to complete its development on a single host individual. Clearly, mycoheterotrophs occupy an extreme position at one end of the mutualism–parasitism continuum occupied by the mycorrhizal symbiosis. Research on plants representing intermediate evolutionary stages between the autotrophic and heterotrophic modes of nutrition is likely to throw new light on those factors that drive mycoheterotrophs in the direction of specificity in their relationships with fungal symbionts. Among the ECM-associated mycoheterotrophs, the helleborine orchids are emerging as a potentially useful target group, while the searches in the AM-associated families Burmanniaceae and Gentianaceae that contain large numbers of putatively autotrophic as well as mycoheterotrophic species could lead to exciting experimental opportunities.

14

Fungal symbioses in lower land plants

Introduction

An unambiguous conclusion from the multiplicity of contemporary phylogenetic studies of the plant kingdom is that the bryophytes were the first green plants to radiate into terrestrial niches (Shaw and Renzaglia, 2004). The earliest confirmed land plant fossils are believed to be from an ancient liverwort dating to the middle Ordovician, about 475 million years ago (Wellman *et al.*, 2003) and it seems likely that the gametophytes of bryophytes will provide critical clues about land plant evolution. As the earliest multicellular autotrophs on land they were likely targets for attack by heterotrophs and they were the first complex organisms to face the challenges associated with the need to acquire nutrients from solid and gaseous substrates. For these reasons, the study of the relationships between extant bryophytes and their fungal symbionts may throw light upon the processes whereby, through time, so many of the descendants of these plants came to form stable associations with mycobionts through which they gained access to nutrients. Of course, it is also likely that fungal symbioses developed with other autotrophs, as shown by the extant *Nostoc*–glomeromycotan association called *Geosiphon* (see Chapter 1). These have not survived in large numbers through the ages, so that the bryophytes remain our best link with early symbiotic evolution.

Among the three phyla of extant bryophytes, the liverworts (Marchantiophyta), the hornworts (Anthocerophyta) and the mosses (Bryophyta), there are records of consistent endophytic associations with fungi only in representatives of the liverworts and hornworts (Duckett *et al.*, 1991; Read *et al.*, 2000; Nebel *et al.*, 2004; Duckett *et al.*, 2006a, b). The occurrence of symbioses between liverworts and fungi has been known for over a century. Schacht *et al.* (1854) provided careful descriptions of fungi in the thalli of *Pellia* and *Priessia* and observed colonization of rhizoids of *Marchantia* and *Lunularia*. Bernard (1909b) used the widespread occurrence of liverwort–fungus associations as the basis for his theory that vascular cryptogams were descended from mycorrhizal liverworts. A period of intensive light microscope analysis (see Stahl, 1949; Boullard, 1988 for reviews) confirmed that these associations are a normal feature of hepatic biology and revealed that the associations were broadly of two kinds, one involving fungi with aseptate and the other with septate hyphae.

More recently, the application of ultrastructural and, to a lesser extent, molecular methods of analysis (see below) has enabled the aseptate fungi to be recognized as belonging to the Glomeromycota, while those with septate hyphae encompass both ascomycetous and basidiomycetous groups.

Hornworts are now thought to be a sister group of the early tracheophytes (Groth-Malonek *et al.*, 2004). Some, for example *Phaeoceros laevis*, can be colonized by arbuscular mycorrhizal (AM) fungi (Boullard, 1988; Ligrone, 1988; Schüßler, 2000), but the distribution of thalli harbouring mycobionts within the group has yet to be investigated.

It is a distinguishing feature of mosses, including the basal lineages *Sphagnum* and *Takakia*, that symbiotic associations of the kinds seen in liverworts are normally absent. Although there are occasional records (Rabatin, 1980; Parke and Linderman, 1980; Mago *et al.*, 1992) of the presence of AM-type fungal associations with mosses, careful scrutiny of many of the reports indicates that the fungi are confined to dead or moribund cells which suggests that they are present as secondary colonists. It thus seems likely that in this group, which is also regarded in current phylogenies (Groth-Malonek *et al.*, 2004) as being sister rather than basal to the tracheophytes, lacked symbiotic associations from the outset.

Fungal symbioses in liverworts

The widespread and consistent occurrence of endophytic fungal colonization in liverworts is of particular interest because molecular studies increasingly point to these plants as being the basal lineage from which tracheophytes arose (Kenrick and Crane, 1997; Davis, 2004; Groth-Malonek *et al.*, 2004; He-Nygren *et al.*, 2006). It is therefore possible that some extant liverwort–fungus associations represent the ancestral forms of the mycorrhizal association that is now so widespread in vascular plants. Recognition of this possibility has led to a renewed interest in the types of association between liverworts and their fungal symbionts and, in particular, to a debate about their functional status and the extent to which they can be regarded as mycorrhizas (Selosse, 2005; Kottke and Nebel, 2005; Duckett *et al.*, 2006). Sadly, while the new studies have enabled progress towards understanding of the extent and diversity of these associations, knowledge of their functions is still lacking. There are reports (Duckett *et al.*, 2006a, b) that some liverwort thalli inoculated with host-specific basidiomycetous endophytes grow more vigorously than their uninoculated counterparts and there is evidence (Bidartondo *et al.*, 2003; see below and Chapter 13) of C transfer to a liverwort by way of its fungal associate. However, there has been no in-depth physiological analysis of the nutrient relations of liverwort–fungus associations. For this reason, and because the fungal colonization occurs in rhizoids, stems and thalli of non-vascular plants rather than in roots (as in tracheophytes), it is misleading, despite some suggestions to the contrary (Brundrett, 2004), to refer to these associations as 'mycorrhizal'. They may, following Kottke and Nebel (2005), be described as 'mycorrhiza-like' but, until such time as essential functional information is available, we prefer to consider them in the broader sense simply as 'symbiotic associations' (Read *et al.*, 2000).

A recently published maximum-likelihood tree of liverwort phylogeny (Davis, 2004), when considered in conjunction with the known symbiotic affinities of these plants and their propensity to occupy terrestrial or epiphytic habits (Kottle and Nebel, 2005), throws new light on the possible evolutionary history of the symbiosis. It emerges (Figure 14.1) that the symbiotic habit may have been gained and lost independently on a number of occasions over time and that the prevalence of liverworts supporting symbiotic associations in terrestrial habitats contrasts with their general absence in epiphytes. While it is tempting to ascribe a functional basis to their apparently specialized correlation with habitat type, the failure of epiphytes to form symbioses could equally well be ascribed to a scarcity of fungal inoculum in arboreal habitats. In this context it is worth noting that the mycorrhizal condition is much more prevalent in terrestrial than epiphytic orchids (see Chapter 12).

The tree places three sister groups, the Complex thalloids (Marchantiales, Monocleales, Blasiales Sphaerocarpales Ricciales), the Haplomitrales-Treubiales and the

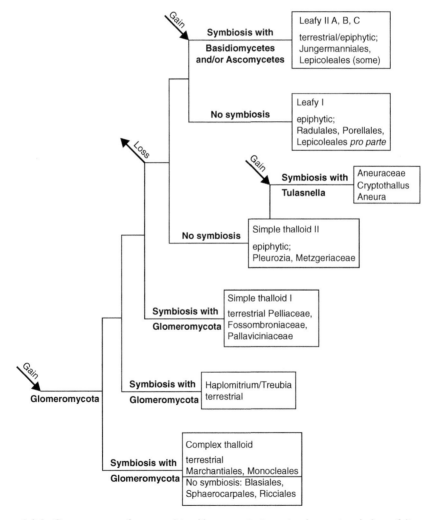

Figure 14.1 Occurrence of mycorrhiza-like associations in the main clades of liverworts. The phylogenetic tree of liverworts is based on the maximum-likelihood tree derived from a twelve gene backbone data matrix by Davis (2004). Modified from Kottke and Nebel (2005).

Simple thalloids (Fossombroniales, Pelliaceae), in basal positions. These primitive groups all have terrestrial habits and are either associated with glomeromycotan fungi (Read *et al.*, 2000; Carafa *et al.*, 2003; Nebel *et al.*, 2004; Duckett *et al.*, 2006b; Ligrone *et al.*, 2007) or, as in the cases of Blasiales, Sphaerocarpales and derived Ricciales, are devoid of fungal associations. Since the last three orders are considered to be sister groups to *Marchantia* and *Monoclea*, which are colonized by members of the Glomeromycota (Davis, 2004), it is likely that those lines which lack symbioses have emerged from a common symbiotic ancestor.

The new taxonomy divides the liverworts with simple thalli in the orders Fossombroniales and Metzgeriales into two groups (Figure 14.1). Those in Simple thalloid 1, which include *Fossombronia, Pellia, Petalophyllum* and *Pallavicinia*, are largely terrestrial and are widely associated with glomeromycotan fungi in nature. In contrast, those in the second group, which are considered to be derived and are mostly of epiphytic habit, including several Metzgeriaceae and *Pleurozia*, appear to have lost the ability to form symbioses. However, in the Aneuraceae, (Simple thalloid 2 group), members of which are either terrestrial (*Aneura*) or subterranean (*Cryptothallus*) (see Colour Plate 14.1c), associations with basidiomycetes of the order Tulasnellales are the norm. Since *Metzgeria* is currently placed in a position basal to *Aneura* (Davis, 2004), it appears that members of the Simple thalloid 2 group first lost their symbiotic condition and later re-established new associations with basidiomycetes.

As indicated (Figure 14.1), the leafy liverworts comprise two sister clades, Leafy 1 and Leafy 2, the dichotomy between them, according to Davis (2004), being ancient. Leafy 1, made up of the orders Porellales, Radullales and some Lepiicoleales, almost all of which are epiphytic, is basal to Leafy 2 and consists of plants that are non-symbiotic. This lends support to the hypothesis (Kottke and Nebel, 2005) that the plant–fungus association was lost by a common ancestor of Simple thalloid 1 and leafy liverworts. In contrast, most members of the derived Leafy 2 group so far examined form symbioses with ascomycetous or basidiomycetous symbionts, sometimes with both. These symbiotic liverworts, most of which are members of the large order Jungermanniales, are all of terrestrial or semi-terrestrial habit, many of the latter being colonists of rotting wood and humus. Within the Jungermanniales are families like the Cephaloziaceae, Cephaloziellaceae and Lepidoziaceae that are primarily colonized by ascomycetous fungi, among which members of the *Hymenoscyphus* aggregate figure prominently (Duckett *et al.*, 1991; Duckett and Read, 1995; Read *et al.*, 2000; Nebel *et al.*, 2004). There are also some families, like the Lophoziaceae and Arnelliaceae, that are most commonly associated with basidiomycetes. Other families have members that can be colonized by both ascomycetes and basidiomycetes, the different classes of fungus sometimes co-occurring in the same thallus. Members of the Jungermanniaceae, Scapaniaceae, Geocalycaceae and Calypogeiaceae fall into this category (Nebel *et al.*, 2004; Duckett *et al.*, 2006a). Among the basidiomycetous fungal associates, members of Subgroup B of the genus *Sebacina* (see Figure 6.5) are common (Kottke *et al.*, 2003). Fungi in this subgroup have now been recognized as symbionts in ericoid mycorrhiza (ERM) and orchid associations (see Chapters 11, 12 and 13). In an elaborate series of cross-inoculation experiments, Duckett *et al.*, 2006a) have demonstrated that, in striking contrast to the wide host ranges seen in many of the liverwort–ascomycete associations, those between basidiomycetes and members of the Lophoziaceae and Arnelliaceae are host-specific.

In summary, it appears that the complexity of mycorrhiza-like associations in hepatics is the result of formation, loss and reformation of different partnerships over evolutionary time.

The structures of liverwort–fungus symbioses

Ultrastructural studies of *Concephalum conicum* (Ligrone and Lopes, 1989) and combined ultrastructural and molecular analysis of *Marchantia foliacea* (Russell and Bulman, 2004) have confirmed that both these terrestrial members of the basal Complex thalloid group of liverworts (Figure 14.1) form symbioses with AM fungi. Colonization takes place primarily through the rhizoids on the ventral surface of the thallus. Distinct zones of colonization are found within the ventral half of the thallus (Figure 14.2a). Within this zone the entirely intracellular hyphae produce arbuscules, consisting of a mass of very finely branched hyphae, coils and vesicles and appear to lack intercellular hyphae (Figure 14.2b). This type of AM colonization conforms to the *Paris*- rather than the *Arum*-type (see Chapter 2).

Amplification of ribosomal sequences of the fungal colonist of *M. foliacea* using PCR confirmed that the thalli were occupied by multiple phenotypes of the genus *Glomus* (Russell and Bulman, 2004). All of these fell into *Glomus* Group A (*sensu* Schüßler *et al.*, 2001), which is considered to have arisen late in glomeromycotan phylogeny. Two of these genotypes were also consistently detected in roots of trees

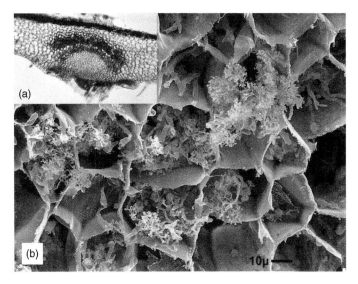

Figure 14.2 Distribution of mycorrhizal colonization in the thallus of *Marchantia foliacea*. (a) Cross-section of thallus stained with fuchsin red, showing distinct zone (dark colouration) surrounding the area of the central rib. (b) Scanning electron micrograph of thallus tissue, showing cells occupied by arbuscules in the main colonization zone. From Russell and Bulman (2004).

in the Podocarpaceae growing with the liverwort, raising a strong possibility that the two types of plant are co-associated with their fungal symbionts in the same New Zealand habitats.

The two remaining basal orders of liverwort, Treubiales and Haplomitriales, both of which are terrestrial, have recently been the subject of intensive ultrastructural analysis (Duckett *et al.*, 2006b; Carafa *et al.*, 2003). Both are extensively colonized by AM-type fungi, but structural aspects of the symbioses in these two orders are quite different.

The lobed green thalli of *Treubia* (see Colour Plate 14.1a) posses a central midrib in which histologically differentiated tissues support extensive and interconnected intra- and intercellular colonization of the glomeromycotan type. Fungal penetration takes place through mucilage-filled clefts in the undersurface of the thallus rather than through rhizoids. The fungi form extensive intracellular *Paris*-type coils (Figure 14.3a, b) and distinctive terminal swellings are produced at the apices of some of the hyphae (Figure 14.3c, d). Above the zone dominated by intracellular colonization is another supporting largely extracellular development, some of which is pseudoparenchymatous in nature. These structures are analogous to those produced by glomeromycotan fungi in the subterranean gametophytes of the primitive tracheophyte *Lycopodium clavatum* (Schmid and Oberwinkler, 1993). Further links with lycopsid-type infections are the frequent presence of intracellular bacterial inhabitants in the fungal hyphae. However, these occur in all glomeromycotan associations in liverworts and in other plants as well (see Chapter 1). While all the structural features of the *Treubia* symbiosis

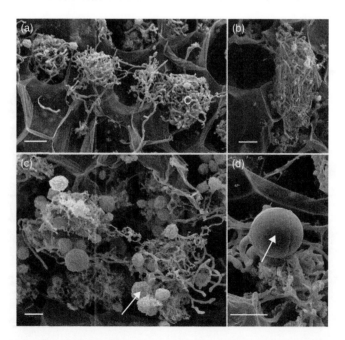

Figure 14.3 Scanning electron micrographs of intracellular fungal colonization in *Treubia lacunosoides*. (a) Early colonization stage showing intracellular coils at different developmental stages; (b) fully developed hyphal coil; (c) more advanced stage of intracellular fungal colonization, showing numerous fungal swellings; (d) young swelling at the end of a short hyphal branch. Scale bars = 10 μm. From Duckett *et al.* (2006b).

suggest that the fungi involved are of the AM type, there is an urgent need for confirmation of their identity using molecular methods.

The axes of *Haplomitrium* are morphologically distinct from those of *Treubia* and superficially resemble those of leafy liverworts (see Colour Plate 14.1b). The plants have sympodial systems of creeping, largely subterranean, axes referred to as 'rhizomes' by Grubb (1970), from which erect leafy shoots emerge (see Colour Plate 14.1d). From the subterranean axes mucilage-covered downward-growing branches emerge that become extensively colonized by aseptate fungi (Carafa *et al.*, 2003) (Figure 14.4). Colonization is restricted to epidermal or adjacent cortical cells and consists, as in *Treubia*, of hyphal coils, being finely branched arbuscule-like structures (arbusculate coils) again typical of *Paris*-type AM (see Chapter 2) (Smith and Smith, 1997). Also, as in *Treubia*, distinctive short-lived terminal swellings develop in colonized cells. Molecular analysis has confirmed that in *H. chilensis* the fungus is in *Glomus* group A (Glomeromycota) and clusters with those from marcantialean and mezgerialean liverworts (Ligrone *et al.*, 2007).

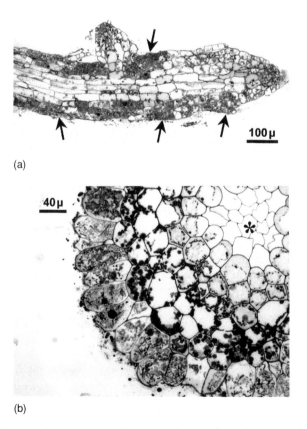

(a)

(b)

Figure 14.4 (a) Longitudinal section of the apical part of a colonized axis in *Haplomitrium ovalifolium*. The arrows indicate colonized areas. A lateral branch is emerging a short distance behind the apex. (b) Transverse section of a colonized axis in *Haplomitrium gibbsiae*. The papillose epidermal cells are heavily colonized while the adjacent cortical cells are fungus free and contain abundant starch grains. The asterisk indicates the central strand. From Carafa *et al.* (2003).

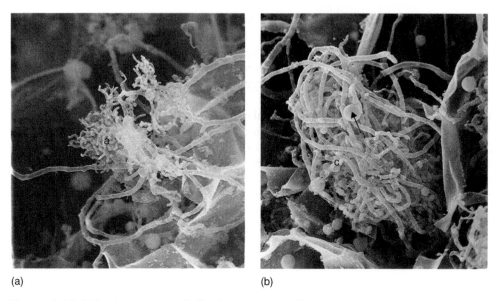

(a)   (b)

Figure 14.5 AM colonization in thalli of the simple thalloid category 1 hepatic *Petalophyllum rolfsii*. (a) Intracellular arbuscule; (b) hyphal coil with terminal vesicles (arrow). Photographs courtesy of J.G. Duckett.

Liverworts in the derived Simple thalloid category 1 (see Figure 14.1) are also normally colonized by fungi of the AM type. These are widespread through the Pelliaceae, Fossombroniaceae, Allisoniaceae and Pallavicinaceae. Cytologically these associations mirror those seen in the Complex thalloids. Studies of *Pellia* were pioneered by Magrou (1925) who reported the production of arbuscules and vesicles in mature tissues of the thallus. Experiments involving the growth of aseptically produced *Pellia* thalli in association with *Glomus*-colonized seedlings of *Plantago lancelolata* have confirmed the ability of this fungus to spread from a higher plant to a liverwort (Read *et al.*, 2000). Thalli of *Fossombronia* and *Petalophyllum* are also colonized by AM-type fungi bearing both arbuscules (Figure 14.5a) and coils (Figure 14.5b).

Whereas the largely epiphtypic members of the Simple thalloid 2 group generally lack fungal symbioses, the derived terrestrial Aneuraceae have been shown by a combination of ultrastructural (Ligrone *et al.*, 1993; Read *et al.*, 2000; Nebel *et al.*, 2004) and molecular (Bidartondo *et al.*, 2003; Kottke *et al.*, 2003) studies to form symbioses with basidiomycetes. The presence of dolipore septa with imperforate parenthosomes in the fungal inhabitants of *Riccardia* (Nebel *et al.*, 2004), *Aneura* (Ligrone *et al.*, 1993) and *Cryptothallus* (Figure 14.6) (Pocock and Duckett, 1984) indicate that the fungi are likely to be heterobasidiomycetes. In the cases of *Aneura* and *Cryptothallus*, this has been confirmed by molecular analyses (Bidartondo *et al.*, 2003; Kottke *et al.*, 2003) revealing the fungi to be members of the genus *Tulasnella* in both cases. The cytology of the fungal colonization in all three genera is similar. Distinct central zones of the thallus are intensively occupied by coils of robust hyphae. As in orchids, there are repeated cycles of intracellular colonization, followed by hyphal degeneration.

Understanding processes of colonization in associations found in many of the terrestrial leafy liverworts of the Leafy 2 group has been facilitated by the fact that the fungi involved can be cultured and the associations resynthesized. Two types of

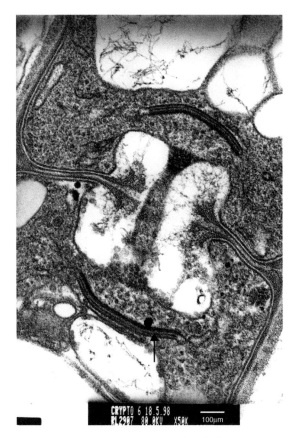

Figure 14.6 Electron micrograph of colonized cell of the mycoheterotrophic liverwort *Cryptothallus mirabilis*, showing dolipore septum with imperforate parenthosome indicative of colonization by a heterobasidiomycete. From Pocock and Duckett (1984).

association are found in this group, each being largely restricted to individual orders (Schuster, 1966; Duckett *et al.*, 1991, 2006; Williams *et al.*, 1994; Pockock and Duckett, 1995). One, seen in many members of the Jungermanniales, involves ascomycetes which are readily identifiable in cytological studies by the presence of simple septa and Woronin bodies (Figure 14.7). The other, involving basidiomycetes, recognized by the production of dolipore septa (see Figure 14.6), is found predominantly in the Arnelliaceae, Scapaniaceae and Lophoziaceae. Inoculation experiments indicate that, whereas the ascomycete associations show little host specificity, the basidiomycetes are more selective in their host preferences (Duckett *et al.*, 2006a). Both ascomycete and basidiomycete types of colonization take place through the rhizoids but, whereas in the former the fungus is restricted to the rhizoids where it causes characteristic swelling of their tips (Figure 14.8a, b) (Duckett *et al.*, 1991; Pocock and Duckett, 1995), in the basidiomycete associations, fungal penetration progresses into the axis tissues of the liverwort (Duckett *et al.*, 2006a). In the case of the ascomycete infections, hyphal penetration of the axes appears to be prevented by overgrowth of host wall materials which produce peg-like structures (see Figure 14.8c).

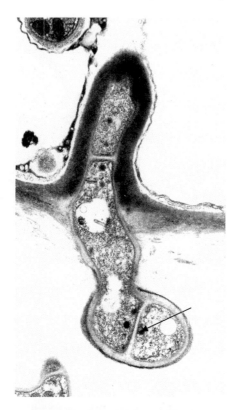

Figure 14.7 Electron micrograph showing penetration of a rhizoid tip of *Cephalozia connivens* by the ascomycete fungus *Hymenoscyphus* (= *Rhizoscyphus*) *ericae*. Note Woronin body (arrow). Photograph courtesy of J.G. Duckett and R. Ligrone.

Provisional identification of the ascomycetous fungi has been enabled by analysis of their culture characteristics following isolation from rhizoids. They were shown to be of the dark septate type and to be similar in appearance to those in the ERM fungal aggregate *Hymenoscyphus ericae* (see Chapter 11) (Williams *et al.*, 1994). When liverworts representative of a number of families of the Jungermanniales were inoculated with strains of *R. ericae* isolated from the mycorrhizas of ericaceous plants, they were shown to develop typical infections (see Figure 14.8b) (Duckett and Read, 1995; Read *et al.*, 2000). Similarly, the fungi isolated from liverwort rhizoids formed typical ERM hyphal complexes in the epidermal cells of the ericaceous plants *Calluna vulgaris* and *Vaccinium oxycoccus* (Figure 14.9). Some degree of specificity in the association was revealed by the fact that another ERM fungus, *Oidiodendron maius*, as well as a number of ectomycorrhizal (ECM) and orchid mycorrhizal fungi failed to colonize liverworts. To date, in only one liverwort of this group, *Cephalozia exiliflora*, has the identity of the fungal symbiont been determined by DNA analysis. This confirmed the fungus to be a member of the *H. ericae* aggregate (Chambers *et al.*, 1999; see Chapter 11).

In the associations formed by basidiomycetes, colonization of the rhizoids, which does not lead to swelling of the tips, is followed by penetration into the liverwort

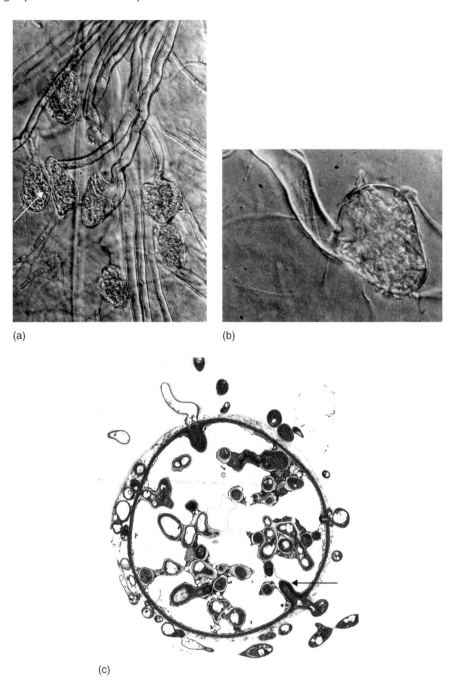

(a)

(b)

(c)

Figure 14.8 Rhizoids of *Cephalozia connivens* showing swollen tips associated with coloniz-ing fungal symbionts. (a) Grown in the wild and naturally colonized by *Rhizoscyphus ericae*; (b) in pure culture inoculated with *H. ericae*; (c) electron micrograph of swollen rhizoid apex showing production of host wall materials around penetrating hyphal tips (arrow). From Duckett and Read (1995).

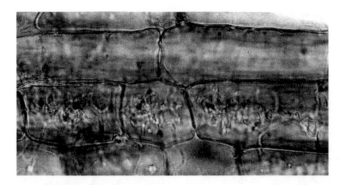

Figure 14.9 Typical ericoid hyphal complexes formed in the epidermal cells of the erica-ceous plant *Vaccinnium oxycoccus* by the fungal endophyte isolated from swollen rhizoid tips of the leafy hepatic *Kurzia*. From Duckett and Read (1995).

axes (Duckett *et al.*, 2006a). There are two patterns. In the Arnelliaceae, the stems have a central core of colonized cells (Figure 14.10a, b), whereas in the Lophoziacea and Scapaniaceae, a mosaic of colonized and uncolonized cells is found. In con-trast to the glomeromycotan infections and those formed by basidiomycetes in the Aneuraceae, there is no evidence of hyphal digestion in the plants (Figure 14.10c). In fact the reverse appears to be the case since, in the mosaic-type stem infections, the colonized liverwort cells die (Duckett *et al.*, 2006a); the sequence therefore being more similar to that seen in ericoid than arbuscular mycorrhizas (see Chapters 2 and 11). In this type of colonization, hyphal tips are often seen to be occluded as they penetrate the walls of the liverwort cells, producing peg-like structures (Figure 14.11).

Little is yet known about the identity of these basidiomycetous associates. However, the presence of flat rather than curved imperforate parenthosomes (see Figure 14.6), together with the results of DNA analysis, indicate that, in some cases, the fungi are attributable to the Sebacinales, including members of the *S. vermifera* group in subgroup B (see Figure 6.5) (Kottke *et al.*, 2003; Nebel *et al.*, 2004). In view of the fact that cross inoculation using basidiomycetous isolates derived from members of these liverwort families has shown considerable host specificity (Duckett *et al.*, 2006a), it seems likely that a number of different fungal genotypes are involved in the associations. Here, as in the case of both the glomeromycotan and ascomycetous symbioses, there is an urgent need for more extensive molecular analysis to enable identification of the taxa associated with each of the liverworts.

Comparisons between features of liverwort symbioses and those of gametophytes in 'lower' tracheophytes

In common with the basal complex thalloid liverworts and members of the families Treubiaceae and Haplomitraceae, the gametophytes of lower tracheophytes, per-haps with the general exception of those in the Equisitopsida (Read *et al.* 2000), are colonized by AM-type fungi. Unlike the liverworts, there are no reports of ascomyc-etous or basidiomycetous associations. The structural features of the AM coloniza-tion are similar to those seen in the Complex and Simple thalloid types of liverwort. Almost all reports indicate the presence of *Paris*-type AM infection; arbuscules are

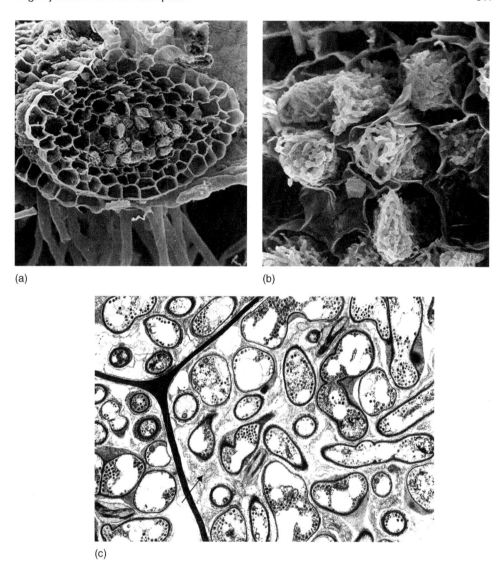

(a)   (b)

(c)

Figure 14.10 Basidiomycetous colonization of the shoot tissues of leafy liverworts of Category 2. (a) Central core of colonized cells in *Southbya* (Arnelliacecae); (b) scanning electron micrograph showing hyphal coils in colonized cells of *Southbya*; (c) transmission electron micrograph of colonized zone of *Southbya* showing structurally intact hyphal colonization. From Duckett *et al.* (2006a).

not recorded. Within the Lycopsida, the colonization found in both the subterranean achlorophyllous gametophytes of *Lycopodium clavatum* (Schmid and Oberwinkler, 1993), *Hupertzia hypogeae* (Winther and Friedman, 2007b) and in those of the green gametophytes of *L. cernuum* (Duckett and Ligrone, 1992) and of this type. Phylogenetic analyses show that symbionts of four *Glomus* phylotypes in a single clade of *Glomus* group A occur in the gametophytes of *L. clavatum* and *H. hypogeae* (Winther and Friedman, 2007b). Ultrastructural analyses of the achlorophyllous

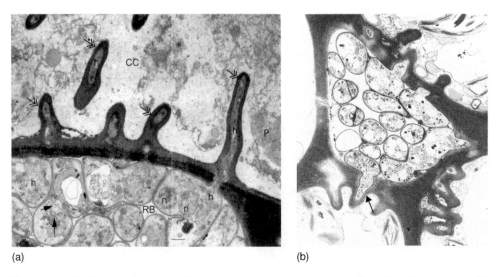

(a)   (b)

Figure 14.11 Encapsulation of hyphae penetrating cortical cells of leafy hepatics. (a) Dense hyphal colonization of the rhizoidal base of *Calypogeia muelleriana* and invasion of these hyphae into adjacent cortical cells in which encapsulation produces ingrowth pegs (arrows). h = hyphae; CC = cortical cell; RB = rhizoid base. From Kottke *et al.* 2003. (b) Ingrowth pegs in tissues surrounding densely colonized cell of *Scapania* spp. From Duckett *et al.* (2006a).

subterranean gametophytes of two members of the Psilotales, *Psilotum nudum* and *Tmesipteris tannensis* have shown these plants also to be colonized by AM fungi (see Chapter 1).

Fungal penetration takes place in a manner similar to that in Marchantiales and ascomycete liverwort associations, but through rhizome hairs rather than rhizoids. Peterson *et al.* (1981) observed dense coils of aseptate hyphae in cortical cells of *P. nudum* thalli. In a detailed comparative anatomical analysis of the gametophytes of both *P. nudum* and *T. tannensis*, Duckett and Ligrone (2005) showed that, unlike the situation in bryophytes, these structures are vascularized. Most of the cortical cells in both species were colonized by AM-type fungi with coarse hyphae producing small vesicle-like structures and dense hyphal coils (Figure 14.12a, b), again indicative of the *Paris*-type association. Fungal structures observed in sporophytes of these species were similar to those seen in gametophytes, indicating the possibility that the achlorophyllous haploid generation might be connected to and supported by the photosynthetic sporophyte (see Chapter 1). The structure of mycorrhizal associations in subterranean gametophytes of *Botrychium* (Ophioglossales) is similar to that in *Psilotum* (Schmid and Oberwinkler, 1994; Winther and Friedman, 2007a; see also Chapter 13).

Conclusions

As is evident from the above account, there is an abundance of structural evidence in favour of the notion that lower tracheophytes, in common with the basal

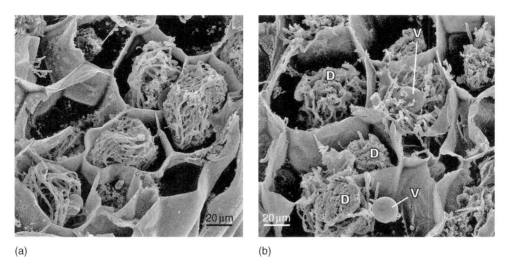

(a)   (b)

Figure 14.12 Scanning electron micrographs of *Tmesipteris* gametophytes. (a) Cortical cells with dense hyphal coils; (b) cortical cells with degenerating coils (D) and hyphae producing vesicles (V). From Duckett and Ligrone (2005).

liverwort groups, are normally colonized by fungi in the Glomeromycota. It is increasingly clear from phylogenetic analyses that in many cases they are placed in a narrow clade of *Glomus* group A. Physiological information relating to the way the symbioses function, and hence their potential significance in nature, is completely lacking. We must presume that mutual benefits are to be gained in the various associations, particularly as the hepatics appear to have explored, in evolutionary terms, a number of different strategies involving symbioses with fungi known to form AM, ECM and ERM relationships with other plants, as well as encompassing full autotrophy to full heterotrophy.

Section 4
Functioning of mycorrhizas in broader contexts

15

The roles of mycorrhizas in successional processes and in selected biomes

Introduction

Studies of the species composition and community structure of assemblages of land plants and their dependent heterotrophs over the last century have enabled the delineation of distinct biomes at the global scale (Odum, 1971). They have also revealed some of the successional dynamics involved in progression from immature and disturbed to mature and stable states in some of these biomes. Over broadly the same time period, commencing with pioneers like Janse (1897) and Gallaud (1905), extensive below-ground surveys have confirmed the presence of the mycorrhizal symbioses in most, but not all, successional stages and in most of the plants that make up the mature biomes. They have also revealed some segregation between plant families, both in the extent to which they are colonized by mycorrhizal fungi and the types of mycorrhiza that they support (Trappe, 1987; Newman and Redell, 1987; Fitter and Moyersoen, 1996; Wang and Qui, 2006).

While it has been relatively easy to achieve the sampling necessary to describe both the species composition of biomes above ground and the nature of mycorrhizal communities below ground, it has proved more demanding to determine relationships between records of the occurrence of mycorrhizas and the possible contributions of the symbioses to the dynamic properties of the biomes in which they occur. Despite these difficulties, some progress has been made. What is emerging is a picture suggesting that the functions of the symbioses go far beyond simple scenarios involving facilitation of mineral nutrient capture by individual plants or of organic C by the associated fungi. It can be hypothesized that while soil and climate have combined to configure the distinctive composition of the autotroph community in each biome, selection will also have favoured mycorrhizal symbioses and mycobionts that are appropriate to that particular set of environmental circumstances. There follows an analysis of the extent to which the emerging evidence supports this hypothesis.

After commencing with a consideration of the roles of mycorrhizas in successional dynamics, attention is turned to arctic-alpine, heathland, boreo-temperate forest and tropical forest biomes, each in turn viewed as a stable climax community occurring along a latitudinal gradient from the poles to equatorial regions.[1] The extent and nature of mycorrhizal colonization in each case is considered with a view to elucidating the functions that might be important under the conditions prevailing in each system. At this stage, there are more questions than there are answers. It is to be hoped that identification of the questions can at least be of assistance to the next generation of researchers who will continue attempts to test hypotheses of the kind presented above.

The roles of mycorrhizal colonization in primary succession

The sequential development of plant communities following major environmental perturbations such as glaciation (Crocker and Major, 1955) and volcanic activity (Simkins and Fiske, 1983) are well documented. It is acknowledged also that scarcity of nutrients in the poorly weathered materials exposed by such events may determine the early stages of the primary succession initiated on them (Gorham et al., 1979). Under these circumstances it seems likely that mycorrhizal fungi would play a role in facilitating the succession but, until recently, there was little direct evidence for such a role in nature. Work of Kazuhide Nara and his group towards the summit of Mount Fuji in Japan has now gone some way to describing their involvement in a volcanic environment. Here, at an altitude of 1500–1600 m, the slopes, which are covered to a depth of up to 10 m with volcanic tephra deposited in the course of an eruption in 1707, support sparse patchy vegetation within which the shrub *Salix reinii* occurs as an ectomycorrhizal pioneer species.

Nara et al. (2003a) began by describing the succession of fungal sporocarp production associated with naturally colonizing plants of *S. reinii*. Two *Laccaria* and one *Inocybe* species were the first colonizers and these were succeeded by other species as hosts grew with age (Figure 15.1). This pattern is not unlike that recorded in secondary successional sites (Last et al., 1983; see Chapter 6), but there was no evidence of replacement of any of the fungi with tree development. The biomass production of the sporocarps was exceptionally large, yields equivalent to as much as 633 kg/ha/ year being recorded in association with the largest *S. reinii* trees. Comparative analysis of above- and below-ground composition of these fungal communities, based upon sequencing of the ITS region of nuclear r-DNA (Nara et al., 2003b), revealed that species dominating the sporocarp population were also the most widely encountered ectomycorrhizal (ECM) associates of *S. reinii* roots (Table 15.1).

While confirming the early presence of ECM fungi and providing evidence of correlations between formation of the symbiosis and performance, these studies do not directly address questions relating to the facilitation of establishment. These were considered in experiments involving seedling transplants. It was first shown by planting seedlings close to healthy or to unhealthy *S. reinii*, in bare ground or in vegetation patches lacking the shrub, that ECM formation occurred prolifically only in association with healthy plants (Nara and Hogetsu, 2004). Furthermore, at the end of the three-month growing season, seedlings associated with healthy willows were of significantly greater biomass as well as N and P contents than those

---

[1]We recognize the importance of grasslands as major terrestrial biomes. They are not discussed in detail here, but aspects of plant interactions in them are considered in Chapter 16.

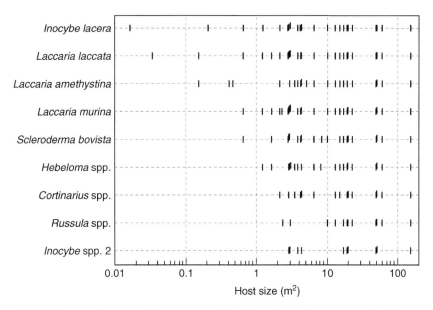

Figure 15.1 The recruitment of ectomycorrhizal fungal species, as revealed by measurements of sporocarp succession, over a period during which the shrubs of the autotroph *Salix reinii* gain size (m² ground area covered) with age. From Nara *et al.* (2003a).

Table 15.1 Species composition in the sporocarp and underground ectomycorrhizal (ECM) communities associated with individuals of *Salix reinii*, in an early successional volcanic desert.

Patch No.	Host size (m²)	Sporocarp abundance rank*					ECM abundance rank†				
		I	2	3	4	5	I	2	3	4	5
Small hosts											
8	0.03	Ll					Ll				
125	0.15	Ll	La				Ll	La			
41	0.21	Il					Il	Tl			
127	0.47	La					La	Lm	Ll		
Middle-sized hosts											
131	2.83	Sb	Ll	Hp	Il	Lm	Il	Lm	Sb	L	Tl
61	3.87	Sb	Ll	Hm	Lm	La	La	Sb	Ll	Lm	Tl
120	4.31	Sb	Ll	La	Il	Lm	Lm	La	Il	Sb	Ll
142	6.58	Il	Sb	Lm	La	I2	Il	Lm	Sb	Ll	La
Large hosts											
89	48.45	Ll	La	Hm	Lm	Sb	La	Ll	Sb	Lm	Ud
83	50.23	Ll	La	Hm	Sb	Il	Il	Ll	Sb	La	Lm
90	59.84	Ll	Sb	Hm	La	Il	La	Lm	Ud	Sb	Ll
139	153.73	Ll	La	Hm	Cd	Sb	Lm	T2	Sb	Il	Ll

*Abbreviations of the ECM fungal species are Cd, *Cortinarius decipiens*; Hm, *Hebeloma mesophaeum*; Hp, *Hebeloma pusillum*; Il, *Inocybe lacera*; I2, *Inocybe* spp.; L, *Laccaria* spp.; La, *Laccaria amethystina*; Ll, *Laccaria laccata*; Lm, *Laccaria murina*; Sb, *Scleroderma bovista*; Tl, *Tomentella* spp.; Ud, Unidentified D morphotype spp. *The data for sporocarps were from Nara *et al.* (2003a). †The ECM abundance rank was based on the relative abundance of ECM root tips in three soil samples collected from each of the small and middle-sized host and in twelve soil samples collected from each large host. From Nara *et al.* (2003b).

planted in bare ground or in patches lacking established willow plants (Table 15.2). Since linear relationships had been shown earlier (Nara *et al.*, 2003a) between photosynthetic rates of established *S. reinii* plants and their tissue N (Figure 15.2a) and P (Figure 15.2b) concentrations, the importance of early colonization by appropriate ECM symbionts was confirmed. The growth of transplant seedlings was most strongly correlated with their N content. Because photosynthetic activity at this site is closely linked to tissue N content (Figure 15.2a), Nara and Hogetsu (2004) suggest that access to this element is the key determinant of survival and growth at this site.

The role of mycelial networks produced by 11 of the commonest mycorrhiza-forming fungi at the site in facilitating growth and nutrient acquisition of pioneer plants was investigated by transplanting seedlings into the field with pre-formed connections to 'mother' plants or lacking any fungal associations (Nara, 2006a). Shared ECM mycelial networks involving fungi isolated from the field site were first established on plants grown in microcosms for some weeks in the laboratory. Alongside these 'mother' plants were then planted seedlings of the same species, so that they became colonized by the shared ECM inoculum. When these microcosms were transplanted intact at the field site so that the roots and mycelia were free to extend into the natural environment, the impacts of the individual inoculant fungi could be assessed and the performance of their plant partners compared with that of non-mycorrhizal transplants. Only eight of fifteen non-mycorrhizal seedlings had survived at the end of the four-month growing period, whereas the majority survived in all other treatments. None of the non-mycorrhizal seedlings survived the ensuing winter (Nara, personal communication). There were significant differences between the biomass, N and P contents of seedlings, depending first upon whether they were mycorrhizal and secondly, upon the fungal species originally inoculated (Figure 15.3). In almost all cases growth, N and P contents were greater in mycorrhizal plants. Plants inoculated with *Laccaria amethystina*, in which no parameter was significantly different from the mycorrhiza-free plants, proved to be exceptions. Seedlings responded particularly strongly to colonization by *Hebeloma leucosarx*, *Russula sororia* and *Inocybe lacera*.

These results provide strong support for the suggestion that early colonization by ECM fungi is essential for survival of the associated willows in this harsh environment. It is likely that the establishment of the first tree seedlings will be dependent upon chance association with spores of pioneer fungi. These plants having established support vigorous mycelial networks that facilitate later recruitment around them. Most of the pioneer fungi found at the Mount Fuji site are generalists in terms of their host preference (see Chapter 6) and so are able also to colonize and facilitate the recruitment of other ectomycorrhizal tree species. Nara and Hogetsu (2004) observed that the later ECM plant colonists, *Betula ermanii* and *Larix kaempferi*, established successfully only in patches containing pre-established *S. reinii* with which they shared ECM symbionts. These fungi are thus effectively driving the succession process in this environment.

A prerequisite for any fungal impacts upon primary succession is that propagules are transported to the newly exposed substrates. Cazares *et al.* (2005) investigated the possible role of propagule availability in determining primary succession along a chronosequence of soils exposed by a retreating glacier. They carried out *in situ* assays of plant mycorrhizal status and also bioassays of 'bait' seedlings grown in

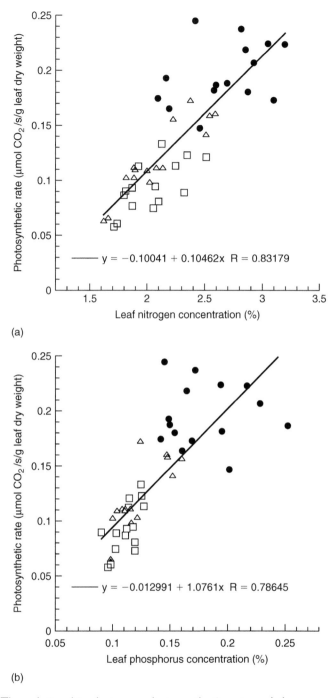

Figure 15.2 The relationships between photosynthetic rate and the concentration of leaf nitrogen (a) and phosphorus (b) of *Salix reinii* with reference to the production of associated ectomycorrhizal sporocarps. ● indicates leaves collected from patches in which ectomycorrhizal sporocarp production was relatively high; △ indicates those from patches with middle productivity; □ indicates those from unproductive patches. The *R*-values in (a) and (b) indicate statistically significant correlation ($P < 0.01$). From Nara *et al.* (2003a).

Table 15.2 Effects of established willow (*Salix reinii*) shrubs on ectomycorrhiza (ECM) formation and the performance of transplanted *S. reinii* seedlings in a volcanic desert on Mt Fuji, Japan.

Transplant sites[†]	No. seedlings[‡] (ECM/total)	No. ECM root tips/seedling[§]	Shoot dry mass[§] (mg/ seedling)	Shoot N amount[§] (μg/seedling)	Shoot P amount[§] (μg/seedling)
Bare ground	0/14	$0.0^a \pm 0.0$	$0.8^a \pm 0.1$	$10.8^a \pm 1.0$	$4.6^a \pm 0.4$
No-willow patch	1/17	$0.2^a \pm 0.2$	$0.8^a \pm 0.1$	$11.0^a \pm 1.2$	$4.5^a \pm 0.3$
Unhealthy willow	10/16	$9.3^b \pm 2.5$	$1.1^a \pm 0.2$	$20.4^b \pm 3.7$	$4.7^a \pm 0.4$
Healthy willow	17/17	$25.5^c \pm 2.9$	$2.2^b \pm 0.3$	$27.7^b \pm 1.6$	$6.4^b \pm 0.5$

From Nara and Hogetsu (2004). [†] Seedlings were transplanted into four habitat types: bare ground, the periphery of vegetation patches lacking willow shrubs (no-willow patch), the periphery of vegetation patches containing normally growing middle-sized ($10 \, m^2$ canopy coverage) willow shrubs (healthy willow) and the periphery of vegetation patches containing apparently unhealthy middle-sized willow shrubs (unhealthy willow). [‡] The number of ECM seedlings followed by the total number of sampled (surviving) seedlings. [§] Mean $\pm$ 1 SE values followed by different letters within a column differ statistically (Tukey's hsd test, $P < 0.05$).

soils collected along the gradient. The soils exposed for the shortest period, between 0–15 years, supported no plants in nature but were shown in the bioassays to contain propagules of dark septate (DS) fungi (see Chapter 7). In soils exposed for longer periods, four distinct groups of plants were recognized. The first, on soils exposed for approximately 25 years, consisted of plants of families generally considered to be non-mycorrhizals, such as Cyperaceae, Juncaceae, Caryophyllaceae and Onagraceae. These were confirmed to be lacking mycorrhizas, but to support low levels of DS colonization. The third and fourth groups consisted of plants, some of which were mycorrhizal. They contained those with AM, ECM and ERM mycorrhizas but, even in the oldest soils sampled, of ~65 years exposure, only 20% of the samples examined supported each of these types of colonization. Such observations provide tantalizing views of the possible roles of propagule availability in determining primary succession, but only combinations of inoculation and transplantation experiments of the kinds described by Nara *et al.* (see above) can establish cause and effect relationships.

We still know too little about the factors determining the transport of fungal propagules to primary successional environments. While wind is almost certainly the primary vector of spores produced by epigeous sporocarps, animals act as vectors of hypogeous species (see also Chapter 6).

Warner *et al.* (1987) demonstrated that spores of AM fungi could be wind blown for up to 2 km, whereas animals were able to transport AM propagules over several miles

Figure 15.3 Effects of mycorrhizal networks of individual ectomycorrhizal fungal species on (a) seedling dry weight, (b) nitrogen content and (c) phosphorus content of current-year *Salix reinii* seedlings in an early successional volcanic desert on Mount Fuji, Japan. C, control; Ll, *Laccaria laccata*; Il, *Inocybe lacera*; La, *Laccaria amethystina*; Lm, *Laccaria murina*; Sb, *Scleroderma bovista*; Hl, *Hebeloma leucosarx*; Hm, *Hebeloma mesophaeum*; Hp, *Hebeloma pusillum*; Rp, *Russula pectinatoides*; Rs, *Russula soraria*; Cg, *Cenococcum geophilum*. Broken lines indicate the mean for *S. reinii* seeds. From Nara (2006a).

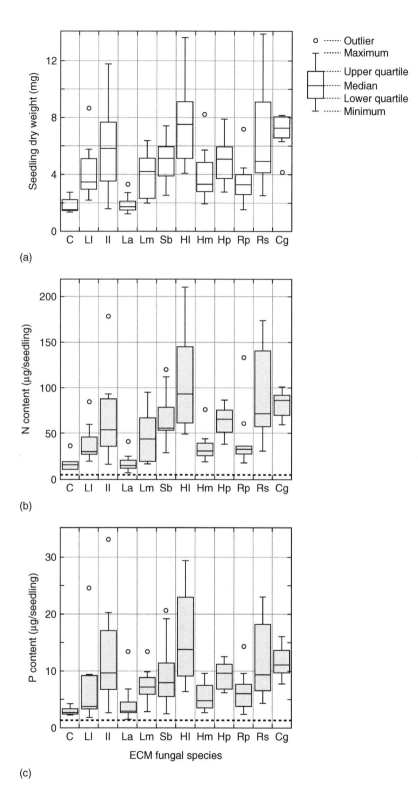

(a)

(b)

(c)

Figure 15.3 (Caption opposite)

of sterile pumice on Mount St Helens after its eruption (Allen, 1988). In all likelihood, the smaller spores of ECM fungi and the microspores of DS fungi, such as *P. fortinii*, will be transported over even greater distances (see Chapters 2 and 7).

The resupinate sporocarps of *Tomentella* spp., that are widespread ECM formers, have been found in recently deglaciated soils of circumpolar regions (Koljalg, 1995). Since these sporocarps are eaten by a wide range of soil invertebrates, there is the possibility that arthropods act as both long- and short-distance vectors of these fungi. Lilleskov and Bruns (2005) observed that the spores of *T. sublilacina* retain viability and can initiate mycorrhiza formation after passage through the intestines of a number of invertebrate species.

The proximity of established vegetation probably frequently complicates the successional process by providing local sources of mycorrhizal propagules. Allen (1988) found such undisturbed patches of the original plant communities in the centre of the main region of pyroclastic flow on Mount St Helens following its eruption. Evidence from species lists compiled for isolated islands formed by volcanic eruptions does little to clarify the picture, though non-mycorrhizal or facultatively mycorrhizal species, the latter often grasses, figured more prominently as early colonists on Krakatoa. These were succeeded by species likely to be more responsive to mycorrhizal fungi including orchids and a *Casuarina* spp. (Simkin and Fiske, 1983). Critical questions concerning the role of mycorrhizal colonization in facilitating establishment of plants in virgin sites can ultimately only be answered by manipulative experimentation, involving addition of plants to such sites and monitoring over a chronosequence the relationship between colonization nutrient capture, growth and survival.

Primary succession on sand-dune ecosystems is a more predictable process in which relationships between soil quality disturbance and mycorrhizal status are apparent (Read, 1989). In the disturbed and nutrient enriched conditions of the drift line, non-mycorrhizal species predominate, particularly those in the families Chenopodiaceae and Brassicaceae. Under more stable conditions, a succession of communities made up of species that are responsive to AM colonization are found, ranging from open grassland dominated by *Ammophila arenaria* on the fore-dunes, to herb-rich closed communities on the more stable dunes. The succession from drift line to stable back-dunes typically covers a gradient of decreasing pH and increasing soil organic matter content (Figure 15.4) over which communities dominated by plants with ectomycorrhiza or ericoid mycorrhiza become increasingly important, forest or heathland replacing grassland as the climax vegetation type.

Some doubts have been expressed about the validity of this model. On the basis of their observation that *Salix repens* retains a predominantly ECM association on calcareous and acidic dunes in the Netherlands, van der Heijden and Vosatka (1999) and van der Heijden (2001) concluded that this gradient-based model may not be widely applicable. There is no doubt that *S. repens* has the ability to colonize calcareous soil, although it is not normally a colonist of the mobile fore-dunes that are occupied by AM grasses. It remains likely that once it has colonized stabilized calcareous sand, the accumulation of its litter in the superficial layers will begin the process of acidification that drives the system in the direction favouring first ECM and then ERM plants. The compatibility of *S. repens* with both AM and ECM fungi (van der Heijden, 2001; see Chapter 6) may contribute to its ability to establish on calcareous sand and so to accelerate successional processes.

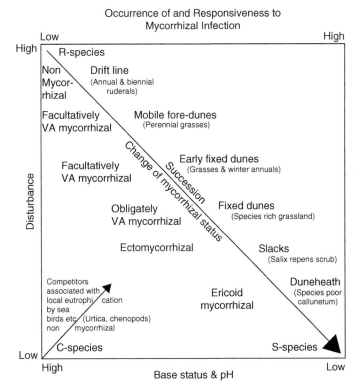

Figure 15.4 Diagrammatic representation of succession of mycorrhizal communities in a coastal sand dune, along an axis representing decreasing disturbance, pH and availability of mineral nutrients and increasing soil organic matter. C = competitor; R = ruderal; S = stress tolerant; VA = arbuscular (AM). From Read (1989), with permission.

The role of mycorrhiza in secondary successions

While primary successions often commence under conditions of nutrient impoverishment, those processes referred to as secondary succession, which follow disturbance of existing vegetation, are normally initiated in an environment of relatively greater enrichment, a pulse of N and P being produced by mineralization of residues left by the previous community (Walker and Syers, 1976). The disturbed soil is characteristically first occupied by weedy annuals, especially of families such as Chenopodiaceae, Brassicaceae and Polygonaceae, which are effective colonizers and which, since the early work of Stahl (1900), have been considered to be largely non-mycorrhizal. The conventional interpretation of the basis of the success of these ruderal plants is that as 'r' strategists they have a high fecundity, short generation time and an ability rapidly to exploit pulses of nutrient availability (Grime, 1979). However, the sensitivity of many such plants to the presence of AM mycelium or to competition from associated AM plants (see Chapter 16) raises the possibility that reduction of inoculum potential of these fungi, which is known to arise from disturbance (see Chapters 2 and 17), is an essential prerequisite for their success.

Decline of nutrient availability as the initial flush of minerals is utilized or lost by leaching leads progressively to competition between plants for resources and to a situation in which mycorrhizal colonization could be expected to provide a nutritional advantage to plants. A possible role for mycorrhizas in determining the trajectory of successional processes was acknowledged by Gorham *et al.* (1979) who proposed that plants characteristic of a particular stage of succession may have a higher 'affinity', through their fungal associates, for nutrients at a particular stage. However, few ecologists have considered mycorrhizas as possibly playing a pivotal role in the successional dynamics. The hypothesis that mycorrhizal colonization might provide hosts with a greater competitive ability and that this leads to acceleration of successional processes was tested by Allen and Allen (1988), who introduced inoculum of AM fungi to a high-altitude soil which had been disturbed by open-cast coal mining and was colonized largely by annual 'non-host' species. The presence of inoculum had the effect of reducing the growth of the ruderals and so, in some plots, led to increases in rates of succession. However, in others, loss of cover provided by the ruderals led to exposure-damage to those species, mostly grasses, that had the potential to respond to colonization. Consequently, the rate of succession declined (Allen, 1989). Experiments of this kind demonstrate the complexity of interacting factors that can influence the successional process and emphasize that above ground, as well as below ground, factors can affect plant response (see Chapter 16).

Much emphasis has been placed by ecologists upon secondary succession in old fields where progressive decrease of availability of N is believed to be the factor driving the process (Odum, 1960; Golley, 1965; Tilman, 1987). Following the ruderal phase, a succession of grass species with increasing ability to compete for N is recognized (Tilman, 1990). Although these grasses are likely to be colonized by AM fungi, the role of mycorrhizas in determining the outcome of competitive interactions between them appear not to have been considered until recently. There is much scope for work which includes the natural symbionts of these organisms. On theoretical grounds, because the decline in N availability arises partly through progressive inhibition of nitrification and replacement of mobile nitrate by relatively immobile ammonium as sources of mineral N (Robertson and Vitousek, 1981), advantages should increasingly accrue to mycorrhizal plants. In this context it is worth noting that the prairie grasses discussed above, were typical of habitats limited by N rather than P availability.

A response seen in some ecosystems to changing N status is the appearance, as transient occupants, of N_2-fixing shrubs and trees (van Cleve and Viereck, 1981). In the succession from grassland to boreal or temperate forest, other trees to appear early are members of the genera *Salix* and *Populus* which, in addition to having lightweight propagules that enhance their capacity for dispersal into successional environments, are characterized by a plasticity which enables them to form both arbuscular and ectomycorrhizas. Compatibility with AM fungi may be a factor facilitating their incorporation into a turf dominated by AM grasses or herbs, while associations with ECM fungi should be advantageous in the situation where a progressively greater proportion of the soil N is present as ammonium or in organic form.

It has been suggested (Read, 1993) that the change in N status of an ecosystem from one in which inorganic N predominates, to the later condition in which accumulating plant residues sequester N largely in organic form, may be the key factor selecting in favour of ECM trees in late stages of succession, be they members of the Fagaceae as in many temperate forests, or of the Pinaceae in boreal forests in the northern hemisphere. By the same logic, where for reasons of climatic stress, for example, at high elevation or latitude, growth of ECM trees is restricted, shrubs with ericoid mycorrhiza that also have the ability to mobilize nutrients from organic sources are favoured (see Chapter 11 and below).

As succession towards ECM forest or ERM heathland proceeds, there are inevitably stages during which cohorts of species with different types of mycorrhiza coexist. Indeed, even in stable forest communities, soil and light conditions may permit the persistence of a herbaceous understorey of plants with arbuscular mycorrhizas beneath a canopy of predominantly ECM trees. However, different patterns of root distribution can provide niche separation. Merryweather and Fitter (1995b) show that seedlings of the herb *Hyacinthoides non-scripta* germinating in the organic matter in ECM *Quercus* woodland are largely non-mycorrhizal. With time, the developing bulb, and the roots produced from it, descend into mineral soil where they develop arbuscular mycorrhizas in isolation from the largely surface-rooting trees. In effect, there are two separate communities, the constituent species of each of which, through their mycorrhiza, are exploiting different resources. Such differentiation can even be seen at the intraspecific level. Reddell and Malajczuk (1984) observed that *Eucalyptus marginata* plants formed AM associations when rooted in mineral soil, but ectomycorrhiza if grown in litter. Plasticity of this kind may be of particular value in fire-susceptible ecosystems of the kind in which *Eucalyptus* spp. occur, these being characterized by a cyclical pattern of accumulation and loss of organic resources due to fire.

If, as proposed by some ecologists (e.g. Clements, 1916; Odum, 1971; MacMahon, 1981), succession is a series of predictable processes, the trajectories of which are primarily influenced by nutritional constraints, a potential clearly exists for mycorrhizal colonization to play a significant role in determining both the rate and direction of the processes. The need, therefore, is for more field-based experiments which investigate the effects of manipulation of mycorrhizal status on the outcome of interaction between species at different stages of the succession. Only by these approaches can the real impact of the symbiosis upon the dynamics of the process be evaluated.

From what has been written earlier in this chapter, it is evident that some pattern can be recognized in the relationship eventually established in stable, climax communities, between biome and predominant mycorrhizal type. On this basis, it was proposed (Read, 1984, 1991) that the combination of climatic and soil factors found at any position along a gradient of latitude or altitude selects in favour of that mycorrhizal type having the functional attributes necessary to enable success of both partners in that environment (Figure 15.5). There is no doubt that, on a global scale, in the absence of disturbance, biome-related segregation of predominant mycorrhizal types can be seen even though a given type rarely if ever occurs to the exclusion of all others. The extent and nature, if any, of the involvement of the mycorrhizal symbiosis in determining these observed patterns remains to be investigated by experiment.

Mycorrhizas in Arctic, Antarctic and alpine biomes

The small number of species occurring, usually as individual plants, in the very high Arctic, in continental and sub-maritime Antarctic and in the nival zone of the Alps are only intermittently snow-covered. They grow in mineral soils that are

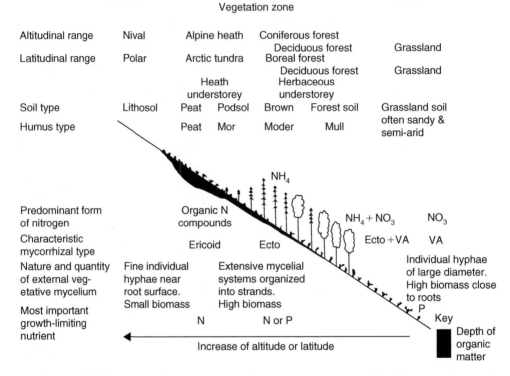

Figure 15.5 Diagrammatic representation of the postulated relationship between latitude or altitude, climate, soil and mycorrhizal type, together with the development of vegetative mycelium associated with mycorrhizas. VA = arbuscular (AM). From Read (1984), with permission.

constantly kept moist by melt water and appear to be largely uncolonized by mycorrhizal fungi. Haselwandter and Read (1982; Haselwandter *et al.*, 1983) found that one of the very few plants to occur in this zone of the Austrian alps, *Ranunculus glacialis*, appeared to be free of all fungal colonization. They concluded that this plant was adequately supplied with N and P contained in melt water during its short growing season and that physical rather than nutritional constraints determined its success in these habitats. Väre *et al.* (1992) reported that none of the six *Ranunculus* spp. examined in the high Arctic at Spitsbergen were colonized by AM fungi. This contrasts with the situation in this genus below the nival zone (Mullen and Schmidt, 1993) and in temperate latitudes (e.g. Harley and Harley, 1987a) where colonization by AM fungi appears to be the norm.

Only two vascular plants are native to mainland Antarctica. These are the graminoid *Deschampsia antarctica* and a member of the Caryophyllaceae, *Colobanthus quitensis*. They occur on thin soils ranging from skeletal gravels to those consisting entirely of acidic organic residues. Extensive analyses of these plants across a latitudinal gradient from 60° to 68°S (Upson, 2006) revealed that they were normally colonized by DS fungi and that, with the very few exceptions of root segments colonized by fungi of the 'fine endophyte' type, AM fungi were absent. This is consistent with the observation of Christie and Nicolson (1983) who could find no AM colonization on *D. antarctica* in the maritime Antarctic. They did, however, find AM colonization at less extreme sub-Antarctic sites in the Falkland Islands and South Georgia, as did Laursen *et al.* (1997) in another sub-Antarctic location, Macquarie Island, where AM colonization was recorded on 18 of the 40 plant species examined.

Below the nival zone of alpine areas and in those regions of the Arctic that have a consistent snow-free growing season, there can be patches of continuous vegetation cover and accumulation of organic matter, at least at the surface, is normal. In their study of *R. adoneus*, carried out at 3500 m in the alpine zone of the Colorado Front Range, Mullen and Schmidt (1993) showed that, whereas the plant was lightly colonized by coarse and fine AM endophytes throughout the year, arbuscules were present only during the short growing season. Their formation was followed by increases of P concentration in the shoots and roots. It was proposed that P acquired in this period was stored for use during growth and flowering the following spring, both of which occur before soils thaw to release nutrients. More studies of this kind, in which the dynamics of colonization and nutrient acquisition are followed through the year in the natural environment of the plant, are much needed. They provide not only ecologically relevant information, but heighten the need for physiological investigations to sort out the relative significances of arbuscules and hyphae in P transport.

Often, especially in those regions of the Arctic where the water table lies permanently near the surface, organic soils support tussock tundra dominated by a preponderance of species in families like the Cyperaceae and Juncaceae, which are typically non-mycorrhizal. These systems contrast with the heath tundra communities on somewhat drier soils that contain a higher proportion of shrubby ERM and ECM species (see below). While the prevalence of cyperaceous plants in tussock tundra may help to explain the small quantities of AM colonization present (Bledsoe *et al.*, 1990), it does not provide a full explanation, since again many species which are 'hosts' to AM fungi at lower altitudes, are uncolonized or only lightly so in Arctic-alpine situations. Even where arbuscular mycorrhizas are observed, they are often, as in the Antarctic, formed by 'fine' rather than 'coarse' endophytes, there being a progressive increase in colonization by *Glomus tenuis* with altitude (Crush, 1973; Haselwandter and Read, 1980; Olsson *et al.*, 2005). Nevertheless, the presence of 'fine endophytes' must not be dismissed because they can play highly significant roles in increasing plant P uptake (Crush, 1973; Smith and Smith, unpublished).

One striking feature to emerge from studies of plants growing at high latitudes and altitudes is the extensive occurrence on their roots of DS fungi (Haselwandter and Read, 1980; Read and Haselwandter, 1981; Christie and Nicolson, 1983; Currah and van Dyke, 1986; Kohn and Stasovski, 1990; Väre *et al.*, 1992; Treu *et al.*, 1996; Jumponnen *et al.*, 1998, Ruotsalainen *et al.*, 2002; Upson, 2006). In a study of 179 vascular plant species of Alberta, Currah and Van Dyke (1986) found that roots of 87% of alpine species were colonized by DS fungi, in contrast with only 9% in non-alpine situations.

The preponderance of fungi of this general type on plants growing in alpine situations is reflected in analyses of soil microflora, which suggest that in Antarctic (Heal *et al.*, 1967), Arctic (Väre *et al.*, 1992) and alpine (Haselwandter and Read, 1980) soils, fungi with DS hyphae dominate the soil microbial community. Recognition of their quantitative importance in these habitats and in other climatically or nutritionally stressed habitats has driven attention towards their possible taxonomic and functional status. Some progress has been made towards determination of their taxonomic positions, most of them being now recognized as ascomycetes of the

Dermateaceae (order Helotiales) (see Chapter 7). Among these, members of the genus *Phialocephala*, particularly *P. fortinii*, were prominent in alpine situations (Read and Haselwandter, 1981; Currah and van Dyke, 1986). Molecular analysis of the most widely occurring DS endophytes of the Antarctic species *Deschampsia. antarctica* and *C. quitensis* have confirmed the importance of taxa within the Dermateaceae in this biome, but have also provided evidence of host preference in the fungi involved (Upson, 2006). Whereas the dominant fungi on the roots of the grass *D. antarctica* had affinities to mollisioid taxa having close relationships with *P. fortinii sensu lato* (see Figure 7.4a), the roots of *C. quitensis* were preferentially colonized by *Cadophora* and *Leptodontidium* species (see Chapter 7).

Much less is known about the possible functions of DS fungi in polar and alpine regions. Haselwandter and Read (1982) isolated DS fungi from healthy, field-collected roots of the alpine sedges *Carex firma* and *C. sempervirens* and obtained a positive growth response in *C. firma* when the plant was inoculated and grown in sand with one of the isolates. However, since neither the substrate nor the climatic condition employed in the experiment reflected those of the alpine habitat, they urged caution in the interpretation of these responses, referring to them as being evidence of an 'association' rather than of a typical mycorrhizal relationship. Upson (2006) carried out an experiment of similar design but under more relevant climatic conditions (day temperature 6°, night temperature 4°C) in which *D. antarctica* was inoculated with DS isolates and grown on perlite supplemented with nutrient solutions containing ammonium or casein hydrolysate as the N source. The results varied according to both N source and fungal inoculant. Yields of the inoculated grass, relative to those of uninoculated controls, were either reduced or unaffected in the ammonium treatment. However, four of the DS isolates produced a significant increase of biomass when the plants were supplied with organic N source (Figure 15.6). These results confirm that there are circumstances in which *D. antarctica* can benefit from the saprotrophic activities of some of its DS associates, but the mechanisms whereby these effects are achieved and the extent to which they would be observed in nature, where the roots would be colonized by a larger number of DS taxa, remain to be elucidated.

Of equally uncertain status are the ECM-like structures which are frequently, but not consistently, found on roots of herbaceous species such as *Kobresia* (Fontana, 1963; Haselwandter and Read, 1980; Kohn and Stasouski, 1990) and *Polygonum viviparum* (Hesselman, 1900; Read and Haselwandter, 1981; Lesica and Antibus, 1986). Where they occur, these associations too are normally formed by fungi with dark mycelia. Among these, *Cenoccum geophilum* appears to be prominent, though the frequent presence of hyphae with clamp connections indicates that basidomycetous fungi may also be involved. There is a suggestion (Väre *et al.*, 1992) that these plants are more frequently colonized in this way when growing with typically ECM plants. Again, there is a need for experimental analysis of the status of these types of colonization.

There is concern that warming global climate will pose particular threats to Arctic and sub-Arctic ecosystems. Predicted temperature increases of up to 5°C in the Arctic (International Panel on Climate Change – IPCC, 2001) can be expected to lead to northward migration of ECM woody shrubs (Bret-Hart *et al.*, 2001; van Wijk *et al.*, 2004; Clemmensen and Michelsen, 2006; Clemmensen *et al.*, 2006). Associated changes in below-ground activities, including more rapid rates of mineralization of organic matter, might in turn lead to increases of fertility and the release of the

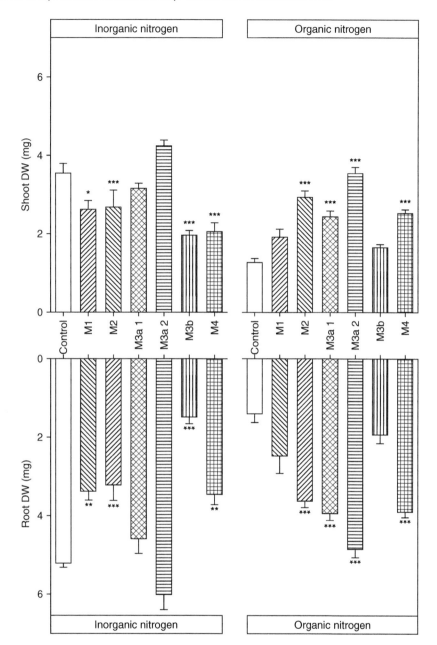

Figure 15.6 Shoot and root dry weight (DW) of *Deschampsia antarctica*, grown in perlite with an inorganic or organic nitrogen form and either (white) uninoculated or inoculated with DS isolates M1, M2, M3a (isolate 1 or 2), M3b or M4. Significant differences between inoculated and uninoculated plants tested by two-way ANOVA (GLM) and Tukey's pairwise comparisons are shown as *$P < 0.05$, ** $P < 0.01$, *** $P < 0.001$. Bars represent the mean (+S.E.) of 5 or 6 replicates. From Upson (2006).

approximately 14% of global soil organic C currently stored in Arctic ecosystems (Mack *et al.*, 2004). Since the ECM mantle and its associated extraradical hyphae occupy the interface between roots and this C reserve, more information on the possible interactions between these compartments is essential. Clemmensen *et al.* (2006) investigated the effects of simulated soil warming and fertilization by N, N and P, or NPK over a 14-year period on the development of ECM mycelium of *Betula nana* growing in tussock and heath tundra. Using ergosterol as a biomarker, they found that fertilizer application increased production of ECM extraradical mycelium at both sites and that temperature increase stimulated its production in the tussock tundra (Figure 15.7). It appears from these results that one consequence of warming and associated eutrophication will be an enhancement of cycling of both C and N in this previously strongly nutrient-limited ecosystem. There is also the possibility that warming will lead to changes in the population structure of ECM fungal communities in the Arctic. In detailed analyses of fungal isolates collected across a gradient extending from the Arctic, through boreal, into temperate forests, it was concluded that *Hebeloma* spp. showed significant physiological adaptation to their local environments (Tibbett *et al.*, 1998a, 1998b). When compared with temperate and boreal forest isolates, the species and races from more northerly latitudes had greater nutrient mobilizing capabilities at low temperatures. These low temperature-adapted fungi may either be displaced northwards as soils warm or, if suitable habitats are not available, driven to extinction.

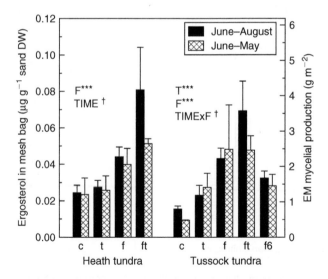

Figure 15.7 Production of external ectomycorrhizal (EM) mycelia associated with *Betula nana* measured as ergosterol concentration in sand-filled ingrowth bags after incubation in the field for one growing season (June–August) or 1 year (June–May) (mean ± 1 SE) in (c), control; (t), temperature enhanced; (f), fertilized; or combined f and t plots after 14-year treatment at a heath tundra in sub-Arctic Sweden and at a tussock tundra in Arctic Alaska. Significant main factor effects (T, temperature enhancement; F, fertilization; TIME, incubation period) and interactions are indicated: ***, $P < 0.0001$; †, $P < 0.10$. At tussock tundra, additional plots fertilized for 6 years (f6) are not included in statistical analyses. From Clemmensen *et al.* (2006).

Mycorrhizas in heathland

Heathlands occur as major biomes under two environmental circumstances. In the first, the upland or maritime heath is found in both continental and island locations at a distinct altitudinal position between the alpine zone and the tree line (Figure 15.8). The second, lowland heath, occupies areas of particularly impoverished acid soil at low elevation. These biomes are characterized by the presence of shrubby, sclerophyllous, evergreen plants of the family Ericaceae, all of which normally have hair roots colonized by ERM fungi (Read, 1983; see Chapter 11).

Analysis of environmental gradients across which plants with ericoid mycorrhizas become increasingly prevalent has shown that such communities arise primarily in response to nutrient impoverishment (Specht, 1981; Rundel, 1988; Read, 1989). Their occurrence in warm Mediterranean climate zones as 'dry-heath' or 'sand-plain' formations, as well as in sub-alpine environments, serves to emphasize the fact that nutritional rather than climatic factors play the primary role in determining the distribution of these communities. The response to low availability of N and P is to allocate increasing proportions of fixed C to the structural components lignin and cellulose, rather than to molecules rich in protein or P, a process which leads directly to sclerophylly (Specht and Rundel, 1990) and to the release of residues of high C:N ratio and considerable recalcitrance. These accumulate at the soil surface to provide the matrix in which ERM roots proliferate.

In northern heaths, the hair roots of dominant plants such as *Calluna vulgaris*, *Erica* spp. and *Vaccinium* spp. are characteristically confined to the top 10 cm or less of the soil profile, where they are closely associated with the litter (Reiners, 1965; Gimingham, 1972; Persson, 1980). Interestingly, when herbaceous species such as *Molinia caerulea*, *Eriophorum vaginatum* and *Carex* spp. coexist with ericaceous shrubs,

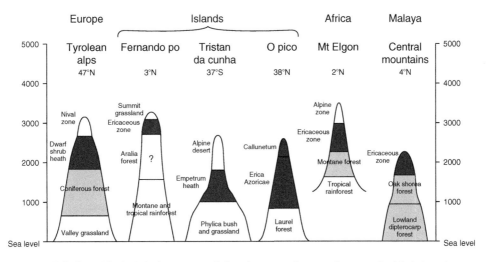

Figure 15.8 Simplified global pattern of distribution of major biomes, highlighting the segregation of predominant mycorrhizal types in association with distinctive types of plant community. Mycorrhizal types: black shading, ericoid; grey shading, ECM; no shading, AM.ecto. Note that in the dipterocarp forests there will be important canopy and understorey plants that are AM (see text). From Read (1993), with permission.

their roots are concentrated at greater depths in the soil profile (Gimingham, 1972), so the two groups of plants are not competing for the same resources. Indeed, the grasses may be colonized by AM fungi, again emphasizing separate strategies of resource acquisition. Read (1993) presented a schematic view of the manner whereby distinctive mutualisms, together with modifications of root distribution and anatomy, might promote species diversity in northern heaths by enabling exploitation of different sources of the critical growth limiting element N. In this (Figure 15.9), the coexistence of ericaceous, leguminous and carnivorous species, typically seen in heaths of moderate acidity, was facilitated by their abilities to use sources of N derived, respectively, from soil organic matter, the atmosphere and captured animals. Structural modifications, in particular production of aerenchyma, enables roots of cyperaceous species like *E. vaginatum* to penetrate water-logged horizons where they exploit N sources untapped by the other groups that are essentially surface rooting.

Questions have been raised concerning the extent to which non-mycorrhizal plants may compete for organic N with co-occurring ERM and ECM fungi in heathland and heath tundra environments. Whereas there is no doubt that some of these non-mycorrhizal plants have the potential to assimilate simple forms of organic N (Chapin *et al.*, 1993; Nasholm *et al.*, 1998), direct comparisons between these and species with ERM and ECM roots suggest that their abilities to do so are relatively

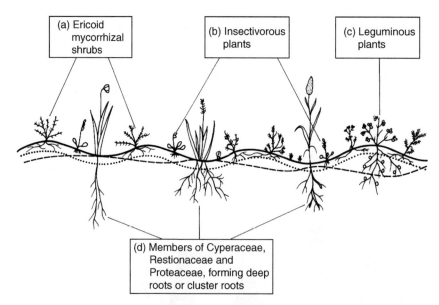

Figure 15.9 Schematic representation of compartmentation of resource acquisition in heathland ecosystems, based on the occurrence of distinctive mutualistic associations or root specializations: (a) Ericoid mycorrhizas occur in dwarf shrubs and play an important role in the mobilization of N in plant litter and microbial protein. (b) Insectivorous plants capture insects, from which they release N. (c) Leguminous plants form nodules which fix N$_2$. (d) Members of the Cyperaceae, Restionaceae and Proteaceae either produce deep roots that tap N in low soil horizons and/or proteoid or cluster roots that are important in capture of nutrients (particularly P, but also possibly N in the surface horizons). From Read (1993), with permission.

limited (Emmerton *et al.*, 2001b). When supplied with ^{15}N-labelled glutamic acid and glycine under aseptic conditions, the ERM plant *Vaccinium vitis-idaea* and the ECM *Betula nana* assimilated these N sources readily, whereas their non-mycorrhizal counterparts *E. vaginatum* and *Luzula wahlenbergii* captured very limited amounts of N from them. It has been confirmed that the ability of *V. macrocarpon* to assimilate amino acids so effectively is expressed only when the plant is growing in the ERM condition (Read *et al.*, 2004). In any event, it must be recognized that because small organic molecules like amino acids can be readily assimilated or mineralized by the assemblage of soil saprotrophs, they will have short half lives in the soil. Under these potentially competitive circumstances, the advantages conferred upon ERM (and some ECM, see Chapter 9) plants by their extensive fine root and extraradical mycelial systems are likely to be important.

Irrespective of differences between non-mycorrhizal plants and those colonized by ERM and ECM fungi in terms of their abilities to compete for and assimilate simple organic N compounds, it is likely that the direct attack on macromolecular organic N sources facilitated by proteolytic activities (see Chapters 9 and 11) will be the distinguishing advantage for ERM and ECM plants when growing in competition with species lacking this attribute (Read *et al.*, 2004).

The extent to which δ^{15}N signatures of plant shoots can provide indications of the extent of organic N use by different mycorrhizal types has been investigated. Comparative analysis of δ^{15}N-enrichment of leaf tissues of ERM (*Vaccinium vitis idaea*), ECM (*Picea mariana*) and AM (*Calamagrostis canadensis*) species all growing in the tundra heath-boreal forest transition zone of Alaska revealed significant differences of enrichment between these species. *P. mariana* had a significantly lower δ^{15}N value (-6.496) than *V. vitis idaea* (-3.837) and *C. canadensis* ($+0.585$) (Schultze *et al.*, 1994). Distinctive rooting depths may have contributed to these differences, but the possibility of discrimination facilitated by ECM, ERM and AM colonization, respectively, was suggested as a possible basis for the observed effects. Some support for the latter possibility was obtained in a field-based study of sub-Arctic plants (Michelsen *et al.*, 1996). The extent of mycorrhizal colonization was examined in fellfield and heathland communities before δ^{15}N enrichment of leaf tissue was determined in plants of known mycorrhizal status, representative of ERM, ECM, AM and non-mycorrhizal categories. In the fellfield, the mean δ^{15}N of the ERM species was -5.5, that of ECM species was -4.1 and of AM or non-mycorrhizal species zero. In the heath, the mean δ^{15}N values of the same groups were -7.6, -6.4 and -1.8, respectively. In all cases the values obtained from the ERM and ECM plants were significantly different from those of the AM or non-mycorrhizal species. Though the differences between ERM and ECM species were not significant ($P = 0.051$ in fellfield and 0.270 in heath), the ECM plants appeared to occupy an intermediate position in the hierarchy. It was concluded that ERM and ECM colonization facilitates access to organic sources of N which were not only the predominant forms of N in the soil but also had significantly lower levels of δ^{15}N enrichment.

More recent work has cast doubt on the extent to which δ^{15}N signatures of shoots can provide information concerning the sources of N assimilated by ERM roots. Using defined N sources of known δ^{15}N enrichment, it has been shown that considerable fractionation of N isotopes occurs in the course of assimilation both by the fungi in pure culture (Emmerton *et al.*, 2001a) and by the mycorrhizal plants (Emmerton *et al.*, 2001b). Of greatest interest in this connection was that root, shoot and whole plant ^{15}N abundance values frequently showed significant differences from those of the source N compounds supplied to the mycorrhizal plants. Furthermore, there were differences between mycorrhizal categories in the extent to which N fractionation occurred in the fungus–root–plant pathway. The interpretation of δ^{15}N data is further discussed in the context of boreal forest biomes later in this chapter.

The ability of ERM fungi to release a wide range of polymer-degrading enzymes and so to facilitate mobilization of the N and P contained in organic resources of various kinds has been emphasized earlier (see Chapter 11). From the biome perspective, the most important attribute of the ericoid mycorrhizas could be the provision of direct access to the detrital remains of the dominant components of the system. Berendse (1994) emphasized the role of litter quality as a determinant of fitness in heathland plants, but envisaged saprotrophs as being the agents facilitating this recycling. Experiments demonstrating the abilities of ERM fungi to release N from residues of their own mycelia and those of ECM fungi (Kerley and Read, 1997), as well as from necromass of their autotrophic associates (Kerley and Read, 1998), have confirmed their potential to effect significant decomposer activities. Whereas ERM plants of *V. macrocarpon* were able to recover 40% of the N contained in *Vaccinium* necromass over 60 days, non-mycorrhizal plants recovered only 5% over the same period (see Chapter 11). Obviously, detrital material produced aseptically by desiccation is not a precise surrogate for that found in nature, but experiments of this kind provide some evidence that ERM fungi can be expected to contribute to the tightly coupled recovery of nutrients from the otherwise recalcitrant residues of their host plants.

In view of the prediction that Arctic and sub-Arctic biomes, including heaths, may be subject to particularly marked temperature increases over the next century (see above), Olsrud *et al.* (2004) have examined the impacts of elevated temperature and CO_2 concentration on aspects of ERM root colonization. They found that the extent of ERM colonization increased in association with greater rates of photosynthesis under conditions of both higher temperature and elevated CO_2. However, in CO_2-enriched plots, leaf N contents of the ericaceous species were reduced. Olsrud *et al.* (2004) propose that declining availability of inorganic N under conditions of elevated CO_2 will limit impacts upon productivity of ericaceous ecosystems.

Whereas plants with ERM are well adapted to the organic N-enriched soils of high latitudes, they have responded negatively to the anthropogenic enhancement of nitrate and ammonium availability seen in industrialized regions further south (Aerts, 2002). The widespread replacement of *Calluna*-dominated heaths by grasslands composed of more productive AM-associated species can be ascribed to a combination of factors, among which the greater competitive ability of the herbs under N-enriched conditions and the failure of ERM systems to assimilate nitrate effectively are likely to be prominent (Aerts and Bobbink, 1999; Aerts, 2002; Read *et al.*, 2004).

Pate and Hopper (1993) point out that the great species diversity of the epacrid-containing heathlands of the southern hemisphere is in striking contrast to the situation seen in northern heaths which are often dominated by monospecific stands of ericaceous species. It appears that with the diversity comes a commensurate increase in the range of root specializations (Lamont, 1982, 1984; Pate, 1994; Lambers *et al.*, 2006), suggesting that niche separation will be even more marked in these systems than it is in less stressed environments. Among the distinctive plants of southern heaths, and some other vegetation types growing on severely P-limited soils, members of the non-mycorrhizal Proteaceae, Restionaceae, as well as mycorrhizal *Casuarina*, are characterized by the production of 'proteoid' or cluster roots. These structures are formed by a highly specialized branching of lateral roots, so that a dense mat of root-lets of narrow diameter is formed (Lamont, 1984; Lambers *et al.*, 2006). The clusters are transitory, undergoing growth, maturation and senescence over a period of around 2–3 weeks, at least under the artificial experimental conditions. The developmental changes are accompanied by distinct changes in excretion of organic acids which have been shown to play a significant role in increasing the availability of P, and probably also of Fe, tightly held in inorganic forms (Dinkelaker *et al.*, 1995; Comerford, 1998; Lambers *et al.*, 2006). It has also been suggested that the location of the cluster roots, near to the soil surface and sometimes in the litter layers, may increase capture of nutrients leached through the soil. Present evidence suggests that formation of cluster roots is an adaptation to extreme soil P limitation. Indeed Lambers *et al.* (2006) show that members of the Proteaceae occur in extremely P-impoverished soils in Western Australia, which apparently do not contain enough P in forms available to plants with AM associations. The latter are found, in the same geographic region, on soils which are still P-limited, but not extremely so. Ericaceous species in the same environments should have the potential to exploit organic residues, via the activities of their associ-ated ERM fungi, whereas insectivorous and N_2-fixing species make use of other adap-tations (see Figure 15.9). It is worthy of note that *Casuarina* seems to have evolved ways of maximizing acquisition of scarce resources in many forms, forming AM and sometimes ECM symbioses, as well as having N_2-fixing nodules and cluster roots.

Stewart *et al.* (1993) provide evidence of some functional segregation with respect to N nutrition in ecosystems able to support plants with a range of strategies. In the fire-prone habitats that they studied, nitrate was relatively abundant immedi-ately after burning, although some ammonium was also present. Sufficient nitrate reductase activity (NR) was observed in shoots and roots of three proteaceous gen-era (*Banksia, Petrophile* and *Stirlingia*) to suggest that they would assimilate nitrate in nature when available. In contrast, two epacrid species, *Astroloma macrocalyx* and *Conostephium pendulum*, both of which have ericoid mycorrhizas, showed barely detectable NR, even after feeding with nitrate via the transpiration stream. These species would be dependent on assimilation of ammonium, which was always present in soil and predominated at sites not burnt for several years, or on utiliza-tion of organic N via ERM associates.

Other plant families with widespread representation in epacridaceous heaths, but which are absent or of little importance in the northern hemisphere, are the Rutaceae, Dilleniaceae and Compositae, most members of which would be expected to be colo-nized by AM fungi. This type of mycorrhiza has been reported, for example, in *Boronia* (Rutaceae), *Hibbertia* (Dilleniacea) and *Helichrysum* (Compositae) (Lamont, 1984). These plants have less fibrous root systems than epacrids and, in addition, penetrate the sandy soils more deeply (Dodd *et al.*, 1984; Pate, 1994), a feature providing spatial sepa-ration of root activity. Further, while AM colonization will enhance their ability to scav-enge for P, members of two of the genera, *Helichrysum* and *Hibbertia*, have been shown to develop significant NR activity (Stewart *et al.*, 1993), again suggesting the likeli-hood of nutritional as well as spatial niche differentiation between these plants and those with ericoid mycorrhizas. However, the demonstrations that AM colonization may increase uptake of ammonium and especially in dry soils of nitrate may be relevant here. It is in just such habitats as these that N acquisition is important (see Chapter 5). The presence of these physiological patterns suggests that selection favouring a range of specializations has been an important factor enabling the coexistence of taxonomically distinct species in Australian heathland systems, where the greatest diversity is found on the least fertile soils (Pate and Hopper, 1993). Tilman's equilibrium model of plant competition (1982, 1988) predicts that, in resource-poor environments, diversity will be low because few species can tolerate extremes of nutrient deprivation. In the sand-plain heaths it appears that, on the contrary, selection of distinctive mutualisms and nutrient acquisition strategies over very long periods of evolution can facilitate coexistence of species in very diverse assemblages. Indeed, recent work, admittedly in artificial systems, suggests that mycorrhizal colonization may be of greatest advantage in competition where resources are poor and plant density low (see Chapter 16).

Mycorrhizas in boreal and temperate forest biomes

Communities of ECM trees, in particular members of the Pinaceae, Fagaceae, Betulaceae and Salicaceae, are the natural dominants of the boreo-temperate biomes of the world. These are the world's largest vegetation system. They stretch as a continuous 1000–2000 km wide circumpolar belt around the Northern Hemisphere (Odum, 1971). As in northern heathlands, the diversity of host species in these ECM forests is characteristically low but, in contrast to the heathland situation, there is a very great diversity of fungal symbionts associated with the plants (see Table 6.2, Chapter 6). The cool climates prevailing across much of the biome lead to low rates of both evapotranspiration and decomposition, with the result that acidic organic residues of plants accumulate either as raw humus in superficial layers of the soil or as peat deposits which can be of considerable depth. Under these circumstances, dwarf shrubs with ericoid mycorrhiza are commonly found as understorey components. However, within this predominantly ECM habitat there are areas in which base enrichment of the soil can provide conditions suitable for the development of plants with AM colonization. These may occur as dominants in the case of forests of *Acer* and *Fraxinus*, as understorey shrubs, or as a herb layer (Brundrett and Kendrick, 1988, 1990a, 1990b; Giesler *et al.*, 1998; Högberg *et al.*, 2003). The changing nutritional circumstances driving change of mycorrhizal type along such gradients are discussed below.

An understanding of the pattern of distribution of roots within the soil profile is important because this determines the nature of the substrates that are accessible to them and, in particular, to their fungal symbionts. In the podzolic and peaty soils, which characterize the boreal zone, as well as in the moder and mull soils of the temperate zone, ECM roots are predominantly found in the superficial organic layers (Meyer, 1973; Harley, 1978; Persson, 1978; George and Marschner, 1996). In the case of the peats which cover so much of the boreal zone, anaerobiosis associated with a high water table and permanent or winter freezing of the deeper layers, restricts the development of mycorrhizal roots, which are essentially aerobic in their requirements, to superficial layers. In black spruce (*Picea mariana*) forests formed on soils of this kind in interior Alaska, almost 100% of first-order fine roots were ECM and 84% of their production occurred within 20 cm of the surface (Ruess *et al.*, 2003). In a sandy podzolic system supporting pine (*Pinus sylvestris*) with an ericaceous understorey,

Persson (1983) observed that the presence of the shrubs influenced the distribution of ECM roots. Within the surface organic (O) horizon, fine roots were largely concentrated in the superficial L (litter) and F (fermentation) layers where the trees grew in the absence of understorey shrubs. However, in areas where they were present, ECM proliferation was depressed by a few centimetres towards the F-H (humus)–mineral soil transition.

An analysis of the distribution of *Picea abies* roots across a latitudinal gradient through Europe from the boreal to the temperate zones (Stober *et al.*, 2000) revealed that the bulk of the fine roots were located in the top 10 cm of the soil profile in every stand. However, there was some evidence of a greater proliferation in the organic horizons of the boreal (93%) than in two temperate sites (75% and 78%). Further, whereas most of the fine roots in the upper 10 cm of the profile were living, those in the deeper layers were mostly dead.

The advance of molecular technologies has now made it possible to combine analysis of fine root distribution with characterization of their fungal associations. Rosling *et al.* (2003), working in a boreal *P. sylvestris–P. abies* forest, confirmed the earlier observations that the superficial organic layers of the podzol are the most intensively exploited by fine roots (Figure 15.10). However, they also showed that considerable numbers of ECM roots occur in the mineral horizons and that half of the identified fungal taxa were associated with these substrates (Table 15.3). The mineral horizons most extensively occupied by ECM roots were the eluviated

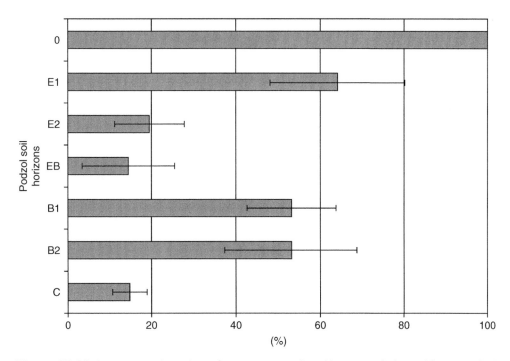

Figure 15.10 Average total number of root tips in each soil horizon of a boreal forest podzol (O = organic; E1, E2, EB = elluviated; B1, B2 = illuviated; C = subsoil) expressed as the percentage of total number of root tips in the organic horizon. From Rosling *et al.* (2003).

Table 15.3 Vertical distribution of ectomycorrhizal fungal taxa in a podzol soil profile.

	O	E1	E2	EB	B1	B2	C
Tylospora spp.	X	X	X	X		X	X
Cortinarius spp.	X		X	X	X	X	X
Piloderma reticulatum	X	X	X	X		X	X
Piloderma spp. JS15686	X	X			X	X	
Piloderma byssiium	X						
Inocybe	X						
Tomentellopsis submollis	X	X					
Piloderma fallax	X	X	X				
Hygrophorus olivaceoalbus	X	X	X				
Russula decolorans	X	X	X	X			
Dermocybe spp.	X	X	X	X			
Tomentelloid		X					
Lactarius utilis		X	X	X	X		
Piloderma spp.2		X	X	X	X		
Piloderma spp.3			X	X	X		
Piloderma spp.1			X	X	X		
Suillus luteus		X	X	X	X		X
unID $\neq$ 15						X	
unID $\neq$ 12						X	
Wilcoxina							X
Russula adusta							X
Tricholoma portentosum							X

Data from Rosling *et al.* (2003). O, organic horizon; E1 and E2, elluviated horizon; B1 and B2, illuviated horizons; C, subsoil; unID, unidentified.

(E1), located immediately below the O horizon and both of the iluviated (B) layers. These are regions from which organic materials are leached (EI), or into which they may be precipitated (B1 and B2). It is of particular interest that the fungal communities occupying the organic and mineral horizons are distinct (Table 15.3). While *Dermocybe* spp., *Tomentellopsis submollis*, and three *Piloderma* spp. were found predominantly in the organic horizon, *Suillus luteus, Lactarius utilis* and three different species of *Piloderma* were associated with the mineral layers.

Further recent studies, one using DNA analysis of hyphal fragments recovered from the L, F, H and B horizons of a pine forest soil (Dickie *et al.*, 2002), another a combination of sequencing with morphotyping in boreal forest substrates of different quality (Tedersoo *et al.*, 2003) and one using morphotyping alone (Koide and Wu, 2003), indicate clustering of distinctive types of ECM fungi in specific niches of coniferous forest soils. In the study of Tedersoo *et al.* (2003), a strong preference of resupinate thelephoroid and atheliod fungi for woody debris was observed.

While proliferation of the mycelia of individual fungi in association with substrates of a particular resource quality have been described previously (Carleton and Read, 1991; Agerer, 1991b; Perez-Moreno and Read, 2000, see below), interspecific differences in occupancy of soil horizons and substrates had not been fully appreciated until recently. It may be of considerable importance if it turns out that these patterns reflect different potentials for exploitation of the different substrates (see below). Lindahl *et al.* (2007) have used DNA-based methods in combination with ^{14}C dating of the organic matter, measurements of C:N ratios and of δ^{15}N natural

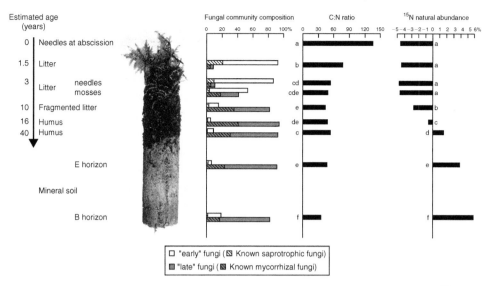

Figure 15.11 Fungal community composition, carbon:nitrogen (C:N) ratio and ^{15}N natural abundance (α^{15}N) throughout the upper soil profile of a Scandinavian *Pinus sylvestris* forest. Different letters in the diagram indicate statistically significant differences between the horizons in C:N ratios and ^{15}N abundance, and the standard error of the mean was <0.3% for ^{15}N natural abundance and <3 for C:N ratio (n = 19–27, for recently abscised needles, n = 3). The age of the organic matter is estimated from the above average Δ^{14}C of three samples from each horizon (five samples of the litter 2 (needles) fraction) and needle abscission age (3 years) is subtracted. Community composition data are expressed as the frequency of total observations. 'Early' fungi are defined as those occurring with a higher frequency in litter samples compared with older organic matter and mineral soil. 'Late' fungi are those occurring with a higher frequency in older organic matter. From Lindahl *et al.* (2007).

abundances to discriminate between different functional groups of fungi involved in C and N cycling down a boreal forest podsol profile. A clear shift in fungal community composition was identified between the surface litter (L horizon) and the underlying F horizon in which the litter had lost its structural integrity (Figure 15.11). The recently-shed litter (less than 4 years old) near the surface of the profile was dominated by saprotrophic fungi, among which non-mycorrhizal ascomycetes of the Helotiales were prominent. This was a zone in which organic C was mineralized and N was retained. ECM fungi, members of a group referred to as 'late fungi' (Figure 15.11), first appeared as a major component of the fungal community in older fragmented litter and humus in the fermentation horizon. The humus zone was characterized by small increases of C:N ratio with age of organic matter, indicating removal of N by organisms using root-derived C. Selective exploitation of N is suggested by the large increases of δ^{15}N at this level. This enrichment of ^{15}N is thought to be driven by fractionation against the heavier isotope during transfer of N from soil through ECM fungi to their host plants (Högberg *et al.*, 1996, 1999; Lindahl *et al.*, 2002; Berg and McClaugherty, 2003; Hobbie and Colpaert, 2003; Hobbie *et al.*, 2005). Among the first of the ECM fungi to appear in the profile were *Cortinarius* spp., which selectively colonized dead moss shoots. Selective exploitation

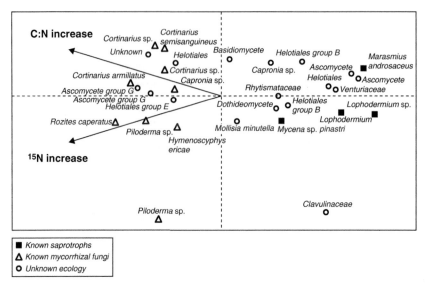

Figure 15.12 Canonical correspondence analysis depicting the relationship between fungal community composition and changes in carbon:nitrogen (C:N) ratio and ^{15}N natural abundance in a boreal forest soil. The arrows represent increasing C:N ratio or ^{15}N abundance with depth and age of the organic matter. ■, known saprotrophic taxa; △, known ectomycorrhizal taxa; ○, taxa with unknown ecology. From Lindahl *et al.* (2007).

of senescent parts of bryophyte shoots by ECM fungi has been reported previously (Carleton and Read, 1991, see Figure 15.16; Genney *et al.*, 2006) and may be particularly important in those boreal systems characterized by a dominance of feather mosses in the forest floor community. As in the case of the Rosling *et al.* study, the species composition of the ECM assemblage changed in the transition from organic to deeper mineral horizons. In a canonical correspondence analysis in which shifts of fungal community composition were compared to changes in the C:N ratio and ^{15}N content of the soil horizons (Figure 15.12), it was shown fungal taxa known to be ectomycorrhizal were most strongly associated with positive changes of C:N and large increases in ^{15}N natural abundance. Saprotrophic fungi, in contrast, were associated with negative changes in C:N and only small changes of ^{15}N natural abundance.

Despite the complexities involved in interpretation of tissue δ^{15}N signatures, data that have emerged from recent studies enable some cautious generalizations to be made about the likely patterns of use of soil N sources both within the ECM functional group and between this group and saprotrophic fungi. These are based upon awareness that δ^{15}N fractionation in the course of mineralization of soil N leads to a depletion of δ^{15}N in the mineral sources of the element and an enrichment in the recalcitrant organic N compounds of the humus (Nadelhoffer and Fry, 1994; Kohzu *et al.*, 1999). Some comparative analyses of the ECM and the saprotrophic groups have suggested that the former are normally enriched in δ^{15}N relative to the latter (Hobbie *et al.*, 1999, 2001; Kohzu *et al.*, 1999), the implication being that ECM fungi are using relatively more of the recalcitrant N source. Deeper insights have been gained from studies in which the $\delta^{15}N$ signatures of the substrates have been directly compared with those of the fungi growing upon them (Gebauer and Dietrich, 1993; Gebauer and Taylor, 1999). These suggest that gross distinctions between N-use patterns of ECM and saprotrophic groups on the basis of their different $\delta^{15}N$ signatures may be oversimplistic. Comparing $\delta^{15}N$ signatures of soil components considered to be supporting mycelia of different fungi with those of their sporocarps, Gebauer and Taylor (1999) identified four functional groups of fungi in a Central European mixed forest environment. ECM fungi known to be capable of using organic N, and saprotrophs capable of using humus N made up two of these groups, both of which had sporocarps enriched in $\delta^{15}N$. The remaining two consisted of ECM fungi lacking the ability to use polymeric N sources and saprotrophs growing on decomposing wood. In both of these, sporocarps showed depletion of $\delta^{15}N$. The suggestion is clearly that while the $\delta^{15}N$ signature reflects that of the N source used, it may not provide an adequate discrimination between ECM and saprotrophs as functional groups.

Further evidence in support of the view that the $\delta^{15}N$ signatures of ECM sporocarps reflected those of the substrate has been provided by comparing relationships between these signatures measured in field-grown specimens and the ability of the same fungi to use organic N when grown in pure culture (Lilleskov et al., 2002a). Taxa such as *Tricholoma inamoenum*, which grew readily on protein and which have been shown to be sensitive to the presence of high levels of inorganic N in soils (Lilleskov et al., 2001), had the highest sporocarp $\delta^{15}N$ contents, while those, including *Laccaria* spp., which are relatively insensitive to high mineral N inputs and failed to use organic N in culture, showed the lowest levels of $\delta^{15}N$ enrichment.

It appears from data presented in Chapters 9 and 10 that ECM fungi isolated from boreal and temperate habitats are little different from saprotrophs in their preferences for and abilities to assimilate the mineral forms of N and P. When mineral forms are present in the soil solution the major function of the mycobiont is to facilitate their capture by providing a dynamic and extensive absorbing system in the form of the extraradical mycelial network. However, it has become increasingly evident that these elements and, in particular, mineral N are often present in such low amounts in the soils of these systems as to limit their productivity (Tamm, 1991). As a consequence of this recognition, increasing attention has been paid over the last two decades to the possibility that ECM fungi of boreal and temperate biomes may have the abilities to degrade some of the macromolecular organic residues present in their immediate natural environments, so gaining access to additional sources of N and P (Read and Perez-Moreno, 2003).

Attempts to deduce the wider functional capabilities of ECM fungi have relied heavily upon analysis of their abilities to mobilize more complex substrates as expressed in pure culture (see Chapters 8, 9 and 10). Because the readily culturable fungi which have been used for these studies represent only a tiny minority among the many species now known to be able to form mycorrhizas, circumspection is required when attempting to interpret the results obtained from them. Nonetheless, it is evident that some abilities to degrade structural polymers representative of those present in plant and fungal residues have been detected. Activities of these kinds can be expected to expose N and P sources that are locked into or protected by the polymeric carbon frameworks. However, the abilities of ECM fungi to perform decomposer functions should not be exaggerated. When comparisons are made between this functional group of fungi and those representative of the ericoid symbiosis (Bending and Read, 1996a, 1996b; see Chapter 11) or saprotrophs (Maijala *et al.*, 1991; Colpaert and van Laere, 1996; Colpaert and van Tichelen, 1996), the abilities of ECM fungi to depolymerize complex C sources are invariably lower than those of the other groups. The failure of most ECM fungi to penetrate the cell walls of their autotrophic partners may be a reflection of their low cellulolytic capabilities. Furthermore, claims that genes indicative of ligninolytic potential have been detected in ECM fungi (Chen *et al.*, 2001) have now been retracted (Cairney *et al.*, 2003). There remains the possibility that some of the so-far unculturable ECM fungi will be shown to express ligninolytic activity, but the likelihood would seem to be that, after making the switch from the saprotrophic to the mycorrhizal habit, these capabilities were lost (see Chapter 6).

On the basis that ECM roots and their associated mycelia proliferate selectively in organic layers present immediately below the surface litter horizons of boreal and temperate forests, several experiments have now investigated the extent of nutrient mobilization from these types of substrate. Entry *et al.* (1991a) compared N and P contents of organic matter colonized by the mat-forming fungus *Hysterangium setchelii* with those of adjacent uncolonized material. They observed a greater than 30% reduction of both elements in the organic residues colonized by the fungus (Table 15.4). Laboratory mesocosm experiments (Bending and Read, 1995a) (see Colour Plate 6.2) investigating the nutrient mobilizing properties of *Suillus bovinus* and *Thelephora terrestris* growing from pine (*P. sylvestris*) to forage in pine litter collected from the FH horizon-derived, revealed that both fungi were able to export some N and P from the organic matrices but that there were interspecific differences between them (Table 15.4). *S. bovinus* exported significantly more of both elements than *T. terrestris*. When these properties were again examined using *Betula pendula–Paxillus involutus* systems and a range of litter types, a marked exploitation of litter P resources but little reduction of their N contents was observed (Perez-Moreno and Read, 2000) (Table 15.4). Such differences between studies may reflect interspecific variation in the abilities of ECM fungi to mobilize N from litter or differences in the incubation times used in the different experiments. They emphasize that generalizations from one mycorrhizal system and substrate type to another must be made with great caution.

Koide and Wu (2003) buried residues of the L, F and H layers in a *Pinus resinosa* plantation and observed that a decrease of C:N ratio was still evident after 16 months of incubation. As they emphasized the duration of the C:N shift and the stage at which the potential for N release is greatest will be influenced by local environmental factors. Among these, availability of moisture, which in some sites can be reduced by the activities of the roots themselves, will be a significant factor affecting the extents and rates of microbial activity. While this activity is likely to be higher under the permanently moist and sometimes elevated temperatures of laboratory incubations, the mesocosm studies confirm the presence of pathways leading both to removal of N and P from litter by ECM fungi and to the increases in C:N and C:P ratios that arise as a consequence. Effective N and P removal during decomposition is indicated by the extremely small residual levels of these elements found in the deeper humic layers of these forest soils (Stevenson, 1982).

These observations leave open to question the identities of the N- and P-enriched substrates that are exploited by the mycorrhizal fungi in and around the FH layer. It is widely accepted that organic sources of N released into the humic soils will be rapidly co-precipitated with polyphenolic materials (Handley, 1954; Northup *et al.*, 1995)

Table 15.4 Nutrient mobilization expressed as percent loss of nitrogen (N) and phosphorus (P) from different organic natural substrates by ectomycorrhizal fungi grown in association with different host plants (bold characters) and in parallel controls with mycorrhizal mycelium absent or very weakly developed.

Type of substrate	Nutrient mobilization (%)			Host plant – ectomycorrhizal fungus combination	Reference
	N	P	Time (days)		
Plant detrital materials					
Douglas fir litter	**32**	**33**	**365**	Pseudotsuga menziesii–mats of Hysterangium setchellii	Entry et al., 1991b
Douglas fir litter	16	19	365	Control (no host plant)–mycorrhizal hyphal mats absent	
Pine FHM	**23**	**22**	**120**	Pinus sylvestris–Suillus bovinus	Bending and Read, 1995a
Pine FHM	**13**	**3**	**120**	Pinus sylvestris–Thelephora terrestris	
Pine FHM	5	0	120	Control (no host plant)–mycorrhizal fungus absent	
Birch FHM	**0**	**40**	**90**	Betula pendula–Paxillus involutus	Perez-Moreno and Read, 2000
Pine FHM	**1**	**35**	90	Betula pendula–Paxillus involutus	
Beech FHM	**14**	**37**	**90**	Betula pendula–Paxillus involutus	
Pine FHM	**25**	**63**	**90**	Betula pendula–Pinus sylvestris linked by Paxillus involutus	Perez-Moreno and Read (unpublished data)
Pine FHM	**25**	**54**	**90**	Betula pendula–Pinus sylvestris linked by Paxillus involutus	
Pollen					
Pine pollen	**76**	**97**	**115**	Betula pendula–Paxillus involutus	Perez-Moreno and Read, 2001a
Pine pollen	42	35	115	Control (non-mycorrhizal plant)–mycorrhizal fungus absent	
Soil animals					
Nematodes	**68**	**65**	**150**	Betula pendula–Paxillus involutus	Perez-Moreno and Read, 2001b
Nematodes	37	25	150	Control (non-mycorrhizal plant)–mycorrhizal fungus absent	

Note: FHM, fermentation-horizon material.

and laboratory based results (Bending and Read, 1996a; Wu *et al.*, 2003) indicate that the ECM fungi tested so far have a very low ability to release N and P from such complexes. As a result, the inherently large requirement of ECM fungi, particularly for N, would be expected to drive their scavenging activities in the direction of N sources that are not co-polymerized or precipitated with polyphenolic materials. In boreal and many temperate forest soils that are heavily loaded with phenolic residues, this must require that the fungi attack N-enriched substrates before they become involved in the immobilization processes.

In recognition of these constraints, a new generation of experiments has investigated the abilities of selected ECM fungi to mobilize N contained in substrates that are both of low phenolic content and likely to be present in quantitatively significant amounts in boreal forest soils. Among the materials used in these studies are dead mycelia (necromass) of the ECM fungi themselves (Andersson *et al.*, 1997), pollen (Perez-Moreno and Read, 2001a), seeds (Tibbett and Sanders, 2002) and quantitatively important representatives of the boreal soil mesofauna, namely nematodes (Perez-Moreno and Read, 2001b) and collembolans (Klironomos and Hart, 2001). The rapidity and intensity with which the mycobionts colonize these nutritionally enriched substrates is striking (see Colour Plate 6.3b). In the case of both pollen and nematodes, the colonization leads to significant removal of both N and P from the source materials relative to that lost from the same materials colonized only by saprotrophic fungi (see Table 15.4).

Broadly based types of characterization which see N mobilization by ECM fungi as a key nutritional feature of boreal forests run the risk of obscuring more subtle patterns associated with distinctive and localized changes of soil condition. Nowhere have such subtleties been better demonstrated than in analyses of local-scale gradients across which, for geological, topographic or hydrological reasons, soil pH, N form and understorey vegetation change dramatically. Along one such gradient of 90 m length at Betsele in northern Sweden a transect revealed a shift from extreme acidity (soil pH 3.5) at one end which was rainwater fed, to near neutrality (soil pH 6.4) at the other, where base-enriched groundwater was discharged (Giesler *et al.*, 1998; Högberg *et al.*, 2003) (Figure 15.13a, b, c). The acidic to neutral pH gradient was also one of markedly decreasing C to N ratio (Högberg *et al.*, 2003, 2006). Along the transect, large changes in the structure and composition of the overstorey and ground flora as well as the mycorrhizal and general microflora occur. These appear to be driven by an N-source gradient, in which organic N dominated at the acidic end, with mineral N, predominantly ammonium, in mid-gradient and nitrate at the discharge end. Pine (*P. sylvestris*) with ericaceous dwarf shrub understorey at the acidic end was progressively replaced by spruce (*P. abies*), first with a short herb and then with a tall herb community in the nitrate-enriched zone. While neither soil respiration nor total soil microbial biomass changed along the gradient, analyses of the phospholipid fatty acid (PFLA) signatures (see Chapters 2 and 6) of this biomass suggested that the physicochemical transition was driving significant qualitative changes in the nature of the microbial populations (Högberg *et al.*, 2003, 2006). These signatures showed drastic decline in fungus to bacteria ratio as pH (Figure 15.14a) and mineral N availability (Figure 15.14b) increased across the gradient. They also indicated a shift in the predominant fungal population from one in which ECM and ERM fungi were likely to be the dominants in the dwarf shrub-pine and short herb environments (Figure 15.15a) to one in which the putative AM fungal

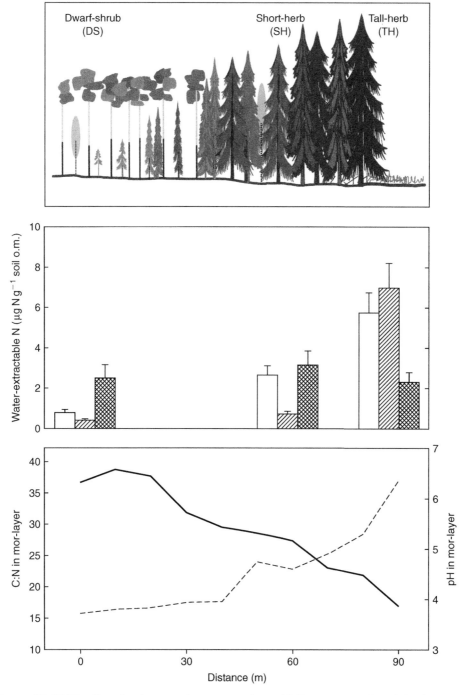

Figure 15.13 The Betsele plant productivity gradient: (a) Forest types: dwarf shrubs understorey (DS) between 0 and 40 m, short herb understorey (SH) between 50 and 80 m and tall herb understorey (TH) at 90 m. (b) water-extractable N forms: open bars, ammonium; hatched bars, nitrate; crossed bars, amino acids. (c) the C:N ratio of the mor-soil (solid line) and pH of the soil solution (dashed line). From Högberg *et al.* (2003).

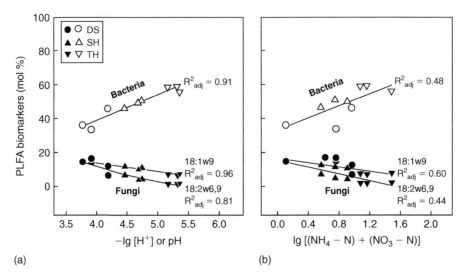

Figure 15.14 Partitioning between major functional microbial groups in forest with three types of understorey. The relationship between the bacterial and the fungal signature lipid biomarkers and (a) $-\log_{10} (H^+)$, i.e. soil pH and (b) $\log_{10}$ (inorganic N). Twelve PLFAs were used as biomarkers for bacteria and two PLFAs for fungi. DS, dwarf shrub; SH, short herb; TH, tall herb. From Högberg et al. (2007).

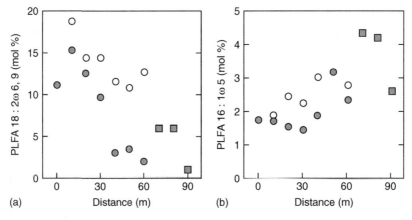

Figure 15.15 Mol percentage of individual phospholipid fatty acids (PLFAs) in the F horizon (O), H horizon (●), and F + H horizon (■) of the mor layer along the Betsele forest gradient. (a) The ERM + ECM fungal indicator 18:2ω6 9; (b) The AM fungal indicator 16:1ω5. From Högberg et al. (2003).

signature 16:1ω5c dominated in the tall herb community (Figure 15.15b). These studies appear to confirm the predominant effect of N availability as a determinant both of the type and functions of mycorrhizas in the boreal biome. Högberg et al. (2003) hypothesize that under conditions of low N supply at the acidic end of the gradient, autotrophs are induced to increase allocation of photosynthate below ground. This, while stimulating the activity of their fungal symbionts, will in turn contribute to

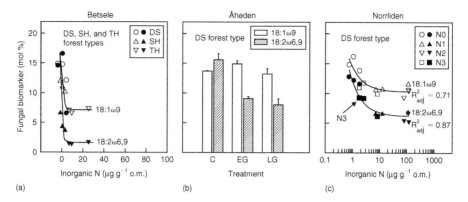

Figure 15.16 The effects of soil N concentration and tree girdling upon the absolute quantities and relative amounts of the two phospholipids fatty acid (PLFA) fungal biomarkers 18:1ω6,9 and 18:2ω6,9 in boreal forests. (a) The marker 18:2ω6,9 declines more strongly than 18:1ω6,9 along the natural gradient of increasing mineral N from dwarf shrub (DS), through short herb (SH) to tall herb (TH) communities at Betsele. (b) Whereas the 18:1ω6,9 signature does not decline in response to early (EG) or late (LG) stem girdling, that of the 18:ω6,9 is reduced by c.50% in the Åheden experiment with dwarf shrub (DS) understorey. (c) Application of mineral N as a fertilizer at three levels (N1, N2, N3-plus control N0) leads to progressively larger reductions of 18:2ω6,9 relative to 18:1ω6,9 in another dwarf shrub forest type at Norrliden. From Högberg et al. (2007).

the production of litter of high C:N ratio. In contrast, it is generally accepted that in boreal forest systems, increases of N loading, however they may arise, lead to a reduction in C allocation belowground (Waring and Running, 1998). This may be to the relatively strong detriment of ECM fungal symbionts with high C demand, while having lesser effects upon bacteria and AM fungi. Such a scenario is one that again fits with determinations of PLFA signatures in this type of ecosystem. Högberg et al. (2006) observed that of the two markers 18:1ω6,9 and 18:2ω6,9 considered to represent the ECM and ERM mycelial biomass, the 18:2ω6,9 was strongly sensitive to availability of mineral N. This was the case not only along the natural gradient of N concentrations at Betsele (Figure 15.16a), but also under three levels of mineral N application in a nearby pine-dwarf ericaceous shrub community (Figure 15.16c). The pattern of differential response of the two markers was repeated following girdling of pine with an ericaceous understorey at Åheden (Figure 15.16b). Since, at this last site, C supply to the ECM community was specifically targeted by the treatment, the indication emerges that the 18:2ω6,9 is a better marker for ECM fungi. The extent to which the 18:1ω6,9 signature encompasses fungi of the ERM and of saprotrophic habit remains to be elucidated. In combination, these observations suggest that under circumstances where C supply from their autotrophic partners is maintained, the structures of the fungal symbiont populations in boreal forests are largely determined by qualitative and quantitative aspects of soil N economies.

Notwithstanding the observation (Figure 15.15b) that under the localized influence of raised pH and N enrichment there can be enhancement of the AM mycelial component of boreal forest soils, there have been few measurements of the extent or function of AM colonization in the herb communities that develop under these circumstances.

Opik *et al.* (2003) observed that *Pulsatilla* spp. growing in soils of pH 8.6 in the boreal region in Estonia were extensively colonized by AM fungi. While there were site-dependent differences in AM fungal community composition, preliminary analysis of different *Pulsatilla* spp. indicated that the associations lacked specificity. When seed of two species of the genus were sown in soil either with or without AMF inoculum, lower establishment was observed in the AM inoculum circumstance (Moora *et al.*, 2004). In this study, an AMF inoculum from grassland was more effective than one from boreal forest soil in colonizing roots, promoting biomass production and elevating P concentration in one *Pulsatilla* species, whereas a boreal forest inoculum was slightly more effective in another. It was concluded here that the distinctive responses of the congeneric species may be attributable to fungal specificity. Further studies of these kinds are required because the herbaceous communities of the boreal forest contain a high proportion of the plant biodiversity seen in the biome.

While P is rarely the primary growth limiting element in such communities, it is of interest that the processes involved in its immobilization and release in litter appear to be the same as those of N (Berg and McClaugherty, 1989). As in the case of N, the onset of energy limitation in the saprotroph population should enable ECM fungi, with their ability to produce a range of phosphomono- and di-esterase (see Chapter 10) to compete effectively for P as well as N. Indeed, the retention of the ECM habit in mull humus forests, in many of which active nitrification occurs (Aber *et al.*, 1989; Ellenberg, 1988), may be attributable to the ability of such fungi to release P from the organic residues. In addition to the general observation that ECM mycelia proliferate most intensively in material from the FH horizon, there is evidence at a finer scale of hyphal growth patterns which are likely to provide intimate contact with resources of a particular quality. Ponge (1990), using the light microscope, observed selective exploitation of pine needles, *Pteridium* leaflets and animal corpses by hyphae of ECM fungi. In an attempt to similate conditions prevailing in the ECM conifer-feather moss communities that cover large area of the boreal forest zone, Carleton and Read (1991) grew mycorrhizal *Pinus* seedlings in association with the feather moss *Pleurozium schreberi*. Hyphae of the fungal symbiont *Suillus bovinus* selectively colonized and formed a sheath-like structure around senescing parts of the moss shoot (Figure 15.17). It was shown that such colonization provided the potential to capture resources from the moss shoot. In nature, this colonization is likely to provide a key link in the nutrient cycle since most of the elements arising at the forest floor in these ecosystems are intercepted by the moss carpet.

At an even finer scale, Agerer (1991b) has demonstrated structural modifications of the hyphal tips of ECM fungi at their point of contact with particular materials in soil. An appressorium-like structure attaches the tip of its substrate (Figure 15.18) providing an enlarged surface of contact for biochemical interaction. It seems that the walls at the tips of ECM hyphae are less hydrophobic than they are behind the tip (Unestam and Sun, 1995; Sun *et al.*, 1999). It is likely to be at the tip that most of the active interactions with the environment occur. In view of the long-distance transport role of the hyphae, which serve as pipe-lines through which movement of C and minerals occurs to and from the growing tip, the need for their walls and membranes to be effectively impermeable at maturity may be paramount.

In addition to interactions with organic materials, the possibility that release of protons, organic anions or CO_2 from hyphal tips might lead to the release of minerals has received increasing attention. The recognition (see above) that in podzolic

Figure 15.17 Observation chamber showing the colonization of senescent parts of the shoot of the feather moss, *Pleurozium schreberi*, by mycelium of *Suillus bovinus* (arrowed, growing from a colonized plant of *Pinus contorta* (not shown). From Carleton and Read (1991), with permission.

soils a significant proportion of ECM roots can occur in the mineral horizons and that some of the fungi involved are localized in this environment has led to the suggestion that they may be directly involved in mineral dissolution and so contribute to the podsolization process (van Breemen *et al.*, 2000b; Smits, 2006; Wallander, 2006). Interest in this aspect of mycorrhizal physiology was strengthened by the description of small (3–10 μm wide) tunnel-like features in hornblende and feldspar grains of Swedish boreal forest podsols (Jongmans *et al.*, 1997). It was postulated that the tunnels were created by mineral dissolution through the activities of low molecular weight organic acids released from ECM hyphae. It was further suggested that the products of dissolution, in particular P from apatite inclusions in feldspars, as well as Ca, Mg and K, if transported to the roots, would enable a bypass of the bulk soil solution (van Breemen *et al.*, 2000a; Landeweert *et al.*, 2001). Further circumstantial evidence in favour of a possible involvement of these hyphae in tunnel formation was provided by the description of a positive linear relationship between tunnel length and density of ECM root tips (Hoffland *et al.*, 2003) and the observation that the tunnels are almost entirely restricted to ECM forests of boreal and temperate biomes (Hoffland *et al.*, 2002). To date, however, there appears to have been no direct observational or experimental verification of the involvement of ECM hyphae in tunnel formation. Hyphae occupy less than 1% of the tunnels observed and there is no confirmation that these are of ECM origin. Further, calculations (Smits, 2006) and models (Sverdrup *et al.*, 2002) indicate that tunnelling

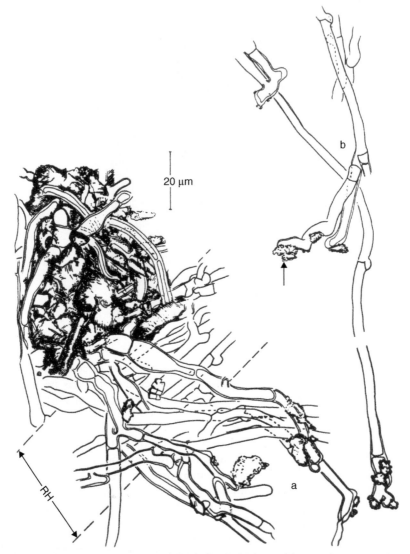

20 µm

b

a

RH

Figure 15.18 Thick-walled hyphae of *Sarcodon imbricatus*: (a) growing into a humus particle. (b) The hyphae in close contact with the soil debris are thick-walled and have somewhat swollen, appressorium-like tips adhering to the soil particles (arrowed). From Agerer (1991), with permission.

of mineral grains contributes very little to total weathering in soils. The possibility remains, however, that in view of the generic property of fungi to release low molecular weight organic anions (Gadd, 1999), as well as to acidify their environments through proton release (Rosling *et al.*, 2004b), the ECM fungal community may be influential in broader soil mineralization processes (see Chapter 10).

The threats posed to boreo-temperate forest biomes by anthropogenically-related increases of atmospheric C and of N deposition are increasingly recognized and the responses of ECM systems to elevation of both elements have been investigated. It has been shown that, under conditions of CO_2 enrichment, the relative abundance of ECM morphotypes changes significantly, there being a shift towards types with greater quantities of rhizomorphs and extramatrical mycelium (Godbold *et al.*, 1997; Godbold and Berntson, 1997). Doubling of CO_2 concentrations around shoots of *Betula pendula* was shown to stimulate growth of ECM mycelium of both *Paxillus involutus* and *Suillus bovinus* in microcosms but, since *P. involutus* was more responsive to the treatment than *S. bovinus*, the need to be aware of interspecific effects on the ECM community was again emphasized (Rouhier and Read, 1998). Through its effects upon availability of C, enhancement of ambient CO_2 can be expected to influence the production and longevity of mycorrhizal roots and associated mycelia. Rygiewicz *et al.* (1997) observed that elevation of CO_2 concentration over microcosms supporting ECM *P. ponderosa* led to enhancement of production but to no impact upon the life span of roots. The observation that abrupt rather than gradual exposure to CO_2 can produce large and probably unrealistic responses (Klironomos *et al.*, 2005; see Chapter 2) should be borne in mind when interpreting results of the experiments reported above.

The responses of ECM systems to elevated CO_2 have been the subject of recent reviews (Gorrissen and Kuyper, 2000; Treseder and Allen, 2000; Rillig *et al.*, 2002a) as well as of meta-analyses (Treseder, 2004; Albertson *et al.*, 2005). Conclusions from the analyses made to date must be made with caution because both the experimental conditions and the responses observed in studies of CO_2 impacts vary greatly. This situation reflects that seen in studies of AM systems (see Chapter 2). Despite the variability, taken overall, the results indicate that both the ECM plants and their fungal partners respond positively to elevation of ambient CO_2 levels. Perhaps not surprisingly, however, in a given symbiont pairing the extent of the response seen in the autotroph and the fungus can be quite different (Albertson *et al.*, 2005). Response ratios calculated on the basis of a large number of morphological and physiological parameters, indicate that the activities of the fungal partners in the ECM symbiosis are particularly stimulated by CO_2 enrichment. The calculations of Albertson *et al.* (2005) indicate that across the experiments ECM fungi showed a 34% response-ratio increase under elevated relative to ambient CO_2 exposure. Since this value is considerably larger than the 21% increase seen in experiments involving AM fungi, it can be predicted that ecosystems dominated by ECM partnerships will be more strongly affected by the predicted global increases of CO_2 than those consisting largely of AM symbionts. ECM plants are generally less CO_2-responsive than their fungal partners being, at +26%, only 1% more so than AM plants. In view of the greater responsiveness, particularly of ECM fungi, to elevated CO_2, Albertson *et al.* (2005) emphasize the need to incorporate mycocentric perceptions in considerations of ecosytem responses to the emerging CO_2 scenarios.

There has been a plethora of studies on the impacts of pollutant N enrichment on mycorrhizas of boreo-temperate forests. These have been extensively reviewed (Jansen and Dighton, 1990; Colpaert and van Tichelen, 1996; Wallenda and Kottke, 1998; Cairney and Meharg, 1999; Avis *et al.*, 2003) and will be only briefly described here. A decline of ECM sporocarp production was described as the first impact of N deposition (Arnolds, 1991). The involvement of N deposition in this decline was strongly implicated by experiments in which N fertilization of forest plots was shown to inhibit sporocarp yields (Rühling and Tyler, 1991; Termoshuizen, 1993). The effects of N enrichment observed on sporocarp production above ground are likely to be a product of impacts of N on below-ground patterns of C allocation to ECM mycelial systems. Experimental analyses of these impacts have produced variable results. Some have shown that N additions to microcosms lead to modest increases of mycelial development (e.g. Arnebrandt, 1994), others to a decrease (e.g. Wallander *et al.*, 1994) and others to no effect (e.g. Wallander *et al.*, 1999). This is not surprising since, as described above, the impacts of N are likely to be mediated at least in part through its effects on canopy photosynthesis and C allocation. They will also be influenced by the initial N status of the system. N enrichment has been increasing slowly across the industrialized world over the last century so, as in the case of CO_2, experiments investigating its effects should reflect the long-term nature of the deposition processes. The cumulative effects of N deposition at the biome level have been examined in spruce (*Picea abies*) forests along a gradient of N deposition, and hence soil mineral N content, from pristine boreal forest environments in Northern Sweden where N mineralization is almost undetectable (Persson *et al.*, 2000) to heavily polluted stands in central Europe (Taylor *et al.*, 2000). As extractable soil N increased in a southerly direction, there was both a decrease in diversity of ECM fungi and a reduction in the proportion of mycobionts that are identified as having proteolytic potential (Figure 15.19). Associated with the enhancement of soil mineral N concentration was an increasing prevalence of the resupinate corticoid fungus *Tylospora fibrillosa* as an ECM symbiont, indicating that the decline of sporophore-producing ECM fungi in these polluted forests is a result of species replacement rather than of complete elimination of symbionts (Taylor *et al.*, 2000).

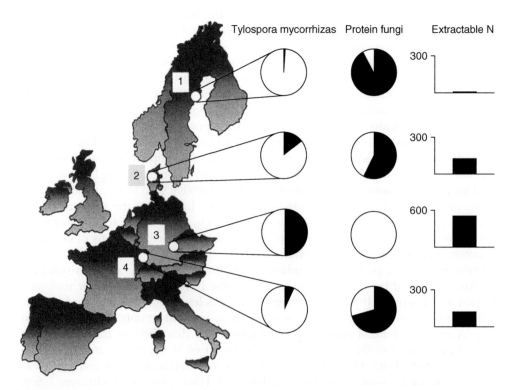

Figure 15.19 (Caption opposite)

An experimental application of N over 16 years in an oak-savanna system dominated by ECM oaks led to loss of diversity and a 50% reduction of overall sporocarp abundance indicating that effects of anthropogenic inputs are similar in boreal and more temperate systems. However, a distinctive feature of the oak savanna was that the treatment ($17 \, \mathrm{g} \, \mathrm{N}/\mathrm{m}^2/\mathrm{year}$) led to a 5-fold increase of sporocarp production by *Russula* spp. (Avis *et al.*, 2003). Such results are indicative of the likelihood that ecosystems may respond in subtly different ways to nutritional perturbations.

Mycorrhizas in tropical and subtropical biomes

Mycorrhizal associations in the wet tropics

Highly productive tropical rainforest ecosystems cover much of the equatorial belt spreading across South-Central America, West Africa and South-East Asia. These forests are widely threatened by anthropogenic activity, both directly through logging activities and indirectly as a result of global climate change. Sadly, despite these threats, we remain largely ignorant of some of the most basic aspects of rainforest biology. The gaps in our knowledge are nowhere better illustrated than in consideration of the mycorrhizal status of these systems. While we recognize that almost all rainforest trees so far examined have mycorrhizal symbionts and have made progress towards understanding of the distribution of mycorrhizal types (Alexander, 1989a), we are still better placed to raise questions about the identity and functions of the fungal symbionts involved than we are to provide answers (Janos, 1996; Alexander and Lee, 2005).

A defining feature of tropical rainforest is the enormous diversity of plant species in the communities. In the tree flora alone, Wright (2002) recorded 1175 species in a $0.52 \, \mathrm{km}^2$ plot of Borneo rainforest. This is more than occur in all of the boreo-temperate forests of the northern hemisphere. Within this diversity it is possible to recognize some patterns with respect to the distribution of mycorrhizal types. While most rainforest tree species associate with AM fungi, there are families of tropical plants which are predominantly or entirely ectomycorrhizal (Alexander, 1989a). These include the Dipterocarpaceae (Lee *et al.*, 1996), Fagaceae (Corner, 1972), many legumes in the Caesalpinoideae (Alexander, 1989b) and members of the subfamily Leptospermoideae in the Myrtaceae (Moyersoen *et al.*, 2001). Circumstantial evidence suggests that there is a relationship between mycorrhizal type and soil conditions. Whereas trees with

---

Figure 15.19 Summary of an analysis of the ectomycorrhizal communities at four Norway spruce (*Picea abies*) forests along a north-south European transect. Pie charts show the proportion of root tips at each site colonized by *Tylospora fibrillosa* and the proportion of species isolates from each site that could utilize protein as a source of nitrogen (protein fungi, *sensu* Abuzinadah and Read, 1986a). None of the isolates from site 3 could utilize protein as an N source. The bar charts show KCl-extractable inorganic N (μg N/g LOI) in the LFH soil horizons (note that the scale at site 3 is twice that at the other three sites). Sites: 1, Åheden, northern Sweden; 2, Klosterhede, western Jutland, Denmark; 3, Waldstein, Fichtelgebirge, Germany; 4, Aubure, Vosges Mountains, NE France. Data from Taylor *et al.* (2000), Persson *et al.* (2000).

AM colonization predominate over large areas of such systems, there are localized occurrences in South American (Singer and Araujo, 1979, 1986; Henkel, 2003), West African (Newbery *et al.*, 1988) and South-East Asian (Whitmore, 1984) forests of communities dominated by ECM species. Although these are often restricted to the most nutrient-poor soils with a surface accumulation of litter and raw humus (Torti *et al.*, 2001; Henkel, 2003), deductions concerning possible ecological relationships between mycorrhizal and soil types must be drawn with caution in view of the background phylogenetic histories of each family. This complication is best illustrated by the generally ectomycorrhizal Dipterocarpaceae which dominate the rainforests of South-East Asia almost irrespective of soil type (Whitmore, 1984), and apparently without particular association with nitrogen-limiting conditions (Alexander and Lee, 2005). This taxonomically-based ECM trait differentiates the fundamentally ECM forests of the Paleotropics from the primarily AM forests of the Neotropics. Despite these reservations, it can be said that, especially in the cases of trees which form uniform groves of the kind seen in many ECM tropical forests, feedbacks between litter quality and quantity, types of mycorrhizal colonization and nutrient supply are likely to be important factors in maintaining the stability of the ecosystem.

Some knowledge of the taxonomic status of the fungal symbionts of tropical rain-forest plants is a prerequisite for understanding mycorrhizal function in these ecosystems. In this context it is clearly important to know whether the diversity of mycobionts below ground reflects in any way that observed in the plants. The picture in this regard is still incomplete, but increasingly intensive sampling of ECM fungi (Henkel *et al.*, 2002; Lee *et al.*, 2003) and the application of molecular methods to analysis of AM fungal associations (Husband *et al.*, 2002a, 2002b; Vandenkoornhuyse, 2002) is beginning to indicate that diversity among these communities is much higher than was previously thought. Husband *et al.* (2002a) analysed the AM fungal communities associated with the roots of emergent seedlings of two tree species co-occurring in the rainforest of Panama. A total of 48 plants were examined using AM fungal-specific primers. Compared with communities of these types sampled from temperate habitats and analysed using the same methods, both the overall diversity and the species richness were higher at the rainforest site. A total of 30 AM fungal types was identified, 17 of which were previously unrecorded (Table 15.5). A shift in the AM community was also detected over time (Husband *et al.*, 2002a,

Table 15.5 The number of AM fungal types and the diversity of the AM fungal community in a number of ecosystems, based on phylogenetic analysis of partial SSU rRNA sequences. The relatively high diversity of AM types in the tropical biome is evident.

Ecosystem	No. of roots sampled	No. of host species	No. of AM fungal types	Diversity[a] (H')	Reference
Tropical forest, Panama	48	2	30	2.33	Husband *et al.*, 2002a
Temperate grassland, UK	47	2	24	1.71	Vandenkoornhuyse *et al.*, 2002
Temperate woodland, UK	49	5	11	1.44	Helgason *et al.*, 1998
Arable fields, UK	79	4	8	1.16	Daniell *et al.*, 2001

From Husband *et al.* (2002a). [a] Shannon-Weiner diversity index.

2002b). Those fungi dominant in newly germinated seedlings were replaced over a year by others and seedlings of different ages sampled at the same time supported significantly different AM communities. Even taking into account some variations in sampling protocols between studies, these results suggest not only that AM diversity is much higher than previously acknowledged, but also that it is far greater than that seen in temperate systems. Bearing in mind that a serious decline of AM diversity has been seen to result from agricultural practices in the temperate zone (Helgason *et al.*, 1998; see Chapter 17) and that logging is already known to reduce AM inoculum potential in the tropics (Alexander *et al.*, 1992), there is every reason to expect that diverse rainforest communities of AM fungi will be seriously threatened by these and other commercially-driven activities in rainforest ecosystems (Janos, 1996). There is an urgent need for more intensive and widespread sampling of AM fungal genotypes of tropical soils, so that these communities can be both characterized and archived for possible use by future generations.

In the absence of similar molecular analyses of ECM associations, the diversity of these fungi can only be inferred from records of occurrence of putatively ECM sporocarps and by morphotyping studies.

Lee *et al.* (2003) measured sporocarp occurrence over 7 years in a 20 ha lowland rainforest dominated by Dipterocarpaceae and Fagaceae in Malaysia. A total of 296 species were recorded of which 66% were new to science. This diversity is greater than the cumulative total of 265 species of putatively ECM sporocarps determined in weekly visits over 21 years to a North temperate forest in Switzerland (Straatsma *et al.*, 2001) and is more than double that recorded over the same 7 year sampling period in an Austrian forest (Straatsma and Krisai-Greilhuber, 2003). Whereas there are now extensive databases describing ECM morphotypes of North temperate ECM forests (Agerer *et al.*, 1987–2002; Goodman *et al.*, 1996–2002), few such records exist for the tropics. However, recent analyses using these approaches to characterize communities on dipterocarp seedlings in South-East Asia (Lee and Alexander, 1996; Lee *et al.*, 1996; Ingleby *et al.*, 1998) indicate that the numbers of morphotypes present in the tropical systems are at least as great or greater than those of temperate systems (Alexander and Lee, 2005).

Studies involving transplantation of non-mycorrhizal seedlings of the dipterocarp *Hopea nervosa* into logged and unlogged areas of Malaysian forest (Lee *et al.*, 1996) revealed that, after 6 months growth, the amount of ECM colonization was the same in both types of system, but seedlings recovered from the unlogged forest supported 40% more morphotypes than their counterparts from the logged system. In addition, there were twice as many uncommon morphotypes in the undisturbed forest, indicating that relatively rare fungi are most likely to be lost as a result of logging activities. Comparisons with the morphotypes present on seedlings of another dipterocarp species, *Shorea leprosula*, indicated that the most common fungi were equally well represented on both host plants, suggesting a lack of host specificity among dipterocarp fungi. In summary, as far as ECM fungal community structure is concerned, the emerging picture is one in which, as in the case of AM fungi, diversity is higher in tropical rainforest than in temperate forest but that this diversity is threatened by the extensive logging activities.

What have we learned of the functions of these obviously widespread mycorrhizal symbioses? There have been attempts using pot experiments to determine whether tropical trees show growth responses to mycorrhizal colonization. As Alexander and

Lee (2005) point out, the results of these must be treated with caution because they have inevitably been carried out over a short time span, have generally used mineral soil low in organic matter and the inoculum has often been either undefined or of a model species. Yazid *et al.* (1994) demonstrated growth increases and improved P uptake in two *Hopea* species following inoculation with a *Pisolithus* isolate. Using the same two species, Lee and Alexander (1994) also showed that application of ECM inoculum, on this occasion collected from the field, increased P uptake. In this case, the growth response to mycorrhizal colonization was greater than that to added P, indicating that factors other than P supply were involved. A similar effect was obtained by Moyersoen *et al.* (1998b) with the widespread AM species of rainforest in Cameroon, *Oubangia alata*. In the same study, it was shown that in the ECM species *Tetraberlinia moreliana* there was increased uptake of P at both high and low levels of P supply, but there was no growth response to the P addition (see Chapters 5 and 10). In an attempt to reproduce more closely the qualitative aspects of the rooting environment, Brearley *et al.* (2003) grew seedlings of three ECM dipterocarp species in pots of mineral soil with and without litter additions. The seedlings grew better in the presence of the litter and showed a negative relationship between their $\delta^{15}N$ signatures and the extent of ECM colonization. Since the litter had a lower $\delta^{15}N$ signature than the soil, this was interpreted as evidence that ECM colonization facilitated uptake of litter-derived N. It can be concluded from these types of experiment that mycorrhizal colonization may facilitate increased access to P and N, but that many, particularly the slow growing shade tolerant species, may not show a growth response as a result of the additional nutrient supplies. This in turn suggests that there may be considerable temporal uncoupling between the absorption and allocation of nutrients, a feature that is further considered below.

In an attempt to determine whether mature stands of tropical trees were limited by P availability, Newbery *et al.* (2002) applied P to a grove of the ECM species *Microberlinia bisulcata* in Cameroon. They found no effects of P addition upon seedling establishment or growth either of seedlings or trees in the fertilized plots, despite the fact that the P was taken up as evidenced by an increased P concentration in the foliage and litter. A fertilizer experiment in lowland ECM dipterocarp forest of Central Malaysia involving factorial application of both P and N produced similar results. While litterfall biomass and P concentrations were increased, indicating that the P at least was taken up, there was no change in tree girth over a 5-year period (Mirmanto *et al.*, 1999). These results do not support the view that P is a widespread limiting element, at least in ECM forests of the lowland tropics. This view is largely supported by studies carried out in the montane forests of the wet tropics (Tanner *et al.*, 1998). However, these authors concluded that there were severe growth limitations in this forest type imposed by low N availability. N was also seen as the critical limiting nutrient in tropical heath forest (Vitousek and Stanford, 1986). Following the contention (Read, 1991b) that the N limitation characteristic of boreal and temperate forests has selected in favour of ECM colonization, it might be suggested that these types of tropical forest would also be dominated by ECM trees. Alexander and Lee (2005) have subjected this hypothesis to scrutiny by examining reports of the mycorrhizal status of trees growing in particularly N-limited tropical environments. They find little evidence to support it. Some of the most intensively examined of these environments in the montane tropics contain few, if any, ECM trees (Tanner, 1977; Tanner *et al.*, 1992; Vitousek *et al.*, 1995).

Similarly, in extremely N-deficient heath forests of Brunei (Moyersoen *et al.*, 2001), Venezuela (Moyersoen, 1993) and Guyana (Bereau *et al.*, 1997), most of the dominant trees associate with AM fungi.

A further possibility, again based upon analysis of stratified soils of the kind found in boreal forest, is that ECM and AM species exploit different horizons, the former occupying predominantly the surface organic layers and the latter the mineral soil. There have been few comparative analyses of ECM and AM root distribution in the tropics. However, Moyersoen *et al.* (1998a) examined the vertical distribution of these root types in a lowland forest of Cameroon and found that fractional colonization of roots by either type was unaffected by horizon. It was later confirmed (Moyersoen *et al.*, 2001), that there were no differences in the relative abundances of ECM and AM roots in organic and mineral layers either in mixed dipterocarp or heath forests of Brunei. It was even apparent that the AM fractional colonization was significantly higher in the organic and acidic soils of the most nutrient stressed heath forests. Observations of these kinds bring into sharp focus the need to investigate the taxonomic and functional status of the AM fungi that dominate such stressed environments. It is also important to view both snap shots of tissue nutrient concentrations and short-term changes in growth (or lack of them) in the context of the whole life span of the trees, not simply the immediate duration of the experiments.

While, when they co-occur, ECM and AM trees may share the same localized soil niches, it remains the case that trees with each of these types of symbiosis are often spatially segregated in tropical forests. Under these circumstances, the structures of the resulting plant communities can be distinctive. Thus, in rainforests of both West Africa (Alexander, 1989a; Hart *et al.*, 1989; Torti *et al.*, 2001) and South America (Henkel, 2003; Mayor and Henkel, 2006), extensive pure stands, often referred to as 'groves', of monodominant ECM trees can be surrounded by species-rich stands of AM trees. On both continents the monodominant ectomycorrhizal species are often caesalpinoid legumes, most of which appear not to be involved in nitrogen fixation. There is little evidence to support the notion that these monodominant stands are associated with particular soil types since they can be found on substrates ranging from leached 'white sand' soils to red tropical oxisols (Nascimento and Proctor, 1997; Henkel *et al.*, 2002; Henkel, 2003). However, there is evidence that they can produce deeper litter layers and greater humic accumulations and that both of these substrates can be intensively exploited by ECM roots (Torti *et al.*, 2001; Henkel, 2003). Working in Guyana on the monodominant ECM caesalpinoid legume *Dicymbe corymbosa*, Henkel (2003) observed extensive root mounds at the base of large trees extending outwards between conspecifics to form forest-wide litter traps with depths of up to 50 cm. These were intensively exploited by ECM mycelia and roots. In a test of the hypothesis that this exploitation would lead to changes of residual litter quality, Mayor and Henkel (2006) examined mass loss from litter bags inserted into *D. corymbosa*-dominated plots without (trenched) and with (untrenched) intact mycorrhizal systems. Leaf litter mass loss was not influenced by the presence of the ectomycorrhizas and the only identifiable change of litter nutrient status was a more effective removal of Ca in the untrenched plots. These results thus argue against the view, originally expressed by Gadgil and Gadgil (1975), that the presence of ECM fungi leads to the suppression of saprotrophic fungi and hence to a reduction of decomposition rate. In the absence of a measurable effect of ECM mycelia on litter decompositon, Mayor and Henkel (2006) concluded that the deep litter layers characteristic of *D. corymbosa* stands may contribute to the maintenance of mono-dominance by inhibiting seedling establishment by AM species. Torti *et al.* (2001) working in monodominant ECM stands of *Gilbertiodendron dewevrei* in Congo came to a similar conclusion. In this context, the abilities of many of the species forming these monodominant stands to produce, often in 'mast' years (see below), very large numbers of seedlings that can be rapidly recruited into the ECM mycelial network, may be important. The evidence collected so far thus points to the likelihood that monodominance of ECM stands in the tropics is maintained by competitor exclusion rather than by nutrient competition, but it is also probable that other as yet unidentified biotic effects are involved.

The difficulties experienced in identifying a direct nutritional benefit to trees from mycorrhizal colonization in the moist tropics suggest that a search for alternative functions might be fruitful. Two hypotheses, both indirectly involving nutrition, are worthy of further investigation. The first, supported by observations of a positive impact of mycorrhizal colonization upon seedling recruitment (Alexander *et al.*, 1992; Onguene and Kuyper, 2002) is that young plants may be sustained by C transfer from illuminated overstorey plants (Read *et al.*, 1985; Read, 1997). The second, that colonization enables the cumulative capture and storage of sufficient reserves to facilitate mast fruiting (Newbery *et al.*, 1997, 2006; Henkel *et al.*, 2005).

The suggestion that organic C might be passed from illuminated 'source' plants to support shaded 'sink' seedlings has been controversial (Robinson and Fitter, 1999; Newbery *et al.*, 2000; see Chapters 4 and 16). However, in addition to experimental evidence of the process gained from ECM ecosystems (see Chapter 16), there is now some indirect support for its occurrence in AM systems in the moist tropics. Bidartondo *et al.* (2002) confirmed that all stages of development of fully mycoheterotrophic plants of a number of families growing on the floor of moist tropical forests were colonized by AM fungi (Chapter 13). On the basis that these fungi are obligate mycorrhizal associates of autotrophs, it is most likely that the mycoheterotrophs are secondary recipients of photosynthate from the autotrophic plants. A pathway for C transfer is therefore demonstrated for these plants as it is for a number of other mycoheterotrophs that are routinely colonized by ECM fungi (see Chapter 13). Clearly, fully mycoheterotrophic plants (whether associated with AM or ECM fungi) represent the end point of an evolutionary process enabling them to survive, often as sole occupants of the forest floor, under conditions of extremely low irradiance. Since in the Orchidaceae partially mycoheterotrophic green plants associated with ECM fungi are common, there is reason to believe that intermediate physiological conditions will be found in seedlings that are recruited into ECM networks in the tropics.

Fruiting is a process involving distinct supra-annual bursts of seed production, evidenced in most cases by large peak years separated by periods during which fruiting may not occur. The process is best recognized in ECM tree families such as Dipterocarpaceae, Fagaceae and in the caesalpinoid legumes of tropical forests, but is also seen in temperate zone Fagaceae as well as in Pinaceae and Betulaceae. While the association between mast fruiting and trees of ECM habit is not exclusive, the event being occasionally observed in AM families (Alexander and Högberg, 1986; Newman and Reddell, 1987), it is predominantly a feature of taxa with the former type of symbiosis. Two non-exclusive theories have been proposed to explain the habit. The first, the resource limitation hypothesis (Isagi *et al.*, 1997; Newbery *et al.*, 1997) proposes that the demands which fruiting places upon the C and mineral nutrient resources of the tree are so great that annual cycles of seed production are not possible. The second, the predator satiation hypothesis (Janzen, 1974; Kelly and Sork, 2002), suggests that irregular seed production leads to reduction of herbivore numbers and so increases seedling survivorship in masting relative to non-masting species.

There is no doubt that mast fruiting produces a heavy demand on the nutrient resources of the tree. In the monodominant ECM stands of *Dicymba corymbosa* in the neotropics, the resource allocation to seed production in a mast year was 3.0 t/ha (Henkel *et al.*, 2005). The African ECM caesalp, *Microberlinia bisulcata*, was likewise shown to invest the equivalent of 55% of the annual dry weight of leaves in seed and pod production during a masting event, the relative amounts of N and P being 13 and 21% respectively (Green and Newbery, 2002). In their phenology and climate ECM response (PACER) hypothesis, Newbery *et al.* (1997) proposed that it was the ability of ECM systems to capture and accumulate C, P and N during unusually moist seasons preceding a mast year that sustained the demands of fruiting on this scale. As an indication of the P stress suffered in association with mast fruiting, it was observed that in years following these events leaves shed from the canopy had lower P concentrations than in non-mast years. In general, removal of P from leaves of ECM species prior to leaf fall was approximately half that from those of non-ECM species (Chuyong *et al.*, 2000). Clearly, while some of the arguments in favour of a functional link between ECM associations and mast fruiting are persuasive, the evidence in their favour remains largely circumstantial and further testing of the suggested relationships are required. Nevertheless, as we have pointed out elsewhere (see Chapter 16), experiments only investigate a snap shot of the whole lives of plants. It may well be that it is our lack of understanding of the roles of enhanced tissue nutrient concentrations (apparent luxury consumption) in the eventual success of plants, that has led to the belief that such accumulations are valueless unless accompanied by growth increases.

Mycorrhizas in the seasonally dry tropics

Outside the wet equatorial forest belt, gradients occur towards biomes with pronounced dry seasons. Vitousek (1984) concluded that the changes of climatic regime across gradients of this kind had pronounced qualitative impacts upon the nutrient status of soils. He concluded that, whereas soils of wet tropical rainforest were most likely to be P-limited, those of the seasonally dry savannas, from which N was frequently lost in fire events, were characteristically N limited. If real, these deficiencies would be expected to select in favour of symbioses that optimized access to the limiting nutrient. These generalizations will of course be moderated by geological history so that some soils, weathered and eroded over millions of years, will be highly deficient in all essential nutrients.

An alternative but still indirect method of assessing the role of mycorrhizas in these challenging ecosystems is to determine the response of native plants to inoculation when grown in forest soil under greenhouse conditions. Extensive studies of normally AM colonized woody plants of the seasonally dry Tibaga River Basin of Southern Brazil have revealed a wide range of responsiveness (Siquera *et al.*, 1998; Zangaro *et al.*, 2000, 2003; Siquera and Saggin-Junior, 2001). While significant numbers of these native species were non-mycorrhizal, many were highly responsive and some apparently dependent upon AM colonization. Pioneer species had small seeds and were found to be both highly susceptible to AM colonization and very responsive in terms of growth. Late successional species with larger seeds were less dependent upon colonization, at least in the early stages of their development (Siqueira *et al.*, 1998). These observations were broadly confirmed in a further study of 80 woody species of the same forest (Zangaro *et al.*, 2003). The responses to AM fungi and the extent of colonization were, respectively, 5.9 and 4.2 times greater in early successional than late successional species. Again, inverse correlations were established between seed weight and responsiveness to AM colonization. Studies of these kinds involving a wide range of native species provide valuable insights into the ecology of hitherto understudied communities and are an important first step towards evaluation of the likely roles of mycorrhiza under conditions of disturbance.

The widespread occurrence of woody N_2-fixing legumes in biomes, notably *Acacia* spp. in Africa, *Prosopis* spp. in America and *Acacia* and *Casuarina* in Australia, might be seen as a response to N limitation of trees in these biomes (Högberg, 1986, 1989, 1992). The legumes, however, comprise only a proportion of the trees in most of the systems in which they occur. In the sub-humid miombo woodlands of Africa, for example, they comprise only 20–25% of the woody species in the community (Högberg and Pearce, 1986). In any event, alleviation of one deficiency is likely to lead to its replacement by another. The dependence of N_2-fixing legumes upon P supplies is demonstrated by experiments showing that application of P to dry subtropical woodlands increased both the yield and nitrogenase activity of leguminous understorey plants in the genera *Acacia* and *Kennedia* in Australia (Hingston *et al.*, 1982). The interdependence of N and P supplies should not be overlooked and may help to explain the observation that N-fixing legumes are generally colonized by AM fungi (Alexander, 1989a, 1989b). The fact that non-nodulated leguminous tree species often successfully co-occur with those that are nodulated in savannah grasslands, suggests, as pointed out by Sprent (1985), that in these relatively arid tropical biomes the production of an extensive root system to enable acquisition of water may be the key determinant of success. This appears to be the strategy of the dominant savannah grasses, the extensive root systems of which are also colonized by AM fungi (Newman *et al.*, 1986).

The absence of large data sets for nutrient status in soils of the dry tropics has led some to measure element concentrations. Högberg and Alexander (1995) examined the leaf N and P concentrations in 98 species and site combinations of Tanzania and north east Zambia, recognizing N-fixing AM species, non-fixing AM species and non-fixing ECM species as three separate functional groups. The concentrations of N and P in each group were expressed in terms of what are widely considered to be their optimal ratios in plant tissues of between 12.5:1 and 10:1. The N-fixing AM species had supra-optimal N:P ratios indicative of P deficiency, non-fixing AM species show suboptimal N:P ratios suggesting P deficiency, whereas ECM species occupied an intermediate category in which the N:P ratio was near optimal. Such a distribution might be considered to support the view that, whereas AM colonization selectively favours the acquisition of P in these systems, the ECM symbiosis provides access to both elements.

Tissue $\delta^{15}N$ signatures have been measured with a view to determining whether the different types of symbiosis found in the dry tropics are providing access to different sources of N (Högberg, 1992; Högberg and Alexander, 1995). The first analyses of nodulated, AM and ECM plants growing in miombo systems of Tanzania suggested that there were indeed differences between the groups. Tissues of ECM species showed greater $\delta^{15}N$ enrichment than those that were nodulated, suggesting that they may have access to organic N sources. AM plants occupied an intermediate position (Högberg, 1992). However, subsequent meaurements of miombo plants in Zambia have failed to confirm this pattern (Högberg and Alexander, 1995). The difficulties inherent in attempts to interpret patterns of nutrient acquisition from measurements of leaf stable isotope enrichment were described earlier. Progress is most likely to arise through studies of the kind employed for heathland and boreal forest plants, which involve direct parallel analyses of nutrient mobilization and stable isotope composition.

Conclusions

Evidence produced from field experiments is beginning to replace speculation and hence to throw new light on the important role played by mycorrhizal fungi in successional regimes at high altitudes and latitudes. They confirm that while the earliest herbaceous pioneers may well be ruderals that lack mycorrhizal colonization, the invasion of woody pioneers, particularly of the ECM kinds, is dependent upon mycorrhizal colonization. The important role played by ecto- and ericoid mycorrhiza in tundra and heathland environments in providing access to the predominantly organic N resources of these biomes is established. However, there is increasing consciousness of the threat to these delicate ecosystems that are posed by global climate change. Again, field experiments have provided valuable documentation of the responses of the mycorrhizal components of these systems both to elevated temperatures and to anthropogenically induced enhancement of N deposition.

Molecular methods have been applied to evaluate the diversity and distribution of ECM fungi in boreal and temperate forest systems. The possibility emerges that individual fungal species selectively exploit well-defined niches in the mineral as well as in the organic soil horizons of podsolic soils, but knowledge of the functional roles of the fungal symbionts lags behind awareness of their spatial distribution. The important role of soil N as a determinant of the structure and activity of mycorrhizal fungal communities has been assessed by careful evaluation of microbial populations along natural gradients of increasing N availability. Such work highlights the sensitivity of mycorrhizal communities to anthropogenic N depositions and provides a mechanistic basis for evaluation of the impacts of these effects.

Although N is still regarded by many as the most important limiting nutrient in terrestrial biomes, there are certainly many regions where P availability limits productivity. In such environments plants characteristically depend on arbuscular mycorrhizas where inorganic P is the predominant form in soil, but a range of other root adaptations, most of them primarily involved with mobilization and assimilation of P, are also found (Table 15.6). AM associations may play key roles in species with N_2-fixing symbioses, enabling the plants to establish on highly weathered and nutritionally impoverished soils. However, in P-limited soils, which, nevertheless, accumulate some P (and N) in organic form, ERM and ECM species will gain greater access to these nutrient sources. As anthropogenic N depositions alter the nutrient balance of ecosystems, it will be important to pay attention to the likely changes in vegetation that may come about as a result of variations in accessibility of forms of N, P and other nutrients brought about by different mycorrhizal associates.

Table 15.6 Diagrammatic representation of the types of P and N resources in soil that can be accessed by plants with different root adaptations, including mycorrhizas of various types. The intensity of shading indicates the extent to which the different adaptations shown are thought to be able to access the nutrients. It is assumed that plants without specialized adaptations access nutrients of all types largely from the soluble inorganic pools in soil, although they may also have some capacity for uptake from soluble organic pools.

Resource \ Strategy	Roots and root hairs	AM	Exudates/clusters	ECM/ERM
Soluble inorganic		P (N)	P	P and N
Insoluble inorganic		P	P	P and N
Labile/soluble organic		(P)	(P)	P and N
Recalcitrant organic				P and N

While understanding of the functional roles of mycorrhiza in tropical biomes still lags behind that obtained in the temperate zone, considerable progress has been made towards an evaluation of the extent of below-ground biodiversity in these systems. Hitherto, unsuspected diversity has been found in AM as well as in ECM fungal assemblages. The challenge now is to bring some of the hitherto undescribed species of both functional groups into culture, so that their functional attributes can be assessed. Research of this kind is a matter of urgency, not only because we need to gain a better appreciation of the roles of these fungi in the nutrient dynamics of tropical forests, but also because widespread devastation of these biomes through human activity is threatening to drive potentially valuable species to extinction before they have even been described.

16

Mycorrhizas in ecological interactions

Introduction

As shown in Chapter 15, mycorrhizal symbioses are prevalent in all major terrestrial biomes. The reasons for this prevalence are not always clear from our current understanding of function of the symbioses determined in pot experiments using single plant species, particularly in relation to whole plant nutrition. In consequence, researchers are now actively addressing the questions posed in the second edition (Smith and Read, 1997), that relate to the biology of mycorrhizal plants in natural environments (see Chapter 15). In particular, we asked whether there are impacts of mycorrhizal colonization on the fitness of individual plants? Do mycorrhizal fungi influence the outcome of competitive interactions? If so, under what circumstances and in what ways are the effects mediated? Investigations have explored the contributions of mycorrhizas to fitness of individuals and species, to outcomes of competitive interactions between species and to the possibility that mycorrhizal status may change both the ability of plants to coexist and the diversity of plant assemblages. With the wider recognition of structural and functional diversity among different mycorrhizal partnerships has come the appreciation that the interdependence of plant and fungal partners may lead to changes in fungal, as well as plant communities.

Experimental approaches to determining which interactions are important in field situations and to understanding the mechanisms that underlie them at physiological, biochemical and molecular levels are by no means easy. In particular, the widespread, indeed universal, occurrence of mycorrhizas means that the appropriate non-mycorrhizal control treatments that are necessary to satisfy experimental designs, require elimination or marked reduction of mycorrhizal colonization. As yet no wholly satisfactory field treatments have been found (Hartnett and Wilson, 2002). The commonly-utilized fungicide applications are not specific to mycorrhizal fungi, with consequent problems of determining whether the outcomes are related to reduction in activity of mycorrhizas or other susceptible groups, such as plant pathogens or saprotrophs. The consequence has been that many physiological ecologists, interested in unravelling details of the interactions and their underlying mechanisms, have necessarily relied on pot or 'microcosm' experiments to increase their understanding of natural communities or the outcomes of field manipulation. The failure of some pot experiments to demonstrate benefits in terms of plant growth and nutrition, coupled with field experiments which showed no increases in the efficiency with which roots took up nutrients, led to the suggestion that there may be other benefits of the symbioses that partially explain their persistence in natural vegetation. Even though it is now appreciated that mycorrhizal fungi may make marked contributions to nutrient uptake in the absence of growth or nutrient responses (see Chapters 4, 5 and 15), one of the consequences of this research has been increased understanding of multitrophic interactions involving soil microflora and fauna, herbivores, plant pathogens and mycorrhizal plants. Moreover, ubiquity of mycorrhizal colonization does not necessarily imply universal benefit or increased fitness. It might simply mean that there is no strong selection against the symbiosis.

This chapter will consider some of the observations and experiments that have been carried out in the field and in pots with the aim of increasing our understanding of mechanisms by which mycorrhizas influence plant and fungal communities. It will also briefly cover multitrophic interactions that are even more difficult to study experimentally, but are likely to have significant impacts in natural ecosystems.

Those interested in following the extensive literature covering development of ideas and investigations relating to the parts played by mycorrhizas in ecological interactions may find three books (Allen, 1991, 1992; van der Heijden and Sanders, 2002) and a number of key reviews particularly valuable. The work of Mike and Edith Allen in the USA played an important part in broadening ecological thought to include the roles of mycorrhizal associations in plant succession and competitive interactions (Allen and Allen, 1984; Allen, 1991, 1992). Major contributions, particularly emphasizing the importance of mycorrhizas in facilitating access to recalcitrant sources of nutrients and hence diversifying pathways of nutrient cycling, have been made by David Read and Jonathan Leake (Leake *et al.*, 2004a; Leake and Read, 1997; Read *et al.*, 2004), while important roles in recognizing non-nutritional benefits of mycorrhizas and in questioning received wisdom generally have been played by Alistair Fitter and collaborators (Fitter, 1991, 2001, 2005; Robinson and Fitter, 1999). The most recent compilation of a wide range of topics is to be found in 'Mycorrhizal Ecology' (van der Heijden and Sanders, 2002), a multi-author volume that reflects much current thought in this area.

Roles of mycorrhizas in mediating effects at the level of single plant species

Overview

Searches of the literature have revealed that experimental approaches to infer or determine mycorrhizal function in the field have most often been applied to communities in which many of the plant species form arbuscular mycorrhizas, rather than one of the other major types. Several factors may have contributed to this apparent bias. First, AM-dominated communities, such as grasslands, prairies, meadows, tropical forests and some temperate woodlands, are often more floristically diverse than those forests and heathlands where most of the species form ectomycorrhizal or ericoid mycorrhizas (see Chapter 15). In consequence, the 'AM communities' hold more fascination for plant ecologists interested in mechanisms underlying interspecific plant interactions and consequences for plant diversity. It is also true that these plant communities are often dominated by grasses and forbes that are easier to manipulate in the field and in pots than the large woody perennials typical of 'ECM communities'. Nevertheless, large numbers of experiments have been done with economically important ECM trees, directed towards improving forestry production (see Chapter 17). A third reason for prevalence of research on roles of arbuscular mycorrhizas is that it is in these symbioses that major variations in occurrence of colonization and plant response have been highlighted and hence questions relating to evolutionary advantage of 'facultative' mycorrhizal status have been raised and explanations sought. In contrast, it is generally accepted that the ECM or ERM states are an essential requirement if the plants are to access the predominant sources of nutrients, particularly organic N, in those ecosystems that they inhabit (see Chapters 9, 11 and 15).

The occurrence (and presumed activity) of mycorrhizas along successional gradients has provided some insights into their possible roles, without resort to manipulation to eliminate the fungal symbionts. Allen and Allen (1984) reviewed available literature and concluded that arbuscular mycorrhizas were likely to be relatively unimportant in disturbed and early successional stages characterized by high availability of nutrients. In contrast, in low nutrient habitats and late successional stages, all plants were found to be colonized, with the implication that the symbioses played significant roles in the success and persistence of species found there. The authors also highlighted the possibility that AM symbioses may be very important determinants of the outcome of plant competition in habitats with high nutrient and water availability because the resultant high biomass production would lead to more intense competition, as suggested by Grime (1979 and see Chapter 15).

Selection appears to have favoured the prevalence of AM rather than ECM colonization in many ecosystems which are primarily P limited. However, it has proved quite difficult to demonstrate experimentally that plants growing in the field under natural conditions benefit from enhanced access to P in terms of whole plant uptake (Fitter, 1985, 1990), leading to the suggestions that impacts of AM colonization in the field are lower than would be expected from pot experiments. One problem stems from the earlier presumption that it is necessary to show either increased tissue P concentrations or efficiency of root uptake (e.g. inflow) in order to demonstrate that arbuscular mycorrhizas play a role in P nutrition. As discussed in Chapter 5, we now know that operation of the AM uptake pathway does not necessarily lead to increased P uptake. Accordingly, failure to demonstrate increased P inflow due to mycorrhizas (based on whole plant measurements) in the field should no longer be taken as evidence that the fungal pathway makes no contribution to plant nutrient acquisition. Indeed, studies employing radioactive tracers to track hyphal P uptake from root-free hyphal compartments buried in field soil have shown considerable AM contributions to P uptake, even in non-responsive crops (Schweiger and Jakobsen, 2000) (see Chapters 5 and 17). The approach could usefully be applied in natural environments. It remains true that the contribution of AM fungi may have no effect on overall growth or whole plant P uptake and that for positively responsive plants effects observed in pots may be greater than those seen in the field (Fitter and Merryweather, 1992). It is a major challenge of current and future research to elucidate what benefits operation of the AM nutrient uptake pathway may have under these conditions.

Several other explanations have been put forward to explain the discrepancy between field and pot experiments. Rates of growth of many plants in nature may be limited by environmental factors other than P deficiency, for example water shortage. In stress-tolerant species, growth rates may be inherently low, so that their P requirements can be satisfied by diffusion processes without involvement of mycorrhizal hyphae. However, even this idea may be challenged on the basis that a slow growing and long-lived root may develop a considerable zone of depletion as a consequence of prolonged nutrient uptake from the same soil zone (Smith FA et al., 2003b). Grazing of extraradical hyphae by arthropods, particularly collembolans, has been demonstrated to eliminate responses to AM colonization in some circumstances (Warnock et al., 1982; Fitter and Sanders, 1992). However, damage might be offset by enhanced nutrient turnover as senescent mycelium is grazed, with consequently increased nutrient availability (see below). It is important not to be defensive about apparent lack of field-based effects. Interactions are extremely complex and difficult to unravel. Increasingly, where appropriate experiments have been carried out, the importance of mycorrhizal fungi in mediating plant interactions, productivity and diversity are being revealed. In any event, the evolutionary persistence of mycorrhizal fungi as symbionts that make considerable demands on their plant partners for C, argues for benefits in the field that confer selective advantages.

Elimination or suppression of mycorrhiza development and functioning in the field to produce non-mycorrhizal controls has most frequently been attempted using the fungicide benomyl as a soil drench. Its addition to alpine grassland reduced colonization by AM fungi in a number of species, but had no effect on tissue P concentrations (Fitter, 1986). One interpretation of results such as this is, indeed, that AM colonization has little impact upon P nutrition of plants in the field. It is necessary, however, to recognize first that tissue concentration may not reflect whole-plant nutrient uptake and even the latter does not reveal the full contribution of the fungal symbionts (see also Chaper 15). Secondly, these studies suffer from the same essential weakness as do those with potted plants, in that most of them probe only a small proportion of the full life cycle of the plant. The extent of this weakness was highlighted by a study of bluebell, *Hyacinthoides non-scripta*, carried out in a deciduous woodland (Merryweather and Fitter, 1995a, 1995b). This work clearly demonstrated, apparently for the first time in a natural population of field-grown plants, that the rate of P uptake necessary to maintain a positive annual P budget can only be achieved in the AM condition. *H. non-scripta* is a perennial, vernal geophyte which characteristically dominates the herb layer of deciduous woodland throughout north-west Europe. It has a coarse root system made up of thick (0.5–10 mm) unbranched elements which are produced annually from the base of the bulb. A new bulb and root system are produced every year. The relationship between AM colonization and P uptake was explored by regularly sampling undisturbed plants of *H. non-scripta* throughout their annual life cycle (Merryweather and Fitter, 1995a). There was a rapid increase in the proportion of root length colonized by AM fungi over the period from root emergence in September (Autumn), to a maximum of over 70% in January and February (Figure 16.1a), even before the shoots appeared above ground. Thereafter, as evidenced by declining numbers of entry points (Figure 16.1b), new colonization slowed. From the time of root emergence, P inflow increased rapidly at a similar rate to that of colonization, although until December values were negative, indicating that net loss of P was occurring (Figure 16.1c).

Maximum inflows were reached during the photosynthetic phase, but these subsequently declined at the same rate as that of colonization. When curves were fitted to data for P inflow and per cent root length colonized, they demonstrated a very similar pattern, with significant correlation between the two variables (Figure 16.1d).

The individual plants of bluebell lose significant amounts of P, particularly at the end of the growing season, in seeds, old leaves and roots as they are shed. Glasshouse grown plants, lacking AM colonization, are unable to capture sufficient P from the soil to balance their P budget. They therefore end the season with a large P deficit, which could not have been sustained in the field. In a subsequent experiment, otherwise undisturbed colonies of *H. non-scripta* growing in the field were drenched with benomyl at two-monthly intervals over two years, a treatment which greatly reduced AM colonization without having any effect on P availability in soil (Merryweather and Fitter, 1996). This led to a large reduction of the P concentration of all vegetative parts, relative to that in colonies drenched with water. However, the flowers and seeds of the benomyl-treated plants had the same P concentration as the controls after the first season, reduction in their P status being observed only after two years (Figure 16.2). This suggests that when P uptake is restricted, *H. non-scripta* protects reproductive structures by selectivity allocating P to them.

Selective allocation to reproductive structures of P acquired by AM plants from P-deficient soils has also been observed in pot-grown plants of wild oats (*Avena fatua*) and tomato (*Solanum lycopersicum*) (Bryla and Koide, 1990; Koide *et al.*, 1988a),

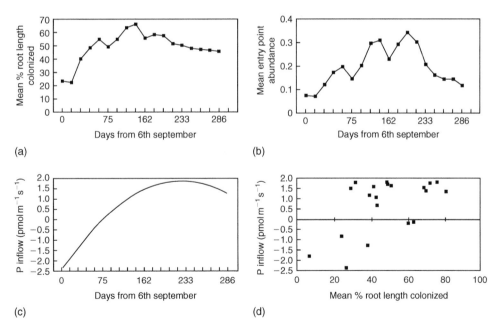

Figure 16.1 Seasonal pattern of AM colonization and P inflow in *Hyacinthoides non-scripta* growing in the field. (a) Mean per cent colonization; (b) mean abundance of entry points; the points on the curves are three-harvest running means; (c) fitted values for P inflow in the field; (d) correlation between P inflow and per cent colonization. Linear regression: $r^2 = 28.8, P = 0.15$. From Merryweather and Fitter (1995a), with permission.

wheat (Li *et al.*, 2005) and *Campanula rotundifolia* (Nuortila *et al.*, 2004). In the last case, although seed output was reduced in the AM plants, the seeds that were produced had higher relative growth rates in the next season than non-mycorrhizal counterparts. Detailed investigations of both male and female function in tomato have demonstrated considerable positive effects on total flower production, fruit mass, seed number and pollen production per plant and per flower, as well as *in vitro* pollen tube growth rates, which were again related to improved P nutrition (Poulton *et al.*, 2001, 2002). By these means, AM colonization may significantly influence fecundity of plants and so play a direct role in determination of fitness in the field, but more work on wild plants in natural environments is required. An extreme example of this effect is seen in the mast fruiting of ECM trees which may possibly be dependent on the apparently luxury accumulation of paid N and as a result of mycorrhizal activity (see Chapter 15).

Plants such as bluebell, with very coarse root systems, can be predicted on theoretical grounds to be responsive to AM colonization, but questions remain as to the role played by the symbiosis in plants such as grasses which, despite the fibrosity of their root systems, retain high levels of colonization in nature. As in the forbs, there are many glasshouse experiments with grasses demonstrating that enhancement of P capture can lead to increases of productivity, but such effects have been difficult to observe in natural communities. Hetrick *et al.* (1988, 1990), examining the responsiveness of two grass species that dominate the tall grass prairies of the USA, found evidence in the C3 species *Bromus inermis*, that, despite greater P acquisition in the AM condition, there was little or no growth response. In nature, *B. inermis* makes most of its growth in the cool seasons of autumn and spring and it is then that arbuscule production is at a maximum. Hetrick *et al.* suggested that early season growth enabled the plant to avoid competition with the other dominant species, *Andropogon gerardii*, a C4 plant which grows in the warm season and is extremely responsive to AM colonization. The authors proposed that benefits of P acquisition may only be expressed at a later growth stage in terms of increased fecundity and improved offspring performance. As the experiments were carried out with *B. inermis* alone in pots, it is likely that competitive advantages were missed and, in this case, it could be that luxury consumption of P early in the growing season enhanced the

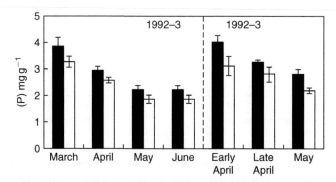

Figure 16.2 Effect of benomyl drench on the concentration of P in the leaves of *Hyacinthoides non-scripta*, measured at intervals over two growing seasons in the field. Black columns, control; white columns, benomyl. From Merryweather and Fitter (1996), with permission.

competitive ability of this grass by pre-empting availability of the element to other species. It cannot be emphasized too strongly that responsiveness to AM colonization determined in single plants in pots, whether positive, neutral or negative, is likely to be strongly modified by both inter- and intra-specific interactions in the field.

Density-dependent effects

For a plant population, density and pot size have effects on the productivity of individuals. A number of ecological studies indicate that the biomass (or other measure of success) of an individual plant is greatest when it is grown alone and declines as the planting density increases. Although the mycorrhizal status of the plants was not recorded in many of these investigations, the assumption must be that potential mycorrhizal hosts were colonized in untreated field soil and results should be interpreted accordingly. Experiments investigating the effects of AM colonization on intraspecific interactions of positively responsive species consistently confirm that AM plants perform best (as individuals) at the lowest planting densities and compete severely for soil resources at higher densities (Koide, 1991b; Allsopp and Stock, 1992b; Hetrick *et al.*, 1994a; West, 1996; Facelli *et al.*, 1999; Schroeder and Janos, 2004). In contrast, there is apparently little or no competition between individuals of the same species when grown without AM inoculum. These plants grow relatively poorly (as individuals), even at the lowest planting densities when non-mycorrhizal and their success (again as individuals) is not changed as density increases (Figure 16.3). In contrast, the negative responses of some plants to AM colonization, apparent with single plants or at low density, is sometimes lessened at high density (Hartnett *et al.*, 1993; Schroeder and Janos, 2004; Li *et al.*, unpublished), not because AM plants grow better at high densities but because the negative effects of density are greater for the larger, non-mycorrhizal plants. This can be explained by greater overlap of roots and hence zones of resource acquisition among the larger non-mycorrhizal plants, leading to greater competition.

Effects of density interact not only with AM colonization but also with P availability and the identity of the AM fungal symbionts, with outcomes varying again with responsiveness of the plants to AM colonization and P supply when grown

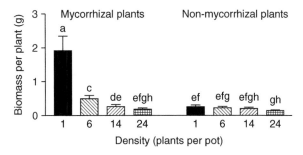

Figure 16.3 Effects of plant density on biomass of plants of *Trifolium subterraneum*, inoculated with *Gigaspora margarita*, or not inoculated. Bars with different letters are significantly different. Results from E. Facelli. See Facelli *et al.* (1999).

singly (Schroeder and Janos, 2004; Li *et al.*, unpublished). Rather surprisingly Facelli *et al.* (1999) found no influence of reduced irradiance on the effects of density on AM responsiveness. In their investigations, intraspecific competition intensity was increased by AM colonization at both low and high irradiance, due to an increase in capture of soil P. The increase in competition was reflected in the greater size-inequality of low-density AM treatments. They came to the important conclusion that the influence of AM colonization at the level of individual plants cannot be expected to be apparent at the population level, because of the large modifying effects of density-dependent processes.

Results such as these carry several implications for effects of density on responsive plants:

1 potentially AM species may be unable to use soil resources effectively in the absence of colonization, so that increased population size increases root length density and the acquisition of nutrients by the population, but not the individuals, until a plateau is reached
2 colonized plants have a combined root and hyphal length density which permits effective use of a much greater volume of soil than non-colonized plants, with the consequence that interplant competition is potentially strong
3 responses to AM colonization will be greatest at the lowest planting densities, both on an individual and a population basis.

Applying similar arguments to negatively responsive species, explains why increasing density can minimize the extent of the negative response.

Most investigations have involved plants of the same age, but the situation when seedlings establish in competition with adult plants may be quite different. In non-competitive treatments, seedlings of *Hypericum perforatum* were more responsive to AM inoculation than adults, although the effect was positive in both cases. When adults and seedlings were in competition, the responses were reduced, more so in seedlings (Moora and Zobel, 1997). The authors concluded that the outcome was influenced by the fact that the negative effect of competition on the (small) seedlings was stronger than the positive effect of AM colonization. These results contrast with reports of benefits of AM on seedlings germinating in established plant communities (Grime *et al.*, 1987; Read, 1991; Gange *et al.*, 1993; Francis and Read, 1994; van der Heijden, 2004). The discrepancies are likely to relate to the nutrient uptake strategies of the competing species and their mycorrhizal responses in mixtures, both of which also influence the impact of AM colonization on plant biodiversity (see below).

A recent investigation of the importance of ECM links in facilitating establishment of seedlings has highlighted similar benefits (Nara, 2006a). The investigation of willow (*Salix reinii*) seedling establishment on a nutrient poor, volcanic desert on the slopes of Mt Fuji, Japan capitalized on the lack of spore-based inoculum at the site. Details of the procedures have been described in Chapter 15. The salient features of the results were that spore-based inoculum had an insignificant role in colonization at the site, common mycelial networks (CMNs) growing from each mother tree facilitated ECM development and that the CMNs appeared to reduce competition for N between the large and small plants, regardless of the identity of the fungal partner and number of ECM root tips. The significance of CMNs for establishing seedlings and in mediating enhanced nutrient uptake and reduced competition has thus been demonstrated without resort to trenching, which has confounded previous investigations.

Competitive interactions in mixed-species assemblages

AM colonization certainly affects the relative performance of adults of different species in mixtures, altering the species diversity and growth of individuals. In one of the earliest published examples, outcomes of competition between the pasture grass *Lolium perenne* and the legume *Trifolium repens* were shown to be markedly influenced by AM colonization. *L. perenne* became extensively colonized but did not respond to colonization, presumably because its roots were relatively efficient at extracting P from soil. In contrast, the highly responsive *T. repens* only performed well in mixtures with *L. perenne* when mycorrhizal (Hall, 1978). Similarly, AM inoculation in pots changed the relative productivity and survivorship of a number of grassland species grown in mixtures in low P soil (Grime *et al.*, 1987). Analagous findings are commonly reported for groups and pairs of species, indicating how AM symbionts can markedly influence the co-occurrence of species, their competitive interactions and, in consequence, the biodiversity of ecosystems (Hetrick *et al.*, 1994a; West, 1996; van der Heijden *et al.*, 2003).

Unravelling the mechanisms by which mycorrhizal fungi alter relative plant performance in interspecific competitive situations is extremely difficult, in part because of the differences in nutrient acquisition and growth strategies of the competing plants and in part because of differences in effectiveness of different fungi. Responsiveness can be strongly modified by potentially competitive interactions and there are considerable dangers of predicting likely outcomes of competition from responsiveness of single plants. This is borne out by experiments with wild species. Growth responses of single species from tall-grass prairie were useful in predicting outcomes when the species were grown in mixtures. In this case, as mentioned above, the warm-season grass *Andropogon gerardii* was highly responsive and this character aligned with its dominance in competition. However, Marler *et al.* (1999), investigating the mechanisms underlying invasiveness of *Centaurea maculata* in grasslands in Montana, observed no positive growth response in this species or in *Festuca idahoensis* with which it successfully competes. However, when paired with large plants of *F. idahoensis,* to reduce the size imbalance of the two species, *C. maculata* was highly responsive to AM inoculation and outcompeted the grass, as it does in the field. The authors suggested that C transfer between the species might have been the basis for the effect in mixtures, but the same group later examined this idea in more detail and found no unequivocal evidence to support it (Zabinski *et al.*, 2002; Carey *et al.*, 2004). The most likely conclusion is that AM colonization conferred on *C. maculata* an enhanced ability to scavenge P which was, in the experiments of Zabinski *et al.* (2002), apparent as increased P concentrations in plant tissues. This may be another example of higher competitive ability, based on pre-emption of limiting resources.

Some insights into mechanisms were obtained from a pot experiment in which wild-type tomato (normally AM, but non-responsive in the soil used) was paired with a mycorrhiza-defective mutant derived from it (Cavagnaro *et al.*, 2004a). When non-inoculated, the two plants (both non-mycorrhizal) grew at the same rate and showed the same responses to P fertilization and to density of planting, whether as single genotypes or in competition with each other in a replacement series (Cavagnaro *et al.*, 2004a). Thus, the only strategy likely to confer enhanced nutrient uptake from low P soil was the ability of the wild type to form AM when inoculated. Despite potential differences in C balance between the wild type (AM) and the mutant (non-mycorrhizal), the outcome of competition in inoculated treatments in the replacement series was a clear growth benefit to the AM wild type, based on an increased ability to compete for P in the low nutrient soil. The growth response was only apparent in competition; the wild-type tomato showed either a neutral or negative response to colonization when grown alone.

The examples provided so far involve AM plants, which are presumed to access the same sources of inorganic nutrients as their fungal symbionts (see Chapter 5). Formation of ectomycorrhizas also allowed *Pinus elliottii* to take up more P from inorganic sources in competition with non-mycorrhizal plants of the grass *Panicum chamaelonche* than when competing with another pine (Pedersen *et al.*, 1999). The authors cautioned that, in nature, the outcome might be quite different because *P. chamaelonche* would then form arbuscular mycorrhizas which would improve its ability to access the available inorganic P.

A recent field experiment has demonstrated significant below-ground interactions between ECM pinyon pine (*Pinus edulis*) and co-occuring AM shrubs. During drought, field performance and root biomass of pine was lower in the presence of shrubs, suggesting below-ground competition for resources (Figure 16.4a). Furthermore, when shrubs were removed experimentally, pine growth both above and below ground was increased. At the same time, ECM colonization doubled (Figure 16.4b), although the diversity of the fungal assemblages was unaffected (McHugh and Gehring, 2006). These results support several previous observations of effects of such processes as trenching to remove competition on performance of target species. The authors suggested that effects were particularly marked in their experiments because of water-limitation at the site.

However, for many plant species, organic N and P are increasingly recognized as major sources of nutrients. This is especially the case in plants forming ECM or ERM (see Chapters 9, 10, 11 and 15), where the fungal symbionts increase the diversity of available resources to include both inorganic and organic forms. The versatility is of great importance in nutrient-poor habitats such as heathlands, allowing the plants to exploit nutrient pools that are unavailable to AM or non-mycorrhizal species, permitting them to coexist with them without direct competition for N or P. Aerts (1999) has highlighted the changes that may occur as inorganic N in soil rises as a consequence of atmospheric N deposition. The consequence could well be that AM grasses like *Molinia caerulea* and *Deschampsia flexuosa*, dependent on inorganic N, would increase in abundance at the expense of members of the Ericaceae which currently dominate the heaths (see also Chapter 15).

There have been rather few experimental investigations of the way ectomycorrhizas influence plant competition for natural sources of nutrients. Perry *et al.* (1989b) used a replacement series to investigate the way both specific and generalist ECM fungi influenced competitive interactions between *Pseudostuga menziesii* and *Pinus ponderosa*, which co-occur in forests of south-west Oregon. They used soil from the litter and humus layers of such forests, which was sterilized before plants, inoculated or not with particular ECM fungi, were transplanted into the

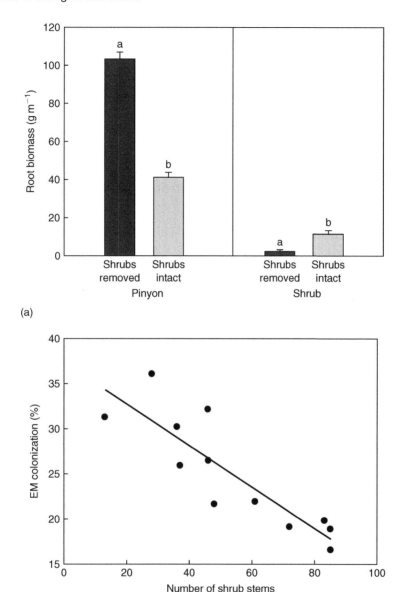

(a)

(b)

Figure 16.4 Effects of shrub removal on growth and ectomycorrhizal colonization of pinyon pine (*Pinus edulis*). (a) Above-ground shrub removal decreased shrub root biomass. In response, pinyon pines with associated shrubs removed had three times greater root biomass than pines with intact associated shrubs. Bars are means and standard errors. a and b indicate significant differences. (b) Relationship between ectomycorrhizal colonization (%) of mature pinyon pine trees and number of shrub stems remaining in the pinyon pine zone. Redrawn from McHugh and Gehring (2006).

pots. Without inoculation (but with mycorrhizas formed by contaminant *Thelephora terrestris*), the two tree species inhibited one another, but this inhibition disappeared when pots were inoculated with two ECM fungi specific for either *P. menziesii* or *P. ponderosa* (*Rhizopogon vinicolor* and *R. ochraceorubens*, respectively). When four fungi were inoculated (the two specific fungi plus two generalists), *P. menziesii* grew significantly better in mixtures than alone and growth of *P. ponderosa* did not decline. Results indicated that the generalist fungus *Laccaria laccata* enhanced N and P uptake by *P. menziesii* and reduced luxury consumption by *P. ponderosa*, but not sufficiently to reduce biomass production. The effect on growth appeared not to be the result of increasing numbers of fungi but rather depended on the presence of effective fungal symbionts. This point is important in relation to the discussion below.

Effects of AM on plant diversity

The influence of mycorrhizas on plant competition can be safely expected to feed through to changes in plant coexistence and biodiversity. However, the outcomes of experiments have been quite varied, so that it is important to analyse the reasons for the variations and, where possible, determine the mechanisms. Grime *et al.* (1987), using microcosms, reconstructed a plant assemblage representative of that occurring in nutrient-deficient calcareous soils in north-west Europe. The plants were a mixture of grasses and herbs, all but two of which (*Arabis hirsuta* and *Rumex acetosa*, from typically non-host families) were known to be heavily colonized by AM fungi in the field. The plants were grown from seed for one year in a sward of the dominant grass *Festuca ovina*, which had been pre-sown either in the non-mycorrhizal condition or as AM individuals which provided a natural source of inoculum for the subordinate plants. Survivorship was monitored throughout the year and the impact of colonization on biomass and on final structure of the community was determined at a single harvest. Among the grasses only *Holcus lanatus* showed a positive response to AM colonization in terms of dry weight; the others were either unaffected or reduced in growth (Table 16.1). Those forbs that are normally colonized by AM fungi showed significantly higher growth in the inoculated than the non-inoculated microcosms (Table 16.1). In contrast, *A. hirsuta* was more productive in the non-mycorrhizal community. Survivorship can provide a more direct index of fitness than productivity, especially in nutrient stressed habitats such as calcareous grasslands, as long as the survivors produce sufficient viable seed. Absence of AM inoculum led to major reductions of survivorship in the AM forbs (Table 16.2), but the reverse was observed with *A. hirsuta* and *R. acetosa*. Only small numbers of individuals of these plants were still alive after six months in AM microcosms. This study provides some insights into the roles of AM colonization in inhibiting the establishment of seedlings in closed turf. The non-host species, together with others, have been shown to be 'turf incompatible' and, in consequence, are relegated to ruderal situations (Grubb, 1976, 1977; Fenner, 1978). Similar negative effects of AM colonization on non-host species have been observed both in microcosms and in field situations (e.g. O'Connor *et al.*, 2002; van der Heijden *et al.*, 1998b).

It has been suggested that sensitivity to the presence of external AM mycelium is a factor determining growth and survivorship of non-hosts. *R. acetosa* and *A. hirsuta*, as well as a range of hosts and non-hosts, were grown in the same pots as AM 'nurse

Table 16.1 Effects of AM inoculation on shoot dry matter production per plant of species grown together in microcosms for 1 year.

	Minus AM inoculation	Plus AM inoculation	Effects of infection
Grasses			
Anthoxanthum odoratum	5.02	5.40	—
Briza media	13.55	54.26	*
Dactylis glomerata	160.19	286.84	NS
Festuca ovina+	922.68	609.43	***
Festuca ovina	24.14	22.02	*
Festuca rubra	27.67	25.37	*
Poa pratensis	15.07	19.34	NS
Forbs			
Arabis hirsuta[1]	0.26	0.13	—
Campanula rotundifolia	0.73	4.20	**
Centaurea nigra	1.70	10.90	***
Centaurium erythraea	0.23	7.08	—
Galium verum	1.87	9.39	**
Hieracium pilosella	0.93	7.63	***
Leontodon hispidus	0.83	3.72	**
Plantago lanceolata	3.62	33.96	***
Rumex acetosa[1]	9.72	8.77	NS
Sanguisorba minor	5.06	17.14	***
Scabiosa columbaria	2.46	10.19	***
Silene nutans	16.85	44.89	***

Data from Grime et al. (1987). [1]Non-host species; not colonized when inoculated. Festuca ovina was introduced to the microcosms as small plants (+) or as seed.

Table 16.2 Survisorship (%) of forbs after 6 months in mycorrhizal and non-mycorrhizal microcosms.

Species	Mycorrhizal	Non-mycorrhizal
Centaurium erythraea	64	2
Galium verum	58	11
Hieraceum pilosella	49	6
Leontodon hispidus	42	13
Plantago lanceolata	71	10
Sanguisorba minor	53	6
Scabiosa columbaria	84	16
Arabis hirsuta	8	42
Rumex acetosa	11	60

Significant increases of survivorship were obtained in most forbs grown in the mycorrhizal condition, the exceptions being the non-hosts Arabis hirsuta and Rumex acetosa, which show the reverse trend.

plants', with the roots separated by mesh that allowed passage of AM mycelium. The mycelium could therefore interact with the roots of the test species, which showed very different responses to its presence (Francis and Read, 1994, 1995). The 'non-hosts' (e.g. *A. hirsuta*) grew poorly and showed a reduction in survivorship, but hosts such as *P. lanceolata* and *C. erythraea* responded positively. The basis of the antagonistic effect of AM fungi upon non-host ruderals remains to be elucidated. In some cases, adverse effects upon root development have been observed in the absence of colonization by the fungus, suggesting that there may be a chemical interaction (Allen *et al.*, 1989; Francis and Read, 1994), whereas in others, inhibition is associated with penetration of the root and prolific production of vesicles which might induce a significant C drain (Francis and Read, 1995). However, the explanation may be even simpler. Although the roots of the plants remained separate, it seems most likely that hyphal exploitation of the 'extra' soil within the mesh conferred a nutritional advantage on the hosts and reduced growth of the non-hosts was the result of competition. An interaction of this type has been observed in a similar experimental system, with the tomato mutant (*rmc*) growing with *Allium porrum* as the nurse plant (Cavagnaro *et al.*, 2004b). Regardless of the mechanisms of interaction, which certainly require more detailed cytological and physiological exploration, it is clear that AM fungi can be seen as major determinants both of the structure and biodiversity of plant communities.

The microcosm experiment of Grime and co-workers provided one of the first clear demonstrations that AM fungi are potentially major determinants both of the structure and diversity of plant assemblages. The effect of arbuscular mycorrhizas in increasing diversity was the result of marked reduction in growth of the dominant species, *F. ovina* and of the increased survival of some very AM responsive species, such as *C. erythraea*, although species richness was not actually changed. The relative performance of canopy dominants and subordinate species is crucial to the outcome. In this plant assemblage, *F. ovina* showed a negative growth response to AM inoculation despite high colonization.

Similar results have been obtained more recently, again using grassland or old-field plants grown in microcosms. van der Heijden *et al.* (1998b) showed increased diversity and, in one experiment, increased productivity which was not observed in the experiments of Grime *et al.* (1987). However, the two investigations were similar in that the response of the dominant grass species to AM inoculum was either neutral or negative. Thus in both these examples plant diversity was shown to increase with AM colonization and this was the result of improved growth and survivorship of AM subordinates, associated with dominants that actually had their own competitive ability reduced by AM colonization. Thus, there are two possible contributors to the effects: increased ability of the subordinates to access nutrients, by virtue of the nutrient scavenging ability of the AM mycelium growing from their roots, and reduced vigour and hence reduced competition by the dominants.

Two field investigations, backed by pot experiments, have given results that at first sight conflict with these microcosm experiments. As we have seen (above and Chapters 4 and 5), tall-grass prairie communities are dominated by warm-season, C4 grasses that are highly responsive to AM colonization. The plant assemblages also include cool-season, C3 grasses that are somewhat less responsive, and a wide range of forbs that show quite variable responses, at least in pot experiments. The interactions between the species were investigated in the field by selective removal of dominant C4 grasses and by suppression of AM fungi over two seasons by application of the fungicide benomyl (Smith *et al.*, 1999). Both treatments increased plant species richness and the benomyl treatment also increased Shannon diversity index. The dominant and highly responsive *A. gerardii* and the subdominant and responsive *Bouteloua curtipendua* both declined when AM colonization was suppressed in the field, whereas the forbs and grasses (with variable responsiveness in pot experiments) increased in abundance. The similar effects of removal of dominants and benomyl application suggests that competition from the dominants is a major factor influencing floristic diversity in this system.

Similarly, the semi-arid herbland in South Australia investigated by O'Connor *et al.* (2002) was dominated by *Medicago minima*, an annual weed that was shown in pots to be highly responsive to AM colonization. Two other weeds, *Carrichtera annua* (Crucifereae, annual non-host) and *Salvia verbenaca* (Labiatae, short-lived perennial and non-responsive host), also made important contributions to overall biomass and there were minor contributions from a range of native species of variable responsiveness. Application of benomyl over one season effectively suppressed AM colonization compared with controls, either unwatered or receiving water only. Application of water as an additional control was necessary in this environment where rainfall is low, highly variable and unpredictable. Comparison of the two control treatments showed that watering increased productivity, but there were no differences from the unwatered control in terms of relative numbers of plants or biomass of each species. Suppression of AM activity markedly reduced the aboveground biomass and numbers of plants of the dominant, *M. minima*. *C. annua* and *S. verbenaca* both showed large increases in biomass, accompanied in the latter by greatly increased seedling survivorship. There were small differences in the other (native) species (*Velleia arguta*, *Erodium crinitum* and *Vittadinia gracilis*) between the benomyl and watered treatments (Figure 16.5). These alterations in biomass and abundance were not reflected in any net change in species richness, but diversity (+29%, Shannon *H'*) and evenness (+32%, Shannon *J*) both increased in mycorrhiza-suppressed plots. The differences in diversity came from changes in contribution of the three major species (Figure 16.6). It appeared that when AM activity was suppressed, the competitive ability of the dominant *M. minima* was strongly reduced, with consequent advantages to *C. annua* and *S. verbenaca* that did not depend on their host status but rather on their lack of marked responsiveness. It is likely, however, that *S. verbenaca* utilizes the AM nutrient uptake pathway when colonized, most probably permitting this species to coexist successfully with *M. minima*.

We must conclude that the effects of AM colonization on plant diversity are not absolute and are strongly influenced by the responsiveness of the species in the community as predicted by Bergelson and Crawley (1988). Attempts have been made to construct conceptual models to aid understanding (Hartnett and Wilson, 2002; Urcelay and Diaz, 2003). It seems clear that in the investigations undertaken so far it is the responsiveness and hence competitiveness of the dominants that has the largest influence. When these are positively responsive then suppression of AM colonization is likely to increase diversity because the subordinates are released from competition. In contrast, when dominants are negatively responsive, AM suppression will decrease diversity because the subordinates will lose the competitive advantage conferred by AM symbioses. The balance of experimental evidence suggests that mycorrhizal responsiveness of the dominants, rather than subordinates, plays a very significant role (see Urcelay and Diaz, 2003), but investigations should be expanded to encompass a wider range of vegetation types.

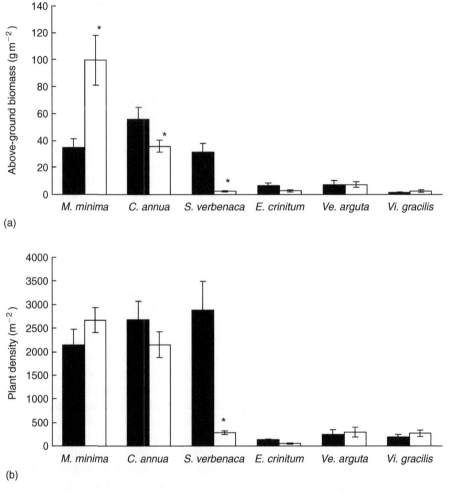

Figure 16.5 Response of major plant species in field plots in a semi-arid herbland to suppression of AM colonization with benomyl (black bars) compared with watered plots (open bars). (a) Above-ground biomass; (b) plant density. An asterisk above the bar indicates that the control is significantly different ($P < 0.05$) from the fungicide-treated plots for that species, determined by Tukey's HSD test. Species shown accounted for >90% of plot biomass ($n = 7$). Species are: *Medicago minima, Carrictera annua, Salvia verbenaca, Erodium crinitum, Vellea arguta, Vittadinia gracilis*. Reproduced from O'Connor *et al.* (2002), with permission of the *New Phytologist*.

Rather little work of this type has been carried out in ECM communities. However, observations in some tropical forest ecosytems suggest that ECM symbioses are of importance in promoting the ability of some tree species to develop monodominant stands, which contrast with highly diverse rainforest often present on similar soils in close proximity and associated with AM fungi (Connell and Lowman, 1989). This effect has been discussed in Chapter 15.

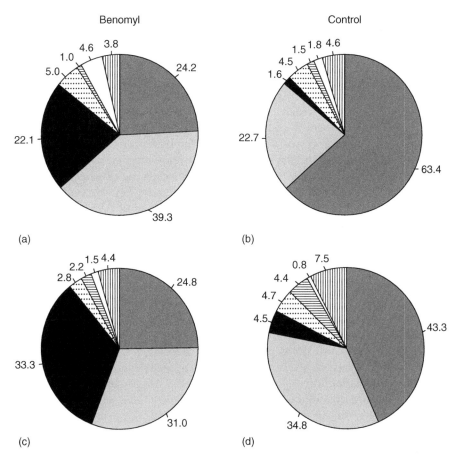

Figure 16.6 Species composition (%) of total aboveground biomass (a & b) and total number of individuals (c & d) in plots with AM fungi suppressed (Benomyl a, c) or AM fungi active (Control b, d). Species included are *M. minima* ▫; *C. annua* ▤; *S. verbenaca* ■; *Ve. arguta* ▨; *Vi. gracilis* ▤; *E. crinitum* ▫; and Others ▥. Data represent the mean of seven replicates.

Effects of the fungal assemblage

In the discussion so far, the possible effects of differences in composition of the AM fungal assemblages to competitive outcomes and hence biodiversity and productivity have been ignored. However, the large diversity of function between different plant–fungus combinations, as well as selectivity in choice of partners, means that changes in the fungal community with respect to both numbers and identity of species could change plant interactions (see Chapters 1 and 4). In their microcosm experiment, van der Heijden *et al.* (1998b) compared the effects of adding four species of AM fungi separately and in combination, on the growth of the component plants and on outcomes in terms of measures of plant productivity and biodiversity. They found, not unexpectedly, that there were large differences in response of individual plants to the fungi inoculated separately and that the mix of four fungi had different effects again. Multiple inoculation was sometimes more effective in producing a response than single inoculation and the authors suggested that plant diversity and also productivity would increase with increasing numbers of AM fungi in the soil assemblage, although in this experiment there were actually no changes in productivity.

Somewhat more persuasive evidence was provided by a second, field experiment which used up to 14 randomly selected AM fungal species inoculated onto a mix of 15 plant species. In this case, productivity in terms of plant biomass and hyphal length density did increase with increasing numbers of fungi and there was also an increase in diversity, all of which parameters reached an asymptote at about eight fungal species per microcosm. P in the plants increased and P in soil fell, indicating that increasing exploitation of soil P was one of the bases for the outcome. Johnson *et al.* (2004) provided some support for this finding, showing that increased P concentrations in the shoots of *Plantago lanceolata* were related to the diversity of AM fungi that developed in the roots, after culture in microcosms with different plant treatments. However, the interpretation that increased diversity of AM fungi is a driver for increased plant diversity and productivity has been criticized on the basis that the results of many experiments are confounded by 'sampling effect' (SE), where the chances of including a fungus conferring a large benefit to a plant increases as the number of AM fungi included in the test increases (Wardle, 1999). This effect was observed in the experiment of Perry *et al.* (1989b; see above), in which increases in productivity of the two tree seedlings when inoculated with four ECM fungi were due to the positive effects of *L. laccata* on *P. menziesii*, and not to the ECM fungal diversity.

Other investigations have failed to show a consistent link between mycorrhizal fungal species richness and plant productivity. The effects of ECM assemblage on growth and productivity of seedlings of two trees were examined in two soils of different nutrient availabilities (Jonsson *et al.*, 2001). Mycorrhizal effects on plant growth varied among the eight ECM fungi and between soils, when the plants were inoculated with one fungus alone. Only in one case (*Betula pendula* on low fertility soil) was there a clear positive relationship between fungal species richness and plant productivity that could be attributed to effects not related to SE. A negative relationship was observed between fungal species richness and productivity of *Pinus silvestris* on high fertility soil; in the other two cases no effects related to fungal species richness were observed.

The observations that productivity is not necessarily related to fungal diversity should not come as a surprise. As with the effects of presence or absence of AM fungi (see above), the outcome will be strongly influenced not only by the ability of the fungi in the assemblage to colonize the roots, but also by the responsiveness of the plant species that make the largest contributions to the overall biomass. In the microcosm experiment of van der Heijden *et al.* (1998b), there was no change in productivity as the number of AM fungi increased, because the plant species that made the largest contributions to biomass were either non-responsive or negatively responsive and this did not change when they were inoculated with four species rather than one. Unfortunately, data for root colonization by AMF were not presented and, in any event, it would have been hard, using the techniques then available, to determine the contributions of the different fungi in mixtures to the overall colonization in the different species. Differences between fungi in type and infectivity of inoculum, as well as their competitive abilities, which are poorly understood (see Chapter 2), are likely to be significant in determining root occupancy and hence the ability to obtain resources for effective exploitation of the soil. In contrast to most investigations in which fungal species richness refers to the numbers of fungi inoculated at the start of the experiment, Jonsson *et al.* (2001) showed that all inoculated ECM fungi did survive to the end of the experiment. Nevertheless, in some cases the proportions of root tips colonized by different fungi showed that they were not present in equal proportions, so that some competitive reductions had occurred.

The significant feedback in the way plants influence AM fungal communities, and *vice versa*, has mostly been studied at the level of spore production. Several investigations have shown positive relationships between numbers of AM fungal spores and numbers of plant species in a community (Burrows and Pfleger, 2002; Chen *et al.*, 2004). These findings support the observation that plant species-richness is sometimes positively correlated with AM fungal biomass (Hedlund *et al.*, 2003). Landis *et al.* (2004) analysed the species composition of AM spore assemblages in an oak wood, taking advantage of natural gradients in plant species composition. They showed a strong correlation between plant species-richness and AM fungal species-richness. These studies for the most part did not measure plant fitness or productivity in relation to the characteristics of the AM fungal species assemblage. However, Hedlund *et al.* (2003) did observe a negative correlation between AM biomass (determined from the signature fatty acid 16:1ω5) and plant biomass, suggesting that extensive root and soil colonization by the fungi was not a driver for high plant productivity.

Bever (2002a, 2003) examined the community dynamics of co-occurring plant and AM fungal species at a grassland site, in terms of growth and spore production as surrogates for 'fitness'. He provided considerable evidence for the existence of asymmetric fitness relationships and negative feedback between plants and AM fungi. He showed that, generally, a fungus delivers the greatest benefit to one plant species, but grows better (produces more spores) on another and expressed the relationships in terms of feedback models (Figure 16.7). He did not find evidence for positive feedback, in which a fungus that delivers the greatest growth benefit to a plant also receives the greatest benefit from it, a situation that would lead to the evolution of specificity and 'best friend partnerships' (see Chapter 1). These findings are highly significant, because they help to explain several of the general observations about the specificity, occurrence and function of AM symbioses in complex

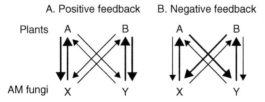

Figure 16.7 Representation of fitness sets that would produce positive and negative feedback within the interaction of plants and AM fungi. In these figures, the thickness and direction of the arrows represent the relative benefit that two plant types (A and B) and two AM fungal types (X and Y) receive from their association. (a) In the case of a highly symmetric fitness relationship between plants and AM fungi, an initial abundance of plant A will result in an increase in representation of AM fungus X, which will increases the growth rate of plant A and thereby generate a positive feedback that can lead to the loss of diversity. (b) Alternatively, in the case of a highly asymmetric fitness relationship, an initial abundance of plant A will increase the representation of AM fungus Y and thereby boost the performance of plant B, resulting in a negative feedback on plant A. Reproduced from Bever (2002), with permission.

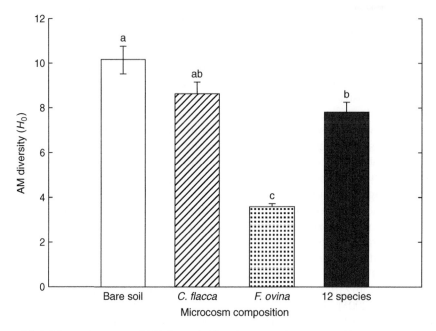

Figure 16.8 Mean diversity of AM fungi (*Ho*) colonizing the roots of bioassay seedlings of *Plantago lanceolata* removed from grassland microcosms after 12 weeks growth. The microcosms had bare soil (open bars), *Carex flacca* (hatched bars), *Festuca ovina* (stippled bars) or a mixture of 12 plant species (black bars). Means ± standard errors of means, $n = 3$. Bars with the same letter are not significantly different ($P > 0.05$). From Johnson *et al.* (2004), with permission of the *New Phytologist*.

communities. Under negative feedback, specific advantageous symbioses will not be able to evolve, fluctuating dynamics in plant and fungal success will occur and co-occurrence of competing species will be maintained, leading to species-richness in both plant and fungal communities. As we have seen, this is the pattern that best fits our present knowledge of AM symbioses as in most cases non-specific associations with considerable functional diversity, that are characteristic of floristically rich ecosystems (Chapter 1).

Bever's analyses are based on spore assemblages, which do not provide information on the AM associations developed with roots and therefore potentially significant for plant growth and nutrition. However, advances in techniques to identify AM sequence variants have allowed the diversity of the AM communities to be followed in roots of *Plantago lanceolata* growing in microcosms containing 12 species of plants from limestone grassland, compared with those in microcosms containing bare soil or monocultures of the non-host *Carex flacca* or the highly mycorrhizal *Festuca ovina* (Johnson *et al.*, 2004, 2005). AM fungal diversity was significantly influenced by floristic composition (Figure 16.8) but, surprisingly, the greatest species richness (Shannon H_o) was found in the bare soil and *C. flacca* microcosms and the least with *F. ovina*. The authors suggested that colonization from soil without plants or supporting a non-host was initiated from spores of a large number of species that had survived the 3-year pretreatment without plants. In contrast, active mycelium in soil was most likely to have initiated colonization in those microcosms supporting host species. If this explanation is correct, colonization in *P. lanceolata* would then reflect the assemblage of fungi that occupied the roots of these host plants and, in consequence, any pre-exisiting plant–fungus selectivity. Even if spores were present in these soils, colonization from them would most likely have been slower and result in lower root occupancy than that from the mycelium. Multifaceted investigations of this type will increasingly be needed to understand the highly complex interactions and feedbacks between assemblages of plants and fungi that are gradually being revealed. Moreover, the length of the experiments will be an important consideration, especially with respect to determining any temporal differences in colonization from different sources of inoculum and changes in fungal assemblages which might possibly be related to the occurrence of *r*- and *K*-selected species.

The role of mycelial networks in distributing nutrients among members of a plant assemblage

As outlined in earlier chapters, there is considerable evidence that plants may be linked into common mycelial networks (CMNs), one for each of the mycorrhizal fungi present at a particular site. There seems little doubt that CMNs are an important means by which seedlings become mycorrhizal when establishing in the neighbourhood of colonized plants. The roles of the networks in redistributing nutrients among the linked plants has been extensively discussed, but there is still no general consensus on their quantitative significance or indeed on mechanisms underlying any transfers. It seems quite likely that the processes facilitated by CMNs vary among different mycorrhizal types. Available data have been thoroughly summarized and reviewed in recent years (Robinson and Fitter, 1999; Simard *et al.*, 2002; Simard and Durall, 2004).

   One of the first demonstrations of movement of ^{14}C-labelled compounds between AM species was obtained in the microcosm experiment of Grime *et al.* (1987), discussed above. Immediately prior to harvesting the microcosms, $^{14}CO_2$ was fed to shoots of *Festuca* and the pattern of distribution of label was determined. The transfer of radioactivity occurred almost exclusively in the microcosms containing AM fungi and only the plant species which were actually colonized contained high amounts of ^{14}C. These observations provide evidence that plants in mixed assemblages are interconnected by their AM fungi. It was also suggested that net movement of organic C between plants was an important contributor to the outcomes in the microcosms in terms of plant growth. However, Bergelson and Crawley (1988) pointed out that a more likely explanation was that subordinates were released from competition. Numerous studies have followed up this observation, some of them investigating possible source-sink effects by experimental manipulations such as shading. In many cases, ^{14}C was certainly transferred from the shoots of one plant to the AM roots of another, but this does not demonstrate transfer to the plant cells of the receiver nor the quantitative contribution of transfers to overall C budgets. It does show that the plants are linked by a common AM fungus and it does indicate that there is the potential for plants to share the expense of maintaining the fungal network and its nutrient uptake activities, which might have important consequences for the relative benefits of AM colonization to the plants involved (see Chapter 4). The extent of contributions to the CMN by different plants might very well be influenced by changes in source-sink relationships.

In a study of the possible roles of AM colonization in mediating interactions between the invasive weed *Centaurea maculosa* and the native grass *Festuca idahoensis*, Marler *et al.* (1999) showed that neither species responded positively to AM inoculation when grown singly in pots. However, when grown together, *C. maculosa* exerted much greater competitive effects in inhibiting the growth of *F. idahoensis* when the plants were inoculated with AM fungi than when they were not. The researchers then used ^{13}C labelling and pots compartmented with mesh, to explore the possibility that C transfer from *F. idahoensis* to *C. maculata* was the basis for successful competition (Zabinski *et al.*, 2002; Carey *et al.*, 2004). The experiments of Zabinski *et al.* (2002) found no evidence for C transfer in either direction between the plants, despite the fact that growth differences were the same as those previously reported. They also demonstrated that *C. maculosa* had significantly higher P concentrations when grown in pots with a grass, with their root systems separated by mesh, and concluded that the weed was more effective at exploiting its AM symbiosis for P acquisition than the native grasses. Carey *et al.* (2004) re-examined the possibility of net C transfer but, despite obtaining some evidence that it occurred, on the basis of distribution of ^{13}C, they were not able completely to discount the involvement of differences in water-use efficiency in altering ^{13}C distribution patterns. Differences in obtaining P from a common mycelial network, or indeed from separate networks formed by different AM fungal species, remain the most parsimonious explanation for competitive outcomes in this case.

The two investigations outlined above studied plants growing at the same time, in pots. Lerat *et al.* (2002) explored the possibility of seasonal differences in C transfer between the deciduous tree, *Acer saccharum* (sugar maple) and the geophyte *Erythronium americanum* (trout lily), with the plants grown in fibre pots located in a maple forest. *E. americanum* produces its leaves in spring, around the time of maple bud-burst, but before the maple leaves are fully photosynthetic. Later, as the maple canopy closes, *E. americanum* dies back and survives as a corm, which produces new roots in early autumn. The potential for these two species, both of which form arbuscular mycorrhizas, to provide ^{14}C to each other is therefore offset between seasons. Labelling with ^{14}CO$_2$ was carried out in both spring and autumn. The results were suggestive of some transfer from *E. americanum* to *A. saccharum* in spring, including to the expanding leaves, which implies transfer from fungus to plant and redistribution to the shoot. No label was found in adjacent ECM *Betula alleghaniensis*, arguing for transfer through the AM mycelial network. In autumn there was also a suggestion that ^{14}C moved from *A. saccharum* to roots of some of the *E. americanum* plants. There was no labelling in the corms, so in this case it is possible that ^{14}C was retained in the fungal structures within the *E. americanum* roots.

Despite some intriguing findings, it is probably fair to say that there is still little unequivocal evidence for transfer of ^{14}C-labelled compounds in nutritionally relevant amounts from one plant species to another based on transfer in an AM mycelial network. The mechanistic bases for the transfers have not been explored and, as outlined in Chapter 5, it appears unlikely that C moves from fungus to plant as the C skeletons of amino acids or other organic N molecules (Pfeffer *et al.*, 2004; Govindarajulu *et al.*, 2005; Jin *et al.*, 2005).

Evidence for transfer of organic nutrients between plants linked by an ECM network is much more persuasive, but significance in terms of outcomes of competitive interactions between plants has not been established either in pots or in the field.

However, the very existence of mycoheterotrophic species such as *Monotropa hypopitys* depends on C transfer between autotroph and heterotroph via common mycelial links (see Chapter 13). The ability of ECM mycelial links to function as potential conduits for the transfer of organic C was demonstrated by autoradiography after feeding $^{14}CO_2$ to the shoots of 'donor' plants (Finlay and Read, 1986a; see Chapter 8). Label was transported through the interconnecting mycelium and accumulated in the colonized roots of 'receiver' plants. The fragility of these hyphal links is such that it is difficult to trace them in natural soil. However, indirect evidence for the existence of such links in the field was provided by the observation that $^{14}CO_2$ fed to an adult plant of *Pinus contorta* was subsequently detected in roots of neighbouring plants of the same or of different species, providing they were also ectomycorrhizal (Read *et al.*, 1985). Only small amounts of activity were found in neighbouring AM species. The greatest amounts of radioactivity were detected in a number of 'receiver' plants that had been subjected to artificial shading during and after the period of isotope feeding to the 'donor'. This suggested that carbon transfer might be influenced by sink strength.

Experiments of this type have, together with similar investigations of AM transfer, been justifiably criticized on the grounds that they are only capable of revealing unidirectional transport and hence do not take into account the possibility of C transfer in the opposite direction (Newman, 1988; Jakobsen, 1991; Robinson and Fitter, 1999). In order to establish unequivocally that net transfer is occurring, it is necessary to demonstrate that one of the interconnected plants gains more material than the other in an exchange. In an enlightening study using double labelling methods, Simard *et al.* (1997a, 1997c) demonstrated that significant net transfer of carbon can occur in interspecific combinations of plants colonized by shared ECM symbionts. The two ECM plant species used in the study, birch (*Betula papyrifera*) and Douglas fir (*Pseudotsuga menziesii*), co-occur naturally in mixed wet forests of British Columbia, where they share seven of 11 ECM morphotypes, which occupy 90% of the root tips of both species (Simard *et al.*, 1997b). The likelihood of the occurrence of interconnection between the species was therefore great. For comparative purposes a further co-occurring species, *Thuja occidentalis* forming arbuscular mycorrhizas, was included in the study. Individual one-year-old plants of each species were planted 50 cm apart in triangular arrangements. The triangular groups were allowed to grow for either one or two years before experimental procedures were imposed. Four to 6 weeks prior to labelling, *P. menziesii* plants were subjected to three shading treatments; deep shade, partial shade and ambient light. *P. menziesii* or *B. papyrifera* in each group were then pulse-labelled (2 h) with either $^{13}CO_2$ or $^{14}CO_2$. The labelling pulse was followed by a 9-day chase under the same shading regimes.

Transfer of organic C occurred in both directions, with significant net transfer from *B. papyrifera* to *P. menziesii* (Figure 16.9a, b). Redistribution from roots to shoots in amounts that were much higher than previously observed in ECM systems was also found (Finlay and Read, 1986a). This is an important point of difference from experiments with plants linked by AM fungi and confirms that transfer of C, possibly as the C skeletons of N compounds, occurs from fungus to plant (see Chapter 9). Nevertheless, the recent suggestion that N transfer occurs as inorganic ammonium/ ammonia throws some doubt on this mechanism (Chalot *et al.*, 2006).

In the experiment carried out one year after planting, net transfer to the conifer was found only when it was grown in full sun. The amount transferred represented 2% of the total isotope fixed by both species, 4% of the isotope assimilated by

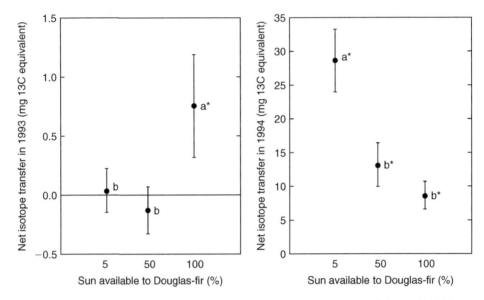

Figure 16.9 Net isotope transfer between *Betula* and *Pseudotsuga*, in full sun (100%) or in 50% or 5% of full sun. (a) 1993; (b) 1994. Means denoted by the same letter do not differ significantly (*P* = 0.01). See text for explanation. Data from Simard (1995), reproduced with permission. See also Simard *et al.* (1997c).

B. papyrifera and 7% of that assimilated by *P. menziesii*. In the 50% and 5% of full sun treatments, isotope transfer to *P. menziesii* was balanced by transfer to *B. papyrifera* (i.e. there was zero net transfer). In the second year after planting, net isotope transfer represented on average 6% of the total isotope fixed, averaged over all shading treatments, and in the deepest shaded it was double that in the 50% or full sun treatments. This amount of transfer represents a substantial carbon gain by *P. menziesii* and is similar to that which has been considered sufficient to improve growth and survival of connected ramets in some clonal plants (Hutchings and Bradbury, 1986; Alpert *et al.*, 1991). No net transfer to the AM *Thuja* occurred, implicating the ECM networks in the transfers between plants.

Both spatial and temporal factors were considered to have contributed to the increased net transfer and to the effect of shading upon C transfers in the second year. In the spatial context, below-ground proximity of roots of the potentially interacting species, as well as the numbers of fungal interconnections between them, were likely to be larger two years after planting than one year. Of perhaps greater importance was the fact that the application of isotope was carried out in August of the second year, fully one month after cessation of shoot elongation in *P. menziesii*, whereas the first year feeding experiment took place in July when shoot activity was ongoing. The likely importance of seasonal effects upon carbon allocation below ground has been stressed earlier (see Chapter 8).

Net transfer from *B. papyrifera* to *P. menziesii* coincided with whole seedling net photosynthetic rates which were 1.5 and 4.3 times greater for *B. papyrifera* than *P. menziesii* in full sun and full shade, respectively, and occurred where foliar N concentrations were 1.2–6.7 times higher in *B. papyrifera* than *P. menziesii*. It is therefore possible that transfer was influenced by gradients of assimilate and/or nutrient concentration between the two species. Since rates of photosynthesis of *B. papyrifera* in full sun were so much greater than those of *P. menziesii* in shade, it is likely that carbon supply to the colonized roots of *Betula* would be greater than to those of *Pseudotsuga*. This, combined with the fact that the *Betula* plants contained significantly more N than those of *P. menziesii*, may have contributed to the sink effect observed.

Simard *et al.* (1997a) examined the roles of the CMN in more detail in a pot experiment in which the same two plant species were grown in separate mesh pouches that allowed the development of extensive mycelial links between them. Direct transfer through the mycelium was prevented in half the microcosms by severing the links just before the labelling pulse was applied. As in the field experiments, both *B. papifera* and *P. menziesii* received isotope from their neighbours and there was an indication that, again, *P. menziesii* received more C from *B. papifera* than *vice versa*, although differences were not significant. Surprisingly, severing mycelial links between the plants had no significant effect on transfer between the plants, so that the relative importance of mycelial links, *vis à vis* transfer through soil must remain in doubt.

In the experiments described by Simard, transfer of C in some form of combination with N, perhaps in the form of amino compounds, is a possibility. This was examined by Arnebrant *et al.* (1993), who investigated transfer of the products of $^{14}CO_2$ and $^{15}N_2$ fixation by ECM and actinorhizal *Alnus glutinosa* to neighbouring ECM *Pinus contorta* in laboratory microcosms. They showed in two out of three experiments that material labelled with both isotopes moved between the plants and appeared not only in the ECM roots of the *Pinus* receivers, but also in the shoots. Transfer in the reverse direction was not investigated, so that net transfer could not be calculated. The plants were not separated by mesh and there were no nonmycorrhizal or AM control plants, so it remains possible that the transfers observed took place indirectly through soil, bypassing the mycelial network. Evidence for the operation of both transfer pathways was found in a field study by He *et al.* (2006), following labelling of shoots of donor plants of *P. sabiniana* with $K^{15}NO_3$. ^{15}N appeared most rapidly in the roots of AM annuals growing nearby (*Cynosurus echinatus*, *Torilis arvensis* and *Trifolium hirsutum*) but, after 4 weeks, was also detectable in their shoots as well as in the roots and shoots of woody species which formed ecto- and/or arbuscular mycorrhizas (*Quercus douglasii*, *Ceanothus cuneatus* and *P. sabiniana*). These results indicate that the CMN is not the only pathway through which N transfers between plants occur in the field and confirm, as shown earlier for AM systems (McNeill and Wood, 1990), that an indirect pathway involving N cycling in soil can result in considerable N movement.

All in all, the studies with C and N isotopes indicate that there is a much greater likelihood of net movement via CMNs formed between ECM species than AM species and that movement through the soil must not be ignored in some environments. There is still much to be learned about the biochemical pathways associated with transfer of C compounds between symbionts but, where gradients of C and N co-occur, there is a strong possibility that organic N compounds may be involved, at least in ECM systems. There remains a major gap between demonstration of nutrient transfers between plants via CMNs and elucidation of their significance both in plant growth and in competitive interactions between individuals and species. The whole plant and community level effects will be even harder to unravel, but of great importance in understanding vegetation dynamics.

Multitrophic interactions

The mycorrhizosphere

The recognition of stimulatory effects of some classes of bacteria on the processes of ECM colonization of roots has heightened awareness of the complexities of microbial interactions in soil. The mantles of ECM roots have long been known to provide a significant habitat for bacteria (Foster and Marks, 1966; Nurmiaho-Lassila *et al.*, 1997) (Figure 16.10), with assemblages distinct from those associated with uncolonized roots (Rambelli, 1973). As techniques enabling more detailed examination have been developed, a great deal more about the bacterial associates of ectomycorrhizas has been revealed. Although we are still a very long way from understanding the diversity of interactions, it is now certain that there can be considerable differences in bacterial populations influenced by identity of plants and ECM fungi, as well as by soils. Interactions of ECM roots and mycelium with saprotrophic fungi has received much less attention, probably in part because of the long-held view that bacteria dominate activities in the rhizosphere (de Boer *et al.*, 2006). Nevertheless, recent work is beginning to demonstrate both that saprotrophic fungi have important roles in rhizosphere processes (Mougel *et al.*, 2006) and that inter-actions between them and mycorrhizal fungal symbionts should not be ignored.

Scanning and transmission electron microscopy by Nurmiaho-Lassila *et al.* (1997) revealed a large number of morphologically different types of bacteria in habitats provided by intact ectomycorrhizas formed between *Pinus sylvestris* and *Suillus bovinus* or *Paxillus involutus* grown in natural forest soil. With *S. bovinus*, not only were bacteria found on the mantle surfaces, but also in inter- and intracellular locations

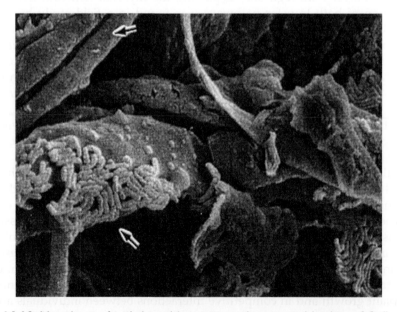

Figure 16.10 Monolayer of rod-shaped bacteria on the external hyphae of *Suillus bovinus*. Bar = 5 μm. Reproduced from Nurmiaho-Lassila *et al.* (1997), with permission.

in the mantle and Hartig net. Mycelial strands of the fungi had relatively sparse bacterial populations, whereas extensive monolayers of bacteria developed on the fine hyphal branches. In contrast, ECM root tips of *P. involutus* harboured few bacteria whether on the surface or internally, but the intact external mycelium supported both bacterial colonies and solitary bacteria. In a subsequent investigation, Timonen *et al.* (1998) showed differences in metabolic activity between the bacteria associated with the different fungal symbionts, although these were not as great as differences associated with the two soils used. Use of PCR and RFLPs enabled the detection of sequences belonging to the Archaea associated with external mycelium of both ECM fungi and uncolonized humus, but not from non-mycorrhizal short roots (Bomberg *et al.*, 2003). Such differences in microbial populations were carried through to the populations of protozoans dependent on them, with numbers of protozoa higher in *S. bovinus* mycorrhizospheres than in those of *P. involutus* (Figures 16.11 and 16.12; see also Colour Plate 16.1) (Timonen *et al.*, 2004).

The roles of bacterial communities in association with ECM fungi have not received much concerted attention, but could include activities that increase nutrient availability in various ways. One such possibility is that bacteria may enhance the production of low molecular weight organic acids and that this may increase the capacity of ECM mycelium to weather minerals (Leyval and Berthelin, 1991), with consequent release

(a) (b) (c)

Figure 16.11 Roots of *Pinus sylvestris* used in studies of the protozoan populations associated with (a) non-mycorrhizal short roots (NM), (b) ectomycorrhizal short roots (MR) and (c) extraradical mycelium (EH) of *Paxillus involutus* (see Figure 16.12). Reproduced from Timonen *et al.* (2004) with permission. See also Colour Plate 16.1.

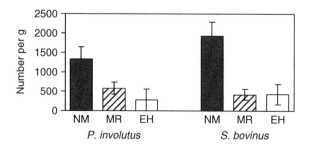

Figure 16.12 Numbers of culturable protozoa on uncolonized short roots (NM), mycorrhizal root tips (MR) and extraradical mycelium (EH) of roots of *Pinus sylvestris* mycorrhizal with either *Paxillus involutus* or *Suillus bovinus* (see Figure 16.11). Reproduced from Timonen *et al.* (2004), with permission.

of a range of nutrients such as P, K and Ca which could then be transferred to the plant symbionts (Griffiths *et al.*, 1991, 1994; Wallander, 2000; Landeweert *et al.*, 2001; Hoffland *et al.*, 2004; see Chapters 10 and 15). The quantitative contribution of mineral-weathering to nutrient acquisition may be rather low, but could become increasingly significant as increased anthropogenic N depositions change the nutrient balance of some forest ecosystems. Another possibility is that bacteria in the ectomycorrhizo-sphere or more closely associated with ECM mantles might have sufficient N_2 fixing capabilities to make significant contributions to forest N balance. This, as pointed out by Izumi *et al.* (2006) and Timonen *et al.* (1998) requires further investigation.

The suggestion that bacteria in ECM rhizospheres might be directly involved in the dynamics of mycorrhiza-formation was made by Bowen and Theodorou (1979). They showed, *in vitro*, that the ability of *Rhizopogon luteolus* to colonize roots of *Pinus radiata* was enhanced by the presence of some bacterial isolates, but inhibited by others. Subsequently, Garbaye and Bowen (1987) examined the process of mycorrhiza formation by three fungal symbionts in different steam-sterilized soils, inoculated with microbial populations from one of the soils. There were differences in responses, dependent on both soil and fungus and positive effects outnumbered negative ones. Subsequent isolation of bacteria from surface-sterilized *P. radiata–R. luteolus* mycorrhizas showed that most were fluorescent pseudomonads and that 80% of them had positive effects on ECM formation by this plant–fungus combin-ation (Garbaye and Bowen, 1989a, 1989b). The term mycorrhiza-helper-bacterium (MHB) was coined to describe these microorganisms. The populations of fluorescent pseudomonads have been further characterized for different ectomycorrhizas. In one case, functional characterization *in vitro*, including a range of attributes such as production of HCN and antibiotics, N_2-fixing and phosphate-solubilizing activ-ity, led Frey-Klett *et al.* (2005) to suggest that the ectomycorrhizosphere controls *P. fluorescens* populations in such as way as to select strains potentially beneficial to the symbiosis. In reviewing the topic, Garbaye (1994) suggested five hypoth-eses to explain the beneficial effects of MHBs. These include facilitation of coloni-zation either by production of cell wall softening enzymes or enhancement of the plant–fungus recognition processes. More work is needed to establish clear relation-ships between effectiveness of MHBs, enzyme production and wall softening using a functionally relevant test system. Nutritional enhancement of fungal growth by MHBs is also a possibility but, in view of the apparent specificity of the effect, it must be of a highly specialized type. Detoxification of compounds present in soil by MHBs could, indirectly, provide a nutritional effect and it is noteworthy that many of the responses to MHBs have been observed in artificial media or nursery soils which have been subjected to sterilization treatments. These have the poten-tial to release toxins and so influence subsequent microbial interactions. Changes in the rhizosphere, possibly leading to the production of chelating agents such as siderophores or stimulants that enhance germination of fungal propagules, might also contribute to the effectiveness of MHBs. There remains considerable scope for experimental analysis of the roles of these bacteria. New molecular methods will help to reveal details of the rather specific interactions and subsequently their underlying mechanisms.

The significance of interactions between ECM fungi and fast growing and destruc-tive wood and litter decomposing saprotrophs, which together dominate the micro-bial communities in highly organic forest soils, has been highlighted by Leake *et al.*

(2002). They point out how few investigations in ecologically relevant situations have been carried out, despite the potential for interactions between the two fungal groups to have major consequences for nutrient cycling. The early field investigations of Gadgil and Gadgil (1971, 1975), which showed that exclusion of ECM fungi by trenching increased litter decomposition by saprotrophs, was found to be difficult to repeat. Recently, the sharp demarcation between ECM, monospecific stands of *Dicymbe corymbosum* and AM-dominated, species-rich rainforest provided a useful system in which to investigate the influences of ECM on litter decomposition (Mayor and Henkel, 2006). The ECM stands naturally have much higher litter accumulation than the AM forest. Trenching in these stands showed no effect whatever of exclusion of ECM fungi on either the fraction of litter remaining after 12 months or in concentrations of major nutrients, N, P and Mg. There were, however, significant effects on potassium and calcium concentrations. Likewise, there were no marked differences in decomposition of litter samples transferred between the two forest types, although again Ca concentrations were higher when ECM fungi had no access to the litter bags. The results of this study strongly argue against suppression of saprotrophs by ECM fungi in this environment (see Chapter 15).

Despite these negative findings, evidence for direct interactions has been forthcoming from experiments using microcosms containing natural soil. It appears that spatial redistribution of nutrients such as P in the mycelial systems of saprotrophs such as *Hypholoma fasciculare* can lead to increased nutrient availability to ECM fungi, either via natural turnover of the mycelium or even via antagonistic effects in the zones where two mycelia interact (Lindahl *et al.*, 1999, 2001). The review by Leake *et al.* (2002) presents data on the interactions between two ECM fungi and the wood-destroying fungus *Phanaerochaete velutina*. Using a combination of $^{14}CO_2$ feeding to the shoots of *Betula pendula* to follow C allocation in the mycelium of associated *Paxillus involutus* or *Suillus bovinus*, and measurements of the extent of mycelial development of *P. velutina* growing from wood blocks, they were able to show that growth of both types of fungi was inhibited in the zones where the mycelia interacted. Although this work did not extend to providing information on redistribution of N or P, it did reveal the fierce competition between the two types of fungi, which is almost certain to feed back into changes in nutrient availability and litter decomposition shown by the 'Gadgil effect'. These effects now deserve re-examination, taking into account the potential changes in below-ground litter as a result of ECM formation itself (Langley and Hungate, 2003). Modification of litter quality and abundance will influence decomposition, nutrient cycling and, hence, plant establishment and succession (Facelli and Facelli, 1993; see Chapter 15).

Interest in the rhizospheres of AM plants has increased in recent years (Azcón-Aguilar and Barea, 1992) and it is clear that AM development can have important effects on soil microflora and fauna, although results are very variable (see Wamberg *et al.*, 2003). Formation of arbuscular mycorrhizas influences both the quality and quantity of root exudation and rhizodeposition (Graham *et al.*, 1981; Marschner *et al.*, 1997; Johnson *et al.*, 2002; Jones *et al.*, 2004). Marschner and Crowley (1996) used *Pseudomonas fluorescens* strain 2-79RL to follow effects of AM formation in *Capsicum* on bacterial activity. This *P. fluorescens* strain harbours a ribosomal promoter coupled to a lux gene cassette and it therefore emits light when in exponential growth phase. Results showed reduced bacterial activity in the rhizospheres of plants colonized by two *Glomus* species. Analysis of the bacterial rhizosphere communities of maize using PCR-Denaturing Gradient Gel Electrophoresis (DGGE) led to the conclusion that AM colonization changed the bacterial community structure both on the root surface and in the soil at greater distances from the root (Marschner and Baumann, 2003). The latter had presumably been invaded by external AM mycelium. Not unexpectedly, differences have also been revealed between the rhizospheres of plants inoculated with different AM fungi (Marschner and Timonen, 2005). In two of these investigations, some effects appeared to be transmitted via the plant between separate compartments of split pots. Negative effects of AM fungi on growth of saprotrophic fungi in the rhizospheres of three plant species were suggested by some data of Smith SE *et al.* (2004a), but have not yet been substantiated. Despite these findings, the functional relevance of changes in bacterial and fungal inhabitants of rhizospheres remains to be revealed.

Grazing by protozoans, collembolans and mites significantly influences bacterial and fungal numbers in density- and time-dependent ways, so that much of the variation in bacterial communities between different experiments may be explained on these bases. For example, Rønn *et al.* (2002) found no effect of mycorrhiza-formation by *Pisum sativum* on bacterial numbers in the rhizosphere (determined from plate counts), but did show marked increases in protozoa that were grazing on and presumably controlling the bacterial numbers. In a more detailed analysis, Wamberg *et al.* (2003), using both plate counts and DGGE, showed that bacterial numbers were unaffected by AM colonization, but specific bacterial groups grew preferentially in the rhizospheres of non-mycorrhizal *P. sativum*. The effects were mediated by changes in rhizodeposition and grazing by protozoans. Moreover, fungivorous members of the soil fauna have the potential to damage the external mycorrhizal mycelium with consequent effects on function, as well as on nutrient cycling (see below).

Interactions between mycorrhizas, pathogens and endophytes

There are many reports of the interactions between mycorrhizal colonization of plants and the incidence and severity of diseases caused by plant pathogens. The effects are variable and influenced by plant nutrition, relative density of the inoculum of the pathogen and mycorrhizal fungus, and whether or not the plants were mycorrhizal before being challenged with propagules of the pathogen (Harley and Smith, 1983; Graham, 1988; Linderman, 1992; Fitter and Garbaye, 1994). The main instances of arbuscular mycorrhizas reducing disease are for root-infecting fungi and nematodes. Shoot pathogens are usually unaffected or their effects increased in AM plants. A number of mechanisms to explain lower disease losses in AM plants have been suggested. These include: competition for colonization sites, so that prior occupancy by a mycorrhizal fungus reduces the opportunities for colonization by pathogens; mobilization of plant defence mechanisms during AM colonization, discussed in Chapters 3 and 6; and improved nutrient status, which increases the resistance of the plants to attack by disease organisms and increased tolerance of disease, particularly root damage. Indeed, direct negative effects of AM hyphae on *Fusarium* within mycorrhizal transformed carrot roots have been observed (Benhamou *et al.*, 1994) and there are reports of decreased development of the pathogen *Phytophthora parasitica* as a result of both localized and systemic induced resistance in mycorrhizal and non-mycorrhizal parts of AM root systems of tomato (Cordier *et al.*, 1998).

Although these interactions have been most closely studied in relation to crops, they are highly relevant to the influence of mycorrhizas in natural ecosystems where pests and pathogens may exert considerable influence on community structure. This was shown to be the explanation for the apparently beneficial impact of AM fungal colonization upon the annual grass *Vulpia ciliata* (Newsham *et al.*, 1994, 1995). Using benomyl to reduce AM colonization of the roots in the field, West *et al.* (1993a) showed that there was no relationship between extent of AM colonization and the rates of P uptake. However, the fungicide also controlled weakly pathogenic fungi such as the cosmopolitan *Fusarium oxysporum*, which was known to reduce fecundity of the plant. The benefits of AM fungi were suggested to be the result of their ability to protect the plant from pathogens (Newsham *et al.*, 1994). This possibility was examined further in field-grown populations of *V. ciliata* exposed to different concentrations of benomyl so as to control the extent of colonization by both AM and pathogenic fungi (Newsham *et al.*, 1995). Fecundity was shown to be largely unchanged by fungicide application, despite the fact that benomyl significantly reduced the abundance of all fungi in roots. However, the abundance of root pathogenic fungi, especially *F. oxysporum*, was negatively correlated with fecundity, even though plants displayed no disease symptoms. The poor relationship between fecundity and benomyl application contrasted markedly with the effects of benomyl on AM and pathogenic fungi and with the negative effects of root pathogens on fecundity. These effects could be explained if the two groups of fungi interacted, so that when both were reduced in abundance by fungicides the positive and negative effects on the plants cancelled each other out, so that the net effect on fecundity was slight.

The interaction between AM fungi and root pathogens was further investigated using a transplant approach (Newsham *et al.*, 1995). Seedlings of *V. ciliata* were grown in a growth chamber with a factorial combination of inoculum of *F. oxysporum* or a *Glomus* sp., both isolated from *V. ciliata* at the field site. They were planted into a natural population of the grass in the field. After 62 days of growth, clear evidence was obtained that colonization by *Glomus* gave a protective effect. Plants inoculated with *Glomus* performed as well as control plants, even when simultaneously inoculated with *F. oxysporum*, whereas those inoculated with *F. oxysporum* alone grew less well (Table 16.3). The *Glomus* sp. had no net effect on the performance of the plants in the absence of the pathogen. Analysis of P status of the tissues showed that there was no correlation between shoot P concentration and the abundance of either type of fungus in roots. Rather, the differences between treatments

Table 16.3 Effects of a factorial combination of *Fusarium oxysporum* (F) and a *Glomus* spp. (G) on shoot biomass and root length of *Vulpia ciliata* plants grown in the laboratory, transplanted into the field and sampled from the field after 62 days' growth.

Variable	Treatment				Main effects		Interaction
	−G −F	+G −F	−G +F	+G +F	G	F	G × F
Log (ln) shoot biomass (mg)	2.4a	2.2a	1.4b	2.2a	$F = 3.5$	$F = 17.3^{***}$	$F = 4.8^{*}$
Root length (cm)	217a	203a	111b	228a	$F = 9.4^{*}$	$F = 9.0^{**}$	$F = 9.0^{**}$

Data from Newsham *et al.* (1995). Means are of 16 replicates; where followed by different letters they differ at $P < 0.05$. Significant main and interaction effects in ANOVA are indicated by: $^{*} P < 0.05$, $^{**} P < 0.01$, $^{***} P < 0.001$

seem to have been due to a reduction in the frequency of pathogenic hyphae within roots brought about by AM colonization. Of course, we now know that an additional contributor to the interaction could have been P uptake via the AM pathway, leading to enhanced competitive acquisition of P by AM plants despite the fact that they showed no net benefits compared with those in which AM colonization was suppressed (see above and Chapter 5).

Suppression of other organisms by AM fungi has been implicated in increased plant survival in nature. The root-feeding nematode, *Pratelenchus penetrans*, plays a significant role in decline of the dune grass *Ammophila arenaria* as dune succession proceeds. Results of de la Peña *et al.* (2006) demonstrate that colonization of *A. arenaria* roots by a suite of AM fungi isolated from dune systems significantly reduced nematode numbers and multiplication rates. The effects were not dependent on pre-establishment of the mutualists in the roots and operated locally. It was suggested that AM colonization was important in facilitating the persistence of *A. arenaria* in stable parts of the dunes, where burial of the roots no longer afforded protection from the nematodes.

Effects of ECM fungi in reducing disease have been known for some time. Both *Pisolothus tinctorius* and *Thelephora terrestris* have been shown to reduce the impacts of the root pathogen *Phytophthora cinnamomi* on *Pinus* spp. (Marx, 1969, 1973), whereas inoculation with *Laccaria laccata* reduced disease caused by *Fusarium oxysporum* in *Pseudotsuga menziesii* (Sylvia and Sinclair, 1983a), *Picea abies* (Sampangi and Perrin, 1985) and *P. sylvestris* (Chakravarty and Unestam, 1987a, 1987b). Mechanisms proposed to explain such effects include protection of the root by the physical barrier imposed by the fungal mantle, production of phenolic compounds in the plant tissues induced by the ECM fungus (Sylvia and Sinclair, 1983b) and antibiotic production by ECM fungi themselves. The last effect may operate before any root colonization has occurred for Duchesne *et al.* (1988a, 1988b) observed that inoculation of *Pinus resinosa* with *Paxillus involutus* reduced pathogenicity of *Fusarium oxysporum* and that increases in seedling survival were associated with a sixfold reduction in sporulation of the pathogen in the *Pinus* rhizosphere. Ethanol-soluble compounds in the rhizosphere exerted fungi-toxic effects within 3 days of ECM inoculation. Such suppressive effects may be of considerable importance in seedling survival, both in natural environments and in forestry plantations. Little is known about the chemical bases for the antibiotic activity, although Duchesne *et al.* (1989a) showed that production of oxalic acid was related to suppression and Kope *et al.* (1991) isolated two antifungal compounds (benzoylformic and mandelic acids) from liquid medium in which *Pisolithus tinctorius* had been growing. It is regrettable that most of the experiments on disease suppressive and antibiotic effects of ECM fungi have been carried out under rather unrealistic conditions. Epidemiological studies under natural conditions are necessary to determine if, or at what stages, colonization by ECM fungi can reduce disease impacts and potentially influence plant communities.

A group of interactions that has received rather little attention is those between mycorrhizas and fungal shoot endophytes. Many plants, including a number of common grasses, harbour seed-transmitted, fungal endophytes from the Clavicipitaceae (e.g. *Neotyphodium* and *Epichloë*). The infections are either symptomless or cause suppression of flowering (Clay and Schardl, 2002). The fungi inhabit the intercellular spaces of the shoots and derive organic C from their photosynthetic partners; the latter receive benefits in terms of reduced herbivory. Several investigations have shown that the presence of fungal endophytes reduces the extent of AM colonization (Chu-Chou *et al.*, 1992; Müller, 2003; Omacini *et al.*, 2006), so that any mycorrhizal influences on the plants are potentially moderated by the endophytes. The mechanism of reduction has not been explored, but might be attributable to competition for organic C or production of toxic metabolites.

Although experiments have only been carried out on grasses, the widespread occurrence of shoot endophytes and their interactions with mycorrhizal colonization mean that it is risky to ignore them in investigations of the effects of either on plant communities. Significantly, benomyl reduces or eliminates shoot endophytes as well as mycorrhizal fungi, so that application to create non-mycorrhizal controls will also remove these symbionts and hence their modifying effects (Omacini *et al.*, 2006).

Interactions with soil fauna and above- and below-ground herbivores

Interactions of mycorrhizal fungi with the vast diversity of animals that inhabit the soil has received rather limited attention, probably because of the enormous difficulties of studying the highly complex interactions, effects of which can only be detected by combining sensitive experimental design with careful data analysis. Pot experiments involving additions of the organisms under investigation risk being too simplistic, whereas field experiments utilizing biocides are difficult to interpret because of the lack of specificity of the chemicals applied to suppress different groups. Recent reviews have addressed some of these complexities in relation to herbivores (Gehring and Whitham, 2002) and soil invertebrates (Gange and Brown, 2002).

It is well established that soil animals, such as collembolans, mites and nematodes play very significant roles in nutrient turnover, due to their effects in fragmenting litter and grazing on decomposer organisms such as fungi and bacteria. The grazing may also stimulate growth and reproduction of the fungi (Lussenhop, 1992). Analysis of effects of below-ground grazing on ectomycorrhizas has concentrated on fungivory, because most of the youngest root tissue is enveloped in mycelium. Ek *et al.* (1994a) examined the effects of different densities of the collembolan *Onychiurus armatus* on ecotymycorrhizas of *Pinus contorta* formed by *Paxillus involutus*. Impacts of fungivory on nutrient uptake by the extramatrical mycelium were examined by placing cups, containing $^{15}NH_4$ or phytin, to which the fungus alone had access, in the soil. Low densities of the collembolan induced greater development of ECM mycelium and increased uptake of and transfer of ^{15}N to the plants. Mycelial growth was reduced only at high densities of *O. armatus*. Collembolan populations did not increase in ECM treatments compared with non-mycorrhizal ones, possibly suggesting that alternative, preferred, food sources were present. In another investigation, ECM birch and pine plants were exposed to either a naturally complex microfaunal assemblage or a highly simplified one. The complex animal assemblage reduced colonization after 57 weeks, but actually increased shoot growth and N and P uptake compared to the plants that had received the simple assemblage (Setala, 1995).

In the case of AM fungi, considerable emphasis has been placed on the possibility that the grazing fauna may damage the fungal networks and hence reduce their capacity to deliver nutrients to associated plants (Fitter and Garbaye, 1994). It was thought that such effects might partially explain the apparent lack of AM responses under field situations. McGonigle and Fitter (1988a) applied the broad-spectrum insecticide chlorfenvinphos to a semi-natural, species-rich grassland in northern Britain and monitored the response of a constituent grass of the plant assemblage, *Holcus lanatus*. In the presence of the insecticide, which reduced density of collembolans to one third of the original population, there was a large increase in P uptake by the grass and a significantly greater shoot biomass. Clearly, the lack of specificity of the insecticide makes interpretation of these responses difficult, but increases of P uptake per unit root length or area is hard to explain by any mechanism other than one involving increased AM function when insects were suppressed.

Pot experiments with collembolans have sometimes suggested that the insects damage AM mycelial networks and consequently reduce the potential for enhanced P delivery to the plants (Warnock *et al.*, 1982). This finding was essentially confirmed by Finlay (1985) whose results did, however, indicate that the density of collembolans was important (as shown also for ectomycorrhizas). At low densities, AM responses were actually increased, possibly because increased nutrient mobilization was more important than mycelial damage. Increase in collembolan numbers did result in increased damage to AM hyphal networks of *Glomus intraradices* in an agar culture experiment (Klironomos and Ursic, 1998), but these authors also showed that AM fungal hyphae were not the preferred food source of *Folsomia candida*. This collembolan selectively grazed on *Alternaria alternata* or *Trichoderma harzianum*, in preference to *G. intraradices* and was more fecund when feeding on the conidial fungi, regardless of density. Using compartmented pots to separate roots from AM hyphae and collembolans, Larsen and Jakobsen (1996) again showed that *F. candida* probably did not graze on hyphae of *G. caledonium* or *G. intraradices*. Furthermore, at densities of the collembolan similar to those found in the field, there were no significant effects on the delivery of ^{32}P to plants of *Trifolium subterraneum* via the mycorrhizal pathway. Later investigations have confirmed that experimental outcomes can be highly variable, emphasizing the complexities of the interactions and their underlying mechanisms (Harris and Boerner, 1990; Kaiser and Lussenhop, 1991; Gange, 2000; Lussenhop and BassiriRad, 2005). Lussenhop and BassiriRad (2005) again confirmed that hyphae of *G. intraradices* were a minor food source for *F. candida*, as only about 5% of gut contents were fungal hyphae and there were no effects of the collembolan on length density of the external mycelium. This work also showed that the greatest N uptake by seedlings of *Fraxinus pennsylvanica* was in AM plants with moderate collembolan numbers. The effects of collembolans in damaging AM fungal mycelium and hence reducing nutrient uptake may well have been given undue emphasis, but the interactions and their outcomes in realistic field situations deserve more consideration. Recently, Johnson *et al.* (2005) demonstrated in a natural grassland that the collembolan *Protaphorura armata* at natural densities decreased C flux through soil measured after pulse labelling the sward with $^{13}CO_2$. Other data presented were strongly indicative of disruption of AM mycelium in soil by the collembolans.

Interactions between AM plants and the fungal-feeding nematode *Aphelenchus avenae* provide an interesting comparison with the work with collembolans (Bakhtiar *et al.*, 2001). This organism is also fungivorous and AM fungi do appear to support its growth. Populations declined when *A. avenae* was inoculated onto non-mycorrhizal plants of *T. subterraneum*, but increased markedly when the plants were colonized by either *G. coronatum* or *Gi. margarita*, with final populations much higher with *G. coronatum*. Presence of the nematode decreased root colonization by both fungi to

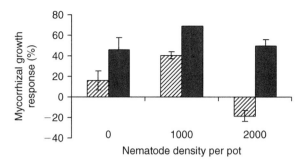

Figure 16.13 Mycorrhizal growth response (%), relative to non-mycorrhizal controls, of *Trifolium subterraneum* colonized by *Gigaspora margarita* (shaded bars) or *Glomus coronatum* (black bars) grown with different densities of the fungivorous nematode *Aphelenchus avenae*. Values are means, ±standard errors of means (*n* = 3). Data of Bakhtiar *et al.* (2001).

the same extent and decreased the percentage of spores with contents, markedly so for *G. coronatum*. It therefore appeared that *G. coronatum* was either more palatable or more accessible to the nematode. Despite this damage, mycorrhizal growth responses were increased at moderate nematode densities (Figure 16.13), again suggesting that increased hyphal turnover as a result of nematode grazing had beneficial effects that offset damage to the hyphae if nematode populations were not exceptionally high. Effects on the two fungi were again different, highlighting the diversity of interactions to be expected in complex soil environments.

Interactions between mycorrhizal fungi and above-ground herbivores are at least as complex as those with soil animals, but our knowledge is even more fragmentary. By surveying published literature, Gehring and Whitham (1994, 2002), have shown that both ECM and AM colonization is reduced by herbivory in about two-thirds of the investigations and either unaffected or very infrequently increased in the remainder. The predominance of negative effects has been attributed to a reduced capacity of the plants to support their fungal symbionts with organic C. The authors highlighted the facts that again most work has been carried out on AM plants and that investigations focused mainly on the extent of colonization and rarely explored outcomes in terms of functioning of the symbioses or consequences for community interactions.

Effects of mycorrhizal colonization on the herbivores has also been investigated, again with a predominance of AM investigations. In these, the generalist herbivores were more often negatively affected by AM colonization than the specialists, an effect attributed to the greater sensitivity of generalists to defensive compounds produced by their hosts and by the possibility that production of these compounds is likely to be enhanced in mycorrhizal plants.

AM hyphae and soil structure

It has long been recognized that hyphae of AM fungi are important binding agents in soil (Miller and Jastrow, 2002). This was implicit in the use of weight of adhering soil to estimate the length of external hyphae (Graham *et al.*, 1982b) and anyone who has washed roots out of soil knows that the mycorrhizal treatments are much harder work than uninoculated ones! Aggregation of sand grains by AM hyphae in dunes has been repeatedly confirmed (Koske, 1975; Koske et al., 1975; Clough and Sutton, 1976; Forster, 1979; Forster and Nicolson, 1979; Graham et al., 1982b). In these ecosystems, AM fungi are particularly important in stabilization, because they use recent photosynthate and, unlike soil saprotrophs, do not depend on readily available organic substrates from soil, which may be in relatively short supply. In many soils, it is also evident that roots and hyphae play a major role, together with other organic components in stabilization of aggregates.

Oades (1993) reviewed the contributions of biological processes to the development and stabilization of soil structure and showed that they vary with soil texture. Aggregate formation and stability depend more on organic matter and the activities of organisms in soils with relatively low clay contents. In many soils, there is evidence for the existence of a hierarchy of aggregates of different sizes, with the smaller ones progressively packaged together to form the larger ones. This hierarchy covers a range of sizes over many orders of magnitude (<2 to >2000 μm), with the smaller aggregates held together by stronger forces than the larger aggregates. Roots and fungal hyphae certainly stabilize macroaggregates (>250 μm) acting as temporary binding agents which hold together smaller particles (Tisdall and Oades, 1979, 1982; Tisdall, 1994, 1995). Hyphae are also probably important in stabilizing microaggregates. Clay particles (<2 μm) adhere to mucilage on the surface of hyphae. The importance of AM hyphae as part of this complex binding network was demonstrated by Tisdall and Oades (1979), who showed that the effectiveness of roots of *Lolium perenne* (ryegrass) in stabilizing aggregates >2000 μm was related to the length of hyphae in those aggregates. Roots of *Trifolium repens* had considerably less effect than those of *L. perenne*, despite the fact that AM colonization of roots was 50% and 13% respectively after 14 weeks' growth. This illustrates that it is the root and hyphal length densities in soil and not the per cent colonization of the roots that are important in this context.

The involvement of roots and hyphae in aggregate formation and stabilization has also been followed in a chronosequence of tall grass prairie restoration (Cook et al., 1988; Miller and Jastrow, 1990; Jastrow et al., 1998). In this case, the development of mycorrhizas was related to different root size classes. The analysis showed that fine roots and hyphae had significant direct effects on the geometric mean diameter (GMD) of water stable aggregates, while very fine roots had no direct effects. The indirect effects of both types of root were assumed to be related to their colonized lengths and hence to production of extraradical hyphae. Of the plant species in the communities, prairie grasses and members of the Compositae, both of which produced extensive fine roots, were the best predictors of high GMD, whereas other species were apparently less important.

Undoubtedly, both roots and hyphae play a part in the stabilization processes (Miller and Jastrow, 2002). Thomas et al. (1986) showed that the contribution of roots of *Allium cepa* to the process was stronger than that of the associated AM hyphae but, unfortunately, their results were confounded by the difference in sizes of AM and non-mycorrhizal root systems. In following up this work, they used a single plant of *Glycine max*, with a split root system and mesh dividers to produce four treatments in a single pot: no roots plus saprotrophic hyphae; hyphae alone; roots plus hyphae (mycorrhizal roots); and roots plus saprotrophic fungi (non-mycorrhizal roots). Although experimental procedures resulted in lower water-stable aggregates at the end of the experiment than in the initial soil, the differences between treatments were significant, in descending order of aggregation: mycorrhizal roots plus hyphae > mycorrhizal hyphae = non-mycorrhizal roots > control.

The contribution of roots and hyphae to aggregate stability increases as the concentration of organic matter in soils increases under pasture. The organic binding agents are relatively transient compared with the more persistent agents that cement the smaller particles. Consequently, the larger aggregates, which are so important in determining the occurrence of free-draining pores, are not only relatively temporary but also fragile and subject to disruption by tillage. Miller and Jastrow (1990) emphasize that, because of the transitory nature of the bonds, stabilization by roots and hyphae depends on their continued production. These points are obviously relevant to the management of plant–mycorrhizal populations to increase and maintain the structural stability of soil and may, as outlined in Chapter 5, have significant consequences for plant water relations as well as sequestration of organic C in soils.

Exudates from hyphae of quite specialized types have been suggested to be important contributors to stabilization of soil structure by AM fungi. There is an increasing body of work investigating the role of glycoproteins produced by AM fungi (termed glomalin or glomalin-related soil protein – GRSP) in stabilizing soil structure. The proteins are poorly characterized biochemically, with difficulties emanating at least partly because they are very stable and extraction from soil requires such treatments as autoclaving in 50 mM citrate buffer (Wright and Jawson, 2001). A recent investigation has shown that 'glomalin' is almost certainly a heat-shock protein which is retained in the AM fungal mycelium (Driver *et al.*, 2005; Gadkar and Rillig, 2006; see Chapter 5). Although it was initially thought that AM fungi secreted the GRSPs, it now appears more likely that accumulation in soil is a result of hyphal turnover, linked to the recalcitrant nature of the glycoprotein. More work is required both to determine the roles of GRSPs in AM fungal biology, as well as quantitative relationships between GRSPs, hyphal length density and turnover and soil structural stability and organic C content (Rillig and Mummey, 2006). Until such work has been carried out it will be advisable to adopt a critical approach to use of GRSPs either as measures of AM activity in soil or as major contributors to soil stabilization and C sequestration.

Conclusions

It is clear that mycorrhizas markedly influence plant interactions in nature and the extent to which they do so is beginning to be revealed. Ecologists are now focusing consistently on the importance of below-ground processes in plant coexistence and competitive interactions, as well as invasiveness, productivity and biodiversity. Consideration of the mechanisms that underlie the effects must in future be informed by experiments and, importantly, take the most parsimonious interpretation of results.

The difficulties associated with experimentation in the field are large and, in most cases, involve application of relatively unspecific biocides, so that outcomes need to be interpreted with caution and conclusions backed up by careful observations and controlled experiments. The modifying effects of plant pathogens, shoot endophytes and soil animals of various types are unlikely to be characterized and understood until we have more effective and specific methods to create effective control treatments. Carefully designed experiments that exclude confounding problems such as sampling effect are also required to unravel mechanisms underlying demonstrations that mycorrhizas can influence productivity and also plant coexistence and diversity. The ways in which they do so appear to be partly dependent on the extent to which the members of the plant assemblages rely on their fungal symbionts for effective nutrient acquisition. Sufficient information is now available to show that broad generalizations are unwise; mycorrhizas do not always increase plant diversity, nor do increasing numbers of fungal symbionts always increase productivity. Likewise, the outcomes of multitrophic interactions are variable and mechanisms even less well understood. They are, however, likely to play significant roles in nature.

Plant ecologists necessarily take a phytocentric view of mycorrhizal symbioses. It is, however, critical to remember that the fungi are also engaged in a multiplicity of interactions, the outcomes of which will influence the composition of fungal assemblages and hence colonization of co-occurring plants. Now that it is clearly appreciated that there is enormous diversity in symbiotic outcome of interactions between different plants and fungi, it should be obvious that the mycorrhizal condition cannot be treated as a 'black box'. A much more 'fungicentric' view of mycorrhizal symbioses and their contribution to vegetation dynamics will be warranted in future.

17

Mycorrhizas in agriculture, horticulture and forestry

Introduction

The widespread occurrence of mycorrhizal fungi of all types on crops and trees in natural ecosystems, together with effects on their mineral nutrition and growth, led to the early recognition that the different mycorrhizal symbioses might be manipulated to increase crop yields in different types of primary production systems. Most plants used in agriculture and horticulture, as well as some forest species, form arbuscular mycorrhizas (AM), but other mycorrhizal types are important in particular situations: ectomycorrhizas (ECM) for forest production and in reafforestation programmes, ericoid mycorrhizas (ERM) for fruit crops such as blueberries and orchid mycorrhizas for enhanced propagation particularly for conservation. As components of the soil biota, all mycorrhizal types are potentially important in restoration of sites degraded by mining or by forestry operations. In consequence, the effects of such disturbance on communities of mycorrhizal fungi and outcomes for productivity are receiving increasing attention. Furthermore, the multifaceted roles of mycorrhizas in soil aggregation and stabilization, in disease tolerance and in mobilizing forms of nutrients that are not directly available to roots (see Chapters 5, 9, 10 and 15) have attracted attention in the areas of biological farming and sustainable management of production systems. Techniques to enhance the yields of edible fruit bodies of ectomycorrhizal fungi, many of which command very high prices, are being actively pursued. This chapter will review selected examples of the application of arbuscular and ectomycorrhizas in managed environments and discuss possible avenues for future work.

Arbuscular mycorrhizas in agriculture and horticulture

There are many instances where crop productivity is influenced by AM symbioses, but there are still rather few examples where inoculation or management to increase AM colonization are carried out as part of normal commercial practice. Attempts have received most publicity in developed and highly mechanized agricultural and horticultural systems. In these, the application of large amounts of fertilizers and pesticides is often routine (Abbott and Robson, 1982, 1991; Menge, 1982; Miller *et al.*, 1986; Hall, 1988; Larsen *et al.*, 2007), which may reduce the potential of the fungal symbionts to have a significant net effect on plant nutrition and growth. However, increasing emphasis on more sustainable and ecologically oriented production systems may at least partially reduce this trend. In recent years, more attention has been paid to the possibility that arbuscular mycorrhizas may increase tolerance of plants to toxic elements or reduce disease (Plenchette *et al.*, 2005; Larsen *et al.*, 2007). Furthermore, the potential of making money from sale of inoculum continues to attract interest from biotechnology companies (Gianinazzi and Vosátka, 2004). The difficulties and costs of producing high quality inoculum of the obligate symbionts, both in terms of AM infectivity and absence of pathogens, must not be underestimated and, in consequence, application may well be effectively limited to high value crops well into the future.

The emphasis of commercial applications of either inoculation with selected fungi or management of indigenous populations has generally been on increasing yields, with relatively less attention being paid to establishment or maintenance of production systems that are 'sustainable' in preserving (or improving) soil resources (Bethlenfalvay and Linderman, 1992; Gianinazzi and Schüepp, 1994; Ryan and Graham, 2002). In these environments, and in less developed systems with low inputs (Sieverding, 1987, 1991; Plenchette *et al.*, 2005), naturally occurring AM populations may play important roles in crop nutrition that is not always appreciated. Indeed, demonstrations that the AM pathway of P uptake operates in colonized plants in the field, even when there are no net benefits in terms of yield, net nutrient uptake or food quality (Chapter 5), means that the AM fungi are indeed integral to the function of root systems and must not be ignored in programmes aimed at maximizing use of nutrients in soil.

Wherever costs of production and application of fertilizers are important, or minimum input or organic agriculture is practised, the contribution of biological processes and organisms (including AM fungi) in nutrient dynamics deserves to receive more attention (Oberson *et al.*, 1993; Plenchette *et al.*, 2005; Jakobsen *et al.*, 2005). Recognition that rock phosphate deposits are limited and that reserves of high quality for fertilizer manufacture could run out before the end of this century, must lend increased urgency to the search to harness plant adaptations that maximize the use of P accumulated in soil. The 'non-nutritional' effects of mycorrhizas in modifying water relations (Chapter 5), reducing the severity of some plant diseases and in stabilizing soil structure are also potentially important (Chapter 16). These have received less emphasis than increases in production, probably because the environmental and economic benefits are less easily quantified. The functions and activities of AM symbioses have rarely been included in integrated systems for pest management, but there is an argument that this should be considered. There is an even stronger argument for their inclusion in programmes for the management of nutrient or water use, particularly where soils are highly P-fixing, fragile or subject to erosion or leaching of nutrients. The possible economic benefits of managing AM populations in agriculture and horticulture need to be critically assessed in the context of the ecology of the systems, not simply the growth of the crops (Miller *et al.*, 1994).

Mycorrhizal involvement in crop nutrition and growth in the field

Unequivocal proof that AM colonization contributes to crop nutrition, growth or yield in the field is difficult to obtain, because roots are normally colonized and appropriate non-mycorrhizal controls are hard to produce (see Chapters, 4, 5 and 16) (Abbott and Robson, 1982; Fitter, 1985, 1990; Hall, 1988). Chemical or heat treatments of soil that eliminate the AM population may alter the levels of available nutrients or release toxic compounds. They also eliminate other members of the soil biota which may themselves have direct positive or negative effects on the plants or interact with AM fungi.

Despite these problems, there are many situations where the effects of soil treatments on plant growth and yield are best explained by activities of AM fungi, because either fumigation decreased growth or inoculation increased it. McGonigle (1988) evaluated 78 field trials with AM fungi and found that inoculation (either in sterilized or untreated field soil) resulted in an average yield increase of 37%. At the time he was doubtful that nutrient uptake via AM fungi was involved, because the magnitude of the increases were not correlated with per cent colonization. We now know that the extent of colonization is not necessarily a good predictor of nutrient uptake via AM fungi (see Chapter 5). However, the possibilities remain that benefits stemmed, at least in part, from lower disease or other non-nutritional effects.

Soil fumigation or steam treatments are often used for high value crops, to reduce losses caused by plant pathogens. These processes sometimes lead to 'stunting', poor yields and variable growth of many plant species including: *Citrus*, *Persea* (avocado), *Capsicum*, *Cassava*, *Vitis*, *Allium*, *Malus* (apple), *Prunus* (peach), *Tamarillo*, *Liquidambar*, *Liriodendron*, *Elaeis* (oil palm), *Manihot* (cassava), *Cacao* and many woody ornamentals. Sometimes stunting can be reversed or reduced by applications of fertilizer, but many species that are highly AM responsive (see Chapter 4) are unable to make effective use of P and other immobile nutrients unless their roots are colonized by AM fungi. This effect is most noticeable in P-fixing soils and varies with plant species (Yost and Fox, 1979; Haas *et al.*, 1987; Li *et al.*, 2005).

Plenchette *et al.* (1983a, 1983b) investigated the effects of fumigation on the growth of 22 species grown in the field under temperate conditions, in a soil of relatively high P availability. The crops fell into three major groups: those which formed arbuscular mycorrhizas and grew better in non-fumigated soil comprised 16 species, including corn (*Zea mays*), carrot (*Daucus carota*), tomato (*Solanum esculentum*), potato (*S. tuberosum*) and a number of legumes; those which formed arbuscular mycorrhizas, but whose growth was unaffected by fumigation, including oats (*Avena sativa*) and wheat (*Triticum aestivum*); and those non-host species, cabbage (*Brassica oleracea*) and garden beet (*Beta vulgaris*), which actually grew better in fumigated soil. These groups would reasonably have been predicted from the discussion on variations in AM responsiveness (see Chapters 4 and 5) and confirm earlier findings of pot experiments with many species. Yost and Fox (1979) explored the mycorrhizal response and P uptake of a number of tropical crops grown in an acid, P-fixing soil. Ten levels of P were applied to establish solution P concentrations in the range 0.012–1.0 µg P/ml. Fumigation had little effect on bicarbonate-extractable P and caused a small but insignificant increase in inorganic N. *Brassica chinensis* grew better on fumigated than unfumigated plots and was non-mycorrhizal in both situations. The other crops all formed arbuscular mycorrhizas in non-fumigated soil and grew much better in this treatment, although the levels of soil P at which fumigation ceased to exert an effect on P concentration in the tissues differed. The order of responsiveness (together with the critical solution P concentration) was *Manihot esculenta* and *Stylosanthes hamata* (>1.6 µg/ml), *Leucaena leucocephala* (1.6 µg/ml), *Allium cepa* (0.8 µg/ml), *Vigna unguiculata* (0.2 µg/ml) and *Glycine max* (0.1 µg/ml).

As highlighted by Menge (1982) and shown subsequently in many situations, the influence of soil type, particularly with respect to P supply and P-fixation, is also important in determining responses.

Fumigation as a technique to eliminate AM fungi for experimental purposes has been criticized because it eliminates other soil organisms (including pathogens) and may also release nutrients, with consequent difficulties in interpretation of data simply in terms of AM involvement (see Chapter 16). Jakobsen discussed these difficulties in the context of the growth of several temperate field crops, including *Pisum* and *Hordeum*, and came to the conclusion that, in the cropping system he was investigating, the AM contributions to P uptake and growth were important and were underestimated by the fumigation technique (Jakobsen and Neilsen, 1983; Jakobsen, 1987). It is worth noting that some cultivars of barley and of wheat respond to AM colonization, with those that are responsive to fertilizer application also being responsive to colonization (Baon *et al.*, 1993; Hetrick *et al.*, 1993a, 1993b; Zhu *et al.*, 2003; Li *et al.*, 2005). Thus, not all cereals fall into the non-responder group (Plenchette *et al.*, 1983a) and, again, there will be significant effects of identity of fungal symbionts and soil conditions (Graham and Abbott, 2000).

Application of more selective biocidal treatments can also reduce AM populations, root colonization and cause stunting of plants (Menge, 1982). Although there are few examples of this causing major problems in production, usually because high levels of fertilizer are used simultaneously, the risk is significant in situations where there is an AM contribution to net nutrient uptake. The systemic fungicide benomyl certainly reduces AM colonization in both pots and in the field (Fitter, 1985; Fitter and Nichols, 1988; Koide *et al.*, 1988b; Sukarno *et al.*, 1993; West *et al.*, 1993b). Together with other fungicides, benomyl has been used experimentally to eliminate or reduce colonization by AM symbionts (Schweiger and Jakobsen, 1999b; Schweiger *et al.*, 2001). In controlled conditions and low P soil, the result is much reduced plant growth that can be directly attributed to lack of AM P uptake (Sukarno *et al.*, 1993; Schweiger and Jakobsen, 1999b; Schweiger *et al.*, 2001). However, in field situations, the occurrence of fungal pathogens, which might also be eliminated, confuses the picture in relation to AM effects on nutrition. Nevertheless, the technique certainly highlighted the possible importance of AM fungi in reducing the effects of pathogens and hence acting as biocontrol agents (West *et al.*, 1993a; Fitter and Garbaye, 1994; Hooker *et al.*, 1994; Graham, 2001). With the phase-out of the fumigant methyl bromide, alternative, low input approaches to eliminate pathogens have been sought. In Florida, Sylvia *et al.* (2001) investigated a strip tillage system in which tomatoes were transplanted into bahaigrass (*Paspalum notatum*) pasture. The system was designed to reduce effects of pathogens and improve soil conservation, but tests showed that competition between crop and grass actually reduced tomato yields. Pot experiments revealed that tomato was highly responsive to AM colonization in the low P and low pH soil used and bahaigrass was less so. AM colonization increased the ability of the tomato to compete with the grass, when they were grown together. However, as shown in a second experiment, application of P reduced the relative competitive ability of tomato. The authors stressed that development of low input systems, such as the one investigated, requires careful development of strategies to maximize nutrient uptake by the crop. They suggested that reduced P application might allow tomatoes to take advantage of their relatively high mycorrhizal responsiveness in this soil type. The involvement of mycorrhizas in productivity during intercropping has scarcely been explored, but deserves investigation.

Severe soil disturbance by tillage, mining or natural causes can also reduce plant nutrient uptake and hence crop yield. A major effect in this case is the disruption of the network of mycorrhizal hyphae in soil, with consequent reductions in colonization, nutrient acquisition and growth. In the field, this has been shown to be important for growth and nutrition of *Zea mays*, particularly with relatively low fertilizer P applications (Evans and Miller, 1988, 1990; McGonigle and Miller, 1993) and a number of species from native vegetation in soil disturbed by mining (Powell, 1980; Jasper *et al.*, 1989, 1992). Variations in susceptibility of different AM fungi to such damage and consequent effects on plant nutrition have been observed (McGonigle *et al.*, 2003).

Crop rotations involving long periods of bare fallow (1 year or more) have sometimes been adopted, with the aim of accumulating moisture and mineral N in the soil profile. In Queensland, Australia, this practice led to severe stunting and P and Zn deficiency in a wide range of taxonomically unrelated crops (Thompson, 1987). This so-called 'Long-fallow Disorder' was tracked down to a deficiency in AM propagules in the soil, with consequent decreases in the rate and extent of AM colonization and uptake of nutrients (Thompson, 1990, 1994). The problem was overcome by eliminating long fallow periods from rotations and adopting management practices that maintain AM fungal populations. In this case, as in many other examples, the data from soil tests used to determine appropriate rates of P and Zn application were obtained in situations where the fungal symbionts normally made substantial contributions to P and Zn uptake by the plants. Reduction in AM colonization, by whatever means, reduced or eliminated the fungal contribution to nutrition, with the consequence that the response to fertilizer was also reduced and the data from the soil tests were no longer good predictors of crop response to P application (Haas *et al.*, 1987; Thompson, 1987). A note of caution is required here, because field experiments on similar soils in south-eastern Australia found that increasing AM colonization by management of rotations rarely resulted in any increases in yield or nutrition of wheat. The discrepancy was attributed to differences in P fertilizer applications between the two regions (Ryan *et al.*, 2002).

A novel experimental approach to determining the potential contribution of AM fungi to tomato production in the field was recently adopted by Cavagnaro *et al.* (2006). AM and non-mycorrhizal treatments were established without soil fumigation or fungicide application, by using a mutant tomato with reduced AM colonization (*rmc*) and its wild-type progenitor, planted in an organically managed field at the same time as a normal commercial tomato crop. Naturally-occurring AM fungi colonized the wild-type tomato normally (24% of root length), but were only able to invade the epidermal cells of the mutant over 4% of root length, with no cortical development. This pattern of colonization does not support any P uptake via the AM pathway (Poulsen *et al.*, 2005). At the final harvest, there were no significant differences in shoot or fruit biomass between the genotypes, a similar finding to several pot experiments with these plants grown in neutral or slightly alkaline soils (Cavagnaro *et al.*, 2004a; Poulsen *et al.*, 2005). However, the wild-type (AM) plants had considerably higher concentrations of both P and Zn in shoots and fruits than the mutant (non-mycorrhizal). Not only do these data indicate that the AM fungal populations were contributing to nutrition of the colonized wild type, but they highlight an increase in food quality in terms of nutrient densities. Similar results have occasionally been observed in highly colonized wheat, compared with wheat grown in field rotations that resulted in low AM colonization (Ryan *et al.*, 2002). Effects of arbuscular mycorrhizas on nutritional value of crops deserve ongoing research, particularly in the light of concerns that use of highly purified fertilizers and other modern agricultural practices are reducing micronutrient densities below those required for human health (Welch, 2002; Welch and Graham, 2002). Possibilities that AM colonization may increase concentrations of nutritionally beneficial compounds like antioxidants are currently being actively explored.

Positive effects of high AM colonization on yields are by no means always observed. Growth depressions have most often been attributed to C drain caused by AM colonization, as discussed for citrus (see Chapter 4). Graham and Abbott (2000) reported a wide range of responses of wheat in a pot experiment, with growth depressions associated with rapid and high colonization by what they referred to as 'aggressive' AM fungi. In the field, high populations of *Glomus macrocarpum* have been linked to tobacco stunt disease, with the poor growth alleviated by soil fumigation that effectively eliminated AM fungi (Modjo and Hendrix, 1986; Hendrix *et al.*, 1992). The mechanism of inhibition of growth was not fully worked out, but reduced root growth of AM plants may have been a contributory factor (Jones and Hendrix, 1987). More recently, Ryan *et al.* (2002) carried out a series of five experiments, examining the effects of cropping sequence on AM colonization, P and N nutrition, growth and yield of wheat. The soils were alkaline vertisols, similar to those on which reduced AM fungal populations had been shown to be associated with Long-fallow Disorder. Low AM colonization following brassicas did not affect early crop growth, P and Zn uptake or yield. In only one of the experiments was there a positive correlation between AM colonization and grain P and Zn concentrations. Findings such as these led Ryan *et al.* (2002) to conclude that farmers need not consider the effects of cropping sequence on extent to which wheat becomes colonized on these soils. They went further to question more generally any positive roles of AM fungi in crop production and suggest that, for some normally non-responsive crops like wheat, the C drain to the fungal symbionts may have significant effects in reducing yields. However, these suggestions may require re-evaluation in the light of demonstration of the involvement of AM fungi in nutrition, independent of net growth benefits (see Chapters 4 and 5).

The work discussed so far has evaluated potential AM effects in terms of net positive or negative effects on plant growth and nutrition. The demonstration that the AM pathway of nutrient uptake may replace that of roots has not yet received much attention in cropping situations. However, several field experiments have demonstrated considerable uptake of ^{32}P into crops and pastures from buried hyphal compartments (HCs) which successfully prevented P uptake by plant roots and root hairs from inside the HCs. In the experiment of Schweiger and Jakobsen (1999b) with wheat, the rate of P uptake per unit length of AM hyphae was similar to that observed in pot experiments. Application of the fungicide carbendazim to the HCs, at concentrations known to inhibit AM fungal P transport, effectively eliminated P uptake via the fungal pathway. The results demonstrate the considerable contribution of native AM fungi to overall P uptake of field-grown winter wheat, even at typical field soil fertility levels of $28\,\mu g$ $NaHCO_3$-extractable P/g. A similar approach was used to investigate the AM contribution to nutrition of peas also grown in a relatively fertile soil (Schweiger *et al.*, 2001). Again, the native AM fungi made a considerable contribution to plant P uptake, which was reduced or eliminated at high levels of application of commercial fungicides. Interestingly, at recommended field application rates, hyphal P uptake and transfer were slightly increased. This effect was attributed to relatively greater negative effects on other members of the soil microbial community and highlights the importance of establishing appropriate protocols that not only control unwanted pathogens but also maintain the function of potentially beneficial organisms. AM contributions to nutrition of grassland and pasture plants growing in undisturbed soil cores have also been assessed using $^{32/33}$P-labelled soil buried in HCs. Johnson *et al.* (2001) showed that there was considerable movement of ^{33}P from soil in the HC to a mixture of grassland species growing in the main pot compartment, which was greatly reduced if the external mycelium of the AM fungal assemblage was severed by rotating the cores. Jakobsen *et al.* (2001) used a similar approach to show that the contribution of native AM fungi to P transfer to *Trifolium subterraneum* was very variable, depending on soil type and on the growth of external mycelium into the labelled soil cores. Inhibition of AM fungal activity with the benomyl applied to the HCs resulted in reduced P concentrations and reduced ^{32}P uptake into plant shoots in several of the soils. However, in one soil, plants grew very badly regardless of fungicide application, the external AM hyphae did not grow into the HCs and there was little or no ^{32}P transfer to the plants. Taken together these experiments clearly demonstrate the importance of the external mycelium of field assemblages of AM fungi and the AM pathway of P uptake to a range of plants of varying responsiveness in several soils of different P status.

It is highly likely that AM fungi make similar contributions to uptake of P, Zn and possibly other nutrients in a variety of crops and pastures in the field, even when net benefits in terms of yield or nutritional quality are not apparent. It may be thought that these 'hidden' contributions are unimportant. However, crop improvement for enhanced efficiency of nutrient uptake increasingly depends on understanding the details of the uptake processes and their genetic control. If crops are normally and unavoidably colonized in the field, then P uptake via AM fungi must be taken into account in research programmes. Modern crop varieties have certainly been selected and bred without taking AM symbiosis into account and the lack of responsiveness of many crop plants may be a consequence of this (Smith *et al.*, 1992; Hetrick *et al.*, 1996). At least two investigations indicate that AM responsiveness may be an inherited trait in wheat and barley and closely related to P responsiveness (Baon *et al.*, 1993; Hetrick *et al.*, 1996). In the main, plant breeders have ignored potentially beneficial AM symbioses, but future genetic research and breeding could usefully capitalize on these findings, particularly bearing in mind that AM fungi are integral to root function of both responsive and non-responsive AM plant species and that colonization in the field and contribution to P uptake may be very high (Ryan *et al.*, 2002; Li *et al.*, 2005, 2006) and hard to eliminate without drastic management. We need to understand fully the pathways and mechanisms of nutrient uptake, including coordinated regulation of direct and AM symbiotic P uptake pathways in order to manipulate them to improve P efficiency of crops either by breeding, genetic manipulation or management.

Evaluation and management of AM fungal populations

Most of the examples given above show that nutrient uptake and hence crop production in the field can involve the naturally occurring assemblages of AM fungi.

The composition of these communities, in terms of species and propagule densities, is not usually well known and the way in which differences in species richness may affect plant productivity is not at all clear. Indeed, the fungal attributes that result in a large contribution to plant nutrient uptake and growth are not fully worked out, so that selection of 'efficient' fungi still remains largely empirical. It is known that an extensive external mycelium is necessary and that different species of fungi vary in the way that this mycelium develops (see Chapters 2 and 5). Rapid and early colonization of the roots and the production of numerous arbuscules are also important and may be functions both of the innate characteristics of the fungal species, the propagule density and other conditions in the soil. The picture is further complicated by differences in the way different plant–fungus combinations function in terms of nutrient transfer and by difficulties in relating physiological effects (e.g. nutrient uptake, growth response) to per cent colonization of the root system, which is frequently measured rather late in the growth period when the density and activity of arbuscules may be relatively low.

In the field, many different AM fungal species are likely to be present and will colonize crops to varying extents, depending on plant–fungus selectivity, as well as relative propagule densities (see Chapter 1). We now recognize that the efficiency of particular fungi may vary when associated with different plant species, but we have little idea whether those fungi which are most efficient with respect to nutrient uptake are the same as those which play a major role in reducing the effects of pathogenic organisms or stabilizing soil. There are essentially two approaches to establishing and maintaining high infectivity of AM fungi in soils used for agriculture or horticulture. These are inoculation (and subsequent management) of selected AM fungi and adoption of field practices which increase the infectivity of indigenous assemblages. The relative merits of these approaches in different situations have been reviewed many times (Abbott and Robson, 1982, 1991; Menge, 1983, 1984; Hall, 1988; Gianinazzi et al., 1990b; Sieverding, 1991; Bethlenfalvay, 1992b; Bethlenfalvay and Linderman, 1992; Wood and Cummings, 1992; Dodd and Thomson, 1994; Brundrett et al., 1996; Ryan and Graham, 2002; Larsen et al., 2007).

There is general consensus that before any approach is adopted a number of key factors need to be evaluated. These include: the responsiveness of the crops to be grown, the assemblage of indigenous AM fungi present, particularly with respect to their infectivity and effectiveness, the possible effects of soil management (e.g. tillage, P and N application) on their AM assemblages and the characteristics of the soil, as they affect both nutrient availability, crop responsiveness and fungal survival. To these need to be added information on the incidence of pathogens and methods that are used to reduce their effects (e.g. fumigation, application of fungicides) and an evaluation of whether AM mycelium in soil may have a significant effect on the establishment and stabilization of soil structure (Tisdall, 1994; Miller and Jastrow, 2002; Rillig and Mummey, 2006) (see Chapter 16). The latter may be very important in fragile environments that are being farmed more intensively in attempts to maintain outputs (Plenchette et al., 2005). Finally, the economic costs and benefits of any management practices are vital considerations and need to be incorporated into long-term plans for sustainable use of soil resources (Miller et al., 1994).

The effects of agricultural practices on AM fungi have recently been comprehensively reviewed (Jansa et al., 2006; Larsen et al., 2007). The factors rank (in decreasing order) crop rotation, soil tillage, fungicide application and application of fertilizers

Table 17.1 Positive and negative influences on arbuscular mycorrhizal assemblages and colonization of subsequent crops by different agricultural management practices.

Management factor	Positive influence	Negative influence
Plant species	Host species High colonization High spore production High mycorrhizal root length density	Non-host species
Bare fallow	None	Reduces populations
Pasture	Increased propagule densities	
Disturbance–tillage–rotation	Minimum tillage pasture phase	Conventional tillage compaction
Management	Organic, biodynamic	Conventional
Fertilizer application	Drip feeding Slow release rock phosphate	High applications of soluble P and N
Fumigation	None	Reduces propagules
Fungicides	Variable effects	Variable effects
Low light (glasshouse)	None	Colonization and growth decreased

in importance in modifying or reducing AM fungal populations. Other factors such as irrigation, burning and grazing, pollutants and topsoil removal are also relevant (Table 17.1). Of course, the most extreme examples of low populations result from soil sterilization, fumigation or use of soil-less media for production of high value crops. Here, the greatest potential for successful inoculation with AM fungi exists, including the production of micropropagated plants (Vestberg and Estaún, 1994; Lovato *et al.*, 1995, 1996). However, few, if any, commercial production systems use inoculation because of the difficulties of producing and applying inoculum and of introducing modifications in cultural practices (Menge, 1984; Wood and Cummings, 1992; Lovato *et al.*, 1995; Gianinazzi and Vosátka, 2004).

Many high value glasshouse crops are increasingly being grown in soil-less media, such as rockwool. The systems have advantages in that such rooting media are light and nutrient supplies can be closely controlled. However, it has been found that pathogen attack can produce significant losses and some research is now being directed towards introducing an element of biological balance to improve plant health into these systems. In one recent investigation of tomato and cucumber production in Denmark, Ravnskov and Larsen (2005) tested the effects of application of commercial AM inoculants in relation to nutrient supply. Tomatoes responded positively to application of P fertilizer over the whole range tested, up to 100% of concentrations recommended to growers. AM colonization was relatively low and none of the inoculants gave any benefits in terms of vegetative biomass under experimental conditions. In contrast, non-mycorrhizal cucumber showed increased growth only up to 50% of recommended P fertilizer concentrations and also showed a positive response to inoculation. A grower trial showed no benefits of inoculation of cucumber in terms of growth or yield, but one of the cultivars tested produced almost 4% more first class fruit. This increase may seem rather small, but it was significant and represented an increased profit to the grower involved of €50000 (£33500) in spring production (2005 prices). In evaluating the findings, Ravnskov and Larsen (2005) concluded that there was potential for ongoing investigations of AM applications for cucumber production, but not for tomatoes. The plants did not suffer from disease either under experimental conditions or in the grower trial, so potential benefits of AM inoculation in disease tolerance could not be assessed in this case, but would be beneficial in future.

In field situations, evaluation of composition of AM fungal assemblages requires both identification of the species present and quantification of propagule densities and infectivity (see Chapter 2). The assemblages are usually described in terms of spore types and numbers, whereas bioassays of various types are used to evaluate the infectivity of the soil based on all the functional propagules. Spore isolation, unfortunately, cannot show which fungi are active in roots and soil. The limitations are clearly appreciated and a number of different methods are being developed to identify AM fungi in the absence of spores. Techniques used with varying success and sensitivity include DNA-based fingerprinting, isozyme banding patterns and fatty acid methyl ester profiles. DNA-based methods, in particular, have the potential to be sufficiently precise to distinguish different strains of a single species but will need further development to make them satisfactorily quantitative. Until we have precise and rapid methods of identification and evaluation of AM fungi which are present as vegetative stages and contribute to soil–plant processes, we are not likely to make much progress in understanding and managing the fungi in agroecosystems. Even then uncertainties in relating taxonomic position or extent of colonization to function are likely to remain significant.

If populations are low or ineffective, then inoculation or management to increase propagule densities can be considered. Conversely, and as cautiously advocated by Ryan and Graham (2002), there may be advantages in managing soils to reduce AM populations for specific crops and on particular soils, if negative effects on yield are such as to have significant effects on crop profitability. This could be achieved by extensive tillage or rotations with non-hosts such as brassicas, but advantages would have to be very carefully offset against the advantages of minimum tillage, which is frequently and effectively adopted to minimize soil erosion. Furthermore, until we know more about the ways in which plant and fungal nutrient uptake processes are integrated and operate in field situations, the practice certainly should not be widely adopted. Additionally, there are many who would argue that conserving biodiversity of potentially beneficial soil microbial populations, including AM fungi, is crucial to soil health.

Inoculation

As Wood and Cummings (1992) predicted, the unculturabilty of the fungi continues to be a major barrier to the development of cheap and easy inoculation techniques. AM inoculum currently has to be grown in symbiosis with plants. Production costs are high, the product often bulky and quality control to maintain infectivity and exclude pathogens is a significant concern (Menge, 1984; Gianinazzi and Vosátka, 2004). Nevertheless, over 20 companies worldwide are reportedly producing AM inoculum (Gianinazzi and Vosátka, 2004), capitalizing in part on the desire of many producers to use 'biological' methods. Symbiotic production does have the important advantage of automatically monitoring the ability of the fungi to colonize roots, at least during the production stage.

The plant-based inocula now available are quite diverse and require different methods of application. As outlined by Gianinazzi and Vosátka (2004), they are produced in different ways ranging from nursery plots to *in vitro* monoxenic root organ cultures. The resulting materials (spores, hyphae, root fragments etc.) are added to different carriers, resulting in a wide range of formulations. The different systems have a range of advantages and disadvantages in terms of ease of use, quality and cost. The suitability of these inocula for different applications depends on the identity of the main AM propagules and on their ability to retain infectivity during storage and to persist in soil or roots from year to year, as well as on the methods available for application. One promising approach is encapsulation of AM roots, containing high densities of fungal vesicles, in alginate beads. The resulting inoculum retained infectivity for at least 3 years (Plenchette and Strullu, 2003). It is a matter for discussion whether 'generic products', containing several AM fungi and potentially suitable for a range of applications, are more appropriate for the market than those with precise formulations and AM fungi specifically tuned to particular end-uses.

Quality control criteria have not yet been set for AM inoculum, but are urgently required so that the products meet reasonable standards. At the very least, the inoculum must initiate colonization in the root systems of plant species that are able to form mycorrhizas, at the doses recommended by the suppliers, it must not contain pathogens or other agents that could reduce plant growth and it must have a reasonable shelf-life when stored under recommended conditions. Although outcomes are difficult to prove without direct experimentation, the products should also meet the expectations and requirements of purchasers including decreasing the need for fertilizer applications, increasing plant growth, flowering, yields or tolerance to disease or pollutants. Not all available inocula meet these criteria, an important issue both with respect to maintaining confidence of the market and meeting critical quarantine standards. No doubt the DNA fingerprinting methods now under development will find application both at the level of identification of wanted and unwanted organisms in the inoculum and survival in roots and soil following application. Further considerations and progress towards regulation of products are documented in two recent reviews (Gianinazzi and Vosátka, 2004; Plenchette *et al.*, 2005).

At present, routine inoculation in broad scale, highly developed farming systems is not realistic, because of the expense of production and application and uncertainties relating to the competitive ability of inoculant fungi in field situations. Management of indigenous populations is the only currently viable option. This was demonstrated to have some potential in bell pepper (*Capsicum annuum* L.) production in Australia (Olsen *et al.*, 1999). The field site was subjected to two cycles of crops which are highly mycorrhizal, establishing a well-developed mycelial network with very high density of infective propagules. AM fungal populations were effectively eliminated by fumigation to provide non-mycorrhizal control plots and P was applied as milled superphosphate at five levels (P0–P5), from zero to 135 kg P/ha. *Capsicum* seedlings were transplanted into the plots, according to normal horticultural practices. Phosphate application did not affect the percentage colonization of the plants. Phosphate nutrition was monitored in the youngest mature leaves and showed that plants in the AM treatments had adequate P concentrations at all P applications except P0, but the non-mycorrhizal plants only achieved adequate nutrition at P5. At harvest, the yields of AM plants were higher than controls, except at P5. Gross margin of non-mycorrhizal plants at P5 was AUD 3440/ha (1999 prices). All AM treatments except P1 showed similar profitability. The authors concluded that AM had only limited potential as a substitute for P at that time, because fertilizer made a low contribution to the overall costs of production. However, this situation will change if P fertilizer costs rise significantly.

In relatively small-scale, high value operations such as nursery production, routine inoculation is certainly feasible and likely to be highly advantageous in increasing growth rates and uniformity of the product. In these situations, potting media are frequently sterilized to eliminate pathogens, so that reintroduction of AM fungi specifically selected for the particular application has potential. Work in this high-tech area continues actively, including production of high quality inoculum for research purposes (Gianinazzi et al., 1990a; Lovato et al., 1995; Gianinazzi and Vosátka, 2004).

Inoculation of seedlings is potentially a good method for establishing selected fungi in roots, before potting on or planting-out into the field. It is appropriate where transplanting is part of the normal production system. An advantage is that the inoculant fungus is established in the root systems at an early stage and may consequently have a competitive advantage over soil-borne species. The method has been tested with a number of different crop species in the field. Sasa et al. (1987) inoculated *Allium porrum* in pots, so that at transplanting the roots were about 80% colonized by a mixture of AM fungi. These plants grew better than uninoculated controls after transplanting, with 5.7 and 1.5 fold increases, in fumigated and unfumigated soil, respectively. Similar increases in growth have been observed for such diverse species as chilli (Bagyaraj and Sreeramulu, 1982), apple (Plenchette et al., 1981) and guayule (*Parthenium argentatum*, Bloss and Pfeiffer, 1984). Snellgrove and Stribley (1986) adapted normal commercial methods in their work with *A. cepa*, but had difficulties establishing colonization in the peat modules used for transplanting. The approach deserves more extensive evaluation and the results highlight the need to fit inoculation procedures to acceptable production methods, as well as to soil-type and plant species. Inoculation with two or more fungi needs to be considered as it could reduce the variation in response that might be expected with different soils, plant species and growing conditions, following inoculation with single species (Sieverding, 1990; Bethlenfalvay and Linderman, 1992).

At the other end of the scale, manufacture of 'home-grown' inoculum of highly colonized roots and soil that is applied to plots immediately before planting a crop could make a valuable contribution to food production in many relatively small, low input systems. Its potential should not be under-rated because it does not make large profits in monetary terms. In most cases, experiments have been carried out with annual crops grown in monoculture. However, tree crops are also important and, particularly in developing countries, are sometimes cultivated in plantations or gardens with considerable species diversity. Examples include coffee and cacao, which are grown with mycorrhiza-responsive shade trees (Wibawa et al., 1995) and many tropical fruits, grown in mixed agroforestry systems (Janos, 1980). It is known that both coffee and cacao, as well as citrus, cashew and many other tree crops of tropical origin, are responsive to AM colonization (Janos, 1987; Alexander, 1988; Sieverding, 1991; Smits, 1992; Smith et al., 1998). Both cultivation and monoculture appear to change the species composition of the fungal populations and reduce their diversity, but the impact of these changes on crop production has not been adequately evaluated (Black and Tinker, 1979; Johnson *et al.*, 1992; Johnson, 1993; Allen *et al.*, 1995; Hendrix *et al.*, 1995; Helgason *et al.*, 1998).

In those situations where the cost and availability of phosphate fertilizers is very significant, AM fungi may increase the accessibility of relatively cheap fertilizers, such as rock phosphate (RP) when used directly on acid soils. Again, the responses will vary with the plant under consideration. In the work of Wibawa *et al.* (1995) on shade trees used in coffee and cacao plantations, *Sesbania grandiflora* appeared to be relatively efficient at acquiring P from deficient soil and did not respond significantly to triple superphosphate (TSP) or RP, regardless of AM inoculation. Two other species grew poorly when unfertilized, responded to TSP but not RP in the absence of inoculation and responded to RP when inoculated with *Gigaspora margarita*. Colonization of the roots was not determined, but spore production was significantly influenced by treatment for all species. In general, inoculation increased spore production to a much greater extent with RP than with TSP, an important consideration in the context of management of AM fungal populations.

The same practices that are used to manage AM fungi for crop production can be used to enhance AM effects on soil structure and structural stability and have the potential to make an important contribution to agricultural and horticultural ecosystems, particularly where erosion is a serious concern. It has also been suggested that an extensive mycelial network that ensures effective removal of P from soil will prevent off-site losses to streams, rivers and groundwater (e.g. Jakobsen *et al.*, 2005b). This possibility has not been extensively tested, but it was recently shown that the main effect of AM colonization in reducing the mobility of P in repacked soil columns was associated with increases in plant growth in low P soils. When available soil P was increased, P losses in leachate were much higher and not reduced if plants were AM (Asghari *et al.*, 2005).

Management

The continuing (though not continuous) presence of host plants is essential for inoculum build-up. Growth responses of plants to colonization are not important; rather the mycorrhizal root length-density in soil and the production of extraradical hyphae and spores will contribute most to the population of propagules. Pasture, which combines production of colonized root and low disturbance, has a high potential for inoculum build-up, as well as production of water-stable soil aggregates, stabilized by hyphae of AM fungi (see Chapter 16). Other management strategies that might maintain fungal populations include sequential cropping, where two or more crops are grown each year, or intercropping, where two crops are grown simultaneously. In either case, if at least one of the crops is potentially mycorrhizal, an adequate inoculum level is likely to be maintained (Andrews and Kassam, 1976; Tisdall and Adem, 1990). Both bare fallow and cropping with non-hosts will, sooner or later, reduce mycorrhizal populations, or may delay re-establishment of a pool of infective propagules (Ocampo and Hayman, 1980; Thompson, 1987; Ryan *et al.*, 2002).

Tillage, other types of disturbance and stockpiling of soil reduce the populations of viable propagules (Jasper *et al.*, 1987, 1989, 1992; Miller and Jastrow, 1992b, 2002) and should be minimized to allow maintenance or build-up of AM populations, as well as to maintain soil structural stability. Soil compaction also has negative effects on root colonization, as well as on root growth itself. Mulligan *et al.* (1985) reported reduced per cent colonization as bulk density increased due to trafficking by agricultural vehicles. It might be expected that restricted root growth would result in lower mycorrhizal root length density in soil, even if there were no effects on the ability of the fungi to colonize the roots. In fact, reported effects of increasing compaction on colonization are variable, with both increases and decreases recorded (Nadian *et al.*, 1996, 1997; Li *et al.*, 1997; Yano *et al.*, 1998). The findings are likely to have been influenced by the extent to which root growth was reduced under compaction, in conjunction with infectivity of the soils. Hyphal extension in soils from colonized roots has also been shown to be reduced, but not to the same extent as the roots (Nadian *et al.*, 1996). Mycorrhizal plants grew better than non-mycorrhizal at all levels of compaction, but there were consistent decreases in responsiveness as compaction increased. Different AM fungi have been shown to respond differently to compaction and to variations in pore size in soil, both with respect to root colonization and ability to extend out into the compacted soil and absorb P or colonize new plants (Drew *et al.*, 2003, 2005).

Biodynamic and organic farm management results in higher per cent colonization of roots of pasture and annual crops than conventional management (Ryan *et al.*, 1994, 2000; Ryan and Ash, 1999). The effects are at least partly the result of lower P applications and use of less available fertilizer sources. The interactions between application of P and N, growth responses and maintenance of soil populations of propagules are of major importance. So far as growth responses are concerned, the form and timing of fertilizer application, as well as the sensitivity of the particular plant–fungus combinations needs to be taken into account. For most crops, P (frequently as superphosphate) is applied once, before sowing, and may have large effects in reducing per cent colonization and growth response. Other forms of P, such as RP, do not have the same effects on colonization and may be much more compatible with maximizing the contribution of arbuscular mycorrhizas in plant nutrition. Drip feeding of nutrients in irrigation water (fertigation) is sometimes practised in intensive vegetable production. In one field investigation with *Capsicum* on a highly P-fixing soil, the practice maintained soil solution P at a relatively low concentration, permitting both extensive mycorrhizal colonization and good growth and yield (Haas *et al.*, 1987). Application of slow-release fertilizers might be expected to have the same effect. Selection of P-tolerant fungi has also been canvassed, with the possibility of using them in inoculation programmes where high P application is practised.

Biological control of pathogens is now an accepted component of pest management programmes. Success in this area will be reflected in the reduced use of pesticides, including fungicides. The consequences in terms of AM function are likely to be positive. Damage to non-target AM fungi will be minimized and the effects of these on plant nutrition, on soil structure and on root-infecting pathogens themselves maximized. The potential for including AM fungi in pest control packages as biocontrol agents has not been widely explored, although some potential certainly exists. In the horticultural industry in particular, it is possible to envisage an integrated package in which AM fungi (possibly in association with other beneficial microorganisms) are applied with the aim of making most effective use of fertilizer and minimizing losses due to disease. Such strategies are likely to become more attractive as the use of chemicals for fumigation and disease control is progressively discouraged and fertilizer becomes a proportionally higher component of the cost of production. Future management of soil microbial populations to maximize the effective use of P reserves stored in various soil fractions will need to take all potential benefits and interactions into account (Jakobsen *et al.*, 2005b).

Ectomycorrhizas and forest production

In native forests managed for timber production, ectomycorrhizas are an accepted part of the ecosystem and their potential significance in tree nutrition has been recognized since the earliest experiments (Frank, 1885). Less attention has been paid to arbuscular mycorrhizas, despite the fact that a number of significant timber trees form this mycorrhizal type. Timber extraction, and especially clear-cut logging, have major impacts on forest ecosystems, not only on the plant species that regenerate in clear cuts, but also on the communities of ECM fungi that survive and colonize regenerating or transplanted seedlings. The composition and diversity of these fungal assemblages is attracting increasing attention because of their potential impacts on forest productivity and development of site management strategies (Jones *et al.*, 2003).

Plantation forestry is increasing in many parts of the world as the demand for timber and pulp increases and in response to calls to increase sequestration of CO_2 through increases in forest cover. Although outplanted or naturally regenerating seedlings can develop ECM associations from any naturally occurring mycorrhizal inoculum, the potential advantages of nursery stock being mycorrhizal before outplanting are clear. Grove and Le Tacon (1993) provide a comprehensive outline of the imperatives to developing successful strategies to maximize the benefits of ectomycorrhizas in forestry. Many commercial practices, particularly those employed to improve hygiene in tree nurseries, are inimical to the growth of all but a few ruderal species of ECM fungi. In consequence, special techniques have been developed which enable selected fungi to colonize plants prior to outplanting. The application of these techniques has facilitated superior performance in tree crops in many parts of the world, particularly those which lack natural local sources of inoculum (see Marx *et al.*, 2002). At the same time, interest has grown in the possibility of harvesting edible fruit bodies of ECM fungi which have been used as commercial inoculum both to supplement diet and revenue.

Inoculation can have benefits at two stages of the timber production systems; in the nursery itself and after outplanting to the field. Much of the experimental work has focused on advantages to be gained from the production of well-developed seedlings that, with their fungal symbionts, will become successfully established in the field. However, rapid growth in nurseries also results in direct savings by increasing the rate of throughput. Experience of the use of inoculated seedlings has indicated that responses to ECM colonization are often greatest under the most extreme conditions, particularly those involving exposure to drought, metal contamination and pathogens. Such observations have led to analysis of the functional basis of the ameliorative effects of ECM fungi. In addition to the well-documented effects on plant nutrition and growth, considerable advances have been made towards understanding the roles of the symbioses in providing resistance to these stresses which, though they also occur in natural ecosystems, are often locally increased by previous land-use practices or by the afforestation process itself. Indeed, many of the discrepancies between nursery and field performance following inoculation may stem from different impacts of soil and other environmental conditions on the individual mycorrhizal partnerships established by inoculation. Forest productivity is threatened by aspects of global change such as direct inputs to soils of nitrogen and sulphur and concomitant decreases in pH, which may have adverse effects on both symbionts. Such effects are thought to be contributory factors in the forest decline syndrome experienced in Europe and north-eastern USA. There are also indications that base-cation availability may come to limit forest productivity, so that any positive effects of ectomycorrhizas on uptake will be beneficial (see Chapter 10).

Inoculum production and inoculation practice

The use of defined inoculum consisting of ECM fungi that were physiologically and ecologically appropriate for the planting site, with a view to improving performance of the crop, was pioneered by Moser (1958) in Austria, Takacs (1967) in Argentina and Theodorou and Bowen (1973) in Australia. Prerequisites for the widespread use of ECM inoculation programmes are the selection of fungal symbionts and the development of methods for the large-scale production of inoculum (Grove and Le Tacon, 1993; Brundrett *et al.*, 1996). The requirements are interrelated because, in addition to being functionally compatible with and enhancing performance of the inoculated crop, the selected fungus must be able to withstand the physical, chemical and biological stresses involved in the production and storage of the inoculum, as well as those imposed by the soil and other site characteristics, first of forest nurseries and later at the sites to which the seedlings are outplanted (Kropp and Langlois, 1990; Grove and Le Tacon, 1993).

One of the most widely used and successful early inoculation programmes employed *Pisolithus tinctorius*. Interest in this fungus was prompted by its apparently wide geographic distribution, broad host range (Marx, 1977) and the knowledge that it became prominent on adverse sites, particularly those subject to drought, high temperature or contamination (Schramm, 1966). It is a striking feature of the programme of inoculum production using *P. tinctorius* that only one vegetative isolate of the fungus was used throughout. This strain was originally obtained from a sporophore found under *Pinus taeda* in Georgia, USA (Marx and Bryan, 1969). Its aggressive traits and ability to enhance growth of plant partners have apparently been enhanced by annual re-isolation over 30 years from seedlings growing in inoculation trials.

In order to eliminate weeds, pathogens and other symbiotic fungi which are potential competitors, seed beds or potting mixes are routinely fumigated before inoculation. Even so, re-invasion of fumigated soil by spores of naturally occurring ECM fungi, particularly *Thelephora terrestris*, normally occurs within days and it is a requirement of the inoculant fungus that it has the ability to colonize roots quickly. *T. terrestris* appears to be the dominant ECM fungus of forest nursery soils worldwide (Mikola, 1970; Ivory, 1980; Marx *et al.*, 1984a) and, whether as a result of re-invasion after fumigation or natural occurrence, its presence as a potential colonist of roots must be recognized in the majority of nursery studies. Because of the ubiquitous occurrence of *T. terrestris*, experiments designed to evaluate the influence of an inoculant fungus are complicated by the fact that most of the uninoculated 'control' plants, as well as some of those in the inoculated treatments, are frequently colonized from natural sources. There may also be other 'casual' colonists, among which E-strain fungi (see Chapter 7) and *Laccaria* species are common. Such trials are therefore comparisons of performance between *T. terrestris* (or other contaminant fungus) and the inoculated symbiont. One exception to this generalization (Xu *et al.*, 2001) is discussed below. Experience with *P. tinctorius* as the introduced organism strongly suggests that large numbers of mycorrhizas must be produced consistently on the roots of the seedlings if maximum promotion of growth is to be achieved when they are outplanted to reforestation sites. In these situations, Marx *et al.* (1976, 1988) showed that, if less then half of all mycorrhizas are formed by *P. tinctorius*, no growth promotion occurs, relative to that seen in *Thelephora*-colonized plants. Outplanting trials in several regions of the world indicate that increased growth in nurseries may not be correlated with improved performance in the field and that inoculant fungi may persist only a few years after outplanting, before being supplanted by naturally occurring fungi. Nevertheless, it is quite likely that early benefits accrue from dependence of young trees on uptake of nutrients from soil via their ECM symbionts. Later, internal recycling of nutrients within large trees plays an increasingly important role, while direct uptake declines (Grove and Le Tacon, 1993).

Various commercial inoculum formulations and inoculation techniques have been developed for use in seedling production systems (Marx and Bryan, 1975; Marx, 1980, 1991; Marx and Kenney, 1982; Sieverding, 1990; Bethlenfalvay and Linderman, 1992; Brundrett *et al.*, 1996). Spore-based inoculum, including pellets or seed coating, is appropriate and convenient for fungi that produce sporocarps with copious spores. Mycelial inoculum types include slurries and alginate bead formulations (Brundrett *et al.*, 1996). Some of the most successful have involved the growth of vegetative mycelium in vermiculite-peat mixtures moistened with liquid nutrient medium (Marx and Kenney, 1982). Vermiculite provides a well-aerated laminated substrate within which the mycelium is protected and addition of peat in different ratios enables adjustment of pH to the required range, usually 4.8 to 5.5. Nutrient solutions commonly have a C:N ratio of between 50 and 60, added in amounts sufficient to ensure that all free C is utilized by the mycorrhizal fungus in the course of its development in the medium. The presence of available C at the time of inoculation leads to competitive exclusion of the mycorrhizal fungus by saprotrophs. Details of inoculum production procedures and methods for screening isolates are available in Brundrett *et al.* (1996).

Development of inoculation procedures commercially presents a number of major challenges. One is the scale of the operation required. Even in 1985, Marx reported that 1.5 billion seedlings of *Pinus* spp. were produced per year in nurseries of the southern USA (Marx, 1985). Since then, worldwide increases in plantation forestry as well as in revegetation programmes have increased the requirement for stock of a wide range of broadleaf and conifer species. In Australia, the forestry department in the state of New South Wales alone produced 7.5 million seedlings and 2 million cuttings in 2006 (S. Sullivan, Forests New South Wales, personal communication). Problems with the use of solid substrates for inoculum production include the large space required for storage, difficulties in maintaining homogeneity of conditions within and between batches and the inability to control physicochemical conditions in the medium in the absence of water. Because of these difficulties, there have been various attempts to use liquids or gels as culture media. The main advantages of submerged, liquid culture are the homogeneity of the medium and the control which can be obtained over physical and chemical conditions. Vessels suitable for large-scale axenic production of fungal inoculum have been developed for other purposes in the chemical and pharmaceutical industries. They are designed to facilitate careful regulation and optimization of culture conditions for particular organisms, reducing the period of culture compared with solid substrates (Le Tacon *et al.*, 1985; Boyle *et al.*, 1987). Inoculum produced in this way can be applied directly as a slurry (Boyle *et al.*, 1987; Gagnon *et al.*, 1988), requiring some form of fragmentation. Unfortunately, this treatment greatly reduces the vigour of many ECM fungi. Attempts have been made to retain viability of fragmented inoculum by incorporation in a protective carrier medium. Sodium alginate has been successfully used, either applied as a gel directly to the bare roots (Deacon and Fox, 1988), granulated (Kropacek *et al.*, 1989) or as beads (Le Tacon *et al.*, 1985; Mauperin *et al.*, 1987). The susceptibility of many fungi to fragmentation damage, even when protected in this way, has led to a search for alternative culture methods. One approach which has considerable promise involves the production of the inoculum inside hydrogel beads, which can be applied directly, circumventing the fragmentation phase (Jeffries and Dodd, 1991; Kuek *et al.*, 1992). Several ECM fungi, including species of *Descocolea*, *Hebeloma*, *Laccaria* and *Pisolithus*, have been successfully grown as inoculum in this way and it has been shown that viability can be retained after storage for up to 7 months at low temperature (Kuek *et al.*, 1992).

Basidiospore inoculum of *P. tinctorius*, either as a suspension sprayed on the soil or encapsulated on seed with clay, has been used on an experimental basis in the USA and elsewhere. This can yield growth responses, but rarely produces as many mycorrhizas per plant as does the 'super-strain' of vegetative inoculum and so is less effective (Marx *et al.*, 1984b, 1991). A delay in production of mycorrhizas from spores might be expected because, as described in Chapter 6, colonization would not normally take place from monokaryotic mycelia. Only after hyphal fusion and the formation of dikaryons does ECM colonization occur.

Good though the responses to inoculation with *P. tinctorius* have been in warmer and more drought susceptible parts of the world, this fungus has proved less successful in cooler climates. In the Pacific north-west of the USA, for example, the 'super-strain' of *P. tinctorius* performed less well that did local isolates of the fungus (Perry *et al.*, 1987). In this region, the US Forest Service developed a spore inoculation programme based upon the use of ECM fungi known to be important in local ecosystems, including species of *Laccaria*, *Hebeloma*, *Rhizopogon* and *Suillus* (Castellano and Molina, 1989). Spores were applied to seed beds through the nursery irrigation system or to container-grown plants using mist-propagation units. Poor colonization was obtained with *Rhizopogon* and *Suillus* spp. (Perry *et al.*, 1987). In contrast, several strains of *Laccaria* produced abundant mycorrhizas in container-grown plants (Molina, 1982). One strain, subsequently referred to as *L. bicolor* S238, was found to have particular promise as an inoculant. Some *Laccaria* and *Hebeloma* strains have been developed as commercial inoculum, producing high levels of colonization on *Pseudostuga menziesii* in containers and under nursery conditions (Hung and Molina, 1986).

It appears, however, that despite success in achieving colonization by vigorous strains of fungi, outplanting performance of the seedlings improved rather little (Perry *et al.*, 1987). This was also the experience in Europe. Le Tacon *et al.* (1988, 1992) describe a number of experiments in France, Spain and Britain in which the performance of nursery inoculated plants of *P. menziesii* and *Pinus sylvestris* has been followed for several years after outplanting. The fungi used were mostly strains of *Thelephora, Hebeloma* and *Laccaria*, including the vigorous Oregon strain S238 of *L. bicolor*, originally isolated by Molina. The extent of success in obtaining colonization by the inoculant fungus varied from nursery to nursery, being apparently determined largely by the rate of re-invasion and vigour of indigenous *Thelephora* strains. Even where high levels of colonization by inoculant fungi were achieved, improvements in performance of the outplanted trees were rarely observed. Inoculation of *P. menziesii* with *L. bicolor* S238 provided significant increases of height growth and a doubling of wood volume at one site in central France six years after outplanting, but at the remaining sites, differences between control plants colonized by *T. terrestris* and those that were inoculated were small. Jackson *et al.* (1995) reported a similar experience with container-grown *P. menziesii* and *Picea sitchensis* which were inoculated with a wider range of fungi and transplanted, after colonization, to six nursery sites across the UK.

A second major challenge is the selection of appropriate fungi for inoculation programmes, based on their performance as symbionts for the plant species in nursery production and on their likely survival and competitiveness at field sites. The huge diversity of fungi now shown to occur in different forest ecosystems makes this a daunting task and may well contribute to the observation that extensive work on methods of inoculum production has not always been matched by demonstration of efficacy in forest productivity. As a result, while there are theoretical studies of the economic advantages to be gained from use of the new inoculation technologies (Kuek, 1994), prospects for their extensive application are not particularly promising.

Nevertheless, a number of targeted programmes have had success, especially in reafforestation projects on degraded sites mainly in SE Asia. In China, eucalypt plantations have been established on sites that are both low in available P and lack appropriate ECM fungi; in consequence establishment of early plantations was poor. Xu *et al.* (2001) investigated the potential of both P fertilizer application and ECM inoculation in the nursery to improve growth and after outplanting. They found that application of superphosphate generally improved seedling survival and stand volume three years after outplanting. The effect of nursery inoculation with ECM fungi was also tested. In these experiments, all inoculated seedlings formed ectomycorrhizas and the non-inoculated controls remained non-mycorrhizal, so that comparisons with seedlings inoculated with several different ECM fungi could be attributed to the effects of these fungi and not to weedy fungal species. There were no differences in growth of inoculated and non-inoculated seedlings at the time of outplanting. Inoculation had variable effects on growth in plantations, depending on P application and the identity of the inoculant fungus. As shown in Figure 17.1, after three years growth in the field, seedlings of *Eucalyptus europhylla* inoculated with *Hebeloma westraliense* showed decreases in stand volume relative to controls, whereas those inoculated with either *Laccaria lateritia* or *Pisolithus albus* showed increases, at least at some levels of P application. The negative effect of *Hebeloma* was related to very poor seedling survival, possibly because of poor survival of the fungus after outplanting. The conclusions were that optimum growth of plantations could be achieved by appropriate fertilizer management, coupled with inoculation with effective ECM fungi. The results bring into focus the need for careful site evaluation with respect to nutrient status and requirement for nutrient applications, as

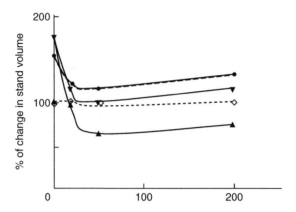

Figure 17.1 Effect of fungal inoculation in the nursery with three ectomycorrhizal fungi on stand volume of *Eucalyptus europhylla* averaged across three years of growth after outplanting. Values are per cent stand volume in relation to uninoculated control at each level of P fertilizer application. Control, ◇; *Pisolithus*, ▼; *Laccaria*, ●; *Hebeloma*, ▲. Redrawn from Xu *et al.* (2001).

well as for testing of inoculant fungi with respect to survival and compatibility with the tree species, as well as effects on timber production (see Chapters 9 and 10).

Unfortunately, it appears that many of those fungi selected to achieve optimal colonization in the nursery are often poor competitors in the field, especially when outplanting sites contain indigenous populations of mycorrhizal fungi. McAfee and Fortin (1986) preinoculated seedlings of *Pinus banksiana* with *L. bicolor*, *P. tinctorius* and *R. rubescens* before transplanting them to denuded, burned or natural pine stands. After two months in the natural stand, colonization by *L. laccata* and *P. tinctorius* had declined significantly, whereas that of *Rhizopogon rubescens* showed modest increase. *P. tinctorius* showed an ability to colonize new roots in a denuded site which lacked competition from an indigenous mycorrhizal population. These are the circumstances in which the greatest successes have been achieved in inoculation programmes involving this fungus. The failure of *L. bicolor* to compete with indigenous fungi is in line with the observation of Bledsoe *et al.* (1982) that the closely related *L. laccata* failed to persist on seedlings of *P. menziesii* when challenged by native fungi on outplanting sites in Washington.

There are a number of possible explanations for the common failure of inoculation to produce beneficial effects at outplanting sites. Probably among the most important of these is the inability of introduced inoculum to persist on the roots of planting stock after transfer from the nursery to the field, as shown for *Hebeloma* in the study outlined above. In addition to the fact that soil conditions experienced by nursery and container-grown plants are very different from those in most outplanting sites, the lifting, storage and transport of seedlings, especially those raised in bare-root nurseries, can be expected to reduce the vigour of fine roots and their fungal associates. These treatments are likely to favour replacement of introduced fungi by those resident in soil of the replant site. It is noteworthy in this context that the most strongly beneficial effects of inoculation have been observed where plants are transferred to disturbed or treeless sites in which inoculum potential of any indigenous fungi is likely to be low. Here, in contrast to the situation so often reported in

Table 17.2 Per cent increase in survival and volume growth of pine seedlings after 2 to 4 years with *Pisolithus tinctorius* ectomycorrhizas over controls with naturally occurring ectomycorrhizas on various adverse sites.

Pinus species	Site	Adversity	% Increase in seedling	
			Survival	Volume
P. resinosa	Coal spoil	pH 3.0	214	60
P. echinata	Coal spoil	pH 4.1	5	400
P. virginiana	Coal spoil	pH 3.1	87	444
P. virginiana	Coal spoil	pH 3.8	480	422
P. rigitaeda	Coal spoil	pH 3.8	0	420
P. rigida	Coal spoil	pH 3.4	57	215
P. rigida	Coal spoil	pH 4.3	8	180
P. taeda	Coal spoil	pH 3.3	20	415
P. taeda	Coal spoil	pH 3.4	14	750
P. taeda	Coal spoil	pH 4.1	41	400
P. taeda	Coal spoil	pH 3.4	96	800
P. taeda	Coal spoil	pH 4.3	16	380
P. taeda	Kaolin spoil	Low fertility	0	1100
P. taeda	Fullers' earth	Low fertility	0	47
P. taeda	Copper basin	Eroded	0	45
P. virginiana	Copper basin	Eroded	0	88
P. taeda	Borrow pit	Droughty	17	412

From Marx, 1975; Marx *et al.*, 1989.

soils with a pre-existing vegetation cover, responses to inoculation can be quite dramatic (Table 17.2) involving improvements in survival as well as increases in yield (Marx, 1991) and they appear to be most marked where the soil is contaminated with metal ions. In this context, the ability of ectomycorrhizal *Betula* spp. to spontaneously colonize mine spoils is widely recognized.

The natural formation of ectomycorrhizas in nurseries means that some forest production systems do not consider inoculation necessary or worthwhile. Furthermore, in some parts of the world, forest regeneration following clear cut is allowed to occur naturally, sometimes after site preparation that may disturb the soil or remove the organic horizons. There has been considerable interest in the effects of such practices on the composition and diversity of assemblages of ECM fungi and the possible impacts of any changes on the ECM-development on and performance of the regenerating seedlings. Jones *et al.* (2003) classify changes into those which result in shifts in the amount or type of inoculum and those which are related to changes in the soil environment. They come to the conclusions that the second group of changes are at least as important as the first and, consequently, that the fungi that do colonize the seedlings are likely to be better adapted to cope with the changed environmental conditions. They do, however, stress that data for diversity are at present based on taxonomic and genetic criteria and that future work on changes in community composition and effects on tree growth must be based on physiological and ecological attributes. In any event, the conclusions are important in informing policy on the way clear-cut sites are prepared and managed for maximum future productivity.

Edible mycorrhizal fungi

While most of the emphasis in applied research on ectomycorrhizas has concentrated on improvement of tree production, there is an increasing awareness of the uses to which fruit bodies of ECM fungi are put as food and medicines. In many countries, fungi, many of them ectomycorrhizal, are gathered in the wild and used directly as food and also to provide an important source of income (Figure 17.2a, b; Colour Plate 17.1). Interestingly, the subsistence uses of edible fungi in general have often been ignored in the context of development projects, probably because harvesting has largely been perceived as for personal use (Boa, 2004). In Mexico, unofficial estimates suggest that hundreds of tonnes of fresh mushrooms are sold annually in markets and that, in rural areas with a traditional knowledge of wild mushrooms, as much as 100% of family income comes from mushroom sales in the season (Jesus Perez-Moreno, personal communication). This contrasts with strong interest in development of industries based on the luxury trade in edible ECM fungi, such as truffles (*Tuber* spp.) and matsutake (*Tricholoma matsutake*), as well as species with perceived medicinal potential

(a)   (b)   (c)

Figure 17.2 Edible mushroom production in Mexico. (a) Girl selling *Boletus edulis* in a market. (b) and (c) Baskets of ECM fungi of which the main ones are: *Gomphus flocculosus, Lactarius salmonicolor, Helvella crispa* and *Amanita aspera* var *franchetii*. Photographs courtesy Jesus Perez-Moreno. See also Colour Plate 17.1.

(Hall and Wang, 1998; Boa, 2004). Harvesting highly sought-after species in the wild can add substantially to the income of rural communities, such that 60% of porcini (*Boletus edulis*) imported into Italy come from China. Exports of matsutake from China, Bhutan, Korea and both USA and Canada help to supply the huge demand for this delicacy in Japan, where availability in the forests where it was traditionally gathered has declined substantially. Such is the demand for and potential contribution of ECM fruit bodies that there are serious dangers that environmental damage will result from harvesting activities. Furthermore, in some places, loss of native woodlands and replacement with plantation crops (such as introduced eucalypts colonized by non-edible ECM fungi) has reduced the production and ease of harvesting of traditional fungal crops, with impact on traditional gatherers (Boa, 2004).

The possibilities of exploiting the commercial value of the fruit bodies produced by ECM fungi, particularly truffles, is being explored in a number of countries. At present, a small number of mycorrhizal species are particularly prized for their gastronomic quality and are hence of very high value (Table 17.3). In 2005, the price for black truffles (*Tuber melanosporum*) reached £1100 per kg. Edible fungi of this type are still mostly collected from natural stands and make only a small contribution (~4%) of the total global production of edible fungi (Figure 17.3), most of which are saprotrophs grown relatively easily under controlled conditions. Production of ECM fruit bodies has apparently declined in native ecosystems for a range of social and environmental reasons (Wang and Hall, 2004). In order to increase supply of ECM fruit bodies, the current demands for which far outstrip supply, three approaches are being adopted: extension of harvesting from the wild to non-traditional regions (see above), attempts to develop management strategies that will enhance or restore production in traditional harvesting areas, and development of new plantations of tree species, specifically to produce fruit bodies rather than timber. Numerous

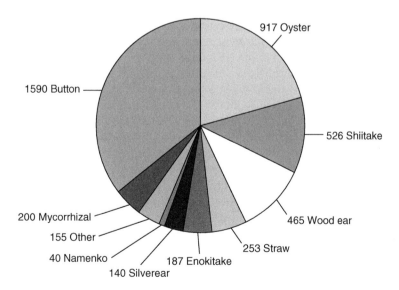

Figure 17.3 The contribution of mycorrhizal fungi to the approximate world production of edible mushrooms in 1991; Values are tonnes × 1000. From Hall *et al.* (1994), with permission.

Table 17.3 Prices of selected edible fungi for the luxury market.

Common name	Latin binomial	How sold	£ per kg	USD per lb
Black truffle	*Tuber melanosporum*	Fresh	425	900
Bianchetto truffles	*Tuber albidum*	Fresh	700	
	T. magnatum	Fresh	1700	
Summer truffles	*T. aestivum*	Fresh	105	
Porcini	*Boletus edulis*	Sliced and dried	95	
Morel	*Morcella elata*	Sliced and shredded	255	
Chanterelle	*Cantharellus cibarius*		15	
Matsutake	*Tricholoma matsutaki*	Dried		76

Source: various web pages, August 2006.

commercial organizations are involved in planting trees which have been pre-colonized by inoculation with appropriate fungi. Particular emphasis has been placed on truffles (*Tuber* spp.) because of their extremely high economic value, the most important of these being the black truffle, *T. melanosporum*. Indeed, according to Wang and Hall (2004), this species, together with *T. uncinatum*, is among the few to have been successfully cultivated commercially. Others include desert truffles, *Lactarius deliciosus* and the tubers of the mycoheterotrophic orchid *Gastrodia elata* (Chapter 13).

Techniques for the germination of the ascospores of *T. melanosporum* and for the aseptic production of mycorrhizas by a number of *Tuber* spp. pioneered in France (Grente *et al.*, 1972; Chevalier and Desmas, 1975; Chevalier and Grente, 1978) and Italy (Palenzona, 1969; Fontana and Bonfante-Fasolo, 1971) are widely used and remain the most popular (Wang and Hall, 2004). *T. melanosporum* has a broad host range and can be successfully grown on calcareous soils with the hardwood genera *Corylus*, *Quercus*, *Carpinus* and *Castanea*, as well as softwoods such as *Pinus*. Commercial production of colonized seedlings, particularly of *Quercus* and *Corylus*, now takes place in a number of centres in both the northern and southern hemispheres. In France alone, about 160 000 plants colonized by *T. melanosporum* were earlier reported to be produced annually, some being exported to the USA (Hall *et al.*, 1994).

Truffières have been established, usually as mixed plantings of *Quercus* and *Corylus* on potentially favourable sites, often at some distance from any other ECM communities to reduce competition between pre-existing and introduced fungi. The possibility of replacement of the truffle fungi with 'weedy' ECM fungal species has given some impetus for the development of molecular markers to monitor establishment and maintenance of truffle fungi and their competitors in plant roots. Many truffières were developed with *T. melanosporum* in the USA in the 1980s and began producing fruit bodies around 8–10 years later. Truffle production was commenced somewhat later in New Zealand in a programme pioneered by Hall, involving the introduction of both fungus and plant as exotic species. These truffières again began production after around 5 or more years, with the successful sites being in the warmer regions north of Christchurch. A second southern hemisphere initiative in Tasmania, Australia, is apparently adopting similar strategies and the establishment of truffières is extending into southern parts of the mainland of Australia, with about 90 or more hectares planted altogether (Wang and Hall, 2004). These ventures aim to capture the benefits of production in the southern hemisphere for sale when traditional supplies from the northern hemisphere are seasonally unavailable.

Commercial truffle production appears set to expand to appropriate sites, with ventures reported from Israel and several Asian countries including Taiwan.

Efforts have been made for some time to establish commercially productive systems for other valuable edible fungi such as *Boletus edulis*, *Cantharellus cibarius*, *Tricholoma matsutake* and *Tuber magnatum*. Seedlings have been successfully colonized under sterile laboratory conditions, but no success in the field has been reported (Wang and Hall, 2004). Despite advances in science and technology, which provide the prospect of large-scale production of a number of edible ECM fungi, the commercial success of the ventures is not yet fully assured. To a large extent the value of the commodities (especially of truffles) is based upon their limited availability, so that prices will certainly drop if large-scale production is achieved. However, especially in crops that can be used for timber, harvesting of edible fruit bodies could be an additional source of revenue or food, especially in developing countries. The large-scale establishment of eucalypt plantations in China, using planting stock pre-inoculated with edible fungi has considerable potential to provide an important dietary supplement (B. Dell and N. Malajczuk, personal communication).

In addition to fruit bodies, symbiotic mycorrhizal orchid tubers have been used in traditional medicine for over 2000 years (Xu and Mu, 1990; Kim and Ko, 1995; Xu and Guo, 2000). These practices continue to the present, so that there is an increasing demand which may threaten natural populations of some species. In order to maintain supplies, considerable research has been directed to the propagation of *Gastrodia elata*, which has as its fungal symbionts *Mycena osmundicola* in early protcorm stages, followed by *Armillaria mellea*. The latter fungus supports growth of the orchid with organic C derived saprotrophically from large logs of wood. This forms the basis for the current production systems (see Chapter 13).

Conclusions

Biological activities in soil are widely recognized as playing a vital part in nutrient cycling and in developing and maintaining soil structure and contributing to 'soil health'. Sustainable land-use, either in highly developed or subsistence economies requires that soil degradation ceases and that soil management practices are adopted to conserve and augment soil resources. Mycorrhizal fungi comprise just one of the functional groups of organisms that are important in soil ecosystems, but their position in forming direct links between roots of plants and the soil fabric means that they play key roles in soil–plant interactions, particularly in nutrient uptake and deposition of organic C. This applies as much to crops as it does to plants in undisturbed ecosystems and the difficulties in experimental demonstration of 'benefits' of mycorrhizas in the field are very similar. In both situations it is clear that mycorrhizas influence nutrient uptake, but benefits of crop inoculation will always depend on the resident mycorrhizal microflora and soil nutrient status.

Enormous efforts have been made to harness the potential benefits of mycorrhizal symbioses in commercial production systems, whether in horticulture, agriculture or forestry. The effects of mycorrhizal associations on production are almost all potentially beneficial, with only a few reports of growth depressions in agricultural field situations that remain imperfectly explained. There is an increasing awareness that mycorrhizal symbionts contribute to nutrient uptake in all types of crops, even when direct benefits of inoculation or management are difficult to demonstrate in field situations. Mycorrhizal inocula have been successfully developed for both AM and ECM fungi. Direct inoculation of AM fungi is currently limited to relatively small-scale, high value systems or subsistence farming. This picture may change when inoculum can be readily produced in a form that is convenient for wider application and contains fungi that are both effective and persist in the highly modified horticultural or agricultural environments. Similarly, in forestry it has often been the experience, particularly in sites with a well-established resident flora of ECM fungi, that inoculant species have had little impact. It thus appears that natural selection, operating over many generations, has produced stable populations of symbionts that are resistant to invasion by alien organisms. The assemblages of fungi may be modified by the agricultural and forestry practices applied, but are, nevertheless, adapted to the production systems in which they are now found. This may, in fact, be a desirable feature otherwise inoculant fungi, artificially selected for their vegetative vigour and now introduced with little legislative restriction to so many parts of the world, would come to dominate natural ecosystems. The consequences of the loss of biodiversity which would result from such invasion cannot as yet be quantified but, clearly, until such time as the roles of an extensive gene pool in sustaining ecosystems is evaluated, it would be prudent to protect and return as much diversity as possible.

It is therefore important to view mycorrhizal associations as integral components of the complex soil ecosystems and to manage those systems in order to maximize the contributions that mycorrhizas most certainly make to soil processes and growth of plants. Research directed towards understanding the activities of mycorrhizal fungi in production systems is valuable both in determining appropriate management strategies and as a background against which effective inoculation techniques may be developed in future. There remains enormous scope for investigation of both fundamental and applied aspects of mycorrhizal symbioses. These studies, if pursued at the cellular, whole plant and community levels, will enrich our understanding of the role of mycorrhizas in extant ecosystems both natural and man made. The need for reliable data on the impacts of disturbance, fragmentation of natural ecosystems and climate change introduces an element of urgency to the quest for knowledge.

References

Abbott LK 1982 Comparative anatomy of vesicular-arbuscular mycorrhizas formed on subterranean clover. *Australian Journal of Botany* **30**, 485–499.

Abbott LK and Robson AD 1978 Growth of subterranean clover in relation to the formation of endomycorrhizas by introduced and indigenous fungi in a field soil. *New Phytologist* **81**, 575–585.

Abbott LK and Robson AD 1979 A quantitative study of the spores and anatomy of mycorrhizas formed by a species of *Glomus*, with reference to its taxonomy. *Australian Journal of Botany* **27**, 363–375.

Abbott LK and Robson AD 1982 The role of vesicular arbuscular mycorrhizal fungi in agriculture and the selection of fungi for inoculation. *Australian Journal of Agricultural Research* **33**, 389–408.

Abbott LK and Robson AD 1991 Field management of VA mycorrhizal fungi. In *The Rhizosphere and Plant Growth*. Eds DL Keister and PB Cregan pp. 355–362. Kluwer Academic Publishers, Dordrecht, The Netherlands.

Abbott LK, Robson AD, Parker CA 1979 Double symbiosis in legumes – the role of mycorrhizas. In *Soil Microbiology and Plant Nutrition*. Eds WJ Broughton, ZK John, JC Bajarao, L Beda pp. 176–181. Penerbit University Malaya, Kuala Lumpur.

Abbott LK, Robson AD, De Boer G 1984 The effect of phosphorus on the formation of hyphae in soil by the vesicular-arbuscular mycorrhizal fungus, *Glomus fasciculatum*. *New Phytologist* **97**, 437–446.

Abbott LK, Robson AD, Jasper D, Gazey C 1992 What is the role of VA mycorrhizal hyphae in soil? In *Mycorrhizas in Ecosystems*. Eds DJ Read, DH Lewis, AH Fitter, IJ Alexander pp. 37–41. CAB International, Wallington, UK.

Abe JP 2005 An arbuscular mycorrhizal genus in the Ericaceae. *Inoculum* **56**, 6.

Aber JD, Nadelhoffer KJ, Steudler P, Melillo JM 1989 Nitrogen saturation in northern forest systems. *Biological Science* **39**, 378–386.

Abras K, Bilger I, Martin F, le Tacon F, Lapeyrie F 1988 Morphological and physiological changes in ectomycorrhizas of spruce (*Picea excelsa* (Lam.) Link) associated with ageing. *New Phytologist* **110**, 535–540.

Abuarghub SM and Read DJ 1988a The biology of mycorrhizas in the Ericaceae. XI. The distribution of nitrogen in the soil of a typical upland Callunetum with special reference to the 'free' amino acids. *New Phytologist* **108**, 425–431.

Abuarghub SM and Read DJ 1988b The biology of mycorrhiza in the Ericaceae. XII. Quantitative analysis of individual 'free' amino acids in relation to time and depth in the soil profile. *New Phytologist* **108**, 433–441.

Abuzinadah RA and Read DJ 1986a The role of proteins in the nitrogen nutrition of ectomycorrhizal plants. I. Utilization of peptides and proteins by ectomycorrhizal fungi. *New Phytologist* **103**, 481–493.

Abuzinadah RA and Read DJ 1986b The role of proteins in the nitrogen nutrition of ectomycorrhizal plants. III. Protein utilisation by *Betula, Picea* and *Pinus* in mycorrhizal association with *Hebeloma crustuliniforme*. *New Phytologist* **103**, 507–514.

Abuzinadah RA and Read DJ 1988 Amino acids as nitrogen sources for ectomycorrhizal fungi: utilisation of individual amino acids. *Transactions of the British Mycological Society* **91**, 473–479.

Abuzinadah RA and Read DJ 1989a Carbon transfer associated with assimilation of organic nitrogen sources by silver birch *Betula pendula* Roth. *Trees* **3**, 17–23.

Abuzinadah RA and Read DJ 1989b The role of proteins in the nitrogen nutrition of ectomycorrhizal plants IV. The utilization of peptides by birch *Betula pendula* L. infected with different mycorrhizal fungi. *New Phytologist* **112**, 55–60.

Abuzinadah RA and Read DJ 1989c The role of proteins in the nitrogen nutrition of ectomycorrhizal plants. V. Nitrogen transfer in birch *Betula pendula* grown in association with mycorrhizal and non-mycorrhizal fungi. *New Phytologist* **112**, 61–68.

Abuzinadah RA, Finlay RD, Read DJ 1986 The role of proteins in the nitrogen nutrition of ectomycorrhizal plants. II Utilisation of proteins by mycorrhizal plants of *Pinus contorta*. *New Phytologist* **103**, 495–506.

Addoms RM and Mounce FC 1931 Notes on the nutrient requirements and histology of the cranberry *V. macrocarpon* with special reference to mycorrhiza. *Plant Physiology* **6**, 563–568.

Addy HD, Schaffer G, Miller MH, Peterson RL 1994 Survival of the external mycelium of a VAM fungus in frozen soil over winter. *Mycorrhiza* **5**, 1–5.

Addy HD, Miller MH, Peterson RL 1997 Infectivity of the propagules associated with extraradical mycelia of two AM fungi following winter freezing. *New Phytologist* **135**, 745–753.

Addy HD, Boswell EP, Koide RT 1998 Low temperature acclimation and freezing resistance of extraradical VA mycorrhizal hyphae. *Mycological Research* **102**, 582–586.

Addy HD, Piercey MM, Currah RS 2005 Microfungal endophytes in roots. *Canadian Journal of Botany-Revue Canadienne De Botanique* **83**, 1–13.

Adriaensen K, Vrålstad T, Noben JP, Vangronsveld J, Colpaert JV 2005 Copper-adapted *Suillus luteus*, a symbiotic solution for pines colonizing Cu mine spoils. *Applied and Environmental Microbiology* **71** 11, 7279–7284.

Aerts R 1999 Interspecific competition in natural plant communities: mechanisms, trade-offs and plant-soil feedbacks. *Journal of Experimental Botany* **50**, 29–37.

Aerts R 2002 The role of various types of mycorrhizal fungi in nutrient cycling and plant competition. In *Mycorrhizal Ecology*. Eds MGA van der Heijden and I Sanders. pp. 117–133. Springer, Berlin, Germany.

Aerts R and Bobbink R 1999 The impact of atmospheric nitrogen deposition on vegetation processes in terrestrial, non-forest ecosystems. In *The Impact of Nitrogen Deposition on Natural and Semi-natural Systems*. Ed. SJ Langan pp. 85–122. Kluwer, Dordrecht, The Netherlands.

Agerer R 1986 Studies on ectomycorrhizae 3. Mycorrhizae formed by 4 fungi in the genera *Lactarius* and *Russula* on spruce. *Mycotaxon* **27**, 1–59.

Agerer R 1987 Studies in ectomycorrhizae 5. Mycorrhizae formed by *Dermocybe cinnamomea* and *Dermocybe sanguinea* on spruce. *Nova Hedwigia* **44**, 68–89.

Agerer R 1987–2002 *Colour Atlas of Ectomycorrhizae*. Einhorn-Verlag, Schwäbisch Gmünd D-72525, Germany.

Agerer R 1990 Studies on ectomycorrhizas. XXIV. Ectomycorrhizas of *Chroogomphus helveticus* and *C. rutilus* Gomphidiaceae, Basidiomycetes and their relationship to those of *Suillus* and *Rhizopogon*. *Nova Hedwigia* **50**, 1–63.

Agerer R 1991a Characterisation of ectomycorrhiza. *Methods in Microbiology* **23**, 25–73.

Agerer R 1991b Ectomycorrhizas of *Sarcodon imbricatus* on Norway spruce and their chlamydospores. *Mycorrhiza* **1**, 21–30.

Agerer R 1991c Comparison of the ontogeny of hyphal and rhizoid strands of *Pisolithus tinctorius* and *Polytrichium juniperinum*. *Cryptogam Botany* **2/3**, 85–92.

Agerer R 1992 Ectomycorrhizal rhizomorphs: organs of contact. In *Mycorrhizas in Ecosystems*. Eds DJ Read, DH Lewis, AH Fitter, IJ Alexander pp. 84–90. CAB International, Wallingford, UK.

Agerer R 1995 Anatomical characteristics of identified ectomycorrhizas: an attempt towards a natural classification. In *Mycorrhiza Structure, Function, Molecular Biology and Biotechnology*. Eds AK Varma and B Hock pp. 685–734. Springer Verlag, Berlin, Germany.

Agerer R 2001 Exploration types of ectomycorrhizae – A proposal to classify ectomycorrhizal mycelial systems according to their patterns of differentiation and putative ecological importance. *Mycorrhiza* **11**, 107–114.

Agerer R 2006 Fungal relationships and structural identity of their ectomycorrhizae. *Mycological Progress* **5**, 67–107.

Ahlich K and Sieber TN 1996 The profusion of dark septate endophytic fungi in non-ectomycorrhizal fine roots of forest trees and shrubs. *New Phytologist* **132**, 259–270.

Ahmad I, Carleton TJ, Malloch DW, Hellebust JA 1990 Nitrogen metabolism in the ectomycorrhizal fungus *Laccaria bicolor* R. Mre. Orton. *New Phytologist* **116**, 431–441.

Ahonen-Jonnarth U, van Hees PAW, Lu S, Finlay RD 2000 Organic acids produced by mycorrhizal *Pinus sylvestris* exposed to elevated aluminium and heavy metal concentrations. *New Phytologist* **146**, 557–567.

Ahonen-Jonnarth U, Goransson A, Finlay RD 2003 Growth and nutrient uptake of ectomycorrhizal *Pinus sylvestris* seedlings in a natural substrate treated with elevated Al concentrations. *Tree Physiology* **23**, 157–167.

Aitchison PA and Butt VS 1973 The relationship between synthesis of inorganic polyphosphate and phosphate uptake by *Chlorella vulgaris*. *Journal of Experimental Botany* **24**, 197–510.

Akiyama K, Matsuoka H, Hayashi H 2002 Isolation and identification of a phosphate deficiency-induced *C*-glycosylflavonoid that stimulates arbuscular mycorrhiza formation in melon roots. *Molecular Plant-Microbe Interactions* **15**, 334–340.

Akiyama K, Matsuzaki K, Hyashi H 2005 Plant sesquiterpenes induce hyphal branching in arbuscular mycorrhizal fungi. *Nature* **435**, 824–827.

Albertson O, Kuyper TW, Gorissen A 2005 Taking mycocentrism seriously: mycorrhizal fungi and plant responses to elevated CO_2. *New Phytologist* **167**, 859–868.

Albrecht C, Burgess T, Dell B, Lapeyrie F 1994 Chitinase and peroxidase activities are induced in *Eucalyptus* roots according to aggressiveness of Australian ectomycorrhizal strains of *Pisolithus* sp. *New Phytologist* **127**, 217–222.

Albrecht C, Guerts R, Lapeyrie F, Bisseling T 1998 Endomycorrhizae and rhizobial Nod factors both require SYM8 to induce the expression of the early nodulin genes *PsENOD5* and *PsENOD12A*. *The Plant Journal* **15**, 605–614.

Alexander C and Alexander IJ 1984 Seasonal changes in populations of the orchid *Goodyera repens* Br. and its mycorrhizal development. *Transactions of the Botanical Society of Edinburgh* **44**, 219–227.

Alexander C and Hadley G 1984 The effect of mycorrhizal infection of *Goodyera repens* and its control by fungicide. *New Phytologist* **97**, 391–400.

Alexander C and Hadley G 1985 Carbon movement between host and mycorrhizal endophyte during development of the orchid *Goodyera repens* Br. *New Phytologist* **101**, 657–665.

Alexander C, Alexander IJ, Hadley G 1984 Phosphate uptake by *Goodyera repens* in relation to mycorrhizal infection. *New Phytologist* **97**, 401–411.

Alexander IJ 1981 *Picea sichensis* and *Lactarius rufus* mycorrhizal association and its effects on seedling growth and development. *Transactions of the British Mycological Society* **76**, 417–423.

Alexander IJ 1983 The significance of ectomycorrhizas in the nitrogen cycle. In *Nitrogen as an Ecological Factor*. Eds JA Lee, S McNeill, IH Rorison pp. 69–94. Blackwell, Oxford.

Alexander IJ 1988 Mycorrhizas of indigenous tropical forest trees: some research priorities. In *Proceeedings of the Asian Seminar*. Ed. FSP Ngi pp. 79–89. Forest Research Institute, Kuala Lumpur, Malaysia.

Alexander IJ 1989a Mycorrhizas in tropical forest. In *Mineral Nutrients in Tropical Forest and Savannah Ecosystems*. Ed. J Proctor pp. 169–188. Blackwell Scientific Publications, Oxford.

Alexander IJ 1989b Systematics and ecology of ectomycorrhizal legumes. *Monographs in Systematic Botany from the Missouri Botanic Garden* **29**, 607–624.

Alexander IJ 2006 Ectomycorrhizas – out of Africa? *New Phytologist* **172**, 589–591.

Alexander IJ and Bigg WL 1981 Light microscopy of ectomycorrhizas using glycol methacrylate. *Transactions of the British Mycological Society* **77**, 425–429.

Alexander IJ and Fairley RI 1983 Effects of N fertilisation on populations of fine roots and mycorrhizas in spruce humus. *Plant and Soil* **74**, 49–53.

Alexander IJ and Hardy K 1981 Surface phosphatase activity of Sitka spruce mycorrhizas from a serpentine site. *Soil Biology and Biochemistry* **13**, 301–305.

Alexander IJ and Högberg P 1986 Ectomycorrhizas of tropical angiospermous trees. *New Phytologist* **102**, 541–549.

Alexander IJ and Lee SS 2005 Mycorrhizas and ecosystem processes in tropical rain forest: implications for diversity. In *Biotic Interactions in the Tropics*. Eds DFRP Burslem, MA Pinard, SE Hartley pp. 165–203. Cambridge University Press, Cambridge.

Alexander IJ, Ahmad N, Lee SS 1992 The role of mycorrhizae in the regeneration of some Malaysian forest trees. *Philosophical Transactions of the Royal Society B* **335**, 379–388.

Alexander T, Meier R, Toth R, Weber HC 1988 Dynamics of arbuscule development and degeneration in mycorrhizas of *Triticum aestivum* L. and *Avena sativa* L. with reference to *Zea mays* L. *New Phytologist* **110**, 363–370.

Alexander T, Toth R, Meier R, Weber HC 1989 Dynamics of arbuscule development and degeneration in onion, bean and tomato with reference to vesicular-arbuscular mycorrhizae with grasses. *Canadian Journal of Botany* **67**, 2505–2513.

Ali NA and Jackson RM 1988 Effects of plant roots and their exudates on germination of spores of ectomycorrhizal fungi. *Transactions of the British Mycological Society* **91**, 253–260.

Allaway WG and Ashford AE 1996 Structure of the hair roots in *Lysinema ciliatum* R. Br. and its implications for their water relations. *Annals of Botany* **77**, 383–388.

Allaway WG and Ashford A 2001 Motile tubular vacuoles in extramatrical mycelium and sheath hyphae of ectomycorrhizal systems. *Protoplasma* **215**, 218–225.

Allaway WG, Carpenter JL, Ashford AE 1985 Amplification of inter-symbiont surface by root epidermal transfer cells in the *Pisonia* mycorrhiza. *Protoplasma* **128**, 227–231.

Allen EB 1984 VA mycorrhizae and colonising annuals: implications for growth, competition and succession. In *VA Mycorrhizae and Reclamation of Arid and Semiarid Lands*. Eds SE Williams and MF Allen pp. 41–51. University of Wyoming Agricultural. Experiment Station, Science Report No. SA1261. University of Wyoming, Laramie, Wyoming, USA.

Allen EB and Allen MF 1984 Competition between plants of different successional stages: mycorrhizae as regulators. *Canadian Journal of Botany* **62**, 2625–2629.

Allen EB and Allen MF 1988 Facilitation of succession by the nonmycotrophic coloniser *Salsola kali* Chenopodiaceae on a harsh site: effects of mycorrhizal fungi. *American Journal of Botany* **75**, 257–266.

Allen EB, Allen MF, Helm DJ, Trappe JM, Molina R, Rincon E 1995 Patterns and regulation of mycorrhizal plant and fungal diversity. *Plant and Soil* **170**, 47–62.

Allen MF 1982 Influence of vesicular-arbuscular mycorrhizae on water movement through *Bouteloua gracilis* H.B.K. Lag ex Steud. *New Phytologist* **91**, 191–196.

Allen MF 1988 Re-establishment of VA mycorrhizae following severe disturbance: comparative patch dynamics of a shrub desert and a sub-alpine volcano. *Proceedings of the Royal Society of Edinburgh* **94B**, 63–71.

Allen MF 1989 Mycorrhizae and rehabilitation of disturbed arid soils: processes and practices. *Arid Soil Research and Rehabilitation* **3**, 229–241.

Allen MF 1991 *The Ecology of Mycorrhizae*. Cambridge University Press, Cambridge, UK.

Allen MF 1992 *Mycorrhizal Functioning*. Routledge: Chapman and Hall, New York, USA.

Allen MF and McMahon JA 1988 Direct VA mycorrhizal inoculation of colonising plants by pocket gophers *Thomomys talpoides* on Mount St. Helens. *Mycologia* **80**, 754–756.

Allen MF, Moore TS, Christensen M 1980 Phytohormone changes in *Bouteloua gracilis* infected by vesicular-arbuscular mycorrhizae. I. Cytokinin increases in the host plant. *Canadian Journal of Botany* **58**, 371–374.

Allen MF, Smith WK, Moore TS, Christensen M 1981 Comparative water relations and photosynthesis of mycorrhizal and non-mycorrhizal *Bouteloua gracilis* H.B.K. Lag ex Steud. *New Phytologist* **88**, 683–693.

Allen MF, Moore TS, Christensen M 1982 Phytohormone changes in *Bouteloua gracilis* infected by vesicular-arbuscular mycorrhizae. II. Altered levels of gibberellin-like substances and abscisic acid in the host plant. *Canadian Journal of Botany* **60**, 468–471.

Allen MF, Allen EB, Friese CF 1989 Responses of the non-mycotrophic plant *Salsola kali* to invasion by vesicular-arbuscular mycorrhizal fungi. *New Phytologist* **111**, 45–49.

Allen TR, Millar T, Berch SM, Berbee ML 2003 Culturing and direct DNA extraction find different fungi from the same ericoid mycorrhizal roots. *New Phytologist* **160** 1, 255–272.

Allsopp N, Stock WD 1992a Mycorrhizas, seed size and seedling establishment in a low nutrient environment. In *Mycorrhizas in Ecosystems*. Eds DJ Read, DH Lewis, AH Fitter, IJ Alexander pp. 59–64. CAB International, Wallingford, UK.

Allsopp N and Stock WD 1992b Density dependent interactions between VA mycorrhizal fungi and even-aged seedlings of two perennial Fabaceae species. *Oecologia* **91**, 281–287.

Alpert P, Warembourg FR, Roy J 1991 Transport of carbon among connected ramets of *Eichhornia crassipes* Pontederiaceae at normal and high levels of CO_2. *American Journal of Botany* **78**, 1459–1466.

Alvarez MR 1968 Quantitative changes in nuclear DNA accompanying post-germination embryonic development in *Vanda* Orchidaceae. *American Journal of Botany* **55**, 1036–1041.

Amaranthus MP and Perry DA 1989 Interaction effects of vegetation type and Pacific madrone soil inocula on survival, growth, and mycorrhiza formation of Douglas-fir. *Canadian Journal of Forest Research* **19**, 550–556.

Amaranthus MP, Molina R, Perry DA 1990 Soil organisms, root growth and forest regeneration. *Proceedings of the Society of the American Foresters*, Spokane, Washington. pp. 89–93.

Ames RN, Reid CPP, Porter L, Cambardella C 1983 Hyphal uptake and transport of nitrogen from two ^{15}N-labelled sources by *Glomus mosseae*, a vesicular-arbuscular mycorrhizal fungus. *New Phytologist* **95**, 381–396.

Amijee F and Stribley DP 1987 Soluble carbohydrates of vesicular-arbuscular mycorrhizal fungi. *The Mycologist* **1**, 20–21.

Amijee F, Tinker PB, Stribley DP 1989 The development of endomycorrhizal root systems. VII. A detailed study of the effects of soil phosphorus on colonization. *New Phytologist* **111**, 435–446.

An Z-Q, Hendrix JW, Hershman DE, Henson GT 1990 Evaluation of the 'Most Probable Number' MPN and wet-sieving methods for determining soil-borne populations of endogonaceous mycorrhizal fungi. *Mycologia* **82**, 576–581.

Andersen TF 1996 A comparative taxonomic study of *Rhizoctonia sensu lato* employing morphological, ultrastructural and molecular methods. *Mycological Research* **100**: 1117–1128.

Anderson IC, Chambers SM, Cairney JWG 1999 Patterns of organic nitrogen utilisation by three Australian *Pisolithus* species. *Mycological Research* **103**, 1579–1587.

Anderson IC, Chambers SM, Cairney JWG 2001 ITS-RFLP and ITS sequence diversity in *Pisolithus* from central and eastern Australian sclerophyll forests. *Mycological Research*, **105**, 1304–1312.

Andersson S, Ek H, Soderstrom B 1997 Effects of liming on the uptake of organic and inorganic nitrogen by mycorrhizal (*Paxillus involutus*) and non-mycorrhizal *Pinus sylvestris* plants. *New Phytologist*, **135**, 763–771.

Andrade SAL, Abreu CA, Abreu MF, Silveira APD 2003 Interaction between lead, soil base saturation rate, and mycorrhiza on soybean development and mineral nutrition. *Revista Brasileira De Ciencia Do Solo* **27**, 945–954.

Andrade SAL, Abreu CA, de Abreu MF, Silveira APD 2004 Influence of lead additions on arbuscular mycorrhiza and *Rhizobium* symbioses under soybean plants. *Applied Soil Ecology* **26**, 123–131.

Andrews DJ and Kassam AH 1976 The importance of multiple cropping in increasing world food supplies. In *Multiple Cropping*. Ed. M Stelly pp. 1–10. American Society of Agronomy: Madison, Wisconsin, USA.

Andrews JA, Harrison KG, Matamala R, Schlesinger WH 1999 Separation of root respiration from total soil respiration using carbon-13 labeling during Free-Air Carbon Dioxide Enrichment (FACE). *Soil Science Society of America, Journal* **63** 5, 1429–1435.

Ané JM, Kiss GB, Reily BK *et al.* 2004 *Medicago truncatula DMI1* required for bacterial and fungal symbioses in legumes. *Science* **303**, 1364–1367.

Antibus RK, Kroehler CJ, Linkins AE 1986 The effects of external pH, temperature, and substrate concentration on acid phosphatase activity of ectomycorrhizal fungi. *Canadian Journal of Botany* **64**, 2383–2387.

Antibus RK, Sinsabaugh RL, Linkins AE 1992 Phosphatase activities and phosphorus uptake from inositol phosphate by ectomycorrhizal fungi. *Canadian Journal of Botany* **70**, 794–801.

Antoniolli ZI 1999 Arbuscular mycorrhizal community in a permanent pasture and development of species-specific primers for detection and quantification of two AM fungi. PhD thesis. The University of Adelaide, Adelaide, Australia.

Antoniolli Z, Schachtman DP, Ophel-Keller K, Smith SE 2000 Variation in rDNA ITS sequences in *Glomus mosseae* and *Gigaspora margarita* spores from a permanent pasture. *Mycological Research* **104**, 708–715.

Aono T, Maldonado-Mendoza IE, Dewbre GR, Harrison MJ, Saito M 2004 Expression of alkaline phosphatase genes in arbuscular mycorrhizas. *New Phytologist* **162**, 525–534.

Arditti J 1979 Aspects of the physiology of orchids. *Advances in Botanical Research* **7**, 421–655.

Arditti J 1992 Mycorrhiza. In *Fundamentals of Orchid Biology*. Ed. J. Arditti pp. 419–451. John Wiley and Sons, New York, USA.

Arditti J and Ghani AKA 2000 Numerical and physical properties of orchid seeds and their biological implications. *New Phytologist* **146** 3, 569–569.

Arines J, Palma JM, Vilarino A 1993 Comparison of protein patterns in non-mycorrhizal and vesicular-arbuscular mycorrhizal roots of red clover. *New Phytologist* **123**, 763–768.

Armstrong L and Peterson RL 2002 The interface between the arbuscular mycorrhizal fungus *Glomus intraradices* and root cells of *Panax quinquefolius*: a *Paris*-type mycorrhizal association. *Mycologia* **94**, 587–595.

Arnebrant K 1994 Nitrogen amendments reduce the growth of extramatrical ectomycorrhizal mycelium. *Mycorrhiza* **5**, 7–15.

Arnebrant K, Ek H, Finlay RD, Söderström B 1993 Nitrogen translocation between *Alnus glutinosa* L. Gaertn.seedlings inoculated with *Frankia* sp. and *Pinus contorta* Doug. ex Loud seedlings connected by a common ectomycorrhizal mycelium. *New Phytologist* **124**, 231–242.

Arnolds E 1991 Decline of ectomycorrhizal fungi in Europe. *Agriculture Ecosystems and Environment* **35**, 209–244.

Aroca R, Porcel R, Ruiz-Lozano JM 2007 How does arbuscular mycorrhizal symbiosis regulate root hydraulic properties and plasma membrane aquaporins in *Phaseolus vulgaris* under drought, cold or salinity stresses? *New Phytologist* **173**, 808–816.

Arocena, JM and Glowa KR 2000 Mineral weathering in ectomycorrhizosphere of subalpine fir *Abies lasiocarpa* Hook. Nutt. as revealed by soil solution composition. *Forest Ecology and Management* **133** 1–2, 61–70.

Arocena JM, Glowa KR, Massicotte HB, Lavkulich L 1999 Chemical and mineral composition of ectomycorrhizosphere soils of subalpine fir (*Abies lasiocarpa* (Hook.) Nutt.) in the Ae horizon of a Luvisol. *Canadian Journal of Soil Science* **79**, 25–35.

Arocena JM, Glowa KR, Massicotte HB 2001 Calcium-rich hypha encrustations on *Piloderma*. *Mycorrhiza* **10**, 209–215.

Asai T 1944 Über die Mykorrhizenbildung der leguminosen Pflanzen. *Japanese Journal of Botany* **13**, 463–485.

Asghari HR, Chittleborough DJ, Smith FA, Smith SE 2005 Influence of arbuscular mycorrhizal AM symbiosis on phosphorus leaching through soil cores. *Plant and Soil* **275**, 181–193.

Ashford AE 1998 Dynamic pleiomorphic vacuole systems: are they endosomes and transport compartments in fungal hyphae? *Advances in Botanical Research* **28**, 119–159.

Ashford AE and Allaway WG 1982 A sheathing mycorrhiza on *Pisonia grandis* R.Br. Nyctaginaceae with development of transfer cells rather than a Hartig net. *New Phytologist* **90**, 511–519.

Ashford AE and Allaway WG (2002) The role of the motile tubular vacuole system in mycorrhizal fungi. *Plant and Soil* **244**, 177–187.

Ashford AE and Orlovich DA 1994 Vacuole transport, phosphorus, and endosomes in the growing tips of fungal hyphae. In *Pollen-Pistil Interactions and Pollen Tube Growth*. Eds AG Stephenson and TH Kao pp. 135–149. American Society of Plant Physiologists, Rockville, Maryland, USA.

Ashford AE, Ling-Lee M, Chilvers G 1975 Polyphosphate in eucalypt mycorrhizas: a cytochemical demonstration. *New Phytologist* **74**, 447–453.

Ashford AE, Peterson CA, Carpenter JL, Cairney JWG, Allaway WG 1988 Structure and permeability of the fungal sheath in the *Pisonia* mycorrhiza. *Protoplasma* **147**, 149–161.

Ashford AE, Allaway WG, Peterson CA, Cairney JWG 1989 Nutrient transfer and the fungus root interface. *Australian Journal of Plant Physiology* **16**, 85–97.

Ashford AE, Ryde S, Barrow KD 1994 Demonstration of a short chain polyphosphate in *Pisolithus tinctorius* and the implications for phosphorus transport. *New Phytologist* **126**, 239–247.

Ashford AE, Allaway WG, Reed ML 1996 A novel role for thick-walled epidermal cells in the mycorrhizal hair roots of *Lysinema ciliatum* R.Br and other Epacridaceae. *Annals of Botany* **77**, 375–382.

Ashkannejhad S and Horton TR 2006 Ectomycorrhizal ecology under primary succession on coastal sand dunes: interactions involving *Pinus contorta*, suilloid fungi and deer. *New Phytologist* **169**, 345–354.

Ashton PS 1982 Dipterocarpaceae. In *Flora Malesiana Series 1, Spermatophyta*. Ed. CGGJ van Steenis pp. 237–552. Martinus-Nijhoff Publications, The Hague, The Netherlands.

Åström H, Giovannetti M, Raudaskoski M 1994 Cytoskeletal components in the arbuscular mycorrhizal fungus *Glomus mosseae*. *Molecular Plant–Microbe Interactions* **7**, 309–312.

Atkinson MA 1975 The fine structure of mycorrhizas. DPhil Thesis, Oxford, UK.

Augé RM 2001 Water relations, drought and vesicular-arbuscular mycorrhizal symbiosis. *Mycorrhiza* **11**, 3–42.

Augé RM 2004 Arbuscular mycorrhizae and soil/plant water relations. *Canadian Journal of Soil Science* **84**, 373–381.

Augé RM, Sylvia DM, Park S *et al.* 2004 Partitioning mycorrhizal influence on water relations of *Phaseolus vulgaris* into soil and plant components. *Canadian Journal of Botany* **82**, 503–514.

Avis PG, McLaughlin DJ, Dentinger BC, Reich PB 2003 Long-term increase in nitrogen supply alters above- and belowground ectomycorrhizal communities and increases the dominance of *Russula* spp. in a temperate oak savanna. *New Phytologist* **160**, 239–253.

Avis PG, Dickie IA, Mueller GM 2006 A 'dirty' business: testing the limitations of terminal restriction fragment length polymorphism (TRFLP) analysis of soil fungi. *Molecular Ecology* **15**, 873–882.

Axelrod DI 1986 Cenozoic history of some western American pines. *Annals of the Missouri Botanical Garden* **73**, 565–641.

Ayling SM, Smith SE, Smith FA 2000 Transmembrane electric potential difference of germ tubes of arbuscular mycorrhizal fungi responds to external stimuli. *New Phytologist* **147**, 631–639.

Azcón R 1987 Germination and hyphal growth of *Glomus mosseae in vitro*: effects of rhizosphere bacteria and cell-free culture media. *Soil Biology and Biochemistry* **19**, 417–419.

Azcón-Aguilar C, Diaz-Rodriguez RM, Barea JM 1986a Effect of free-living fungi on the germination of *G. mosseae* on soil extract. In *Physiological and Genetical Aspects of Mycorrhizae*. Eds V Gianinazzi-Pearson and S Gianinazzi pp. 515–519. INRA, Paris, France.

Azcón-Aguilar C, Diaz-Rodriguez RM, Barea JM 1986b Effect of soil micro-organisms on spore germination and growth of the vesicular-arbuscular mycorrhizal fungus *Glomus mosseae*. *Transactions of the British Mycological Society* **86**, 337–340.

Azcón-Aguilar C and Barea JM 1992 Interactions between mycorrhizal fungi and other rhizosphere microorganisms. In *Mycorrhizal Functioning*. Ed. MF Allen pp. 163–198. Chapman and Hall, London, UK.

Azcón-Aguilar C, Bago B, Barea JM 1999 Saprophytic growth of arbuscular mycorrhizal fungi. In *Mycorrhiza; Structure, Function, Molecular Biology and Biotechnology* 2nd edn. Eds A Varma and B Hock pp. 557–569. Springer, Berlin, Germany.

Azul M, Agerer R, Freitas H 2006 *Quercirhiza dendrohyphidiomorpha* on *Quercus suber*. *Descriptions of Ectomycorrhizae* **9/10**, 87–91.

Baar J, Horton TR, Kretzer AM, Bruns TD (1999) Mycorrhizal colonization of *Pinus muricata* from resistant propagules after a stand-replacing wildfire. *New Phytologist* **143**, 409–418.

Baas R and Kuiper D 1989 Effects of vesicular-arbuscular mycorrhizal infection and phosphate on *Plantago major* ssp. *pleiosperma* in relation to internal cytokinin concentration. *Physiologia Plantarum* **76**, 211–215.

Bååth E and Söderström B 1979 Fungal biomass and fungal immobilisation of plant nutrients in Swedish coniferous forest soils. *Revue Ecologie Biologie des Sols* **16**, 477–489.

Bago B and Azcón-Aguilar C 1997 Changes in the rhizospheric pH induced by arbuscular mycorrhiza formation in onion *Allium Cepa* L. *Zeitschrift fur Pflanzenernahrung und Bodenkunde* **160**, 333–339.

Bago B, Azcón-Aguilar C, Goulet A, Piché Y 1998a Branched adsorbing structures BAS: a feature of the extraradical mycelium of symbiotic arbuscular mycorrhizal fungi. *New Phytologist* **139**, 375–388.

Bago B, Azcón-Aguilar C, Piché Y 1998b Architecture and developmental dynamics of the external mycelium of the arbuscular mycorrhizal fungus *Glomus intraradices* grown under monoxenic conditions. *Mycologia* **90**, 52–62.

Bago B, Pfeffer PE, Douds DDJ, Brouillette J, Bécard G, Shachar-Hill Y 1999a Carbon metabolism in spores of the arbuscular mycorrhizal fungus *Glomus intraradices* as revealed by nuclear magnetic resonance spectroscopy. *Plant Physiology* **121**, 263–271.

Bago B, Zipfel W, Williams RM, Piché Y 1999b Nuclei of symbiotic arbuscular mycorrhizal fungi as revealed by in vivo two-photon microscopy. *Protoplasma* **209**, 77–89.

Bago B, Pfeffer PE, Shachar-Hill Y 2000 Carbon metabolism and transport in arbuscular mycorrhizas. *Plant Physiology* **124**, 949–958.

Bago B, Pfeffer PE, Zipfel W, Lammers PJ, Shachar-Hill Y 2002a Tracking metabolism and imaging transport in arbuscular mycorrhizal fungi. *Plant and Soil* **244**, 189–197.

Bago B, Zipfel W, Williams RM *et al.* 2002b Translocation and utilization of fungal storage lipid in the arbuscular mycorrhizal symbiosis. *Plant Physiology* **128**, 108–124.

Bago B, Pfeffer PE, Abubaker J *et al.* 2003 Carbon export from arbuscular mycorrhizal roots involves the translocation of carbohydrate as well as lipid. *Plant Physiology* **131**, 1496–1507.

Bagyaraj DJ and Sreeramulu KR 1982 Preinoculation with VA mycorrhiza improves growth and yield of chilli transplanted in the field and saves phosphatic fertiliser. *Plant and Soil* **69**, 375–381.

Bagyaraj DJ, Manjunath A, Patil RB 1979 Occurrence of vesicular-arbuscular mycorrhizas in some tropical aquatic plants. *Transactions of the British Mycological Society* **72**, 164–167.

Bain HF 1937 Production of synthetic mycorrhiza in the cultivated cranberry. *Journal of Agricultural Research* **55**, 811–835.

Bajwa R and Read DJ 1985 The biology of mycorrhiza in the Ericaceae. IX. Peptides as nitrogen sources for the ericoid endophyte and for mycorrhizal and non-mycorrhizal plants. *New Phytologist* **101**, 459–467.

Bajwa R and Read DJ 1986 Utilisation of mineral and amino N sources by the ericoid mycorrhizal endophyte *Hymenoscyphus ericae* and by mycorrhizal and non-mycorrhizal seedlings of *Vaccinium*. *Transactions of the British Mycological Society* **87**, 269–277.

Bajwa R, Abuarghub S, Read DJ 1985 The biology of mycorrhiza in the Ericaceae. X. The utilization of proteins and the production of proteolytic enzymes by the mycorrhizal endophyte and by mycorrhizal plants. *New Phytologist* **101**, 469–486.

Bakhtiar Y, Miller D, Cavagnaro TR, Smith SE 2001 Interactions between two arbuscular mycorrhizal fungi and fungivorous nematodes and control of the nematode with fenamifos. *Applied Soil Ecology* **17**, 107–117.

Bakken LR and Olsen RA 1983. Buoyant densities and dry matter contents of microorganisms–conversion of a measured biovolume into biomass. *Applied and Environmental Microbiology* **45**, 1188–1195.

Balestrini R and Bonfante P 2005 The interface compartment in arbuscular mycorrhizae: a special type of plant cell wall? *Plant Biosystems* **139**, 8–15.

Balestrini R, Berta G, Bonfante P 1992 The plant nucleus in mycorrhizal roots: positional and structural modifications. *Biology of the Cell* **75**, 235–243.

Baon JB 1994 Phosphorus uptake and growth of barley as affected by soil temperature and mycorrhizal infection. *Journal of Plant Nutrition* **17**, 2–3.

Baon JB, Smith SE, Alston AM, Wheeler RD 1992 Phosphorus efficiency of three cereals as related to indigenous mycorrhizal infection. *Australian Journal of Agricultural Research* **43**, 479–491.

Baon JB, Smith SE, Alston AM 1993 Mycorrhizal responses of barley cultivars differing in P efficiency. *Plant and Soil* **157**, 97–105.

Barea JM and Azcón-Aguilar C 1983 Mycorrhizas and their significance in nodulating nitrogen-fixing plants. *Advances in Agronomy* **36**, 1–54.

Barea JM, Azcón R, Hayman DS 1975 Possible synergistic interactions between *Endogone* and phosphate solubilising bacteria in low phosphate soils. In *Endomycorrhizas*. Eds FE Sanders, B Mosse, PB Tinker pp. 409–417. Academic Press, London, UK.

Barea JM, Azcón-Aguilar C, Azcón R 1987 Vesicular-arbuscular mycorrhiza improve both symbiotic N_2 fixation and N uptake from soil as assessed with a ^{15}N technique under field conditions. *New Phytologist* **106**, 717–725.

Barea JM, Azcón R, Azcón-Aguilar C 1992 Vesicular-arbuscular mycorrhizal fungi in nitrogen-fixing systems. *Methods in Microbiology* **24**, 391–416.

Barker SJ, Stummer B, Gao L, Dispain I, O'Connor PJ, Smith SE 1998a A mutant in *Lycopersicon esculentum* Mill. with highly reduced VA mycorrhizal colonization: isolation and preliminary characterisation. *The Plant Journal* **15**, 791–797.

Barker SJ, Tagu D, Delp G 1998b Regulation of root and fungal morphogenesis in mycorrhizal symbioses. *Plant Physiology* **116**, 1201–1207.

Barker SJ, Duplessis S, Tagu D 2002 The application of genetic approaches for investigations of mycorrhizal symbioses. *Plant and Soil* **244**, 85–95.

Barker SJ, Edmonds-Tibbett T, Forsyth L *et al.* 2005 Root infection of the reduced mycorrhizal colonization *rmc* mutant of tomato reveals genetic interaction between symbiosis and disease. *Physiogical and Molecular Plant Pathology* **67**, 277–283.

Barkman A and Sverdrup H 1996 *Critical Loads of Acidity and Nutrient Imbalance for Forest Ecosystems in Skåne*, Report 1. Department of Chemical Technology II, Lund University, Lund, Sweden.

Barmicheva KM 1989 Ultrastructure of *Neottia nidus-avis* mycorrhizas. *Agriculture, Ecosystems and Environment* **29**, 23–27.

Barroso J and Pais MSS 1885 Cytochimie-characterisation cytochimique de l'interface hote/ endophyte des endomycorhizes d'*Ophrys lutea*. Role de l'hôte dans la synthese des poly-saccharides. *Annales des Sciences Naturelles. Botanique et Biologie Végétale* **13**, 237–244.

Barrow JR 2003 Atypical morphology of dark septate fungal root endophytes of *Bouteloua* in arid southwestern USA rangelands. *Mycorrhiza* **13**, 239–247.

Barrow JR and Aaltonen RE 2001 Evaluation of the internal colonization of *Atriplex canescens* (Pursh) Nutt. roots by dark septate fungi and the influence of host physiological activity. *Mycorrhiza* **11**, 199–205.

Barrow JR and Osuna P 2002 Phosphorus solubilization and uptake by dark septate fungi in fourwing saltbush, *Atriplex canescens* (Pursh) Nutt. *Journal of Arid Environments* **51**, 449–459.

Barrow NJ, Malajczuk N, Shaw TC 1977 A direct test of the ability of vesicular-arbuscular mycorrhiza to help plants take up fixed soil phosphate. *New Phytologist* **78**, 269–276.

Bartholdy BA, Berreck M, Haselwandter K 2001 Hydroxamate siderophore synthesis by *Phialocephala fortinii*, a typical dark septate fungal root endophyte. *Biometals* **14**, 33–42.

Bartlett EM and Lewis DH 1973 Surface phosphatase activity of mycorrhizal roots of beech. *Soil Biology and Biochemistry* **5**, 249–257.

Baylis GTS 1959 Effect of vesicular-arbuscular mycorrhizas on growth of *Griselinia littoralis* Cornaceae. *New Phytologist* **58**, 274–280.

Baylis GTS 1961 The significance of mycorrhizas and root nodules in New Zealand vegeta-tion. *Proceedings of the Royal Society of New Zealand* **89**, 45–50.

Baylis GTS 1962 Rhizophagus. The catholic symbiont. *Australian Journal of Science* **25**, 195–200.

Baylis GTS 1967 Experiments on the ecological significance of phycomycetous mycorrhizas. *New Phytologist* **66**, 231–243.

Baylis GTS 1972 Fungi, phosphorus and the evolution of root systems. *Search* **3**, 257–258.

Bayman P, Lebron LL, Tremblay RL, Lodge DJ 1997 Variation in endophytic fungi from roots and leaves of *Lepanthes* (Orchidaceae). *New Phytologist* **135**, 143–149.

Beau C 1920 Sur le role trophique des endophytes d'orchidées. *Comptes Rendus de l'Academie des Sciences* **171**, 675–677.

Bécard G and Fortin JA 1988 Early events of vesicular-arbuscular mycorrhiza formation in Ri T-DNA transformed roots. *New Phytologist* **108**, 211–218.

Bécard G and Pfeffer PE 1993 Status of nuclear division in arbuscular mycorrhizal fungi during *in vitro* development. *Protoplasma* **174**, 62–68.

Bécard G and Piché Y 1989a Fungal growth stimulation by CO_2 and root exudates in vesicular-arbuscular mycorrhizal symbiosis. *Applied and Environmental Microbiology* **55**, 2320–2325.

Bécard G and Piché Y 1989b New aspects on the aquisition of biotrophic status by a vesicular-arbuscular mycorrhizal fungus, *Gigaspora margarita*. *New Phytologist* **112**, 77–83.

Bécard G, Doner LW, Rolin DB, Douds DD, Pfeffer PE 1991 Identification and quantification of trehalose in vesicular-arbuscular mycorrhizal fungi by *in vivo* ^{13}C NMR and HPLC analyses. *New Phytologist* **118**, 547–552.

Bécard G, Douds DD, Pfeffer PE 1992 Extensive *in vitro* hyphal growth of vesicular-arbuscular mycorrhizal fungi in the presence of CO_2 and flavonols. *Applied and Environmental Microbiology* **58**, 821–825.

Bécard G, Taylor LP, Douds DDJ, Pfeffer PE, Doner LW 1995 Flavonoids are not necessary plant signal compounds in arbuscular mycorrhizal symbioses. *Molecular Plant–Microbe Interactions* **8**, 252–258.

Bécard G, Kosuta S, Tamasloukht M, Séjalon-Delmas N, Roux C 2004 Partner communication in the arbuscular mycorrhizal interaction. *Canadian Journal of Botany* **82**, 1186–1197.

Becker WN and Gerdemann JW 1977 Colorimetric quantification of vesicular-arbuscular mycorrhizal infection in onion. *New Phytologist* **78**, 289–295.

Beever RE and Burns DJW 1980 Phosphorus uptake, storage and utilisation by fungi. *Advances in Botanical Research* **8**, 128–219.

Beguiristain TR and Lapeyrie F 1997 Host plant stimulates hypaphorine accumulation in *Pisolithus tinctorius* hyphae during ectomycorrhizal infection while excreted fungal hypaphorine controls root hair development. *New Phytologist* **136**, 525–532.

Beguiristain T, Cote R, Rubini P, Jayallemand C, Lapeyrie F (1995) Hypaphorine accumulation in hyphae of the ectomycorrhizal fungus, *Pisolithus tinctorius*. *Phytochemistry* **40**, 1089–1091.

Beilby JP 1980 Fatty acid and sterol composition of ungerminated spores of the vesicular-arbuscular mycorrhizal fungus *Acaulospora laevis*. *Lipids* **15**, 949–952.

Beilby JP and Kidby DK 1980 Biochemistry of ungerminated and germinated spores of the vesicular-arbuscular mycorrhizal fungus, *Glomus caledonium*: changes in neutral and polar lipids. *Journal of Lipid Research* **21**, 739–750.

Bell TL and Adams MA 2004 Ecophysiology of ectomycorrhizal fungi associated with *Pinus* spp. in low rainfall areas of Western Australia. *Plant Ecology* **171** 1–2, 35–52.

Bell TL and Pate JS 1996 Nitrogen and phosphorus nutrition in mycorrhizal Epacridaceae of south-west Australia. *Annals of Botany* **77**, 389–397.

Bell TL, Pate JS, Dixon KW 1996 Relationships between fire response, morphology, root anatomy and starch distribution in south-west Australian Epacridaceae. *Annals of Botany* **77**, 357–364.

Bellion M, Courbot M, Jacob C, Blaudez D, Chalot M 2006 Extracellular and cellular mechanisms sustaining metal tolerance in ectomycorrhizal fungi. *FEMS Microbiology Letters*, **254**, 173–181.

Bending GD and Read DJ 1995a The structure and function of the vegetative mycelium of ectomycorrhizal plants. V. The foraging behaviour of ectomycorrhizal mycelium and the translocation of nutrients from exploited organic matter. *New Phytologist* **130**, 401–409.

Bending GD and Read DJ 1995b The structure and function of the vegetative mycelium of ectomycorrhizal plants VI. Activities of nutrient mobilising enzymes in birch litter colonised by *Paxillus involutus* Fr. Fr. *New Phytologist* **130**, 411–417.

Bending GD and Read DJ 1996a Effects of the soluble polyphenol tannic acid on the activities of ericoid and ectomycorrhizal fungi. *Soil Biology and Biochemistry* **28**, 1595–1602.

Bending GD and Read DJ 1996b Nitrogen mobilization from protein-polyphenol complex by ericoid and ectomycorrhizal fungi. *Soil Biology and Biochemistry* **28**, 1603–1612.

Bending GD and Read DJ 1997 Lignin and soluble phenolic degradation by ectomycorrhizal and ericoid mycorrhizal fungi. *Mycological Research* **101**, 1348–1354.

Benedetto A, Magurno F, Bonfante P, Lanfranco L 2005 Expression profiles of a phosphate transporter *GmosPT* from the endomycorrhizal fungaus *Glomus mosseae*. *Mycorrhiza* **15**, 620–627.

Benhamou N, Fortin JA, Hamel C, St-Arnaud M, Shatilla A 1994 Resistance responses of mycorrhizal Ri T-DNA-transformed carrot roots to infection by *Fusarium oxysporum* f. sp. *chrysanthemi*. *Phytopathology* **84**, 958–968.

Benjdia M, Rikirsch E, Muller T *et al.* 2006 Peptide uptake in the ectomycorrhizal fungus *Hebeloma cylindrosporum*: characterization of two di- and tripeptide transporters (HcPTR2A and B). *New Phytologist* **170**, 401–410.

Bentivenga SP and Hetrick BAD 1992 Seasonal and temperature effects on mycorrhizal activity and dependence of cool-season and warm-season tallgrass prairie grasses. *Canadian Journal of Botany* **70**, 1596–1602.

Berbee ML and Taylor JW 1993 Dating the evolutionary radiations of the true fungi. *Canadian Journal of Botany* **71**, 1114–1127.

Berbee ML and Taylor JW 2001 Fungal molecular evolution; gene trees and geologic time. In *The Mycota: a comprehensive treatise on fungi as experimental systems for basic and applied research in systematics and evolution, Part B*. Eds DJ McLaughlin, EG McLaughlin, PA Lemke pp. 229–245. Springer-Verlag, Berlin, Germany.

Berch SM, Allen TR, Berbee ML 2002 Molecular detection, community structure and phylogeny of ericoid mycorrhizal fungi. *Plant and Soil* **244**, 55–66.

Berch SM, Massicotte HB, Tackaberry LE 2005 Re-publication of a translation of The vegetative organs of *Monotropa hypopitys* L. published by F Kamienski in 1882, with an update on *Monotropa* mycorrhizas. *Mycorrhiza* **15**, 323–332.

Bereau M, Gazel M, Garbaye J 1997 Mycorrhizal symbiosis in trees of the tropical rainforest of French Guiana. *Canadian Journal of Botany* **75**, 711–716.

Berendse F (1994) Litter decomposability – a neglected component of plant fitness. *Journal of Ecology* **82**, 187–190.

Berg B and Staaf H 1981 Leaching, accumulation and release of nitrogen in decomposing forest litter. In *Terrestrial Nitrogen Cycles*. Eds FE Clark and T Rosswall. *Ecological Bulletins* **33**, 163–178.

Berg B and McClaugherty CA 1989 Nitrogen and phosphorus release from decomposing litter in relation to the disappearance of lignin. *Canadian Journal of Botany* **67**, 1148–1156.

Berg B, McClaugherty C 2003 *Plant litter – Decomposition, Humus Formation, Carbon Sequestration*. Springer, Berlin, Germany.

Bergelson JM, Crawley MJ 1988 Mycorrhizal infection and plant species diversity. *Nature* **334**, 202.

Bergero R, Perotto S, Girlanda M, Vidano G, Luppi AM 2000 Ericoid mycorrhizal fungi are common root associates of a Mediterranean ectomycorrhizal plant (*Quercus ilex*). *Molecular Ecology* **9**, 1639–1649.

Bermudes D and Benzing DH 1989 Fungi in neotropical epiphyte roots. *Biosystems* **23**, 65–73.

Bernard N 1899 Sur la germination du *Neottia nidus-avis*. *Comptes Rendus de l'Academie des Sciences* **128**, 1253–1255.

Bernard N 1901 Études sur la tuberisation. These presenté a la Faculté des Sciences Paris. 103 pp.

Bernard N 1904a Recherches experimentales sur les orchidées I-III. Methodes de culture; champignon endophyte; La germination des orchidées. *Revue Générale de Botanique* **16**, 405–451.

Bernard N 1904b Récherches experimentales sur les orchidées IV. Les semis horticoles. *Revue Générale de Botanique* **16**, 458–476.

Bernard N 1904c Le champignon endophyte des orchidées. *Comptes Rendus de l'Academie des Sciences* **138**, 828–830.

Bernard N 1905 Nouvelles especes d'endophytes d'orchidées. *Comptes Rendus de l'Academie des Sciences* **140**, 1272–1273.

Bernard N 1909a L'evolution dans la symbiose. Les orchidées et leur champignons commenseux. *Annales des Sciences Naturelles; Botanique. Paris* **9**, 1–196.

Bernard N 1909b Remarques sur l'immunité chez les plantes. *Bulletin de l'Institut Pasteur* **7**, 369–386.

Bersoult A, Camut S, Perhald A, Kereszt A, Kiss GB, Cullimore JV 2005 Expression of the *Medicago truncatula* DMI2 gene suggests roles of the symbiotic receptor kinase in nodules and during early nodule development. *Molecular Plant-Microbe Interactions* **18**, 869–876.

Berta G and Bonfante-Fasolo P 1983 Apical meristems in mycorrhizal and uninfected roots of *Calluna vulgaris* L. Hull. *Plant and Soil* **71**, 285–291.

Berta G, Gianinazzi-Pearson V, Gay G, Torri G 1988 Morphogenetic effects of endomycorrhiza formation on the root system of *Calluna vulgaris* L. Hull. *Symbiosis* **5**, 33–44.

Berta G, Fusconi A, Trotta A, Scannerini S 1990a Morphogenetic modifications induced by the mycorrhizal fungus *Glomus* strain E_3 in the root system of *Allium porrum* L. *New Phytologist* **114**, 207–215.

Berta G, Sgorbati S, Soler V *et al.* 1990b Variations in chromatin structure in host nuclei of a vesicular arbuscular mycorrhiza. *New Phytologist* **114**, 199–205.

Berta G, Tagliasacchi AM, Fusconi A, Gerlero D, Trotta A, Scannerini S 1991 The mitotic cycle in root apical meristems of *Allium porrum* L. is controlled by the endomycorrhizal fungus *Glomus* sp. strain E_3. *Protoplasma* **161**, 12–16.

Berta G, Fusconi A, Trotta A 1993 VA mycorrhizal infection and the morphology and function of root systems. *Environmental and Experimental Botany* **33**, 159–173.

Berta G, Trotta A *et al.* 1995 Arbuscular mycorrhizal induced changes to plant growth and root system morphology in *Prunus cerasifera*. *Tree Physiology* **15**, 281–293.

Berta G, Fusconi A, Sampo S, Lingua G, Perticone S, Repetto O 2000 Polyploidy in tomato roots as affected by arbuscular mycorrhizal colonization. *Plant and Soil* **226**, 37–44.

Besserer A, Puech-Pages V, Kiefer P *et al.* 2006 Strigolactones stimulate arbuscular mycorrhizal fungi by activating mitochondria. *PLoS Biology* **4**, 1239–1247.

Bestel-Corre G, Dumas-Gaudot E, Gianinazzi S 2004 Proteomics as a tool to monitor plant-microbe endosymbioses in the rhizosphere. *Mycorrhiza* **14**, 1–10.

Bethlenfalvay GJ 1992a Vesicular-arbuscular mycorrhizal fungi in nitrogen-fixing legumes: problems and prospects. *Methods in Microbiology* **24**, 375–389.

Bethlenfalvay GJ 1992b Mycorrhizae and crop production. In *Mycorrhizae in Sustainable Agriculture*. Eds GJ Bethlenfalvay and RG Linderman pp. 1–27. ASA Special publication No. 54: Madison, Wisconsin, USA.

Bethlenfalvay GJ and Linderman RG (eds) 1992 *Mycorrhizae in Sustainable Agriculture*. ASA special publication No. 54, Madison, Wisconsin, USA.

Bethlenfalvay GJ and Newton WE 1991 Agro-ecological aspects of the mycorrhizal, nitrogen-fixing legume symbiosis. *Beltsville Symposia in Agricultural Research* **14**, 349–354.

Bethlenfalvay GJ and Pacovsky RS 1983 Light effects in mycorrhizal soybeans. *Plant Physiology* **73**, 969–972.

Bethlenfalvay GJ, Pacovsky RS, Brown MS 1982a Parasitic and mutualistic asssociations between a mycorrhizal fungus and soybean: development of the endophtye. *Phytopathology* **72**, 894–897.

Bethlenfalvay GJ, Brown MS, Pacovsky RS 1982b Parasitic and mutalistic associations between a mycorrhizal fungus and soybean: development of the host plant. *Phytopathology* **72**, 889–893.

Bethlenfalvay GJ, Bayne HC, Pacovsky RS 1983 Parasitic and mutualistic association between a mycorrhizal fungus and soybean. *Physiologia Plantarum* **57**, 543–549.

Bethlenfalvay GJ, Schreiner RP, Mihara KL 1997 Mycorrhizal fungi effects on nutrient composition and yield of soybean seeds. *Journal of Plant Nutrition* **20**, 581–591.

Bevege DI and Bowen GD 1975 Endogone strain and host plant differences in development of vesicular-arbuscular mycorrhizas. In *Endomycorrhizas*. Eds FE Sanders, B Mosse, PB Tinker pp. 77–86. Academic Press, London, UK.

Bevege DI, Bowen GD, Skinner MF 1975 Comparative carbohydrate physiology of ecto- and endo-mycorrhizas. In *Endomycorrhizas*. Eds FE Sanders, B Mosse, PB Tinker pp. 149–174. Academic Press, London, UK.

Bever JD 1999 Dynamics within mutualism and the maintenance of diversity: inference from a model of interguild frequency dependence. *Ecology Letters* **2**, 52–62.

Bever JD 2002a Host-specificity of AM fungal population growth rates can generate feedback on plant growth. *Plant and Soil* **244**, 281–290.

Bever JD 2002b Negative feedback within a mutualism: host-specific growth of mycorrhizal fungi reduces plant benefit. *Proceedings of the Royal Society of London, Series B, Biological Sciences* **269**, 2595–2601.

Bever JD 2003 Soil community feedback and the coexistence of competitors: conceptual frameworks and empirical tests. *New Phytologist* **157** 3, 465–473.

Bever JD and Morton J 1999 Heritable variation and mechanisms of inheritance of spore shape within a population of *Scutellospora pellucida*, an arbuscular mycorrhizal fungus. *American Journal of Botany* **86**, 1209–1216.

Bever JD, Morton JB, Antonovics J, Schultz PA 1996 Host-dependent sporulation and species diversity of arbuscular mycorrhizal fungi in a mown grassland. *Journal of Ecology* **84**, 71–82.

Bever JD, Westover KM, Antonovics J 1997 Incorporating the soil community into plant population dynamics: the utility of the feedback approach. *Journal of Ecology* **85**, 561–573.

Bever JD, Schultz PA, Pringle A, Morton JB 2001 Arbuscular mycorrhizal fungi: more diverse than meets the eye, and the ecological tale of why. *Bioscience* **51**, 923–931.

Bever JD, Pringle A, Schultz PA 2002 Dynamics within the plant-arbuscular mycorrhizal fungal mutualism: testing the nature of community feedback. In *Mycorrhizal Ecology*. Eds MGA van der Heijden and IR Sanders pp. 268–292. Springer Verlag, Berlin, Germany.

Beyrle H, Penningsfeld F, Hock B 1991 The role of nitrogen concentration in determining the outcome of the interaction between *Dactylorhiza incarnata* L. Soo and *Rhizoctonia* sp. *New Phytologist* **117**, 665–672.

Beyrle HF, Smith SE, Peterson RL, Franco CMM 1995 Colonisation of *Orchis morio* protocorms by a mycorrhizal fungus: effects of nitrogen nutrition and glyphosate in modifying the responses. *Canadian Journal of Botany* **73**, 1128–1140.

Bhupinderpal S, Nordgren A, Lofvenius MO, Högberg MN, Mellander PE, Högberg P 2003 Tree root and soil heterotrophic respiration as revealed by girdling of boreal Scots pine forest: extending observations beyond the first year. *Plant Cell and Environment* **26**, 1287–1296.

Bianciotto V and Bonfante P 1992 Quantification of the nuclear DNA content of two arbuscular mycorrhizal fungi. *Mycological Research* **96**, 1071–1076.

Bianciotto V and Bonfante P 1993 Evidence of DNA replication in an arbuscular mycorrhizal fungus in the absence of the host plant. *Protoplasma* **176**, 100–105.

Bianciotto V, Barbiero G, Bonfante P 1995 Analysis of the cell-cycle in an arbuscular mycorrhizal fungus by flow-cytometry and bromodeoxyuridine labeling. *Protoplasma* **188**, 161–169.

Bianciotto V, Bandi C, Minerdi D, Sironi M, Tichy HV, Bonfante P 1996 An obligately endosymbiotic mycorrhizal fungus itself harbors obligately intracellular bacteria. *Applied and Environmental Microbiology* **62**, 3005–3010.

Bianciotto V, Genre A, Jargeat P, Lumini E, Bécard G, Bonfante P 2004 Vertical transmission of endobacteria in the arbuscular mycorrhizal fungus *Gigaspora margarita* through generation of vegetative spores. *Applied and Environmental Microbiology* **70**, 3600–3608.

Bidartondo MI 2005 The evolutionary ecology of myco-heterotrophy. *New Phytologist* **167** 2, 335–352.

Bidartondo MI and Bruns TD 2001 Extreme specificity in epiparasitic Monotropoideae Ericaceae: widespread phylogenetic and geographical structure. *Molecular Ecology* **10** 9, 2285–2295.

Bidartondo MI and Bruns TD 2002 Fine-level mycorrhizal specificity in the Monotropoideae Ericaceae: specificity for fungal species groups. *Molecular Ecology* **11** 3, 557–569.

Bidartondo MI and Bruns TD 2005 On the origins of extreme mycorrhizal specificity in the Monotropoideae (Ericaceae): performance trade-offs during seed germination and seedling development. *Molecular Ecology* **14**, 1549–1560.

Bidartondo MI, Kretzer AM, Pine EM, Bruns TD 2000 High root concentration and uneven ectomycorrhizal diversity near *Sarcodes sanguinea* (Ericaceae): a cheater that stimulates its victims? *American Journal of Botany* **87**, 1783–1788.

Bidartondo MI, Redecker D, Hijri I *et al.* 2002 Epiparasitic plants specialized on arbuscular mycorrhizal fungi. *Nature* **419**, 389–392.

Bidartondo MI, Bruns TD, Weiss M, Sergio C, Read DJ 2003 Specialized cheating of the ectomycorrhizal symbiosis by an epiparasitic liverwort. *Proceedings of the Royal Society of London Series B, Biological Sciences* **270**, 835–842.

Bidartondo MI, Burghardt B, Gebauer G, Bruns TD, Read DJ 2004 Changing partners in the dark: isotopic and molecular evidence of ectomycorrhizal liaisons between forest orchids and trees. *Proceedings of the Royal Society of London Series B, Biological Sciences* **271**, 1799–1806.

Biermann B, Linderman RG 1983 Use of vesicular-arbuscular mycorrhizal roots, intraradical vesicles and extraradical vesicles as inoculum. *New Phytologist* **95**, 97–105.

Bildusas IJ, Dixon RK, Pfleger FL, Stewart EL 1986 Growth, nutrition and gas exchange of *Bromus inermis* inoculated with *Glomus fasciculatum*. *New Phytologist* **102**, 303–311.

Binder M and Hibbett DS 2006 Molecular systematics and biological diversification in Boletales. *Mycologia* **98**, 917–925.

Birch CPD 1986 Development of VA mycorrhizal infection in seedlings in semi-natural grassland turf. In *Physiological and Genetical Aspects of Mycorrhizae*. Eds V Gianinazzi-Pearson and S Gianinazzi pp. 233–237. INRA, Paris, France.

Björkman E 1960 *Monotropa hypopitys* L. an epiparasite on tree roots. *Physiologia Plantarum* **13**, 308–327.

Björkman E 1970 Mycorrhiza and tree nutrition in poor forest soils. *Studia Forestalia Suecia* **83**, 1–24.

Black R and Tinker PB 1979 The development of endomycorrhizal root systems II. Effect of agronomic factors and soil conditions on the development of vesicular-arbuscular mycorrhizal infection in barley and on the endophyte spore density. *New Phytologist* **83**, 401–413.

Blair DA, Peterson RL, Bowley SR 1988 Nuclear DNA content in the root cells of *Lotus* and *Trifolium* colonized by the VAM fungus, *Glomus versiforme*. *New Phytologist* **109**, 167–170.

Blakeman JP, Mokahel MA, Hadley G 1976 The effect of mycorrhizal infection on respiration and activity of some oxidase enzymes of orchid protocorms. *New Phytologist* **77**, 697–704.

Blancaflor EB, Zhao LM, Harrison MJ 2001 Microtubule organization in root cells of *Medicago truncatula* during development of an arbuscular mycorrhizal symbiosis with *Glomus versiforme*. *Protplasma* **217**, 154–165.

Blaudez D, Chalot M, Dizengremel P, Botton B 1998 Structure and function of the ectomycorrhizal association between *Paxillus involutus* and *Betula pendula* – II. Metabolic changes during mycorrhiza formation. *New Phytologist* **138**, 543–552.

Blaudez D, Jacob C, Turnau K *et al.* 2000 Differential responses of ectomycorrhizal fungi to heavy metals *in vitro*. *Mycological Research* **104**, 1366–1371.

Blaudez D, Botton B, Dizengremel P, Chalot M 2001 The fate of C-14 glutamate and C-14 malate in birch roots is strongly modified under inoculation with Paxillus involutus. *Plant Cell and Environment* **24**, 449–457.

Bledsoe CS, Tennyson K, Lopushinsky W 1982 Survival and growth of outplanted Douglas-fir seedlings inoculated with mycorrhizal fungi. *Canadian Journal of Forestry Research* **12**, 720–723.

Bledsoe C, Klein P, Bliss LC 1990 A survey of mycorrhizal plants on Truelove Lowland, Devon Island, N.W.T. Canada. *Canadian Journal of Botany* **68**, 1848–1856.

Blee KA and Anderson AJ 1996 Defense-related transcript accumulation in *Phaseolus vulgaris* L. colonized by the arbuscular mycorrhizal fungus *Glomus intraradices* Schenck and Smith. *Plant Physiology* **110**, 675–688.

Blee KA and Anderson AJ 2002 Transcripts for genes encoding soluble acid invertase and sucrose synthase accumulate in root tip and cortical cells containing mycorrhizal arbuscules. *Plant Molecular Biology* **50**, 197–211.

Blilou I, Bueno P, Ocampo JA, Garcia-Garrido JM 2000 Induction of catalase and ascorbate peroxidase activities in tobacco roots inoculated with the arbuscular mycorrhizal *Glomus mosseae*. *Mycological Research* **104**, 722–725.

Bloss HE and Pfeiffer CM 1984 Latex content and biomass increase in mycorrhizal guayule *Parthenium argentatum* under field conditions. *Annals of Applied Biology* **104**, 175–183.

Boa E 2004 Wild edible fungi: a global overview of their use and importance to people. FAO Corporate Document Depository: Non-wood Forest Products 17.

Boddington CL and Dodd JC 2000 The effect of agricultural practices on the development of indigenous arbuscular mycorrhizal fungi. I. Field studies in an Indonesian ultisol. *Plant and Soil* **218**, 137–144.

Bolan NS 1991 A critical review on the role of mycorrhizal fungi in the uptake of phosphorus by plants. *Plant and Soil* **134**, 189–207.

Bolan NS, Robson AD, Barrow NJ 1983 Plant and soil factors including mycorrhizal infection causing sigmoidal response of plants to applied phosphorus. *Plant and Soil* **73**, 187–201.

Bolan NS, Robson AD, Barrow NJ 1984 Increasing phosphorus supply can increase the infection of plant roots by vesicular arbuscular mycorrhizal fungi. *Soil Biology and Biochemistry* **16**, 419–420.

Bolan NS, Robson AD, Barrow NJ 1987 Effects of vesicular-arbuscular mycorrhiza on the availability of iron phosphates to plants. *Plant and Soil* **99**, 401–410.

Bomberg M, Jurgens G, Saano A, Sen R, Timonen S 2003 Nested PCR detection of archaea in defined compartments of pine mycorrhizospheres developed in boreal forest humus microcosms. *FEMS Microbiology Ecology* **43**, 163–171.

Bonanomi A, Wiemken A, Boller T, Salzer P 2001 Local induction of a mycorrhiza-specific class III chitinase gene in cortical root cells of *Medicago truncatula* containing developing or mature arbuscules. *Plant Biology* **3**, 194–199.

Bonello P, Bruns TD, Gardes M 1998 Genetic structure of a natural population of the ectomycorrhizal fungus *Suillus pungens*. *New Phytologist* **138**, 533–542.

Bonfante P, Bergero R, Uribe X, Romera C, Rigau J, Puigdomenech P 1996 Transcriptional activation of a maize α-tubulin gene in mycorrhizal maize and transgenic tobacco plants. *The Plant Journal* **9**, 737–743.

Bonfante P, Genre A, Faccio A *et al.* 2000 The *Lotus japonicus LjSym4* gene is required for the successful symbiotic infection of root epidermal cells. *Molecular Plant-Microbe Interactions* **13**, 1109–1120.

Bonfante-Fasolo P 1980 Occurrence of a basidiomycete in living cells of mycorrhizal hair roots of *Calluna vulgaris*. *Transactions of the British Mycological Society* **75**, 320–325.

Bonfante-Fasolo P 1988 The role of the cell wall as a signal in mycorrhizal associations. In *Cell to Cell Signals in Plant, Animal and Microbial Symbiosis*. Ed. S Scannerini pp. 219–235. Springer, Berlin, Germany.

Bonfante-Fasolo P and Fontana A 1985 VAM fungi in *Ginkgo biloba* roots: their interactions at cellular level. *Symbiosis* **1**, 53–67.

Bonfante-Fasolo P and Gianinazzi-Pearson V 1979 Ultrastructural aspects of endomycorrhiza in the Ericaceae. I. Naturally infected hair roots of *Calluna vulgaris* L. Hull. *New Phytologist* **83**, 739–744.

Bonfante-Fasolo P and Gianinazzi-Pearson V 1982 Ultrastructural aspects of endomycorrhiza in the Ericaceae. III. Morphology of the dissociated symbionts and modifications occurring during their reassociation in axenic culture. *New Phytologist* **91**, 691–704.

Bonfante-Fasolo P and Grippiolo R 1984 Cytochemical and biochemical observations on the cell wall of the spore of *Glomus epigaeum*. *Protoplasma* **123**, 140–151.

Bonfante-Fasolo P and Perotto S 1992 Plants and endomycorrhizal fungi: the cellular and molecular basis of their interaction. In *Molecular Signals in Plant-Microbe Communications*. Ed. DPS Verma pp. 445–470. CRC Press, Boca Raton, Florida, USA.

Bonfante-Fasolo P, Gianinazzi-Pearson V, Martinengo L 1984 Ultrastructural aspects of endomycorrhiza in the Ericaceae IV. Comparison of infection by *Pezizella ericae* in host and non-host plants. *New Phytologist* **98**, 329–333.

Bonfante-Fasolo P, Berta G, Fusconi A 1987a Distribution of nuclei in a VAM fungus during its symbiotic phase. *Transactions of the British Mycological Society* **88**, 263–266.

Bonfante-Fasolo P, Perotto S, Testa B, Faccio A 1987b Ultrastructural localisation of cell surface sugar residues in ericoid mycorrhizal fungi by gold-labelled lectins. *Protoplasma* **139**, 25–35.

Bonfante-Fasolo P, Peretto R, Perotto S 1992 Cell surface interactions in endomycorrhizal symbiosis. In *Perspectives in Plant Cell Recognition*. Eds JA Callow and JR Green pp. 239–255. Cambridge University Press, Cambridge, UK.

Borchers SL and Perry DA 1990 Growth and ectomycorrhiza formation of Douglas-fir seedlings grown in soils collected at different distances from pioneering hardwoods in southwest Oregon clear-cuts. *Canadian Journal of Forest Research* **20**, 717–721.

Botton B and Chalot M 1995 Nitrogen assimilation: enzymology in ectomycorrhizas. In *Mycorrhizas. Structure, Function, Molecular Biology and Biotechnology*. Eds A Varma and B Hock pp. 325–363. Springer-Verlag, Berlin, Germany.

Bougher NL, Grove TS, Malajczuk N 1990 Growth and phosphorus acquisition of karri *Eucalyptus diversicolor* F. Muell. seedlings inoculated with ectomycorrhizal fungi in relation to phosphorus supply. *New Phytologist* **114**, 77–85.

Bougoure DS and Cairney JWG 2005a Assemblages of ericoid mycorrhizal and other root-associated fungi from *Epacris pulchella* (Ericaceae) as determined by culturing and direct DNA extraction from roots. *Environmental Microbiology* **7** 6, 819–827.

Bougoure DS and Cairney JWG 2005b Fungi associated with hair roots of *Rhododendron lochiae* (Ericaceae) in an Australian tropical cloud forest revealed by culturing and culture-independent molecular methods. *Environmental Microbiology* **7** 11, 1743–1754.

Bougoure JJ, Bougoure DS, Cairney JWG, Dearnaley JDW 2005c ITS-RFLP and sequence analysis of endophytes from *Acianthus, Caladenia* and *Pterostylis* (Orchidaceae) in southeastern Queensland. *Mycological Research* **109**, 452–460.

Bougoure JJ, Ludwig M, Brundett M, Grierson PF 2006 The fungal endophyte of the mycoheterotrophic *Rhizanthella gardneri* (Orchidaceae). Proceedings of the 5th International Conference on Mycorrhizae, Granada, Spain.

Boullard B 1951 Champignons endophytes de quelques Fougères indigènes et observations relatives à *Ophioglossum vulgatum* L. *Le Botaniste* **XXXV**, 257–281.

Boullard B 1958 La mycotrophie chez les pteridophytes; sa frequence, ses charactères, sa signification. *Botaniste* **41**, 1–185.

Boullard B 1988 Observations on the co-evolution of fungi with hepatics. In *Coevolution of Fungi with Plants and Animals*. Eds KA Pirozynski and DL Hawksworth pp. 107–124. Academic Press, London, UK.

Bowen GD 1968 Phosphate uptake by mycorrhizas and uninfected roots of *Pinus radiata* in relation to root distribution. *Proceedings of 9th International Congress of Soil Science* **2**, 219–228.

Bowen GD 1973 Mineral nutrition of mycorrhizas. In *Ectomycorrhizas*. Eds GC Marks and TT Kozolowski pp. 151–201. Academic Press, New York, USA.

Bowen GD 1987 The biology and physiology of infection and its development. In *Ecophysiology of VA Mycorrhizal Plants*. Ed. GR Safir pp. 27–70. CRC Press, Boca Raton, Florida, USA.

Bowen GD and Smith SE 1981 The effects of mycorrhizas on nitrogen uptake by plants. *Ecological Bulletins* **33**, 237–247.

Bowen GD and Theodorou C 1967 Studies on phosphate uptake by mycorrhizas. 14th IUFRO Congress, Munich **5**, 116–138.

Bowen GD and Theodorou C 1979 Interactions between bacteria and ectomycorrhizal fungi. *Soil Biology and Biochemistry* **11**, 119–126.

Bowen GD, Skinner MF, Bevege DI 1974 Zinc uptake by mycorrhizal and uninfected roots of *Pinus radiata* and *Araucaria cunninghamii*. *Soil Biology and Biochemistry* **6**, 141–144.

Boyd R, Furbank RT and Read DJ 1986 Ectomycorrhiza and the water relations of trees. Physiological and genetical aspects of mycorrhizae. In *Physiological and genetical aspects of mycorrhizae*. Eds V Gianinazzi-Pearson and S Gianinazzi pp. 689–693. INRA, Paris, France.

Boyle CD, Robertson WJ, Salonius PO 1987 Use of mycelial slurries of ectomycorrhizal fungi as inoculum for commercial tree seedling nurseries. *Canadian Journal of Forestry Research* **17**, 1480–1486.

Bradbury SM, Peterson RL, Bowley SR 1991 Interactions between three alfalfa nodulation genotypes and two *Glomus* species. *New Phytologist* **119**, 115–120.

Bradbury SM, Peterson RL, Bowley SR 1993 Further evidence for a correlation between nodulation genotypes in alfalfa *Medicago sativa* L. and mycorrhiza formation. *New Phytologist* **124**, 665–673.

Bradley R, Burt AJ, Read DJ 1981 Mycorrhizal infection and resistance to heavy metal toxicity in *Calluna vulgaris*. *Nature* **292**, 335–337.

Bradley R, Burt AJ, Read DJ 1982 The biology of mycorrhiza in the Ericaceae. VIII. The role of mycorrhizal infection in heavy metal resistance. *New Phytologist* **91**, 197–209.

Brand F 1989 Studies on ectomycorrhizae XXI – Beech ectomycorrhizas and rhizomorphs of *Xerocomus chrysenteron* (Boletales). *Nova Hedwigia* **48**, 469–483.

Brand F 1991 Ektomykorrhizen an *Fagus sylvatica*. Charakterisierung und Identifizierung, ökologische Kennzeichnung und unsterile Kultivierung. *Libri Botanici* vol. 2, IHW-Verlag, Eching, Germany.

Brand F and Agerer R 1986 Studies on ectomycorrhizae VIII – Mycorrhizae formed by *Lactarius subdulcis*, *L. vellereus* and *Laccaria amethystina* on beech. *Zeitschrift für Mykologie* **52**, 287–320.

Brand F, Gronbach E, Taylor AFS 1992 *Picerhiza bicolorata*. In *Colour Atlas of Ectomycorrhizae*. Ed. R Agerer. Schwabisch Gmund. Munich, Germany.

Braunberger PG, Miller MH, Peterson RL 1991 Effect of phosphorus nutrition on morphological characteristics of vesicular-arbuscular mycorrhizal colonization of maize. *New Phytologist* **119**, 107–113.

Braunberger PG, Abbott LK, Robson AD 1994 The effect of rain in the dry-season on the formation of vesicular-arbuscular mycorrhizas in the growing season of annual clover-based pastures. *New Phytologist* **127**, 107–114.

Brearley F, Press MC, Scholes JD 2003 Nutrients obtained from leaf litter can improve the growth of dipterocarp seedlings. *New Phytologist* **160**, 101–110.

Brechenmacher L, Weidmann S, van Tuinen D *et al.* 2004 Expression profiling of up-regulated plant and fungal genes in early and late stages of *Medicago truncatula-Glomus mosseae* interactions. *Mycorrhiza* **14**, 253–262.

Bret-Harte MS, Shaver GR, Zoerner JP *et al.* 2001 Developmental plasticity allows *Betula nana* to dominate tundra subjected to an altered environment. *Ecology* **82**, 18–32.

Breuninger M and Requena N 2004 Recognition events in AM symbiosis: analysis of fungal gene expression at the early appressorium stage. *Fungal Genetics and Biology* **41**, 794–804.

Brook JP 1952 Mycorrhiza of *Pernettya macrostigma*. *New Phytologist* **51**, 388–397.

Brown MS and Bethlenfalvay GJ 1988 The Glycine, Glomus, Rhizobium symbiosis. VII. Photosynthetic nutrient use efficiency in nodulated, mycorrhizal soybeans. *Plant Physiology* **86**, 1292–1297.

Brown MT and Wilkins DA 1985 Zinc tolerance of mycorrhizal *Betula*. *New Phytologist* **99**, 101–106.

Brownlee C, Duddridge JA, Malibari A, Read DJ 1983 The structure and function of mycelial systems of ectomycorrhizal roots with special reference to their role in assimilate and water transport. *Plant and Soil* **71**, 433–443.

Bruce A, Smith SE, Tester M 1994 The development of mycorrhizal infection in cucumber: effects of P supply on root growth, formation of entry points and growth of infection units. *New Phytologist* **127**, 507–514.

Brun A, Chalot M, Botton B, Martin F 1992 Purification and characterisation of glutamine synthetase and NADP-glutamate dehydrogenase from the ectomycorrhizal fungus *Laccaria laccata*. *Plant Physiology* **99**, 938–944.

Brundrett MC 2002 Coevolution of roots and mycorrhizas of land plants. *New Phytologist* **154**, 275–304.

Brundrett M 2004 Diversity and classification of mycorrhizal associations. *Biological Reviews* **79**, 473–495.

Brundrett MC and Kendrick B 1988 The mycorrhizal status, root anatomy and phenology of plants in a sugar maple forest. *Canadian Journal of Botany* **66**, 1153–1173.

Brundrett MC and Kendrick B 1990a The roots and mycorrhizas of herbaceous woodland plants. I. Quantitative aspects of morphology. *New Phytologist* **114**, 457–468.

Brundrett M and Kendrick B 1990b The roots and mycorrhizas of herbaceous woodland plants. II. Structural aspects of morphology. *New Phytologist* **114**, 469–479.

Brundrett MC, Piché Y, Peterson RL 1984 A new method for observing the morphology of vesicular-arbuscular mycorrhizae. *Canadian Journal of Botany* **62**, 2128–2134.

Brundrett MC, Piché Y, Peterson RL 1985 A developmental study of the early stages in vesicular-arbuscular mycorrhiza development. *Canadian Journal of Botany* **63**, 184–194.

Brundrett MC, Murase G, Kendrick B 1990 Comparative anatomy of roots and mycorrhizae of common Ontario trees. *Canadian Journal of Botany* 68, 551–578.

Brundrett MC, Melville L, Peterson RL 1994 *Practical Methods in Mycorrhizal Research.* Mycologue Publications, Waterloo, Canada.

Brundrett M, Bougher N, Dell B, Grove T, Malajczuk N 1996 *Working with Mycorrhizas in Forestry and Agriculture.* Australian Centre for International Agricultural Research, Canberra, Australia.

Bruns TD 1995 Thoughts on the processes that maintain local species-diversity of ectomycorrhizal fungi. *Plant and Soil,* 170, 63–73.

Bruns TD and Gardes M 1993 Molecular tools for the identification of ectomycorrhizal fungi – taxon-specific oligonucleotide probes for suilloid fungi. *Molecular Ecology,* 2, 233–242.

Bruns TD and Read DJ 2000 *In vitro* germination of nonphotosynthetic, myco-heterotrophic plants stimulated by fungi isolated from the adult plants. *New Phytologist* 148, 335–342.

Bruns TD and Shefferson RP 2004 Evolutionary studies of ectomycorrhizal fungi: recent advances and future directions. *Canadian Journal of Botany,* 82, 1122–1132.

Bruns TD, Szaro TM, Gardes M *et al.* 1998 A sequence database for the identification of ectomycorrhizal basidiomycetes by phylogenetic analysis. *Molecular Ecology* 7, 257–272.

Bruns TD, Bidartondo MI, Taylor DL 2002 Host specificity in ectomycorrhizal communities: what do the exceptions tell us? *Integrative and Comparative Biology* 42, 352–359.

Bryla DR and Koide RT 1990 Regulation of reproduction in wild and cultivated *Lycopersicon esculentum* Mill. by vesicular arbuscular mycorrhizal infection. *Oecologia* 84, 74–81.

Bücking H and Heyser W 1999 Elemental composition and function of polyphosphates in ectomycorrhizal fungi – an X-ray microanalytical study. *Mycological Research* 103, 31–39.

Bücking H and Heyser W 2000 Subcellular compartmentation of elements in non-mycorrhizal and mycorrhizal roots of *Pinus sylvestris*: an X-ray microanalytical study. I. The distribution of phosphate. *New Phytologist* 145, 311–320.

Bücking H, Kuhn AJ, Schroder WH, Heyser W 2002 The fungal sheath of ectomycorrhizal pine roots: an apoplastic barrier for the entry of calcium, magnesium, and potassium into the root cortex? *Journal of Experimental Botany* 53, 1659–1669.

Bücking H and Shachar-Hill Y 2005 Phosphate uptake, transport and transfer by the arbuscular mycorrhizal fungus *Glomus intraradices* is stimulated by increased carbohydrate availability. *New Phytologist* 165, 899–912.

Buee M, Rossignol M, Jauneau A, Ranjeva R, Bécard G 2000 The pre-symbiotic growth of arbuscular mycorrhizal fungi is induced by a branching factor partially purified from plant root exudates. *Molecular Plant-Microbe Interactions* 13, 693–698.

Bultel G 1926 Les orchidées germinées sans champignons ont des plantes normales. *Revue Horticulturale* 98, 125.

Burgeff H 1909 *Die wurzelpiltze der orchideen.* Gustav Fischer, Jena, Germany.

Burgeff H 1911 *Die ansucht tropischer orchideen aus samen.* Gustav Fischer, Jena, Germany.

Burgeff H 1932 *Saprophytismus und Symbiose.* Gustav Fischer, Jena, Germany.

Burgeff H 1936 *Samenkeimung der Orchideen.* Gustav Fisher, Jena, Germany.

Burgeff H 1959 Mycorrhiza of orchids. In *The Orchids.* Ed. K. Withner pp. 361–395. The Ronald Press Company, New York, USA.

Burgeff H 1961 *Mikrobiologie des Hochmores.* Gustav Fischer Verlag, Stuttgart, Germany.

Burges A 1939 The defensive mechanism in orchid mycorrhizas. *New Phytologist* 38, 273–283.

Burgess TI, Malajczuk N, Grove TS 1993 The ability of 16 ectomycorrhizal fungi to increase growth and phosphorus uptake of *Eucalyptus globulus* Labill. and *E. diversicolor* F. Muell. *Plant and Soil* 153, 155–164.

Burgess T, Dell B, Malajczuk N 1994 Variation in mycorrhizal development and growth stimulation of 20 isolates of *Pisolithus* inoculated onto *Eucalyptus grandis* W. Hill ex Maiden. *New Phytologist* 127, 731–739.

Burgess T, Laurent P, Dell B, Malajczuk N, Martin F 1995 Effect of fungal-isolate aggressivity on the biosynthesis of symbiosis-related polypeptides in differentiating eucalypt ectomycorrhizas. *Planta* **195**, 408–417.

Burke RM and Cairney JWG 1997a Carbohydrolase production by the ericoid mycorrhizal fungus *Hymenoscyphus ericae* under solid-state fermentation conditions. *Mycological Research* **101**, 1135–1139.

Burke RM and Cairney JWG 1997b Purification and characterization of a β-1,4-endoxylanase from the ericoid mycorrhizal fungus *Hymenoscyphus ericae*. *New Phytologist* **139**, 637–645.

Burke RM and Cairney JWG 1998a Carbohydrate oxidases in ericoid and ectomycorrhizal fungi: a possible source of Fenton radicals during the degradation of lignocellulose. *New Phytologist* **139**, 637–645.

Burke RM and Cairney JWG 1998b Measuring production and activity of plant cell wall degrading enzymes in ectomycorrhizal fungi. In *Laboratory Manual for Mycorrhizal Research*. Ed. A Varma pp. 173–185. Springer-Verlag: Berlin, Germany.

Burke RM and Cairney JWG 2002 Laccases and other polyphenol oxidases in ecto- and ericoid mycorrhizal fungi. *Mycorrhiza* **12** 3, 105–116.

Bürkert B and Robson A 1994 ^{65}Zn uptake in subterranean clover *Trifolium subterraneum* L. by three vesicular-arbuscular mycorrhizal fungi in a root-free sandy soil. *Soil Biology and Biochemistry* **26**, 1117–1124.

Burleigh SH and Harrison MJ 1999 The down-regulation of *Mt4*-like genes by phosphate fertilization occurs systemically and involves phosphate translocation to the shoots. *Plant Physiology* **119**, 241–248.

Burleigh SH, Cavagnaro TR, Jakobsen I 2002 Functional diversity of arbuscular mycorrhizas extends to the expression of plant genes involved in P nutrition. *Journal of Experimental Botany* **53**, 1593–1601.

Burleigh SH, Kristensen BK, Bechmann IE 2003 A plasma membrane zinc transporter from *Medicago truncatula* is up-regulated in roots by Zn fertilization, yet down-regulated by arbuscular mycorrhizal colonization. *Plant Molecular Biology* **52**, 1077–1088.

Burns RG 1986 Interaction of enzymes with soil mineral and organic colloids. In *Interactions of Soil Minerals with Natural Organics and Microbes*. Soil Science Society of America Special Publication 17. pp. 429–451.

Burrows RL and Pfleger FL 2002 Arbuscular mycorrhizal fungi respond to increasing plant diversity. *Canadian Journal of Botany* **80**, 120–130.

Burt AJ, Hashem AR, Shaw G, Read DJ 1986 Comparative analysis of metal tolerance in ericoid and ectomycorrhizal fungi. In *Physiological and Genetical Aspects of Mycorrhizae*. Eds V Gianinazzi-Pearson and S Gianinazzi pp. 683–687. INRA, Paris, France.

Busse MD and Ellis JR 1985 Vesicular-arbuscular mycorrhizal *Glomus fasciculatum* influence on soybean drought tolerance in high phosphorus soil. *Canadian Journal of Botany* **63**, 2290–2294.

Bütehorn B, Gianinazzi-Pearson V, Franken P 1999 Quantification of β-tubulin RNA expression during asymbiotic and symbiotic development of the arbuscular mycorrhizal fungus *Glomus mosseae*. *Mycological Research* **103**, 360–364.

Butler EJ 1939 The occurences and systematic position of the vesicular-arbuscular type of mycorrhizal fungi. *Transactions of the British Mycological Society* **22**, 274–301.

Butler GM 1958 The development and behaviour of mycelial strands in *Merulius lacrymans* (Wulf.) Fr. II. Hyphal behaviour during strand formation. *Annals of Botany* **22**, 219–236.

Buwalda JG and Goh KM 1982 Host-fungus competition for carbon as a cause of growth depressions in vesicular-arbuscular mycorrhizal ryegrass. *Soil Biology and Biochemistry* **14**, 103–106.

Buwalda JG, Stribley DP, Tinker PB 1984 The development of endomycorrhizal root systems. V. The detailed pattern of development of infection and the control of infection level by host in young leek plants. *New Phytologist* **96**, 411–427.

Cairney JWG 2002 *Pisolithus* – death of the pan-global super fungus. *New Phytologist* **153** 2, 199–201.

Cairney JWG and Alexander IJ 1992 A study of spruce (*Picea sitchensis* (Bong.) Carr.) ectomycorrhizas. II. Carbohydrate allocation in ageing *Picea sitchensis/Tylospora fibrillosa* (Burt.) Donk ectomycorrhizas. *New Phytologist* **122**, 153–158.

Cairney JWG and Ashford AE 2002 Biology of mycorrhizal associations of epacrids Ericaceae. *New Phytologist* **154**, 305–326.

Cairney JWG and Burke RM 1998 Extracellular enzyme activities of the ericoid mycorrhizal endophyte *Hymenoscyphus ericae* Read Korf and Kernan: their likely roles in decomposition of moribund plant tissue in soil. *Plant and Soil* **205**, 181–192.

Cairney JWG and Burke RM 1994 Fungal enzymes degrading plant cell walls: their possible significance in the ectomycorrhizal symbiosis. *Mycological Research* **98**, 1345–1356.

Cairney JWG and Meharg AA 1999 Influences of anthropogenic pollution on mycorrhizal fungal communities. *Environmental Pollution* **106**, 169–182.

Cairney JWG, Ashford AE, Allaway WG 1989 Distribution of photosynthetically fixed carbon within root systems of *Eucalyptus pilularis* plants ectomycorrhizal with *Pisolithus tinctorius*. *New Phytologist* **112**, 495–500.

Cairney JWG, Jennings DH, Agerer R 1991 The nomenclature of fungal multi-hyphal linear aggregates. *Cryptogamic Botany* **2/3**, 246–251.

Cairney JWG, Rees BJ, Allaway WG, Ashford AE 1994 A basidiomycete isolated from a *Pisonia* mycorrhiza forms sheathing mycorrhizas with transfer cells on *Pisonia grandis*. *New Phytologist* **126**, 91–98.

Cairney JWG, Sawyer NA, Sharples JM, Meharg AA 2000 Intraspecific variation in nitrogen source utilisation by isolates of the ericoid mycorrhizal fungus *Hymenoscyphus ericae* (Read) Korf and Kernan. *Soil Biology and Biochemistry* **32**, 1319–1322.

Cairney JWG, Van Leerdam DM, Chen DM 2001 Metal insensitivity in ericoid mycorrhizal endophytes from *Woollsia pungens* (Epacridaceae). *Australian Journal of Botany* **49**, 571–577.

Cairney JWG, Taylor AFS, Burke RM 2003 No evidence for lignin peroxidase genes in ectomycorrhizal fungi. *New Phytologist* **160**, 461–462.

Calantzis C, Morandi D, Arnould C, Gianinazzi-Pearson V 2001 Cellular interactions between *G. mosseae* and a Myc-dmi2 mutant in *Medicago truncatula*. *Symbiosis* **30**, 97–108.

Caldwell BA, Jumpponen A, Trappe JM 2000 Utilization of major detrital substrates by dark-septate, root endophytes. *Mycologia* **92**, 230–232.

Caldwell MM and Richards JH 1989 Hydraulic lift – water efflux from upper roots improves effectiveness of water uptake by deep roots. *Oecologia* **79** 1, 1–5.

Caldwell MM, Dawson TE, Richards JH 1998 Hydraulic lift: Consequences of water efflux from the roots of plants. *Oecologia* **113**, 151–161.

Calleja M, Mousain D, Lecouvreur B, D'Auzac J 1980 Influence de la carence phosphatée sur les activités phosphatases acides de trois champignons mycorhiziens: *Hebeloma edurum* Metrod, *Suillus granulatus* L. and Fr. O. Kuntze et *Pisolithus tinctorius* Pers. Coker and Couch. *Physiologie Végetale* **18**, 489–504.

Callow JA, Capaccio LCM, Parish G, Tinker PB 1978 Detection and estimation of polyphosphate in vesicular-arbuscular mycorrhizas. *New Phytologist* **80**, 125–134.

Camel SB, Reyes-Solis MG, Ferrera-Cerrato R, Franson RL, Brown MS, Bethlenfalvay GJ 1991 Growth of vesicular-arbuscular mycorrhizal mycelium through bulk soil. *Soil Science Society of America Journal* **55**, 389–393.

Cameron DD, Leake JR, Read DJ 2006 Mutualistic mycorrhiza in orchids: evidence from the plant-fungus carbon and nitrogen transfers in the green-leaved terrestrial orchid *Goodyera repens*. *New Phytologist* **171**, 405–416.

Cameron DD, Johnson I, Leake JR, Read DJ 2007 Mycorrhizal acquisition of inorganic phosphorus by the green-leaved terrestrial orchid *Goodyera repens*. *Annals of Botany* **99**, 831–834.

Campbell EO 1962 The mycorrhiza of *Gastrodia cunninghamii*. *Transactions of the Royal Society of New Zealand, Botany* **1**, 289–296.

Campbell EO 1971 Notes on the fungal associations to two *Monotropa* sp. in Michigan. *Michigan Botany* **10**, 63–67.

Canovas FM, Dumas-Gaudot E, Recorbet G, Jorrin J, Mock HP, Rossignol M 2004 Plant proteome analysis. *Proteomics* **4**, 285–298.

Capaccio LCM and Callow JA 1982 The enzymes of polyphosphate metabolism in vesicular-arbuscular mycorrhizas. *New Phytologist* **91**, 81–91.

Carafa A, Duckett JG, Ligrone R. (2003) Subterranean gametophytic axes in the primitive liverwort *Haplomitrium* harbour a unique type of endophytic association with aseptate fungi. *New Phytologist* **160**, 185–197.

Carey EV, Marler MJ, Callaway RM 2004 Mycorrhizae transfer carbon from a native grass to an invasive weed: evidence from stable isotopes and physiology. *Plant Ecology* **172**, 133–141.

Carleton TJ and Read DJ 1991 Ectomycorrhizas and nutrient transfer in conifer-feather moss ecosystems. *Canadian Journal of Botany* **69**, 778–785.

Carling DE 1996 Grouping in *Rhizoctonia solani* by hyphal anastomosis reaction. In *Rhizoctonia spp: taxonomy, molecular biology, ecology, pathology and disease control*. Eds B Sneh, S Jabaji-Hari, S Neate, G Dijst pp. 37–47. Kluwer, Dordrecht, The Netherlands.

Carr GR, Hinkley MA, Le Tacon F, Hepper CM, Jones MGK, Thomas E 1985 Improved hyphal growth of two species of vesicular-arbuscular mycorrhizal fungi in the presence of suspension cultured plant cells. *New Phytologist* **101**, 417–426.

Carrodus BB 1966 Absorption of nitrogen by mycorrhizal roots of beech. I. Factors affecting assimilation of nitrogen. *New Phytologist* **65**, 358–371.

Carrodus BB 1967 Absorption of nitrogen by mycorrhizal roots of beech. 2. Ammonium and nitrate as sources of nitrogen. *New Phytologist* **66**, 1–4.

Carvalho LM, Correia PM, Martins-Loucao MA 2004 Arbuscular mycorrhizal fungal propagules in a salt marsh. *Mycorrhiza* **14**, 165–170.

Casarin V, Plassard C, Hinsinger P, Arvieu JC 2004 Quantification of ectomycorrhizal fungal effects on the bioavailability and mobilization of soil P in the rhizosphere of *Pinus pinaster*. *New Phytologist* **163**, 177–185.

Casselton LA and Kües U 1994 Mating-type genes in homobasidiomycetes. In *The Mycota I. Growth, Differentiation and Sexuality*. Eds JGH Wessels and F Meinhardt pp. 307–321. Springer-Verlag, Berlin, Germany.

Castellano MA and Trappe JM 1985 Mycorrhizal associations of five species of Mono-tropoideae in Oregon. *Mycologia* **77**, 499–502.

Castellano MA and Molina R 1989 Mycorrhizae. In *The Container Tree Nursery Manual, Agriculture Handbook 674*. Eds TD Landis, RW Tinus, SE McDonald, JP Barnett pp. 101–167. USDA Forest Service, Washington, DC, USA.

Cavagnaro TR, Gao L-L, Smith FA, Smith SE 2001a Morphology of arbuscular mycorrhizas is influenced by fungal identity. *New Phytologist* **151**, 469–475.

Cavagnaro TR, Smith FA, Kolesik P, Ayling SM, Smith SE 2001b Arbuscular mycorrhizas formed by *Asphodelus fistulosus* and *Glomus coronatum*: three-dimensional analysis of plant nuclear shift using laser scanning confocal microscopy. *Symbiosis* **30**, 109–121.

Cavagnaro TR, Smith FA, Lorimer MF, Haskard KA, Ayling SM, Smith SE 2001c Quantitative development of *Paris*-type arbuscular mycorrhizas formed between *Asphodelus fistulosus* and *Glomus coronatum*. *New Phytologist* **149**, 105–113.

Cavagnaro TR, Smith FA, Ayling SM, Smith SE 2003 Growth and phosphorus nutrition of a *Paris*-type arbuscular mycorrhizal symbiosis. *New Phytologist* **157**, 127–134.

Cavagnaro TR, Smith FA, Hay G, Carne-Cavagnaro VL, Smith SE 2004a Inoculum type does not affect overall resistance of an arbuscular mycorrhiza-defective tomato mutant to colonisation but inoculation does change competitive interactions with wild-type tomato. *New Phytologist* **161**, 485–494.

Cavagnaro TR, Smith FA, Smith SE 2004b Interactions between arbuscular mycorrhizal fungi and a mycorrhiza-defective mutant tomato: does a non-infective fungus alter the ability of an infective fungus to colonise the roots – and vice versa? *New Phytologist* **164**, 485–491.

Cavagnaro TR, Smith FA, Smith SE, Jakobsen I 2005 Functional diversity in arbuscular mycorrhizas: exploitation of soil patches with different phosphate enrichment differs among fungal species. *Plant Cell and Environment* **28**, 642–650.

Cavagnaro TR, Jackson LE, Six J *et al.* 2006 Arbuscular mycorrhizas, microbial communities, nutrient availability, and soil aggregates in organic tomato production. *Plant and Soil* **282**, 209–225.

Cazares E, Trappe JM, Jumpponen A 2005 Mycorrhiza-plant colonization patterns on a subalpine glacier forefront as a model system of primary succession. *Mycorrhiza* **15**, 405–416.

Chabaud M, Venard C, Defaux-Petras A, Bécard G, Barker DG 2002 Targeted inoculation of *Medicago truncatula in vitro* root cultures reveals *MtENOD11* expression during early stages of infection by arbuscular mycorrhizal fungi. *New Phytologist* **156**, 265–273.

Chabot S, Bel-Rhlid R, Chênevert R, Piché Y 1992 Hyphal growth promotion *in vitro* of the VA mycorrhizal fungus, *Gigaspora margarita* Becker and Hall, by the activity of structurally specific flavonoid compounds under CO_2-enriched conditions. *New Phytologist* **122**, 461–467.

Chakravarty P and Unestam T 1987a Differential influence of ectomycorrhizae on plant growth and disease resistance of *Pinus sylvestris* seedlings. *Journal of Phytopathology* **120**, 104–120.

Chakravarty P and Unestam T 1987b Mycorrhizal fungi prevent disease in stressed pine seedlings. *Journal of Phytopathology* **118**, 335–340.

Chalot M and Brun A 1998 Physiology of organic nitrogen acquisition by ectomycorrhizal fungi and ectomycorrhizas. *FEMS Microbiology Reviews* **22** 1, 21–44.

Chalot M, Brun A, Botton B 1990 Occurrence and distribution of the fungal NADP-dependent glutamate dehydrogenase in spruce and beech ectomycorrhizas. In *Fast Growing Trees and Nitrogen Fixing Trees*. Eds D Werner and P Muller pp. 324–327. Gustsav Fischer, Stuttgart, Germany.

Chalot M, Brun A, Khalid A, Dell B, Rohr R, Botton B 1991 Occurrence and distribution of aspartate aminotransferases in spruce and beech ectomycorrhizas. *Canadian Journal of Botany* **68**, 1756–1762.

Chalot M, Brun A, Finlay RD, Söderström B 1994a Metabolism of [^{14}C]glutamate and [^{14}C]glutamine by the ectomycorrhizal fungus *Paxillus involutus*. *Microbiology* **140**, 1641–1649.

Chalot M, Brun A, Finlay RD, Söderström B 1994b Respiration of [^{14}C]alanine by the ectomycorrhizal fungus *Paxillus involutus*. *FEMS Microbiology Letters* **121**, 87–92.

Chalot M, Kytövita MM, Brun A, Finlay RD, Söderström B 1995 Factors affecting amino acid uptake by the ectomycorrhizal fungus *Paxillus involutus*. *Mycological Research* **99**, 1131–1138.

Chalot M, Brun A, Botton B, Söderström B 1996 Kinetics, energetics and specificity of a general amino acid transporter from the ectomycorrhizal fungus *Paxillus involutus*. *Microbiology-UK* **142**, 1749–1756.

Chalot M, Javelle A, Blaudez D *et al.* 2002 An update on nutrient transport processes in ectomycorrhizas. *Plant and Soil* **244**, 165–175.

Chalot M, Blaudez D, Brun A 2006 Ammonia: a candidate for nitrogen transfer at the mycorrhizal interface. *Trends in Plant Science* **11**, 263–266.

Chambers SM, Sharples JM, Cairney JWG 1998 Towards a molecular identification of the *Pisonia* mycobiont. *Mycorrhiza* **7**, 319–321.

Chambers SM, Williams PG, Seppelt RD, Cairney JWG 1999 Molecular identification of *Hymenoscyphus* sp. from rhizoids of the leafy liverwort *Cephaloziella exiliflora* in Australia and Antarctica. *Mycological Research* **103**, 286–288.

Chambers SM, Liu G, Cairney JWG 2000 ITS rDNA sequence comparison of ericoid mycorrhizal endophytes from *Woollsia pungens*. *Mycological Research* **104**, 168–174.

Chapin FS III, Moilanen L, Kielland K 1993 Preferential use of organic nitrogen for growth by a nonmycorrhizal arctic sedge. *Nature* **63**, 150–153.

Charvet-Candela V, Hitchin S, Reddy MS, Cournoyer B, Marmeisse R, Gay G 2002a Characterization of a *Pinus pinaster* cDNA encoding an auxin up-regulated putative peroxidase in roots. *Tree Physiology* **22**, 231–238.

Charvet-Candela V, Hitchin S, Ernst D, Sandermann H Jr, Marmeisse R, Gay G 2002b Characterization of an Aux/AA cDNA upregulated in *Pinus pinaster* roots in response to colonization by the ectomycorrhizal fungus *Hebeloma cylindrosporum. New Phytologist* **154**, 769–777.

Chase MW 2001 The origin and biogeography of Orchidaceae. In *Genera Orchidacearum Vol 2.* Eds AM Pridgeon, PJ Cribb, MW Chase, F Rasmussan pp. 1–5. Oxford University Press, Oxford, UK.

Chase MW 2004 Monocot relationships: an overview. *American Journal of Botany* **91** 10, 1645–1655.

Chase MW, Cameron KM, Barrett RL, Freudenstein JV 2003 DNA data and Orchidaceae systematics: a new phylogenetic classification. In *Orchid Conservation.* Eds KW Dixon, SP Kell, RL Barrett, PJ Cribb pp. 69–90. Natural History Publications, Kota Kinabalu, Malaysia.

Chase MW, Fay MF, Devey DS *et al.* 2006 Multi-gene analysis of monocot relationships. In *Monocots: Comparative Biology and Evolution.* Eds JT Columbus, EA Friar, CW Hamilton, JM Porter, LM Prince, MG Simpson pp. 63–75. Rancho Santa Ana Botanical Garden, Claremont, California, USA.

Chen A, Chambers SM, Cairney JWG 1999 Utilisation of organic nitrogen and phosphorus sources by mycorrhizal endophytes of *Woollsia pungens* Cav. F. Muell. *Epacridaceae. Mycorrhiza* **8**, 181–187.

Chen BD, Roos P, Borggaard OK, Zhu YG, Jakobsen I 2005a Mycorrhiza and root hairs in barley enhance acquisition of phosphorus and uranium from phosphate rock but mycorrhiza decreases root to shoot uranium transfer. *New Phytologist* **165**, 591–598.

Chen BD, Zhu YG, Smith FA 2005b Effects of arbuscular mycorrhizal inoculation on uranium and arsenic accumulation by Chinese brake fern *Pteris vittata* L. from a uranium mining-impacted soil. *Chemosphere* **62**, 1404–1473.

Chen DM, Taylor AFS, Burke RM, Cairney JWG 2001 Identification of genes for lignin peroxidases and manganese peroxidases in ectomycorrhizal fungi. *New Phytologist* **152**, 151–158.

Chen X, Tang JJ, Fang ZG, Shimizu K 2004 Effects of weed communities with various species numbers on soil features in a subtropical orchard ecosystem. *Agriculture Ecosystems and Environment* **102**, 377–388.

Chen X-Y. and Hampp R 1993 Sugar uptake by protoplasts of the ectomycorrhizal fungus *Amanita muscaria. New Phytologist* **125**, 601–608.

Chevalier G and Desmas C 1975 Synthése axénique des mycorhizes de *Tuber melanosporum, T. uncinatum* et *T. rufum* sur *Pinus sylvestris* à partir de cultures pures du champignon. *Annales de Phytopathologie* **7**, 338.

Chevalier G and Grente J 1978 Application pratique de la symbiose ectomycorrhizienne: production à grande echelle de plantes mycorhizes par la truffe *Tuber melanosporum* Vitt. *Mushroom Science* **10**, 483–505.

Chilvers GA 1968a Some distinctive types of eucalypt mycorrhiza. *Australian Journal of Botany* **26**, 49–70.

Chilvers GA 1968b Low power electronmicroscopy of the root cap region of eucalypt mycorrhiza. *New Phytologist* **67**, 663–665.

Chilvers GA and Gust LW 1982 The development of mycorrhizal populations on pot-grown seedlings of *Eucalyptus st-johnii. New Phytologist* **90**, 677–699.

Chilvers GA and Harley JL 1980 Visualisation of phosphate accumulation in beech mycorrhizas. *New Phytologist* **84**, 319–326.

Chilvers GA and Pryor LD 1965 The structure of eucalypt mycorrhizas. *Australian Journal of Botany* **13**, 245–259.

Chiou T-J, Liu H, Harrison MJ 2001 The spatial expression patterns of a phosphate transporter MtPT1 from *Medicago truncatula* indicate a role in phosphate transport at the root/soil interface. *The Plant Journal* **25**, 281–293.

Christie P and Nicolson TH 1983 Are mycorrhizas absent from the Antarctic? *Transactions of the British Mycological Society* **80**, 557–560.

Christie P, Li XL, Chen BD 2004 Arbuscular mycorrhiza can depress translocation of zinc to shoots of host plants in soils moderately polluted with zinc. *Plant and Soil* **261**, 209–217.

Christoph H 1921 Untersuchungen über die mycotrophen Verhältnisse der Ericales und die Keimung von Pyrolaceen. *Beihefte Botanisches Centralblatt* **38**, 115–117.

Chu-Chou M, Guo B, An Z-Q *et al.* 1992 Suppression of mycorrhizal fungi in fescue by the *Acremonium coenophialum* endophyte. *Soil Biology and Biochemistry* **24** 7, 633–637.

Chuyong GB, Newbery DM, Songwe NC (2002) Litter breakdown and mineralization in a central African rain forest dominated by ectomycorrhizal trees. *Biogeochemistry* **61**, 73–94.

Citterio S, Santagostino A, Fumagalli P, Prato N, Ranalli P, Sgorbati S 2003 Heavy metal tolerance and accumulation of Cd, Cr and Ni by *Cannabis sativa* L. *Plant and Soil* **256**, 243–252.

Claassen N, Barber SA 1976 Simulation model for nutrient uptake from soil by a growing plant root system. *Agronomy Journal* **68**, 961–964.

Clapp JP, Young JPW, Merryweather J, Fitter AH 1995 Diversity of fungal symbionts in arbuscular mycorrhizas from a natural communitiy. *New Phytologist* **130**, 259–265.

Clapp JP, Helgason T, Daniell T, Young JPW 2002 Genetic studies of the structure and diversity of arbuscular mycorrhizal fungal communities. In *Mycorrhizal Ecology*. Eds MGA van der Heijden and IR Sanders pp. 202–224. Springer-Verlag, Berlin, Germany.

Clarkson DT 1985 Factors affecting mineral nutrient acquisition by plants. *Annual Review of Plant Physiology* **36**, 77–115.

Clay K, Schardl CL 2002 Evolutionary origins and ecological consequences of endophyte symbiosis with grasses. *American Naturalist* **160**, 99–127.

Clayton JS, Bagyaraj DJ 1984 Vesicular-arbuscular mycorrhizas in submerged aquatic plants in New Zealand. *Aquatic Botany* **19**, 251–262.

Clements FE 1916 Plant succession: an analysis of the development of vegetation. Carnegie Institute of Washington Publication No. 242, 1–512.

Clements MA 1981 The germination of Australian orchid seed. *Proceedings of the Orchid Symposium*. 13th International Botanical Congress Sydney. Orchid Society of New South Wales. pp. 5–8.

Clements MA 1982a Australian native orchids epiphytic and terrestrial. In *Orchid Biology: Reviews and Perspectives II*. Ed. J Arditti pp. 295–303. Cornell University Press, Ithaca, New York.

Clements MA 1982b Development in the symbiotic germination of Australian terrestrial orchids. In *Proceedings of the 10th World Orchid Conference*. Eds J Stewart and CN van der Merwe.

Clements MA 1988 Orchid mycorrhizal associations. *Lindleyana* **3**, 73–86.

Clemmensen KE, Michelsen A, Jonasson S, Shaver GR 2006 Increased ectomycorrhizal fungal abundance after long-term fertilization and warming of two arctic tundra ecosystems. *New Phytologist* **171**, 391–404.

Clough KS and Sutton JC 1976 Direct observation of fungal aggregates in sand-dune soil. *Canadian Journal of Microbiology* **24**, 326–333.

Clowes FAL 1951 The structure of mycorrhizal roots of *Fagus sylvatica*. *New Phytologist* **53**, 525–529.

Codignola A, Verotta L, Spanu P, Maffei M, Scannerini S, Bonfante-Fasolo P 1989 Cell wall bound phenols in roots of vesicular-arbuscular mycorrhizal plants. *New Phytologist* **112**, 221–228.

Coleman MD, Bledsoe CS Lopushinsky W 1989 Pure culture response of ectomycorrhizal fungi to imposed water stress. *Canadian Journal of Botany* **67**, 29–39.

Colpaert JV and van Assche JA 1987 Heavy metal tolerance in some ectomycorrhizal fungi. *Functional Ecology* **1**, 415–421.

Colpaert JV and van Assche JA 1992 Zinc toxicity in ectomycorrhizal *Pinus Sylvestris*. *Plant and Soil* **143**, 201–211.

Colpaert JV and van Assche JA 1993 The effects of cadmium on ectomycorrhizal *Pinus sylvestris* L. *New Phytologist* **123**, 325–333.

Colpaert JV and van Laere A 1996 A comparison of the extracellular enzyme activities of two ectomycorrhizal and a leaf-saprotrophic basidiomycete colonizing beech leaf litter. *New Phytologist* **134**, 133–141.

Colpaert JV and van Tichelen KK 1996 Decomposition, nitrogen and phosphorus mineralization from beech leaf litter colonized by ectomycorrhizal or litter-decomposing basidiomycetes. *New Phytologist* **134**, 123–132.

Colpaert JV, van Laere A, van Tichelen KK, van Assche JA 1997 The use of inositol hexaphosphate as a phosphorus source by mycorrhizal and non-mycorrhizal Scots Pine (*Pinus sylvestris*). *Functional Ecology* **11**, 407–415.

Colpaert JV, van Tichelen KK, van Assche JA, van Laere A 1999 Short-term phosphorus uptake rates in mycorrhizal and non-mycorrhizal roots of intact *Pinus sylvestris* seedlings. *New Phytologist* **143**, 589–597.

Colpaert JV, Vandenkoornhuyse P, Adriaensen K, Van Gronsveld J 2000 Genetic variation and heavy metal tolerance in the ectomycorrhizal basidiomycete *Suillus luteus*. *New Phytologist* **147**, 367–379.

Comerford NB 1998 Soil phosphorus bioavailabiilty. In *Phosphorus in Plant Biology*. Eds JP Lynch and J Deikman pp. 136–147. American Society of Plant Physiologists, Rockville, Maryland, USA.

Conjeaud C, Scheromm P, Mousain D 1996 Effects of phosphorus fertilisation and ectomycorrhizal infection on the carbon balance in maritime pine seedlings *Pinus pinaster* Soland. in Ait. *New Phytologist* **133**, 345–351.

Connell JH and Lowman MD 1989 Low-diversity tropical rainforests: some possible mechanisms for their existance. *American Naturalist* **134**, 88–119.

Cook BD, Jastrow JD, Miller RM 1988 Root and mycorrhizal endophyte development in a chronosequence of restored tallgrass prairie. *New Phytologist* **110**, 355–362.

Cooke JC, Gemma JN, Koske RE 1987 Observations of nuclei in vesicular-arbuscular mycorrhizal fungi. *Mycologia* **79**, 331–333.

Cooke MA, Widden P, O'Halloran I 1993 Development of vesicular-arbuscular mycorrhizae in sugar maple *Acer saccharum* and effects of base-cation ammendment on vesicle and arbuscule formation. *Canadian Journal of Botany* **71**, 1421–1426.

Cooper KM 1976 A field survey of mycorrhizas in New Zealand ferns. *New Zealand Journal of Botany* **14**, 169–181.

Cooper KM and Lösel D 1978 Lipid physiology of vesicular-arbuscular mycorrhiza. I. Composition of lipids in roots of onion, clover and ryegrass infected with *Glomus mosseae*. *New Phytologist* **80**, 143–151.

Cooper KM and Tinker PB 1978 Translocation and transfer of nutrients in vesicular-arbuscular mycorrhizas. II. Uptake and translocation of phosphorus, zinc and sulphur. *New Phytologist* **81**, 43–52.

Cooper KM and Tinker PB 1981 Translocation and transfer of nutrients in vesicular-arbuscular mycorrhizas. IV. Effect of environmental variables on movement of phosphorus. *New Phytologist* **88**, 327–339.

Cordier C, Pozo MJ, Barea JM, Gianinazzi S, Gianinazzi-Pearson V 1998 Cell defence responses associated with localized and systemic resistance to *Phytophthora parasitica* induced in tomato by an arbuscular mycorrhizal fungus. *Molecular Plant-Microbe Interactions* **11**, 1017–1028.

Corner EJH 1972 *Boletus in Malaysia*. Botanic Gardens, Singapore.

Corradi N, Hijri M, Fumagalli L, Sanders IR 2004 Arbuscular mycorrhizal fungi Glomeromycota harbour ancient fungal tubulin genes that resemble those of the chytrids Chytridiomycota. *Fungal Genetics and Biology* **41**, 1037–1045.

Cosgrove DJ 1967 Metabolism of organic phosphates in soil. In *Soil Biochemistry, vol. I*. Eds AD McLaren and GH Peterson pp. 216–228. M. Decker, New York, USA.

Courbot M, Diez L, Ruotolo R, Chalot M, Leroy P 2004 Cadmium-responsive thiols in the ectomycorrhizal fungus *Paxillus involutus*. *Applied and Environmental Microbiology* **70**, 7413–7417.

Coutts MP and Nicholl BC 1990a Growth and survival of shoots, roots and mycorrhizal mycelium in clonal Sitka spruce during the first growing season after planting. *Canadian Journal of Forest Research* **20**, 861–868.

Coutts MP and Nicholl BC 1990b Waterlogging tolerance of roots of Sitka-spruce clones and of strands from *Thelephora terrestris* mycorrhizas. *Canadian Journal of Forest Research* **20**, 1896–1899.

Cox G, Sanders F 1974 Ultrastructure of the host-fungus interface in a vesicular-arbuscular mycorrhiza. *New Phytologist* **73**, 901–912.

Cox G and Tinker PB 1976 Translocation and transfer of nutrients in vesicular-arbuscular mycorrhizas. I. The arbuscule and phosphorus transfer: a quantitative ultrastructural study. *New Phytologist* **77**, 371–378.

Cox G, Sanders FE, Tinker PB, Wild JA 1975 Ultrastructural evidence relating to host-endophyte transfer in a vesicular-arbuscular mycorrhiza. In *Endomycorrhizas*. Eds FE Sanders, B Mosse, PB Tinker pp. 297–312. Academic Press, London, UK.

Cox G, Moran KJ, Sanders FE, Nockolds C, Tinker PB 1980 Translocation and transfer of nutrients in vesicular-arbuscular mycorrhizas. III. Polyphosphate granules and phosphorus translocation. *New Phytologist* **84**, 649–659.

Cribb PJ, Kell SP, Dixon KW, Barrett RL 2003 Orchid conservation – a global perspective. In *Orchid Conservation*. Eds KW Dixon, SP Kell, RL Barrett, PJ Cribb pp. 1–24. Natural History Publication, Kota Kinabalu, Malaysia.

Crocker RL and Major J 1955 Soil development in relation to vegetation and surface age at Glacier Bay, Alaska. *Journal of Ecology* **43**, 427–448.

Cromack K, Sollins P, Granstein WC *et al.* 1979 Calcium oxalate accumulation and soil weathering in mats of the hypogeous fungus *Hysterangium crassum*. *Soil Biology and Biochemistry* **11**, 463–468.

Cromer DAN 1935 The significance of the mycorrhiza of *Pinus radiata*. *Bulletin of Forest Bureau of Australia* **16**, 1–19.

Cronquist A 1981 *An Integrated System of Classification of Flowering Plants*. Columbia University Press, New York, USA.

Crossett RN and Loughman BC 1966 The absorption and translocation of phosphorus by seedlings of *Hordeum vulgare* L. *New Phytologist* **65**, 459–468.

Crush JR 1973 Significance of endomycorrhizas in tussock grassland in Otago, New Zealand. *New Zealand Journal of Botany* **11**, 645–660.

Cui M, Caldwell MM 1996a Facilitation of plant phosphate acquisition by arbuscular mycorrhizas from enriched soil patches. I. Roots and hyphae exploiting the same soil volume. *New Phytologist* **133**, 453–460.

Cui M, Caldwell MM 1996b Facilitation of plant phosphate acquisition by arbuscular mycorrhizas from enriched soil patches. II. Hyphae exploiting root-free soil. *New Phytologist* **133**, 461–467.

Cullings K 1994 Molecular phylogeny of the Monotropoideae Ericaceae with a note on the placement of the Pyroloideae. *Journal of Evolutionary Biology* **7** 4, 501–516.

Cullings KW 1996 Single phylogenetic origin of ericoid mycorrhizae within the Ericaceae. *Canadian Journal of Botany* **74**, 1896–1909.

Cullings KW, Szaro TM, Bruns TD 1996 Evolution of extreme specialisation with a lineage of ectomycorrhizal parasites. *Nature* **379**, 63–66.

Cumming JR and Weinstein LH 1990 Aluminum-mycorrhizal interactions in the physiology of pitch pine seedlings. *Plant Soil* **125**, 7–18.

Cummings MP and Welschmeyer NA 1998 Pigment composition of putatively achlorophyllous angiosperms. *Plant Systematics and Evolution* **210**, 105–111.

Currah RS and Van Dyke M 1986 A survey of some perennial vascular plant species native to Alberta for occurrence of mycorrhizal fungi. *Canadian Field-Naturalist* **100**, 330–342.

Currah RS, Siegler L, Hambleton S 1987 New records and new taxa from the mycorrhizae of terrestial orchids of Alberta. *Canadian Journal of Botany* **65**, 2473–2482.

Currah RS, Hambleton S, Smreciu EA 1988 The mycorrhizae and mycorrhizal fungi of *Calypso bulbosa* (Orchidaceae). *American Journal of Botany* **75**, 737–750.

Currah RS, Smreciu EA, Hambleton S 1990 Mycorrhizae and mycorrhizal fungi of boreal species of *Platanthera* and *Coeloglossum* (Orchidaceae). *Canadian Journal of Botany* **68**, 1171–1181.

Currah RS, Tsuneda A, Murakami S 1993 Morphology and ecology of *Phialocephala fortinii* in roots of *Rhododendron brachycarpos*. *Canadian Journal of Botany* **71**, 1639–1644.

Currah RS, Zettler LW, McInnis TM 1997 *Epulorhiza inquilina* sp. nov. from *Platanthera* (Orchidaceae) and a key to *Epulorhiza* species. *Mycotaxon* **61**, 335–342.

Curtis JT 1939 The relation of specificity of orchid mycorrhizal fungi to the problem of symbiosis. *American Journal of Botany,* **26**, 390–398.

Daft MJ and El Giahmi AA 1978 Effects of arbuscular mycorrhiza on plant growth. VIII. Effects of defoliation and light on selected hosts. *New Phytologist* **80**, 365–372.

Daft MJ and Hogarth BG 1983 Competitive interactions amongst four species of *Glomus* on maize and onion. *Transactions of the British Mycological Society* **80**, 339–345.

Daft MJ and Nicolson TH 1966 Effect of *Endogone* mycorrhiza on plant growth. *New Phytologist* **65**, 343–350.

Daft MJ and Nicolson TH 1969a Effect of *Endogone* mycorrhiza on plant growth. II. Influence of soluble phosphate on endophyte and host in maize. *New Phytologist* **68**, 945–952.

Daft MJ and Nicolson TH 1969b Effect of *Endogone* mycorrhiza on plant growth. III. Influence of inoculum concentration on growth and infection in tomato. *New Phytologist* **68**, 953–963.

Daft MJ and Nicolson TH 1972 Effect of *Endogone* mycorrhiza on plant growth. IV. Quantitative relationships between the growth of the host and development of the endophyte in tomato and maize. *New Phytologist* **71**, 287–295.

Daft MJ, Hacskaylo E, Nicolson TH 1975 Arbuscular mycorrhizas in plants colonising coal spoils in Scotland and Pennsylvania. In *Endomycorrhizas*. Eds FE Sanders, B Mosse, PB Tinker pp. 561–580. Academic Press, London, UK.

Daft MJ, Chilvers MT, Nicolson TH 1980 Mycorrhizas of the Lilliflorae. I. Morphogenesis of *Endymion non-scriptus* L. Garcke and its mycorrhizas in nature. *New Phytologist* **85**, 181–189.

Dahlberg A and Stenlid J 1990 Population structure and dynamics in *Suillus bovinus* as indicated by spatial distribution of fungal clones. *New Phytologist* **115**, 487–493.

Dalal RC 1977 Soil organic phosphorus. *Advances in Agronomy* **29**, 83–117.

Dalpé Y 1986 Axenic synthesis of ericoid mycorrhiza in *Vaccinium angustifolium* Ait. by *Oidiodendron* species. *New Phytologist* **103**, 391–396.

Dalpé Y, Litten W, Sigler L 1989 *Scytalidium vaccinii* a new species, an ericoid endophyte of *Vaccinium angustifolium* roots. *Mycotaxon* **35**, 371–378.

Daniell T, Husband R, Fitter AH, Young JPW 2001 Molecular diversity of arbuscular mycorrhizal fungi colonising arable crops. *FEMS Microbiology Ecology* **36**, 203–209.

Daniels BA and Duff DM 1978 Variation in germination and spore morphology among four isolates of *Glomus mosseae*. *Mycologia* **70**, 1261–1267.

Daniels BA and Menge JA 1980 Secondary sporocarp formation by *Glomus epigaeus*, a vesicular-arbuscular mycorrhizal fungus, in long-term storage. *Mycologia* **72**, 1235–1238.

Daniels BA and Trappe JM 1980 Factors affecting spore germination of the vesicular-arbuscular mycorrhizal fungus, *Glomus epigaeus*. *Mycologia* **72**, 457–471.

Daniels BA, McCool PA, Menge JA 1981 Comparative inoculum potential of spores of six vesicular-arbuscular mycorrhizal fungi. *New Phytologist* **89**, 385–391.

Daniels Hetrick BA 1984 Ecology of VA mycorrhizal fungi. In *V.A. Mycorrhiza*. Eds CL Powell and DJ Bagyaraj pp. 35–55. CRC Press, Boca Raton, Florida, UK.

Daniels Hetrick BA and Bloom J 1986 The influence of host plant on production and colonization ability of vesicular-arbuscular mycorrhizal spores. *Mycologia* **78**, 32–36.

Daniels Hetrick BA and Wilson GWT 1989 Supression of mycorrhizal fungus spore germination in non-sterile soil: relationship to mycorrhizal growth response in big bluestem. *Mycologia* **81**, 382–390.

Danielson RM 1984 Ectomycorrhizal associations in jack pine stands in north-eastern Alberta. *Canadian Journal of Botany* **62**, 932–939.

Danielson RM 1982 Taxonomic affinities and criteria for identification of the common ectendomycorrhizal symbiont of pines. *Canadian Journal of Botany* **60**, 7–18.

Danielson RM and Pruden M 1989 The ectomycorrhizal status of urban spruce. *Mycologia* **81**, 335–341.

Danielson RM and Visser S 1989a Effects of forest soil acidification on ectomycorrhizal and vesicular-arbuscular mycorrhizal development. *New Phytologist* **112**, 41–47.

Danielson RM and Visser S 1989b Host response to inoculation and behaviour of introduced and indigenous ectomycorrhizal fungi of jack pine grown on soil-sand tailings. *Canadian Journal of Forest Research* **19**, 1412–1421.

Danneberg G, Latus C, Zimmer W, Hundeshagen B, Schneider-Poetsch H, Bothe H 1993 Influence of vesicular-arbuscular mycorrhiza on phytohormone balances in maize *Zea mays* L. *Journal of Plant Physiology* **141**, 33–39.

David-Schwartz R, Badani H, Smadar W, Levy AA, Galili G, Kapulnik Y 2001 Identification of a novel genetically controlled step in mycorrhizal colonization: plant resistance to infection by fungal spores but not extra-radical hyphae. *The Plant Journal* **27**, 561–569.

David-Schwartz R, Gadkar V, Wininger S *et al.* 2003 Isolation of a premycorrhizal infection *pmi2* mutant of tomato, resistant to arbuscular mycorrhizal fungal colonization. *Molecular Plant-Microbe Interactions* **16**, 382–388.

Davis EC 2004 A molecular phylogeny of leafy liverworts Jungermannidae: Marchantriophyta. In *Molecular Systematics of Bryophytes*. Eds B Goffinet, V Hollowell, R Magill pp. 61–86. Missouri Botanical Garden Press, St Louis, Missouri, USA.

de Bary A 1887 *Comparative Morphology and Biology of the Fungi, Mycetozoa and Bacteria*. English translation. Clarendon Press, Oxford, UK.

de Boer GA, Kowalchuk GA, Van Veen JA 2006 Root-food and the rhizosphere microbial community composition. *New Phytologist* **170**, 3–6.

de la Peña E, Rodriguez Echeverria S, van der Putten WH, Freitas H, Moens M 2006 Mechanism of control of root-feeding nematodes by mycorrhizal fungi in the dune grass *Ammophila arenaria*. *New Phytologist* **169**, 829–840.

de la Providencia IE, de Souza FA, Fernandez G, Delmas NS, Declerck S 2005 Arbuscular mycorrhizal fungi reveal distinct patterns of anastomosis formation and hyphal healing mechanisms between different phlogenetic groups. *New Phytologist* **165**, 261–271.

de Miranda JCC and Harris PJ 1994 The effect of soil phosphorus on the external mycelium growth of arbuscular mycorrhizal fungi during the early stages of mycorrhiza formation. *Plant and Soil* **166**, 271–280.

de Silva, RL and Wood RKS 1964 Infection of plants by *Corticium solani* and *C. praticola* – effect of plant exudates. *Transactions of the British Mycological Society* **47**, 15–24.

Deacon JW and Fox FM 1988 Delivery of microbial inoculants into the root zone of transplant crops. In *Brighton Crop Protection Conference – Pests and Diseases*. pp. 645–653. Association of Applied Biologists, Wellesbourne Park, Warwick, UK.

Deacon JW, Donaldson SJ, Last FT 1983 Sequences and interactions of mycorrhizal fungi on birch. *Plant and Soil* **71**, 257–262.

Dearnaley JDW and McGee PA 1996 An intact microtubule cytoskeleton is not necessary for interfacial matrix formation in orchid protocorm mycorrhizas. *Mycorrhiza* **6** 3, 175–180.

Debaud JC, Pepin R, Bruchet G 1981 Etude des ectomycorrhizes de *Dryas octopetala*. Obtention de syntheses mycorhiziennes et de carpophores d'*Hebeloma alpinum* et *H. marginatum*. *Canadian Journal of Botany* **59**, 1014–1020.

Debaud JC, Gay G, Bruchet G 1986 Intra-specific variability in an ectomycorrhizal fungus: *Hebeloma cylindrosporum*. 1. Preliminary studies on *in vitro* fruiting, spore germination and sexual comportment. In *Physiological and Genetical Aspects of Mycorrhizae*. Eds V Gianinazzi-Pearson and S Gianinazzi pp. 581–588. INRA, Paris, France.

Debaud JC, Marmeisse R, Gay F 1995 Intra-specific genetic variation in ectomycorrhizal fungi. In *Mycorrhiza: Structure, Function, Molecular Biology and Biotechnology*. Eds A Varma and B Hock pp. 79–113. Springer-Verlag, Berlin, Germany.

Declerck S, de Boulois HD, Bivort C, Delvaux B 2003 Extraradical mycelium of the arbuscular mycorrhizal fungus *Glomus lamellosum* can take up, accumulate and translocate radiocaesium under root-organ culture conditions. *Environmental Microbiology* **5**, 510–516.

Dell B, Botton B, Martin F, Le Tacon F 1989 Glutamate dehydrogenases in ectomycorrhizas of spruce *Picea excelsa* L. and beech *Fagus sylvatica* L. *New Phytologist* **111**, 683–692.

Dell B, Malajczuk N, Thomson G 1990 Ectomycorrhiza formation in *Eucalyptus*. V. A tuberculate ectomycorrhiza of *Eucalyptus pilularis*. *New Phytologist* **114**, 633–640.

Dell J, Malajczuk N, Bougher NL, Thomson G 1994 Development and function of *Pisolithus* and *Scleroderma* ectomycorrhizas formed *in vivo* with *Allocasuarina*, *Casuarina* and *Eucalyptus*. *Mycorrhiza* **5**, 129–138.

Delp G, Timonen S, Rosewarne G, Barker SJ, Smith SE 2003 Differential expression of *Glomus intraradices* genes in external mycelium and mycorrhizal roots of tomato and barley. *Mycological Research* **107**, 1083–1093.

Demchenko K, Winzer T, Stougaard J, Parniske M, Pawlowski K 2004 Distinct roles of *Lotus japonicus* SYMRK and SYM15 in root colonization and arbuscule formation. *New Phytologist* **163**, 381–392.

Denny HJ and Wilkins DA 1987 Zinc tolerance in *Betula* spp. IV. The mechanism of ectomycorrhizal amelioration of zinc toxicity. *New Phytologist* **106**, 545–553.

Dettmann ME 1992 Structure and floristic composition of Cretaceous vegetation of southern Gondwana: implications for angiosperm biogeography. *Palaeobotanist* **41**, 224–233.

Dexheimer J and Serrigny J 1983 Étude ultrastruturale des endomycorhizes d'une orchidée tropicale: *Epidendrum ibaguense* H.B.K.I. Localisation des activité phosphatasiques acides *et* alkalines. *Bulletin Société Botanique de France* **130**, 187–194.

Dexheimer J, Gianinazzi S, Gianinazzi-Pearson V 1979 Ultrastructural cytochemistry of the host-fungus interface in the endomycorrhizal association *Glomus mosseae/Allium cepa*. *Zeitschrift für Pflanzenphysiologie* **92**, 191–206.

Dexheimer J, Marx C, Gianinazzi-Pearson V, Gianinazzi S 1985 Ultracytological studies on plasmalemma formations produced by host and fungus in vesicular-arbuscular mycorrhizae. *Cytologia* **50**, 461–471.

Diaz S 1996 Effects of elevated [CO_2] at the community level mediated by root symbionts. *Plant and Soil* **187**, 309–320.

Dickie IA, Koide RT, Steiner KC 2002 Influences of established trees on mycorrhizas, nutrition, and growth of *Quercus rubra* seedlings. *Ecological Monographs* **72**, 505–521.

Dickson S 2004 The *Arum-Paris* continuum of mycorrhizal symbioses. *New Phytologist* **161**, 187–200.

Dickson S and Kolesik P 1999 Visualisation of mycorrhizal fungal structures and quantification of their surface area and volume using laser scanning confocal microscopy. *Mycorrhiza* **9**, 205–213.

Dickson S and Smith SE 2001 Cross walls in arbuscular trunk hyphae form after loss of metabolic activity. *New Phytologist* **151**, 735–742.

Dickson S, Schweiger PF, Smith FA, Söderström B, Smith S 2003 Paired arbuscules in the *Arum*-type arbuscular mycorrhizal symbiosis with *Linum usitatissimum* L. *Canadian Journal of Botany* **81**, 457–463.

Dickson S, Smith FA, Smith SE 2007 Structural differences in arbuscular mycorrhizal symbioses: more than 100 years after Gallaud, where next? *Mycorrhiza* **17**, 375–393.

Dighton J and Mason PA 1985 Mycorrhizal dynamics during forest tree development. In *Development Biology of Higher Fungi*. Eds D Moore, LA Casselton, DA Wood, JC Frankland pp. 117–139. Cambridge University Press, Cambridge, UK.

Dighton J, Poskitt JM, Howard DM 1986 Changes in occurrence of basidiomycete fruit bodies during forest stand development: with specific reference to mycorrhizal species. *Transactions of the British Mycological Society* **87**, 163–171.

Dighton J, Thomas ED, Latter PM 1987 Interactions between tree roots, mycorrhizas, a saprophytic fungus and the decomposition of organic substrates in a microcosm. *Biology and Fertility of Soils* **4**, 145–150.

Dinkelaker B and Marschner H 1992 *In vivo* demonstration of acid phosphatase activity in the rhizosphere of soil-grown plants. *Plant and Soil* **144**, 199–205.

Dinkelaker B, Hengeler C, Marschner H 1995 Distribution and function of proteoid roots and other root clusters. *Botanica Acta* **108**, 183–200.

Ditengou FA and Lapeyrie F 2000 Hypaphorine from the ectomycorrhizal fungus *Pisolithus tinctorius* counteracts activities of indole-3-acetic acid and ethylene but not synthetic auxins in eucalypt seedlings. *Molecular Plant-Microbe Interactions* **13** 2, 151–158.

Ditengou FA, Beguiristain T, Lapeyrie F 2000 Root hair elongation is inhibited by hypaphorine, the indole alkaloid from the ectomycorrhizal fungus *Pisolithus tinctorius*, and restored by indole-3-acetic acid. *Planta* **211**, 722–728.

Dixon KW, Kell SP, Barrett RL, Cribb PJ 2003 *Orchid Conservation*. Natural History Publications, Kota Kinabalu, Malaysia.

Dixon RK, Pallardy SG, Garrett HE, Cox GS 1983 Comparative water relations of container-grown and bare-root ectomycorrhizal and non-mycorrhizal *Quercus velutina* seedlings. *Canadian Journal of Botany* **61**, 1559–1565.

Dixon RK, Garrett HE, Cox GS, Marx DH, Sander IL 1984 Inoculation of three *Quercus* species with eleven isolates of ectomycorrhizal fungi. I. Inoculation success and seedling growth relationships. *Forest Science* **30**, 364–372.

Dixon RK, Garrett HE, Cox GS 1987 Cytokinins in the root pressure exudate of *Citrus jambhiri* Lush. colonised by vesicular-arbuscular mycorrhizae. *Tree Physiology* **4**, 9–18.

Doak KD 1928 The mycorrhizal fungus of *Vaccinium*. *Phytopathology* **18**, 101–108.

Dodd JC 1994 Approaches to the study of the extraradical mycelium of arbuscular mycorrhizal fungi. In *Impact of Arbuscular Mycorrhizas on Sustainable Agriculture and Natural Ecosystems*. Eds S Gianinazzi and H Schüepp pp. 147–166. Birkhauser, Basel, Switzerland.

Dodd JC and Thomson BD 1994 The screening and selection of inoculant arbuscular-mycorrhizal and ectomycorrhizal fungi. *Plant and Soil* **159**, 149–158.

Dodd J, Hedle EM, Pate JS, Dixon KW 1984 Rooting patterns of sandplain plants and their functional significance. In *Kwongan. Plant life of the sandplain*. Eds JS Pate and JS Beard pp. 146–177. University of Western Australia Press, Nedlands, Australia.

Dodd JC, Burton CC, Burns RG, Jeffries P 1987 Phosphatase activity associated with the roots and the rhizosphere of plants infected with vesicular-arbuscular mycorrhizal fungi. *New Phytologist* **107**, 163–172.

Dominik T 1969 Key to ectotrophic mycorrhizas. *Folia Forestalia Polonica Seria A Lesnictwo* **15**, 309–328.

Dorr I and Kollmann R 1969 Fine structure of mycorrhiza in *Neottia nidus-avis*. *Planta* **89**, 372–375.

Dosskey MG, Linderman RG, Boersma L 1990 Carbon-sink stimulation of photosynthesis in Dougas fir seedlings by some ectomycorrhizas. *New Phytologist* **115**, 269–274.

Doudrick RL and Anderson NA 1989 Incompatibility factors and mating competence of two *Laccaria* spp. Agaricales associated with black spruce in northern Minnesota. *Phytopathology* **79**, 694–700.

Doudrick RL, Furnier GR, Anderson NA 1990 The number and distribution of incompatibility alleles in *Laccaria laccata* var. *moelleri* (Agaricales). *Phytopathology* **80**, 869–872.

Douds DDJ, Schenck NC 1990 Increased sporulation of vesicular-arbuscular mycorrhizal fungi by manipulation of nutrient regimens. *Applied and Environmental Microbiology* **56**, 413–418.

Douds DDJ, Galvez L, Bécard G, Kapulnik Y 1998 Regulation of arbuscular mycorrhizal development by plant host and fungus species in alfalfa. *New Phytologist* **138**, 27–35.

Douds DDJ, Johnson CR, Koch KE 1988 Carbon cost of the fungal symbiont relative to net leaf P accumulation in a split-root VA mycorrhizal symbiosis. *Plant Physiology* **86**, 491–496.

Douds DDJ, Pfeffer PE, Shachar-Hill Y 2000 Carbon partitioning, cost and metabolism of arbuscular mycorrhizas. In *Arbuscular Mycorrhizas: Physiology and Function*. Eds Y Kapulnik and DDJ Douds pp. 107–129. Kluwer Academic Publishers, Dordrecht, The Netherlands.

Douglas AE 1998 Host benefit and the evolution of specialisation in symbiosis. *Heredity* **81**, 599–603.

Douglas GC, Heslin MC, Reid C 1989 Isolation of *Oidiodendron maius* from *Rhododendron* and ultrastructal characterization of synthesized mycorrhizas. *Canadian Journal of Botany* **67**, 2206–2212.

Downes GM, Alexander IJ, Cairney JWG 1992 A study of spruce *Picea sitchensis* Bong Carr. ectomycorrhizas. I. Morphological and cellular changes in mycorrhizas formed by *Tylospora fibrillosa* Burt Donk and *Paxillus involutus* Batsch ex Fr. Fr. *New Phytologist* **122**, 141–152.

Drew EA, Murray RS, Smith SE, Jakobsen I 2003 Beyond the rhizosphere: growth and function of arbuscular mycorrhizal external hyphae in sands of varying pore sizes. *Plant and Soil* **251**, 105–114.

Drew EA, Murray RS, Smith SE 2005 Functional diversity of external hyphae of AM fungi: ability to colonise new hosts is influenced by fungal species, distance and soil conditions. *Applied Soil Ecology* **32**, 350–365.

Driver JD, Holben WE, Rillig MC 2005 Characterization of glomalin as a hyphal wall component of arbuscular mycorrhizal fungi. *Soil Biology and Biochemistry* **37**, 101–106.

Druge U and Schonbeck F 1993 Effect of vesicular-arbuscular mycorrhizal infection on transpiration, photosynthesis and growth of flax *Linum usitatissimum* L. in relation to cytokinin levels. *Journal of Plant Physiology* **141**, 40–48.

Duc G, Trouvelot A, Gianinazzi-Pearson V, Gianinazzi S 1989 First report of non-mycorrhizal plant mutants Myc- obtained in pea *Pisum sativum* L. and fababean *Vicia faba* L. *Plant Science* **60**, 215–222.

Duchesne LC, Peterson RL, Ellis BE 1988a Interaction between the ectomycorrhizal fungus *Paxillus involutus* and *Pinus resinosa* induces resistance to *Fusarium oxysporum*. *Canadian Journal of Botany* **66**, 558–562.

Duchesne LC, Peterson RL, Ellis BE 1988b Pine root exudate stimulates the synthesis of antifungal compounds by the ectomycorrhizal fungus *Paxillus involutus*. *New Phytologist* **108**, 471–476.

Duchesne LC, Ellis BE, Peterson RL 1989a Disease suppression by the ectomycorrhizal fungus *Paxillus involutus*: contribution of oxalic acid. *Canadian Journal of Botany* **67**, 2726–2730.

Duchesne LC, Peterson RL, Ellis BE 1989b The time-course of disease suppression and antibiosis by the ectomycorrhizal fungi *Paxillus involutus*. *New Phytologist* **111**, 693–698.

Duckett JG and Ligrone R 1992 A light and electron-microscope study of the fungal endophytes in the sporophyte and gametophyte of *Lycopodium cernuum* with observations on the gametophyte-sporophyte junction. *Canadian Journal of Botany*, **70**, 58–72.

Duckett JG and Ligrone R 2005 A comparative cytological analysis of fungal endophytes in the sporophyte rhizomes and vascularized gametophytes of *Tmesipteris* and *Psilotum*. *Canadian Journal of Botany* **83** 11, 1443–1456.

Duckett JG, Russell J, Ligrone R 2006 Basidiomycetous endophytes in jungermannialean leafy liverworts have novel cytologies and species-specific host ranges: a cytological and experimental study. *Canadian Journal of Botany* **84**, 1075–1093.

Duckett JG and Read DJ 1995 Ericoid mycorrhizas and rhizoid-ascomycete associations in liverworts share the same mycobiont – isolation of the partners and resynthesis of the associations *in vitro*. *New Phytologist* **129**, 439–477.

Duckett JG, Renzaglia KS, Pell K 1991 A light and electron-microscope study of rhizoid ascomycete associations and flagelliform axes in British hepatics with observations on the effects of the fungi on host morphology. *New Phytologist* **118**, 233–257.

Duckett JG, Carafa A, Ligrone R 2006b A highly differentiated glomeromycotean association with the mucilage-secreting, primitive antipodean liverwort *Treubia* (Treubiaceae): clues to the origins of mycorrhizas. *American Journal of Botany* **93**, 797–813.

Ducousso M, Bena G, Bourgeois C *et al.* 2004 The last common ancestor of Sarcolaenaceae and Asian dipterocarp trees was ectomycorrhizal before the India-Madagascar separation, about 88 million years ago. *Molecular Ecology* **13**, 231–236.

Duddridge JA 1980 *A Comparative Ultrastructural Analysis of a Range of Mycorrhizal Associations*. PhD thesis, University of Sheffield, Sheffield, UK.

Duddridge JA 1986a The development and ultrastructure of ectomycorrhizas. III. Compatible and incompatible interactions between *Suillus grevillei* (Klotzsch) Sing. and 11 species of ectomycorrhizal hosts *in vitro* in the absence of exogenous carbohydrates. *New Phytologist* **103**, 457–464.

Duddridge JA 1986b The development and ultrastructure of ectomycorrhizas. IV. Compatible and incompatible interactions between *Suillus grevillei* (Klotzsch) Sing. and a number of ectomycorrhizal hosts *in vitro* in the presence of exogenous carbohydrates. *New Phytologist* **103**, 465–471.

Duddridge JA 1987 Specificity and recognition in ectomycorrhizal associations. In *Fungal Infection of Plants*. Eds GF Pegg and PG Ayres pp. 25–44. Cambridge University Press, Cambridge, UK.

Duddridge JA and Read DJ 1982a An ultrastructural analysis of the development of mycorrhizas in *Monotropa hypopitys*. L. *New Phytologist* **92**, 203–214.

Duddridge JA and Read DJ 1982b Ultrastructural analysis of the development of mycorrhizas in *Rhododendron ponticum*. *Canadian Journal of Botany* **60**, 2345–2356.

Duddridge JA and Read DJ 1984a The development and ultra-structure of ectomycorrhizas. I. Ectomycorrhizal development on pine in the field. *New Phytologist* **96**, 565–573.

Duddridge JA and Read DJ 1984b The development and ultra-structure of ectomycorrhizas. II. Ectomycorrhizal development on pine *in vitro*. *New Phytologist* **96**, 575–582.

Duddridge JA and Read DJ 1984c Modification of the host-fungus interface in mycorrhizas synthesised between *Suillus bovinus* Fr O. Kuntz and *Pinus sylvestris* L. *New Phytologist* **96**, 583–588.

Duddridge JA, Malibari A, Read DJ 1980 Structure and function of mycorrhizal rhizomorphs with special reference to their role in water transport. *Nature* **287**, 834–836.

Dueck TA, Visser P, Ernst WHO, Schat H 1986 Vesicular-arbuscular mycorrhizae decrease zinc-toxicity to grasses growing in zinc-polluted soil. *Soil Biology and Biochemistry* **18**, 331–333.

Dumas-Gaudot E, Valot B *et al.* 2004 Proteomics as a way to identify extra-radicular fungal proteins from *Glomus intraradices*-RiT-DNA carrot root mycorrhizas. *FEMS Microbiology Ecology* **48**, 401–411.

Dunham SM, O'Dell TE, Molina R 2003 Analysis of nrDNA sequences and microsatellite allele frequencies reveals a cryptic chanterelle species *Cantharellus cascadensis* sp nov from the American Pacific Northwest. *Mycological Research* **107**, 1163–1177.

Dunne MJ and Fitter AH 1989 The phosphorus budget of a field-grown strawberry *Fragaria x ananassa* cv. Hapil crop: evidence for a mycorrhizal contribution. *Annals of Applied Biology* **114**, 185–193.

Duplessis S, Courty PE, Tagu D, Martin F 2005 Transcript patterns associated with ectomycorrhiza development in *Eucalyptus globulus* and *Pisolithus microcarpus*. *New Phytologist* **165**, 599–611.

Duponnois R, Diedhiou S, Chotte JL, Sy MO 2003 Relative importance of the endomycorrhizal and or ectomycorrhizal associations in *Allocasuarina* and *Casuarina* genera. *Canadian Journal of Microbiology* **49**, 281–287.

Durall DM, Jones MD, Tinker PB 1994 Allocation of C-14 carbon in ectomycorrhizal willow. *New Phytologist* **128**, 109–114.

Eason WR and Newman EI 1990 Rapid cycling of nitrogen and phosphorus from dying roots of *Lolium perenne*. *Oecologia* **82**, 432–436.

Eason WR, Newman EI, Chuba PN 1991 Specificity of interplant cycling of phosphorus: the role of mycorrhizas. *Plant and Soil* **137**, 267–274.

Egerton-Warburton LM and Allen EB 2000 Shifts in arbuscular mycorrhizal communities along an anthropogenic nitrogen deposition gradient. *Ecological Applications* **10**, 484–496.

Egger KN 1986 Substrate hydrolysis patterns of postfire ascomycetes (Pezizales). *Mycologia* **78**, 771–780.

Egger KN 1996 Molecular systematics of E-strain mycorrhizal fungi: *Wilcoxina* and its relationship to *Tricharina* Pezizales. *Canadian Journal of Botany* **74**, 773–779.

Egger KN and Fortin JA 1990 Identification of taxa of E-strain mycorrhizal fungi by restriction fragment analysis. *Canadian Journal of Botany* **68**, 1482–1488.

Egger KN and Paden JW 1986 Biotrophic associations between lodgepole pine seedlings and postfire ascomycetes Pezizales in monoxenic culture. *Canadian Journal of Botany* **64**, 2719–2725.

Egger KN and Sigler L 1993 Relatedness of the ericoid endophytes *Scytalidium vaccinii* and *Hymenoscyphus ericae* inferred from analysis of ribosomal DNA. *Mycologia* **85**, 219–230.

Egger KN, Danielson RM, Fortin JA 1991 Taxonomy and population structure of E-strain mycorrhizal fungi inferrred from ribosomal and mitochondrial DNA polymorphisms. *Mycological Research* **95**, 866–872.

Eissenstat DM 1990 A comparison of phosphorus and nitrogen transfer between plants of different phosphorus status. *Oecologia* **82**, 342–347.

Eissenstat DM, Graham JH, Syvertsen JP, Drouillard DL 1993 Carbon economy of sour orange in relation to mycorrhizal colonization and phosphorus status. *Annals of Botany* **71**, 1–10.

Ek H, Andersson S, Arnebrant K, Söderström B 1994 Growth and assimilation of NH_4^+ and NO_3^- by *Paxillus involutus* in association with *Betula pendula* and *Picea abies* as affected by substrate pH. *New Phytologist* **128**, 629–637.

Ekblad A, Wallander H, Carlsson R, HussDanell K 1995 Fungal biomass in roots and extramatrical mycelium in relation to macronutrients and plant biomass of ectomycorrhizal *Pinus sylvestris* and *Alnus incana*. *New Phytologist* **131**, 443–451.

El Abyad MSH and Webster J 1968 Studies on pyrophilous discomycetes, I. Comparative physiological studies. *Transactions of the British Mycological Society* **51**, 353–367.

El-Badaoui K and Botton B 1989 Production and characterisation of exocellular proteases in ectomycorrhizal fungi. *Annales des Sciences Forestières* **46**, 728–730.

El-Kherbawy M, Angle JS, Heggo A, Chaney RL 1989 Soil pH, rhizobia, and vesicular-arbuscular mycorrhizae inoculation effects on growth and heavy metal uptake of alfalfa *Medicago sativa* L. *Biology and Fertility of Soils* **8**, 61–65.

Elfstrand M, Feddermann N, Ineichen K *et al.* 2005 Ectopic expression of the mycorrhiza-specific chitinase gene *Mtchit 3-3* in *Medicago truncatula* root-organ cultures stimulates spore germination of glomalean fungi. *New Phytologist* **167**, 557–570.

Elias KS and Safir GR 1987 Hyphal enlongation of *Glomus fasciculatus* in response to root exudates. *Applied and Environmental Microbiology* **53**, 1928–1933.

Ellenberg H 1988 *Vegetation Ecology of Central Europe.* Cambridge University Press, Cambridge, UK.

Elliott GC, Lynch J, Lauchli A 1984 Influx and efflux of P in roots of intact maize plants. *Plant Physiology* **76**, 336–341.

Eltrop L and Marschner H 1996 Growth and mineral nutrition of non-mycorrhizal and mycorrhizal Norway spruce *Picea abies* seedlings grown in semi-hydroponic sand culture. 1. Growth and mineral nutrient uptake in plants supplied with different forms of nitrogen. *New Phytologist* **133** 3, 469–478.

Emmerton KS, Callaghan TV, Jones HE, Leake JR, Michelsen A, Read DJ 2001a Assimilation and isotopic fractionation of nitrogen by mycorrhizal fungi. *New Phytologist* **151**, 503–511.

Emmerton KS, Callaghan TV, Jones HE, Leake JR, Michelsen A, Read DJ 2001b Assimilation and isotopic fractionation of nitrogen by mycorrhizal and nonmycorrhizal subarctic plants. *New Phytologist* **151**, 513–524.

Enstone DE, Peterson CA, Ma FS 2002 Root endodermis and exodermis: structure, function, and responses to the environment. *Journal of Plant Growth Regulation* **21**, 335–351.

Entry JA, Donnelly PK, Cromack K Jr 1991a Influence of ectomycorrhizal mat soils on lignin and cellulose degradation. *Biology and Fertility of Soils* **11**, 75–78.

Entry JA, Rose CL, Cromack K Jr 1991b Litter decomposition and nutrient release in ectomycorrhizal mat soils of a Douglas fir ecosystem. *Soil Biology and Biochemistry* **23**, 285–290.

Ericsson T 1995 Growth and shoot-root ratio of seedlings in relation to nutrient availability. *Plant and Soil* **169**, 205–214.

Ernst R 1967 The effect of carbohydrate selection on the growth rate of freshly germinated *Phalaenopsis* and *Dendrobium* seed. *American Orchid Society Bulletin* **36**, 1068–1073.

Ernst R, Arditti J, Healey PL 1971 Carbohydrate physiology of orchid seedlings II. Hydrolysis and effects of oligosaccharides. *American Journal of Botany* **58**, 827–835.

Esnault A-L, Masuhara G, McGee PA 1994 Involvement of exodermal passage cells in mycorrhizal infection of some orchids. *Mycological Research* **98**, 672–676.

Evans DG and Miller MH 1988 Vesicular-arbuscular mycorrhizas and the soil-disturbance-induced reduction of nutrient absorption in maize. I. Causal relations. *New Phytologist* **110**, 67–74.

Evans DG and Miller MH 1990 The role of the external mycelial network in the effect of soil disturbance upon vesicular-arbuscular mycorrhizal colonization of maize. *New Phytologist* **114**, 65–71.

Ezawa T and Yoshida T 1994 Acid phosphatase specific to arbuscular mycorrhizal infection in marigold and possible role in symbiosis. *Soil Science and Plant Nutrition* **40**, 655–665.

Ezawa T, Kuwahara S-Y, Sakamoto K, Yoshida T, Saito M 1999 Specific inhibitor and substrate specificity of alkaline phosphatase expressed in the symbiotic phase of the arbuscular mycorrhizal fungus, *Glomus etunicatum*. *Mycologia* **91**, 636–641.

Ezawa T, Smith SE, Smith FA 2001 Enzyme activity involved in glucose phosphorylation in two arbuscular mycorrhizal fungi: indication that polyP is not the main phosphagen. *Soil Biology and Biochemistry* **33**, 1279–1281.

Ezawa T, Smith SE, Smith FA 2002 P metabolism and transport in AM fungi. *Plant and Soil* **244**, 221–230.

Ezawa T, Cavagnaro TR, Smith SE, Smith FA, Ohtomo R 2004 Rapid accumulation of polyphosphate in extraradical hyphae of an arbuscular mycorrhizal fungus as revealed by histochemistry and a polyphosphate kinase/luciferase system. *New Phytologist* **161**, 387–392.

Ezawa T, Hayatsu M, Saito M 2005 A new hypothesis on the strategy for acquisition of phosphorus in arbuscular mycorrhiza: up-regulation of secreted acid phosphatase gene in the host plant. *Molecular Plant-Microbe Interactions* **18**, 1046–1053.

Faber BA, Zasoski RJ, Burau RG, Uriu K 1990 Zinc uptake by corn as affected by vesicular-arbuscular mycorrhizae. *Plant and Soil* **129**, 121–130.

Faber BA, Zasoski RJ, Munns DN, Shackel K 1991 A method for measuring hyphal nutrient and water uptake in mycorrhizal plants. *Canadian Journal of Botany* **69**, 87–94.

Facelli JM and Facelli E 1993 Interactions after death: plant litter controls priority effects in a successional plant community. *Oecologia* **95**, 277–282.

Facelli E and Facelli JM 2002 Soil phosphorus heterogeneity and mycorrhizal symbiosis regulate plant intra-specific competition and size distribution. *Oecologia* **133**, 54–61.

Facelli E, Facelli JM, Smith SE, McLaughlin MJ 1999 Interactive effects of arbuscular mycorrhizal symbiosis, intraspecific competition and resource availability on *Trifolium subterraneum* cv. Mt. Barker. *New Phytologist* **141**, 535–547.

Fairchild GL and Miller MH 1988 Vesicular-arbuscular mycorrhizas and the soil-disturbance-induced reduction of nutrient absorption in maize. II. Development of the effect. *New Phytologist* **110**, 75–84.

Fay P, Mitchell DT, Osborne BA 1996 Photosynthesis and nutrient-use efficiency of barley in response to low arbuscular mycorrhizal colonization and addition of phosphorus. *New Phytologist* **132**, 425–433.

Federspiel A, Schuler R, Haselwandter K 1991 Effect of pH, L-ornithine and L-proline on the hydroxamate siderophore production by *Hymenoscyphus ericae*, a typical ericoid mycorrhizal fungus. *Plant and Soil* **130**, 259–261.

Fedorova M, van de Mortel J, Matsumoto PA *et al.* 2002 Genome-wide identification of nodule-specific transcripts in the model legume *Medicago truncatula*. *Plant Physiology* **130**, 519–537.

Fenner M 1978 A comparison of the abilities of colonisers and closed turf species to establish from seed in artificial swards. *Journal of Ecolgy* **66**, 953–965.

Ferdinandsen C and Winge O 1925 *Cenococcum* Fr. A monographic study. *Kungl Veterinaer og Landbohoskol Aarskrift* 332–382.

Fernando AA and Currah RS 1995 *Leptodontidium orchidicola–Mycelium radicis atrovirens* complex – aspects of its conidiogenesis and ecology. *Mycotaxon* **54**, 287–294.

Fernando AA and Currah RS 1996 A comparative study of the effects of the root endophytes *Leptodontidium orchidicola* and *Phialocephala fortinii* Fungi Imperfecti on the growth of some subalpine plants in culture. *Canadian Journal of Botany* **74** 7, 1071–1078.

Ferrol N, Barea JM, Azcón-Aguilar C 2000 The plasma membrane H$^+$-ATPase gene family in the arbuscular mycorrhizal fungus *Glomus mosseae*. *Current Genetics* **37**, 112–118.

Fester T, Strack D, Hause B 2001 Reorganization of tobacco root plastids during arbuscule development. *Planta* **213**, 864–868.

Fester T, Hause B, Schmidt D *et al.* 2002a Occurrence and localization of apocarotenoids in arbuscular mycorrhizal plant roots. *Plant and Cell Physiology* **43**, 256–265.

Fester T, Schmidt D, Lohse S *et al.* 2002b Stimulation of carotenoid metabolism in arbuscular mycorrhizal roots. *Planta* **216**, 148–154.

Ficus EL and Markhart AH 1979 Relationships between root system water transport properties and plant size in *Phaseolus*. *Plant Physiology* **64**, 770–773.

Finlay RD 1985 Interactions between soil micro-arthropods and endomycorrhizal associations of higher plants. In *Ecological Interactions in Soil*. Eds AH Fitter, D Atkinson, DJ Read, MB Usher pp. 319–331. Blackwell Scientific Publications, Oxford, UK.

Finlay RD 1989 Functional aspects of phosphorus uptake and carbon translocation in incompatible ectomycorrhizal associations between *Pinus sylvestris* and *Suillus grevillei* and *Boletus cavipes*. *New Phytologist* **112**, 185–192.

Finlay R 1992 Uptake and translocation of nutrients by ectomycorrhizal fungal mycelia. In *Mycorrhizas in Ecosystems*. Eds DJ Read, DH Lewis, AH Fitter, IJ Alexander pp. 91–97. CAB International, Wallingford, UK.

Finlay RD and Read DJ 1986a The structure and function of the vegetative mycelium of ectomycorrhizal plants. I. Translocation of ^{14}C-labelled carbon between plants interconnected by a common mycelium. *New Phytologist* **103**, 143–156.

Finlay RD and Read DJ 1986b The structure and function of the vegetative mycelium of ectomycorrhizal plants. II. The uptake and distribution of phosphorus by mycelium interconnecting host plants. *New Phytologist* **103**, 157–165.

Finlay RD and Read DJ 1986c The uptake and distribution of phosphorus by ectomycorrhizal mycelium. In *Physiological and Genetical aspects of Mycorrhizae*. Eds V Gianinazzi-Pearson and S Gianinazzi pp. 351–355. INRA, Paris, France.

Finlay RD and Söderström B 1989 Mycorrhizal mycelia and their role in soil and plant communities. In *Developments in Plant and Soil Sciences, Vol 39, Ecology of Arable Land, Perspectives and Challenges*. Eds M Clarholm and L Bergström pp. 139–148. Kluwer Academic Publishers, Dordrecht, The Netherlands.

Finlay RD and Söderström B 1992 Mycorrhiza and carbon flow to the soil. In *Mycorrhiza Functioning*. Ed. M Allen pp. 134–160. Chapman and Hall, London, UK.

Finlay RD, Ek H, Odham G, Söderström B 1988 Mycelial uptake, translocation and assimilation of nitrogen from ^{15}N labelled ammonium by *Pinus sylvestris* plants infected with four different ectomycorrhizal fungi. *New Phytologist* **110**, 59–66.

Finlay RD, Ek H, Odham G, Söderström B 1989 Uptake, translocation and assimilation of nitrogen from ^{15}N-labelled ammonium and nitrate sources by intact ectomycorrhizal systems of *Fagus sylvatica* infected with *Paxillus involutus*. *New Phytologist* **113**, 47–55.

Finlay RD, Frostegård Å, Sonnerfeldt A.-M 1992 Utilisation of organic and inorganic nitrogen sources by ectomycorrhizal fungi in pure culture and in symbiosis with *Pinus contorta* Dougl. ex Loud. *New Phytologist* **120**, 105–115.

Finn RF 1942 Mycorrhizal inoculation in soil of low fertility. *Black Rock Forest Papers* **1**, 116–117.

Fitter AH 1985 Functioning of vesicular-arbuscular mycorrhizas under field conditions. *New Phytologist* **99**, 257–265.

Fitter AH 1986 Effect of benomyl on leaf phosphorus concentration in alpine grasslands: a test of mycorrhizal benefit. *New Phytologist* **103**, 767–776.

Fitter AH 1988 Water relations of red clover, *Trifolium pratense* L., as affected by VA mycorrhizal infection and phosphorus supply before and during drought. *Journal of Experimental Botany* **39**, 595–604.

Fitter AH 1990 The role and ecological significance of vesicular-arbuscular mycorrhizas in temperate ecosystems. *Agriculture, Ecosystems and Environment* **29**, 137–151.

Fitter AH 1991 Costs and benefits of mycorrhizas: implications for functioning under natural conditions. *Experientia* **47**, 350–355.

Fitter AH 2001 Specificity, links and networks in the control of diversity in plant and microbial communities. In *Ecology: Achievement and Challenge*. Eds MC Press, NJ Hontley, S Levin pp. 95–111. Blackwell Science, Oxford, UK.

Fitter AH 2005 Darkness visible: reflections on underground ecology. *Journal of Ecology* **93**, 231–243.

Fitter AH and Garbaye J 1994 Interactions between mycorrhizal fungi and other soil organisms. *Plant and Soil* **159**, 123–132.

Fitter AH and Merryweather JW 1992 Why are some plants more mycorrhizal than others? an ecological enquiry. In *Mycorrhizas in Ecosystems*. Eds DJ Read, DH Lewis, AH Fitter, I Alexander pp. 26–36. CAB International, Wallingford, UK.

Fitter AH and Moyersoen B 1996 Evolutionary trends in root-microbe symbioses. *Philosophical Transactions of the Royal Society of London, Series B Biological Sciences* **351**, 1367–1375.

Fitter AH and Nichols R 1988 The use of benomyl to control fungal infection by vesicular-arbuscular mycorrhizal fungi. *New Phytologist* **110**, 201–206.

Fitter AH and Sanders IR 1992 Interactions with soil fauna. In *Mycorrhizal Functioning: An Integrative Plant-Fungal Process*. Ed. MF Allen pp. 333–354. Chapman and Hall, London, UK.

Fitter AH, Graves JD, Watkins NK, Robinson D, Scrimgeour C 1998a Carbon transfer between plants and its control in networks of arbuscular mycorrhizas. *Functional Ecology* **12**, 406–412.

Fitter AH, Wright WJ, Williamson L, Belshaw M, Fairclough J, Meharg AA 1998b The phosphorus nutrition of wild plants and the paradox of arsenic tolerance: does leaf phosphate concentration control flowering? In *Phosphorus in Plant Biology: Regulatory Roles in Molecular, Cellular, Organismic and Ecosystem Processes*. Eds JP Lynch and J Deikman pp. 39–51. American Society of Plant Physiologists, Rockville, Maryland, USA.

Fitter AH, Heinemeyer A, Staddon P 2000 The impact of elevated CO_2 and global climate change on arbuscular mycorrhizas: a mycocentric approach. *New Phytologist* **147**, 179–187.

Fitter AH, Heinemeyer A, Husband R, Olsen E, Ridgway KP, Staddon PL 2004 Global environmental change and the biology of arbuscular mycorrhizas: gaps and challenges. *Canadian Journal of Botany* **82**, 1133–1139.

Fleming LV 1983 Succession of mycorrhizal fungi on birch: infection of seedlings planted around mature trees. *Plant and Soil* **71**, 263–267.

Fleming LV 1984 Effects of soil trenching and coring on formation of ectomycorrhizas on birch seedlings grown around mature trees. *New Phytologist* **98**, 143–153.

Fleming LV 1985 Experimental study of sequences of ectomycorrhizal fungi on birch *Betula* sp. seedling root systems. *Soil Biology and Biochemistry* **17**, 591–600.

Fleming LV, Deacon JW, Last FT, Donaldson SJ 1984 Influence of propagating soil on the mycorrhizal succession of birch seedlings transplanted to a field site. *Transactions of the British Mycological Society* **82**, 707–712.

Fogel R and Trappe JM 1978 Fungus consumption (mycophagy) by small animals. *North West Science* **52**, 1–31.

Fontana A 1963 Micorrize ectotriche in una Ciperacea: *Kobresia belliardi* Degl. *Giornale Botanico Italiano* **70**, 639–641.

Fontana A and Bonfante Fasolo P 1971 Sintesi micorizica di *Tuber brumale* Vitt. con *Pinus nigra* Arnold. *Allionia* **17**, 15–18.

Ford L, Forbert L Burger JA, Miller OK 1985 Comparative effects of four mycorrhizal fungi on loblolly pine seedlings growing in a greenhouse in a Piedmont soil. *Plant and Soil* **83**, 215–221.

Forster SM 1979 Microbial aggregation of sand in an embryo dune system. *Soil Biology and Biochemistry* **11**, 537–543.

Forster SM and Nicolson TH 1979 Microbial aggregation of sand in a maritime dune succession. *Soil Biology and Biochemistry* **13**, 205–208.

Fortin JA, Piché Y, Lalonde M 1980 Technique for observation of early morphological changes during ectomycorrhiza formation. *Canadian Journal of Botany* **58**, 361–365.

Foster RC and Marks GC 1966 The fine structure of the mycorrhizas of *Pinus radiata* D. Don. *Australian Journal of Biological Science* **19**, 1027–1038.

Fox, FM 1986 Groupings of ectomycorrhizal fungi of birch and pine, based on establishment of mycorrhizas on seedlings from spores in unsterile soils. *Transactions of the British Mycological Society* **87**, 371–380.

France RC and Reid CPP 1983 Interactions of nitrogen and carbon in the physiology of ectomycorrhizas. *Canadian Journal of Botany* **61**, 964–984.

France RC and Reid CPP 1984 Pure culture growth of ectomycorrhizal fungi on inorganic nitrogen sources. *Microbial Ecology* **10**, 187–195.

Francis R and Read DJ 1984 Direct transfer of carbon between plants connected by vesicular-arbuscular mycorrhizal mycelium. *Nature* **307**, 53–56.

Francis R and Read DJ 1994 The contributions of mycorrhizal fungi to the determination of plant community structure. *Plant and Soil* **159**, 11–25.

Francis R and Read DJ 1995 Mutualism and antagonism in the mycorrhizal symbiosis, with special reference to impacts on plant community structure. *Canadian Journal of Botany* **73** (Suppl. 1), 1301–1309.

Francis R, Finlay RD, Read DJ 1986 Vesicular-arbuscular mycorrhizas in natural vegetation systems. IV. Transfer of nutrients in inter- and intra-specific combinations of host plants. *New Phytologist* **102**, 103–111.

Francke HL 1934 Beitrage zur Kenntnis der Mykorrhiza von *Monotropa hypopitys* L. Analyse und Synthese der Symbiose. *Flora* Jena **129**, 1–52.

Frank A 1877 Über die biologishen Verhältnisse des Thalles einiger Krustenflechten. *Beitrage zur biologie der Pflanzen* **2**, 123–200.

Frank AB 1894 Die bedeutung der Mykorrhizapilze fur die gemeine Kiefer. *Forshwissenchaftliche Centralblatt* **16**, 1852–1890.

Frank AB 1985 Über die auf Wurzelsymbiose berhende Ernährung gewiser Bäume durch unterirdische Pilze. *Berichte der Deutschen Botanishen Gesellschaft* **3**, 128–145.

Franke T 2002 The myco-heterotrophic *Voyria flavescens* Gentianaceae and its associated fungus. *Mycology Progress* **1**, 367–376.

Franke T, Beenker L, Döring M, Kocyan A, Agerer R 2006 Arbuscular mycorrhizal fungi of the Glomus-group A lineage Glomerales; Glomeromycota detected in mycoheterotrophic plants from tropical Africa. *Mycology Progress* **5**, 24–31.

Franken P and Gnädinger F 1994 Analysis of parsely arbuscular endomycorrhiza: infection development and mRNA levels of defence related genes. *Molecular Plant-Microbe Interactions* **7**, 612–620.

Franken P, Lapopin L, Meyer-Gauen G, Gianinazzi-Pearson V 1997 RNA accumulation and genes expressed in spores of the arbuscular mycorrhizal fungus, *Gigaspora rosea*. *Mycologia* **89**, 293–297.

Fredeen AL and Terry N 1988 Influence of vesicular-arbuscular mycorrhizal infection and soil phosphorus level on growth and carbon metabolism of soybean. *Canadian Journal of Botany* **66**, 2311–2316.

Freisleben R 1933 Über experimentelle Mycorrhizabildung bei Ericaceen. *Berichte Deustche Botanishe Gesellschaft* **60**, 315–356.

Freisleben R 1934 Zur Frage der Mykotrophia in der gattung *Vaccinium* L. *Jachbuch für Wissenschaftliche Botanik* **80**, 421–456.

Freisleben R 1935 Wietere Untersuchungen über die Mykotrophie der Ericaceen. *Jahrbuch für Wissencschlaftliche Botanik* **82**, 413–459.

Freisleben R 1936 Wieterer Untersunchungen über die Mykotrophie der Ericaceen. *Jachbuch für Wissenschaftliche Botanik* **82**, 413–459.

Frenot Y, Bergstrom DM, Gloaguen JC, Tavenard R, Strullu DG 2005 The first record of mycorrhizae on sub-Antarctic Heard Island: a preliminary examination. *Antarctic Science* **17**, 205–210.

Frey B and Schüepp H 1993 Acquisition of nitrogen by external hyphae of arbuscular mycorrhizal fungi associated with *Zea mays* L. *New Phytologist* **124**, 221–230.

Frey B, Zierold K, Brunner I 2000 Extracellular complexation of Cd in the Hartig net and cytosolic Zn sequestration in the fungal mantle of *Picea abies-Hebeloma crustuliniforme* ectomycorrhizas. *Plant Cell and Environment* **23**, 1257–1265.

Frey-Klett P, Chavatte M, Clausse ML *et al.* 2005 Ectomycorrhizal symbiosis affects functional diversity of rhizosphere fluorescent pseudomonads. *New Phytologist* **165**, 317–328.

Fries N 1983 Spore germination, homing reaction, and intersterility groups in *Laccaria laccata* Agaricales. *Mycologia* **75**, 221–227.

Fries N 1987 Ecological and evolutionary aspects of spore germination in the higher Basidiomycetes. *Transactions of the British Mycological Society* **88**, 1–7.

Fries N and Mueller GM 1984 Incompatibility systems, cultural features and special circumscriptions in the ectomycorrhizal genus *Laccaria* Agaricales. *Mycologia* **76**, 633–642.

Fries N and Newman W 1990 Sexual incompatibility in *Suillus luteus* and *S. granulatus*. *Mycological Research* **94**, 64–70.

Fries N and Sun P 1992 The mating system of *Suillus bovinus*. *Mycological Research* **96**, 237–238.

Friese CF and Allen MF 1991 The spread of VA mycorrhizal fungal hyphae in the soil: inoculum types and external hyphal architecture. *Mycologia* **83**, 409–418.

Friese CF and Allen MF 1993 The interaction of harvester ants and vesicular-arbuscular mycorrhizal fungi in a patchy semi-arid environment: the effects of mound structure on fungal dispersion and establishment. *Functional Ecology* **7**, 13–20.

Fuchs A and Ziegenspeck H 1924 Aus der Monographie des *Orchis traunsteineri* Saut. III. Entwicklungsgechichte einiger deutscher Orchideen. *Botanisches Archiv* **5**, 120–132.

Fuchs A and Ziegenspeck H 1927 Entwicklungsgeschichte der Axen der einheimischen Orchideen und ihre Physiologie und Biologie III. *Botanisches Archiv* **18**, 378–475.

Fujimura KF, Smith JE, Horton TR, Weber NS, Spatafora JW 2005 Pezizalean mycorrhizas and sporocarps in ponderosa pine (*Pinus ponderosa*) after prescribed fires in eastern Oregon, USA. *Mycorrhiza* **15**, 79–86.

Furlan V, Fortin JA 1977 Effects of light intensity on the formation of vesicular-arbuscular endomycorrhizas on *Allium cepa* by *Gigaspora calospora*. *New Phytologist* **79**, 335–340.

Furman TE 1966 Symbiotic relationship of *Monotropa*. *American Journal of Botany* **53**, 627.

Furman TE and Trappe JM 1971 Phylogeny and ecology of mycotrophic achlorophyllous angiosperms. *Quarterly Review of Biology* **46**, 214–225.

Fusconi A and Bonfante-Fasolo P 1984 Ultrastructural aspects of host-endophyte relationships in *Arbutus unedo* mycorrhizas. *New Phytologist* **96**, 397–410.

Gadd GM 1993 Interactions of fungi with toxic metals. *New Phytologist* **124**, 25–60.

Gadd GM 1999 Fungal production of citric and oxalic acid: importance in metal speciation, physiology and biogeochemical processes. *Advances in Microbial Physiology* **41**, 47–92.

Gadgil RL and Gadgil PD 1971 Mycorrhiza and litter decomposition. *Nature* **233**, 133.

Gadgil RL and Gadgil PD 1975 Suppression of litter decomposition by mycorrhizal roots of *Pinus radiata*. *New Zealand Journal of Forest Science* **5**, 35–41.

Gadkar V and Rillig M 2006 The arbuscular mycorrhizal fungal protein glomalin is a putative homolog of heat shock protein 60. *FEMS Microbiological Letters* **263**, 93–101.

Gadkar V, David-Schwartz R, Kunik T, Kapulnik Y 2001 Arbuscular mycorrhizal fungal colonization. Factors involved in host recognition. *Plant Physiology* **127**, 1493–1499.

Gadkar V, David-Schwartz R, Nagahashi G, Douds DD, Wininger S, Kapulnik Y 2003 Root exudate of *pmi* tomato mutant M161 reduces AM fungal proliferation in vitro. *FEMS Microbiology Letters* **223**, 193–198.

Gagnon J, Langlois CG, Fortin JA 1988 Growth and ectomycorrhiza formation of containerised black spruce seedlings as affected by nitrogen fertilisation, inoculum type and symbiont. *Canadian Journal of Forest Research* **18**, 922–929.

Gallaud I 1905 Études sur les mycorrhizes endotrophes. *Revue Générale de Botanique* **17**, 5–48, 66–83, 123–135, 223–239, 313–325, 425–433, 479–500.

Gams W 1963 Mycelium radicis atrovirens in forest soils, isolation from soil microhabitats and identification. In *Soil organisms* Eds J Docksen and J van der Drift pp. 176–182. North-Holland, Amsterdam, The Netherlands.

Gandolfi A, Sanders IR, Rossi V, Menozzi P 2003 Evidence of recombination in putative ancient asexuals. *Molecular Biology and Evolution* **20**, 754–761.

Gange AC 2000 Arbuscular mycorrhizal fungi, collembola and plant growth. *Trends in Ecology and Evolution* **15**, 369–372.

Gange AC and Brown VK 2002 Actions and interactions of soil invertebrates and arbuscular mycorrhizal fungi in affecting structure of plant communities. In *Mycorrhizal Ecology*. Eds MGA van der Heijden and I Sanders pp. 321–344. Springer, Berlin, Germany.

Gange AC, Brown VK, Sinclair GS 1993 Vesicular-arbuscular mycorrhizal fungi: a determinant of plant community structure in early succession. *Functional Ecology* **7**, 616–622.

Gao L-L, Delp G, Smith SE 2001 Colonization patterns in a mycorrhiza-defective mutant tomato vary with different arbuscular-mycorrhizal fungi. *New Phytologist* **151**, 477–491.

Gao L-L, Knogge W, Delp G, Smith FA, Smith SE 2004 Expression patterns of defense-related genes in different types of arbuscular mycorrhizal development in wild-type and mycorrhiza-defective mutant tomato. *Molecular Plant-Microbe Interactions* **17**, 1103–1113.

Gao L-L, Smith FA, Smith SE 2006 The *rmc* locus does not affect plant interactions or defence-related gene expression when tomato *Solanum lycopersicum* is infected with the root fungal parasite, *Rhizoctonia*. *Functional Plant Biology* **33**, 289–296.

Garbaye J 1994 Helper bacteria: a new dimension to the mycorrhizal symbiosis. *New Phytologist* **128**, 197–210.

Garbaye J and Bowen GD 1987 Effect of different microflora on the success of ectomycorrhizal inoculation of *Pinus radiata*. *Canadian Journal of Forest Research* **17**, 941–943.

Garbaye J and Bowen GD 1989a Ectomycorrhizal infection of *Pinus radiata* by *Rhizopogon luteolus* is stimulated by microorganisms naturally present in the mantle of ectomycorrhizas. *New Phytologist* **112**, 383–388.

Garbaye J and Bowen GD 1989b Stimulation of ectomycorrhizal infection of *Pinus radiata* by some microorganisms associated with the mantle of ectomycorrhizas. *New Phytologist* **112**, 383–388.

García-Garrido JM, García-Romera I, Ocampo JA 1992 Cellulase activity in lettuce and onion plants colonized by the vesicular-arbuscular mycorrhizal fungus *Glomus mosseae*. *Soil Biology and Biochemistry* **25**, 503–504.

García-Garrido JM, Tribak M, Rejon-Palomares A, Ocampo JA, García-Romera I 2000 Hydrolytic enzymes and ability of arbuscular mycorrhizal fungi to colonize roots. *Journal of Experimental Botany* **51**, 1443–1448.

García-Romera I, García-Garrido JM, Martinez-Molina E, Ocampo JA 1990 Possible influence of hydrolytic enzymes on vesicular-arbuscular mycorrhizal infections of alfalfa. *Soil Biology and Biochemistry* **22**, 149–152.

García-Romera I, García-Garrido JM, Ocampo JA 1991 Pectolytic enzymes in the vesicular-arbuscular mycorrhizal fungus *Glomus mosseae*. *FEMS Microbiology Letters* **78**, 343–346.

Gardes M and Bruns TD 1993 ITS primers with enhanced specificity for basidiomycetes – application to the identification of mycorrhizae and rusts. *Molecular Ecology* **2**, 113–118.

Gardes M and Bruns TD 1996 Community structure of ectomycorrhizal fungi in a *Pinus muricata* forest: above- and below-ground views. *Canadian Journal of Botany* **74**, 1572–1583.

Gardes M, Mueller GM, Fortin JA, Kropp BR 1991a Mitochondrial DNA polymorphism in *Laccaria bicolor, L. laccata, L. proxima* and *L. amethystina*. *Mycological Research* **95**, 206–216.

Gardes M, Mueller G, Fortin JA, Kropp BR 1991b Restriction fragment length polymorphisms in ribosomal DNA from our species for *Laccaria*. *Phytopathology* **80**, 1312–1317.

Gardner IC 1986 Mycorrhizae of actinorhizal plants. *Mircen Journal* **2**, 147–160.

Garnier A, Berredjem A, Botton B 1997 Purification and characterization of the NAD-dependent glutamate dehydrogenase in the ectomycorrhizal fungus *Laccaria bicolor* (Maire) Orton. *Fungal Genetics and Biology* **22**, 168–176.

Garrec JP and Gay G 1978 Influence des champignons mycorrhiziens sur l'accumulation des elements mineraux dans les racines de pin d'alep. Analyse directe par microsonde electronique. In *Root Physiology and Symbiosis*. Eds A Reidacker and MJ Gagnaire-Michard pp. 486–488. Proceedings of the IUFROP Symposium, Nancy, France.

Garrett SD 1963 *Soil Fungi and Soil Fertility*. Pergamon Press, Oxford, UK.

Garriock ML, Peterson RL, Ackerley CA 1989 Early stages in colonization of *Allium porrum* (leek) roots by the vesicular-arbuscular mycorhizal fungus, *Glomus versiforme*. *New Phytologist* **112**, 85–92.

Gaudinski JB, Trumbore SE, Davidson EA, Zheng SH 2000 Soil carbon cycling in a temperate forest: radiocarbon-based estimates of residence times, sequestration rates and partitioning of fluxes. *Biogeochemistry* **51**, 33–69.

Gaudinski JB, Trumbore SE, Davidson EA, Cook AC, Markewitz D, Richter DD 2001 The age of fine-root carbon in three forests of the eastern United States measured by radiocarbon. *Oecologia* **129**, 420–429.

Gavito ME and Olsson PA 2003 Allocation of plant carbon to foraging and storage in arbuscular mycorrhizal fungi. *FEMS Microbiology Ecology* **45**, 181–187.

Gavito ME, Schweiger P, Jakobsen I 2003 P uptake by arbuscular mycorrhizal hyphae: effect of soil temperature and atmospheric CO_2 enrichment. *Global Change Biology* **9**, 106–116.

Gavito ME, Olsson PA, Rouhier H *et al.* 2005 Temperature constraints on the growth and functioning of root organ cultures with arbuscular mycorrhizal fungi. *New Phytologist* **168**, 179–188.

Gay G and Debaud JC 1987 Genetic study on indole-3-acetic acid production by ectomycorrhizal *Hebeloma* species: inter- and intraspecific variability in homo- and di-karyotic mycelia. *Applied Microbiology and Biotechnology* **26**, 141–146.

Gay G, Marmeisse R, Fouillet P, Buntertreau M, Debaud JC 1993 Genotype/nutrition interactions in the ectomycorrhizal fungus *Hebeloma cylindrosporum* Romagnési. *New Phytologist* **123**, 335–343.

Gay G, Normand L, Marmeisse R, Sotta B, Debaud JC 1994b Auxin overproducer mutants of *Hebeloma cylindrosporum* Romagnési have increased mycorrhizal activity. *New Phytologist* **128**, 645–657.

Gazey C, Abbott LK, Robson AD 1992 The rate of development of mycorrhizas affects the onset of sporulation and production of external hyphae by two species of *Acaulospora*. *Mycological Research* **96**, 643–650.

Gea L, Normand L, Vian B, Gay G 1994 Structural aspects of ectomycorrhiza of *Pinus pinaster* (Ait.) Sol. Formed by an IAA-overproducer mutant of *Hebeloma cylindrosporum* Romagnési. *New Phytologist* **128**, 659–670.

Gebauer G and Dietrich P 1993 Nitrogen isotope ratios in different compartments of a mixed stand of spruce, larch and beech trees and of understorey vegetation including fungi. *Isotopen Praxus* **29**, 35–44.

Gebauer G and Meyer M 2003 N-15 and C-13 natural abundance of autotrophic and mycoheterotrophic orchids provides insight into nitrogen and carbon gain from fungal association. *New Phytologist* **160** 1, 209–223.

Gebauer G and Taylor AFS 1999 N-15 natural abundance in fruit bodies of different functional groups of fungi in relation to substrate utilization. *New Phytologist* **142** 1, 93–101.

Gehrig H, Schüßler A, Kluge M 1996 *Geosiphon pyriforme*, a fungus forming endocytobiosis with *Nostoc* cyanobacteria, is an ancestral member of the Glomales: evidence by SSU rRNA analysis. *Journal of Molecular Evolution* **43**, 71–81.

Gehring CA and Whitham TG 1994 Interactions between aboveground herbivores and mycorrhizal mutualists of plants. *Trends in Ecology and Evolution* **9**, 251–255.

Gehring CA and Whitham TG 2002 Mycorrhizae-herbivore interactions: population and community consequences. In *Mycorrhizal Ecology*. Eds MGA van der Heijden and I Sanders pp. 295–320. Springer, Berlin, Germany.

Gemma JN and Koske RE 1988 Seasonal variation in spore abundance and dormancy of *Gigaspora gigantea* and in mycorrhizal inoculum potential of a dune soil. *Mycologia* **80**, 211–216.

Gemma JN and Koske RE 1992 Are mycorrhizal fungi present in early stages of primary succession? In *Mycorrhizas in Ecosystems*. Eds DJ Read, BG Lewis, AH Fitter, IJ Alexander pp. 183–189. CAB International, Wallingford, UK.

Genetet I, Martin F, Stewart GR 1984 Nitrogen assimilation in mycorrhizas. Ammonium assimilation in the N-starved ectomycorrhizal fungus *Cenococcum graniforme*. *Plant Physiology* **76**, 395–399.

Genney DR, Anderson IC, Alexander IJ 2006 Fine-scale distribution of pine ectomycorrhizas and their extramatrical mycelium. *New Phytologist* **170**, 381–390.

Genre A and Bonfante P 1997 A mycorrhizal fungus changes microtubule orientation in tobacco root cells. *Protoplasma* **199**, 30–38.

Genre A and Bonfante P 1998 Actin versus tubulin configuration in arbuscule-containing cells from mycorrhizal tobacco roots. *New Phytologist* **140**, 745–752.

Genre A and Bonfante P 1999 Cytoskeleton-related proteins in tobacco mycorrhizal cells: gamma-tubulin and clathrin localisation. *European Journal of Histochemistry* **43**, 105–111.

Genre A and Bonfante P 2002 Epidermal cells of a symbiosis-defective mutant of *Lotus japonicus* show altered cytoskeleton organisation in the presence of a mycorrhizal fungus. *Protoplasma* **219**, 43–50.

Genre A and Bonfante P 2005 Building a mycorrhizal cell: how to reach compatibility between plants and arbuscular mycorrhizal fungi. *Journal of Plant Interactions* **1**, 3–13.

Genre A, Chabaud M, Timmers T, Bonfante P, Barker DG 2005 Arbuscular mycorrhizal fungi elicit a novel intracellular apparatus in *Medicato truncatula* root epidermal cells before infection. *The Plant Cell* **17**, 3489–3499.

George E and Marschner H 1996 Nutrient and water uptake by roots of forest trees. *Zeitschrift fur Pflanzenernährung und Bodenkunde* **159**, 11–21.

George E, Haussler K-U, Vetterlein D, Gorgus E, Marschner H 1992 Water and nutrient translocation by hyphae of *Glomus mosseae*. *Canadian Journal of Botany* **70**, 2130–2137.

Gerdemann JW 1955 Relation of a large soil-borne spore to phycomycetous mycorrhizal infections. *Mycologia* **47**, 619–632.

Gerdemann JW 1964 The effect of mycorrhiza on the growth of maize. *Mycologia* **56**, 342–349.

Gerdemann JW 1965 Vesicular-arbuscular mycorrhizae formed on maize and tuliptree by *Endogone fasciculata*. *Mycologia* **57**, 562–575.

Gerdemann JW 1968 Vesicular-arbuscular mycorrhiza and plant growth. *Annual Review of Phytopathology* **6**, 397–418.

Gerdemann JW 1975 Vesicular-arbuscular mycorrhizae. In *The Development and Function of Roots*. Eds JG Torrey and DT Clarkson pp. 575–591. Academic Press, London, UK.

Gerdemann JW and Nicolson TH 1963 Spores of mycorrhizal *Endogone* species extracted from soil by wet sieving and decanting. *Transactions of the British Mycological Society* **46**, 235–244.

Gerdemann JW and Trappe JM 1974 The *Endogonaceae* in the Pacific Northwest. *Mycologia Memoir* **5**, 1–76.

Gerdemann JW and Trappe JM 1975 Taxonomy of the Endogonaceae. In *Endomycorrhizas*. Eds FE Sanders, B Mosse, PB Tinker pp. 35–51. Academic Press, London, UK.

Gerlitz TGM and Gerlitz A 1997 Phosphate uptake and polyphosphate metabolism of mycorrhizal and nonmycorrhizal roots of pine and of *Suillus bovinus* at varying external pH measured by in vivo ^{31}P-NMR. *Mycorrhiza* **7** 2, 101–106.

Gianinazzi S 1991 Vesicular arbuscular endo-mycorrhizas: cellular, biochemical and genetic aspects. *Agriculture, Ecosystems and Environment* **35**, 105–119.

Gianinazzi S and Schüepp H (eds) 1994 *Impact of Arbuscular Mycorrhizas on Sustainable Agriculture and Natural Ecosystems*. Advances in Life Sciences. Birkhaüser, Basel, Switzerland.

Gianinazzi S and Vosátka M 2004 Inoculum of arbuscular mycorrhizal fungi for production systems: science meets business. *Canadian Journal of Botany* **82**, 1264–1271.

Gianinazzi S, Gianinazzi-Pearson V, Dexheimer J 1979 Enzymatic studies on the metabolism of vesicular-arbuscular mycorrhiza. III. Ultrastructural localization of acid and alkaline phosphatase in onion roots infected by *Glomus mosseae* Nicol. and Gerd. *New Phytologist* **82**, 127–132.

Gianinazzi S, Trouvelot A, Gianinazzi-Pearson V 1990a Conceptual approaches for the rational use of VA endomycorrhizae in agriculture: possibilities and limitations. *Agriculture, Ecosystems and Environment* **29**, 153–161.

Gianinazzi S, Trouvelot A, Gianinazzi-Pearson V 1990b Role and use of mycorrhizas in horticultural crop production. In *XXIII International Horticultural Congress ISHS – International Society for Horticultural Science* Firenze, Italy pp. 25–30. Parretti Grafiche, Firenze, Italy.

Gianinazzi-Pearson V 1984 Host-fungus specifity, recognition and compatibility in mycorrhizae. In *Genes Involved in Microbe–Plant Interactions*. Eds DPS Verma and TH Hohn pp. 225–253. Springer-Verlag, Vienna, Austria.

Gianinazzi-Pearson V 1996 Plant cell responses to arbuscular mycorrhizal fungi: getting to the roots of the symbiosis. *The Plant Cell* **8**, 1871–1883.

Gianinazzi-Pearson V and Bonfante-Fasolo P 1986 Variability in wall structure and mycorrhizal behaviour of ericoid fungal isolates. In *Physiological and Genetical Aspects of Mycorrhizae*. Eds V Gianinazzi-Pearson and S Gianinazzi pp. 563–567. INRA, Paris, France.

Gianinazzi-Pearson V and Brechenmacher L 2004 Functional genomics of arbuscular mycorrhiza: decoding the symbiotic cell programme. *Canadian Journal of Botany* **82**, 1228–1234.

Gianinazzi-Pearson V and Denarie J 1997 Red carpet genetic programmes for root endosymbioses. *Trends in Plant Science* **2**, 371–372.

Gianinazzi-Pearson V and Gianinazzi S 1978 Enzymatic studies on the metabolism of vesicular-arbuscular mycorrhiza. II. Soluble alkaline phosphatase specific to mycorrhizal infection in onion roots. *Physiological Plant Pathology* **12**, 45–53.

Gianinazzi-Pearson V and Gianinazzi S 1983 The physiology of vesicular-arbuscular mycorrhizal roots. *Plant and Soil* **71**, 197–209.

Gianinazzi-Pearson V, Fardeau J-C, Asimi S, Gianinazzi S 1981a Source of additional phosphorus absorbed from soil by vesicular-arbuscular mycorrhizal soybeans. *Physiologie Végétale* **19**, 33–43.

Gianinazzi-Pearson V, Morandi D, Dexheimer J, Gianinazzi S 1981b Ultrastructural and ultracytochemical features of a *Glomus tenuis* mycorrhiza. *New Phytologist* **88**, 633–639.

Gianinazzi-Pearson V, Gianinazzi S, Trouvelot A 1985 Evaluation of the infectivity and effectiveness of indigenous vesicular-arbuscular fungal populations in some agricultural soils in Burgundy. *Canadian Journal of Botany* **63**, 1521–1524.

Gianinazzi-Pearson V, Bonfante-Fasolo P, Dexheimer J 1986 Ultrastructural studies of surface interactions during adhesion and infection by ericoid endomycorrhizal fungi. In *Recognition in Microbe-Plant Symbiotic and Pathogenic Interactions*. Ed. B Lugtenberg pp. 273–282. Nato ASI Series H, Cell Biology 4.

Gianinazzi-Pearson V, Branzanti B, Gianinazzi S 1989 *In vitro* enhancement of spore germination and early hyphal growth of a vesicular-arbuscular mycorrhizal fungus by host root exudates and plant flavonoids. *Symbiosis* **7**, 243–255.

Gianinazzi-Pearson V, Gianinazzi S, Brewin NJ 1990 Immunocytochemical localisation of antigenic sites on the perisymbiotic membrane of vesicular-arbuscular endomycorrhiza using monoclonal antibodies reacting against the peribacteroid membrane of nodules. In *Endocytobiology IV*. Eds P Nardon, V Gianinazzi-Pearson, AH Grenier, L Margulis, DC Smith pp. 127–131. INRA, Paris, France.

Gianinazzi-Pearson V, Smith SE, Gianinazzi S, Smith FA 1991a Enzymatic studies on the metabolism of vesicular-arbuscular mycorrhizas. V. Is H^+-ATPase a component of ATP-hydrolysing enzyme activities in plant-fungus interfaces? *New Phytologist* **117**, 61–74.

Gianinazzi-Pearson V, Gianinazzi S, Guillemin JP, Trouvelot A, Duc G 1991b Genetic and cellular analysis of resistance to vesicular arbuscular VA mycorrhizal fungi in pea mutants. In *Advances in Molecular Genetics of Plant-Microbe Interactions*. Eds H Hennecke and DPS Verma pp. 336–342. Kluwer Academic Publishers, Dordrecht, The Netherlands.

Gianinazzi-Pearson V, Gianinazzi S, Guillemin JP, Trouvelot A, Duc G 1994a Genetic and cellular analysis of resistance to vesicular-arbuscular VA mycorrhizal fungi in pea mutants. *Advances in Molecular Genetics of Plant-Microbe Interactions* **1**, 336–342.

Gianinazzi-Pearson V, Lemoine M-C, Arnould C, Gollotte A, Morton JB 1994b Localization of β 1–3 glucans in spore and hyphal walls of fungi in the Glomales. *Mycologia* **86**, 478–485.

Gianinazzi-Pearson V, Dumas-Gaudot E, Gollotte A, Tahiri-Alaoui A, Gianinazzi S 1996 Cellular and molecular defence-related root responses to invasion by arbuscular mycorrhizal fungi. *New Phytologist* **133**, 45–57.

Gianinazzi-Pearson V, Arnould C, Oufattole M, Arango M, Gianinazzi S 2000 Differential activation of H^+-ATPase genes by an arbuscular mycorrhizal fungus in root cells of transgenic tobacco. *Planta* **211**, 609–613.

Gibson BR and Mitchell DT 2004 Nutritional influences on the solubilization of metal phosphate by ericoid mycorrhizal fungi. *Mycological Research* **108**, 947–954.

Gibson BR and Mitchell DT 2005 Phosphatases of ericoid mycorrhizal fungi: kinetic properties and the effect of copper on activity. *Mycological Research* **109**, 478–486.

Gibson F and Deacon JW 1988 Experimental study of establishment of ectomycorrhizas in different regions of birch root systems. *Transactions of the British Mycological Society* **91**, 239–251.

Gibson F and Deacon JW 1990 Establishment of ectomycorrhizas in asceptic culture: effects of glucose, nitrogen and phosphorus in relation to successions. *Mycological Research* **94**, 166–172.

Giesler R, Högberg M, Högberg P 1998 Soil chemistry and plants in Fennoscandian boreal forest as exemplified by a local gradient. *Ecology* **79**, 119–137.

Gifford RM and Evans LT 1981 Photosynthesis, carbon partitioning, and yield. *Annual Reviews of Plant Physiology* **32**, 485–509.

Gil F and Gay JL 1977 Ultrastructural and physiological properties of the host interfacial components of haustoria of *Erisiphe pisi in vivo* and *in vitro*. *Physiological Plant Pathology* **10**, 1–12.

Gildon A, Tinker PB 1983 Interactions of vesicular-arbuscular mycorrhizal infections and heavy metals in plants. II. The effects of infection on uptake of copper. *New Phytologist* **95**, 263–268.

Gill RA and Jackson RB 2000 Global patterns of root turnover for terrestrial ecosystems. *New Phytologist* **147** 1, 13–31.

Gilmore AE 1971 The influence of endotrophic mycorrhizae on the growth of peach seedlings. *Journal of the American Society for Horticultural Science* **96**, 35.

Gimingham CH 1960 Biological flora of the British Isles. *Calluna vulgaris* L. Hull. *Journal of Ecology* **48**, 455–483.

Gimingham CH 1972 *Ecology of Heathlands*. Chapman and Hall, London, UK.

Giollant M, Guillot J, Damez M, Dusser M, Didier P, Didier E 1993 Characterization of a lectin from *Lactarius deterrimus*: research on the possible involvement of the fungal lectin in recognition between mushroom and spruce during the early stages of mycorrhizae formation. *Plant Physiology* **10**, 513–522.

Giovannetti G and Fontana A 1982 Mycorrhizal synthesis between *Cistaceae* and *Tuberaceae*. *New Phytologist* **92**, 533–537.

Giovannetti M 1985 Seasonal variations of vesicular-arbuscular mycorrhizas and endogonaceous spores in a maritime sand dune. *Transactions of the British Mycological Society* **84**, 679–684.

Giovannetti M 1997 Host signals dictating growth direction, morphogenesis and differentiation in arbuscular mycorrhizal symbionts. In *Eukaryotism and Symbiosis: Intertaxonomic Combination versus Symbiotic Adaptation*. Eds HEA Schenk, R Herrmann, KW Jeon, NE Müller, W Schwemmler. Springer-Verlag, Berlin, Germany.

Giovannetti M 2000 Spore germination and pre-symbiotic mycelial growth. In *Arbuscular Mycorrhizas: Physiology and Function*. Eds Y Kapulnik and DD Douds pp. 47–68. Kluwer Academic Press, Dordrecht, The Netherlands.

Giovannetti M and Citernesi AS 1993 Time-course of appressorium formation on host plants by arbuscular mycorrhizal fungi. *Mycological Research* **97**, 1140–1142.

Giovannetti M and Hepper CM 1985 Vesicular-arbuscular mycorrhizal infection in *Hedysarum coronarium* and *Onobrychis viciifolia*: host-endophyte specificity. *Soil Biology and Biochemistry* **17**, 899–900.

Giovannetti M and Lioi L 1990 The mycorrhizal status of *Arbutus unedo* in relation to compatible and incompatible fungi. *Canadian Journal of Botany* **68**, 1239–1244.

Giovannetti M and Mosse B 1980 An evaluation of techniques for measuring vesicular-arbuscular mycorrhizal infection in roots. *New Phytologist* **84**, 489–500.

Giovannetti M and Sbrana C 1998 Meeting a non-host: the behaviour of AM fungi. *Mycorrhiza* **8**, 123–130.

Giovannetti M and Sbrana C 2001 Self and non-self responses in hyphal tips of arbuscular mycorrhizal fungi. In *Cell Biology of Plant and Fungal Tip Growth*. Ed. A Geitmann pp. 221–231. IOS Press, Amsterdam, The Netherlands.

Giovannetti M, Sbrana C, Avio L, Citernesi AS, Logi C 1993a Differential hyphal morphogenesis in arbuscular mycorrhizal fungi during pre-infection stages. *New Phytologist* **125**, 587–593.

Giovannetti M, Avio L, Sbrana C, Citernesi AS 1993b Factors affecting appressorium development in the vesicular-arbuscular mycorrhizal fungus *Glomus mosseae* Nicol. and Gerd. Gerd. and Trappe. *New Phytologist* **123**, 115–122.

Giovannetti M, Sbrana C, Citernesi AS *et al.* 1994 Recognition and infection process, basis for host specificity of arbuscular mycorrhizal fungi. In *Impact of Arbuscular Mycorrhizas on Sustainable Agriculture and Natural Ecosystems*. Eds S Gianinazzi and H Schüepp pp. 61–72. Birkhäuser Verlag, Basel, Switzerland.

Giovannetti M, Azzolini D, Citernesi AS 1999 Anastomosis formation and nuclear and protoplasmic exchange in arbuscular mycorrhizal fungi. *Applied and Environmental Microbiology* **65**, 5571–5575.

Giovannetti M, Sbrana C, Logi C 2000 Microchambers and video-enhanced light microscopy for monitoring cellular events in living hyphae of arbuscular mycorrhizal fungi. *Plant and Soil* **226**, 153–159.

Giovannetti M, Fortuna P, Citernesi AS, Morini S, Nuti MP 2001 The occurrence of anastomosis formation and nuclear exchange in intact arbuscular mycorrhizal networks. *New Phytologist* **151**, 717–724.

Giovannetti M, Sbrana C, Strani P, Agnolucci M, Rinaudo V, Avio L 2003 Genetic diversity of isolates of *Glomus mosseae* from different geographic areas detected by vegetative compatibility testing and biochemical and molecular analysis. *Applied and Environmental Microbiology* **69**, 616–624.

Giovannetti M, Sbrana C, Avio L, Strani P 2004 Patterns of below-ground plant interconnections established by means of arbuscular mycorrhizal networks. *New Phytologist* **164**, 175–181.

Girlanda M, Selosse MA, Cafasso D *et al.* 2006 Inefficient photosynthesis in the Mediterranean orchid *Limodorum abortivum* is mirrored by specific association to ectomycorrhizal Russulaceae. *Molecular Ecology* **15**, 491–504.

Givinish TJ and Renner SS 2004 Tropical intercontinetnal disjunctions: Gondwana breakup, immigration from the boreotropics, and transoceanic dispersal. *International Journal of Plant Science* **165**, S1–S6.

Glassop D, Godwin RMC, Smith SE, Smith FW 2007 Rice phosphate transporters associated with phosphate uptake in rice roots colonised with arbuscular mycorrhizal fungi. *Canadian Journal of Botany* **85**, 644–651.

Glassop D, Smith SE, Smith FW 2005 Cereal phosphate transporters associated with the mycorrhizal pathway of phosphate uptake into roots. *Planta* **222**, 688–698.

Glenn MG, Chew FS, Williams PH 1988 Influence of glucosinolate content of *Brassica* (Cruciferae) on germination of vesicular-arbuscular fungi. *New Phytologist* **110**, 217–225.

Glen M, Tommerup IC, Bougher NL, O'Brien PA 2002 Are Sebacinaceae common and widespread ectomycorrhizal associates of *Eucalyptus* species in Australian forests? *Mycorrhiza* **12**, 243–247.

Glerum C and Balatinecz JJ 1980 Formation and distribution of food reserves during autumn and their subsequent utilisation in jack pine. *Canadian Journal of Botany* **58**, 40–54.

Glowa KR, Arocena JM, Massicotte HB 2003 Extraction of potassium and/or magnesium from selected soil minerals by *Piloderma*. *Geomicrobiology Journal* **20**, 99–111.

Gnekow MA, Marschner H 1989 Role of VA-mycorrhiza in growth and mineral nutrition of apple *Malus pumila* var. *domestica* rootstock cuttings. *Plant and Soil* **119**, 285–293.

Gobert A and Plassard C 2002 Differential NO_3-dependent patterns of NO_3-uptake in *Pinus pinaster*, *Rhizopogon roseolus* and their ectomycorrhizal association. *New Phytologist* **154** 2, 509–516.

Godbold DL and Berntson GM 1997 Elevated atmospheric CO_2 concentration changes ectomycorrhizal morphotype assemblages in *Betula papyrifera*. *Tree Physiology* **17**, 347–350.

Godbold DL, Berntson GM, Bazzaz FA 1997 Growth and mycorrhizal colonization of three North American tree species under elevated atmospheric CO_2. *New Phytologist* **137**, 433–440.

Godbold DL, Hoosbeek MR, Lukac M, Cotrufo F *et al*. 2006 Mycorrhizal hyphal turnover as a dominant process for carbon input into soil organic matter. *Plant and Soil* **281**, 15–24.

Godbout C and Fortin JA 1983 Morphological features of synthesised ectomycorrhizae on *Alnus crispa* and *A. rugosa*. *New Phytologist* **94**, 249–262.

Godbout C and Fortin JA 1985 Synthesised ectomycorrhizas of aspen: fungal genus level of structural characterisation. *Canadian Journal of Botany* **63**, 252–262.

Gogala N 1991 Regulation of mycorrhizal infection by hormonal factors produced by hosts and fungi. *Experientia* **47**, 331–340.

Goh CJ, Sim AA, Lim G 1992 Mycorrhizal associations in some tropical orchids. *Lindleyana* **7**, 13–17.

Golley FB 1965 Structure and function of an old-field broomsedge community. *Ecological Monographs* **35**, 113–137.

Gollotte A, Gianinazzi-Pearson V, Giovannetti M, Sbrana C, Avio L, Gianinazzi S 1993 Cellular localization and cytochemical probing of resistance reactions to arbuscular mycorrhizal fungi in the 'locus *a*' myc⁻ mutant of *Pisum sativum* L. *Planta* **191**, 112–122.

Gollotte A, Gianinazzi-Pearson V, Gianinazzi S 1994 Etude immunocytochimique des interfaces plante-champignon endomycorhizien à arbuscules chez des pois isogeneiques myc⁺ ou resistant à endomycorrhization (myc⁻ PB). *Acta Botanica Gallica* **141**, 449–454.

Gonzalez-Chavez C, Harris PJ, Dodd J, Meharg AA 2002 Arbuscular mycorrhizal fungi confer enhanced arsenate resistance on *Holcus lanatus*. *New Phytologist* **155**, 163–171.

Gonzalez-Guerrero M, Azcón-Aguilar C, Mooney M *et al*. 2005 Characterization of a *Glomus intraradices* gene encoding a putative Zn transporter of the cation diffusion facilitator family. *Fungal Genetics and Biology* **42**, 130–140.

Goodman D, Durall D, Trofymow T, Berch S 1998 A manual of concise descriptions of North American ectomycorrhizae. *Mycorrhiza* **8**, 57–59.

Goodman DM, Durall DM, Trofymow JA, Berch SM (eds) 1996–2002 *A Manual of Concise Descriptions of North American Ectomycorrhizae*. Mycologue Publications, Sidney, BC, Canada.

Gordon JC and Larson PR 1968 Seasonal course of photosynthesis, respiration and distribution of ¹⁴C in young *Pinus resinosa* trees as related to wood formation. *Plant Physiology* **43**, 1617–1624.

Gorham E, Vitousek PM, Reiners WA 1979 The regulation of chemical budgets over the course of terrestrial ecosystem succession. *Annual Review of Ecological Systems* **10**, 53–84.

Gorissen A and Kuyper TW 2000 Fungal species-specific response of ectomycorrhizal Scots pine (*Pinus sylvestris*) to elevated [CO₂]. *New Phytologist* **146**, 163–168.

Goss RW 1960 Mycorrhiza of ponderosa pine in Nebraska grassland soils. *University of Nebraska. Agricultural Experiment Station Research Bulletin* **192**, 47.

Govindarajulu M, Pfeffer P, Jin H *et al*. 2005 Nitrogen transfer in the arbuscular mycorrhizal symbiosis. *Nature* **435**, 819–823.

Grace C and Stribley DP 1991 A safer procedure for routine staining of vesicular-arbuscular mycorrhizal fungi. *Mycological Research* **95**, 1160–1162.

Graham JH 1982 Effect of citrus root exudates on germination of chlamydospores of the vesicular-arbuscular mycorrhizal fungus, *Glomus epigaeum*. *Mycologia* **74**, 831–835.

Graham JH 1986 Citrus mycorrhizae: potential benefits and interactions with pathogens. *HortScience* **21**, 1302–1306.

Graham JH 1988 Interactions of mycorrhizal fungi with soilborne plant pathogens and other organisms: an introduction. *Phytopathology* **78**, 365–366.

Graham JH 2001 What do root pathogens see in mycorrhizas? *New Phytologist* **149**, 357–359.

Graham JH and Abbott LK 2000 Wheat responses to aggressive and non-aggressive arbuscular mycorrhizal fungi. *Plant and Soil* **220**, 207–218.

Graham JH and Eissenstat DM 1994 Host genotype and the formation and function of VA mycorrhizae. *Plant and Soil* **159**, 179–185.

Graham JH and Syvertsen JP 1984 Influence of vesicular-arbuscular mycorrhiza on the hydraulic conductivity of root of two citrus rootstocks. *New Phytologist* **97**, 277–284.

Graham JH and Syvertsen JP 1985 Host determinants of mycorrhizal dependency of citrus rootstock seedlings. *New Phytologist* **101**, 667–676.

Graham JH, Leonard RT, Menge JA 1981 Membrane-mediated decrease in root exudation responsible for inhibition of vesicular-arbuscular mycorrhiza formation. *Plant Physiology* **68**, 548–552.

Graham JH, Leonard RT, Menge JA 1982a Interaction of light-intensity and soil-temperature with phosphorus inhibition of vesicular arbuscular mycorrhiza formation. *New Phytologist* **91**, 683–690.

Graham JH, Linderman RG, Menge JA 1982b Development of external hyphae by different isolates of mycorrhizal *Glomus* spp. in relation to root colonization and growth of Troyer citrange. *New Phytologist* **91**, 183–189.

Graham JH, Syvertsen JP, Smith ML 1987 Water relations of mycorrhizal and phosphorus fertilized non-mycorrhizal citrus under drought stress. *New Phytologist* **105**, 411–419.

Graham JH, Eissenstat DM, Drouillard DL 1991 On the relationship between a plant's mycorrhizal dependency and rate of vesicular-arbuscular mycorrhizal colonization. *Functional Ecology* **5**, 773–779.

Graham JH, Duncan LW, Eissenstat DM 1997 Carbohydrate allocation patterns in citrus genotypes as affected by phosphorus nutrition, mycorrhizal colonization and mycorrhizal dependency. *New Phytologist* **135**, 335–343.

Grandmaison J, Benhamou N, Furlan V, Visser SA 1988 Ultrastructural localization of *N*-acetylglucosamine residues in the cell wall of *Gigaspora margarita* throughout its life cycle. *Biology of the Cell* **63**, 89–100.

Graw D, Moawad M, Rehm S 1979 Untersuchungen zur wirts und wirkungs spezifitat der VA mykorrhiza. *Zeitschrift für Aker und Pflanzenbau* **148**, 85–98.

Green NE, Graham SO, Schenk NC 1976 The influence of pH on the germination of vesicular-arbuscular mycorrhizal spores. *Mycologia* **68**, 929–934.

Green JJ and Newbery DM 2002 Reproductive investment and seedling survival of the mast-fruiting rainforest tree, *Microberlinia bisulcata*. A. Chev. *Plant Ecology* **162**, 169–183.

Grellet GA, Meharg AA, Alexander IJ 2005 Carbon availability affects nitrogen source utilisation by *Hymenoscyphus ericae*. *Mycological Research* **109**, 469–477.

Grente J, Chevalier C, Pollacsek A 1972 La germination de l'ascospore de *Tuber melanosporum* et la synthese sporale de mycorrhizes. *Compte Rendu Hebdomadaire des Sciences de l'Academie des Sciences, Series D* **275**, 743–746.

Grey WE 1991 Influence of temperature on colonization of spring barleys by vesicular arbuscular mycorrhizal fungi. *Plant and Soil* **137**, 181–190.

Griffith W 1845 On the root-parasites referred by authors to Rhizantheae; and on various plants related to them. *Transactions of the Linnean Society of London* **19**, 303–347.

Griffiths RP and Caldwell BA 1992 Mycorrhizal mat communities in forest soils. In *Mycorrhizas in Ecosystems*. Eds DJ Read, DH Lewis, AH Fitter, IJ Alexander pp. 98–105. CAB International, Wallingford, UK.

Griffiths RP, Ingham ER, Caldwell BA, Castellano MA, Cromack KJ 1991 Microbial characteristics of ectomycorrhizal mat communities in Oregon and Californa. *Biology and Fertility of Soils* **11**, 196–202.

Griffiths RP, Baham JE, Caldwell BA 1994 Soil solution chemistry of ectomycorrhizal mats in forest soil. *Soil Biology and Biochemistry* **26**, 331–337.

Grime JP 1979 *Plant Strategies and Vegetation Processes*. John Wiley, New York, USA.

Grime JP, Mackey JML, Hillier SH, Read DJ 1987 Floristic diversity in a model system using experimental microcosms. *Nature* **328**, 420–422.

Grimoldi AA, Kavanova M, Lattanzi FA, Schnyder H 2005 Phosphorus nutrition-mediated effects of arbuscular mycorrhiza on leaf morphology and carbon allocation in perennial ryegrass. *New Phytologist* **168**, 435–444.

Grogan P, Baar J, Bruns TD 2000 Below-ground ectomycorrhizal community structure in a recently burned bishop pine forest. *Journal of Ecology* **88**, 1051–1062.

Gronbach E and Agerer R 1986 Charakterisierung und Inventur der Fichtenmykorrhizen im Höglwald und dern Reaktionen auf saure Beregnung. *Forstwissenschaftliche Centralblatt* **105**, 329–335.

Groom P 1895 Contribution to the knowledge of monocotyledonous saprophytes. *Journal of the Linnean Society of London* **31**, 149–215.

Groth-Malonek M, Pruchner D, Grewe F, Knoop V 2004 Ancestors of trans-splicing mitochondrial introns support serial sister relationships of hornworts and mosses with vascular plants. *Molecular Biology and Evolution* **22**, 117–125.

Grove TS and Le Tacon F 1993 Mycorrhiza in plantation forestry. *Advances in Plant Pathology* **9**, 191–228.

Grubb PJ 1970 Observations on structure and biology of *Haplomitrium* and *Takakia*, hepatics with roots. *New Phytologist* **69** 2, 303.

Grubb PJ 1976 A theoretical background to the conservation of ecologically distinct groups of annuals and biennials in the chalk grassland ecosystem. *Biological Conservation* **10**, 53–76.

Grubb PJ 1977 The maintenance of species richness in plant communities: the importance of the regeneration niche. *Biological Reviews* **52**, 107–145.

Grubisha LC, Trappe JM, Molina R, Spatafora JW 2002 Biology of the ectomycorrhizal genus *Rhizopogon*. VI. Re-examination of infrageneric relationships inferred from phylogenetic analyses of ITS sequences. *Mycologia* **94**, 607–619.

Grunig CR and Sieber TN 2005 Molecular and phenotypic description of the widespread root symbiont *Acephala applanata* gen. et sp nov., formerly known as dark-septate endophyte Type 1. *Mycologia* **97** 3, 628–640.

Grunig CR, Duo A, Sieber TN 2006 Population genetic analysis of *Phialocephala fortinii* s.l. and *Acephala applanata* in two undisturbed forests in Switzerland and evidence for new cryptic species. *Fungal Genetics and Biology* **43**, 410–421.

Grunwald U, Nyamsuren O, Tamasloukht M *et al.* 2004 Identification of mycorrhiza-regulated genes with arbuscule development-related expression profile. *Plant Molecular Biology* **55**, 553–566.

Grunze N, Willmann M, Nehls U 2004 The impact of ectomycorrhiza formation on mono-saccharide transporter gene expression in poplar roots. *New Phytologist* **164**, 147–155.

Guaragnella N and Butow RA 2003 ATO3 encoding a putative outward ammonium transporter is an RTG-independent retrograde responsive gene regulated by GCN4 and the Ssy1-Ptr3-Ssy5 amino acid sensor system. *Journal of Biological Chemistry* **278** 46, 45882–45887.

Guerin-Laguette A, Shindo Matsushita N, Suzuki K, Lapeyrie F 2004 The mycorrhizal fungus *Tricholoma matsutake* stimulates *Pinus densiflora* seedling growth *in vitro*. *Mycorrhiza* **14**, 397–400.

Guidot A, Debaud JC, Effosse A, Marmeisse R 2004 Below-ground distribution and persistence of an ectomycorrhizal fungus. *New Phytologist* **161**, 539–547.

Güimil S, Chang HS *et al.* 2005 Comparative transcriptomics of rice reveals an ancient pattern of response to microbial colonization. *Proceedings of the National Academy of Sciences of the United States of America* **102**, 8066–8070.

Gunning BES and Robards AW 1976 *Intercellular Communication in Plants. Studies on Plasmodesmata*. Springer-Verlag, Berlin, Germany.

Güttenberger M 1995 The protein complement of ectomycorrhizas. In *Mycorrhiza: Structure, Molecular Biology and Function*. Eds AK Varma and B Hock pp. 59–77. Springer, Berlin, Germany.

Güttenberger M 2000 Arbuscules of vesicular-arbuscular mycorrhizal fungi inhabit an acidic compartment within plant roots. *Planta* **211**, 299–304.

Güttenberger M and Hampp R 1992 Ectomycorrhizins-symbiosis-specific or artifactual polypeptides from ectomycorrhizas? *Planta* **188**, 129–136.

Haas JH, Bar-Yosef B, Krikun J, Barak R, Markovitz T, Kramer S 1987 Vesicular-arbuscular mycorrhizal-fungus infestation and phosphorus fertigation to overcome pepper stunting after methyl bromide fumigation. *Agronomy Journal* **79**, 905–910.

Hadley G 1969 Cellulose as a carbon source for orchid mycorrhiza. *New Phytologist* **68**, 933–939.

Hadley G 1970 Non-specificity of symbiotic infection in orchid mycorrhiza. *New Phytologist* **69**, 1015–1023.

Hadley G 1975 Organisation and fine structure of orchid mycorrhizas. In *Endomycorrhizas*. Eds FE Sanders, B Mosse, PB Tinker pp. 335–351. Academic Press, London, UK.

Hadley G 1982 Orchid mycorrhiza. In *Orchid Biology; Reviews and Perspectives II*. Ed. J Arditti pp. 81–118. Cornell University Press, Ithaca, New York, USA.

Hadley G and Ong SH 1978 Nutritional requirements of orchid endophytes. *New Phytologist* **81**, 561–569.

Hadley G and Purves S 1974 Movement of [14]carbon from host to fungus in orchid mycorrhiza. *New Phytologist* **73**, 475–482.

Hadley G and Williamson B 1971 Analysis of post infection growth stimulus in orchid mycorrhiza. *New Phytologist* **70**, 445–455.

Hadley G and Williamson B 1972 Features of mycorrhizal infection in some malayan orchids. *New Phytologist* **71**, 1111–1118.

Hadley G, Johnson RPC and John DA 1971 Fine structure of the host-fungus interface in orchid mycorrhiza. *Planta* **100**, 191–199.

Hagerberg D, Thelin G, Wallander H 2003 The production of ectomycorrhizal mycelium in forests: relation between forest nutrient status and local mineral sources. *Plant and Soil* **252**, 279–290.

Hahn A and Hock B 1994 Immunochemical detection of arbuscular mycorrhizae. *Experientia* **50**, 913.

Hahn MG, Buchell P, Gervone F *et al.* 1989 Roles of cell wall constituents in plant-pathogen interactions. In *Plant Microbe Interactions: Molecular and Genetic Perspectives*. Eds T Kosuge and EW Nester pp. 131–181. McGraw-Hill, New York, USA.

Hahn A, Gianinazzi-Pearson V, Hock B 1994 Characterization of arbuscular mycorrhizal fungi by immunochemical methods. In *Impact of Arbuscular Mycorrhizas on Sustainable Agriculture and Natural Ecosystems*. Eds S Gianinazzi and H Schüepp pp. 25–39. Burkhäuser Verlag, Basel, Switzerland.

Hahn A, Wright S, Hock B 2001 Immunochemical characterization of mycorrhizal fungi. In *The Mycota – Fungal Associations*. Ed. B Hock pp. 29–44. Springer, Berlin, Germany.

Haines BL, Best GR 1976 *Glomus mosseae*, endomycorrhizal with *Liquidambar styraciflua* L. seedlings retards NO_3, NO_2 and NH_4 nitrogen loss from a temperate forest soil. *Plant and Soil* **45**, 257–261.

Hall IR 1976 Response of *Coprosma robusta* to different forms of endomycorrhizal inoculum. *Transactions of the British Mycological Society* **67**, 409–411.

Hall IR 1978 Effects of endomycorrhizas on the competitive ability of white clover. *New Zealand Journal of Agricultural Research* **21**, 509–515.

Hall IR 1988 Potential for exploiting vesicular-arbuscular mycorrhizas in agriculture. In *Biotechnology in Agriculture* pp. 141–174. Alan R. Liss Inc., New York, USA.

Hall IR and Wang Y 1998 Methods for cultivating edible ectomycorrhizal mushrooms. In *Mycorrhiza Manual*. Ed. A Varma pp. 99–114. Springer, Berlin, Germany.

Hall IR, Brown G, Byars J 1994 *The Black Truffle: its History, Uses and Cultivation*. Crop and Food Research, Lincoln, New Zealand.

Hamada M 1939 Studien uber die Mykorrhiza von *Galeola septentrionalis* Reichb. f.-Ein neuer Fall der Mykorrhiza-Bildung durch intraradicale Rhizomorpha. *Japan Journal of Botany* **10**, 151–212.

Hambleton S and Currah RS 1997 Fungal endophytes from the roots of alpine and boreal Ericaceae. *Canadian Journal of Botany* **75** 9, 1570–1581.

Hambleton S and Sigler L 2005 *Meliniomyces*, a new anamorph genus for root-associated fungi with phylogenetic affinities to *Rhizoscyphus ericae* (*Hymenoscyphus ericae*), Leotiomycetes. *Studies in Mycology* **53**, 1–27.

Hambleton S, Egger KN, Currah RS 1998 The genus *Oidiodendron*: species delimitation and phylogenetic relationships based on nuclear ribosomal DNA analysis. *Mycologia* **90**, 854–868.

Hambleton S, Huhtinen S, Currah RS 1999 *Hymenoscyphus ericae*: a new record from western Canada. *Mycological Research* **103**, 1391–1397.

Hamel C and Smith DL 1991 Interspecific N-transfer and plant development in a mycorrhizal field-grown mixture. *Soil Biology and Biochemistry* **23**, 661–666.

Handley LL, Daft MJ, Wilson J, Scrimgeour CM, Ingleby K, Sattar MA 1993 Effect of the ecto- and VA- mycorrhizal fungi *Hydnagium carneum* and *Glomus clarum* on the $\delta^{15}N$ and $\delta^{13}C$ values of *Eucalyptus globulus* and *Ricinus communis*. *Plant, Cell and Environment* **16**, 375–382.

Handley WRC 1954 *Mull and Mor Formation in Relation to Forest Soils*. Forestry Commission Bulletin No. 23. HMSO, London, UK.

Hans J, Hause B, Strack D, Walter MH 2004 Cloning, characterization, and immunolocalization of a mycorrhiza-inducible 1-deoxy-D-xylulose 5-phosphate reductoisomerase in arbuscule-containing cells of maize. *Plant Physiology* **134**, 614–624.

Hansen J *et al.* 1997 Conifer carbohydrate physiology: updating classical views. In *Ecophysiology of Coniferous Forests*. Eds H Rennenberg, W Eschrich, H Ziegler pp. 97–108. Backhuys, Leiden, Germany.

Hardie K 1985 The effect of removal of extraradical hyphae on water uptake by vesicular-arbuscular mycorrhizal plants. *New Phytologist* **101**, 677–684.

Hardie K and Leyton L 1981 The influence of vesicular-arbuscular mycorrhiza on growth and water relations of red clover I. In phosphate deficient soil. *New Phytologist* **89**, 599–608.

Harley JL 1936 Mycorrhiza of *Fagus sylvatica*. DPhil thesis, Oxford University, Oxford, UK.

Harley JL 1964 Incorporation of carbon dioxide into excised beech mycorrhizas in the presence and absence of ammonia. *New Phytologist* **63**, 203–208.

Harley JL 1969 *The Biology of Mycorrhiza*. Leonard Hill, London, UK.

Harley JL 1973 Symbiosis in the ecosystem. *Journal of the Natural Science Council Sri Lanka* **1**, 31–48.

Harley JL 1978 Mycorrhizas as nutrient absorbing organs. *Proceedings of the Royal Society of London Series B* **203**, 1–21.

Harley JL 1985 Specificity and penetration of tissues by mycorrhizal fungi. *Proceedings of the Indian Academy of Sciences (Plant Sciences)* **94**, 99–109.

Harley JL 1991 The history of research on mycorrhiza and the part played by Professor Beniamino Peyronel. In *Fungi, Plants and Soil. A review of forty years' research at the Soil Mycology Centre on the occasion of the centenary of the birth of its founder Beniamino Peyronel*. Ed. A Fontana pp. 31–47. Centro di Studio sulla Micologica del Torreno del CNR, Torino, Italy.

Harley JL and Harley EL 1987a A check-list of mycorrhiza in the British Flora. *New Phytologist* **105**, 1–102.

Harley JL and Harley EL 1987b A check list of mycorrhiza in the British Flora – addenda, errata and index. *New Phytologist* **107**, 741–749.

Harley JL and Jennings DH 1958 The effect of sugars on the respiratory responses of beech mycorrhiza to salts. *Proceedings of the Royal Society of London Series B* **148**, 403–418.

Harley JL and Loughman BC 1963 The uptake of phosphate by excised mycorrhizal roots of the beech. IX. The nature of the phosphate compounds passing to the host. *New Phytologist* **62**, 350–359.

Harley JL and McCready CC 1950 Uptake of phosphate by excised mycorrhizas of beech. I. *New Phytologist* **49**, 388–397.

Harley JL and McCready CC 1952a Uptake of phosphate by excised mycorrhiza of the beech. II. Distribution of phosphate between host and fungus. *New Phytologist* **51**, 56–64.

Harley JL and McCready CC 1952b Uptake of phosphate by excised mycorrhizas of the beech. III. The effect of the fungal sheath on the availability of phosphate to the core. *New Phytologist* **51**, 343–348.

Harley JL and McCready CC 1981 Phosphate accumulation in *Fagus* mycorrhizas. *New Phytologist* **89**, 75–80.

Harley JL and Smith SE 1983 *Mycorrhizal Symbiosis*. Academic Press, London, UK.

Harley JL and Waid JS 1955 A method of studying active mycelia on living roots and other surfaces in the soil. *Transactions of British Mycological Society* **38**, 104–118.

Harley JL and Wilson JM 1959 The absorption of potassium by beech mycorrhizas. *New Phytologist* **58**, 281–298.

Harley JL, McCready CC, Brierley JK 1953 Uptake of phosphate by excised mycorrhizal roots of beech. IV. The effect of oxygen concentrations upon host and fungus. *New Phytologist* **52**, 124–132.

Harley JL, Brierley JK, McCready CC 1954 The uptake of phosphate by excised mycorrhizal roots of the beech. V. The examination of possible sources of misinterpretation of the quantities of phosphorus passing into the host. *New Phytologist* **53**, 92–98.

Harley JL, McCready CC, Brierley JK, Jennings DH 1956 The salt respiration of excised beech mycorrhizas. II. The relationship between oxygen consumption and phosphate absorption. *New Phytologist* **55**, 1–28.

Harley JL, McCready CC, Brierley JK 1958 The uptake of phosphorus by excised mycorrhizal roots of beech. VII. Translocation of phosphorus in mycorrhizal roots. *New Phytologist* **57**, 353–362.

Harold FM 1966 Inorganic polyphosphates in biology: structure, metabolism and function. *Bacteriological Reviews* **30**, 772–794.

Harrier LA, Wright F, Hooker JE 1998 Isolation of the 3-phosphoglycerate kinase gene of the arbuscular mycorrhizal fungus *Glomus mosseae* Nicol. and Gerd. Gerdemann and Trappe. *Current Genetics* **34**, 386–392.

Harrington TC and McNew DL 2003 Phylogenetic analysis places the *Phialophora*-like anamorph genus *Cadophora* in the Helotiales. *Mycotaxon* **87**, 141–151.

Harris D and Paul EA 1987 Carbon requirements of vesicular-arbuscular mycorrhizae. In *Ecophysiology of VA mycorrhizae*. Ed. GR Safir pp. 93–105. CRC Press, Boca Raton, Florida, USA.

Harris KK and Boerner REJ 1990 Effects of belowground grazing by collembola on growth, mycorrhizal infection and P uptake of *Geraneum robertianum*. *Plant and Soil* **129**, 203–210.

Harrison AF 1983 *Soil Organic Phosphorus – a review of World Literature*. CAB International, Wallingford, UK.

Harrison CR 1977 Ultrastructural and histochemical changes during germination of *Cattleya aurantica* (Orchidaceae). *Botanical Gazette* **138**, 41–45.

Harrison MJ 1996 A sugar transporter from *Medicago truncatula*: altered expression pattern in roots during vesicular-arbuscular (VA) mycorrhizal associations. *The Plant Journal* **9**, 491–503.

Harrison MJ 1999 Molecular and cellular aspects of the arbuscular mycorrhizal symbiosis. *Annual Review of Plant Physiology and Plant Molecular Biology* **50**, 361–389.

Harrison MJ 2005 Signaling in the arbuscular mycorrhizal symbiosis. *Annual Review of Microbiology* **59**, 19–42.

Harrison MJ and van Buuren ML 1995 A phosphate transporter from the mycorrhizal fungus *Glomus versiforme*. *Nature* **378**, 626–632.

Harrison MJ and Dixon RA 1993 Isoflavonoid accumulation and expression of defense gene transcripts during the establishment of vesicular-arbuscular mycorrhizal associations in roots of *Medicago truncatula*. *Molecular Plant-Microbe Interactions* **6**, 643–654.

Harrison MJ and Dixon RA 1994 Spatial patterns of expression of flavonoid/isoflavonoid pathway genes during interactions between roots of *Medicago truncatula* and the mycorrhizal fungus *Glomus versiforme*. *The Plant Journal* **6**, 9–20.

Harrison MJ, Dewbre GR, Liu JY 2002 A phosphate transporter from *Medicago truncatula* involved in the acquisition of phosphate released by arbuscular mycorrhizal fungi. *The Plant Cell* **14**, 2413–2429.

Hart MM and Reader RJ 2002a Taxonomic basis for variation in the colonization strategy of arbuscular mycorrhizal fungi. *New Phytologist* **153**, 335–344.

Hart MM and Reader RJ 2002b Does percent root length colonization and soil hyphal length reflect the extent of colonization for all AMF? *Mycorrhiza* **12**, 297–301.

Hart MM and Reader RJ 2002c Host plant benefit from association with arbuscular mycorrhizal fungi: variation due to differences in size of mycelium. *Biology and Fertility of Soils* **36**, 357–366.

Hart TB, Hart JA, Murphy PG 1989 Monodominant and species-rich forests of the humid tropics – causes for their co-occurrence. *American Naturalist* **133**, 613–633.

Hartnett DC and Wilson GWT 1999 Mycorrhizae influence plant community structure and diversity in tall grass prairie. *Ecology* **80**, 1187–1195.

Hartnett DC and Wilson GWT 2002 The role of mycorrhizas in plant community structure and dynamics: lessons from grasslands. *Plant and Soil* **244**, 319–331.

Hartnett DC, Hetrick BAD, Wilson GWT, Gibson DJ 1993 Mycorrhizal influence on intra- and interspecific neighbour interactions among co-occuring prairie grasses. *Journal of Ecology* **81**, 787–795.

Harvais G and Hadley G 1967a The development of *Orchis purpurella* in asymbiotic and inoculated cultures. *New Phytologist* **66**, 217–230.

Harvais G and Hadley G 1967b The relation between host and endophyte in orchid mycorrhiza. *New Phytologist*, **66**, 205–215.

Haselwandter K 1995 Mycorrhizal fungi – siderophore production. *Critical Reviews in Biotechnology* **15**, 287–291.

Haselwandter K and Read DJ 1980 Fungal association of roots of dominant and sub-dominant plants in high-alpine vegetation systems with special reference to mycorrhiza. *Oecologia* **45**, 57–62.

Haselwandter K and Read DJ 1982 The significance of a root-fungus association in two *Carex* species of high-alpine plant communities. *Oecologia* **53**, 352–354.

Haselwandter K, Hoffmann A, Holzmann HP, Read DJ 1983 Availability of nitrogen and phosphorus in the nival zone of the Alps. *Oecologia* **57**, 266–269.

Haselwandter K, Bobleter O, Read DJ 1990 Degradation of ^{14}C-labelled lignin and dehydropolymer of coniferyl alcohol by ericoid and ectomycorrhizal fungi. *Archives of Microbiology* **153**, 352–354.

Haselwandter K, Dobernigg B, Beck W, Jung G, Cansier A, Winkelmann G 1992 Isolation and identification of hydroxamate siderophores of ericoid mycorrhizal fungi. *BioMetals* **5**, 51–56.

Hatch AB 1937 The physical basis of mycotrophy in the genus *Pinus*. *Black Rock Forest Bulletin* **6**, 168.

Hatch AB and Doak KD 1933 Mycorrhizal and other features of the root system of *Pinus*. *Journal of the Arnold Arboretum* **14**, 85–99.

Hattingh MJ 1975 Uptake of ^{32}P-labelled phosphate by endomycorrhizal roots in soil chambers. In *Endomycorrhizas*. Eds FE Sanders, PB Tinker, B Mosse pp. 289–295. Academic Press, London, UK.

Hattingh MJ, Gray LE, Gerdemann JW 1973 Uptake and translocation of ^{32}P-labeled phosphate to onion roots by endomycorrhizal fungi. *Soil Science* **116**, 383–387.

Haug I, Weber R, Oberwinkler F, Tschen J 1991 Tuberculate mycorrhizas of *Castanopsis borneensis* King and *Engelhardtia roxburghiana* Wall. *New Phytologist* **117**, 25–35.

Haugen LM, Smith SE 1992 The effect of high temperature and fallow period on infection of mung bean and cashew roots by the vesicular-arbuscular mycorrhizal fungus *Glomus intraradices*. *Plant and Soil* **145**, 71–80.

Häussling M and Marschner H 1989 Organic and inorganic soil phosphates and acid phosphatase activity in the rhizosphere of 80-year-old Norway spruce *Picea abies* L. Karst. trees. *Biology and Fertility of Soils* **8**, 128–133.

Hawkins HJ, Johansen A, George E 2000 Uptake and transport of organic and inorganic nitrogen by arbuscular mycorrhizal fungi. *Plant and Soil* **226**, 275–285.

Hayman DS 1970 Endogone spore numbers in soil and vesicular-arbuscular mycorrhiza in wheat as influenced by season and soil treatment. *Transactions of the British Mycological Society* **54**, 53–63.

Hayman DS 1974 Plant growth responses to vesicular-arbuscular mycorrhiza. VI. Effect of light and temperature. *New Phytologist* **73**, 71–80.

Hayman DS 1983 The physiology of vesicular-arbuscular endomycorrhizal symbiosis. *Canadian Journal of Botany* **61**, 944–963.

Hayman DS and Mosse B 1971 Plant growth responses to vesicular-arbuscular mycorrhiza. I. Growth of *Endogone*-inoculated plants in phosphate-deficient soils. *New Phytologist* **70**, 19–27.

Hayman DS and Mosse B 1972 Plant growth responses to vesicular-arbuscular mycorrhiza. III. Increased uptake of labile P from soil. *New Phytologist* **71**, 41–47.

Hayman DS, Johnson AM, Ruddlesdin I 1975 The influence of phosphate and crop species on Endogone spores and vesicular-arbuscular mycorrhiza under field conditions. *Plant and Soil* **43**, 489–495.

Haystead A, Malajczuk N, Grove TS 1988 Underground transfer of nitrogen between pasture plants infected with VA mycorrhizal fungi. *New Phytologist* **108**, 417–423.

He XH, Bledsoe C, Zasoski RJ, Southworth D, Horwath WR 2006 Rapid nitrogen transfer from ectomycorrhizal pines to adjacent ectomycorrhizal and arbuscular mycorrhizal plants in a California oak woodland. *New Phytologist* **170**, 143–151.

He-Nygren X, Juslen A, Ahonen I, Glenny D, Piippo S 2006 Illuminating the evolutionary history of liverworts (Marchantiophyta) – towards a natural classification. *Cladistics* **22**, 1–31.

Heal OW, Bailey AD, Latter PM 1967 Bacteria, fungi and protozoa in Signy Island soils compared with those from a temperate moorland. *Philosophical Transactions of the Royal Society* B **252**, 191–197.

Heath MC 1981 A generalized concept of host-parasite specificity. *Phytopathology* **71**, 1121–1123.

Heath MC and Heath IB 1975 Ultrastructural changes associated with the haustorial mother cell during haustorium formation in *Uromyces phaseoli* var. *vignae*. *Protoplasma* **84**, 297–314.

Heckman DS, Geiser DM, Eidell BR, Stauffer RL, Kardos NL, Hedges SB 2001 Molecular evidence for the early colonization of land by fungi and plants. *Science* **293**, 1129–1133.

Hedlund K, Regina IS *et al.* 2003 Plant species diversity, plant biomass and responses of the soil community on abandoned land across Europe: idiosyncracy or above-belowground time lags. *Oikos* **103**, 45–58.

Heinemeyer A, Ridgway KP, Edwards EJ, Benham DG, Young JPW, Fitter AH 2004 Impact of soil warming and shading on colonization and community structure of arbuscular mycorrhizal fungi in roots of a native grassland community. *Global Change Biology* **10**, 52–64.

Heinemeyer A, Ineson P, Ostle N, Fitter A 2006 Respiration of the external mycelium in the arbuscular mycorrhizal symbiosis shows strong dependence on recent photosynthesis and acclimation to temperature. *New Phytologist* **171**, 159–170.

Heinrich PA and Patrick JW 1986 Phosphorus acquisition in the soil-root system of *Eucalyptus pilularis* Smith seedlings. II. The effect of ectomycorrhiza on seedling phosphorus and dry weight acquisition. *Australian Journal of Botany* **34**, 445–454.

Helgason T, Daniell TJ, Husband R, Fitter AH, Young JPW 1998 Ploughing up the wood-wide web? *Nature* **394**, 431.

Helgason T, Fitter AH, Young JPW 1999 Molecular diversity of arbuscular mycorrhizal fungi colonising *Hyacinthoides non-scripta* (bluebell) in a seminatural woodland. *Molecular Ecology* **8**, 659–666.

Helgason T, Merryweather JW, Denison J, Wilson P, Young JPW, Fitter AH 2002 Selectivity and functional diversity in arbuscular mycorrhizas of co-occurring fungi and plants from a temperate deciduous woodland. *Journal of Ecology* **90**, 371–384.

Helgason T, Watson I, Young JPW 2003 Phylogeny of the Glomerales and Diversisporales (Fungi): Glomeromycota from actin and elongation factor 1-alpha sequences. *FEMS Microbiology Letters* **229**, 127–132.

Hendricks JJ, Mitchell RJ, Kuehn KA, Pecot SD, Sims SE 2006 Measuring external mycelia production of ectomycorrhizal fungi in the field: the soil matrix matters. *New Phytologist* **171**, 179–186.

Hendrix JW, Jones KJ, Nesmith WC 1992 Control of pathogenic mycorrhizal fungi in maintenance of soil productivity by crop-rotation. *Journal of Production Agriculture* **5**, 383–386.

Hendrix JW, Guo BZ, An ZQ 1995 Divergence of mycorrhizal fungal communities in crop production systems. *Plant and Soil* **170**, 131–140.

Henkel TW 2003 Monodominance in the ectomycorrhizal *Dicymbe corymbosa* (Caesalpinaceae) from Guyana. *Journal of Tropical Ecology* **19**, 417–437.

Henkel TW, Terborgh J, Vilgalys RJ 2002 Ectomycorrhizal fungi and their leguminous hosts in the Pakaraima Mountains of Guyana. *Mycological Research* **106**, 515–531.

Henkel TW, Mayor JR, Woolley LP 2005 Mast fruiting and seedling survival of the ectomycorrhizal, monodominant *Dicymbe corymbosa* (Caesalpinaceae) in Guyana. *New Phytologist* **167**, 543–556.

Henrion B, Letacon F, Martin F 1992 Rapid identification of genetic-variation of ectomycorrhizal fungi by amplification of ribosomal-RNA genes. *New Phytologist* **122**, 289–298.

Hentschel E, Godbold DL, Marschner P, Schlegel H, Jentschke G 1993 The effect of *Paxillus involutus* Fr on aluminum sensitivity of Norway spruce seedlings. *Tree Physiology* **12**, 379–390.

Hepper CM 1977 A colorimetric method for estimating vesicular-arbuscular mycorrhizal infection in roots. *Soil Biology and Biochemistry* **9**, 15–18.

Hepper CM 1979 Germination and growth of *Glomus caledonius* spores: the effects of inhibitors and nutrients. *Soil Biology and Biochemistry* **11**, 269–277.

Hepper CM 1981 Techniques for studying the infection of plants by vesicular-arbuscular mycorrhizal fungi under axenic conditions. *New Phytologist* **88**, 641–647.

Hepper CM 1983a Effect of phosphate on germination and growth of vesicular-arbuscular mycorrhizal fungi. *Transactions of the British Mycological Society* **80**, 487–490.

Hepper CM 1983b Limited independent growth of a vesicular-arbuscular mycorrhizal fungus *in vitro*. *New Phytologist* **93**, 537–542.

Hepper CM 1984a Inorganic sulphur nutrition of the vesicular-arbuscular mycorrhizal fungus *Glomus caledonium*. *Soil Biology and Biochemistry* **16**, 669–671.

Hepper CM 1984b Isolation and culture of VA mycorrhizal (VAM) fungi. In *VA Mycorrhizae*. Eds CL Powell and DJ Bagjaraj pp. 95–112. CRC Press, Boca Raton, Florida, USA.

Hepper CM and Jakobsen I 1983 Hyphal growth from spores of the mycorrhizal fungus *Glomus caledonius*: effect of amino acids. *Soil Biology and Biochemistry* **15**, 55–58.

Hepper CM and Mosse B 1975 Techniques used to study the interaction between *Endogone* and plant roots. In *Endomycorrhizas*. Eds FE Sanders, B Mosse, PB Tinker pp. 65–75. Academic Press, London, UK.

Hepper CM and Smith GA 1976 Observations on the germination of *Endogone* spores. *Transactions of the British Mycological Society* **66**, 189–194.

Hepper CM and Warner A 1983 Role of organic matter in growth of a vesicular-arbuscular mycorrhizal fungus in soil. *Transactions of the British Mycological Society* **81**, 155–156.

Herold A 1980 Regulation of photosynthesis by sink activity – the missing link. *New Phytologist* **86**, 131–144.

Hesselman H 1990 Om mykorrhizabildingar hos arktiska växter. *Bilhang Till Vetenskakadamie Handlingar* **26**, 1–46.

Hetrick BAD, Wilson GT, Kitt DG, Schwab AP 1988 Effects of soil microorganisms on mycorrhizal contribution to growth of big bluestem grass in non sterile soil. *Soil Biology and Biochemistry* **20**, 501–507.

Hetrick BAD, Wilson GWT, Hartnett DC 1989 Relationship between mycorrhizal dependence and competitive ability of two tallgrass prairie grasses. *Canadian Journal of Botany* **67**, 2608–2615.

Hetrick BAD, Wilson GWT, Todd TC 1990 Differential responses of C_3 and C_4 grasses to mycorrhizal symbiosis, phosphorus fertilization, and soil microorganisms. *Canadian Journal of Botany* **68**, 461–467.

Hetrick BAD, Wilson GWT, Leslie JF 1991 Root architecture of warm- and cool- season grasses: Relationship to mycorrhizal dependence. *Canadian Journal of Botany* **69**, 112–118.

Hetrick BAD, Wilson GWT, Todd TC 1992 Relationships of mycorrhizal symbiosis, rooting strategy, and phenology among tallgrass prairie forbs. *Canadian Journal of Botany* **70**, 1521–1528.

Hetrick BAD, Wilson GWT, Cox TS 1993a Mycorrhizal dependance of modern wheat cultivars and ancestors: a synthesis. *Canadian Journal of Botany* **71**, 512–518.

Hetrick BAD, Wilson GWT, Cox TS 1993b Mycorrhizal dependence of modern wheat varieties and ancestors. *Canadian Journal of Botany* **70**, 2032–2040.

Hetrick BAD, Hartnett DC, Wilson GWT, Gibson DJ 1994a Effects of mycorrhizae, phosphorus availability and plant density on yield relationships among competing tallgrass prairie grasses. *Canadian Journal of Botany* **72**, 168–176.

Hetrick BAD, Wilson GWT, Schwab AP 1994b Mycorrhizal activity in warm- and cool-season grasses: variation in nutrient uptake strategies. *Canadian Journal of Botany* **72**, 1002–1008.

Hetrick BAD, Wilson GWT, Todd TC 1996 Mycorrhizal response in wheat cultivars: relationship to phosphorus. *Canadian Journal of Botany* **74**, 19–25.

Hibbett, DS and Donoghue MJ 1995 Progress toward a phylogenetic classification of the Polyporaceae through parsimony analysis of mitochondrial ribosomal DNA-sequences. *Canadian Journal of Botany* **73**, S853–S861.

Hibbett DS, Gilbert LB, Donoghue MJ 2000 Evolutionary instability of ectomycorrhizal symbioses in basidiomycetes. *Nature* **407**, 506–508.

Hibbett DS, Nilsson M, Snyder M, Fonseca J, Costanzo J, Schonfeld M 2005 Automated phylogenetic taxonomy: an example in the Homobasidiomycetes (mushroom forming fungi). *Systematic Biology* **54**, 660–668.

Hietala AM 1997 The mode of infection of a pathogenic uninucleate *Rhizoctonia* sp. in conifer seedling roots. *Canadian Journal of Forest Research-Revue Canadienne De Recherche Forestiere* **27**, 471–480.

Hietala AM and Sen R 1996 *Rhizoctonia* associates with forest trees. In *Rhizoctonia* spp: *Taxonomy, Molecular Biology, Ecology, Pathology and Disease Control*. Eds R Sneh *et al.* pp. 351–358. Kluwer, Dordrecht, The Netherlands.

Hijri M and Sanders IR 2004 The arbuscular mycorrhizal fungus *Glomus intraradices* is haploid and has a small genome size in the lower limit of eukaryotes. *Fungal Genetics and Biology* **41**, 253–261.

Hijri M and Sanders IR 2005 Low gene copy number shows that arbuscular mycorrhizal fungi inherit genetically different nuclei. *Nature* **433**, 160–163.

Hilbert JL and Martin F 1988 Regulation of gene expression in ectomycorrhizas. I. Protein changes and the presence of ectomycorrhiza specific polypeptides in the *Pisolithus-Eucalyptus* symbiosis. *New Phytologist* **110**, 339–346.

Hilbert JL, Costa G, Martin F 1991 Ectomycorrhizin synthesis and polypeptide changes during the early stage of eucalypt mycorrhiza development. *Plant Physiology* **97**, 977–984.

Hildebrandt U, Schmelzer E, Bothe H 2002 Expression of nitrate transporter genes in tomato colonized by an arbuscular mycorrhizal fungus. *Physiologia Plantarum* **115**, 125–136.

Hildebrandt U, Ouziad F, Marner F-J, Bothe H 2006 The bacterium *Paenibacillus validus* stimulates growth of the arbuscular mycorrhizal fungus *Glomus intraradices* up to the formation of fertile spores. *FEMS Microbiology Letters* **254**, 258–267.

Hingston FJ, Malajczuk N, Grove TS 1982 Acetylene reduction (N_2-fixation) by Jarrah forest legumes following fire and phosphate application. *Journal of Applied Ecology* **19**, 631–646.

Hirrel MC and Gerdemann JW 1979 Enhanced carbon transfer between onions infected with a vesicular-arbuscular mycorrhizal fungus. *New Phytologist* **83**, 731–738.

Ho I and Trappe JM 1973 Translocation of ^{14}C from *Festuca* plants to their endomycorrhizal fungi. *Nature* **244**, 30–31.

Ho I and Trappe JM 1975 Nitrate reducing capacity of two vesicular-arbuscular mycorrhizal fungi. *Mycologia* **67**, 886–888.

Ho I and Trappe JM 1987 Enzymes and growth substances of *Rhizopogon* species in relation to mycorrhizal host and infrageneric taxonomy. *Mycologia* **79**, 553–558.

Hobbie EA and Colpaert JV 2003 Nitrogen availability and colonization by mycorrhizal fungi correlate with nitrogen isotope patterns in plants. *New Phytologist* **157**, 115–126.

Hobbie EA, Macko SA, Shugart HH 1999 Insights into nitrogen and carbon dynamics of ectomycorrhizal and saprotrophic fungi from isotopic evidence. *Oecologia* **118**, 353–360.

Hobbie EA, Weber NS, Trappe JM 2001 Mycorrhizal *vs* saprotrophic status of fungi: the isotopic evidence. *New Phytologist* **150**, 601–610.

Hobbie EA, Tingey DT, Rygiewicz PT, Johnson MG, Olszyk DM 2002 Contributions of current year photosynthate to fine roots estimated using a C-13-depleted CO_2 source. *Plant and Soil* **247**, 233–242.

Hobbie EA, Jumpponen A, Trappe J 2005 Foliar and fungal [15]N:[14]N ratios reflect development of mycorrhizae and nitrogen supply during primary succession: testing analytical models. *Oecologia* **146**, 258–268.

Hodge A 1996 Impact of elevated CO_2 on mycorrhizal associations and implications for plant growth. *Biology and Fertility of Soils* **23**, 388–398.

Hodge A 2001 Arbuscular mycorrhizal fungi influence decomposition of, but not plant nutrient capture from, glycine patches in soil. *New Phytologist* **151**, 725–734.

Hodge A, Campbell CD, Fitter AH 2001 An arbuscular mycorrhizal fungus accelerates decomposition and acquires nitrogen directly from organic material. *Nature* **413**, 297–299.

Hoffland E, Giesler R, Jongmans T, van Breemen N 2002 Increasing feldspar tunneling by fungi across a north Sweden podzol chronosequence. *Ecosystems* **5**, 11–22.

Hoffland E, Giesler R, Jongmans AG, van Breemen N 2003 Feldspar tunneling by fungi along natural productivity gradients. *Ecosystems* **6**, 739–746.

Hoffland E, Kuyper TW, Wallender H 2004 The role of fungi in weathering. *Frontiers in Ecology and the Environment* **2**, 258–264.

Högberg MN and Högberg P 2002 Extramatrical ectomycorrhizal mycelium contributes one-third of microbial biomass and produces, together with associated roots, half the dissolved organic carbon in a forest soil. *New Phytologist* **154**, 791–795.

Högberg MN, Bååth E, Nordgren A, Arnebrant K, Högberg P 2003 Contrasting effects of nitrogen availability on plant carbon supply to mycorrhizal fungi and saprotrophs – a hypothesis based on field observations in boreal forest. *New Phytologist* **160**, 225–238.

Högberg MN, Högberg P, Myrold DD 2007 Is microbial community composition in boreal forest soils determined by pH, C-to-N ratio, the trees or all three? *Oecologia* **150**, 590–601.

Högberg MN, Myrold DD, Giesler R, Högberg P 2006 Contrasting patterns of soil N-cycling in model ecosystems of Fennoscandian boreal forests. *Oecologia* **147**, 96–107.

Högberg P 1982 Mycorrhizal associations in some woodland and forest trees and shrubs in Tanzania. *New Phytologist* **92**, 407–415.

Högberg P 1986 Soil nutrient availability, root symbioses and tree composition in tropical Africa – a review. *Journal of Tropical Ecology* **2**, 359–372.

Högberg P 1989 Root symbioses of trees in savannas. In *Mineral Nutrients in Tropical Forest and Savanna Ecosystems*. Ed. J Proctor pp. 123–136. Blackwell, Oxford, UK.

Högberg P 1992 Root symbioses of trees in African dry tropical forests. *Journal of Vegetation Science* **3**, 393–400.

Högberg P and Alexander IJ 1995 Roles of root symbioses in African woodland and forest: evidence from ^{15}N abundance and foliar analysis. *Journal of Ecology* **83**, 217–224.

Högberg P and Pearce GD 1986 Mycorrhizas in Zambian trees in relation to taxonomy, vegetation communities and successional patterns. *Journal of Ecology* **74**, 775–785.

Högberg P and Read DJ 2006 Towards a more plant physiological perspective on soil ecology. *Trends in Ecology and Evolution* **21**, 544–548.

Högberg P, Nasholm T, Hogbom L, Stahl L 1994 Use of ^{15}N labelling and ^{15}N natural abundance to quantify the role of mycorrhizas in N uptake by plants: importance of seed N and of changes in the ^{15}N labelling of available N. *New Phytologist* **127**, 515–519.

Högberg P, Hogbom L, Schinkel H, Högberg M, Johannisson C, Wallmark H 1996 N-15 abundance of surface soils, roots and mycorrhizas in profiles of European forest soils. *Oecologia* **108**, 207–214.

Högberg P, Högberg MN, Quist ME, Ekblad A, Nasholm T 1999 Nitrogen isotope fractionation during nitrogen uptake by ectomycorrhizal and non-mycorrhizal *Pinus sylvestris*. *New Phytologist* **142**, 569–576.

Högberg P, Nordgren A, Buchmann N *et al.* 2001 Large-scale forest girdling shows that current photosynthesis drives soil respiration. *Nature* **411**, 789–792.

Hohnjec N, Vieweg ME, Puhler A, Becker A, Kuster H 2005 Overlaps in the transcriptional profiles of *Medicago truncatula* roots inoculated with two different *Glomus* fungi provide insights into the genetic program activated during arbuscular mycorrhiza. *Plant Physiology* **137**, 1283–1301.

Holevas CD 1966 The effect of vesicular-arbuscular mycorrhiza on the uptake of soil phosphorus by strawberry *Fragaria* sp. var. Cambridge Favourite. *The Journal of Horticultural Science* **41**, 57–64.

Holley JD, Peterson RL 1979 Development of a vesicular-arbuscular mycorrhiza in bean roots. *Canadian Journal of Botany* **57**, 1960–1978.

Hooker JE, Jaizme-Vega M, Atkinson D 1994 Biocontrol of plant pathogens using arbuscular mycorrhizal fungi. In *Impact of Arbuscular Mycorrhizas on Sustainable Agriculture and Natural Ecosystems*. Eds S Gianinazzi and H Schüepp pp. 191–200. Birkhäuser Verlag, Basel, Switzerland.

Horan DP and Chilvers GA 1990 Chemotropism – the key to ectomycorrhizal formation. *New Phytologist* **116**, 297–301.

Horan DP, Chilvers GA, Lapeyrie FF 1988 Time sequence of the infection process in eucalypt ectomycorrhizas. *New Phytologist* **109**, 451–458.

Horton TR and Bruns TD 1998 Multiple-host fungi are the most frequent and abundant ectomycorrhizal types in a mixed stand of Douglas fir (*Pseudotsuga menziesii*) and bishop pine (*Pinus muricata*). *New Phytologist* **139** 2, 331–339.

Horton JL and Hart SC 1998 Hydraulic lift: a potentially important ecosystem process. *Trends in Ecology and Evolution* **13** 6, 232–235.

Horton TR, Cazares E, Bruns TD 1998 Ectomycorrhizal, vesicular-arbuscular and dark septate fungal colonization of bishop pine (*Pinus muricata*) seedlings in the first 5 months of growth after wildfire. *Mycorrhiza* **8**, 11–18.

Hosny M and Dulieu H 1994 Organisation of the genome of *Scutellospora castanea*, an endomycorrhizal fungus. In *Abstracts of the 4th European Symposium on Mycorrhizas* p. 90. Estacion del Zaidin, Granada, Spain.

Hosny M, Gianinazzi-Pearson V, Dulieu H 1998 Nuclear DNA content of 11 fungal species in Glomales. *Genome*, 422–428.

Howe R, Evans RL, Ketteridge SW 1997 Copper-binding proteins in ectomycorrhizal fungi. *New Phytologist* **135**, 123–131.

Huang RS, Smith WK, Yost RS 1985 Influence of vesicular-arbuscular mycorrhiza on growth, water relations, and leaf orientation in *Leucaena leucocephala* Lam. De Wit. *New Phytologist* **99**, 229–243.

Humpert AJ, Muench EL, Giachini AJ, Castellano MA, Spatafora JW 2001 Molecular phylogenetics of *Ramaria* and related genera: evidence from nuclear large subunit and mitochondrial small subunit rDNA sequences. *Mycologia* **93**, 465–477.

Hung L-L and Molina R 1986 Temperature and time in storage influence the efficacy of selected isolates of fungi in commercially produced ectomycorrhizal inoculum. *Forest Science* **32**, 534–545.

Husband R, Herre EA, Turner SL, Gallery R, Young JPW 2002a Molecular diversity of arbuscular mycorrhizal fungi and patterns of host association over time and space in a tropical forest. *Molecular Ecology* **11**, 2669–2678.

Husband R, Herre EA, Young JPW 2002b Temporal variation in the arbuscular mycorrhizal communities colonising seedlings in a tropical forest. *FEMS Microbiology Ecology* **42**, 131–136.

Hutchings MJ and Bradbury SM 1986 Ecological perspectives on clonal perennial herbs. *BioScience* **36**, 178–182.

Hutchinson L 1990 Studies on the systematics of ectomycorrhizal fungi in axenic culture. II. The enzymatic degradation of selected carbon and nitrogen compounds. *Canadian Journal of Botany* **68**, 1522–1530.

Hutton BJ, Dixon KW, Sivasithamparam K 1994 Ericoid endophytes of Western Australian heaths (*Epacridaceae*). *New Phytologist* **127**, 557–566.

Imaizumi-Anraku H, Takeda N *et al.* 2005 Plastid proteins crucial for symbiotic fungal and bacterial entry into plant roots. *Nature* **433**, 527–531.

Imhof S 1997 Root anatomy and mycotrophy of the achlorophyllous *Voyria tenella* Hook. (Gentianaceae). *Botanica Acta* **110**, 298–305.

Imhof S 1998 Subterranean structures and mycotrophy of the achlorophyllous *Triuris hyalina* (Triuridaceae). *Canadian Journal of Botany* **76**, 2011–2019.

Imhof S 1999a Anatomy and mycotrophy of the achlorophyllous *Afrothismia winkleri*. *New Phytologist* **144**, 533–540.

Imhof S 1999b Root morphology, anatomy and mycotrophy of the achlorophyllous *Voyria aphylla* Jacq. Pers. (Gentianaceae). *Mycorrhiza* **9**, 33–39.

Imhof S 1999c Subterranean structures and mycorrhiza of the achlorophyllous *Burmannia tenella* (Burmanniaceae). *Canadian Journal of Botany* **77**, 637–643.

Imhof S 2001 Subterranean structures and mycotrophy of the achlorophyllous *Dictyostega orobanchoides* (Burmanniaceae). *Revista de Biologia Tropical* **49**, 239–247.

Imhof S 2003 A dorsiventral mycorrhizal root in the achlorophyllous *Sciaphila polygyna* (Triuridaceae). *Mycorrhiza* **13**, 327–332.

Imhof S 2006 Two distinct fungi colonize roots and rhizomes of the myco-heterotrophic *Afrothismia gesnerioides* (Burmanniaceae). *Canadian Journal of Botany* **84** 5, 852–861.

Imhof S and Weber HC 1997 Root anatomy and mycotrophy AM of the achlorophyllous *Voyria truncata* (Standley) Standley and Steyermark (Gentianaceae). *Botanica Acta* **110**, 127–134.

Imhof S and Weber HC 2000 Root structures and mycorrhiza of the achlorophyllous *Voyria obconica* (Gentianaceae). *Symbiosis* **29** 3, 201–211.

Ingestad T, Arveby A, Kähr M 1986 The influence of ectomycorrhiza on nitrogen nutrition and growth of *Pinus sylvestris* seedlings. *Physiologia Plantarum* **68**, 575–582.

Ingleby K, Mason PA, Last FT, Fleming LV 1990 Identification of ectomycorrhizas. Institute of Terrestrial Ecology Research Publication No. 5 HMSO, London, UK.

Ingleby K, Munro RC, Noor M, Mason PA, Clearwater MJ 1998 Ectomycorrhizal populations and growth of *Shorea parvifolia* (Dipterocarpaceae) seedlings regenerating under three different forest canopies following logging. *Forest Ecology and Management* **111**, 171–179.

IPCC 2001 *Climate Change 2001, Intergovernmental Panel on Climate Change Third Assessment Report*. Cambridge University Press, New York, USA.

Isagi Y, Sugimura K, Sumida A, Ito H 1997 How does masting happen and synchronize? *Journal of Theoretical Biology* **187**, 231–239.

Isayenkov S, Fester T, Hause B 2004 Rapid determination of fungal colonization and arbuscule formation in roots of *Medicago truncatula* using real-time RT PCR. *Journal of Plant Physiology* **161**, 1379–1383.

Ivory MH 1980 Ectomycorrhizal fungi of lowland tropical pines in natural forests and exotic plantations. In *Tropical Mycorrhiza Research*. Ed. P Mikola pp. 110–117. Clarendon Press, Oxford, UK.

Izumi H, Anderson IC, Alexander IJ, Killham K, Moore ERB 2006 Endobacteria in some ectomycorrhiza of Scots pine *(Pinus sylvestris)*. *FEMS Microbiological Ecology* **56**, 34–43.

Jackson RM, Walker C, Luff S, McEvoy C 1995 Inoculation and field testing of Sitka spruce and Douglas fir with ectomycorrhizal fungi in the United Kingdom. *Mycorrhiza* **5**, 165–173.

Jackucs E, Bratek Z, Agerer R 1998 *Genea verrucosa* Vitt. + *Quercus* spec. *Descriptions of Ectomycorrhizae* **3**, 19.

Jacob C, Courbot M, Brun A *et al.* 2001 Molecular cloning, characterization and regulation by cadmium of a superoxide dismutase from the ectomycorrhizal fungus *Paxillus involutus*. *European Journal of Biochemistry* **268**, 3223–3232.

Jacob C, Courbot ML, Martin F, Brun A, Chalot M 2004 Transcriptomic responses to cadmium in the ectomycorrhizal fungus *Paxillus involutus*. *FEBS Letters* **576**, 423–427.

Jacobs PE, Peterson RL, Massicotte HB 1989 Altered fungal morphogenesis during early stages of ectomycorrhiza formation in *Eucalyptus pilularis*. *Scanning Microscopy* **3**, 249–255.

Jacquelinet-Jeanmougin S, Gianinazzi-Pearson V 1983 Endomycorrhizas in the Gentianaceae. I. The fungi associated with *Gentiana lutea* L. *New Phytologist* **95**, 663–666.

Jacquelinet-Jeanmougin S, Gianinazzi-Pearson V, Gianinazzi S 1987 Endomycorrhizas in the Gentianaceae. II. Ultrastructural aspects of symbiont relationships in *Gentiana lutea* L. *Symbiosis* **3**, 269–286.

Jakobsen I 1986 Vesicular-arbuscular mycorrhiza in field grown crops. III. Mycorrhizal infection and rates of phosphorus inflow in pea plants. *New Phytologist* **104**, 573–581.

Jakobsen I 1987 Effects of V.A. mycorrhiza on yield and harvest index of field-grown pea. *Plant and Soil* **98**, 407–415.

Jakobsen I 1991 Carbon metabolism in mycorrhiza. In *Methods in Microbiology*. Eds JR Norris, DJ Read, A Varma. Academic Press, London, UK.

Jakobsen I 1995 Transport of phosphorus and carbon in VA mycorrhizas. In *Mycorrhiza, Structure, Function, Molecular Biology and Biotechnology*. Eds A Varma and B Hock pp. 297–324. Springer-Verlag, Berlin, Germany.

Jakobsen I 2004 Hyphal fusion to plant species connections – giant mycelia and community nutrient flow. *New Phytologist* **164**, 4–7.

Jakobsen I and Neilsen NE 1983 Vesicular-arbuscular mycorrhiza in field-grown crops. I. Mycorrhizal infection in cereals and peas at various times and soil depths. *New Phytologist* **93**, 401–413.

Jakobsen I and Rosendahl L 1990 Carbon flow into soil and external hyphae from roots of mycorrhizal cucumber plants. *New Phytologist* **115**, 77–83.

Jakobsen I, Abbott LK, Robson AD 1992a External hyphae of vesicular-arbuscular mycorrhizal fungi associated with *Trifolium subterraneum* L. 1. Spread of hyphae and phosphorus inflow into roots. *New Phytologist* **120**, 371–380.

Jakobsen I, Abbott LK, Robson AD 1992b External hyphae of vesicular-arbuscular mycorrhizal fungi associated with *Trifolium subterraneum* L. 2. Hyphal transport of ^{32}P over defined distances. *New Phytologist* **120**, 509–516.

Jakobsen I, Gazey C, Abbott IK 2001 Phosphate transport by communities of arbuscular mycorrhizal fungi in intact soil cores. *New Phytologist* **149**, 95–103.

Jakobsen I, Smith SE, Smith FA 2002 Function and diversity of arbuscular mycorrhizae in carbon and mineral nutrition. In *Mycorrhizal Ecology*. Eds MGA van der Heijden and IR Sanders pp. 75–92. Springer-Verlag, Berlin, Germany.

Jakobsen I, Chen BD, Munkvold L, Lundsgaard T, Zhu YG 2005a Contrasting phosphate acquisition of mycorrhizal fungi with that of root hairs using the non-mycorrhizal root hairless barley mutant. *Plant Cell and Environment* **28**, 928–938.

Jakobsen I, Leggett ME, Richardson AE 2005b Rhizosphere microorganisms and plant phosphorus uptake. In *Phosphorus: Agriculture and the Environment* pp. 437–494. American Society of Agronomists, Crop Science Society of America, Soil Science Society of America, Madison, USA.

Jalal MAF and Read DJ 1983 The organic-acid composition of Calluna heathland soil with special reference to phyto-toxicity and fungitoxicity.1. Isolation and identification of organic-acids. *Plant and Soil* **70** 2, 257–272.

Jalal MAF, Read DJ, Haslam E 1982 Phenolic composition and its seasonal variation in *Calluna vulgaris*. *Phytochemistry* **21**, 1397–1401.

Jambois A, Dauphin A, Kawano T *et al.* 2005 Competitive antagonism between IAA and indole alkaloid hypaphorine must contribute to regulate ontogenesis. *Physiologia Plantarum* **123**, 120–129.

James TY, Kauf F, Schoch CL *et al.* 2006 Reconstructing the early evolution of fungi using a six-gene phylogeny. *Nature* **443,** 818–822.

Janos DP 1980 Vesicular-arbuscular mycorrhizae affect lowland tropical rain forest plant growth. *Ecology* **61**, 151–162.

Janos DP 1987 VA mycorrhizas in humid tropical ecosystems. In *Ecophysiology of VA Mycorrhizal Plants*. Ed. GR Safir pp. 107–134. CRC Press, Boca Raton, Florida, USA.

Janos DP 1996 Mycorrhizas succession and the rehabilitation of deforested land in the humid tropics. In *Fungi and Environmental Change*. Eds J Frankland, N Magan, G Gadd pp. 129–162. Cambridge University Press, Cambridge, UK.

Janouskova M and Vosatka M 2005 Response to cadmium of *Daucus carota* hairy roots dual cultures with *Glomus intraradices* or *Gigaspora margarita*. *Mycorrhiza* **15**, 217–224.

Janouskova M, Pavlikova D, Macek T, Vosatka M 2005 Arbuscular mycorrhiza decreases cadmium phytoextraction by transgenic tobacco with inserted metallothionein. *Plant and Soil* **272**, 29–40.

Jansa J, Mozafar A, Anken T, Ruh R, Sanders IR, Frossard E 2002a Diversity and structure of AMF communities as affected by tillage in a temperate soil. *Mycorrhiza* **12**, 225–234.

Jansa J, Mozafar A, Banke S, McDonald BA, Frossard E 2002b Intra- and intersporal diversity of ITS rDNA sequences in *Glomus intraradices* assessed by cloning and sequencing, and by SSCP analysis. *Mycological Research* **106**, 670–681.

Jansa J, Mozafar A, Frossard E 2003 Long-distance transport of P and Zn through the hyphae of an arbuscular mycorrhizal fungus in symbiosis with maize. *Agronomie* **23**, 481–488.

Jansa J, Weimken A, Frossard E 2006 The effects of agricultural practices on arbuscular mycorrhizal fungi. In *Function of Soils for Human Societies and the Environment*. Eds E Frossard, W Blum, B Warkentin. Geological Society, London, UK.

Jansa J, Smith FA, Smith SE 2007 Are there benefits of simultaneous root colonization by different arbuscular mycorrhizal fungi? *New Phytologist* online early Doi: I0:1111/j.1469-8137.2007.02294.x

Janse JM 1897 Les endophytes radicaux de quelques plantes Javanaise. *Annales du Jardin Botanique de Buitenzorg* **14**, 53–212.

Jansen AE and Dighton J 1990 Effects of air pollutants on ectomycorrhiza. A review. *Air Pollution Research Report* **30**, 1–58.

Jany JL, Martin F, Garbaye J 2003 Respiration activity of ectomycorrhizas from *Cenococcum geophilum* and *Lactarius* sp. in relation to soil water potential in five beech forests. *Plant and Soil* **255**, 487–494.

Janzen DH 1974 Tropical backwater rivers, animals and mast fruiting by the Dipterocarpaceae. *Biotropica* **6**, 69–103.

Jargeat P, Gay G, Debaud JC, Marmeisse R 2000 Transcription of a nitrate reductase gene isolated from the symbiotic basidiomycete fungus *Hebeloma cylindrosporum* does not require induction by nitrate. *Molecular and General Genetics* **263**, 948–956.

Jargeat P, Rekangalt D, Verner MC *et al.* 2003 Characterisation and expression analysis of a nitrate transporter and nitrite reductase genes, two members of a gene cluster for nitrate assimilation from the symbiotic basidiomycete *Hebeloma cylindrosporum*. *Current Genetics* **43**, 199–205.

Jasper DA, Robson AD, Abbott LK 1979 Phosphorus and the formation of vesicular-arbuscular mycorrhizas. *Soil Biology and Biochemistry* **11**, 501–505.

Jasper DA, Robson AD, Abbott LK 1987 V.A. mycorrhizal fungi in revegetation after soil disturbance by mining. In *Mycorrhizae in the Next Decade*. Eds DM Sylvia, LL Hung, JH Graham pp. 152. Institute of Food and Agricultural Sciences, University of Florida, Gainsville, Florida, USA.

Jasper DA, Robson AD, Abbott LK 1988 Revegetation in an iron-ore mine – nutrient requirements for plant growth and the potential role of vesicular-arbuscular (VA) mycorrhizal fungi. *Australian Journal of Soil Research* **26**, 497–507.

Jasper DA, Abbott LK, Robson AD 1989 Soil disturbance reduces the infectivity of external hyphae of vesicular-arbuscular mycorrhizal fungi. *New Phytologist* **112**, 93–99.

Jasper DA, Abbott LK, Robson AD 1991 The effect soil disturbance on vesicular-arbuscular mycorrhizal fungi in soils from different vegetation types. *New Phytologist* **118**, 471–476.

Jasper DA, Abbott LK, Robson AD 1992 Soil disturbance in native ecosystems – the decline and recovery of infectivity of VA mycorrhizal fungi. In *Mycorrhizas in Ecosystems*. Eds DJ Read, DH Lewis, AH Fitter, IJ Alexander pp. 151–155. CAB International, Wallingford, UK.

Jasper DA, Abbott LK, Robson AD 1993 The survival of infective hyphae of vesicular-arbuscular mycorrhizal fungi in dry soil: an interaction with sporulation. *New Phytologist* **124**, 473–479.

Jastrow JD, Miller RM, Lussenhop J 1998 Contributions of interacting biological mechanisms to soil aggregate stabilization in restored prairie. *Soil Biology and Biochemistry* **30**, 905–916.

Javelle A, Chalot M, Söderström B, Botton B 1999 Ammonium and methylamine transport by the ectomycorrhizal fungus *Paxillus involutus* and ectomycorrhizas. *FEMS Microbiology Ecology* **30**, 355–366.

Javelle A, Rodriguez-Pastrana BR, Jacob C *et al.* 2001 Molecular characterization of two ammonium transporters from the ectomycorrhizal fungus *Hebeloma cylindrosporum*. *FEBS Letters* **505**, 393–398.

Javelle A, Andre B, Marini AM, Chalot M 2003a High-affinity ammonium transporters and nitrogen sensing in mycorrhizas. *Trends in Microbiology* **11**, 53–55.

Javelle A, Morel M, Rodriguez-Pastrana BR *et al.* 2003b Molecular characterization, function and regulation of ammonium transporters (Amt) and ammonium-metabolizing enzymes (GS, NADP-GDH) in the ectomycorrhizal fungus *Hebeloma cylindrosporum*. *Molecular Microbiology* **47**, 411–430.

Javelle A, Chalot M, Brun A, Botton B 2004 Nitrogen transport and metabolism in mycorrhizal fungi and mycorrhizas. In *Plant Surface Microbiology*. Eds A Varma, L Abbott, D Werner, R Hampp pp. 393–429. Springer, Berlin, Germany.

Javot H, Penmetsa RV, Terzaghi N, Cook DR, Harrison MJ 2007 A *Medicago truncatula* phosphate transporter indispensable for the arbuscular mycorrhizal symbiosis. *Proceedings of the National Academy of Sciences of the United States of America* **104**, 1720–1725.

Javot H, Pumplin N, Harrison MJ 2007 Phosphate in the arbuscular mycorrhizal symbiosis: transport properties and regulatory roles. *Plant, Cell and Environment* **30**, 310–322.

Jayachandran K, Schwab AP, Hetrick BAD 1989 Mycorrhizal mediation of phosphorus availability: synthetic iron chelate effects on phosphorus solubilization. *Soil Science Society of America Journal* **53**, 1701–1706.

Jayachandran K, Schwab AP, Hetrick BAD 1992 Mineralization of organic phosphorus by vesicular-arbuscular mycorrhizal fungi. *Soil Biology and Biochemistry* **24**, 897–903.

Jeffries P and Dodd JC 1991 The use of mycorrhizal inoculants in forestry and agriculture. In *Handbook of Applied Mycology Volume 1: Soil and Plants*. Eds DK Arora, B Rai, KG Mukerji, GR Knudsen pp. 155–185. Marcel Dekker, New York, USA.

Jennings DH 1995. *The Physiology of Fungal Nutrition*. Cambridge University Press, Cambridge, UK.

Jentschke G, Fritz E, Godbold DL 1991 Distribution of lead in mycorrhizal and nonmycorrhizal Norway spruce seedlings. *Physiologia Plantarum* **81**, 417–422.

Jentschke G, Marschner P, Vodnik D *et al.* 1998 Lead uptake by *Picea abies* seedlings: effects of nitrogen source and mycorrhizas. *Journal of Plant Physiology* **153**, 97–104.

Jentschke G, Winter S, Godbold DL 1999 Ectomycorrhizas and cadmium toxicity in Norway spruce seedlings. *Tree Physiology* **19**, 23–30.

Jentschke G, Brandes B, Kuhn AJ, Schroder WH, Becker JS, Godbold DL 2000 The mycorrhizal fungus *Paxillus involutus* transports magnesium to Norway spruce seedlings. Evidence from stable isotope labeling. *Plant and Soil* **220**, 243–246.

Jentschke G, Brandes B, Kuhn AJ, Schroder WH, Godbold DL 2001a Interdependence of phosphorus, nitrogen, potassium and magnesium translocation by the ectomycorrhizal fungus *Paxillus involutus*. *New Phytologist* **149**, 327–337.

Jentschke G, Godbold DL, Brandes B 2001b Nitrogen limitation in mycorrhizal Norway spruce (*Picea abies*) seedlings induced mycelial foraging for ammonium: implications for Ca and Mg uptake. *Plant and Soil* **234**, 109–117.

Jifon JL, Graham JH, Drouillard DL, Syvertsen JP 2002 Growth depression of mycorrhizal Citrus seedlings grown at high phosphorus supply is mitigated by elevated CO_2. *New Phytologist* **153**, 133–142.

Jin H, Pfeffer P, Douds D, Piotrowski E, Lammers P, Shachar Hill Y 2005 The uptake, metabolism, transport and transfer of nitrogen in an arbuscular mycorrhizal symbiosis. *New Phytologist* **168**, 687–696.

Johansen A 1999 Depletion of soil mineral N by roots of *Cucumis sativus* L. colonized or not by arbuscular mycorrhizal fungi. *Plant and Soil* **209**, 119–127.

Johansen A and Jensen ES 1996 Transfer of N and P from intact or decomposing roots of pea to barley interconnected by an arbuscular mycorrhizal fungus. *Soil Biology and Biochemistry* **28**, 73–81.

Johansen A, Jakobsen I, Jensen ES 1992 Hyphal transport of ^{15}N-labelled nitrogen by a vesicular-arbuscular mycorrhizal fungus and its effect on depletion of inorganic soil N. *New Phytologist* **122**, 281–288.

Johansen A, Jakobsen I, Jensen ES 1993a External hyphae of vesicular-arbuscular mycorrhizal fungi associated with *Trifolium subterraneum* L.3. Hyphal transport of ^{32}P and [15]N. *New Phytologist* **124**, 61–68.

Johansen A, Jakobsen I, Jensen ES 1993b Hyphal transport by a vesicular-arbuscular mycorrhizal fungus of N applied to the soil as ammonium or nitrate. *Biology and Fertility of Soils* **16**, 66–70.

Johansen A, Jakobsen I, Jensen ES 1994 Hyphal N transport by a vesicular-arbuscular mycorrhizal fungus associated with cucumber grown at three nitrogen levels. *Plant and Soil* **160**, 1–9.

Johansen A, Finlay RD, Olsson PA 1996 Nitrogen metabolism of external hyphae of the arbuscular mycorrhizal fungus *Glomus intraradices*. *New Phytologist* **133**, 705–712.

Johansson T, Le Quere A, Ahren D *et al.* 2004 Transcriptional responses of *Paxillus involutus* and *Betula pendula* during formation of ectomycorrhizal root tissue. *Molecular Plant-Microbe Interactions* **17**, 202–215.

Johnson CN 1996 Interactions between mammals and ectomycorrhizal fungi. *Trends in Ecology and Evolution* **11**, 503–507.

Johnson D, Leake JR, Read DJ 2001 Novel in-growth core system enables functional studies of grassland mycorrhizal mycelial networks. *New Phytologist* **152**, 555–562.

Johnson D, Leake JR, Ostle N, Ineson P, Read DJ 2002a *In situ* [13]CO_2 pulse-labelling of upland grassland demonstrates a rapid pathway of carbon flux from arbuscular mycorrhizal mycelia to the soil. *New Phytologist* **153**, 327–334.

Johnson D, Leake JR, Read DJ 2002b Transfer of recent photosynthate into mycorrhizal mycelium of an upland grassland: short-term respiratory losses and accumulation of C-14. *Soil Biology and Biochemistry* **34**, 1521–1524.

Johnson D, Vandenkoornhuyse PJ, Leake JR *et al.* 2004 Plant communities affect arbuscular mycorrhizal fungal diversity and community composition in grassland microcosms. *New Phytologist* **161**, 503–515.

Johnson D, Ijdo M, Genney DR, Anderson IC, Alexander IJ 2005 How do plants regulate the function, community structure, and diversity of mycorrhizal fungi? *Journal of Experimental Botany* **56**, 1751–1760.

Johnson D, Krsek M, Wellington EMH *et al.* 2005 Soil invertebrates disrupt carbon flow through fungal networks. *Science* **309**, 1047.

Johnson NC 1993 Can fertilization of soil select less mutualistic mycorrhizae? *Ecological Applications* **3**, 749–757.

Johnson NC 1996 Interactions between mammals and ectomycorrhizal fungi. *Trends in Ecology and Evolution* **11** 12, 503–507.

Johnson NC, Zak DR, Tilman D, Pfleger FL 1991 Dynamics of vesicular-arbuscular mycorrhizae during old field succession. *Oecologia* **86**, 349–358.

Johnson NC, Tilman D, Wedin D 1992 Plant and soil controls on mycorrhizal fungal communities. *Ecology* **73**, 2034–2042.

Johnson NC, Graham JH, Smith FA 1997 Functioning of mycorrhizal associations along the mutualism-parasitism continuum. *New Phytologist* **135**, 575–586.

Jolicoeur M, Germette S, Gaudette M, Perrier M, Bécard G 1998 Intracellular pH in arbuscular mycorrhizal fungi – a symbiotic physiological marker. *Plant Physiology* **116**, 1279–1288.

Joner EJ 1994 Arbuscular mycorrhiza and plant utilization of organic phosphorus in soil. Doctor Scientiarum thesis, Agricultural University of Norway.

Joner EJ and Jakobsen I 1994 Contribution by two arbuscular mycorrhizal fungi to P uptake by cucumber *Cucumis sativus* L. from ^{32}P-labelled organic matter during mineralization in soil. *Plant and Soil* **163**, 203–209.

Joner EJ and Johansen A 2000 Phosphatase activity of external hyphae of two arbuscular mycorrhizal fungi. *Mycological Research* **104**, 81–86.

Joner EJ and Leyval C 1997 Uptake of ^{109}Cd by roots and hyphae of a *Glomus mosseae/Trifolium subterraneum* mycorrhiza from soil amended with high and low concentrations of cadmium. *New Phytologist* **135**, 353–360.

Joner EJ and Leyval C 2001 Time-course of heavy metal uptake in maize and clover as affected by root density and different mycorrhizal inoculation regimes. *Biology and Fertility of Soils* **33**, 351–357.

Joner EJ, Briones R, Leyval C 2000 Metal-binding capacity of arbuscular mycorrhizal mycelium. *Plant and Soil* **226**, 227–234.

Jones DL, Hodge A, Kuzyakov Y 2004 Plant and mycorrhizal regulation of rhizodeposition. *New Phytologist* **163**, 459–480.

Jones K and Hendrix JW 1987 Inhibition of root extension in tobacco by the mycorrhizal fungus *Glomus macrocarpus* and its prevention by benomyl. *Soil Biology and Biochemistry* **19**, 297–299.

Jones MD and Hutchinson TC 1986 The effect of mycorrhizal infection on the response of *Betula papyrifera* to nickel and copper. *New Phytologist* **102**, 429–442.

Jones MD and Hutchinson TC 1988a Nickel toxicity in mycorrhizal birch seedlings infected with *Lactarius rufus* or *Scleroderma flavidum*. Effects on growth, photosynthesis, respiration and transpiration. *New Phytologist* **108**, 451–459.

Jones MD and Hutchinson TC 1988b Nickel toxicity in mycorrhizal birch seedlings infected with *Lactarius rufus* or *Scleroderma flavidum*. 2 Uptake of nickel, calcium, magnesium, phosphorus and iron. *New Phytologist* **108**, 461–470.

Jones MD and Smith SE 2004 Exploring functional definitions of mycorrhizas: are mycorrhizas always mutualisms? *Canadian Journal of Botany* **82**, 1089–1109.

Jones MD, Durall DM, Tinker PB 1990 Phosphorus relationships and production of extramatrical hyphae by two types of willow ectomycorrhizas at different soil phosphorus levels. *New Phytologist* **115**, 259–267.

Jones MD, Durall DM, Tinker PB 1991 Fluxes of carbon and phosphorus between symbionts in willow ectomycorrhizas and their changes with time. *New Phytologist* **119**, 99–106.

Jones MD, Durall DM, Tinker PB 1998 A comparison of arbuscular and ectomycorrhizal *Eucalyptus coccifera*: growth response, phosphorus uptake efficiency and external hyphal production. *New Phytologist* **140**, 125–134.

Jones MD, Durall DM, Cairney JWG 2003 Ectomycorrhizal fungal communities in young forest stands regenerating after clearcut logging. *New Phytologist* **157**, 399–422.

Jongmans AG, van Breemen N, Lundstrom U *et al.* 1997 Rock-eating fungi. *Nature* **389**, 682–683.

Jonsson LM, Nilsson MC, Wardle DA, Zackrisson O 2001 Context dependent effects of ectomycorrhizal species richness on seedling productivity. *Oikos* **93**, 353–364.

Jonsson U, Rosengren U, Thelin G, Nihlgard B 2003 Acidification-induced chemical changes in coniferous forest soils in southern Sweden 1988–1999. *Environmental Pollution* **123**, 75–83.

Journet EP, Carreau V, Gouzy J *et al.* 2001 The model legume *Medicago truncatula*: recent advances and perspectives in genomics. *OCL-Oleagineux Corps Gras Lipides* **8**, 478–484.

Journet EP, van Tuinen D, Gouzy J *et al.* 2002 Exploring root symbiotic programs in the model legume *Medicago truncatula* using EST analysis. *Nucleic Acids Research* **30**, 5579–5592.

Judson OP and Normark BB 1996 Ancient asexual scandals. *Trends in Ecology and Evolution* **11**, 41–46.

Julou T, Burghardt B, Gebauer G, Berveiller D, Damesin C, Selosse MA 2005 Mixotrophy in orchids: insights from a comparative study of green individuals and nonphotosynthetic individuals of *Cephalanthera damasonium*. *New Phytologist* **166**, 639–653.

Jumpponen A 2001 Dark septate endophytes – are they mycorrhizal? *Mycorrhiza* **11** 4, 207–211.

Jumpponen A and Trappe JM 1998 Dark septate endophytes: a review of facultative biotrophic root-colonizing fungi. *New Phytologist* **140** 2, 295–310.

Jumpponen A, Mattson KG, Trappe JM 1998 Mycorrhizal functioning of *Phialocephala fortinii* with *Pinus contorta* on glacier forefront soil: interactions with soil nitrogen and organic matter. *Mycorrhiza* **7**, 261–265.

Jun J, Abubaker J, Rehrer C, Pfeffer PE, Shachar-Hill Y, Lammers PJ 2002 Expression in an arbuscular mycorrhizal fungus of genes putatively involved in metabolism, transport, the cytoskeleton and the cell cycle. *Plant and Soil* **244**, 141–148.

Jungk A, Asher CJ, Edwards DG, Meyer D 1990 Influence of phosphate status on phosphate uptake kinetics of maize *Zea mays* and soybean *Glycine max*. *Plant and Soil* **124**, 175–182.

Juniper S and Abbott LK 1993 Vesicular-arbuscular mycorrhizas and soil salinity. *Mycorrhiza* **4**, 45–57.

Juniper S and Abbott LK 2004 A change in the concentration of NaCl in soil alters the rate of hyphal extension of some arbuscular mycorrhizal fungi. *Canadian Journal of Botany* **82**, 1235–1242.

Kaiser PA and Lussenhop J 1991 Collembolan effects on establishment of vesicular-arbuscular mycorrhizae in soybean *Glycine max*. *Soil Biology and Biochemistry* **23**, 307–308.

Kaldorf M, Zimmer W, Bothe H 1994 Genetic evidence for the occurrence of assimilatory nitrate reductase in arbuscular mycorrhizal and other fungi. *Mycorrhiza* **5**, 23–28.

Kaldorf M, Schmelzer E, Bothe H 1998 Expression of maize and fungal nitrate reductase genes in arbuscular mycorrhiza. *Molecular Plant-Microbe Interactions* **11**, 439–448.

Kaldorf M, Renker C, Fladung M, Buscot F 2004 Characterization and spatial distribution of ectomycorrhizas colonizing aspen clones released in an experimental field. *Mycorrhiza* **14**, 295–306.

Kamienski F 1881 Die Vegetationsorgane der *Monotropa hypopitys* L. *Botanische Zeitung* **29**, 458.

Kamienski F 1882 Les organs végétatifs du *Monotropa hypopitys* L. *Mémoires de la Sociéte national des Sciences Naturelles et Mathématiques de Cherbourg* **26**, 1–40.

Kapulnik Y and Kushnir U 1992 Growth dependency of wild primitive and modern cultivated wheat lines on vesicular-arbuscular mycorrhizal fungi. *Euphytica* **56**, 27–36.

Kapulnik Y, Volpin H *et al.* 1996 Suppression of defence responses in mycorrhizal alfalfa and tobacco roots. *New Phytologist* **133**, 59–64.

Karagiannidis N and Hadjisavva-Zinoviadi S 1998 The mycorrhizal fungus *Glomus mosseae* enhances growth, yield and chemical composition of a durum wheat variety in 10 different soils. *Nutrient Cycling in Agroecosystem* **52**, 1–7.

Karandashov V, Bucher M 2005 Symbiotic phosphate transport in arbuscular mycorrhizas. *Trends in Plant Science* **10**, 22–29.

Karandashov V, Nagy R, Wegmüller S, Amrhein N, Bucher M 2004 Evolutionary conservation of a phosphate transporter in the arbuscular mycorrhizal symbiosis. *Proceedings of the National Academy of Sciences of the United States of America* **101**, 6285–6290.

Karen O, Högberg N, Dahlberg A, Jonsson L, Nylund JE 1997 Inter- and intraspecific variation in the ITS region of rDNA of ectomycorrhizal fungi in Fennoscandia as detected by endonuclease analysis. *New Phytologist* **136**, 313–325.

Kasuya MCM, Masaka K, Igarashi T 1995 Mycorrhizae of *Monotropastrum globosum* growing in a *Fagus crenata* forest. *Mycoscience* **36**, 461–464.

Keller G 1996 Utilization of inorganic and organic nitrogen sources by high-subalpine ectomycorrhizal fungi of *Pinus cembra* in pure culture. *Mycological Research* **100**, 989–998.

Kelly D and Sork VL 2002 Mast seeding in perennial plants: Why, how, where? *Annual Review of Ecology and Systematics* **33**, 427–447.

Kenrick P and Crane PR 1997 The origin and early evolution of plants on land. *Nature* **389**, 33–39.

Kerley SJ and Read DJ 1995 The biology of mycorrhiza in the Ericaceae 18. Chitin degradation by *Hymenoscyphus ericae* and transfer of chitin-nitrogen to the host plant. *New Phytologist* **131**, 369–375.

Kerley SJ and Read DJ 1997 The biology of mycorrhiza in the Ericaceae 19. Fungal mycelium as a nitrogen source for the ericoid mycorrhizal fungus *Hymenoscyphus ericae* and its host plants. *New Phytologist* **136**, 691–701.

Kerley SJ and Read DJ 1998 The biology of mycorrhiza in the Ericaceae 20. Plant and mycorrhizal necromass as nitrogenous substrates for the ericoid mycorrhizal fungus *Hymenoscyphus ericae* and its host. *New Phytologist* **139**, 353–360.

Kernan MJ and Finocchio AF 1983 A new Discomycete associated with the roots of *Monotropa uniflora* Ericaceae. *Mycologia* **75**, 916–920.

Kerner A 1894 *The Natural History of Plants*, vol. 1. English edn Translated by FW Oliver. Blackie and Sons, London, UK.

Khan AG, Belik M 1995 Occurrence and ecological significance of mycorrhizal symbiosis in aquatic plants. In *Mycorrhiza, Structure, Function, Molecular Biology and Biotechnology*. Eds A Varma and B Hock pp. 627–666. Springer-Verlag, Berlin, Germany.

Kidston R, Lang WH 1921 On the old red sandstone plants showing structure from the Rhynie chert bed, Aberdeenshire. Part V. The thallophyta occurring in the peat bed; the succession of the plants through a vertical section of the bed, and the conditions of accumulation and preservation of the deposit. *Transactions of the Royal Society of Edinburgh* **52**, 855–902.

Kielland K 1994 Amino acid absorption by arctic plants: implications for plant nutrition and nitrogen cycling. *Ecology* **75**, 2373–2383.

Kielland K 1997 Role of free amino acids in the nitrogen economy of arctic cryptogams. *Ecoscience* **4** 1, 75–79.

Kiers ET and van der Heijden MGA 2006 Mutualistic stability in the arbuscular mycorrhizal symbiosis: exploring hypotheses of evolutionary cooperation. *Ecology* **87**, 1627–1636.

Killham K and Firestone MK 1983 Vesicular-arbuscular mycorrhizal mediation of grass response to acidic and heavy-metal depositions. *Plant and Soil* **72**, 39–48.

Kim HJ and Ko MK 1995 Mycelial running method for *Armillaria mushroom* logs in *Gastrodia elata* cultivation. *Forest Research Institute Korea Journal of Forest Science* **51**, 89–95.

Kinden DA and Brown MF 1975a Electron microscopy of vesicular-arbuscular mycorrhizae of yellow poplar. I. Characterization of endophytic structures by scanning electron stereoscopy. *Canadian Journal of Microbiology* **21**, 989–993.

Kinden DA and Brown MF 1975b Electron microscopy of vesicular-arbuscular mycorrhizae of yellow poplar. II. Intracellular hyphae and vesicles. *Canadian Journal of Microbiology* **21**, 1768–1780.

Kishino H and Hasegawa M 1989 Evaluation of the maximum likelihood estimate of the evolutionary tree topologies from DNA sequence data, and the branching order in Hominoidea. *Journal of Molecular Evolution* **29**, 170–179.

Kistner C and Parniske M 2002 Evolution of signal transduction in intracellular symbiosis. *Trends in Plant Science* **7**, 511–518.

Kistner C, Winzer T *et al.* 2005 Seven *Lotus japonicus* genes required for transcriptional reprogramming of the root during fungal and bacterial symbiosis. *Plant Cell* **17**, 2217–2229.

Kjøller R 2006 Disproportionate abundance between ectomycorrhizal root tips and their associated mycelia. *FEMS Microbiology and Ecology* **58**, 214–224.

Kjøller R and Bruns TD 2003 Rhizopogon spore bank communities within and among California pine forests. *Mycologia* **95** 4, 603–613.

Kling M, Gianinazzi-Pearson V, Lherminier J, Jakobsen I 1996 The development and functioning of mycorrhizas in pea mutants. In First International Conference on Mycorrhizas ICOM 1. Berkeley, California p. 71.

Klironomos JN 2003 Variation in plant response to native and exotic arbuscular mycorrhizal fungi. *Ecology* **84**, 2292–2301.

Klironomos JN and Hart MM 2001 Food-web dynamics – animal nitrogen swap for plant carbon. *Nature* **410** 6829, 651–652.

Klironomos JN and Hart MM 2002 Colonization of roots by arbuscular mycorrhizal fungi using different sources of inoculum. *Mycorrhiza* **12**, 181–184.

Klironomos JN and Ursic M 1998 Density-dependent grazing on the extraradical hyphal network of the arbuscular mycorrhizal fungus, *Glomus intraradices*, by the collembolan, *Folsomia candida*. *Biology and Fertility of Soils* **26**, 250–253.

Klironomos JN, Allen MF, Rillig MC *et al.* 2005 Abrupt rise in atmospheric CO_2 overestimates community response in a model plant-soil system. *Nature* **433**, 621–624.

Knöbel M and Weber HC 1988 Vergleichende Untersuchungen zur Mycotrophie bei *Gentiana verna* L. und *Voyria truncata* (Stand.) Stand. and Stey. (Gentianaceae). *Beitrage Biologie der Pflanzen* **63**, 463–477.

Knudson L 1929 Seed germination and growth of *Calluna vulgaris*. *New Phytologist* **32**, 127–155.

Knudson L 1930 Flower production by orchid grown non-symbiotically. *Botanical Gazette* **89**, 192–199.

Koch AM, Kuhn G, Fontanillas P, Fumagalli L, Goudet J, Sanders IR 2004 High genetic variability and low local diversity in a population of arbuscular mycorrhizal fungi. *Proceedings of the National Academy of Sciences of the United States of America* **101**, 2369–2374.

Koch KE and Johnson CR 1984 Photosynthate partitioning in split-root citrus seedlings with mycorrhizal and nonmycorrhizal root systems. *Plant Physiology* **75**, 26–30.

Koch R 1912 *Complete Works*, col. I, pp. 650–660. George Thieme, Leipzig, Germany.

Kohn LM and Stasovski E 1990 The mycorrhizal status of plants at Alexandra Fiord, Ellesmere island. Canada, a high arctic site. *Mycologia* **82**, 23–35.

Kohzu A, Yoshioka T, Ando T, Takahashi M, Koba K, Wada E 1999 Natural ^{13}C and ^{15}N abundance of field-collected fungi and their ecological implications. *New Phytologist* **144**, 323–330.

Koide R 1985a The nature of growth depressions in sunflower caused by vesicular-arbuscular mycorrhizal infection. *New Phytologist* **99**, 449–462.

Koide R 1985b The effect of VA mycorrhizal infection and phosphorus status on sunflower hydraulic and stomata properties. *Journal of Experimental Botany* **36**, 1087–1098.

Koide RT 1991a Nutrient supply, nutrient demand and plant response to mycorrhizal infection. *New Phytologist* **117**, 365–386.

Koide RT 1991b Density-dependent response to mycorrhizal infection in *Abutilon theophrasti* Medic. *Oecologia* **85**, 389–395.

Koide RT 1993 Physiology of the mycorrhizal plant. In *Advances in Plant Pathology*. Eds DS Ingram and PH Williams pp. 33–54. Academic Press, London, UK.

Koide RT 2000 Functional complementarity in the arbuscular mycorrhizal symbiosis. *New Phytologist* **147**, 233–235.

Koide RT and Elliott G 1989 Cost, benefit and efficiency of vesicular-arbuscular mycorrhizal symbiosis. *Functional Ecology* **3**, 4–7.

Koide RT and Kabir Z 2000 Extraradical hyphae of the mycorrhizal fungus *Glomus intraradices* can hydrolyse organic phosphate. *New Phytologist* **148**, 511–517.

Koide RT and Lu X 1992 Mycorrhizal infection of wild oats: maternal effects on offspring growth and reproduction. *Oecologia* **90**, 218–226.

Koide R and Mooney HA 1987 Spatial variation in inoculum potential of vesicular-arbuscular mycorrhizal fungi caused by formation of gopher mounds. *New Phytologist* **107**, 173–182.

Koide RT and Mosse B 2004 A history of research on arbuscular mycorrhiza. *Mycorrhiza* **14**, 145–163.

Koide RT and Schreiner RP 1992 Regulation of the vesicular-arbuscular mycorrhizal symbiosis. *Annual Review of Plant Physiology and Plant Molecular Biology* **43**, 557–581.

Koide RT and Wu T 2003 Ectomycorrhizas and retarded decomposition in a *Pinus resinosa* plantation. *New Phytologist* **158**, 401–407.

Koide RT, Li M, Lewis J, Irby C 1988a Role of mycorrhizal infection in the growth and reproduction of wild vs cultivated plants. 1. Wild vs. cultivated oats. *Oecologia* **77**, 537–542.

Koide R, Huenneke LF, Hamberg SP, Mooney HA 1988b Effects of applications of fungicide, phosphorus and nitrogen on the structure and productivity of an annual serpentine plant community. *Functional Ecology* **2**, 335–344.

Koide RT, Dickie IA, Goff MD 1999 Phosphorus deficiency, plant growth and the phosphorus efficiency index. *Functional Ecology* **13**, 733–736.

Koide RT, Goff MD, Dickie IA 2000 Component growth efficiencies of mycorrhizal and nonmycorrhizal plants. *New Phytologist* **148**, 163–168.

Koide RT, Xu B, Sharda J, Lekberg Y, Ostiguy N 2005 Evidence of species interactions within an ectomycorrhizal fungal community. *New Phytologist* **165**, 305–316.

Koljalg U 1995 *Tomentella* (Basiodiomycota) and related genera in temperate Eurasia. *Synopsis Fungorum 9*. Fungiflora, Oslo, Norway.

Koljalg, U, Dahlberg A, Taylor AFS *et al.* 2000 Diversity and abundance of resupinate thelephoroid fungi as ectomycorrhizal symbionts in Swedish boreal forests. *Molecular Ecology* **9**, 1985–1996.

Koljalg U, Larsson KH, Abarenkov K *et al.* 2005 UNITE: a database providing web-based methods for the molecular identification of ectomycorrhizal fungi. *New Phytologist* **166**, 1063–1068.

Kope HH, Tsantrizos YS, Fortin JA, Ogilvie KK 1991 *p*-Hydroxybenzoylformic acid and (R)-(-)-*p*-hydroxymandelic acid, two antifungal compounds isolated from the liquid culture of the ectomycorrhizal fungus *Pisolithus arhizus*. *Canadian Journal of Microbiology* **37**, 258–264.

Koske RE 1975 Endogone spores in Australian sand dunes. *Canadian Journal of Botany* **53**, 668–672.

Koske RE 1981 Multiple germination by spores of *Gigaspora gigantea*. *Transactions of the British Mycological Society* **76**, 328–330.

Koske RE and Gemma JN 1990 VA mycorrhizae in strand vegetation of Hawaii: evidence for long-distance codispersal of plants and fungi. *American Journal of Botany* **77**, 466–474.

Koske R, Gemma JN 1992 Fungal reactions to plants prior to mycorrhizal formation. In *Mycorrhizal Functioning: An Integrative Plant-Fungal Process*. Ed. MF Allen pp. 3–36. Chapman and Hall, London, UK.

Koske RE, Sutton JC, Sheppard BR 1975 Ecology of *Endogone* in Lake Huron sand dunes. *Canadian Journal of Botany* **53**, 87–93.

Kosuta S, Chabaud M, Lougnon G *et al.* 2003 A diffusible factor from arbuscular mycorrhizal fungi induces symbiosis-specific MtENOD11 expression in roots of *Medicago truncatula*. *Plant Physiology* **131**, 952–962.

Kothari SK, Marschner H, George E 1990 Effect of VA mycorrhizal fungi and rhizosphere microorganisms on root and shoot morphology, growth and water relations in maize. *New Phytologist* **116**, 303–311.

Kothari SK, Marschner H, Römheld V 1991 Contribution of the VA mycorrhizal hyphae in acquisition of phosphorus and zinc by maize grown in a calcareous soil. *Plant and Soil* **131**, 177–185.

Kottke I and Nebel M 2005 The evolution of mycorrhiza-like associations in liverworts: an update. *New Phytologist* **167**, 330–334.

Kottke I and Oberwinkler F 1987 The cellular structure of the Hartig net: coenocytic and transfer cell-like organisation. *Nordic Journal of Botany* **7**, 85–95.

Kottke I, Beiter A, Weiss M, Haug I, Oberwinkler F, Nebel M 2003 Heterobasidiomycetes form symbiotic associations with hepatics: Jungermanniales have sebacinoid mycobionts while *Aneura pinguis* (Metzgeriales) is associated with a Tulasnella species. *Mycological Research* **107**, 957–968.

Kramer PJ and Wilbur KM 1949 Absorption of radioactive phosphorus by mycorrhizal roots of pine. *Science* **110**, 8–9.

Kretzer AM and Bruns TD 1999 Use of atp6 in fungal phylogenetics: an example from the Boletales. *Molecular Phylogenetics and Evolution* **13**, 483–492.

Kretzer A, Li YN, Szaro T, Bruns TD 1996 Internal transcribed spacer sequences from 38 recognized species of *Suillus sensu* lato: phylogenetic and taxonomic implications. *Mycologia* **88**, 776–785.

Kretzer A, Li Y, Szaro T, Bruns TD 1998 Internal transcribed spacer sequences from 38 recognized species of Suillus sensu lato: Phylogenetic and taxonomic implications. *Mycologia* **90**, 464–464.

Kretzer AM, Bidartondo MI, Grubisha LC, Spatafora JW, Szaro TM, Bruns TD 2000 Regional specialization of *Sarcodes sanguinea* (Ericaceae) on a single fungal symbiont from the *Rhizopogon ellenae* (Rhizopogonaceae) species complex. *American Journal of Botany* **87**, 1778–1782.

Kretzer AM, Dunham S, Molina R, Spatafora JW 2004 Microsatellite markers reveal the below ground distribution of genets in two species of *Rhizopogon* forming tuberculate ectomycorrhizas on Douglas fir. *New Phytologist* **161**, 313–320.

Kristiansen KA, Taylor DL, Kjøller R, Rasmussen HN, Rosendahl S 2001 Identification of mycorrhizal fungi from single pelotons of *Dactylorhiza majalis* (Orchidaceae) using single-strand conformation polymorphism and mitochondrial ribosomal large subunit DNA sequences. *Molecular Ecology* **10**, 2089–2093.

Kristiansen KA, Freudenstein JV, Rasmussen FN, Rasmussen HN 2004 Molecular identification of mycorrhizal fungi in *Neuwiedia veratrifolia* (Orchidaceae). *Molecular Phylogenetics and Evolution* **33**, 251–258.

Kron KA, Judd WS, Stevens PF *et al.* 2002 Phylogenetic classification of Ericaceae: molecular and morphological evidence. *Botanical Review* **68**, 335–423.

Kropacek K, Budlin P, Majstrik V 1989 The use of granulated ectomycorrhizal inoculum for reforestation of deteriorated regions. *Agriculture, Ecosystems and Environment* **28**, 263–269.

Kropp BR 1990 Variation in acid phosphatase activity among progeny from controlled crosses in the ectomycorrhizal fungus *Laccaria bicolor*. *Canadian Journal of Botany* **68**, 864–866.

Kropp BR, Castellano MA, Trappe JM 1985 Performance of outplanted western hemlock *Tsuga heterophylla* Raf. Sarg. seedlings inoculated with *Cenococcum geophilum*. *Tree Plant Not* **36**, 13–16.

Kropp BR and Fortin JA 1988 The incompatibility system and relative ectomycorrhizal performance of monokaryons and reconstituted dikaryons of *Laccaria bicolor*. *Canadian Journal of Botany* **66**, 289–294.

Kropp BR and Langlois CG 1990 Ectomycorrhizae in reforestation. *Canadian Journal of Forest Research* **20**, 432–451.

Kropp BR, McAfee BJ, Fortin JA 1987 Variable loss of ectomycorrhizal ability in monokaryotic and dikaryotic cultures of *Laccaria bicolor*. *Canadian Journal of Botany* **65**, 500–504.

Kropp BR and Trappe JM 1982 Ectomycorrhizal fungi of *Tsuga heterophylla*. *Mycologia* **74**, 479–488.

Kubota M, McGonigle TP, Hyakumachi M 2001 *Clethra barbinervis*, a member of the order Ericales, forms arbuscular mycorrhizae. *Canadian Journal of Botany* **79**, 300–306.

Kucey RMN 1987 Increased phosphorus uptake by wheat and field beans inoculated with a phosphorus-solubilizing *Penicillium bilaji* strain and with vesicular-arbuscular mycorrhizal fungi. *Applied and Environmental Microbiology* **53**, 2699–2703.

Kucey RMN and Janzen HH 1987 Effects of VAM and reduced nutrient availability on growth and phosphorus and micronutrient uptake of wheat and field beans under greenhouse conditions. *Plant and Soil* **104**, 71–78.

Kucey RMN and Paul EA 1982 Biomass of mycorrhizal fungi associated with bean roots. *Soil Biology and Biochemistry* **14**, 413–414.

Kuek C 1994 Issues concerning the production and use of inocula of ectomycorrhizal fungi in increasing the economic productivity of plantations. *Plant and Soil* **159**, 221–230.

Kuek C, Tommerup IC, Malajczuk N 1992 Hydrogel bead inocula for the production of ectomycorrhizal eucalypts for plantations. *Mycological Research* **96**, 273–277.

Kües U and Cassleton L 1992 Fungal mating-type genes – regulators of sexual development. *Mycological Research* **96**, 993–1006.

Kuhn AJ, Schroder WH, Bauch J 2000 The kinetics of calcium and magnesium entry into mycorrhizal spruce roots. *Planta* **210**, 488–496.

Kuhn G, Hijri M, Sanders IR 2001 Evidence for the evolution of multiple genomes in arbuscular mycorrhizal fungi. *Nature* **414**, 745–748.

Kuiters AT and Denneman CAJ 1987. Water-soluble phenolic substances in soils under several coniferous and deciduous tree species. *Soil Biology and Biochemistry* **19** 6, 765–769.

Kusano S 1911a Preliminary note on *Gastrodia elata* and its mycorrhiza. *Annals of Botany* **25**, 521–523.

Kusano S 1911b *Gastrodia elata* and its symbiotic association with *Armillaria mellea*. *Journal of the College of Agriculture, University of Tokyo* **4**, 1–66.

Lagrange H, Jay-Allgmand C, Lapeyrie F 2001 Rutin, the phenolglycoside from eucalyptus root exudates, stimulates Pisolithus hyphal growth at picomolar concentration. *New Phytologist* **149**, 349–355.

Laiho O 1965 Further studies on the ectendotrophic mycorrhiza. *Acta Forestalia Fennica* **79**, 1–35.

Laiho O 1970 *Paxillus involutus* as a mycorrhizal symbiont of forest trees. *Acta Forestalia Fennica* **106**, 1–65.

Laiho O and Mikola P 1964 Studies on the effects of some eradicants on mycorrhizal development in forest nurseries. *Acta Forestalia Fennica* **77**, 1–34.

Lambais MR and Mehdy MC 1993 Suppression of endochitinase, β-1,3-endoglucanase, and chalcone isomerase expression in bean vesicular-arbuscular mycorrhizal roots under different soil phosphate conditions. *Molecular Plant-Microbe Interactions* **6**, 75–83.

Lambers H, Chapin FSI, Pons TL 1998 *Plant Physiological Ecology*. Springer-Verlag, Berlin, Germany.

Lambers H, Shane MW, Cramer MD, Pearse SJ, Veneklaas EJ 2006 Root structure and functioning for efficient acquisition of phosphorus: matching morphological and physiological traits. *Annals of Botany* **98**, 693–713.

Lambert DH, Baker DE, Cole H Jr 1979 The role of mycorrhizae in the interactions of phosphorus with zinc, copper, and other elements. *Soil Science Society of America Journal* **43**, 976–980.

Lamhamedi MS and Fortin JA 1991 Genetic variations of ectomycorrhizal fungi: extramatrical phase of *Pisolithus* sp. *Canadian Journal of Botany* **69**, 1927–1934.

Lamhamedi MS, Fortin JA, Kope HH, Kropp BR 1990 Genetic variation in ectomycorrhiza formation by *Pisolithus arhizus* on *Pinus pinaster* and *Pinus banksiana*. *New Phytologist* **115**, 689–697.

Lamhamedi MS, Bernier PY, Fortin JA 1992 Hydraulic conductance and soil water potential at the soil root interface in *Pinus pinaster* seedlings inoculated with different dikaryons of *Pisolithus* sp. *Tree Physiology* **10**, 231–244.

Lamhamedi MS, Godbout C, Fortin JA 1994 Dependence of *Laccaria bicolor* basidiome development on current photosynthesis of *Pinus strobus* seedlings. *Canadian Journal of Forest Research* **14**, 412–415.

Lammers PJ, Jun J *et al.* 2001 The glyoxylate cycle in arbuscular mycorrhizal fungi. Carbon flux and gene expression. *Plant Physiology* **127**, 1–12.

Lamont BB 1981 Specialised roots of non-symbiotic origin in heathlands. In *Heathlands and Related Shrublands of the World.B. Analytical Studies*. Ed. RL Specht pp. 183–195. Elsevier, Amsterdam, The Netherlands.

Lamont BB 1982 Mechanisms for enhancing nutrient uptake in plants, with particular reference to mediterranean, South Africa and Western Australia. *Botanical Reviews* **48**, 597–689.

Lamont BB 1984 Specialised modes of nutrition. In *Kwongan. Plant Life in the Sandplain*. Eds JS Pate and JS Beard pp. 326–245. University of Western Australia Press, Nedlands, Western Australia.

Lan J, Xu JT, Li JS 1994 Study on symbiotic relation between *Gastrodia elata* and *Armillariella mellea* by autoradiography. *Acta Mycologica Sinica* **13**, 219–222.

Landeweert R, Hoffland E, Finlay R, Kuyper TW, van Breemen N 2001 Linking plants to rocks: ectomycorrhizal fungi mobilise nutrients from minerals. *Trends in Ecology and Evolution* **16**, 248–254.

Landis FC, Gargas A, Givnish TJ 2004 Relationships among arbuscular mycorrhizal fungi, vascular plants and environmental conditions in oak savannas. *New Phytologist* **164**, 493–504.

Lanfranco L, Delpero M, Bonfante P 1999 Intrasporal variability of ribosomal sequences in the endomycorrhizal fungus *Gigaspora margarita*. *Molecular Ecology* **8**, 37–45.

Lanfranco L, Novero M, Bonfante P 2005 The mycorrhizal fungus *Gigaspora margarita* possesses a CuZn superoxide dismutase that is up-regulated during symbiosis with legume hosts. *Plant Physiology* **137**, 1319–1330.

Langley JA, Hungate BA 2003 Mycorrhizal controls on belowground litter quality. *Ecology* **84**, 2302–2312.

Langley JA, Chapman SK, Hungate BA 2006 Ectomycorrhizal colonisation slows root decomposition: the post-mortem fungal legacy. *Ecology Letters* **9**, 955–959.

Langlois CG and Fortin JA 1984 Seasonal variations in the uptake of [^{32}P]phosphate ions by excised ectomycorrhizae and lateral roots of *Abies balsamea*. *Canadian Journal of Forest Research* **14**, 412–415.

Lapeyrie FF and Chilvers GA 1985 An endomycorrhiza-ectomycorrhizal succession associated with enhanced growth by *Eucalyptus dumosa* seedlings planted in a calcarious soil. *New Phytologist* **100**, 93–104.

Lapeyrie F and Mengden K 1993 Quantitative estimation of surface carbohydrates of ectomycorrhizal fungi in pure culture and during *Eucalyptus* root infection. *Mycological Research* **97**, 603–609.

Lapeyrie F, Ranger J, Vairelles D 1991 Phosphate-solubilizing activity of ectomycorrhizal fungi in vitro. *Canadian Journal of Botany* **69**, 342–346.

Lapopin L and Franken P 2000 Modification of plant gene expression. In *Arbuscular Mycorrhizas: Physiology and Function*. Eds Y Kapulnik and DD Douds pp. 69–84. Kluwer Academic Publishers, Dordrecht, The Netherlands.

Largent DL, Sugihara N, Wishner G 1980 Occurrence of mycorrhizae on ericaceous and pyrolaceous plants in Northern California. *Canadian Journal of Botany* **58**, 2274–2279.

Larsen J and Jakobsen I 1996 Interactions between a mycophagous collembola, dry yeast and the external mycelium of an arbuscular mycorrhizal fungus. *Mycorrhiza* **6**, 259–264.

Larsen J, Thingstrup I, Jakobsen I, Rosendahl S 1996 Benomyl inhibits phosphorus transport but not fungal alkaline phosphatase activity in a *Glomus*-cucumber symbiosis. *New Phytologist* **132**, 127–133.

Larsen J, Ravnskov S, Sorensen J 2007 Capturing the benefits of arbuscular mycorrhiza in horticulture. In *Mycorrhizae and Crop Production*. Eds C Hamel and C Plenchette pp. 123–150. The Haworth Press, Inc., Binghamton, New York, USA.

Last FT, Mason PA, Wilson J, Deacon JW 1983 Fine roots and sheathing mycorrhizas: their formation, function and dynamics. *Plant and Soil* **71**, 9–21.

Last FT, Dighton J, Mason PA 1987 Successions of sheathing mycorrhizal fungi. *Trends in Ecology and Evolution* **2**, 157–161.

Laurent P, Voiblet C, Tagu D *et al.* 1999 A novel class of ectomycorrhiza-regulated cell wall polypeptides in *Pisolithus tinctorius. Molecular Plant-Microbe Interactions* **12**, 862–871.

Laursen GA, Treu R, Seppelt RD, Stephenson SL 1997 Mycorrhizal assessment of vascular plants from subantarctic Macquarie Island. *Arctic and Alpine Research* **29**, 483–491.

Law R and Lewis DH 1983 Biotic environments and the maintenance of sex – some evidence from mutualistic symbioses. *Biological Journal of the Linnean Society* **20**, 249–276.

Le Quere A, Wright DP, Söderström B, Tunlid A, Johansson T 2005 Global patterns of gene regulation associated with the development of ectomycorrhiza between birch (*Betula pendula* Roth.) and *Paxillus involutus* (Batsch) Fr. *Molecular Plant-Microbe Interactions* **18**, 659–673.

Le Page BA, Currah RS, Stockey RA, Rothwell GW 1997 Fossil ectomycorrhizae from the Middle Eocene. *American Journal of Botany* **84**, 410–412.

Le Tacon F, Skinner FA, Mosse B 1983 Spore germination and hyphal growth of a vesicular-arbuscular mycorrhizal fungus, *Glomus mosseae* Gerdemann and Trappe, under decreased oxygen and increased carbon dioxide concentrations. *Canadian Journal of Microbiology* **29**, 1280–1285.

Le Tacon F, Jung G, Mugnier J, Michelot P, Mauperin C 1985 Efficiency in a forest nursery of an ectomycorrhizal fungus inoculum produced in a fermentor and entrapped in polymeric gels. *Canadian Journal of Botany* **63**, 1664–1668.

Le Tacon F, Garbaye J, Bouchard D *et al.* 1988 Field results from ectomycorrhizal inoculation in France. In *Canadian Workshop on Mycorrhizae in Forestry*. Eds M Lalonde and Y Piché pp. 51–74. CRBF Faculté de Foresterie et de Géodésie Université Laval, Ste-Foy, Québec, Canada.

Le Tacon F, Alvarez I, Bouchard D *et al.* 1992 Variations in field response of forest trees of nursery ectomycorrhizal inoculation in Europe. In *Mycorrhizas in Ecosystems*. Eds DJ Read, DH Lewis, AH Fitter, IJ Alexander pp. 119–134. CAB International, Wallingford, UK.

Leake JR 1994 The biology of myco-heterotrophic 'saprophytic' plants. *New Phytologist* **127**, 171–216.

Leake JR 2004 Myco-heterotroph/epiparasitic plant interactions with ectomycorrhizal and arbuscular mycorrhizal fungi. *Current Opinion in Plant Biology* **7**, 422–428.

Leake JR 2005 Plants parasitic on fungi: unearthing the fungi in myco-heterotrophs and debunking the 'saprophytic' plant myth. *Mycologist* **19**, 113–122.

Leake JR and Miles W 1996 Phosphodiesters as mycorrhizal P sources. I. Phosphodiesterase production and the utilisation of DNA as a phosphorus source by the ericoid mycorrhizal fungus *Hymenoscyphus ericae. New Phytologist* **132**, 435–444.

Leake JR and Read DJ 1989a The biology of the Ericaceae. XIII. Some characteristics of the extracellular proteinase activity of the ericoid endophyte *Hymenoscyphus ericae. New Phytologist* **112**, 69–76.

Leake JR and Read DJ 1989b The effects of phenolic compounds on nitrogen mobilisation by ericoid mycorrhizal systems. *Agriculture, Ecosystems and Environment* **29**, 225–236.

Leake JR and Read DJ 1990a Proteinase activity in mycorrhizal fungi. I. The effect of extracellular pH on the production and activity of proteinase by ericoid endophytes from soils of contrasted pH. *New Phytologist* **115**, 243–250.

Leake JR and Read DJ 1990b Proteinase activity in mycorrhizal fungi. II. The effects of mineral and organic nitrogen sources on induction of extracellular proteinase in *Hymenoscyphus ericae* (Read) Korf and Kernan. *New Phytologist* **116**, 123–128.

Leake JR and Read DJ 1990c Chitin as a nitrogen source for mycorrhizal fungi. *Mycological Research* **94**, 993–995.

Leake JR and Read DJ 1990d The effects of phenolic compounds on nitrogen mobilisation by ericoid mycorrhizal systems. *Agriculture, Ecosystems and Environment* **29**, 225–236.

Leake JR and Read DJ 1991 Proteinase activity in mycorrhizal fungi. 3. Effects of protein, protein hydrolysate, glucose and ammonium on production of extracellular proteinase by *Hymenoscyphus ericae* (Read) Korf and Kernan. *New Phytologist* **117** 2, 309–317.

Leake JR and Read DJ 1997 Mycorrhizal fungi in terrestrial habitats. In *The Mycota IV. Environmental and Microbial Relationships*. Eds DT Wicklow and B Söderström pp. 281–301. Springer, Berlin, Germany.

Leake JR, Shaw G, Read DJ 1990a The biology of mycorrhiza in the Ericaceae. XVI. mycorrhiza and iron uptake in *Calluna vulgaris* (L.) Hull in the presence of two calcium salts. *New Phytologist* **114**, 651–657.

Leake JR, Shaw G, Read DJ 1990b The role of ericoid mycorrhiza in the ecology of ericaceous plants. *Agriculture, Ecosystems and Environment* **29**, 237–251.

Leake JR, Donnelly DP, Boddy L 2002 Interactions between ectomycorrhizal and saprotrophic fungi. In *Mycorrhizal Ecology*. Eds MGA van der Heijden and IR Sanders pp. 346–372. Springer-Verlag, Berlin, Germany.

Leake JR, Johnson D, Donnelly DP, Muckle GE, Boddy L, Read DJ 2004a Networks of power and influence: the role of mycorrhizal mycelium in controlling plant communities and agroecosystem functioning. *Canadian Journal of Botany* **82**, 1016–1045.

Leake JR, McKendrick SL, Bidartondo M, Read DJ 2004b Symbiotic germination and development of the myco-heterotroph *Monotropa hypopitys* in nature and its requirement for locally distributed *Tricholoma* spp. *New Phytologist* **163**, 405–423.

Leake JR, Ostle NJ, Rangel-Castro JI, Johnson D 2006 Carbon fluxes from plants through soil organisms determined by field (CO_2)-C-13 pulse-labelling in an upland grassland. *Applied Soil Ecology* **33**, 152–175.

Lee SS and Alexander IJ 1994 The response of seedlings of two dipterocarp species to nutrient additions and ectomycorrhizal infection. *Plant and Soil* **163**, 299–306.

Lee SS, Alexander IJ, Moura-Costa P, Yap SW 1996 Mycorrhizal infection of dipterocarp seedlings in logged and undisturbed forests. *Proc 5th Round Table Conference on Dipterocarps*. Eds S Appaneh, EC Khoo. Forest Research Institute of Malaysia, Kuala Lumpur, Malaysia.

Lee SS, Watling R, Turnbull E 2003 Diversity of putative ectomycorrhizal fungi in Pasoh Forest Reserve. In *Pasoh: Ecology of a Lowland Rain Forest in Southeast Asia*. Eds T Okuda, N Manokaran, Y Matsumoto, K Niiyama, SC Thomas, PS Ashton pp. 149–159. Springer, Tokyo.

Lee Y-J and George E 2005 Contribution of mycorrhizal hyphae to uptake of metal cations by cucumber plants at two levels of phosphorus supply. *Plant and Soil* **278**, 361–370.

Lefebvre DD and Glass ADM 1982 Regulation of phosphate influx in barley roots: effects of phosphate deprivation and reduction of influx with provision of orthophosphate. *Physiologia Plantarum* **54**, 199–206.

Lefevre C and Müller W 1998 *Tricholoma nagnivelare* Peck Redhead. In *A Manual of Concise Descriptions of North American Ectomycorrhizae*. Eds DM Goodman, DM Durall, JA Trofymow, SM Berch. Mycologue Publications, Victoria, BC, Canada: co-published by BC Ministry of Forests, Canadian Forest Service, Victoria, BC.

Lei J and Dexheimer J 1988 Ultrastructural localisation of ATPase activity in *Pinus sylvestris/ Laccaria laccata* mycorrhizal association. *New Phytologist* **108**, 329–334.

Lei J, Ding H, Lapeyrie F, Piché Y, Malajczuk N, Dexheimer J 1990a Ectomycorrhiza formation on the roots of *Eucalyptus globulus* and *Pinus caribaea* with two isolates of *Pisolithus tinctorius*: structural and cytochemical observations. In *Endocytobiology* vol. IV. Ed. P Nardon pp. 123–126. INRA, Paris, France.

Lei J, Lapeyrie F, Malajczuk N, Dexheimer J 1990 Infectivity of pine and eucalypt isolates of *Pisolithus tinctorius* Pers. Coker and Couch on roots of *Eucalyptus urophylla* S.T. Blake *in vitro*. *New Phytologist* **116**, 115–122.

Lei J, Bécard G, Catford JG, Piché Y 1991 Root factors stimulate ^{32}P uptake and plasmalemma ATPase activity in vesicular-arbuscular mycorrhizal fungus, *Gigaspora margarita*. *New Phytologist* **118**, 289–294.

Lemmon AR and Milinkovitch MC 2002 The metapopulation genetic algorithm: an efficient solution for the problem of large phylogeny estimation. *Proceedings of the National Academy of Sciences of the USA* **99**, 10516–10521.

Lemoine MC, Gianinazzi-Pearson V, Gianinazzi S, Straker CJ 1992 Occurence and expression of acid phosphatase of *Hymenoscyphus ericae* (Read) Korf and Kernan, in isolation or associated with plant roots. *Mycorrhiza* **1**, 137–146.

Lemoine MC, Gollotte A, Gianinazzi-Pearson V 1995 β 1-3 glucan localization in walls of the endomycorrhizal fungi *Glomus mosseae* and *Acaulospora laevis* during colonization of host roots. *New Phytologist* **129**, 97–105.

Lerat S, Gauci R, Catford JG, Vierheilig H, Piché Y, Lapointe L 2002 C-14 transfer between the spring ephemeral *Erythronium americanum* and sugar maple saplings via arbuscular mycorrhizal fungi in natural stands. *Oecologia* **132**, 181–187.

Lerat S, Lapointe L, Gutjahr S, Piché Y, Vierheilig H 2003a Carbon partitioning in a split-root system of arbuscular mycorrhizal plants is fungal and plant species dependent. *New Phytologist* **157**, 589–595.

Lerat S, Lapointe L, Piché Y, Vierheilig H 2003b Variable carbon-sink strength of different *Glomus mosseae* strains colonizing barley roots. *Canadian Journal of Botany* **81**, 886–889.

Lesica P and Antibus RK 1986 Mycorrhizae of alpine fell-field communities on soils derived from chrystalline and calcareous parent materials. *Canadian Journal of Botany* **64**, 1691–1697.

Lesica P and Antibus RK 1990 The occurrence of mycorrhizae in vascular epiphytes of two Costa Rican rain forests. *Biotropica* 250–258.

Levy J, Bres C *et al.* 2004 A putative Ca^{2+} and calmodulin-dependent protein kinase required for bacterial and fungal symbioses. *Science* **303**, 1361–1364.

Levy Y and Krikun J 1980 Effect of vesicular-arbuscular mycorrhiza on *Citrus jambhiri* water relations. *New Phytologist* **85**, 25–31.

Levy Y, Syvertsen JP, Nemec S 1983 Effect of drought stress and vesicular-arbuscular mycorrhiza on citrus transpiration and hydraulic conductivity of roots. *New Phytologist* **93**, 61–66.

Lewis DH 1963 Uptake and utilisation of substances by beech mycorrhiza. DPhil thesis, Oxford University, UK.

Lewis DH 1973 Concepts in fungal nutrition and the origin of biotrophy. *Biological Reviews* **48**, 261–278.

Lewis DH 1976 Interchange of metabolites in biotrophic symbioses between angiosperms and fungi. In *Perspectives in Experimental Biology*, Vol 2. Botany. Ed. N Sutherland pp. 207–219. Pergamon Press, Oxford, UK.

Lewis DH and Harley JL 1965a Carbohydrate physiology of mycorrhizal roots of beech. I. Identity of endogenous sugars and utilisation of exogenous sugars. *New Phytologist* **64**, 224–237.

Lewis DH and Harley JL 1965b Carbohydrate physiology of mycorrhizal roots of beech. II. Utilisation of exogenous sugars by uninfected and mycorrhizal roots. *New Phytologist* **64**, 238–256.

Lewis DH and Harley JL 1965c Carbohydrate physiology of mycorrhizal roots of beech. III. Movement of sugars between host and fungus. *New Phytologist* **64**, 256–269.

Leyval C and Berthelin J 1989 Interactions between *Laccaria laccata*, *Agrobacterium radiobacter* and beech roots: influence on P, K, Mg, and Fe mobilization from minerals and plant growth. *Plant and Soil* **117**, 103–110.

Leyval C and Berthelin J 1991 Weathering of a mica by roots and rhizospheric microorganisms of pine. *Soil Science Society of America, Journal* **55**, 1009–1016.

Li HY, Zhu YG, Marschner P, Smith FA, Smith SE 2005 Wheat responses to arbuscular mycorrhizal fungi in a highly calcareous soil differ from those of clover, and change with plant development and P supply. *Plant and Soil* **277**, 221–232.

Li HY, Smith SE, Holloway RE, Zhu YG, Smith FA 2006 Arbuscular mycorrhizal AM fungi contribute to phosphorus uptake by wheat grown in a phosphorus-fixing soil even in the absence of positive growth responses. *New Phytologist* **172**, 536–543.

Li XL and Christie P 2001 Changes in soil solution Zn and pH and uptake of Zn by arbuscular mycorrhizal red clover in Zn-contaminated soil. *Chemosphere* **42**, 201–207.

Li XL, George E, Marschner H 1991a Extension of the phosphorus depletion zone in VA-mycorrhizal white clover in a calcareous soil. *Plant and Soil* **136**, 41–48.

Li XL, George E, Marschner H 1991b Phosphorus depletion and pH decrease at the root-soil and hyphae-soil interfaces of VA mycorrhizal white clover fertilized with ammonium. *New Phytologist* **119**, 307–404.

Li XL, Marschner H, George E 1991c Acquisition of phosphorus and copper by VA-mycorrhizal hyphae and root-to-shoot transport in white clover. *Plant and Soil* **136**, 49–57.

Li XL, George E, Marschner H, Zhang JL 1997 Phosphorus acquisition from compacted soil by hyphae of a mycorrhizal fungus associated with red clover (*Trifolium pratense*). *Canadian Journal of Botany* **75**, 723–729.

Lian CL, Narimatsu M, Nara K, Hogetsu T 2006 *Tricholoma matsutake* in a natural *Pinus densiflora* forest: correspondence between above- and below-ground genets, association with multiple host trees and alteration of existing ectomycorrhizal communities. *New Phytologist* **171**, 825–836.

Ligrone R 1988 Ultrastructure of a fungal endophyte in *Phaeoceros laevis* (L) Prosk (Anthocerotophyta). *Botanical Gazette* **149**, 92–100.

Ligrone R and Lopes C 1989 Cytology and development of a mycorrhiza-like infection in the gametophyte of *Conocephalum conicum* (L) Dum. (Marchantiales, Hepatophyta). *New Phytologist* **111**, 423–433.

Ligrone R, Pocock K, Duckett JG 1993 A comparative ultrastructural study of endophytic basidiomycetes in the parasitic achlorophyllous hepatic Cryptothallus mirabilis and the closely allied photosynthetic species Aneura pinguis (Metzgeriales). *Canadian Journal of Botany* **71**, 666–679.

Ligrone R, Carafa A, Lumini E, Bianciotti V, Bonfante P, Dockett JG 2007 Glomeromycotean associations in liverworts: a molecular, cellular and taxonomic analysis. *American Journal of Botany* **94**, 1756–1777.

Lihnell D 1942 *Cenococcum graniforme* aus Mykorrizabildner von Walkdbäumen. *Symbolae Botanicae Upsaliensis* **5**, 1–8.

Lilleskov EA and Bruns TD 2003 Root colonization of two ectomycorrhizal fungi of contrasting life history strategies are mediated by addition of organic nutrient patches. *New Phytologist* **159**, 141–151.

Lilleskov EA and Bruns TD 2005 Spore dispersal of a resupinate ectomycorrhizal fungus, *Tomentella sublilacina*, via soil food webs. *Mycologia* **97**, 762–769.

Lilleskov EA, Fahey TJ, Lovett GM 2001 Ectomycorrhizal fungal aboveground community change over an atmospheric nitrogen deposition gradient. *Ecological Applications* **11**, 397–410.

Lilleskov EA, Fahey T, Horton TR, Lovett GM 2002a Belowground ectomycorrhizal fungal community change over a nitrogen deposition gradient in Alaska. *Ecology* **83**, 104–115.

Lilleskov EA, Hobbie EA, Fahey TJ 2002b Ectomycorrhizal fungal taxa differing in response to nitrogen deposition also differ in pure culture organic nitrogen use and natural abundance of nitrogen isotopes. *New Phytologist* **154**, 219–231.

Lim LL, Fineran BA, Cole ALJ 1983 Ultrastructure of intrahyphal hyphae of *Glomus fasiculatum* (Thaxter) Gerdemann and Trappe in roots of white clover (*Trifolium repens* L.) *New Phytologist* **95**, 231–239.

Lindahl B, Stenlid J, Olsson S, Finlay R 1999 Translocation of ^{32}P between interacting mycelia of a wood-decomposing fungus and ectomycorrhizal fungi in microcosm systems. *New Phytologist* **144**, 183–193.

Lindahl B, Finlay R, Olsson S 2001 Simultaneous, bidirectional translocation of ^{32}P and ^{33}P between wood blocks connected by mycelial cords of *Hypholoma fasciculare*. *New Phytologist* **150**, 189–194.

Lindahl B, Taylor AFS, Finlay RD 2002 Defining nutritional constraints on carbon cycling in boreal forests – towards a less 'phytocentric' perspective. *Plant and Soil* **242**, 123–135.

Lindahl B, Ihrmark K, Boberg J *et al.* 2007 Spatial separation of litter decomposition and mycorrhizal nitrogen uptake in boreal forest. *New Phytologist* **173**, 611–620.

Lindeberg G and Lindeberg M 1977 Pectinolytic ability of some mycorrhizal and saprophytic Hymenomycetes. *Archives of Microbiology* **115**, 9–12.

Linder S and Axelsson B 1982 Changes in carbon uptake and allocation patterns as a result of irrigation and fertilisation in a young *Pinus sylvestris* stand. In *Carbon Uptake and Allocation in Subalpine Ecosystems as a Key to Management*. Ed. RH Waring pp. 38–44. Forest Research Laboratory, Oregon State University, USA.

Linderman RG 1992 Vesicular-arbuscular mycorrhizae and soil microbial interactions. In *Mycorrhizae in Sustainable Agriculture*. Eds GJ Bethlenfalvay and RG Linderman pp. 45–70. ASA Special Publication 54, Madison, Wisconsin, USA.

Ling-Lee M, Chilvers GA, Ashford AE 1975 Polyphosphate granules in three different kinds of tree mycorrhiza. *New Phytologist* **75**, 551–554.

Linkins AE and Antibus RK 1981 Mycorrhizae of *Salix rotundifolia* in coastal arctic tundra. In *Arctic and Alpine Mycology: Proceedings of the First International Symposium on Arctic-Alpine Mycology*, Fairbanks, Alaska. Eds GA Laursen and FF Ammirati pp. 509–531. University of Washington Press, Seattle, USA.

Littke WR, Bledsoe CS, Edmonds RL 1984 Nitrogen uptake and growth *in vitro* by *Hebeloma crustuliniforme* and other Pacific Northwest mycorrhizal fungi. *Canadian Journal of Botany* **62**, 647–652.

Liu JY, Blaylock LA, Endre G *et al.* 2003 Transcript profiling coupled with spatial expression analyses reveals genes involved in distinct developmental stages of an arbuscular mycorrhizal symbiosis. *The Plant Cell* **15**, 2106–2123.

Liu JY, Blaylock LA, Harrison MJ 2004 cDNA arrays as a tool to identify mycorrhiza-regulated genes: identification of mycorrhiza-induced genes that encode or generate signaling molecules implicated in the control of root growth. *Canadian Journal of Botany* **82**, 1177–1185.

Liu Y, Zhu YG, Chen BD, Christie P, Li XL 2005a Influence of the arbuscular mycorrhizal fungus *Glomus mosseae* on uptake of arsenate by the As hyperaccumulator fern *Pteris vittata* L. *Mycorrhiza* **15**, 187–192.

Liu Y, Zhu YG, Chen BD, Christie P, Li XL 2005b Yield and arsenate uptake of arbuscular mycorrhizal tomato colonized by *Glomus mosseae* BEG167 in As spiked soil under glasshouse conditions. *Environment International* **31**, 867–873.

Lloyd-MacGilp SA, Chambers SM, Dodd JC, Fitter AH, Walker C, Young JPW 1996 Diversity of the ribosomal internal transcribed spacers within and among isolates of *Glomus mosseae* and related mycorrhizal fungi. *New Phytologist* **33**, 103–111.

LoBuglio KF and Taylor JW 2002 Recombination and genetic differentiation in the mycorrhizal fungus *Cenococcum geophilum* Fr. *Mycologia* **94**, 772–780.

LoBuglio KF, Berbee ML, Taylor JW 1996 Phylogenetic origins of the asexual mycorrhizal symbiont *Cenococcum geophilum* Fr. and other mycorrhizal fungi among the ascomycetes. *Molecular Phylogenetics and Evolution* **6**, 287–294.

Lodge DJ and Wentworth TR 1990 Negative associations among VA-mycorrhizal fungi and some ectomycorrhizal fungi inhabiting the same root-system. *Oikos* **57**, 347–356.

Lodge DJ, McDowell WH, Swiney CP 1994 The importance of nutrient pulses in tropical forests. *Tree* **9**, 384–387.

Loewe A, Einig W, Shi L, Dizengremel P, Hampp R 2000 Mycorrhiza formation and elevated CO_2 both increase the capacity for sucrose synthesis in source leaves of spruce and aspen. *New Phytologist* **145**, 565–574.

Loftus BJ, Fung E, Roncaglia P *et al.* 2005 The genome of the basidiomycetous yeast and human pathogen *Cryptococcus neoformans*. *Science* **307**, 1321–1324.

Logi C, Sbrana C, Giovannetti M 1998 Cellular events involved in survival of individual arbuscular mycorrhizal symbionts growing in the absence of the host. *Applied and Environmental Microbiology* **64**, 3473–3479.

Lorillou S, Botton B, Martin F 1996 Nitrogen source regulates the biosynthesis of NADP-glutamate dehydrogenase in the ectomycorrhizal basidiomycete *Laccaria bicolor*. *New Phytologist* **132**, 289–296.

Louis I and Lim G 1987 Spore density and root colonization of vesicular-arbuscular mycorrhizas in tropical soil. *Transactions of the British Mycological Society* **88**, 207–212.

Louis I and Lim G 1988 Effect of storage of inoculum on spore germination of a tropical isolate of *Glomus clarum*. *Mycologia* **80**, 157–161.

Lovato PE, Schüepp H, Trouvelot A, Gianinazzi S 1995 Application of arbuscular mycorrhizal fungi (AMF) in orchard and ornamental plants. In *Mycorrhiza, Structure, Function, Molecular Biology and Biotechnology*. Eds A Varma and B Hock pp. 443–467. Springer-Verlag, Berlin, Germany.

Lovato PE, Gianinazzi Pearson V, Trouvelot A, Gianinazzi S 1996 The state of art of mycorrhizas and micropropagation. *Advances in Horticultural Science* **10**, 46–52.

Lu S and Miller MH 1989 The role of VA mycorrhizae in the absorption of P and Zn by maize in field and growth chamber experiments. *Canadian Journal of Soil Science* **69**, 97–109.

Lück R 1940 Zur Biologie der heimischen *Pyrola* arten. *Physikalish-Ökonomische Gesellschaft zu Königsberg Schriften* **71**, 300–334.

Lundeberg G 1970 Utilisation of various nitrogen sources, in particular bound soil nitrogen, by mycorrhizal fungi. *Studia Forestalia Suecica* **79**, 1–95.

Lussenhop J 1992 Mechanisms of microarthropod-microbial interactions in soil. *Advances in Ecological Research* **23**, 1–33.

Lussenhop J and BassiriRad H 2005 Collembola effects on plant mass and nitrogen acquisition by ash seedlings *Fraxinus pennsylvanica*. *Soil Biology and Biochemistry* **37**, 645–650.

Lutz RW and Sjolund RD 1973 *Monotropa uniflora*: ultrastructural details of its mycorrhizal habit. *American Journal of Botany* **60**, 339–345.

Lux HB and Cumming JR 1999 Effect of aluminum on the growth and nutrition of tulip-poplar seedlings. *Canadian Journal of Forest Research* **29** 12, 2003–2007.

Lux HB and Cumming HR 2001 Mycorrhizae confer aluminium resistance to tulip-poplar seedlings. *Canadian Journal of Forest Research* **31**, 694–702.

Lyr H and Hoffman G 1967 Growth rates and growth periodicity of tree roots. *International Review of Forest Research* **2**, 81.

Ma M, Tan TK, Wong SM 2003 Identification and molecular phylogeny of *Epulorhiza* isolates from tropical orchids. *Mycological Research* **107**, 1041–1049.

Maas PJM and Ruyters P 1986 *Voyria* and *Voyriella* saprophytic Gentianaceae. *Flora Neotropica 41.* New York Botanical Gardens, New York, USA.

Mabberley DJ 1997 *The Plant Book*, 2nd edn. Cambridge University Press, Cambridge, UK.

MacDougal DT 1899 Symbiotic saprophytism. *Annals of Botany* **13**, 1–47.

Mack MC, Schuur EAG, Bret-Harte MS, Shaver GR, Chapin FS 2004 Ecosystem carbon storage in arctic tundra reduced by long-term nutrient fertilization. *Nature* **431**, 440–443.

MacMahon JA 1981 Successional processes: comparisons among biomes with special reference to probable roles of and influences on animals. In *Forest Succession: Concept and Application*. Eds D West, H Shugart, D Botkin pp. 277–304. Springer-Verlag, New York, USA.

Madan R, Pankhurst C, Hawke B, Smith S 2002 Use of fatty acids for identification of AM fungi and estimation of the biomass of AM spores in soil. *Soil Biology and Biochemistry* **34**, 125–128.

Mader P, Vierheilig H, Streitwolf-Engel R *et al.* 2000 Transport of ^{15}N from a soil compartment separated by a polytetrafluoroethylene membrane to plant roots via the hyphae of arbuscular mycorrhizal fungi. *New Phytologist* **146**, 155–161.

Madhani HD and Fink GR 1998 The control of filamentous differentiation and virulence in fungi. *Trends in Cell Biology* **8** 9, 348–353.

Maeda D, Ashida K, Iguchi K *et al.* 2006 Knockdown of an arbuscular mycorrhiza-inducible phosphate transporter gene of *Lotus japonicus* supresses mutualistic symbiosis. *Plant and Cell Physiology* **47**, 807–817.

Mago P, Agnes CA, Mukerji KJ 1992 VA mycorrhizal status of some Indian bryophytes. *Phytomorphology* **42**, 231–239.

Magrou MJ 1925 La symbiose chez les Hepatiques. Le *Pellia epiphylla* et son champignon commensal. *Annales des Sciences Naturelles; Botanique* **7**, 725–780.

Magrou MJ 1936 Culture et inoculation du champignon symbiotique d'*Arum maculatum*. *Comptes Rendues hebd Seanc Acad. Sci* **203**, 887–888.

Magrou MJ 1946 Sur la culture de quelques champignons de mycorhizes à arbuscules et à vesicules. *Revue Generale de Botanique* **53**, 49–77.

Mahmood S, Finlay RD, Wallander H, Erland S 2002 Ectomycorrhizal colonisation of roots and ash granules in a spruce forest treated with granulated wood ash. *Forest Ecology and Management* **160**, 65–74.

Maia LC, Yano AM, Kimbrough JW 1996 Species of Ascomycota forming ectomycorrhizae. *Mycotaxon* **57**, 371–390.

Maijala P, Fagerstedt K, Raudaskoski M 1991 Detection of extracellular cellulolytic and proteolytic activity in ectomycorrhizal fungi and *Heterobasidium annosum* Fr. Bref. *New Phytologist* **117**, 643–648.

Malajczuk N, Molina R, Trappe JM 1984 Ectomycorrhiza formation in *Eucalyptus*. II. The ultrastructure of compatible and incompatible mycorrhizal fungi and associated roots. *New Phytologist* **96**, 43–53.

Malajczuk N, Lapeyrie F, Garbaye J 1990 Infectivity of pine and eucalypt isolates of *Pisolithus tinctorius* on the roots of *Eucalyptus urophylla in vitro*. *New Phytologist* **114**, 627–631.

Maldonado-Mendoza IE, Dewbre GR, Harrison MJ 2001 A phosphate transporter gene from the extraradical mycelium of an arbuscular mycorrhizal fungus *Glomus intraradices* is regulated in response to phosphate in the environment. *Molecular Plant Microbe Interactions* **14**, 1140–1148.

Maldonado-Mendoza IE, Dewbre GR, van Buuren ML, Versaw W, Harrison MJ 2002 Methods to estimate the proportion of plant and fungal RNA in an arbuscular mycorrhizal fungus. *Mycorrhiza* **12**, 67–74.

Mandyam K and Jumpponen A 2005 Seeking the elusive function of the root-colonising dark septate endophytic fungi. *Studies in Mycology* **53**, 173–189.

Mangin L 1910 *Introduction a l'Etude des Mycorrhizes des Arbres Forestières*. Nouvelles Archives Museum National d'Histoire Naturelle, Paris, France.

Manjarrez-Martinez M 2007 Characterization of the life cycle and cellular interactions of AM fungi with the reduced mycorrhizal colonization (*rmc*) mutant of tomato (*Solanum lycopersicum* L.). PhD thesis, The University of Adelaide, Adelaide, Australia.

Manjunath A and Habte M 1988 Development of vesicular-arbuscular mycorrhizal infection and uptake of immobile nutrients in *Leucena leucocephala*. *Plant and Soil* **106**, 97–103.

Manning JC and Van Standen J 1987 The development and mobilisation of seed reserves in some African orchids. *Australian Journal of Botany* **35**, 343–353.

Manske CGB 1989 Genetical analysis of the efficiency of VA mycorrhiza with spring wheat. *Agriculture, Ecosystems and Environment* **29**, 273–280.

Marjanovic Z, Uehlein N, Kaldenhoff R *et al.* 2005a Aquaporins in poplar: what a difference a symbiont makes! *Planta* **222**, 258–268.

Marjanovic Z, Uwe N, Hampp R 2005b Mycorrhiza formation enhances adaptive response of hybrid poplar to drought. In *Biophysics from Molecules to Brain: In Memory of Radoslav K. Andjus* **1048**, 496–499.

Marks GC and Foster RC 1967 Succession of mycorrhizal associations on individual roots of radiata pine. *Australian Forestry* **31**, 194–201.

Marler MJ, Zabinski CA, Callaway RM 1999 Mycorrhizae indirectly enhance competitive effects of an invasive forb on a native bunchgrass. *Ecology* **80**, 1180–1186.

Marmeisse R, Debaud JC, Casselton LA 1992a DNA probes for species and strain identification in ectomycorrhizal fungus *Hebeloma*. *Mycological Research* **96**, 161–165.

Marmeisse R, Gay G, Debaud JC, Casselton LA 1992b Genetic transformation of the symbiotic basidiomycete fungus *Hebeloma cylindrosporum*. *Current Genetics* **22**, 41–45.

Marmeisse R, Jargeat P, Wagner F, Gay G, Debaud JC 1998 Isolation and characterization of nitrate reductase deficient mutants of the ectomycorrhizal fungus *Hebeloma cylindrosporum*. *New Phytologist* **140**, 311–318.

Marschner H 1995 *Mineral Nutrition of Higher Plants*. Academic Press, London, UK.

Marschner H and Dell B 1994 Nutrient uptake in mycorrhizal symbiosis. *Plant and Soil* **159**, 89–102.

Marschner H, Römheld V, Horst WJ, Martin P 1986 Root induced changes in the rhizosphere: Importance for the mineral nutrition of plants. *Zeitschrift fur Pflanzenernahrung und Bodenkunde* **149**, 441–456.

Marschner P and Baumann K 2003 Changes in bacterial community structure induced by mycorrhizal colonisation in split-root maize. *Plant and Soil* **251**, 279–289.

Marschner P and Crowley DE 1996 Root colonization of mycorrhizal and non-mycorrhizal pepper (*Capsicum annuum*) by *Pseudomonas fluorescens* 2–79RL. *New Phytologist* **134**, 115–122.

Marschner P and Timonen S 2005 Interactions between plant species and mycorrhizal colonization on the bacterial community composition in the rhizosphere. *Applied Soil Ecology* **28**, 23–36.

Marschner P, Godbold DL, Jentschke G 1996 Dynamics of lead accumulation in mycorrhizal and non-mycorrhizal Norway spruce (*Picea abies* (L) Karst). *Plant and Soil* **178**, 239–245.

Marschner P, Crowley DE, Higashi RM 1997 Root exudation and physiological status of a root-colonizing fluorescent pseudomonad in mycorrhizal and non-mycorrhizal pepper (*Capsicum annuum* L.). *Plant and Soil* **189**, 11–20.

Marsh JF and Schultze M 2001 Analysis of arbuscular mycorrhizas using symbiosis-defective plant mutants. *New Phytologist* **150**, 525–532.

Martin CA and Stutz JC 2004 Interactive effects of temperature and arbuscular mycorrhizal fungi on growth, P uptake and root respiration of *Capsicum annuum* L. *Mycorrhiza* **14**, 241–244.

Martin F 1985 Nitrogen-[15] NMR studies on nitrogen assimilation and amino acid biosynthesis in the ectomycorrhizal fungus *Cenococcum graniforme*. *FEBS Letters* **182**, 350–354.

Martin F 2007 Fair trade in the Underworld: the Ectomycorrhizal Symbiosis. In *The Mycota VIII Biology of the Fungal Cell* Edn 2. Eds RT Howard and NAR Gow pp. 291–308.

Martin F and Botton B 1993 Nitrogen metabolism of ectomycorrhizal fungi and ectomycorrhiza. *Advances in Plant Pathology* **9**, 83–102.

Martin F and Canet D 1986 Biosynthesis of amino acids during ^{13}C glucose utilisation by the ectomycorrhizal ascomycete *Cenococcum geophilum* monitored by ^{13}C nuclear magnetic resonance. *Physiology Végétale* **24**, 209–218.

Martin F and Tagu D 1995 Ectomycorrhiza development: a molecular perspective. In *Mycorrhiza: Structure, Function, Molecular Biology and Biotechnology*. Eds A Varma and B Hock pp. 29–58. Springer-Verlag, Berlin, Germany.

Martin F, Marchal JP, Timinska A, Canet D 1985 The metabolism and physical state of polyphosphates in ectomycorrhizal fungi. A ^{31}P nuclear magnetic resonance study. *New Phytologist* **101**, 275–290.

Martin F, Stewart G, Genetet I, Le Tacon F 1986 Assimilation of $^{15}NH_4^+$ by beech (*Fagus sylvatica*) ectomycorrhizas. *New Phytologist* **102**, 85–94.

Martin F, Ramstedt M, Söderhäll K, Canet D 1988a Carbohydrate and aminoacid metabolism in the ectomycorrhizal ascomycete *Sphaerosporella brunnea* during glucose utilization. *Plant Physiology* **86**, 935–940.

Martin F, Stewart GR, Genetet I, Mourot B 1988b The involvement of glutamate dehydrogenase and glutamine synthetase in ammonia assimilation by the rapidly growing ectomycorrhizal ascomycete, *Cenococcum geophilum* Fr. *New Phytologist* **110**, 541–550.

Martin F, Chalot M, Brun A, Lorillou S, Botton B, Dell B 1992 Spatial distribution of nitrogen assimilation pathways in ectomycorrhizas. In *Mycorrhizas in Ecosystems*. Eds DJ Read, DH Lewis, AH Fitter, IJ Alexander pp. 311–315. CAB International, Wallingford, UK.

Martin F, Coté R, Canet D 1994 NH_4^+ assimilation in the ectomycorrhizal basidiomycete *Laccaria bicolor* Maire Orton, a ^{15}N-NMR study. *New Phytologist* **128**, 479–485.

Martin F, Diez J, Dell B, Delaruelle C 2002 Phylogeography of the ectomycorrhizal *Pisolithus* species as inferred from nuclear ribosomal DNA ITS sequences. *New Phytologist* **153**, 345–357.

Martin F, Tuskan GA, DiFazio SP, Lammers P, Newcombe G, Podila GK 2004 Symbiotic sequencing for the *Populus* mesocosm. *New Phytologist* **161**, 330–335.

Martin JF 1986 Mycorhization de *Monotropa uniflora* L. par des Russulaceae. *Bulletin de la Société Mycologique de France* **102**, 155–159.

Martinez D, Larrondo LF, Putnam N *et al.* 2004 Genome sequence of the lignocellulose degrading fungus *Phanerochaete chrysosporium* strain RP78. *Nature Biotechnology* **22**, 695–700.

Martino E, Coisson JD, Lacourt I, Favaron F, Bonfante P, Perotto S 2000 Influence of heavy metals on production and activity of pectinolytic enzymes in ericoid mycorrhizal fungi. *Mycological Research* **104**, 825–833.

Martino E, Perotto S, Parsons R, Gadd GM 2003 Solubilization of insoluble inorganic zinc compounds by ericoid mycorrhizal fungi derived from heavy metal polluted sites. *Soil Biology and Biochemistry* **35**, 133–141.

Marulanda A, Azcón R, Ruiz-Lozano JM 2003 Contribution of six arbuscular mycorrhizal fungal isolates to water uptake by *Lactuca sativa* plants under drought stress. *Physiologia Plantarum* **119**, 526–533.

Marx C, Dexheimer J, Gianinazzi-Pearson V, Gianinazzi S 1982 Enzymatic studies on the metabolism of vesicular-arbuscular mycorrhizas. IV. Ultracytoenzymological evidence ATPase for active transfer processes in the host-arbuscular interface. *New Phytologist* **90**, 37–43.

Marx DH 1969 The influence of ectotrophic mycorrhizal fungi on the resistance of pine roots to pathogenic infections. I. Antagonism of mycorrhizal fungi to root pathogenic fungi and soil bacteria. *Phytopathology* **59**, 153–163.

Marx DH 1973 Mycorrhizae and feeder root diseases. In *Ectomycorrhizae*. Eds GC Marks and TT Kozlowski pp. 351–382. Academic Press, New York, USA.

Marx DH 1975 Mycorrhizas and establishment of trees on strip-mined land. *Ohio Journal of Science* **75**, 288–297.

Marx DH 1977 Tree host range and world distribution of the ectomycorrhizal fungus *Pisolithus tinctorius*. *Canadian Journal of Microbiology* **23**, 217–223.

Marx DH 1979a Synthesis of *Pisolithus* ectomycorrhizae on White Oak seedlings in fumigated nursery soil. Forest Service Research Note, USDA, SE 280.

Marx DH 1979b Synthesis of ectomycorrhizae by different fungi in Northern Red Oak seedlings. Forest Service Research Note, USDA, SE 282.

Marx DH 1980 Role of mycorrhizae in forestation of surface mines. In *Proceedings of Symposium on Trees for Reclamation*. pp. 109–116. Compact Commission and USDA Forest Service, Lexington, Kentucky, USA.

Marx DH 1985 Trials and tribulations of an ectomycorrhizal fungus inoculation program. In *Proceedings of the 6th North American Conference on Mycorrhizae*. Ed. R Molina pp. 62–63. Forest Research Laboratory, Corvallis, Oregon, USA.

Marx DH 1991 The practical significance of ectomycorrhizae in forest establishment. In *The Marcus Wallenberg Foundation Symposia 7: Ecophysiology of Ectomycorrhizae of Forest Trees.* pp. 54–83. The Marcus Wallenberg Foundation, Stockholm, Sweden.

Marx DH and Bryan WC 1969 Studies on ectomycorrhizae of pine in an electronically air-filtered, air-conditioned, plant growth room. *Canadian Journal of Botany* **47**, 1903–1909.

Marx DH and Bryan WC 1971 Formation of ectomycorrhizae on half-sib progenies of slash pine in aseptic culture. *Forest Science* **17**, 488–492.

Marx DH and Bryan WC 1975 Growth and ectomycorrhizal development of loblolly pine seedlings in fumigated soil infected with the fungal symbiont *Pisolithus tinctorius*. *Forest Science* **21**, 245–254.

Marx DH and Kenney DS 1982 Production of ectomycorrhizal fungus inoculum. In *Methods and Principles of Mycorrhizal Research.* Ed. NC Schenck pp. 131–146. American Phytopathological Society, St Paul, Minnesota, USA.

Marx DH and Zak B 1965 Effect of pH on mycorrhizal formation of slash pine in aseptic culture. *Forest Science* **11**, 66–75.

Marx DH, Bryan WC, Cordell CE 1976 Growth and ectomycorrhizal development of pine seedlings in nursery soils infested with the fungal symbiont *Pisolithus tinctorius*. *Forest Science* **22**, 91–100.

Marx DH, Hatch AB, Mendicino JF 1977 High fertility decreases sucrose content and susceptibility of Loblolly pine roots to ectomycorrhizal infection by *Pisolithus tinctorius*. *Canadian Journal of Botany* **55**, 1569–1574.

Marx DH, Morris WG, Mexal JG 1978 Growth and ectomycorrhizal development of loblolly pine seedlings in fumigated and nonfumigated nursery soil infested with different fungal symbionts. *Forest Science* **24**, 193–203.

Marx DH, Cordell CE, Kenney DS *et al.* 1984a Commercial vegetative inoculum of *Pisolithus tinctorius* and inoculation techniques for development of ectomycorrhizae on bare-root tree seedlings. *Forest Science* **30**, 1–101.

Marx DH, Jarl K, Ruehle JL, Bell W 1984b Development of *Pisolithus tinctorius* ectomycorrhizae on pine seedlings using basidiospore encapsulated seeds. *Forest Science* **30**, 897–907.

Marx DH, Cordell CE, Clark AI 1988 Eight-year performance of loblolly pine with *Pisolithus* ectomycorrhizae on a good-quality forest site. *Southern Journal of Applied Forestry* **12**, 275–280.

Marx DH, Maul SB, Cordell CE 1989 Application of specific ectomycorrhizal fungi in world forestry. In *Frontiers in Industrial Mycology.* pp. 78–98. American Mycological Society, Chapman Hall, New York, USA.

Marx DH, Ruehle JL, Cordell CE 1991 Methods for studying nursery and field responses of trees to specific ectomycorrhiza. *Methods in Microbiology* **23**, 383–412.

Marx DH, Marrs LF, Cordell CE 2002 Practical use of mycorrhizal fungal technology in forestry, reclamation, arboriculture, agriculture and horticulture. *Dendrobiology* **47**, 29–42.

Mason PA, Last FT, Pelham J, Ingleby K 1982 Ecology of some fungi associated with an ageing stand of birches (*Betula pendula* and *B. pubescens*). *Forest Ecology and Management* **4**, 19–39.

Mason PA, Wilson J, Last FT 1983 The concept of succession in relation to the spread of sheathing mycorrhizal fungi in inoculated tree seedlings growing in unsterile soils. *Plant and Soil* **71**, 247–256.

Massicotte HB, Peterson RL, Ackerley CA, Piché Y 1986 Structure and ontogeny of *Alnus crispa - Alpova diplophloeus* ectomycorrhizae. *Canadian Journal of Botany* **64**, 177–192.

Massicotte HB, Melville LH, Peterson RL 1987a Scanning electron microscopy of ectomycorrhizae, potential and limitations. *Scanning Microscopy* **1**, 1439–1454.

Massicotte HB, Peterson RL, Ackerley CA, Ashford AE 1987b Ontogency of *Eucalyptus pilularis – Pisolithus tinctorius* ectomycorrhizae. II. Transmission electron microscopy. *Canadian Journal of Botany* **65**, 1940–1947.

Massicotte HB, Peterson RL, Ashford AE 1987c Ontogeny of *Eucalyptus pilularis – Pisolithus tinctorius* ectomycorrhizae. I. Light microscopy and scanning electron microscopy. *Canadian Journal of Botany* **65**, 1927–1939.

Massicotte HB, Peterson RL, Melville LH 1989 Ontogeny of *Alnus diplophloeus* ectomycorrhizae I. Light microscopy and scanning electron microscopy. *Canadian Journal of Botany* **67**, 191–200.

Massicotte HB, Peterson RL, Ackerley CA, Melville LH 1990 Structure and ontogeny of *Betula alleghaniensis-Pisolithus tinctorius* ectomycorrhizae. *Canadian Journal of Botany* **68**, 579–593.

Massicotte HB, Melville LH, Li CY, Peterson RL1992 Structural aspects of Douglas-fir (*Pseudotsuga menziesii* (Mirb) Franco) tuberculate ectomycorrhizae. *Trees* **6**, 137–146.

Massicotte HB, Melville LH, Molina R, Peterson RL 1993 Structure and histochemistry of mycorrhizae synthesized between *Arbutus menziesii* (Ericaceae) and two basidiomycetes, *Pisolithus tinctorius* Pisolithaceae and *Piloderma bicolor* Corticiaceae. *Mycorrhiza* **3**, 1–11.

Massicotte HB, Melville LH, Peterson RL, Molina R 1999 Biology of the ectomycorrhizal fungal genus, Rhizopogon - IV. Comparative morphology and anatomy of ectomycorrhizas synthesized between several *Rhizopogon* species on Ponderosa pine (*Pinus ponderosa*). *New Phytologist* **142**, 355–370.

Masuhara G and Katsuya K 1991 Fungal coil formation of *Rhizoctonia repens* in seedlings of *Galeola septentrionalis* (Orchidaceae). *The Botanical Magazine of Tokyo* **104**, 275–281.

Masuhara G and Katsuya K 1992 Mycorrhizal differences between genuine roots and tuberous roos of adult plants of *Sprianthes sinensis* var *amoena* (Orchidaceae). *Botanical Magazine, Tokyo* **105**, 453–460.

Masuhara G and Katsuya K 1994 *In situ* and *in vitro* specificity between *Rhizoctonia* spp. and *Spiranthes sinensis* (Persoon) Ames. var *amoena* (M. Bieberstein) Hara (Orchidaceae). *New Phytologist* **127**, 711–718.

Masuhara G, Kimura S, Katsuya K 1988 Seasonal changes in the mycorrhizae of *Bletilla striata* (Orchidaceae). *Transactions of the Mycological Society of Japan* **29**, 25–31.

Masuhara G, Katsuya K, Yamaguchi K 1993 Potential for symbiosis of *Rhizoctonia solani* and binucleate *Rhizoctonia* with seeds of *Spiranthes sinensis* var *amoena in vitro*. *Mycological Research* **97**, 746–752.

Matamala R, Gonzalez-Meler MA, Jastrow JD, Norby RJ, Schlesinger WH 2003 Impacts of fine root turnover on forest NPP and soil C sequestration potential. *Science* **302**, 1385–1387.

Mauperin CH, Mortier F, Garbaye J, Le Tacon F, Garr G 1987 Viability of an ectomycorrhizal inoculum produced in a liquid medium and entrapped in a calcium alginate gel. *Canadian Journal of Botany* **65**, 2326–2329.

Maurel C and Chrispeels MJ 2001 Aquaporins. A molecular entry into plant water relations. *Plant Physiology* **125** 1, 135–138.

May RM 1981 Patterns in multi-species communities. In *Theoretical Ecology. Principles and Applications*. Ed. RM May pp. 197–227. Blackwell Scientific, Oxford, UK.

Mayo K, Davis RE, Motta J 1986 Stimulation of germination of spores of *Glomus versiforme* by spore associated bacteria. *Mycologia* **78**, 426–431.

Mayor JR and Henkel TW 2006 Do ectomycorrhizas alter leaf-litter decomposition in monodominant tropical forests of Guyana? *New Phytologist* **169**, 579–588.

Mazzola M, Wong OT, Cook RJ 1996 Virulence of *Rhizoctonia oryzae* and *R. solani* AG-8 on wheat and detection of *R. oryzae* in plant tissue by PCR. *Phytopathology* **86**, 354–360.

McAfee BJ and Fortin A 1986 Competitive interactions of ectomycorrhizal mycobionts under field conditions. *Canadian Journal of Botany* **64**, 848–852.

McAfee BJ and Fortin JA 1988 Comparative effects of the soil microflora on ectomycorrhizal inoculation of conifer seedlings. *New Phytologist* **108**, 443–449.

McComb AL 1938 The relation between mycorrhizae and the development and nutrient absorption of pine seedlings in a prairie nursery. *Journal of Forestry* **36**, 1148–1154.

McComb AL and Griffith JE 1946 Growth stimulation and phosphorus absorption of mycorrhizal and non-mycorrhizal White Pine and Douglas Fir seedlings in relation to fertiliser treatment. *Plant Physiology* **21**, 11–17.

McCormick MK, Whigham DF, O'Neill J 2004 Mycorrhizal diversity in photosynthetic terrestrial orchids. *New Phytologist* **163**, 425–438.

McGee PA 1985 Lack of spread of endomycorrhizas of *Centaurium* (Gentianaceae). *New Phytologist* **101**, 451–458.

McGee PA 1986 Mycorrhizal associations of plant species in a semiarid community. *Australian Journal of Botany* **34**, 585–593.

McGee PA 1989 Variation in propagule numbers of vesicular-arbuscular mycorrhizal fungi in a semi-arid soil. *Mycological Research* **92**, 28–33.

McGee PA and Baczocha N 1994 Sporocarpic Endogonales and Glomales in the scats of *Rattus* and *Perameles*. *Mycological Research* **98**, 246–249.

McGonigle TP 1988 A numerical analysis of published field trials with vesicular-arbuscular mycorrhizal fungi. *Functional Ecology* **2**, 473–478.

McGonigle TP and Fitter AH 1988a Ecological consequences of arthropod grazing on VA mycorrhizal fungi. *Proceedings of the Royal Society of Edinburgh* **94B**, 25–32.

McGonigle TP and Fitter AH 1988b Growth and phosphorus inflows of *Trifolium repens* L. with a range of indigenous vesicular-arbuscular mycorrhizal infection levels under field conditions. *New Phytologist* **108**, 59–65.

McGonigle TP and Fitter AH 1990 Ecological specificity of vesicular-arbuscular mycorrhizal associations. *Mycological Research* **94**, 120–122.

McGonigle TP and Miller MH 1993 Responses of mycorrhizae and shoot phosphorus of maize to the frequency and timing of soil disturbance. *Mycorrhiza* **4**, 63–68.

McGonigle TP and Miller MH 2000 The inconsistent effect of soil disturbance on colonization of roots by arbuscular mycorrhizal fungi: a test of the inoculum density hypothesis. *Applied Soil Ecology* **14**, 147–155.

McGonigle TP, Miller MH, Evans DG, Fairchild GL, Swan JA 1990 A new method which gives an objective measure of colonization of roots by vesicular-arbuscular mycorrhizal fungi. *New Phytologist* **115**, 495–501.

McGonigle TP, Yano K, Shinhama T 2003 Mycorrhizal phosphorus enhancement of plants in undisturbed soil differs from phosphorus uptake stimulation by arbuscular mycorrhizae over non-mycorrhizal controls. *Biology and Fertility of Soils* **37**, 268–273.

McHugh TA, Ghering CA 2006 Below-ground interactions with arbuscular mycorrhizal shrubs decrease the performance of pinyon pine and the abundance of its ectomycorrhizas. *New Phytologist* **171**, 171–178.

McIlveen WD and Cole H 1976 Spore dispersal of Endogonaceae by worms, ants, wasps and birds. *Canadian Journal of Botany* **54**, 1486–1489.

McKendrick SL, Leake JR, Read DJ 2000a Symbiotic germination and development of myco-heterotrophic plants in nature: transfer of carbon from ectomycorrhizal *Salix repens* and *Betula pendula* to the orchid *Corallorhiza trifida* through shared hyphal connections. *New Phytologist* **145**, 539–548.

McKendrick SL, Leake JR, Taylor DL, Read DJ 2000b Symbiotic germination and development of myco-heterotrophic plants in nature: ontogeny of *Corallorhiza trifida* and characterization of its mycorrhizal fungi. *New Phytologist* **145**, 523–537.

McKendrick SL, Leake JR, Taylor DL, Read DJ 2002 Symbiotic germination and development of the myco-heterotrophic orchid *Neottia nidus-avis* in nature and its requirement for locally distributed *Sebacina* spp. *New Phytologist* **154**, 233–247.

McKercher RB and Anderson G 1986 Characterisation of the inositol penta- and hexaphosphate fractions of a number of Canadian and Scottish soils. *Journal of Soil Science* **19**, 302–310.

McLean CB, Cunnington JH, Lawrie AC 1999 Molecular diversity within and between ericoid endophytes from the Ericaceae and Epacridaceae. *New Phytologist* **144**, 351–358.

McNabb RFR 1961 Mycorrhiza in New Zealand Ericales. *Australian Journal of Botany* **9**, 57–61.

McNaughton SJ and Oesterheld M 1990 Extramatrical mycorrhizal abundance and grass nutrition in a tropical grazing ecosystem, the Serengeti National Park, Tanzania. *Oikos* **59**, 92–96.

McNeill A and Unkovich M 2006 The nitrogen cycle in terrestrial ecosystems. In *Nutrient Cycling in Terrestrial Ecosystems*. Eds P Marschner and Z Rengel pp. 37–64. Springer-Verlag, Berlin Heidelberg, Germany.

McNeill A and Wood M 1990 Fixation and transfer of nitrogen by white clover to rygrass. *Soil Use and Management* **6**, 84–86.

Meharg AA 1994 Integrated tolerance mechanisms: constitutive and adaptive plant responses to elevated metal concentrations in the environonment. *Plant, Cell and Environment* **17**, 989–993.

Meharg AA 2003 The mechanistic basis of interactions between mycorrhizal associations and toxic metal cations. *Mycological Research* **107**, 1253–1265.

Meharg AA 2004 Arsenic in rice – understanding a new disaster for South-East Asia. *Trends in Plant Science* **9**, 415–417.

Meharg AA and Cairney JWG 1999 Co-evolution of mycorrhizal symbionts and their hosts to metal contaminated environments. *Advances in Ecological Research* **30**, 70–112.

Meharg AA and Hartley-Whitaker J 2002 Arsenic uptake and metabolism in arsenic resistant and nonresistant plant species. *New Phytologist* **154**, 29–43.

Meharg AA and MacNair MR 1992 Supression of the high-affinity phosphate-uptake system – a mechanism of arsenate tolerance in *Holcus lanatus* L. *Journal of Experimental Botany* **43**, 519–524.

Meharg AA, Bailey K, Breadmore K, MacNair MR 1994 Biomass allocation, phosphorus nutrition and vesicular-arbuscular mycorrhizal infection in clones of Yorkshire Fog, *Holcus lanatus* L. (Poaceae) that differ in their phosphate uptake kinetics and tolerance to arsenate. *Plant and Soil* **160**, 11–20.

Meixner C, Ludwig-Muller J, Miersch O, Greshoff P, Staehelin C, Vierheilig H 2005 Lack of mycorrhizal autoregulation and phytohormonal changes in the supernodulating soybean mutant *nts1007*. *Planta* **222**, 709–715.

Mejstrik V 1970 The anatomy of roots and mycorrhizae of the orchid *Dendrobium cunnighamii* Lindl. *Biologia Plantarum Praha* **12**, 105–109.

Mejstrik VK and Krause HH 1973 Uptake of ^{32}P by *Pinus radiata* roots inoculated with *Suillus luteus* and *Cenococcum graniforme* from different sources of available phosphate. *New Phytologist* **72**, 137–140.

Melin E 1917 Studier över de Norrlandska Myrmarkernas Vegetation med Särskildhänsyn till deras Skogsvegetation efter torrläggning. *Norrlands Handbibliotek* **7**, Uppsala, Sweden.

Melin E 1923 Experimentelle Untersuchungen über die Konstitution and Ökologie der Mykorrhizen von *Pinus sylvestris* und *Picea abies* L Karst. *Mykologische Untersuchungen und Berichte* **2**, 72–331.

Melin E 1925 *Untersuchungen über die Bedeutung der Baummykorriza*. Gustav Fischer, Jena, Germany.

Melin E 1927 Studier över barrträdsplatans utveckling i rahumus. II. Mykorrhizans utbildning hos tallplantan i olika rahumus former. *Meddelanden fran Statens Skogsforskningsinstitut* **23**, 433–494.

Melin E and Nilsson H 1950 Transfer of radioactive phosphorus to pine seedlings by means of mycorrhizal hyphae. *Physiologia Plantarum* **3**, 88–92.

Melin E and Nilsson H 1953a Transfer of labelled nitrogen from glutamic acid to pine seedlings through the mycelium of *Boletus variegatus* S.W. Fr. *Nature, London* **171**, 434.

Melin E and Nilsson H 1953b Transport of labelled phosphorus to pine seedlings through the mycelium of *Cortinarius glaucopus* Shaeff ex. Fr. Fr. *Svensk Botanisk Tidskrift* **48**, 555–558.

Melin E and Nilsson H 1955 ^{45}Ca used as indicator of transport of cations to pine seedlings by means of mycorrhizal mycelium. *Svensk Botanisk Tidskrift* **49**, 119–121.

Melin E and Nilsson H 1957 Transport of C^{14}-labelled photosynthate to the fungal associate of pine mycorrhiza. *Svensk Botanisk Tidskrift* **51**, 166–186.

Mellersh D and Parniske M 2006 Common symbiosis genes of *Lotus japonicus* are not required for intracellular accommodation of the rust fungus *Uromyces loti*. *New Phytologist* **170**, 641–644.

Menge JA 1982 Effects of soil fumigants and fungicides on vesicular-arbuscular fungi. *Phytopathology* **72**, 1125–1132.

Menge JA 1983 Utilization of vesicular-arbuscular mycorrhizal fungi in agriculture. *Canadian Journal of Botany* **61**, 1015–1024.

Menge JA 1984 Inoculum production. In *VA Mycorrhizae*. Eds CL Powell and DJ Bagyaraj pp. 187–201. CRC Press, Boca Raton, Florida, USA.

Menge JA, Jarrell WM, Labanauskas CK *et al.* 1982 Predicting mycorrhizal dependency of Troyer Citrange on *Glomus fasciculatus* in California citrus soils and nursery mixes. *Soil Science Society America Journal* **46**, 762–768.

Menkis A, Allmer J, Vasiliauskas R, Lygis V, Stenlid J, Finlay R 2004 Ecology and molecular characterization of dark septate fungi from roots, living stems, coarse and fine woody debris. *Mycological Research* **108**, 965–973.

Merryweather JW and Fitter AH 1991 A modified method for elucidating the structure of the fungal partner in a vesicular-arbuscular mycorrhiza. *Mycological Research* **95**, 1435–1437.

Merryweather J and Fitter A 1995a Phosphorus and carbon budgets: mycorrhizal contribution in *Hyacinthoides non-scripta* (L.) Chouard ex Rothm. under natural conditions. *New Phytologist* **129**, 619–627.

Merryweather J and Fitter A 1995b Arbuscular mycorrhiza and phosphorus as controlling factors in the life history of *Hyacinthoides non-scripta* (L.) Chouard ex Rothm. *New Phytologist* **129**, 629–636.

Merryweather J and Fitter A 1996 Phosphorus nutrition of an obligately mycorrhizal plant treated with the fungicide benomyl in the field. *New Phytologist* **132**, 307–311.

Merryweather J and Fitter A 1998a The arbuscular mycorrhizal fungi of *Hyacinthoides non-scripta* I. Diversity of fungal taxa. *New Phytologist* **138**, 117–129.

Merryweather J and Fitter A 1998b The arbuscular mycorrhizal fungi of *Hyacinthoides non-scripta* II. Seasonal and spatial patterns of fungal populations. *New Phytologist* **138**, 131–142.

Merryweather JW and Fitter AH 1998c Patterns of arbuscular mycorrhizal colonisation of the roots of *Hyacinthoides non-scripta* after disruption of soil mycelium. *Mycorrhiza* **8**, 87–92.

Mexal J and Reid CPP 1973 Growth of selected mycorrhizal fungi in response to induced water stress. *Canadian Journal of Botany* **51** 9, 1579–1588.

Meyer FH 1973 Distribution of ectomycorrhizae in native and man-made forests. In *Ectomycorrhizae*. Eds GC Marks and TT Kozlowski pp. 79–105. Academic Press, New York, USA.

Meysselle JP, Gay G, Debaud JC 1991 Intraspecific genetic variation in acid phosphatase activity in monokaryotic and dikaryotic population of the ectomycorrhizal fungus *Hebeloma cylindrosporum*. *Canadian Journal of Botany* **69**, 808–813.

Michalenko GO, Hohl HR, Rast D 1976 Chemistry and architecture of mycelial wall of *Agaricus bisporus*. *Journal of General Microbiology* **92**, 251–262.

Michelsen A, Schmidt IK, Jonasson S, Quarmby C, Sleep D 1996 Leaf ^{15}N abundance of subarctic plants provides field evidence that ericoid, ectomycorrhizal and non- and arbuscular mycorrhizal species access different sources of soil nitrogen. *Oecologia* **105**, 53–63.

Midgley DJ, Chambers SM, Cairney JWG 2002 Spatial distribution of fungal endophyte genotypes in a *Woollsia pungens* (Ericaceae) root system. *Australian Journal of Botany* **50**, 559–565.

Midgley DJ, Chambers SM, Cairney JWG 2004a Distribution of ericoid mycorrhizal endophytes and root-associated fungi in neighbouring Ericaceae plants in the field. *Plant and Soil* **259**, 137–151.

Midgley DJ, Chambers SM, Cairney JWG 2004b Inorganic and organic substrates as sources of nitrogen and phosphorus for multiple genotypes of two ericoid mycorrhizal fungal taxa from *Woollsia pungens* and *Leucopogon parviflorus* (Ericaceae). *Australian Journal of Botany* **52**, 63–71.

Midgley DJ, Chambers SM, Cairney JWG 2004c Utilisation of carbon substrates by multiple genotypes of ericoid mycorrhizal fungal endophytes from eastern Australian Ericaceae. *Mycorrhiza* **14**, 245–251.

Miflin BJ and Lea PJ 1976 The path of ammonia assimilation on the plant kingdom. *Trends in Biochemical Sciences* **1**, 103–106.

Mikola P 1948 On the physiology and ecology of *Cenococcum graniforme*. *Communicationes Instituti Forestalis Fenniae* **36**, 1–104.

Mikola P 1965 Studies on ectendotrophic mycorrhiza of pine. *Acta Forestalia Fennica* **79**, 1–56.

Mikola P 1970 Mycorrhizal inoculation in afforestation. *International Review of Forest Research* **3**, 123–196.

Mikola P 1998 Ectendomycorrhiza of conifers. *Silva Fennica* **22**, 19–27.

Miller JC, Rajapakse S, Garber RK 1986 Vesicular-arbuscular mycorrhizae in vegetable crops. *HortScience* **21**, 974–984.

Miller M, McGonigle T, Addy H 1994 An economic approach to evaluate the role of mycorrhizas in managed ecosystems. *Plant and Soil* **159**, 27–35.

Miller RM and Jastrow JD 1990 Hierarchy of root and mycorrhizal fungal interactions with soil aggregation. *Soil Biology and Biochemistry* **22**, 579–584.

Miller RM and Jastrow JD 1992a The application of VA mycorrhizae to ecosystem restoration and reclamation. In *Mycorrhizal Functioning*. Ed. MF Allen pp. 438–467. Chapman and Hall, London, UK.

Miller RM and Jastrow JD 1992b The role of mycorrhizal fungi in soil conservation. In *Proceedings of a Symposium on Mycorrhizae in Sustainable Agriculture*. Eds GJ Bethlenfalvay and RG Linderman pp. 29–44. ASA Special Publication No. 54, Madison, Wisconsin, USA.

Miller RM and Jastrow JD 2002 Mycorrhizal influence on soil structure. In *Arbuscular Mycorrhizae: Molecular Biology and Physiology*. Eds Y Kapulnik and DD Douds pp. 3–18. Kluwer Academic Publishers, Dordrecht, The Netherlands.

Miller RM and Lodge DJ 1997 Fungal responses to disturbance: agriculture and forestry. In *Mycota IV Environmental and Microbial Relationships*. Eds Wicklow and B Söderström pp. 65–84. Springer-Verlag, Berlin, Germany.

Miller SL and Allen EB 1992 Mycorrhizae, nutrient translocation and interactions between plants. In *Mycorrhizal Functioning*. Ed. MF Allen pp. 301–332. Chapman and Hall, London, UK.

Miller SL, Koo CD, Molina R 1991 Characterization of red alder ectomycorrhizae – a preface to monitoring belowground ecological responses. *Canadian Journal of Botany* **69**, 516–531.

Minerdi D, Bianciotto V, Bonfante P 2002 Endosymbiotic bacteria in mycorrhizal fungi: from their morphology to genomic sequences. *Plant and Soil* **244**, 211–219.

Mirmanto E, Proctor J, Green J, Nagy L, Suriantata 1999 Effects of nitrogen and phosphorus fertilization in a lowland evergreen rainforest. *Philosophical Transactions of the Royal Society of London Series B, Biological Sciences* **354**, 1825–1829.

Mitchell DT and Read DJ 1981 Utilisation of inorganic and organic phosphates by the mycorrhizal endophytes of *Vaccinium macrocarpon* and *Rhododendron ponticum*. *Transactions of the British Mycological Society* **76**, 255–260.

Mitchell DT, Sweeney M, Kennedy A 1992 Chitin degradation by *Hymenoscyphus ericae* and the influence of *H. ericae* on the growth of ectomycorrhizal fungi. In *Mycorrhizas in Ecosystems*. Eds DJ Read, DH Lewis, AH Fitter, IJ Alexander pp. 246–251. CAB International, Wallingford, UK.

Mitchell HL, Finn RF, Rosendahl RO 1937 The growth and nutrition of white pine (*Pinus strobus* L.) seedlings in culture. *Black Rock Forest Papers* **1**, 58–73.

Mitchell RJ, Garrett HE, Cox GS, Atalay A 1986 Boron and ectomycorrhizal influences on indole-3-acetic acid levels and indole-3-acetic acid oxidase and peroxidase activities of *Pinus echinata* roots. *Tree Physiology* **1**, 1–8.

Modjo HS and Hendrix JW 1986 The mycorrhizal fungus *Glomus macrocarpum* as a cause of tobacco stunt disease. *Phytopathology* **76**, 668–691.

Modjo HS, Hendrix JW, Nesmith WC 1987 Mycorrhizal fungi in relation to control of tobacco stunt disease with soil fumigants. *Soil Biology and Biochemistry* **19**, 289–295.

Molina R 1979 Pure culture synthesis and host specificity of red alder mycorrhizae. *Canadian Journal of Botany* **57**, 1223–1228.

Molina R 1981 Ectomycorrhizal specificity in the genus *Alnus*. *Canadian Journal of Botany* **59**, 325–334.

Molina R 1982 Use of the ectomycorrhizal fungus *Laccaria laccata* in forestry. 1. Consistency between isolates in effective colonisation of containerised conifer seedlings. *Canadian Journal of Botany* **12**, 469–473.

Molina R and Chamard J 1983 Use of the ectomycorrhizal fungus *Laccaria laccata* in forestry. II. Effects of fertiliser forms and levels on ectomycorrhizal development and growth of container-grown Douglas-fir and Ponderosa pine. *Canadian Journal of Forest Research* **13**, 89–95.

Molina R and Trappe JM 1982a Lack of mycorrhizal specificity in the ericaceous hosts *Arbutus menziesii* and *Arctostaphylos uva-ursi*. *New Phytologist* **90**, 495–509.

Molina R and Trappe JM 1982b Patterns of ectomycorrhizal host specificity and potential among Pacific Northwest conifers and fungi. *Forest Science* **28**, 423–458.

Molina R, Massicotte H, Trappe JM 1992 Specificity phenomena in mycorrhizal symbiosis: community ecological consequences and practical application. In *Mycorrhizal Functioning*. Ed. MF Allen pp. 357–423. Chapman and Hall, New York, USA.

Molina R, Massicotte HB, Grubisha LC, Spatafora JW 1999 *Rhizopogon*. In *Ectomycorrhizal Fungi: Key Genera in Profile*. Eds JWG Cairney and SM Chambers pp. 129–161. Springer-Verlag, Heidelberg, Germany.

Mollison JE 1943 *Goodyera repens* and its endophyte. *Transactions of the Botanical Society of Edinburgh* **33**, 391–403.

Moncalvo JM, Vilgalys R, Redhead SA *et al.* 2002 One hundred and seventeen clades of euagarics. *Molecular Phylogenetics and Evolution* **23**, 357–400.

Monreal M, Berch SM, Berbee M 1999 Molecular diversity of ericoid mycorrhizal fungi. *Canadian Journal of Botany* **77**, 1580–1594.

Montanini B, Moretto N, Soragni E, Percudani R, Ottonello S 2002 A high-affinity ammonium transporter from the mycorrhizal ascomycete *Tuber borchii*. *Fungal Genetics and Biology* **36**, 22–34.

Moora M and Zobel M 1997 Can arbuscular mycorrhiza change the effect of root competition between conspecific plants of different ages? *Canadian Journal of Botany* **76**, 613–619.

Moora M, Opik M, Sen R, Zobel M 2004 Native arbuscular mycorrhizal fungal communities differentially influence the seedling performance of rare and common *Pulsatilla* spp. *Functional Ecology* **18**, 554–562.

Moore-Parkhurst S and Englander L 1982 Mycorrhizal status of *Rhododendron* spp. in commercial nurseries in Rhode Island. *Canadian Journal of Botany* **60**, 2342–2344.

Moore RT 1987 The genera of Rhizoctonia-like fungi – *Ascorhizoctonia*, *Ceratorhiza* Gen-Nov, *Epulorhiza* Gen-Nov, *Moniliopsis*, and *Rhizoctonia*. *Mycotaxon* **29**, 91–99.

Morandi D, Bailey JA, Gianinazzi-Pearson V 1984 Isoflavonoid accumulation in soybean roots infected with vesicular-arbuscular mycorrhizal fungi. *Physiological Plant Pathology* **24**, 357–364.

Morandi D, Gollotte A, Camporota P 2002 Influence of an arbuscular mycorrhizal fungus in the interaction of a binucleate *Rhizoctonia* species with Myc$^+$ and Myc$^-$ pea roots. *Mycorrhiza* **12**, 97–102.

Morandi D, Prado E, Sagan M, Duc G 2005 Characterisation of new symbiotic *Medicago truncatula* Gaertn. mutants, and phenotypic or genetic complementary information on previously described mutants. *Mycorrhiza* **15**, 283–289.

Morel M, Jacob C, Kohler A *et al.* 2005 Identification of genes differentially expressed in extraradical mycelium and ectomycorrhizal roots during *Paxillus involutus-Betula pendula* ectomycorrhizal symbiosis. *Applied and Environmental Microbiology* **71**, 382–391.

Morrison TM 1957a Mycorrhiza and phosphorus uptake. *Nature* **179**, 907.

Morrison TM 1957b Host-endophyte relationship in mycorrhiza of *Pernettya macrostigma*. *New Phytologist* **56**, 247–257.

Morrison TM 1962a Absorption of phosphorus from soils by mycorrhizal plants. *New Phytologist* **61**, 10–20.

Morrison TM 1962b Uptake of sulphur by mycorrhizal plants. *New Phytologist* **61**, 21–27.

Morte A, Diaz G, Rodriguez P, Alarcon JJ, Sanchez-Blanco MJ 2001 Growth and water relations in mycorrhizal and nonmycorrhizal *Pinus halepensis* plants in response to drought. *Biologia Plantarum* **44**, 263–267.

Morton J, Franke M, Cloud G 1992 The nature of fungal species in Glomales (Zygomycetes). In *Mycorrhizas in Ecosystems*. Eds DJ Read, DH Lewis, AH Fitter, IJ Alexander pp. 65–73. CAB International, Wallingford, UK.

Morton JB 1990a Evolutionary relationships among arbuscular mycorrhizal fungi in the Endogonaceae. *Mycologia* **82**, 192–207.

Morton JB 1990b Species and clones of arbuscular mycorrhizal fungi (Glomales, Zygomycetes): their role in macro- and microevolutionary processes. *Mycotaxon* **37**, 493–515.

Morton JB and Benny GL 1990 Revised classification of arbuscular mycorrhizal fungi (Zygomycetes): a new order, Glomales, two new suborders, Glominae and Gigasporineae, and two new families, Acaulosporaceae and Gigasporaceae, with an emendation of Glomaceae. *Mycotaxon* **37**, 471–491.

Mosca AML and Fontana A 1975 Sull utilizzazione dell azoto proteico da paste del micelio di *Boletus luteus* L. *Allionia* **20**, 47–52.

Moser M 1958 Die künstliche Mykorrhizaimpfung an Forstpflanzen. I. Erfahrungen bei der Reinkultur von Mykorrhizapilzen. *Forstwissenschaftlich Centralblatt* **77**, 32–40.

Mosse B 1953 Fructifications associated with mycorrhizal strawberry roots. *Nature* **171**, 974.

Mosse B 1957 Growth and chemical composition of mycorrhizal and non-mycorrhizal apples. *Nature* **179**, 922–924.

Mosse B 1959 The regular germination of resting spores and some observations on the growth requirements of an *Endogone* sp. causing vesicular-arbuscular mycorrhiza. *Transactions of the British Mycological Society* **42**, 273–286.

Mosse B 1973 Advances in the study of vesicular-arbuscular mycorrhiza. *Annual Review of Phytopathology* **11**, 171–196.

Mosse B and Hayman DS 1971 Plant growth responses to vesicular-arbuscular mycorrhiza. II. In unsterilized field soils. *New Phytologist* **70**, 29–34.

Mosse B and Hepper C 1975 Vesicular-arbuscular mycorrhizal infections in root organ cultures. *Physiological Plant Pathology* **5**, 215–223.

Mosse B and Phillips JM 1971 The influence of phosphate and other nutrients on the development of vesicular-arbuscular mycorrhiza in culture. *Journal of General Microbiology* **69**, 157–166.

Mosse B, Hayman DS, Arnold DJ 1973 Plant growth responses to vesicular-arbuscular mycorrhiza. V. Phosphate uptake by three plant species from P-deficient soils labelled with ^{32}P. *New Phytologist* **72**, 809–815.

Motta JJ 1969 Cytology and morphogenesis in the rhizomorph of *Armillaria mellea*. *American Journal of Botany* **56**, 610–619.

Mougel C, Offre P, Ranjard L *et al.* 2006 Dynamic of the genetic structure of bacterial and fungal communities at different developmental stages of *Medicago truncatula* Gaertn. cv. Jemalong line J5. *New Phytologist* **170**, 165–175.

Moyersoen B 1993 Ectomicorrizas and micorrhizas del sur de Venezuela. *Scientia Guaianae* **3**, Caracas, Venezuela.

Moyersoen B 2006 *Pakaraimaea dipterocarpaceae* is ectomycorrhizal, indicating an ancient Gondwanaland origin of the ectomycorrhizal habit in Dipterocarpaceae *New Phytologist* **172**, 753–762.

Moyersoen B, Fitter AH, Alexander IJ 1998a Spatial distribution of ectomycorrhizas and arbuscular mycorrhizas in Korup National Park rain forest, Cameroon, in relation to edaphic parameters. *New Phytologist* **139**, 311–320.

Moyersoen B, Alexander IJ, Fitter AH 1998b Phosphorus nutrition of ectomycorrhizal and arbuscular mycorrhizal tree seedlings from a lowland tropical rain forest in Korup National Park, Cameroon. *Journal of Tropical Ecology* **14**, 47–61.

Moyersoen B, Becker P, Alexander IJ 2001 Are ectomycorrhizas more abundant than arbuscular mycorrhizas in tropical heath forests? *New Phytologist* **150**, 591–599.

Mueller WC, Tessier BJ, Englander L 1986 Immunocytochemical detection of fungi in the roots of *Rhododendron*. *Canadian Journal of Botany* **64**, 718–723.

Mueller GM 1991 *Laccaria laccata* complex in North America and Sweden – intercollection pairing and morphometric analyses. *Mycologia* **83**, 578–594.

Muhsin TM and Zwiazek JJ 2002 Ectomycorrhizas increase apoplastic water transport and root hydraulic conductivity in *Ulmus americana* seedlings. *New Phytologist* **153** 1, 153–158.

Mullen RB and Schmidt SK 1993 Mycorrhizal infection, phosphorus uptake, and phenology in *Ranunculus adoneus*: implications for the functioning of mycorrhizae in alpine systems. *Oecologia* **94**, 229–234.

Muller J 2003 Artificial infection by endophytes affects growth and mycorrhizal colonisation of *Lolium perenne*. *Functional Plant Biology* **30** 4, 419–424.

Muller WH, Stalpers JA, van Aelst AC, van der Krift TP, Boekhout T 1998 Field emission gun-scanning electron microscopy of septal pore caps of selected species in the *Rhizoctonia* s.l. complex. *Mycologia* **90**, 170–179.

Mulligan MF, Smucker AJM, Safir GS 1985 Tillage modifications of dry edible bean root colonization by VAM fungi. *Agronomy Journal* **77**, 140–142.

Munkvold L, Kjøller R, Vestberg M, Rosendahl S, Jakobsen I 2004 High functional diversity within species of arbuscular mycorrhizal fungi. *New Phytologist* **164**, 357–364.

Münzenberger, B 1991 Lösliche und Zellwandgebundense Phenole in Mykorrhizen und nicht mykorrhizierten Wurzeln der Fichte (*Picea abies* L. Karst) und des Erdbeerbaumes (*Arbutus unedo* L.) und ihre Bedeutung in der Pilz-Wurzel-Interaktion. Dissertation. University of Tübingen, Germany.

Münzenberger B, Kottke I, Oberwinkler F 1992 Ultrastructural investigations of *Arbutus unedo-Laccaria amethystea* mycorrhiza synthesised *in vitro*. *Trees* **7**, 40–47.

Murdoch CL, Jackobs JA, Gerdemann JW 1967 Utilization of phosphorus sources of different availability by mycorrhizal and non-mycorrhizal maize. *Plant and Soil* **27**, 329–334.

Murphy PJ, Langridge P, Smith SE 1997 Cloning plant genes differentially expressed during colonization of roots of *Hordeum vulgare* by the vesicular-arbuscular mycorrhizal fungus *Glomus intraradices*. *New Phytologist* **135**, 291–301.

Myers MD and Leake JR 1996 Phosphodiesters as mycorrhizal P sources II. Ericoid mycorrhiza and the utilisation of nuclei as a phosphorus source by *Vaccinium macrocarpon*. *New Phytologist* **132**, 445–451.

Nadelhoffer KJ and Fry B 1994 Nitrogen isotope studies in forest ecosystems. In *Stable Isotopes in Ecology and Environmental Science*. Eds K Lajtha and R Michener pp. 22–44. Blackwell, London, UK.

Nadian H, Smith SE, Alston AM, Murray RS 1996 The effect of soil compaction on growth and P uptake by *Trifolium subterraneum*: interactions with mycorrhizal colonisation. *Plant and Soil* **182**, 39–49.

Nadian H, Smith SE, Alston AM, Murray RS 1997 Effects of soil compaction on plant growth, phosphorus uptake and morphological characteristics of vesicular-arbuscular mycorrhizal colonization of *Trifolium subterraneum*. *New Phytologist* **135**, 303–311.

Nadian H, Smith SE, Alston AM, Murray RS, Siebert BD 1998 Effects of soil compaction on phosphorus uptake and growth of *Trifolium subterraneum* colonized by four species of vesicular-arbuscular mycorrhizal fungi. *New Phytologist* **139**, 155–165.

Nagahashi G and Douds DD 1997 Appressorium formation by AM fungi on isolated cell walls of carrot roots. *New Phytologist* **136**, 299–304.

Nagy R, Karandashov V, Chague W *et al.* 2005a The characterization of novel mycorrhiza-specific phosphate transporters from *Lycopersicon esculentum* and *Solanum tuberosum* uncovers functional redundancy in symbiotic phosphate transport in solanaceous species. *The Plant Journal* **42**, 236–250.

Nagy R, Vasconcelos MJV, Zhao S *et al.* 2005b Differential regulation of five Pht1 phosphate transporters from maize *Zea mays* L. *Plant Biology* **8**, 186–197.

Nair MG, Safir GR, Siqueira JO 1991 Isolation and identification of vesicular-arbuscular mycorrhiza-stimulatory compounds from clover (*Trifolium repens*) roots. *Applied and Environmental Microbiology* **57**, 434–439.

Nakamura SJ 1982 Nutritional conditions required for non-symbiotic culture of an achlorophyllous orchid, *Galeola septentrionalis*. *New Phytologist* **90**, 701–715.

Nara K 2006a Ectomycorrhizal networks and seedling establishment during early primary succession. *New Phytologist* **169**, 169–178.

Nara K 2006b Pioneer dwarf willow may facilitate tree succession by providing late colonizers with compatible ectomycorrhizal fungi in a primary successional volcanic desert. *New Phytologist* **171** 1, 187–198.

Nara K and Hogetsu T 2004 Ectomycorrhizal fungi on established shrubs facilitate subsequent seedling establishment of successional plant species. *Ecology* **85** 6, 1700–1707.

Nara K, Nakaya H, Hogetsu T 2003a Ectomycorrhizal sporocarp succession and production during early primary succession on Mount Fuji. *New Phytologist* **158**, 193–206.

Nara K, Nakaya H, Wu BY, Zhou ZH, Hogetsu T 2003b Underground primary succession of ectomycorrhizal fungi in a volcanic desert on Mount Fuji. *New Phytologist* **159**, 743–756.

Nascimento MT and Proctor J 1997 Soil and plant changes across a monodominant rain forest boundary on Maraca Island, Brazil. *Global Ecology and Biogeography Letters* **6**, 387–395.

Nasholm T and Persson J 2001 Plant acquisition of organic nitrogen in boreal forests. *Physiologia Plantarum* **111** 4, 419–426.

Nasholm T, Ekblad A, Nordin A, Giesler R, Högberg M, Högberg P 1998 Boreal forest plants take up organic nitrogen. *Nature* **392**, 914–916.

Nebel M, Kreier H-P, Preussung M, Weiss M 2004 Symbiotic fungal associations of liverworts are the possible ancestors of mycorrhizae. In *Frontiers in Basidiomycote Mycology*. Eds R Agerer, M Piepenbring, P Blanz pp. 339–360. HIW Verlag, Ecking, Germany.

Nehls U, Wiese J, Guttenberger M, Hampp R 1998 Carbon allocation in ectomycorrhizas: identification and expression analysis of an *Amanita muscaria* monosaccharide transporter. *Molecular Plant-Microbe Interactions* **11**, 167–176.

Nehls U, Kleber R, Wiese J, Hampp R 1999 Isolation and characterization of a general amino acid permease from the ectomycorrhizal fungus *Amanita muscaria*. *New Phytologist* **144**, 343–349.

Nehls U, Mikolajewski S, Magel E, Hampp R 2001a Carbohydrate metabolism in ectomycorrhizas: gene expression, monosaccharide transport and metabolic control. *New Phytologist* **150**, 533–541.

Nehls U, Bock A, Ecke M, Hampp R 2001b Differential expression of the hexose-regulated fungal genes *AmPAL* and *AmMst1* within *Amanita/Populus* ectomycorrhizas. *New Phytologist* **150**, 583–589.

Nehls U, Bock A, Einig W, Hampp R 2001c Excretion of two proteases by the ectomycorrhizal fungus *Amanita muscaria*. *Plant Cell and Environment* **24**, 741–747.

Nelsen CE 1987 The water relations of vesicular-arbuscular mycorrhizal systems. In *Ecophysiology of VA Mycorrhizal Plants*. Ed. GR Safir pp. 71–91. CRC Press, Boca Raton, Florida, USA.

Nelsen CE and Safir GR 1982 The water relations of well watered, mycorrhizal and non-mycorrhizal onion plants. *Journal of the American Society for Horticultural Science* **107**, 271–274.

Nemeth K, Bartels H, Vogel M, Mengel K 1987 Organic compounds extracted from arable and forest soils by electro-ultrafiltration and recovery rates of amino acids. *Biology and Fertility of Soils* **5**, 271–275.

Neninger A and Heyser W 1998 The pH conditions in the interface of pine ectomycorrhizae. In *Botanikertagung*. Eds E Beifluss, H Bucking, A Mathews, W Heyser. Deutsche Botanische Gesellschaft, Bremen, Germany.

Neumann E and George E 2004 Colonisation with the arbuscular mycorrhizal fungus *Glomus mosseae* Nicol. and Gerd. enhanced phosphorus uptake from dry soil in *Sorghum bicolor* L. *Plant and Soil* **261**, 245–255.

Newbery DM 2005 Ectomycorrhizas and mast fruiting in trees: linked by climate-driven tree resources? *New Phytologist* **167** 2, 324–326.

Newbery DM, Alexander IJ, Thomas DW, Gartlan JS 1988 Ectomycorrhizal rain-forest legumes and soil phosphorus in Korup National Park, Cameroon. *New Phytologist* **109**, 433–450.

Newbery DM, Alexander IJ, Rother JA 1997 Phosphorus dynamics in a lowland African rainforest: the influence of ectomycorrhizal trees. *Ecological Monographs* **67**, 367–409.

Newbery DM, Alexander IJ, Rother JA 2000 Does proximity to conspecific adults influence the establishment of ectomycorrhizal trees in rain forest? *New Phytologist* **147**, 401–409.

Newbery DM, Chuyong GB, Green JJ, Songwe NC, Tchuenteu F, Zimmermann L 2002. Does low phosphorus supply limit seedling establishment and tree growth in groves of ectomycorrhizal trees in a central African rainforest? *New Phytologist* **156**, 297–311.

Newbery DM, Chuyong GB, Zimmermann L 2006 Mast fruiting of large ectomycorrhizal African rain forest trees: importance of dry season intensity, and the resource-limitation hypothesis. *New Phytologist* **170**, 561–579.

Newman EI 1966 A method of estimating the total length of root in a sample. *Journal of Applied Ecology* **3**, 139–145.

Newman EI 1988 Mycorrhizal links between plants: their functioning and ecologial significance. *Advances in Ecological Research* **18**, 243–270.

Newman EI and Reddell P 1987 The distribution of mycorrhizas among families of vascular plants. *New Phytologist* **106**, 745–751.

Newman EI and Ritz K 1986 Evidence on the pathways of phosphorus transfer between vesicular-arbuscular mycorrhizal plants. *New Phytologist* **104**, 77–87.

Newman EI, Child RD, Patrick CM 1986 Mycorrhizal infection in grasses of Kenyan savanna. *Journal of Ecology* **74**, 1179–1183.

Newman EI and Reddell P 1987 The distribution of mycorrhizas among families of vascular plants. *New Phytologist* **106**, 745–751.

Newman EI, Devoy CLN, Easen NJ, Fowles KJ 1994 Plant species that can be linked by VA mycorrhizal fungi. *New Phytologist* **126**, 691–693.

Newsham KK 1999 *Phialophora graminicola*, a dark septate fungus, is a beneficial associate of the grass Vulpia ciliata ssp ambigua. *New Phytologist* **144** 3, 517–524.

Newsham KK, Fitter AH, Watkinson AK 1994 Root pathogenic and arbuscular mycorrhizal fungi determine fecundity of asymptomatic plants in the field. *Journal of Ecology* **82**, 805–814.

Newsham KK, Fitter AH, Watkinson AR 1995 Arbuscular mycorrhiza protect an annual grass from root pathogenic fungi in the field. *Journal of Ecology* **83**, 991–1000.

Newton AC 1991 Mineral nutrition and mycorrhizal infection of seedling oak and birch. III. Epidemiological aspects of ectomycorrhizal information, and the relationship to seedling growth. *New Phytologist* **117**, 53–60.

Newton AC 1992 Towards a functional classification of ectomycorrhizal fungi. *Mycorrhiza* **2**, 75–79.

Nicolson TH 1959 Mycorrhiza in the Graminae I. Vesicular-arbuscular endophytes, with special reference to the external phase. *Transactions of the British Mycological Society* **42**, 421–438.

Nicolson TH 1967 Vesicular-arbuscular mycorrhiza – a universal plant symbiosis. *Science Progress* **55**, 561–581.

Nicolson TH 1975 Evolution of vesicular-arbuscular mycorrhizas. In *Endomycorrhizas*. Eds FE Sanders, B Mosse, PB Tinker pp. 25–34. Academic Press, London, UK.

Nielsen JS, Joner EJ, Declerck S, Olsson S, Jakobsen I 2002 Phospho-imaging as a tool for visualization and noninvasive measurements of P transport dynamics in arbuscular mycorrhizas. *New Phytologist* **154**, 809–819.

Nieuwdorp PJ 1969 Some investigations on the mycorrhiza of *Calluna, Erica* and *Vaccinium*. *Acta Botanic Neerlandica* **18**, 180–196.

Nilsen P, Borja I, Knutsen H, Brean R 1998 Nitrogen and drought effects on ectomycorrhizae of Norway spruce (*Picea abies* L. (Karst.)). *Plant and Soil* **198**, 179–184.

Nilsson LO and Wallander H 2003 Production of external mycelium by ectomycorrhizal fungi in a Norway spruce forest was reduced in response to nitrogen fertilization. *New Phytologist* **158** 2, 409–416.

Norkrans B 1950 Studies in growth and cellulolytic enzymes of *Tricholoma*. *Symbolae Botanicae Upsaliensis* **11**, 126.

Norris JR, Read D, Varma AK eds 1994 *Mycorrhizal Research. Methods in Microbiology*. Academic Press, London, UK.

Northup RR, Yu ZS, Dahlgren RA, Vogt KA 1995 Polyphenol control of nitrogen release from pine litter. *Nature* **377**, 227–229.

Novero M, Faccio A, Genre A *et al.* 2002 Dual requirement of the *LjSym4* gene for mycorrhizal development in epidermal and cortical cells of *Lotus japonicus* roots. *New Phytologist* **154**, 741–749.

Nuesch J 1963 Defense reactions in orchid bulbs. *Symposium of the Society for General Microbiology* **13**, 335–343.

Nuortila C, Kytoviita MM, Tuomi J 2004 Mycorrhizal symbiosis has contrasting effects on fitness components in *Campanula rotundifolia*. *New Phytologist* **164**, 543–553.

Nurmiaho-Lassila EL, Timonen S, Haahtela K, Sen R 1997 Bacterial colonization patterns of intact *Pinus sylvestris* mycorrhizospheres in dry pine forest soil: an electron microscopy study. *Canadian Journal of Microbiology* **43**, 1017–1035.

Nylund J-E 1981 The formation of ectomycorrhiza in conifers: structural and physiological studies with special reference to the microbiont *Piloderma croceum* Erikss. and Agorts J. PhD thesis, Faculty of Science, University of Uppsala, Sweden.

Nylund JE 1987 The ectomycorrhizal information zone and its relation to acid polysaccharides of cortical cell walls. *New Phytologist* **106**, 505–516.

Nylund J-E 1988 The regulation of mycorrhiza formation – carbohydrate and hormone theories reviewed. *Scandinavian Journal of Forest Research* **3**, 465–479.

Nylund J-E and Unestam T 1982 Structure and physiology of ectomycorrhizae I. The process of mycorrhiza formation in Norway spruce *in vitro*. *New Phytologist* **91**, 63–79.

Nylund J-E and Wallander H 1989 Effects of ectomycorrhiza on host growth and carbon balance in a semi-hydroponic cultivation system. *New Phytologist* **112**, 389–398.

Nylund JE and Wallander H 1992 Ergosterol analysis as a means of quantifying mycorrhizal biomass. *Methods in Microbiology* **24**, 77–88.

Nylund JE, Kasimir A, Strandberg-Arveby A 1982a Cell wall penetration and papilla formation in senescent cortical cells during ectomycorrhiza synthesis *in vitro*. *Physiological Plant Pathology* **21**, 71–73.

Nylund JE, Kasimir A, Strandberg-Arveby A, Unestam T 1982b Simple diagnosis of ectomycorrhiza formation and demonstration of the architecture of the Hartig net by means of a clearing technique. *European Journal of Forest Pathology* **12**, 103–107.

O'Connor P 1994 A comparison of the development and symbiotic efficiency of two isolates of a vesicular-arbuscular mycorrhizal fungus. BSc (Hons) Soil Science Thesis, University of Adelaide, Australia.

O'Connor PJ, Smith SE, Smith FA 2001 Arbuscular mycorrhizal associations in the southern Simpson Desert. *Australian Journal of Botany* **49**, 493–499.

O'Connor PJ, Smith SE, Smith EA 2002 Arbuscular mycorrhizas influence plant diversity and community structure in a semiarid herbland. *New Phytologist* **154**, 209–218.

O'Dell TE, Massicotte HB, Trappe JM 1993 Root colonisation of *Lupinus latifolius* Agardh. and *Pinus contorta* Dougl. by *Phialocephala fortinii* Wang and Wilcox. *New Phytologist* **124**, 93–100.

Oades JM 1993 The role of biology in the formation, stabilization and degradation of soil structure. *Geoderma* **56**, 377–400.

Oaks A1994 Primary N assimilation in higher plants and its regulation. *Canadian Journal of Botany* **72**, 739–750.

Oberson A, Fardeau JC, Besson JM, Sticher H 1993 Soil phosphorus dynamics in cropping systems managed according to conventional and biological agricultural methods. *Biology and Fertility of Soils* **16**, 111–117.

Ocampo JA 1986 Vesicular-arbuscular mycorrhizal infection of 'host' and 'non-host' plants: effect on the growth responses of the plants and competition between them. *Soil Biology and Biochemistry* **18**, 607–610.

Ocampo JA and Hayman DS 1980 Influence of plant interactions on vesicular-arbuscular mycorrhizal infections. II. Crop rotations and residual effects of non-host plants. *New Phytologist* **84**, 27–35.

Odum EP 1960 Organic matter production and turnover in old field succession. *Ecology* **41**, 34–49.

Odum EP 1971 Fundamentals of Ecology. W.B. Saunders, Philadelphia, USA.

Ogawa M 1985 Ecological characters of ectomycorrhizal fungi and their mycorrhizae. *Japanese Annual Research Quarterly* **18**, 305–314.

Ogura-Tsujita Y and Yukawa T 2008 High mycorrhizal specificity in a widespread myco-heterotrophic plant, *Eulophic zollingeri* (Orchidaceae). *American Journal of Botany* **95**, 1–6.

Ohtomo R, Sekiguchi Y, Mimura T, Saito M, Ezawa T 2004 Quantification of polyphosphate: different sensitivities to short-chain polyphosphate using enzymatic and colorimetric methods as revealed by ion chromatography. *Analytical Biochemistry* **328**, 139–146.

Oldroyd GED, Harrison MJ, Udvardi M 2005 Peace talks and trade deals. Keys to long-term harmony in legume-microbe symbioses. *Plant Physiology* **137**, 1205–1210.

Oliver AJ, Smith SE, Nicholas DJD, Wallace W, Smith FA 1983 Activity of nitrate reductase in *Trifolium subterraneum*: effects of mycorrhizal infection and phosphate nutrition. *New Phytologist* **94**, 63–79.

Olsen JK, Schaefer JT, Edwards DG, Hunter MN, Galea VJ, Muller LM 1999 Effects of a network of mycorrhizae on capsicum (*Capsicum annuum* L.) grown in the field with five rates of applied phosphorus. *Australian Journal of Agricultural Research* **50**, 239–252.

Olsrud M, Melillo JM, Christensen TR, Michelsen A, Wallander H, Olsson PA 2004 Response of ericoid mycorrhizal colonization and functioning to global change factors. *New Phytologist* **162**, 459–469.

Olsson PA 1999 Signature fatty acids provide tools for determination of the distribution and interactions of mycorrhizal fungi in soil. *FEMS Microbiology Ecology* **29**, 303–310.

Olsson PA and Wilhelmsson P 2000 The growth of external AM fungal mycelium in sand dunes and in experimental systems. *Plant and Soil* **226**, 161–169.

Olsson PA, Thingstrup I, Jakobsen I, Bååth E 1999 Estimation of biomass of arbuscular mycorrhizal fungi in a linseed field. *Soil Biology and Biochemistry* **31**, 1879–1887.

Olsson PA, Jakobsen I, Wallander H 2002 Foraging and resource allocation strategies of mycorrhizal fungi in a patchy environment. In *Mycorrhizal Ecology*. Eds MGA van der Heijden and IR Sanders pp. 93–115. Springer-Verlag, Berlin, Germany.

Olsson PA, Larsson L, Bago B, Wallander H, van Aarle I 2003 Ergosterol and fatty acids for biomass estimation of mycorrhizal fungi. *New Phytologist* **159**, 7–9.

Olsson PA, Burleigh SH, van Aarle I 2005 The influence of external nitrogen on carbon allocation to *Glomus intraradices* in monoxenic arbuscular mycorrhiza. *New Phytologist* **168**, 677–686.

Omacini M, Eggers T, Bonowski M, Gange A, Jones TH 2006 Leaf endophytes affect mycorrhizal status and growth of co-infected and neighboring plants. *Functional Ecology* **20**, 226–232.

Onguene NA and Kuyper TW 2002 Importance of the ectomycorrhizal network for seedling survival and ectomycorrhiza formation in rain forests of south Cameroon. *Mycorrhiza* **12** 1, 13–17.

Opik M, Moora M, Liira J, Koljalg U, Zobel M, Sen R 2003 Divergent arbuscular mycorrhizal fungal communities colonize roots of *Pulsatilla* spp. in boreal Scots pine forests and grassland soils. *New Phytologist* **160**, 581–593.

Orlov A 1957 Observations on absorbing roots of spruce (*Picea excelsa* Link.) in natural conditions. *Botanicheskii Zhurna SSSR* **42**, 1172–1180.

Orlov AY 1960 Further observations on the growth of absorbing roots of spruce (*Picea excelsa* Link.) in natural conditions. *Botanicheskii Zhurna SSSR* **45**, 888–896.

Orlovich DA and Ashford AE 1993 Polyphosphate granules are an artefact of specimen preparation in the ectomycorrhizal fungus *Pisolithus tinctorius*. *Protoplasma* **173**, 91–102.

Otero JT, Ackerman JD, Bayman P 2002 Diversity and host specificity of endophytic *Rhizoctonia*-like fungi from tropical orchids. *American Journal of Botany* **89**, 1852–1858.

Ott T, Fritz E, Polle A, Schutzendubel A 2002 Characterisation of antioxidative systems in the ectomycorrhiza-building basidiomycete Paxillus involutus (Bartsch) Fr. and its reaction to cadmium. *FEMS Microbiology Ecology* **42**, 359–366.

Owusu-Bennoah E and Wild A 1979 Autoradiography of the depletion zone of phosphate around onion roots in the presence of vesicular-arbuscular mycorrhiza. *New Phytologist* **82**, 133–140.

Pacioni G and Comandini O 1999 *Tuber*. In *Ectomycorrhizal Fungi: Key Genera in Profile*. Eds JWG Cairney and SM Chambers pp. 163–186. Springer, Berlin, Germany.

Pacovsky RS 1989 Carbohydrate, protein and amino acid status of *Glycine-Glomus-Bradyrhizobium* symbioses. *Physiologia Plantarum* **75**, 346–354.

Pairunan AK, Robson AD, Abbott LK 1980 The effectiveness of vesicular-arbuscular mycorrhizas in increasing growth and phosphorus uptake of subterranean clover from phosphorus sources of different solubilities. *New Phytologist* **84**, 327–338.

Pais MSS and Barroso J 1983 Localisation of polyphenol oxidase during the establishment of *Ophrys lutea* endomycorrhizas. *New Phytologist* **95**, 219–222.

Palenzona A 1969 Sintesi micorrizica tra *Tuber aestivum* Vitt. e semenzali di *Corylus avellana* L. *Allionia* **15**, 121–131.

Paris F, Botton B, Lapeyrie F 1996 In vitro weathering of phlogopite by ectomycorrhizal fungi .2. Effect of K^+ and Mg_2^+ deficiency and N sources on accumulation of oxalate and H^+. *Plant and Soil* **179**, 141–150.

Parke JL and Linderman RG 1980 Association of vesicular-arbuscular mycorrhizal fungi with the moss *Funaria hygrometrica*. *Canadian Journal of Botany* **58**, 1898–1904.

Paszkowski U 2006 Mutualism and parasitism: the yin and yang of plant symbioses. *Current Opinion in Plant Biology* **9**, 1–7.

Paszkowski U and Boller T 2002 The growth defect of *lrt1*, a maize mutant lacking lateral roots, can be complemented by symbiotic fungi or high phosphate nutrition. *Planta* **214**, 584–590.

Paszkowski U, Kroken S, Roux C, Briggs SP 2002 Rice phosphate transporters include an evolutionarily divergent gene specifically activated in arbuscular mycorrhizal symbiosis. *Proceedings of the National Academy of Sciences of the United States of America* **99**, 13324–13329.

Paszkowski U, Jakovleva L, Boller T 2006 Maize mutants affected at distinct stages of the arbuscular mycorrhizal symbiosis. *Plant Journal* **47**, 165–173.

Pate JS 1994 The mycorrhizal association: just one of many nutrient acquiring specializations in natural ecosystems. *Plant and Soil* **159**, 1–10.

Pate JS and Hopper SD 1993 Rare and common plants in ecosystems, with special reference to the south-west Australian flora. In *Biodiversity and Ecosystem Function. Ecological Studies* **99**. Eds ED Schultze and HA Mooney pp. 293–325. Springer-Verlag, Heidelberg, Germany.

Pate JS, Stewart GR, Unkovich M 1993 ^{15}N natural abundance of plant and soil components of a *Banksia* woodland ecosystem in relation to nitrate utilisation, life form, mycorrhizal status and N$_2$-fixing abilities of component species. *Plant, Cell and Environment* **16**, 365–373.

Patrick JW 1989 Solute efflux from the host at plant-microorganism interfaces. *Australian Journal of Plant Physiology* **16**, 53–67.

Pattinson GS, Smith SE, Doube BM 1997 Earthworm *Apporectodea trapezoides* had no effect on dispersal of a vesicular-arbuscular mycorrhizal fungus, *Glomus intraradices*. *Soil Biology and Biochemistry* **29**, 1079–1088.

Paul LR, Chapman BK, Chanway CP 2007 Nitrogen fixation associated with *Suillus tomentosus* tuberculate ectomycorrhizae on *Pinus contorta* var. *latifolia*. *Annals of Botany* **99**, 1101–1109.

Pawlowska TE, Taylor JW 2004 Organization of genetic variation in individuals of arbuscular mycorrhizal fungi. *Nature* **427**, 733–737.

Pearson JN and Jakobsen I 1993a Symbiotic exchange of carbon and phosphorus between cucumber and three arbuscular mycorrhizal fungi. *New Phytologist* **124**, 481–488.

Pearson JN and Jakobsen I 1993b The relative contribution of hyphae and roots to phosphorus uptake by arbuscular mycorrhizal plants measured by dual labelling with ^{32}P and ^{33}P. *New Phytologist* **124**, 489–494.

Pearson JN, Abbott LK, Jasper DA 1994 Phosphorus, soluble carbohydrates and the competition between two arbuscular mycorrhizal fungi colonizing subterranean clover. *New Phytologist* **127**, 101–106.

Pearson V and Read DJ 1973a The biology of mycorrhiza in the Ericaceae. I. The isolation of the endophyte and synthesis of mycorrhizas in aseptic culture. *New Phytologist* **72**, 371–379.

Pearson V and Read DJ 1973b The biology of mycorrhiza in the Ericaceae. II The transport of carbon and phosphorus by the endophyte and the mycorrhiza. *New Phytologist* **72**, 1325–1331.

Pearson V and Read DJ 1975 The physiology of the mycorrhizal endophyte of *Calluna vulgaris*. *Transactions of the British Mycological Society* **64**, 1–7.

Pearson V and Tinker PB 1975 Measurement of phosphorus fluxes in the external hyphae of endomycorrhizas. In *Endomycorrhizas*. Eds FE Sanders, B Mosse, PB Tinker pp. 277–287. Academic Press, London, UK.

Pedersen CT, Sylvia DM, Shilling DG 1999 *Pisolithus arhizus* ectomycorrhiza affects plant competition for phosphorus between *Pinus elliottii* and *Panicum chamaelonche*. *Mycorrhiza* **9**, 199–204.

Peng S, Eissenstat DM, Graham JH, Williams K, Hodge NC 1993 Growth depression in mycorrhizal citrus at high-phosphorus supply: analysis of carbon costs. *Plant Physiology* **101**, 1063–1071.

Pennington RT, Cronk QCB and Richardson JA 2004 Introduction and synthesis: plant phylogeny and the origins of major biomes. *Philosophical Transactions of the Royal Society of London Series B, Biological Sciences* **359,** 1455–1464.

Percudani R, Trevisi A, Zambonelli A, Ottonello S 1999 Molecular phylogeny of truffles (Pezizales: Terfeziaceae, Tuberaceae) derived from nuclear rDNA sequence analysis. *Molecular Phylogenetics and Evolution* **13**, 169–180.

Perez-Moreno J and Read DJ 2000 Mobilization and transfer of nutrients from litter to tree seedlings via the vegetative mycelium of ectomycorrhizal plants. *New Phytologist* **145** 2, 301–309.

Perez-Moreno J and Read DJ 2001a Exploitation of pollen by mycorrhizal mycelial systems with special reference to nutrient recycling in boreal forests. *Proceedings of the Royal Society of London Series B, Biological Sciences* **268** 1474, 1329–1335.

Perez-Moreno J and Read DJ 2001b Nutrient transfer from soil nematodes to plants: a direct pathway provided by the mycorrhizal mycelial network. *Plant Cell and Environment* **24** 11, 1219–1226.

Perkins AJ and McGee PA 1995 Distribution of the orchid mycorrhizal fungus, *Rhizoctonia solani*, in relation to its host, *Pterostylis acuminata*, in the field. *Australian Journal of Botany* **43** 6, 565–575.

Perkins AJ, Masuhara G, Mcgee PA 1995 Specificity of the associations between *Microtis parviflora* (Orchidaceae) and its mycorrhizal fungi. *Australian Journal of Botany* **43**, 85–91.

Perombelon M and Hadley G 1965 Production of pectic enzymes by pathogenic and symbiotic *Rhizoctonia* strains. *New Phytologist* **64**, 144–151.

Perotto R, Perotto S, Faccio A, Bonfante P 1990 Cell surface in *Calluna vulgaris* L. hair roots *in situ* localization of polysaccharide components. *Protoplasma* **155**, 1–18.

Perotto R, Bettini V, Bonfante P 1993 Evidence of two polygalacturonases produced by a mycorrhizal ericoid fungus during its saprotrophic growth. *FEMS Microbiology Letters* **114**, 85–92.

Perotto S, Peretto R, Faccio A, Schubert A, Varma A, Bonfante P 1995 Ericoid mycorrhizal fungi: cellular and molecular bases of their interactions with the host plant. *Canadian Journal of Botany* **73** Suppl. 1, 557–568.

Perotto S, Coisson JD, Perugini I, Cometti V, Bonfante P 1997 Production of pectin degrading enzymes by ericoid mycorrhizal fungi. *New Phytologist* **135**, 151–162.

Perry DA, Molina R, Amaranthus MP 1987 Mycorrhizae, mycorrhizospheres, and reforestation: current knowledge and research needs. *Canadian Journal of Forest Research* **17**, 929–940.

Perry DA, Amaranthus MP, Borchers JC, Borchers SL, Brainerd RE 1989a Bootstrapping in ecosystems. *BioScience* **39**, 230–237.

Perry DA, Margolis H, Choquette C, Molina R, Trappe JM 1989b Ectomycorrhizal mediation of competition between coniferous tree species. *New Phytologist* **112**, 501–511.

Persson H 1978 Root dynamics in a young Scots pine stand in central Sweden. *Oikos* **30**, 508–519.

Persson H 1979 Fine root production, mortality and decomposition in forest ecosystems. *Vegetatio* **4**, 101–109.

Persson H 1980 Spatial distribution of fine-root growth, mortality, and decomposition in a young Scots Pine stand. *Oikos* **34**, 77–87.

Persson HA 1983 The status and function of fine roots in boreal forests. *Plant and Soil* **71**, 87–101.

Persson H, van Oene H, Harrison AF *et al.* 2000 Experimental sites in the NIPHYS/CANIF project. In *Carbon and Nitrogen cycling in European Forest Ecosystems. Ecological Studies*, vol 142. Ed. E-D Schulze pp. 14–46. Springer, Berlin, Germany.

Peters NK and Verma DPS 1990 Phenolic compounds as regulators of gene expression in plant-microbe interactions. *Molecular Plant-Microbe Interactions* **3**, 4–8.

Peterson CA 1988 Exodermal Casparian bands: their significance for ion uptake by roots. *Physiologia Plantarum* **72**, 204–208.

Peterson RL and Bonfante P 1994 Comparative structure of vesicular-arbuscular mycorrhizas and ectomycorrhizas. *Plant and Soil* **159**, 79–88.

Peterson RL and Bradbury SM 1995 Use of plant mutants, intraspecific variants and non-hosts in studying mycorrhiza formation and function. In *Mycorrhiza, Structure, Function, Molecular Biology and Biotechnology*. Eds A Varma and B Hock pp. 157–180. Springer-Verlag, Berlin, Germany.

Peterson RL and Currah RS 1990 Synthesis of mycorrhizae between protocorms of *Goodyera repens* (Orchidaceae) and *Ceratobasidium cereale*. *Canadian Journal of Botany* **68**, 1117–1125.

Peterson RL and Guinel FC 2000 The use of plant mutants to study regulation of colonisation by AM fungi. In *Arbuscular Mycorrhizas: Physiology and Function*. Eds Y Kapulnik and DDJ Douds pp. 147–171. Kluwer Academic Publishers, Dordrecht, The Netherlands.

Peterson RL and Massicotte HB 2004 Exploring structural definitions of mycorrhizas, with emphasis on nutrient-exchange interfaces. *Canadian Journal of Botany* **82**, 1074–1088.

Peterson RL, Howarth MJ, Whittier DP 1981 Interactions between a fungal endophyte and gametophyte cells in *Psilotum nudum*. *Canadian Journal of Botany* **59**, 711–720.

Peterson RL, Bonfante P, Faccio A, Uetake Y 1996 The interface between fungal hyphae and orchid protocorm cells. *Canadian Journal of Botany* **74**, 1861–1870.

Peterson RL, Massicotte HB, Melville L 2004 *Mycorrhizas: anatomy and cell biology*. NRC Research Press and CABI Publishing, Ottawa and Wallingford.

Peterson TA, Mueller WC, Englander RL 1980 Anatomy and ultrastructure of a *Rhododendron* root-fungus association. *Canadian Journal of Botany* **58**, 2421–2433.

Peyronel B 1923 Fructification de l'endophyte à arbuscules et à vesicules des mycorhizes endotrophes. *Bulletin de la Societie Mycologique* **39**, 119–126.

Peyronel B 1924 Prime ricerche sulle micorize endotrofiche e sulla micoflora radicicola normal delle fanerogame. *Rivista di Biologia* **6**, 17–53.

Pfeffer PE, Douds DD, Bécard G, Shachar-Hill Y 1999 Carbon uptake and the metabolism and transport of lipids in an arbuscular mycorrhiza. *Plant Physiology* **120**, 587–598.

Pfeffer PE, Bago B, Shachar-Hill Y 2001 Exploring mycorrhizal function with NMR spectroscopy. *New Phytologist* **150**, 543–553.

Pfeffer PE, Douds DD, Bücking H, Schwartz DP, Shachar-Hill Y 2004 The fungus does not transfer carbon to or between roots in an arbuscular mycorrhizal symbiosis. *New Phytologist* **163**, 617–627.

Phillips JM and Hayman DS 1970 Improved procedures for clearing roots and staining parasitic and vesicular-arbuscular mycorrhizal fungi for rapid assessment of infection. *Transactions of the British Mycological Society* **55**, 158–161.

Phipps CJ and Taylor TN 1996 Mixed arbuscular mycorrhizae from the Triassic of Antarctica. *Mycologia* **88**, 707–714.

Piché Y, Fortin JA, Peterson RL, Posluszny U 1982 Ontogeny of dichotomizing apices in mycorrhizal short roots of *Pinus strobus*. *Canadian Journal of Botany* **60**, 1523–1528.

Piché Y, Peterson RL, Ackerley CA 1983a Early development of ectomycorrhizal short roots of pine. *Scanning Electron Microscopy* 1467–1474.

Piché Y, Peterson RL, Howarth MJ, Fortin JA 1983b A structural study of the interaction between the ectomycorrhizal fungus *Pisolithus tinctorius* and *Pinus strobus* roots. *Canadian Journal of Botany* **61**, 1185–1193.

Piché Y, Ackerley CA, Peterson RL 1986 Structural characteristics of ectendomycorrhizas synthesized between roots of *Pinus resinosa* and the E-strain fungus, *Wilcoxina mikolae* var *mikolae*. *New Phytologist* **104**, 447.

Pichot J and Binh T 1976 Actions des endomyocorhizes sur la croissance et la nutrition phosphatée de *l'Agrostis* en vase de vegetation et sur le phosphore isotopiqument diluable du sol. *Agronomie Tropicale* **4**, 375–378.

Piercey MM, Graham SW, Currah RS 2004 Patterns of genetic variation in *Phialocephala fortinii* across a broad latitudinal transect in Canada. *Mycological Research* **108**, 955–964.

Pigott CD 1982 Survival of mycorrhiza formed by *Cenococcum geophilum* Fr. in dry soils. *New Phytologist* **92**, 513–517.

Pijl L Van der 1934 Die Mykorrhiza von *Burmannia* und *Epirrhizanthes* und die Fortflanzung ihres Endophyten. *Recueil des Travaux Botaniques Néerlandais* **31**, 761–779.

Pirazzi R and Di Gregoria A 1987 Accrescimento di conifere micorizate con specie diverse di *Tuber* spp. *Micologia Italica* **16**, 49–62.

Pirozynski CA and Malloch DW 1975 The origin of land plants: a matter of mycotrophism. *Biosystems* **6**, 153–164.

Pirozynski KA and Dalpé Y 1989 Geological history of the Glomaceae with particular reference to mycorrhizal symbiosis. *Symbiosis* **7**, 1–36.

Plassard C, Scheromm P, Llamas H 1986 Nitrate assimilation by maritime pine and ectomycorrhizal fungi in pure culture. In *Physiological and Genetical Aspects of Mycorrhizae.* Eds V Gianinazzi-Pearson and S Gianinazzi pp. 383–388. INRA editions, Paris, France.

Plassard C, Scheromm P, Mousain D, Salsac L 1991 Assimilation of mineral nitrogen and ion balance in the two partners of ectomycorrhizal symbiosis: Data and hypothesis. *Experientia* **47**, 340–349.

Plassard C, Barry D, Eltrop L, Mousain D 1994 Nitrate uptake in maritime Pine (*Pinus pinaster*) and the ectomycorrhizal fungus *Hebeloma cylindrosporum* – effect of ectomycorrhizal symbiosis. *Canadian Journal of Botany* **72**, 189–197.

Plassard C, Guerin-Laguette A, Very AA, Casarin V, Thibaud JB 2002 Local measurements of nitrate and potassium fluxes along roots of maritime pine. Effects of ectomycorrhizal symbiosis. *Plant Cell and Environment* **25**, 75–84.

Plenchette C and Strullu DG 2003 Long-term viability and infectivity of intraradical forms of *Glomus intraradices* vesicles encapsulated in alginate beads. *Mycological Research* **107**, 614–616.

Plenchette C, Furlan V, Fortin JA 1981 Growth stimulation of apple trees in unsterilised soil under field conditions with VA mycorrhizal inoculation. *Canadian Journal of Botany* **59**, 2003–2008.

Plenchette C, Fortin JA, Furlan V 1983a Growth responses of several plant species to mycorrhizae in a soil of moderate P-fertility. I. Mycorrhizal dependency under field conditions. *Plant and Soil* **70**, 199–209.

Plenchette C, Fortin JA, Furlan V 1983b Growth responses of several plant species to mycorrhizae in a soil of moderate P-fertility. II. Soil fumigation induced stunting of plants corrected by reinoculation of the wild endomycorrhiza flora. *Plant and Soil* **70**, 211–217.

Plenchette C, Clermont-Dauphin C, Meynard JM, Fortin JA 2005 Managing arbuscular mycorrhizal fungi in cropping systems. *Canadian Journal of Plant Science* **85**, 31–40.

Pocock K and Duckett JG 1984 A comparative ultrastructural analysis of the fungal endophytes in *Cryptothallus mirabilis* Hulm. and other British thalloid hepatics. *Journal of Bryology* **13**, 227–233.

Pocock K and Duckett JG 1995 On the occurrence of branched and swollen rhizoids in British hepatics: their relationship with the substratum and association with fungi. *New Phytologist* **99**, 281–304.

Podila GK, Zheng J, Balasubramanian S *et al.* 2002 Fungal gene expression in early symbiotic interactions between *Laccaria bicolor* and red pine. *Plant and Soil* **244**, 117–128.

Polidori E, Agostini D, Zeppa S *et al.* 2002 Identification of differentially expressed cDNA clones in *Tilia platyphyllos-Tuber borchii* ectomycorrhizae using a differential screening approach. *Molecular Genetics and Genomics* **266**, 858–864.

Ponge JF (1990) Ecological study of a forest humus by observing a small volume. I. Penetration of pine litter by mycorrhizal fungi. *European Journal of Forest Pathology* **20**, 290–303.

Pons F, Gianinazzi-Pearson V 1984 Influence du phosphore, du potassium, de l'azote et du pH sur le comportement *in vitro* de champignons endomyorhizogenes à vésicules et arbuscules. *Cryptogamie* **5**, 87–100.

Porcel R, Aroca R, Azcón R, Ruiz-Lozano JM 2006 PIP aquaporin gene expression in arbuscular mycorrhizal *Glycine max* and *Lactuca sativa* plants in relation to drought stress tolerance. *Plant Molecular Biology* **60**, 389–404.

Porter WM 1979 The 'Most Probable Number' method for enumerating infective propagules of vesicular-arbuscular mycorrhizal fungi in soil. *Australian Journal of Soil Research* **17**, 515–519.

Porter WM, Abbott LK, Robson AD 1978 Effect of rate of application of superphosphate on populations of vesicular arbscular endophytes. *Australian Journal of Experimental Agriculture and Animal Husbandry* **18**, 573–578.

Posada D and Crandall KA 1998 MODELTEST: testing the model of DNA substitution. *Bioinformatics, Oxford* **14**, 817–818.

Possingham JV and Groot Obbink J 1971 Endotrophic mycorrhiza and the nutrition of grape vines. *Vitis* **10**, 120–130.

Poulsen KH, Nagy R, Gao L-L *et al.* 2005 Physiological and molecular evidence for Pi uptake via the symbiotic pathway in a reduced mycorrhizal colonisation mutation in tomato associated with a compatible fungus. *New Phytologist* **168**, 445–453.

Poulton JL, Koide RT, Stephenson AG 2001 Effects of mycorrhizal infection, soil phosphorus availability and fruit production on the male function in two cultivars of *Lycopersicon esculentum*. *Plant, Cell and Environment* **24**, 841–849.

Poulton JL, Bryla D, Koide RT, Stephenson AG 2002 Mycorrhizal infection and high soil phosphorus improve vegetative growth and the female and male functions in tomato. *New Phytologist* **154**, 255–264.

Powell CL 1975 Plant growth responses to vesicular-arbuscular mycorrhiza. VIII. Uptake of P by onion and clover infected with different *Endogone* spore types in ^{32}P labelled soil. *New Phytologist* **75**, 563–566.

Powell CL 1976 Development of mycorrhizal infections from *Endogone* spores and infected root segments. *Transactions of the British Mycological Society* **66**, 439–445.

Powell CL 1977 Mycorrhizas in hill country soils. I. Spore bearing mycorrhizal fungi in thirty seven soils. *New Zealand Journal of Agricultural Research* **20**, 53–57.

Powell CL 1979 Spread of mycorrhizal fungi through soil. *New Zealand Journal of Agricultural Research* **22**, 335–339.

Powell CL 1980 Mycorrhizal infectivity of eroded soils. *Soil Biology and Biochemistry* **12**, 247–250.

Preston CD, Pearman DA, Dines TD 2002 *New Atlas of the British Flora*. Oxford University Press, Oxford, UK.

Pringle A, Moncalvo JM, Vilgalys R 2000 High levels of variation in ribosomal DNA sequences within and among spores of a natural population of the arbuscular mycorrhizal fungus *Acaulospora colossica*. *Mycologia* **92**, 259–268.

Purves S and Hadley G 1975 Movement of carbon compounds between the partners in orchid mycorrhizas. In *Endomycorrhizas*. Eds FE Sanders, B Mosse, PB Tinker pp. 173–194. Academic Press, London, UK.

Queloz V, Grunig CR, Sieber TN, Holdenrieder O 2005 Monitoring the spatial and temporal dynamics of a community of the tree-root endophyte *Phialocephala fortinii* s.l. *New Phytologist* **168**, 651–660.

Querejeta JI, Egerton-Warburton LM, Allen MF 2003 Direct nocturnal water transfer from oaks to their mycorrhizal symbionts during severe soil drying. *Oecologia* **134**, 55–64.

Quick WP and Schaffer AA 1996 Sucrose metabolism in sources and sinks. In *Photoassimilate distribution in plants and crops*. Eds E Zamski and A Schaffer pp. 115–156. Marcel Dekker, New York, USA.

Quoreshi AM, Ahmad I, Malloch D, Hellebust JA 1995 Nitrogen metabolism in the ectomycorrhizal fungus *Hebeloma crustuliniforme*. *New Phytologist* **131**, 263–271.

Rabatin SC 1980 The occurrence of the vesicular-arbuscular mycorrhiza fungus *Glomus tenuis* with moss. *Mycologia* **72**, 191–194.

Raghothama KG 1999 Phosphate acquisition. *Annual Review of Plant Physiology and Plant Molecular Biology* **50**, 665–693.

Rains KC, Nadkarni NM, Bledsoe CS 2003 Epiphytic and terrestrial mycorrhizas in a lower montane Costa Rica cloud forest. *Mycorrhiza* **13**, 257–264.

Rambelli A 1973 The rhizosphere of mycorrhizae. In *Ectomycorrhizae*. Eds GL Marks and TT Koslowski pp. 299–343. Academic Press, New York, USA.

Ramsay RR, Dixon KW, Sivasithamparam K 1986 Patterns of infection and endophytes associated with Western Australian orchids. *Lindleyana* **1**, 203–214.

Ramsay RR, Sivasithamparam K, Dixon KW 1987 Anastomosis groups among Rhizoctonia-like endophytic fungi in South-western Australian *Pterostylis* species (Orchidaceae). *Lindleyana* **2**, 161–166.

Randall BL and Grand LF 1986 Morphology and possible mycobiont (*Suillus pictus*) of a tuberculate ectomycorrhiza on *Pinus strobus. Canadian Journal of Botany* **64**, 2182–2191.

Rasmussen HN 1995 *Terrestrial Orchids: from seed to mycotrophic plant.* Cambridge University Press, Cambridge, UK.

Rasmussen HN and Whigham DF 1994 Seed ecology of dust seeds *in situ*: a new study technique and its application in terrestrial orchids. *American Journal of Botany* **80**, 1374–1378.

Rasmussen HN and Whigham DF 1998 The underground phase: a special challenge in studies of terrestrial orchid populations. *Botanical Journal of the Linnean Society* **126**, 49–64.

Rasmussen N, Lloyd DC, Ratcliffe RG, Hansen PE, Jakobsen I 2000 ^{31}P NMR for the study of P metabolism and translocation in arbuscular mycorrhizal fungi. *Plant and Soil* **226**, 245–253.

Rausch C, Daram P, Brunner S *et al.* 2001 A phosphate transporter expressed in arbuscule-containing cells in potato. *Nature* **414**, 462–466.

Raven JA, Smith FA 1976 Nitrogen assimilation and transport in vascular land plants in relation to intracellular pH regulation. *New Phytologist* **76**, 415–431.

Ravnskov S and Jakobsen I 1995 Functional compatibility in arbuscular mycorrhizas measured as hyphal P transport to the plant. *New Phytologist* **129**, 611–618.

Ravnskov S and Larsen J 2005 Implementation of arbuscular mycorrhizas in greenhouse-grown vegetables. In *Implementation of Biocontrol in Practice in Temperate Regions – Present and Near Future.* Research Centre Flakkebjerg, Denmark pp. 193–198. Research Centre Flakkebjerg, Denmark.

Ravnskov S, Larsen J, Olsson PA, Jakobsen I 1999 Effects of various organic compounds on growth and phosphorus uptake of an arbuscular mycorrhizal fungus. *New Phytologist* **141**, 517–524.

Ravnskov S, Wu Y, Graham JH 2003 Arbuscular mycorrhizal fungi differentially affect expression of genes coding for sucrose synthases in maize roots. *New Phytologist* **157**, 539–545.

Rawald W 1963 Untersuchungen zur Stickstoffernährung der höheren Pilze. In *Mykorrhiza.* Eds W Rawald and H Lyr pp. 67–83. Internal Mykorrhiza Symposium Weimar, 1960. Gustav Fisher, Jena, Germany.

Rayner MC 1915 Obligate symbiosis in *Calluna vulgaris. Annals of Botany, London* **29**, 97–133.

Rayner MC 1927 Mycorrhiza. *New Phytologist* Reprint **15**.

Rayner ADM, Powell KA, Thompson W, Jennings DH 1985 Morphogensis of vegetative organs. In *Developmental Biology of the Higher Fungi.* Eds D Moore, LA Casselton, DA Wood, JC Frankland pp. 249–279. Cambridge University Press, Cambridge, UK.

Read DJ 1974 *Pezizella ericae* sp.nov. the perfect state of a typical mycorrhizal endophyte of Ericaceae. *Transactions of the British Mycological Society* **63**, 381–383.

Read DJ 1983 The biology of mycorrhiza in the Ericales. *Canadian Journal of Botany* **61**, 985–1004.

Read DJ 1984 The structure and function of the vegetative mycelium of mycorrhizal roots. In *The Ecology and Physiology of the Fungal Mycelium.* Eds DH Jennings and ADM Rayner pp. 215–240. Cambridge University Press, Cambridge, UK.

Read DJ 1989 Mycorrhizas and nutrient cycling in sand dune ecosystems. *Proceedings of the Royal Society, Edinburgh* **96b**, 80–110.

Read DJ 1990 Ecological integration by mycorrhizal fungi. *Endocytobiology* **4**, 99–107.

Read DJ 1991a Mycorrhizas in ecosystems. *Experientia* **47**, 376–391.

Read DJ 1991b Mycorrhizas in ecosystems – nature's response to the 'Law of the Minimum'. In *Frontiers in Mycology.* Ed. DL Hawksworth pp. 101–130. CAB International, Wallingford, UK.

Read DJ 1991c Mycorrhizal fungi in natural and semi-natural plant communities. In *Ecophysiology of Ectomycorrhizae of Forest Trees.* M Wallenberg Foundation, Stockholm, Sweden. *Wallenberg Foundation Symposium Proceedings* Vol **7** pp. 27–53.

Read DJ 1992 The mycorrhizal mycelium. In *Mycorrhizal Functioning* Ed. MF Allen pp. 102–133. Chapman and Hall, London.

Read DJ 1993 Mycorrhiza in plant communities. *Advances in Plant Pathology* **9**, 1–31.

Read DJ 1996 The structure and function of the ericoid mycorrhizal root. *Annals of Botany* **77**, 365–374.

Read DJ 1997 The ties that bind. *Nature* **388,** 517–518.

Read DJ and Boyd R 1986 Water relations of mycorrhizal fungi and their host plants. In *Water, Fungi and Plants.* Eds P Ayres and L Boddy pp. 287–303. Cambridge University Press, Cambridge, UK.

Read DJ and Haselwandter K 1981 Observations on the mycorrhizal status of some alpine plant communities. *New Phytologist* **88**, 341–352.

Read DJ and Kerley SJ 1995 The status and function of ericoid mycorrhizal systems. In *Mycorrhiza, Structure, Function Molecular Biology and Biotechnology.* Eds A Varma and B Hock pp. 499–520. Springer-Verlag, Berlin, Germany.

Read DJ and Perez-Moreno J 2003 Mycorrhizas and nutrient cycling in ecosystems – a journey towards relevance? *New Phytologist* **157** 3, 475–492.

Read DJ and Stribley DP 1973 Effect of mycorrhizal infection on nitrogen and phosphorus nutrition of ericaceous plants. *Nature, London* **244**, 81.

Read DJ and Stribley DP 1975 Some mycological aspects of the biology of mycorrhiza in the Ericacease. In *Endomycorrhizas.* Eds FE Sanders, B Mosse, PB Tinker pp. 105–117. Academic Press, London, UK.

Read DJ, Francis R, Finlay RD 1985 Mycorrhizal mycelia and nutrient cycling in plant communities. In *Ecological Interactions in Soil.* Eds AH Fitter, D Atkinson, DJ Read, MB Usher pp. 193–217. Blackwell Scientific Publications, Oxford, UK.

Read DJ, Leake JR, Langdale AR 1989 The nitrogen nutrition of mycorrhizal fungi and their host plants. In *Nitrogen, Phosphorus and Sulphur Utilisation by Fungi.* Eds LL Boddy, R Marchant, DJ Read pp. 181–204. Cambridge University Press, Cambridge, UK.

Read DJ, Duckett JG, Francis R, Ligrone R, Russell A 2000 Symbiotic fungal associations in 'lower' land plants. *Philosophical Transactions of the Royal Society of London Series B, Biological Sciences* **355**, 815–830.

Read DJ, Leake JR, Perez-Moreno J 2004 Mycorrhizal fungi as drivers of ecosystem processes in heathland and boreal forest biomes. *Canadian Journal of Botany* **82**, 1243–1263.

Reddell P and Malajczuk N 1984 Formation of mycorrhizae by jarrah (*Eucalyptus marginata* Donn ex Smith) in litter and soil. *Australian Journal of Botany* **32**, 511–520.

Reddell P and Spain AV 1991 Earthworms as vectors of viable propagules of mycorrhizal fungi. *Soil Biology and Biochemistry* **23**, 767–774.

Reddell P, Spain AV, Hopkins M 1997a Dispersal of spores of mycorrhizal fungi in scats of native mammals in tropical forests of northeastern Australia. *Biotropica* **29**, 184–192.

Reddell P, Yun Y, Shipton WA 1997b Cluster roots and mycorrhizae in *Casuarina cunninghamiana* – their occurrence and formation in relation to phosphorus supply. *Australian Journal of Botany* **45**, 41–51.

Reddy SM, Pandey AK, Melayah D, Marmeisse R, Gay G 2003 The auxin responsive gene-*C61* is up-regulated in *Pinus pinaster* roots following inoculation with ectomycorrhizal fungi. *Plant, Cell and Environment* **26**, 681–691.

Redecker D, Kodner R, Graham LE 2000 Glomalean fungi from the Ordovician. *Science* **289**, 1920–1921.

Redecker D, Szaro TM, Bowman RJ, Bruns TD 2001 Small genets of *Lactarius xanthogalactus, Russula cremoricolor* and *Amanita francheti* in late-stage ectomycorrhizal successions. *Molecular Ecology* **10**, 1025–1034.

Reed ML 1987 Ericoid mycorrhiza of Epacridaceae in Australia. In *Mycorrhizae in the Next Decade. Practical Applications and Research Priorities.* Eds DM Sylvia, LL Hung, JH Graham p. 335. Institute of Food and Agricultural Sciences, University of Florida, Gainesville, USA.

Rees B, Shepherd VA, Ashford AE 1994 Presence of a motile tubular vacuole system in different phyla of fungi. *Mycological Research* **97**, 985–992.

Reeves FB, Wagner D, Moorman T, Kiel J 1979 The role of endomycorrhizae in revegetation practices in the semi-arid west. I. A comparison of incidence of mycorrhizae in severely disturbed vs. natural environments. *American Journal of Botany* **66**, 6–13.

Reid CPP, Kidd FA, Ekwebelam SA 1983 Nitrogen nutrition, photosynthesis and carbon allocation in ectomycorrhizal pine. *Plant and Soil* **71**, 415–432.

Reinecke T and Kindle H 1994 Inducible enzymes of the 9,10-dihydro-phenanthrene pathway. Sterile orchid plants responding to fungal infection. *Molecular Plant Microbe Interactions* **7**, 449–454.

Reiners WA 1965 Ecology of a heath-shrub synusia in the pine barrens of Long Island, New York. *Bulletin Torrey Bot Club* **92**, 448–464.

Remy W, Taylor TN, Hass H, Kerp H 1994 Four hundred-million-year-old vesicular arbuscular mycorrhizae. *Proceedings of the National Academy of Sciences of the United States of America* **91**, 11841–11843.

Requena N, Fuller P, Franken P 1999 Molecular characterization of *GmFOX2*, an evolutionarily highly conserved gene from the mycorrhizal fungus *Glomus mosseae*, down-regulated during interaction with rhizobacteria. *Molecular Plant-Microbe Interactions* **12**, 934–942.

Requena N, Mann P, Hampp R, Franken P 2002 Early developmentally regulated genes in the arbuscular mycorrhizal fungus *Glomus mosseae*: identification of *GmGIN1*, a novel gene with homology to the C-terminus of metazoan hedgehog proteins. *Plant and Soil* **244**, 129–139.

Requena N, Breuninger M, Franken P, Ocón A 2003 Symbiotic status, phosphate, and sucrose regulate the expression of two plasma membrane H^+-ATPase genes from the mycorrhizal fungus *Glomus mosseae*. *Plant Physiology* **132**, 1540–1549.

Reynolds HL, Hartley AE, Vogelsang KM, Bever JD, Schultz PA 2005 Arbuscular mycorrhizal fungi do not enhance nitrogen acquisition and growth of old-field perennials under low nitrogen supply in glasshouse culture. *New Phytologist* **167**, 869–880.

Rhodes LH and Gerdemann JW 1975 Phosphate uptake zones of mycorrhizal and non-mycorrhizal onions. *New Phytologist* **75**, 555–561.

Rhodes LH and Gerdemann JW 1978a Hyphal translocation and uptake of sulfur by vesicular-arbuscular mycorrhizae of onions. *Soil Biology and Biochemistry* **10**, 355–360.

Rhodes LH and Gerdemann JW 1978b Influence of phosphorus nutrition on sulphur uptake by vesicular-arbuscular mycorrhizae of onions. *Soil Biology and Biochemistry* **10**, 361–364.

Rhodes LH and Gerdemann JW 1978c Translocation of calcium and phosphate by external hyphae of vesicular-arbuscular mycorrhizae. *Soil Science* **126**, 125–126.

Richard C and Fortin JA 1973 The identification of *Mycelium radicis atrovirens* (*Phialocephala dimorphospora*). *Canadian Journal of Botany* **51**, 2247–2248.

Richard C and Fortin J-A 1974 Geographic distribution, ecology, physiology and pathogenicity of *Mycelium radicis atrovirens*. *Phytoprotection* **55**, 67–88.

Richardson KA, Peterson RL, Currah RS 1992 Seed reserves and early symbiotic protocorm development of *Plantanthera hyperborea* (Orchidaceae). *Canadian Journal of Botany* **70**, 291–300.

Richardson KA, Currah RS, Hambleton SM 1993 Basidiomycetous endophytes from the roots of neotropical epiphytic Orchidaceae. *Lindleyana* **8**, 127–137.

Richardson MJ 1970 Studies on *Russula emetica* and other agarics in a Scots pine plantation. *Transactions of the British Mycological Society* **55**, 217–229.

Rillig MC and Mummey DL 2006 Mycorrhizas and soil structure. *New Phytologist* **171**, 41–53.

Rillig MC, Treseder KK, Allen MF 2002a Global change and mycorrhizal fungi. In *Mycorrhizal Ecology*. Eds MGA van der Heijden and IR Sanders. Springer-Verlag, Berlin, Germany.

Rillig MC, Wright SF, Shaw MR, Field CB 2002b Artificial climate warming positively affects arbuscular mycorrhizae but decreases soil aggregate water stability in an annual grassland. *Oikos* **97**, 52–58.

Ritz K and Newman EI 1985 Evidence for rapid cycling of phosphorus from dying roots to living plants. *Oikos* **45**, 174–180.

Rivett M 1924 The root tubercles in *Arbutus unedo*. *Annals of Botany* **28**, 661–677.

Roberts P 1993 *Exidiopsis* species from Devon, including the new segregate genera *Ceratosebacina*, *Endoperplexa*, *Microsebacina* and *Serendipita*. *Mycological Research* **97**, 467–478.

Roberts P 1999 *Rhizoctonia*-forming *Fungi – a Taxonomic Guide*. The Royal Botanic Garden, Kew, UK.

Robertson DC and Robertson JA 1982 Ultrastructure of *Pterospora andromedea* Nutall and *Sarcodes sanguinea* Torrey mycorrhizas. *New Phytologist* **92**, 539–551.

Robertson DC and Robertson JA 1985 Ultrastructural aspects of *Pyrola* mycorrhizae. *Canadian Journal of Botany* **63**, 1089–1098.

Robertson GP and Vitousek PM 1981 Nitrification potentials in primary and secondary succession. *Ecology* **62**, 376–386.

Robertson NF 1954 Studies on the mycorrhiza of *Pinus sylvestris*. *New Phytologist* **53**, 253–283.

Robinson D 1994 The responses of plants to non-uniform supplies of nutrients. *New Phytologist* **127**, 635–674.

Robinson D and Fitter AH 1999 The magnitude and control of carbon transfer between plants linked by a common mycorrhizal network. *Journal of Experimental Botany* **50**, 9–13.

Robinson SA, Slade AP, Fox GG, Phillips R, Ratcliffe RG, Stewart GR 1991 The role of glutamate dehydrogenese in plant nitrogen metabolism. *Plant Physiology* **95**, 509–516.

Rodriguez A, Dougall T, Dodd JC, Clapp JP 2001 The large subunit ribosomal RNA genes of *Entrophospora infrequens* comprise sequences related to two different glomalean families. *New Phytologist* **152**, 159–167.

Rommell LG 1938 A trenching experiment in spruce forest and its bearing on the problems of mycotrophy. *Svensk Botanisk Tidskrift* **32**, 89–99.

Rommell LG 1939a The ecological problem of mycotrophy. *Ecology* **20**, 163–167.

Rommell LG 1939b Barrskogens marksvamphar och deras roll i skogens liv. *Svensk Botanisk Tidskrift* **37**, 348–375.

Rønn R, Gavito M, Larsen J, Jakobsen I, Frederiksen H, Christensen S 2002 Response of free-living soil protozoa and microorganisms to elevated atmospheric CO_2 and presence of mycorrhiza. *Soil Biology and Biochemistry* **34** 7, 923–932.

Rose SL 1980 Mycorrhizal associations of some actinomycete nodulated nitrogen-fixing plants. *Canadian Journal of Botany* **58**, 1449–1454.

Rose SL and Youngberg CT 1981 Tripartite associations in snow-brush *Ceanothus velutinus*: effect of vesicular-arbuscular mycorrhizae on growth, nodulation and nitrogen fixation. *Canadian Journal of Botany* **59**, 34–39.

Rosendahl S and Stukenbrock EH 2004 Community structure of arbuscular mycorrhizal fungi in undisturbed vegetation revealed by analyses of LSU rDNA sequences. *Molecular Ecology* **13**, 3179–3186.

Rosendahl S and Taylor JW 1997 Development of multiple genetic markers for studies of genetic variation in arbuscular mycorrhizal fungi using AFLP^{Tm}. *Molecular Ecology* **6**, 821–829.

Rosewarne GM, Barker SJ, Smith SE 1997 Production of near-synchronous fungal colonization in tomato for developmental and molecular analyses of mycorrhiza. *Mycological Research* **101**, 966–970.

Rosling A, Landeweert R, Lindahl BD *et al.* 2003 Vertical distribution of ectomycorrhizal fungal taxa in a podzol soil profile. *New Phytologist* **159**, 775–783.

Rosling A, Lindahl BD, Finlay RD 2004a Carbon allocation to ectomycorrhizal roots and mycelium colonising different mineral substrates. *New Phytologist* **162**, 795–802.

Rosling A, Lindahl BD, Taylor AFS, Finlay RD 2004b Mycelial growth and substrate acidification of ectomycorrhizal fungi in response to different minerals. *FEMS Microbiology Ecology* **47**, 31–37.

Ross JP 1971 Effect of phosphate fertilization on yield of mycorrhizal and nonmycorrhizal soybeans. *Phytopathology* **61**, 1400–1403.

Rost FWD, Shepherd VA, Ashford AE 1995 Estimation of vacuolar pH in actively growing hyphae of the fungus *Pisolithus tinctorius*. *Mycological Research* **99**, 549–553.

Rouhier H and Read DJ 1998 Plant and fungal responses to elevated atmospheric carbon dioxide in mycorrhizal seedlings of *Pinus sylvestris*. *Environmental and Experimental Botany* **40** 3, 237–246.

Rousseau JVD and Reid CPP 1990 Effects of phosphorus and ectomycorrhizas on the carbon balance of loblolly pine seedlings. *Forest Science* **36**, 101–112.

Rousseau JVD and Reid CPP 1991 Effects of phosphorus fertilisation and mycorrhizal development on phosphorus nutrition and carbon balance of loblolly pine. *New Phytologist* **117**, 319–326.

Rousseau JVD, Sylvia DM, Fox AJ 1994 Contribution of ectomycorrhiza to the potential nutrient-absorbing surface of pine. *New Phytologist* **128**, 639–644.

Roussel H, van Tuinen D, Franken P, Gianinazzi S, Gianinazzi-Pearson V 2001 Signalling between arbuscular mycorrhizal fungi and plants: identification of a gene expressed during early interactions by differential RNA display analysis. *Plant and Soil* **232**, 13–19.

Rudawska M 1986 Sugar metabolism of ectomycorrhizal Scots pine seedlings as influenced by different nitrogen forms and levels. In *Physiological and Genetical Aspects of Mycorrhizae*. pp. 389–394. INRA, Paris, France.

Ruess RW, Hendrick RL, Burton AJ *et al*. 2003 Coupling fine root dynamics with ecosystem carbon cycling in black spruce forests of interior Alaska. *Ecological Monographs* **73**, 643–662.

Rufyikiri G, Huysmans L, Wannijn J, Van Hees M, Leyval C, Jakobsen I 2004 Arbuscular mycorrhizal fungi can decrease the uptake of uranium by subterranean clover grown at high levels of uranium in soil. *Environmental Pollution* **130**, 427–436.

Rühling A and Tyler G 1991 Effects of simulated nitrogen deposition to the forest floor on the macrofungal flora of a beech forest. *Ambio* **20** 6, 261–263.

Ruinen J 1953 Epiphytosis – a second view of epiphytism. *Annales Bogoriensis* **1**, 101–157.

Ruiz-Lozano JM 2003 Arbuscular mycorrhizal symbiosis and alleviation of osmotic stress. New perspectives for molecular studies. *Mycorrhiza* **13**, 309–317.

Ruiz-Lozano JM, Porcel R, Aroca R 2006 Does the enhanced tolerance of arbuscular mycorrhizal plants to water deficit involve modulation of drought-induced plant genes? *New Phytologist* **171**, 693–698.

Rundel PW 1988 Leaf structure and nutrition in mediterranean-climate sclerophylls. In *Mediterranean-type Ecosystems*. Ed. RL Specht pp. 157–167. Kluwer, Dordrecht, The Netherlands.

Ruotsalainen AL, Tuomi J, Vare H 2002 A model for optimal mycorrhizal colonization along altitudinal gradients. *Silva Fennica* **36**, 681–694.

Rusca TA, Kennedy PG, Bruns TD 2006 The effect of different pine hosts on the sampling of *Rhizopogon* spore banks in five Sierra Nevada forests. *New Phytologist* **170**, 551–560.

Russell A and Bulman S 2004 The liverwort *Marcantia foliacea* forms a specialized symbiosis with arbuscular mycorrhizal fungi in the genus *Glomus*. *New Phytologist* **165**, 567–579.

Ryan EA and Alexander IJ 1992 Mycorrhizal aspects of improved growth of spruce when grown in mixed stands on heathlands. In *Mycorrhizas in Ecosystems*. Eds DJ Read, DH Lewis, AH Fitter, IJ Alexander pp. 237–245. CAB International, Wallingford, UK.

Ryan M and Ash J 1999 Effects of phosphorus and nitrogen on growth of pasture plants and VAM fungi in SE Australian soils with contrasting fertiliser histories conventional and biodynamic. *Agriculture Ecosystems and Environment* **73**, 51–62.

Ryan MH and Graham JH 2002 Is there a role for arbuscular mycorrhizal fungi in production agriculture? *Plant and Soil* **244**, 263–271.

Ryan MH, Chilvers GA, Dumaresq DC 1994 Colonisation of wheat by VA-mycorrhizal fungi was found to be higher on a farm managed in an organic manner than on a conventional neighbour. *Plant and Soil* **160**, 33–40.

Ryan MH, Small DR, Ash JE 2000 Phosphorus controls the level of colonisation by arbuscular mycorrhizal fungi in conventional and biodynamic irrigated dairy pastures. *Australian Journal of Experimental Agriculture* **40**, 663–670.

Ryan MH, Norton RM, Kirkegaard JA, McCormick KM, Knights SE, Angus JF 2002 Increasing mycorrhizal colonisation does not improve growth and nutrition of wheat on vertisols in south-eastern Australia. *Australian Journal of Agricultural Research* **53**, 1173–1181.

Ryan MH, McCully ME, Huang CX 2003 Location and quantification of phosphorus and other elements in fully hydrated, soil-grown arbuscular mycorrhizas: a cryo-analytical scanning electron microscopy study. *New Phytologist* **160**, 429–441.

Rygiewicz PT and Anderson CP 1994 Mycorrhizae alter quality and quantity of carbon allocated below ground. *Nature* **369**, 58–60.

Rygiewicz PT, Bledsoe CS, Zasoski RJ 1984 Effects of ectomycorrhizae and solution pH on [15N]ammonium uptake by coniferous seedlings. *Canadian Journal of Forest Research* **14**, 885–892.

Rygiewicz PT, Johnson MG, Ganio LM, Tingey DT, Storm MJ 1997 Lifetime and temporal occurrence of ectomycorrhizae on ponderosa pine (*Pinus ponderosa* Laws) seedlings grown under varied atmospheric CO_2 and nitrogen levels. *Plant and Soil* **189**, 275–287.

Safir GR, Boyer JS, Gerdemann JW 1971 Mycorrhizal enhancement of water transport in soybean. *Science* **172**, 581–583.

Safir GR, Boyer JS, Gerdemann JW 1972 Nutrient status and mycorrhizal enhancement of water transport in soybean. *Plant Physiology* **49**, 700–703.

Safir GR, Coley SC, Siqueira JO, Carlson PS 1990 Improvement and synchronization of VA mycorrhiza fungal spore germination by short-term cold storage. *Soil Biology and Biochemistry* **22**, 109–111.

Salzer P and Hager A 1991 Sucrose utilisation of the ectomycorrhizal fungi *Amanita muscaria* and *Hebeloma crustuliniforme* depends on the cell wall-bound invertase activity of their host *Picea abies*. *Botanica Acta* **104**, 439–445.

Salzer P and Hager A 1993 Characteristization of wall-bound invertase isoforms of *Picea abies* cells and regulation of ectomycorrhizal fungi. *Physiologia Plantarum* **88**, 52–59.

Salzer P, Bonanomi A *et al.* 2000 Differential expression of eight chitinase genes in *Medicago truncatula* roots during mycorrhiza formation, nodulation, and pathogen infection. *Molecular Plant-Microbe Interactions* **13**, 763–777.

Sampangi R and Perrin R 1985 Attempts to elucidate the mechanisms involved in the protective effect of *Laccaria laccata* against *Fusarium oxysporum*. In *Physiological and Genetical Aspects of Mycorrhizae*. Eds V Gianinazzi Pearson and S Gianninazzi pp. 807–810. INRA, Paris, France.

Sancholle M and Dalpé Y 1993 Taxonomic relevance of fatty acids of arbuscular mycorrhizal fungi and related species. *Mycotaxon* **69**, 187–193.

Sanders FE and Sheikh NA 1983 The development of vesicular-arbuscular mycorrhizal infection in plant root systems. *Plant and Soil* **71**, 223–246.

Sanders FE and Tinker PB 1971 Mechanism of absorption of phosphate from soil by *Endogone* mycorrhizas. *Nature* **233**, 278–279.

Sanders FE and Tinker PB 1973 Phosphate flow into mycorrhizal roots. *Pesticide Science* **4**, 385–395.

Sanders FE, Mosse B, Tinker PB Eds 1975 *Endomycorrhizas*. Academic Press, London, UK.

Sanders FE, Tinker PB, Black RLB, Palmerley SM 1977 The development of endomycorrhizal root systems. I. Spread of infection and growth-promoting effects with four species of vesicular-arbuscular endophyte. *New Phytologist* **78**, 257–268.

Sanders IR 2002 Specificity in the arbuscular mycorrhizal symbiosis. In *Mycorrhizal Ecology*. Eds MGA van der Heijden and IR Sanders pp. 416–437. Springer-Verlag, Berlin, Germany.

Sanders IR and Fitter AH 1992a Evidence for differential responses between host-fungus combinations of vesicular-arbuscular mycorrhizas from a grassland. *Mycological Research* **96**, 415–419.

Sanders IR and Fitter AH 1992b The ecology and functioning of vesicular-arbuscular mycorrhizas in co-existing grassland species II. Nutrient uptake and growth of vesicular-arbuscular mycorrhizal plants in a semi-natural grassland. *New Phytologist* **120**, 525–533.

Sanders IR, Alt M, Groppe K, Boller T, Wiemken A 1995 Identification of ribosomal DNA polymorphisms among and within spores of the Glomales: application to studies on the genetic diversity of arbuscular mycorrhizal fungal communities. *New Phytologist* **130**, 419–427.

Sanders IR, Koch A, Kuhn G 2003 Arbuscular mycorrhizal fungi: genetics of multigenomic, clonal networks and its ecological consequences. *Biological Journal of the Linnean Society* **79**, 59–60.

Sangwanit V and Bledsoe SC 1987 Mycorrhizal fungi enhance uptake and storage of amino acids in mycorrhizal Douglas-Fir and western hemlock seedlings. In *Mycorrhizae in the Next Decade*. Eds DM Sylvia, LL Hung, JH Graham p. 262. Institute of Food and Agriculture Sciences, University of Florida, Gainsville, Florida, USA.

Santantonio D and Santantonio E 1987 Effect of thinning on production and mortality of fine roots in a *Pinus radiata* plantation on a fertile site in New Zealand. *Canadian Journal of Forest Research* **17**, 919–928.

Sasa M, Zahka G, Jakobsen I 1987 The effect of pretransplant inoculation with VA mycorrhizal fungi on the subsequent growth of leeks in the field. *Plant and Soil* **97**, 279–283.

Sawaki H, Saito M 2001 Expressed genes in the extraradical hyphae of an arbuscular mycorrhizal fungus *Glomus intraradices*, in the symbiotic phase. *FEMS Microbiology Letters* **195**, 109–113.

Sawyer NA, Chambers SM, Cairney JWG 2003 Distribution of *Amanita* spp. genotypes under eastern Australian sclerophyll vegetation. *Mycological Research* **107**, 1157–1162.

Scales P and Peterson RL 1991a Structure and development of *Pinus banksiana-Wilcoxina* ectendomycorrhizae. *Canadian Journal of Botany* **69**, 2135–2148.

Scales P and Peterson RL 1991b Structure of ectomycorrhizae formed by *Wilcoxima mikolae* var. *mikolae* with *Picea mariana* and *Betula alleghaniensis*. *Canadian Journal of Botany* **69**, 2149–2157.

Scannerini S 1968 Sull' ultrastruttura delle ectomicorrize. II. Ultrastruttura di una micorriza di ascomycete: *Tuber albidum X Pinus strobus*. *Allionia* **14**, 77–95.

Scannerini S and Bonfante-Fasolo P 1983 Comparative ultrastructural analysis of mycorrhizal associations. *Canadian Journal of Botany* **61**, 917–943.

Schacht H 1854 Uber Pilzfäden im Inneren der Zellen und Stärkemehlkörner. *Flora* **7**, 618–624.

Schachtman DP, Reid RJ, Ayling SM 1998 Phosphorus uptake by plants: from soil to cell. *Plant Physiology* **116**, 447–453.

Schadt CW, Mullen RB, Schmidt SK 2001 Isolation and phylogenetic identification of a dark-septate fungus associated with the alpine plant *Ranunculus adoneus*. *New Phytologist* **150**, 747–755.

Schaeffer C, Wallenda T, Güttenberger M, Hampp R 1995 Acid invertase in mycorrhizal and non-mycorrhizal roots of Norway spruce (*Picea abies* [L.] Karst.) seedlings. *New Phytologist* **129**, 417–424.

Schellenbaum L, Gianinazzi S, Gianinazzi-Pearson V 1993 Comparison of acid soluble protein synthesis in roots of endomycorrhizal wild type *Pisum sativum* and corresponding isogenic mutants. *Journal of Plant Physiology* **141**, 2–6.

Scheltema MA, Abbott LK, Robson AD 1987a Seasonal variation in the infectivity of VA mycorrhizal fungi in annual pastures in a mediterranean environment. *Australian Journal of Agricultural Research* **38**, 707–715.

Scheltema MA, Abbott LK, Robson AD, De'Ath G 1987b The spread of mycorrhizal infection by *Gigaspora calospora* from a localized inoculum. *New Phytologist* **106**, 727–734.

Schenck NC 1982 *Methods and Principles of Mycorrhizal Research*. American Phytopathological Society, St Paul, Minnesota, USA.

Schenck NC, Graham SO, Green NE 1975 Temperature and light effect on contamination and spore germination of vesicular-arbuscular mycorrhizal fungi. *Mycologia* **67**, 1189–1192.

Scheromm P, Plassard C, Salsac L 1990a Nitrate nutrition of maritime pine (*Pinus pinaster* Soland *in* Ait.) ectomycorrhizal with *Hebeloma cylindrosporum* Romagn. *New Phytologist* **114**, 93–98.

Scheromm P, Plassard C, Salsac L 1990b Regulation of nitrate reductase in the ectomycorrhizal basidiomycete, *Hebeloma cylindrosporum* Romagn., cultured on nitrate or ammonium. *New Phytologist* **114**, 441–447.

Schier GA and McQuattie CJ 1996 Response of ectomycorrhizal and nonmycorrhizal pitch pine (*Pinus rigida*) seedlings to nutrient supply and aluminum: growth and mineral nutrition. *Canadian Journal of Forest Research* **26** 12, 2145–2152.

Schmid E and Oberwinkler F 1993 Mycorrhiza-like interaction between the achlorophyllous gametophyte of *Lycopodium clavatum* L and its fungal endophyte studied by light and electron-microscopy. *New Phytologist* **124** 1, 69–81.

Schmid E and Oberwinkler F 1994 Light and electron-microscopy of the host-fungus interaction in the achlorophyllous gametophyte of *Botrychium lunaria*. *Canadian Journal of Botany* **72** 2, 182–188.

Schmid E and Oberwinkler F 1996 Light and electron microscopy of a distinctive VA mycorrhiza in mature sporophytes of O*phioglossum reticulatum*. *Mycological Research* **100**, 843–849.

Schramm JR 1966 Plant colonisation studies on black wastes from anthracite mining in Pennsylvania. *Transactions of the American Philosophical Society* **56**, 1–94.

Schreiner RP and Koide RT 1993a Antifungal compounds from the roots of mycotrophic and non-mycotrophic plant species. *New Phytologist* **123**, 99–105.

Schreiner RP and Koide RT 1993b Mustards, mustard oils and mycorrhizas. *New Phytologist* **123**, 107–113.

Schreiner RP and Koide RT 1993c Stimulation of vesicular-arbuscular mycorrhizal fungi by mycotrophic and nonmycotrophic plant root systems. *Applied and Environmental Microbiology* **59**, 2750–2752.

Schroeder MS and Janos DP 2004 Phosphorus and intraspecific density alter plant responses to arbuscular mycorrhizas. *Plant and Soil* **264**, 335–348.

Schubert A, Marzachi C, Mazzitelli M, Cravero MC, Bonfante-Fasolo P 1987 Development of total and viable extraradical mycelium in the vesicular-arbuscular mycorrhizal fungus *Glomus clarum* Nicol. and Schenck. *New Phytologist* **107**, 183–190.

Schubert A, Wyss P, Wiemken A 1992 Occurrence of trehalose in vesicular-arbuscular mycorrhizal fungi and in mycorrhizal roots. *Journal of Plant Physiology* **140**, 41–45.

Schubert A, Allara P, Morte A 2004 Cleavage of sucrose in roots of soybean (*Glycine max*) colonized by an arbuscular mycorrhizal fungus. *New Phytologist* **161**, 495–501.

Schüepp H, Miller DD, Bodmer M 1987 A new technique for monitoring hyphal growth of vesicular arbuscular mycorrhizal fungi through soil. *Transactions of the British Mycological Society* **89**, 429–435.

Schuler R and Haselwandter K 1988 Hydroxamate siderophore production by ericoid mycorrhizal fungi. *Journal of Plant Nutrition* **11**, 907–913.

Schultze ED 1989 Air-pollution and forest decline in a spruce (*Picea abies*) Forest. *Science* **244** 4906, 776–783.

Schultze ED, Chapin FS III, Gebauer F 1994 Nitrogen nutrition and isotope differences among life forms at the northern tree-line of Alaska. *Oecologia* **100**, 406–412.

Schüßler A 2000 *Glomus claroideum* forms an arbuscular mycorrhiza-like symbiosis with the hornwort *Anthoceros punctatus*. *Mycorrhiza* **10**, 15–21.

Schüßler A 2002 Molecular phyology, taxonomy and evolution of *Geosiphon pyriformis* and arbuscular mycorrhizal fungi. *Plant and Soil* **244**, 75–83.

Schüßler A and Wolf E 2005 *Geosiphon pyriformis* – a glomeromycotan soil fungus forming endosymbiosis with Cyanobacteria. In *In Vitro Culture of Mycorrhizas*. Eds S Declerck, DG Strullu, JA Fortin pp. 272–289. Springer, Berlin, Germany.

Schüßler A, Mollenhauer D, Schnepf E, Kluge M 1994 *Geosiphon pyriforme*, an endosymbiotic association of fungus and cyanobacteria: the spore structure resembles that of arbuscular mycorrhizal AM fungi. *Botanica Acta* **107**, 36–45.

Schüßler A, Schwarzott D, Walker C 2001 A new fungal phylum, the *Glomeromycota*: phylogeny and evolution. *Mycological Research* **105**, 1413–1421.

Schuster RM 1966 *The Hepaticae and Anthocerotae of North America*. Vol 1. Columbia University Press, New York, USA.

Schwarzott D, Walker C, Schüßler A 2001 *Glomus*, the largest genus of the arbuscular mycorrhizal fungi (Glomales), is nonmonophyletic. *Molecular Phylogenetics and Evolution* **21**, 190–197.

Schweiger PF and Jakobsen I 1999a The role of mycorrhizas in plant P nutrition: fungal uptake kinetics and genotype variation. In *Plant Nutrition – Molecular Biology and Genetics*. Eds G Gissel-Nielsen and A Jensen pp. 277–289. Kluwer Academic Publishers, Dordrecht, The Netherlands.

Schweiger PF and Jakobsen I 1999b Direct measurement of arbuscular mycorrhizal phosphorus uptake into field-grown winter wheat. *Agronomy Journal* **91**, 998–1002.

Schweiger P and Jakobsen I 2000 Laboratory and field methods for measurement of hyphal uptake of nutrients in soil. *Plant and Soil* **226**, 237–244.

Schweiger P, Spliid NH, Jakobsen I 2001 Fungicide application and phosphorus uptake by hyphae of arbuscular mycorrhizal fungi into field-grown peas. *Soil Biology and Biochemistry* **33**, 1231–1237.

Schwob I, Ducker M, Sallanon H, Coudret A 1998 Growth and gas exchange responses of *Hevea brasiliensis* seedlings to inoculation with *Glomus mosseae*. *Trees* **12**, 236–240.

Scott GD 1969 *Plant Symbiosis*. Edward Arnold, London, UK.

Selle A, Willmann M, Grunze N, Gessler A, Weiss M, Nehls U 2005 The high-affinity poplar ammonium importer PttAMT1.2 and its role in ectomycorrhizal symbiosis. *New Phytologist* **168**, 697–706.

Selosse MA 2005 Are liverworts imitating mycorrhizas? *New Phytologist* **165**, 345–349.

Selosse MA, Weiss M, Jany JL, Tillier A 2002a Communities and populations of sebacinoid basidiomycetes associated with the achlorophyllous orchid *Neottia nidus-avis* (L.) LCM Rich. and neighbouring tree ectomycorrhizae. *Molecular Ecology* **11**, 1831–1844.

Selosse MA, Bauer R, Moyersoen B 2002b Basal hymenomycetes belonging to the Sebacinaceae are ectomycorrhizal on temperate deciduous trees. *New Phytologist* **155**, 183–195.

Selosse MA, Faccio A, Scappaticci G, Bonfante P 2004 Chlorophyllous and achlorophyllous specimens of *Epipactis microphylla* (Neottieae, Orchidaceae) are associated with ectomycorrhizal septomycetes, including truffles. *Microbial Ecology* **47**, 416–426.

Selosse MA, Setaro S, Glatard F, Richard F, Urcelay C, Weiss M 2007 Sebacinales are common mycorrhizal associates of Ericaceae. *New Phytologist* **174**, 864–878.

Sen R, Hietala A, Zelmer A 1999 Common anastomosis and ITS-RFLP groupings among binuculate *Rhizoctonia* isolates representing root endophytes of *Pinus sylvestris*, *Ceratorhiza* spp from orchid mycorrhizas and a phytopathogenic anastomosis group. *New Phytologist* **144**, 331–341.

Serrigny J and Dexheimer J 1985 Étude ultrastructurale des endomycorhizes d'une orchidée tropicale: *Epidendrum ibaguense*. II. Localisation des ATPases et des nucleosides diphosphatases. *Cytologia* **50**, 779–788.

Serrigny J and Dexheimer J 1986 Endomycorrhize d'une orchidée tropicale: *Epidendrum ibaguense*. Étude comparative des activités phosphatasiques acides entre le champignon symbiote associe et isole. In *Physiological and Genetical Aspects of Mycorrhizae*. Eds V Gianinazzi-Pearson and S Gianinazzi pp. 271–275. INRA, Paris, France.

Setala H 1995 Growth of birch and pine seedlings in relation to grazing by soil fauna on ectomycorrhizal fungi. *Ecology* **76**, 1844–1851.

Setaro S, Weiss M, Oberwinkler F, Kottke I 2006 Sebacinales form ectendomycorrhizas with *Cavendishia nobilis*, a member of the Andean clade of Ericaceae, in the mountain rain forest of southern Ecuador. *New Phytologist* **169**, 355–365.

Seviour RJ, Willing RR, Chilvers GA 1973 Basidiocarps associated with ericoid mycorrhizas. *New Phytologist* **72**, 381–385.

Sgorbati S, Berta G, Trotta A *et al.* 1993 Chromatin structure variation in successful and unsuccessful arbuscular mycorrhizas of pea. *Protoplasma* **175**, 1–8.

Shachar-Hill Y, Pfeffer PE, Douds D, Osman SF, Doner LW, Ratcliffe RG 1995 Partioning of intermediary carbon metabolism in vesicular-arbuscular mycorrhizal leek. *Plant Physiology* **108**, 7–15.

Sharples JM, Chambers SM, Meharg AA, Cairney JWG 2000a Genetic diversity of root-associated fungal endophytes from *Calluna vulgaris* at contrasting field sites. *New Phytologist* **148**, 153–162.

Sharples JM, Meharg AA, Chambers SM, Cairney JWG 2000b Symbiotic solution to arsenic contamination. *Nature* **404**, 951–952.

Sharples JM, Meharg AA, Chambers SM, Cairney JWG 2000c Mechanism of arsenate resistance in the ericoid mycorrhizal fungus *Hymenoscyphus ericae*. *Plant Physiology* **124**, 1327–1334.

Shaw G, Leake JR, Baker AJM, Read DJ 1990 The biology of mycorrhiza in the Ericaceae. XVII. The role of mycorrhizal infection in the regulation of iron uptake by ericaceous plants. *New Phytologist* **115**, 251–258.

Shaw J and Renzaglia K 2004 Phylogeny and diversification of bryophytes. *American Journal of Botany* **91** 10, 1557–1581.

Shefferson RP, Weiss M, Kull T, Taylor DL 2005 High specificity generally characterizes mycorrhizal association in rare lady's slipper orchids, genus *Cypripedium*. *Molecular Ecology* **14**, 613–626.

Shepherd VA, Orlovich DA, Ashford AE 1993a Cell-to-cell transport via motile tubules in growing hyphae of a fungus. *Journal of Cell Science* **105**, 1173–1178.

Shepherd VA, Orlovich DA, Ashford AE 1993b A dynamic continuum of pleiomorphic tubules and vacuoles in growing hyphae of a fungus. *Journal of Cell Science* **104**, 495–507.

Shi LB, Güttenberger M, Kottke I, Hampp R 2002 The effect of drought on mycorrhizas of beech (*Fagus sylvatica* L.): changes in community structure, and the content of carbohydrates and nitrogen storage bodies of the fungi. *Mycorrhiza* **12**, 303–311.

Shiroya T, Lister GR, Slankis V, Krotkov G, Nelson CD 1966 Seasonal changes in respiration, photosynthesis and translocation of the ^{14}C labelled products of photosynthesis in young *Pinus strobus* L. plants. *Annals of Botany* **30**, 81–91.

Shishkoff N 1986 The dimorphic hypodermis of plant roots: its distribution in the angiosperms, staining properties and interaction with root invading fungi. PhD thesis, University of Cornell.

Sieber TN 2002 Fungal root endophytes. In *Plant Roots: The Hidden Half*. Eds Y Waisel, A Eshel, U Kafkafi pp. 887–917. Marcel Dekker, New York, USA.

Sieber TN and Grunig CR 2006 Biodiversity of fungal root-endophyte communities and populations in particular of the dark septate endophyte *Phialocephala fortinii*. In *Microbial Root Endophytes*. Eds B Schulz, C Boyle, TN Sieber pp. 107–132. Springer, Berlin, Germany.

Sieverding E 1987 On-farm inoculum production of V.A.M. inoculum. In *Mycorrhizae in the Next Decade*. Eds DM Sylvia, LL Hung, JH Graham pp. 284–285. Institute of Food and Agricultural Sciences, University of Florida, Gainsville, Florida, USA.

Sieverding E 1990 Should VAM inocula contain single or several fungal species? *Agriculture, Ecosystems and Environment* **29**, 391–396.

Sieverding E 1991 *Vesicular-Arbuscular Mycorrhiza Management in Tropical Agrosystems*. Deutsche Gesellschaft fur Technische Zusammenarbeit GTZ GmbH, Eschborn, Germany.

Sieverding E, Toro S, Mosquera O 1989 Biomass production and nutrient concentrations in spores of VA mycorrhizal fungi. *Soil Biology and Biochemistry* **21**, 69–72.

Simard SW 1995 Interspecific carbon transfer in ectomycorrhizal tree species mixtures. PhD Thesis, Oregon State University.

Simard SW and Durall DM 2004 Mycorrhizal networks: a review of their extent, function, and importance. *Canadian Journal of Botany* **82**, 1140–1165.

Simard SW, Jones MD, Durall DM, Perry DA, Myrold DD, Molina R 1997a Reciprocal transfer of carbon isotopes between extomycorrhizal *Betula papyrifera* and *Pseudotsuga menziesii*. *New Phytologist* **137**, 529–542.

Simard SW, Molina R, Smith JE, Perry DA, Jones MD 1997b Shared compatibility of ectomycorrhizae on *Pseudotsuga menziesii* and *Betula papyrifera* seedlings grown in mixture in soils from southern British Columbia. *Canadian Journal of Forest Research* **27**, 331–342.

Simard SW, Perry DA, Jones MD, Myrold DD, Durall DM, Molina R 1997c Net transfer of carbon berween ectomycorrhizal tree species in the field. *Nature* **388**, 579–582.

Simard SW, Jones MD, Durall DM 2002 Carbon and nutrient fluxes within and between mycorrhizal plants. In *Mycorrhizal Ecology*. Eds MGA van der Heijden and IR Sanders pp. 33–74. Springer-Verlag, Berlin, Germany.

Siminovitch D, Singh J, De la Roche IA 1975 Studies on membrane in plant cells resistant to extreme freezing. I. Augmentation of phospholipid and membrane substance without change in unsaturation of fatty acids during hardening of black locust bark. *Cryobiology* **12**, 144–153.

Simkin T and Fiske RS 1983 *Krakatau 1883 – the Volcanic Eruption and its Effects*. Smithsonian Institution Press, Washington, DC, USA.

Simon L, Bousquet J, Lévesque RC, Lalonde M 1993 Origin and diversification of endomycorrhizal fungi and coincidence with vascular land plants. *Nature* **363**, 67–69.

Simoneau P, Viemont JD, Moreau JC, Strullu DG 1993 Symbiosis-related polypeptides associated with the early stages of ectomycorrhiza organogenesis in birch (*Betula pendula* Roth). *New Phytologist* **124**, 495–504.

Simpson D and Daft MJ 1990 Spore production and mycorrhizal development in various tropical crop hosts infected with *Glomus clarum*. *Plant and Soil* **121**, 171–178.

Singer R and Araujo I 1979 Litter decomposition and ectomycorrhiza in Amazonian forest. I. A comparison of litter decomposing and ectomycorrhizal basidiomycetes in latosol-terra-firme forest and white podzol campinarana. *Acta Amazonica* **9**, 25–41.

Singh KG 1965 Comparison of techniques for isolation of root-infecting fungi. *Nature* **206** 4989, 1169.

Singh KG 1974 Mycorrhiza in the Ericaceae with particular reference of *Calluna vulgaris*. *Svensk Botanisk Tidskrift* **68**, 1–16.

Siqueira JO and Saggin-Junior OJ 2001 Dependency on arbuscular mycorrhizal fungi and responsiveness of some Brazilian native woody species. *Mycorrhiza* **11**, 245–255.

Siqueira JO, Hubbell DH, Schenck NC 1982 Spore germination and germ tube growth of a vesicular-arbuscular mycorrhizal fungus *in vitro*. *Mycologia* **74**, 952–959.

Siqueira JO, Sylvia DM, Gibson J, Hubbell DH 1985 Spores, germination and germ tubes of vesicular-arbuscular mycorrhizal fungi. *Canadian Journal of Microbiology* **31**, 965–972.

Siqueira JO, Safir GR, Nair MG 1991 Stimulation of vesicular-arbuscular mycorrhiza formation and growth of white clover by flavonoid compounds. *New Phytologist* **118**, 87–93.

Siqueira JO, Carneiro MAC, Curi N, Rosado SCS, Davide AC 1998 Mycorrhizal colonization and mycotrophic growth of native woody species as related to successional groups in southeastern Brazil. *Forest Ecology and Management* **107**, 241–252.

Sivak MN and Walker DA 1986 Photosynthesis *in vivo* can be limited by phosphate supply. *New Phytologist* **102**, 499–512.

Skinner MF and Bowen GD 1974a The uptake and translocation of phosphate by mycelial strands of pine mycorrhizas. *Soil Biology and Biochemistry* **6**, 53–56.

Skinner MF and Bowen GD 1974b The penetration of soil by mycelial strands of ectomycorrhizal fungi. *Soil Biology and Biochemistry* **6**, 57–81.

Smernik RJ, Dougherty W 2007 Identification of phytate in phosphorus-31 nuclear magnetic resonance spectra: the need for spiking. *Soil Science Society of America Journal* **71**, 1045–1050.

Smith AH and Zeller SM 1966 A preliminary account of the North American species of *Rhizopogon*. *Memoirs of the New York Botanical Gardens* **14**, 1–178.

Smith DC and Douglas AE 1987 *The Biology of Symbiosis*. Edward Arnold, London, UK.

Smith DC, Muscatine L, Lewis DH 1969 Carbohydrate movement from autotrophs to heterotrophs in parasitic and mutualistic symbiosis. *Biological Reviews* **44**, 17–90.

Smith FA 2000 Measuring the influence of mycorrhizas. *New Phytologist* **148**, 1–6.

Smith FA and Smith SE 1981 Mycorrhizal infection and growth of *Trifolium subterraneum*: comparison of natural and artificial inocula. *New Phytologist* **88**, 311–325.

Smith FA and Smith SE 1990 Solute transport at the interface: ecological implications. *Agriculture, Ecosystems and Environment* **28**, 475–478.

Smith FA and Smith SE 1996 Mutualism and parasitism: diversity in function and structure in the 'arbuscular' (VA) mycorrhizal symbiosis. *Advances in Botanical Research* **22**, 1–43.

Smith FA and Smith SE 1997 Structural diversity in vesicular-arbuscular mycorrhizal symbiosis. *New Phytologist* **137**, 373–388.

Smith FA, Smith SE, St John BJ, Nicholas DJD 1986 Inflow of N and P into roots of mycorrhizal and non-mycorrhizal onions. In *Physiological and Genetical Aspects of Mycorrhizae*. Eds V Gianinazzi-Pearson and S Gianinazzi pp. 371–375. INRA, Paris, France.

Smith FA, Dickson S, Morris C, Reid RJ, Tester M, Smith SE 1995 Phosphate transfer in VA mycorrhizas. Special mechanisms or not? In *Structure and Function of Roots*. Eds F Baluska, M Ciamporová, O Gaspariková, PW Barlow pp. 155–161. Kluwer Academic Publishers, Dordrecht, The Netherlands.

Smith FA, Jakobsen I, Smith SE 2000 Spatial differences in acquisition of soil phosphate between two arbuscular mycorrhizal fungi in symbiosis with *Medicago truncatula*. *New Phytologist* **147**, 357–366.

Smith FA, Timonen S, Smith SE 2001 Mycorrhizae. In *Root Ecology*. Eds CWPM Blom and EJW Visser. Springer, Berlin, Germany.

Smith FA, Smith SE, Timonen S 2003b Mycorrhizas. In *Root Ecology*. Eds H de Kroon and EJW Visser pp. 257–295. Springer-Verlag, Berlin, Germany.

Smith FW, Mudge SR, Rae AL, Glassop D 2003a Phosphate transport in plants. *Plant and Soil* **248**, 71–83.

Smith HF, O'Connor PJ, Smith SE, Smith FA 1998 Vesicular-arbuscular mycorrhizas of durian and other plants of forest gardens in West Kalmantan, Indonesia. In *Soils of Tropical Forest Ecosystems*. Eds A Schulte and D Ruhiyat pp. 192–198. Springer-Verlag, Berlin, Germany.

Smith MD, Hartnett DC, Wilson GWT 1999 Interacting influence of mycorrhizal symbiosis and competition on plant diversity in tallgrass prairie. *Oecologia* **121**, 574–582.

Smith N, Mori SA, Henderson A, Stevenson DW, Heald SV 2004b *Flowering Plants of the Neotropics*. Princeton University Press, New Jersey, USA.

Smith SE 1966 Physiology and ecology of orchid mycorrhizal fungi with reference to seedling nutrition. *New Phytologist* **65**, 488–499.

Smith SE 1967 Carbohydrate translocation in orchid mycorrhizal fungi. *New Phytologist* **66**, 371–378.

Smith SE 1973 Asymbiotic germination of orchid seeds on carbohydrates of fungal origin. *New Phytologist* **72**, 497–499.

Smith SE 1980 Mycorrhizas of autotrophic higher plants. *Biological Reviews* **55**, 475–510.

Smith SE 1982 Inflow of phosphate into mycorrhizal and non-mycorrhizal plants of *Trifolium subterraneum* at different levels of soil phosphate. *New Phytologist* **90**, 293–303.

Smith SE 1995 Discoveries, discussions and directions in mycorrhizal research. In *Mycorrhiza: Structure, Function, Molecular Biology and Biotechnology*. Eds A Varma and B Hock pp. 3–24. Springer-Verlag, Berlin, Germany.

Smith SE and Daft MJ 1977 Interactions between growth, phosphate content and N_2 fixation in mycorrhizal and non-mycorrhizal *Medicago sativa*. *Australian Journal of Plant Physiology* **4**, 403–413.

Smith SE and Dickson S 1991 Quantification of active vesicular-arbuscular mycorrhizal infection using image analysis and other techniques. *Australian Journal of Plant Physiology* **18**, 637–648.

Smith SE and Gianinazzi-Pearson V 1988 Physiological interactions between symbionts in vesicular-arbuscular mycorrhizal plants. *Annual Review of Plant Physiology and Plant Molecular Biology* **39**, 221–244.

Smith SE and Gianinazzi-Pearson V 1990 Phosphate uptake and arbuscular activity in mycorrhizal *Allium cepa* L.: effects of photon irradiance and phosphate nutrition. *Australian Journal of Plant Physiology* **17**, 177–188.

Smith SE and Read DJ 1997 *Mycorrhizal Symbiosis*, 2nd edn. Academic Press Ltd, London, UK.

Smith SE and Smith FA 1973 Uptake of glucose, trehalose and mannitol by leaf slices of the orchid *Bletilla hyacinthina*. *New Phytologist* **72**, 957–964.

Smith SE and Smith FA 1990 Structure and function of the interfaces in biotrophic symbioses as they relate to nutrient transport. *New Phytologist* **114**, 1–38.

Smith SE and Walker NA 1981 A quantitative study of mycorrhizal infection in *Trifolium*: separate determination of the rates of infection and of mycelial growth. *New Phytologist* **89**, 225–240.

Smith SE, Smith FA, Nicholas DJD 1981 Effects of endomycorrhizal infection on phosphate and cation uptake by *Trifolium subterraneum*. *Plant and Soil* **63**, 57–64.

Smith SE, St John BJ, Smith FA, Nicholas DJD 1985 Activity of glutamine synthetase and glutamate dehydrogenase in *Trifolium subterraneum* L. and *Allium cepa* L.: effects of mycorrhizal infection and phosphate nutrition. *New Phytologist* **99**, 211–227.

Smith SE, Walker NA, Tester M 1986 The apparent width of the rhizosphere of *Trifolium subterraneum* L. for vesicular-arbuscular mycorrhizal infection: effects of time and other factors. *New Phytologist* **104**, 547–558.

Smith SE, Long CM, Smith FA 1989 Infection of roots with a dimorphic hypodermis: possible effects on solute uptake. *Agriculture, Ecosystems and Environment* **29**, 403–407.

Smith SE, Robson AD, Abbott LK 1992 The involvement of mycorrhizas in assessment of genetically dependent efficiency of nutrient uptake and use. *Plant and Soil* **146**, 169–179.

Smith SE, Dickson S, Morris C, Smith FA 1994 Transport of phosphate from fungus to plant in VA mycorrhizas: calculation of the area of symbiotic interface and of fluxes of P from two different fungi to *Allium porrum* L. *New Phytologist* **127**, 93–99.

Smith SE, Smith FA, Jakobsen I 2003c Mycorrhizal fungi can dominate phosphate supply to plants irrespective of growth responses. *Plant Physiology* **133**, 16–20.

Smith SE, Smith FA, Jakobsen I 2004a Functional diversity in arbuscular mycorrhizal AM symbioses: the contribution of the mycorrhizal P uptake pathway is not correlated with mycorrhizal responses in growth or total P uptake. *New Phytologist* **162**, 511–524.

Smith SE, Barker SJ, Zhu YG 2006 Fast moves in arbuscular mycorrhizal symbiotic signalling. *Trends in Plant Science* **11**, 369–371.

Smits M 2006 Mineral tunnelling by fungi. In *Fungi in Biogeochemical Cycles*. Ed. GM Gadd pp. 311–327. Cambridge University Press, Cambridge, UK.

Smits WTM 1992 Mycorrhizal studies in dipterocarp forests in Indonesia. In *Mycorrhizas in Ecosystems*. Eds DJ Read, DH Lewis, AH Fitter, IJ Alexander. pp. 283–292. CAB International, Wallingford, UK.

Smreciu EA and Currah RS 1989 Symbiotic germination of seeds of terrestrial orchids of North America and Europe. *Lindleyana* **4**, 6–15.

Sneh B, Burpee L, Ogoshi A 1991 *Identification of* Rhizoctonia *species*. APS Press, St. Paul, USA.

Snellgrove RC and Stribley DP 1986 Effects of pre-inoculation with a vesicular-arbuscular mycorrhizal fungus on growth on onions transplanted to the field as multi-seeded peat modules. *Plant and Soil* **92**, 387–397.

Snellgrove RC, Splittstoesser WE, Stribley DP, Tinker PB 1982 The distribution of carbon and the demand of the fungal symbiont in leek plants with vesicular-arbuscular mycorrhizas. *New Phytologist* **92**, 75–87.

Söderström B 1979 Seasonal fluctuations of active fungal biomass in the horizons of a podzolized pine forest soil in Central Sweden. *Soil Biology and Biochemistry* **11**, 149–154.

Söderström B and Read DJ 1987 Respiratory activity of intact and excised ectomycorrhizal mycelial systems growing in unsterilised soil. *Soil Biology and Biochemistry* **19**, 231–236.

Söderström B, Finlay RD, Read DJ 1988 The structure and function of the vegetative mycelium of ectomycorrhizal plants. IV. Qualitative analysis of carbohydrate contents of mycelium interconnecting host plants. *New Phytologist* **109**, 163–166.

Sokolovski SG, Meharg AA, Maathuis FJM 2002 *Calluna vulgaris* root cells show increased capacity for amino acid uptake when colonized with the mycorrhizal fungus *Hymenoscyphus ericae*. *New Phytologist* **155**, 525–530.

Solaiman MZ and Saito M 1997 Use of sugars by intraradical hyphae of arbuscular mycorrhizal fungi revealed by radiorespirometry. *New Phytologist* **136**, 533–538.

Solaiman MZ and Saito M 2001 Phosphate efflux from intraradical hyphae of *Gigaspora margarita in vitro* and its implication for phosphorus translocation. *New Phytologist* **151**, 525–533.

Solaiman MZ, Ezawa T, Kojima T, Saito M 1999 Polyphosphates in intraradical and extraradical hyphae of an arbuscular mycorrhizal fungus, *Gigaspora margarita*. *Applied and Environmental Microbiology* **65**, 5604–5606.

Soltis PS, Soltis DE, Doyle JJ 1992 *Molecular Systematics of Plants*. Chapman and Hall, London, UK.

Son CL, Smith SE 1988 Mycorrhizal growth responses: interactions between photon irradiance and phosphorus nutrition. *New Phytologist* **108**, 305–314.

Sonnewald U, Lerchl J, Zenner R, Frommer W 1994 Manipulation of sink-source relations in transgenic plants. *Plant, Cell and Environment* **17**, 649–658.

Spanu P and Bonfante-Fasolo P 1988 Cell-wall-bound peroxidase activity in roots of mycorrhizal *Allium porrum*. *New Phytologist* **109**, 119–124.

Spanu P, Boller T, Ludwig A, Wiemken A, Faccio A, Bonfante-Fasolo P 1989 Chitinase in roots of mycorrhizal *Allium porrum*: regulation and localization. *Planta* **177**, 447–455.

Specht RL 1979 Heathlands and related shrublands of the world. In *Ecosystems of the World. Heathlands and Related Shrublands. Descriptive Studies*. Ed RL Specht pp. 1–18. Elsevier, Amsterdam, The Netherlands.

Specht RL 1981 Mallee ecosystems in southern Australia. In *Ecosystems of the World. Vol. 11. Mediterranean-type Shrublands. Descriptive Studies*. Eds F di Castri, DW Goodall, RL Sprecht pp. 203–231. Elsevier, Amsterdam, The Netherlands.

Specht RL and Rundel PW 1990 Sclerophylly and foliar nutrient status of Mediterranean-climate plant communities in southern Australia. *Australian Journal of Botany* **38**, 459–474.

Sprent JI 1985 Nitrogen fixation in arid environments. In *Plants for Arid Lands*. Eds GE Wickens, JR Goodin, DV Field pp. 215–229. George Allen and Unwin, London, UK.

St John TV, Coleman DC, Reid CPP 1983a Association of vesicular-arbuscular mycorrhizal hyphae with soil organic particles. *Ecology* **64**, 957–959.

St John TV, Coleman DC, Reid CPP 1983b Growth and spatial distribution of nutrient absorbing organs – selective exploitation of soil heterogeneity. *Plant and Soil* **71**, 487–493.

Staddon PL and Fitter AH 1998 Does elevated atmospheric carbon dioxide affect arbuscular mycorrhizas? *Trends in Ecology and Evolution* **13**, 455–458.

Staddon PL, Heinemeyer A, Fitter AH 2002 Mycorrhizas and global environment change: research at different scales. *Plant and Soil* **244**, 253–261.

Stahl E 1900 Der sinn der mycorhizenbildung. *Jahbuch fur Wissenschaftliche Botanik* **34**, 539–668.

Stahl M 1949 Die mycorrhiza de lebermoose mit besonderer beruchsichtigung der thallosen formen. *Planta* **37**, 103–148.

Stendell ER, Horton TR, Bruns TD 1999 Early effects of prescribed fire on the structure of the ectomycorrhizal fungus community in a Sierra Nevada ponderosa pine forest. *Mycological Research* **103**, 1353–1359.

Stevens PF 2004 Angiosperm phylogeny website. <http://www.mobot.org/MOBOT/research/AP/>

Stevenson FJ 1982 *Humus Chemistry*. John Wiley, New York, USA.

Stewart GR, Mann AF, Fentem PA 1980 Enzymes of glutamate formation: glutamate dehydrogenase, glutamine synthetase and glutamate synthase. In *The Biochemistry of Plants, a Comprehensive Treatise*, vol 5. Ed. BJ Miflin pp. 271–327. Academic Press, New York, USA.

Stewart GR, Pate JS, Unkovich M 1993 Characteristics of inorganic nitrogen assimilation of plants in fire-prone Mediterranean type vegetation. *Plant, Cell and Environment* **16**, 351–363.

Stober C, George E, Persson H 2000 Root growth and response to nitrogen. In *Carbon and Nitrogen Cycling in European Forest Ecosystems. Ecological Studies* **142**, 99–121.

Stoessl A and Arditti J 1984 Orchid phytoalexins. In *Orchid Biology: Reviews and Perspectives III*. Ed. J Arditti pp. 153–175. Cornell University Press, Ithaca, New York, USA.

Stone EL 1950 Some effects of mycorrhizae on the phosphorus nutrition of Monterey pine seedlings. *Proceedings of the Soil Science Society of America* **14**, 340–345.

Stoyke G and Currah RS 1991 Endophytic fungi from the mycorrhizae of alpine ericoid plants. *Canadian Journal of Botany* **69** 2, 347–352.

Stoyke G and Currah RS 1993 Resynthesis in pure culture of a common sub-alpine fungus-root association using *Phialocephala fortinii* and *Menziesia ferruginea*, Ericaceae. *Arctic and Alpine Research* **25** 3, 189–193.

Stoyke G, Egger KN, Currah RS 1992 Characterization of sterile endophytic fungi from the mycorrhizae of sub-alpine plants. *Canadian Journal of Botany* **70**, 2009–2016.

Straatsma G and Krisai-Greilhuber I 2003 Assemblage structure, species richness, abundance, and distribution of fungal fruit bodies in a seven year plot-based survey near Vienna. *Mycological Research* **107**, 632–640.

Straatsma G, Ayer F, Egli S 2001 Species richness, abundance, and phenology of fungal fruit bodies over 21 years in a Swiss forest plot. *Mycological Research* **105**, 515–523.

Straker CJ and Mitchell DT 1986 The activity and characterization of acid phosphatases in endomycorrhizal fungi of the Ericaceae. *New Phytologist* **104**, 243–256.

Straker CJ and Mitchell DT 1987 Kinetic characterisation of a dual phosphate uptake system in the endomycorrhizal fungus of *Erica hispidula*. *New Phytologist* **106**, 129–137.

Straker CJ, Gianinazzi-Pearson V, Gianinazzi S, Cleyet-Marel J-C, Bousquet N 1989 Electrophoretic and immunological studies on acid phosphatase from a mycorrhizal fungus of *Erica hispida*. *New Phytologist* **111**, 215–222.

Strauss SH and Martin F 2004 Poplar genomics comes of age. *New Phytologist* **164**, 1–4.

Streitwolf-Engel R, Boller T, Wiemken A, Sanders IR 1997 Clonal growth traits of two *Prunella* species are determined by co-occurring arbuscular mycorrhizal fungi from a calcareous grassland. *Journal of Ecology* **85**, 181–191.

Stribley DP and Read DJ 1974a The biology of mycorrhiza in the Ericaceae. III. Movement of carbon-[14] from host to fungus. *New Phytologist* **73**, 731–741.

Stribley DP and Read DJ 1974b The biology of mycorrhiza in the Ericaceae. IV. The effect of mycorrhizal infection on uptake of ^{15}N from labelled soil by *Vaccinium macrocarpon* Ait. *New Phytologist* **73**, 1149–1155.

Stribley DP and Read DJ 1980 The biology of mycorrhiza in the Ericaceae. VII. The relationship between mycorrhizal infection and the capacity to utilise simple and complex organic nitrogen sources. *New Phytologist* **86**, 365–371.

Stribley DP, Read DJ, Hunt R 1975 The biology of mycorrhiza in the Ericaceae. V. The effect of mycorrhizal infection, soil type and partial soil sterilisation by γ irradiation on growth of cranberry *Vaccinium macrocarpon* Ait. *New Phytologist* **75**, 119–130.

Stribley DP, Tinker PB, Rayner JH 1980a Relation of internal phosphorus concentration and plant weight in plants infected by vesicular-arbuscular mycorrhizas. *New Phytologist* **86**, 261–266.

Stribley DP, Tinker PB, Snellgrove RC 1980b Effect of vesicular-arbuscular mycorrhizal fungi on the relations of plant growth, internal phosphorus concentration and soil phosphate analyses. *Journal of Soil Science* **31**, 655–672.

Struble JE and Skipper HD 1988 Vesicular-arbuscular mycorrhizal fungal spore production as influenced by plant species. *Plant and Soil* **109**, 277–280.

Strullu DG 1978 Cytologie des endomycorrhizes. *Physiologie Végétale* **16**, 657–669.

Strullu DG and Gerault A 1974 Ultrastructure et évolution du champignon symbiotique des racines de *Dactylorchis maculata*. *Journal de Microscopie, Paris* **20**, 285–294.

Strullu DG and Gerault A 1977 Études des ectomycorrhizes à basidiomycèetes et à ascomycètes du *Betula pubescens* Ehr. en microscopie electronique. *Comptes Rendus d'Academie des Sciences, Paris*, Série D **284**, 2243–2246.

Strullu DG, Gourret JP, Garrec JP 1981a Microanalyse des granules vacuolaires des ectomycorhizes, endomycorhizes et endomycothalles. *Physiologie Végétale* **19**, 367–378.

Strullu DG, Gourret JP, Garrec JP, Fourcy A 1981b Ultrastructure and electron-probe microanalysis of the metachromatic vacuolar granules occurring in *Taxus* mycorrhizas. *New Phytologist* **87**, 537–545.

Strullu DG, Harley JL, Gourret JP, Garrec JP 1982 Ultrastructure and microanalysis of polyphosphate granules of the ectomycorrhiza of *Fagus sylvatica*. *New Phytologist* **92**, 417–423.

Stubblefield SP, Taylor TN, Trappe JM 1987a Antarctic VAM fossils. *American Journal of Botany* **74**, 1904–1911.

Stubblefield SP, Taylor TN, Trappe JM 1987b Fossil mycorrhizae: a case for symbiosis. *Science* **237**, 59–60.

Stukenbrock EH and Rosendahl S 2005a Clonal diversity and population genetic structure of arbuscular mycorrhizal fungi (*Glomus* spp.) studied by multilocus genotyping of single spores. *Molecular Ecology* **14**, 743–752.

Stukenbrock EH and Rosendahl S 2005b Development and amplification of multiple co-dominant genetic markers from single spores of arbuscular mycorrhizal fungi by nested multiplex PCR. *Fungal Genetics and Biology* **42**, 73–80.

Subin VIN, Ronkonen NL, Saukkonen AV 1977 Effects of fertilizers on the fructification of macromycetes on young birch trees. *Mikologiya I Fitopatologia* **11**, 294–303.

Sukarno N, Smith SE, Scott ES 1993 The effect of fungicides on vesicular-arbuscular mycorrhizal symbiosis I. The effects on vesicular-arbuscular mycorrhizal fungi and plant growth. *New Phytologist* **25**, 139–147.

Sukarno N, Smith FA, Smith SE, Scott ES 1996 The effect of fungicides on vesicular-arbuscular mycorrhizal symbiosis. II. The effects on area of interface and efficiency of P uptake and transfer to plant. *New Phytologist* **132**, 583–592.

Summerbell RC 2005 From Lamarckian fertilizers to fungal castles: recapturing the pre-1985 literature on endophytic and saprotrophic fungi associated with ectomycorrhizal root systems. *Studies in Mycology* **53**, 191–256.

Sun YP, Unestam T, Lucas SD, Johanson KJ, Kenne L, Finlay R 1999 Exudation-reabsorption in a mycorrhizal fungus, the dynamic interface for interaction with soil and soil microorganisms. *Mycorrhiza* **9**, 137–144.

Sutton JC 1973 Development of vesicular-arbuscular mycorrhizae in crop plants. *Canadian Journal of Botany* **51**, 2487–2493.

Sutton JC, Barron CL 1972 Population dynamics of *Endogone* spores in soil. *Canadian Journal of Botany* **50**, 1909–1914.

Sverdrup H, Hagen-Thorn A, Holmqvist J *et al.* 2002 Biogeochemical processes and mechanisms. In *Developing Principles for Sustainable Forestry in Southern Sweden*. Eds H Sverdrup and I Stjernquist pp. 91–196. Kluwer Academic Publishers, Dordrecht, The Netherlands.

Swaminathan V 1979 Nature of the inorganic fraction of soil phosphate fed on by vesicular-arbuscular mycorrhizae of potatoes. *Proceedings of the Indian Academy of Science* **88B**, 423–433.

Swaty RL, Gehring CA, Van Ert M, Theimer TC, Keim P, Whitham TG 1998 Temporal variation in temperature and rainfall differentially affects ectomycorrhizal colonization at two contrasting sites. *New Phytologist* **139**, 733–739.

Swaty RL, Deckert RJ, Whitham TG, Gehring CA 2004 Ectomycorrhizal abundance and community composition shifts with drought: predictions from tree rings. *Ecology* **85**, 1072–1084.

Sylvia DM and Schenck NC 1983 Germination of chlamydospores of the three *Glomus* species as affected by soil matric potential and fungal contamination. *Mycologia* **75**, 30–35.

Sylvia DM and Sinclair WA 1983a Supressive influence of *Laccaria laccata* on *Fusarium oxysporum* and on Douglas fir seedlings. *Phytopathology* **73**, 384–389.

Sylvia DM and Sinclair WA 1983b Phenolic compounds and resistance to fungal pathogens induced in primary roots of Douglas fir by the ectomycorrhizal fungus *Laccaria laccata*. *Phytopathology* **73**, 390–397.

Sylvia DM, Alagely AK, Chellemi DO, Demchenko LW 2001 Arbuscular mycorrhizal fungi influence tomato competition with bahiagrass. *Biology and Fertility of Soils* **34**, 448–452.

Taber WA and Taber RA 1987 Carbon nutrition and respiration of *Pisolithus tinctorius*. *Transactions of the British Mycological Society* **89**, 13–26.

Tagu D, Rampant PF *et al.* 2001 Variation in the ability to form ectomycorrhizas in the F1 progeny of an interspecific poplar *Populus* spp. cross. *Mycorrhiza* **10** 5, 237–240.

Tagu D, Python M, Crétin C, Martin F 1993 Cloning symbiosis-related cDNAs from eucalypt ectomycorrhizas by PCR-assisted differential screening. *New Phytologist* **125**, 339–343.

Tagu D, Nasse B, Martin F 1996 Cloning and characterization of hydrophobins-encoding cDNAs from the ectomycorrhizal Basidiomycete *Pisolithus tinctorius*. *Gene* **168**, 93–97.

Tagu D, Rampant PF, Lapeyrie F, Frey-Klett P, Vion P, Villar M 2001 Variation in the ability to form ectomycorrhizas in the F1 progeny of an interspecific poplar (*Populus* spp.) cross. *Mycorrhiza* **10**, 237–240.

Tagu D, Lapeyrie F, Martin F 2002 The ectomycorrhizal symbiosis: genetics and development. *Plant and Soil* **244**, 97–105.

Tahiri-Alaoui A and Antoniw JF 1996 Cloning of genes associated with the colonization of tomato roots by the arbuscular mycorrhizal fungus *Glomus mosseae*. *Agronomie* **16**, 699–707.

Takacs EA 1967 Produccion de cultivos puros de hongos micorrhizógenos en el Centro Nacional de Investigaciones Agropecuarias, Castelar. *IDIA Supplement of Forestry* **4**, 83.

Tallis JH 1991 *Plant Community History*. Chapman and Hall, London, UK.

Tamasloukht M, Séjalon-Delmas N, Kluever A *et al.* 2003 Root factors induce mitochondrial-related gene expression and fungal respiration during the developmental switch from asymbiosis to presymbiosis in the arbuscular mycorrhizal fungus *Gigaspora rosea*. *Plant Physiology* **131**, 1468–1478.

Tamm CO 1985 The Swedish optimum nutrition experiments in forest stands. Aims, methods, yield, results. *Journal Royal Swedish Academy of Agriculture and Forestry Supplement* **17**, 9–29.

Tamm CO 1991 *Nitrogen in Terrestrial Ecosystems*. Springer-Verlag, Berlin, Germany.

Tanner EVJ 1977 Four montane rain forests of Jamaica: a quantitative characterisation of the floristics, the soils and the foliar mineral levels, and a discussion of the inter-relationships. *Journal of Ecology* **65**, 883–918.

Tanner EVJ, Kapos V, Franco W 1992 Nitrogen and phosphorus fertilization effects on Venezuelan montane forest trunk growth and litterfall. *Ecology* **73**, 78–86.

Tanner EVJ, Vitousek PM, Cuevas E 1998 Experimental investigation of nutrient limitation of forest growth on wet tropical mountains. *Ecology* **79**, 10–22.

Tarafdar JC and Marschner H 1994 Phosphatase activity in the rhizosphere and hyphosphere of VA mycorrhizal wheat supplied with inorganic and organic phosphorus. *Soil Biology and Biochemistry* **26**, 387–395.

Tarkka MT, Schrey S, Nehls U 2006 The alpha-tubulin gene *AmTuba1*: a marker for rapid mycelial growth in the ectomycorrhizal basidiomycete *Amanita muscaria*. *Current Genetics* **49**, 294–301.

Tashima Y, Terashita T, Umata H, Matsumoto M 1978 *In vitro* development from seed to flower in *Gastrodia verrucosa* under fungal symbiosis. *Transactions of the Mycological Society of Japan* **19**, 449–453.

Tawaraya K 2003 Arbuscular mycorrhizal dependency of different plant species and cultivars. *Soil Science and Plant Nutrition* **49**, 655–668.

Tawaraya K, Watanabe S, Yoshida E, Wagatsuma T 1996 Effect of onion *Allium cepa* root exudates on the hyphal growth of *Gigaspora margarita*. *Mycorrhiza* **6**, 57–59.

Tawaraya K, Hashimoto K, Wagatsuma T 1998 Effect of root exudate fractions from P-deficient and P-sufficient onion plants on root colonisation by the arbuscular mycorrhizal fungus *Gigaspora margarita*. *Mycorrhiza* **8**, 67–70.

Tawaraya K, Tokairin K, Wagatsuma T 2001 Dependence of *Allium fistulosum* cultivars on the arbuscular mycorrhizal fungus, *Glomus fasciculatum*. *Applied Soil Ecology* **17**, 119–124.

Tawaraya K, Takaya Y, Turjaman M *et al.* 2003 Arbuscular mycorrhizal colonization of tree species grown in peat swamp forests of Central Kalimantan, Indonesia. *Forest Ecology and Management* **182**, 381–386.

Taylor AFS and Alexander IJ 1989 Demography and population dynamics of ectomycorrhizas of Sitka spruce fertilised with N. *Agriculture, Ecosystems and Environment* **28**, 493–496.

Taylor AFS and Alexander IJ 2005 The ectomycorrhizal symbioses: life in the real world. *Mycologist* **19**, 102–112.

Taylor AFS, Högbom L, Högberg M, Lyon AJE, Näsholm T, Höberg P 1997 Natural ^{15}N abundance in fruit bodies of ectomycorrhizal fungi from boreal forests *New Phytologist* **136**, 713–720.

Taylor AFS, Martin F, Read DJ 2000 Fungal diversity in ectomycorrhizal communities of Norway spruce (*Picea abies* L. Karst.) and beech (*Fagus sylvatica* L.) along north-south transects in Europe. In Carbon and nitrogen cycling in European forest ecosystems. *Ecological Studies* **142**, 343–365.

Taylor AFS, Gebauer G, Read DJ 2004 Uptake of nitrogen and carbon from double-labelled (^{15}N and ^{13}C) glycine by mycorrhizal pine seedlings. *New Phytologist* **164**, 383–388.

Taylor DL 1997 The evolution of mycoheterotrophy and specificity in some North American orchids. PhD thesis. University of California, Berkeley, California, USA.

Taylor DL and Bruns TD 1997 Independent, specialized invasions of ectomycorrhizal mutualism by two nonphotosynthetic orchids. *Proceedings of the National Academy of Sciences of the United States of America* **94**, 4510–4515.

Taylor DL and Bruns TD 1999a Community structure of ectomycorrhizal fungi in a *Pinus muricata* forest: minimal overlap between the mature forest and resistant propagule communities. *Molecular Ecology* **8**, 1837–1850.

Taylor DL and Bruns TD 1999b Population, habitat and genetic correlates of mycorrhizal specialization in the 'cheating' orchids *Corallorhiza maculata* and *C. mertensiana*. *Molecular Ecology* **8** 10, 1719–1732.

Taylor DL, Bruns TD, Leake JR, Read DJ 2002 Mycorrhizal specificity and function in myco-heterotrophic plants. In *The Ecology of Mycorrhizas*. Eds MGA van der Heijden and IR Sanders pp. 375–413. Springer-Verlag, Berlin, Germany.

Taylor DL, Bruns TD, Szaro TM, Hodges SA 2003 Divergence in mycorrhizal specialization within *Hexalectris spicata* (Orchidaceae), a nonphotosynthetic desert orchid. *American Journal of Botany* **90**, 1168–1179.

Taylor JW, Jacobson DJ, Fisher MC 1999 The evolution of asexual fungi: reproduction, speciation and classification. *Annual Review of Phytopathology* **37**, 197–246.

Taylor TN, Remy W, Hass H, Kerp H 1995 Fossil arbuscular mycorrhizae from the Early Devonian. *Mycologia* **87**, 560–573.

Taylor TN, Klavins SD, Krings M, Taylor EL, Kerp H, Hass H 2004 Fungi from the Rhynie chert: a view from the dark side. *Transactions of the Royal Society of Edinburgh-Earth Sciences* **94**, 457–473.

Taylor TN, Kerp H, Haas H 2005 Life history biology of early land plants: deciphering the gametophyte phase. *Proceedings of the National Academy of Sciences of the United States of America* **102**, 5892–5897.

Tedersoo L, Koljalg U, Hallenberg N, Larsson KH 2003 Fine scale distribution of ectomycorrhizal fungi and roots across substrate layers including coarse woody debris in a mixed forest. *New Phytologist* **159**, 153–165.

Tedersoo L, Hansen K, Perry BA, Kjoller R 2006 Molecular and morphological diversity of pezizalean ectomycorrhiza. *New Phytologist* **170**, 581–596.

Tedersoo L, Pellet P, Koljalg U, Selosse MA 2007 Parallel evolutionary paths to mycoheterotrophy in understorey Ericaceae and Orchidaceae: ecological evidence for mixotrophy in Pyrolae. *Oecologia* **151**, 206–217.

Terashita T 1982 Fungi inhabiting wild orchids in Japan II. Isolation of symbionts from *Spiranthes sinensis* var. *amoena*. *Transactions of the Mycological Society of Japan* **15**, 121–133.

Terashita T 1985 Fungi inhabiting wild orchids in Japan III. A symbiotic experiment with *Armillarea mellea* and *Galeola septentrionalis*. *Transactions of the Mycological Society of Japan* **26**, 47–53.

Terashita T 1996 Biological species of *Armillaria* symbiotic with *Galeola septentrionalis*. *Nippon Kingakukai Kaiho* **37**, 45–49.

Terashita T and Chuman S 1987 Fungi inhabiting wild orchids in Japan IV: *Armillariella tabescens*, a new symbiont of *Galeola septentrionalis*. *Transactions of the Mycological Society of Japan* **28**, 145–154.

Termorshuizen AJ 1993 The influence of nitrogen fertilizers on ectomycorrhizas and their fungal carpophores in young stands of *Pinus sylvestris*. *Forest Ecology and Management* **57** 1–4, 179–189.

Tester M, Smith SE, Smith FA, Walker NA 1986 Effects of photon irradiance on the growth of shoots and roots, on the rate of initiation of mycorrhizal infection and on the growth of infection units in *Trifolium subterraneum* L. *New Phytologist* **103**, 375–390.

Theodorou C 1968 Inositol phosphate in needles of *Pinus radiata* D. Don and the phytase activity of mycorrhizal fungi. *Proceedings of the 9th International Congress of Soil Science* **3**, 483–493.

Theodorou C 1971 The phytase activity of the mycorrhiza fungus *Rhizopogon roseolus*. *Soil Biology and Biochemistry* **3**, 89–90.

Theodorou C 1978 Soil moisture and the mycorrhizal association of *Pinus radiata* D. Don. *Soil Biology and Biochemistry* **10**, 33–37.

Theodorou C and Bowen GD 1973 Inoculation of seeds and soil with basidiospores of mycorrhizal fungi. *Soil Biology and Biochemistry* **5**, 765–771.

Thomas GW and Jackson RM 1982 *Complexipes moniliformis* – ascomycete or zygomycete? *Transactions of the British Mycological Society* **79**, 149–186.

Thomas RS, Dakessian S, Ames RN, Brown MS, Bethlenfalvay GJ 1986 Aggregation of a silty clay loam soil by mycorrhizal onion roots. *Soil Society of America, Journal* **50**, 1494–1499.

Thompson JP 1987 Decline of vesicular-arbuscular mycorrhizae in Long Fallow Disorder of field crops and its expression in phosphorus deficiency of sunflower. *Australian Journal of Agricultural Research* **38**, 847–867.

Thompson JP 1990 Soil sterilization methods to show VA-mycorrhizae aid P and Zn nutrition of wheat in vertisols. *Soil Biology and Biochemistry* **22**, 229–240.

Thompson JP 1994 Inoculation with vesicular-arbuscular mycorrhizal fungi from cropped soil overcomes long-fallow disorder of linseed *Linum usitatissimum* L. by improving P and Zn uptake. *Soil Biology and Biochemistry* **26**, 1133–1143.

Thomson BD, Robson AD, Abbott LK 1990a Mycorrhizas formed by *Gigaspora calospora* and *Glomus fasciculatum* on subterranean clover in relation to carbohydrate concentrations in roots. *New Phytologist* **114**, 217–225.

Thomson BD, Clarkson DT, Brain P 1990b Kinetics of phosphorus uptake by the germ-tubes of the vesicular-abuscular mycorrhizal fungus, *Gigaspora margarita*. *New Phytologist* **116**, 647–653.

Thomson BD, Grove TS, Malajczuk N, Hardy GSTJ 1994 The effectiveness of ectomycorrhizal fungi in increasing the growth of *Eucalyptus globulus* Labill. in relation to root colonisation and hyphal development in soil. *New Phytologist* **126**, 517–524.

Thomson J, Melville LH, Peterson RL 1989 Interaction between the ectomycorrhizal fungus *Pisolithus tinctorius* and root hairs of *Picea mariana* (Pinaceae). *American Journal of Botany* **76**, 632–636.

Thomson J, Matthes-Sears U, Peterson RL 1990 Effect of seed provenance and fungal species on bead formation in roots of *Picea mariana* seedlings. *Canadian Journal of Forest Research* **20**, 1746–1752.

Tibbett M and Sanders FE 2002 Ectomycorrhizal symbiosis can enhance plant nutrition through improved access to discrete organic nutrient patches of high resource quality. *Annals of Botany* **89**, 783–789.

Tibbett M, Sanders FE, Cairney JWG 1998a The effect of temperature and inorganic phosphorus concentration on acid phosphatase production and growth rate in arctic and temperate isolates of the ectomycorrhizal fungi *Hebeloma* spp. in axenic culture. *Mycological Research* **102**, 129–135.

Tibbett M, Sanders FE, Minto SJ, Dowell M, Cairney JWG 1998b Utilization of organic nitrogen by ectomycorrhizal fungi (*Hebeloma* spp.) of arctic and temperate origin. *Mycological Research* **102**, 1525–1532.

Tibbett M, Sanders FE, Cairney JWG 2002 Low-temperature-induced changes in trehalose, mannitol and arabitol associated with enhanced tolerance to freezing in ectomycorrhizal basidiomycetes *Hebeloma* spp. *Mycorrhiza* **12**, 249–255.

Tilman D 1982 *Resource Competition and Community Structure*. Princetown University Press, NJ, USA.

Tilman D 1987 Secondary succession and the pattern of plant dominance along experimental nitrogen gradients. *Ecological Monographs* **57**, 189–214.

Tilman D 1988 *Plant Strategies and the Dynamics and Structure of Plant Communities*. Princetown University Press, NJ, USA.

Tilman D 1990 Mechanisms of plant competition for nutrients: the elements of a predictive theory of competition. In *Perspectives on Plant Competition*. Eds JB Grace and D Tilman pp. 117–142. Academic Press, New York, USA.

Timmer LW and Leyden RF 1980 The relationship of mycorrhizal infection to phosphorus-induced copper deficiency in sour orange seedlings. *New Phytologist* **85**, 15–23.

Timonen S and Peterson RL 2002 Cytoskeleton in mycorrhizal symbiosis. *Plant and Soil* **244**, 199–210.

Timonen S and Smith SE 2005 Effect of the arbuscular mycorrhizal fungus *Glomus intraradices* on expression of cytoskeletal proteins in tomato roots. *Canadian Journal of Botany* **83**, 176–182.

Timonen S, Jørgensen KS, Haahtela K, Sen R 1998 Bacterial community structure at defined locations of *Pinus sylvestris-Suillus bovinus* and *Pinus sylvestris-Paxillus involutus* mycorrhizospheres in dry pine forest soil and nursery peat. *Canadian Journal of Microbiology* **44**, 499–513.

Timonen S, Smith FA, Smith SE 2001 Microtubules of the mycorrhizal fungus *Glomus intraradices* in symbiosis with tomato roots. *Canadian Journal of Botany* **79**, 307–313.

Timonen S, Christensen S, Ekelund F 2004 Distribution of protozoa in Scots pine mycorrhizospheres. *Soil Biology and Biochemistry* **36**, 1087–1093.

Tinker PB 1975a Effects of veiscular-arbuscular mycorrhizas on higher plants. *Symposium of the Society for Experimental Biology* **29**, 325–349.

Tinker PB 1975b Soil chemistry of phosphorus and mycorrhizal effects on plant growth. In *Endomycorrhizas*. Eds FE Sanders, B Mosse, PB Tinker pp. 353–371. Academic Press, London, UK.

Tinker PB 1978 Effects of vesicular-arbuscular mycorrhizas on plant nutrition and plant growth. *Physiologie Végétale* **16**, 743–751.

Tinker PBH and Nye PH 2000 *Solute Movement in the Rhizosphere*. Oxford University Press, Oxford, UK.

Tinker PB, Durall DM, Jones MD 1994 Carbon use efficiency in mycorrhizas: theory and sample calculations. *New Phytologist* **128**, 115–122.

Tisdall JM 1994 Possible role of soil micro-organisms in aggregation of soils. *Plant and Soil* **159**, 115–121.

Tisdall JM 1995 Formation of soil aggregates and accumulation of soil organic matter. In *Structure and Organic Matter Storage in Agricultural Soils*. Eds MR Carter and BA Stewart pp. 57–96. *Advances in Soil Science*, Lewis Publishers, CRC Press Inc., Boca Raton, FL, USA.

Tisdall JM and Adem HH 1990 Mechanized relay-cropping in an irrigated red-brown earth in south-eastern Australia. *Soil Use and Management* **6**, 21–28.

Tisdall JM and Oades JM 1979 Stabilization of soil aggregates by the root systems of ryegrass. *Australian Journal of Soil Research* **17**, 429–441.

Tisdall JM and Oades JM 1982 Organic matter and water-stable aggregates in soils. *Journal of Soil Science* **33**, 141–163.

Tisserant B, Gianinazzi-Pearson V, Gianinazzi S, Gollotte A 1993 *In planta* histochemical staining of fungal alkaline phosphatase activity for analysis of efficient arbuscular mycorrhizal infections. *Mycological Research* **97**, 245–250.

Tobar RM, Azcon R, Barea JM 1994 Improved nitrogen uptake and transport from ^{15}N-labelled nitrate by external hyphae of arbuscular mycorrhiza under water-stressed conditions. *New Phytologist* **126**, 119–122.

Tommerup IC 1983 Spore dormancy in vesicular-arbuscular mycorrhizal fungi. *Transactions of the British Mycological Society* **81**, 37–45.

Tommerup IC 1984a Effect of soil water potential on spore germination by vesicular-arbuscular mycorrhizal fungi. *Transactions of the British Mycological Society* **83**, 193–202.

Tommerup IC 1984b Persistence of infectivity by germinated spores of vesicular-arbuscular mycorrhizal fungi in soil. *Transactions of the British Mycological Society* **82**, 275–282.

Tommerup IC 1984c Development of infection by a vesicular-arbuscular mycorrhizal fungus in *Brassica napus* and *Trifolium subterraneum*. *New Phytologist* **98**, 487–495.

Tommerup IC and Abbott LK 1981 Prolonged survival and viability of V.A. mycorrhizal hyphae after root death. *Soil Biology and Biochemistry* **13**, 431–433.

Torti SD, Coled PD, Kursar TA 2001 Causes and consequences of monodominance in tropical lowland forests. *American Naturalist* **157**, 141–153.

Toth R 1992 The quantification of arbuscules and related structures using morphometric cytology. In *Methods in Microbiology*. Eds JR Norris, DJ Read, AK Varma pp. 275–299. Academic Press, London, UK.

Toth R and Toth D 1982 Quantifying vesicular-arbuscular mycorrhizae using a morphometric technique. *Mycologia* **74**, 182–187.

Toth R, Doane C, Bennett E, Alexander T 1990 Correlation between host-fungal surface areas and percent colonization in VA mycorrhizae. *Mycologia* **82**, 519–522.

Toth R, Miller RM, Jarstfer AG, Alexander T, Bennett EL 1991 The calculation of intraradical fungal biomass from percent colonization in vesicular-arbuscular mycorrhizae. *Mycologia* **83**, 553–558.

Toussaint JP, St-Arnaud M, Charest C 2004 Nitrogen transfer and assimilation between the arbuscular mycorrhizal fungus *Glomus intraradices* Schenck and Smith and Ri T-DNA roots of *Daucus carota* L. in an in vitro compartmented system. *Canadian Journal of Microbiology* **50**, 251–260.

Tranvan H, Habricot Y, Jeannette E, Gay G, Sotta B 2000 Dynamics of symbiotic establishment between an IAA-overproducing mutant of the ectomycorrhizal fungus *Hebeloma cylindrosporum* and *Pinus pinaster*. *Tree Physiology* **20**, 123–129.

Trappe JM 1967 Pure culture synthesis of Douglas fir mycorrhizae with species of *Hebeloma, Suillus, Rhizopogon* and *Astraeus*. *Forest Science* **13**, 121–130.

Trappe JM 1977 Selection of fungi for ectomycorrhizal inoculation in nurseries. *Annals Review of Phytopathology* **15**, 203–222.

Trappe JM 1987 Phylogenetic and ecologic aspects of mycotrophy in the angiosperms from an evolutionary standpoint. In *Ecophysiology of VA mycorrhizal Plants*. Ed. GR Safir pp. 5–25. CRC Press, Boca Raton, Florida, USA.

Trappe JM and Berch SM 1985 The prehistory of mycorrhizae: AB Frank's predecessors. In *Proceedings of the 6th North America Conference on Mycorrhizae College of Forestry*. Oregon State University Corvallis, Oregon, USA.

Treseder KK 2004 A meta-analysis of mycorrhizal responses to nitrogen, phosphorus, and atmospheric CO_2 in field studies. *New Phytologist* **164**, 347–355.

Treseder KK and Allen MF 2000 Mycorrhizal fungi have a potential role in soil carbon storage under elevated CO_2 and nitrogen deposition. *New Phytologist* **147**, 189–200.

Treu R, Laursen GA, Stephenson SL, Landolt JC, Densmore R 1996 Mycorrhizae from Denali National Park and Preserve, Alaska. *Mycorrhiza* **6**, 21–29.

Trojanowski J, Haider K, Hüttermann A 1984 Decomposition of ^{14}C-labelled lignin, holocellulose and lignocellulose by mycorrhizal fungi. *Archives of Microbiology* **139**, 202–206.

Trouvelot A, Kough JL, Gianinazzi-Pearson V 1986 Mesure du taux de mycorhization VA d'un système radiculaire. Recherche de méthodes d'estimation ayant une signification fonctionelle. In *Physiological and Genetical Aspects of Mycorrhizae*. Eds V Gianinazzi-Pearson and S Gianinazzi pp. 217–221. INRA, Paris, France.

Trudell SA, Rygiewicz PT, Edmonds RL 2003 Nitrogen and carbon stable isotope abundances support the myco-heterotrophic nature and host-specificity of certain achlorophyllous plants. *New Phytologist* **160**, 391–401.

Tsai SM, Phillips DA 1991 Flavonoids released naturally from alfalfa promote development of symbiotic *Glomus* spores *in vitro*. *Applied and Environmental Microbiology* **57**, 1485–1488.

Tullio M, Pierandrei F, Salerno A, Rea E 2003 Tolerance to cadmium of vesicular arbuscular mycorrhizae spores isolated from a cadmium-polluted and unpolluted soil. *Biology and Fertility of Soils* **37**, 211–214.

Turnau K, Kottke I, Oberwinkler F 1993 Element localization in mycorrhizal roots of *Pteridium aquilinum* L. Kuhn collected from experimental plots treated with cadmium dust. *New Phytologist* **123**, 313–324.

Turnau K, Mleczko P, Blaudez D, Chalot M, Botton B 2002 Heavy metal binding properties of *Pinus sylvestris* mycorrhizas from industrial wastes. *Acta Societatis Botanicorum Poloniae* **71**, 253–261.

Turnbull MH, Goodall R, Stewart GR 1995 The impact of mycorrhizal colonisation upon nitrogen source utilisation and metabolism of seedlings of *Eucalyptus grandis* Hill ex Maiden and *Eucalyptus maculata* Hook. *Plant Cell and Environment* **18**, 1386–1394.

Uchiumi T, Shimoda Y *et al.* 2002 Expression of symbiotic and nonsymbiotic globin genes responding to microsymbionts on *Lotus japonicus*. *Plant and Cell Physiology* **43**, 1351–1358.

Uebel E and Heinsdorf D 1997 Results of long-term K and Mg fertilizer experiments in afforestation. *Forest Ecology and Management* **91**, 47–52.

Uetake Y and Peterson RL 1997 Changes in actin filament arrays in protocorm cells of the orchid species *Spiranthes sinensis* induced by the symbiotic fungus *Ceratobasidium cornigerum*. *Canadian Journal of Botany* **75**, 1661–1669.

Uetake Y and Peterson RL 1998 Association between microtubules and symbiotic fungal hyphae in protocorm cells of the orchid species, *Spiranthes sinensis*. *New Phytologist* **140**, 715–722.

Uetake Y, Farquhar ML, Peterson RL 1997 Changes in microtubule arrays in symbiotic orchid protocorms during fungal colonization and senescence. *New Phytologist* **135**, 701–709.

Uetake Y, Kojima T, Ezawa T, Saito M 2002 Extensive tubular vacuole system in an arbuscular mycorrhizal fungus, *Gigaspora margarita*. *New Phytologist* **154**, 761–768.

Umata H 1995 Seed germination of *Galeola altissima*, an achlorophyllous orchid, with aphyllophorales fungi. *Mycoscience* **36**, 369–372.

Umata H 1998a *In vitro* symbiotic association of an achlorophyllous orchid, *Erythrorchis ochobiensis* with orchid and non-orchid fungi. *Memoirs of the Faculty of Agriculture, Kagoshima University* **34**, 97–107.

Umata H 1998b A new biological function of shiitake mushroom, *Lentinula edodes*, in a mycoheterotrophic orchid, *Erythrorchis ochobiensis. Mycoscience* **39**, 85–89.

Unestam T 1991 Water replellency, mat formation, and leaf-stimulated growth of some ectomycorrhizal fungi. *Mycorrhiza* **1**, 13–20.

Unestam T and Sun YP 1995 Extramatrical structures of hydrophobic and hydrophilic ectomycorrhizal fungi. *Mycorrhiza* **5**, 201–311.

Untereiner WA and Malloch D 1999 Patterns of substrate utilization in species of Capronia and allied black yeasts: ecological and taxonomic implications. *Mycologia* **91** 3, 417–427.

Upson R 2006 *The Status of Dark Septate Root Fungal Endophytes in Antarctic Plant Communities.* PhD Thesis, University of Sheffield, Sheffield, UK.

Urban A, Weiss M, Bauer R 2003 Ectomycorrhizas involving sebacinoid mycobionts. *Mycological Research* **107**, 3–14.

Urcelay C and Diaz S 2003 The mycorrhizal dependence of subordinates determines the effect of arbuscular mycorrhizal fungi on plant diversity. *Ecology Letters* **6**, 388–391.

Ursic M and Peterson RL 1997 Morphological and anatomical characterization of ectomy-corrhizas and ectendomycorrhizas on *Pinus strobus* seedlings in a southern Ontario nursery. *Canadian Journal of Botany* **75** 12, 2057–2072.

Ursino DJ, Nelson CD, Krotkov G 1968 Seasonal changes in the distribution of photo-assimilated ^{14}C in young pine plants. *Plant Physiology* **43**, 845–852.

Vallorani L, Polidori E, Sacconi C *et al.* 2002 Biochemical and molecular characterization of NADP glutamate dehydrogenase from the ectomycorrhizal fungus *Tuber borchii. New Phytologist* **154**, 779–790.

van Aarle IM, Olsson PA, Söderström B 2002a Arbuscular mycorrhizal fungi respond to the substrate pH of their extraradical mycelium by altered growth and root colonization. *New Phytologist* **155**, 173–182.

van Aarle IM, Rouhier H, Saito M 2002b Phosphatase activities of arbuscular mycorrhizal intraradical and extraradical mycelium, and their relation to phosphorus availability. *Mycological Research* **106**, 1224–1229.

van Aarle IM, Cavagnaro TR, Smith SE, Smith FA, Dickson S 2005 Metabolic activity of *Glomus intraradices* in *Arum-* and *Paris*-type arbuscular mycorrhizal colonization. *New Phytologist* **166**, 611–618.

van Breemen N, Finlay R, Lundstrom U, Jongmans AG, Giesler R, Olsson M 2000a Mycorrhizal weathering: a true case of mineral plant nutrition? *Biogeochemistry* **49**, 53–67.

van Breemen N, Lundstrom US, Jongmans AG 2000b Do plants drive podzolization via rock-eating mycorrhizal fungi? *Geoderma* **94**, 163–171.

Van Cleve K and Viereck LA 1981 Forest succession in relation to nutrient cycling in the boreal forest of Alaska. In *Forest Succession*. Eds DC West, HH Shugart, BD Botkin pp. 185–212. Springer, Berlin, Germany.

van der Heijden EW 2000 Mycorrhizal symbiosis of *Salix repens*: diversity and functional significance. PhD thesis, Wageningen University, The Netherlands.

van der Heijden EW 2001 Differential benefits of arbuscular mycorrhizal and ectomycorrhizal infection of *Salix repens. Mycorrhiza* **10** 4, 185–193.

van der Heijden EW and Kuyper TW 2001a Laboratory experiments imply the conditionality of mycorrhizal benefits for *Salix repens*: role of pH and nitrogen to phosphorus ratios. *Plant and Soil* **228**, 275–290.

van der Heijden EW and Kuyper TW 2001b Does origin of mycorrhizal fungus or mycorrhizal plant influence effectiveness of the mycorrhizal symbiosis? *Plant and Soil* **230**, 161–174.

van der Heijden EW and Kuyper TW 2003 Ecological strategies of ectomycorrhizal fungi of *Salix repens*: root manipulation versus root replacement. *Oikos* **103**, 668–680.

van der Heijden EW and Vosatka M 1999 Mycorrhizal associations of *Salix repens* L. communities in succession of dune ecosystems. II. Mycorrhizal dynamics and interactions of ectomycorrhizal and arbuscular mycorrhizal fungi. *Canadian Journal of Botany* **77** 12, 1833–1841.

van der Heijden MGA 2004 Arbuscular mycorrhizal fungi as support systems for seedling establishment in grassland. *Ecology Letters* **7**, 293–303.

van der Heijden MGA, Boller T, Wiemken A, Sanders IR 1998a Different arbuscular mycorrhizal fungal species are potential determinants of plant community structure. *Ecology* **79**, 2082–2091.

van der Heijden MGA, Klironomos JN, Ursic M *et al.* 1998b Mycorrhizal fungal diversity determines plant biodiversity, ecosystem variability and productivity. *Nature* **396**, 69–72.

van der Heijden MGA and Sanders IR 2002 Mycorrhizal ecology: synthesis and perspectives. In *Mycorrhizal Ecology*. Eds MGA van der Heijden and IR Sanders pp. 441–456. Springer-Verlag, Berlin, Germany.

van der Heijden MGA, Wiemken A, Sanders IR 2003 Different arbuscular mycorrhizal fungi alter coexistence and resource distribution between co-occurring plants. *New Phytologist* **157**, 569–578.

van Hees PAW, Lundstrom US, Giesler R 2000 Low molecular weight organic acids and their Al-complexes in soil solution – composition, distribution and seasonal variation in three podzolized soils. *Geoderma* **94**, 173–200.

van Hees PAW, Jones DL, Jentschke G, Godbold DL 2004 Mobilization of aluminium, iron and silicon by *Picea abies* and ectomycorrhizas in a forest soil. *European Journal of Soil Science* **55**, 101–111.

van Kessel C, Singleton PW, Hoben HJ 1985 Enhanced N transfer form soybean to maize by vesicular-arbuscular VAM fungi. *Plant Physiology* **79**, 562–563.

van Rhijn P, Fang Y *et al.* 1997 Expression of early nodulin genes in alfalfa mycorrhizae indicates that signal transduction pathways used in forming arbuscular mycorrhizae and *Rhizobium*-induced nodules may be conserved. *Proceedings of the National Academy of Sciences of the United States of America* **94**, 5467–5472.

van Schöll L 2006 Ectomycorrhizal fungi and *Pinus sylvestris*: aluminium toxicity, base cation deficiencies and exudation of organic anions. PhD Thesis. University of Wageningen, The Netherlands.

van Schöll L, Keltjens WG, Hoffland E, van Breemen N 2005 Effect of ectomycorrhizal colonization on the uptake of Ca, Mg and Al by *Pinus sylvestris* under aluminium toxicity. *Forest Ecology and Management* **215**, 352–360.

van Schöll L, Hoffland E, van Breemen N 2006a Organic anion exudation by ectomycorrhizal fungi and *Pinus sylvestris* in response to nutrient deficiencies. *New Phytologist* **170**, 153–163.

van Schöll L, Smits MM, Hoffland E 2006b Ectomycorrhizal weathering of the soil minerals muscovite and hornblende. *New Phytologist* **171**, 805–814.

van Tichelen KK and Colpaert JV 2000 Kinetics of phosphate absorption by mycorrhizal and non-mycorrhizal Scots pine seedlings. *Physiologia Plantarum* **110** 1, 96–103.

van Tichelen KK, Vanstraelen T, Colpaert JV 1999 Nutrient uptake by intact mycorrhizal *Pinus sylvestris* seedlings: a diagnostic tool to detect copper toxicity. *Tree Physiology* **19**, 189–196.

van Tuinen D, Jacquot E, Zhao B, Gollotte A, Gianinazzi-Pearson V 1998 Characterization of root colonization profiles by a microcosm community of arbuscular mycorrhizal fungi using 25S rDNA-targeted nested PCR. *Molecular Ecology* **7**, 879–887.

van Wijk MT, Clemmensen KE, Shaver GR *et al.* 2004 Long-term ecosystem level experiments at Toolik Lake, Alaska, and at Abisko, Northern Sweden: generalizations and differences in ecosystem and plant type responses to global change. *Global Change Biology* **10**, 105–123.

Vandenkoornhuyse P, Leyval C, Bonnin I 2001 High genetic diversity in arbuscular mycorrhizal fungi: evidence for recombination events. *Heredity* **87**, 243–253.

Vandenkoornhuyse P, Husband R, Daniell TJ, Watson IJ, Duck JM, Fitter AH, Young JPW 2002 Arbuscular mycorrhizal community composition associated with two plant species in a grassland ecosystem. *Molecular Ecology* **11**, 1555–1564.

Vandenkoornhuyse P, Ridgway KP, Watson IJ, Fitter AH, Young JPW 2003 Co-existing grass species have distinctive arbuscular mycorrhizal communities. *Molecular Ecology* **12**, 3085–3095.

Vanderplank JE 1978 *Genetic and Molecular Basis of Plant Pathogenesis*. Springer-Verlag, Berlin, Germany.

Väre H, Vestberg M, Eurola S 1992 Mycorrhiza and root-associated fungi in Spitzbergen. *Mycorrhiza* **1**, 93–104.

Varma AK and Bonfante P 1994 Utilisation of cell-wall related carbohydrates by ericoid mycorrhizal endophytes. *Symbiosis* **16**, 301–313.

Vegh I, Fabre E, Gianinazzi-Pearson V 1979 Présence en France de *Pezizella ericae* Read, champignon endomycorhizogène des Ericacées horticoles. *Phytopathologische Zeitschrift* **96**, 231–243.

Vermeulen P 1946 *Studies on Dactylorchis*. Utrecht, The Netherlands.

Vestberg M, Estaún V 1994 Micropropagated plants, an opportunity to positively manage mycorrhizal activities. In *Impact of Arbuscular Mycorrhizas on Sustainable Agriculture and Natural Ecosystems*. Eds S Gianinazzi and H Schüepp pp. 217–226. Birkhäuser Verlag, Basel, Switzerland.

Vezina L-P, Margolis HA, Ouimet R 1988 The activity, characterisation and distribution of the nitrogen assimilation enzyme, glutamine synthetase, in jack pine seedlings. *Tree Physiology* **4**, 9–118.

Vezina L-P, Margolis HA, Delaney S 1989 Changes in the activity of enzymes involved with primary nitrogen metabolism due to ectomycorrhizal symbiosis on jack pine seedlings. *Physiologia Plantarum* **75**, 55–62.

Viera A and Glenn MG 1990 DNA content of vesicular-arbuscular mycorrhizal fungal spores. *Mycologia* **82**, 263–267.

Vierheilig H 2004a Further root colonization by arbuscular mycorrhizal fungi in already mycorrhizal plants is suppressed after a critical level of root colonization. *Journal of Plant Physiology* **161**, 339–341.

Vierheilig H 2004b Regulatory mechanisms during the plant-arbuscular mycorrhizal fungus interaction. *Canadian Journal of Botany* **82**, 1166–1176.

Vierheilig H and Ocampo JA 1990a Role of root extract and volatile substances of non-host plants on vesicular-arbuscular mycorrhizal spore germination. *Symbiosis* **9**, 199–202.

Vierheilig H and Ocampo JH 1990b Effects of isothiocyanate on germination of spores of *G. mosseae*. *Soil Biology and Biochemistry* **22**, 1161–1162.

Vierheilig H, Alt M, Mohr U, Boller T, Wiemken A 1994 Ethylene biosynthesis and activities of chitinase and β-1,3-glucanase in the roots of host and non-host plants of vesicular-arbuscular mycorrhizal fungi after inoculation with *Glomus mosseae*. *Journal of Plant Physiology* **143**, 337–343.

Vierheilig H, Alt M, Lange J, Gut-Rella M, Wiemken A, Boller T 1995 Colonization of transgenic tobacco constitutively expressing pathogenesis-related proteins by the vesicular-arbuscular mycorrhizal fungus *Glomus mosseae*. *Applied and Environmental Microbiology* **61**, 3031–3034.

Vierheilig H, Alt-Hug M, Engel-Streitwolf R, Mäder P, Wiemken A 1998a Studies on the attractional effect of root exudates on hyphal growth of an arbuscular mycorrhizal fungus in a soil compartment-membrane system. *Plant and Soil* **203**, 137–144.

Vierheilig H, Coughlan AP, Wyss U, Piché Y 1998b Ink and vinegar, a simple staining technique for arbuscular-mycorrhizal fungi. *Applied and Environmental Microbiology* **64**, 5004–5007.

Vieweg MF, Fruhling M, Quandt HJ *et al.* 2004 The promoter of the *Vicia faba* L. leghemoglobin gene *VfLb29* is specifically activated in the infected cells of root nodules and in the arbuscule-containing cells of mycorrhizal roots from different legume and nonlegume plants. *Molecular Plant-Microbe Interactions* **17**, 62–69.

Vieweg MF, Hohnjec N, Kuster H 2005 Two genes encoding different truncated hemoglobins are regulated during root nodule and arbuscular mycorrhiza symbioses of *Medicago truncatula*. *Planta* **220**, 757–766.

Vilarino A, Sainz MJ 1997 Treatment of *Glomus mosseae* propagules with 50% sucrose increases spore germination and inoculum potential. *Soil Biology and Biochemistry* **29**, 1571–1573.

Villarreal-Ruiz L, Anderson IC, Alexander IJ 2004 Interaction between an isolate from the *Hymenoscyphus ericae* aggregate and roots of *Pinus* and *Vaccinium*. *New Phytologist* **164**, 183–192.

Visser S 1995 Ectomycorrhizal fungal succession in jack pine stands following wildfire. *New Phytologist* **129**, 389–401.

Vitousek PM 1984 Litterfall, nutrient cycling and nutrient limitation in tropical forests. *Ecology* **65**, 285–298.

Vitousek PM and Stanford RL 1986 Nutrient cycling in moist tropical forest. *Annual Review of Ecology and Systematics* **17**, 137–167.

Vitousek PM, Shearer G, Kohl DH 1989 Foliar ^{15}N natural abundance in Hawaiian rainforest: patterns a possible mechanisms. *Oecologia* **78**, 383–388.

Vitousek PM, Gerrish G, Turner DR, Walker LR, Muellerdombois D 1995 Litterfall and nutrient cycling in 4 Hawaiian montane rain-forests. *Journal of Tropical Ecology* **11**, 189–203.

Vogt KA, Edmunds RL, Grier CC 1981 Biomass and nutrient concentrations of sporocarps produced by mycorrhizal and decomposer fungi in *Abies amabilis* stands. *Oecologia* **50**, 170–175.

Vogt KA, Grier CC, Meier CE, Edmonds RL 1982 Mycorrhizal role in net primary production and nutrient cycling in *Abies amabilis* ecosystems in western Washington. *Ecology* **63**, 370–380.

Vogt KA, Publicover DA, Vogt DJ 1991 A critique of the role of ectomycorrhizas in forest ecology. *Agriculture, Ecosystems and Environment* **35**, 171–190.

Voiblet C, Duplessis S, Encelot N, Martin F 2001 Identification of symbiosis-regulated genes in *Eucalyptus globulus-Pisolithus tinctorius* ectomycorrhiza by differential hybridization of arrayed cDNAs. *The Plant Journal* **25**, 181–191.

Volpin H, Elkind Y, Okon Y, Kapulnik Y 1994 A vesicular-arbuscular mycorrhizal fungus *Glomus intraradix* induces a defense response in alfalfa roots. *Plant Physiology* **104**, 683–689.

Volpin H, Philipps DA, Okon Y, Kapulnik Y 1995 Suppression of an isoflavonoid phytoalexin defense response in mycorrhizal alfalfa roots. *Plant Physiology* **108**, 1449–1454.

Vrålstad T, Holst-Jensen A, Schumacher T 1998 The postfire discomycete *Geopyxis carbonaria* (Ascomycota) is a biotrophic root associate with Norway spruce (*Picea abies*) in nature. *Molecular Ecology* **7**, 609–616.

Vrålstad T, Fossheim T, Schumacher T 2000 *Piceirhiza bicolorata* – the ectomycorrhizal expression of the *Hymenoscyphus ericae* aggregate? *New Phytologist* **145**, 549–563.

Vrålstad T, Myhre E, Schumacher T 2002a Molecular diversity and phylogenetic affinities of symbiotic root-associated ascomycetes of the Helotiales in burnt and metal polluted habitats. *New Phytologist* **155**, 131–148.

Vrålstad T, Schumacher T, Taylor AFS 2002b Mycorrhizal synthesis between fungal strains of the *Hymenoscyphus ericae* aggregate and potential ectomycorrhizal and ericoid hosts. *New Phytologist* **153**, 143–152.

Wagner F, Gay G, Debaud JC 1988 Genetic variability of glutamate dehydrogenase activity in monokaryotic and dikaryotic mycelia of the ectomycorrhizal fungus *Hebeloma cylindrosporum*. *Applied Microbiology and Biotechnology* **28**, 566–576.

Wagner F, Gay G, Debaud JC 1989 Genetic variation of nitrate reductase activity in mono- and dikaryotic populations of the ectomycorrhizal fungus *Hebeloma cylindrosporum* Romagnési. *New Phytologist* **113**, 259–264.

Wahrlich W 1886 Beitrage zur Kenntnis der Orchideen Wurzelpilze. *Botanische Zeitung* **44**, 480–488.

Walker C 1979 *Complexipes moniliformis*: a new genus and species tentatively placed in Endogonaceae. *Mycotaxon* **10**, 99–104.

Walker CM 1992 Systematics and taxonomy of the arbuscular endomycorrhizal fungi Glomales – a possible way forward. *Agronomie* **12**, 887–897.

Walker C, Blaszkowski J, Schwarzott D, Schussler A 2004 *Gerdemannia* gen. nov., a genus separated from *Glomus*, and *Gerdemanniaceae* fam. nov., a new family in the Glomeromycota. *Mycological Research* **108**, 707–718.

Walker NA and Smith SE 1984 The quantitative study of mycorrhizal infection. II. The relation of rate of infection and speed of fungal growth to propagule density, the mean length of the infection unit and the limiting value of the fraction of the root infected. *New Phytologist* **96**, 55–69.

Walker RF, West DC, McLaughlin SB 1982 *Pisolithus tinctorius* ectomycorrhizae reduce moisture stress of Virginia pine on a southern Appalachian coal spoil. In *Proceedings of the Seventh North American Forest Biology Workshop*. Ed. BA Thielges pp. 374–383. University of Kentucky, Lexington, USA.

Walker RF, West DC, McLaughlin SB, Amundsen CC 1989 Growth, xylem pressure potential, and nutrient absorption of loblolly pine on a reclaimed surface mine as affected by an induced *Pisolithus tinctorius* infection. *Forest Science* **35** 2, 569–581.

Walker TW and Syers JK 1976 The fate of phosphorus during pedogenesis. *Geoderma* **15**, 1–19.

Wallace A, Mueller RT, Alexander GV 1978 Influence of phosphorus on zinc, iron, manganese, and copper uptake in plants. *Soil Science* **126**, 336–341.

Wallace GD 1975 Studies of the Monotropoideae (Ericaceae) taxonomy and distribution. *Wasmann Journal of Biology* **33**, 1–88.

Wallander H 2000 Uptake of P from apatite by *Pinus sylvestris* seedlings colonised by different ectomycorrhizal fungi. *Plant and Soil* **218**, 249–256.

Wallander H 2006 External mycorrhizal mycelia – the importance of quantification in natural ecosystems. *New Phytologist* **171** 2, 240–242.

Wallander H and Hagerberg D 2004 Do ectomycorrhizal fungi have a significant role in weathering of minerals in forest soil? *Symbiosis* **37** 1–3, 249–257.

Wallander H and Nylund J-E 1992 Effects of excess nitrogen and phosphorus starvation on the extramatrical mycelium of ectomycorrhizas of *Pinus sylvestris* L. *New Phytologist* **120**, 495–503.

Wallander H, Nylund J-E, Sundberg B 1994 The influence of IAA, carbohydrate and mineral concentration in host tissue on ectomycorrhizal development on *Pinus sylvestris* L. in relation to nutrient supply. *New Phytologist* **127**, 521–528.

Wallander H, Arnebrant K, Dahlberg A 1999 Relationships between fungal uptake of ammonium, fungal growth and nitrogen availability in ectomycorrhizal *Pinus sylvestris* seedlings. *Mycorrhiza* **8**, 215–223.

Wallander H, Nilsson LO, Hagerberg D, Bååth E 2001 Estimation of the biomass and seasonal growth of external mycelium of ectomycorrhizal fungi in the field. *New Phytologist* **151**, 753–760.

Wallander H, Goransson H, Rosengren U 2004 Production, standing biomass and natural abundance of ^{15}N and ^{13}C in ectomycorrhizal mycelia collected at different soil depths in two forest types. *Oecologia* **139**, 89–97.

Wallenda T and Kottke I 1998 Nitrogen deposition and ectomycorrhizas. *New Phytologist* **139**, 169–187.

Wallenda T and Read DJ 1999 Kinetics of amino acid uptake by ectomycorrhizal roots. *Plant Cell and Environment* **22** 2, 179–187.

Wamberg C, Christensen S, Jakobsen I, Muller AK, Sorensen SJ 2003 The mycorrhizal fungus (*Glomus intraradices*) affects microbial activity in the rhizosphere of pea plants (*Pisum sativum*). *Soil Biology and Biochemistry* **35**(10), 1349–1357.

Wang B and Qiu Y-L 2006 Phylogenetic distribution and evolution of mycorrhizae in land plants. *Mycorrhiza* **16**, 299–636.

Wang CJK and Wilcox HE 1985 New species of ectomycorrhizal and pseudomycorrhizal fungi: *Phialophora finlandia*, *Chloridium paucisporum*, and *Phialocephala fortinii*. *Mycologia* **77**, 951–958.

Wang G, Coleman D, Freckman D *et al.* 1989 Carbon partitioning patterns of mycorrhizal versus non-mycorrhizal plants: real-time dynamic measurements using $^{11}CO_2$. *New Phytologist* **112**, 489–493.

Wang H, Wang Z, Zhang F, Liu J, He X 1997 A cytological study on the nutrient-uptake mechanism of a saprophytic orchid *Gastrodia elata*. *Acta Botanica Sinica* **39**, 500–504.

Wang Y and Hall IR 2004 Edible ectomycorrhizal mushrooms: challenges and achievements. *Canadian Journal of Botany* **82**, 1063–1073.

Warcup JH 1971 Specificity of mycorrhizal associations in some Australian terrestrial orchids. *New Phytologist* **70**, 41–46.

Warcup JH 1975 Factors affecting symbiotic germination of orchid seeds. In *Endomycorrhizas*. Eds FE Sanders, B Mosse, PB Tinker pp. 87–104. Academic Press, London, UK.

Warcup JH 1980 Ectomycorrhizal associations of Australian indigenous plants. *New Phytologist* **85**, 531–535.

Warcup JH 1981 The mycorrhizal relationships of Australian orchids. *New Phytologist* **87**, 371–381.

Warcup JH 1985 *Rhizanthella gardneri* (Orchidaceae), its *Rhizoctonia* endophyte and close association with *Melaleuca uncinata* (Myrtaceae) in Western Australia. *New Phytologist* **99**, 273–280.

Warcup JH 1988 Mycorrhizal associations of isolates of *Sebacina vermifera*. *New Phytologist* **110**, 227–231.

Warcup JH 1991 The *Rhizoctonia* endophytes of *Rhizanthella* (Orchidaceae). *Mycological Research* **95**, 656–659.

Warcup JH and Talbot PHB 1967 Perfect states of Rhizoctonias associated with orchids. *New Phytologist* **66**, 631–641.

Warcup JH and Talbot PHB 1970 Perfect states of Rhizoctonias associated with orchids II. *New Phytologist* **70**, 35–40.

Warcup JH and Talbot PHB 1980 Perfect states of Rhizoctonias associated with orchids III. *New Phytologist* **86**, 267–272.

Wardle DA 1999 Is 'sampling effect' a problem for experiments investigating biodiversity – ecosystem function relationships? *Oikos* **87**, 403–407.

Waring RH and Running SW 1998 *Forest Ecosystems-analysis at Multiple Scales*. p. 369. Academic Press, San Diego, USA.

Warner A 1984 Colonization of organic matter by vesicular-arbuscular mycorrhizal fungi. *Transactions of the British Mycological Society* **82**, 352–354.

Warner A and Mosse B 1983 Spread of vesicular-arbuscular mycorrhizal fungi between separate root systems. *Transactions of the British Mycological Society* **80**, 353–354.

Warnock AJ, Fitter AH, Usher MB 1982 The influence of a springtail, *Folsoma candida* (Insecta, Collembola) on the mycorrhizal association of leek *Allium porrum* and the vesicular arbuscular mycorrhizal endophyte *Glomus fasiculatus*. *New Phytologist* **90**, 285–292.

Warren Wilson J and Harley JL 1983 The development of mycorrhiza on seedlings of *Fagus sylvatica* L. *New Phytologist* **95**, 673–695.

Weber HC 1981 Orchideen auf weg zum Parasitismus? Uber die moglichkeit einer phylogenetischen umkonstruktion der infektionsorgane von *Corallorhiza trifida* Chat (Orchidaceae) zu kontaktorganen parasitischer blutenpflanzen. *Berichte der Deutschen Botanischen Gesellschaft* **94**, 275–286.

Wegel E, Schauser L, Sandal N, Stougaard J, Parniske M 1998 Mycorrhiza mutants of *Lotus japonicus* define genetically independent steps during symbiotic infection. *Molecular Plant-Microbe Interactions* **11**, 933–936.

Weidmann S, Sanchez L, Descombin J, Chatagnier O, Gianinazzi S, Gianinazzi-Pearson V 2004 Fungal elicitation of signal transduction-related plant genes precedes mycorrhiza establishment and requires the *dmi3* gene in *Medicago truncatula*. *Molecular Plant-Microbe Interactions* **17**, 1385–1393.

Weiss M and Oberwinkler F 2001 Phylogenetic relationships in Auriculariales and related groups – hypotheses derived from nuclear ribosomal DNA sequences. *Mycological Research* **105**, 403–415.

Weiss M, Selosse MA, Rexer KH, Urban A, Oberwinkler F 2004 Sebacinales: a hitherto overlooked cosm of heterobasidiomycetes with a broad mycorrhizal potential. *Mycological Research* **180**, 1003–1010.

Weissenhorn I, Glashoff A, Leyval C, Berthelin J 1994 Differential tolerance to Cd and Zn of arbuscular mycorrhizal (AM) fungal spores isolated from heavy metal-polluted and unpolluted soils. *Plant and Soil* **167** 2, 189–196.

Weissenhorn I, Leyval C, Belgy G, Berthelin J 1995 Arbuscular mycorrhizal contribution to heavy metal uptake by maize (*Zea mays* L.) in pot culture with contaminated soil. *Mycorrhiza* **5**, 245–251.

Welch RM 2002 The impact of mineral nutrients in food crops on global human health. *Plant and Soil* **247**, 83–90.

Welch RM and Graham RD 2002 Breeding crops for enhanced micronutrient content. *Plant and Soil* **245**, 205–214.

Wellings NP, Wearing AH, Thompson JP 1991 Vesicular-arbuscular mycorrhizae (VAM) improve phosphorus and zinc nutrition and growth of pigeonpea in a vertisol. *Australian Journal of Agricultural Research* **42**, 835.

Wellman CH, Osterloff PL, Mohiuddin U 2003 Fragments of the earliest land plants. *Nature* **425**, 282–285.

Werner NJ, Allen MF, MacMahon JA 1987 Dispersal agents of vesicular-arbuscular mycorrhizal fungi in a disturbed arid ecosystem. *Mycologia* **79**, 721–730.

Wessels JGH 1992 Gene expression during fruiting in *Schizophyllum commune*. *Mycological Research* **96**, 609–620.

West HM 1996 Influence of arbuscular mycorrhizal infection on competition between *Holcus lanatus* and *Dactylis glomerata*. *Journal of Ecology* **84**, 429–438.

West HM, Fitter AH, Watkinson AR 1993a Response of *Vulpia ciliata* ssp. *ambigna* to removal of mycorrhizal infection and to phosphate application under natural conditions. *Journal of Ecology* **81**, 351–358.

West HM, Fitter AH, Watkinson AR 1993b The influence of three biocides on the fungal associates of the roots of *Vulpia ciliata* ssp *ambigna* under natural conditions. *Journal of Ecology* **81**.

Wheeler CT, Tilak M, Scrimgeour CM, Hooker JE, Handley LL 2000 Effects of symbiosis with *Frankia* and arbuscular mycorrhizal fungus on the natural abundance of ^{15}N in four species of Casuarina. *Journal of Experimental Botany* **51**, 287–297.

Whitmore TC 1984 *Tropical Rainforests of the Far East*, 2nd edn. Clarendon Press, Oxford, UK.

Whittingham J and Read DJ 1982 Vesicular-arbuscular mycorrhiza in natural vegetation systems. III. Nutrient transfer between plants with mycorhizal interconnections. *New Phytologist* **90**, 277–284.

Wibawa A, Baon JB, Nurkholis 1995 Growth of shade trees for coffee and cacao as affected by mycorrhizal and rhizobial inoculation. In *Proceedings of the 2nd Symposium on Biology and*

Biotechnology of Mycorrhizae and 3rd Asian Conference on Mycorrhizae. Biotrop Special Publication 56. Eds Supriyanto and JT Kartana pp. 209–214. Seameo Biotrop, Bogor, Indonesia.

Wiese J, Kleber R, Hampp R, Nehls U 2000 Functional characterization of the *Amanita muscaria* monosaccharide transporter, AmMst1. *Plant Biology* **2**, 278–282.

Wikström N, Savolainen V and Chase MW 2001 Evolution of the angiosperms: calibrating the family tree. *Philosophical Transactions of the Royal Society of London Series B* **268**, 2211–2220.

Wilcox HE 1964 Xylem in roots of *Pinus resinosa* Ait. In relation to heterorhyzy and growth activity. In *Formation of Wood in Forest Trees.* Ed. M. Zimmerman pp. 459–478. Academic Press, New York, USA.

Wilcox HE 1968a Morphological studies of the root of red pine, *Pinus resinosa.* I. Growth characteristics and branching patterns. *American Journal of Botany* **55**, 247–254.

Wilcox HE 1968b Morphological studies of the roots of red pine, *Pinus resinosa.* II. Fungal colonisation of roots and development of mycorrhizae. *American Journal of Botany* **55**, 686–700.

Wilcox HE 1971 Morphology of ectendomycorrhizae in *Pinus resinosa.* In *Proceedings of the 1st North American Conference on Mycorrhizae. University of Illinois, 1969* pp. 54–67. US Forestry Service, Washington, DC, USA.

Wilcox HE and Wang CJK 1987a Ectomycorrhizal and endomycorrhizal association of *Phialophora finlandia* with *Pinus resinosa, Picea rubens,* and *Betula alleghaniensis. Canadian Journal of Forest Research* **17**, 976–990.

Wilcox HE and Wang CJK 1987b Mycorrhizal and pathological associations of dematiaceous fungi in roots of 7-month-old tree seedlings. *Canadian Journal of Forest Research* **17**, 884–889.

Wilcox HE, Ganmore-Neumann R, Wang CJK 1974 Characteristics of two fungi producing ectendomycorrhizae in *Pinus resinosa. Canadian Journal of Botany* **52**, 2279–2282.

Williams PG, Roser DJ, Seppelt RD 1994 Mycorrhizas of hepatics in continental Antarctica. *Mycological Research* **98**, 34–36.

Williamson B 1970 Induced DNA synthesis in orchid mycorrhiza. *Planta* **92**, 347–354.

Williamson B 1973 Acid phosphatase and esterase activity in orchid mycorrhiza. *Planta* **112**, 149–158.

Williamson B and Alexander IJ 1975 Acid phosphatases localised in the sheath of beech mycorrhiza. *Soil Biology and Biochemistry* **7**, 195–198.

Williamson B and Hadley G 1969 DNA content of nucleii in orchid protocorms symbiotically infected with *Rhizoctonia. Nature* **222**, 582–583.

Williamson B and Hadley G 1970 Penetration and infection of orchid protocorms by *Thanatephorus cucumeris* and other *Rhizoctonia* isolates. *Phytopathology* **60**, 1092–1096.

Willis KJ and McElwain JC 2002 *The Evolution of Plants.* Oxford University Press, Oxford, UK.

Wilson BJ, Addy HD, Tsuneda A, Hambleton S, Currah RS 2004 *Phialocephala sphaeroides* sp nov., a new species among the dark septate endophytes from a boreal wetland in Canada. *Canadian Journal of Botany* **82**, 607–617.

Wilson GWT and Hartnett DC 1997 Effects of mycorrhizae on plant growth and dynamics in experimental tallgrass prairie microcosms. *American Journal of Botany* **84**, 478–482.

Wilson GWT and Hartnett DC 1998 Interspecific variation in plant responses to mycorrhizal colonization in tallgrass prairie. *American Journal of Botany* **85**, 1732–1738.

Wilson GWT, Daniels Hetrick BA, Gerschefske Kitt D 1988 Suppression of mycorrhizal growth response of big bluestem grass by non sterile soil. *Mycologia* **80**, 338–343.

Wilson JM 1984 Competition for infection between vesicular-arbuscular mycorrhizal fungi. *New Phytologist* **97**, 427–435.

Wilson JM and Tommerup IC 1992 Interactions between fungal symbionts: VA mycorrhizae. In *Mycorrhizal Functioning.* Ed. MF Allen pp. 199–248. Chapman and Hall, London, UK.

Wilson JM and Trinick MJ 1982 Factors affecting the estimation of numbers of infective propagules of vesicular arbuscular mycorrhizal fungi by the most probable number method. *Australian Journal of Soil Research* **21**, 73–81.

Wilson JM and Trinick MJ 1983 Infection development and interactions between vesicular-arbuscular mycorrhizal fungi. *New Phytologist* **93**, 543–553.

Wilson KH, Wilson WJ, Radosevich JL *et al.* 2002 High-density microarray of small-subunit ribosomal DNA probes. *Applied and Environmental Microbiology* **68**, 2535–2541.

Winther JL and Friedman WE 2007a Arbuscular mycorrhizal symbionts in *Botrychium* (Ophioglossaceae). *American Journal of Botany* **94**, 1248–1255.

Winther JL and Friedman WE 2007b Arbuscular mycorrhizal associations in hycopodiaceae. *New Phytologist* online early doi:10.1111/j.1469-8137.2007.02276.x

Wipf D, Benjdia M, Tegeder M, Frommer WB 2002 Characterization of a general amino acid permease from *Hebeloma cylindrosporum. FEBS Letters* **528**, 119–124.

Wong KKY and Fortin JA 1990 Root colonisation and intraspecific mycobiont variation in ectomycorrhiza. *Symbiosis* **8**, 197–231.

Wong KKY, Piché Y, Montpetit D, Kropp BR 1989 Differences in the colonisation of *Pinus banksiana* roots by sib-monokaryotic and dikaryotic strains of ectomycorrhizal *Laccaria bicolor. Canadian Journal of Botany* **67**, 1717–1726.

Wong KKY, Piché Y, Fortin JA 1990. Differential development of root colonisation among four closely related genotypes of ectomycorrhizal *Laccaria bicolor. Mycological Research* **94**, 876–884.

Wood T and Cummings B 1992 Biotechnology and the future of VAM commercialization. In *Mycorrhizal Functioning.* Ed. MF Allen pp. 468–487. Chapman and Hall, London, UK.

Worley JF and Hacskaylo E 1959 The effect of available soil moisture on the mycorrhizal association of Virginian pine. *Forest Science* **5**, 267–268.

Wright DP, Read DJ, Scholes JD 1998a Mycorrhizal sink strength influences whole plant carbon balance of *Trifolium repens* L. *Plant, Cell and Environment* **21**, 881–891.

Wright DP, Scholes JD, Read DJ 1998b Effects of VA mycorrhizal colonization on photosynthesis and biomass production of *Trifolium repens* L. *Plant, Cell and Environment* **21**, 209–216.

Wright DP, Scholes JD, Read DJ, Rolfe SA 2000 Changes in carbon allocation and expression of carbon transporter genes in *Betula pendula* Roth. colonized by the ectomycorrhizal fungus *Paxillus involutus* (Batsch) Fr. *Plant Cell and Environment* **23**, 39–49.

Wright DP, Johansson T, Le Quere A, Söderström B, Tunlid A 2005 Spatial patterns of gene expression in the extramatrical mycelium and mycorrhizal root tips formed by the ectomycorrhizal fungus *Paxillus involutus* in association with birch (*Betula pendula*) seedlings in soil microcosms. *New Phytologist* **167**, 579–596.

Wright SJ 2002 Plant diversity in tropical forests: a review of mechanisms of species coexistence. *Oecologia* **130**, 1–14.

Wright SF and Jawson L 2001 A pressure cooker method to extract glomalin from soils. *Soil Science Society of America Journal* **65**, 1734–1735.

Wright SF and Upadhyaya A 1998 A survey of soils for aggregate stability and glomalin, a glycoprotein produced by hyphae of arbuscular mycorrhizal fungi. *Plant and Soil* **198**, 97–107.

Wright SJ 2002 Plant diversity in tropical forests: a review of mechanisms of species coexistence. *Oecologia* **130** 1, 1–14.

Wright W, Fitter A, Meharg A 2000 Reproductive biomass in *Holcus lanatus* clones that differ in their phosphate uptake kinetics and mycorrhizal colonization. *New Phytologist* **146**, 493–501.

Wu TH, Sharda JN, Koide RT 2003 Exploring interactions between saprotrophic microbes and ectomycorrhizal fungi using a protein-tannin complex as an N source by red pine (*Pinus resinosa*). *New Phytologist* **159**, 131–139.

Wulf A, Manthey K *et al.* 2003 Transcriptional changes in response to arbuscular mycorrhizal development in the model plant *Medicago truncatula. Molecular Plant-Microbe interactions* **16**, 306–314.

Wyss P, Boller T, Wiemken A (1991) Phytoalexin response is elicited by a pathogen (*Rhizoctonia solani*) but not by a mycorrhizal fungus (*Glomus mosseae*) in soybean roots. *Experientia* **47**, 395–399.

Xavier LJC and Germida JJ 2003 Bacteria associated with *Glomus clarum* spores influence mycorrhizal activity. *Soil Biology and Biochemistry* **35**, 471–478.

Xiao G and Berch SM 1992 Ericoid mycorrhizal fungi of *Gaultheria shallon*. *Mycologia* **84**, 470–471.

Xiao G and Berch SM 1995 The ability of known ericoid mycorrhizal fungi to form mycorrhizae with *Gaultheria shallon*. *Mycologia* **87** 4, 467–470.

Xiao GP and Berch SM 1999 Organic nitrogen use by salal ericoid mycorrhizal fungi from northern Vancouver Island and impacts on growth *in vitro* of *Gaultheria shallon*. *Mycorrhiza* **9** 3, 145–149.

Xu D, Dell B, Malajczuk N, Gong M 2001 Effects of P fertilisation and ectomycorrhizal fungal inoculation on early growth of eucalypt plantations in southern China. *Plant and Soil* **233**, 47–57.

Xu J and Guo S 2000 Retrospect on the research of the cultivation of Gastrodia elata B1, a rare Chinese traditional medicine. *Chinese Medical Journal* **113**, 686–692.

Xu J-T and Mu C 1990 The relation between growth of *Gastrodia elata* protocorms and fungi. *Acta Botanica Sinica* **32**, 26–33.

Yagame T, Yamato M, Mii M, Suzuki A, Iwase K 2007 Developmental processes of achlorophyllous orchid, *Epipogium roseum*: from seed germination to flowering under symbiotic cultivation with mycorrhizal fungus. *Journal of Plant Research* **120**, 229–236.

Yamada A, Maeda K, Kobayashi H, Murata H 2006 Ectomycorrhizal symbiosis *in vitro* between *Tricholoma matsutake* and *Pinus densiflora* seedlings that resembles naturally occurring 'shiro'. *Mycorrhiza* **16**, 111–116.

Yamato M 2001 Identification of a mycorrhizal fungus in the roots of achlorophyllous *Sciaphila tosaensis* Makino, Triuridaceae. *Mycorrhiza* **11** 2, 83–88.

Yamato M, Yagame T, Suzuki A, Iwase K 2005 Isolation and identification of mycorrhizal fungi associating with an achlorophyllous plant, *Epipogium roseum* (Orchidaceae). *Mycoscience* **46**, 73–77.

Yanai RD, Blum JD, Hamburg SP, Arthur MA, Nezat CA, Siccama TG 2005 New insights into calcium depletion in northeastern forests. *Journal of Forestry* **103**, 14–20.

Yang CS and Korf RP 1985 A monograph of the genus *Tricharina* and of a new, segregate genus, *Wilcoxina* Pezizales. *Mycotaxon* **24**, 467–531.

Yang CS and Wilcox HE 1984 An E-strain ectomycorrhiza formed by a new species *Tricharina mikolae*. *Mycologia* **76**, 675–684.

Yano K, Yamauchi A, Iijima M, Kono Y 1998 Arbuscular mycorrhizal formation in undisturbed soil counteracts compacted soil stress for pigeon pea. *Applied Soil Ecology* **10**, 95–102.

Yawney WJ and Schultz RC 1990 Anatomy of a vesicular-arbuscular endomycorrhizal symbiosis between sugar maple *Acer saccharum* Marsh and *Glomus etunicatum* Becker and Gerdemann. *New Phytologist* **114**, 47–58.

Yazid SM, Lee SS, Lapeyrie F 1994 Growth-stimulation of *Hopea* spp. (Dipterocarpaceae) seedlings following ectomycorrhizal inoculation with an exotic strain of *Pisolithus tinctorius*. *Forest Ecology and Management* **67**, 339–343.

Yokoyama J, Fukuda T, Tsukaya H 2005 Molecular identification of the mycorrhizal fungi of the epiparasitic plant *Monotropastrum humile* var. *glaberrimum* (Ericaceae). *Journal of Plant Research* **118**, 53–56.

Yost RS and Fox RL 1979 Contribution of mycorrhizae to P nutrition of crops growing on an Oxisol. *Agronomy Journal* **71**, 903–908.

Young BW, Massicotte HB, Tackaberry LE, Baldwin QF, Egger KN 2002 *Monotropa uniflora*: morphological and molecular assessment of mycorrhizae retrieved from sites in the sub-boreal Spruce biogeoclimatic zone in central British Columbia. *Mycorrhiza* **12**, 75–82.

Yu T, Nassuth A, Peterson RL 2001b Characterization of the interaction between the dark septate fungus *Phialocephala fortinii* and *Asparagus officinalis* roots. *Canadian Journal of Microbiology* **47**, 741–753.

Yu TE, Egger KN, Peterson RL 2001a Ectendomycorrhizal associations – characteristics and functions. *Mycorrhiza* **11**, 167–177.

Zabinski CA, Quinn L, Callaway RM 2002 Phosphorus uptake, not carbon transfer, explains arbuscular mycorrhizal enhancement of *Centaurea maculosa* in the presence of native grassland species. *Functional Ecology* **16**, 758–765.

Zak B 1971 Characterization and classification of mycorrhizae of Douglas fir. II. *Pseudotsuga menzieii* + *Rhizopogon vinicolor*. *Canadian Journal of Botany* **49**, 1079–1084.

Zak B 1973 Classification of Ectomycorrhizae. In *Ectomycorrhizae*. Eds GC Marks and TT Kozlowski pp. 43–78. Academic Press, New York, USA.

Zak B 1974 Ectendomycorrhiza of Pacific madrone *Arbutus menziessi*. *Transactions of the British Mycological Society* **62**, 202–204.

Zak B 1976a Pure culture synthesis of bearberry mycorrhizae. *Canadian Journal of Botany* **54**, 1297–1305.

Zak B 1976b Pure culture synthesis of Pacific madrone ectendomycorrhizae. *Mycologia* **68**, 362–369.

Zangaro W, Bononi VLR, Trufen SB 2000 Mycorrhizal dependency, inoculum potential and habitat preference of native woody species in South Brazil. *Journal of Tropical Ecology* **16**, 603–621.

Zangaro W, Nisizaki SMA, Domingos JCB, Nakano EM 2003 Mycorrhizal response and successional status in 80 woody species from south Brazil. *Journal of Tropical Ecology* **19**, 315–324.

Zelmer CD and Currah RS 1995 Evidence for a fungal liason between *Corallorhiza trifida* (Orchidaceae) and *Pinus contorta* (Pinaceae). *Canadian Journal of Botany* **73**, 862–866.

Zelmer CD, Cuthbertson L, Currah RS 1996 Fungi associated with terrestrial orchid mycorrhizas, seeds and protocorms. *Mycoscience* **37**, 439–448.

Zerova MY 1955 Mykorrhiza formation in forest trees of the Ukrainian SSR. In *Mycotrophy in Plants*. Ed A Imshenetskii (translated from Russian by Israel Programme for Scientific Translocations, 1967). USDA and NSF, Washington DC, USA.

Zhang YH and Zhuang WY 2004 Phylogenetic relationships of some members in the genus *Hymenoscyphus* (Ascomycetes, Helotiales). *Nova Hedwigia* **78**, 475–484.

Zhu H, Guo D-C, Dancik BP 1990 Purification and characterisation of an extracellular acid proteinase from the ectomycorrhizal fungus *Hebeloma crustuliniforme*. *Applied and Environmental Microbiology* **56**, 837–843.

Zhu H, Dancik BP, Higginbotham KO 1994 Regulation of extracellular proteinase production in an ectomycorrhizal fungus *Hebeloma crustuliniforme*. *Mycologia* **86**, 227–234.

Zhu YG, Christie P, Laidlaw AS 2001 Uptake of Zn by arbuscular mycorrhizal white clover from Zn-contaminated soil. *Chemosphere* **42**, 193–199.

Zhu YG, Smith SE, Barritt AR, Smith FA 2001 Phosphorus P efficiencies and mycorrhizal responsiveness of old and modern wheat cultivars. *Plant and Soil* **237**, 249–255.

Zhu YG, Smith FA, Smith SE 2003 Phosphorus efficiencies and responses of barley (*Hordeum vulgare* L.) to arbuscular mycorrhizal fungi grown in highly calcareous soil. *Mycorrhiza* **13**, 93–100.

Zimmer K, Hymson NA, Gebauer G, Allen EB, Allen MF, Read DJ 2007 Wide geographic and ecological distribution of nitrogen and carbon gains from fungi in pyroloids and monotropoids (Ericaeae) and in orchids. *New Phytologist* **175**, 166–175.

Index

Page numbers in **bold** refers to tables and page numbers in *italics* refers to figures.

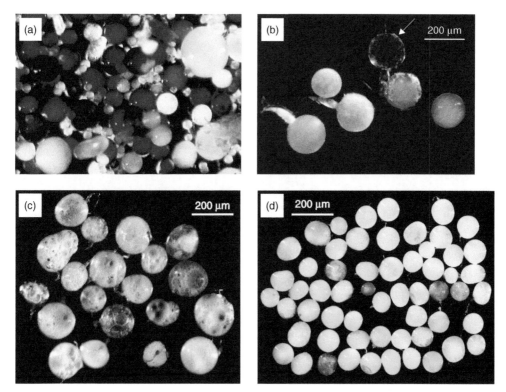

Plate 1.1 (a) Collection of AM spores of nine species of AM fungi isolated from a grassland that had developed from an abandoned agricultural site. Reproduced from Bever *et al.* (2001), Arbuscular mycorrhizal fungi: more diverse than meets the eye, and the ecological tale of why. *Bioscience* **51**, 923–931, Copyright, American Institute of Biological Sciences, with permission. (b) Spores of *Acaulospora laevis*. A translucent sporiferous saccule can be seen attached to one of the spores (arrowed). (c) A group of spores of *Glomus mosseae*. (d) A group of spores of *Gigaspora gigantea*. Images (b), (c) and (d) are reproduced courtesy of Joe Morton.

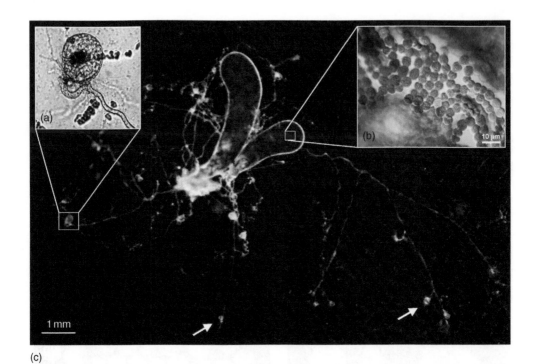

(c)

Plate 1.2 *Geosiphon pyriformis*, a glomeromycotan soil fungus forming endosymbiosis with a cyanobacterium and representing an *in-vitro* culture of a mycorrhiza-like association. (a) Spore and subtending hypha of *G. pyriformis*; (b) cyanobacterial partner; (c) bladders of *G. pyriformis*, with emanating external mycelium. From Schüßler and Wolf (2005), *Geosiphon pyriformis* – a glomeromycotan soil fungus forming endosymbiosis with Cyanobacteria. In *In Vitro Culture of Mycorrhizas*. Eds S Declerck, DG Strullu, JA Fortin pp. 272–289, with kind permission of Springer Science and Business Media.

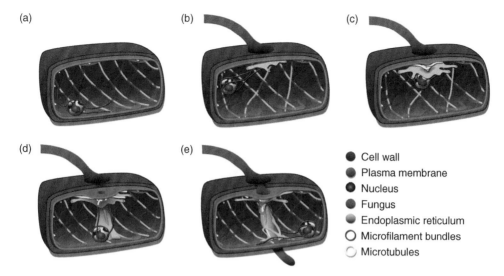

(a)
(b)
(c)
(d)
(e)

- ● Cell wall
- ● Plasma membrane
- ● Nucleus
- ● Fungus
- ● Endoplasmic reticulum
- ○ Microfilament bundles
- ◌ Microtubules

Plate 3.1 Formation of the prepenetration apparatus (PPA) that facilitates passage of AM fungi through root epidermal cells. (a) An epidermal cell prior to fungal contact, showing peripheral position of the plant cell nucleus and arrangement of cytoskeletal elements and endoplasmic reticulum (ER). (b) Appressorium formation on the outer surface of the host cell results in important rearrangements of microtubules, microfilament bundles and ER, associated with initial nuclear movement toward the surface appressorium. (c, d) Assembly of the transient PPA within the cytoplasmic column and subsequent transcellular nuclear migration. (e) AM infection hypha penetrates the cell wall and crosses the epidermal cell through the apoplastic compartment constructed within the cytoplasmic column. Coding of different structures as in legend. Courtesy Andrea Genre.

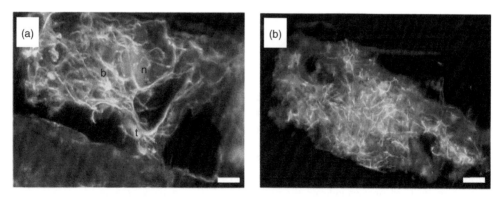

Plate 3.2 Microtubule arrangement in cells of *Nicotiana tabacum* containing arbuscules of *Gigaspora margarita*. (a) Bundles can be seen running through the arbuscule branches (b), along the arbuscular trunk (t) and around the nucleus (n). (b) Shorter bundles are visible among the arbuscule branches, connecting them together. Bar = 10 μm. From Genre and Bonfante (1998), with permission.

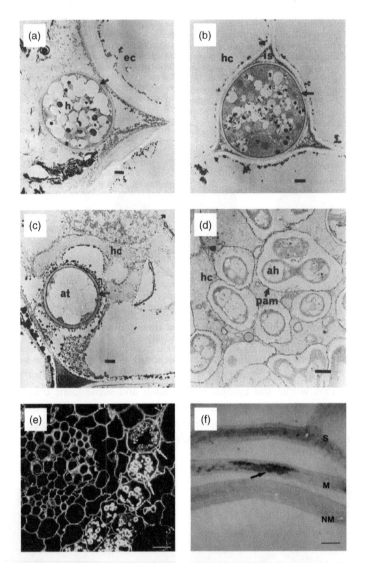

Plate 4.1 (a–d) Transmission electron micrographs of details of the arbuscular mycorrhizal interaction between *Allium cepa* and *Glomus intraradices*. Sections of roots fixed and stained to demonstrate ATPase activity by the deposition of lead phosphate which appears as fine electron-dense precipitates. (a), (b), (c) no inhibitors; (d) in the presence of molybdate, which inhibits non-specific phosphatases but not H^+-ATPases. (a), Extraradical hypha at the root surface; (b), intercellular hypha; (c), arbuscular trunk hypha in a cortical cell; (d), fine arbuscular branches within a cortical cell. Arrows indicate the presence of ATPase activity on fungal (a–c) and plant (d) membranes. Bars, 0.5 μm. The plate is reproduced from Smith and Smith, 1996, with permission. Origins of figures as follows, (a)–(c), V. Gianinazzi-Pearson, unpublished; (d), from Gianinazzi-Pearson *et al.* 1991. (e) Transverse section of a root of *Nicotiana tabacum* colonized by *Glomus fasciculatum* showing light microscopic immunolocalization of H^+-ATPase (bright silver signal) in arbuscule-containing root cortical cells (solid arrows). No signal can be seen in non-colonized cells or in the stele (open arrows). Bar = 15 μm. (f) GUS activity in arbuscule containing cells of a recently colonized root segment (M) of *N. tabacum* transformed with a *pma4-gusA* reporter construct (blue staining arrowed). No GUS activity is evident in non-mycorrhizal roots (NM) or in roots containing senescent arbuscules (S). Bar = 0.5 mm.

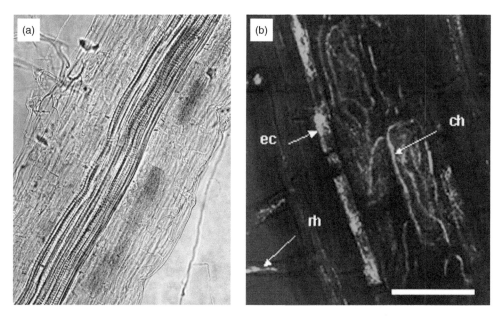

Plate 5.1 Cell-type-specific promoter activity of mycorrhiza-induced Pht1 transporters. (a) Histochemical staining for GUS activity in *Medicago truncatula* roots carrying an *MtPT4* promoter-UidA fusion. Blue staining indicates GUS activity associated with *Arum*-type arbuscules of *Glomus versiforme*. Image courtesy Maria Harrison. (b) Confocal image of a hairy root of *Solanum tuberosum*, transformed with an *StPT3* promoter-Fluorescent Timer chimeric gene. Green fluorescence originating from Fluorescent-Timer in cells colonized by *Paris*-type coils formed by *Gigaspora margarita*. rh, root hairs; ec, epidermal cells; ch, coiled hyphae. Reproduced from Karandashov *et al. Proceedings of the National Academy of Sciences of the United States of America*, **101**, 6285–6290. Copyright (2002) National Academy of Sciences, USA.

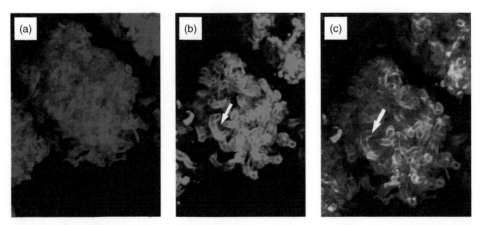

Plate 5.2 Confocal images showing immunolocalization of MtPT4 in roots of *Medicago truncatula* colonized by *Glomus versiforme*. The roots were probed with anitibodies specific for MtPT4 and these were visualized with a secondary antibody conjugated with AlexaFluor488 (green fluorescence). Roots were counterstained with WGA-Texas red to visualize fungal structures (red fluorescence). (a) WGA-Texas red visualization of arbuscules; (b) green fluorescence showing immunolocalization of MtPT4 surrounding branches of the arbuscule (arrowed); (c) merged images showing both red and green fluorescence. Images courtesy Maria Harrison.

Plate 6.1 Ectomycorrhizal roots formed by different fungal symbionts. (a) Mycorrhiza formed by *Russula ochroleuca* on spruce (*Picea abies*) showing diagnostic greenish yellow patches (arrowed) ×44. From Agerer (1987–2002), with permission. (b) Irregularly branched cluster of mycorrhizal roots of beech (*Fagus sylvatica*) formed by *Lactarius subdulcis*. ×10. From Agerer (1987–2002), with permission. (c) Mycorrhizal root of the *Piceirhiza bicolorata* type on spruce (*Picea abies*) ×20. From Agerer (1987–2002), with permission. (d) Cluster of mycorrhizal roots formed by *Sebacina* spp. on cork oak *Quercus suber*. ×5. From Azul *et al.* (2006). (e) Dichotomous mycorrhizal roots formed by *Rhizopogon idahoensis* on pine (*Pinus* spp.) ×6. Photograph courtesy R. Molina. (f) Section across two tuberculate mycorrhizas of *Pinus* spp. formed by *Rhizopogon* showing dense clusters of individual ectomycorrhizal roots encased in a rind of mycelium ×3. Photograph courtesy B. Zak.

Plate 6.1 (Caption opposite)

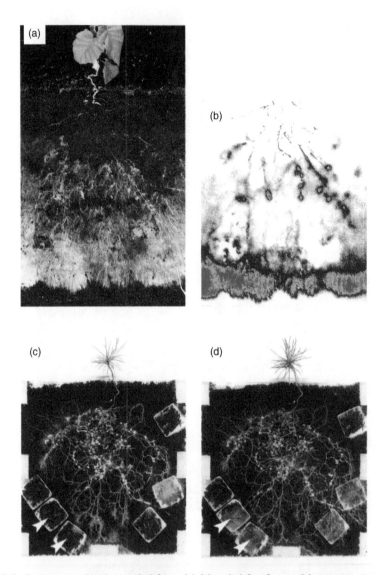

Plate 6.2 Ectomycorrhizal mycelial fans. (a) Mycelial fan formed by vegetative mycelium of *Paxillus involutus* growing across a homogenized peat substrate in a microcosm from roots of birch (*Betula pendula*) showing advancing hyphal front at the lower part of the system. Photograph courtesy J.R. Leake. (b) Colour laser scanning autoradiogram of the microcosm shown in (a) taken 48 hours after feeding shoots of the *Betula* plant with $^{14}CO_2$ showing selective allocation and accumulation of labelled carbon (red colouration) in individual mycorrhizal roots and in the advancing hyphal front. Photograph courtesy J.R. Leake. (c) Mycelial fan formed by *Suillus bovinus* growing across homogenized peat from roots of *Pinus sylvestris*. The advancing hyphal front is making contact with trays of relatively nutrient rich organic matter collected from the fermentation horizon (FH) of a forest soil (arrows lower left). From Bending and Read (1995a). (d) The same microcosm as in (c) after a further 40 days showing intensive exploitation of the FH by mycelium of *S. bovinus* (arrows). From Bending and Read (1995a).

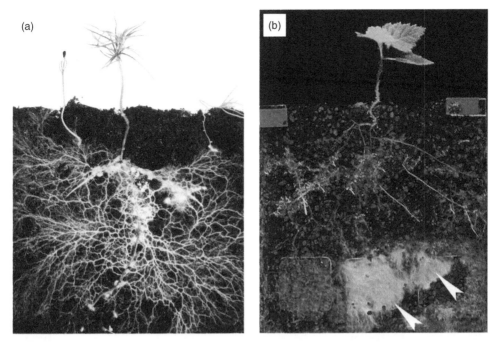

Plate 6.3 (a) Extensive foraging system formed by rhizomorphs of *Suillus bovinus* gowing in symbiosis with *Pinus sylvestris*. The advancing hyphal front has made contact with developing pine seedlings to the left and right of the more mature plant and has initiated mycorrhiza formation so incorporating them into the common mycelial network. From Read (1991c) with permission. (b) Mycorrhizas formed on birch (*Betula pendula*) by the fungal symbiont *Paxillus involutus*. The upper part of the microcosm contains homogenized peat mixed with inert plastic beads, while to the lower half are added trays of the same material either supplemented with mesofaunal necromass as an additional nutrient source (two trays arrowed) or peat only. Note intensive proliferation of *Paxillus* hyphae in the necromass-containing trays. From Perez-Moreno and Read (2001b).

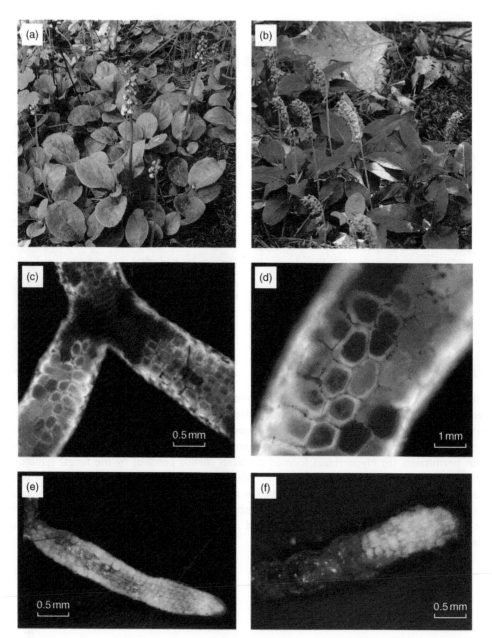

Plate 7.1 (a) Shoots and flowering spikes of *Pyrola minor* (Pyrolae-Ericaceae) growing on the forest floor under a canopy of *Pinus sylvestris*. Photograph courtesy of K. Zimmer. (b) Shoots and flowering spikes of *Orthilia secunda* (Pyrolae-Ericaceae) growing under mixed conifer–deciduous canopy. Photograph courtesy of K. Zimmer. (c) Light micrograph of mycorrhizal root of *P. minor* showing epidermal cells densely packed with intracellular hyphal complexes and lack of fungal mantle. Photograph courtesy of K. Zimmer. (d) As (c) but root of *O. secunda*. Photograph courtesy of K. Zimmer. (e) Seedling of *O. secunda* emerging from seed coat (brown extreme right), recovered from a buried seed packet. Dense white clumps in epidermal cells are hyphal complexes formed by a *Sebacina* spp. Photograph and fungal identification courtesy of M. Bidartondo. (f) Seedling of *Pyrola chlorantha* emerging from seed coat (brown) showing epidermal cells heavily colonized by a *Sebacina* spp. Photograph and fungal identification courtesy of M. Bidartondo.

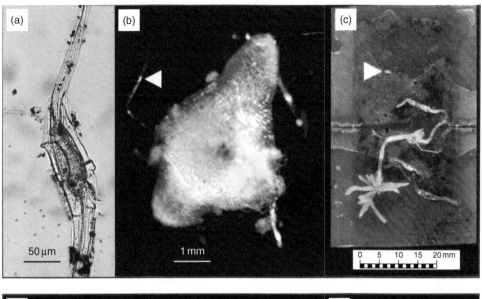

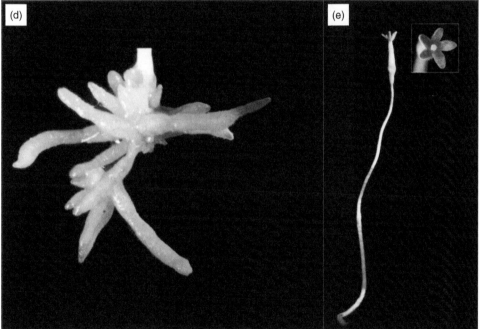

Plate 13.1 Stages in the germination and development of the mycoheterotroph *Voyria tenella* (Gentianaceae) in symbiosis with a *Glomus* spp. as revealed after burial of seed packets in tropical rainforest (French Guyana, South America). (a) Ungerminated seed with filiform appendages to provide buoyancy. This plant has the smallest seeds known in any higher plant. (b) Seedling at an early stage of development. The highly reflective filaments are ungerminated seeds. The horizontal arrow points to an embryo. (c) Fully developed plant with roots, shoot and unopened flower after seed had been buried for 12 months in the natural habitat of the plant. The horizontal arrow points to the seedling shown in (b). (d) Detail of the root system of the plant. (e) Fully developed shoot with flower (inset). Note the complete absence of leaves. Photographs courtesy of J.R. Leake and S.L. McKendrick.

Plate 13.2 (a) Flowering scape of the mycoheterotrophic plant *Sarcodes sanguinea* (Monotropoideae) emerging from the floor of a mixed conifer forest. Photograph courtesy T Bruns. Inset: seedling of *S. sanguinea* after germination *in vitro* in association with its highly specific fungal partner *Rhizopogon ellenae*. From Bruns and Read (2000). (b) Flowering scape of the mycoheterotrophic plant *Pterospora andromedea* (Monotropoideae). Photograph courtesy of V. L. Wong. (c). Flowering scapes of the mycoheterotrophic plant *Monotropa hypopitys* growing under *Pinus sylvestris* in the UK. Photograph by D. J. Read. (d) Seedlings of *P. andromedea* contained in the seed packet in which they were buried in nature. Seedlings are enveloped in mycelium of the specific mycorrhizal symbiont *Rhizopogon salebrosus*. Extensive rhizomorphs of the fungus are evident. Photograph courtesy M. Bidartondo. (e) Seedling of *M. hypopitys* recovered from a seed packet after 21 months burial under the ECM shrub *Salix repens* showing extensive development of monotropoid mycorrhizal roots associated with *Tricholoma cingulatum* and the production of a shoot bud (arrowed and inset). Photograph courtesy S. L. McKendrick. (f) Seedling of *M. uniflora* recovered from a seed packet showing extensive development of monotropoid mycorrhizal roots and the formation of a shoot bud (square and inset). Arrow indicates ungerminated seed. Photograph courtesy M. Bidartondo.

Plate 13.3 (a) Whole plants of the mycoheterotrophic 'coralroot orchid' *Corallorhiza trifida*, showing the reduced and leafless largely achlorophyllous shoots and coarse prolifically branched roots system. Photograph by D.J. Read. (b) Seedlings of *C. trifida* recovered from a single seed packet buried under the ECM shrub *Salix repens*. The fungal symbiont in all cases was a member of the Thelephoraceae. (c) A mature seedling from a seed packet showing initiation of a shoot bud (extreme left). (d) Individual introduced seedlings (arrowed) of *C. trifida* developing in a microcosm in association with the autotrophic ECM tree *Betula pendula*. (e) Seedling of *C. trifida* (right) connected to ECM roots of *B. pendula* by individual hyphae (single arrows) and rhizomorph (double arrow) of its thelephoroid mycorrhizal associate. (f) Ectomycorrhizal roots of birch formed by the thelephoroid associate of *C. trifida* in the vicinity of the introduced orchid seedlings. Plates 13.3 b–c from McKendrick *et al.* (2000a) and Plates 13.3 d–f from McKendrick *et al.* (2006b).

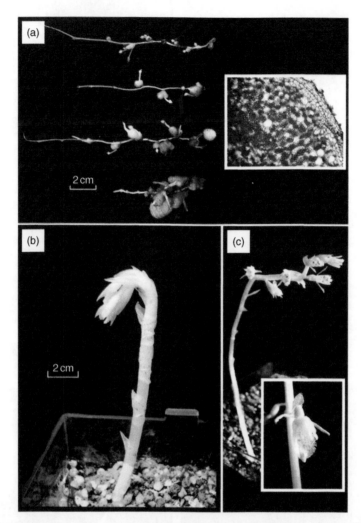

Plate 13.4 Stages in the *in vitro* development of the fully mycoheterotrophic orchid *Epipogium roseum*. (a) Underground organs of *E. roseum* developing in a lava pumice–saw dust mixture 16 weeks after introduction of seed and inoculation with a fungus closely related to *Coprinus disseminatus*. Slender filiform horizontally-running stolons support the development of swollen tubers from which white inflorescence stalks arise. Cells of the tubers are packed with starch grains (inset). (b) Emergence of an inflorescence stalk 26 weeks after seed sowing. (c) Fully developed flowering spike at 27 weeks after sowing with detail of flower (inset). From Yagame *et al.* (2007).

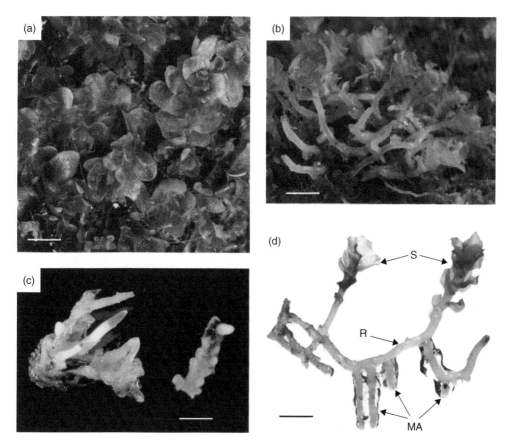

Plate 14.1 (a) Gametophyte thalli of the AM-colonized basal leafy liverwort *Treubia lacuno-soides* (Bar = 1 cm). From Duckett *et al.* (2006b). (b) Gametophyte thalli of the AM-colonized basal leafy liverwort *Haplomitrium ovalifolium* (Bar = 0.5 cm). From Carafa *et al.* (2003). (c) Gametophytes of the mycoheterotrophic subterranean liverwort *Cryptothallus mirabilis* showing stages in sporophyte development (Bar = 0.5 cm). Photograph by D.J. Read. (d) Individual thalli of *H. ovalifolium* showing leafy shoots (S), largely achlorophyllous basal runners (R), and descending 'mycorrhizal axes' (MA) in which the AM fungal symbiont is primarily located (Bar = 0.5 cm). From Carafa *et al.* (2003).

 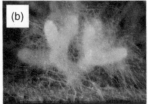 

Plate 16.1 Roots of *Pinus sylvestris* used in studies of the protozoan populations associated with (a) non-mycorrhizal short roots (NM), (b) ectomycorrhizal short roots (MR) and (c) extraradical mycelium of *Paxillus involutus* (see Figure 16.11).

Plate 17.1 Edible mushroom production in Mexico. (a) Girl selling *Boletus edulis* in a market. (b) and (c) Baskets of ECM fungi of which the main ones are: *Gomphus flocculosus*, *Lactarius salmonicolor*, *Helvella crispa* and *Amanita aspera* var *franchetii*. Photographs courtesy Jesus Perez-Moreno.

Printed and bound by CPI Group (UK) Ltd, Croydon, CR0 4YY

12/05/2025

01869445-0001